The Genetic Code

First position	Second position				Third position
	U	**C**	**A**	**G**	
U	Phe	Ser	Tyr	Cys	U
	Phe	Ser	Tyr	Cys	C
	Leu	Ser	Stop	Stop	A
	Leu	Ser	Stop	Trp	G
C	Leu	Pro	His	Arg	U
	Leu	Pro	His	Arg	C
	Leu	Pro	Gln	Arg	A
	Leu	Pro	Gln	Arg	G
A	Ile	Thr	Asn	Ser	U
	Ile	Thr	Asn	Ser	C
	Ile	Thr	Lys	Arg	A
	Met	Thr	Lys	Arg	G
G	Val	Ala	Asp	Gly	U
	Val	Ala	Asp	Gly	C
	Val	Ala	Glu	Gly	A
	Val	Ala	Glu	Gly	G

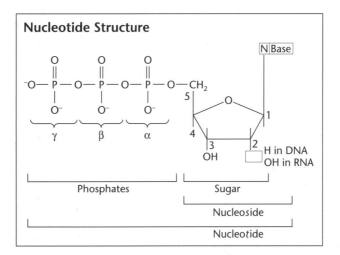

Names of Nucleic Acid Subunits

Base	Nucleoside	Nucleotide	Abbreviation RNA	DNA
Adenine	Adenosine	Adenosine triphosphate	ATP	dATP
Guanine	Guanosine	Guanosine triphosphate	GTP	dGTP
Cytosine	Cytidine	Cytidine triphosphate	CTP	dCTP
Thymine	Thymidine	Thymidine triphosphate		dTTP
Uracil	Uridine	Uridine triphosphate	UTP	

THIRD EDITION

molecular genetics of bacteria

THIRD EDITION
molecular genetics of bacteria

Larry Snyder and Wendy Champness

Department of Microbiology and Molecular Genetics
Michigan State University
East Lansing, Michigan

ASM PRESS

WASHINGTON, D.C.

Address editorial correspondence to ASM Press, 1752 N St. NW, Washington, DC
20036-2904, USA

Send orders to ASM Press, P.O. Box 605, Herndon, VA 20172, USA
Phone: (800) 546-2416 or (703) 661-1593
Fax: (703) 661-1501
E-mail: books@asmusa.org
Online: estore.asm.org

Library of Congress Cataloging-in-Publication Data

Snyder, Larry.
  Molecular genetics of bacteria / Larry Snyder and Wendy Champness.—3rd ed.
    p. ; cm.
  Includes bibliographical references and index.
  ISBN-13: 978-1-55581-399-4 (hardcover)
  ISBN-10: 1-55581-399-2 (hardcover)
 1. Bacterial genetics. 2. Bacteriophages—Genetics. 3. Molecular genetics. I.
Champness, Wendy. II. Title.
  [DNLM: 1. Bacteria—genetics. 2. Bacteriophages—genetics. 3. Chromosomes,
Bacterial. 4. Genetics, Microbial—methods. 5. Molecular Biology—methods. QW 51
S6747m 2007]

  QH434.S59 2007
  572.8′293—dc22

                                                    2006101363

10 9 8 7 6 5 4 3 2 1

Cover photo: Localization of the cell division protein FtsZ in growing *Bacillus subtilis*
cells. FtsZ (green) is fused to green fluorescent protein, and cellular membranes are
stained with FM 4-64 (false-colored purple). FtsZ localizes as a ring at the future
division site that constricts during septal biogenesis. Courtesy of Jamie Gregory and
Kit Pogliano, University of California, San Diego.

Cover and interior design: Susan Brown Schmidler
Cover illustration: Terese Winslow

Contents

CHAPTER **3**

Bacterial Genetic Analysis: Forward and Reverse 139

CHAPTER 11

DNA Repair and Mutagenesis   459

CHAPTER 12

Regulation of Gene Expression: Operons   499

Preface

Thanks to the continued popularity of the textbook *Molecular Genetics of Bacteria*, we undertook a third edition. Much progress has been made in bacterial molecular genetics since the second edition. Some of it has come in the exciting new field of bacterial cell biology. Technical advances that have allowed the visualization of proteins moving within bacterial cells in real time have begun to reveal how incomprehensibly complex even these relatively simple cells are. A number of cellular phenomena once thought to be unique to eukaryotes are now known to have had their origins in bacteria, and this list continues to grow. These phenomena will be much easier to study with bacteria than with eukaryotes. Therefore, rather than having exhausted its promise, as has been predicted numerous times in the last quarter century, the approach of using bacteria as model systems for higher eukaryotes is apparently helping to usher in a new era in cell biology in which new principles could be discovered that are common to all organisms.

Discoveries with bacteria also continue to inform research in many other areas. The field of DNA repair and mutagenesis has long been such a source. The counterparts of bacterial genes for mismatch repair systems, repair systems for oxygen damage, excision repair systems, and mutagenic polymerases have all been found in the human genome, and deficiencies in them have been implicated in hereditary dispositions to some types of cancer. Insights gained from studying how recombination systems promote replication bypass of DNA damage in bacteria promise to be useful in understanding similar functions in eukaryotes, again with implications for human cancer and other diseases. Regulation by riboswitches, recently discovered in bacteria, is increasingly being implicated in gene regulation in eukaryotes.

Bacterial molecular genetics has also meshed nicely with structural biology and related biophysical experiments to broaden our understanding of proteins and structures in cells. Structural homologs of protein chaperonins, discovered in bacteria by using a combination of molecular genetic and structural approaches, are now known to exist in the eukaryotic cytoplasm, where they play a role in actin filament and microtubule formation.

Detailed genetic and biophysical studies on how RNA polymerase recognizes promoters and initiates transcription are now being confirmed and augmented by structural studies. This has also led to a much deeper understanding of how individual transcriptional activators and repressors work, since almost every step of the initiation of transcription is the target of some type of activator or repressor. Our picture of how the universally conserved SecYEG channel transports proteins into and through the cytoplasmic membrane has also begun to take shape due to a combination of bacterial molecular genetics and structural biology approaches. Molecular genetics and structural biology have also joined forces to further our understanding of the mechanism of action of transposases and recombinases and how these remarkable enzymes switch and rejoin DNA strands.

Bacterial genomics and the related techniques of microarray analysis of transcriptional profiles and proteomics have also built on the results of bacterial molecular genetics to provide many surprising insights. Genomics has revealed that protein secretion systems which transport proteins through the outer membrane of gram-negative bacteria are related to transformation and conjugation systems. Genomics has also revealed that there are a limited number of families of transcriptional regulators and recombinases and that the members of a family all seem to share an overall mechanism of action. Another surprising insight from genomics is the extent to which related bacteria that cause very different types of diseases, or occupy very different ecological niches, differ only in their exchangeable DNA elements, including prophages and genetic islands. This insight promises to be very useful for a number of practical applications, including designing therapeutic regimens for bacterial diseases. Genomics has also greatly expanded the known repertoire of small noncoding regulatory RNAs, the first of which were found by using molecular genetic techniques.

Yet another reason to keep abreast of developments in bacterial molecular genetics, independent of the field of study in biology, is that many of the most powerful techniques being used in other fields of biology are based on bacterial systems. Some examples are phage display, protein purification by affinity tags, gateway cloning to test the solubility of fusions of a protein to many different affinity tags to facilitate purification of the protein, seamless cloning to avoid adding extraneous DNA sequences to translational fusions, and recombineering to make site-directed mutational changes in cloned genes. Effective use of such technologies requires a good understanding of their basis.

The third edition follows roughly the same outline as the earlier editions. We have introduced more material on gram-positive bacteria, especially *Bacillus subtilis* but also other organisms such as *Streptomyces* and *Staphylococcus*, where appropriate. Whenever possible, we have further integrated the insights obtained from genomics and structural biology into those obtained from molecular genetics. As before, the first two chapters review DNA replication and gene expression, especially as they pertain to bacteria. They also include some of the concepts and techniques of molecular genetics, subjects that are also taught in biochemistry and molecular biology classes. But these are not merely review chapters, since they also include some of the most exciting new developments, including the latest on how bacterial chromosomes replicate, segregate, and partition; how RNA polymerase recognizes promoters and initiates transcription; and how protein export systems transfer proteins into and through membranes. Some earlier sections of these chapters could be assigned as review for students with a previous background in biochemistry and molecular biology, but later sections should be treated as novel material. The third chapter still contains the fundamentals of genetic analysis, but it has been substantially rewritten to be more pertinent to bacterial genetic analysis and now includes information on how genetic mapping data in bacteria are obtained and analyzed. In earlier editions, this information was distributed among other chapters that dealt with the particular means of DNA exchange, whether conjugation, transformation, or transduction. However, the basic concepts, such as selected versus unselected markers, are common to mapping with all of them, and, in our experience, the processes unique to genetic mapping in bacteria can be understood before exposure to the molecular details of the mechanism of gene exchange being used—as, in fact, they were before our understanding of the means of gene exchange reached its present molecular levels. As before, this chapter ends with a section outlining many of the techniques of reverse genetics in bacteria, including methods for gene knockouts, etc; this section has been substantially revised and updated. This topic is further developed throughout the book.

Later chapters cover the same specific topics in the same order as before, but they are substantially rewritten and updated to include new information on each of the topics. The final chapter is now devoted to compartmentalization in bacteria, including a detailed treatment of protein secretion systems in both gram-negative and gram-positive bacteria and a detailed discussion of what has been learned from *B. subtilis* sporulation about how different cell compartments communicate with each other during development. Throughout the book, we emphasize the experiments that underpin our understanding of the principles that we present, and we include examples from a wide array of bacteria: gram-negative and gram-positive, model organisms, and bacteria that are important to medicine and biotechnology. As in earlier editions, we do not mention the names of most investigators who have made major contributions to bacterial molecular genetics. We include only those names that have become icons in the field because they are associated with certain seminal experiments (e.g., Meselson and Stahl or Luria and Delbrück), models (e.g., Jacob and Monod), or a structure (e.g., Watson and Crick). Many other names are available in the suggested readings, where we give some of the original references to the developments under discussion, and in the credit lines for sources of figures and tables, which are now given at the end of the book.

We are indebted to a number of people who helped us in various ways. Some read sections of the book at our request and made valuable suggestions. Some, who have used the book for teaching, have pointed out ways to make it more useful for them and their students. Others have noticed factual errors or errors of omission and have pointed out references that helped us check our facts. Yet others furnished original figures that we could incorporate into the text. The list includes Cindy Arvidson, Dennis Arvidson, Nora Ausmees, Melanie Berkmen, Tom Bernhardt, Helmut Bertrand, Rob Britton, Bill Burkholder, Mark Buttner, Rich Calendar, Allan Campbell, Don Court, Keith Derbyshire, Alan Derman, Marie Elliot, Jeff Errington, Kim Findlay, Peter Geiduschek, Jim Golden, Sue Golden, Sue Gottesman, Gabriel Guarneros, Tina Henkin, Mike Kahn, Ken Kreuzer, Lee Kroos, Beth Lazazzera, Bebe Magee, Pete Magee, Ian Molineaux, Justin Nodwell, Greg Pettis, Patrick Piggot, Joe Pogliano, Kit Pogliano, Larry Reitzer, Bill Reznikoff, June Scott, Maggie Smith, Linc Sonenshein, Valley Stewart, Lynn Thomason, and Joanne Willey. However, in the end, any mistakes and omissions were all ours.

As with the first two editions, it was a great pleasure to work with the professionals at ASM Press. For the first edition, as neophyte authors, we depended on the expert advice of the director of ASM Press at the time, Patrick Fitzgerald. In preparing the second and third editions, we have been indebted to the current director, Jeff Holtmeier, for his unstinting enthusiasm, encouragement, and patience. We have also had the good fortune to work again with a number of the same professionals who did a masterful job with the first two editions, including Susan Birch, then the ASM Press production manager; Yvonne Strong, who copyedited the manuscript; Susan Brown Schmidler, who created the book and cover design; and Terese Winslow, who created the cover illustration. For the third edition, we especially thank the current production manager, Kenneth April, who directed the entire project and who worked with us with extraordinary professionalism, dedication, and patience. We also thank Patrick Lane of ScEYEnce Studios for bringing an attractive aestheticism to the rendering of our hand-drawn illustrations into the final figures.

Larry Snyder
Wendy Champness

Introduction

The goal of this textbook is to introduce the student to the field of bacterial molecular genetics. Bacteria are relatively simple organisms, and some are quite easy to manipulate in the laboratory. For these reasons, many methods in molecular biology and recombinant DNA technology have been developed around bacteria, and these organisms often serve as model systems for understanding cellular functions and developmental processes in more complex organisms. Much of what we know about the basic molecular mechanisms in cells, such as translation and replication, has originated with studies of bacteria. This is because such central cellular functions have remained largely unchanged throughout evolution. Ribosomes have a similar structure in all organisms, and many of the translation factors are highly conserved. The DNA replication apparatus of all organisms contains features in common such as sliding clamps and editing functions, which were first described in bacteria and their phages. Chaperones that help other proteins fold and topoisomerases that change the topology of DNA were first discovered in bacteria and their viruses, called phages. Studies of repair of DNA damage and mutagenesis in bacteria have also led the way to an understanding of such pathways in eukaryotes. Excision repair systems, mutagenic polymerases, and mismatch repair systems are remarkably similar in all organisms and have recently been implicated in some types of human cancers.

Also, recent evidence indicates that the cell biology of bacteria might be much more complex and more like that of eukaryotes than previously thought. For a long time it has been possible to observe the seemingly purposeful movement of constituents on the cytoskeleton within eukaryotic cells. However, bacterial cells, being much smaller, were thought to be merely "bags of enzymes" that could rely only on passive diffusion to move

1

their cellular constituents around. Now new technologies make it possible to observe movement within bacterial cells, revealing, for example, that some proteins involved in cell division and partitioning oscillate from one end of the cell to the other in a helical pattern during the cell cycle (see chapter 1), as though they were moving on mysterious helical tracks. Bacteria even have many of the structural proteins related to the proteins of the cytoskeleton, once thought to be limited to eukaryotes. For example, their cell division protein, FtsZ, is very similar structurally to the tubulins that make up microtubules, and forms similar dynamic tubules. Other proteins (called the Mre proteins), which help give bacterial cells their shape and structure, form actin-like filaments. Even intermediate filaments, which give eukaryotic cells some of their structure, have been found in some bacteria (see, for example, Ausmees et al., Daniel and Errington, Thanedar and Margolin, and van den Ent et al., Suggested Reading). We seem to be entering another era in biology similar to the early days of molecular genetics, when studies of bacteria led the way to the discovery of new principles of cell biology that are common to all organisms. The historical parallels are striking. Before the advent of molecular genetics, bacteria were not thought to have genetics like other organisms. When some clever people showed otherwise, the relative simplicity of bacteria allowed the development of molecular genetics and molecular biology, some of the most important scientific advances ever. Then bacteria were thought to be only bags of enzymes with little internal structure. Now they are being shown to have a complex dynamic cellular structure that has much in common with the cells of higher organisms. Again, studies of relatively simple bacteria, with their malleable genetic systems, might uncover basic principles of cell biology that are common to all organisms and that we can now only just imagine.

However, bacteria are not just important as laboratory tools to understand higher organisms; they are important and interesting in their own right. For instance, they play an essential role in the ecology of Earth. They are the only organisms that can "fix" atmospheric nitrogen, that is, convert N_2 to ammonia, which can be used to make nitrogen-containing cellular constituents such as proteins and nucleic acids. Without bacteria, the natural nitrogen cycle would be broken. Bacteria are also central to the carbon cycle because of their ability to degrade recalcitrant natural polymers such as cellulose and lignin. Bacteria and some types of fungi thus prevent Earth from being buried in plant debris and other carbon-containing material. Toxic compounds including petroleum, many of the chlorinated hydrocarbons, and other products of the chemical industry can also be degraded by bacteria. For this reason,

these organisms are essential in water purification and toxic waste cleanup. Moreover, bacteria produce most of the naturally occurring so-called greenhouse gases, such as methane and carbon dioxide, which are in turn used by other types of bacteria. This cycle helps maintain climate equilibrium. Bacteria have even had a profound effect on the geology of Earth, being responsible for some of the major iron ores and other types of deposits in Earth's crust.

Another unusual feature of bacteria and archaea (see below) is their ability to live in extremely inhospitable environments, many of which are devoid of life except for bacteria. These organisms are the only ones living in the Dead Sea, where the salt concentration in the water is very high. Some types of bacteria live in hot springs at temperatures close to the boiling point of water, and others survive in atmospheres devoid of oxygen, such as eutrophic lakes and swamps.

Bacteria that live in inhospitable environments sometimes enable other organisms to survive in those environments through symbiotic relationships. For example, symbiotic bacteria make life possible for tubular worms next to hydrothermal vents on the ocean floor, where living systems must use hydrogen sulfide in place of oxygen. In this symbiosis, the bacteria fix carbon dioxide by using the reducing power of the hydrogen sulfide given off by the hydrothermal vents, thereby furnishing food in the form of high-energy carbon compounds for the worms. Symbiotic cyanobacteria allow fungi to live in the Arctic tundra in the form of lichens. The bacterial partners in the lichens fix atmospheric nitrogen and make carbon-containing molecules through photosynthesis to allow their fungal partners to grow on the tundra in the absence of nutrient-containing soil. Symbiotic nitrogen-fixing *Rhizobium* and *Azorhizobium* spp. in the nodules on the roots of legumes and some other types of higher plants allow plants to grow in nitrogen-deficient soils. Other types of symbiotic bacteria digest cellulose to allow cows and other ruminant animals to live on a diet of grass. Chemiluminescent bacteria even generate light for squid and other marine animals, allowing individuals to find each other in the darkness of the deep ocean.

Bacteria are also worth studying because of their role in disease. They cause many human, plant, and animal diseases, and new diseases are continuously appearing. Knowledge gained from the molecular genetics of bacteria will help in the development of new ways to treat or otherwise control old diseases, as well as new ones.

Some bacteria also benefit us directly. The role of our commensal bacteria in human health is only beginning to be appreciated. It has been estimated that of the 10^{14} cells in a human body, only 10% are human! Of course bacterial cells are much smaller, but this shows how our bodies are adapted to live with an extensive bacterial flora, which help us digest food and avoid disease among other roles, many of which are yet to be uncovered.

Bacteria have also long been used to make many useful compounds such as antibiotics and chemicals such as benzene and citric acid. Bacteria and their phages are also the source of many of the useful enzymes used in molecular biology.

In spite of substantial progress, we have only begun to understand the bacterial world around us. Bacteria are the most physiologically diverse organisms on Earth, and the importance of bacteria to life on Earth and the potential uses to which bacteria can be put can only be guessed at. Thousands of different types of bacteria are known, and new insights into their cellular mechanisms and their applications constantly emerge from research with bacteria. Moreover, it is estimated that less than 1% of the types of bacteria living in the soil and other environments have ever been isolated; the undiscovered bacteria may have all manner of interesting and useful functions. Clearly, studies of bacteria will continue to be essential to our future efforts to understand, control, and benefit from the biological world around us, and bacterial molecular genetics will be an essential tool in these efforts. However, before discussing this field, we must first briefly discuss the evolutionary relationship of the bacteria to other organisms.

The Biological Universe

The Eubacteria

According to the current view, all organisms on Earth belong to three major divisions: the eubacteria, the archaea (formerly archaebacteria), and the eukaryotes. Figure 1 shows the microbiologists' view of the living world, where microbes provide most of the variety and eukaryotes occupy a relatively small niche. This is not so far-fetched a concept. Recent sequence data show that we differ from chimpanzees by only about 1.5% of our DNA sequence while 25 to 50% of the genes in a typical bacterium are unique to the species. Furthermore, the gene order is so similar among higher organisms that the order obtained from one species, for example the gene order of the human from the Human Genome Project, can be used to predict the order of genes in dogs, or even, to a lesser extent, in chickens. The order of genes in one species of a bacterium offers no such clue to the order of genes in another species of bacterium, unless the two are very closely related.

Most of the familiar bacteria such as *Escherichia coli*, *Streptococcus pneumoniae*, and *Staphylococcus aureus* are **eubacteria**. These organisms can differ greatly in

their physical appearance. Although most are single celled and rod shaped or spherical, some are multicellular and undergo complicated developmental cycles. The cyanobacteria (formerly called blue-green algae) are eubacteria, but they have chlorophyll and can be filamentous, which is why they were originally mistaken for algae. The antibiotic-producing actinomycetes, which include *Streptomyces* spp., are also eubacteria, but they form hyphae and stalks of spores, making them resemble fungi. Another eubacterial group, the *Caulobacter* spp., have both free-swimming and sessile forms that attach to surfaces through a holdfast structure. One of the most dramatic-appearing eubacteria of all is the genus *Myxococcus*, which can exist as free-living single-celled organisms but can also aggregate to form fruiting bodies much like slime molds. Eubacterial cells are usually much smaller than the cells of higher organisms, but a eubacterium that is 1 mm long, longer than even most eukaryotic cells, has been found. Many multiply by simple division, but one very large eubacterium, *Epulopiscium*, which lives in surgeonfish on the Great Barrier Reef, gives birth to a large number of live progeny. Because eubacteria come in so many shapes and sizes, they cannot be distinguished by their physical appearance but only by biochemical criteria such as the sequence of their ribosomal RNAs (rRNAs) and the absence of organelles.

GRAM-NEGATIVE AND GRAM-POSITIVE EUBACTERIA

The eubacteria can be further divided into two major subgroups, the **gram-negative** and **gram-positive** eubacteria. This division is based on the response to a test called the Gram stain. Gram-negative eubacteria retain little of the dye and are pink after this staining procedure, whereas gram-positive bacteria retain more of the dye and turn deep blue. The difference in staining reflects the fact that gram-negative eubacteria are surrounded by a thinner structure composed of both an inner and an outer membrane while the structure surrounding gram-positive bacteria is much thicker, consisting of a single membrane surrounded by a thicker wall. However, the difference between these groups seems to be more fundamental than the possession of an outer membrane. Individual types of gram-negative bacteria are in general more closely related to other gram-negative bacteria than they are to gram-positive bacteria, suggesting that the eubacteria separated into these two groups long before modern bacterial species arose.

The Archaea

The **archaea** (formerly called archaebacteria) are single-celled organisms that resemble eubacteria but are very different biochemically. The archaea are mostly

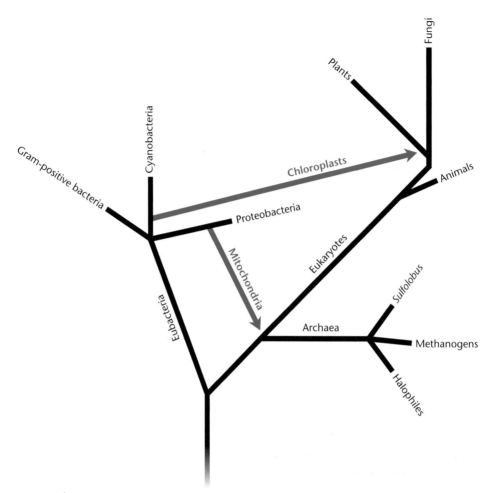

Figure 1 The evolutionary tree showing the points of divergence of eubacteria, archaea, and eukaryotes. The points of transfer of the eubacterial symbiotic organelles (mitochondria and chloroplasts) to eukaryotes are shown in purple.

represented by extremophiles (or "extreme-condition-loving" organisms) that, as their name implies, live under extreme conditions where other types of organisms cannot survive, such as at the very high temperatures in sulfur springs, at high pressures on the ocean floor, and at very high osmolality such as in the Dead Sea. Some of the archaea also perform relatively unusual biochemical functions such as making methane.

The separate classification of archaea from the true bacteria or eubacteria is fairly recent and is based mostly on the sequence of their rRNAs and the structures of their RNA polymerase and lipids (see Olsen et al., Suggested Reading). In fact, some evidence obtained by comparing the sequences of translation factors and membrane ATPases suggests that the archaea may be more closely related to eukaryotes than they are to eubacteria (Figure 1) (see Iwabe et al., Suggested Reading). The archaea themselves form a very diverse group of organisms and are sometimes divided into two kingdoms.

Although substantial progress is being made, much less is known about the archaea than about the eubacteria. The examples in this book come mostly from the eubacteria, which are referred to as just "bacteria" throughout.

The Eukaryotes

The **eukaryotes** are members of the third kingdom of organisms on Earth. They include organisms as seemingly diverse as plants, animals, and fungi. The name "eukaryotes" is derived from their nuclear membrane. They usually have a nucleus (the genus *Giardia* is a known exception), and the word "karyon" in Greek means "nut"—which is what the nucleus must have resembled to early cytologists. The eukaryotes can be unicellular like yeasts and protozoans and some types of algae, or they can be multicellular like plants and animals. In spite of their widely diverse appearances, lifestyles, and relative complexity, however, all eukaryotes are remarkably similar at the biochemical level, particularly in their pathways for macromolecular synthesis.

The Prokaryotes and the Eukaryotes

Organisms on Earth are also sometimes divided into prokaryotes and eukaryotes. This classification is based on whether the organism has a nucleus and some other organelles. In contrast to eukaryotes, both archaea and eubacteria lack a nuclear membrane, which caused them to be lumped together as **prokaryotes**, which means "before the nucleus." They were given this name because they were thought to be the most primitive organisms, existing before the development of "higher organisms," or eukaryotes, which have a nucleus.

The presence or absence of a nuclear membrane greatly influences the mechanisms available to make proteins in the cell. Messenger RNA (mRNA) synthesis and translation can occur simultaneously in prokaryotes, since no nuclear membrane separates the ribosomes (which synthesize proteins) from the DNA. However, in most eukaryotes, the DNA is physically separated from the ribosomes. Therefore, mRNA made in the nucleus must be transported through the nuclear membrane before it can be translated into protein in the cytoplasm, and transcription and translation cannot occur simultaneously.

Besides lacking a nucleus, prokaryotic cells lack many other cellular constituents common to eukaryotes, including mitochondria and chloroplasts, which is not surprising since they are the origin of mitochondria and chloroplasts (see below). They also lack such visible organelles as the Golgi apparatus and the endoplasmic reticulum. The absence of mitochondria, chloroplasts, and most organelles generally gives prokaryotic cells a much simpler appearance under the microscope.

MITOCHONDRIA AND CHLOROPLASTS

All eukaryotic cells contain mitochondria. In addition, plant cells and some unicellular eukaryotic cells contain chloroplasts. The mitochondria of eukaryotic cells are the sites of efficient ATP generation through respiration, and the chloroplasts are the sites of photosynthesis.

Recent evidence indicates that the mitochondria and chloroplasts of eukaryotes are descended from free-living eubacteria that formed a symbiosis with eukaryotes. In fact, these organelles resemble bacteria in many ways. For instance, they contain DNA that encodes the components of oxidative phosphorylation and photosynthesis as well as rRNAs, and transfer RNAs (tRNAs). Even more striking, the mitochondrial and chloroplast rRNA proteins, as well as the membranes of the organelles, more closely resemble those of the eubacteria than they do those of eukaryotes. In any case, the similarities are too great to leave any doubt that these organelles were originally derived from eubacteria.

Not only is there little doubt of the eubacterial origin of mitochondria and chloroplasts, but also it is possible to guess the eubacterial families to which they are most closely related. Comparisons of the sequences of highly conserved organelle genes, such as those for the rRNAs, with those of eubacteria suggest that mitochondria are descended from the proteobacteria and that chloroplasts are descended from the cyanobacteria (Figure 1).

Mitochondria and chloroplasts may have come to be associated with early eukaryotic cells, perhaps a type of archaea, when these cells engulfed eubacteria to take advantage of their superior energy-generating systems or the ability to obtain energy from light through photosynthesis. The eukaryotic cell contributed its ability to manage large DNAs to the symbiosis. The engulfed bacteria eventually lost many of their own genes, which moved to the chromosome, from where they are transported back into the organelle. They had then lost their autonomy and become permanent symbionts of the eukaryotic cells. This process may still be going on; some modern-day eukaryotes called dinoflagellates are known to engulf cyanobacteria when they are in the light during the day, allowing the dinoflagellates to photosynthesize, and then discard them at night, when they have no use for them.

What Is Genetics?

Genetics can be simply defined as the manipulation of DNA to study cellular and organismal functions. Since DNA encodes all of the information needed to make the cell and the complete organism, the effects of changing this molecule can give clues to the normal functions of the cell and organism.

Before the advent of methods for manipulating DNA in the test tube, the only genetic approaches available for studying cellular and organismal functions were those of classical genetics. In this type of analysis, mutants (i.e., individuals that differ from the normal, or wild-type, members of the species by a certain observable attribute, or phenotype) that are altered in the function being studied are isolated. The changes in the DNA, or mutations, responsible for the altered function are then localized in the chromosome by genetic crosses. The mutations are then grouped into genes by allelism tests to determine how many different genes are involved. The functions of the genes can then sometimes be deduced from the specific effects of the mutations on the organism. The ways in which mutations in genes involved in a biological system can alter the biological system provide clues to the normal functioning of the system.

Classical genetic analyses continue to contribute greatly to our understanding of developmental and cellular biology. A major advantage of the classical genetic approach is that mutants altered in a function can be isolated and characterized without any a priori understanding of the molecular basis of the function. Classical genetic analysis is also often the only way to determine how many gene products are involved in a function and, through suppressor analysis, to find other genes whose products may interact either physically or functionally with the products of these genes.

The development of **molecular genetic techniques** has greatly expanded the range of methods available for studying genes and their functions. These techniques include methods for isolating DNA and identifying the regions of DNA that encode particular functions, as well as methods for altering or mutating DNA in the test tube and then returning the mutated DNA to cells to determine the effect of the mutation on the organism.

The approach of first cloning a gene and then altering it in the test tube before reintroducing it into the cells to determine the effect of the alterations is sometimes called **reverse genetics** and is essentially the reverse of a classical genetic analysis. In classical genetics, a gene is known to exist only because a mutation in it has caused an observable change in the organism. With the molecular genetic approach, a gene can be isolated and mutated in the test tube without any knowledge of its function. Only after the mutated gene has been returned to the organism does its function become apparent.

Rather than one approach supplanting the other, molecular genetics and classical genetics can be used to answer different types of questions, and the two approaches often complement each other. In fact, the most remarkable insights into biological functions have sometimes come from a combination of classical and molecular genetic approaches.

Bacterial Genetics

In bacterial genetics, genetic techniques are used to study bacteria. Applying genetic analysis to bacteria is no different in principle from applying it to other organisms. However, the methods that are available differ greatly. Some types of bacteria are relatively easy to manipulate genetically. As a consequence, more is known about some bacteria than is known about any other type of organism. Some of the properties of bacteria that facilitate genetic experiments are listed below.

Bacteria Are Haploid

One of the major advantages of bacteria for genetic studies is that they are **haploid.** This means that they have only one copy or **allele** of each gene. This property makes it much easier to identify cells with a particular type of mutation.

In contrast, most higher organisms are diploid, with two alleles of each gene, one on each homologous chromosome. Most mutations are recessive, which means that they do not cause a phenotype in the presence of a normal copy of the gene. Therefore, in diploid organisms, most mutations have no effect unless both copies of the gene in the two homologous chromosomes have the mutation. Backcrosses between different organisms with the mutation are usually required to produce offspring with the mutant phenotype, and even then only some of the progeny of the backcross have the mutated gene in both homologous chromosomes. With a haploid organism such as a bacterium, however, most mutations have an immediate effect and there is no need for backcrosses.

Short Generation Times

Another advantage of some bacteria for genetic studies is that they have very short generation times. The **generation time** is the length of time the organism takes to reach maturity and produce offspring. If the generation time of an organism is too long, it can limit the number of possible experiments. Some strains of the bacterium *E. coli* can reproduce themselves every 20 min under ideal conditions. With such rapid multiplication, cultures of the bacteria can be started in the morning and the progeny can be examined later in the day.

Asexual Reproduction

Another advantage of bacteria is that they multiply asexually, by cell division. Sexual reproduction, in which individuals of the same species must mate with each other to give rise to progeny, can complicate genetic experiments because the progeny are never identical to their parents. To achieve purebred lines of a sexually reproducing organism, a researcher must repeatedly cross the individuals with their relatives. However, if the organism multiplies asexually by cell division, all the progeny are genetically identical to their parent and to each other. Genetically identical organisms are called **clones**. Some lower eukaryotes such as yeasts and some types of plants such as water hyacinths can also multiply asexually to form clones. Identical twins, formed from the products of division of an egg after it has been fertilized, are clones of each other. Recently, some mammals have been cloned by transplanting a somatic cell into the ovary, where its surroundings rarely cause it to revert to an egg cell and multiply to form a clone of the organism. However, bacteria form clones of themselves every time they divide.

Colony Growth on Agar Plates

Genetic experiments often require that numerous individuals be screened for a particular property. Therefore, it helps if large numbers of individuals of the species being studied can be propagated in a small space.

With some types of bacteria, thousands, millions, or even billions of individuals can be screened on a single agar-containing petri plate. Once on an agar plate, these bacteria divide over and over again, with all the progeny remaining together on the plate until a visible lump or colony has formed. Each colony is composed of millions of bacteria, all derived from the original bacterium and hence all clones of the original bacterium.

Colony Purification

The ability of some types of bacteria to form colonies through the multiplication of individual bacteria on plates allows colony purification of bacterial strains and mutants. If a mixture of bacteria containing different mutants or strains is placed on an agar plate, individual mutant bacteria or strains in the population each multiply to form colonies. However, these colonies may be too close together to be separable or may still contain a mixture of different strains of the bacterium. If the colonies are picked and the bacteria are diluted before replating, discrete colonies that result from the multiplication of individual bacteria may appear. No matter how crowded the bacteria were on the original plate, a pure strain of the bacterium can be isolated in one or a few steps of colony purification.

Serial Dilutions

To count the number of bacteria in a culture or to isolate a pure culture, it is often necessary to obtain discrete colonies of the bacteria. However, because bacteria are so small, a concentrated culture contains billions of bacteria per milliliter. If such a culture is plated directly on a petri plate, the bacteria all grow together and discrete colonies do not form. Serial dilutions offer a practical method for diluting solutions of bacteria before plating to obtain a measurable number of discrete colonies. The principle is that if smaller dilutions are repeated in succession, they can be multiplied to produce the total dilution. For example, if a solution is diluted in three steps by adding 1 ml of the solution to 99 ml of water, followed by adding 1 ml of this dilution to another 99 ml of water and finally by adding 1 ml of the second dilution to another 99 ml of water, the final dilution is $10^{-2} \times 10^{-2} \times 10^{-2} = 10^{-6}$, or one in a million. To achieve the same dilution in a single step, 1 ml of the original solution would have to be added to 1,000 liters (about 250 gallons) of water. Obviously, it is more convenient to handle three solutions of 100 ml each than to handle a solution of 250 gallons, which weighs about 1,000 lb!

Selections

Probably the major advantage of bacterial genetics is the opportunity to do selections, by which very rare mutants and other types of strains can be isolated. To select a rare strain, billions of the bacteria are plated under conditions where only the desired strain, not the bulk of the bacteria, can grow. In general, these conditions are called the selective conditions. For example, a nutrient may be required by most of the bacteria but not by the strain being selected. Agar plates lacking the nutrient then present selective conditions for the strain, since only the strain being selected multiplies to form a colony in the absence of the nutrient. In another example, the desired strain may be able to multiply at a temperature that would kill most of the bacteria. Incubating agar plates at that temperature would provide the selective condition. After the strain has been selected, a colony of the strain can be picked and colony purified away from other contaminating bacteria under the same selective conditions.

The power of selections with bacterial populations is awesome. Using a properly designed selection, a single bacterium can be selected from among billions placed on an agar plate. If we could apply such selections to humans, we could find one individual in the entire human population of Earth.

Storing Stocks of Bacterial Strains

Most types of organisms must be continuously propagated; otherwise they age and die off. Propagating organisms requires continuous transfers and replenishing of the food supply, which can be very time-consuming. However, many types of bacteria can be stored in a dormant state and therefore do not need to be continuously propagated. The conditions used for storage depend on the type of bacteria. Some bacteria sporulate and so can be stored as dormant spores. Others can be stored by being frozen in glycerol or being dried. Storing organisms in a dormant state is particularly convenient for genetic experiments, which often require the accumulation of large numbers of mutants and other strains. The strains remain dormant until the cells are needed, at which time they can be revived.

Genetic Exchange

Genetic experiments with an organism usually require some form of exchange of DNA or genes between members of the species. Most types of organisms on Earth are known to have some means of genetic exchange, which presumably accelerates evolution and increases the adaptability of a species.

Exchange of DNA from one bacterium to another can occur in one of three ways. In transformation, DNA released from one cell enters another cell of the same species. In **conjugation**, plasmids, which are small autonomously replicating DNA molecules in bacterial cells, transfer DNA from one cell to another. Finally, in **transduction**, a bacterial virus accidentally picks up DNA from a cell it has infected and injects this DNA into another cell. The ability to exchange DNA between strains of a bacterium makes possible genetic crosses and complementation tests as well as the tests essential to genetic analysis.

Phage Genetics

Some of the most important discoveries in genetics have come from studies with viruses that infect bacteria; these viruses are called **bacteriophages, or phages** for short. Phages are not alive; instead, they are just genes wrapped in a protective coat of protein and/or membrane, as are all viruses. Because phages are not alive, they cannot multiply outside a bacterial cell. However, if a phage encounters a type of bacterial cell that is sensitive to phages, the phage, or at least its DNA or RNA, enters the cell and directs it to make more phage.

Phages are usually identified by the holes, or **plaques**, they form in layers of sensitive bacteria. In fact, the name "phage" (Greek for "eat") derives from these plaques, which look like eaten-out areas. A plaque can form when a phage is mixed with large numbers of susceptible bacteria and the mixture is placed on an agar plate. As the bacteria multiply, one may be infected by the phage, which multiplies and eventually breaks open or **lyses** the bacterium, releasing more phage. As the surrounding bacteria are infected, the phage spread, even as the bacteria multiply to form an opaque layer called a **bacterial lawn**. Wherever the original phage infected the first bacterium, the plaque disrupts the lawn, forming a clear spot on the agar. Despite its empty appearance, this spot contains millions of the phage.

Phages offer many of the same advantages for genetics as bacteria. Thousands or even millions of phages can be put on a single plate. Also, like bacterial colonies, each plaque contains millions of genetically identical phage. By analogy to the colony purification of bacterial strains, individual phage mutants or strains can be isolated from other phages through **plaque purification**.

Phages Are Haploid

Phages are, in a sense, haploid, since they usually have only one copy of each gene. As with bacteria, this property makes isolation of phage mutants relatively easy, since all mutants immediately exhibit their phenotypes without the need for backcrosses.

Selections with Phages

Selection of rare strains of a phage is possible; as with bacteria, it requires conditions under which only the desired phage strain can multiply to form a plaque. For phage, these selective conditions may be a bacterial host in which only the desired strain can multiply or a temperature at which only the phage strain being selected can multiply. Note that the bacterial host must be able to multiply under the same selective conditions; otherwise, a plaque cannot form.

As with bacteria, selections allow the isolation of very rare strains or mutants. If selective conditions can be found for the strain, millions of phages can be mixed with the bacterial host and only the desired strain multiplies to form a plaque. A pure strain can then be obtained by picking the phage from the plaque and plaque purifying the strain under the same selective conditions.

Crosses with Phages

Phage strains can be crossed very easily. The same cells are infected with different mutants or strains of the phage. The DNA of the two phages is then in the same cell, where the molecules can interact genetically with each other, allowing genetic manipulations such as gene mapping and allelism tests.

A Brief History of Bacterial Molecular Genetics

Because of the ease with which they can be handled, bacteria and their phages have long been the organisms of choice for understanding basic cellular phenomena, and their contributions to this area of study are almost countless. The following chronological list should give a feeling for the breadth of these contributions and the central position that bacteria have occupied in the development of modern molecular genetics. Some original references are given at the end of the chapter under Suggested Reading.

Inheritance in Bacteria

In the early part of this century, biologists agreed that inheritance in higher organisms follows Darwinian principles. According to Charles Darwin, changes in the hereditary properties of organisms occur randomly and are passed on to the progeny. In general, the changes that happen to be beneficial to the organism are more apt to be passed on to subsequent generations.

With the discovery of the molecular basis for heredity, Darwinian evolution now has a strong theoretical foundation. The properties of organisms are determined by the sequence of their DNA, and as the organisms multiply, changes in this sequence sometimes occur randomly and without regard to the organism's environment. However, if a random change in the DNA happens to be beneficial in the situation in which the organism finds itself, the organism has an improved chance of surviving and reproducing.

As late as the 1940s, many bacteriologists thought that inheritance in bacteria was different from inheritance in other organisms. It was thought that rather than enduring random changes, bacteria could adapt to their environment by some sort of "directed" change and that the adapted organisms could then somehow pass on the change to their offspring. Such opinions were encouraged by the observations of bacteria growing under selective conditions. For example, in the presence of an antibiotic, all the bacteria in the culture soon become resistant to the antibiotic. It seemed as though the resistant bacterial mutants appeared in response to the antibiotic.

One of the first convincing demonstrations that inheritance in bacteria follows Darwinian principles was made in 1943 by Salvador Luria and Max Delbrück (see Suggested Reading). Their work demonstrated that particular phenotypes, in their case resistance to a virus, occur randomly in a growing population, even in the absence of the virus. By the directed-change or adaptive-mutation hypothesis, the resistant mutants should have appeared only in the presence of the virus.

The demonstration that inheritance in bacteria follows the same principles as inheritance in higher organisms set the stage for the use of bacteria in studies of basic genetic principles common to all organisms.

Transformation

As discussed at the beginning of the Introduction, most organisms exhibit some mechanism for exchanging genes. The first demonstration of genetic exchange in bacteria was made by Fred Griffith in 1928. He was studying two variants of pneumococci, now called *Streptococcus pneumoniae*. One variant formed smooth-appearing colonies on plates and was pathogenic in mice. The other variant formed rough-appearing colonies on plates and did not kill mice. Only live, and not dead, smooth-colony-forming bacteria could cause disease, since the disease requires that the bacteria multiply in the infected mice. However, when Griffith mixed dead smooth-colony formers with live rough-colony formers and injected the mixture into mice, the mice became sick and died. Moreover, he isolated live smooth-colony formers from the dead mice. Apparently, the dead smooth-colony formers were "transforming" some of the live rough-colony formers into the pathogenic, smooth-colony-forming type. The "transforming principle"

given off by the dead smooth-colony formers was later shown to be DNA, since addition of purified DNA from the dead smooth-colony formers to the live rough-colony formers in a test tube transformed some members of the rough type to the smooth type (see Avery et al., Suggested Reading). This method of exchange is called transformation, and this experiment provided the first direct evidence that genes are made of DNA. Later experiments by Alfred Hershey and Martha Chase in 1952 (see Suggested Reading) showed that phage DNA alone is sufficient to direct the synthesis of more phages.

Conjugation

In 1946, Joshua Lederberg and Edward Tatum (see Suggested Reading) discovered a different type of gene exchange in bacteria. When they mixed some strains of *E. coli* with other strains, they observed the appearance of recombinant types that were unlike either parent. Unlike transformation, which requires only that DNA from one bacterium be added to the other bacterium, this means of gene exchange requires direct contact between two bacteria. It was later shown to be mediated by plasmids and is called conjugation.

Transduction

In 1953, Norton Zinder and Joshua Lederberg discovered yet a third mechanism of gene transfer between bacteria. They showed that a phage of *Salmonella enterica* serovar Typhimurium could carry DNA from one bacterium to another. This means of gene exchange is called transduction and is now known to be quite widespread.

Recombination within Genes

At the same time, experiments with bacteria and phages were also contributing to the view that genes were linear arrays of nucleotides in the DNA. By the early 1950s, recombination had been well demonstrated in higher organisms, including fruit flies. However, recombination was thought to occur only between mutations in different genes and not between mutations in the same gene. This led to the idea that genes were like "beads on a string" and that recombination is possible between the "beads," or genes, but not within a gene. In 1955, Seymour Benzer disproved this hypothesis by using the power of phage genetics to show that recombination is possible within the *r*II genes of phage T4. He mapped numerous mutations in the *r*II genes, thereby demonstrating that genes are linear arrays of mutable sites in the DNA. Later experiments with other phage and bacterial genes showed that the sequence of nucleotides in the DNA directly determines the sequence of amino acids in the protein product of the gene.

Semiconservative DNA Replication

In 1953, James Watson and Francis Crick published their structure of DNA. One of the predictions of this model is that DNA replicates by a semiconservative mechanism, in which specific pairing occurs between the bases in the old and the new DNA strands, thus essentially explaining heredity. In 1958, Matthew Meselson and Frank Stahl used bacteria to confirm that DNA replicates by this semiconservative mechanism.

mRNA

The existence of mRNA was also first indicated by experiments with bacteria and phage. In 1961, Sydney Brenner, François Jacob, and Matthew Meselson used phage-infected bacteria to show that ribosomes are the site of protein synthesis and confirmed the existence of a "messenger" RNA that carries information from the DNA to the ribosome.

The Genetic Code

Also in 1961, phages and bacteria were used by Francis Crick and his collaborators to show that the genetic code is unpunctuated, three lettered, and redundant. These researchers also showed that not all possible codons designated an amino acid and that some were nonsense. These experiments laid the groundwork for Marshall Nirenberg and his collaborators to decipher the genetic code, in which a specific three-nucleotide set encodes one of 20 amino acids. The code was later verified by the examination of specific amino acid changes due to mutations in the lysozyme gene of phage T4.

The Operon Model

François Jacob and Jacques Monod published their operon model for the regulation of the lactose utilization genes of *E. coli* in 1961 as well. They proposed that a repressor blocks RNA synthesis on the *lac* genes unless the inducer, lactose, is bound to the repressor. Their model has served to explain gene regulation in other systems, and the *lac* genes and regulatory system continue to be used in molecular genetic experiments, even in systems as far removed from bacteria as animal cells and viruses.

Enzymes for Molecular Biology

The early 1960s saw the start of the discovery of many interesting and useful bacterial and phage enzymes involved in DNA and RNA metabolism. In 1960, Arthur Kornberg demonstrated the synthesis of DNA in the test tube by an enzyme from *E. coli*. The next year, a number of groups independently demonstrated the synthesis of RNA in the test tube by RNA polymerases from bacteria. From this time on, other interesting and useful enzymes for molecular biology were isolated from bacteria and their phages, including polynucleotide kinase, DNA ligases, topoisomerases, and many phosphatases.

From these early observations, the knowledge and techniques of molecular genetics exploded. For example, in the early 1960s, techniques were developed for detecting the hybridization of RNA to DNA and DNA to DNA on nitrocellulose filters. These techniques were used to show that RNA is made on only one strand in specific regions of DNA, which later led to the discovery of promoters and other regulatory sequences. By the late 1960s, restriction endonucleases had been discovered in bacteria and shown to cut DNA in specific places (see Linn and Arber, Suggested Reading). By the early 1970s, these restriction endonucleases were being exploited to introduce foreign genes into *E. coli* (see Cohen et al., Suggested Reading), and by the late 1970s, the first human gene had been expressed in a bacterium. Also in the late 1970s, methods to sequence DNA by using enzymes from phages and bacteria were developed.

In 1988, a thermally stable DNA polymerase from a thermophilic bacterium was used to invent the technique called the polymerase chain reaction (PCR). This extremely sensitive technique allows the amplification of genes and other regions of DNA, facilitating their cloning and study.

These examples illustrate that bacteria and their phages have been central to the development of molecular genetics and recombinant DNA technology. Contrast the timing of these developments with the timing of comparable major developments in physics (early 1900s) and chemistry (1920s and 1930s), and you can see that molecular genetics is arguably the most recent major conceptual breakthrough in the history of science.

What's Ahead

This textbook emphasizes how molecular genetic approaches can be used to solve biological problems. As an educational experience, the methods used and the interpretation of experiments are at least as important as the conclusions drawn. Therefore, whenever possible, the experiments that led to the conclusions are presented. The first two chapters, of necessity, review the concepts of macromolecular synthesis that are essential to understanding bacterial molecular genetics. However, they also introduce more current material including some of the most interesting recent advances in bacterial cell biology. Chapter 1, besides reviewing the basics of DNA replication and the techniques of molecular biology, presents some recent advances in understanding the processes of chromosome segregation and partitioning and their coordination with cell division and also discusses recent advances in determining the bacterial cytoskeleton. Chapter 2, in addition to reviewing the basics of protein synthesis, presents more current developments concerning protein folding and transport. We recommend that the review sections of these chapters be assigned as background reading, especially for students with an adequate biochemistry background, and that more time be spent on the more current topics. Chapter 3, similarly, reviews basic genetic principles, but with a special emphasis on bacterial genetics. Students are not likely to get some of this material in more general genetics courses, at least not in the same depth. The chapter also includes more current applications, including gene knockouts, reverse genetics, and saturation genetics. Chapters 4 through 14 deal with more specific topics and the techniques that can be used to study them, with particular emphasis on recent evidence concerning the relatedness of seemingly disparate topics. The last two chapters on global regulation and cell compartmentalization deal with the exciting current topics of cell-cell communication and protein secretion out of bacterial cells and into eukaryotic cells, and they finish with the paradigm for bacterial development, sporulation in *Bacillus subtilis*. We hope that this textbook will help put modern molecular genetics in an historical perspective, bring the reader up to date on current advances in bacterial molecular genetics, and position the reader to understand future developments in this exciting and rapidly progressing field of science.

SUGGESTED READING

Ausmess, N., J. R. Kuhn, and C. Jacobs-Wagner. 2003. The bacterial cytoskeleton: an intermediate filament-like function in cell shape. *Cell* **115**:705–713.

Avery, O. T., C. M. MacLeod, and M. McCarty. 1944. Studies on the chemical nature of the substance inducing transformation of pneumococcal types. I. Induction of transformation by a desoxyribonucleic acid fraction isolated from pneumococcus type III. *J. Exp. Med.* **79**:137.

Brenner, S., F. Jacob, and M. Meselson. 1961. An unstable intermediate carrying information from genes to ribosomes for protein synthesis. *Nature* (London) **190**:576.

Cairns, J., G. S. Stent, and J. D. Watson. 1966. *Phage and the Origins of Molecular Biology.* Cold Spring Harbor Laboratory Press, Cold Spring Harbor, N.Y.

Cohen, S. N., A. C. Y. Chang, H. W. Boyer, and R. B. Helling. 1973. Construction of biologically functional bacterial plasmids *in vitro. Proc. Natl. Acad. Sci. USA* **70**:3240–3244.

Crick, F. H. C., L. Barnett, S. Brenner, and R. J. Watts-Tobin. 1961. General nature of the genetic code for proteins. *Nature* (London) **192**:1227–1232.

Daniel, R. A., and J. Errington. 2003. Control of cell morphogenesis in bacteria: two distinct ways to make a rod shaped cell. *Cell* **113**:767–776.

Hershey, A. D., and M. Chase. 1952. Independent functions of viral protein and nucleic acids in growth of bacteriophage. *J. Gen. Physiol.* **36**:393.

Iwabe, N., K. Kuma, M. Hasegawa, S. Osawa, and T. Miyata. 1989. Evolutionary relationship of archaebacteria, eubacteria and eukaryotes inferred from phylogenetic trees of duplicated genes. *Proc. Natl. Acad. Sci. USA* **86**:9355–9359.

Jacob, F., and J. Monod. 1961. Genetic regulatory mechanisms in the synthesis of proteins. *J. Mol. Biol.* **3**:3183.

Lederberg, J., and E. L. Tatum. 1946. Gene recombination in *E. coli. Nature* (London) **158**:558.

Linn, S., and W. Arber. 1968. Host specificity of DNA produced by *Escherichia coli*. X. *In vitro* restriction of phage fd replicative form. *Proc. Natl. Acad. Sci. USA* **59**:1300–1306.

Luria, S., and M. Delbrück. 1943. Mutations of bacteria from virus sensitivity to virus resistance. *Genetics* **28**:491–511.

Meselson, M., and F. W. Stahl. 1958. The replication of DNA in *Escherichia coli. Proc. Natl. Acad. Sci. USA* **44**:671.

Nirenberg, M. W., and J. H. Matthei. 1961. The dependence of cell-free protein synthesis in *E. coli* upon naturally occurring or synthetic polynucleotides. *Proc. Natl. Acad. Sci. USA* **47**: 1588–1602.

Olby, R. 1974. *The Path to the Double Helix.* Macmillan Press, London, United Kingdom.

Olsen, G. J., C. R. Woese, and R. Overbeek. 1994. The winds of (evolutionary) change: breathing new life into microbiology. *J. Bacteriol.* **176**:1–6.

Schrodinger, E. 1944. *What is Life? The Physical Aspect of the Living Cell.* Cambridge University Press, Cambridge, United Kingdom.

Thaneder, S., and W. Margolin. 2004. FtsZ exhibits rapid movement and oscillation waves in helix-like patterns in *Escherichia coli. Curr. Biol.* **14**:1167–1173.

van den Ent, F., L. A. Amos, and J. Löwe. 2001. Prokaryotic origin of the actin cytoskeleton. *Nature* (London) **413**: 39–44.

Watson, J. D. 1968. *The Double Helix.* Atheneum, New York, N.Y.

Zinder, N. D., and J. Lederberg. 1952. Genetic exchange in *Salmonella. J. Bacteriol.* **64**:679–699.

CHAPTER **1**

The Bacterial Chromosome: DNA Structure, Replication, and Segregation

DNA Structure

The science of molecular genetics began with the determination of the structure of DNA. Experiments with bacteria and phages (i.e., viruses that infect bacteria) in the late 1940s and early 1950s, as well as the presence of DNA in chromosomes of higher organisms, had implicated this macromolecule as the hereditary material (see Introduction). In the 1930s, biochemical studies of the base composition of DNA by Erwin Chargaff established that the amount of guanine always equals the amount of cytosine and that the amount of adenine always equals the amount of thymine, independent of the total base composition of the DNA. In the early 1950s, X-ray diffraction studies by Rosalind Franklin and Maurice Wilkins showed that DNA is a double helix. Finally, in 1953, Francis Crick and James Watson put together the chemical and X-ray diffraction information in their famous model of the structure of DNA. This story is one of the most dramatic in the history of science and has been the subject of many historical treatments, some of which are listed at the end of this chapter.

Figure 1.1 illustrates the Watson-Crick structure of DNA, in which two strands wrap around each other to form a double helix. These strands can be extremely long, even in a simple bacterium, extending up to 1 mm—a thousand times longer than the bacterium itself. In a human cell, the strands that make up a single chromosome (which is one DNA molecule) are hundreds of millimeters, or many inches, long.

The Deoxyribonucleotides

If we think of DNA strands as chains, deoxyribonucleotides form the links. Figure 1.2 shows the basic structure of deoxyribonucleotides, called deoxynucleotides for short. Each is composed of a **base**, a **sugar**, and a

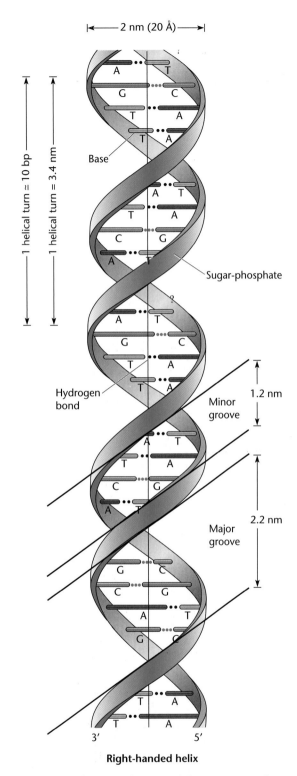

Figure 1.1 A schematic drawing of the Watson-Crick structure of DNA, showing the helical sugar-phosphate backbones of the two strands held together by hydrogen bonding between the bases. Also shown are the major and minor grooves and the dimensions of the helix.

phosphate group. DNA bases are **adenine** (A), **cytosine** (C), **guanine** (G), and **thymine** (T), which have either one or two rings, as shown in Figure 1.2. The bases with two rings (A and G) are the purines, and those with only one ring (T and C) are pyrimidines. A third pyrimidine, uracil (U), replaces thymine in RNA. The carbons and nitrogens making up the rings of the bases are numbered sequentially, as shown in the figure. All four DNA bases are attached to the five-carbon sugar deoxyribose. This sugar is identical to ribose, which is found in RNA, except that it does not have an oxygen attached to the second carbon, hence the name deoxyribose. The carbons in the sugar of a nucleotide are also numbered 1, 2, 3, and so on, but they are primed to distinguish them from the carbons in the bases (Figure 1.2). The nucleotides also have one or more phosphate groups attached to a carbon of the deoxyribose sugar, as shown. The carbon to which the phosphate group is attached is indicated, although if the group is attached to the 5′ carbon (the usual situation), the carbon to which it is attached is often not stipulated.

The components of the deoxynucleotides have special names. A **deoxynucleoside** (rather than -*tide*) is a base attached to a sugar but lacking a phosphate. Without phosphates, the four deoxynucleosides are called **deoxyadenosine, deoxycytidine, deoxyguanosine,** and **deoxythymidine.** As shown in Figure 1.2, the deoxynucleotides have one, two, or three phosphates attached to the sugar and are known as deoxynucleoside *mono*phosphates, *di*phosphates, or *tri*phosphates, respectively. The individual deoxynucleoside monophosphates, called deoxyguanosine monophosphate, etc., are often abbreviated dGMP, dAMP, dCMP, and dTMP, where the d stands for *deoxy*; the G, A, C, or T stands for the base; and the MP stands for *mono*phosphate. In turn, the *di*phosphates and *tri*phosphates are abbreviated dGDP, dADP, dCDP, dTDP, and dGTP, dATP, dCTP, dTTP, respectively. Collectively, the four deoxynucleoside triphosphates are often referred to as dNTPs.

The DNA Chain

Phosphodiester bonds join each deoxynucleotide link in the DNA chain. As shown in Figure 1.3, the phosphate attached to the last (5′) carbon of the deoxyribose sugar of one nucleotide is attached to the third (3′) carbon of the sugar of the next nucleotide, thus forming one strand of nucleotides connected 5′ to 3′, 5′ to 3′, etc.

The 5′ and 3′ Ends

Clearly, at the ends of DNA will be nucleotides that are linked to only one other nucleotide. At one end of the DNA chain, a nucleotide will have a phosphate attached to its 5′ carbon that does not connect it to another

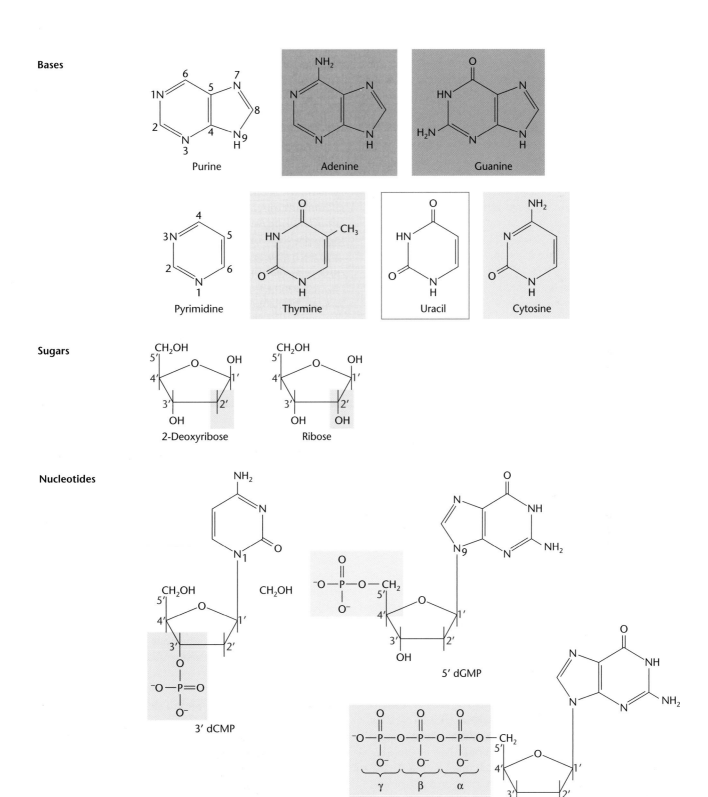

Figure 1.2   The chemical structures of deoxyribonucleotides, showing the bases and sugars and how they are assembled into a deoxyribonucleotide.

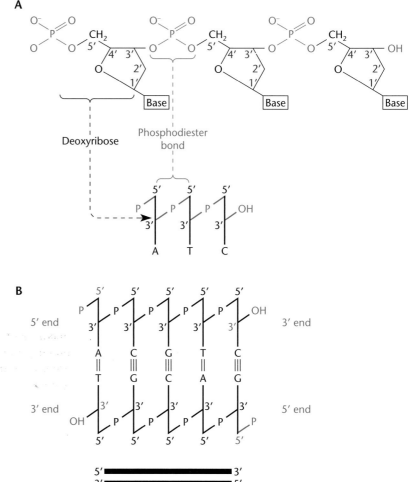

Figure 1.3 (A) Schematic drawing of a DNA chain showing the 3′-to-5′ attachment of the phosphates to the sugars, forming phosphodiester bonds. (B) Two strands of DNA bind at the bases in an antiparallel arrangement of the phosphate-sugar backbones.

nucleotide. This end of the strand is called the **5′ end** or the **5′ phosphate end** (Figure 1.3B). On the other end, the last nucleotide lacks a phosphate at its 3′ carbon. Because it has only a hydroxyl group (the OH in Figure 1.3B), this end is called the **3′ end** or the **3′ hydroxyl end**.

Base Pairing

The sugar and phosphate groups of DNA form what is often called a **backbone** to support the bases, which jut out from the chain. This structure allows the bases to form hydrogen bonds with each other, thereby holding together two separate nucleotide chains (Figure 1.3B). This role of the four bases was first suggested by their ratios in DNA.

First, Erwin Chargaff found that no matter the source of the DNA, the concentration of guanine (G) always equals the concentration of cytosine (C) and the concentration of adenine (A) always equals the concentration of thymine (T). These ratios, named Chargaff's

rules, gave Watson and Crick one of the essential clues to the structure of DNA. They proposed that the two strands of the DNA are held together by specific hydrogen bonding between the bases in opposite strands, as shown in Figure 1.4. Thus, the amounts of A and T and of C and G are always the same because A's pair only with T's and G's pair only with C's to hold the DNA strands together. Each A-and-T pair and each G-and-C pair in DNA is called a **complementary base pair,** and the sequences of two strands of DNA are said to be **complementary** if one strand always has a T where there is an A in the other strand and a G where there is a C in the other strand.

It did not escape the attention of Watson and Crick that the complementary base-pairing rules essentially explain heredity. If A pairs only with T and G pairs only with C, then each strand of DNA can replicate to make a complementary copy of itself, so that the two replicated DNAs will be exact copies of each other. Offspring

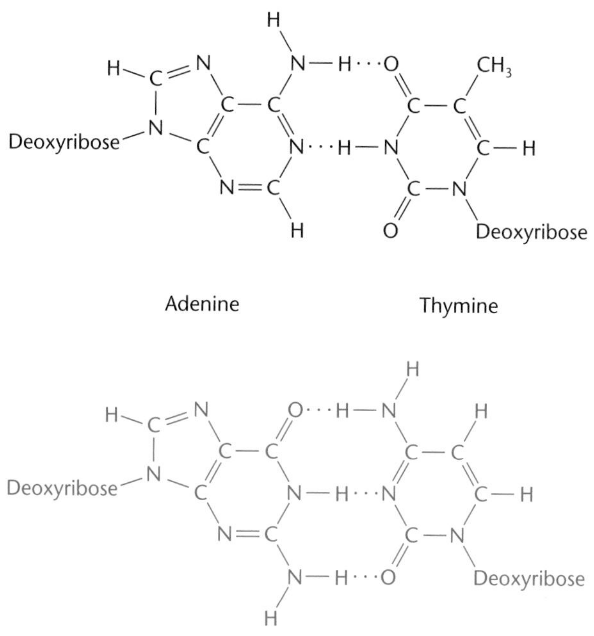

Figure 1.4 The two complementary base pairs found in DNA. Two hydrogen bonds form in adenine-thymine base pairs. Three hydrogen bonds form in guanine-cytosine base pairs.

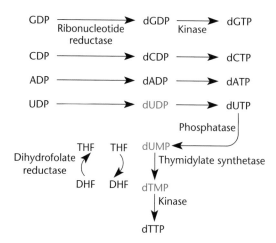

Figure 1.5 The pathways for synthesis of the deoxynucleotides from the ribonucleotides. Some of the enzymes referred to in the text are identified. THF, tetrahydrofolate; DHF, dihydrofolate.

containing these DNAs would have the same sequence of nucleotides in their DNAs as their parents and so would be exact copies of their parents.

Antiparallel Construction

As mentioned at the beginning of this section, the complete DNA molecule consists of two long chains wrapped around each other in a double helix (Figure 1.1). The double-stranded molecule can be thought of as being like a circular staircase, with the alternating phosphates and deoxyribose sugars forming the railings and the bases connected to each other forming the steps. However, the two chains run in opposite directions, with the phosphates on one strand attached 5′ to 3′, 5′ to 3′, etc., to the sugars and those on the other strand attached 3′ to 5′, 3′ to 5′, etc. This arrangement is called **antiparallel**. In addition to phosphodiester bonds running in opposite directions, the antiparallel construction causes the 5′ phosphate end of one strand and the 3′ hydroxyl end of the other to be on the same end of the double-stranded DNA molecule (Figure 1.3B).

The Major and Minor Grooves

Because the two strands of the DNA are wrapped around each other to form a double helix, the helix has two grooves between the two strands (Figure 1.1). One of these grooves is wider than the other, so it is called the **major groove**. The other, narrower groove is called the **minor groove**. Most of the modifications to DNA that are discussed in this and later chapters occur in the major groove of the helix.

The Mechanism of DNA Replication

The molecular details of DNA replication are probably similar in all organisms on Earth. The basic process of replication involves **polymerizing, or linking, the nucleotides of DNA into long chains or strands, using the sequence on the other strand as a guide**. Because the nucleotides must be made before they can be put together into DNA, the nucleotides are called the **precursors** of DNA synthesis.

Deoxyribonucleotide Precursor Synthesis

The precursors of DNA synthesis are the four deoxyribonucleoside triphosphates, dATP, dGTP, dCTP, and dTTP. The triphosphates are synthesized from the corresponding ribose nucleoside diphosphates by the pathway shown in Figure 1.5. In the first step, the enzyme **ribonucleotide reductase** reduces (i.e., removes an oxygen from) the ribose sugar to produce the deoxyribose sugar by changing the hydroxyl group at the 2′ position (the second carbon) of the sugar to a hydrogen. Then an enzyme known as a **kinase** adds a phosphate to the deoxynucleoside diphosphate to make the deoxynucleoside triphosphate precursor.

The deoxynucleoside triphosphate dTTP is synthesized by a somewhat different pathway from the other three. The first step is the same. Ribonucleotide reductase synthesizes the nucleotide dUDP (deoxyuridine diphosphate) from the ribose UDP. However, from then on, the pathway differs. A phosphate is added to make dUTP, and the dUTP is converted to dUMP by a phosphatase that removes the phosphates. This molecule is then converted to dTMP by the enzyme thymidylate synthetase, using tetrahydrofolate to donate a methyl group. Kinases then add two phosphates to the dTMP to make the precursor dTTP.

Deoxyribonucleotide Polymerization

The complex process of DNA replication involves many enzymes and other cellular components. In the end, complementary copies of each of the extremely long strands of a double-stranded DNA must be made. In this section, we discuss the obstacles that DNA replication must overcome and how the various functions involved overcome these obstacles.

DNA POLYMERASES

The properties of the DNA polymerases, the enzymes that actually join the deoxynucleotides together to make the long chains, are the best guides to an understanding of the replication of DNA. These enzymes make DNA by linking one deoxynucleotide to another to generate a long chain of DNA. This process is called DNA polymerization, hence the name DNA polymerases.

Figure 1.6 shows the basic process of DNA polymerization by DNA polymerase. The DNA polymerase attaches the first phosphate (called α) of one deoxynucleoside triphosphate to the 3′ carbon of the sugar of the next deoxynucleoside triphosphate, in the process releasing the last two phosphates (called the β and γ phosphates) of the first deoxynucleoside triphosphate to produce energy for the reaction. Then the α phosphate of another deoxynucleoside triphosphate is attached to the 3′ carbon of this deoxynucleotide, and the process continues until a long chain is synthesized.

DNA polymerases also need a template strand to direct the synthesis of the new strand. As mentioned in the base-pairing section, complementary base pairing dictates that wherever there is a T in the template strand, an A is inserted in the strand being synthesized, and so forth according to the base-pairing rules. The DNA polymerase can move only in the 3′-to-5′ direction on the template strand, linking deoxynucleotides in the new strand in the 5′-to-3′ direction. When replication is completed, the product is a new double-stranded DNA with antiparallel strands, one strand of which is the old template strand and one strand of which is the newly synthesized strand.

There are two DNA polymerases which participate in normal DNA replication in E. coli; they are called DNA polymerase III and DNA polymerase I (see Table 1.1).

DNA polymerase III is a large complex composed of the enzyme which polymerizes the nucleotides in a complex with accessory proteins, which make new DNA at the replication fork. DNA polymerase I is one of the enzymes responsible for removing RNA primers in the lagging strand and filling the gaps between Okazaki fragments (see below). It also plays a role in DNA repair, as discussed in chapter 11.

NUCLEASES

Enzymes which degrade DNA strands by breaking the phosphodiester bonds are just as important in replication as the enzymes which polymerize DNA by forming phosphodiester bonds between the nucleotides. These bond-breaking enzymes, called nucleases, can be grouped into two major categories. One type can initiate breaks in the middle of a DNA strand and so are called endonucleases, from a Greek word meaning "within," and the other type can remove nucleotides only from the ends of DNA strands and so are called exonucleases, from a Greek word meaning "outside." Exonucleases can in turn be divided into two groups. Some exonucleases can degrade only from the 3′ end of a DNA strand, degrading DNA in the 3′-to-5′ direction. These are called 3′ exonucleases; one example of their activity is their role in the editing function associated with DNA polymerase I and III, which is discussed below. Other exonucleases, called 5′ exonucleases, degrade DNA strands only from the 5′ end, an example being the 5′ exonuclease activity of DNA polymerase I, which removes RNA primers during replication. Nucleases can also leave either 3′ or 5′ phosphate deoxynucleotides, depending on which side of the phosphodiester bond they cut.

DNA LIGASES

DNA ligases are enzymes which form phosphodiester bonds between the ends of chains of DNA, another reaction that DNA polymerases cannot perform for themselves. During replication, these enzymes join the 5′ phosphate at the end of one DNA chain to the 3′ hydroxyl at the end of another chain to make a longer continuous chain.

PRIMASES

Another type of enzyme which performs a reaction that DNA polymerases cannot perform for themselves is the group of primases. These are enzymes that make RNA primers to initiate the synthesis of new strands of DNA. DNA polymerases cannot start the synthesis of a new strand of DNA; they can only attach deoxynucleotides to a preexisting 3′ OH group. The 3′ OH group to which DNA polymerase adds a deoxynucleotide is called the primer (Figure 1.7).

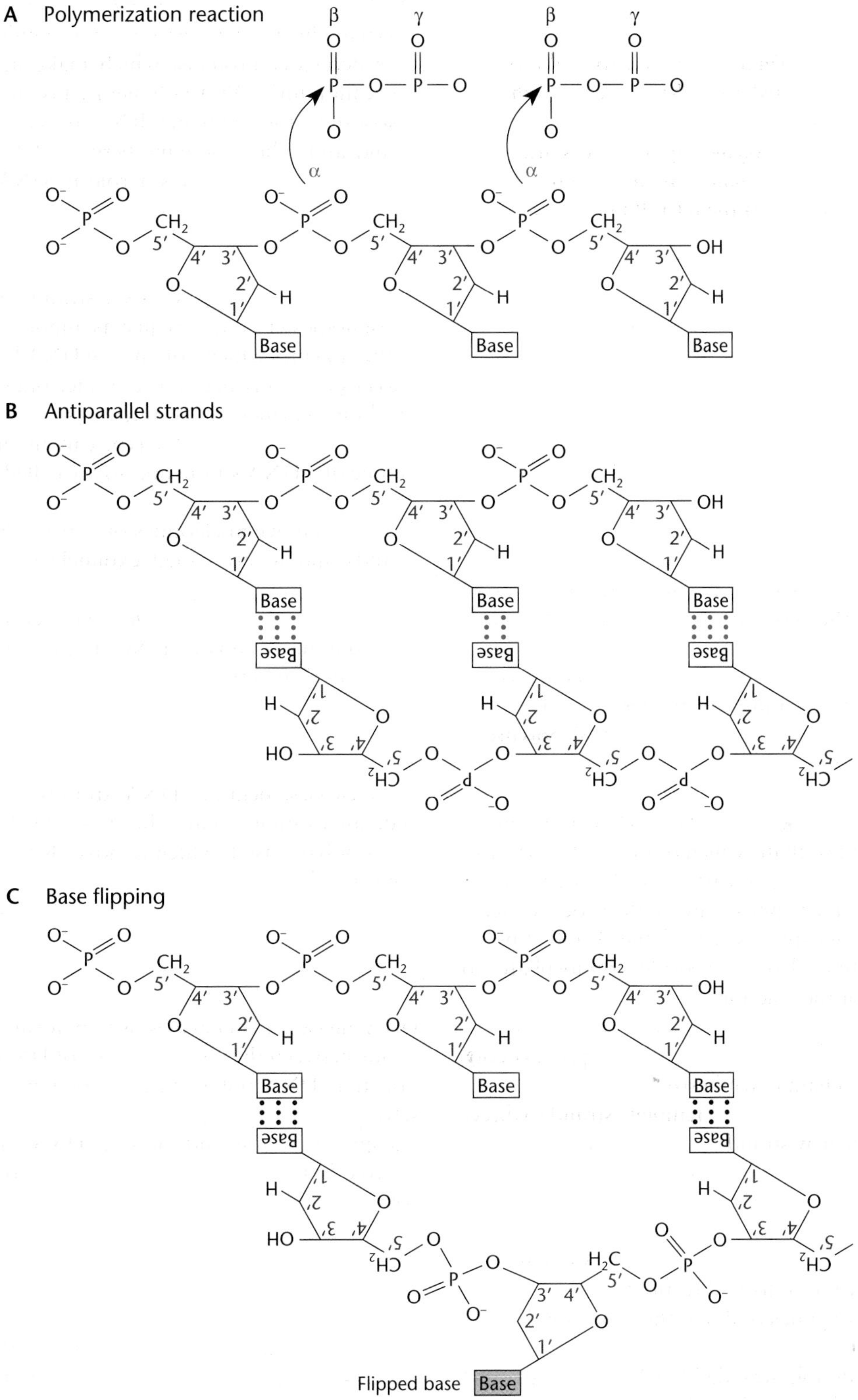

A   Polymerization reaction

B   Antiparallel strands

C   Base flipping

Flipped base   Base

Figure 1.6   Features of DNA. (A) Polymerization of the deoxynucleotides during DNA synthesis. The β and γ phosphates of each deoxynucleoside triphosphate are cleaved off to give energy for the polymerization reaction. (B) The strands of DNA are antiparallel. (C) A single base can be flipped out from the double helix, which could be important in recombination and repair.

TABLE 1.1	Proteins involved in *E. coli* DNA replication	
Protein	**Gene**	**Function**
DnaA	*dnaA*	Initiator protein; primosome (priming complex) formation
DnaB	*dnaB*	DNA helicase
DnaC	*dnaC*	Delivers DnaB to replication complex
SSB	*ssb*	Binding to single-stranded DNA
Primase	*dnaG*	RNA primer synthesis
DNA ligase	*lig*	Sealing DNA nicks
DNA gyrase		Supercoiling
α	*gyrA*	Nick closing
β	*gyrB*	ATPase
DNA Pol I	*polA*	Primer removal; gap filling
DNA Pol III (holoenzyme, contains 11 polypeptides)		
α	*dnaE*	Polymerization
ε	*dnaQ*	3′-to-5′ editing
RNase H	*rnhA*	Removes RNA primers
θ	*holE*	Present in core (αεθ)
β	*dnaN*	Sliding clamp
τ*ᵃ*	*dnaX*	Organizes complex; joins leading and lagging DNA Pol III
γ*ᵇ*	*dnaX*	Binds clamp loaders and SSB protein
δ	*holA*	Clamp loading
δ′	*holB*	Clamp loading
χ	*holC*	Binds SSB
ψ	*holD*	Binds SSB

ᵃ Full-length product of the *dnaX* gene.

ᵇ Shorter product of the *dnaX* gene produced by translational frameshifting (see Box 2.3).

The requirement of a primer for DNA polymerase creates an apparent dilemma in DNA replication. When a new strand of DNA is synthesized, there is no DNA upstream (i.e., on the 5′ side) to act as a primer. The cell usually solves this problem by using RNA as the primer to initiate the synthesis of new strands. Unlike DNA polymerase, RNA polymerase does not require a primer to initiate the synthesis of new strands. The RNA primers that are used to initiate the synthesis of new strands of DNA are made either by the RNA polymerase (which makes all the other RNAs, including mRNA, tRNA, and rRNA) or by the special enzymes called primases. During DNA replication, special enzymes recognize and remove the RNA primer (see below).

ACCESSORY PROTEINS

Ten different proteins travel with the DNA polymerase as part of a **DNA replication complex** as the polymerase moves along the template strand. These proteins are called **DNA polymerase accessory proteins**; together with the polymerizing activity, they form the **DNA polymerase III**

Figure 1.7 Functions of the primer and template in DNA replication. (A) The DNA polymerase adds deoxynucleotides to the 3′ end of the primer by using the template strand to direct the selection of each base. (B) Simple illustration of 5′-to-3′ DNA synthesis. The dotted line indicates the primer.

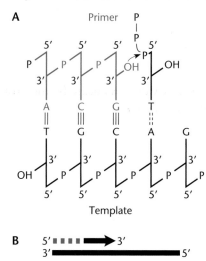

holoenzyme. One of these accessory proteins forms a clamp that helps keep the DNA polymerase from falling off the template strand. In bacteria, this clamp, called the β clamp, consists of two polypeptide products of the *dnaN* gene that form a ring around the DNA. By forming a ring around the DNA, the β clamp cannot come off easily and release the DNA polymerase; this allows replication to continue for long distances without the DNA polymerase being released. However, the DNA polymerase must also replicate the lagging strand, where it periodically comes off and hops ahead to synthesize another Okazaki fragment (see below). Other proteins therefore make up the **clamp loader** that is required to help the β clamp come off periodically and then reload on the DNA. The clamp loader is a complicated structure that consists of three γ (or two τ and one γ) proteins and one each of δ, δ′, and ψ, which form a five-sided ring. Yet other proteins which travel with the DNA polymerase are exonucleases that serve an **editing function** to correct mistakes made by the DNA polymerase (see the section on editing, below) and other proteins of unknown function. Table 1.1 lists many of the DNA replication proteins, the gene encoding them, and their functions.

Semiconservative Replication

The process of replication described above is called **semiconservative replication** because each time a DNA molecule replicates, the two old strands are *conserved* to become part of two new DNA molecules. Each of the new molecules consists of one old conserved strand and one newly synthesized strand, so that each new molecule is only partly conserved (or semiconserved). The next time the DNA molecule replicates, each strand serves as a template, becoming the conserved strand in a new double-stranded molecule.

THE MESELSON-STAHL EXPERIMENT

The semiconservative mechanism was suggested by the structure of DNA. However, other mechanisms of DNA replication are also possible. Soon after Watson and Crick published their structure for DNA, Matthew Meselson and Frank Stahl performed an experiment that showed that DNA does replicate by the proposed semiconservative mechanism.

One prediction of the semiconservative-replication hypothesis is that after replication, each DNA molecule should have one newly synthesized strand and one old strand from the original molecule. If it could be demonstrated that newly synthesized DNA had one new strand and one old strand, this prediction would have been fulfilled. Figure 1.8 illustrates the details of their experiment. First, they chose the heavy isotopes of nitrogen, carbon, and hydrogen, three of the types of atoms contained

in DNA, as markers for new strands. DNA synthesized from these heavy isotopes is more dense than DNA synthesized from the normal atoms.

To incorporate the heavy isotopes into newly synthesized DNA, they grew *E. coli* for about the time the cells took to divide on a medium containing the heavy isotopes instead of one containing the normal isotopes. Then they extracted the DNA from the cells and analyzed its density by density gradient equilibrium centrifugation. As shown in Figure 1.8, the density of the DNA is reflected by the position of the DNA in the gradient; heavier DNA, composed of atoms of the heavy isotopes, bands farther down. If the DNA replicates by a semiconservative mechanism, then after one cycle of replication in the presence of the heavy isotopes, one strand should be made from precursors with the heavy isotopes and the newly replicated molecule should then consist of one heavy strand and one light strand. Consequently, its density is intermediate between those of light DNA, with no heavy isotopes, and heavy DNA, in which both new strands contain heavy isotopes. After a short time of replication, Meselson and Stahl did observe DNA of intermediate density, composed of one light strand and one heavy strand; thus, their results supported the semiconservative mechanism of replication.

Replication of Double-Stranded DNA

The replication of most long DNA molecules such as bacterial DNA begins at one point and moves in both directions from there. In the process, the two old strands of DNA are separated and used as templates to synthesize new strands. The structure where the two strands are separated and the new synthesis is occurring is referred to as the **replication fork.** The DNA polymerase enzyme cannot separate the two strands of a bacterial chromosome, replicate them, and separate the daughter DNAs by itself. Many other proteins are required for replication, as discussed in this section.

HELICASES AND HELIX-DESTABILIZING PROTEINS

One task the DNA polymerase cannot perform in the replication of double-stranded DNA is the separation of the strands of DNA at the replication fork. The strands of DNA must be separated for the two strands to serve as templates. The bases of the DNA are inside of the double helix, where they are not available to pair with the incoming deoxynucleotides, to direct which nucleotide will be inserted at each step. The strands of DNA are separated by proteins called **DNA helicases** (see Singleton and Wigley, Suggested Reading). Some of these proteins form a ring around one strand of the DNA and suck the strand through the ring; others "snowplow" along the DNA, separating the strands as they go. It takes a lot of

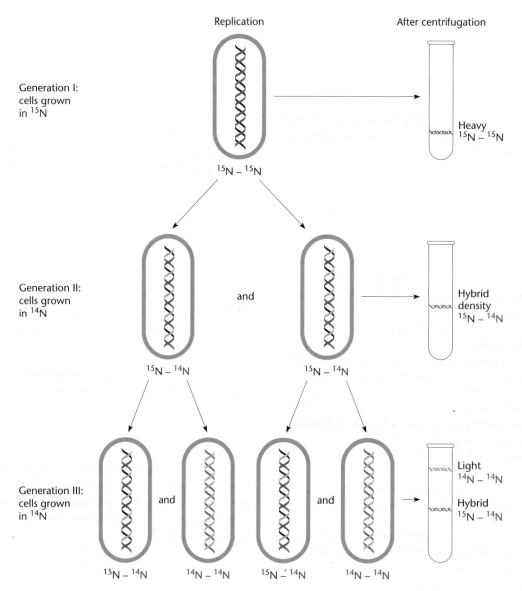

Figure 1.8   The Meselson-Stahl experiment. Synthesis of a "light" complementary strand gives a double-stranded DNA of hybrid light and heavy density. The light-light, heavy-light, and heavy-heavy DNAs can be separated by equilibrium density centrifugation.

energy to separate the strands of DNA, and helicases cleave a lot of ATP to form ADP in the process. There are about 20 different helicases in *E. coli*, and each helicase works in only one direction, either the 3′-to-5′ or the 5′-to-3′ direction. The large hexameric DnaB helicase that normally separates the strands of DNA ahead of the replication fork is a large doughnut-shaped protein composed of six polypeptide products of the *dnaB* gene. It sucks one strand through its hole in the 5′-to-3′ direction as it opens the strands of DNA ahead of the replication fork (see Figure 1.12). Because it forms a ring around the DNA at the fork, another protein, DnaC, is required to

associate it with the replication fork. Some other helicases are discussed in later chapters in connection with recombination and repair.

Once the strands of DNA have been separated, they also must be prevented from coming back together (or from annealing to themselves if they happen to be complementary over short regions). Separation of the strands is maintained by proteins called **helix-destabilizing proteins** or single-strand-binding proteins (SSB). These are proteins that bind preferentially to single-stranded DNA and prevent double-stranded helical DNA from reforming prematurely.

OKAZAKI FRAGMENTS AND THE REPLICATION FORK

A major problem in replicating double-stranded DNA is created because the two strands are antiparallel. As already discussed, in one strand the phosphates connect the sugars 3' to 5' and in the other strand they connect the sugars 5' to 3'. However, DNA polymerase can move only in the 3'-to-5' direction on the template strand, synthesizing the new DNA in the 5'-to-3' direction. How can the replication fork move in one direction on a double-stranded DNA and make complementary copies of both strands at the same time? Because of the antiparallel structure, the DNA polymerase on one of the two strands would have to be moving in the wrong direction overall.

This problem is overcome by replicating the two strands differently (Figure 1.9). On one template strand, DNA polymerase III initiates synthesis from an RNA primer and moves along the template DNA in the 3'-to-5' direction. This newly synthesized strand is referred to as the **leading strand** (Figure 1.9). To replicate the other strand, the DNA polymerase must wait until the DNA strands have been separated by the DnaB helicase before it can load on the DNA, which is why this is called the **lagging-strand** synthesis (Figure 1.9). This DNA polymerase makes short pieces called **Okazaki fragments** in the opposite direction to that in which the fork is moving. Synthesis of each Okazaki fragment requires a new RNA primer, about 10 to 12 nucleotides long. In *E. coli*, these primers are synthesized by DnaG primase, which produces a new RNA primer about once every 2 kb, recognizing the sequence 3'-GTC-5' and beginning synthesis opposite the T. These RNA primers are then used to prime DNA synthesis by DNA polymerase III, which continues until it encounters another piece being synthesized further along the DNA (Figure 1.10). However, before these short pieces can be joined to make a long, continuous strand of DNA, the short RNA primers must be removed. Most of the RNA primer is removed by an enzyme called RNase H, which removes the RNA strand of a DNA:RNA double helix (Table 1.1). Then DNA polymerase I comes into play. Using its concerted 5' exonuclease and DNA polymerase activities (Figure 1.11), DNA polymerase I removes what remains of each RNA primer and replaces it with DNA, using the upstream Okazaki fragment as a primer. The Okazaki fragments are then joined together by DNA ligase as the replication fork moves on, as shown in Figure 1.9. By using RNA rather than DNA to prime the synthesis of Okazaki fragments, the cell lowers the mistake rate (see below).

What actually happens at the replication fork is more complicated than suggested by this simple picture. For one thing, this picture ignores the overall topological restraints on the DNA that is replicating. The **topology** of a molecule refers to its position in space. Because the circular DNA is very long and its strands are wrapped around each other, pulling the two strands apart introduces stress into other regions of the DNA in the form of **supercoiling.** Unless the two strands of DNA were free to rotate around each other, supercoiling would cause the chromosome to look like a telephone cord wound up on itself. To relieve this stress, enzymes called **topoisomerases** undo the supercoiling ahead of the replication fork. DNA supercoiling and topoisomerases are discussed later in the chapter.

THE TROMBONE MODEL OF CHROMOSOME REPLICATION

The picture of the two strands of DNA replicating independently as it is usually drawn is too simple. Rather than replicating independently, the two DNA polymerase III holoenzymes replicating the leading and lagging strands are joined to each other through their τ subunits (Table 1.1). To accommodate the fact that the two DNA polymerases must move in opposite directions and still remain joined, the lagging-strand template loops out as the Okazaki fragment is synthesized. The loop is then relaxed as the sliding clamp releases the lagging-strand polymerase, allowing the DNA polymerase to hop ahead to the next RNA primer to begin synthesizing the next Okazaki fragment (Figure 1.12). According to the current model, the old β clamp is left behind and a new one is assembled by the clamp loader at the site of the new RNA primer. The DNA polymerase is released when it gets to a nick, which tells it that it has completely synthesized the fragment. This model has been referred to as the "trombone" model of replication because the loops forming and contracting at the replication fork resemble the "oom-paa-oom-paa" of the musical instrument. It has not been proven for bacterial replication but has been proven for T4 phage (see chapter 7). The situation is probably similar in all bacteria and even higher organisms, although in some other bacteria including *Bacillus subtilis* and in eukaryotes, different DNA polymerases are used to replicate the leading and lagging strands (see Dervyn et al., Suggested Reading).

THE GENES FOR REPLICATION PROTEINS

Most of the genes for replication proteins have been found by isolating mutants defective in DNA replication but not RNA or protein synthesis. Since a mutant cell that cannot replicate its DNA will die, any mutation (for definitions of mutants and mutations, see "Replication Errors" below and chapter 3) that inactivates a gene whose product is required for DNA replication will kill the cell. Therefore, for experimental purposes, only a

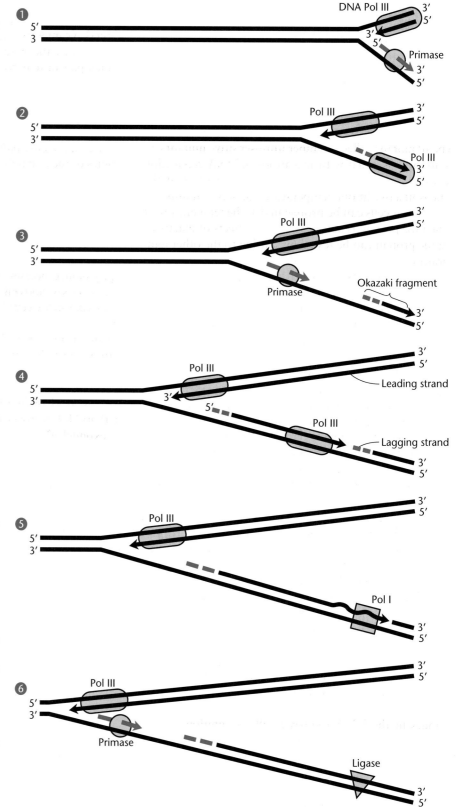

Figure 1.9 Discontinuous synthesis of one of the two strands of DNA during chromosome replication. (1) DNA Pol III replicates one strand, and the primase synthesizes RNA on the other strand in the opposite direction. (2) Pol III extends the RNA primer to synthesize an Okazaki fragment. (3) The primase synthesizes another RNA primer. (4) Pol III extends this primer until it reaches the previous primer. (5) Pol I removes the first RNA primer and replaces it with DNA. (6) DNA ligase seals the nick to make a continuous DNA strand, and the process continues. The strand that is synthesized continuously is the leading strand; the strand that is synthesized discontinuously is the lagging strand.

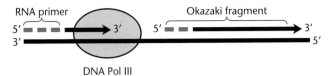

Figure 1.10   Synthesis of short Okazaki fragments from an RNA primer.

type of mutant called a **temperature-sensitive mutant** can be usefully isolated with mutations in DNA replication genes. These are mutants in which the product of the gene is inactive at one temperature but active at another. The mutant cells can be propagated at the temperature at which the protein is active. Then, the effects of inactivating the protein can be tested by shifting to the other temperature. The molecular basis of temperature-sensitive mutants is discussed in more detail in chapter 3.

The immediate effect of a temperature shift on a mutant with a mutation in a DNA replication gene depends on whether the product of the gene is continuously required for replication at the replication forks or is involved only in the initiation of rounds of replication. For example, if the mutation is in a gene for DNA polymerase III or in the gene for the DnaG primase, replication ceases immediately. However, if the temperature-sensitive mutation is in a gene whose product is required only for initiation of DNA replication, for example, the gene for DnaA or DnaC (see the section on initiation of chromosome replication, below), the replication rate for the population will slowly decline. Unless the cells have been somehow synchronized in their cell cycle, each cell is at a different stage of replication, with some cells having just finished a round of replication and other cells having just begun a new round. Cells in which rounds of chromosome replication were under way at the time of the temperature shift will complete their replication but not start a new round. Therefore, the rate of replication decreases until the rounds of replication in all the cells are completed.

Replication Errors

To maintain the stability of a species, replication of the DNA must be almost free of error. Otherwise, changes in the DNA sequence called **mutations** occur, and these changes are passed on to subsequent generations. Depending on where these changes occur, they can severely alter the protein products of genes or other cellular functions. To avoid such instability, the cell has mechanisms that reduce the error rate.

As DNA replicates, the wrong base is sometimes inserted into the growing DNA chain. For example,

Figure 1.13 shows the incorrect incorporation of a T opposite a G. Such a base pair in which the bases are paired wrongly is called a **mismatch**. Mismatches can occur when the bases take on forms called **tautomers**, which pair differently from the normal form of the base (see chapter 3). After the first replication in Figure 1.13, the mispaired T is usually in its normal form and pairs correctly with an A, causing a GC-to-AT change in the sequence of one of the two progeny DNAs and thus changing the base pair at that position on all subsequent copies of the mutated DNA molecule.

Editing

One way the cell reduces mistakes during replication is through **editing** functions. Sometimes these functions are performed by separate proteins, and sometimes they are part of the DNA polymerase itself. Editing proteins are aptly named because they go back over the newly replicated DNA looking for mistakes, recognizing and removing incorrectly inserted bases (Figure 1.14). If the last nucleotide inserted in the growing DNA chain creates a mismatch, the editing function stops the replication until the offending nucleotide is removed. The replication then continues, inserting the correct nucleotide. Because the DNA chain grows in the 5'-to-3' direction, the last nucleotide added is at the 3' end. The enzyme activity that removes this nucleotide is therefore called a **3' exonuclease.** The editing proteins probably recognize a mismatch because the mispairing (between T and G in the example) causes a minor distortion in the structure of the double-stranded helix of the DNA.

In some DNA polymerases, for example, DNA polymerase I, the 3' exonuclease editing activity is part of the DNA polymerase itself. However, in the DNA polymerase that replicates the bacterial chromosome, the editing functions are accessory proteins encoded by separate genes whose products travel along the DNA with the DNA polymerase during replication. In *E. coli*, the 3' exonuclease editing function is encoded by the *dnaQ* gene (Table 1.1), and *dnaQ* mutants, also called *mutD* mutants (i.e., cells with a mutation in this gene that inactivates the 3' exonuclease function), show much higher rates of spontaneous mutagenesis than do cells containing the **wild-type,** or normally functioning, *dnaQ* gene product. Because of their high spontaneous mutation rates, *mutD* mutants of *E. coli* are often used to introduce random mutations into plasmids and bacteriophages.

RNA PRIMERS AND EDITING

The importance of the editing functions in lowering the number of mistakes during replication may explain why DNA replication is primed by RNA rather than by DNA.

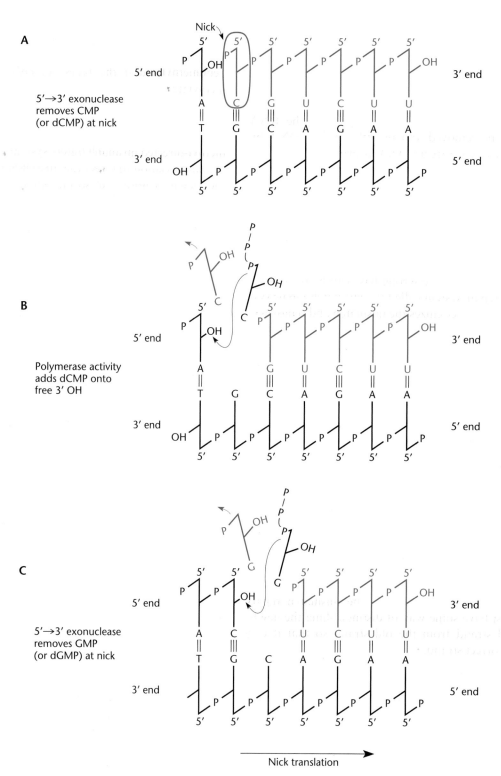

Figure 1.11 DNA polymerase I can remove the nucleotides of an RNA primer by using its "nick translation" activity. (A) A break, or nick, in the DNA strand occurs after the DNA polymerase III holoenzyme has incorporated the last deoxynucleotide before it encounters a previously synthesized RNA primer. (B) In the example, the 5'-to-3' exonuclease activity of DNA polymerase I removes the CMP at the nick and its DNA polymerase activity incorporates a dCMP onto the free 3' hydroxyl. (C) This process continues, moving the nick in the 5'-to-3' direction.

When the replication of a DNA chain has just initiated, the helix may be too short for distortions in its structure to be easily recognized by the editing proteins. The mistakes may then go uncorrected. However, if the first nucleotides inserted in a growing chain are ribonucleotides rather than deoxynucleotides, an RNA primer is synthesized rather than a DNA primer. The RNA primer can be removed and resynthesized as DNA by using preexisting upstream DNA as primer. Under these conditions, the editing functions are active and mistakes are avoided.

Methyl-Directed Mismatch Repair

Sometimes the wrong base pair is inserted into DNA in spite of the vigilance of the editing functions. However, the cell still has another chance to prevent a permanent mistake or mutation: the wrong base can be recognized by another repair system called the **mismatch repair system.** This system recognizes the mismatch and removes it as well as DNA in the same strand around the mismatch, leaving a gap in the DNA that is refilled by the action of DNA polymerase, which inserts the correct nucleotide.

The mismatch repair system is very effective at removing mismatches from DNA. However, by itself, it would not lower the rate of spontaneous mutagenesis unless it repaired the correct strand of DNA at the mismatch. In the example shown in Figure 1.13, a T was mistakenly incorporated opposite a G. If the mismatch repair system changes the T to the correct C in the *newly* replicated DNA, a GC base pair will be restored at this position and no change in the sequence or mutation will have occurred. However, if it repairs the G in the *old* DNA in the mismatch to an A, the mismatch will have been removed and replaced by an AT base pair with correct pairing, but a GC-to-AT change would have occurred in the DNA sequence at the site of the mismatch, creating a mutation. To prevent mutations, the mismatch repair system must have some way of distinguishing the newly synthesized strand from the old strand so that it can repair the correct strand.

Different organisms seem to use different mechanisms to distinguish the new and old strand for mismatch repair. In *E. coli*, it is the state of methylation of the DNA strands that allows the mismatch repair system to distinguish the new strand from the old strand after replication. In *E. coli* and other enteric bacteria, the A's in the symmetric sequence GATC/CTAG are methylated at the 6′ position of the larger of the two rings of the adenine base. These methyl groups are added to the bases by the enzyme **deoxyadenosine methylase (Dam methylase)**, but this occurs only *after* the nucleotides have been incorporated into the DNA. Since DNA replicates by a semiconservative mechanism, the "A" in the GATC/CTAG

sequence in the newly synthesized strand remains temporarily unmethylated after replication of a region containing this sequence. The DNA at this site is said to be **hemimethylated** if the bases on only one strand are methylated. Figure 1.15 shows that a hemimethylated GATC/CTAG sequence tells the mismatch repair system which strand is newly synthesized and should be repaired. For this reason, the repair system is called the **methyl-directed mismatch repair system.** The use of Dam hemimethylation to direct the mismatch repair system to the newly synthesized strand seems to be restricted mostly to enteric bacteria, since most bacteria and eukaryotes do not have a Dam methylase. Nevertheless, all organisms do possess a mismatch repair system, making it seem likely that other mechanisms exist to distinguish the new from the old strand immediately following DNA replication.

Methylation of GATC sequences in *E. coli* plays other roles as well, including helping to time the initiation of chromosomal DNA replication (see below).

Role of Editing and Mismatch Repair in Maintaining Replication Fidelity

It is possible to estimate how much each of the repair systems lowers the mistake rate of replication. On the basis of normal mutation rates, the *E. coli* genome probably contains a mistake in about 1 in 10^{10} nucleotides after replication. DNA polymerase III of *E. coli* makes a mistake about once every 10^5 times it incorporates a nucleotide. Proofreading corrects 99% of these before the replication apparatus moves on. Mismatch repair corrects an additional 99.9% of the remaining mistakes. Overall, since there are about 4.7×10^6 nucleotides in each strand of *E. coli* DNA, the cell contains $1/10^{10} \times 4.7 \times 10^6 = 4.7 \times 10^{-4}$ mistakes every time it replicates a chromosome. In other words, approximately 1 in every 2,000 progeny bacteria will have a mistake in its DNA, since the entire bacterial DNA must replicate once every time the cell divides. Mistake levels like these are apparently low enough to be tolerated, and some mistakes may even be desirable because they increase diversity in the population and speed up evolution.

In contrast, mutant bacteria that lack either the editing functions or the mismatch repair system have unacceptably high mistake rates. An *E. coli* bacterium that lacks the editing function makes a mistake on average once every 10^8 times a nucleotide is added during replication, or 100 times more frequently than the wild-type bacteria. This means that 1 in every 20 cells will have a mistake after one round of DNA replication. If the mismatch repair system is also inactivated, the mistake rate is increased another 1,000-fold or more, for an average of 50 or more mistakes every time the DNA replicates

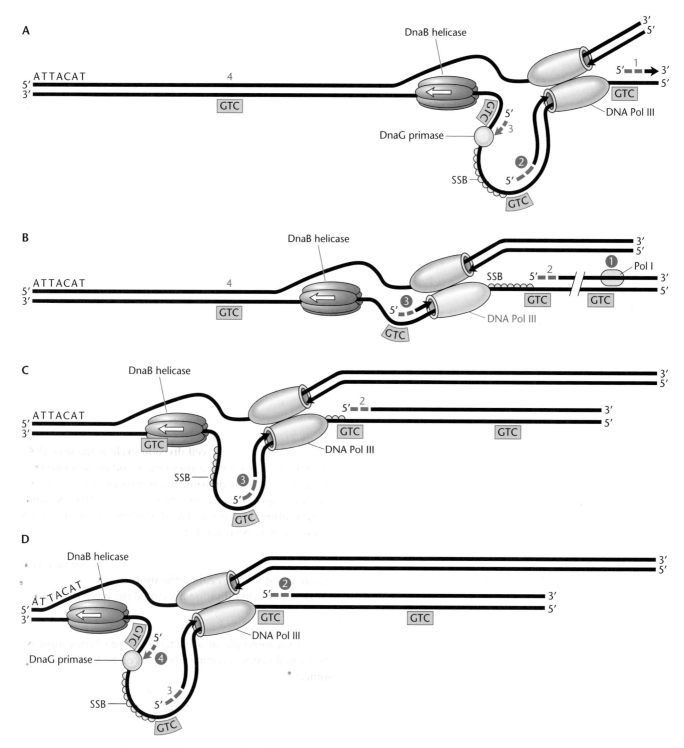

Figure 1.12 The "trombone" model for how both the leading and lagging strands of the DNA helix might be simultaneously replicated at the replication fork. SSB, single-strand binding protein. RNA primers are shown in purple. (A) Pol III holoenzyme is synthesizing lagging-strand DNA from primer 2 (circled) and has just run into primer 1. (B) Lagging-strand Pol III (indicated by shading in purple) has been released from the template at primer 1 and hopped ahead, reassembling on the DNA at primer 3 (circled) to synthesize an Okazaki fragment. Both the leading- and lagging-strand Pol III enzymes, except for the

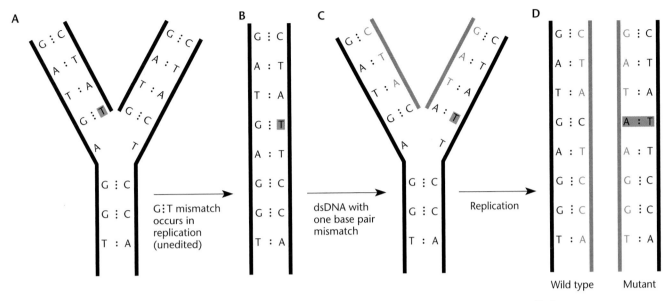

Figure 1.13   Mistakes in base pairing can lead to changes in the DNA sequence called mutations. If a T is mistakenly put opposite a G during replication (A), it can lead to an AT base pair replacing a GC base pair in the progeny DNA (B to D).

and the cell divides; i.e., every bacterium would have about 50 new mutations. It would not be very long before these *E. coli* bacteria were severely compromised by mutations and bore little resemblance to their ancestors. By lowering the spontaneous mutation rate, DNA correction systems are important for maintaining replication fidelity and the stability of the species.

Replication of the Bacterial Chromosome and Cell Division

So far we have discussed the details of DNA replication, but we have not discussed how the bacterial DNA as a whole replicates, nor have we discussed how the replication process is coordinated with division of the bacterial cell. To simplify the discussion, we consider only bacteria that grow as individual cells and divide by binary fission to form two cells of equal size, even though this is far from the only type of multiplication observed among bacteria.

The replication of the bacterial DNA occurs during the cell division cycle. The **cell division cycle** is the time during which a cell is born, grows larger, and divides into two progeny cells. **Cell division** is the process by which the larger cell splits into the two new cells. The **division time**, or **generation time**, is the time that elapses from the point when a cell is born until it divides. This time is usually approximately the same for all the individuals in the population under certain growth conditions. The original cell before cell division is called the **mother cell**, and the two progeny cells after division are called the **daughter cells**.

Structure of the Bacterial Chromosome

The DNA molecule of a bacterium that carries most of its normal genes is commonly referred to as its **chromosome**, by analogy to the chromosomes of higher organisms.

---

clamp loaders (not shown), have remained bound to each other during the release and reassembly process. Meanwhile, Pol I is removing primer 1 (circled) and replacing it with DNA. (C) Pol III is continuing lagging-strand synthesis from primer 3 (circled). (D) Pol III has completed the Okazaki fragment and has run into primer 2 (circled). Pol III holoenzyme will now hop to primer 4 (circled). In all panels, the arbitrary sequence ATTACAT shows the progress of the replication fork. The sequence 3'GTC5' boxed in purple shows where the primer synthesis initiates. The length of primers and Okazaki fragments is not drawn to scale.

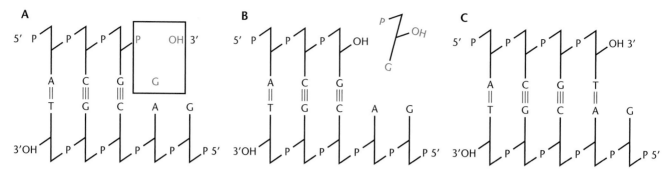

Figure 1.14 The editing function of DNA polymerase. (A) A G is mistakenly put opposite an A while the DNA is replicating. (B and C) The DNA polymerase stops while the G is removed and replaced by a T before the replication continues.

This name distinguishes the molecule from plasmid DNA, which can be almost as large as chromosomal DNA but usually carries genes that are not always required for growth of the bacterium (see chapter 4).

Most bacteria have only one chromosome; in other words, there is only one *unique* DNA molecule per cell that carries most of the normal genes. There are exceptions, including *Vibrio cholerae*, the bacterium responsible for cholera, although even in this case the second chromosome shows more characteristics of a plasmid than a chromosome in how it initiates replication (see chapter 4). Just because bacteria have one chromosome does not mean that there is necessarily only one copy of this chromosomal DNA in each bacterial cell. Bacterial cells whose chromosomes have replicated, but for some reason have not divided, have more than one copy of this chromosomal DNA per cell. Also, as discussed below, when bacteria such as *E. coli* are reproducing very rapidly, new rounds of replication initiate before others are completed, increasing the DNA content of the cells. However, these individual chromosomal DNAs are not unique and therefore do not represent new chromosomes, since they are derived from each other by replication.

The structure of bacterial DNA differs significantly from that of the chromosomes of higher organisms. For example, DNA in the chromosomes of most bacteria is circular (for exceptions, see Box 1.1), with a circumference of approximately 1 mm. In contrast, eukaryotic chromosomes are linear with free ends. As discussed in Box 1.1, the circularity of bacterial chromosomal DNA allows it to replicate in its entirety without using telomeres, as eukaryotic chromosomes do, or terminally redundant ends, as some phages do. Even in cases where bacterial chromosomes are linear, they do not use the same mechanism, involving telomerases to replicate their ends, that is used by eukaryotic chromosomes (Box 1.1).

Another difference between the DNA of bacteria and eukaryotes is that the DNA in eukaryotes is wrapped around proteins called histones to form nucleosomes. Bacteria contain histone-like proteins including HU, HN-S, Fis, and IHF, around which DNA is often wrapped, and archaea do have rudimentary histones related to those of eukaryotes. However, in general, DNA is much less structured in bacteria than in eukaryotes.

Replication of the Bacterial Chromosome

The replication of the circular bacterial chromosome initiates at a unique site in the DNA called the *origin of chromosomal replication*, or *oriC*, and proceeds in both directions around the circle. On the *E. coli* chromosome, *oriC* is located at 84.3 min. At the positions where polymerases add the nucleotides, the double-stranded DNA splits and forms two new double-stranded DNAs. As mentioned above, the place in DNA at which replication is occurring is known as the replication fork. The two replication forks proceed around the circle until they meet and **terminate** chromosomal replication. As discussed below in the section on termination, many bacteria do not terminate replication at a unique site in the DNA but, rather, terminate replication where the two replication forks meet. Each time the two replication forks proceed around the circle and meet, a **round of replication** has been completed and two new DNAs, called the **daughter DNAs**, are created.

Initiation of Chromosome Replication

Much has been learned about the molecular events occurring during the initiation of replication. Some of this information has a bearing on how the initiation of chromosome replication is regulated and serves as a model for the interaction of proteins and DNA.

Two types of functions are involved in the initiation of chromosome replication. One consists of the sites or

A   Hemimethylated DNA

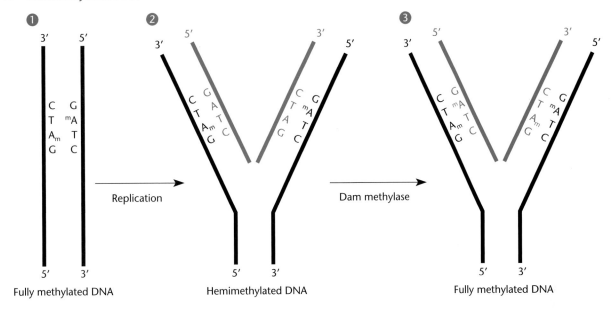

B   Methyl-directed mismatch repair

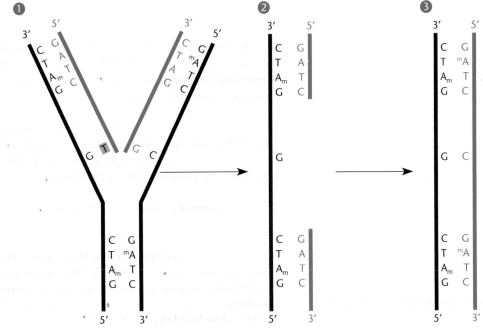

Figure 1.15   The methyl-directed mismatch repair system and how replication creates hemimethylated DNA sequences. (A) The A in the sequence GATC is methylated on both strands (1). After replication, the A in GATC in the new strand is not immediately methylated by the Dam methylase (2, 3). (B) The newly replicated DNA contains a GT mismatch (1). The newly synthesized strand is recognized because it is not methylated at the nearby GATC sequence, and the T in the mismatch is removed along with neighboring sequences (2). The sequence is resynthesized, replacing the T with the correct C. The neighboring GATC sequence is then methylated by the Dam methylase (3). The newly synthesized strand is shown in purple.

BOX 1.1

Linear Chromosomes in Bacteria

Not all bacteria have circular chromosomes. Some, including *Borrelia burgdorferi* (the causative agent of Lyme disease), *Streptomyces* spp., and *Rhodococcus fascians,* have linear chromosomes. As mentioned in the text, the replication of the ends of linear DNAs presents special problems because DNA polymerases cannot prime their own replication. This means that they cannot replicate all the way to a 3′ end in a linear DNA. If the last Okazaki fragment is primed by RNA at the end of a linear DNA and the RNA primer is then removed, there is no upstream primer DNA to prime its replacement with DNA as there is in a circular DNA. Eukaryotic chromosomes, which are linear, solve this problem by having junk DNA called telomeres at their ends. This junk DNA does not need complementary sequences to be synthesized from the template during replication. They use an enzyme called telomerase, which contains an RNA that is complementary to the repeated sequences at their ends. This enzyme makes reiterated copies of the repeated telomeric sequences at the ends. When the linear chromosome replicates, some of these repeated sequences at the 3′ end are lost, but this is not a problem because they will be resynthesized by the telomerase before the DNA replicates again.

Linear bacterial chromosomes seem to solve the linear-chromosome replication problem in different ways. The mechanism used by *Borrelia* to replicate its linear chromosome is best understood and is illustrated in the figure. The 5′ phosphate and 3′ OH at each end of its linear chromosome are joined to each other to form hairpins, as shown. The chromosome replicates not from these ends but from an *ori* sequence somewhere in the middle of the chromosome; it replicates bidirectionally from this origin, with each replication fork having only a leading strand. When this leading-strand replication gets to the ends, it replicates right around the hairpins to form a dimerized chromosome, with two copies of the chromosome forming a circle containing two copies of the chromosome linked end to end as shown. An enzyme called telomere resolvase protein (ResT) (in analogy to the telomerase of eukaryotes even though it does not work in the same way) then re-creates the original hairpin ends from these double circles by making a staggered break in the two strands where the original ends were and then rejoining the 3′ end of one strand to the 5′ end of the other strand to form a hairpin. The ResT enzyme works somewhat like some topoisomerases and Y recombinases (see chapter 9) in that the breakage and rejoining process goes through a 3′ phosphoryltyrosine intermediate where the 3′ phosphate end is covalently joined to a tyrosine (Y [see inside cover]) before it is joined to the 5′ hydroxyl end.

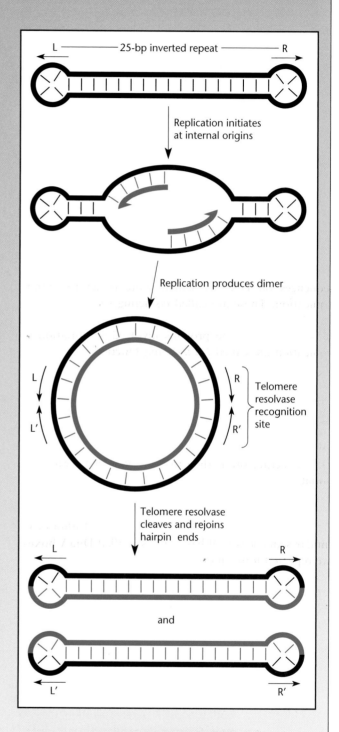

The very large linear chromosome of *Streptomyces* seems to solve this problem in a very different way from that of *Borrelia.* It has inverted-repeated sequences at its ends and a protein, terminal protein (TP), attached to the 5′ ends.

BOX 1.1 (continued)

Linear Chromosomes in Bacteria

Replication to the 3′ ends of the chromosome is thought to involve both these inverted repeats and the TP, in a process called "patching." After the linear DNA replicates, the 3′ ends of each DNA remain single stranded, which allows the inverted-repeat sequences to form hairpins. Replication of these hairpins, combined with some sort of slippage, then allows complete replication of the ends by a process not completely understood. Interestingly, bacteria that can have linear chromosomes also often contain linear plasmids, and these plasmids often replicate by a mechanism similar to that used by the chromosome (see chapter 4). Apparently, once the problem of replicating a linear DNA to the end is solved in a particular type of cell, other DNAs in the cell adopt the same mechanism. However, the relative advantages to having a circular versus a linear chromosome is the subject of speculation.

References

Kobryn, K., and G. Chaconas. 2002. ResT, a telomere resolvase encoded by the Lyme disease spirochete. *Mol. Cell* **9**:195–201.

Yang, C. C., C. H. Huang, C. Y. Li, Y. G. Tsay, S. C. Lee, and C. W. Chen. 2002. The terminal proteins of linear *Streptomyces* chromosome and plasmids: a novel class of replication priming proteins. *Mol. Microbiol.* **43**:297–305.

sequences on DNA at which proteins act to initiate replication. These are called *cis*-acting sites. The prefix *cis* means "on this side of," and these sites act only on the same DNA. The proteins involved in initiation of replication are called *trans*-acting functions. The prefix *trans* means "on the other side of," and these functions can act on any DNA in the same cell, not just the DNA from which they were made. These concepts are used again later in the book.

ORIGIN OF CHROMOSOMAL REPLICATION

One *cis*-acting site is the *oriC* site, at which replication initiates. The sequence of *oriC* is well defined and is similar in most bacteria. Figure 1.16 shows the structure of the origin of replication of *E. coli*. Less than 260 bp of DNA is required for initiation at this site. Within *oriC*, similar sequences of 9 bases, the so-called **DnaA boxes**, are repeated four times. In addition, regions 13 bp long with a higher than average AT base pair frequency are repeated three times. These two types of repeated sequences are thought to be very important for the initiation of chromosome replication.

INITIATION PROTEINS

Many *trans*-acting proteins are also required for the initiation of DNA replication, including the DnaA, DnaB, and DnaC proteins. DnaA is required only for initiation, but both DnaB and DnaC are also required for primer synthesis once DNA replication is under way. Many proteins used in other cellular functions are also involved, such as the primase (DnaG) and the normal RNA polymerase that makes most of the RNA in the cell.

Figure 1.17 outlines how DnaA, DnaB, DnaC, and other proteins participate in the initiation of chromosome replication. In the first step, 10 to 12 molecules of the DnaA protein bind to the DnaA boxes in the *oriC* region. This has the effect of wrapping the DNA around the aggregated DnaA proteins, as shown in the figure. The bending helps separate the strands of the DNA in the region of the bend as shown.

Once the strands are partially opened, the DnaB protein binds to the *oriC* region with the help of the DnaG protein. This binding is also aided by supercoiling at the origin (see the section on supercoiling, below) and by the helix-destabilizing protein, or single-strand binding

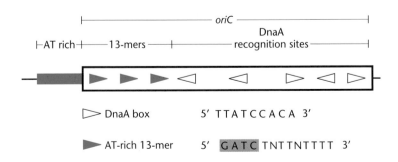

Figure 1.16 Structure of the *oriC* region of *E. coli*. Shown are the position of the AT-rich 13-mers and the position of some of the DnaA binding sequences (DnaA boxes). Also shown is an additional important 12-bp AT-rich region.

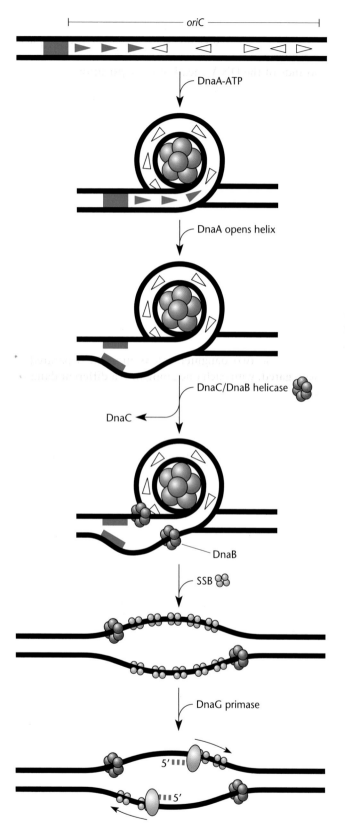

protein (SSB in Figure 1.17), which helps keep the helix from re-forming. The DnaB protein is a helicase that opens the strands further for priming and replication. The DnaC protein may then leave. This structure, with many copies of DnaA as well as DnaB and other proteins, is called a **primosome**. The primosome may help the DnaG primase or another RNA polymerase to synthesize an RNA primer to start replication.

RNA PRIMING OF INITIATION

Initiation of DNA replication requires RNA primers, but which RNA polymerase makes the primer for initiating leading-strand synthesis is not completely clear. The RNA polymerase that synthesizes most of the RNA molecules, including mRNA, in the cell (see chapter 2) is needed to initiate rounds of replication. However, its role may be to help separate the strands of DNA in the *oriC* region by transcribing through this region, because the strands may have to be separated before the DnaA protein can bind. In this case, the RNA primers themselves may be synthesized by the DnaG protein, the same RNA polymerase that makes RNA primers for lagging-strand synthesis at the replication fork. Alternatively, the normal RNA polymerase may make the primer that initiates leading-strand synthesis, while the DnaG protein makes only the primers that initiate lagging-strand synthesis. More experiments are needed to answer this question.

Termination of Chromosome Replication

After the replication of the chromosome initiates in the *oriC* region and proceeds around the circular chromosome in both directions, the two replication forks must meet somewhere on the other side of the chromosome and the two daughter chromosomes must separate. Do they meet and terminate replication at a certain well-defined site in the DNA, or do they terminate replication wherever they happen to meet? Also, are specific proteins required for termination of chromosome replication?

As with most cellular processes, the process of termination of chromosome replication is especially well understood in *E. coli*. In this bacterium, chromosome replication usually terminates in a certain region but not at a well-defined unique site. This termination region, called *ter*, contains clusters of sites called *ter* sequences,

Figure 1.17 Initiation of replication at the *E. coli* origin (*oriC*) region. About a dozen DnaA-ATP proteins bind to the origin, wrapping the DNA around themselves and opening the helix. DnaC helps the DnaB helicase to bind. The DnaG primase synthesizes RNA primers, initiating replication. SSB, single-strand binding protein.

which are only 22 bp long. These sites act somewhat like the one-way gates in an automobile parking lot, allowing the replication forks to pass through in one direction but not in the other.

Figure 1.18 shows how the one-way nature of *ter* sequences causes replication to terminate in the *ter* region. In the illustration, two *ter* site clusters called *terA* and *terB* bracket the termination region. Replication forks can pass site *terA* in the clockwise direction but not in the counterclockwise direction. The opposite is true for *terB*. Thus, the clockwise-moving replication fork can pass through *terA*, but if it gets to *terB* before it meets the counterclockwise-moving fork, it stalls because it cannot move clockwise through *terB*. Similarly, the replication fork moving in the counterclockwise direction stalls at site *terA* and waits for the clockwise-moving replication fork. When the counterclockwise and clockwise replication forks meet, at *terA*, *terB*, or somewhere between them, the two forks terminate replication, releasing the two daughter DNAs. This picture is somewhat oversimplified because each replication fork passes through a gauntlet of many *ter* sites, each of which only slows it down so that it eventually stops. While one replication fork is slowing down at a succession of *ter* sites, the other replication fork has time to make it around the chromosome to meet it at a position more or less opposite the replication origin.

Encountering a *ter* DNA sequence, by itself, is not sufficient to stop the replication fork. Proteins are also required to terminate replication at *ter* sites. These proteins, called terminus utilization substance (Tus) in *E. coli*

and replication terminator protein (RTP) in *B. subtilis*, are thought to bind to the *ter* sites and stop the replicating helicase (DnaB in *E. coli*) that is separating the strands of the DNA ahead of the replication fork. These proteins may also assist in the orderly separation of the two newly synthesized daughter DNAs to prevent any free 3′ ends from being used as primers to continue replication. However, in spite of these seeming advantages, at least some types of bacteria do not seem to absolutely need their *ter* sites. As evidence, the entire *ter* region of the *B. subtilis* chromosome can be deleted and the bacteria can still multiply, albeit more slowly. Apparently, the chromosome can terminate replication fairly reliably wherever the forks happen to meet. There are probably subtle advantages to terminating chromosome replication in the *ter* region opposite the *oriC* origin of replication, but, as is often the case, these advantages are not apparent in a laboratory situation.

Chromosome Segregation

Once the DNA has replicated, and the cells are ready to divide, the two daughter DNAs must be separated, or segregated, with each one going into a different daughter cell. Otherwise, one cell gets both chromosomes and the other cell gets neither. Chromosome segregation encounters a number of obstacles. The very long daughter chromosomes might have been joined by recombination or might have become interlinked or otherwise tangled during replication. Even if they are not physically joined, their separation would be very difficult if they were both spread out throughout the cell. Bacteria have a number

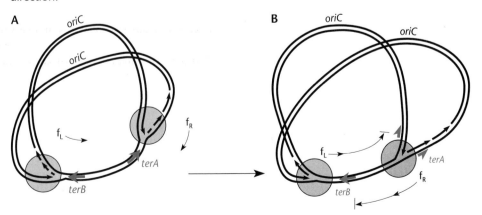

Figure 1.18   Termination of chromosome replication in *E. coli*. (A) The replication forks that start at *oriC* can traverse *terA* and *terB* in only one direction, opposite to that indicated by the purple arrows. (B) When they meet, between or at one of the two clusters, chromosome replication terminates. f_L is the fork that initiated to the left and moved in the counter-clock-wise direction. f_R is the fork that initiated to the right and moved in the clockwise direction.

of systems to ensure that their chromosomes segregate properly into the daughter cells during cell division. These systems are discussed separately in this section.

RESOLUTION OF DIMERIZED CHROMOSOMES

Sometimes the two circular daughter DNAs have become joined in a chromosome dimer, in which they are joined end to end to form a double-length circle. Such dimerized chromosomes are fairly common and arise by recombination between the two daughter DNAs as a result of replication restarts at DNA damage (Box 1.2). Such dimers obviously cause problems for chromosome segregation because the two daughter chromosomes cannot be separated if they are part of the same larger molecule.

If dimerized chromosomes can be created by recombination, they can also be resolved into the individual chromosomes by a second recombination. The general recombination system resolves the dimers by recombination anywhere within the repeated chromosomes. However, the general recombination system can both create and resolve dimers depending on how many crossovers occur between the daughter DNAs. An odd number of crossovers occurring between any two sequences on the two daughter DNAs in the dimer will resolve the dimer, but an even number of crossovers will recreate a dimer.

In *E. coli*, where it is best understood, and probably in other bacteria as well, a very clever mechanism called the Xer recombination system is used to resolve chromosome dimers. Rather than using the general recombination system, the cell uses a system involving a site-specific recombinase (see chapter 9) called the XerC,D recombinase to resolve chromosome dimers. This system is designed so that it resolves dimers into the individual chromosomes but does not create new dimers or make even larger multimers. It is also designed so that its action is coordinated with division of the cells. The Xer recombination system consists of two proteins called XerC and XerD and a specific site in the chromosome called *dif*. If two copies of the *dif* site occur on the same DNA, such as occurs when the chromosome is dimerized, the Xer proteins promote a recombination between the two *dif* sites, resolving the dimer into the individual chromosomes. To ensure that there is normally only one *dif* site in the cell until just before cell division, the *dif* site is located close to the *ter* region, so it is not replicated until just before the chromosome has completed replication and just before the cell divides (see below). As added insurance, the activity of the Xer site-specific recombination system is also made dependent on the formation of the division septum. Only when a division septum begins to form and a protein called FtsK is recruited to the division septum is this recombination system activated (see

Aussel et al., Suggested Reading). The FtsK protein is also a DNA translocase that may pump the DNA through the division septum before the cell divides, to avoid guillotining the dimerized chromosomal DNA.

Models have been presented to use these observations to explain how the XerC,D recombinase promotes recombination only between *dif* sites that are on the same molecule to resolve dimers and not between *dif* sites on different molecules, which would create new dimers. According to one model, when the cell tries to segregate a dimerized chromosome before it divides, the dimerized chromosome is being pulled toward both ends of the cell rather than toward one end or the other (see the section on partitioning, below). Therefore, when the septum begins to form, the dimerized chromosome is still spread through the division septum. When the FtsK protein binds to the division septum, it also binds to the XerC,D recombinase and to one *dif* site in the dimer. It then pumps the DNA through the septum until it gets to the other *dif* site, where it promotes site-specific recombination to separate the two linked chromosomes on opposite sides of the division septum. This model is supported by results from studies of a *B. subtilis* protein related to FtsK, named SpoIIIE (see Liu et al., Suggested Reading). This protein is a DNA translocase that pumps one copy of the chromosome into the forming spore during sporulation. In the absence of this protein, the chromosome is guillotined as the spore forms and the spore receives only part of the chromosome. The same protein is used to help segregate chromosomes during normal cell division.

DECATENATION

Replicating DNAs can also become joined to each other through the formation of catenenes, in which the daughter DNAs become interlinked like the links on a chain. These interlinks could form as the natural result of terminating a round of chromosome replication or could be caused by topoisomerases passing the strands of the two DNAs through each other (see "Topoisomerases" below). Once such interlinks are formed, the only way to unlink them is to break both strands of one of the two DNAs and pass the two strands of the other DNA through the break. The break must then be resealed. This double-strand passage, called decatenation, is one of the reactions performed by type II topoisomerases (see Figure 1.25). A type II topoisomerase called topoisomerase IV (topo IV) is thought to be responsible for removing most of the interlinks between the daughter DNAs in *E. coli* after replication. More recent evidence suggests that topo IV also removes positive supercoils ahead of the replication fork (see Khodursky et al., Suggested Reading).

BOX 1.2

Restarting Replication Forks

Once replication forks have been created, they proceed around the chromosome, making the two daughter DNAs as they go. In an ideal world, a replication fork, once formed, should be able to replicate the entire chromosome,

oriC-dependent replication

DNA lesion

DNA nick

Gap created

Double-strand break created

Stalled replication forks

D-loop formation

Pri proteins assemble the replisome, using 3′ OH as a primer, and recombination functions process the DNA structure

Replication continues

making two complete copies of the chromosome. However, DNA is not static in the cell and is constantly being altered, either by chemical damage from inside and outside influences or by participation in other cellular functions such as recombination. When a replication fork encounters such an alteration in the DNA, it is blocked and can fall off, in much the same way as damage to a railroad track or repairs to the track can block the passage of a train and cause it to be derailed. This is not a trivial problem for the cell because if it cannot finish the round of replication it will die. It has been estimated that *E. coli* replication forks encounter some obstacle in almost every round of chromosome replication.

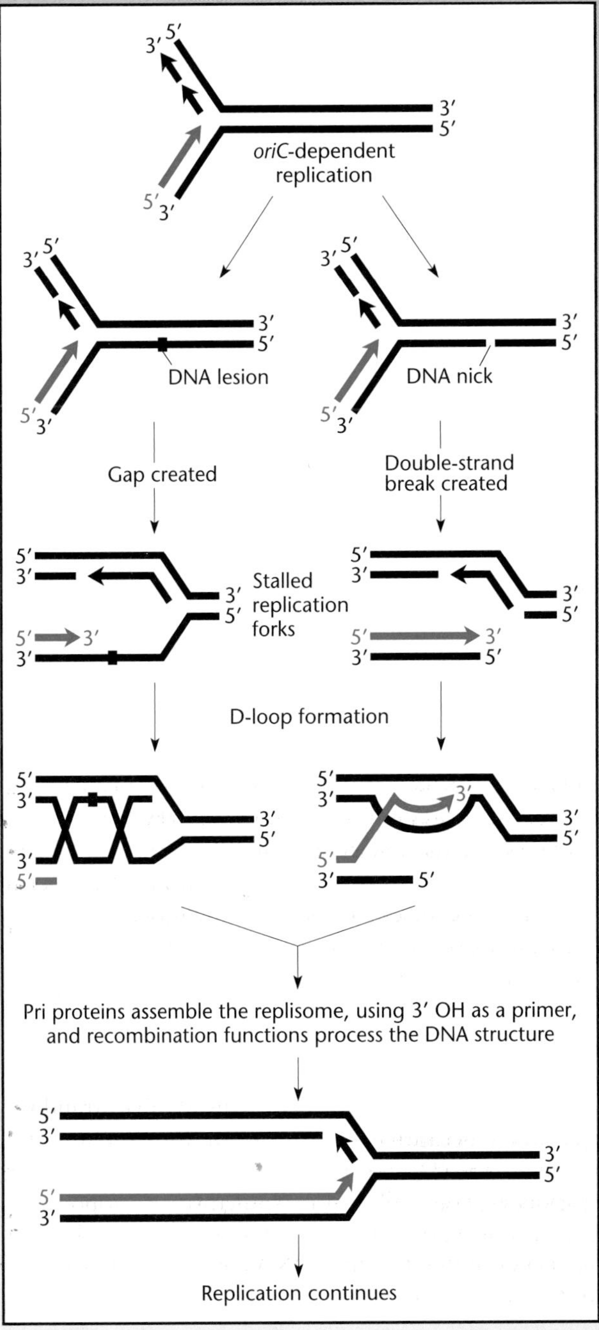

What happens when the replication fork encounters an alteration in the DNA depends on the nature of the alteration, as shown in the figure. If the alteration involves chemical damage to a base that prevents proper pairing with the complementary base, the leading-strand replication fork might stall but the DnaB helicase continues on the lagging strand, leaving a gap opposite the leading strand (see the figure). Such a gap, in which the DNA is single-stranded for a stretch, could be repaired by recombination with the other daughter DNA, as described in chapter 11. Alternatively, if the alteration is a break in one of the two strands, the replication fork may proceed past the break, causing both strands in one of the two daughter DNAs to be broken. This type of damage can also be repaired, using the other daughter DNA, by a different recombination pathway called double-strand break repair, which is also outlined in chapters 10 and 11. Both of these ways of bypassing the damage involve recombination between the two daughter chromosomes that can lead to dimerization of the chromosome, which must be resolved by the XerC,D recombinase before the cell can divide (see the text). In fact, recombination-mediated bypass of damage to the DNA seems to be the major reason for dimerization of the chromosome and may even be the major reason for the existence of the recombination systems. A problem comes when the replication machinery has to reassemble and reinitiate synthesis after the damage has been repaired. Normally, the replication apparatus can initiate synthesis of DNA only at the *oriC* region, which is somewhere else in the chromosome. Proteins called primosome proteins PriA, PriB, and PriC, as well as another protein named DnaT, may cooperate with the recombination functions in reinitiating replication at the block. These proteins were first found because they are required to initiate the replication of some phages (see chapter 7). The PriA, PriB, PriC, and DnaT proteins are thought to help reassemble the replication apparatus, using

(continued)

BOX 1.2 (continued)

Restarting Replication Forks

the single-stranded 3′ OH end of DNA that invades the double-stranded DNA to form the "D-loop" recombination intermediate as the primer to initiate new DNA synthesis (see the figure).

Reference
Cox, M. M., M. F. Goodman, K. N. Kreuzer, D. J. Sherratt, S. J. Sandler, and K. J. Marians. 2000. The importance of repairing stalled replication forks. *Nature* 404:37–41.

CONDENSATION

By itself, just passing the two daughter DNAs through each other would not necessarily have the effect of separating the two interlinked DNAs. New interlinks would be created as fast as old ones were removed, since the topo IV has no way of knowing whether it is removing or creating interlinks when it passes two DNAs through each other. To completely decatenate the two DNAs, it is necessary to simultaneously separate them as their strands are being passed through each other, much like fishing line should be reeled in as it is untangled, to separate the untangled part from the part which is still tangled.

One way to separate the two daughter DNAs from each other is to **condense** them in different parts of the cell. If they are more condensed, they do not overlap as much in the cell and so are less apt to become interlinked. Condensation of chromosomes prior to mitosis has been known for a long time to occur in eukaryotic cells, where the chromosomes are only clearly visible just before mitosis. We now know that bacteria also condense their daughter DNAs to make them easier to separate prior to division, even though it is more difficult to visualize the condensation of bacterial chromosomes because of their smaller size.

Supercoiling

One way bacteria condense DNAs is through **supercoiling** (see Figure 1.24). In bacteria, all DNAs are negatively supercoiled, which means that DNA is twisted in the opposite direction to the Watson-Crick helix, creating underwinds. As discussed in more detail below, the underwinds introduce stress into the DNA, causing it to wrap up on itself, much like a rope wraps up on itself if the two ends are rotated in opposite directions. This twisting occurs in loops in the DNA, causing the DNA to be condensed into a smaller space.

Condensins

Proteins called **condensins** also help condense the DNA in the cell, making the daughter DNA molecules easier to separate. These proteins were first discovered in eukaryotes, where they help to condense DNA in chromosomes and were named SMC proteins (for *structural maintenance of chromosome*). Condensins are long, dumbbell-shaped proteins with globular domains at the ends and a long coiled-coil region holding them together. The long coiled-coil region has a hinge so that it can fold back on itself and the two globular domains can bind together. The condensins bind to DNA through their globular domains and hold it in large loops. They can bind to DNA by themselves but also bind to other proteins called **kleisins**, which may hold them into a network-like array and condense the DNA into even smaller spaces.

For a long time condensins were thought to occur only in eukaryotes, where the condensation of the large chromosomes is clearly evident. The condensin of *E. coli*, called MukB, was found because mutations in its gene interfere with chromosome segregation. MukB was suspected of condensing the DNA because this protein and supercoiling of the DNA can compensate for each other in allowing proper segregation of the daughter chromosomes into daughter cells. While conditions that either inactivate MukB or decrease supercoiling cause only minor defects, conditions that both inactivate MukB and remove negative supercoiling cause more major defects in chromosome segregation. This was interpreted to mean that MukB and supercoiling could compensate for each other, suggesting that they both do the same thing, i.e., condense the daughter DNAs, making them easier to segregate during cell division (for a review, see Holmes and Cozzarelli, Suggested Reading). Once MukB was found to be a condensin, other proteins (MukF and maybe MukE) that interact with it were found to be related structurally to the eukaryotic kleisins, although they do not share amino acid sequences (see Fennel-Fezzie et al., Suggested Reading). *B. subtilis* also has a condensin, which is more similar in amino acid sequence to the eukaryotic condensins and so was also named SMC protein (see Britton et al., Suggested Reading). It also has kleisins named ScpA and ScpB, which are also more closely related to the eukaryotic kleisins. It is curious that some bacteria have condensins related to MukB of *E. coli* while others have condensins more closely related to the eukaryotic SMC proteins. Usually proteins that perform such an essential role in cells as condensins do are closely related to each other, even when they are present in distantly related species of bacteria.

PARTITIONING

Not only must the two daughter chromosomes be segregated after replication, but also they must be segregated in such a way that each daughter cell gets only one of the two copies of the chromosome. Otherwise, one daughter cell would get two chromosomes and the other would be left with no chromosome and eventually would die. Since chromosomeless cells appear so infrequently, there must be some process that directs the two daughter chromosomes to different daughter cells. This apportionment of one daughter chromosome to each of the two daughter cells is called **partitioning**. In spite of extensive study, the process of chromosome partitioning is still not understood very well. However, some of the genes whose products play roles in this process are now being identified, and a lot of relevant information is being accumulated. This is one of the most exciting areas of current research in bacterial cell biology, and a number of surprises have arisen from this work.

The Par Proteins

Early work has concentrated on the functions of the so-called partitioning proteins, the products of the *par* genes. The Par functions were first discovered in plasmids, which are small DNA molecules that are found in bacterial cells and that replicate independently of the chromosome (see chapter 4). Because they exist independently of the chromosome, plasmids must also have a system for partitioning, otherwise they would often be lost from cells when the cells divide. The Par systems of plasmids are known to fall into two families, one represented by the Par system of plasmid R1 and the other, much larger, represented by plasmids P1, F, and many others. It is the second of these families to which the known Par functions of chromosomes belong. Because plasmids are much smaller than the chromosome, it has been easier to do experiments with them, and we know much more about how plasmid Par systems work than we do about how the corresponding chromosomal systems work. However, we can assume that the functioning of the plasmid systems gives clues to the functioning of the homologous chromosomal functions.

Plasmid Par functions usually consist of two proteins and a site on the DNA (often called *parS*) at which they act. One of these proteins, often called ParB, binds in many copies to the site on the DNA. This binding of ParB to the site often occurs only after the plasmid has replicated; it may require pairing of the two newly replicated daughter plasmid DNAs. The binding of ParB to the site allows the binding of the other protein, often called ParA, to the complex in a process called nucleation. Once bound, the ParA protein polymerizes to form dynamic filaments that extend across the cytoplasm and may push (or pull) the daughter plasmids toward opposite ends of the cell, allowing their segregation or partitioning. The ParA proteins have ATPase activity, and the cleavage of ATP affects their polymerization and depolymerization (see Fig. 4.18).

Filament formation. Par proteins can form various types of filaments. The type of dynamic filament formed seems to depend on the family to which the partitioning system belongs. In the case of the R1 partitioning system, a single long filament forms that extends to the ends of the cell. Interestingly, the Par protein that forms these filaments, called ParM, is related to eukaryotic actin and to *E. coli* MreB, which play similar roles in moving cellular constituents around in cells (see chapter 4 and Box 1.3). The other type of Par system, and the one to which the known chromosomal systems are related, forms a number of shorter filaments that radiate out from the site on the plasmid, forming a flower-like structure reminiscent of the mitotic spindle. For this reason, the site on the plasmid has been compared to the centromere in eukaryotic chromosomes and the partitioning system has been compared to the mitotic spindle that pushes chromosomes apart. Plasmid partitioning systems are discussed in more detail in chapter 4.

Par functions and bacterial chromosomes. As mentioned, some types of bacteria also have Par functions that are encoded by the bacterial chromosome and that seem to perform a similar role in partitioning the chromosomes during cell division. The situation is clearest in *B. subtilis* and *Caulobacter crescentus*, two species of bacteria in which this system has been studied most extensively. Their Par functions are like the plasmid Par functions in that one is an ATPase and binds to the other, which in turn binds to specific sites on the chromosome. In *B. subtilis*, the proteins analogous to the ParA and ParB proteins of plasmids are called Soj and Spo0J, respectively. These names come from early genetic studies of *B. subtilis* sporulation, where *spo0J* was identified as a gene required for sporulation and *soj* was a suppressor of *spo0J*. There are also a number of *parS*-like sites close to the origin of chromosome replication; Spo0J has been shown to bind to these sites, as expected if it is analogous to ParB. In addition, plasmids lacking their own Par system but containing the *parS*-like site of the *B. subtilis* chromosome are more faithfully partitioned into daughter cells, but only if both Spo0J and Soj are

BOX 1.3

The Bacterial Cytoskeleton and Bacterial Cell Biology

It has long been possible to observe the movement of proteins and other cellular components within living eukaryotic cells. Such seemingly purposeful movement of cellular constituents such as occurs during cytoplasmic streaming can be observed even under low-power light microscopes. In eukaryotic cells, movement is directed by three types of filaments, i.e., actin, tubulin, and intermediate filaments, and these make up what is called the cytoskeleton. However, bacterial cells were not thought to require such directed movement. Accordingly, bacteria were generally considered to be merely "bags of enzymes" with little cellular organization. Because of their smaller size, it was generally thought that random diffusion is sufficient to move constituents from one place to another in the bacterial cell in a timely fashion so that they reach their destination in time to perform their function. Without the need for directed movement within cells, bacteria had no need for a cytoskeleton.

B   MreB and Mbl role in cell shape determination

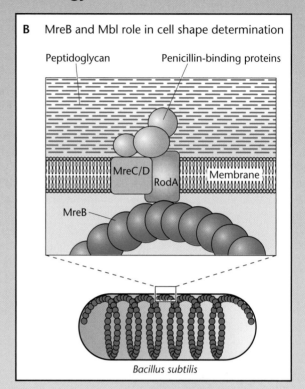

Bacillus subtilis

A   MreB roles in *Caulobacter crescentus*

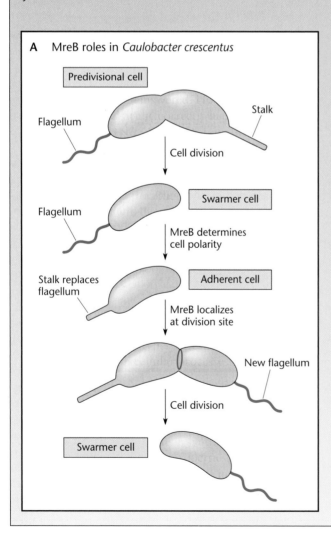

C   FtsZ role in division site selection

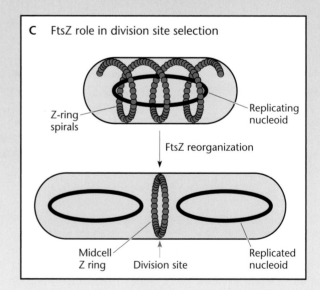

Recent advances in technology have altered our whole view of the bacterial cell. Not only have all three types of cytoskeletal filaments been found in bacteria, but also these filaments have been found to direct the movement of many cellular constituents including the chromosome, the enzymes for cell wall biosynthesis, and the division apparatus. While not homologous to their eukaryotic equivalents at the amino acid

BOX 1.3 (continued)

The Bacterial Cytoskeleton and Bacterial Cell Biology

sequence level, these filaments share remarkable structural and functional features with their eukaryotic equivalents.

The first actin-like filament detected in bacteria was due to polymers formed by the MreB protein in *E. coli.* The role of this and other actin-like filaments in bacteria is to determine cell shape. The gene for the MreB protein of *E. coli* was found because mutations in this gene cause the cell to become round rather than rod shaped. Later, the MreB protein was shown to be structurally very similar to eukaryotic actin and to form eukaryotic-cell-like dynamic filaments that lengthen on one end and shorten on the other, cleaving ATP in the process. These actin-like filaments are helical and lie along the inner face of the membrane. Actin-like proteins are now known to exist in many types of bacteria. *Bacillus subtilis* has at least three actin-like proteins, at least two of which, MreB and Mbl, also help determine cell shape (see Carballido-Lopez and Errington, below). It is an attractive idea that the enzymes that synthesize the cell wall, and therefore give the cell its shape, are directed to their site of synthesis by their association with the actin-like filaments. There is also evidence that MreB is involved in the segregation of the bacterial chromosome, at least in some types of bacteria. This, combined with the similarity of some Par functions to actin, suggests that chromosome separation can be directed by actin-like filaments.

Caulobacter presents a particularly interesting case that demonstrates how actin-like filaments in bacteria help direct the placement of cellular constituents (see the figure, panel A). Unlike *E. coli* and *B. subtilis,* this bacterium is visually asymmetric, with one end of the cell forming a narrower stalk that holds it to solid surfaces and the other end forming a flagellum that allows it to swim. After cell division, one cell becomes a swarmer cell that is able to swim to another location before it forms a stalk on the same end as the flagella had formed previously, and the cycle is repeated. Interestingly, MreB is required to determine on which end the stalk forms (see Wagner et al., below, and the figure, panel B). Using methods similar to those described later in the book, these researchers found that if the cell is depleted of MreB, the stalk does not form at either end, but that when MreB is

restored, the stalk can form at both ends and sometimes even in the middle. Other researchers found that other gene products, and even the origins of DNA replication, are usually restricted to one end of the cell or the other and show the same loss of orientation on depletion of MreB (see Gitai, below). In the absence of MreB, the cell has forgotten which end was which! These results point out the importance of a preexisting cytoskeleton to direct synthesis of a new cell.

Tubulin is another type of filament-forming protein that also exists in bacteria, in the form of FtsZ. The FtsZ protein forms a ring at the site of septum formation during cell division (see the text). It is structurally very similar to eukaryotic tubulin and forms similar filaments, called microtubules in eukaryotes. Both use GTP to drive filament formation, both form dynamic multistranded filaments, and both move by the assembly and disassembly of shorter protofilaments. The major difference is that the FtsZ filaments are composed of only one protein, FtsZ, while eukaryotic microtubules are composed of two proteins, tubulin α and tubulin β. Recent work has shown that FtsZ also forms helical structures in the cell that oscillate in the cell and then concentrate at the site of septum formation to form a ring, just before cell division (see the figure, panel C, and Thanedar and Margolin and Michie et al., below). In *E. coli,* MinC and MinD also oscillate in a helical pattern from one end of the cell to the other (see the figure panel D) and inhibit FtsZ ring formation at the poles. It is not clear whether these proteins create their own helical filaments or whether they are moving on helical structures created by others.

The final eukaryotic-cell-like filament to be found in bacteria is crescentin in *Caulobacter crescentus* (see Ausmees et al., below). This protein forms helical bundles of filaments in the inside curvature of the cell that gives *C. crescentus* its characteristic crescent shape (see the figure, panel E) and its name. In eukaryotes, these filaments are also responsible for maintaining cell shape. Crescentin shares many features with the proteins that form eukaryotic intermediate filaments. They are also long proteins containing coiled-coiled domains in

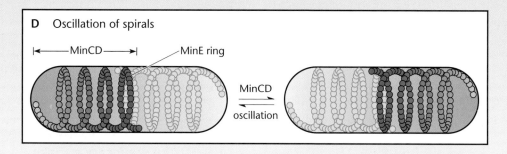

D   Oscillation of spirals

(continued)

BOX 1.3 (continued)

The Bacterial Cytoskeleton and Bacterial Cell Biology

their middle. In vitro, they can be denatured in strong detergents, and when the detergent is removed they reassemble spontaneously into long filaments without any requirement for energy or even divalent cations. It seems likely that other filaments related to the intermediate filaments of eukaryotes will be found in bacteria. Like so many other cellular constituents that we once thought were unique to eukaryotes, even the cytoskeleton of eukaryotes has its origin in microorganisms.

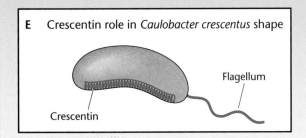

E   Crescentin role in *Caulobacter crescentus* shape

Flagellum

Crescentin

References

Ausmees, N., J. R. Kuhn, and C. Jacobs-Wagner. 2003. The bacterial cytoskeleton: an intermediate filament-like function in cell shape. *Cell* **115**:705–713.

Carballido-López, R., and J. Errington. 2003. A dynamic bacterial cytoskeleton. *Trends Cell Biol.* **13**:577–583.

Gitai, Z. 2005. The new bacterial cell biology: moving parts and sub-cellular architecture. *Cell* **120**:577–586.

Grantcharova, N., U. Lustig, and K. Flardh. 2005. Dynamics of FtsZ assembly during sporulation in *Streptomyces coelicolor* A3(2). *J. Bacteriol.* **187**:3227–3237.

Michie, K. A., L. G. Monahan, P. L. Beech, and E. J. Harry. 2006. Trapping of a spiral-like intermediate of the bacterial cytokinetic protein FtsZ. *J. Bacteriol.* **188**:1680–1690.

Raskin, D. M., and P. A. deBoer. 1999. Rapid pole-to-pole oscillation of a protein required for directing division to the middle of *Escherichia coli*. *Proc. Natl. Acad. Sci. USA* **96**:4971–4976.

Thanedar, S., and W. Margolin. 2004. FtsZ exhibits rapid movement and oscillation waves in helix-like patterns in *Escherichia coli*. *Curr. Biol.* **14**:1167–1173.

Wagner, J. K., C. D. Galvani, and Y. V. Brun. 2005. *Caulobacter crescentus* requires RodA and MerB for stalk synthesis and prevention of ectopic pole formation. *J. Bacteriol.* **187**:544–553.

present in the cell, suggesting that both proteins help in the partitioning of plasmids. However, while Spo0J is required for normal partitioning of the chromosome, Soj is not. This suggests either that Soj does not perform the same role in chromosome partitioning that ParA performs in plasmid partitioning or that some other protein can substitute for Soj in chromosome partitioning.

The situation is clearer in *C. crescentus*, which has two functions named ParA and ParB, based on their similarity to the ParA and ParB functions of plasmids. In *C. crescentus*, the ParA and ParB functions are required for proper chromosome partitioning. The ParB function of *Caulobacter* binds close to the origin of replication, which is expected if its role is to pull or push the *ori* regions toward opposite ends of the cell before cell division (see below).

Surprisingly, *E. coli* seems to lack Par functions, at least ones closely enough related to those of plasmids to be identifiable. However, it does have a site called *migS*, which lies near the origin of chromosome replication and seems to play a centromere-like role in being the site at which the chromosomes are pulled apart before cell division (see Yamaichi and Niki, Suggested Reading). The only missing components are the functions that bind to this site and pull the daughter chromosomes apart. It seems possible that the functions that bind this site and pull the DNAs toward the ends of the cell are normal

proteins of the cytoskeleton rather than proteins identifiable as Par functions (see below and Box 1.3).

Where Are the Replication Forks?

The location where chromosome replication occurs in the cell might also give clues to how the chromosomes segregate. This is a very active area of research, and the results are often contradictory. Chromosome DNA replication in bacterial cells is performed by the replicative DNA polymerase with the help of all of its accessory proteins that make up the replication fork. Therefore, if we knew where the replicating DNA polymerase was in the cell, we would know where replication was occurring. In an early experiment, the position of the replicative DNA polymerase in *B. subtilis* (called PolC) during the cell cycle was determined by genetically fusing the DNA polymerase to green fluorescent protein (GFP) and then seeing where the fluorescence was located (see Lemon and Grossman, Suggested Reading). The experimental results were consistent with the model shown in Figure 1.19. The two replication forks stay together at the midpoint of the cell until replication is complete, and then they move apart to the one-quarter and three-quarters positions of the cell, where the next midpoints will occur.

The situation seems to be somewhat different in *C. crescentus*, where the termini of chromosome

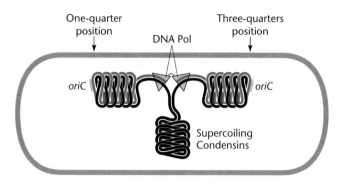

Figure 1.19 A model for replication and partitioning of chromosomes during the *Bacillus subtilis* cell cycle. According to the model, replication occurs at a "factory" in the middle of the cell, and the newly replicated DNA moves out toward the one-quarter and three-quarters positions of the cell, where it is condensed by supercoiling and condensins.

replication remain at one end while the origins are on the opposite ends of the cell, where the stalk and flagella are attached. As they replicate, one daughter origin moves to the other end of the cell so that, when replication is complete, the two chromosomes are mirror images of each other with the termini together in the middle and the origins at the far ends, one of which has a stalk and one of which has a flagellum (Box 1.3). Intermediate genes move to their respective positions in the nucleoids as they are synthesized, so that the pattern of genes in the nucleoids is recreated as they are synthesized (see Viollier et al., Suggested Reading). Recent work with *E. coli* also suggests a similar pattern of recreating the left-right organization of genes in the nucleoids as they are synthesized but with a very different spatial organization in the cell (see Wang et al., Suggested Reading). This work involved fluorescent probes that emit at different wavelengths so that the locations of two different genes could be visualized in the same cell. According to this work, rather than the origins of replication in *E. coli* immediately moving toward opposite ends, or poles, of the cell after they replicate, the newly replicated origins remain in the middle of the cell. As the two replication forks replicate the DNA in opposite directions from these origins, the newly replicated DNA fans out to opposite ends of the developing daughter nucleoids so that genes, replicated by the left and right forks, retain their left and right positions in the daughter nucleoids. As a result, the position of genes in the nucleoids roughly corresponds to their position on the genetic map, with the origin at the top and the terminus at the bottom and the genes to the left and right of the origin in the genetic map going to the left and right sides of the nucleoid. Because they are

progressively layered on the ends of the nucleoids as they are synthesized, the genes closest to the origin, which are replicated first, end up on the inside and the genes farther away end up on the outside. As a result, the order of genes in the nucleoid may recapitulate the genetic map (see the *E. coli* genetic map in Figure 3.30). This is interesting because it suggests that *E. coli* cells are not symmetrical, as has been commonly assumed, but, rather, have right and left ends that are somehow imprinted, perhaps by the cytoskeleton (Box 1.3). It is somewhat surprising that different types of bacteria seem to spatially organize their nucleoids so differently during replication. However, all of this work is very preliminary and subject to unknown artifacts, and the conclusions may have to be substantially revised as more evidence accumulates. What is increasingly likely, however, is that the different regions of the nucleoid have their own spatial location in the cell, to which they move as soon as they are synthesized, and that the organization is not nearly as haphazard as once believed.

Cell Division

Much has also been learned about how the bacterial division septum forms. The most important protein in this process is a protein called FtsZ, which forms a ring around the midpoint of the cell. Before the cell is ready to divide, the FtsZ protein exists as helical filaments that traverse the cell. When the cell is about to divide, these filaments converge on the middle of the cell and form a ring at the site of the future septum (see Box 1.3). The FtsZ ring then attracts other proteins, including the DNA translocase FtsK discussed above, which help form the **division septum** and which eventually squeezes the cell apart into the two daughter cells. The following major questions may be asked: why does the septum form only in the middle of the cell, and why does the forming septum not guillotine the bacterial nucleoid as it forms? The answers to these questions lie, at least in part, in two systems: the Min system and the nucleoid occlusion system.

THE Min PROTEINS

In *E. coli*, three proteins called MinC, MinD, and MinE are known to be involved in selecting the site for the division septum to form. The *min* genes of *E. coli* were found because mutations in these genes can cause the division septa to form in the wrong places, sometimes pinching off smaller cells called minicells. Apparently, in the absence of the Min proteins, division septa can form in places other than in the middle of the cell, for example, at the one-quarter and three-quarters positions, the sites of future division septa. When this happens, the smaller minicells are pinched off that lack a chromosome, hence the name Min proteins, for minicell-producing. It was predicted that the Min proteins would be localized in the ends of the cell, where they could prevent FtsZ from forming a division septum anywhere but the middle of the cell. However, when the localization of the Min proteins in the cell was studied, using GFP fusions to the Min proteins, a very surprising result was in store: the Min proteins oscillate in a helical pattern from one pole of the cell to the other during the cell cycle as though they are moving on an invisible track (Box 1.3). The MinC and MinD proteins oscillate the most, collecting at one end of the cell and then all moving to the other end. MinD may drive the oscillation of MinC, which may in turn inhibit the formation of the FtsZ ring. It has been hypothesized that the purpose of the oscillation may be to ensure that the concentration of the MinC division inhibitor is highest at the poles, where it needs to inhibit FtsZ, and lowest in the middle of the cell, where it is just passing through. The MinE protein forms a spiral ring which oscillates back and forth in the middle of the cell, apparently driving the oscillation of MinC and MinD. The significance of these MinE rings is unclear since they do not seem to have anything directly to do with FtsZ ring formation. It is almost as though the Min proteins play the role of "division site policemen," constantly scanning the cell to make sure that FtsZ does not loiter and form a septum somewhere it is not supposed to. In fact, they are more like "Keystone Kops" in this role, chasing each other from one pole to the other as the cell goes through its cycle.

Proteins similar to the MinC and MinD proteins of *E. coli* have also been found in *B. subtilis*, although MinE seems to be lacking in this bacterium. Instead, *B. subtilis* has two other proteins, named DivIVA and EzrA, which play roles in this process. Mutations in the *min* genes of *B. subtilis* also allow division septa to form at the ends. However, the Min proteins of *B. subtilis* do not oscillate but, rather, just gather at the poles. Instead, it is the Par functions of *B. subtilis*, Soj and Spo0J, that seem to oscillate, although with a slower periodicity than the Min proteins of *E. coli* (see above and Box 1.3). *C. crescentus* seems to lack Min proteins altogether, although it is possible that their role could be played by other, unrelated proteins. Thus, there are differences, still poorly understood, in how bacterial chromosomes replicate and segregate prior to division. The specific proteins involved may differ in other bacteria.

NUCLEOID OCCLUSION

As mentioned, the FtsZ ring should also not initiate the assembly of a division septum while the nucleoid is still occupying the center of the cell or it might guillotine the chromosome. In fact, it was observed in *E. coli* that FtsZ rings never formed in the center of the cell when it was still occupied by the nucleoid, which had not yet segregated. Proteins that inhibit FtsZ ring formation in the presence of the nucleoid have been detected in both *E. coli* and *B. subtilis* at about the same time and were named **nucleoid occlusion** (NO) proteins. Both proteins were found because they are essential only if the Min system is inactivated. The reasoning is that the NO and Min systems can at least partially substitute for each other in localizing the division septum. If one or the other is missing, the FtsZ protein still forms a ring in the middle of the cell, which usually is not occupied by the nucleoid by this time. However, if both are missing, the division septum is apt to occur anywhere, even in regions that are occupied by the nucleoid. The protein in *B. subtilis*, named Noc, was found serendipitously, because its gene, *noc*, is adjacent to the genes for the Par functions, *soj* and *spo0J*, and it was observed that mutations in this gene could not be combined with mutations in the *minD* gene without making the cells very sick (see Wu and Errington, Suggested Reading). The reason they were sick is because they were forming long ropes and not dividing properly. The NO protein in *E. coli*, named SlmA, was found by directly looking for genes whose products were required only if one of the *min* genes was also inactivated by a synthetic lethal selection (see Bernhardt and de Boer, Suggested Reading). A synthetic lethal screen is designed to isolate mutations in genes whose products are required only if another gene product is absent, in this case the products of the *min* genes. The investigators expressed the *min* genes from an inducible promoter and looked for mutants that were sick and failed to form colonies only in the absence of inducer. Some of these mutants had mutations in a gene that was named *slmA* by the investigators. While mutants deficient in Min proteins had more Z rings, these were never over the nucleoids. However, mutants that lacked both the Min proteins and SlmA often formed Z rings over the nucleoids, as expected for a mutant deficient in nucleoid occlusion. The use of inducible promoters and other examples of synthetic phenotypes are discussed in more detail in later chapters. It is not clear how NO systems work. One idea is that the Noc and SlmA proteins are FtsZ inhibitors that are bound all over the DNA, so they will inhibit FtsZ ring formation when the DNA in the nucleoid is nearby.

Coordination of Cell Division with Replication of the Chromosome

It is not sufficient to know how chromosomes replicate and are segregated into the daughter cells prior to division. Something must coordinate the replication of the chromosome with division of the cells. If the cells divided before the replication of the chromosome was completed, there would not be two complete chromosomes to segregate into the daughter cells and one cell would end up without a complete chromosome. The mechanism by which cell division is coordinated with replication of the DNA is still not understood, but there is a lot of relevant information.

TIMING OF REPLICATION IN THE CELL CYCLE

It is important to know when replication occurs during the cell cycle. Experiments were designed to determine the relationship between the time of chromosome replication and the cell cycle in *E. coli* (see Helmstetter and Cooper, Suggested Reading). The conclusions are still generally accepted, so it is worth going over them in some detail.

These scientists recognized that if the DNA content of cells at different stages in the cell cycle could be measured, it would be possible to determine how far chromosome replication had proceeded at that time in the cell cycle. Since bacterial cells are too small to allow observation of DNA replication in a single cell, it was necessary to measure DNA replication in a large number of cells. However, cells growing in culture are all at different stages in their cell cycles. Therefore, to know how far replication had proceeded at a certain stage in the cell cycle, it was necessary to synchronize cells in the population so that all were the same age or point in their life cycle at the same time.

Helmstetter and Cooper accomplished this by using what they called a bacterial "baby machine." Their idea was to first label the DNA of a growing culture of bacterial cells by adding radioactively labeled nucleosides and then fix the bacterial cells on a membrane. When the cells on the filter divided, one of the two daughter cells would no longer be attached and would be released into the medium. All of the daughter cells released at a given time would be newborns and so would be the same age. This means that cells that divided to release the daughter cells at a given time would also be the same age and would have DNA in the same replication state. The amount of radioactivity in the released cells is then a measure of how much of the chromosome had replicated in cells of this age. This experiment was done under different growth conditions to show how the timing of replication and the timing of cell division are coordinated under different growth conditions.

Figure 1.20 shows the results of these experiments. For convenience, the following letters were assigned to each of the intervals during the cell cycle. The letter I denotes the time from when the last round of chromosome replication initiated until a new round begins. The letter C is the time it takes to replicate the entire chromosome, and the letter D is the time from when a round of chromosome replication is completed until cell division occurs. The top of the figure shows the relationship of I, C, and D when the cells are growing very slowly with a generation time of 70 min. Under these conditions, I is 70 min, C is 40 min, and D is 20 min. However, when the cells are growing in a richer medium and are dividing more rapidly with a generation time of only 30 min, the pattern changes. The C and D intervals remain about the same, but the I interval is much shorter, only about 30 min.

Some conclusions may be drawn from these data. One conclusion is that the C and D intervals remain about the same independent of the growth rate. At 37°C, the time it takes the chromosome to replicate is always about 40 min, and it takes about 20 min from the time a round of replication terminates until the cell divides. However, the I interval gets shorter when the cells are growing faster and have shorter generation times. In fact, the I interval is approximately equal to the generation time—the time it takes a newborn cell to grow and divide. This makes sense because, as discussed later, initiation of chromosome replication occurs every time the cells reach a certain size. They reach this size once every generation time, independent of how fast they are growing.

Another point apparent from the data is that in cells growing rapidly with a short generation time, the I interval can be shorter than the C interval. If I is shorter than C, a new round of chromosomal DNA replication will begin before the old one is completed. This explains the higher DNA content of fast-growing cells than of slow-growing cells. It also explains the observation that genes closer to the origin of replication are present in more copies than are genes closer to the replication terminus.

Despite providing these important results, this elegant analysis does not allow us to tell whether division is coupled to initiation or termination of chromosomal DNA replication. The fact that the I interval always equals the generation time suggests that the events leading up to division are set in motion at the time a round of chromosome replication is initiated and are completed 60 min later independent of how fast the cells are growing. However, it is also possible that the act of termination of a round of chromosome replication sets in motion a cell division 20 min later. More experiments are needed to resolve these issues.

Timing of Initiation of Replication

A new round of replication must be initiated each time the cell divides, or the amount of DNA in the cell would increase until the cells were stuffed full of it or decrease until no cell had a complete copy of the chromosome. Clearly, initiation of replication is exquisitely timed. In

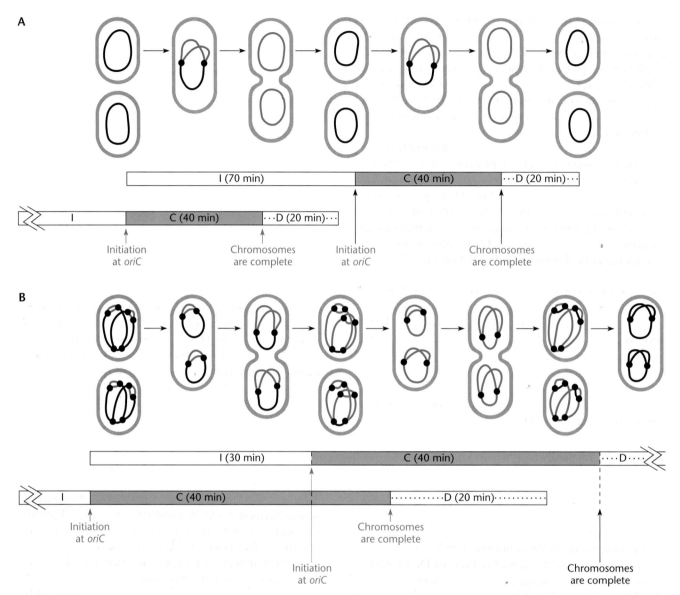

Figure 1.20   The timing of DNA replication during the cell cycle, with two different generation times. The time between initiations (I) is the only time that changes. See the text for definitions of I, C, and D.

cells growing very rapidly, in which the next rounds of replication initiate before the last ones are completed, so that the cells contain a number of origins of replication, all of the origins in a cell "fire" simultaneously, indicating tight control.

A number of attempts have been made to correlate the timing of initiation of chromosome replication with other cellular parameters during the cell cycle. Most such evidence points to initiation of replication being tied to cell mass. After cells divide, their mass or weight continuously increases until they divide again. The initiation of chromosome replication occurs each time the cell achieves a certain mass, the **initiation mass.** If cells are

growing faster in richer medium, they are larger and achieve the initiation mass sooner than do smaller, slower-growing cells, explaining why new rounds of chromosome replication occur before the termination of previous rounds in faster-growing cells but not in slower-growing cells. However, these experiments by themselves do not explain what it is about the cell mass that triggers initiation.

ROLE OF THE DnaA PROTEIN
Most evidence indicates that the timing of initiation of chromosome replication is tied to the intracellular concentration of DnaA protein. This makes sense. The

DnaA protein is the first protein to bind to each origin during initiation. Many copies of DnaA bind to each origin, thereby opening the DNA strands and allowing the DnaB helicase protein to bind and the replication fork to form (Figure 1.17). Therefore, initiation of chromosome replication can occur only when there are sufficient copies of DnaA in the cell to allow many copies of DnaA to bind to each origin. However, what matters is not the absolute amount of DnaA in the cell. What matters is the *ratio* of the amount of DnaA protein to the number of origins of replication in the cell. According to this model, as the cell grows, the amount of the DnaA protein in the cell progressively increases but the number of origins of replication stays the same since no new initiations are occurring. Finally, the ratio of DnaA protein to origins reaches the critical number and initiation occurs.

We can add more detail to this model. There are three DnaA binding sequences, "DnaA boxes," in each *oriC* region (Figure 1.17). These sites can be recognized because they have the nucleotide sequence TT(A/T)TNCACA reading in the 5′-to-3′ direction, where (A/T) means either A or T and N means any of the four bases. Because their nucleotide sequences are somewhat different, these three sites bind DnaA protein with different affinities (i.e., tightness). Initiation occurs only if all three boxes are occupied by DnaA. Then and only then will additional copies of DnaA "pile on," as shown in Figure 1.17, and initiation will occur.

Let us see what this model predicts will happen as the cells go through the cell cycle. Initially, after a round of chromosome replication has just initiated, the ratio of DnaA proteins to origins of replication is low and at most only one or two of the DnaA boxes is occupied. As DnaA protein accumulates but the number of origins of replication remains the same, the ratio of DnaA protein to origins steadily increases. Finally, there is enough DnaA that all three DnaA boxes are occupied and initiation occurs. This model is consistent with evidence that either artificially decreasing the amount of DnaA in the cell or increasing the number of copies of the DnaA boxes will delay initiation of rounds of chromosome replication.

This model also accounts for why initiation does not reoccur immediately after it has occurred. Once initiation occurs, there are twice as many origins as previously, so the ratio of DnaA protein to origins drops to half of what it was and the process repeats itself. Also, there is a site close to the origin called *datA* that can bind 300 to 400 copies of DnaA. After chromosome replication has initiated, there will also be more of these sites to bind DnaA and further lower its availability. There also may be other factors at play. DnaA binds either ATP or ADP, but only the ATP-bound form can initiate replication.

The ATP/ADP ratio is higher in faster-growing cells, which may help ensure that initiation occurs only in growing cells. Also, the assembly of the sliding clamp on the DNA polymerase activates the ATP-to-ADP activity of DnaA, so that it can no longer initiate replication. The sliding clamp does this by interacting with a relative of DnaA called Hda, which then activates the ATPase activity of DnaA (see Kurz et al., Suggested Reading).

HEMIMETHYLATION AND SEQUESTRATION

At least some types of bacteria, including *E. coli* and *C. orescentus*, have yet another means of delaying the initiation of new rounds of chromosome replication until the cell divides. In these bacteria at least, methylation of the DNA plays a role in delaying initiation. In *E. coli*, in which this is best understood, the methylation is due to the Dam methylase, the same enzyme involved in mismatch repair (see "Methyl-Directed Mismatch Repair" above). The Dam methylase methylates the two A's in the DNA sequence GATC/CTAG. The methylation occurs only after the DNA is synthesized, so that the A in a newly synthesized strand of this sequence is not methylated immediately. Intriguingly, the sequence GATC/CTAG appears 11 times in only 245 bp in the *oriC* region of the chromosome, much more often than would be expected by chance alone. Furthermore, the promoter region of the *dnaA* gene, the region in which mRNA synthesis initiates for the DnaA protein, also has GATC/CTAG sequences, and no DnaA protein is synthesized unless these sequences are fully methylated.

Figure 1.21 depicts a model of how the Dam methylase may sequester *oriC* regions after initiation. Immediately after an *oriC* region has been used to initiate replication, the GATC/CTAG sequences in the *oriC* region are hemimethylated; only the A in the old strand of the sequence is fully methylated. According to the model, the hemimethylated *oriC* region is sequestered by binding to the membrane, a process that renders it nonfunctional for the initiation of new rounds of replication and prevents it from being further methylated by the Dam methylase. In addition, a protein called SeqA (*seq*uestration protein A) may be involved in binding hemimethylated *oriC* regions to the membrane.

There is direct evidence to support this role of methylation in *oriC* sequestration and regulation of DnaA protein synthesis after initiation. For instance, the GATC/CTAG sequences in the *oriC* region and in the promoter region for the *dnaA* gene remain hemimethylated much longer after replication than do GATC/CTAG sequences elsewhere in the chromosome. Also, increasing the amount of Dam methylase in the cell causes premature initiation of replication, as might be expected if higher than normal levels of Dam methylase fully

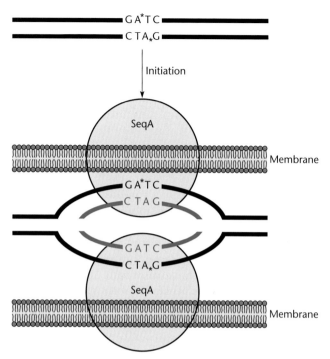

Figure 1.21   Model of the sequestration of the *oriC* region of the *E. coli* chromosome after initiation of chromosome replication. Before initiation, the *oriC* region is methylated in both strands. After initiation, only one of the two strands is methylated; hence, the region is hemimethylated. A protein called SeqA helps bind the hemimethylated *oriC* to the membrane, thereby sequestering it and preventing further initiation and methylation. The newly synthesized strand is shown in purple, and the methylated bases are starred. Only 1 of the 11 GATC/CTAG sequences in the *oriC* region is shown.

methylate the GATC/CTAG sequences, which would unsequester *oriC* sooner than the hemimethylated sequences would allow. Finally, hemimethylated *oriC* regions bind more readily to membranes than do fully methylated *oriC* regions in the presence of SeqA (see Slater et al., Suggested Reading).

More recent work has shown that SeqA does not remain with the origin of replication but, rather, remains in clusters at the center of the cell with the replication forks. There is speculation that SeqA may play a number of roles in replication, including helping drive the newly replicated DNA out of the center of the cell and toward the poles. It is known to help distinguish the new and old strands for mismatch repair (see above and chapter 11).

The Bacterial Nucleoid

As mentioned at the beginning of this chapter, the DNA of even a simple bacterium is approximately 1 mm long, while bacteria themselves measure only micrometers in

0.5 µm

Figure 1.22   Thin section of *E. coli* showing condensed DNA.

length. Therefore, the DNA is about 1,000 times longer than the bacterium itself and must be condensed to fit in the cell, but it also must be folded in such a way that it is available for transcription, recombination, and other functions.

Figure 1.22 shows a picture of a thin section of an *E. coli* cell. The chromosome is not spread all over the cell but is condensed in only one part. As the chromosome replicates, this condensed mass becomes larger until it finally separates into two masses of DNA, just before cell division.

Condensed bacterial DNA isolated from bacteria is shown in Figure 1.23. This condensed structure is called the bacterial **nucleoid**. The nucleoid is composed of 30 to 50 loops of DNA emerging from a more condensed region, or **core**. Seeing this tangle of loops, it is difficult to imagine that the DNA in this complicated structure is actually one long, continuous circular molecule.

Supercoiling in the Nucleoid

One of the most noticeable features of the nucleoid is that most of the DNA loops are twisted up on themselves. This twisting is the result of supercoiling of the DNA, as discussed above.

Figure 1.24 illustrates supercoiling. In this example, the ends of a DNA molecule have been rotated in opposite directions and the DNA has become twisted up on

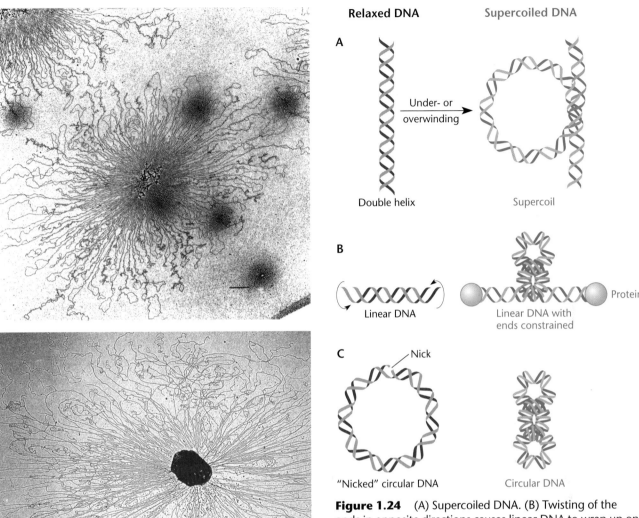

Figure 1.23 Electron micrographs of bacterial nucleoids.

Figure 1.24 (A) Supercoiled DNA. (B) Twisting of the ends in opposite directions causes linear DNA to wrap up on itself. The supercoiling is lost if the ends of the DNA are not somehow constrained. (C) A break, or nick, in one of the two strands of a circular DNA will relax the supercoils.

itself to relieve the stress. The DNA remains supercoiled as long as its ends are constrained and so cannot rotate, and a circular DNA has no free ends that can rotate. Therefore, a supercoiled circular DNA remains supercoiled but a linear DNA immediately loses its supercoiling unless the ends are somehow constrained.

Even a circular DNA loses its supercoiling if one of the strands of the DNA is cut, thereby allowing the strands to rotate around each other. The phosphodiester bond connecting the two deoxyribose sugars on the other strand serves as a swivel and rotates, resulting in relaxed (i.e., not supercoiled) DNA. A DNA with a phosphodiester bond broken in one of the two strands is said to be nicked. When nucleoids are prepared, some of the loops are usually relaxed, probably by nicks introduced during the extraction process. The fact that only some, and not all, of the loops of DNA in the nucleoid are relaxed tells us something about the structure of the nucleoid, i.e., that there are periodic barriers to rotation of the DNA. A break or nick in a circular DNA should relax the whole DNA unless portions of the molecule are periodically attached to barriers that prevent rotation of the strands, such as condensins and their associated kleisins, mentioned above.

SUPERCOILING OF NATURAL DNAs

It is possible to estimate the supercoiling of natural DNAs. According to the Watson-Crick structure, the two strands are wrapped around each other about once every 10.5 bp to form the double helix. Therefore, in a DNA of 2,100 bp, the two strands should be wrapped around each other about 2,100/10.5, or 200, times. In a supercoiled DNA of this size, however, the two strands are wrapped around each other either more or less than 200 times. If they are wrapped around each other more than once every 10.5 bp, the DNA is said to be **positively supercoiled;** if less than once every 10.5 bp, it is **negatively supercoiled.**

Most DNA in bacteria is negatively supercoiled, with an average of one negative supercoil for every 300 bp, although there are localized regions of higher or lower negative supercoiling. Also, in some regions, such as ahead of a transcribing RNA polymerase, the DNA may be positively supercoiled.

SUPERCOILING STRESS

Some of the stress due to supercoiling of the DNA, which causes it to twist up on itself, can be relieved if the DNA is wrapped around something else such as proteins. Sailors know about this effect: if you twist a rope as you roll it up to store it, it does not try to unroll itself again when you are finished. Wrapping DNA around proteins in the cell is called constraining the supercoils. Unconstrained supercoils cause stress in the DNA, which can be relieved by twisting the DNA up on itself, as shown in Figure 1.24, and making the DNA more compact. The stress due to unconstrained supercoils can have other effects as well, for example, helping to separate the strands of DNA during reactions such as replication, recombination, and initiation of RNA synthesis at promoters.

Topoisomerases

The supercoiling of DNA in the cell is modulated by topoisomerases, (see Wang, Suggested Reading). All organisms have these proteins, which manage to remove the supercoils from a circular DNA without permanently breaking either of the two strands. They perform this feat by binding to DNA, breaking one or both of the strands, and passing the DNA strands through the break before resealing it. As long as the enzyme holds the cut ends of the DNA so that they do not rotate, this process, known as **strand passage,** either introduces or removes supercoils from DNA.

The topoisomerases are classified into two groups, type I and type II (Figure 1.25). These two types differ in

Figure 1.25 Action of the two types of topoisomerases. The type I topoisomerases break one strand of DNA and pass the other strand through the break, removing one supercoil at a time. The type II topoisomerases break both strands and pass another part of the same DNA through the breaks, introducing or removing two supercoils at a time.

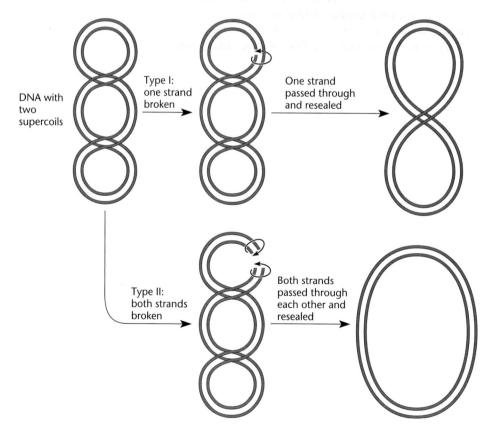

how many strands are cut and how many strands pass through the cut. The type I topoisomerases cut one strand and pass the other strand through the break before resealing the cut. The type II topoisomerases cut both strands and pass two other strands from somewhere else in the DNA or even another DNA through the break before resealing it. Hence, type I topoisomerases change DNA one supercoil at a time whereas type II topoisomerases change DNA two supercoils at a time, as shown in Figure 1.25.

TYPE I TOPOISOMERASES

Bacteria have several type I topoisomerases. The major bacterial type I topoisomerase removes negative supercoils from DNA. In E. coli and Salmonella enterica serovar Typhimurium, the topA gene encodes this type I topoisomerase. As expected, DNA isolated from E. coli with a topA mutation is more negatively supercoiled than normal.

TYPE II TOPOISOMERASES

Bacteria also have more than one type II topoisomerase. Because type II topoisomerases can break both strands and pass two other DNA strands through the break, they can either separate two linked circular DNA molecules or link them up. Linkage sometimes happens after replication or recombination. The major type II topoisomerase in bacteria is called gyrase instead of topoisomerase II because rather than removing negative supercoils like most type II topoisomerases, this enzyme adds them. Gyrase acts by first wrapping the DNA around itself and then cutting the two strands before passing another part of the DNA through the cuts, thereby introducing two negative supercoils. Adding negative supercoils increases the stress in the DNA and so requires energy; hence, gyrase needs ATP for this reaction.

The gyrase of E. coli is made up of four polypeptides, two of which are encoded by the gyrA gene and two of which are encoded by gyrB. These genes were first identified by mutations that make the cell resistant to antibiotics that affect gyrase (see Table 1.2). The GyrA subunits seem to be responsible for breaking the DNA and holding it as the strands pass through the cuts. The GyrB subunits have the ATP site that furnishes the energy for the supercoiling.

As mentioned above, the other major type II topoisomerase in E. coli, topo IV, decatenates daughter chromosomes after infection, allowing them to be segregated into the daughter cells.

The Bacterial Genome

The discussion of the replication and structure of the bacterial genome ignores the complexity of the sequences and the functions they encode. Box 1.4 discusses some of the features of bacterial chromosomal sequences that have been discovered from sequencing bacterial genomes. To date, more than 500 bacterial genomes have been sequenced. Most of these were chosen for sequencing because they had medical or ecological importance or are model systems for molecular genetics research. Bacteria are in some ways the ideal subjects for genome analysis because their genomes are relatively small, ranging from ca. 0.5 Mb with only about 500 genes for some obligate parasites to ca. 10 Mb with 10,000 genes for some free-living bacteria. They contain few introns and much less repetitive DNA than eukaryotic genomes. One interesting outcome of the sequencing of various genomes is the information about how some bacteria from the same genus, but with very different lifestyles, differ mostly in the prophages and genetic islands they carry. For example, the harmless E. coli K-12, which is used extensively for molecular genetic analysis, and the extremely pathogenic E. coli O157:H7, which causes major outbreaks of food poisoning, differ mostly in their prophages and genetic islands but otherwise are very similar in sequence. Knowledge such as this is important to our understanding of how bacteria adapt to different environments and also may suggest ways to combat bacterial diseases, particularly emergent diseases. We discuss prophages in chapter 8 and genetic islands in chapter 9.

TABLE 1.2	Antibiotics that block replication	
Antibiotic	**Source**	**Target**
Trimethoprim	Chemically synthesized	Dihydrofolate reductase
Hydroxyurea	Chemically synthesized	Ribonucleotide reductase
5-Fluorodeoxyuridine	Chemically synthesized	Thymidylate synthetase
Nalidixic acid	Chemically synthesized	gyrA subunit of gyrase
Novobiocin	Streptomyces sphaeroides	gyrB subnit of gyrase
Mitomycin C	Streptomyces caespitosus	Cross-links DNA

BOX 1.4

Features of Bacterial Genomes

Once a genome is sequenced, it can be "annotated," which means that it is analyzed for features including open reading frames, RNA-encoding genes, repeat sequences, and regulatory sequences. Ongoing advances in bioinformatics are allowing more and more information to be derived from such sequence information. A fundamental tool, the **Basic Local Alignment Search Tool (BLAST)**, can find regions of similarity between different genomes or within a single genome. BLAST can be used for many types of similarity searches (see References, below, and Box 2.7). Nucleotide-nucleotide searches can be analyzed with **blastn.**

Bacterial genomes are densely packed with coding information. For example, the *E. coli* genome contains at least 4,400 genes in approximately 4.6 Mb of DNA. The average gene density in bacterial genomes is 1.1 kb; that is, on average, one gene is found in every 1.1 kb. In contrast, the mouse and human genomes have average gene densities of one gene every 60 to 80 kb. The major types of coding information evaluated at this time encode for proteins, rRNAs, or tRNAs, but more and more regions encoding small RNAs and small peptides are being discovered. Recent research on such gene products is discussed in later chapters.

The sequences of more and more genomes have been entered into databases, and proteins with the same function often have sequences, called domains and motifs, in common (see Box 2.7). Therefore, from the sequence of a gene alone, it is possible to search the databases to find similar proteins of known function. This makes it possible to guess the functions of many of the gene products of the organism.

By sequencing genomes, a number of insights into evolution may emerge. For example, there may be interesting patterns in how genes of similar function are organized in different species. An important outcome of comparing genome sequences has been the observation of the high degree of conservation in genetic linkage, called **synteny.** For example, two *Bacillus* species, *Bacillus subtilis*, which has been

a model genetic organism, and *Bacillus lichenformis,* which is used in biotechnology because of its highly efficient secretory system, share large syntenic regions interspersed with unique regions composed of prophages, insertion sequence elements, and metabolic gene clusters called genetic islands.

Genome sequencing has revealed why *E. coli* strains can be either "intestinal friends" or "intestinal foes." Since the significance of *E. coli* as a model organism is discussed in the text, it is important to clarify what distinguishes the disease-causing *E. coli* strains, such as the O157:H7 strain that is the causative agent of deadly infections worldwide, from the *E. coli* laboratory strains such as K-12. Both *E. coli* K-12 and *E. coli* O157:H7 share 4.1 Mb of DNA sequence homology, but scattered throughout the genome of O157:H7 are long DNA regions that encode virulence characteristics. This additional DNA, approximately 1 Mb, is almost completely composed of prophages and other horizontally transferable genetic "islands" (see chapter 8).

The minimum number of genes required to make an organism might be determined by comparing the total genomes of many different organisms. Only the genes shared by all types of organisms would be absolutely required to make a living organism. Another reason to sequence the entire DNA of a living organism is to make its identification as unambiguous as possible, for example in epidemiology so as to trace the source of a disease. Repetitive DNA sequences are especially useful for such analyses.

Bacterial genomes have relatively little repetitive DNA compared to eukaryotic genomes. For example, less than 1% of the *E. coli* genome sequence is repetitive whereas almost 50% of the human genome consists of repetitive sequences. However, certain classes of repeated sequences are found in many bacteria. Repetitive extragenic palindromic (REP) elements, which are imperfect palindromes of 30 to 40 bp in length, and enterobacterial repetitive intergenic consensus (ERIC) sequences, which are several hundred base pairs long,

Size range of genomes		
Organism	**Size (Mbp)**	**Comments**
Mycoplasma genitalium	0.58	Smallest cell genome
Treponema pallidum	1.14	Causes syphilis
Helicobacter pylori	1.67	Causes duodenal ulcers
Sulfolobus solfataricus	2.25	Found in hot springs in Yellowstone National Park
Bacillus subtilis	4.20	Soil bacterium; "model" for development
Escherichia coli	4.64	Intestinal bacterium; "model" for genetics and physiology
Pseudomonas aeruginosa	6.26	Causes respiratory infections
Streptomyces coelicolor	8.4	Antibiotic-producing soil bacterium

BOX 1.4 (continued)

Features of Bacterial Genomes

are important examples. Both of these sequences are found in extragenic regions and are dispersed throughout the bacterial genomes. The biological importance of these sequences is not understood, but they have proven useful in **strain typing**, the process of identifying specific bacterial strains from medical or environmental samples, because these elements contain conserved sequences that can be used to design primers for an application of the PCR technique called REP-PCR (see Figure 1.33). Other repetitive sequences characterize specific bacteria. For example, the STAR sequence is a *Staphylococcus aureus* repeat signature (see Box 2.7). CRISPER (clustered regularly interspaced palindromic repeats) sequences are another example of repeat sequence found in gram-positive pathogens.

The sizes of the genomes of some common bacteria are shown in the table.

References
Blattner, F. R., G. Plunkett III, C. A. Bloch, N. T. Perna, V. Burland, M. Riley, J. Collado-Vides, J. D. Glasner, C. K. Rode, G. F. Mayhew, et al. 1997. The complete genome sequence of *Escherichia coli* K-12. *Science* **277**:1453–1462.

Earl, A. M., R. Losick, and R. Kolter. 2007. *Bacillus subtilis* genome diversity. *J. Bacteriol.* **189**:1163–1170. (Supplemental material at jb.asm.org.)

Perna, N. T., G. Plunkett III, V. Burlande, B. Mau, J. D. Glasner, D.-J. Rose, G. F. Mayhew, P. S. Evans, J. Gregor, H. A. Kirkpatrick, G. Posfai, J. Hackett, S. Klink, A. Boutin, Y. Shao, L. Miller, E. J. Grotbeck, N. W. Davis, A. Lim, E. T. Dimalanta, K. D. Potamousis, J. Apodaca, T. S. Anantharaman, J. Lin, G. Yen, D. C. Schwartz, R. A. Welch, and F. R. Blattner. 2001. Genome sequence of enterohaemorrhagic *Escherichia coli* O157:H7. *Nature* **409**:529–533.

Public Access Genomics Resources
Comprehensive Microbial Resource:
www.tigr.org/tigrscripts/CMR2/CMRHomePage.spl

E. coli-based databases for *E. coli* and other microbes: ecocyc.org

European Bioinformatics Institute: www.ensembl.org

France: www.genoscope.cns.fr

Japan: www.ddbj.nig.ac.jp

United Kingdom: www.ebi.ac.uk

United States (Department of Energy Joint Genomes Institute): www.jgi.doe.gov

United States (Lawrence Berkeley Laboratory): www.microbesonline.org

United States (National Center for Biotechnology Information database; contains extensive user information and assistance): www.ncbi.nlm.nih.gov and, for sequence alignments, www.ncbi.nlm.nih.gov/BLAST

Antibiotics That Affect Replication and DNA Structure

Antibiotics are substances that block the growth of cells. Many antibiotics are naturally synthesized chemical compounds made by soil microorganisms, especially actinomycetes, to help them compete with other soil microorganisms. Consequently, this group of compounds has a broad spectrum of activity and target specificity. Antibiotics have proven useful in enhancing our understanding of cellular functions as well as in treating diseases. Many antibiotics stop the growth of bacteria by specifically blocking DNA replication or by changing the molecule's structure. Table 1.2 lists a few representative antibiotics that affect DNA, along with their targets in the cell and their sources. Because some parts of the replication machinery have remained relatively unchanged throughout evolution, many of these antibiotics work against essentially all types of bacteria. Some even work against eukaryotic cells and so are used as antifungal agents and in tumor chemotherapy.

Antibiotics That Block Precursor Synthesis

As discussed above, DNA is polymerized from the deoxynucleoside triphosphates. Any antibiotic that blocks the synthesis of these deoxynucleotide precursors will block DNA replication.

INHIBITION OF DIHYDROFOLATE REDUCTASE

Some of the most important precursor synthesis blockers are antibiotics that inhibit the enzyme dihydrofolate reductase. One such compound, trimethoprim, works very effectively in bacteria, and the antitumor drug methotrexate (amethopterin) inhibits the dihydrofolate reductase of eukaryotes. Methotrexate is used as an antitumor agent.

Antibiotics, like trimethoprim, that inhibit dihydrofolate reductase kill the cell by depleting it of tetrahydrofolate, which is needed for many biosynthetic reactions. This inhibition is overcome, however, if the cell lacks the enzyme thymidylate synthetase, which synthesizes dTMP; therefore, most mutants that are resistant to trimethoprim have mutations that inactivate the *thyA* thymidylate synthetase gene. The reason is apparent from the pathway for dTMP synthesis shown in Figure 1.5. Thymidylate synthetase is solely responsible for converting tetrahydrofolate to dihydrofolate when it transfers a methyl group from tetrahydrofolate to dUMP to make dTMP. The dihydrofolate reductase is the only enzyme in the cell that can restore the tetrahydrofolate needed for other biosynthetic reactions. However, if the cell lacks thymidylate synthetase, there is no need for a dihydrofolate reductase to restore tetrahydrofolate. Therefore, inhibition of the dihydrofolate reductase by trimethoprim has no effect, thus making *thyA* mutant cells resistant to the antibiotic. Of course, if the cell lacks a thymidylate synthetase, it cannot make its own dTMP and must be provided with thymidine in the medium so that it can replicate its DNA.

There is more than one mechanism by which cells can achieve trimethoprim resistance. They can have an altered dihydrofolate reductase to which trimethoprim cannot bind, or they can have more copies of the gene so that they make more enzyme. Some plasmids and transposons carry genes for resistance to trimethoprim. These genes encode dihydrofolate reductases that are much less sensitive to trimethoprim and so can act even in the presence of high concentrations of the antibiotic.

INHIBITION OF RIBONUCLEOTIDE REDUCTASE
The antibiotic hydroxyurea inhibits the enzyme ribonucleotide reductase, which is required for the synthesis of all four precursors of DNA synthesis (Figure 1.5). The ribonucleotide reductase catalyzes the synthesis of the deoxynucleoside diphosphates dCDP, dGDP, dADP, and dUDP from the ribonucleoside diphosphates, an essential step in deoxynucleoside triphosphate synthesis. Mutants resistant to hydroxyurea have an altered ribonucleotide reductase.

COMPETITION WITH dUMP
5-Fluorodeoxyuridine and the related 5-fluorouracil have monophosphate forms resembling dUMP, the substrate for the thymidylate synthetase. By competing with the natural substrate for this enzyme, they inhibit the synthesis of deoxythymidine monophosphate. Mutants resistant to these compounds have an altered thymidylate synthetase. These are useful antibiotics for the treatment of fungal as well as bacterial infections.

Antibiotics That Block Polymerization of Deoxynucleotides

The polymerization of deoxynucleotide precursors into DNA would also seem to be a tempting target for antibiotics. However, there seem to be surprisingly few

antibiotics that directly block this process. Most antibiotics that block polymerization do so indirectly, by binding to DNA or by mimicking the deoxynucleotides and causing chain termination rather than by inhibiting the DNA polymerase itself.

DEOXYNUCLEOTIDE PRECURSOR MIMICS
Dideoxynucleotides are similar to the normal deoxynucleotide precursors except that they lack a hydroxyl group on the 3′ carbon of the deoxynucleotide. Consequently, they can be incorporated into DNA, but then replication stops because they cannot link up with the next deoxynucleotide. These compounds are not useful antibacterial agents, probably because they are not phosphorylated well in bacterial cells. However, this property of prematurely terminating replication has made them the basis for DNA sequencing (see below).

CROSS-LINKING
Mitomycin C blocks DNA synthesis by cross-linking the guanine bases in DNA to each other. Sometimes the cross-linked bases are in opposing strands. If the two strands are attached to each other, they cannot be separated during replication. Even one cross-link in DNA that is not repaired prevents replication of the chromosome. This antibiotic is also a useful antitumor drug, probably for the same reason.

Antibiotics That Affect DNA Structure

ACRIDINE DYES
The acridine dyes include proflavine, ethidium, and chloroquine. These compounds insert between the bases of DNA and thereby cause frameshift mutations and inhibit DNA synthesis. Their ability to insert themselves between the bases in DNA has made acridine dyes very useful in genetics and molecular biology. Some of these applications are discussed in later chapters. In general, acridine dyes are not useful as antibiotics because of their toxicity due to their ability to block DNA synthesis in the mitochondria of eukaryotic cells. Some members of this large family of antibiotics have long been used as antimalarial drugs because of their ability to block DNA synthesis in the mitochondria (kinetoplasts) of trypanosomes. This is the basis for the antimalarial activity of the tonic water in a gin and tonic.

THYMIDINE MIMIC
5-Bromodeoxyuridine (BUdR) is similar to thymidine and is efficiently phosphorylated and incorporated in its place. However, BUdR incorporated into DNA often mispairs and increases replication errors. DNA containing BUdR is also more sensitive to some wavelengths of ultraviolet (UV) light (which makes BUdR useful in

enrichment schemes for isolating mutants; see chapter 3). Moreover, DNA containing BUdR has a different density from DNA containing exclusively thymidine (another feature of BUdR that is useful in experiments).

Antibiotics That Affect Gyrase

Many antibiotics and antitumor drugs affect topoisomerases. The type II topoisomerase, gyrase, in bacteria is a target for many different antibiotics. These antibiotics kill the bacterial cell because gyrase is required for bacterial growth. Because this enzyme is similar among all bacteria, these antibiotics have a broad spectrum of activity and kill many types of bacteria.

GyrA INHIBITION

Nalidixic acid specifically binds to the GyrA subunit, which is involved in cutting the DNA and in strand passage. This activity makes nalidixic acid and its many derivatives, including oxolinic acid and chloromycetin, very useful antibiotics. Another antibiotic that binds to the GyrA subunit, ciprofloxacin (Cipro), is used for treating gonorrhea, anthrax, and bacterial dysentery. However, because these antibiotics can induce prophages, they may actually make some diseases worse (see chapter 8).

The mechanism of killing by these antibiotics is not completely understood. They are known to cause degradation of the DNA and can cause the DNA to become covalently linked to gyrase, presumably by trapping it in an intermediate state in the process of strand passage. Bacteria resistant to nalidixic acid have an altered *gyrA* gene.

GyrB INHIBITION

Novobiocin and its more potent relative coumermycin bind to the GyrB subunit, which is involved in ATP binding. These antibiotics do not resemble ATP, but by binding to the gyrase they somehow prevent ATP cleavage, perhaps by changing the conformation of the enzyme. Mutants resistant to novobiocin have an altered *gyrB* gene.

Molecular Biology Manipulations with DNA

In addition to their basic science interest, the meticulous studies of the mechanism of DNA replication in bacteria and the enzymes involved in DNA replication discussed above have led to many practical applications in molecular biology. These applications have had profound effects on many aspects of our everyday lives including medicine, agriculture, and even law enforcement. We review some of these applications in this section.

Restriction Endonucleases

Among the most useful enzymes that alter DNA are the restriction endonucleases. These are enzymes that recognize specific sequences in DNA and cut the DNA in or close to the recognition sequence. They are usually accompanied by methylating activities that modify DNA by methylating the DNA in the recognition sequence, making it immune to cutting by the endonuclease activity. These enzymes are made exclusively by bacteria, and their role seems to be to defend against phages by cutting incoming unmodified phage DNA. Some of them may also play a role in preventing the loss of plasmids by killing the cell if it is cured of the plasmid (see Box 4.3).

The restriction endonucleases are classified into three groups, types I, II, and III. The enzymes in these groups differ mostly in the relationship between their methylating and cutting activities. The type II enzymes have proven to be most useful because the methylating activity can be separated from the cutting activity. Hundreds of type II enzymes are known, and many of them can be purchased from biochemical supply companies. What makes them so useful is that they each recognize their own specific sequence in DNA and then cut the DNA at or close to the recognition sequence. Also, because the sequences recognized by many of them are palindromic by making staggered breaks in these sequences, they leave complementary (sticky) ends that can be used for DNA cloning (see below). The recognition sequences are often 4, 6, or 8 bp long. The sequences recognized by some restriction endonucleases are shown in Table 1.3.

USING RESTRICTION ENDONUCLEASES TO CREATE RECOMBINANT DNA

As mentioned, one of the properties of some type II restriction endonucleases that make them so useful is that the sequences they recognize read the same in the 5'-to-3' direction on both strands. Such a sequence is said to have a twofold rotational symmetry or be a **palindrome** because it reads the same if you rotate it through 180°

TABLE 1.3	Recognition sequences of restriction endonucleases
Enzyme	**Recognition sequence**[a]
Sau3A	*GATC/CTAG*
BamHI	G*GATCC/CCTAG*G
EcoRI	G*AATTC/CTTAA*G
PstI	CTGCA*G/G*ACGTC
HindIII	A*AGCTT/TTCGA*A
SmaI	CCC*GGG/GGG*CCC
NotI	GC*GGCCGC/CGCCGG*CG

[a]Asterisks indicate where the endonucleases cut.

and read the other strand. The word comes from a palindrome in English, where letters read the same in both directions, as in MADAM I'M ADAM. Because the sequence reads the same on both strands, the restriction endonuclease binds to the identical sequence on both strands and then cuts the two strands at the same place in the sequence. For example, the restriction endonuclease HindIII (so called because it was the third restriction endonuclease found in *Haemophilus influenzae*) recognizes the 6-bp sequence 5'AAGCTT3'/3'TTCGAA5' and cuts between the two A's on each strand (Table 1.3), which is in the same place in the sequences of the two strands read in the 5'-to-3' direction. Such a break is called a "staggered break" because the breaks in the two strands are not exactly opposite each other in the DNA, which has the effect of leaving short single-stranded ends on both ends of the broken DNA. Because the original sequence which was cut had a twofold rotational symmetry, both of these single-stranded ends have the same sequence read from 5' to 3', in this case 5'AGCT3'. The two single-stranded ends are complementary to each other and so can pair with each other. More importantly, each can pair with the single-stranded end of any other DNA that had been cut with the same restriction endonuclease, because they all have the single-stranded ends with the same sequence. These single-stranded ends are called **sticky ends** because they can pair with ("stick" to) any other single-stranded ends with the complementary sequence. Other restriction nucleases might leave the same sticky ends even if they recognize a somewhat different sequence. Such restriction endonucleases are said to be **compatible**. When two sticky ends pair with each other, they leave a double-stranded DNA with staggered nicks in the two strands, which can then be sealed by DNA ligase, as illustrated in Figure 1.26. The new DNA that has been created this way is called **recombinant DNA** because two DNAs have been recombined into new sequence combinations. While other methods can be used to make recombinant DNA, this method involving restriction endonucleases has been and continues to be the most generally applicable.

CLONING AND CLONING VECTORS

There is only one molecule of a recombinant DNA when it first forms. In order for it to be useful, many copies of the recombinant DNA molecule are needed. This is the function of **cloning vectors.** A cloning vector is a DNA that has its own origin of replication and is capable of independent replication in the cell, for example, plasmids. DNAs which have an *ori* sequence that makes them capable of independent replication in the cell are called **replicons.** The process of cloning a piece of DNA into a circular plasmid with its own *ori* sequence is illustrated

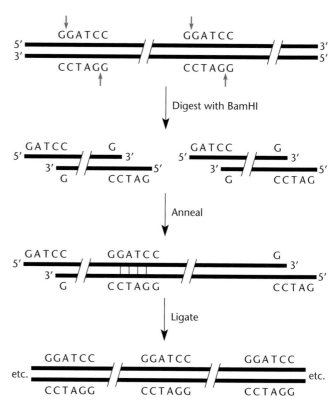

Figure 1.26 Creation of sticky complementary ends by cutting with a restriction endonuclease. The two single-stranded ends can pair with each other, and the nicks can be sealed by DNA ligase.

in Figure 1.27. Once it has been joined to another piece of DNA and introduced into a cell, the cloning vector replicates itself, along with the piece of DNA to which it is joined, making many exact copies of the original DNA molecule. These exact replicas of the piece of DNA are called **DNA clones** in analogy to the genetic replicas of an organism that are made when an organism replicates itself asexually. Phages also are capable of independent replication in their bacterial hosts, and some of these have been modified to serve as convenient cloning vectors. Some examples of cloning vectors and their relative advantages are discussed in subsequent chapters.

DNA LIBRARIES

A DNA library is a collection of DNA clones that includes all, or at least almost all, the DNA sequences of an organism. One way to make a DNA library is with restriction endonucleases. The entire DNA of an organism is cut with restriction endonucleases, and the pieces are ligated into a cloning vector cut with a compatible enzyme. The mixture is then transformed or transfected into cells, and the transformations or plaques are pooled. If the collection is large enough, every DNA sequence of

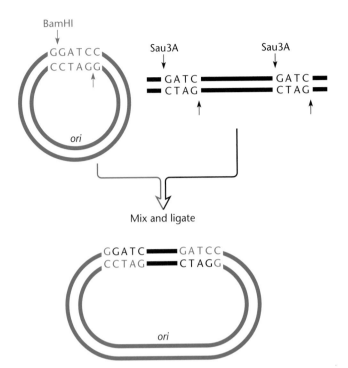

Figure 1.27 DNA cloning. The compatible restriction endonucleases Sau3A and BamHI were used to clone a piece of DNA into a cloning vector. The DNA to be cloned was cut with Sau3A and ligated into a cloning vector cut with BamHI. The piece of DNA inserted into the cloning vector cannot replicate by itself, since it lacks an *ori* region; however, once it is inserted into the cloning vector, it will replicate each time the cloning vector replicates.

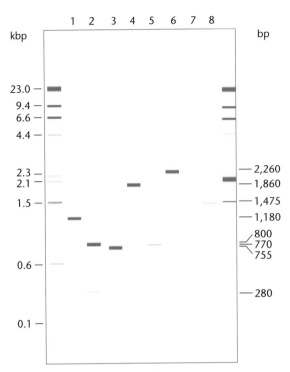

Figure 1.28 Agarose gel electrophoresis of fragments of DNA. Smaller fragments migrate faster on the gel and so move farther in the same amount of time. The outside lanes are marker DNAs of known size for comparison.

the organism is represented somewhere in the pooled clones, and the library is complete. The trick then is to find the clone you want out of all the clones in the library; some methods to do this are mentioned in this section and other chapters.

The number of clones required to make a complete DNA library of an organism depends on the complexity of the DNA of that organism. For example, if we make a library of *E. coli* DNA cut with an enzyme that recognizes six base pairs (a six-hitter) like EcoRI, we need about $4.5 \times 10^6/4 \times 10^3 \approx 1,100$ different clones, since *E. coli* DNA contains approximately 4.5×10^6 bp and a six-hitter like EcoRI cuts the DNA about once every 4,000 bp. In contrast, a library of λ DNA should require only about $5 \times 10^4/4 \times 10^3 \approx 13$ clones, since λ DNA has only about 50,000 bp. An important point is that these are minimum estimates of the number of clones required to make a library of the DNA of the organism; not all clones are equally represented in the library because of random statistical fluctuation. Also, some pieces may be easier to clone, for example because they are smaller or contain no genes whose products are toxic to the cell.

PHYSICAL MAPPING

Another important use of restriction endonucleases is in **physical mapping.** In analogy to genetic mapping, where the approximate position of mutations is determined by genetic crosses (see the discussion of genetic analysis in chapter 3), physical mapping is the process of determining the exact position of particular sequences in the nucleotide sequence of the DNA. Because restriction endonucleases cut the DNA only at specific sequences, they create unique-sized pieces for each DNA molecule. From the sizes of these pieces, it is possible to determine where the recognition sequences for the restriction endonuclease must have been on the original DNA. By comparing the sizes of the pieces left by a number of different restriction endonucleases, it is possible to order the restriction sites with respect to each other and construct a physical map of the DNA. Figures 1.28 and 1.29 illustrate the reasoning behind the physical mapping of restriction sites in a DNA.

RESTRICTION SITE POLYMORPHISMS

While all the members of a particular species are very similar genetically, there are minor differences called polymorphisms between individuals. These minor genetic differences are reflected in differences in the sequence of

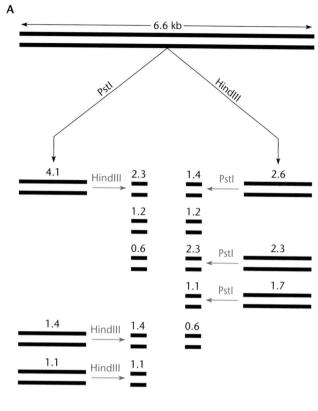

Figure 1.29   A method for mapping the restriction sites on a DNA fragment. The original fragment of 6.6 kb (6.6 × 10³ bp) contains two recognition sites for the restriction endonuclease HindIII and two sites for PstI. (A) The fragment is digested with HindIII and PstI separately, and the isolated fragments are digested with the other restriction endonuclease (in purple). (B) From the size of the fragments, the order of the sites must have been as shown.

DNA from each organism, in particular in differences in the locations of restriction sites. These differences are called **restriction fragment length polymorphisms (RFLPs)** and can be due to deletions or insertions between the positions of the sites or inversions of DNA containing the sites. RFLPs in the DNA can be useful for determining the ancestry of a particular individual, mapping genetic diseases, and, in forensic science, identifying the person who committed a crime and left behind some blood or other material containing their DNA.

Hybridizations

Many applications in molecular genetics have come from our knowledge of the structure of DNA and how it is synthesized and held together. The two strands of DNA

are held together in the double helix by hydrogen bonds between the complementary bases. Heating double-stranded DNA or treating it at high pH disrupts these hydrogen bonds and causes the two strands to separate. If the temperature is then lowered or the pH is returned

Figure 1.30   Hybridization methods. (A) Method of Southern blot hybridization. In step 1, DNA is isolated and digested with a restriction endonuclease. In steps 2 and 3, after electrophoresis (step 2), the DNA is transferred and fixed to a filter (step 3). In step 4, the filter is hybridized with a probe. Only bands complementary to the probe appear as dark bands because the signal detection procedure reveals the radioactivity or reactive chemical in the probe. (B) Plate hybridizations. The colonies or plaques on a plate are transferred to a membrane filter, and the DNA is denatured. The DNA on the filters is then hybridized to a labeled probe as in a Southern blot hybridization to identify the colonies or plaques that contain DNA complementary to the probe.

A   Step 1   DNA isolation and digestion

Step 2   DNA electrophoresis

Step 3   DNA transfer

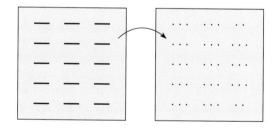

Step 4   DNA hybridization

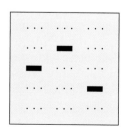

B

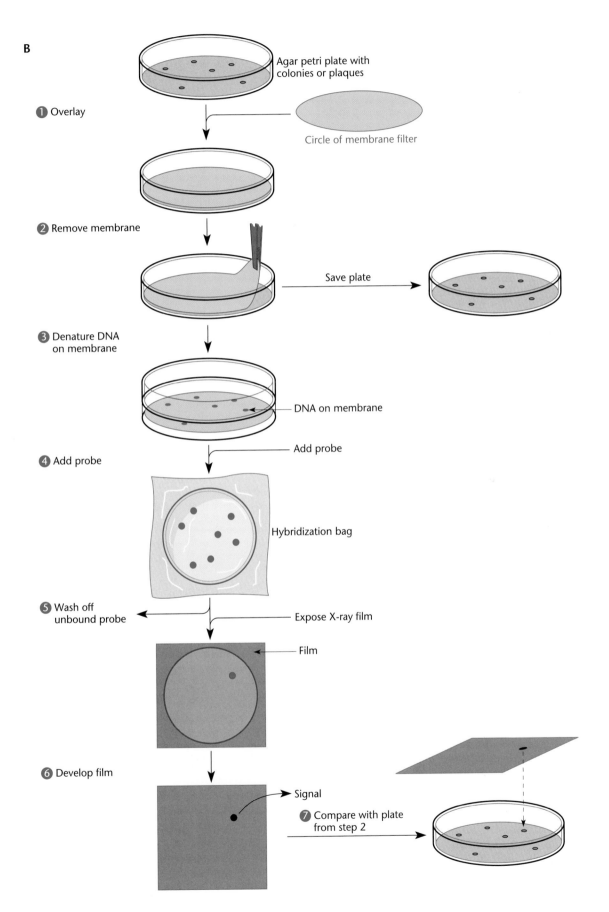

to neutral, the complementary sequences eventually find each other and a new double helix is formed. Two strands of RNA or a strand of DNA and a strand of RNA can also be held together by such a double-stranded helix, provided that their sequences are complementary. This process is called **hybridization.** Under optimal conditions, two strands of DNA or RNA form a double-stranded helix only if their sequences are almost perfectly complementary, making hybridization very specific and sensitive and allowing the detection of RNA or DNA of a particular sequence among thousands of other sequences.

BLOTS AND PLATE HYBRIDIZATIONS

Most methods of hybridization utilize a membrane filter made of nitrocellulose or some other related compound. First, a single-stranded DNA or RNA is fixed to the membrane. Then a solution containing another DNA or RNA is added to the membrane. If the second RNA or DNA hybridizes to the RNA or DNA fixed to the membrane, it too becomes fixed to the membrane. The hybridization can be detected provided that the second DNA or RNA has been labeled somehow, for example with radioactivity or fluorescent chemicals. Similar techniques can be used to detect the binding of specific antibodies to proteins fixed on a filter or even the binding of proteins to DNA fixed on a filter and vice versa.

Some of the most useful techniques in molecular biology involve filter hybridization (Figure 1.30), in which

the membrane filter receives a replica of the molecules on a gel or of the colonies or plaques on a petri plate: the filter is layered on the gel or plate, and the macromolecules (DNA, RNA, or proteins) are transferred to their same position on the filter as they were on the gel or petri plate. The transfer can be by diffusion, capillary action, or use of an electric field, depending on the application. Transfer of DNA, RNA, or proteins from a gel to a filter is called **blotting,** and the filter containing the replica is a **blot.** The blot can then be hybridized to a labeled probe to determine the location of particular DNA or RNA sequences or proteins on the original gel or plate.

The first such procedure for transferring DNA bands from a gel to a filter was named **Southern blotting** because it was developed by Ed Southern. When similar procedures were developed for blotting RNA and proteins, they were whimsically given the names of other directions on the compass: **Northern blotting** for RNA blots, **Western blotting** for protein blots, and so on.

The principle behind Southern blotting and an example of such a blot are shown in Figures 1.30 and 1.31. In the left-hand panel of Figure 1.31, a mixture of DNA fragments has been applied to an agarose gel and subjected to an electric field. The fragments separate on the basis of their size, with the smaller fragments moving faster. After electrophoresis, the fragments can be stained with ethidium bromide and the gel can be photographed to show the positions of all of the bands (Figure 1.31,

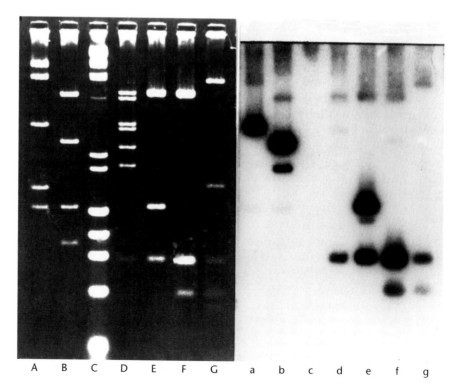

Figure 1.31 Results of a Southern blot hybridization. Lanes A to G show the total DNA in each lane. Lanes a to g show the bands that hybridize to a specific probe. Markers are shown in lane C.

lanes A to G). If a blot is made of this gel and the blot is hybridized to DNA of a particular sequence, the only bands that appear are the bands which contain DNA complementary to the probe (Figure 1.31, lanes a to g). For detailed protocols of this and other types of blotting, consult a cloning manual (see Suggested Reading).

Applications of the Enzymes Used in DNA Replication

As mentioned above, meticulous work on the properties of the enzymes involved in DNA replication has not only increased our understanding of DNA replication but also led directly to many important practical applications. As discussed in the introductory chapter, many of these enzymes were first detected and purified from bacteria and phage-infected bacteria, but their applications extend to the molecular genetics of all organisms. A few of the more prominent applications are discussed here.

DNA POLYMERASES

The properties of DNA polymerases have been exploited in many applications in molecular genetics. As discussed earlier in this chapter, these enzymes all extend a primer polynucleotide chain by attaching the 5′ phosphate of an incoming deoxynucleotide to the 3′ end of the growing primer chain. They can synthesize DNA only by extending primers which are hybridized to a template DNA, and the choice of which deoxynucleotide to add at each step is determined by complementary base pairing between the incoming deoxynucleotide and the template DNA, leading to synthesis of DNA which is a complementary copy of the template.

DNA Sequencing

One important application of DNA polymerases is in DNA sequencing, the process of determining the sequence of deoxynucleotides in DNA. This technology has received much publicity recently with the sequencing of the entire almost 1-m-long human DNA, containing many billions of deoxynucleotides, the so-called Human Genome Project. Sequencing of entire genomes required a tremendous amount of technical ingenuity, but the principle behind it is quite simple, involving knowledge of the properties of DNA polymerases.

The DNA-sequencing methods used today are based on a method developed by Fred Sanger, who received a Nobel Prize for this work (his second; the first was for sequencing a protein). This method is called the dideoxy method and is based on the chain-terminating property of the dideoxynucleotides. As mentioned, the dideoxynucleotides are like the normal deoxynucleotides, except that they have hydrogens at the 3′ positions of the deoxyribose sugar. The dideoxynucleotides can be phosphorylated to give the dideoxynucleoside triphosphates, which are then incorporated into DNA by DNA polymerases that are unable to tell the difference between them and the normal deoxynucleotides. However, because it lacks a hydroxyl group at the 3′ position, an incorporated dideoxynucleotide cannot be joined to the 5′ phosphate of the next deoxynucleotide, and so the growing DNA chain terminates. From the length of the DNA chain that is made, we can deduce which base must have been next in the template DNA, because the base in the chain-termininating dideoxynucleotide which was last added was the complement of this base. In the original method on which all the later variations are based, four separate polymerizing reactions were run, with the mixture for each reaction containing a small amount of one of the four dideoxynucleoside triphosphates (ddTTP, ddGTP, ddATP, or ddCTP) mixed with the normal deoxynucleoside triphosphates. A short DNA primer complementary to a known sequence adjacent to the unknown DNA sequence is hybridized to the DNA, and DNA polymerase is added. The DNA polymerase extends the primer, making a DNA which is complementary to the template. Each time the DNA polymerase encounters a base in the template DNA that is complementary to the dideoxynucleotide used in that reaction, there is a chance that the dideoxynucleotide instead of the normal deoxynucleotide will be incorporated into the chain and the growth of that chain will be terminated. Since each reaction mixture contains a different dideoxynucleoside triphosphate, each of the four reactions produces a set of shortened DNA chains of different lengths determined by the positions of the complementary nucleotides in the template DNA. Therefore, the sequence of the template DNA can be read by determining which of the four reactions yielded the next longest DNA chain.

Random Shotgun Sequencing of Bacterial Genomes

Special techniques are required to sequence entire genomes because they are so long. A single sequencing reaction reveals the sequence of only hundreds of base pairs, while even bacterial genomes are on the order of millions of base pairs long. Box 1.5 describes commonly used procedures to sequence bacterial genomes. One way that works particularly well for bacterial genomes is **random shotgun sequencing**. The genome is first broken into random smaller pieces that can be sequenced in their entirety. These sequences can then be entered into a computer, where available software will look for overlapping sequences between the pieces. This allows the ordering of

Bacterial Genome Sequencing

Many efforts, known as genome projects, are under way to sequence the DNA of entire organisms. The most famous of these is the Human Genome Project, as a result of which the sequence of the billions of deoxynucleotides making up the almost 1-meter (or 1-yard)-long DNA of a human is now known. The genomic DNA of phage λ was the first to be sequenced, and now many other virus and bacterial genomes have been sequenced. At the time of this writing, more than 500 microbial genomes have been sequenced and published, and several hundred more will be finished in the near future. Microbes of particular medical or environmental importance or ones which serve as model systems for cell and molecular biology were the first to be chosen for sequencing. Advances in automation of sequencing technology have greatly increased the pace of sequence data accumulation.

The method used for most bacterial genomic sequencing projects is random shotgun sequencing. This is because bacterial genomes are relatively small and have relatively little repetitive DNA compared to eukaryotes. This approach was used for *Haemophilus influenzae* (see Fleischmann et al., below) and is illustrated in the flowchart. First, random pieces of the genome are cloned to produce libraries of the total DNA. These pieces are then sequenced, typically until the same sequence is obtained eight times on average, to provide an eightfold coverage of the genome. The sequencing is automated. Computerized analysis of the sequences allows their assembly into so-called "contigs," which are overlapping sequences. The sequences can then be put in order. Finally, sequencing of unrepresented regions that have no contigs on one side or the other is usually necessary to "close the gaps."

Once the genome is sequenced, it can be "annotated," which means that it is analyzed for features such as open reading frames, RNA-encoding genes, repeat sequences, consensus sequences, etc. (see Box 1.4). Ongoing advances in

Flowchart for genomic sequencing

1 **Isolate genomic DNA**
   DNA should be in pieces of >20 kb
2 **Shear DNA**
   DNA fragments will be of random lengths
3 **Size fractionate DNA**
   Collect fragments in size range from 1.5 to 2 kb
4 **Construct plasmid library**
   Inserts are the 1.5- to 2-kb genomic DNA
5 **Randomly sequence inserts**
   Sequencing process is highly automated
   Need ~15,000 sequence runs of ~500 to 600 bases each per megabase of genome DNA
6 **Assemble randomly generated sequence information in contiguous segments ("contigs")**
7 **Close gaps with directed sequencing**
   Several hundred reactions needed
8 **Analyze sequence**
   Bioinformatics allows "annotation" of open reading frames, etc.

bioinformatics are allowing more and more information to be derived from such sequence information.

References
Fleischmann, R. D., M. D. Adams, O. White, R. A. Clayton, E. F. Kirkness, A. R. Kerlavage, C. J. Bult, J.-F. Tomb, B. A. Dougherty, J. M. Merrick, K. McKenney, G. Sutton, W. Fitzhugh, C. Fields, J. D. Gocayne, J. Scott, R. Shirley, L.-I. Liu, A. Glodek, J. M. Kelley, J. F. Weldman, C. A. Phillips, T. Spriggs, E. Hedblom, M. D. Cotton, T. R. Utterback, M. C. Hanna, D. T. Nguyen, D. M. Saudek, R. C. Brandon, L. D. Fine, J. L. Fritchman, J. L. Fuhrmann, N. S. M. Geoghagen, C. L. Gnehm, L. A. McDonald, K. V. Small, C. M. Fraser, H. O. Smith, and J. C. Venter. 1995. Whole-genome random sequencing and assembly of *Haemophilus influenzae* Rd. *Science* 269:496–512. www.ornl.gov/microbialgenomes/.

Franguel, L., K. E. Nelson, C. Buchrueserm, A. Danchin, P. Glaser, and F. Kunst. 1999. Cloning and assembly strategies in bacterial genome projects. *Microbiology* 145:2625–2634.

the sequences because bacterial genomes have very little repetitive DNA. Some sequences are difficult to obtain, perhaps because they are particularly GC rich, have extensive secondary structure, or express toxic gene products. This can lead to gaps in the sequence that must be filled in by more laborious techniques. Bacterial genome sequencing has been automated to the point where a task that once took dozens of researchers months to accomplish can now be done by a few researchers in days.

Site-Specific Mutagenesis

Another exploitation of the properties of DNA polymerases is in site-specific mutagenesis. These methods allow the investigator to make a desired change at a particular site in the sequence of DNA rather than relying on more traditional methods of mutagenesis which more or less randomly cause mutations and do not target them to a particular site. Most methods for site-specific mutagenesis rely on synthetic DNA primers which are mostly complementary to the sequence of the DNA being mutagenized except for the desired mutational change. When this primer is hybridized to the DNA and used to prime the synthesis of new DNA by the DNA polymerase, the synthesized DNA has the same sequence as the template except for the change in the attached primer. This method of site-specific mutagenesis can be used only to make minor changes, such as single-base-pair changes, in the DNA sequence, because if the sequence of the primer is altered too much, it no longer hybridizes to the template DNA being mutagenized.

The difficult part of site-specific mutagenesis is in replicating the mutated DNA and selecting it from among the myriad of DNA molecules that have not been mutated. Many methods have been developed to accomplish this, one of which is the "two-primer" method illustrated in Figure 1.32. The DNA to be mutagenized is cloned into a cloning vector with a unique restriction site as shown. The entire cloning vector containing the clone is then replicated using two primers. One primer is complementary to the sequence being mutagenized, except that it contains the desired mutational change. The other

Figure 1.32 Use of two primers to eliminate the wild-type sequence after site-specific mutagenesis. See the text for details.

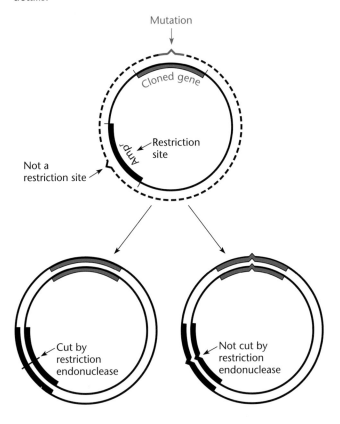

primer is complementary to the region of the cloning vector containing the unique restriction site, except that it has a single-base change in the recognition sequence for the restriction endonuclease. Both these primers are present at high concentrations, so that usually the replication of the DNA uses both primers rather than only one or the other. After the DNA polymerase has made a copy of the DNA and the DNA has been introduced into cells and allowed to replicate, the DNA is isolated and cut with the restriction nuclease. The DNAs not descended from a DNA that had replicated from the primers are cut and eliminated. The DNAs descended from a DNA replicated from the two primers are not cut and will survive. No such selection is 100% effective, and a few of the surviving DNA clones might still have to be sequenced to find one with the desired mutation.

Polymerase Chain Reaction

One of the most useful technical applications involving DNA polymerases is the **polymerase chain reaction** (**PCR**). This technology makes it possible to selectively amplify regions of DNA out of much longer DNAs. It is called the polymerase chain reaction because each newly synthesized DNA serves as the template for more DNA synthesis in a sort of chain reaction until large amounts of DNA have been amplified from a single DNA molecule. The power of this method is that it can be used to detect and amplify sequences from just a few molecules of DNA from any biological specimen, for example a drop of blood or a single hair; this has made it very useful in criminal investigations to identify the perpetrators of crimes on the basis of DNA typing. However, for our purposes here, it also has many other applications including the physical mapping of DNA, gene cloning, mutagenesis, and DNA sequencing.

The principles behind the use of PCR to amplify a region of DNA are outlined in Figure 1.33. PCR takes advantage of the same properties of DNA polymerases that are important in other applications, i.e., their ability to make a complementary copy of a DNA template starting from the 3' hydroxyl of a primer DNA. PCR uses two primers complementary to sequences on either side of the region to be amplified. One primer has the same sequence in the 5'-to-3' direction as one of the strands on one side of the region to be amplified, and the other has the complementary sequence to this strand on the other side of the region to be amplified, but written in the opposite direction. Thus, the two primers will prime the synthesis of DNA in opposite directions over the region to be amplified. The DNA is denatured to separate the strands and hybridized to the primers. One primer will prime the synthesis of DNA over the region to be amplified and

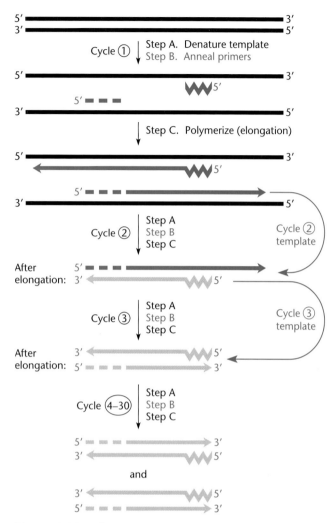

Figure 1.33   The steps in a PCR. In the first cycle, the template is denatured by heating. Primers are added; they hybridize to the separated strands for the synthesis of the complementary strand. The strands of the DNA are separated by the next heating cycle, and the process is repeated. The DNA polymerase survives the heating steps because it is from a thermophilic bacterium. The DNA sequence is amplified approximately 10^9-fold.

will continue polymerizing past the region. If the two strands are again separated by heating and the temperature is again lowered, the other primer can then hybridize to this newly synthesized strand and prime replication back over the region. Now, however, the DNA polymerase will run off the end of the template DNA when it reaches the end of the first primer sequence, leading to the synthesis of a short piece of DNA with the primer sequences on both ends. If this DNA is then heated to separate the strands, this shorter piece can bind another primer DNA, which can then prime the synthesis of the complementary strand of this

shorter DNA. This process of heating and cooling can be repeated 30 or 40 times until large numbers of copies of the particular DNA region have accumulated, beginning from one or very few longer DNA molecules containing the region.

In principle, any DNA polymerase could be used to perform such an amplification. However, most DNA polymerases would be inactivated by the high temperatures required to separate the strands of double-stranded DNA, making it necessary to add fresh DNA polymerase after each heating step. This is where the DNA polymerases from thermophilic bacteria such as *Thermus aquaticus* come to the rescue. These bacteria normally live at very high temperatures, and so their DNA polymerase, called the *Taq* polymerase, can survive the high temperatures needed to separate the strands of DNA, obviating the need to add new DNA polymerase at each step. We can just mix the primers, a tiny amount of biological material containing DNA with the region to be amplified, and the *Taq* polymerase; set a thermocycler to heat and cool over and over again; and come back a few hours later. Voilà, we should have large amounts of the region of the amplified DNA, which we can detect on a gel.

PCR MUTAGENESIS
PCR can also be used either to make specific changes in a DNA sequence or to randomly mutagenize a region of DNA. Making specific changes by PCR is similar to the other means of site-specific mutagenesis. A complementary primer is made, but with the desired change in the sequence. When the polymerase uses the primer to amplify the region, the specific change is made in the sequence.

PCR can be used to make random changes in a sequence because the *Taq* polymerase makes many mistakes, particularly in the presence of manganese ions, because it lacks an editing function. In fact, the mistake level during normal amplification by *Taq* polymerase is so high that clones made from PCR fragments should usually be sequenced to be certain that no unwanted mutations have been introduced.

CLONING OF PCR-AMPLIFIED FRAGMENTS
PCR is also useful for adding sequences, such as restriction sites for cloning, to the ends of the amplified fragment. Although the primers used for PCR amplification must be complementary to the sequence being amplified at the 3′ end, they need not be complementary at their 5′ end. Therefore, the 5′ end of the primer sequence can include, for example, the recognition sequence for a specific restriction endonuclease, making it easier to clone the PCR amplified fragment (see above). Figure 1.34 illustrates how we can use PCR amplification to introduce

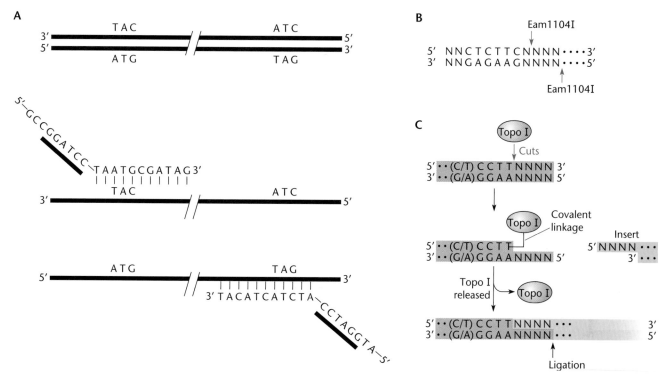

Figure 1.34 Methods for cloning PCR-amplified fragments. (A) Use of PCR to add convenient restriction sites to the ends of an amplified fragment. The primers contain sequences at their 5′ ends that are not complementary to the gene to be cloned but, rather, contain the sequence of the cleavage site for BamHI (underlined). The amplified fragment contains a BamHI cleavage site at both ends. See the text for details. Extra random bases are added at the 5′ ends, as shown, so that the site can be cut in the amplified fragment. Some restriction endonucleases do not cut a site that begins at the end of the DNA. (B) The recognition site of restriction endonuclease Eam1104I, showing where it cuts the DNA relative to the position of the sequence it recognizes (purple). The letter N means any of the four bases. (C) The recognition site of topoisomerase I (Topo I) from vaccinia virus. It breaks the phosphodiester bond 3′ of the second T in its recognition sequence (purple), transferring the phosphate bond to a tyrosine in the Topo I enzyme. This causes the dissociation of the end of the DNA, leaving a single-stranded 5′ overhang as shown. Any other DNA with a complementary 5′ single-stranded overhang can pair with this DNA and be ligated to it by the Topo I enzyme. (C/T) means that the base can be either C or T in the recognition sequence, and N means any base. Details are given in the text.

restriction sites at the ends of an amplified fragment, for example to create a fusion in an expression vector (see Figure 2.45). PCR fragments can also be cloned as blunt-end fragments into specialized plasmid cloning vectors; this is described in chapter 4.

Seamless Cloning
The above technology of introducing restriction sites into a PCR-amplified fragment allows the cloning of almost any piece of DNA into any cloning vector without depending on the presence of preexisting restriction sites. However, for some applications it has the disadvantage that it introduces extra base pairs, in the form of the

restriction site, between the vector and in the cloned piece of DNA. This can be a problem if a region internal to a coding sequence is being replaced or in making some types of translational fusions (see chapter 2). A variation of this method, called **seamless cloning,** overcomes this limitation. It depends on the properties of a type II restriction endonuclease called Eam1104I. This endonuclease recognizes a specific 6-bp sequence but cuts outside the sequence rather than within the sequence itself. The site it recognizes and the cuts it makes are illustrated in Figure 1.34B. It makes a staggered break 1 bp from its recognition site in the 3′ direction and 4 bp in the 5′ direction as shown. Because it makes a staggered break, it leaves a 5′ overhang of 3 bases. However, unlike endonucleases that cut in their recognition site, the sequences of the overhangs left by this endonuclease are unique and depend on the sequence of bases next to the recognition site, shown as N in the figure. To use this technology to fuse a known sequence to a site in the cloning vector, both the cloning vector and the piece to be inserted are amplified by PCR, using primers that introduce an Eam1104I site. In addition, 5′ of the Eam1104I site, the primer used to amplify the cloning vector has sequences complementary to the sequence of the piece to be cloned so that when they are amplified, both pieces have the same sequence next to the recognition site. As a consequence, when both of the PCR-amplified DNAs are cut by Eam1104I, their overhangs will be complementary and can base pair with each other. When they are joined by ligation, the sequence on the cloning vector and the sequence on the piece being cloned are joined without introducing any extraneous base pairs between them; i.e., the joining is seamless.

Topo I Cloning

One problem with the above cloning methods is that they all involve ligation steps. Even under optimal conditions, ligation is inefficient, which limits the numbers of clones that can be obtained. This is a particular problem in some applications such as making libraries for genomic sequencing. A more efficient system for doing cloning that does not rely on ligation is called **Topo cloning** because it relies on the type I topoisomerase of vaccinia virus. Like other type I topoisomerases, this enzyme makes a single-stranded break in one strand of DNA and holds the broken ends while the other strand passes through the break, thereby introducing or removing supercoils in the DNA one at a time (see above). However, unlike most topoisomerases, topo I of vaccinia virus has a strong sequence specificity and only creates breaks next to the 5-bp sequence shown in Figure 1.34C. It makes a break next to the 3′ T in the sequence and remains attached to it through the phosphate bond to one of its tyrosines to form a 3′ phosphoribosyltyrosine bond, much like Y recombinases (see Figure 9.32). Normally, the DNA would now rotate around the other strand and the topoisomerase would religate the strands of DNA to remove a supercoil. However, if the topoisomerase has

cut the DNA too close to the end (within 10 bp of the end), the DNA falls apart and the topoisomerase remains joined to the T with a 5′ single-stranded overhang, as shown. Note that this resembles the overhang left after a restriction endonuclease makes a staggered cut in the DNA. If another DNA with the complementary 5′ overhang pairs with this DNA, the topoisomerase will ligate the two ends, joining the two DNAs.

To use this technology, a plasmid cloning vector is constructed that has two recognition sites for the topoisomerase. This vector is cut on the 3′ side of the recognition sites, for example with a restriction endonuclease, so that the desired overhang sequences remain after the topoisomerase acts. The topoisomerase is then added; it cuts the DNA, leaves the desired overhangs, and remains attached to the DNA. This activated vector can then be mixed with any DNA fragment with the same overhang sequences, and the bound topoisomerases insert the fragment into the vector with high efficiency. Once the activated vector has been prepared, it can be used to clone many fragments with high efficiency. One particularly useful application depends on the fact that the *Taq* polymerase that is commonly used in PCR amplifications naturally leaves the single base, A, as a 5′ overhang in the amplified fragment because the *Taq* polymerase also has a terminal transferase activity. If the activated vector has been constructed so that it has a single base, T, as a 5′ overhang, any PCR-amplified fragment can be effectively cloned into it.

Topo cloning has the disadvantage that preparing the activated vector is time-consuming and technical, limiting the choice of vectors that can be used. In some applications, it is necessary to clone a fragment into a number of vectors, for example to try fusing a protein encoded by the fragment to a number of different affinity tags to see which one works best to purify that particular protein. The original vector used for Topo cloning can be engineered so that the cloned fragment is bracketed by convenient restriction sites that can be used to move the cloned fragment into other vectors. Alternatively, the original vector used for Topo cloning can be engineered to use with other technologies such as Gateway technology, which is based on the integrase and exisase of lysogenic phage (see chapter 8), which efficiently transfer the cloned fragment into other vectors.

SUMMARY

1. DNA consists of two strands wrapped around each other in a double helix. Each strand consists of a chain of nucleotides held together by phosphates joining their deoxyribose sugars. Because the phosphate joins the third carbon of one sugar to the fifth carbon of the next sugar, the DNA strands have a directionality, or polarity, and have distinct 5' phosphate and 3' hydroxyl ends. The two strands of DNA are antiparallel, so that the 5' end of one is on the same end as the 3' end of the other.

2. DNA is synthesized from the precursor deoxynucleoside triphosphates by DNA polymerase. The first phosphate of each nucleotide is attached to the 3' hydroxyl of the next deoxynucleotide, giving off the terminal two phosphates to provide energy for the reaction.

3. DNA polymerases require both a primer and a template strand. The pairing of the bases between the incoming deoxynucleotide and the base on the template strand dictates which deoxynucleotide will be added at each step, with A always pairing with T and G always pairing with C. The DNA polymerase synthesizes DNA in the 5'-to-3' direction, moving in the 3'-to-5' direction on the template.

4. DNA polymerases cannot put down the first deoxynucleotide, so RNA is usually used to prime the synthesis of a new strand. Afterward, the RNA primer is removed and replaced by DNA using upstream DNA as a primer. The use of RNA primers helps reduce errors by allowing editing.

5. DNA polymerase does not synthesize DNA by itself but needs other proteins to help it replicate DNA. These other proteins are helicases that separate the strands of the DNA, ligases to join two DNA pieces together, primases to synthesize RNA primers, and other accessory proteins to keep the DNA polymerase on the DNA and reduce errors.

6. Both strands of double-stranded DNA are usually replicated from the same end, so that the overall direction of DNA replication is from 5' to 3' on one strand and from 3' to 5' on the other strand. Because DNA polymerase can polymerize only in the 5'-to-3' direction, it must replicate one strand in short pieces and ligate these afterward to form a continuous strand. The short pieces are called Okazaki fragments. The two DNA polymerases replicating the leading and lagging strands remain bound to each other in a process called the trombone model of replication.

7. The DNA in a bacterium that carries most of the genes is called the bacterial chromosome. The chromosome of most bacteria is a long, circular molecule that replicates in both directions from a unique origin of replication, *oriC*. Replication of the chromosomes initiates each time the cells reach a certain size. For fast-growing cells, new rounds of replication initiate before old ones are completed. This accounts for the fact that fast-growing cells have a higher DNA content than slower-growing cells.

8. Chromosome replication terminates, and the two daughter DNAs separate, when the replication forks meet. Multiple *ter* sites that act as "one-way gates" delay movement of the replication forks on the chromosome. Proteins that are inhibitors of the DnaB helicase stop replication at these sites.

9. To separate the daughter DNAs after replication, dimerized chromosomes, created by recombination between the daughter DNAs, are resolved by XerC and XerD, a site-specific recombination system which promotes recombination between the *dif* sites on the daughter chromosomes. The FtsK protein is a DNA translocase that promotes XerC,D recombination at *dif* sites to prevent dimerized chromosomal DNA from being guillotined by the forming septum. Topo IV decatenates the intertwined daughter DNAs by passing the double-stranded DNAs through each other.

10. The daughter chromosomes are segregated by condensing the DNAs through supercoiling by DNA gyrase and by condensins and kleisins that hold the DNA in large supercoiled loops.

11. The FtsZ protein forms a ring at the midpoint of the cell, attracting other proteins, which form the division septum.

12. The Min proteins prevent the formation of FtsZ rings anywhere in the cell other than in the middle. Nucleoid occlusion proteins prevent the formation of FtsZ rings over nucleoids.

13. Initiation of a round of chromosome replication occurs once every time the cell divides. Initiation occurs when the ratio of DnaA protein to origins of replication reaches a critical number. After replication initiates, the DnaA protein is diluted out by having more DNA to bind to and its ATPase is activated by interaction with the β clamp to prevent reinitiation. In some bacteria including *E. coli*, related enterics, and *Caulobacter crescentus*, new initiations are prevented by hemimethylation of the newly replicated DNA at the origin and by sequestration until the replication fork has left the origin.

14. The chromosomal DNA of bacteria is usually one long, continuous circular molecule about 1,000 times as long as the cell itself. This long DNA is condensed in a small part of the cell called the nucleoid. In this structure, the DNA loops out of a central condensed core region. Some of these loops of DNA are negatively supercoiled. In *E. coli*, most DNAs have one supercoil about every 300 bases.

15. The enzymes that modulate DNA supercoiling in the cell are called topoisomerases. There are two types of topoisomerases in cells. Type I topoisomerases remove supercoils one at a time by breaking only one strand and passing the other strand through the break. Type II topoisomerases remove or add supercoils two at a time by breaking both

(continued)

SUMMARY (continued)

strands and passing another region of the DNA through the break. The enzyme responsible for adding the negative supercoils to DNA in bacteria is a type II topoisomerase called gyrase. Topo IV decatenates daughter DNAs after replication. It may also remove positive supercoils ahead of the replication fork.

16. Type II restriction endonucleases recognize defined sequences in DNA and cut at a defined position in or near the recognition sites, which has made them very useful in physical mapping of DNA and in DNA cloning. Most type II restriction endonucleases cut at the same position in both strands of symmetric sequences. If the cuts are not immediately opposite each other, single-stranded sticky ends will form that can hybridize to any other sticky end cut with the same or a compatible enzyme. This property has made these enzymes very useful for DNA cloning and for DNA manipulations in vitro.

17. The physical map of a DNA shows the actual location of sequences in DNA including the position of recognition sequences for restriction endonucleases.

18. Cloned DNA consists of multiple copies of a DNA sequence descended from a single molecule. If a piece of DNA is inserted into a cloning vector, the DNA will replicate along with the vector, making millions of clones of the original DNA.

19. A DNA library is a collection of clones that, among themselves, contain all the DNA sequences of the organism.

20. Knowledge of the properties of the enzymes involved in DNA replication has led to many applications including DNA sequencing, site-specific mutagenesis, and PCR.

21. Some antibiotics block DNA replication or affect the structure of DNA. The most useful of these inhibit the synthesis of deoxynucleotides or inhibit the gyrase enzyme. Antibiotics that inhibit deoxynucleotide synthesis include inhibitors of dihydrofolate reductase (trimethoprim and methotrexate) and inhibitors of ribonucleotide reductase (hydroxyurea). Inhibitors of gyrase include novobiocin, nalidixic acid and ciprofloxacin. Acridine dyes also affect the structure of DNA by intercalating between the bases and are used as antimalarial drugs as well as in molecular biology.

QUESTIONS FOR THOUGHT

1. Some viruses, such as adenovirus, avoid the problem of lagging-strand synthesis by replicating the individual strands of the DNA in the leading-strand direction simultaneously from both ends so that eventually the entire molecule is replicated. Why do bacterial chromosomes not replicate in this way?

2. Why are DNA molecules so long? Would it not be easier to have many shorter pieces of DNA? What are the advantages and disadvantages of a single long DNA molecule?

3. Why do cells have DNA as their hereditary material instead of RNA like some viruses?

4. What effect would shifting a temperature-sensitive mutant with a mutation in the *dnaA* gene for initiator protein DnaA have on the rate of DNA synthesis? Would the rate drop linearly or exponentially? Would the slope of the curve be affected by the growth rate of the cells at the time of the shift? Explain.

5. The gyrase inhibitor novobiocin inhibits the growth of almost all types of bacteria. What would you predict about the gyrase of the bacterium *Streptomyces sphaeroides*, which makes this antibiotic? How would you test your hypothesis?

6. How do you think chromosome replication and cell division are coordinated in bacteria like *E. coli*? How would you go about testing your hypothesis?

7. Why is termination of chromosome replication so sloppy that the *ter* region is nonessential for growth and there has to be more than one *ter* site in each direction to completely stop the replication fork? What are the advantages to not having a definite site on the chromosome at which replication always terminates?

PROBLEMS

1. You are synthesizing DNA on the template 5'ACCTTAC-CGTAATCC3' from an upstream primer. You add three of the deoxynucleotides but leave out the fourth deoxynucleotide, deoxycytosine triphosphate, from the reaction mixture. What DNA would you make? Draw a picture.

2. You are synthesizing DNA from the same template and with the same upstream primer, but instead of just deoxy-

thymidine triphosphate you add an equal mixture of the inhibitor of replication, dideoxythymidine triphosphate, and the normal nucleotide deoxythymidine triphosphate in addition to the other three deoxynucleoside triphosphates. What DNAs would you make? Draw a picture.

3. You are growing *E. coli* with a generation time of only 25 min. How long will the I periods, C periods, and D periods be? Draw a picture showing when the various events occur during the cell cycle.

4. You are growing *E. coli* with a generation time of 90 min. How long will the I, C, and D periods be? Draw a picture.

5. You are measuring the supercoiling of a plasmid from a *topA* mutant of *E. coli* which lacks the major type I topoisomerase and comparing it with the same plasmid from *E. coli* without the *topA* mutation. Would you expect there to be more or fewer negative supercoils in the plasmid from the mutant? Why?

6. Design a downstream PCR primer to amplify the sequences upstream of the sequence 5′CGATCTTAAT3′ and add an EcoRI restriction site for cloning.

7. Design PCR primers to introduce Eam1104I sites to seamlessly fuse a sequence that ends with 5′ . . . GCGACGTACGA3′ on the cloning vector to a sequence that begins with 5′ **TACGA**AGCTCT3′ on the insert. The overlapping sequence is bolded.

SUGGESTED READING

Aussel, L., F.-X. Barre, M. Aroyo, A. Stasiak, A. J. Stasiak, and D. Sherratt. 2002. FtsK is a DNA motor protein that activates chromosome dimer resolution by switching the catalytic state of the XerC and XerD recombinases. *Cell* **108:**195–205.

Bernhardt, T. G., and P. A. J. de Boer. 2005. SlmA, a nucleoid-associated, FtsZ binding protein required for blocking septal ring assembly over chromosomes in *E. coli. Mol. Cell* **18:**555–564.

Blakely, G., G. May, R. McCulloch, L. K. Arciszewska, M. Burke, S. T. Lovett, and D. J. Sherratt. 1993. Two related recombinases are required for site-specific recombination at *dif* and *cer* in *E. coli. Cell* **75:**351–361.

Britton, R. A., D. C. Lin, and A. D. Grossman. 1998. Characterization of a prokaryotic SMC protein involved in chromosome partitioning. *Genes Dev.* **12:**1254–1259.

Carballido-López, R. 2006. The bacterial actin-like cytoskeleton. *Microbiol. Mol. Biol. Rev.* **70:**888–909.

Cohen, S. N., A. C. Y. Chang, H. W. Boyer, and R. B. Helling. 1973. Construction of biologically functional bacterial plasmids in vitro. *Proc. Natl. Acad. Sci. USA* **70:**3240–3244.

Christensen, B. B., T. Atlung, and F. G. Hansen. 1999. DnaA boxes are important elements in setting the initiation mass of *Escherichia coli. J. Bacteriol.* **181:**2683–2688.

Danna, K. J., and D. Nathans. 1971. Specific cleavage of simian virus 40 DNA by restriction endonuclease of *Haemophilus influenzae. Proc. Natl. Acad. Sci. USA* **68:**2913–2917.

Dervyn, E., C. Suski, R. Daniel, C. Bruand, J. Chapuis, J. Errington, L. Janniere, and S. D. Ehrich. 2001. Two essential DNA polymerases at the bacterial replication fork. *Science* **294:**1716–1719.

Fennell-Fezzie, R., S. D. Gradia, D. Akey, and J. M. Berger. 2005. The MukF subunit of *Escherichia coli* condensin: architecture and functional relationship to kleisins. *EMBO J.* **24:**1921–1930.

Helmstetter, C. E., and S. Cooper. 1968. DNA synthesis during the division cycle of rapidly growing *Escherichia coli* B/r. *J. Mol. Biol.* **31:**507–518.

Holmes, V. F., and N. R. Cozzarelli. 2000. Closing the ring: links between SMC proteins and chromosome partitioning, condensation, and supercoiling. *Proc. Natl. Acad. Sci. USA* **97:**1322–1324.

Khodursky, A. B., B. J. Peter, M. B. Schmid, J. DeRisi, D. Botstein, P. O. Brown, and N. R. Cozzarelli. 2000. Analysis of topoisomerase function in bacterial replication fork movement: use of DNA microarrays. *Proc. Natl. Acad. Sci. USA* **97:**9419–9424.

Kurz, M., B. Dalrymple, G. Wijffels, and K. Kongsuwan. 2004. Interaction of the sliding clamp β-subunit and Hda, a DnaA-related protein. *J. Bacteriol.* **186:**3508–3515.

Lemon, K. P., and A. D. Grossman. 1998. Localization of bacterial DNA polymerase: evidence for a factory model of replication. *Science* **282:**1516–1519.

Linn, S. 1996. The DNases, topoisomerases, and helicases of *Escherichia coli* and *Salmonella*, p. 764–772. *In* F. C. Neidhardt, R. Curtiss III, J. L. Ingraham, E. C. C. Lin, K. B. Low, B. Magasanik, W. S. Reznikoff, M. Riley, M. Schaechter, and H. E. Umbarger (ed.), Escherichia coli *and* Salmonella: *Cellular and Molecular Biology*, 2nd ed. ASM Press, Washington, D.C.

Linn, S., and W. Arber. 1968. Host specificity of DNA produced by *Escherichia coli* X: *in vitro* restriction of phage fd replicative form. *Proc. Natl. Acad. Sci. USA* **59:**1300–1306.

Liu, N.-J., R. J. Dutton, and K. Pogliano. 2006. Evidence that the SpoIIIE DNA translocase participates in membrane fusion during cytokinesis and engulfment. *Mol. Microbiol.* **59:**1097–1113.

Olby, R. 1974. *The Path to the Double Helix*. Macmillan Press, London, United Kingdom.

Sambrook, J., and D. Russell. 2001. *Molecular Cloning: a Laboratory Manual*, 3rd ed. Cold Spring Harbor Laboratory Press, Cold Spring Harbor, N.Y.

Singleton, M. R., and D. B. Wigley. 2002. Modularity and specialization in superfamily 1 and 2 helicases. *J. Bacteriol.* **184:**1819–1826.

Shih, Y.-L., and L. Rothfield. 2006. The bacterial cytoskeleton. *Microbiol. Mol. Biol. Rev.* **70:**729–754.

Slater, S., S. Wold, M. Lu, E. Boye, K. Skarsted, and N. Kleckner. 1995. The *E. coli* SeqA protein binds *oriC* in two different methyl-modulated reactions appropriate to its roles in DNA replication initiation and origin sequestration. *Cell* **82:**927–936.

Viollier, P. H., M. Thanbichler, P. T. McGrath, L. West, M. Meewan, H. H. McAdams, and L. Shapiro. 2004. Rapid and sequential movement of individual chromosomal loci to specific subcellular locations during bacterial DNA replication. *Proc. Natl. Acad. Sci. USA* **101:**9257–9262.

Wang, J. C. 1996. DNA topoisomerases. *Annu. Rev. Biochem.* **65:**635–692.

Wang, X., X. Liu, C. Possoz, and D. J. Sherratt. 2006. The two *Escherichia coli* chromosome arms locate to separate cell halves. *Genes Dev.* **20:**1727–1731.

Watson, J. D. 1968. *The Double Helix.* Atheneum, New York, N.Y.

Watson, J. D., and F. H. C. Crick. 1953. Molecular structure of nucleic acids. *Nature* (London) **171:**737–738.

Wu, L. J., and J. Errington. 2004. Coordination of cell division and chromosome segregation by a nucleoid occlusion protein in *Bacillus subtilis. Cell* **117:**915–925.

Yamaichi, Y., and H. Niki. 2004. *migS,* a *cis*-acting site that affects bipolar positioning of *oriC* on the *Escherichia coli* chromosome. *EMBO J.* **23:**221–233.

CHAPTER **2**

Bacterial Gene Expression: Transcription, Translation, and Protein Folding

Uncovering the mechanism of protein synthesis, and therefore of gene expression, was one of the most dramatic accomplishments in the history of science. The process of protein synthesis is sometimes called the **central dogma** of molecular biology, which states that information in DNA is copied into RNA to be translated into protein. We now know of many exceptions to the central dogma. For example, information does not always flow from DNA to RNA but sometimes in the reverse direction, from RNA to DNA. The information in RNA is often changed after it has been copied from the DNA. Moreover, the information in DNA may be translated differently depending on where it is in a gene. Despite these exceptions, however, the basic principles of the central dogma remain sound.

This chapter outlines the process of protein synthesis and gene expression. The discussion is meant to be only a broad overview but with special emphasis on topics essential to an understanding of the chapters that follow and on subjects unique to bacteria. For more detailed treatments, consult any modern biochemistry textbook.

Overview

DNA carries the information for the synthesis of RNA and proteins in regions called **genes**. The first step in expressing a gene is to **transcribe**, or copy, an RNA from one strand in that region. The word *transcription* is descriptive because the RNA is copied in the same language as DNA, a language written in a sequence of nucleotides. If the gene carries information for a protein, this RNA transcript is called **messenger RNA (mRNA)**. An mRNA is a messenger because it carries the gene's message to a **ribosome**. Once on the ribosome, the information in the mRNA can be **translated** into the protein. *Translation* is another descriptive word because one language—the sequence of nucleotides in DNA and RNA—is translated into a different language—a sequence of amino acids in a protein. The mRNA is translated as it moves along the ribosome, 3 nucleotides at a time. Each 3-nucleotide sequence, called a **codon,** carries information for a specific amino acid. The assignment of each of the possible codons to amino acids is called the genetic code.

The actual translation from the language of nucleotide sequences to the language of amino acid sequences is performed by small RNAs called transfer RNAs (tRNAs) and enzymes called aminoacyl-tRNA synthetases. The enzymes attach specific amino acids to each tRNA. Then a tRNA specifically pairs with a codon in the mRNA as it moves through the ribosome, and the amino acid it carries is inserted into the growing protein. The tRNA pairs with the codon in the mRNA through a complementary three-nucleotide sequence called the **anticodon.** The base-pairing rules for codons and anticodons are basically the same as the base-pairing rules for DNA replication, and the pairing is antiparallel. The only differences are that RNA has uracil (U) rather than thymine (T) and that the pairing between the last of the three bases in the codon and the first base in the anticodon is less stringent.

This basic outline of gene expression leaves many important questions unanswered. How does mRNA synthesis begin and end at the correct places and on the correct strand in the DNA? Similarly, how does translation start and stop at the correct places on the mRNA? What actually happens to the tRNA and ribosomes during translation? The answers to these questions and many others are important for the interpretation of genetic experiments, so we will discuss the structure of RNA and proteins and the processes by which they are synthesized in much more detail.

The Structure and Function of RNA

In this section, we review the basic components of RNA and how it is synthesized. We also review how structure varies among different types of cellular RNAs and the role each type plays in cellular processes.

Types of RNA

There are many different types of RNA in cells. Some of these, including mRNA, ribosomal RNA (rRNA), and tRNA, are involved in protein synthesis. Each of these types of RNAs has special properties, which are discussed later in the section. Others are involved in regulation and replication.

RNA Precursors

RNA is similar to DNA in that it is composed of a chain of nucleotides. However, RNA nucleotides contain the sugar ribose instead of deoxyribose. These five-carbon sugars differ in the second carbon, which is attached to a hydroxyl group in ribose rather than the hydrogen found in deoxyribose (see chapter 1). Figure 2.1A shows the structure of a **ribonucleoside triphosphate,** so named because of the different sugar.

The only other difference between RNA and DNA chains when they are first synthesized is in the bases. Three of the bases—adenine, guanine, and cytosine—are the same, but RNA has uracil instead of the thymine found in DNA (Figure 2.1B). The RNA bases can also be modified later, as discussed below.

Figure 2.1C shows the basic structure of an RNA polynucleotide chain. As in DNA, RNA nucleotides are held together by phosphates that join the 5′ carbon of one ribose sugar to the 3′ carbon of the next. This arrangement ensures that, as with DNA chains, the two ends of an RNA polynucleotide chain will be different from each other, with the 5′ end terminating in a phosphate group and the 3′ end terminating in a hydroxyl group. When it is first synthesized, the 5′ end of an RNA chain has three phosphates attached to it, although two of the phosphates are usually removed soon after.

According to convention, the sequence of bases in RNA is given from the 5′ end to the 3′ end, which is actually the direction in which the phosphates are attached 3′ to 5′, 3′ to 5′, etc. Also, by convention, regions in RNA that are closer to the 5′ end in a given sequence are referred to as **upstream** and regions that are 3′ are referred to as **downstream** because RNA is both made and translated in the 5′-to-3′ direction.

RNA Structure

Except for the sequence of bases and minor differences in the pitch of the helix, little distinguishes one DNA molecule from another. However, RNA chains generally have more structural properties than DNA, tending to be folded into complex structures, and can be extensively modified.

PRIMARY STRUCTURE

All RNA is created equal. No matter what their function, all RNA transcripts are made the same way, from a DNA template. Only the sequences of their nucleotides and their lengths are different. The sequence of nucleotides in RNA is the **primary structure** of the RNA. In some cases, the primary structure of an RNA is changed after it is transcribed from the DNA (see "RNA Processing and Modification" below).

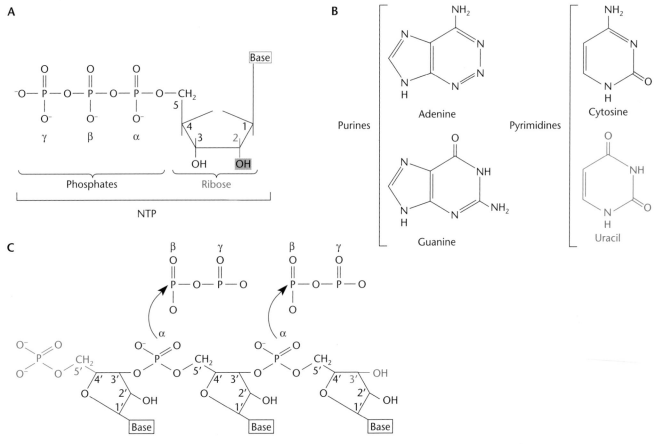

Figure 2.1 RNA precursors. (A) A ribonucleoside triphosphate (NTP) contains a ribose sugar, a base, and three phosphates. (B) The four bases in RNA. (C) An RNA polynucleotide chain with the 5′ and 3′ ends shown in purple.

SECONDARY STRUCTURE

Unlike DNA, RNA is usually single stranded. However, pairing between the bases in some regions of the molecule may cause it to fold up on itself to form double-stranded regions. Such double-stranded regions are called the **secondary structure** of the RNA. All RNAs, including mRNAs, probably have extensive secondary structure.

Figure 2.2 shows an example of RNA secondary structure, in which the sequence 5′AUCGGCA3′ has paired with the complementary sequence 5′UGCUGAU3′ somewhere else in the molecule. As in double-stranded DNA, the paired strands of RNA are antiparallel; i.e., pairing occurs only when the two sequences are complementary when read in opposite directions (5′ to 3′ and 3′ to 5′) and the double-stranded RNA forms a helix, although in the slightly different A form rather than the B form of a DNA:DNA helix. However, the pairing rules for double-stranded RNA are slightly different from the pairing rules for DNA. In RNA, guanine can pair with uracil as well as with cytosine. Because these GU pairs do not share

Figure 2.2 Secondary structure in an RNA. (A) The RNA folds back on itself to form a hairpin loop. The presence of a GU pair (in parentheses) does not disrupt the structure. (B) Different regions of the RNA can also pair with each other to form a pseudoknot. In the example, the loop of the hairpin is pairing with another region of the RNA. The purple dashes show the phosphate-ribose backbone; the black dashes show the hydrogen bonds. Details are given in the text.

A Hairpin

```
      U
   G     C
  C       A
  A — U
  C — G
  G — C
 (G   U)
  C — G
  U — A
5′ A–C–U–C–A–U–A–U–C–C–G–G–C 3′
```

B Pseudoknot

```
3′ U–C–A–C–A
                U     G
             G     C
             C       A
             C     A
             A — U
             C — G
             G — C
            (G   U)
             C — G
             U — A
5′ A–C–U–C–A–U–A–U
```

hydrogen bonds, as indicated in the figure, they do not contribute to the stability of the double-stranded RNA. Thus, the GU pair shown in Figure 2.2 does not disrupt the helix, although it does not help hold the structure together.

Each base pair that forms in the RNA makes the secondary structure of the RNA more stable. Consequently, the RNA generally folds so that the greatest number of continuous base pairs can form. The stability of a structure can be predicted by adding up the energy of all its hydrogen bonds that contribute to the structure. By eye, it is very difficult to predict which regions of a long RNA will pair to give the most stable structure. However, computer software is available that, given the sequence of bases (primary structure) of the RNA, can predict the most stable secondary structure.

TERTIARY STRUCTURE

Double-stranded regions of RNAs created by base pairing are stiffer than single-stranded regions. As a result, an RNA that has secondary structure will have a more rigid shape than one without double-stranded regions. Also, the intermingled paired regions cause the RNA to fold back on itself extensively. One type of tertiary structure occurs when the unpaired region in a hairpin such as that shown in Figure 2.2 pairs with another region of the same RNA molecule to form a knot. Such a structure is called a pseudoknot, rather than a real knot, because it is held together only by hydrogen bonds, which are more easily broken than covalent chemical bonds. Together, these effects give many RNAs a well-defined three-dimensional shape, called its **tertiary structure.** Proteins or other cellular constituents recognize RNA forms by their tertiary structure, which also gives ribozymes their enzymatic activity (see below).

RNA Processing and Modification

The folding of the RNA molecule as a result of secondary and tertiary structure are examples of **noncovalent changes,** because only hydrogen bonds, not chemical (**covalent**) bonds, are formed or broken. However, once the RNA is synthesized, covalent changes can occur during **RNA processing** and **RNA modification.**

RNA processing involves forming or breaking phosphate bonds in the RNA after it is made. For example, the terminal phosphates at the 5′ end may be removed, or the RNA may be cut into smaller pieces and even religated into new combinations, requiring the breaking and making of many phosphate bonds. In one of the most extreme cases of RNA processing, called RNA editing, nucleotides can be excised or added to mRNA after it has been transcribed from DNA.

RNA modification, by contrast, involves altering the bases or sugars of RNA. Examples include methylation of the bases or sugars of rRNA and enzymatic alteration of the bases of tRNA. In eukaryotes, "caps" of inverted methylated nucleotides are added to the 5′ ends of some types of mRNA. In bacteria, mRNAs are not capped and only the stable rRNAs and tRNAs are extensively modified.

Transcription

Transcription is the synthesis of RNA on a DNA template. The process of transcription is probably fairly similar in all organisms, but it is best understood in bacteria.

Structure of Bacterial RNA Polymerase

The transcription of DNA into RNA is the work of **RNA polymerase.** In bacteria, the same RNA polymerase makes all the cellular RNAs, including rRNA, tRNA, and mRNA. There are approximately 2,000 molecules of this RNA polymerase in each bacterial cell. Only the primer RNAs of Okazaki fragments are made by a different RNA polymerase. In contrast, eukaryotes have three nuclear RNA polymerases, as well as a mitochondrial RNA polymerase, which make their cellular RNAs.

Figure 2.3 shows a schematic structure of *E. coli* RNA polymerase, which has six subunits and a molecular weight of more than 400,000, making it one of the largest enzymes in the *E. coli* cell. The core enzyme consists of five subunits: two identical α subunits, two very large subunits called β and β′, and the ω subunit. The α, β, and β′ subunits are essential parts of the RNA polymerase, and the ω subunit helps in its assembly. A sixth subunit, the σ factor, is required only for initiation and cycles off the enzyme after initiation of transcription. Without the σ factor, the RNA polymerase is called the core enzyme; with it, it is called the **holoenzyme.**

Crystal structures of *Thermus aquaticus* RNA polymerase, which is known to be very similar to the *E. coli* enzyme but easier to crystallize, have revealed the structure shown in Figure 2.3. The overall shape has what has been described as a crab claw appearance. Regions of the β and β′ subunits form the pincers of the crab claw, as shown. The two α subunits, αI and αII, are on the opposite end from the claw, making up the rear end when the enzyme is transcribing DNA. The carboxy-terminal regions of the α polypeptides hang out from the enzyme, where they can contact other proteins or regions upstream of the promoter and stimulate transcription initiation (see below). The ω subunit is also on this end of the RNA polymerase, wrapped around the β′ subunit. When the σ factor is bound to form the holoenzyme, it wraps around the front end of the core enzyme in such a way that it can contact the DNA as it enters the open claw. One of its domains, domain σ_2, contacts the

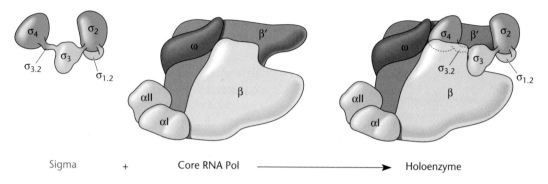

Figure 2.3   The structure of bacterial RNA polymerase, showing the sigma factor σ (shown in purple), the core enzyme without σ, and the holoenzyme with σ attached. The functions of the domains of σ, σ_1, and $\sigma_{3.2}$ are discussed later in this chapter. One α subunit, αI, contacts the β subunit, and the other, αII, contacts the β′ subunit.

β′ pincer and is in position to bind to the −10 region of the promoter on the DNA while two other domains, σ_3 and σ_4, contact the β subunit further upstream in the active-center channel in such a way that the σ_4 domain is in position to contact the −35 region of the promoter (see below). The RNA polymerases of all eubacteria are probably very similar in sequence and composition. The only known difference is that in some types of bacteria the β and β′ subunits are attached to each other to form an even larger polypeptide. The eukaryotic and archaeal core RNA polymerases have more subunits and seem more complex, and have very different sequences, but their basic overall structure is very similar to that of bacterial RNA polymerase.

Overview of Transcription

TRANSCRIPTION INITIATION

Much like DNA polymerase (see chapter 1), the RNA polymerase makes a complementary copy of a DNA template, building a chain of RNA by attaching the 5′ phosphate of a ribonucleotide to the 3′ hydroxyl of the one preceding it (Figure 2.4). However, in contrast to DNA polymerases, RNA polymerases do not need a preexisting primer to initiate the synthesis of a new chain of RNA. To begin transcription, the RNA polymerase binds to the promoter sequence and separates the strands of the DNA, exposing the bases. Unlike DNA polymerases, which require helicases to separate the strands, the RNA polymerase can complete this step by itself. Then a ribonucleoside triphosphate complementary to the nucleotide at the transcription start site bonds to the template. The second ribonucleoside triphosphate comes in, pairing with the next complementary base in the DNA template. RNA polymerase catalyzes the reaction in which the α phosphate of the second nucleotide joins with the

3′ hydroxyl of the first nucleotide. Then the third nucleotide comes in and bonds to the second, and so forth. The RNA polymerase makes a complementary copy, i.e., **transcribes** the sequence of one strand of DNA into RNA. As shown in Figure 2.5, the strands of DNA in a region that is transcribed are named to reflect the sequence of the RNA made from that region. The template strand of DNA that is copied is called the **transcribed strand.** The other strand, which has the same sequence as the RNA copy, is called the **coding strand.**

Figure 2.4   RNA transcription. (A) The polymerization reaction in which the newly synthesized RNA pairs with the template strand of DNA during transcription. (B) RNA polymerase synthesizes RNA in the 5′-to-3′ direction, moving 3′ to 5′ on the template. RNA is shown as a wavy line, and both strands of DNA are shown as straight lines.

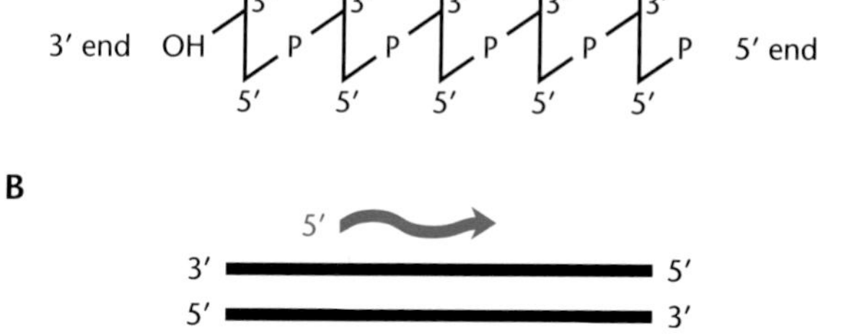

A

DNA { 5′ A T T A C G A C C T A C G C A T 3′ Coding strand
      3′ T A A T G C T G G A T G C G T A 5′ Template strand

RNA 5′ A U U A C G A C C U A C G C A U 3′

B

DNA { 5′ ▬▬▬▬▬▬▬▬▬▬▬▬ 3′
      3′ ▬▬▬▬▬▬▬▬▬▬▬▬ 5′

RNA 5′ ～～～～～～～➤

Figure 2.5 RNA polymerase transcribes only one strand of DNA. (A) The coding strand has the same sequence as the mRNA. The template strand is the strand to which the mRNA is complementary if read in the 3′-to-5′ direction. (B) Schematic illustration of the DNA and RNA strands.

since it has the same sequence as the mRNA that encodes the protein, even if the RNA that is made does not actually encode a protein. The sequence of a gene is usually written as the sequence of the coding strand.

PROMOTERS

RNA transcripts are copied from only selected regions of the DNA, rather than from the whole molecule, so the RNA polymerase can only start making an RNA chain from a double-stranded DNA at certain sites. These DNA regions are called **promoters,** and the RNA polymerase recognizes a particular T or C in the promoter region as a **transcription start site,** shown as +1 in Figure 2.6. Thus, usually the first base in the chain is an A or a G laid down opposite to a T or C, respectively. Sequences 5′ of the start site on the coding strand are said to be upstream, and sequences 3′ of the start site are said to be downstream.

The RNA polymerase recognizes different types of promoters on the basis of which type of σ factor is attached. The most common promoters are those recognized by the RNA polymerase with the vegetative σ

called σ⁷⁰ in *E. coli.* The σ factors are often named for their size, and this one has a molecular weight of 70,000 (or a molecular mass of 70 kDa).

Even promoters of the same type are not identical to each other, but they do share certain sequences known as **consensus sequences** by which they can be distinguished. Figure 2.6 shows the consensus sequence of the σ⁷⁰ promoter in *E. coli.* The promoter sequence has two important regions: a short AT-rich region about 10 bp upstream of the transcription start site, known as the **−10 sequence,** and a region about 35 bp upstream of the start site, called the **−35 sequence.** With some exceptions, the σ⁷⁰ factor must bind to both sequences to start transcription.

THE STEPS OF TRANSCRIPTION

Figure 2.7 shows an overview of the steps of transcription. The RNA polymerase core, together with the σ factor, recognizes a promoter and begins transcription with a nucleoside triphosphate. As the RNA chain begins to grow, the RNA polymerase holoenzyme releases its σ factor, and the five-subunit core enzyme continues moving along the transcribed DNA strand in the 3′-to-5′ direction, polymerizing RNA in the 5′-to-3′ direction. Inside an opening in the DNA helix approximately 17 bases long, called a **transcription bubble,** the elongating RNA and the complementary strand of DNA pair with each other to form a DNA-RNA hybrid of approximately 8 or 9 bp, which has a double-helix structure similar to that of a double-stranded DNA molecule. In this way, the RNA polymerase moves along the DNA template until it reaches a terminator. The RNA polymerase and the RNA **transcript** are then released.

Details of Transcription

It was once assumed that after initiation occurs, the RNA polymerase moves along the DNA at a uniform rate, polymerizing nucleotides into RNA. However, it is

Figure 2.6 (A) Typical structure of a σ⁷⁰ bacterial promoter with −35 and −10 regions. (B) The consensus sequences of such promoters. (C) The relationship of these sequences with respect to the start site of transcription on the two strands. RNA synthesis typically starts with an A or a G, and no primer is required. N, any nucleotide.

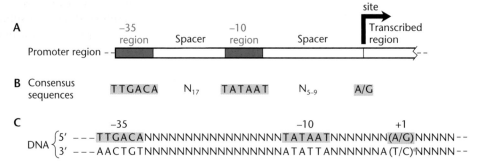

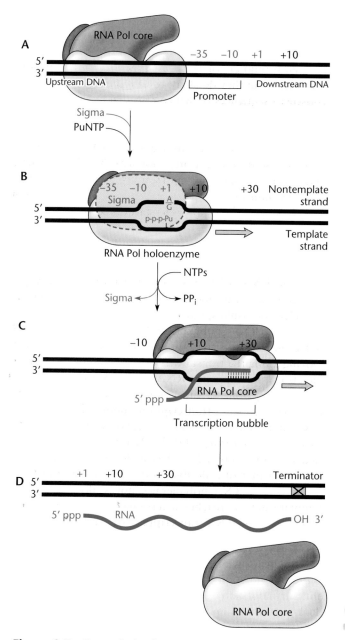

Figure 2.7 Transcription begins at a promoter and ends at a transcription terminator. (A and B) RNA polymerase core (RNA Pol) may be randomly bound to DNA until it binds σ factor and recognizes a promoter. Transcription begins when the strands of DNA are opened at the promoter. The β and γ phosphates are removed as pyrophosphate (PP$_i$). (C) As the RNA polymerase moves along the DNA, polymerizing ribonucleotides, it forms a transcription bubble containing an RNA-DNA double-stranded hybrid, which helps hold the RNA polymerase on the DNA. The sigma factor is released. (D) The RNA polymerase encounters a transcription terminator and comes off the DNA, releasing the newly synthesized RNA.

now known that the RNA polymerase often starts making RNA and then repeatedly aborts, synthesizing a number of short RNAs before finally leaving the promoter. Even after transcription is under way, the polymerase often pauses and sometimes even backs up (backtracks) before continuing. In this section we discuss in detail each of the steps in transcription (Figure 2.8), which have been established over many years by a large number of researchers. We discuss these steps one at a time because each of them is the basis for regulatory mechanisms that are discussed in later chapters.

BINDING
In the first step (Figure 2.9), the RNA polymerase core enzyme binds to a σ factor to form the holoenzyme. The bound sigma factor then directs it to the correct promoter, in a process called promoter recognition (Figure 2.10). The σ factor must be able to recognize the promoter even though the DNA in the promoter is still in a double-stranded state. Sigma factors consist of a number of domains held together by flexible linkers. One domain of the bound σ, σ$_4$, recognizes the −35 sequence when it is still in the double-stranded state, while another σ domain, σ$_2$, first binds to double-stranded DNA at the AT-rich −10 sequence in what is called a closed complex (RP$_c$). Later this region will open (see below). The σ$_3$ domain is close to the β subunit in the active-site channel, while σ$_2$ is bound to the β' subunit. Most σ factors are related to σ^{70}, and their domains play similar roles in recognizing their specific promoters. Figure 2.11 shows the conserved regions of the σ^{70} family of sigma factors and the role played by some of the conserved domains in promoter recognition and initiation of transcription.

Not all σ^{70} promoters have the same features. Figure 2.12 shows some additional features of some σ^{70} promoters, and Figure 2.13 shows which regions of the RNA polymerase recognize these features. The strength of a promoter (how strongly the RNA polymerase binds) can be enhanced by sequences upstream of the promoter called UP (for upstream) elements to which the carboxyl terminus of the α subunits, called αCTD (for carboxyl terminal domain α subunits) in the figure, can also bind and help stabilize the binding. Also, some promoters lack a −35 sequence and instead have what is called an extended −10 sequence. This sequence is recognized not by σ$_4$ but, rather, by σ$_3$.

ISOMERIZATION
When the RNA polymerase first binds to the promoter, the DNA is double stranded. This is called the closed complex, because the DNA strands are still "closed." In the next step, the β' pincer of the crab claw closes around the DNA to form the active-site channel around the

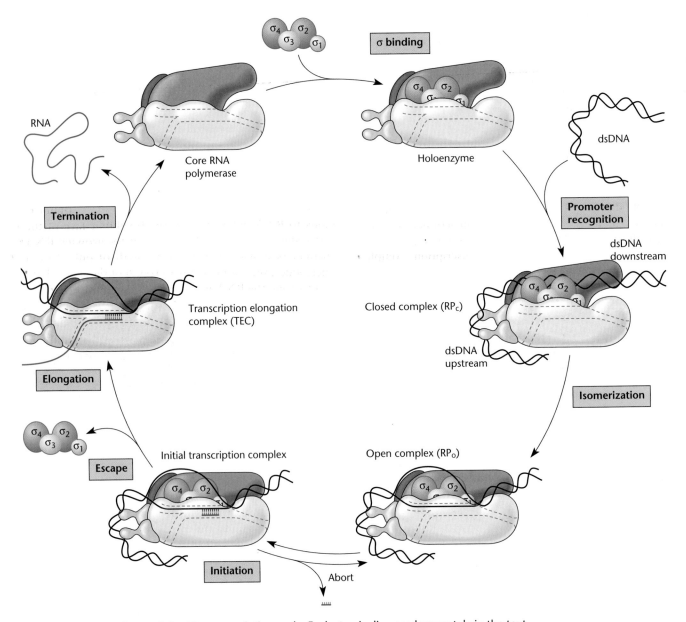

Figure 2.8   The transcription cycle. Each step is discussed separately in the text. The shape and placement of the RNA polymerase subunits are schematized for ease of illustration. dsDNA, double-stranded DNA.

template strand of the DNA. This allows the σ_2 region to separate the strands of DNA at the -10 region and bind to the nontemplate strand, in a process called isomerization (Figures 2.8 and 2.10). Recall that AT base pairs are less stable than GC pairs, so the AT-rich -10 sequence is relatively easy to melt. The complex is now called the open complex (RP_o) since the strands of DNA at the -10 region of the promoter are now "open." The $+1$ nucleotide (Figure 2.10) of the template strand is being

held in the active-site channel, where the polymerization reaction is about to occur.

INITIATION

In the initiation process, a single nucleoside triphosphate (usually an ATP or GTP), enters through the secondary channel and pairs with nucleotide $+1$ (usually a T or C, respectively) in the template strand in the active center. Then a second nucleoside triphosphate enters, and a

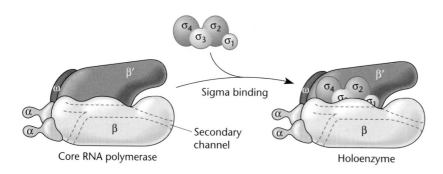

Figure 2.9 Sigma binding. Sigma (σ) is shown in purple. The secondary channel is discussed below.

phosphodiester bond forms between its α phosphate and the 3′ hydroxyl of the first nucleotide, releasing two phosphates in the form of pyrophosphate. This is called the **initiation** or **initial transcription complex** and is the step at which the antibiotic rifampin can block transcription (see the section on "Antibiotic inhibitors of transcription" below). As shown in Figure 2.14, rifampin binds to RNA polymerase in the β subunit face of the active-site channel in such a way that the growing RNA encounters it when it reaches a length of only 2 or 3 nucleotides, preventing further growth of the RNA chain and freezing the RNA polymerase on the promoter. This

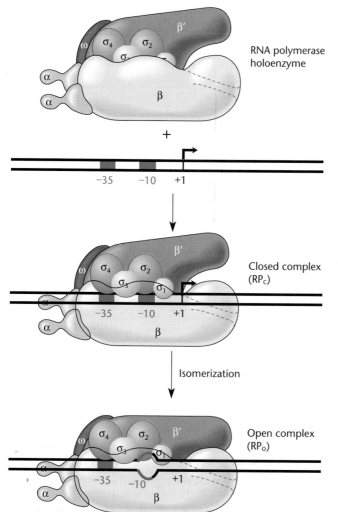

Figure 2.10 Promoter recognition. Sigma factor binds the promoter region as discussed in the text. The closed complex isomerizes to an open complex (see below). In the open complex, σ_2 binds the nontemplate single-stranded DNA of the −10 sequence.

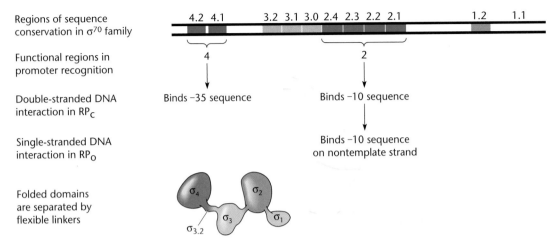

Figure 2.11   Functional regions of σ⁷⁰. There are four major regions of sequence conservation in the σ⁷⁰ family; these are divided into subregions as shown. Region 1.1 is not conserved in alternative sigma factors. Also shown are regions functional in promoter recognition.

explains why rifampin blocks only the initiation of transcription.

Even in the absence of the antibiotic rifampin, the RNA polymerase is not yet free to continue transcription. When the RNA chain grows to a length of about 10 nucleotides, it encounters the σ$_{3.2}$ loop in the active-site channel blocking the exit, called the exit channel (Figure 2.15). This causes transcription to stop, often releasing a short transcript about 10 nucleotides in length. This is called abortive transcription and occurs to various degrees on many promoters, for reasons that are not understood. Eventually a growing transcript pushes the σ$_{3.2}$ loop aside and enters the exit channel, causing the σ factor to be released from the core RNA

polymerase. At long last, the RNA polymerase has escaped the promoter and entered the elongation phase.

ELONGATION PHASE

Figure 2.16 shows the **transcription elongation complex** in the process of elongating the RNA transcript. Most of the features are mentioned above, including the approximately 17-bp transcription bubble where the two strands of DNA are separated and the approximately 8- to 9-bp RNA:DNA hybrid that forms in the active site before the DNA template strand and the newly synthesized RNA strands separate and the newly synthesized RNA exits through the RNA exit channel. The RNA polymerase is capable of synthesizing RNA at a rate of 30 to 100

Figure 2.12   Variations on the basic σ⁷⁰ promoter. (A) The consensus −10 and −35 regions; (B) location of the UP element; (C) location of the extended −10 sequence.

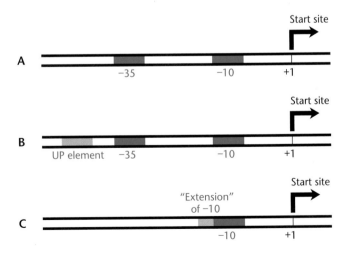

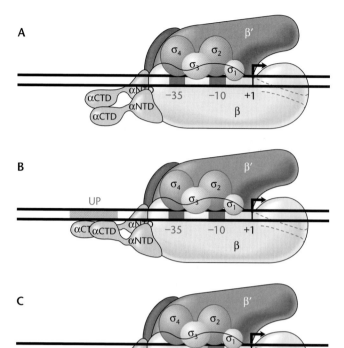

Figure 2.13 Flexibility of the α subunit carboxyl-terminal domains (αCTD). (A) Interaction of σ⁷⁰ at consensus −10 and −35 regions. (B) Flexible linkers, which allow binding of αCTDs to UP elements. (C) σ₃ binding to an extended −10 region.

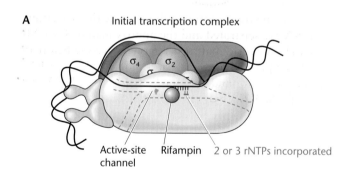

Figure 2.14 Initiation of transcription and action of the antibiotic rifampin. (A) Two or three ribonucleoside triphosphates (rNTPs) are polymerized and are incorporated in the active site. Rifampin binds to the wall of the active-site channel, preventing further elongation of RNA. (B) Structure of the antibiotic rifampin. By convention, most carbon atoms are not shown.

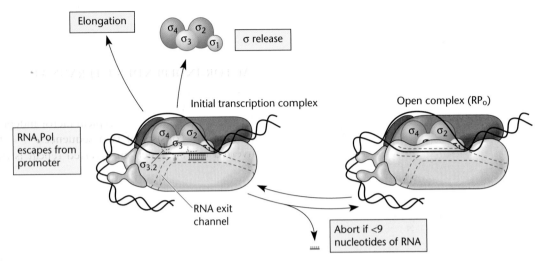

Figure 2.15   Abortive transcription and RNA polymerase (RNAP) escape from the promoter. RNAP can escape from the promoter only if more than 10 or 11 nucleotides are polymerized. At 12 nucleotides, the RNA transcript displaces the $\sigma_{3.2}$ region which blocks the active-site channel. With RNAP escape, σ is released.

nucleotides per second. However, it often pauses and even backtracks. This phenomenon often occurs when hairpins form in the RNA as it exits the RNA exit channel because the newly synthesized RNA contains inverted-repeated sequences. It is not clear why hairpins cause pausing and backtracking, but they may pull the RNA polymerase backward or bind to it and change its conformation. Backtracking creates special problems for the TEC. When the RNA polymerase is forced backward, it pushes the 3′ hydroxyl end of the newly synthesized RNA forward, shoving it into the secondary channel through which the nucleotides enter, as shown in Figure 2.17. It would remain this way, permanently blocked, except for the action of two proteins called GreA and GreB. These proteins insert their N terminus, which contains a ribonuclease (RNase) activity, into the secondary channel (Figure 2.17) and degrade the 3′ end of the RNA in the channel back until it is now in its

Figure 2.16   The transcription elongation complex. During elongation, NTPs enter through the secondary channel (shown in purple) and are polymerized at the active site; then nascent RNA exits through the RNA exit channel.

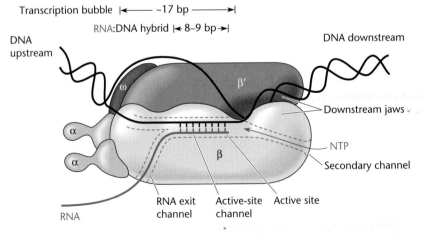

Transcription elongation complex
(TEC)

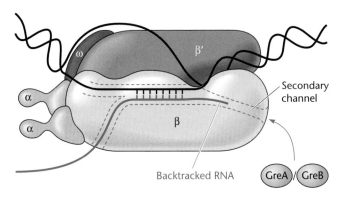

Figure 2.17 Backtracked transcription elongation complexes. GreA and GreB can enter the secondary channel to degrade RNA.

proper place in the active center so that transcription can continue.

It is not clear why RNA polymerase pausing and backtracking are tolerated, since they reduce the rate of transcription overall and create the necessity for the Gre proteins. One possibility is that selective pausing helps the folding of the RNA or the protein being translated from the RNA. Some genes whose products must be made in large amounts, such as the rRNA genes (see below), have special mechanisms to reduce pausing and backtracking. The rRNAs have sequences called antitermination sites, which bind to the RNA polymerase and preempt the binding of hairpins that appear further along in the emerging rRNA. The rRNA genes may have an additional motivation for reducing pausing; to avoid ρ-dependent termination of transcription (see below). The ρ factor can terminate transcription if the RNA polymerase pauses and if the RNA is not being translated, and the rRNAs are not translated.

TERMINATION

Once the RNA polymerase has initiated transcription at a promoter, it continues along the DNA, polymerizing ribonucleotides, until it encounters a transcription termination site in the DNA. These sites are not necessarily at the end of individual genes. In bacteria, more than one gene is often transcribed into a single RNA, so that a transcription termination site does not occur until the end of the cluster of genes being transcribed. Even if only a single gene is being transcribed, the transcription termination site may occur far downstream of the gene.

Bacterial DNA has two basic types of transcription termination sites: **factor independent** and **factor dependent.** As their names imply, these types are distinguished

by whether they work with just RNA polymerase and DNA alone or need other factors before they can terminate transcription.

FACTOR-INDEPENDENT TERMINATION

The factor-independent transcription terminators are easy to recognize because they have similar properties. As shown in Figure 2.18, a typical factor-independent terminator site consists of two sequences. The first is an inverted repeat. When an inverted repeat is transcribed

Figure 2.18 Transcription termination at a factor-independent termination site. (A) Sequence of a typical site. (B) The U-rich RNA causes RNA polymerase to pause, allowing a hairpin loop to form, dissociating RNA polymerase and RNA.

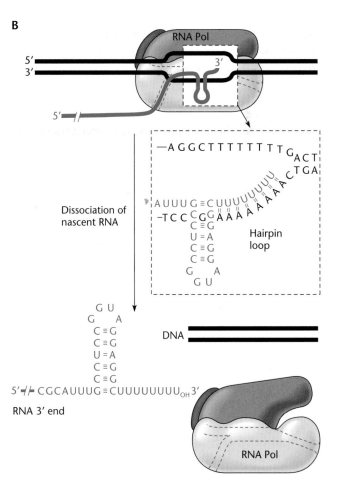

into RNA, the RNA can form a **hairpin** (Figure 2.18B). The inverted repeat is followed by a short string of A's. Transcription then terminates somewhere in the string of A's in the DNA, leaving a string of U's at the end of the RNA.

Figure 2.18B also shows how a factor-independent transcription terminator might work. The transcription of the U-rich RNA from the A-rich template might cause the RNA polymerase to pause and allow time for the GC-rich hairpin to form. The hairpin then causes the RNA polymerase to be released by an unknown mechanism, possibly influenced by the fact that the AU base pairs that form are less stable, so the RNA polymerase and RNA spontaneously fall off the DNA, terminating transcription.

FACTOR-DEPENDENT TERMINATION

While factor-independent terminators are easily recognizable, the factor-dependent transcription terminators have very little sequence in common with each other and so are not readily apparent. *E. coli*, in which factor-dependent termination is best understood, has three transcription termination factors, Rho (ρ), Tau (τ), and NusA. Since τ and NusA are not as specific and their role is less well understood, we concentrate on the ρ factor. The ρ factor probably exists in all types of bacteria, and so this type of termination is probably universal.

Any model for how the ρ factor terminates transcription at ρ-dependent termination sites has to incorporate the following facts about ρ-dependent termination. First, ρ usually causes the termination of RNA synthesis only if the RNA is not being translated. In bacteria, which lack a nuclear membrane, transcription and translation can occur at the same time, so this is an important criterion (see Introduction). Second, ρ is an RNA-dependent ATPase and so will cleave ATP to get energy, but this occurs only if RNA is present. Finally, ρ is also an RNA-DNA helicase. It is similar to the DNA helicases that separate the strands of DNA during replication, but it unwinds only a double helix with RNA in one strand and DNA in the other.

Figure 2.19 illustrates a current model for how ρ terminates transcription. Recent structural evidence shows that ρ forms a hexameric (six-sided) ring made up of six subunits encoded by the *rho* gene. This ring binds to a sequence in the mRNA called the *rut* site (for *rho utilization* site). However, ρ can bind to a *rut* site in the mRNA only if the mRNA in this region is not being translated and occupied by ribosomes, for example, if translation has terminated upstream at a nonsense codon as shown in Figure 2.19. The *rut* sites are not very distinctive but are about 40 nucleotides long and have many C's and not much secondary structure. Once ρ has bound to a *rut* site

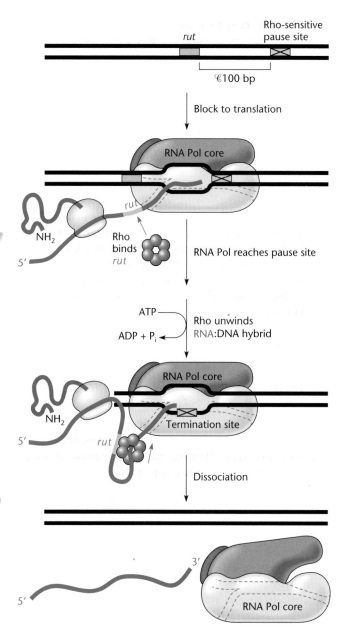

Figure 2.19 A model for factor-dependent transcription termination at a ρ-sensitive pause site. The ρ factor attaches to the mRNA at a *rut* site if the mRNA is not being translated and forms a ring around it. It then moves along the mRNA with the cleavage of ATP until it catches up with paused RNA polymerase at a ρ-sensitive pause site. The helicase activity of the ρ factor then dissociates the RNA-DNA hybrid in the transcription bubble, causing the RNA polymerase and the RNA to be released.

through the outside of its ring, the mRNA downstream of the *rut* site can enter the hole in the ring, perhaps because one of the ρ subunits is displaced to open the ring and then replaced to re-form the ring around the mRNA. The ring then moves along the mRNA in the 5'-to-3' direction, chasing the RNA polymerase. Energy for this movement is provided by the cleavage of ATP to ADP by the ATPase activity of ρ as shown in Figure 2.19. The ring may move because the mRNA binds sequentially to each of the subunits in the ring. Cleavage of ATP by the bound subunit may change the conformation of that subunit so that the mRNA is transferred to the next subunit. In this way, the ρ factor ring rotates down the mRNA behind the RNA polymerase at a speed of about 60 nucleotides per s. However, the RNA polymerase is capable of transcribing at 100 nucleotides per s, so the ρ factor can catch up only if the RNA polymerase pauses at a ρ-dependent termination site. Then the ρ factor can catch up to the RNA polymerase and its RNA-DNA helicase activity disrupts the RNA-DNA helix in the transcriptional bubble in the RNA polymerase, stopping transcription and releasing the RNA polymerase from the DNA template. While this model accounts for most of the known activities of ρ, it leaves unanswered the question of how ρ can access the DNA-RNA helix, which is in the inside of the RNA polymerase. Perhaps the RNA polymerase partially opens up when it is paused at a ρ-dependent termination site. The coupling of transcription termination to translation blockage ensures that transcription of the gene will not stop unless translation has terminated. However, ρ-dependent termination is not very efficient, and transcription continues through a ρ-sensitive pause site as much as 50% of the time. ρ-dependent termination not only occurs at the end of transcribed regions but also accounts for ρ-dependent polarity (see the section on polarity, below).

rRNAs and tRNAs

Transcription of the genes for all the RNAs of the cell is basically the same. However, rRNAs and tRNAs play special roles in protein synthesis, so their fate after transcription differs from that of mRNAs.

The ribosomes are some of the largest organelles in bacterial cells and are composed of both proteins and RNA. Bacterial ribosomes contain three types of rRNA: 16S, 23S, and 5S. S (from Svedberg, the name of the person who pioneered this way of measuring the sizes of molecules) is a measure of how fast a molecule sediments in an ultracentrifuge. In general, the higher the S value, the larger the RNA. This designation has persisted even though this method of measuring molecular size is seldom used nowadays.

The rRNAs are among the most highly evolutionarily conserved of all the cellular constituents, as indeed are many of the components of translation. For this reason, they have formed the basis for molecular phylogeny (Box 2.1). Comparisons of the sequences of rRNAs and

other constituents of the translation apparatus from different species permit estimates to be made of how long ago these constituents separated evolutionarily.

In some strains of bacteria, the 23S rRNA is broken into two pieces. Nevertheless, the ribosomes containing the broken rRNA still function because the overall structure of the ribosome holds the pieces of rRNA together. The breaks occur because the rRNA genes contain parasitic DNAs whose sequences are cut out of the rRNA (see below). However, unlike better-behaved RNA introns, these introns are not spliced back together again after they leave the RNA.

In addition to their structural role in the ribosome, the rRNAs play a direct role in translation. The 23S rRNA is the peptidyltransferase enzyme, which joins amino acids into protein on the ribosome, making it a ribozyme (see below). The 16S RNA is directly involved in both initiation and termination of translation.

The rRNAs and tRNAs make up the bulk of RNA in cells for two reasons. In a rapidly growing bacterial cell, about half of the total RNA synthesis is devoted to making these RNAs. Also, the rRNAs and tRNAs are far more stable than mRNA. They are not usually degraded until a long time after they are synthesized. With this combination of high synthesis rate and high stability, the rRNAs and tRNAs together amount to more than 95% of the total RNA in a bacterial cell.

Not only do the rRNAs physically associate in the ribosome, but also they are synthesized together as long precursor RNAs containing all three species, or forms, of rRNAs separated by so-called spacer regions. The precursors often contain one or more tRNAs as well (Figure 2.20). After the precursor RNA is synthesized, the individual rRNAs and tRNAs are cut from it. At some point during the processing, the RNAs are modified to make the mature rRNAs and tRNAs.

The faster a cell makes proteins, the faster it grows. Ribosomes are the site of protein synthesis; therefore, cells can increase their growth rate by increasing the number of their ribosomes. In many bacteria, the coding sequences for the rRNAs are repeated in 7 to 10 different places around the genome. Duplication of these genes leads to higher rates of rRNA synthesis in these bacteria. However, although the precursor RNAs encoded by these different regions have identical rRNAs, they each encode different tRNAs and spacer regions.

MODIFICATION OF tRNA

Many of the RNAs in cells are modified after they are made. For example, the rRNAs are methylated after they are made, and this sometimes confers resistance to some antibiotics (see below). The tRNAs are probably the most highly processed and modified RNAs in cells

BOX 2.1

Molecular Phylogeny

Of all the cellular components, the translation apparatus is the most highly conserved. The structures of ribosomes, translation factors, aminoacyl transferases, tRNAs, and the genetic code itself have changed remarkably little in billions of years of evolution. This is why these components have been used extensively in molecular phylogeny. By comparing the sequences of the rRNAs and other components of the translation apparatus and determining how much they have diverged, it has been possible to establish phylogenetic trees that include all organisms on Earth. The high level of conservation probably also explains why so many different antibiotics target the translation apparatus as opposed to other cellular components. An antibiotic designed to inhibit translation in one type of bacteria will probably inhibit translation in many other types of bacteria.

The conservation of components of the translation apparatus is so high that "rooted" evolutionary trees can be made that include eukaryotes and archaea (see the introduction to the book). Such trees are usually not too different from what has been obtained from physiological and other comparisons, but there are sometimes surprises. For example, a 1-mm-long organism found in sea clams around thermal vents in the sea floor was shown to be a bacterium on the basis of its 16S rRNA sequence. Most bacteria are thousands of times smaller than this. Also, the sequence of the translation elongation factors led to the suggestion that the archaea are more closely related to eukaryotes than they are to other bacteria, prompting the change of their name to archaea.

Many of the initiation and elongation factors in eubacteria have their counterparts in archaea and eukaryotes. Nevertheless, the major differences in the translation apparatus come in the translation initiation factors. While eubacteria have only three initiation factors (some of them have more than one form), archaea and eukaryotes have many more. As is the case with other cellular functions, archaea share more of their initiation factors with eukaryotes than they do with the eubacteria. Also, some of the initiation factors, while conserved, seem to have somewhat different functions in the three kingdoms of life. These differences may reflect differences in the initiation sites for translation.

References

Ganoza, M. C., M. C. Kiel, and H. Aoki. 2002. Evolutionary conservation of reactions in translation. *Microbiol. Mol. Biol. Rev.* **66:**460–485.

Ibawe, N., K.-I. Kuma, M. Hasegawa, S. Osawa, and T. Migata. 1989. Evolutionary relationships of archaebacteria, eubacteria and eukaryotes inferred from phylogenetic trees of duplicated genes. *Proc. Natl. Acad. Sci. USA* **86:**9355–9359.

Pace, N. R. 1997. A molecular view of microbial diversity and the biosphere. *Science* **276:**734–740.

Woese, C. R. 1987. Bacterial evolution. *Microbiol. Rev.* **51:**221–271.

Figure 2.20 The precursor of rRNA. The long molecule contains the 16S, 23S, and 5S rRNAs, as well as one or more tRNAs. Nucleases cut the individual RNAs out of the long precursor after it is synthesized.

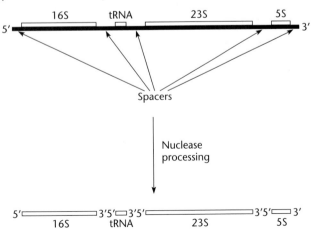

(see Björk and Hagervall, Suggested Reading). Figure 2.21 shows a "mature" tRNA that was originally cut out of a much longer molecule that may also have included the rRNAs. Then, some of the bases were modified by specific enzymes, creating altered bases such as pseudouracil and thiouracil. Finally, an enzyme called CCA transferase added the sequence CCA to the 3′ end. Clearly, much had to be done to this molecule after it was synthesized before it could become a functional, mature tRNA.

Proteins

The proteins do most of the work of the cell. While there are a few RNA enzymes, most of the enzymes that make and degrade energy sources and make cell constituents are proteins. Also, proteins make up much of the structure of the cell. Because of these diverse roles, there are

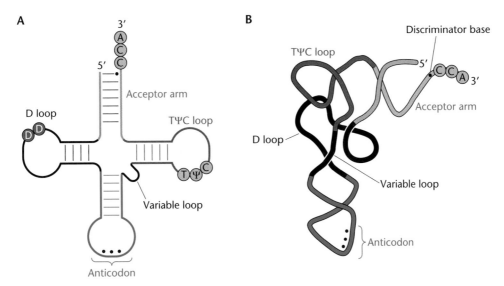

Figure 2.21 The structure of mature tRNAs, showing the secondary structure and some modifications. (A) The standard clover leaf representation of tRNA, showing the base pairing that holds the molecule together and some of the standard modifications. Y is the modified base pseudouracil. tRNAs also contain thymine (T) and pseudouracil (Ψ) among other modifications. The CCA end is where an amino acid attached. (B) The actual folding of the tRNA. Discriminator base: the position of a base important in correct aminoacylation.

many more types of proteins than there are types of other cell constituents. Even in a relatively simple bacterium, there are thousands of different types of proteins, and most of the DNA sequences in bacteria are dedicated to genes encoding proteins.

Protein Structure

Unlike DNA and RNA, which consist of a chain of nucleotides held together by phosphodiester bonds between the sugars and phosphates, proteins consist of chains of 20 different amino acids held together by **peptide bonds**. Figure 2.22 shows the formation of a peptide bond between two amino acids. The peptide bond is formed by joining the amino group (NH$_2$) of one amino acid to the carboxyl group (COOH) of another. These amino acids in turn are joined to other amino acids by the same type of bond, making a chain. A short chain of amino acids is called an **oligopeptide**, and a long one is called a **polypeptide.**

Like RNA and DNA, polypeptide chains have direction. However, in polypeptides, the direction is defined by their carboxyl and amino groups. One end of the chain, the **amino terminus or N terminus**, has an unattached amino group. The amino acid at this end is called the **N-terminal amino acid.** On the other end of the polypeptide, the unattached carboxyl group is called the **carboxyl terminus or C terminus**, and the amino acid is called the **C-terminal amino acid.** As we shall see, proteins are synthesized from the N terminus to the C terminus.

Protein structure terminology is the same as that for RNA structures. Proteins have primary, secondary, and tertiary structures, as well as quaternary structures. All of these are shown in Figure 2.23.

Figure 2.22 Two amino acids joined by a peptide bond. The bond connects the amino group on one amino acid to the carboxylic acid (carboxyl) group on the one preceding it. R is the side group of the amino acid that differs in each type of amino acid.

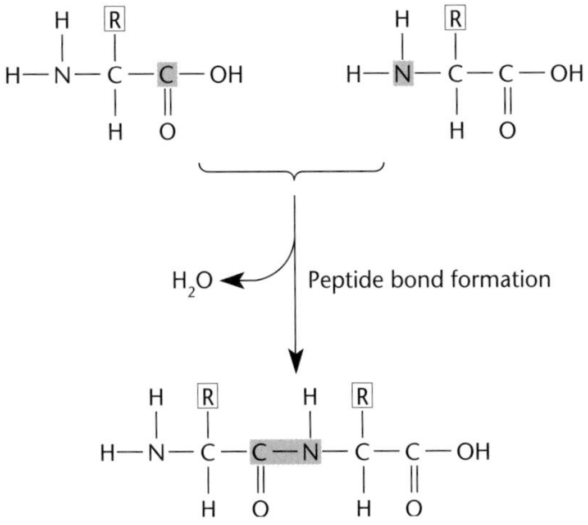

Primary structure

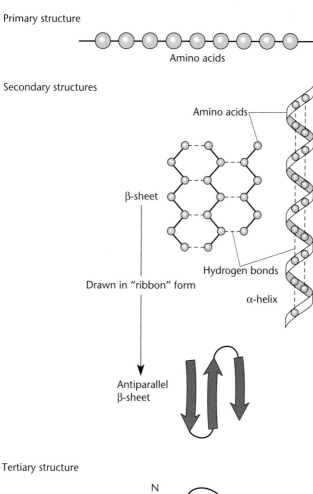

Amino acids

Secondary structures

Amino acids

β-sheet

Hydrogen bonds

Drawn in "ribbon" form

α-helix

Antiparallel
β-sheet

Tertiary structure

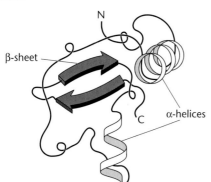

N

β-sheet

C

α-helices

Quaternary structure

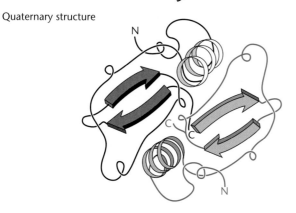

N

C

C

N

PRIMARY STRUCTURE

Primary structure refers to the sequence of amino acids and the length of a polypeptide. Because polypeptides are made up of 20 amino acids instead of just four nucleotides, as in RNA, many more primary structures are possible for polypeptides than for RNA chains.

SECONDARY STRUCTURE

Also like RNA, polypeptides can have a **secondary structure,** in which parts of the chain are held together by hydrogen bonds. However, because many more types of pairings are possible between amino acids than between nucleotides, the secondary structure of a polypeptide is more difficult to predict. The two basic forms of secondary structures in polypeptides are α-helices, where a short region of the polypeptide chain forms a helix owing to the pairing of each amino acid with the one before and the one after it, and β-sheets, in which stretches of amino acids pair with other stretches to form loops or sheetlike structures (Figure 2.23). Computer software is available to help predict which secondary structures of a polypeptide are possible on the basis of its primary structure. However, none of these programs are entirely reliable, and a technique called X-ray crystallography is the only way to be certain of the secondary structure of a polypeptide.

TERTIARY STRUCTURE

Polypeptides usually also have a well-defined **tertiary structure,** in which they fold up on themselves with hydrophobic amino acids such as leucine and isoleucine, which are not very soluble in water, on the inside and charged amino acids such as glutamate and lysine, which are more water soluble, on the outside. We discuss the structure of proteins in more detail in the section on protein folding (below).

QUATERNARY STRUCTURE

Proteins made up of more than one polypeptide chain also have **quaternary structure.** Such proteins are called **multimeric proteins.** When the polypeptides are the same, the protein is a **homomultimer.** When they are different, the protein is a **heteromultimer.** Other names reflect the number of polypeptides composing the protein. For example, the term **homodimer** describes a protein composed of two identical polypeptides whereas **heterodimer** describes a protein made of two different chains encoded by different genes. The names trimer, tetramer, and so on refer to increasing numbers of

Figure 2.23 Primary, secondary, tertiary, and quaternary structures of proteins.

polypeptides. Hence, ρ factor is a homohexamer (see above).

The polypeptide chains in a protein are usually held together by hydrogen bonds. The only covalent chemical bonds in most proteins are the peptide bonds holding adjacent amino acids together to form the polypeptide chains. As a result, if the protein is heated, it falls apart into its individual polypeptide chains. However, some proteins are unusually stable; these include extracellular enzymes, which must be able to function in the harsh environment outside the cell. Such proteins are often also held together by **disulfide bonds** between cysteine amino acids in the protein. We discuss how disulfide bonds are formed and broken later in the chapter.

Even though the secondary, tertiary, and quaternary structures are usually specific to a protein, determining the structure of even simple proteins is laborious and generally requires analysis of X-ray diffraction patterns of the crystallized protein.

Translation

The translation of the sequence of nucleotides in mRNA to the sequence of amino acids in a protein occurs on the ribosome.

As mentioned in the section on RNA, the ribosome is one of the largest and most complicated structures in cells, consisting of both three different RNAs and over 50 different proteins. It is also one of the major constituents of the cell, and much of the cell's capacity goes to making ribosomes. Each cell contains thousands of ribosomes, with the actual number depending on the growth conditions. It is also one of the most evolutionarily highly conserved structures in cells, having remained largely unchanged in shape and structure from bacteria to humans. For this reason the sequence of the rRNAs is often used in molecular phylogeny, to classify species (Box 2.1).

The ribosome is actually an enormous enzyme which performs the complicated role of polymerizing specific amino acids into polypeptide chains, using the information in mRNA as a guide. As such, a better name for it might have been amino acid polymerase, in analogy to DNA and RNA polymerases. It was given the historical name "ribosome" because it is large enough to have been visualized under the electron microscope, and so it was called a "some" (for body), and because it contains ribonucleotides. The recent determination of the structure of ribosomes has led to insights into how it performs its function of polymerizing amino acids.

RIBOSOMAL SUBUNITS
Figure 2.24 shows the components of a ribosome. The complete ribosome, called the 70S ribosome, consists of two subunits, the 30S subunit, which contains 16S

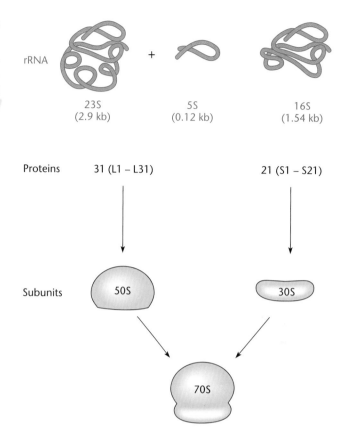

Figure 2.24   The composition of a ribosome containing one copy each of the 16S, 23S, and 5S rRNAs as well as many proteins. The proteins of the large 50S subunit are designated L1 to L31. The proteins of the small 30S subunit are designated S1 to S21. The simple subunit shapes shown here are used as icons to represent ribosomes in illustrations throughout the textbook.

rRNA, and the 50S subunit, which contains both 23S and 5S rRNA. Each subunit also contains **ribosomal proteins;** the 30S subunit contains 21 different proteins, while the 50S subunit contains 31 different proteins. Like the different terms for rRNA, the names of ribosomal subunits are derived from their sedimentation rates. The 30S and 50S subunits normally exist separately in the cell; only when they are translating an mRNA do they come together to form the complete 70S ribosome.

The two ribosomal subunits play very different roles in translation. To initiate translation, the 30S subunit binds to the mRNA. Then the 30S ribosome binds to the 50S subunit, to make the 70S ribosome. From this point on, the 30S subunit mostly helps select the correct tRNA for each codon while the 50S subunit does most of the work of forming the peptide bonds and translocating the tRNAs from one site on the ribosome to another (see below). The 70S ribosome moves along the mRNA, allowing tRNA anticodons to pair with the mRNA codons and translate the nucleic acid chain into a polypeptide. After the polypeptide chain is completed, the ribosome separates again into the 30S and 50S subunits. The role of the subunits is discussed in more detail below in the section on initiation of translation.

Details of Protein Synthesis

In this section, we return to the process of translation in more detail. First, we discuss reading frames. Then we discuss translation elongation, or what happens as the 70S ribosome moves along the mRNA, translating its nucleotides into amino acids. Finally, we discuss how translation is initiated and terminated.

READING FRAMES

As discussed in the overview of translation, each three-nucleotide sequence, or codon, in the mRNA encodes a specific amino acid, and the assignment of the codons is known as the genetic code. Because there are three nucleotides in each codon, an mRNA can be translated in three different frames in each region. Usually, initiation of translation at a initiator codon establishes the **reading frame of translation.** Once translation has begun, the ribosome moves three nucleotides at a time through the coding part of the mRNA. If the translation is occurring in the proper frame for protein synthesis, we say the translation is in the **zero frame** for that protein. If the translation is occurring in the wrong reading frame, it can be displaced either back by one nucleotide in each codon (the −1 frame) or forward by one nucleotide (+1 frame). In a few instances, translational frameshifts occur that change the reading frame even after translation has initiated.

TRANSLATION ELONGATION

Before translation can begin, a specific amino acid is attached to each tRNA by its **cognate aminoacyl-tRNA synthetase** (Figure 2.25). Each of these enzymes specifically recognizes only one type of tRNA, hence the name *cognate*. How each cognate tRNA-synthetase recognizes its own tRNA varies, but the anticodon (i.e., the three tRNA nucleotides that base pair with the complementary mRNA sequence [see "Overview" above]) is not the only determinant. Often, if the anticodon changes in a given tRNA, the cognate synthetase still attaches the amino acid for the original tRNA, and that amino acid is inserted for a different codon in the mRNA. This is the basis of nonsense suppression, which is discussed in chapter 3. Finally, the tRNA with its amino acid becomes bound to a protein called **translation elongation factor Tu (EF-Tu).**

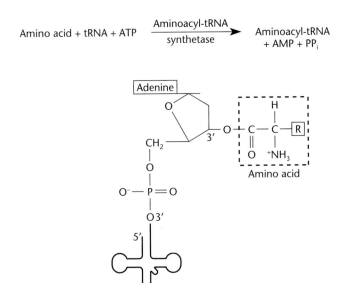

Figure 2.25   Aminoacylation of a tRNA by its cognate aminoacyl-tRNA synthetase.

During translation, the ribosome moves three nucleotides at a time along the mRNA in the 5′-to-3′ direction, allowing tRNAs carrying amino acids (aminoacylated tRNAs) to pair with the larger mRNA. Which tRNA can enter the ribosome depends on the sequence of the mRNA codon occupying the ribosome at that time. At a particular place in one of its loops, the tRNA must have three nucleotides that are complementary to the mRNA bases, so that the tRNA and mRNA can pair with each other. As mentioned above, this tRNA sequence is called the anticodon (Figure 2.26). To pair, the two RNA sequences must be complementary when read in opposite directions. In other words, the 3′-to-5′ sequence of the anticodon must be complementary to the 5′-to-3′ sequence of the codon.

Figure 2.26   Complementary pairing between a tRNA anticodon and an mRNA codon.

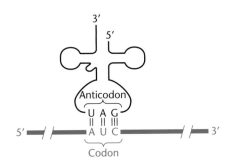

If the anticodon is complementary to the mRNA codon passing through the ribosome, the complex of tRNA plus EF-Tu can enter, binding to a site called the A site (see below) on the ribosome (Figure 2.27). The pairing of only three bases is enough to direct the right tRNA to the A site on the ribosome; in fact, sometimes the pairing of only two bases is sufficient to direct the anticodon-codon interaction (see the section on wobble, below). However, the hydrogen bonding in base pairs is not strong enough to stably hold the tRNA at this site. Apparently, the presence of the bound EF-Tu and the structure of the ribosome at the A site help to stabilize the binding. The tRNA is bound between the 30S and 50S subunits of the ribosome such that the anticodon loop is in communication with the mRNA in the 30S subunit and the acceptor end of the aminoacylated tRNA containing the bound amino acid is in communication with the 23S rRNA in the 50S subunit. After the aminoacylated tRNA is bound to the A site, the GTP on EF-Tu is cleaved to GDP and the EF-Tu is released from the ribosome.

A number of enzymes and sites on the ribosome then participate in the elongation of the polypeptide chain. A simple schematic view of this process is as follows. After the aminoacylated tRNA is bound at the **A site** (for acceptor site) of the ribosome, the **peptidyltransferase,** which is actually the ribozyme 23S rRNA (28S in eukaryotes) (see below), catalyzes the formation of a peptide bond between the incoming amino acid at the A site and the growing polypeptide at an adjacent site called the **P site.** RNAs can be enzymes as well as proteins, and when an RNA is an enzyme it is called a **ribozyme.** Another enzyme, **translation elongation factor G (EF-G)** or the **translocase,** then enters the ribosome and moves or translocates the polypeptide-containing tRNA from the A site to the P site, displacing the tRNA at the P site and making room for another aminoacylated tRNA to enter the A site. The tRNA which has been displaced then moves to yet another site, the **E site,** before it exits the ribosome. In the meantime, the mRNA is moving through the ribosome three nucleotides at a time and each tRNA can remain in contact with its own codon through its anticodon sequence as it marches through the ribosome, thereby discouraging frameshifting.

According to one attractive model, there are distinct A and P sites on *both* the 30S and 50S subunits of the ribosome. The anticodon end of the tRNA binds to the sites on the 30S subunit, while the acceptor CCA end, to which the amino acid or polypeptide is attached, binds to the sites on the 50S subunit. A tRNA bound to the A site on the 30S subunit and the corresponding A site on the 50S subunit is said to be bound to the A/A site, while one bound to the A site on the 30S subunit but the P site on the 50S subunit is bound to the A/P site, etc. The incoming aminoacylated tRNA first binds to the A site on the 30S subunit through its anticodon end. The CCA end of tRNA is still bound to EF-Tu and is therefore "masked." Once EF-Tu cycles off, the CCA end is free to bind to the A site on the 50S ribosome, so the aminoacylated tRNA is now bound to the A/A sites. The peptide bond then forms, concomitant with the movement of the CCA end of the tRNA to the P site on the 50S ribosome, and so it is now bound to the A/P sites. EF-G then moves the anticodon end of the tRNA to the P site on the 30S subunit, so the tRNA is now bound to the P/P site. There may be other dual sites, called E/E sites, to which the tRNA temporarily binds before it exits the ribosome. One attractive feature of this model is that it allows the growing polypeptide chain to stay fixed at the P site on the 50S subunit and exit through the channel in the 50S subunit as it grows while a progression of tRNAs "sashay" through the ribosome, making contacts with the different sites, like folk dancers on a promenade.

Interestingly, recent structural studies have indicated that the translation factors EF-Tu and EF-G and even the release factors may mimic each other in their various states, allowing them to bind to the same limited A and P sites on the ribosome to perform their very different roles in translation (Box 2.2).

The translation of even a single codon in an mRNA requires a lot of energy. First, ATP must be cleaved for an aminoacyl-tRNA synthetase to attach an amino acid to a tRNA (Figure 2.25). Also, EF-Tu requires that a GTP be cleaved to GDP before it can be released from the ribosome after the tRNA is bound (Figure 2.27A). Yet another GTP must be cleaved to GDP for the EF-G to move the tRNA with the attached polypeptide to the P site (Figure 2.27C). In all, the energy of three or possibly four nucleoside triphosphates is required for each step of translation.

STRUCTURE OF THE RIBOSOME

A variety of physical techniques, combined with much indirect information accumulated over the years from genetics and biochemistry, have revealed many details of the overall structure of the ribosome. The crystal structures of the individual subunits and the entire 70S ribosome have been determined and correlated with the earlier indirect information. A number of laboratories participated in this project, and this awesome achievement will go down in history as one of the major milestones in molecular biology. We can review only a few of the most salient features here.

The two subunits of the ribosome are rather round, with a flat side that binds to the other subunit, leaving a gap between them. It is through this gap that the aminocylated tRNAs enter and pass through the ribosome,

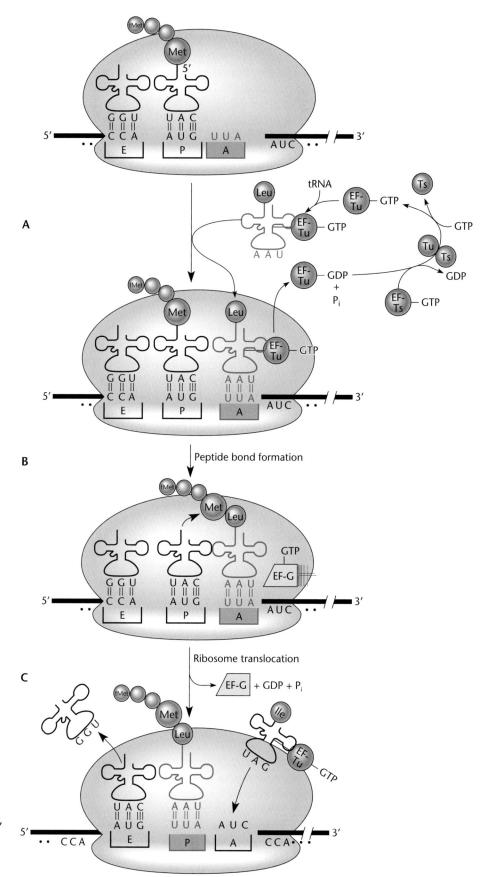

Figure 2.27 Overview of translation. (A) The tRNA bound to its amino acid and complexed with EF-Tu comes into the A site on the 30S ribosome. (B) Peptidyl-transferase, 23S rRNA in the 50S ribosome, bonds the next amino acid to the growing polypeptide. (C) The tRNA is moved to the P site by EF-G, making room at the A site for another tRNA. Finally, the tRNA, now stripped of its amino acid, moves to the E site before exiting the ribosome.

BOX 2.2

Mimicry in Translation

The ribosome is a very busy place during translation, with numerous factors and tRNAs cycling quickly through the A and P sites. Different factors have to enter the ribosome for each of the steps and then leave when they have finished their function. Somehow the sequences of rRNA and proteins in the A and P sites are able to accommodate and specifically bind all of these factors. One way the complexity of the system seems to be reduced is by having the various factors and tRNAs mimic each other, which allows them to bind to the same sites on the ribosome. For example, the translation factor G (EF-G) seems to be roughly the same shape as the translation factor Tu (EF-Tu) bound to an aminoacylated tRNA. This may allow EF-G to enter the A site, displace the tRNA (now attached to the growing polypeptide), and move it to the P site. Another example is the mimicry between the tRNAs and the release factors. Not only do the release factors resemble tRNAs in their shape, but also they seem to bind to specific terminator codons through amino acids in the release factors and nucleotide bases in the nonsense codon, rather than through base paring between the codon and the anticodon on a tRNA. When the peptidyltransferase attempts to transfer the polypeptide to the release factor in the A site, it sets in motion the string of events that cause translation to be terminated and the polypeptide and mRNA to be released from the ribosome (see the text). It is an attractive idea that the release factors replaced what were once terminator tRNAs that responded to these terminator codons. Perhaps in the earliest forms of life everything in translation was done by RNA; now RNA is used to make proteins and the proteins, being more versatile, play many of the roles previously played by RNA.

References

Clark, B. F. C., S. Thirup, M. Kjeldgaard, and J. Nyborg. 1999. Structural information for explaining the molecular mechanism of protein biosynthesis. *FEBS Lett.* **452**:41–46.

Nyborg, J., P. Nissen, M. Kjeldgaard, S. Thirup, G. Polekhina, B. F. C. Clark, and L. Reshetnikova. 1996. Structure of the ternary complex of Ef-Tu: macromolecular mimicry in translation. *Trends Biochem. Sci.* **21**:81–82.

Song, H., P. Mugnier, A. K. Das, H. M. Webb, D. R. Evans, M. F. Tuit, R. A. Hemmings and D. Barford. 2000. The crystal structure of human eukaryotic release factor eRF1–mechanism of stop codon recognition and peptidyl-tRNA hydrolysis. *Cell* **100**:311–321.

contributing their amino acid to the growing polypeptide chain. The polypeptide chain being synthesized passes out through a channel running through the 50S subunit. This channel is long enough to hold a chain of about 70 amino acids, and so a polypeptide of this length must be synthesized before the N-terminal end of a protein first emerges from the ribosome. The 50S ribosomal subunit is rather rigid, with no moveable parts, but the 30S subunit has three domains or regions that can move relative to each other during translation.

The rRNAs play many of the most important roles in the ribosome, and the ribosomal proteins seem to be present mostly to give rigidity to the structure, helping cement the rRNAs in place. This has contributed to speculation that RNAs were the primordial enzymes and that proteins came along later in the earliest stages of life on Earth. The 23S rRNA rather than a ribosomal protein also performs the enzymatic function which forms the peptide bonds. As mentioned above, a region of the 23S rRNA is the peptidyltransferase enzyme, which forms the peptide bonds between the carboxyl end of the growing polypeptide and the amino group of the incoming amino acid. Thus, 23S rRNA is an enzyme or ribozyme. The 23S rRNA also forms most of the channel in the 50S subunit through which the growing polypeptide passes. The 16S RNA has a region close to its 3′ end that base pairs with the Shine-Dalgarno region in the translation initiation region in the mRNA to initiate translation (see below). A structure of the ribosome illustrating some of the features discussed is shown in Figure 2.28.

The Genetic Code

As mentioned in the overview, the **genetic code** determines which amino acid will be inserted into a protein for each three-nucleotide set, or **codon**, in the mRNA. More precisely, the genetic code is the assignment of each possible combination of three nucleotides to one of the 20 amino acids. The code is universal, with a few minor exceptions (Box 2.3), meaning that it is the same in all organisms from bacteria to humans. The assignment of each codon to its amino acid appears in Table 2.1.

REDUNDANCY

In the genetic code, more than one codon often encodes the same amino acid. This feature of the code is called **redundancy**. There are $4 \times 4 \times 4 = 64$ possible codons that can be made of four different nucleotides taken three at a time. Thus, without redundancy, there would be far too many codons for only 20 amino acids.

A tRNA

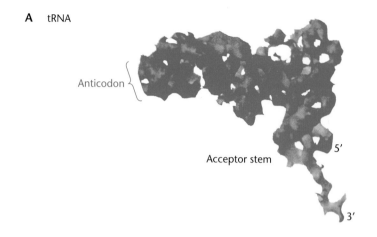

Anticodon

Acceptor stem

5′

3′

B

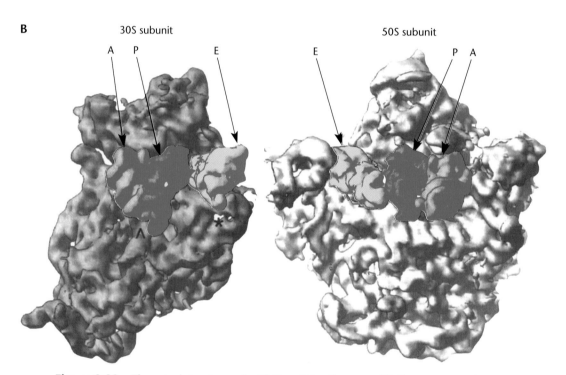

30S subunit

50S subunit

A P   E

E   P A

Figure 2.28 The actual structures of a tRNA and the ribosome. (A) The structure of a tRNA bound to the P site of the ribosome. The anticodon loop is on the left, and the 3′ acceptor end where the polypeptide is attached is at the bottom. (B) The two subunits of the ribosome separated and rotated to show the channel between them through which the tRNAs move. The 30S subunit is on the left, and the 50S subunit is on the right. The tRNAs bound at the A, P, and E sites are indicated in purple.

WOBBLE

Codons that encode the same amino acid often differ only by their third base, which is why they tend to be together in the same column when the code is presented as in Table 2.1. This pattern of redundancy in the code is due to less stringent pairing or **wobble** between the first base in the anticodon on the tRNA and the last base in the codon on the mRNA (remember that RNA sequences are always given 5′ to 3′ and the pairing of strands of RNA, like DNA, is antiparallel). As a consequence of wobble, the same tRNA can pair with more than one of the codons for a particular amino acid, so there can be fewer types of tRNA than there are codons. For example, even though there are two codons for lysine, AAA and AAG,

BOX 2.3

Exceptions to the Code

One of the greatest scientific discoveries of the 20th century was the discovery of the universal genetic code. Whether human or bacterial or plant, for the most part all organisms on Earth use the same three bases in nucleic acids to designate each of the amino acids. However, although the code is mostly universal, there are exceptions to this general rule. In some situations a codon can mean something else. We gave the example of initiation codons that encode different amino acids when internal to a gene than they do at the beginning of a gene, where they invariably encode methionine (see the text). Also, some organelles and primitive microorganisms use different code words for some amino acids. Mitochondria are notorious for using different code words for some amino acids and for termination. For example, in mammalian mitochrondria the normally nonsense codon UGA designates tryptophan. Also, some protozoans use the nonsense codons UAA and UAG for glutamine. In these organisms, UGA is the only nonsense codon. Some yeasts of the genus *Candida*, the causative agent of thrush, ringworm, and vaginal yeast infections, recognize the codon CUG as serine instead of the standard leucine. In bacteria, the only known exceptions to the universal code involve the codon UGA, which encodes the amino acid glutamine in some bacteria of the genus *Mycoplasma*, which are responsible for some plant and animal diseases.

Some exceptions to the code occur only at specific sites in the mRNA. For example, UGA encodes the rare amino acid selenocysteine in some contexts. This amino acid exists at one or a very few positions in certain bacterial and eukaryotic proteins. It has its own unique aminoacyl-synthetase, translation elongation factor (EF-Tu), and tRNA, to which the amino acid serine is added and then converted into selenocysteine. This tRNA then inserts the amino acid selenocysteine for the codon UGA, but only at a very few, unique positions in proteins and not every time a UGA appears in frame. But how does the tRNA know that these are sites to insert selenocysteine and not the numerous other UGAs, which usually signify the end of a polypeptide? The answer seems to be that the specific selenocystyl EF-Tu has extra sequences that recognize the mRNA sequences around the selenocystyl codon and only if the UGA codon is flanked by these particular sequences will this EF-Tu allow its tRNA to enter the ribosome. It is a mystery why the cell goes to so much trouble to insert selenocysteine in a specific site in only a very few proteins. In some instances where selenocysteine was replaced by cysteine, the mutated protein still functioned, albeit less efficiently. However, it may be required in the active center of some anaerobic metabolism enzymes, and this amino acid has persisted throughout evolution, existing in organisms from bacteria to humans.

A striking deviation from the code was discovered recently in the methanogenic archaea (archaea that make natural gas). These bacteria insert pyrrolysine for the normally nonsense codon UAG. Unlike selenocysteine, which is chemically derived from serine already on its tRNA, pyrrolysine is loaded on its own tRNA, as such, by its own aminoacyl transferase. It therefore qualifies as the 22nd amino acid (after formylmethionine). Its aminoacylated tRNA uses the normal EF-Tu and is inserted whenever the codon UAG appears in the mRNA (see Blight et al., below).

Other exceptions violate the rule that the code is read three bases at a time until a nonsense codon is encountered. This happens with high-level frameshifting and readthrough of nonsense codons. In high-level frameshifting, the ribosome can back up one base or go forward one base before continuing translation. High-level frameshifting usually occurs where there are two cognate codons next to each other in the RNA, for example, in the sequence *UUUUC*, where both UUU and UUC are phenylalanine codons that are presumably recognized by the same tRNA through wobble. Then the ribosome with the tRNA bound can slip back one codon before it continues translating, creating a frameshift. Sites at which high-level frameshifting occurs, "shifty sequences," usually have common features. They have a secondary structure such as a pseudoknot in the RNA just downstream of the frameshifted region, which causes the ribosome to pause (Figure 2.2). They also have a Shine-Dalgarno sequence just upstream of the frameshifted site, to which the ribosome then binds through its 16S rRNA, shifting the ribosome one nucleotide on the mRNA and causing the frameshift. Sometimes both the normal protein and the frameshifted protein, which has a different carboxyl end, can function in the cell. Examples are the *E. coli* DNA polymerase accessory proteins γ and τ which are both the products of the *dnaX* gene (Table 1.1) but differ by a frameshift. Frameshifting can also allow the readthrough of nonsense codons to make "polyproteins," as occurs in many retroviruses such as human immunodeficiency virus (the AIDS virus). Moreover, high-level frameshifting can play a regulatory role, for example, in the regulation of the RF2 gene in *E. coli*. The RF2 protein causes release of the ribosome at the nonsense codons UGA and UAA (see "Translation termination"). The gene for RF2 in *E. coli* is arranged so that its function in translation termination can be used to regulate its own synthesis through frameshifting. How long the ribosome pauses at a UGA codon depends on the amount of RF2 in the

(continued)

Exceptions to the Code

cell. If there is a lot of RF2 in the cell, the pause is brief and the polypeptide is quickly released by RF2. If there is less RF2, the ribosome will pause for longer, allowing time for a −1 frameshift. The RF2 protein is translated in the −1 frame, so this is the correct frame for translation of RF2 and more RF2 will be made if there is not enough for rapid termination.

In the most dramatic cases of frameshifting, the ribosome can hop over large sequences in the mRNA and then continue translating. This is known to occur in gene *60* of bacteriophage T4 and the *trpR* gene of *E. coli.* Somehow, the ribosome quits translating the mRNA at a certain codon and "hops" to the same codon further along. Presumably, the secondary and tertiary structures of the mRNA between the two codons cause the ribosome to hop. In the case of gene *60* of T4, the hopping occurs almost 100% of the time and the protein that results is the normal product of the gene. In the *E. coli trpR* gene, the hopping is less efficient and the physiological significance of the hopped form is unknown. .

High-level readthrough of nonsense codons can also give rise to more than one protein from the same ORF. Instead of stopping at a particular nonsense codon, the ribosome sometimes continues making a longer protein in addition to the shorter one. Examples are the synthesis of the head

proteins in the RNA phage Qβ and the synthesis of Gag and Pol proteins in some retroviruses. Many plant viruses also make readthrough proteins. Again, it seems to be the sequence around the nonsense codon that destines it for high-level readthrough. However, it is important to emphasize that these are all exceptions and normally the codons on an mRNA are translated faithfully one after the other from the translational initiation region until a nonsense codon is encountered.

References
Alam, S. L., J. F. Atkins, and R. F. Gesteland. 1999. Programmed ribosomal frameshifting: much ado about knotting! *Proc. Natl. Acad. Sci. USA* **96:**14177–14179.

Blight, S. K., R. C. Larue, A. Mahapatra, D. G. Longstaff, E. Chang, G. Zhao, P. T. Kang, K. B. Green-Church, M. K. Chan, and J. A. Krzycki. 2004. Direct charging of tRNA$_{CAU}$ with pyrrolysine *in vitro* and *in vivo*. *Nature* **431:**333–335.

Bock, A., K. Forchhammer, J. Heider, W. Leinfelder, G. Sawers, B. Veprek, and F. Zinoni. 1991. Selenocysteine: the 21st amino acid. *Mol. Microbiol.* **5:**515–520.

Gesteland, R. F., and J. F. Atkins. 1996. Recoding: dynamic reprocessing of translation. *Annu. Rev. Biochem.* **65:**741–768.

Maldonada, R., and A. J. Herr. 1998. Efficiency of T4 gene 60 translational bypassing. *J. Bacteriol.* **180:**1822–1830.

TABLE 2.1	The genetic code				
First position (5′ end)	**Second position**				**Third position (3′ end)**
	U	**C**	**A**	**G**	
U	Phe	Ser	Tyr	Cys	U
	Phe	Ser	Tyr	Cys	C
	Leu	Ser	Stop	Stop	A
	Leu	Ser	Stop	Trp	G
C	Leu	Pro	His	Arg	U
	Leu	Pro	His	Arg	C
	Leu	Pro	Gln	Arg	A
	Leu	Pro	Gln	Arg	G
A	Ile	Thr	Asn	Ser	U
	Ile	Thr	Asn	Ser	C
	Ile	Thr	Lys	Arg	A
	Met	Thr	Lys	Arg	G
G	Val	Ala	Asp	Gly	U
	Val	Ala	Asp	Gly	C
	Val	Ala	Glu	Gly	A
	Val	Ala	Glu	Gly	G

E. coli makes do with only one tRNA for lysine, which, because of wobble, can respond to both lysine codons.

The binding at the third position is not totally random, however, and certain rules apply (Figure 2.29). For example, a G in the first position of the anticodon might pair with either a C or a U in the third position of the codon but not with an A or a G, explaining why UAU and UAC, but not UAA or UAG, are codons for tyrosine. Looking at the figure, we might predict the existence of a tyrosine tRNA with the anticodon GUA. The rules for wobble are difficult to predict, however, because the bases in tRNA are sometimes modified, and a modified base in the first position of an anticodon can have altered pairing properties.

NONSENSE CODONS

Not all codons stipulate an amino acid; of the 64 possible, only 61 nucleotide combinations actually encode an amino acid. The other three, UAA, UAG, and UGA, are **nonsense codons** in most organisms. The nonsense codons are usually used to terminate translation at the end of genes (see the section on termination of translation below).

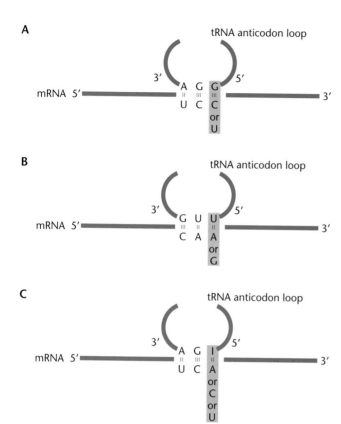

Figure 2.29 Wobble pairing between the anticodon on the tRNA and the codon in the mRNA. Many pairing interactions are possible in the third position of the codon. Alternative pairings for the anticodon base are shown: guanine (A), uracil (B), and inosine (a purine base found only in tRNAs) (C).

AMBIGUITY

In general, each codon specifies a single amino acid, but some can specify a different amino acid depending on where they are in the mRNA. For example, the codons AUG and GUG encode formylmethionine if they are at the beginning of the coding region but encode methionine or valine, respectively, if they are internal to the coding region. The codons CUG, UUG, and even AUU also sometimes encode formylmethionine if they are at the beginning of a coding sequence.

UGA is another exception. This codon is usually used for termination but encodes the amino acid selenocysteine in a few positions in genes (Box 2.3) and encodes tryptophan in some types of bacteria.

CODON USAGE

Just because more than one codon can encode an amino acid does not mean that all the codons are used equally in all organisms. The same amino acid may be preferentially encoded by different codons in different organisms.

This codon preference may reflect higher concentrations of certain tRNAs or may be related to the base composition of the DNA of the organism. While mammals and other higher eukaryotes have an average G+C content of about 50% (so that there are about as many AT base pairs in the DNA as there are GC base pairs), some bacteria and their viruses have very high or very low G+C contents. How the G+C content can influence codon preference is illustrated by some members of the genera *Pseudomonas* and *Streptomyces*. These organisms have G+C contents of almost 75%. To maintain such high G+C contents, the codon usage of these bacteria favors the codons that have the most G's and C's for each amino acid.

Translation Initiation

The process of initiating the synthesis of a new polypeptide chain is very different from the process of translation once it is under way. For example, the 30S, but not the 50S, ribosomal subunit works with other factors unique to initiation. Initiation of translation in bacteria is somewhat different from initiation in eukaryotic organisms, and we shall point out some of these differences as we go along.

TRANSLATIONAL INITIATION REGIONS

In the chain of thousands of nucleotides that make up an mRNA, the ribosome must bind and initiate translation at the correct site. If the ribosome starts working at the wrong initiation codon, the protein will have the wrong N-terminal amino acids or the mRNA will be translated out of frame and all of the amino acids will be wrong. Hence, mRNA have sequences called **translational initiation regions (TIRs)** that flag the correct first codon for the ribosome. In spite of extensive research, it is still not possible to predict with 100% accuracy whether a sequence is a TIR. However, some general features of TIRs are known.

Initiation Codons

All TIRs have an **initiation codon,** which codes for a definable amino acid. The three bases in these codons are usually AUG or GUG but in rare cases are UUG or AUA. There is even one known case of an *E. coli* gene with the initiation codon AUU. This initiation codon is used to initiate translation of one of the initiation factors for translation and is used to regulate its translation (see below).

The initiation codon does not have to be the first sequence in the mRNA chain. In fact, the 5′ end of the mRNA may be some distance from the TIR and the initiation codon; this region is called the **5′ untranslated region (5′ UTR)** or **leader sequence.**

Regardless of which amino acid these sequences call for in the genetic code (Table 2.1), if they are serving as initiation codons, they encode methionine (actually formylmethionine [see below]) as the N-terminal amino acid. After translation, this methionine is usually cut off (see the section on removal of the methionine, below). Notice that for the initiation codons, there seems to be "wobble in reverse," with the first position being the one that can wobble instead of the third position. The significance of this is unknown but might relate to the fact that these codons are recognized at the P site on the ribosome rather than the A site.

Shine-Dalgarno Sequences

Given that the initiation codons code for amino acids other than methionine when internal to a coding region, the presence of one codon is clearly not enough to define a TIR. These sequences may also occur out of frame, in which case they would not read as an amino acid. They could even appear in an mRNA sequence that is not translated at all. Obviously, other regions around these three bases must help define them as a place to begin translation.

Many bacterial genes have 5 to 10 nucleotides on the 5′ side (upstream) of the initiation codon that define a TIR. These sequences, named the **Shine-Dalgarno (S-D) sequence** after the two scientists who first noticed them, are complementary to short sequences within certain regions of the 16S RNA. Figure 2.30 shows an example of a typical bacterial TIR with a characteristic S-D sequence. By pairing with their complementary sequences on the 16S rRNA, S-D sequences help define TIRs by properly aligning the mRNA on the ribosome. However, these sequences are not always easy to identify because they can be very short and do not have a distinct sequence, being complementary to different regions of

the 16S rRNA. Moreover, not all bacterial genes have S-D sequences. The initiation codon sometimes resides at the extreme 5′ end of the mRNA, leaving no room for an S-D sequence. In such cases, the sequence that interacts with the 16S rRNA of the ribosome may be downstream of the initiation codon.

Because of this lack of universality, often the only way to be certain that translation is initiated at a particular initiation codon is to sequence the N terminus of the polypeptide to see if the N-terminal amino acids correspond to the codons immediately adjacent to the putative initiation codon.

INITIATOR tRNA

Translation initiation requires a unique aminoacylated tRNA, the formylmethionine tRNA ($fMet-tRNA_f^{Met}$). This unique aminoacyl-tRNA has a formyl group attached to the amino group of the methionine (Figure 2.31), making it resemble a peptidyl-tRNA rather than a normal aminoacyl-tRNA. This causes it to bind to the P site rather than the A site of the ribosome, which is an important step in initiation, as discussed below. The initiator $fMet-tRNA_f^{Met}$ is synthesized somewhat differently from the other aminoacyl-tRNAs. Unlike other tRNAs, this special tRNA does not have its own aminoacyltransferase, and it uses the aminoacyltransferase of the normal $tRNA^{Met}$ to attach methionine to the $tRNA_f^{Met}$. Then an enzyme called **transformylase** adds a formyl group to the amino group of the methionine on the $tRNA_f^{Met}$ to form $fMet-tRNA_f^{Met}$.

STEPS IN INITIATION OF TRANSLATION

The current accepted view of the steps in the initiation of translation at a TIR are outlined in Figure 2.32. It can be seen that the process of initiation of translation at a TIR is very different from the process of elongation and, in

Figure 2.30 Structure of a typical bacterial translation initiation region (TIR) showing the pairing between the S-D sequence in the mRNA and a short sequence close to the 3′ end of the 16S rRNA. The initiator codon, typically AUG or GUG, is 5 to 10 bases downstream of the S-D sequence. N designates any base.

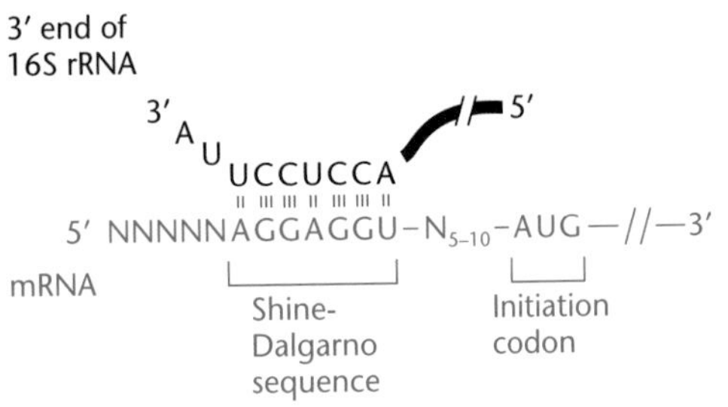

Figure 2.31 Comparison between methionine (Met) and N-formylmethionine (fMet).

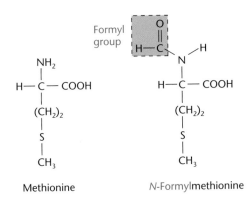

Methionine   N-Formylmethionine

addition to fMet-tRNA$_f^{Met}$, requires three different initiation factors, IF1, IF2, and IF3. These initiation factors interact mostly with the P site of the ribosome rather than the A site like the elongation factors discussed above.

For initiation to occur, the 70S ribosome must first be separated or dissociated into its smaller 30S and 50S subunits. This dissociation occurs after the termination step of translation (see below) and requires the IF3 initiation factor, which binds to the 30S subunit and helps keep the subunits dissociated. Therefore, ribosomes are continuously cycling between the 70S ribosome and the 30S and 50S subunits depending on whether they have initiated translation. This is called the **ribosome cycle.**

Once the subunits are dissociated, IF1 binds to the A site on the 30S ribosome to prevent the fMet-tRNA$_f^{Met}$ from inadvertently binding to this site and perhaps also to help IF3 keep the ribosome subunits apart during the initiation process. Then the other three components, the TIR site on an mRNA, an fMet-tRNA$_f^{Met}$, and IF2, all bind to the P site of the ribosome, perhaps in any order. The initial binding of fMet-tRNA$_f^{Met}$ is anticodon independent, and the tRNA binds no matter what mRNA codon occupies the P site. However, IF2, with the help of

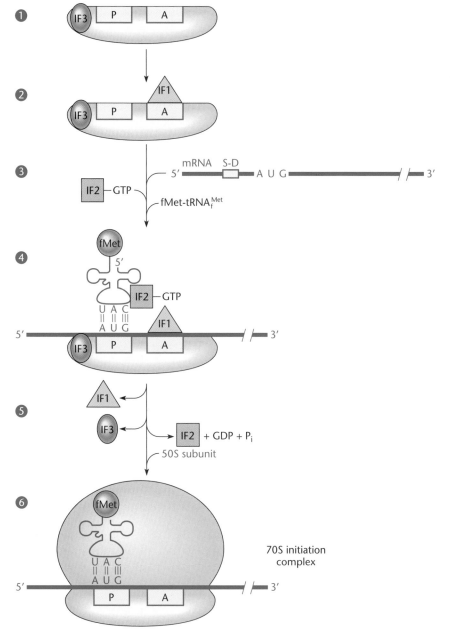

Figure 2.32 Initiation of translation. (1) The IF3 factor binds the 30S subunit to keep it dissociated from the 50S subunit during initiation. (2) IF1 binds to the A site to block this site. (3 and 4) The P site of the 30S subunit then binds the mRNA TIR site, the fMet-tRNA$_f^{Met}$, and IF2-GTP, probably in a random order. (5) IF1 and IF3 are released, the cleavage of GTP on IF2 correctly positions the fMet-tRNA$_f^{Met}$ on the P site, and the 50S subunit binds. (6) The 70S ribosome is ready to accept another aminoacyl-tRNA at the A site.

IF3, adjusts the fMet-tRNA$_f^{Met}$ and the mRNA initiator codon so that the binding becomes codon specific (see below). IF1 and IF3 are then ejected, and IF2 promotes the association of this initiation complex with the 50S large subunit of the ribosome. IF2 is then released, with the cleavage of GTP to GDP. The newly formed 70S ribosome is now ready for translation, and another aminoacylated tRNA can enter the A site. The peptidyltransferase reaction then joins this incoming amino acid to the fMet amino acid at the P site as shown in Figure 2.33. IF2 therefore plays a role similar to EF-Tu in that it helps position the fMet-tRNA$_f^{Met}$ on the ribosome by using the energy of GTP cleavage. However, unlike EF-Tu, it does not seem to accompany the aminoacylated tRNA into the ribosome and it has the additional function of helping promote the association of the two ribosomal subunits.

Figure 2.33   The peptidyltransferase reaction.

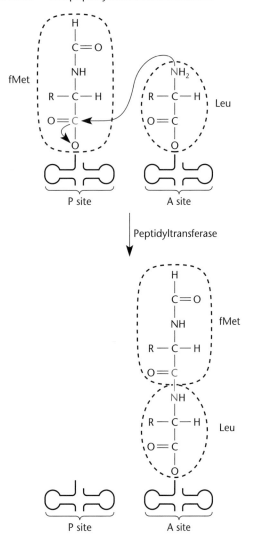

Regulation of IF3 Synthesis

As mentioned above, the initiator codon is usually AUG but can also be GUG or rarely UUG, AUA, or even AUU. Interestingly, the AUU initiator codon is used only to initiate the translation of a single mRNA, the IF3 initiation factor, at least in *E. coli*. This helps regulate the synthesis of IF3, so that the cell has no more or no less than it needs to initiate translation. As mentioned above, IF3 helps discriminate against initiator codons so that the fMet-tRNA$_f^{Met}$ binds only to the correct initiation codons. This causes it to discriminate against its own initiator codon, AUU, and help regulate its own translation. The more IF3 there is in the cell, the less translation will initiate at AUU and the less IF3 will be made, while the less IF3 there is, the more its own mRNA will be translated.

TRANSLATION INITIATION FROM LEADERLESS mRNAs

As mentioned, a few mRNAs in bacteria do not have standard TIRs with leader sequences containing S-D sequences. In these rare mRNAs, the initiator codon can be right at the 5′ end or very close to it. It is not understood how the ribosome recognizes such an initiator codon and initiates translation, but the mechanism seems to be very different from that of initiation at a more normal TIR. There is some evidence that a complex first forms between fMet-tRNA$_f^{Met}$, IF2, and the small subunit of the ribosome. This complex may then help recognize the initiation codon, in the absence of upstream sequences to help distinguish the initiation codon. Other evidence suggests that the 70S ribosome itself recognizes the leaderless initiator codon. It is intriguing to think that the process of initiation of translation at initiator codons without leader sequences may resemble more closely the process used in eukaryotes (see below) and may be the remnants of a process used before these kingdoms of life separated.

REMOVAL OF THE FORMYL GROUP AND THE N-TERMINAL METHIONINE

Normally, polypeptides do not have a formyl group attached to their N terminus. In fact, they usually do not even have methionine as their N-terminal amino acid. The formyl group is removed from the polypeptide after it is synthesized by a special enzyme called **peptide deformylase** (Figure 2.34). The N-terminal methionine is also usually removed by an enzyme called **methionine aminopeptidase**.

TRANSLATION INITIATION IN ARCHAEA AND EUKARYOTES

Translation initiation in the archaea is similar to that in the eubacteria. Like bacteria, archaea use well-defined ribosome-binding sites with leader sequences and

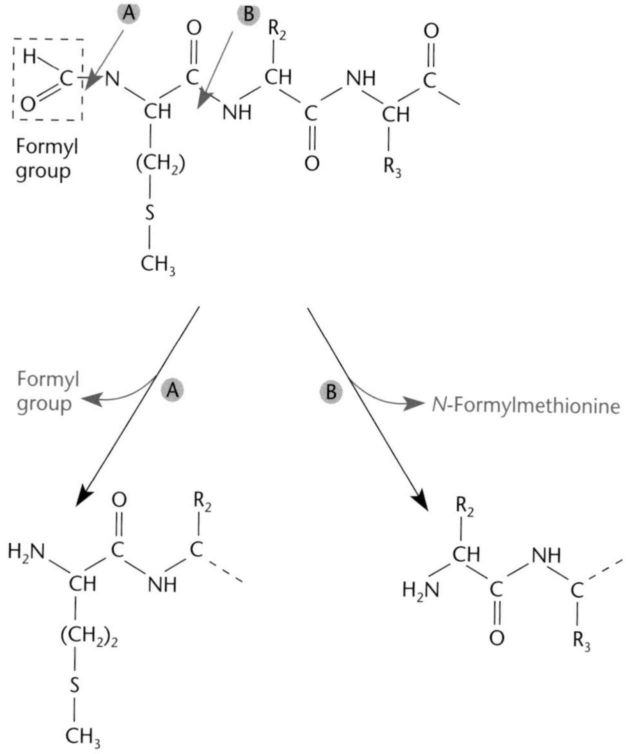

Figure 2.34 Removal of the N-terminal formyl group by peptide deformylase (A) and of the N-terminal methionine by methionine aminopeptidase (B).

formylmethionine for initiation of translation. In contrast, eukaryotes do not seem to use special ribosome-binding sites but usually use the first AUG from the 5′ end of the mRNA as the initiation codon. This does not mean, however, that sequences around this initiator AUG are not important for its recognition. Also, secondary structure in the mRNA may mask other AUG sequences that could potentially be used as initiator codons. Also, although eukaryotes have a special methionine tRNA that responds to the first AUG codon, called Met-tRNA$_i$, the methionine attached to the eukaryotic initiator tRNA is never formylated. The first methionine is, however, usually removed by an aminopeptidase after the protein is synthesized. Eukaryotes and archaea also seem to use many more initiation factors and elongation factors than do bacteria. Although the exact role of most of these initiation factors is unknown, many are obviously related to the initiation and translation factors of eubacteria. Table 2.2 shows some of the translation factors in eubacteria and their counterparts in archaea and eukaryotes. It is interesting that the mechanism of translation initiation in the archaea is sort of a hybrid between that in eubacteria and that in eukaryotes. The archaea use formylated methionine and S-D sequences like eubacteria, but their initiation factors are more akin to those in eukaryotes.

TABLE 2.2	Translation factors		
Eubacteria	**Archaea**	**Eukaryotes**	**Function in eubacteria**
IF1	+	eIF1A	Block A site
	+	eIF1	Bind mRNA
IF2	+	eIF2γ, eIF5B	Initiate tRNA binding
	++	eIF2α, eIF2β	
	I, II, III	eIF2Bα, eIF2Bβ, eIF2Bδ Bγ, BBε	
IF3		eIF3	Subunit dissociation
	+	eIF3A	
	+	eIF4A	
		eIF4B–eIF4H	
		eIF5	
	++	eIF5A, eIF5B	
		eIF6	
EF-Tu	+	eEF1α	tRNA binding
EF-G	+	eEF2	Translocation
RF1	+	eRF1	Termination
RF2			Termination
RF3		eRF3	Release RF1, RF2

Translation Termination

Once initiated, translation proceeds along the mRNA, one codon at a time, until the ribosome encounters one of the nonsense codons, UAA, UAG, or UGA. These codons do not encode an amino acid, so they have no corresponding tRNA (Table 2.1). When a ribosome comes to a nonsense codon, translation stops. Similar to the positioning of translation initiators, the nonsense codon that terminates translation may not be at the end of the mRNA molecule. The region between this last codon and the 3′ end of the mRNA is called the **3′ untranslated region.**

RELEASE FACTORS

In addition to a codon for which there is no tRNA, termination of translation requires **release factors.** These proteins recognize the nonsense codons and promote the release of the polypeptide from the tRNA and the ribosome from the mRNA. In *E. coli*, there are two translation release factors, called RF1 and RF2. The two release factors respond to specific nonsense codons: RF1 responds to UAA and UAG, whereas RF2 responds to UAA and UGA. Another factor called RF3 helps to release these factors from the ribosome after termination. The release factors may pair with the nonsense codons in the mRNA directly, through pairing between specific amino acids in the RF and the nonsense codon in the mRNA, much like base pairing between nucleotide tRNA and a codon. As evidence, mutations in RF1 can allow it to respond to all three nonsense codons. In fact, eukaryotes have only one RF, which responds to all three nonsense codons. Some other types of bacteria and mitochondria also only have one RF, but those that do, generally use UGA to encode an amino acid and not as a nonsense codon (Box 2.3).

RELEASE OF THE POLYPEPTIDE

Figure 2.35 outlines the process of translation termination. After translation stops at the nonsense codon, the A site is left unoccupied because there is no tRNA to pair with the nonsense codon. The release factors bind to the A site of the ribosome instead. They then somehow cooperate with EF-G and ribosome release factor (RRF) to cleave the polypeptide chain from the tRNA and release it and the mRNA from the ribosome. An attractive model to explain how this could happen is suggested by the observation that the release factors mimic aminoacylated tRNA (Box 2.2). If the release factor is occupying the A site, then the peptidyltransferase might try to transfer the polypeptide chain to the release factor rather than to the amino acid on a tRNA normally occupying the A site. When EF-G then tries to translocate the

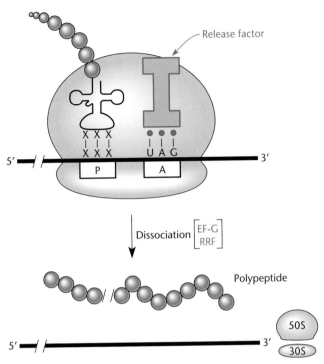

Figure 2.35 Termination of translation at a nonsense codon. A specific release factor interacts with the ribosome stalled at the nonsense codon, possibly through specific pairing between amino acids in the release factor and the nonsense codon (purple dots). Translocation by EF-G causes dissociation of the ribosome from the mRNA, possibly with the assistance of ribosome release factor (RRF).

release factor with the polypeptide attached to the P site, it may trigger a series of reactions that release the polypeptide. The role of RRF in this process is uncertain but it might be involved in releasing the mRNA after termination. Termination is more efficient when it occurs in the proper **context,** that is, when a nonsense codon is surrounded by certain sequences. This is probably one of the reasons why some types of cells can tolerate nonsense suppressors.

A problem occurs when the ribosome reaches the 3′ end of an mRNA without encountering a nonsense codon. Then the release factors do not release the ribosome from the mRNA and the ribosome is jammed on the mRNA. A special mechanism involving a hybrid tRNA and mRNA called **tmRNA** is then used to release the ribosome from the mRNA and degrade the defective protein (Box 2.4).

Previously, it was thought that an enzyme called **peptidyl-tRNA hydrolase** is involved in removing the polypeptide from the last tRNA. The finding that this enzyme is essential in *E. coli* encouraged such speculation. However, this enzyme is not associated with the

BOX 2.4

Traffic Jams on mRNA: Removing Stalled Ribosomes with tmRNA

When a ribosome reaches a nonsense codon in frame, the release factors release it, along with the finished polypeptide, from the mRNA. But what happens if the ribosome gets to the end of an mRNA before it encounters a nonsense codon? This might happen fairly often, because mRNA is constantly being degraded and transcription often terminates prematurely. The release factors can function only at a nonsense codon, and so the ribosome should stall on the mRNA. Not only would this cause a traffic jam and use up ribosomes, but also the protein that is being made will be defective because it is shorter than normal, and accumulation of defective proteins may cause problems for the cell. This is where a small RNA called tmRNA comes to the rescue. As the name implies, tmRNA is both a tRNA and an mRNA, as shown in the figure. It can be loaded with alanine like a tRNA but also contains a short ORF terminating in a nonsense codon like an mRNA. If the ribosome reaches the end of an mRNA without encountering a nonsense codon, the tmRNA enters the A site of the stalled ribosome, and alanine is inserted as the next amino acid of the polypeptide. Then, by a process that is not well understood, the ribosome shifts from translating the ORF on the mRNA to translating the ORF on the tmRNA, where it soon encounters the nonsense codon. The release factors then release the ribosome and the truncated polypeptide fused to a short "tag" sequence of only about 10 amino acids encoded by the tmRNA. The tag sequence which has been attached to the carboxy end of the truncated polypeptide is recognized by the Clp protease (see the text), which degrades the entire defective polypeptide so that it cannot cause problems for the cell. In some cases, tmRNA-mediated degradation may play a regulatory role, allowing the degradation of proteins until they are needed (see Abo et al. and Withey and Friedman, below).

References

Abo, T., T. Inada, K. Ogawa, and H. Aiba. 2000. SsrA-mediated tagging and proteolysis of LacI and its role in the regulation of the *lac* operon. *EMBO J.* **19:**3762–3769.

Gillet, R., and B. Felden. 2001. Emerging views on the tmRNA-mediated protein tagging and ribosome rescue. *Mol. Microbiol.* **42:**879–885.

Keller, K. C., P. R. Waller, and R. T. Sauer. 1996. Role of a peptide tagging system in degradation of proteins synthesized from damaged messenger RNA. *Science* **271:**990–993.

Withey, J., and D. Friedman. 1999. Analysis of the role of *trans*-translation in the requirement of tmRNA for λ*imm*^P22 growth in *Escherichia coli*. *J. Bacteriol.* **181:**2148–2157.

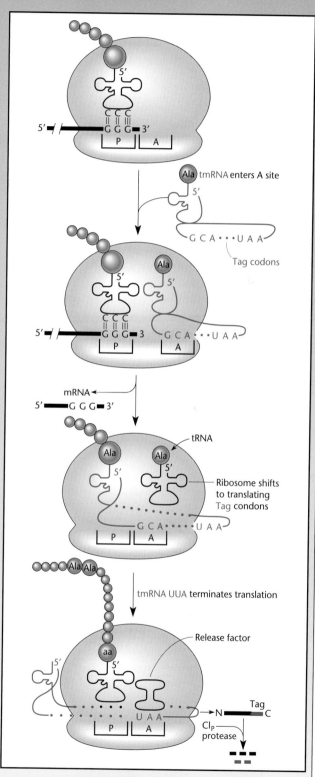

ribosome, where normal release is thought to occur. Rather, the peptidyl-tRNA hydrolase is apparently a scavenger enzyme, releasing polypeptides from tRNA following aberrant termination and allowing reuse of the tRNA. This could be especially important in the case of rare tRNAs, explaining why the peptidyl-tRNA hydrolase is essential for growth, at least in *E. coli*.

Polycistronic mRNA

In bacteria and archaea, the same mRNA can encode more than one polypeptide. Such mRNAs, called **polycistronic mRNAs**, must have more than one TIR to allow simultaneous translation of more than one sequence of the mRNA.

The name "polycistronic" is derived from *cistron*, which is the genetic definition of the coding region for each polypeptide, and *poly*, which means many. Figure 2.36 shows a typical polycistronic mRNA, in which the coding sequence for one polypeptide is followed by the coding sequence for another. The space between two coding regions can be very short, and the coding sequences may even overlap. For example, the coding region for one polypeptide may end with the nonsense codon UAA, but the last A may be the first nucleotide of the initiator codon AUG for the next coding region. Even if the two coding regions overlap, the two polypeptides on an mRNA can be translated independently by different ribosomes.

Polycistronic mRNAs do not exist in eukaryotes, in which, as described above, TIRs are much less well defined and translation usually initiates at the AUG

codon closest to the 5′ end of the RNA. In eukaryotes, the synthesis of more than one polypeptide from the same mRNA usually results from differential splicing of the mRNA or from high-level frameshifting during the translation of one of the coding sequences (Box 2.3; see the section on reading frames, above). Polycistronic RNA leads to phenomena unique to bacteria, i.e., polarity and translational coupling, which are described in the following sections.

TRANSLATIONAL COUPLING

Two or more polypeptides encoded by the same polycistronic mRNA can be translationally coupled. Two genes are **translationally coupled** if translation of the upstream gene is required for translation of the gene immediately downstream.

Figure 2.37 shows an example of how two genes could be translationally coupled. The TIR including the AUG initiation codon of the second gene is inside a hairpin on the mRNA, and so it cannot be recognized by a ribosome. However, a ribosome arriving at the UGA stop codon for the first gene can open up this secondary structure, allowing another ribosome to bind and initiate translation on the second gene. Thus, translation of the second gene depends on the translation of the first.

POLAR EFFECTS ON GENE EXPRESSION

Some mutations that affect the expression of a gene in a polycistronic mRNA can have secondary effects on the transcription of downstream genes. Such mutations are said to exert a polar effect on gene expression. Several

Figure 2.36 Structure of a polycistronic mRNA. (A) The coding sequence for each polypeptide is between the TIR and the stop codon. The region 5′ of the first initiation codon is called the leader sequence, and the untranslated region between a stop codon for one gene and the next translational initiation region is known as the intercistronic spacer. (B) The association of the 30S and 50S ribosome at a TIR and their dissociation at a stop codon. A 30S and 50S subunit associate at a downstream TIR.

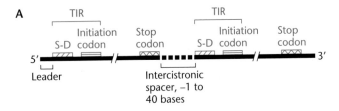
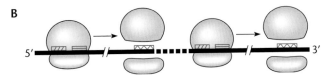

Figure 2.37 Model for translational coupling in a polycistronic mRNA. The secondary structure of the RNA blocks translation of the second polypeptide (A) unless it is disrupted by a ribosome translating the first coding sequence (B).

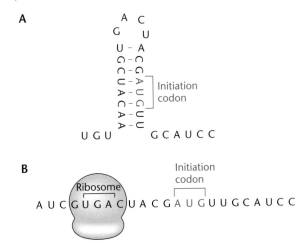

types of mutations can result in polar effects. One type of mutation that can cause a polar effect is an insertion mutation that carries a transcriptional terminator. For example, if a transposon "hops" into a polycistronic transcription unit, the transcriptional terminators on the transposon prevent the transcription of genes downstream in the same polycistronic transcription unit. Likewise, a "knockout" of a gene by insertion of an antibiotic resistance gene with a transcriptional terminator causes a polar effect on the genes downstream in the same transcription unit.

A second type of mutation that can cause a polar effect is a mutation that disrupts translation so that ribosomes dissociate. Within an open reading frame, a change of an amino acid codon to a nonsense codon can cause ribosome dissociation. Then a downstream gene expressed on the same mRNA that is translationally coupled to the upstream gene will not be translated as described above. Additional mutations that can cause this type of effect are frameshift mutations and deletion mutations that shift the reading frame. Ribosomes that are translating out of frame are likely to encounter a nonsense triplet and so to dissociate.

ρ-DEPENDENT POLARITY

Recall that translation of mRNAs in bacteria normally occurs simultaneously with transcription and that the mRNA is translated in the same 5′-to-3′ direction as it is transcribed. Moreover, ribosomes often load onto a TIR as soon as it is vacated by the preceding ribosome, so the mRNA is coated with translating ribosomes. If a nonsense mutation causes dissociation of ribosomes, the abnormally naked mRNA downstream may be targeted by the transcription termination factor ρ, which may find an exposed *rut* sequence in the mRNA and cause transcription termination, as shown in Figures 2.19 and 2.38. The nonsense mutation will then have prevented the expression of the downstream gene by preventing its transcription as shown. Such ρ-dependent polarity effects are relatively rare because the effect occurs only if a *rut* sequence recognizable by ρ and a ρ-dependent terminator lie between the point of the mutation and the next downstream TIR.

Superficially, translational coupling and polarity due to transcription termination have similar effects; in both cases, blocking the translation of one polypeptide affects the synthesis of another polypeptide normally encoded on the same mRNA. However, as we have seen, the molecular bases of the two phenomena are completely different.

RNases and mRNA Processing and Decay

Most bacterial mRNAs have a half-life of only a few minutes, varying from less than 1 to 20 min. The coupling of translation and transcription greatly affects mRNA

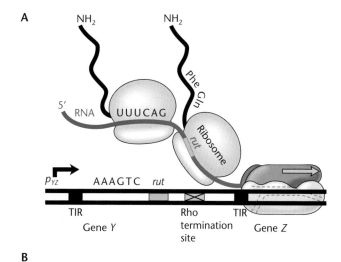

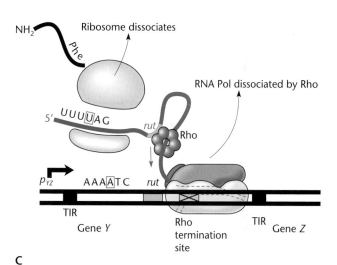

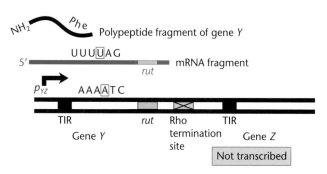

Figure 2.38   Polarity in transcription of a polycistronic mRNA transcribed from p_{YZ}. (A) Normally the *rut* site is masked by ribosomes translating the mRNA of gene Y. (B) If translation is blocked in gene Y by a mutation that changes the codon CAG to UAG (boxed in purple), the ρ factor can cause transcription termination before the RNA polymerase reaches gene Z. (C) Only fragments of the normal gene Y protein and mRNA are produced, and gene Z is not even transcribed into mRNA.

stability (Box 2.5). Although many RNases are known, the details of how they regulate mRNA decay are not fully understood.

Protein Folding

Translating the mRNA into a polypeptide chain is only the first step in making an active protein. To be an active protein, the polypeptide must fold into its final conformation. This is the most stable state of the protein and is determined by the primary structure of its polypeptides. Theoretically, every protein would eventually fold into its most stable structure. However, without the help of other factors, folding might take too long for the protein to be useful.

Protein Chaperones

Some proteins, called **chaperones**, help other proteins fold into their final conformation. Some chaperones are dedicated to the folding of only one other protein, while others are general chaperones which help many different proteins fold. We only discuss general chaperones here.

THE DnaK PROTEIN AND OTHER Hsp70 CHAPERONES

The **Hsp70** family of chaperones is the most prevalent and ubiquitous type of general chaperone, existing in all types of cells with the possible exception of some archaea (see Bukau and Horwich, Suggested Reading). These chaperones are also highly conserved evolutionarily, being almost the same size and having almost the same amino acid sequence whether they come from a human cell or a bacterial cell. These chaperones are called the Hsp70 proteins because they are about 70 kDa in size and because more of them are made, along with many other proteins, if cells are subjected to a sudden increase in temperature or "heat shock" (Hsp70: heat shock protein 70 kDa), although other stresses such as ethanol can have the same effect. More of them are made after such stresses to help refold proteins that have been denatured by the environmental stress, although they also help fold proteins under normal conditions. The Hsp70 type of chaperone was first discovered in *E. coli*, where it was given the name **DnaK** because it is required to assemble the DNA replication apparatus of λ phage and so is required for λ DNA replication. This name for the Hsp70 chaperone in *E. coli* is still widely used in spite of being a misnomer. The chaperone has nothing directly to do with DNA, but functions only in folding proteins. In its role as a heat shock protein, the DnaK protein of *E. coli* also functions as a cellular thermometer, regulating the synthesis of other proteins in response to a heat shock (see chapter 13).

To understand how Hsp70 chaperones including DnaK help fold proteins, it is necessary to understand something about the structure of most proteins. As discussed, proteins are made up of chains of amino acids that are folded up into well-defined structures which are often rounded or globular. The amino acids that make up proteins can be charged, polar, or hydrophobic (see inside book cover for a listing). Amino acids that are charged (either acidic or basic) or polar tend to be more soluble in water and are called hydrophilic (water loving). Amino acids that are not charged or polar are hydrophobic (water hating) and tend to be in the inside of the globular protein among other hydrophobic amino acids and away from the water on the surface. If the hydrophobic amino acids are exposed, they tend to associate with hydrophobic amino acids on other proteins and cause the proteins to precipitate. This is essentially what happens when you cook an egg. High temperatures cause the proteins in the egg to unfold, exposing their hydrophobic regions, which then associate with each other, causing the proteins to precipitate into a hard white mass.

The Hsp70-type chaperones help proteins fold by binding to the hydrophobic regions in denatured proteins and nascent proteins as they emerge from the ribosome and keeping these regions from binding to each other prematurely as the protein folds. The Hsp70 proteins have an ATPase activity which, by cleaving bound ATP to ADP, helps the chaperone periodically bind to, and dissociate from, the hydrophobic regions of the protein they are helping fold. The Hsp70-type chaperones are helped in their protein-folding role by smaller proteins called **cochaperones**. The major cochaperones in *E. coli* were named DnaJ and GrpE, again for historical reasons. The DnaJ cochaperone helps DnaK recognize some proteins and to cycle on and off the proteins by regulating its ATPase activity. It can also sometimes function as a chaperone by itself. The GrpE protein is a nucleotide exchange protein, which helps regenerate the ATP-bound form of DnaK from the ADP-bound form, allowing the cycle to continue.

TRIGGER FACTOR AND OTHER CHAPERONES

Given the prevalence and central role of DnaK in the cell, it came as a surprise that *E. coli* can live without DnaK since *E. coli* mutants that lack DnaK still multiply, albeit slowly. In fact, the only reason they are sick at all is because they are making too many copies of the other heat shock proteins since DnaK also functions as the cellular thermometer (the regulation of heat shock is discussed in chapter 13). One reason why cells lacking DnaK are not dead is that other chaperones can substitute for it. One of these is **trigger factor**. This type of chaperone has so far been found only in bacteria, and

BOX 2.5

Stability and Degradation of mRNA

The mRNAs of bacteria are generally short-lived, and the timing of their degradation is important in regulating the output of genes. Nevertheless, the pathways for mRNA degradation are not completely known, even in *E. coli*. The figure illustrates some of the better characterized enzymatic pathways that process and degrade mRNAs. Part A shows that RNase E is involved in general degradation of mRNAs. Part B shows that RNase III or RNase P process some polycistronic mRNAs. The sequences of recognition sites are not absolutely defined.

The table describes the properties of the above-mentioned enzymes. For example, some enzymes that degrade mRNA also process other RNAs. RNase III also processes rRNA precursors and RNase P also processes tRNA precursors.

About 1 to 2% of the RNAs of *E. coli* and many other bacteria contain 3' poly(A) tails, which are added after transcription. The presence of a poly(A) tail can affect the stability of an mRNA. The table shows the enzymes that produce and process the poly(A) tails.

Reference
Kushner, S. R. 2005. mRNA decay and processing, p. 327–345. *In* N. P. Higgins (ed.), *The Bacterial Chromosome*. ASM Press, Washington, D.C.

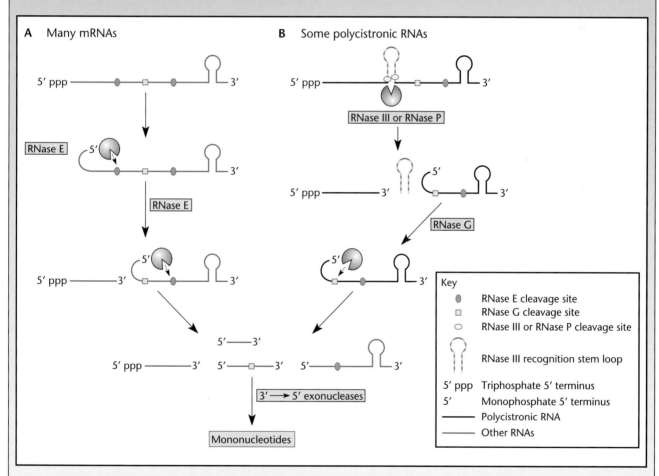

(continued)

Stability and Degradation of mRNA

Enzymes of *E. coli* involved in mRNA processing		
Enzyme	Substrate(s)	Description
RNase E	mRNA, 9S and 16S rRNA, tRNA	Endonuclease, highly conserved in all G$^-$, some G$^+$
RNase III	30S rRNA, polycistronic mRNA	Endonuclease, cleaves double-stranded RNA in some stem-loops; found in both G$^+$ and G$^-$
RNase P	Polycistronic mRNA, tRNA precursors	Ribozyme, necessary to process 5′ end of tRNAs
RNase G	5′ end of 16S rRNA, 9S rRNA, mRNA	Endonuclease; some bacteria have this instead of RNase E, and some have both
Poly(A) polymerase	Any 3′ OH mRNA	Found in both G$^+$ and G$^-$
PNPase	mRNA, poly(A) tails	3′–5′ exonuclease, sometimes also a poly(A) polymerase; found in all G$^+$ and G$^-$

much less is known about it. It is bound more closely to the exit pore of the ribosome and helps proteins fold as they emerge from the ribosome. It is also a **prolyl isomerase.** Of all the amino acids, only proline has an asymmetric carbon, which allows it to exist in two isomers. Trigger factor can convert the prolines in a protein from one isomer to the other. There are many other examples of chaperones being prolyl isomerases.

Besides these chaperones, there are others such as ClpA, ClpB, and ClpX, which form cylinders and unfold misfolded proteins by sucking them through the cylinder. This takes energy, and they cleave a lot of ATP to provide the energy. Some of them, including ClpA and ClpX, feed the unfolded proteins directly to an associated protease called ClpP, which degrades the unfolded protein. Therefore, they are not really chaperones since they do not try to fold misfolded proteins but just recycle their amino acids to be used to make other proteins. ClpB, another cylindrical chaperone, does not associate with a protease but seems to cooperate with the small heat shock proteins IbpA and IbpB to help redissolve precipitated proteins so that they can be refolded by DnaK (see Mogk et al., Suggested Reading).

CHAPERONINS
Besides the relatively simple protein chaperones, cells contain much larger structures that help proteins fold. These large structures are called **chaperonins,** and they

exist in all forms of life, including the archaea and eukaryotes. They are composed of two large cylinders with hollow chambers held together back to back with openings at their ends (Figure 2.39). They help fold a misfolded protein by taking it up in one of the chambers. A cap called a **cochaperonin** is then put on the chamber, and the protein folds within the more hospitable environment of the chamber. A more detailed model for what happens in the chamber and how this helps a protein fold is suggested by their structure (see Wang and Boisvert, Suggested Reading). When the misfolded protein is first taken up, the lining of the chamber consists of mostly hydrophobic amino acids, which bind the exposed hydrophobic regions of the misfolded protein. When the cochaperonin cap is put on the chamber, the lining may switch to being mostly hydrophilic amino acids, driving the hydrophobic regions of the misfolded protein to the interior, where they belong. The cap then comes off, releasing the folded protein. This process takes a lot of energy, and a number of ATP molecules are cleaved to ADP in the process. A different unfolded protein can then enter the other chamber, and the process is repeated in the other chamber. This has been described as a two-stroke engine, where the folding role switches from one chamber to the other. It is a mystery why chaperonins have two chambers and why the folding has to alternate between the two chambers. Chaperonins composed of only one chamber function more poorly, although they

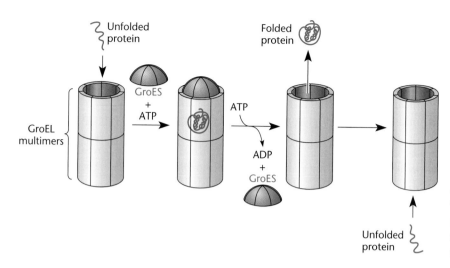

Figure 2.39 Chaperonins. The GroEL (Hsp60)-type chaperonin multimers form two connected cylinders (shown separately). A denatured protein enters the chamber in one of the cylinders, and the chamber is capped by the cochaperonin GroES (Hsp10). The denatured protein can then be helped to fold in the chamber. The other chamber plays a regulatory role but is the chamber that takes up the next unfolded protein in a sort of two-stroke engine. Details are given in the text.

might retain some of their activity (see Sun et al., Suggested Reading).

As is often the case with universal cellular functions, the first chaperonin was discovered in *E. coli*, where it was named GroEL because it helps assemble the E protein of λ phage into the phage head. The GroEL chaperonin, which has served as the model, consists of 14 identical polypeptides (7 making up each cylinder) of 60 kDa. Its cochaperonin cap is called GroES, which is also made up of seven subunits, each 10 kDa in size. Unlike DnaK and the other chaperones, GroEL and GroES are required for *E. coli* growth, even at lower temperatures. Presumably the GroEL chaperonin is required to fold some essential protein or proteins, which cannot fold without its help.

The chaperonins come in two general types called the group I and group II chaperonins. The group I chaperonins are related to GroEL, are composed of 60-kDa subunits, and are found in all the eubacteria and the mitochondria and chloroplasts of eukaryotes, which makes sense since these organelles are derived from eubacteria (see the introductory chapter). These chaperonins are induced by heat shock and other stresses, so they are called the **Hsp60 proteins** for heat shock 60-kDa proteins (see Bukau and Horwich, Suggested Reading). The group II chaperonins are found in the archaea and in the cytoplasm of eukaryotes. They have very little amino acid sequence in common with the group I chaperonins and are not composed of identical subunits (i.e., they are mixed multimers); they might have eight or more polypeptide subunits per cylinder. Furthermore, if they have a cochaperonin cap, it might be attached to the opening of the chamber rather than being detachable like GroES. Nevertheless, the two types form similar cylindrical structures and presumably use a similar mechanism

to help fold proteins. Note that here is yet another example of where the archaea and eukaryotes are similar to each other and different from the eubacteria.

Membrane Proteins and Protein Export

In order to function, proteins not only must be folded properly but also must reach their final destination in the cell. Often this means that they must leave the cytoplasm where they were synthesized and enter the membranes surrounding the cell, or in some instances leave the cell altogether in the form of extracellular proteins. This is a conceptually simpler process in gram-positive bacteria, with only one membrane to pass through, than it is in gram-negative bacteria, with both an inner and outer membrane. Nevertheless, the processes are very similar in gram-negative and gram-positive bacteria. The primary difference is that when a protein to be secreted passes through the inner membrane of gram-negative bacteria, it must also get through the outer membrane before it is outside the cell. The complexity of these specialized mechanisms in gram-negative bacteria is discussed in chapter 14.

Specific words are used to designate the proteins that leave the cytoplasm and the processes which allow them to do this. Proteins that reside in the inner membrane of gram-negative bacteria are called **inner membrane proteins (IMPs)**. Proteins that reside in the outer membrane are **outer membrane proteins (OMPs)**. Proteins that pass through the inner membrane into the periplasm or outer membrane of gram-negative bacteria are **exported proteins**. Proteins that pass completely out of the cell into the external environment are **secreted proteins**. The process of passing a protein through a membrane is **translocation**.

By far the largest group of proteins that leave the cytoplasm are destined for the inner membrane. Inner membrane proteins often extend through the membrane a number of times and have stretches that are in the periplasm and other stretches that are in the cytoplasm. The stretches that traverse the membrane have mostly uncharged, nonpolar (hydrophobic) amino acids, which make them more soluble in the membranes, which are made up of lipids and so are very hydrophobic. A stretch of about 20 amino acids is long enough to extend from one side of the bipolar lipid membrane to the other, and such stretches in proteins are called the **transmembrane domains.** The other stretches between them are called the **cytoplasmic domains** and the **periplasmic domains,** depending on whether they extend into the cytoplasm on one side of the membrane or into the periplasm on the other side. Transmembrane proteins are very important because they allow communication from outside the cell to the cytoplasm; some of them are discussed in chapter 13 in the section on two-component global regulatory systems.

Secreted proteins and even transmembrane proteins contain many amino acids that are either polar or charged (basic or acidic), which make it difficult for them to pass through the membranes. They must be helped in their translocation through the membrane by other specialized proteins that are dedicated to this purpose. Some of these proteins form a channel in the membrane with a hydrophilic core through which hydrophilic regions of proteins can pass. Some transported proteins use their own channel, but most use the more general channel called the **translocase.** The translocase is discussed next.

The Translocase System

Many of the proteins that help other proteins pass through the membranes are part of the **Sec system.** A current picture of the structure of the export channel that helps a protein pass through the inner membrane, as well as how it works, is outlined in Figure 2.40. The genes for this system were first found in a search for *E. coli* mutants defective in transport of proteins into the periplasm. This was an elegantly designed selection and is discussed in detail in chapter 14.

Some of the proteins of the Sec system form the channel in the membrane called the translocase, through which proteins can pass (Figure 2.40A). As mentioned, most proteins have many charged and polar amino acids, both of which are hydrophilic (water loving), and this channel must be able to allow these hydrophilic amino acids to pass through the very hydrophobic membrane. It must also be normally closed, and open only when a protein is passing through it; otherwise other proteins and small molecules would pass in and out of the cell through the channel, which would be disastrous for the cell. The channel is made up of three proteins, SecY, SecE, and SecG, and is therefore called the **SecYEG channel** or **SecYEG translocase.** These three proteins form a heterotrimer, made up of one of each of the three different polypeptides. The SecY protein is by far the largest and forms the major part of the channel, while the other two smaller proteins play more ancillary, albeit important roles. It is not clear how many of each protein contribute to the channel. One heterotrimer, made up of one copy of each of the proteins, can form a large enough channel to let an unfolded protein through (see van den Berg et al., Suggested Reading), but it is still possible that more than one of these heterotrimers called protomers fuse to form a larger channel in the membrane. It is more likely that a number of the protomers join together to form a larger complex called an oligomer, which plays some role in the translocation process.

Much of what we know about the details of how proteins are exported through the SecYEG channel comes from genetic studies, which have been largely confirmed by recent structural studies. Mutations, called *prl* mutations, were isolated that allow the transport of proteins with defective or even missing signal sequences. These mutations can be in *secA, secE,* or *secY.* A region of the SecY protein forms a hydrophobic "plug," which opens

---

Figure 2.40 Protein export. (A) Cutaway view of the export channel. SecY, SecE, and SecG (not shown) form the translocase. SecY forms the channel, ring, and plug. The signal sequence of the exported protein moves the plug toward SecE. (B) Posttranslational export by the SecB-SecA system. SecB keeps the protein unfolded until it binds to SecA, which interacts with SecY. The signal sequence is removed, in this case by Lep protease. The exported protein is folded or may be secreted across the outer membrane in a gram-negative bacterium. (C) Cotranslational export by the SRP system. SRP binds to the first transmembrane domain as it emerges from the ribosome and then binds to the FtsY docking protein, bringing the ribosome to interact with SecY. The protein is translated, driving it into the SecY channel. The transmembrane domains of the protein somehow escape through the side of the channel into the membrane.

A Export channel

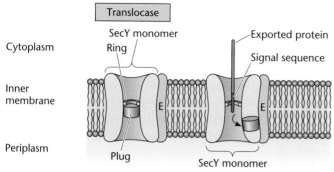

Translocase

Cytoplasm

SecY monomer
Ring
Exported protein
Signal sequence

Inner membrane

Periplasm

Plug
SecY monomer

B Posttranslational export

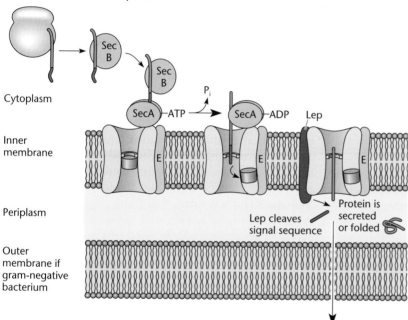

Cytoplasm

Sec B

Sec B

SecA — ATP → SecA — ADP

P$_i$

Lep

Inner membrane

Periplasm

Lep cleaves signal sequence

Protein is secreted or folded

Outer membrane if gram-negative bacterium

C Cotranslational export

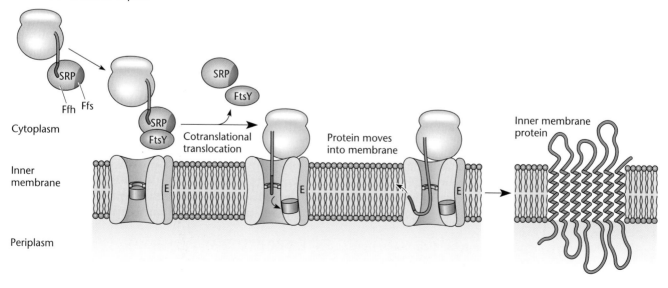

SRP

Ffh Ffs

Cytoplasm

SRP
FtsY

SRP
FtsY

Cotranslational translocation

Protein moves into membrane

Inner membrane protein

Inner membrane

Periplasm

only when an exported protein is passing through (Figure 2.40). The binding of a signal sequence (see below) in a protein to be transported causes the plug to move over toward SecE on the side of the channel, opening the channel as shown in the figure. This prevents the translocation of proteins that do not have a bona fide signal sequence. Apparently, *prl* mutations leave the plug at least partially open, perhaps by making binding of the plug to SecE stronger. This allows proteins without a good signal sequence to pass through. The role of SecG is less clear. It is not absolutely required for protein translocation but seems to stimulate the rate of protein movement through the channel. Two other nonessential proteins, called SecD and SecF, are also bound to the channel and are highly evolutionarily conserved, but their role in transport is not known. The genetic analyses that led to the identification of the *sec* genes and *prl* mutations were very elegant and are discussed in some detail in chapter 14.

The Signal Sequence

As mentioned above, the defining feature of proteins that are to be exported into the inner membrane or beyond by the SecYEG channel is the presence at their N terminus of a short hydrophobic sequence of amino acids called the **signal sequence** (Figure 2.40A). While the amino acids in this short sequence differ from one type of protein to another, it is made up of mostly hydrophobic amino acids. However, the fate of this signal sequence depends on whether the protein is to be exported through the inner membrane or is to remain in the inner membrane as an IMP. In the case of exported proteins, the signal sequence is removed by a protease as the protein passes through the SecYEG channel. The most prevalent of the proteases which clip off signal sequences in *E. coli* is the **Lep protease** (for "leader peptide protease"), but there is at least one other more specialized protease called LspA, which removes the leader sequence from lipoproteins destined for the outer membrane. Proteins that are destined to be exported beyond the inner membrane but have just been synthesized and so still retain their signal sequence are called **presecretory proteins.** When the short signal sequence is removed in the SecYEG channel, the presecretory protein becomes somewhat shorter before it reaches its final destination in the periplasm, the outer membrane, or outside the cell. This shortening of the protein after it is synthesized is easy to detect on sodium dodecyl sulfate-polyacrylamide gels and is often taken as evidence that the protein is exported by the SecYEG channel.

IMPs also have a signal sequence at their N terminus, but this is generally longer and more hydrophobic than the signal sequence of secreted proteins. It becomes the

first transmembrane domain of the protein (see above) and is not cleaved off when the protein enters the membrane. Which type of signal sequence a protein has determines which of the targeting factors directs it to the SecYEG translocon and whether it remains in the inner membrane or is exported through to the periplasm and beyond (see below).

The Targeting Factors

The targeting factors recognize proteins to be transported into or through the inner membrane and help target them to the membrane. Enteric gram-negative bacteria like *E. coli* have at least two separate systems that target proteins to and through the membranes. One of these is the **SecB** system. This targeting system is dedicated to proteins that will be directed through the inner membrane into the periplasm or exported from the cell. The other system is the **signal recognition particle (SRP) system,** which may exist in all organisms including humans. In bacteria, this targeting system seems to be dedicated to proteins that are destined to reside in the inner membrane. Another protein, SecA, participates in both pathways, at least for some proteins; it is found in all bacteria but not in eukaryotes, although in eukaryotes other proteins may play a similar role.

THE SecB PATHWAY

Proteins that have a removable signal sequence and are exported are most often targeted by the SecB system in bacteria. The SecB protein is a specialized chaperone that binds to presecretory proteins even after they are synthesized (**posttranslational translocation**), thereby preventing them from folding prematurely and leaving the signal sequence exposed. The exposed signal sequence may itself also play a role in preventing the premature folding of the presecretory protein. The SecB chaperone then passes the unfolded protein to SecA, which functions somehow to facilitate the association of the protein with the SecYEG channel, perhaps by binding to the signal sequence and to the channel in the membrane simultaneously (Figure 2.40B). After SecA binds to the channel, the ATP on SecA is cleaved to ADP, which drives the protein into the channel. The SecG component of the channel might also help drive the protein into the channel (not shown). SecA then cycles off the translocase channel, and the protein passes through the channel, losing its signal sequence in the process. SecB is not an essential protein, and the cell can use DnaK or other general chaperones as substitutes for SecB to help transport some proteins.

THE SRP PATHWAY

The SRP pathway for protein targeting in bacteria generally targets proteins that are to remain in the inner membrane. It consists of a particle (the signal recognition particle) made up of both a small **4.5S RNA**, encoded by the *ffs* gene, and at least one protein, **Ffh**, as well as a specific receptor on the membrane, called **FtsY** in *E. coli*, to which the SRP binds. FtsY is sometimes referred to as the docking protein because it "docks" proteins targeted by the SRP pathway on the membrane. Its name is another misnomer. Its gene was originally identified through temperature-sensitive mutations that cause *E. coli* to not divide properly and to form long filaments of many cells linked end to end at higher temperatures (filament temperature-sensitive Y; temperature-sensitive mutations are discussed in chapter 3). Presumably, FtsY is required to insert some inner membrane proteins required for cell division into the inner membrane, and these proteins are not being inserted properly at the higher temperatures due to the mutational defect in FtsY. The role of the docking protein is to direct the protein to the SecYEG translocon. The SecA protein might also help target some SRP translocated proteins to the SecYEG translocon, particularly those with long periplasmic domains.

Figure 2.40C illustrates how the SRP system works. The SRP binds to the first hydrophobic transmembrane sequence of an IMP as this region of the protein emerges from the ribosome. It is debatable whether binding of the SRP stops translation of the emerging protein in bacteria as it does in eukaryotes (see below) or whether the particle binds quickly enough so that the complex has time to bind to the FtsY receptor in the membrane before translation continues. It is also not clear whether the FtsY receptor remains on the membrane or can bind to the SRP complex in the cytoplasm and then direct it to the membrane. In any case, once the complex has bound to the membrane, translation of the protein continues, feeding the protein directly into the SecYEG translocon as the inner membrane protein is translated. The energy of translation drives the polypeptide out of the ribosome into the SecYEG translocon, obviating the need for SecA in at least most cases. This is called **cotranslational translocation** because the protein is translated as it is inserted into the membrane. There is a good reason why proteins destined for the inner membrane are cotranslated with their insertion into the translocon in the membrane while exported proteins can first be translated and then inserted into the translocon after they are translated in their entirety. With their long transmembrane domains, the inner membrane proteins are too hydrophobic to remain soluble in the cytoplasm and would precipitate if they were translated in their entirety in the cytoplasm before they were transported. Presecretory proteins are generally less hydrophobic, so they can be held in a partially unfolded state by chaperones like SecB until they

can be transported (see Lee and Bernstein, Suggested Reading).

What happens after an inner membrane protein enters the SecYEG channel is less clear. The transmembrane domains of the protein must escape the SecYEG channel and enter the surrounding membrane while the periplasmic and cytoplasmic domains must stay where they are. Presumably the SecYEG channel opens up and allows the transmembrane domains of the protein to escape into the membrane. Another inner membrane protein called **YidC** might help in this process. The role of YidC in protein translocation is not clear, but it seems to be required for the transport of some proteins but not others. Some inner membrane proteins bypass SecYEG altogether and require only YidC to enter the inner membrane. It seems possible that YidC is required to assemble some complexes of proteins in the membrane such as the membrane ATPase, which is the role it plays in mitochondria.

It is interesting to compare the Sec systems of eukaryotes with those of *E. coli*, the bacterium in which these systems are best understood. Eukaryotes do not have SecB or SecA and use the SRP system to translocate all exported proteins through the translocon into the endoplasmic reticulum. Even though they lack SecA, they may have other systems that help direct already translated proteins to the translocon. The translocon itself is highly conserved and is composed of three proteins in all three kingdoms of life; these proteins form similar structures even though they have different names. The sequences of the SecY and SecE subunits are similar in all three kingdoms (eubacteria, eukaryotes, and archaea). Only the third subunit, called the SecG subunit in eubacteria, is very different in eukaryotes and archaea, where it may have a different function. If SecG plays a role in driving presecretory proteins into the channel (see above), eukaryotes and archaea may not have any need for such a function since most of their protein export is driven by cotranslation (see below). It is also interesting that, while eukaryotes have other such channels, the translocase which helps transported proteins to enter the endoplasmic reticulum of eukaryotic cells is the one most similar to the SecYEG channel of bacteria. This makes sense, since the endoplasmic reticulum plays a role in protein translocation similar to the role played by the inner membrane of bacteria.

The SRP system was first described to occur in eukaryotes, where it is much larger, consisting of a 300-nucleotide RNA and eight proteins, six in the SRP and two in the docking protein, called the SRP receptor (SR). However, some of the proteins in eukaryotes are very similar to those in eubacteria. In fact, the protein in the SRP of eubacteria was named the Ffh protein (for "fifty-four homolog") because it is so similar to the 54-kDa

SRP protein in eukaryotes. Also, the SRP system of eukaryotes targets both membrane and presecretory proteins to the endoplasmic reticulum, the organelle that plays an analogous role in protein transport to the inner membrane of bacteria, and seems to recognize both proteins with a true signal sequence and membrane proteins, which merely have a long hydrophobic N-terminal transmembrane segment. Furthermore, the SRP of eukaryotes is known to stop translation of both secreted and inner membrane proteins as the signal sequence emerges from the ribosome, enforcing cotranslational transport of both types of proteins. The eukaryotic SRP can do this because the extra proteins it contains make it long enough to extend all the way from the exit pore for polypeptides on the large subunit of the ribosome to the A site on the ribosome (see Wild et al., Suggested Reading). When the SRP binds to the signal sequence of a protein as the protein emerges from the exit channel, it extends all the way to the A site on the ribosome, blocking the entrance of aminoacylated tRNAs to the A site and stopping translation. When the SRP-ribosome complex then binds to the SR docking or receptor protein, the SRP is removed and translation continues, feeding the protein directly into the translocon channel. If binding of the SRP to the emerging signal sequence also arrests translation in bacteria, it must be by a different mechanism, considering the much smaller size of the bacterial SRP.

Protein Secretion

Some translocated proteins do not stop when they reach the inner membrane, the periplasmic space, or the outer membrane, but keep going until they are outside the cell. As mentioned above, these are called secreted proteins. They include extracellular enzymes that degrade large molecules such as polysaccharides so that the smaller degradation products can be transported back into the cell to be used as food. Other systems are required to secrete extracellular structures that form on the outside of cells, such as pili and flagella. Pili and flagella stick out of the surface of the cell and help it move on solid and liquid surfaces, respectively. Other examples of secreted proteins include proteins that are secreted directly from the bacterial cell into a eukaryotic host cell to help pathogenic bacteria establish an infection and the relaxosomes that are attached to plasmids which are secreted from one bacterium to another or from a bacterium to a plant cell (see below and chapter 5).

There are basically six known types of secretion systems in gram-negative bacteria: types I through VI. Some of these depend on the SecYEG translocase described above or another system called the Tat system to transport proteins through the inner membrane before they take over to transport it the rest of the way through the outer membrane to the outside of the cell. The type II system, represented by the cholera toxin secretion system of *Vibrio cholerae*, is an example of such a system. The toxin is first transported through the inner membrane by the SecYEG translocon into the periplasmic space. It then uses its own complex structure to pass through the outer membrane. Once the toxin is outside the cell, its B subunit can help its A subunit enter a eukaryotic cell, where it acts as a toxin. Other examples of secreted proteins that use the SecYEG channel are the so-called autotransporter or type V systems, represented by the immunoglobulin A (IgA) protease of *Neisseria gonorrhoeae*, in which the protein carries its own channel to get through the outer membrane. Once in the periplasm, part of the protein folds into a β barrel that inserts into the outer membrane and helps the remainder of the protein through the membrane. Others, such as the components of pili, use chaperones and ushers to get through the outer membrane.

Other transport systems can help proteins through both membranes, without the help of the SecYEG translocon. The type I systems, sometimes called ABC transporters (for "ATP-binding cassette"), are dedicated to secreting only one protein, such as the hemolysin of *E. coli*, directly through both membranes to the outside of the cell. Type IV secretion systems secrete virulence proteins directly into eukaryotic cells and are ancestrally related to plasmid conjugation systems; therefore, they are discussed in chapter 5. The most dramatic type of secretion system that can transport proteins through both membranes are the type III secretion systems. These systems form large, syringe-like multiprotein complexes that inject the protein directly through both membranes into a eukaryotic cell and are represented by the type III secretion system of *Yersinia pestis*, which injects proteins called Yops directly into eukaryotic phagocytes. This disables the phagocyte, whose role is to ingest bacteria and destroy them. Type III secretion systems have attracted much attention recently because of their role in bacterial pathogenesis and their similarity in different types of pathogenic bacteria including both plant and animal pathogens. These and other types of secretion systems are discussed in more detail in chapter 14.

Disulfide Bonds

Another characteristic of proteins that are exported to the periplasm or secreted outside the cell is that many of them have disulfide bonds between cysteines (C) (see inner cover). In other words, two of the cysteines in the protein are held together by covalent bonds between their sulfides. These disulfide bonds can be either between two cysteines in the same polypeptide chain or between cysteines in different polypeptide chains.

Secreted proteins need the covalent disulfide bonds to hold them together in the harsh environments of the periplasm and outside the cell. Failure to form the correct disulfide bonds or formation of disulfide bonds between the wrong cysteines can result in inactivity of the protein. The sulfur atom of a cysteine in a disulfide bond is in its oxidized form because one of its electrons is now shared between the two sulfurs, while the sulfur atom of an unbound cysteine is in its reduced form because it has an extra electron. The disulfide bonds are formed by enzymes called **disulfide oxidoreductases** as the proteins pass through the oxidizing environment in the periplasmic space between the inner and outer membranes of gram-negative bacteria. The oxidoreductases contain the motif C-X-X-C, two cysteines separated by two other amino acids, where X can be any amino acid. This is sometimes called the thioredoxin motif because it is also found in thioredoxin, which plays a role in reducing proteins inside the cytoplasm. However, even though the X's in the motif can be any amino acid, the oxidizing strength of the oxidoreductase (i.e., its ability to form a disulfide bond) is much stronger with some amino acids separating the cysteines than others. For example, in *E. coli*, the major periplasmic enzyme that forms disulfide bonds is **DsbA** (for disulfide bond A). The X-X in the C-X-X-C motif in DsbA is Pro-His (P-H), which makes it a stronger oxidizer than the other oxidizing oxidoreductases in *E. coli* (see Tan et al., Suggested Reading).

The way the oxidoreductases work to create disulfide bonds in the periplasm is illustrated in Figure 2.41. Basically, the disulfide bond forms when an extra electron is passed from a cysteine in the protein being exported to a stronger oxidizer like DsbA in the periplasm. This oxidizes the cysteine in the exported protein, causing it to form a disulfide bond with another cysteine in the exported protein. The electron transferred to DsbA reduces the cysteines in its C-P-H-C motif, destroying their disulfide bond. These cysteines in DsbA are in turn oxidized to form another disulfide bond by passing the extra electron to DsbB, a weaker oxidoreductase in the inner membrane, reducing its C-X-X-C motif. DsbB in turn passes the electron to quinones in the membrane, whose job it is to pass electrons to electron acceptors such as oxygen during electron transport. The disulfide bond in the C-P-H-C motif in DsbA can then be used to form another disulfide bond between two cysteines in another exported protein in the periplasm.

Not only do proteins that are found inside the cell in the cytoplasm lack disulfide bonds, but also this type of bond cannot normally form in the cytoplasm. This is because of the "reducing atmosphere" inside the cytoplasm due to the presence of high concentrations of small reducing molecules, mostly glutathione and

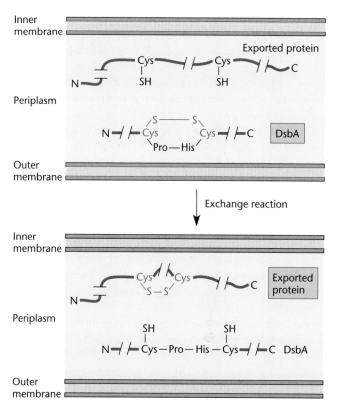

Figure 2.41 Disulfide bond formation in the periplasm. Oxidoreducatases in the periplasm exchange disulfide bonds (in purple) with the protein as it enters the periplasm. Only DsbA is shown. The broken line indicates that the cysteines in the exported protein are different distances apart and can even be on different polypeptides. Details are given in the text.

thioredoxin. In fact, the appearance of disulfide bonds in some cytoplasmic regulatory proteins is taken as a signal by the cell that oxidizing chemicals are accumulating in the cell and that proteins should be made to combat the potentially lethal oxidative chemical stress (see chapter 13).

Regulation of Gene Expression

The previous sections have reviewed how a gene is expressed in the cell, from the time mRNA is transcribed from the gene until the protein product of the gene reaches its final destination in or outside the cell and has its effect (i.e., is expressed). The products of different genes are made in vastly different amounts, depending on how much of the product of the gene is required by the cell. This is sometimes referred to as the copy number of the protein and is determined by many factors including the strength of the promoter and the strength of the TIR on the mRNA. The amount of any particular gene product made by the cell also often varies depending on the state in which the cell finds itself. In general, genes are expressed in the cell only when their products are needed by the cell and then only as much as is required to make the amount of product needed by the cell. This saves energy and prevents the products of different genes from interfering with each other. The process by which the output of genes is changed depending on the state of the cell is called the **regulation of gene expression** and can occur at any stage in the expression of the gene. Genes whose products regulate the expression of other genes are called **regulatory genes.** The product of a regulatory gene can either inhibit or stimulate the expression of a gene. If it inhibits the expression, the regulation is negative; if it stimulates the expression, the regulation is positive. A regulatory protein need not necessarily be one or the other, however; some regulatory gene products are both positive and negative regulators depending on the situation. Sometimes the product of a regulatory gene can regulate the expression of only one other gene, and sometimes it can regulate the expression of many genes. The set of genes regulated by the same regulatory gene product is called a **regulon.** Sometimes a gene can also regulate its own expression as well as the expression of one or more other genes. If a gene product regulates its own expression, it is said to be **autoregulated.** We discuss the molecular mechanisms of regulation of gene expression in much more detail in chapter 12, but in this chapter we briefly review some basic concepts needed to understand the next chapters.

Transcriptional Regulation

Usually the expression of a gene is regulated by controlling the amount of mRNA that is made on the gene. This is called **transcriptional regulation.** It makes sense to regulate gene expression at this level since it is wasteful to make mRNA on a gene if the expression of the gene is going to be blocked at a later stage anyway. Also, bacterial genes are often arranged in an **operon** (see above), so mRNA can be made simultaneously on a number of genes whose products perform related functions. The expanded definition of an operon is all of the genes whose products are translated from the same mRNA plus the promoter and other *cis*-acting sequences required for expression and regulation of the genes of the operon. The regulatory gene is normally not considered part of the operon unless it is cotranscribed with the other genes of the operon. Sometimes a regulatory gene is part of an operon it regulates, in which case the regulatory gene is autoregulated.

The regulation of transcription of an operon usually occurs at the start point of transcription, at the promoter.

Whether a gene is expressed depends on whether the promoter for the gene is used to make mRNA. Transcriptional regulation at the promoter for a gene can be either negative or positive, depending on whether the regulatory gene product is a transcriptional **repressor** or a transcriptional **activator**, respectively. The difference between regulation of transcription by repressors and activators is illustrated in Figure 2.42. A repressor binds to the DNA at an **operator** sequence close to, or even overlapping, the promoter and somehow prevents RNA polymerase from using the promoter, usually either by physically obstructing access to the promoter by the RNA polymerase or by bending the promoter so that the RNA polymerase is unable to bind to it. An activator, in contrast, binds upstream of the promoter at an **upstream activator site (UAS),** where it can help the RNA polymerase bind to the promoter or help open the promoter after the RNA polymerase binds. Sometimes a transcriptional regulator can be a repressor on some promoters and an activator on other promoters depending on where it binds relative to the start site of transcription.

Whether a repressor represses transcription or an activator activates transcription depends on the state of the repressor or activator. Some regulatory proteins bind small molecules called **effectors,** which affects their activity. Effectors are often molecules that can be used by the cell if it turns on the operon or an essential metabolite that does not have to be made by the cell if it is present in the medium. If the small-molecule effector causes transcription of the operon to be turned on, for example by binding to a repressor and changing it so that the repressor can no longer bind to the DNA, the small molecule is called an **inducer.** If, by binding to a repressor, the effector causes the operon to be turned off, it is called a **corepressor.**

The regulatory molecule can also have its activity changed by being phosphorylated by another protein in the cell in response to a certain set of conditions (see Box 13.4). In this type of regulation, a phosphate (PO_4) group is transferred from an amino acid in another protein called a phosphotransferase to an amino acid in the regulatory protein in response to some environmental condition, changing the activity of the regulatory protein. There are even examples of where a protein changes from being a repressor to being an activator (or vice versa) on binding the effector.

Not all transcriptional regulation occurs at the promoter, however. Sometimes transcription starts and then stops prematurely depending on other factors in the cell. Such regulation is called attenuation of transcription. These and other mechanisms of transcriptional regulation are discussed in subsequent chapters.

A

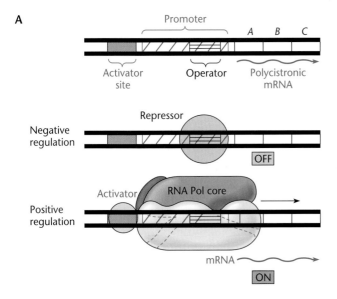

B

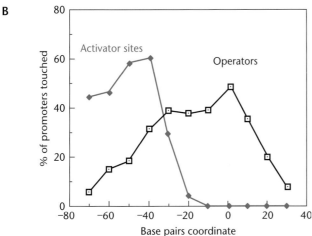

Figure 2.42   (A) The two general types of transcriptional regulation. In negative regulation, a repressor binds to an operator and turns the operon off. In positive regulation, an activator protein binds upstream of the promoter and turns the operon on. (B) Graph showing the usual locations of activator sites relative to operators. Activator sites are usually farther upstream. Each datum point indicates the middle of the known region on the DNA where a regulatory protein binds. Zero on the *x* axis marks the start point of transcription.

Posttranscriptional Regulation

More rarely, expression of a gene can be regulated at later stages in the expression of a gene. For example, translation of the gene may be inhibited even after the mRNA has been made. This is called **translational regulation.** Alternatively, the mRNA may be degraded as soon as it is made, before it can be translated. The protein product of the gene may even have its activity

regulated after it is made. It may be degraded by other proteins called proteases if it is not needed; it may have its activity altered by being phosphorylated, methylated, or adenoribosylated, depending on the conditions in which the cell finds itself; or the product of a pathway may inhibit the activity of an enzyme in the pathway, by a process called **feedback inhibition.** In general, a type of regulation of gene expression which operates after the mRNA for a gene has been made is called **posttranscriptional regulation.** Specific examples of posttranscriptional regulation are also discussed in subsequent chapters.

Introns and Inteins

The simple picture of gene expression in which an RNA copy is made of the DNA and then faithfully translated into a protein is complicated by the existence of parasitic DNA elements called introns or inteins (Box 2.6). These DNA elements can integrate themselves into a gene and disrupt the coding sequence of the gene, so that the sequence of nucleotides in the DNA no longer represents the sequence of amino acids in the final protein product. In order not to inactivate the protein product of the gene and harm their host, these DNA elements often process (splice) themselves out of the mRNA (intron) or out of the protein product after it is made (intein). Introns and inteins are found in all forms of life, but introns in particular are much more common in eukaryotes than they are in bacteria. Furthermore, they often play regulatory roles in eukaryotes through differential splicing, which has not been observed in bacteria.

Useful Concepts

We have introduced a lot of detail in this chapter, so it is worth reviewing some of the most important concepts and words. As with any field, molecular genetics has its own jargon, and in order to follow a paper or seminar that includes some molecular genetics, this jargon must be very familiar.

Figure 2.43 shows a typical gene with a promoter and transcription terminator. The mRNA is transcribed beginning at the promoter and ending at the transcription terminator. The direction on the DNA or RNA is indicated by the direction of the phosphate bonds between the carbons on the ribose or deoxyribose sugars in the backbone of the polynucleotide. These carbons are labeled with a prime (′) to distinguish them from the carbons in the bases of the nucleotides. On one end of the RNA, the 5′ carbon of the terminal nucleotide is not joined to another nucleotide by a phosphate bond. Therefore, this is called the **5′ end.** Similarly, the other end is called the **3′ end** because the 3′ carbon of the last

BOX 2.6

Selfish DNAs: RNA Introns and Protein Inteins

The chromosomal DNA of all organisms abounds with parasitic DNA elements, so named because they cannot replicate themselves but can replicate only when the host DNA replicates. These parasitic DNAs often have few functions except for the ability to move from one DNA to another and thereby parasitize new hosts (see Box 10.1). When such a parasitic DNA integrates into a region of the DNA encoding a protein or RNA, it will disrupt the coding sequence, which is why these elements are sometimes called **intervening sequences.** Like all good parasites, these DNA elements do as little harm to their host as possible, which makes sense since the parasite is dependent on the host for its own survival. Sometimes an intervening sequence inserts into the coding sequence for an essential RNA or protein. This would disrupt the coding sequence and could be lethal in a haploid organism like a bacterium, except that many of the parasitic elements minimize damage to their host by splicing their

sequences out of RNAs and proteins after they are made, restoring the RNAs and proteins to functionality. Intervening sequences that splice themselves out of the RNA are called **introns,** while those that splice themselves out of the protein product of a gene are called **inteins.** The sequences upstream and downstream of the intron or intein in a gene that are rejoined following splicing are called **exons** and **exteins,** respectively.

The two types of RNA introns in bacteria are called group I and group II, based on their mechanism of splicing (see the figure panel A). Both groups of introns are typically self-splicing, meaning that they are capable of splicing themselves out of the RNA without the help of proteins. The RNA of an intron is therefore an enzyme. RNA enzymes are called **ribozymes** to distinguish them from the more common protein enzymes. Other known ribozymes are some RNases and the 23S rRNA peptidyltransferase (see the text). In group

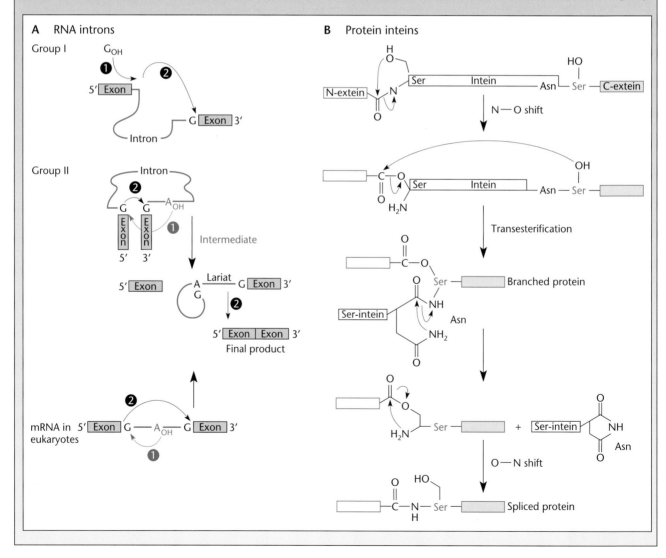

BOX 2.6 (continued)

Selfish DNAs: RNA Introns and Protein Inteins

I introns, a free guanosine nucleoside or nucleotide residue initiates the splicing by breaking the RNA at the 5′ end of the intron, called the 5′ splice site, initiating a series of phosphodiester bond transfers that complete the splicing process. The group I introns are typically found in bacteriophage protein-coding regions and in the tRNA genes of bacteria. Group I introns (and inteins [see below]) typically move by encoding a DNA endonuclease that makes a double-strand break at a specific site in a DNA that lacks them. This initiates a double-strand break repair recombination that inserts the intron in the site on the DNA. In this way the intron can move, but it can move only into the same site on another DNA which lacks an intron at that site. It is essential that they move into the same site because sequences flanking the intron play a role in its splicing and they would not be able to splice themselves out of an mRNA anywhere else, where the flanking sequences would be different. This ability to move, but only into the same place in another DNA, is called **homing,** and the DNA endonuclease they encode is called a **homing endonuclease.** Homing and double-strand break repair recombination are discussed in chapter 10.

Group II intron splicing is similar to group I splicing except that the initiating nucleotide is a specific adenine base internal to the intron, creating a characteristic "lariat" structure of the intron. This type of splicing is more analogous to mRNA splicing in eukaryotes, as shown. While common in lower eukaryotes, group II introns are much rarer in bacteria and are typically found only in other movable elements such as conjugative plasmids and transposons (see chapters 5 and 9).

Even though many of these introns are themselves enzymes and can splice themselves out of the RNA, many encode maturase proteins that help them fold into the structure required for splicing. In the group II introns, these maturase proteins are also a reverse transcriptase and, in combination with the "lariat" intron RNA, form the DNA endonuclease that cuts the target DNA. However, they move by a different process called **retrohoming** because it goes from RNA to DNA, the reverse of the normal direction. Retrohoming is essentially the reverse of splicing in that the intron splices itself *into* a DNA rather than *out* of an RNA as in splicing. In a process somewhat analogous to splicing, some lower eukaryotes such as trypanosomes and nematodes also attach short RNA sequences to the 5′ ends of their mRNA after synthesis, which may help their translation and stability.

Protein inteins are parasitic DNAs like self-splicing introns, except that they splice themselves out of the protein product of the gene rather than out of the mRNA. Inteins probably also exist in all organisms from bacteria to humans. Inteins

self-splice themselves out of a protein by the mechanism shown in the figure (panel B). The first amino acid in the intein is always cysteine or serine. This amino acid can be rearranged so that it is attached to the amino acid upstream through its side chain rather than by a normal peptide bond. Such a bond is called an ester bond or thioester bond depending on whether the first amino acid in the intein is a serine or a cysteine, respectively, and is called an N-O shift because the bond to the nitrogen in the peptide bond is being shifted to the oxygen in the side chain of the serine. In the next step, this (thio)ester bond is attacked by the side chain of the first amino acid just downstream of the intein, which can be a serine, threonine, or cysteine. This replaces the side chain of the first serine or cysteine in the intein with the side chain of the first amino acid in the downstream extein and leads to the formation of a branched protein, in which one branch is the intein, as shown in the figure. This reaction is called a transesterification, because the ester bond is being transferred to a different amino acid. The last amino acid in the intein is now connecting the intein branch to the rest of the protein, as shown. In almost all known inteins, the last amino acid is an asparagine, whose side chain can then cyclize to release the intein branch. The two exteins are now joined to each other, but they are being held together by a (thio)ester bond to the side chain of the first amino acid in the downstream extein rather than a peptide bond. The peptide bond is re-formed by the reversal of the original reaction (called an O-N shift) to restore the normal peptide bond, and the intein has been successfully spliced out of the protein, leaving no trace. Complicated though these reactions seem, they occur spontaneously without the help of any other proteins or energy and therefore can occur in a test tube containing just the purified protein with the intein. They also occur in any type of cell into which the gene containing the intein is inserted, independent of the original source of the intein-containing gene, which has made them very useful for some types of applications.

Not only can intein splicing be used to remove an intein from a protein that contains it, but also it can be used to bring different proteins encoded by different genes together, in a process called *trans* splicing. This phenomenon was first observed with the split *dnaE* gene for the replicative DNA polymerase of a strain of the cyanobacterium *Synechocystis*, where intein splicing brings two widely separated parts of a gene product together to form an active enzyme (see Wu et al. below). The *dnaE* gene for the DNA polymerase in this strain of bacteria is split into two parts separated by 745,226 bp of DNA. Apparently, an intein was once integrated into the

(continued)

Selfish DNAs: RNA Introns and Protein Inteins

gene for the DNA polymerase, where it could splice itself out of the protein. Some time later, another large DNA was inserted into the intein. Even though the two parts of the DNA polymerase gene are now split wide apart, and the DNA polymerase is made in two pieces with one or the other end of the intein attached, the two parts of the intein can still find each other and perform the splicing reaction as shown in the figure, joining the two parts of the DNA polymerase together to make the active enzyme. The intein can even be split into three pieces and splice itself out of the protein! This technology can be exploited to assemble proteins from different clones (see reference to Sun et al. below). RNA introns are known to perform similar *trans*-splicing reactions, and *trans*-splicing has many applications in molecular genetics.

References

Martinez-Abarca, F., and N. Toro. 2000. Group II introns in the bacterial world. *Mol. Microbiol.* **38:**917–926. (Microreview.)

Sun, W., J. Hu, and X.-Q. Liu. 2004. Synthetic two-piece and three-piece split inteins for protein *trans*-splicing. *J. Biol. Chem.* **279:** 35281–35286.

Vepritskiy, A. A., I. A. Vitol, and S. A. Nierzwicki-Bauer. 2002. Novel group I intron in the tRNA^{Leu} (UAA) gene of a proteobacterium isolated from a deep subsurface environment. *J. Bacteriol.* **184:** 1481–1487.

Wu, H., Z. Hu, and X. Q. Liu. 1998. Protein *trans*-splicing by a split intein encoded in a split DnaE gene of *Synechocystis* sp. PCC6803. *Proc. Natl. Acad. Sci. USA* **95:**9226–9231.

Xu, M.-Q., and F. B. Perler. 1996. The mechanism of protein splicing and its modulation by mutation. *EMBO J.* **15:**5146–5153.

nucleotide on this end is not joined to another nucleotide by a phosphate bond. The direction on DNA or RNA from the 5′ end to the 3′ end is called the **5′-to-3′ direction.** An RNA polymerase molecule synthesizes mRNA in the 5′-to-3′ direction, moving 3′ to 5′ on the **transcribed strand** of DNA. The opposite strand of DNA from the transcribed strand, the coding strand, has the same sequence and 5′ to 3′ polarity as the RNA, so it is called the **coding strand.** Sequences of DNA in the region of a gene are usually shown as the sequence of the coding strand if this is known. Also, if a gene product is made from the region of the DNA and if the coding strand is known, the relative positions of sequences are given as though they were in a river flowing in the 5′-to-3′ direction. A sequence in the 5′ direction of another sequence on the coding strand is **upstream** of that sequence, while a sequence in the 3′ direction is **downstream.** Therefore, the promoter for a gene and the S-D sequences are both upstream of the initiation codon, while the termination codon and the transcription termination sites are both downstream.

The positions of nucleotides around a promoter are numbered as shown in Figure 2.42. The position of the first nucleotide in the RNA is called the start point and is given the number +1; the distance in nucleotides from this point to another point is numbered negatively or positively, depending on whether the second site is upstream or downstream of the start point, respectively. We have already used this numbering system in Figure 2.6, which shows a σ^{70} promoter with consensus sequences at −10 and −35 relative to the start point of transcription. Note that these definitions can be used to

describe only a region of DNA which is known to encode an RNA or protein, where we know which is the coding strand and which is the transcribed strand. Otherwise, what is upstream on one strand of DNA is downstream on the other strand.

Because mRNAs are both made and translated in the 5′-to-3′ direction, an mRNA can be translated while it is still being made, at least in prokaryotes, where there is no nuclear membrane separating the DNA from the cytoplasm where the ribosomes reside. We have discussed how this can lead to phenomena unique to bacteria such as ρ-dependent polarity and is used to regulate the synthesis of RNA on some genes in bacteria by a process called attenuation (see chapter 12).

It is important to distinguish promoters from **translational initiation regions (TIRs)** and to distinguish **transcription termination** sites from **translation termination** sites. Figure 2.43 also illustrates this difference. Transcription begins at the 5′ end of the mRNA at the promoter, but the place where translation begins, the TIR, can be some distance from the 5′ end. The untranslated region on the 5′ end of an mRNA upstream of the TIR is called the **5′ untranslated region** or **leader region** and can be quite long. Similarly, a nonsense codon in the reading frame for the protein is not a transcription terminator, only a translation terminator. The transcription terminator and therefore the 3′ end of the mRNA may be some distance from the nonsense codon which terminates translation of the mRNA. The distance from the last nonsense codon to the 3′ end of the mRNA is the **3′ untranslated region.** These distinctions are dramatically illustrated in the case of polycistronic mRNAs, which encode more

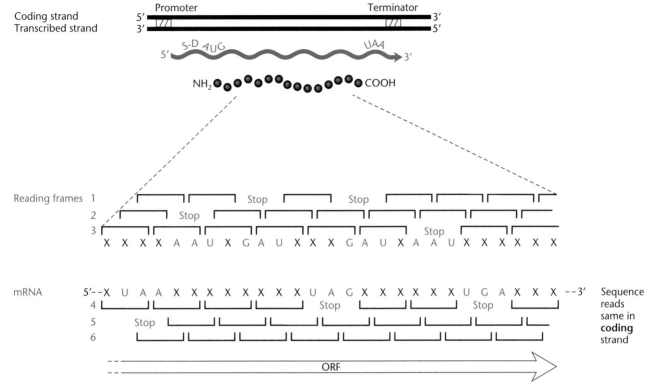

Figure 2.43 Relationship between a gene in DNA and the coding sequence in mRNA. There are a total of six different sequences in the two strands of DNA that may contain ORFs, but generally only one ORF encodes a polypeptide in each region.

than one polypeptide. These mRNAs have a separate TIR and nonsense codon for each gene and can have noncoding or untranslated sequences upstream of, downstream of, and between the genes. Eukaryotes do not seem to have polycistronic mRNAs, possibly because their TIRs are less well defined, so that they cannot be recognized unless they are at the 5′ end of the mRNA.

Open Reading Frames

The concept of an **open reading frame (ORF)** is very important, particularly in this age of genomics. As discussed above, a reading frame in DNA is a succession of nucleotides in the DNA taken three at a time, the same way the genetic code is translated. Each DNA sequence has six reading frames, three on each strand, as illustrated in Figure 2.43. An ORF is a string of potential codons for amino acids in DNA unbroken by nonsense codons in one of these reading frames. Computer software can show where all the ORFs are in a sequence, and most DNA sequences have many ORFs on both strands, although most of these are short. The region shown in Figure 2.43 contains many ORFs, but only the longest, in frame 6, is likely to encode a polypeptide. However, the

presence of even a long ORF in a DNA sequence does not necessarily indicate that the sequence encodes a protein, and fairly long ORFs often occur by chance. More information is usually required to establish which, if any, of the ORFs in a sequence encode(s) a protein.

If an ORF does encode a polypeptide, it will begin with a TIR, but, as discussed above, TIRs are sometimes difficult to identify. Clues to whether an ORF is likely to encode a protein may come from the choice of the third base in the codon for each amino acid in the ORF. Because of the redundancy of the code, an organism has many choices of codons for each amino acid, but each organism prefers to use some codons over others (see the section on codon usage, above, and Table 2.1).

A more direct way to determine if an ORF actually encodes a protein is to ask which polypeptides are made from the DNA in an in vitro transcription-translation system. These systems use extracts of cells, typically of *E. coli*, from which the DNA has been removed but the RNA polymerase, ribosomes, and other components of the translation apparatus remain. When DNA with the ORFs under investigation is added to these extracts, polypeptides can be synthesized from the added DNA. If the size of one of these polypeptides corresponds to the size of an ORF on the DNA, this ORF probably encodes a protein. Another way is to make a translation fusion of a reporter gene to the ORF and determine whether the reporter gene is expressed (see the following section).

Transcriptional and Translational Fusions

Probably the most convenient way to determine which of the possible ORFs on the two strands of DNA in a given region are translated into proteins is to make **transcriptional** and **translational fusions** to the ORFs. These methods make use of **reporter genes** such as *lacZ* (β-galactosidase), *gfp* (green fluorescent protein), *lux* (luciferase), or other genes whose products are easy to detect. Figure 2.44 illustrates the concepts of translational and transcription fusions.

An ORF can be translated only if it is transcribed into RNA. Transcriptional fusions can be used to determine whether this has occurred. To make a transcriptional fusion, a reporter gene with the sequence for a TIR but no promoter of its own is fused immediately downstream of the promoter. If the gene is transcribed into mRNA, the reporter gene will also be transcribed and its product will be detectable in the cell. Transcriptional fusions also offer a convenient way of determining how much mRNA is made on a coding sequence. In general, the more reporter gene product that is made in the transcriptional fusion, the more mRNA was made on the upstream coding sequence. Many examples of the use of transcriptional fusions in studying the regulation of operons are given in subsequent chapters.

In a translational fusion, the two coding sequences are cloned in such a way that they are translated in the same frame and there are no nonsense codons between them. Translation beginning at a TIR upstream of one of the coding sequences will proceed through the other coding sequence, making a **fusion protein** that contains both polypeptide sequences. The coding sequence can be fused either to the amino terminus of the reporter gene product or to its carboxyl terminus. The reporter gene product can then be assayed as before to determine how much of the fusion protein has been made. The reporter gene

Figure 2.44   Transcriptional and translational fusions to express *lacZ* (which encodes β-galactosidase). In both types of fusions, transcription begins at the +1 site at the p_{ORFA} promoter upstream of OrfA. (A) In a transcriptional fusion, both the upstream OrfA coding region and the downstream *lacZ* reporter gene are translated from their own TIRs. Only the TIR for *lacZ* is shown as S-D and ATG (boxed). The translation of the upstream OrfA continues until it encounters a nonsense codon in frame, as indicated by the dashed line. (B) In a translational fusion, a fusion protein is translated from the TIR upstream of OrfA to make a fusion protein containing the LacZ reporter protein fused to the remaining product of the upstream OrfA. The prime (′) symbols indicate that part of each protein may be deleted.

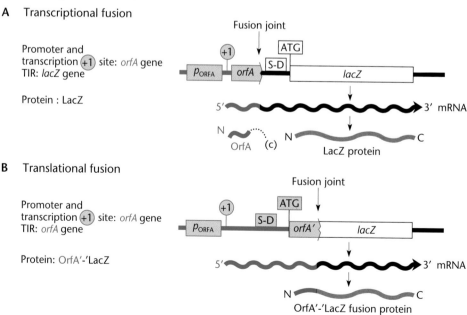

product must retain its activity even when fused to the potential polypeptide encoded by the ORF; otherwise it will not be detectable. Many reporter genes have been chosen because their products remain active even when fused to other polypeptides. Translation fusions are also often used to attach affinity tags to proteins to use in their purification (see the following section).

EXPRESSION VECTORS

One important application of transcriptional and translational fusions is in **expression vectors**. It is often desirable to make large amounts of a protein for biochemical or structural studies. We may have cloned the gene for a protein which we want to purify, but the protein is normally made in small amounts and/or in an organism from which it is difficult to purify proteins. Or we may not have a convenient activity for the protein to help in its purification. To synthesize large amounts of the protein, we can clone the gene into an expression vector, which is a cloning vector designed so that a gene cloned into it will be transcribed and/or translated from the cloning vector. We can then express the protein product of the gene in larger amounts from the expression vector or fuse it to other proteins that are much easier to purify.

Most expression vectors are designed to work in *E. coli*. An example of a basic *E. coli* plasmid expression vector is shown in Figure 2.45. Expression vectors can be divided into two groups: **transcription vectors** and **translation vectors**. Transcription vectors transcribe the cloned gene into mRNA from a promoter on the expression vector, but the cloned gene must be translated from its own TIR; therefore, genes from bacteria distantly related to *E. coli* or genes from eukaryotes, in general, cannot be expressed from a transcription vector. In a translation vector, the ORF to be expressed is translationally fused to the TIR on the vector that is recognized in *E. coli*. Translation vectors will translate essentially any gene, but the gene must be cloned into the cloning vector in such a way that its ORF is translationally fused in frame to the TIR on the cloning vector. To reproduce the protein exactly, it is necessary to know the DNA sequence of the gene and the position where translation of the gene normally begins so that it can be fused to the TIR in such a way that the N-terminal amino acids will remain unchanged. PCR offers a convenient way to make such fusions (see chapter 1). Other problems may also be encountered when expressing foreign proteins in *E. coli*. For example, only self-splicing introns are removed from the mRNA in *E. coli*, which lacks RNA-splicing systems. If the gene contains introns, it might be necessary to perform a PCR amplification of the gene for cloning from the mRNA from which the introns have already been removed (cDNA), using a technique called reverse transcription PCR (RT-PCR). Also, most types of eukaryotic modifications to proteins such as glycosylation do not occur in *E. coli*.

Affinity Tags

Another current application of gene fusions involves attachment of **affinity tags** to proteins. This is a powerful technology because it allows the easy purification of protein products of genes that have been cloned, even if their activity is not known. Affinity tags have some property that makes them very easy to purify. If the coding sequence for a protein we want to purify has been translationally fused to the coding sequence of an affinity tag, we can purify the tag and the protein will come along for the ride. In the example shown in Figure 2.45, the affinity vector attaches a string of six histidines called a His tag to either the amino or carboxyl terminus of the protein whose coding sequence has been cloned into the vector. Histidine binds strongly to nickel, and so the string of histidines binds to a column that contains nickel. The procedure is to break open cells containing the fusion protein and pass the extracts through a column to which nickel is bound. Only the fusion protein containing the attached His tag remains on the column; the other thousands of types of proteins all pass through. The fusion protein can then be eluted by washing with a solution containing high concentrations of imidazole, which also binds to nickel and so will displace the His tag and the protein to which it is fused from the column. In this way, the fusion protein can be separated from most of the other proteins in the extract in a single step. Some of the affinity vectors have been designed so that they also include a sequence of a few amino acids that is the site of cleavage for a specific protease such as thrombin. This sequence is introduced in the fusion between the affinity tag and the protein being purified, so that cleavage of the purified fusion protein with the protease removes the affinity tag from the protein. The purified protein, an almost exact replica of the native protein, can then be used to make specific antibodies or in any other application requiring a purified protein.

There are many affinity tags available. Some widely used ones are glutathione *S*-transferase (GST) and maltose-binding protein, which bind strongly to reduced glutathione and amylose, a polymer of maltose, respectively. However, His tags have the advantage over most affinity tags in that they are smaller and bind to nickel, even if the protein to which they are attached is denatured (unfolded) by detergents or other harsh treatments. Many proteins precipitate into inclusion bodies when they are synthesized in large amounts from an expression vector, and they must be redissolved before they can be applied to an affinity column. The conditions used to

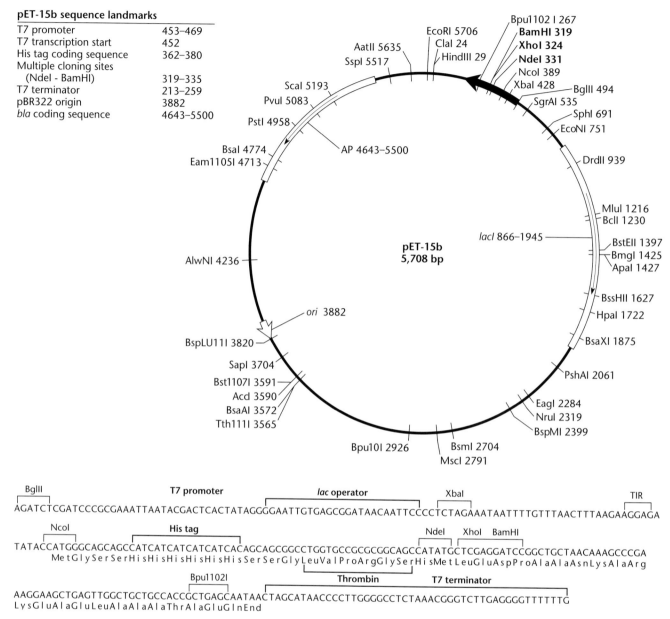

pET-15b sequence landmarks

T7 promoter	453–469
T7 transcription start	452
His tag coding sequence	362–380
Multiple cloning sites	
(NdeI - BamHI)	319–335
T7 terminator	213–259
pBR322 origin	3882
bla coding sequence	4643–5500

Figure 2.45 A cloning vector for attaching a His tag to a protein. If the ORF for a gene is cloned into the BamHI site on the vector in such a way that it will be translated in the right frame from the upstream TIR (RBS in the figure), a chain of six histidines will be attached to the N terminus of the protein. The protein can then be easily purified on nickel-containing columns. The His tag can subsequently be cut off the purified protein by adding the specific protease thrombin. This type of vector uses the T7 phage promoter and a strong TIR to synthesize large amounts of the fusion protein. The use of such phage-derived cloning vectors is discussed in chapter 7.

dissolve inclusion bodies also denature the proteins. A His tag allows the denatured protein to be purified. The protein can then sometimes be renatured after purification or can be used in its denatured state to make antibodies, etc.

Inducible Expression Vectors

Many expression vectors express the cloned gene from a regulated promoter such as the *lac* promoter. The protein product of the cloned gene is then synthesized only when the promoter is turned on. These inducible vectors are particularly useful if the protein product of a cloned gene is toxic to the cell. Even relatively nontoxic proteins can kill the cell if they are made in very large amounts from an expression vector. If the promoter from which the cloned gene is transcribed is regulated, the cells can be grown before the gene is induced. Even if the cells die after the induction, they still contain large amounts of the cloned gene product. The *lac* (p_{LAC}) and L-*ara* promoters are often used as inducible promoters; the regulation of these promoters is discussed in chapter 12.

Bacterial Genome Annotation

Many of the concepts and methods discussed in this chapter are relevant to the process of genome annotation in bacteria (Box 2.7). A genome annotation describes the functional features of a genome including its ORFs and RNAs as well as methods for analyzing the proteome and transcriptome. Such experiments are discussed in chapters 13 and 14.

Antibiotics That Block Transcription and Translation

As in chapter 1, we devote the remainder of this chapter to a discussion of antibiotics because these compounds not only are among the most useful therapeutic agents but also allow mutants to be isolated for genetic studies. Studies of how antibiotics affect transcription and translation have greatly contributed to our understanding of these processes.

Antibiotic Inhibitors of Transcription

Some of the components of the transcription apparatus are the targets of antibiotics used in treatment of bacterial infections and in tumor therapy. Some of these antibiotics are made by soil bacteria and fungi, and some have been synthesized chemically. Table 2.3 lists examples, along with their sources and their targets.

INHIBITORS OF RIBONUCLEOSIDE TRIPHOSPHATE SYNTHESIS

Some antibiotics that inhibit transcription do so by inhibiting the synthesis of the ribonucleoside triphosphates. An example is azaserine, which inhibits purine biosynthesis.

Uses

Azaserine and other antibiotics that block the synthesis of the ribonucleotides are usually not specific to transcription, since the ribonucleotides, including ATP and GTP, have many other uses in the cell. This lack of specificity limits the usefulness of these antibiotics for studying transcription, although some of them have other uses.

INHIBITORS OF RNA SYNTHESIS INITIATION

Rifamycin and its more commonly used derivative, rifampin, block transcription by binding to the β subunit of RNA polymerase and specifically blocking the initiation of RNA synthesis. The antibiotic binds in the active-

---

BOX 2.7

Annotation and Comparative Genomics

Genome sequencing is one way to begin the study of a bacterium. However, this information is most useful in the context of other information about the bacterium. In this book, we show how the methods of genetic analysis and genomic analysis complement each other to permit a more complete understanding of how a bacterium functions. Figure 1 summarizes many of the types of genomic and genetic experiments available. The following text briefly describes these experiments; many of them are more fully discussed and illustrated in upcoming chapters. In some cases websites are included in the list; in other cases Web search terms are given, as websites change rapidly. For a general reference on genome annotation, see Gibson and Muse (below). For a reference on bioinformatics, see Mount (below). Finally, for an example of a journal publication that

describes a comprehensive annotation analysis of *Escherichia coli* K-12, see Riley et al. (below).

Genome sequence
Genome-sequencing methodology is discussed in Box 1.5.

Annotation and comparative genomics
For analysis of a new genome sequence, the use of bioinformatics resources, as described below, can give us a profile of the similarities of DNA sequences and gene products to those of other organisms.

Functional annotation
Genome sequence information is accumulating faster than we can understand it, but tools for analyzing genome sequences are also rapidly increasing in number and sophistication.

(continued)

BOX 2.7 (continued)

Annotation and Comparative Genomics

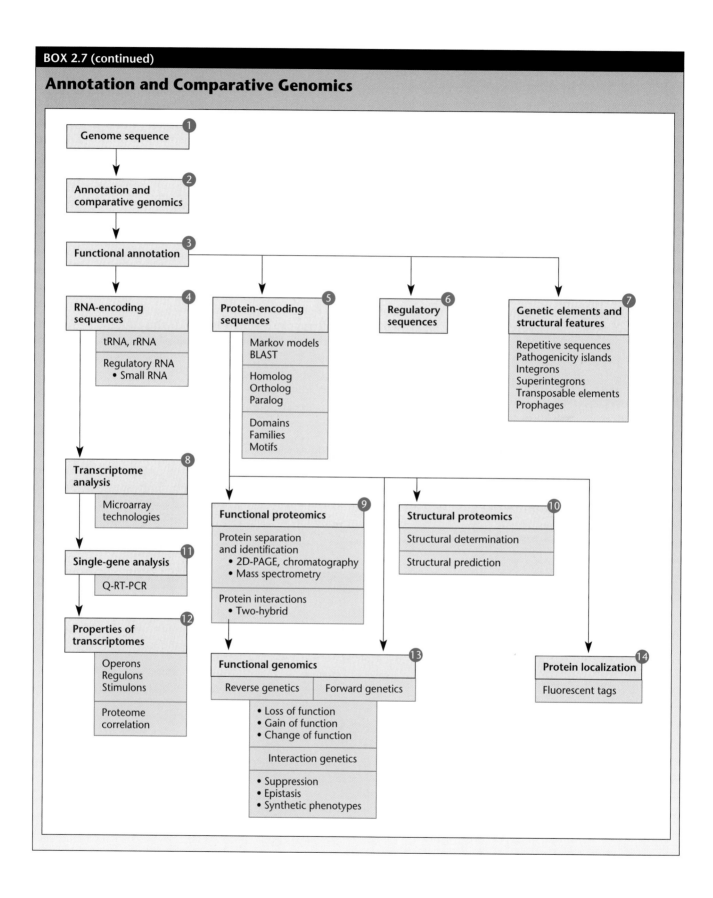

BOX 2.7 (continued)

Annotation and Comparative Genomics

Examples of tools available for public use are included in the text that follows. Newer and improved tools will be continuously developed; often a simple Web browser search can reveal useful tools and databases. However, it is important to note that, inevitably, databases contain errors: the level of inaccuracy in annotation is as high as 5 to 10%.

RNA-encoding sequences

rRNA-encoding sequences are extremely highly conserved (see Box 2.1) and so are easily recognized. For recognition of tRNA sequences, a useful bioinformatics tool is www.genetics. wustl.edu/eddy/tRNAscan-SE/. Methods for identifying small, noncoding, regulatory RNAs are only now being developed (see chapter 13).

Protein-encoding sequences

For gene finding in prokaryotes (i.e., distinguishing coding from noncoding DNA), an especially useful tool is **GLIMMER** (for "**Gene Locator and Interpolated Markov Modeler**") (www.tigr.org/software/). A **Markov model** is a statistical tool useful in a situation in which a system, in this case a protein sequence, undergoes a series of changes in its state (i.e., amino acid substitutions) and a change from one state to the next is independent of the history of the state. A type of Markov model that is especially useful in genome annotation is the **hidden Markov model (HMM).** An HMM uses previous data sets to weight an analysis; in other words, an HMM is able to "train itself" if it is "given" a set of about 50 related sequences. Once trained, the HMM places more value on states that are conserved, e.g., common amino acid substitutions. This allows them to be highly sensitive. In addition, an HMM can be further trained to recognize codons or DNA sequences characteristic of a particular organism; this type of model is called an **interpolated Markov model (IMM).**

An HMM is able to consider all possible combinations of factors such as gaps, matches, and mismatches that could affect the alignment of a set of sequences. Thus, an HMM can pick out amino acid positions that are or are not conserved. The basis for using HMMs is the HMMER statistical tool (http://hmmer.wustl.edu/).

It is interesting that HMMs have uses in applications other than genome annotation: for example, in speech recognition programs! For protein-encoding sequences, the most common sequence analysis tool for predicting function is called **BLAST (Basic Local Alignment Search Tool).** The U.S. National Center for Biotechnology Information (NCBI) has a publicly available website, with step-by-step tutorials (www.ncbi.nlm.nih.gov/BLAST). To use BLAST, a query of a

nucleotide (blastn; see Box 1.4) or amino acid sequence is submitted for comparison to the publicly available databases. Searches sometimes ask that a sequence be submitted in **FASTA** format, which means that a sequence is submitted as an uninterrupted sequence, using the standard amino acid and nucleic acid codes. The website www.broad.mit.edu contains tutorials on FASTA as well as many other genomics topics.

The BLAST algorithm can translate a sequence in all six possible reading frames (see Figure 2.43). Moreover, the BLAST search can be performed in several ways: translated query versus protein database (blastx), protein query versus translated database (tblastn), or translated query versus translated database (tblastx). Numerous additional variations of BLAST are available at the NCBI BLAST website, including protein-protein BLAST (blastp) and position-specific iterated BLAST (**PSI-BLAST**).

After a query is submitted, the BLAST algorithm calculates the statistical significance of any matches found. The significance of a similarity is expressed as an **E-value.** The E-value ("E" for expected) is a term that indicates the significance of an alignment found between two sequences. An E-value is the number of database hits of similar quality that you would expect to find by chance. One sequence would be the query sequence and the other would be a related sequence found in a database, for example, by a BLAST search. The lower the E-value, the closer the similarity found. Generally, an E-value greater than 0.01 to 0.05 is considered to be insignificant.

The relatedness of gene sequences can be categorized as **homologs,** which are genes or sequences that share common ancestry. Homologs can be classified as **orthologs,** which are genes that are similar in sequence and have a common ancestor but are found in different species and (in some cases) have similar functions, and **paralogs,** which are genes that arose by duplication within a given species and may have similar functions. In addition, proteins are categorized into "families," in which the individual members share certain features, as discussed below.

It is important to note that the matches that result from some BLAST searches, such as blastp, are matches to protein domains rather than to genes per se. A protein domain is generally an independently folding element of a protein; thus, proteins are mosaics of domains. For example, see the σ factor domains in Figure 2.11.

The regions of sequence conservation among proteins that are found by an HMM analysis can be used to categorize proteins into families and so can provide information about the function of a protein. The term "family" is used in

(continued)

BOX 2.7 (continued)

Annotation and Comparative Genomics

many contexts and can refer to many types of categories. In a scheme that broadly defines families, proteins are divided into three types of families. One type of family contains sequences based on one or more shared protein domains. A given domain may be found in more than one functional type of protein. A second type of family contains enzyme families, in which the gene products all perform the same biological function. Enzyme names are assigned by the Enzyme Commission (http://ca.expasy.org/enzyme/). A very large collection of metabolic pathways has been compiled at the website http://www.genome.ad.jp/kegg/. This site provides so-called **Kegg maps,** which provide a preliminary suggested pathway in which an enzyme might function; experimental support for the enzyme classification is important. A third family type is the "superfamily," which contains two or more proteins that are related by sequence but have not necessarily been tested for biochemical function. Thus, the importance of experimental study of protein function (see below) cannot be overstated.

Several research groups have combined the information from BLAST analyses into large databases. These include **COGs** (from the NCBI: www.ncbi.nlm.nih.gov/COG/), **Pfams** (from the Sanger Centre in the United Kingdom: www.sanger.ac.uk/Software/Pfam/), and **TIGRFAMS** (from the U.S. public/corporate Institute for Genomic Research: www.tigr.org/TIGRFAMs/).

A **COG** is a **cluster of orthologous genes.** The NCBI has defined approximately 18 COGs by comparing protein-encoding regions of dozens of complete genomes of the major phylogenetic lineages; each COG is defined by proteins from at least three lineages. NCBI COGs are grouped into the following four major categories:

1. Information storage and processing, containing translation, ribosomal structure and biogenesis functions; transcription functions; and DNA replication, recombination, and repair functions

2. Cellular processes, containing functions for cell division and partitioning; posttranslational modification, protein turnover, and chaperones; cell envelope biogenesis and the outer membrane; cell motility and secretion; inorganic ion transport and metabolism; and signal transduction mechanisms

3. Metabolism, containing functions for energy production and conversion; carbohydrate transport and metabolism; amino acid transport and metabolism; nucleotide transport and metabolism; coenzyme metabolism; lipid metabolism; and secondary metabolite biosynthesis, transport, and catabolism

4. "Poorly characterized," containing "general function prediction only" and "function unknown"

Links to COG information can be found at the website www.ncbi.nlm.nih.gov/COG.

The Pfam families are more likely to describe domains than full-length proteins, for example, indicating evidence for an ATP-binding domain in a protein.

The TIGRFAM families are a versatile resource for protein classifications and include superfamilies, which include all proteins with amino acid homology but which may differ in biological function; equivalogs, which include proteins that are conserved in function; and subfamilies, which contain proteins incompletely evaluated for function.

Motifs are conserved patterns of amino acids, often the amino acids comprising the active site of a protein. Thus, a motif can indicate that a protein has biochemical activity similar to that of other proteins with a related motif.

Regulatory sequences

Regulatory sequences are usually determined experimentally, but bioinformatics techniques are becoming more effective. For example, at the NCBI website mentioned above, **ELPH** can find motifs in DNA (or protein) sequences, **TransTerm** can find rho-independent transcriptional terminator sites, and **RBSfinder** can identify potential prokaryotic translational ribosome-binding sites.

Genetic elements and structural features

Subsequent chapters (Chapters 4 through 9) describe the features of DNA elements such as prophages, integrons, etc. Structural features, such as repetitive sequences, are discussed throughout the book.

Transcriptome analysis

Parallel analysis of the expression of thousands of genes in a bacterial genome can be done by **transcriptome** analysis (**microarray** analysis), including the use of cDNA microarrays, oligonucleotide arrays, and Affymetrix GeneChip arrays (see chapter 13). A database for microarray data is GEO (for *Gene Expression Omnibus*). Software named MADAM (for *microarray data management*) can be found at the NCBI website.

Functional proteomics

Functional protein domains and motifs can be predicted by special algorithms and experimental methods (**proteomics**). The HMM type of algorithm described above is very useful in identifying protein domains and motifs. Some examples of useful Web search terms are Swiss-Prot, PIR (Protein Information

BOX 2.7 (continued)

Annotation and Comparative Genomics

Resource), Ensembl UniProt, ProDom, PROSITE, TRANS-FAC (for transcription factors), and SENTRA (for prokaryotic signal transduction proteins).

High-throughput protein identification is now possible, using a combination of a protein separation technique such as **two-dimensional polyacrylamide gel electrophoresis** (**2D-PAGE**) or **chromatography** and fragmentation of proteins into peptides followed by analysis of the peptides by **mass spectrometry.** By using comparisons with genome sequence data, the "mass spectrometry" data can be used to identify individual proteins. In another type of **proteome** analysis, interactions between proteins can be detected using **two-hybrid screens** (see chapter 13).

Structural proteomics

Three-dimensional protein structures are determined from the crystal structure or from nuclear magnetic resonance (NMR) spectroscopy. Development of predictive algorithms has allowed some structures to be predicted from the sequence alone; predicting protein structures from amino acid sequences is a very challenging but active research area. Examples of some Web search terms are "EXPASY,"

"Swiss-model," "Touchstone," and "COILS prediction," to mention just a few. Another sample of search terms— "HMMTOP," "TMHMM," and "TmPred"—are tailored for the prediction of transmembrane helices and topology.

Single-gene analysis

A popular method for single-gene transcriptional analysis is quantitative RT-PCR (**Q-RT-PCR**). The abbreviation RT is used to refer to either **reverse transcriptase PCR** or **real-time PCR,** which includes the use of reverse transcriptase. RT-PCR methods are described in chapter 13.

Properties of transcriptomes

The regulatory units of gene expression include operons, which are genes transcribed into the same mRNA; regulons, which are operons under the control of the same molecule; and stimulons, which are regulons affected by the same environmental stimulus. Correlation of proteome and transcriptome data is important for a full understanding of gene expression and is discussed in chapters 13 and 14.

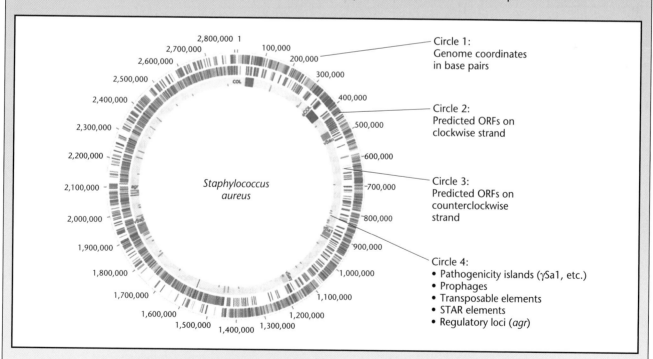

Circle 1:
Genome coordinates in base pairs

Circle 2:
Predicted ORFs on clockwise strand

Circle 3:
Predicted ORFs on counterclockwise strand

Circle 4:
• Pathogenicity islands (γSa1, etc.)
• Prophages
• Transposable elements
• STAR elements
• Regulatory loci (*agr*)

A conventional representation of some of the genome sequence information that can be obtained from steps 1 through 7 of the flowchart, using *Staphylococcus aureus* as an example (see Box 1.4). Interestingly, other pathogenic bacteria of the genus *Staphylococcus* which cause very different diseases, such as *S. epidermidis,* differ only in circle 4 with different prophages, genetic islands, and transposons (see Gill et al., below).

(continued)

BOX 2.7 (continued)

Annotation and Comparative Genomics

Functional genomics
Reverse genetics can be used to determine the function of a gene whose sequence is known. **Forward genetics** can be used to identify and sequence a gene whose function is known. Often this requires a large repertoire of genetic techniques rather than merely knocking out the gene. **Interaction genetics** seeks to elucidate the significant interrelationships and subtle interactions of genes and gene products. Methods for all of these aspects of genetic analysis are discussed throughout this textbook.

Protein localization
Gene fusion techniques using, for example, fluorescent probes can often locate the gene product within the cell. For an example, see the cover of this book.

References
Gibson, G., and S. V. Muse. 2004. *A Primer of Genome Science,* 2nd ed. Sinauer Associates, Inc. Sunderland, Mass.

Gill, S. R., et al. 2005. Insights on evolution of virulence and resistance from the complete genome analysis of an early methicillin-resistant *Staphylococcus aureus* strain and a biofilm-producing methicillin-resistant *Staphylococcus epidermidis* strain. *J. Bacteriol.* **187:**2426–2438.

Mount, D. W. 2004. *Bioinformatics: Sequence and Genome Analysis,* 2nd ed. Cold Spring Harbor Laboratory Press, Cold Spring Harbor, N.Y.

Riley, M., T. Abe, M. B. Arnaud, M. K. B. Berlyn, F. R. Blattner, R. R. Chaudhuri, J. D. Glasner, T. Horiuchi, I. M. Keseler, T. Kosuge, H. Mori, N. T. Perna, G. Plunkett III, K. E. Rudd, M. H. Serres, G. H. Thomas, N. R. Thomson, D. Wishart, and B. L. Wanner. 2006. *Escherichia coli* K-12: a cooperatively developed annotation snapshot—2005. *Nucleic Acids Res.* **34:**1–9.

site channel of RNA polymerase and limits growth of the RNA chain to a few nucleotides (Figure 2.14).

Uses

The property of blocking only initiation of transcription has made these antibiotics very useful in the study of transcription. For example, they have been used to analyze the steps in initiation of RNA synthesis and to study the stability of RNA and proteins in the cell. These antibiotics are useful therapeutic agents in the treatment of tuberculosis and other difficult-to-treat bacterial infections because they inhibit the RNA polymerases of essentially all types of bacteria but not the RNA polymerases of eukaryotes, so that they are not toxic to humans and animals. Accordingly, many derivatives have been made from them.

Resistance

In rifampin-resistant mutants, one or more amino acids in the β subunit of RNA polymerase lining the active-site channel have been changed so that rifampin can no longer bind but the RNA polymerase still functions. Such chromosomal mutations conferring resistance to rifampin and other streptovarycin-type antibiotics are fairly common and have limited the usefulness of these antibiotics somewhat.

INHIBITORS OF RNA ELONGATION AND TERMINATION

Streptolydigin also binds to the β subunit of the RNA polymerase of bacteria but can block RNA synthesis after it is under way. It has a weaker affinity for RNA polymerase than does rifampin, and so it blocks transcription only when added at higher concentrations, which limits its usefulness. Bicyclomycin targets the transcription terminator protein ρ and prevents transcription termination.

INHIBITORS THAT AFFECT THE DNA TEMPLATE

Actinomycin D and bleomycin block transcription by binding to the DNA. After bleomycin binds, it also nicks the DNA. While such drugs have been useful for studying

TABLE 2.3	Antibiotics that block RNA synthesis	
Antibiotic	**Source**	**Target or action**
Streptolydigin	*Streptomyces lydicus*	β subunit of RNA polymerase
Actinomycin D	*Streptomyces antibioticus*	Binds DNA
Rifampin	*Nocardia mediterranei*	β subunit of RNA polymerase
Bleomycin	*Streptomyces verticulus*	Cuts DNA

transcription in bacteria, they are not very useful in antibacterial therapy because they are not specific to bacteria and are very toxic to humans and animals. They are, however, used in antitumor therapy.

Antibiotic Inhibitors of Translation

Because it is somewhat different from the eukaryotic translation apparatus and is highly conserved, the translation apparatus of bacteria is a particularly tempting target for antibacterial drugs. In fact, antibiotics that inhibit translation are among the most useful of all the antibiotics, and some of them are household words. Commonly used antibiotics directed against the translation apparatus of bacteria are listed in Table 2.4, which also lists their target and source. Some of these antibiotics are also very useful in combating fungal diseases and in cancer chemotherapy.

INHIBITORS THAT MIMIC TRNA

Puromycin mimics the 3′ end of tRNA with an amino acid attached (aminoacylated tRNA). It enters the ribosome as does an aminoacylated tRNA, and the peptidyltransferase attaches it to the growing polypeptide. However, it does not translocate properly from the A site to the P site, and the peptide with puromycin attached to its carboxyl terminus is released from the ribosome, terminating translation.

Uses

Studies with puromycin have contributed greatly to our understanding of translation. The model of the A and P sites in the ribosome and the concept that the 50S ribosome contains the enzyme for peptidyl bond formation, which was recently shown to be the 23S rRNA itself, came from studies with this antibiotic. Puromycin is not a very useful antibiotic for treating bacterial diseases,

however, because it also inhibits translation in eukaryotes, making it toxic in humans and animals.

INHIBITORS THAT BIND TO THE 23S rRNA

Chloramphenicol

Chloramphenicol inhibits translation by binding to ribosomes and preventing the binding of aminoacylated tRNA to the A site. It might also inhibit the peptidyltransferase reaction, preventing the formation of peptide bonds. Structural studies have shown that chloramphenicol binds to specific nucleotides in the 23S rRNA, although ribosomal proteins are also part of the binding site.

Uses. Chloramphenicol is effective at low concentrations and therefore has been one of the most useful antibiotics for studying cellular functions. For example, it has been used to determine the time in the cell cycle when proteins required for cell division and for initiation of chromosomal replication are synthesized. It is also quite useful in treating bacterial diseases since it is not very toxic for humans and animals, being fairly specific for the translation apparatus of bacteria. It can also cross the blood-brain barrier, making it useful for treating diseases of the central nervous system such as bacterial meningitis. Whatever toxicity it does have may result from inhibition of the translational apparatus of mitochondria, which is similar to the translation apparatus of bacteria (see the introduction to this book). Chloramphenicol is bacteriostatic, which means that it stops the growth of bacteria without actually killing them. Such antibiotics should not be used in combination with antibiotics that depend on cell growth for their killing activity, such as penicillin, since they neutralize the effect of these other antibiotics.

TABLE 2.4	Antibiotics that block translation	
Antibiotic	**Source**	**Target**
Puromycin	*Streptomyces alboniger*	Ribosomal A site
Kanamycin	*Streptomyces kanamyceticus*	16S rRNA
Neomycin	*Streptomyces fradiae*	16S rRNA
Streptomycin	*Streptomyces griseus*	30S ribosome
Thiostrepton	*Streptomyces azureus*	23S rRNA
Gentamicin	*Micromonospora purpurea*	16S rRNA
Tetracycline	*Streptomyces rimosus*	Ribosomal A site
Chloramphenicol	*Streptomyces venezuelae*	Peptidyltransferase
Erythromycin	*Saccharopolyspora erythraea*	23S rRNA
Fusidic acid	*Fusidium coccineum*	Translation elongation factor G
Kirromycin	*Streptomyces collinus*	Translation elongation factor Tu

Resistance. It takes many mutations in ribosomal proteins to make bacteria resistant to chloramphenicol, so that resistant mutants are very rare. Some bacteria have enzymes that inactivate chloramphenicol. The genes for these enzymes are often carried on plasmids and transposons, interchangeable DNA elements that are discussed in chapters 4 and 9. The best-characterized chloramphenicol resistance gene is the *cat* gene of transposon Tn9, whose product is an enzyme that specifically acetylates (adds an acetyl group to) chloramphenicol, thereby inactivating it. The *cat* gene has been used extensively as a reporter gene to study gene expression in both bacteria and eukaryotes and has been introduced into many plasmid cloning vectors.

Macrolides

Erythromycin is a member of a large group of antibiotics called the macrolide antibiotics, which are large ring structures. These antibiotics may also inhibit translation by binding to the 23S rRNA and blocking the exit channel of the growing polypeptide. This causes the polypeptide to be released prematurely at either the peptidyltransferase reaction or the translocation step, causing the peptidyl-tRNA to dissociate from the ribosome.

Uses. Erythromycin and other macrolide antibiotics have been among the most useful antibiotics. They are effective mostly against gram-positive organisms but are also useful in treating some gram-negative bacterial diseases including *Legionella*, *Mycoplasma*, and *Rickettsia* infections.

Resistance. One of the most foreboding developments in medicine is the extent to which pathogenic bacteria have become resistant to the macrolides through the misuse of these once most useful antibiotics. They achieve this resistance in a number of ways. One way is methylation of a specific adenine base in the 23S rRNA by enzymes called the Erm methylases. Methylation of this base causes a conformational change in the 23S rRNA that might prevent proper binding of the antibiotics. These enzymes are encoded by plasmids and transposons that are exchanged readily between bacteria. Others become resistant by altering preexisting efflux pumps so that they pump macrolides out of the cell or by acquiring them from other resistant bacteria. Some mutational changes in the 23S rRNA can also confer resistance to these antibiotics. Yet others acquire functions that inactivate the antibiotics enzymatically, although these do not seem to be as important clinically. New derivatives of these antibiotics are being made constantly to stay ahead of the advancing bacterial resistance.

Thiostrepton

Thiostrepton and other thiopeptide antibiotics block translation by binding to 23S rRNA in the region of the ribosome involved in the peptidyltransferase reaction and preventing the binding of EF-G. Thiostrepton is specific to gram-positive bacteria; it does not enter gram-negative bacterial cells.

Uses. Thiostrepton has limited usefulness because it is is not very soluble. It is used mostly in veterinary medicine and agriculture.

Resistance. Most thiostrepton-resistant mutants are missing the L11 ribosomal protein from the 50S ribosomal subunit. This protein seems not to be required for protein synthesis but plays a role in guanosine tetraphosphate (ppGpp) synthesis (see chapter 13). Other mutations confer resistance by changing nucleotides 1067 and 1095 in the 23S rRNA; these nucleotides presumably are close to where the antibiotic binds. Plasmids and transposon genes can confer thiostrepton resistance by methylating ribose sugars of the 23S rRNA in certain positions. Eukaryotes may be insensitive to this antibiotic because the analogous ribose sugars of the eukaryotic 28S rRNAs are normally extensively methylated.

INHIBITORS OF BINDING OF AMINOACYLATED tRNA TO THE A SITE

Tetracycline was one of the first antibiotics isolated. Recent evidence suggests that it may inhibit translation by allowing aminoacylated tRNA–EF-Tu complex to bind to the A site of the ribosome and allowing the GTP on EF-Tu to be cleaved to GDP but then inhibiting the next step, causing a futile cycle of binding and release of the aminoacylated tRNA from the A site. It might also inhibit the binding of the release factors to the A site, which is interesting considering that the release factors might mimic aminoacylated tRNA (Box 2.2). Interestingly, it is also one of the few known examples of a naturally produced chlorinated hydrocarbon. Most chlorinated hydrocarbons are human-made and have only been made chemically in recent times.

Uses

Tetracycline has been a very useful antibiotic for treating bacterial diseases, although it is somewhat toxic to humans because it also inhibits the eukaryotic translation apparatus. It is very broad spectrum, acting against both gram-negative and gram-positive bacteria as well as some protozoans such as the one that causes ameobic desentery. It is also used to treat acne. However, this is another case where overuse has led to the spread of resistance, and it is no longer the first choice of antibiotic against many infections.

Resistance

In some types of bacteria, ribosomal mutations confer low levels of resistance to tetracycline by changing protein S10 of the ribosome. However, most clinically important resistance to tetracycline and its derivatives is acquired on plasmids and transposons. One of these genes, *tetM*, carried by the conjugative transposon Tn*916* and its relatives (see chapter 5), encodes an enzyme that confers resistance by methylating certain bases in the 16S rRNA. The *tetM* gene is ubiquitous; related genes occur in both gram-positive and gram-negative bacteria. Other tetracycline resistance genes, such as the *tet* genes carried by transposon Tn*10* and plasmid pSC101 of *E. coli*, encode membrane proteins that confer resistance by pumping tetracycline out of the cell. These tetracycline resistance genes are extensively used as reporter genes and as markers for genetic analysis in *E. coli,* and the *tetA* gene from pSC101 has been introduced into many plasmid cloning vectors (see chapter 4). However, the *tet* genes of Tn*10* and pSC101 are specific for *E. coli* and do not confer tetracycline resistance in many other types of bacteria, which limits their usefulness as genetic markers in bacteria other than *E. coli*. One of the more interesting types of resistance to tetracyclines is due to the so-called ribosome protection proteins, represented by TetO and TetQ. This is the type of resistance exhibited by the soil bacteria that make tetracycline. These proteins bind to the A site of the ribosome and release tetracycline from the A site. They may be able to bind to the A site because they mimic the translation factor EF-G. This is yet another example of molecular mimicry in translation (Box 2.2).

INHIBITORS OF TRANSLOCATION

Aminoglycosides

Kanamycin and its close relatives neomycin and gentamicin are members of a larger group of antibiotics, the aminoglycoside antibiotics, which also includes streptomycin. Their mechanism of action is somewhat obscure, but they seem to affect some aspect of translocation by binding to the A site of the ribosome. They also cause misreading of mRNA by the ribosome, and the high level of translation errors they cause could be a major cause of their lethality.

Uses.   Aminoglycosides have a very broad spectrum of action, and some of them inhibit translation in plants and animal cells as well as in bacteria. For example, the ability of neomycin to block translation in plants and animals has made it very useful in biotechnology, where it is used to select transgenic plants containing bacterial genes that confer resistance to these antibiotics (see the discussion of resistance below). However, their toxicity, especially during sustained use, and high rates of resistance somewhat limits their usefulness as therapeutic agents.

Resistance.   Bacterial mutants resistant to aminoglycosides are quite rare, and multiple mutations are required to confer high levels of resistance. The fact that resistant mutants are rare has contributed to the usefulness of kanamycin and its relatives in biotechnology. However, most of the clinically important resistance is due to genes exchanged on transposons and plasmids. The products of some of these genes inacivate the aminoglycosides by phosphorylating, acetylating, or adenylating (adding adenosine to) them. For example, the *neo* gene for kanamycin and neomycin resistance, from transposon Tn*5*, phosphorylates these antibiotics. The *neo* gene has been very important in genetics and biotechnology because it expresses kanamycin resistance in almost all gram-negative bacteria and even makes plant and animal cells resistant to kanamycin (more accurately, to a derivative, G418), provided that it is transcribed and translated in the plant or animal cells.

Fusidic acid

Fusidic acid specifically inhibits translation elongation factor G (EF-G, called EF-2 in eukaryotes), probably by preventing its dissociation from the ribosome after GTP cleavage. It has been very useful in studies of the function of ribosomes. In *E. coli*, mutations that confer resistance to fusidic acid are in the *fusA* gene, which encodes EF-G. Unexpectedly, some acetyltransferases that confer resistance to chloramphenicol also bind to fusidic acid, inactivating it. It is not clear why these two types of resistance should be associated with the same protein.

SUMMARY

1. RNA is a polymer made up of a chain of ribonucleotides. The bases of the nucleotides—adenine, cytosine, uracil, and guanine—are attached to the five-carbon sugar ribose. Phosphate bonds connect the sugars to make the RNA chain, attaching the third (3′) carbon of one sugar to the fifth (5′) carbon of the next sugar. The 5′ end of the RNA is the nucleotide that has a free phosphate attached to the 5′ carbon of its sugar. The 3′ end has a free hydroxyl group at the 3′ carbon, with no phosphate attached. RNA is both made and translated from the 5′ end to the 3′ end.

2. After they are synthesized, RNAs can undergo extensive processing and modification. Processing occurs when phosphate bonds are broken or new phosphate bonds are formed. Modification occurs when the bases or the sugars of the RNA are chemically altered, for example by methylation. The rRNAs and tRNAs, but not the mRNAs, of bacteria are extensively modified.

3. The primary structure of an RNA is its sequence of nucleotides. The secondary structure is formed by hydrogen bonding between bases in the same RNA to give localized double-stranded regions. The tertiary structure is the three-dimensional shape of the RNA due to the stiffness of the double-stranded regions of secondary structure. All RNAs including mRNA, rRNA, and tRNA probably have secondary and tertiary structure.

4. The enzyme responsible for making RNA is called RNA polymerase. One of the largest enzymes in the cell, the bacterial RNA polymerase has five subunits plus another detachable subunit, the σ factor, which comes off after the initiation of transcription. Another factor, ω, helps in its assembly.

5. Transcription begins at well-defined sites on DNA called promoters. The type of promoter used depends on the type of σ factor bound to the RNA polymerase.

6. Transcription stops at sequences in the DNA called transcription terminators, which can be either factor dependent or factor independent. The factor-independent terminators have a string of A's that follows a symmetric sequence. The symmetry of sequence allows the RNA transcribed from that region to fold back on itself to form a loop or hairpin, which causes the RNA molecule to fall off the DNA template. The factor-dependent terminators do not have such a well-defined sequence. The ρ protein is the best-characterized termination factor in *E. coli*. It forms a ring that encircles the RNA moving toward the RNA polymerase. If the RNA polymerase pauses at a ρ termination site, the ρ factor causes it to dissociate, releasing the mRNA.

7. Most of the RNA in the cell falls into three groups: messenger (mRNA), ribosomal (rRNA), and transfer (tRNA). Of these, mRNA is very unstable, existing for only a few minutes before being degraded. rRNAs in bacteria are further divided into three types: 16S, 23S, and 5S. Both rRNA and tRNA are very stable and account for about 95% of the total RNA. Other RNAs include the primers for DNA replication and small RNAs involved in regulation or RNA processing.

8. Ribosomes, the site of protein synthesis, are made up of two subunits, the 30S subunit and the 50S subunit, as well as many proteins. The 16S rRNA is in the 30S subunit, while the 23S and 5S rRNAs are in the 50S subunit.

9. Polypeptides are chains of the 20 amino acids (rarely 21 or 22), which are held together by peptide bonds between the amino group of one amino acid and the carboxyl group of another. The amino terminus (N terminus) of the polypeptide has the amino acid with an unattached amino group. The carboxyl terminus (C terminus) of a polypeptide has the amino acid with a free carboxyl group.

10. Translation is the synthesis of polypeptides from mRNA. During translation, the mRNA moves in the 5′-to-3′ direction along the ribosome three nucleotides at a time. Three reading frames are possible depending on how the ribosome is positioned at each triplet.

11. The genetic code is the assignment of each possible three-nucleotide codon sequence in mRNA to 1 of 20 amino acids. The code is redundant, with more than one codon sometimes encoding the same amino acid. Because of wobble, the first position of the tRNA anticodon (written 5′ to 3′) does not have to behave by the standard base-pairing complementarity to the third position of the antiparallel codon sequence, and other pairings are possible.

12. Initiation of translation occurs at translation initiation regions (TIRs) on the mRNA that consist of an initiation codon, usually AUG or GUG, and often a Shine-Dalgarno (S-D) sequence, a short sequence that is complementary to part of the 16S rRNA and precedes the initiation codon.

13. The first tRNA to enter the ribosome is a special methionyl-tRNA called fMet-tRNA$_f^{Met}$, which carries the amino acid formylmethionine. After the polypeptide has been synthesized, the formyl group and often the first methionine are removed.

14. Translation termination occurs when one of the terminator or nonsense codons UAA, UAG, or UGA is encountered as the ribosome moves down the mRNA. Proteins called ribosome release factors (RFs) are also required for release of the polypeptide.

15. The primary structure of a polypeptide is the sequence of amino acids in the polypeptide. Proteins can be made up of more than one polypeptide chain, which can be the same as or different from each other. The secondary structure results from hydrogen bonding of the amino acids to form α-helical regions and β-sheets. Tertiary structure refers to how the chains fold up on themselves, and quaternary structure refers to one or more different polypeptide chains folding up on each other.

16. Proteins that help other proteins fold are called chaperones. The most ubiquitous chaperones are the Hsp70 chaperones, called DnaK in *E. coli*, which are almost the same in all types of cells from bacteria to humans. These chaperones bind to the hydrophobic regions of proteins and prevent them from associating prematurely. They are aided by their smaller cochaperones, DnaJ and GrpE, which help in binding to proteins and cycling ADP off the chaperone, respectively. Other proteins, called Hsp60 chaperonins, also help proteins fold, but by a very different mechanism. One, the Hsp60 chaperonin, called GroEL in *E. coli*, forms large cylindrical structures with internal chambers that take up unfolded proteins and help them refold properly. A cochaperonin called GroES forms a cap on the cylinder after the unfolded protein is taken up. Chaperonins like are GoEL are found in bacteria and in the organelles of eukaryotes and are called group I chaperonins. Another type, group II chaperonins, are found in the cytoplasm of eukaryotes and in archaea. They have a similar structure but a very different amino acid sequence.

17. The process of passing proteins through membranes is called transport. Proteins which pass through the inner membrane into the periplasm and beyond are said to be exported. Proteins which pass out of the cell are secreted. The *sec* system, responsible for transporting many proteins into and through the inner membrane, consists of the SecYEG channel in the inner membrane, through which proteins pass. Proteins to be transported through the inner membrane and those whose final destination is the inner membrane are recognized by different targeting factors. The SecB-SecA system recognizes proteins that are to be transported through the membrane by the *sec* system after they are synthesized. These proteins characteristically have a short hydrophobic signal sequence at their N terminus which is cleaved off by a peptidase as the protein passes through the SecYEG channel in the inner membrane. These targeting factors are unique to bacteria. Another targeting system, the signal recognition particle (SRP), specifically recognizes proteins destined to remain in the inner membrane. The SRP binds to the first hydrophobic transmembrane domain as it emerges from the ribosome. Translation can then continue, feeding the protein into the membrane as it is synthesized in a process called cotranslation. The SRP targeting system is more universal, being found in a modified form also in eukaryotes, where it plays a more general role in protein transport, also transporting proteins with removable signal sequences.

18. Proteins can also be held together by disulfide linkages between cysteines in the protein. Generally, only proteins that are exported into the periplasm or out of the cell have disulfide bonds. These disulfide bonds are made by oxidoreductases in the periplasm of gram-negative bacteria.

19. In gram-negative bacteria, proteins which are secreted out of the cell often have specialized structures to help them pass through the outer membrane. Some of these use the SecYEG translocon to get through the inner membrane, whereas others make elaborate structures that pass proteins through both membranes. The most dramatic of these are the type III secretion systems of pathogenic bacteria, which act like syringes to inject proteins through both bacterial membranes and directly into the eukaryotic host cell, either plant or animal.

20. An open reading frame (ORF) is a string of amino acid codons in DNA unbroken by a nonsense codon. In vitro transcription-translation systems or transcriptional and translation fusions are often required to prove that an ORF in DNA actually encodes a protein.

21. The strand of DNA from which the mRNA is made is the transcribed strand. The opposite strand, which has the same sequence as the mRNA, is the coding strand.

22. A sequence 5' on the coding strand of DNA is said to be upstream, whereas a sequence 3' is downstream.

23. The TIR sequence of a gene does not necessarily occur at the beginning of the mRNA. The 5' end of the mRNA is called the 5' untranslated region. Similarly, the sequence downstream of the nonsense codon is the 3' untranslated region.

24. Because mRNA is both transcribed and translated in the 5'-to-3' direction, it can be translated as it is transcribed in bacteria, which have no nuclear membrane.

25. Bacteria often make polycistronic mRNAs with more than one polypeptide coding sequence on an mRNA. This makes possible polarity of transcription and translational coupling, phenomena unique to bacteria.

26. The expression of genes is regulated, depending on the conditions in which the cell is found. This regulation can be either transcriptional or posttranscriptional. Transcriptional regulation can be either negative or positive depending on whether the regulatory protein is a repressor or an activator, respectively. A repressor binds to an operator or operators which are usually close to the promoter and prevents transcription from the promoter. An activator binds to an upstream activator sequence (UAS), upstream of the promoter, and allows transcription from the promoter. Transcriptional regulation can also occur after the RNA polymerase leaves the promoter, as attenuation or antitermination of transcription. Posttranscriptional regulation can occur at the level of translation of the mRNA, stability of the mRNA, or processing and modification of the gene product.

(continued)

27. Gene fusions have many uses in modern molecular genetics. They can be either transcriptional or translational fusions. In a transcriptional fusion, the two coding regions are transcribed into the same mRNA but each is translated from its own TIR. In a translational fusion, the two coding regions are fused to each other so that they are translated in the same frame and with no nonsense codons between them. A translational fusion makes a fusion protein with the two polypeptides fused to each other.

28. Expression vectors are designed to allow the synthesis of the product of a cloned gene in a convenient host such as *E. coli*. They can be either transcription or translation vectors. In transcription vectors, the cloned gene is transcribed from a promoter on the vector but translated from its own TIR. Translation vectors also contain a TIR from which the gene can be translated. Affinity vectors are translation vectors that fuse a polypeptide that is easily purified to the protein product of the cloned gene. Some expression vectors have inducible promoters, so that the cloned gene is expressed only when the inducer is added.

29. Many naturally occurring antibiotics attack components of the transcription and translation apparatuses. Some of the more useful are rifampin, streptomycin, tetracycline, thiostrepton, chloramphenicol, and kanamycin. Besides their uses in treating bacterial infections, tumor chemotherapy, and biotechnology, such antibiotics have also helped us understand the workings of the transcription and translation apparatuses. In addition, the genes that confer resistance to these antibiotics have served as selectable genetic markers and reporter genes in molecular genetic studies of both bacteria and eukaryotes.

QUESTIONS FOR THOUGHT

1. Which do you think came first in the very earliest life on Earth, DNA, RNA, or protein? Why?

2. Why is the genetic code universal?

3. Why do you suppose prokaryotes have polycistronic mRNAs but eukaryotes do not?

4. Why do you suppose mitochondrial genes show differences in their genetic code from chromosomal genes?

5. Why is selenocysteine inserted into proteins of almost all organisms but only into a few sites in a few proteins in these organisms?

6. Why do so many antibiotics inhibit the translation process as opposed to, say, amino acid biosynthesis?

7. Why do you think chaperonins have two linked chambers and alternate the folding of proteins between the two chambers?

8. Why do some proteins have specialized systems of their own for membrane transport instead of using the general secretory *sec* system? Why do not all exported proteins use the *sec* system?

9. How do you suppose the transmembrane domains of an inner membrane protein escape the SecYEG channel into the inner membrane?

10. List all the reasons you can think of why bacteria would regulate the expression of their genes.

PROBLEMS

1. What is the longest open reading frame in the mRNA sequence 5′AGCUAACUGAUGUGAUGUCAACGUCCUAC-UCUAGCGUAGUCUAAAG3′? Remember to look in all three frames.

2. Where do you think translation is most likely to start in the mRNA sequence 5′UAAGUGAAAGAUGUGAAUGAAG-UAGCCACCAAAGUCACUAAUGCUUCCAACA3′? Why?

3. Which of the following is more likely to be a factor-independent transcription termination site? Note that in each case, only the transcribed strand of the DNA is shown in the 3′-to-5′ direction.

a. 3′AACGACTAGTACGACATACTAGTCGTTG-GCAAAAAAAATGCA5′

b. 3′ACTAGCCTAAGCATCTTGCATCAGGCACA-GAAAAAAAAAATCGCA5′

4. Design a 20-nucleotide PCR primer that could be used as the upstream primer to introduce a BamHI restriction site to clone a protein-coding sequence that begins with ATGUUGC-GATTU to fuse it downstream of a His tag in which translation begins with ATGCCG<u>CATCATCATCATCATCAT</u>**GGATCC**T. The six histidine codons on the cloning vector are underlined, and a BamHI restriction endonuclease recognition site on the vector is shown in bold.

5. What would be the effect of a mutation that inactivates the regulatory protein of an operon on the expression of the operon if

a. the regulation is negative?

b. the regulation is positive?

6. Outline how you would use PCR and cloning with fusion to an affinity tag to purify a human protein of unknown function whose gene you have identified in the human genome sequence.

7. Define homolog, ortholog and paralog.

8. Define COG and TIGRFAM equivalogs.

SUGGESTED READING

Agashe, V. R., S. Guha, H.-C. Chang, P. Genevaux, M. Hayer-Hartl, M. Stemp, C. Georgopoulos, F. Ulrich-Hartl, and J. M. Barral. 2004. Function of trigger factor and DnaK in multidomain protein folding: increase in yield at the expense of folding speed. *Cell* **117**:199–209.

Artsimovitch, I. 2005. Control of transcription termination and antitermination: the structure of bacterial RNA polymerase, p. 311–326. *In* N. P. Higgins (ed.), *The Bacterial Chromosome*. ASM Press, Washington, D.C.

Ban, N., P. Nissen, J. C. Jansen, M. Capel, P. B. Moore, and T. A. Steitz. 1999. Placement of protein and RNA structures into a 5Å resolution map of the 50S ribosomal subunit. *Nature* **400**:841–847.

Björk, G. R., and T. G. Hagervall. 25 July 2005, posting date. Chapter 4.6.2, Transfer RNA modification. *In* A. Böck, R. Curtiss III, J. B. Kaper, F. C. Neidhardt, T. Nyström, K. E. Rudd, and C. L. Squires (ed.), EcoSal—*Escherichia coli* and *Salmonella*: cellular and molecular biology. ASM Press, Washington, D.C. [Online.] http://www.ecosal.org.

Brock, J. E., R. L. Paz, P. Cottle, and G. R. Janssen. 2007. Naturally occurring adenines within mRNA coding sequences affect ribosome binding and expression in *Escherichia coli*. *J. Bacteriol.* **189**:501–510.

Bukau, B., and A. L. Horwich. 1998. The Hsp70 and Hsp60 chaperone machines. Review. *Cell* **92**:351–366.

Cate, J. H., M. M. Yusupov, G. Z. Yusupova, T. N. Earnest, and H. F. Noller. 1999. X-ray crystal structures of 70S ribosome functional complexes. *Science* **285**:2095–2104.

Doherty, G. P., D. H. Meredith, and P. J. Lewis. 2006. Subcellular partitioning of transcription factors in *Bacillus subtilis*. *J. Bacteriol.* **188**:4101–4110.

Gabashvili, I. S., R. K. Agrwal, C. M. T. Spahn, R. A. Grassucci, D. I. Svergun, and J. Frank. 2000. Solution structure of the *E. coli* 70S ribosome at 11.5Å resolution. *Cell* **100**:537–549.

Ganoza, M. C., M. C. Kiel, and H. Aoki. 2002. Evolutionary conservation of reactions in translation. *Microbiol. Mol. Biol. Rev.* **66**:460–485.

Geszvain, K., and R. Landick. 2005. The structure of bacterial RNA polymerase, p. 286–295. *In* N. P. Higgins (ed.), *The Bacterial Chromosome*. ASM Press, Washington, D.C.

Han, M.-J., and S. Y. Lee. 2002. The *Escherichia coli* proteome: past, present, and future prospects. *Microbiol. Mol. Biol. Rev.* **66**:460–485.

Lee, H. C., and H. D. Bernstein. 2001. The targeting pathway of *Escherichia coli* presecretory and integral membrane proteins is specified by the hydrophobicity of the targeting signal. *Proc. Natl. Acad. Sci. USA* **98**:3471–3476.

Maguire, B. A., and R. A. Zimmermann. 2001. The ribosome in focus. Minireview. *Cell* **104**:813–816.

Meinnel, T., C. Sacerdot, M. Graffe, S. Blanquet, and M. Springer. 1999. Discrimination by *Escherichia coli* initiation factor IF3 against initiation on non-canonical codons relies on complementarity rules. *J. Mol. Biol.* **290**:825–837.

Mogk, A., E. Deuerling, S. Vorderwulbecke, E. Vierling, and B. Bukau. 2003. Small heat shock proteins, ClpB and the DnaK system form a functional triade in reversing protein aggregation. *Mol. Microbiol.* **50**:585–595.

Sun, Z., D. J. Scott, and P. A. Lund. 2003. Isolation and characterisation of mutants of GroEL that are fully functional as single rings. *J. Mol. Biol.* **332**:715–728.

Tam, P. C., A. P. Maillard, K. K. Chan, and F. Duong. 2005. Investigating the SecY plug movement at the SecYEG translocation channel. *EMBO J.* **24**:3380–3388.

Tan, J., Y. Lu, and J. C. A. Bardwell. 2005. Mutational analysis of the disulfide catalysts DsbA and DsbB. *J. Bacteriol.* **187**:1504–1510.

Van den Berg, B., W. M. Clemons, Jr., I. Collinson, Y. Modis, E. Hartmann, S. C. Harrison, and T. A. Rapoport. 2004. X-ray structure of a protein-conducting channel. *Nature* **427**:36–44.

Wang, J., and D. C. Boisvert 2003. Structural basis for GroEL-assisted protein folding from the crystal structure of (GroEL-KMgATP)14 at 2.0 Å resolution. *J. Mol. Biol.* **327**:843–855.

Wild, K., K. R. Rosendal, and I. Sinning. 2004. A structural step into the SRP cycle. *Mol. Microbiol.* **53**:357–363.

Yusupov, M. M., G. Z. Yusupova, A. Baucom, K. Lieberman, T. N. Earnest, J. H. Cate, and H. F. Noller. 2001. Crystal structure of the ribosome at 5.5Å resolution. *Science* **292**:883–896.

CHAPTER **3**

Bacterial Genetic Analysis: Forward and Reverse

As mentioned in the introductory chapter, the relative ease with which bacteria can be handled genetically has made them very useful model systems for understanding many life processes, and much of the information on basic macromolecular synthesis discussed in the first two chapters came from genetic experiments with bacteria. In this chapter, we introduce the genetic concepts and definitions that are used in later chapters.

Definitions

In genetics, as in any field of knowledge, we need definitions. However, words do not mean much when taken out of context, and so here we define only the most basic terms. We will define other important terms as we go along; these appear in boldface.

Terms Used in Genetics

These words are common to all types of genetic experiments, whether with prokaryotic or eukaryotic systems.

MUTANT

The word **mutant** refers to an organism that is the direct offspring of a normal member of the species (the **wild type**) but is different. Organisms of the same species isolated from nature that have different properties are usually not called mutants but, rather, variants or **strains,** because, even if one of the strains has recently arisen from the other in nature, we have no way of knowing which one is the mutant and which is the wild type.

PHENOTYPE

The phenotypes of an organism are all the observable properties of that organism. Usually in genetics, the term **phenotype** means **mutant phenotype,** or the characteristics of the mutant organism that differ from those of the wild type. The corresponding normal property is sometimes referred to as the **wild-type phenotype.**

GENOTYPE

The **genotype** of an organism refers to the actual sequence of its DNA. If two organisms have the same genotype, they are genetically identical. Identical twins have almost the same genotype. If two organisms differ by only one mutation, they are said to be **isogenic** except for that mutation.

MUTATION

A **mutation** refers to any heritable change in the DNA sequence. Practically every imaginable type of change is possible, and all changes are called mutations. However, the word "heritable" must be emphasized. Changes or damage that are repaired, so that the original sequence is restored, are not inherited. Hence, only a permanent change in the sequence of deoxynucleotides constitutes a mutation.

ALLELE

Different forms of the same gene are called **alleles.** For example, if one form of a gene has a mutation and the other has the wild-type sequence, the two forms of the gene are different alleles of the same gene. In this case, one gene is a mutant allele and the other gene is the wild-type allele. Diploid organisms can have two different alleles of the same gene, one on each homologous chromosome. The term "allele" can also refer to genes with the same or similar sequences that appear at the same chromosomal location in closely related species. However, similar gene sequences occurring in different chromosomal locations are not alleles; rather, they are **copies** of the gene.

USE OF GENETIC DEFINITIONS

The following example illustrates the use of the definitions in the previous section. For an explanation of the methods, see the introductory chapter.

A culture of *Pseudomonas fluorescens* normally grows as bright green colonies on agar plates. However, suppose that one of the colonies is colorless. It probably arose through multiplication of a **mutant** organism. The **mutant phenotype** is "colorless colony," and the corresponding **wild-type phenotype** is "green colony." The mutant bacterium that formed the colorless colony probably had a **mutation** in a gene for an enzyme required to make the green pigment. Perhaps the mutation consists of a base pair change in the gene, causing the insertion of a wrong amino acid in the polypeptide, and the resulting enzyme cannot function. Thus, the mutant and wild-type bacteria have different **alleles** of this gene, and we can refer to the gene in the colorless colony-forming bacteria as the **mutant allele** and the gene in the green-colony-forming bacteria as the **wild-type allele.**

In the example above, we only know that a mutation has occurred because of the lack of color. However, recall that any heritable change in the DNA sequence is a mutation, and so mutations can occur without changing the organism's phenotype. Many such changes, called **silent mutations,** have been found by sequencing the DNA directly.

Genetic Names

There are some commonly accepted rules for naming mutants, phenotypes, and mutations in bacteria, although different publications sometimes use different notations. We use the terms recommended by the American Society for Microbiology.

NAMING MUTANT ORGANISMS

The mutant organism can be given any name as long as the designation does not refer specifically to the phenotype or the gene thought to have been mutated. This rule helps to avoid confusion if the gene with the mutation is introduced into another strain or if other mutations occur or are transferred into the original organism. Quite often, someone who has isolated a mutant names it after himself or herself, giving it his or her initials and a number (e.g., *Escherichia coli* AB2497). This notation is not intended to gratify the person's ego but to inform others where they can obtain the mutant strain and get advice about its properties. If another mutation alters the mutant strain, this new strain is usually given another name, such as *E. coli* AB2498.

NAMING GENES

Bacterial genes are designated by three lowercase italic letters that usually refer to the function of the gene's product, when it is known. For example, the name *his* refers to a gene whose product is an enzyme required to synthesize the amino acid histidine. Sometimes more than one gene encodes a product with the same function, or an enzymatic pathway requires more than one different polypeptide. In these cases, a capital letter designating each individual gene follows the three lowercase letters. For example, the *hisA* and *hisB* genes both encode polypeptides required to synthesize histidine. A mutation that inactivates either gene will make the cell unable to synthesize histidine.

NAMING MUTATIONS

Hundreds of different types of mutations can occur in a single gene, and so all alleles of a particular gene have a specific allele number. For example, *hisA4* refers to the *hisA* gene with mutation number 4, and the *hisA* gene with mutation number 4 is referred to as the *hisA4* allele.

If a mutation is known to inactivate the product of a gene, a superscript minus sign, or simply the word "mutation," may be added to the gene or allele name. For example, *hisA⁻* or a *hisA* mutation inactivates the product of the *hisA* gene. Alternatively, the designation *hisA⁺* refers specifically to the wild-type form of the *hisA* gene, which encodes a functional gene product.

Different nomenclatural rules apply if a mutation is a deletion or insertion. We defer a discussion of these rules until we discuss these types of mutations (see below).

NAMING PHENOTYPES

Phenotypes are also denoted by three-letter names, but the letters are not italicized and the first letter is capitalized. As with genotypes, superscripts are often used to distinguish mutant from wild-type phenotypes. For example, His⁻ describes the phenotype of an organism with a mutated *his* gene that cannot grow without histidine in its environment. The corresponding wild-type organisms grow without histidine, so they are phenotypically His⁺. Another example, Rif^r, describes resistance to the antibiotic rifampin, which blocks RNA synthesis (see chapter 2). A mutation in the *rpoB* gene, which encodes a subunit of the RNA polymerase, makes the cell resistant to this antibiotic. The corresponding wild-type phenotype is rifampin sensitivity, or Rif^s.

Useful Phenotypes in Bacterial Genetics

What phenotypes are useful for genetic experiments depends on the organism being studied. For bacterial genetics, the properties of the colonies formed on agar plates are the most useful phenotypes (see the introductory chapter).

The visual appearance of colonies sometimes provides useful mutant phenotypes, such as the colorless colony discussed above. Colonies formed by mutant bacteria might also be smaller than normal or smooth instead of wrinkled. The mutant bacterium may not multiply to form a colony at all under some conditions, or, conversely, it may multiply when the wild type cannot.

Many mutant phenotypes have been used to study cellular processes such as DNA recombination and repair, mutagenesis, and development. The following sections describe a few of the more commonly used phenotypes. In later chapters, we discuss many more types of mutants and demonstrate how mutations can be used to study life processes.

Auxotrophic Mutants

Some of the most useful bacterial mutants are **auxotrophic mutants,** or auxotrophs. Of the two types of these mutants, one cannot multiply without a particular growth supplement that is not required by the original, wild-type isolate. For example a His⁻ auxotrophic mutant cannot grow unless the medium is supplemented with the amino acid histidine, while the wild type could grow without added histidine. Similarily, a Bio⁻ auxotrophic mutant cannot grow without the vitamin biotin, which is not needed by the wild type.

The other type of auxotrophic mutant cannot use a particular substance for growth that can be used by the wild type. For example, the wild-type bacteria may be able to use the sugar maltose as a sole carbon and energy source but a Mal⁻ auxotrophic mutant must be given another carbon and energy source, such as glucose, in order to grow. Other examples are mutants that cannot use a particular amino acid as a nitrogen source or a particular phosphate-containing compound as a source of phosphate.

Even though these two types of auxotrophs seem opposite, their molecular basis is similar. In both types, a mutation has altered a gene encoding an enzyme of a metabolic pathway, thereby inactivating the enzyme. The only difference is that in the first case, the inactivated enzyme was in a **biosynthetic** pathway, which is required to synthesize a substance, while in the second case, the inactivated enzyme was in a **catabolic** pathway, which is required to degrade a substance to use it as a carbon and energy source, a nitrogen source, or a phosphate source.

ISOLATING AUXOTROPHIC MUTANTS

Figure 3.1 shows a simple method for isolating mutants auxotrophic for histidine and biotin. In this experiment, eight colonies from plate 1, which contains all the nutrients the bacteria need, including histidine and biotin, were picked up with a loop and transferred onto two other plates. These plates are the same as plate 1 except that plate 2 lacks biotin but has histidine and plate 3 lacks histidine but has biotin. The bacteria from most of the colonies can multiply on all three types of plates. However, the bacteria in colony 2 grow only on plate 2; they are mutants that require added histidine, that is, they are His⁻. These mutants do not require biotin, and so they are Bio⁺. Similarly, the bacteria in colony 6 are Bio⁻ but His⁺ since they can grow on plate 3 but not on plate 2. Under real conditions, mutants that require histidine or biotin would not be this frequent, and thousands of colonies would have to be tested to find one mutant

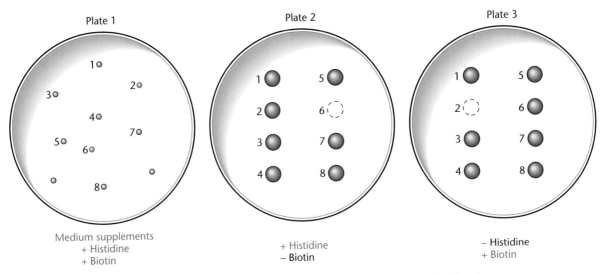

Figure 3.1 Detection of auxotrophic mutants. Colonies were scraped with a loop from plate 1 and transferred to plates 2 and 3. Colony 6 was formed by a bacterium that could not multiply without biotin and so was Bio⁻. The bacteria in colony 2 are descendents of a His⁻ bacterium.

that required histidine, biotin, or indeed any growth supplement not required by the wild type.

In principle, it should be possible to find auxotrophic mutants unable to synthesize any compound required for growth or unable to use any carbon and energy source. However, auxotrophic mutants must be supplied with the compound they cannot synthesize, and these compounds must enter the cell. Yet many bacteria cannot take in some compounds that have a high electrical charge, such as nucleotides, so that some types of auxotrophs are very difficult to isolate.

Conditional Lethal Mutants

As mentioned, auxotrophic mutants can be isolated because they have mutations in genes whose products are required under only certain conditions. The cells can be grown under conditions where the product of the mutated gene is not required and tested under conditions where it is required. However, many gene products of the cell are essential for growth no matter what conditions the bacteria find themselves in. The genes that encode such functions are called **essential genes**. Examples of essential genes include those for RNA polymerase, ribosomal proteins, DNA ligase, and some helicases. Cells with mutations that inactivate essential genes cannot be isolated unless the mutations inactivate the gene under only some conditions. Hence, any mutants that are isolated will have **conditional lethal mutations,** because these DNA changes are lethal only under some conditions.

TEMPERATURE-SENSITIVE MUTANTS

The most generally useful conditional lethal mutations in bacteria are mutations that make the mutant **temperature sensitive** for growth. Usually, such mutations change an amino acid of a protein so that the protein no longer functions at higher temperatures but still functions at lower temperatures. The higher temperatures are called the **nonpermissive temperatures** for the mutant, whereas the temperatures at which the protein still functions are the **permissive temperatures** for the mutant.

Mutations can affect the temperature stability of proteins in various ways. Often, an amino acid required for the protein's stability at the nonpermissive temperature is changed, causing the protein to unfold, or denature, partially or completely. The protein could then remain in the inactive state or be destroyed by cellular proteases that remove abnormal proteins. If the protein remains, it sometimes spontaneously renatures (refolds) when the temperature is lowered; growth can then resume immediately. With other mutations, the protein is irreversibly denatured and must be resynthesized before growth can resume.

The temperature ranges used to isolate temperature-sensitive mutations depend on the organism. Bacteria can be considered poikilothermic (meaning "cold-blooded" when applied to animals) organisms, since their cell temperature varies with the outside temperature. Therefore, their proteins are designed to function over a wide range of temperatures. However, different species of bacteria differ greatly in their preferred temperature range. For example, a "mesophilic" bacterium such as *E. coli* may grow well in a range of temperatures from 20 to 42°C. In contrast, a "thermophilic" bacterium such as *Bacillus stearothermophilus* may grow well only between 42 and 60°C. For *E. coli*, a temperature-sensitive mutation may leave a protein functional at 33°C, the permissive temperature, but not at 42°C, the nonpermissive temperature. For *B. stearothermophilus*, the temperature-sensitive mutation may leave a protein active at the permissive temperature of 47°C but render it nonfunctional at the nonpermissive temperature of 55°C.

Isolating Temperature-Sensitive Mutants

In principle, temperature-sensitive mutants are as easy to isolate as auxotrophic mutants. If a mutation that makes the cell temperature sensitive occurs in a gene whose protein product is required for growth, the cells stop multiplying at the nonpermissive temperature. To isolate such mutants, the bacteria are incubated on a plate at the permissive temperature until colonies appear and then the colonies are transferred to a plate incubated at the nonpermissive temperature. Bacteria that can form colonies at the permissive temperature but not at the nonpermissive temperature are temperature-sensitive mutants. However, temperature-sensitive mutants are usually much rarer than auxotrophic mutants. Many changes in a protein will inactivate it, but very few will make a protein functional at one temperature and nonfunctional at another. The frequency of occurrence of different types of mutations is discussed later in this chapter.

COLD-SENSITIVE MUTANTS

Cells with proteins that fail to function at low temperatures are called specifically **cold-sensitive mutants.** Mutations that make a bacterium cold sensitive for growth are often in genes whose products must form a larger complex such as the ribosome. The increased movement at the higher temperature may allow the mutated protein, despite its altered shape, to enter the complex, but it is unable to do so at lower temperatures. Such mutations often show a phenotype only after a long delay, so they

are generally less useful than mutations that make a protein unstable at higher temperatures. Both mutations are in a sense temperature sensitive, but the name is usually reserved for heat-sensitive mutations.

NONSENSE MUTATIONS

Mutations that change a codon in a gene to one of the three nonsense codons—UAA, UAG, or UGA—can also be conditional lethal mutations. A nonsense mutation causes translation to stop within the gene unless the cell has a "nonsense suppressor" tRNA, as explained later in the chapter. Because nonsense mutations are more generally useful in viral genetics than bacterial genetics, they are discussed in more detail in chapter 7.

Resistant Mutants

Among the most useful types of bacterial mutants to isolate are resistant mutants. If a substance kills or inhibits the growth of a bacterium, mutants resistant to the substance can often be isolated merely by plating the bacteria in the presence of the substance.

The numerous mechanisms of resistance depend on the basis for toxicity and on the options available to prevent the toxicity (examples are given in Table 3.1). For example, the mutation may destroy a cell surface receptor to which the toxic substance must bind to enter the cell. If the substance cannot enter the mutant cell, it cannot kill the cell. Alternatively, a mutation might change the "target" affected by the substance inside the cell. For example, an antibiotic might normally bind to a ribosomal protein and affect protein translation. However, if the antibiotic cannot bind to a mutant (but still functional) protein, it cannot kill the cell. An example of such a resistance mutation is a mutation to streptomycin resistance in *E. coli* (Table 3.1) The antibiotic streptomycin binds to the 16S rRNA in the 30S subunit of the ribosome and blocks translation. However, some mutations in the gene for the S12 protein, *rpsL* (for *r*ibosomal *p*rotein *s*mall-subunit *L*), prevent streptomycin from binding to the ribosome but do not inactivate the S12 protein. These mutations therefore confer streptomycin

TABLE 3.1	Some resistance mutations	
Substance	**Toxicity**	**Resistance mutation**
Bacteriophage T1	Infects and kills	Inactivates *tonB* outer membrane protein; phage cannot absorb
Streptomycin	Binds to ribosomes; inhibits translation	Changes ribosomal protein S12 so that it no longer binds
Chlorate	Converted to chlorite, which is toxic	Inactivates nitrate reductase, which converts chlorate to chlorite
High concentrations of valine, no isoleucine	Feedback inhibits acetolactate synthetase; starves for isoleucine	Activates a valine-insensitive acetolactate synthetase

resistance on *E. coli*. In some cases, the substance added to the cells is not toxic until one of the cell's own enzymes changes it. A mutation inactivating the enzyme that converts the nontoxic substance into the toxic one could make the cell resistant to that substance.

Inheritance in Bacteria

Salvador Luria and Max Delbrück were among the first people to attempt to study bacterial inheritance quantitatively. They published a now-classic paper in the journal *Genetics* in 1943. This paper is still very much worth reading and is listed in the Suggested Reading section at the end of the chapter. As discussed in the introductory chapter, the experiments and reasoning of Luria and Delbrück helped debunk what was then a popular misconception among bacteriologists. At the time of the Luria and Delbrück studies, it was generally thought that bacteria were different from other organisms in their inheritance. It was generally accepted that heredity in higher organisms followed "neo-Darwinian" principles. According to Charles Darwin, random mutations occur, and if one happens to confer a desirable phenotype, organisms with this mutation are selected by the environment and become the predominant members of the population. Undesirable as well as desirable mutations continuously occur, but only the desirable mutations are passed on to future generations.

However, many bacteriologists thought that heredity in bacteria followed different principles. They thought that bacteria, rather than changing as the result of random mutations, somehow "adapt" to the environment by a process of directed change, after which the adapted organism would pass the adaptation on to its offspring. This process is called Lamarckian inheritance, and acceptance of it was encouraged by the observation that all the bacteria in a culture exposed to a toxic substance seem to become resistant to that substance in response (Figure 3.2).

The Luria and Delbrück Experiment

The Luria and Delbrück experiment was designed to test two hypotheses for how mutants arise in bacterial cultures: the **random-mutation** hypothesis and the **directed-change** hypothesis. The random-mutation hypothesis predicts that the mutants appear randomly prior to the addition of the selective agent, whereas the directed-change hypothesis predicts that mutants appear only in response to a selective agent.

One distinction between the two hypotheses is reflected in the distribution of the number of mutants in a series of cultures. If the random-mutation hypothesis is correct, mutations that occur early in the growth of a culture will have a disproportionate effect on the number of mutants in the culture and the fraction of mutants should vary widely from culture to culture. If the directed-change hypothesis is correct, the number of mutants per culture should be approximately equal, subject only to statistical fluctuations. Figure 3.3 illustrates this principle. In culture 1, only one mutation occurred, but this mutation gave rise to eight resistant mutants because it occurred early. In culture 2, two mutations arose, but they gave rise to only six resistant mutants because they occurred later.

Therefore, to determine if mutants appear before or after addition of a selective agent, one can grow several cultures in the absence of the selective agent, add the agent to all the cultures at the same time, and then measure the fraction of bacteria resistant to the selective agent in each culture. If the random-mutation hypothesis is correct, the number of mutant colonies will vary among all the cultures, depending on when the mutations occurred, as illustrated in Figure 3.3. In contrast, if the directed-change hypothesis is correct, each bacterium has the same chance of becoming a mutant, but only after the selective agent is added, and so the same percentage of the bacteria should become resistant in all the cultures. Therefore, a result in which the number of mutants per culture varies greatly will favor the random-mutation hypothesis but a result in which the number of mutants in a series of cultures is about the same will favor the directed-change hypothesis.

In their experiments, Luria and Delbrück used *E. coli* as the bacterium and bacteriophage T1 as the selective agent. As shown in Table 3.1 and Figure 3.4, phage T1 kills wild-type *E. coli*, but a mutation in the gene for an outer membrane protein called TonB can make these cells resistant to killing by the phage. If bacteria are spread on an agar plate with the phage, only those resistant to the phage multiply to form a colony. All the others are killed. The number of colonies on the plate is therefore a measure of the number of bacteria resistant to the bacteriophage in the culture.

Figure 3.5 shows the two experiments that Luria and Delbrück performed and Table 3.2 gives some representative results. It can be seen that the two experiments give very different results, even though they seem superficially similar. In experiment 1, the authors started one culture of bacteria. After incubating it, they took out small aliquots and plated them with and without phage T1 to measure the number of resistant mutants as well as the total number of bacteria in the culture. They then calculated the fraction of resistant mutants. In experiment 2, they started a large number of relatively smaller cultures. After incubating these cultures, they measured the number of resistant mutants and the total number of bacteria

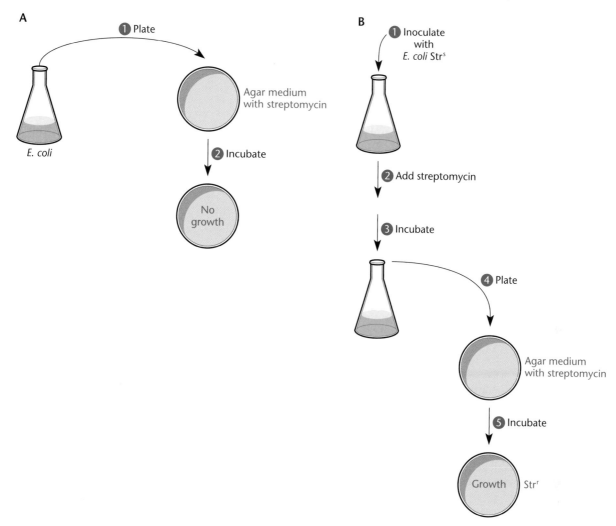

Figure 3.2 Resistant mutants seem to appear in response to the presence of a toxic substance. (A) Sensitivity of wild-type *E. coli* to the antibiotic streptomycin. Plating an *E. coli* culture on streptomycin-containing agar medium results in a lack of colony growth after incubation. (B) Emergence of streptomycin-resistant *E. coli* mutants. A flask containing wild-type *E. coli* is incubated in the presence of streptomycin. When the contents of the flask are transferred to an agar medium also containing streptomycin, streptomycin-resistant mutants form colonies.

in each culture. It can easily be seen that in experiment 1 the number of resistant mutants in each aliquot is almost the same, subject only to sampling errors and statistical fluctuations. However, in experiment 2 a very large variation in the number of resistant bacteria per culture was found. Some cultures had no resistant mutants, while some had many. One culture even had 107 resistant mutants! Luria and Delbrück referred to this and the other mutant-rich cultures as "jackpot" cultures. Apparently, these are cultures in which a mutation to resistance occurred very early. Hence, these results fulfill the predictions of the random-mutation hypothesis. In contrast,

the directed-change hypothesis predicts that the results of the two experiments should be the same, and certainly no jackpot cultures should appear in the second experiment. Box 3.1 presents these predictions in statistical terms.

The Newcombe Experiment

The analysis by Luria and Delbrück was fairly sophisticated mathematically and so was not generally understood. Some people still held to the belief that bacteria were somehow different from other organisms in their inheritance. Consequently, Howard Newcombe in 1951 devised an experiment that was conceptually simpler and

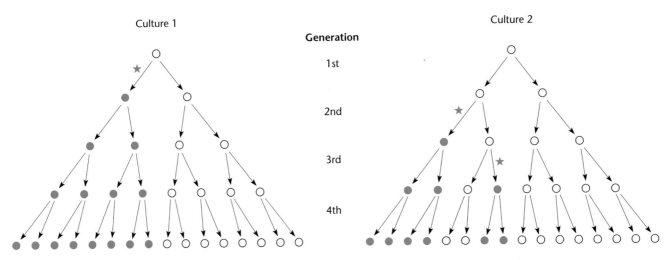

Figure 3.3 Earlier mutations give rise to more mutant progeny in a growing culture. Only one mutation occurred in culture 1, but it gave rise to eight mutant progeny because it occurred in the first generation. In culture 2, two mutations occurred, one in the second generation and one in the third. However, because these mutations occurred later, they gave rise to only six mutant progeny. The mutant cells are shaded.

so convinced many skeptics. In his experiment, he also used *E. coli* mutants resistant to phage T1.

The principle behind Newcombe's experiment is that if the random-mutation hypothesis is correct, mutants

Figure 3.4 When bacteriophage T1 infects wild-type *E. coli*, it binds to a receptor in the outer membrane, protein TonB (Table 3.1). After phage replication, the *E. coli* cell is lysed and new phage are released. A mutation in the *tonB* gene results in an altered (mutant) receptor to which T1 can no longer bind or eliminates the receptor, and so the cells are not infected and survive.

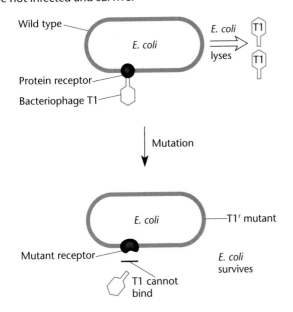

should be **clonal;** that is, one mutant bacterium will give rise to more, even in the absence of the selective agent (see the introductory chapter). However, if the directed-change hypothesis is correct, mutants should not be clonal, because they are not multiplying to form a colony before the selective agent is added. Instead, all the mutants first appear at the time the selective agent is added.

To detect clones of resistant bacteria, Newcombe analyzed cultures grown on agar plates. According to the random-mutation hypothesis, the number of colonies due to resistant mutants on an agar plate varies depending on whether the colonies are left alone or disturbed by having been spread out on the plate. When the colonies on a plate are not disturbed, all the descendants of a particular bacterium remain together in the same colony. However, if the colonies on a plate are disturbed, each resistant bacterium should give rise to a separate colony of resistant bacteria. Consequently, a spread plate has many more resistant colonies than an unspread plate. However, the directed-change hypothesis predicts that the mutants need not arise from each other (i.e., are not clonal), so that the number of resistant colonies on the undisturbed and disturbed plates should be about the same, because the resistant bacteria will have appeared only at the time the phage were added.

How Newcombe did his experiment is illustrated in Figure 3.6, and some of his actual data are given in Table 3.3. He first spread the same number of bacteria (an average of 5.1×10^4) on plates and incubated them. (He actually used many more plates than the six pictured in Figure 3.6, but this number serves as an illustration.)

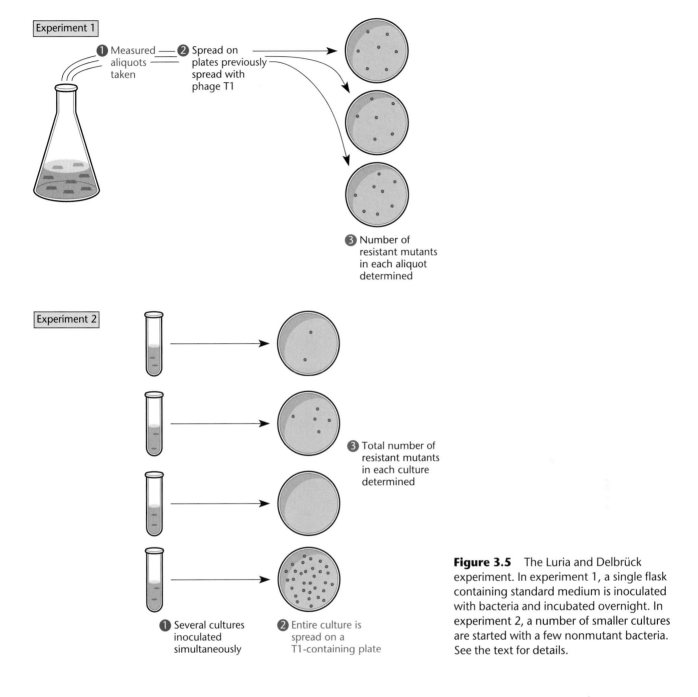

Figure 3.5   The Luria and Delbrück experiment. In experiment 1, a single flask containing standard medium is inoculated with bacteria and incubated overnight. In experiment 2, a number of smaller cultures are started with a few nonmutant bacteria. See the text for details.

After 5 h, he removed three of the plates and sprayed one with the virus without disturbing the bacteria. This is the unsp (*unsp*read) plate in Table 3.3. He treated the second plate in the same way, except that he spread the bacteria around before spraying them with the virus. This is the sp (*sp*read) plate in Table 3.3. He washed the bacteria off the third plate, diluted them, and plated them without the virus to determine the total number of bacteria on the plates at this time. After a 6-h incubation, he took the remaining three plates out of the incubator and subjected

them to the same treatment. He incubated all the plates overnight, and the next day he counted the colonies produced by phage-resistant mutants. The data in Table 3.3 show that the spread plates have many more resistant colonies than the corresponding unspread plates, a result supporting the random-mutation hypothesis. The difference became greater the longer the plates were incubated. After 5 h, only a few resistant colonies had appeared and there was not much difference between the unspread and spread plates (8 and 13 colonies,

TABLE 3.2	The Luria and Delbrück experiment		
Experiment 1		**Experiment 2**	
Aliquot no.	No. of resistant bacteria	Culture no.	No. of resistant bacteria
1	14	1	1
2	15	2	0
3	13	3	3
4	21	4	0
5	15	5	0
6	14	6	5
7	26	7	0
8	16	8	5
9	20	9	0
10	13	10	6
		11	107
		12	0
		13	0
		14	0
		15	1
		16	0
		17	0
		18	64
		19	0
		20	35

respectively) because the mutants had not had much time to multiply, and so each colony contained few resistant bacteria. However, after 6 h, the numbers of unspread and spread colonies were very different (49 and 3,719, respectively). Note also that the number of resistant bacteria, as measured by the resistant colonies on the spread plates, increased faster than the total population. At 5 h, there were 13 resistant bacteria in 2.6×10^8 total bacteria, so the fraction of resistant bacteria was $13/(2.6 \times 10^8)$, or 1 mutant for every 2×10^7 bacteria. By 6 h, there were 3,719 resistant mutants in 2.8×10^9 total bacteria, so that the fraction was $3,719/(2.8 \times 10^9)$, or 1 mutant for every 7×10^5 bacteria. Therefore, the fraction of resistant bacteria rose about 30-fold in just 1 h. In other words, the number of resistant mutants apparently increased about 30 times faster than the total number of

BOX 3.1

Statistical Analysis of the Number of Mutants per Culture

A simple statistical analysis shows that the number of mutants in experiment 2 of Luria and Delbrück does not follow a normal distribution. If it did, the variance would be approximately equal to the mean:

$$\text{Variance} = \sum_{i=1}^{n} \frac{\left(M_i - \bar{M}\right)^2}{n-1} = \text{mean} = \bar{M} = \sum_{i=1}^{n} \frac{M_i}{n}$$

where M_i is the number of mutants in each culture and n is the number of cultures.

In experiment 1 of Luria and Delbrück, the variance was 18.23 and the mean was 16.7, so they are approximately equal and vary owing to statistical fluctuations and pipetting errors. In experiment 2, however, the variance was 752.38 and the mean was 11.35—very different values. Therefore, the number of mutants per culture does *not* follow a normal distribution, and the result is not consistent with the directed-change hypothesis; however, it is consistent with the random-mutation or neo-Darwinian hypothesis.

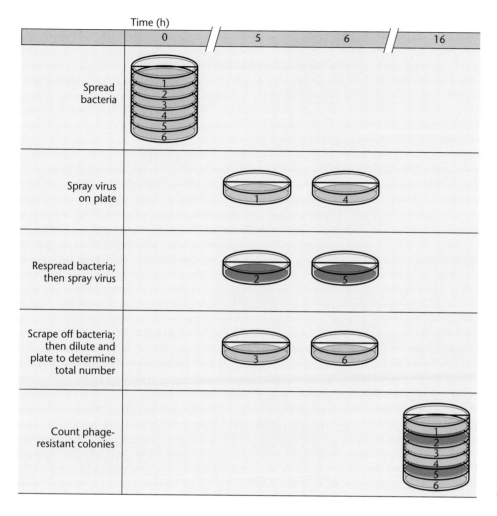

Figure 3.6 The Newcombe experiment. See the text for details.

bacteria during this hour. Much of this increase is not real and is presumably due to phenotypic lag, as discussed below. In any case, the number of resistant mutants does increase faster than the total population, fulfilling one prediction of the random-mutation hypothesis but not of the directed-change hypothesis, as explained below in the section on mutation rates.

The Lederbergs' Experiment

The experiments that really buried the directed-change hypothesis, at least as the sole explanation for some types of resistant bacterial mutants, were the replica-plating

experiments of the Lederbergs (see Lederberg and Lederberg, Suggested Reading). They spread millions of bacteria on a plate without an antibiotic and allowed the bacteria to form a lawn during overnight incubation. This plate was then replicated onto another plate containing the antibiotic. After incubating the antibiotic-containing plate, the Lederbergs could determine where antibiotic-resistant mutants had arisen on the original plate by aligning the two plates and marking the regions on the first plate where antibiotic-resistant mutants had grown on the second. They cut these regions out of the original plate, diluted the bacteria, and repeated the

TABLE 3.3	The Newcombe experiment		No. of resistant colonies	
Incubation time (h)	No. of bacteria plated	Ending no. of bacteria	Unspread	Spread
5	5.1×10^4	2.6×10^8 (plate 3)	8 (plate 1)	13 (plate 2)
6	5.1×10^4	2.8×10^9 (plate 6)	49 (plate 4)	3,719 (plate 5)

experiment. This time, there were many more resistant mutants than previously. Eventually, by repeating this process, they obtained a pure culture of bacteria, all of which were resistant to the antibiotic even though they had never been exposed to it! Therefore, the bacteria must have acquired the resistance independently of exposure to the antibiotic and passed the resistance on to their offspring. Therefore, the experiments of the Lederbergs also contributed to the proof that at least some types of mutants of *E. coli* arise through random mutations.

Mutation Rates

As defined above, a mutation is any heritable change in the DNA sequence of an organism, and we usually know that a mutation has occurred because of a phenotypic change in the organism. The **mutation rate** can be loosely defined as the chance of mutation to a particular phenotype. Mutation rates can differ because mutations to some phenotypes occur much more often than mutations to other phenotypes. When many possible different mutations in the DNA can give rise to a particular phenotype, the chance that a mutation to that phenotype will occur is relatively high. However, if only a very few types of mutations can cause a particular phenotype, the mutation rate for that phenotype is relatively low. For example, the spontaneous mutation rate for the His⁻ phenotype is hundreds of thousands of times higher than the mutation rate for Str^r. Approximately 11 gene products are required for histidine biosynthesis, and each has hundreds of amino acids, many of which are essential for activity. Changing any of these amino acids inactivates the enzyme. By contrast, streptomycin resistance results from a change in one of very few amino acids in a single ribosomal protein, S12, so that the mutation rate for streptomycin resistance is very low. Hence, a mutation to Str^r occurs spontaneously in about 1 in 10^{10} to 10^{11} cells whereas a mutation to His⁻ occurs in about 1 in 10^6 to 10^7 cells.

We can summarize how the mutation rate to a particular phenotype reflects the number and types of mutations that can cause the phenotype. Generally, if the mutation rate for a phenotype is high, the phenotype probably results from inactivation of the product of a gene or genes. If the mutation rate is low, the phenotype probably is due to a subtle change in the properties of a gene product. An extremely high mutation rate for a particular phenotype may indicate not a mutation but rather the loss of a plasmid or prophage or the occurrence of some programmed recombination event such as inversion of an invertible sequence. We discuss plasmids and prophages and other gene rearrangements in subsequent chapters.

Calculating Mutation Rates

To calculate anything, we must first define it. The mutation rate is usually defined as the chance of a mutation each time a cell grows and divides. This is a reasonable definition because, as discussed in chapter 1, DNA replicates once each time the cell divides, and most mutations occur during this process. The number of times a cell grows and divides in a culture is called the number of **cell generations** or **cell divisions**. This is not to be confused with **generation time**, which is the time it takes for a cell to grow and divide. The mutation rate is the number of mutations to a particular phenotype that have occurred in a growing culture divided by the total number of cell generations or cell divisions that have occurred in the culture during the same time.

DETERMINING THE NUMBER OF CELL GENERATIONS
The total number of cell generations that have occurred in an exponentially growing culture is easy to calculate and is simply the the total number of cells in the culture minus the number of cells in the starting innoculum. To understand this, see Figure 3.7. In this illustration, a culture that was started from one cell multiplies to form eight cells in seven cell divisions, or cell generations. This number equals the final number of cells (8) minus the number of cells at the beginning (1). In general, the number of cell generations that have occurred in the culture equals $N_2 - N_1$ if N_2 equals the number of cells at time 2 and N_1 equals the number of cells at time 1.

Figure 3.7   The number of cell generations or divisions (7) equals the total number of cells in an exponentially growing culture (8) minus the number at the beginning (1).

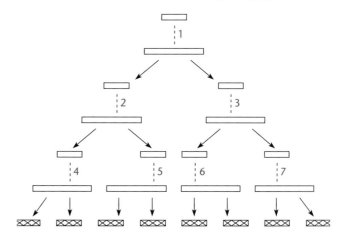

Therefore, from the definition, the mutation rate (a) is given by

$$a = \frac{m_2 - m_1}{N_2 - N_1}$$

where m_2 and m_1 are the number of mutations in the culture at time 2 and time 1, respectively.

Usually, a culture is started with a few cells and ends with many, so we can often ignore the initial cells and just call the number of cell generations N, where N is the total number of cells in the culture. Then the mutation rate equation can be simplified to

$$a = m/N$$

where m is the number of mutations that have occurred in the culture and N is the number of bacteria. This equation assumes that there were either no or at most an insignificant number of mutants in the culture when it was started, which is likely if the culture was started with only a few cells.

DETERMINING THE NUMBER OF MUTATIONS THAT HAVE OCCURRED IN A CULTURE

From the equations above, it looks as though it might be easy to calculate the mutation rate. The total number of mutations is simply divided by the total number of cells. The problem comes in determining the number of mutational events that have occurred in a culture, because mutant cells, not mutational events, are usually what are detected. Recall from Figure 3.3 that one mutant cell resulting from a single mutational event can give rise to many mutant cells, depending on when the mutation occurred during the growth of the culture. Therefore, one cannot determine the number of mutations in a culture merely by counting the number of mutant cells. However, in some cases the number of mutant cells can form the basis of a calculation of the number of mutational events and, by extension, the mutation rate. Some examples of such situations are below.

Using the Data of Luria and Delbrück To Calculate the Mutation Rate

Luria and Delbrück used data like those shown in Table 3.2 to calculate the mutation rate to T1 phage resistance. They assumed that even though the number of mutants per culture does not follow a normal distribution, the number of mutations per culture should do so, because each cell has the same chance of acquiring a mutation to T1 phage resistance each time it grows and divides. For convenience, the **Poisson distribution** can be used to approximate the normal distribution in a case like this. According to the Poisson distribution, if P_i is the probability of having i mutations in a culture, then

$$P_i = \frac{m^i e^{-m}}{i!}$$

where m is the average number of mutations per culture, the number they wanted to know. Therefore, if they knew how many cultures had a certain number of mutations, they could calculate the average number of mutations per culture. However, this is not as obvious as it seems. The data give the number of T1-phage-resistant mutants per culture but do not indicate how many of the cultures had one, two, three, or more mutations. For example, cultures with one mutant probably had one mutation, but others, even the one with 107 mutants, might also have had only one mutation. Only the number of cultures with zero mutations seems clear—those with zero mutants, or 11 of 20. Therefore, the probability of having zero mutations equals 11/20. Applying the formula for the Poisson distribution, the probability of having zero mutations is given by

$$\frac{11}{20} = \frac{m^0 e^{-m}}{0!} = \frac{1 \times e^{-m}}{1} = e^{-m}$$

and $m = -\ln 11/20 = 0.59$. Therefore, in this experiment, an average of 0.59 mutation occurred in each culture. From the equation for mutation rate,

$$a = m/N = 0.59/(5.6 \times 10^8) = 1.06 \times 10^{-9}$$

Therefore, there are 1.06×10^{-9} mutations per cell generation if there were a total of 5.6×10^8 total bacteria per culture. In other words, a mutation for phage T1 resistance occurs about once every 10^9, or every billion, times a cell divides.

There are a number of problems with measuring mutation rates this way, as indeed there are with any way of measuring mutation rates. One problem is phenotypic lag (see below). Some of the cultures with no mutants have presumably had some mutations, but they have not been expressed yet. Another problem is that the method is wasteful in that it ignores most of the data and considers only the cultures with no mutants. In their classic paper, Luria and Delbrück also derived an equation to estimate the mutation rate by using the number of mutants in all of the cultures. Others have subsequently proposed methods to measure mutation rates from such data (see, e.g., Lea and Coulson; and Jones et al., Suggested Reading).

Calculating the Mutation Rate from Newcombe's Data

Newcombe's data can be used more directly to calculate the number of mutations per culture (refer to Figure 3.6). On the unspread plate, each mutation gives rise to only one resistant colony. Therefore, the number of resistant colonies on the unspread plates equals the number of mutational events that have occurred at the time of incubation.

According to Newcombe's data in Table 3.3, from 5 to 6 h there were $49 - 8 = 41$ new resistant colonies on the unspread plates. Therefore, 41 mutations to phage resistance must have occurred during that time interval. During this time, the total number of bacteria went from 2.6×10^8 to 2.8×10^9 based on the total number of bacteria on the plates (Table 3.3). From the equation for mutation rate,

$$a = \frac{m_2 - m_1}{N_2 - N_1} = \frac{49 - 8}{2.8 \times 10^9 - 2.6 \times 10^8}$$

the mutation rate to T1 resistance is $41/(2.54 \times 10^9) = 1.6 \times 10^{-8}$ mutation per cell generation. In other words, Newcombe's data indicate that a mutation to resistance to phage T1 occurs a little more than once every hundred million times a cell divides. Notice that Newcombe's data give a mutation rate about 10 times higher than that derived from the data of Luria and Delbrück. This discrepancy can be explained because both methods are subject to phenotypic lag, as we explain later.

Using the Increase in the Fraction of Mutants To Measure Mutation Rates

As mentioned above, Newcombe's data fulfilled one prediction of the random-mutation hypothesis, i.e., that the number of mutants should increase faster than the total population. In other words, the fraction of mutants in the population should increase as the population grows. At first, it seems surprising that the total number of mutants increases faster than the total population until one thinks about where mutants come from. If the multiplication of old mutants were the only source of mutants, the fraction of mutants would remain constant or even drop if the mutants did not multiply as rapidly as the normal type (which is often the case). However, new mutations occur constantly, and their progeny are also multiplying. Therefore, new mutations are continuously adding to the total number of mutants.

This fact can also be used to measure mutation rates. The higher the mutation rate, the faster the proportion of mutants will increase (Figure 3.8). In fact, if we plot the fraction of mutants (M/N) against time (in doubling

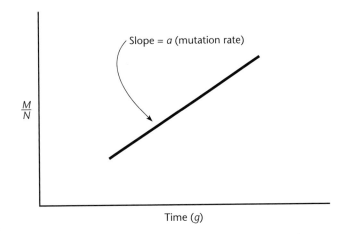

Figure 3.8   The fraction of mutants increases as a culture multiplies, and the slope is the mutation rate. M is the number of mutants, N is the total number of cells, and Time is the time in generation times which is the total time elapsed divided by the time it takes the culture to double in mass (i.e., the doubling time [g]).

times), as in Figure 3.8, the slope of this curve is the mutation rate. In theory, this fact could be used to calculate mutation rates. In practice, however, mutation rates are usually low and the number of bacteria we can conveniently work with are relatively small so each new mutation makes too large a contribution to the number of mutants and we do not get a straight line. To make this method practicable, we would have to work with trillions of bacteria in a large chemostat.

The fact that other mutations are causing some mutants to become wild type again (in a process known as reversion [see later sections]) also affects the results shown in Figure 3.8. However, reversion of mutants becomes significant only when the number of mutants is very large and the number of mutants multiplied by the mutation rate back to the wild type (called the reversion rate) begins to approximate the rate at which new mutants are being formed, which is the forward mutation rate times the number of nonmutant bacteria. At earlier stages of culture growth, the vast majority of bacteria are nonmutant, so the latter product is much larger than the former product. Also, for reasons discussed later in this chapter, reversion rates for many types of mutants are much lower than forward mutation rates. Therefore, the contribution to the mutation rate of the reversion of mutants to the wild type can generally be ignored, at least in the early stages of growing a culture.

PHENOTYPIC LAG

Some of the difficulty in accurately determining mutation rates results from **phenotypic lag**. Most phenotypes are not immediately evident after a mutation but appear some time later. The length of the lag depends on the molecular basis for the phenotype.

Mutations to phage T1 resistance would be expected to show a phenotypic lag. Recall that resistance to phage T1 derives from the alteration or loss of the protein product of a gene, *tonB*. This is an outer membrane protein to which the phage binds to start the infection. The mutant bacteria survive because they lack the wild-type protein in their outer membrane, so they cannot be infected by the virus. However, when the mutation first occurs, the mutant bacteria still have wild-type TonB in their outer membrane and so are T1 sensitive. Only after a few generations is all the wild-type TonB diluted out, so that the progeny cells can no longer absorb virus and are resistant.

It may seem that phenotypic lag would not be a serious problem in measuring mutation rates, because bacteria go through so many generations in a culture. However, in an exponentially growing culture, half of all mutations occur in the last generation time. Therefore, when the bacteria are plated with the selective agent, in our example phage T1, more than half of the mutations are not counted because they have not yet been expressed and the bacteria are still sensitive. Obviously, ignoring more than half of all mutations will introduce a significant error in the mutation rate. Phenotypic lag therefore introduces a significant error into the mutation rate calculated by either the Luria and Delbrück or Newcombe method but not in the rate of increase of the fraction of the mutants in a culture as shown in Figure 3.8, provided that the culture is large enough.

Some methods for measuring mutation rates are also influenced by differences in the growth rate of mutants relative to the original or wild type. Quite often, mutants grow measurably more slowly than the wild type even under nonselective conditions. Note that the two methods that have been described for determining mutation rates are not affected by such differences.

PRACTICAL IMPLICATIONS OF POPULATION GENETICS

The fact that the proportion of mutants increases as the culture grows presents both opportunities and problems in genetics. This fact can be advantageous in the isolation of a rare mutant such as one resistant to streptomycin. If we grow a culture from a few bacteria and plate 10^9 bacteria on agar containing streptomycin, we might not find any resistant mutants, since they occur at a frequency of only about 1 in 10^{11} cell generations. However, if we add a large number of bacteria to fresh broth, grow the broth culture to saturation, and then repeat this process a few times, the fraction of streptomycin-resistant mutants will increase. Then when we plate 10^9 bacteria, we may find many streptomycin-resistant mutants.

Because the fraction of all types of mutants increases as the culture multiplies, if we allow a culture to go through enough generations, it will become a veritable "zoo" of different kinds of mutants—virus resistant, antibiotic resistant, auxotrophic, and so on. To deal with this problem, most researchers store cultures under nongrowth conditions (e.g., as spores or lyophilized cells or in a freezer) that still maintain cell viability. An alternative is to periodically colony purify bacteria in the culture to continuously isolate the progeny of a single cell (see the introductory chapter). The progeny of a single cell are not likely to be mutated in a way that could confound our experiments.

Summary

Two very important points emerge from this discussion of mutations and mutation rates. First, measuring mutation rates is not as simple as one might think. The mutation rate is not simply the number of mutants with a particular phenotype divided by the total number of organisms in the culture. To calculate the mutation rate, we must use special methods to measure the number of mutations or must apply statistical methods to the data. Second, mutants of all kinds accumulate in cultures as we grow them. Consequently, it is best to store bacteria without growing them or to periodically isolate a single cell before mutants have had a chance to become such a significant proportion of the total population that we are apt to pick one.

Types of Mutations

As defined above, any heritable change in the sequence of nucleotides in DNA is a mutation. A single base pair may be changed, deleted, or inserted; a large number of base pairs may be deleted or inserted; or a large region of the DNA may be duplicated or inverted. Regardless of how many base pairs are affected, a mutation is considered to be a **single mutation** if only one error in replication, recombination, or repair has altered the DNA sequence.

As discussed earlier in this chapter, to be considered a mutation, the change in the DNA sequence must be heriditable. Damage to DNA, by itself, is not a mutation, but a mutation can occur when the cell attempts to repair damage or replicate over it and a strand of DNA is synthesized that is not completely complementary to the original sequence. The wrong sequence is then faithfully replicated through subsequent generations and thus becomes a mutation.

Lethal changes in the DNA sequence (as also mentioned earlier) do occur but cannot usually be scored as mutations since the cells do not survive. Ordinarily, to be scored as a mutation, the change must be heritable and so cannot be lethal. For example, deletion of a gene required for growth is usually lethal because bacteria are haploid and usually have only one gene of each type. If the gene is deleted, the organism cannot multiply and will die without leaving progeny. Therefore, such a deletion is not scored as a mutation.

The properties and causes of the different types of mutations are probably not very different in all organisms, but they are more easily studied with bacteria. A geneticist can often make an educated guess about what type of mutation is causing a mutant phenotype merely by observing some of its properties.

One property that distinguishes mutations is whether they are **leaky**. The term "leaky" means something very specific in genetics. It means that in spite of the mutation, the gene product still retains some activity.

Another property of mutations is whether they **revert**. If the sequence has been changed to a different sequence, it can often be changed back to the original sequence by a subsequent mutation. The organism in which a mutation has reverted is called a **revertant,** and the **reversion rate** is the rate at which the mutated sequence in DNA returns to the original wild-type sequence.

Usually, the reversion rate is much lower than the mutation rate that gave rise to the mutant phenotype. As an illustration, consider the previously discussed example, histidine auxotrophy (His⁻). Any mutation that inactivates any of the approximately 11 genes whose products are required to make histidine will cause a His⁻ phenotype. Since thousands of changes can result in this phenotype, the mutation rate for His⁻ is relatively high. However, once a *his* mutation has occurred, the mutation can revert only through a change in the mutated sequence that restores the original sequence. Everything else being equal, the reversion rate to His⁺ revertants would be expected to be thousands of times lower than the forward mutation rate to His⁻.

Some types of revertants are very easy to detect. For example, His⁺ revertants can be obtained by plating large numbers of His⁻ mutants on a plate with all the growth requirements except histidine. Most of the bacteria cannot multiply to form a colony. However, any His⁺ revertants in the population will multiply to form a colony (see Figure 3.21). The appearance of His⁺ colonies when large numbers of a His⁻ mutant are plated would be evidence that the *his* mutation can revert.

Base Pair Changes

A **base pair change** is when one base pair in DNA, for example a GC pair, is changed into another base pair, for example an AT pair. Base pair changes can be classified as **transitions** or **transversions** (Figure 3.9). In a

transition, the purine (A and G) in a base pair is replaced by the other purine and the pyrimidine (C and T) is replaced by the other pyrimidine. Thus, an AT pair would become a GC pair or a CG would become a TA. In a transversion, by contrast, the purines change into pyrimidines and vice versa. For example, a GC could become a TA, or a CG could become an AT.

BASE PAIR CHANGES RESULTING FROM MISPAIRING

Base pair changes can be the result of mistakes in replication, recombination, or repair. Figure 3.10A shows an example of mispairing during replication. In this example, a T instead of the usual C is mistakenly placed opposite a G as the DNA replicates. In the next replication, this T usually pairs correctly with an A, causing a GC-to-AT transition in one of the two daughter DNAs. Mistakes in pairing may occur because the bases are sometimes in a different form called the enol form, which causes them to pair differently (Figure 3.10B).

Mispairing between a purine and a pyrimidine causes a transition, whereas mispairing between two purines or two pyrimidines causes a transversion. Because a pyrimidine in the enol form still pairs with a purine and a purine in the enol form still pairs with a pyrimidine, mispairing during replication usually leads to transition mutations. Furthermore, all four bases can undergo the shift to the enol form, and either the base in the DNA template or the incoming base can be in the enol form and cause mispairing. Thus, the thymine in the enol form pictured in Figure 3.10B might be in the template, in which case the transition would be AT to GC, or it could be the incoming base, resulting in a GC-to-AT transition.

Mistakes during replication leading to mutations are not random, however, and some sites are much more prone to base pair changes than are others. Mutation-prone sites are called **hot spots.** Mispairing occurs fairly often during replication, and it is an obvious advantage for the cell to reduce the number of base pair change mutations that occur during replication. In chapter 1, we discussed some of the mechanisms used by cells to reduce these base pair changes, including editing and methyl-directed mismatch repair.

DEAMINATION OF BASES IN DNA

Deamination, or the removal of an amino group, can also cause base pair changes. Cytosine is particularly susceptible to deamination; it becomes uracil when deaminated, since the only difference between cytosine and uracil is the amino group at the 6 position of the cytosine ring (see chapters 1 and 2 for structures). However, uracil pairs with adenine instead of guanine. Therefore, unless the uracils due to deamination of cytosines are

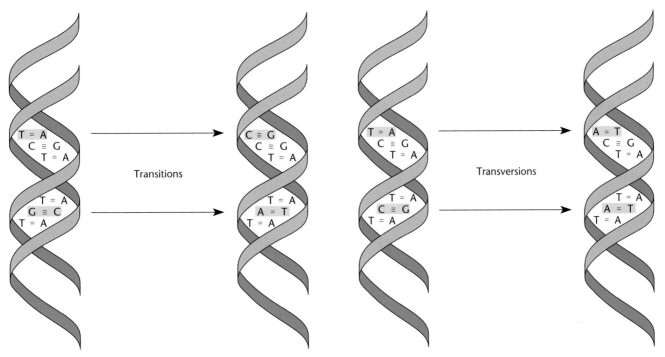

Figure 3.9 Transitions versus transversions. The mutations are shown in purple.

removed from DNA, they cause CG-to-TA transitions the next time the DNA replicates.

Because of the special problems caused by deamination of cytosine, cells have evolved a special mechanism for removing uracil from DNA whenever it appears (Figure 3.11). An enzyme called uracil-*N*-glycosylase, the product of the *ung* gene in *E. coli*, recognizes the uracil as unusual in DNA and removes the uracil base. The DNA strand in the region where the uracil was removed is then degraded and resynthesized, and the correct cytosine is inserted opposite the guanine. As expected, *ung* mutants of *E. coli* show high rates of spontaneous mutagenesis, and most of the mutations are GC-to-AT transitions. All organisms have the problem of cytosine deamination in their DNA, explaining why the testicles of warm-blooded animals including mammals are external where the average temperature is lower and deaminations are less frequent.

OXIDATION OF BASES

Reactive forms of oxygen such as peroxides and free radicals are given off as by-products of oxidative metabolism, and these forms can react with and alter the bases in DNA. A common example is the altered guanine base, 8-oxoG, which sometimes mistakenly pairs with adenine instead of cytosine, causing GC-to-TA or AT-to-CG transversion mutations. Repair systems specific to damage

such as deamination and oxidation are discussed in more detail in chapter 11.

CONSEQUENCES OF BASE PAIR CHANGES

Whether a base pair change causes a detectable phenotype depends, of course, on where the mutation occurs and what the actual change is. Even a change in an open reading frame (ORF) that encodes a polypeptide may not result in an altered protein. If the mutated base is the third in a codon, the amino acid inserted into the protein may not be different because of the degeneracy of the code (see the section on the genetic code in chapter 2). Mutations in the coding region of a gene that do not change the amino acid sequence of the polypeptide product are called silent mutations.

The change may also occur in a region that does not encode a polypeptide but, rather, is a regulatory sequence such as an operator or promoter. Alternatively, the mutation may occur in a region that has no detectable function. We first discuss mutations that change the coding region of a polypeptide.

MISSENSE MUTATIONS

Most base pair changes in bacterial DNA cause one amino acid in a polypeptide to be replaced by another. These mutations are called **missense mutations** (Figure 3.12). However, even a missense mutation that changes an

A

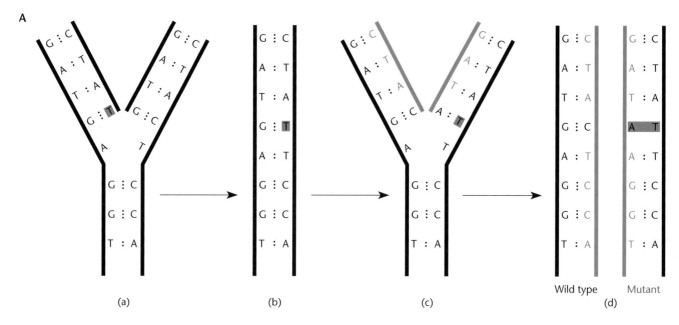

(a)   (b)   (c)   (d)

Wild type   Mutant

B

Deoxyribose

Guanine   Thymine (enol form)

Figure 3.10   (A) A mispairing during replication can lead to a base pair change in the DNA (shown in purple). The purple line shows the second generation of newly replicated DNA. The mispaired base is boxed. (B) Pairing between guanine and an alternate enol form of thymine can cause G-T mismatches.

amino acid in a protein does not always inactivate the protein. If the original and new amino acids have similar properties, the change may have little or no effect on the activity of a protein. For example, a missense mutation changing an acidic amino acid, such as glutamate, into another acidic amino acid, such as aspartate, may have less effect on the functioning of the protein than does a mutation that substitutes a basic amino acid, such as arginine, for an acidic one. The consequences also depend on which amino acid is changed. Certain amino acids in any given protein sequence are more essential to activity than others, and a change at one position can have much more effect than a change elsewhere. Investigators often use this fact to determine which amino acids are essential for activity in different proteins. Some methods to change specific amino acids in a protein, called site-specific mutagenesis, are discussed in chapter 1, and other methods are introduced in later chapters.

NONSENSE MUTATIONS

Instead of changing a codon into one coding for a different amino acid, base pair changes sometimes produce one of the nonsense codons, UAA, UAG, or UGA. These changes are called **nonsense mutations.**

While nonsense mutations are base pair changes and have the same causes as other base pair changes, the consequences are very different. Because the nonsense codons are normally used to signify the end of a gene, these codons are normally recognized by release factors (see chapter 2), which cause release of the translating ribosome and the polypeptide chain. Therefore, if a mutation to one of the nonsense codons occurs in an ORF for a protein, the protein translation terminates prematurely at the site of the nonsense codon and the shortened or truncated polypeptide is released from the ribosome (Figure 3.13). For this reason, nonsense mutations are sometimes called "chain-terminating mutations." These

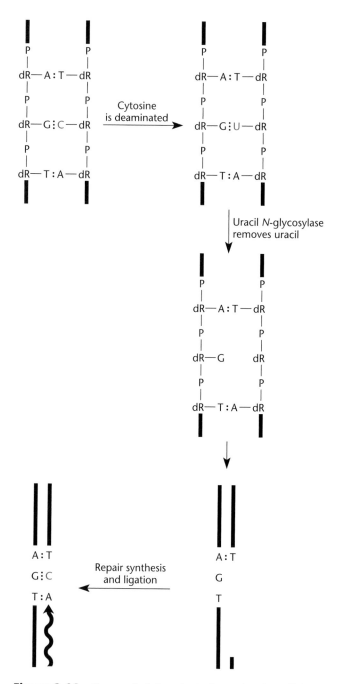

Figure 3.11 Removal of deaminated cytosine (uracil) from DNA by uracil-*N*-glycosylase. The uracil base is cleaved off, and the DNA strand is degraded and resynthesized with cytosine at that position.

mutations almost always inactivate the protein product of the gene in which they occur. If, however, they occur in a noncoding region of the DNA or in a region that encodes an RNA rather than a protein, such as a gene for a tRNA, they are indistinguishable from other base pair changes.

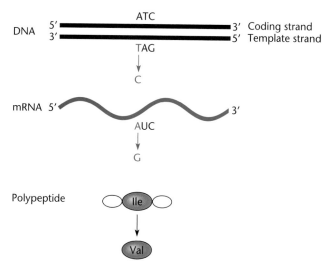

Figure 3.12 Missense mutation. A mutation that changes T to C in the DNA template strand will result in an A-to-G change in the mRNA. The mutant codon GUC is translated as valine instead of isoleucine.

The three nonsense codons—and their corresponding mutations—are sometimes referred to by color designations: **amber** for UAG, **ochre** for UAA, and **opal** for UGA. These names have nothing to do with the effects of the mutation. Rather, when nonsense mutations were

Figure 3.13 Nonsense mutation. Changing the CAA codon, encoding glutamine (Gln), to UAA, a nonsense codon, causes truncation of the polypeptide gene product.

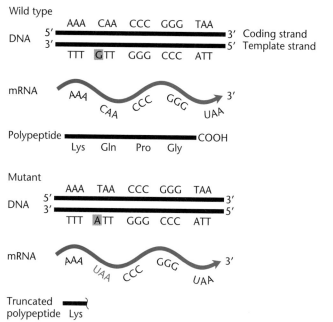

first discovered at the California Institute of Technology, their molecular basis was unknown. The investigators thought that descriptive names might be confusing later on if their interpretations were wrong, so they followed the lead of physicists with their "quarks" and "barns." The first nonsense mutations to be discovered, to UAG, were called amber mutations. Following suit, UAA and UGA mutations were also named after colors—ochre and opal, respectively.

PROPERTIES OF BASE PAIR CHANGE MUTATIONS

Base pair changes are often leaky. A substituted amino acid may not work nearly as well as the original at that position in the chain, but the protein can retain some activity. Even nonsense mutations are usually somewhat leaky because sometimes an amino acid is inserted for a nonsense codon, albeit at a low frequency. In wild-type *E. coli*, UGA tends to be most leaky, followed by UAG; the nonsense codon UAA tends to be the least leaky.

Base pair mutations also revert. If the base pair has been changed to a different base pair, it can also be changed back to the original base pair by a subsequent mutation. Moreover, base pair changes are a type of point mutation because they map to a particular "point" on the DNA, as discussed later.

Frameshift Mutations

A high percentage of all spontaneous mutations are **frameshift mutations** (Figure 3.14). This type of mutation occurs when a base pair or a few base pairs are removed from or added to the DNA, causing a shift in the reading frame if they occur in an ORF encoding a polypeptide. Because the code is three lettered, any addition or subtraction that is not a multiple of 3 causes a frameshift in the translation of the remainder of the gene. For example, adding or subtracting 1, 2, or 4 base pairs

causes a frameshift, but adding or subtracting 3 or 6 base pairs does not. Mutations that remove or add base pairs are usually called frameshift mutations even if they do not occur in an ORF and do not actually cause a frameshift in the translation of a polypeptide.

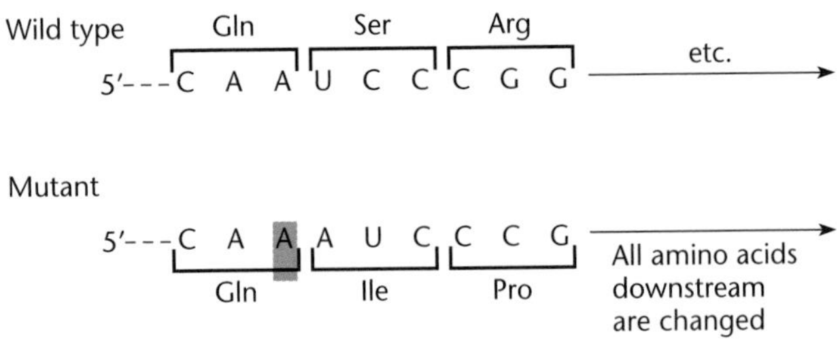

CAUSES OF FRAMESHIFT MUTATIONS

Spontaneous frameshift mutations often occur where there is a short repeated sequence that can slip. As an example, Figure 3.15 shows a string of AT base pairs in the DNA. Since any one of the A's in one strand can pair with any T in the other strand, the two strands could slip with respect to each other, as in the illustration. Slippage during replication could leave one T unpaired, and an AT base pair would be left out on the other strand when it replicates. Alternatively, the slippage could occur before the base was added, and an extra AT base pair could appear in one strand as shown.

PROPERTIES OF FRAMESHIFT MUTATIONS

Frameshift mutations are usually not leaky and almost always inactivate the protein, because every amino acid in the protein past the point of the mutation is wrong. The protein is usually also truncated, because a nonsense codon is usually encountered while the gene is being translated in the wrong frame. Because, in general, 3 of the 64 codons are the nonsense codons, one of these should be encountered by chance about every 20 codons when the region is being translated in the wrong frame.

Another property of frameshift mutations is that they revert. If a base pair has been subtracted, one can be added to restore the correct reading frame and vice versa. More often, frameshift mutations do not revert but are suppressed by the addition or subtraction of a base pair close to the site of the original mutation that restores the original reading frame. This means of frameshift suppression is discussed later in this chapter. Finally, frameshift mutations are a type of point mutation.

Some types of pathogenic bacteria apparently take advantage of the frequency and high reversion rate of frameshift mutations to avoid host immune systems. In such bacteria, genes required for the synthesis of cell surface components that are recognized by host immune systems often have repeated sequences. Consequently, these genes can be turned off and on by frameshift mutations and subsequent reversion. Frameshift mutations may aid in the synthesis of virulence gene products by *Bordetella pertussis*, the causative agent of whooping cough. Frameshift mutations are also used to reversibly inactivate genes of *Haemophilus influenzae* and *Neisseria gonorrhoeae*, which cause spinal meningitis and gonorrhea, repectively.

Figure 3.14 Frameshift mutation. The wild-type mRNA is translated glutamine (Gln)-serine (Ser)-arginine (Arg)-, etc. Addition of an A (boxed) would shift the reading frame, so that the codons would be translated glutamine (Gln)-isoleucine (Ile)-proline (Pro)-, etc., with all downstream amino acids being changed.

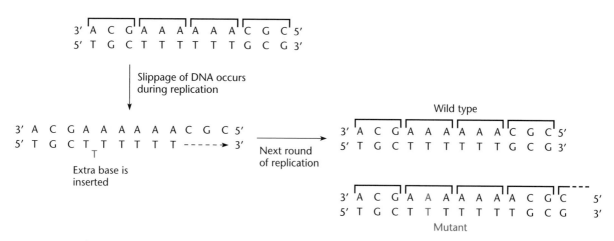

Figure 3.15 Slippage of DNA at a repeated sequence can cause a frameshift mutation.

Deletion Mutations

Deletion mutations may be very long, removing thousands of base pairs and possibly many genes. Often, the only limitation on these mutations in bacterial DNA is that they cannot delete any essential genes, since haploid bacteria generally only have one copy of each gene. Some deletions in bacteria can be quite long, however, since bacterial genomes often possess long stretches of genes that can be deleted without causing a loss of cell viability. Deletion mutations often constitute a high percentage of spontaneous mutations; for example, in *E. coli*, almost 5% of all spontaneous mutations are deletions.

CAUSES OF DELETIONS

Deletions can be caused by recombination between different regions of the DNA. Recombination usually occurs between the same regions in two DNAs because the two DNAs are identical in the same region. However, recombination can sometimes mistakenly occur between two different regions if they are similar enough in sequence. In the latter case, the strands of two DNA molecules are broken and rejoined in new combinations, hence the name **recombination**. This process is discussed later in this chapter and in detail in chapter 10.

As shown in Figure 3.16, recombination can give rise to deletions in two ways. Deletions result from recombination between two sequences that are **direct repeats**, that is, two sequences that are similar or identical when read in the 5′-to-3′ direction on the same strand of DNA. However, these repeated sequences can be on different or the same DNA molecule.

Figure 3.16 shows that a deletion can occur when different copies of a direct repeated sequence in daughter DNAs mistakenly pair with each other. The two regions are then broken and rejoined, removing the sequence between the two direct repeats as shown. This is sometimes called "unequal crossing over," because the two DNAs are not equally aligned during the recombination. Alternatively, the two direct repeats on the same DNA molecule could pair, "looping out" the intervening sequences. Breaking and rejoining the DNA would remove the looped-out sequences as shown. For purposes of illustration, the directly repeated sequences shown in the figure are much shorter than would normally be required for recombination. Depending on the recombination sequence and the organism, these repeated sequences can be quite short, but the frequency of recombination causing deletions is much higher with longer repeated sequences. Usually, direct repeats that promote mistaken recombination are hundreds or even thousands of base pairs long. Bacterial DNA contains several types of repeats, the longest of which include insertion sequence (IS) elements and the rRNA genes, which are often repeated in many places in the DNA. IS elements are transposons and are discussed in chapter 9.

PROPERTIES OF DELETION MUTATIONS

Deletions have very distinctive properties. They are usually not leaky; deleting part or all of a gene usually totally inactivates the gene product. Mutations that inactivate more than one gene simultaneously are most often deletions. Moreover, deletion mutations sometimes fuse one gene to another, sometimes putting one gene under the control of another.

The most distinctive property of long deletion mutations is that they never revert. Every other type of mutation reverts at some frequency, but for a deletion to revert, the missing sequence would somehow have to be found and reinserted. Deletions also behave differently from point mutations in genetic crosses, not mapping to

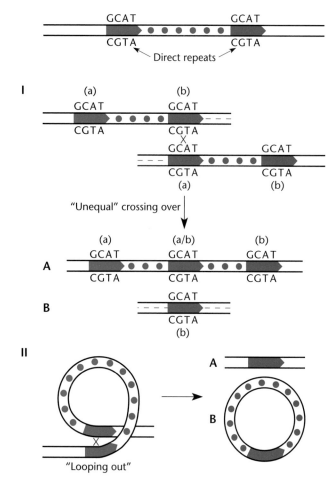

Figure 3.16 Recombination between directly repeated sequences can cause deletion mutations. (I) The recombination can occur between repeated sequences in different DNAs, resulting in a duplication (A) or a deletion (B). (II) Alternatively, it can occur between repeated sequences in the same DNA, resulting in a deletion (A) and the looped-out deleted segment (B).

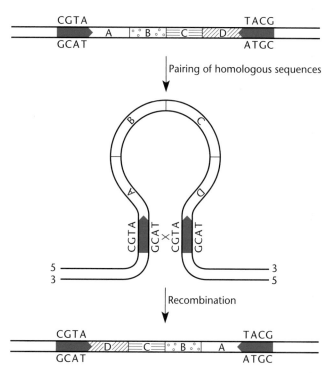

Figure 3.17 Recombination between inverted repeats can cause inversion mutations. The order of genes within the inversion is reversed after the recombination.

a single point. This property can be very useful in some types of genetic mapping, as discussed in later chapters.

NAMING DELETION MUTATIONS
Deletion mutations are named differently from other mutations. The Greek letter for "d," Δ (delta), for *deletion*, is written in front of the gene designation and allele number, e.g., Δ*his8*. Often, deletions remove more than one gene, and so, if known, the deleted regions are shown, followed by a number to indicate the particular deletion. For example, Δ(*lac-proAB*)195 is deletion number 195 extending through the *lac* and *proAB* genes on the *E. coli* chromosome. Often a deletion removes one or more known genes but extends into a region of unknown genes, so that the endpoints of the deletion are not known. In this case, the deletion is often named after the known gene. For example, the Δ*his8* deletion may delete the entire *his* operon but also extend an unknown distance into neighboring genes.

Inversion Mutations

Sometimes a DNA sequence is not removed, as in a deletion, but, rather, is flipped over, or **inverted.** After such an **inversion,** all the genes in the inverted region face in the opposite orientation.

CAUSES OF INVERSIONS
Inversions are caused in the same way as deletions, by recombination between repeats. However, recombination between inverted sequences rather than directly repeated sequences produces inversions. Inverted repeats read almost the same in the 5′-to-3′ direction on opposite strands (see chapter 1). Also unlike deletions, the recombination that produces inversions must occur between two regions on the same DNA (Figure 3.17).

PROPERTIES OF INVERSION MUTATIONS
Unlike deletions, inversion mutations can generally revert. Recombination between the inverted repeats that caused the mutation will "reinvert" the affected sequence,

recreating the original order. However, an inversion might occur between very short inverted repeats or repeats that are not exactly the same. Then the recombination event would have to occur between the exact bases involved in the first recombination to restore the correct sequence. Such a recombination could be a very rare event, and reversions of such a mutation would be very rare.

Inversion mutations often cause no phenotype. If the inversion involves a longer sequence, including many genes, generally the only affected regions are those in the **inversion junctions,** where the recombination occurred. Most of the genes in the inverted region are still intact, although they are present in the reverse order. Consequently, even very long inversion mutations often cause no obvious phenotypes. Like deletions, inversion mutations sometimes fuse one gene to another gene. This property provides a mechanism for detecting them. The occurrence of inversions in evolution is discussed in Box 3.2.

NAMING INVERSIONS

A mutation known to be an inversion is given the letters IN followed by the genes in which the inversion junctions occur, provided that these are known, followed by the number of the mutation. For example, IN (*purB-trpA*) *3* is inversion number 3 in which the inverted region extends from somewhere within the gene *purB* to somewhere within the gene *trpA*.

Tandem Duplication Mutations

In a duplication mutation, a sequence is copied from one region of the DNA to another. The most common, a **tandem duplication,** consists of a sequence immediately followed by its duplicate. Tandem duplications occur frequently and can be very long.

CAUSES OF TANDEM DUPLICATIONS

Like deletions, tandem duplications can result from recombination between directly repeated sequences in DNA. In fact, as shown in Figure 3.18, they are probably often created at the same time as a deletion. Pairing between two directly repeated sequences in different DNAs, followed by recombination, can give rise to a tandem duplication and a deletion as the two products.

PROPERTIES OF TANDEM DUPLICATION MUTATIONS

Although the mechanism by which tandem duplications arise is similar to the mechanism that creates deletions and inversions, the properties of tandem duplications are unique. Tandem duplication mutations that occur within a single gene usually inactivate the gene and are not leaky. However, if the duplicated region is long enough

to include one or more genes, no genes are inactivated, including those in which the recombination occurred—the **duplication junctions.** This conclusion may seem surprising, but consider the example shown in Figure 3.18. Direct repeats in genes A and C on different DNAs pair with each other. The repeats in the two DNAs are then broken and rejoined to each other, creating a duplication in one DNA and a deletion in the other. Only part of gene A exists in the duplicate, but an entire gene A exists upstream. Conversely, only part of gene C exists upstream but the entire gene exists in the duplicate. There are now two copies of gene B, both of which are unaltered. Therefore, intact genes A, B, and C still exist after the duplication, and there would be no indication that a mutation had even occurred unless there happened to be a phenotype associated with the presence of two copies of gene B or potential altered expression of genes A and B. Like deletions and inversions, duplications sometimes fuse two genes to put expression of one gene under the control of a different gene. In the example in Figure 3.18, part of gene A has been fused to gene C, which might put genes A and B under the control of the promoter for gene C.

The most characteristic property of tandem duplications is that they are very unstable and revert at a high frequency. Even though the mistaken recombinations that lead to a duplication are usually rare, recombination anywhere within the duplicated segments can delete them, restoring the original sequence. The instability of tandem duplications is often the salient feature that allows their identification. Later in the chapter, we discuss an actual case of the genetic analysis of duplications in the *his* operon of *Salmonella*.

ROLE OF TANDEM DUPLICATION MUTATIONS IN EVOLUTION

Tandem duplication mutations may play an important role in evolution. Ordinarily, a gene cannot change without loss of its original function, and if the lost function was a necessary one, the organism will not survive. However, when a duplication has occurred, there are two copies of the genes in the duplicated region, and now one of these is free to evolve to a different function. This mechanism would allow organisms to acquire more genes and become more complex. However, how tandem duplications could persist long enough for some of the duplicated genes to evolve is not clear.

Insertion Mutations

Insertion mutations are caused by the insertion of a large piece of DNA into a region, usually by transposons "hopping" into the DNA. **Transposons** are DNA elements that can promote their own movement from

BOX 3.2

Inversions and the Genetic Map

Even a single large inversion mutation causes a dramatic change in the genetic map, or order of genes in the DNA, of an organism. The order of all the genes is reversed between the sites of the recombination that led to the inversion. We would also expect inversions to be fairly frequent because repeated sequences often exist in inverted orientation with respect to each other. In spite of this, inversions seem to have occurred very infrequently in evolution. As evidence, consider the genetic maps of *S. enterica* serovar Typhimurium and *E. coli*. These bacteria presumably diverged billions of generations ago. Nevertheless, the maps are very similar except for one short inverted sequence between about 25 and 27 min on the *E. coli* map. At present, we can only speculate on why the genetic maps are so highly conserved. Perhaps organisms with this gene order have some selective advantage, or perhaps other sequences in the DNA cannot be inverted without disadvantaging the organism.

Termination of chromosome replication after sites like *terA* and *terB* (see chapter 1) may help explain why so few large inversions seem to have occurred in the evolution of bacteria. There may be sequences that resemble *terA* and *terB* distributed around the chromosome, but because they are on the wrong strand, they do not cause termination. However, an inversion mutation would reverse their orientation, so that if the *terA* site were preceded by a *terB*-like site, the DNA between the two sites would not be replicated. This situation would be lethal. However, there are probably other explanations for the rarity of large inversions, and there may be other sequences whose orientation relative to the origin of replication must be conserved.

Reference
Mahan, M. J., and J. R. Roth. 1991. Ability of a bacterial chromosome to invert is dictated by included material rather than flanking sequences. *Genetics* **129**:1021–1032.

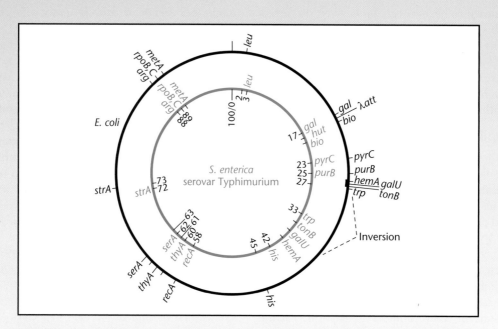

Genetic maps of *S. enterica* serovar Typhimurium and *E. coli*, showing a high degree of conservation. The region from *hemA* to the 40-min position is inverted in *E. coli* relative to that in serovar Typhimurium.

one place in the DNA to another. In doing so, they create insertion mutations. Although these elements are usually thousands of base pairs long, sometimes only part of a transposon moves, or hops, producing a shorter insertion. Indeed, the movement of relatively short trans-

posons, known as **insertion elements**, produces the majority of insertion mutations. These elements, which are only about 1,000 bp long, carry no easily identifiable genes. Most bacteria carry several insertion elements in their chromosome (see chapter 9).

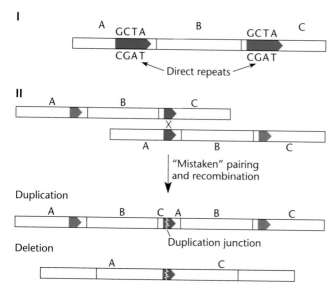

Figure 3.18 Formation of tandem duplication mutations by recombination between directly repeated sequences on different DNA molecules. A deletion, the reciprocal recombinant, is created at the same time as the duplication.

PROPERTIES OF INSERTION MUTATIONS

Insertion of DNA into a gene almost always inactivates the gene; therefore, insertion mutations are usually not leaky. Transposons also contain many transcription termination sites, and so their insertion results in polarity (see chapter 2), which prevents the transcription of genes normally copied onto the same mRNA as the gene with the insertion. Finally, insertion mutations seldom revert, because the inserted DNA must be precisely removed, with no DNA sequences remaining. These last two unusual properties of insertion mutations led to their discovery.

SELECTING INSERTION MUTATIONS

A significant percentage of all spontaneous mutations are insertion mutations, but their phenotypes are difficult to distinguish from those of other types of mutations. However, transposons can serve as useful tools in genetics experiments, because many carry a selectable gene, such as one for antibiotic resistance. The insertion of such a transposon into a cell's DNA makes the mutant cells antibiotic resistant and easy to isolate. Moreover, transposon insertions are relatively easy to map, both genetically and physically. The methods of transposon mutagenesis are central to bacterial molecular genetics and biotechnology and so are discussed in some detail in later chapters.

NAMING INSERTION MUTATIONS

An insertion mutation in a particular gene is represented by the gene name, two colons, and the name of the insertion. For example, *galK*::Tn*5* denotes the insertion of the transposon Tn*5* into the *galK* gene. If more than one Tn*5* insertion exists in *galK*, the mutations can be numbered to distinguish them (e.g., *galK35*::Tn*5*). When insertion mutations are constructed for use in genetic experiments, they are denoted with the capital Greek letter Ω (omega) followed by the name of the insertion. For example, pBR322Ω::*kan* is a kanamycin resistance gene inserted into plasmid pBR322.

Reversion versus Suppression

Reversion mutations are often detected through the restoration of a mutated function. As discussed above, a reversion actually restores the original sequence of a gene. However, sometimes the function that was lost because of the original mutation can be restored by a second mutation elsewhere in the DNA. Whenever one mutation in the DNA relieves the effect of another mutation, that mutation has been **suppressed** and the second mutation is called a **suppressor mutation**. The following sections present some of the mechanisms of suppression.

Intragenic Suppressors

Suppressor mutations in the same gene as the original mutation are called **intragenic suppressors**, from the Latin prefix "intra" meaning "within." These mutations can restore the activity of a mutant protein by many means. For example, the original mutation may have made an unacceptable amino acid change that inactivated the protein, but changing another amino acid somewhere else in the polypeptide could restore the protein's activity. This form of suppression is not uncommon and is often interpreted to indicate an interaction between the two amino acids in the protein.

The suppression of one frameshift mutation by another frameshift mutation in the same gene is another example of intragenic suppression. If the original frameshift resulted from the removal of a base pair, the addition of another base pair close by could return translation to the correct frame. The second frameshift can restore the activity of the protein product, provided that ribosomes, while translating in a different frame, do not encounter any nonsense codons in that frame or insert any amino acids that alter the activity of the protein.

Intergenic Suppressors

Intergenic (or **extragenic**) **suppressors** do not occur in the same gene as the original mutation. The prefix "inter" comes from the Latin for "between." There are many ways in which intergenic suppression can occur. The suppressing mutation may restore the activity of the mutated gene product or provide another gene product to take its place. Alternatively, it may alter another gene product with which the original gene product must interact in a complementary way so that now the two mutated gene products can again interact properly.

One common way an intergenic suppressing mutation may restore the viability of the cell is by preventing the accumulation of a toxic intermediate. If the gene for a step in a biochemical pathway is mutated, a toxic intermediate in that pathway can accumulate, causing cell death. However, a suppressing mutation in another gene of the pathway may prevent the accumulation, allowing the cell to survive even though it still has the original mutation. The suppression of *galE* mutations by *galK* mutations provides an illustration of such an intergenic suppressor. Cells with *galE* mutations are galactose sensitive (galactosemic), and their growth is inhibited by galactose in the medium. The reason is apparent from the pathway for galactose utilization shown in Figure 3.19. Many types of cells use the sugar galactose by first converting it to glucose. In the first step, galactose is phosphorylated by the product of the *galK* gene, a galactose kinase. The second step is the transfer of galactose 1-phosphate to uridine diphosphoglucose (UDPglucose) by the product of the *galT* gene, a transferase. The glucose produced is used as a carbon and energy source. The third step is the isomerization of the galactose on UDPgalactose to UDPglucose by the product of the *galE* gene, an isomerase. The newly synthesized UDPglucose can then cycle back into the pathway to convert more galactose to glucose.

The reason why cells with *galE* mutations are galactose sensitive is that the absence of the GalE epimerase permits the accumulation of both phosphorylated galactose and UDPgalactose, which are toxic to cells when present in high concentrations. Consequently, if we plate large numbers of a *galE* mutant strain on plates containing galactose, most of the cells will be inhibited. However, a few mutants multiply to form colonies. Most of these mutants have not undergone reversion of the *galE* mutation but are double mutants with the original *galE* mutation and a suppressing *galK* mutation. The *galK* mutation blocks the first step of the pathway, so that no toxic intermediates accumulate. Revertants with reversion of the original *galE* mutation could also grow. However, we would expect *galE* reversion mutations to

gal genes | P | E | T | K

Promoter Epimerase Transferase Galactokinase

Pathway

1 Galactose + ATP $\xrightarrow{\text{GalK}}$ Galactose-1-PO$_4$ + ADP

2 Galactose-1-PO$_4$ + UDPglucose $\xrightarrow{\text{GalT}}$ UDPgalactose + glucose

3 UDPgalactose $\xrightarrow{\text{GalE}}$ UDPglucose

Figure 3.19 The pathway to galactose utilization in *E. coli* and most other organisms. *galK* mutations suppress *galE* mutations because they prevent the accumulation of the toxic intermediates galactose 1-phosphate and UDPgalactose.

be much rarer than *galK* suppression mutations because many changes inactivate the *galK* gene but only one base pair change can cause the *galE* mutation to revert. Also, the *galE*$^+$ revertants can be distinguished from *galE galK* double mutants because *galE*$^+$ revertants are Gal$^+$ and grow on galactose as the sole carbon and energy source. In contrast, the *galE galK* double mutants are still Gal$^-$, and so another carbon source such as glucose must be provided in the medium.

Nonsense Suppressors

Nonsense suppressors are another type of intergenic suppressor. A **nonsense suppressor** is usually a mutation in a tRNA gene that changes the anticodon of the tRNA product of the gene, so that it now recognizes a nonsense codon. In Figure 3.20, for example, the gene for a tRNA with the anticodon 3′GUC5′ (so that it normally recognizes the glutamine codon 5′CAG3′) mutates, causing the anticodon to become 3′AUC5′. This altered anticodon can pair with the nonsense codon UAG instead of CAG. However, the anticodon mutation does not significantly change the tertiary shape of the tRNA, which means that the cognate aminoacyl-tRNA synthetase still loads it with glutamine. Therefore, this mutated tRNA binds with the amber codon UAG, allowing insertion

Figure 3.20 Formation of a nonsense suppressor tRNA. (A) Gene *X* and gene *Y* contain CAG codons, encoding glutamine. Other codons are not shown. The bacterium also has two different tRNA genes inserting glutamine for the CAG codon. Only the anticodons of the tRNAs are shown. (B) A mutation occurs in gene *X* (shown in purple), changing the CAG codon to UAG and causing the synthesis of a truncated polypeptide (also shown in purple). (C) A suppressor mutation in the gene for one of the two tRNAs changes its anticodon so that it now pairs with the nonsense codon UAG. The translational machinery now sometimes inserts glutamine for the UAG nonsense codon in gene *X*, allowing synthesis of the complete polypeptide. The anticodon of the other tRNA still pairs with the CAG codon, allowing synthesis of the gene *Y* protein and the products of other genes carrying the CAG codon.

A Wild type

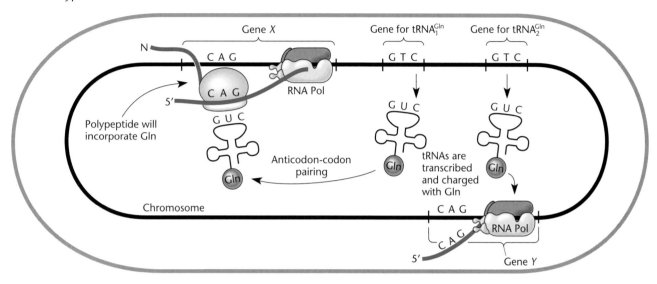

B Mutant$_A$ with nonsense mutation in gene X

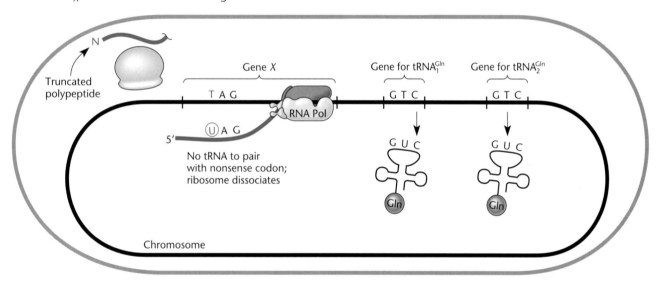

C Mutant$_B$ with nonsense mutation in gene X and nonsense suppressor mutation in gene for tRNA$_1^{Gln}$

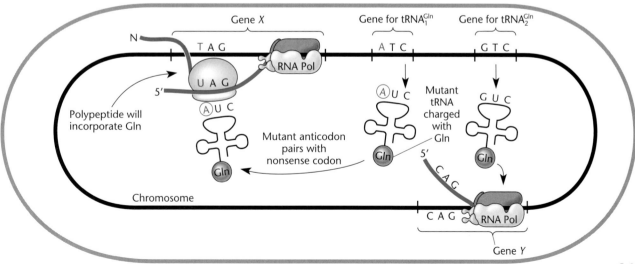

of glutamine into the growing polypeptide instead of translation termination. This can lead to synthesis of the active polypeptide and suppression of the amber mutation.

The mutated tRNA is called a **nonsense suppressor tRNA,** and nonsense suppressors themselves are referred to as amber suppressors, ochre suppressors, or opal suppressors depending on whether they suppress UAG, UAA, or UGA mutations, respectively. Table 3.4 lists several *E. coli* nonsense suppressor tRNAs.

Nonsense suppressors can also be classified as **allele-specific suppressors** because they suppress only one type of allele of a gene, that is, one with a particular type of nonsense mutation. In contrast, the *galK* suppressing mutations discussed above suppress any *galE* mutation and so are not allele specific.

SUPPRESSED POLYPEPTIDES ARE NOT NORMAL

The polypeptide synthesized as the result of a nonsense suppressor is not always fully active. Usually, the amino acid inserted at the site of a nonsense mutation is not the same amino acid that was encoded by the original gene. This changed amino acid sometimes causes the polypeptide to be almost inactive or temperature sensitive.

TYPES OF NONSENSE SUPPRESSORS

Not all tRNA genes can be mutated to form a nonsense suppressor. Generally, if there is only one type of tRNA to respond to a particular codon, the gene for that tRNA cannot be mutated to make a suppressor tRNA. The original codon to which the tRNA responded would be "orphaned," and no tRNA would respond to it wherever it appears in an mRNA. Often when a tRNA can be mutated to a nonsense suppressor tRNA, it is because there is another tRNA that can respond to the same original codons. Cells quite often have more than one tRNA, encoded by different genes, that respond to the same codons. In the example, two different tRNAs encoded by different genes recognize the codon CAG, one of which continues to recognize CAG after the other has been mutated to recognize UAG.

Wobble (see chapter 2) offers the only exception to the rule that a tRNA can be mutated to a nonsense suppressor only if there is another tRNA to respond to the original codon. Because of wobble, the same tRNA can sometimes respond both to its original codon and to one of the nonsense codons. For example, in a particular organism, there may be only one tRNA that recognizes the codon for tryptophan, 5'UGG3'. If the anticodon, 3'ACC5', is mutated to ACU, by wobble the tRNA might be able to recognize both the tryptophan codon UGG and the nonsense codon UGA, so that the suppressor strain could be viable. Wobble also allows the same suppressor tRNA to recognize more than one nonsense codon. In *E. coli*, all naturally occurring ochre suppressors also suppress amber mutations. From the wobble rules (see chapter 2), we know that a suppressor tRNA with the anticodon AUU could recognize both the UAG and UAA nonsense codons in mRNA (Table 3.4). Note that in Table 3.4 anticodons are written 5'-3' even though they pair with the codon 3'-5'.

EFFICIENCY OF SUPPRESSION

Nonsense suppression is never complete, because the nonsense codons are also recognized by release factors, which free the polypeptide from the ribosome (see chapter 2). Therefore, suppression of a nonsense codon and translation of the codon complete protein depends on the outcome of a race between the release factors and the suppressor tRNA. If the tRNA can base pair with the nonsense codon before the release factors terminate translation at that point, translation will continue. Sequences around the nonsense codon influence the outcome of this race and determine the efficiency of suppression of nonsense mutations at particular sites.

NONSENSE SUPPRESSOR STRAINS ARE USUALLY SICK

It would seem that nonsense suppressors would tend to translate through the proper nonsense codons at the ends of genes, resulting in proteins that are longer than normal. However, because nonsense suppressors are never 100% efficient, some of the correct proteins are always synthesized. Moreover, since the efficiency of suppression depends on the sequence of nucleotides in the gene around the nonsense codon, the nonsense codons at the ends of genes presumably have a "context" that favors termination rather than suppression. Also, more than

TABLE 3.4	Some *E. coli* nonsense suppressor tRNAs		
Suppressor name	tRNA	Anticodon change	Suppressor type
supE	tRNA^Gln	CUG-CUA	Amber
supF	tRNA^Tyr	GUA-CUA	Amber
supB	tRNA^Gln	UUG-UUA	Ochre/amber
supL	tRNA^Lys	UUU-UUA	Ochre/amber

one type of nonsense codon often lies in frame at the end of genes, presumably to avoid suppression by any particular tRNA suppressor.

Nevertheless, cells do pay a price for nonsense suppression. Cells with nonsense suppressors usually grow more slowly. Only lower organisms such as bacteria, fungi, and roundworms seem to tolerate nonsense suppressors, which are known to be lethal in higher organisms, including fruit flies and humans.

Genetic Analysis in Bacteria

One of the cornerstones of modern biological research is genetic analysis. Gregor Mendel probably performed the first definitive genetic analysis of a cellular function almost 150 years ago, when he crossed wrinkled peas with smooth peas and counted the number of progeny of each type. The methods of genetic analysis have become considerably more sophisticated since then and are still central to research in cell and developmental biology. The first information about many basic cellular and developmental processes often comes from a genetic analysis of the process. Advantages of the genetic approach are that it requires few assumptions and can be applied to any type of organism, even ones about which little to nothing is known. Now that hundreds of bacterial genomes have been sequenced, we can often tentatively identify the function of a gene product on the basis of similarities in sequence and structure to those that have already been characterized. Genetic analysis, including techniques of reverse genetics, is still the only way to determine how many gene products are involved in a function and to obtain a preliminary idea of the role of each gene product in the function. Suppressor analysis also offers one of the best ways to ascertain which gene products interact with each other in performing the function. Bacteria (and their phages) are ideal for demonstrating basic genetic principles, which is why so many discoveries in basic genetics have been made using these organisms. However, it is important to keep in mind that the basic principles discussed here are universal, applying equally to all organisms including humans, and that only the details of how a genetic analysis is performed differ from one type of organism to another. Genetic analysis is covered in general genetics textbooks, and we review the basic principles here only as they apply particularly to bacteria.

Isolating Mutants

As discussed in the introductory chapter, a classical genetic analysis begins with finding mutants in which the function is altered. This process is called the **isolation of mutants** because the mutant organisms are somehow found and separated or "isolated" from the myriad of normal or nonmutant organisms with which they are associated. As discussed in the introductory chapter, a major reason why bacteria are such excellent genetic subjects is the relative ease with which mutants can be isolated. Bacteria are generally haploid, meaning that they have only one allele of each gene. This makes the effects of even recessive mutations immediately apparent, obviating the need for backcrosses to obtain homozygous individuals that show the effects of the mutation. Bacteria also multiple asexually, not requiring crosses with another organism to make progeny. Generally, no two organisms produced by mating are genetically identical (unless they are identical twins), and so the progeny of a mutant are not identical to the original mutant, necessitating backcrosses between the progeny and the original mutant to try to make them more similar. Such backcrosses are not necessary with bacteria and other asexually reproducing organisms since the progeny of a mutant bacterium are usually identical to the original mutant. To make genetically identical bacteria, we do not need to clone them; they clone themselves when they multiply. Bacteria are also small, and numbers equivalent to the entire human population on Earth can be placed on a single petri plate, facilitating the isolation of even very rare mutants.

TO MUTAGENIZE OR NOT TO MUTAGENIZE?

The first step in obtaining a collection of mutants for a genetic analysis is to decide whether to allow the mutations to occur spontaneously or to deliberately mutagenize the organism. Spontaneous mutations occur normally as mistakes in DNA replication, but the frequency of mutations can be greatly increased by treating the cells with some chemicals or with some types of irradiation. Treatments such as chemicals or UV irradiation, which cause mutations, are said to be **mutagenic,** and agents that cause mutations are **mutagens.** In general, treatments that damage DNA are mutagenic, but it is important to keep in mind that damage to DNA is not a mutation. Mutations are heritable changes in the sequence of normal deoxynucleotides in the DNA (see the definition of a mutation, above). Damaged DNA may mispair more frequently during replication, causing mutations, or mutations may arise during misguided attempts by the cell to repair the damage. Mutagenesis is discussed in much more detail in chapter 11.

Both spontaneous and induced mutations have advantages in a genetic analysis. To decide whether to mutagenize the cells and, if so, which mutagen to use, we must first ask how frequent the mutations are likely to be. Spontaneous mutations are usually much rarer than induced mutations and so are more difficult to isolate. Therefore, to isolate very rare types of mutants or ones for which there is no good selection, we might have to use a mutagen. On the other hand, mutants containing spontaneously arising mutations are less likely to contain more than one different mutation, and the presence of multiple mutations can confuse the analysis later.

One major advantage of inducing mutations by using mutagens is that a particular mutagen often induces only a particular type of mutation. Spontaneous mutations can be base pair changes, frameshifts, duplications, insertions, or deletions. However, the acridine dye mutagens, such as acriflavine, cause only frameshift mutations, and base analogs, such as 2-aminopurine, cause only base pair changes. Therefore, the use of a particular mutagen may make it possible to restrict the mutations to the type desired.

Isolating Independent Mutations

For an effective genetic analysis, mutants defective in a function should have mutations that are as representative as possible of all the mutations that can cause the phenotype. If the strains in a collection of mutants carry many different mutations, we can get a better idea of how many genes can be mutated to give the phenotype and how many types of mutations can cause the phenotype. A general rule is that if some genes are represented by only a single mutation, then, by the Poisson distribution discussed above, there are likely to be many other genes in which mutations could give the same phenotype. However, these other genes have not yet been detected by mutations and have therefore been missed.

There are two ways to ensure that the maximum number of different mutations are represented in a collection of mutants. One way is to avoid picking **siblings,** which are organisms that are descendents of the same original mutant. Two sibling mutants always have the same mutation. The best way to avoid picking siblings is to isolate only one mutant from each of a number of different cultures, all started from nonmutant bacteria. If two mutants arose in different cultures, their mutations must have arisen independently and they could not be siblings. Another way to avoid getting the same mutation is to use more than one mutagen. All mutagens have preferred hot spots and tend to mutagenize some sites more than others (see the discussion of mutational spectra in the *r*II genes of T4 phage in chapter 7). If all the mutants are obtained with the same mutagen, many of them have mutations in the same hot spot, but mutants obtained with different mutagens tend to have different mutations.

Selecting Mutants

Even after mutagenesis, mutants are rare and still must be found among the myriad of individuals that remain normal for the function. The process of finding mutants is called **screening.** Screening for mutants is usually the most creative part of a genetic analysis. One must anticipate the phenotypes that might be caused by mutations in the genes for a particular function. This is where the geneticist earns her or his pay, because predicting what types of mutations are possible and how to select them often requires intuition as well as rational thinking, but it is one of the more enjoyable aspects of genetics. For example, what do you imagine would be the phenotype of mutants defective in protein transport through the membrane? Specific examples of screening for this and other types of mutants are discussed in later chapters.

Screening for mutant bacteria usually involves finding **selective conditions** to distinguish the mutants from the original type. These are usually conditions under which either the mutant or the wild type cannot multiply to form a colony. Agar plates and media with selective conditions are called selective plates and selective media, respectively.

Selections can be either positive or negative. In a **positive selection,** selective conditions are chosen under which the mutant but not the original wild type can multiply. Figure 3.21 shows an example of a positive selection for His$^+$ revertants of a *his* mutation. A *galK* mutation is another example of a type of mutation that can be selected by a positive selection, by plating a *galE* mutant on medium containing galactose plus another carbon source (see above). In a **negative selection,** selective conditions are used under which the wild type but not the mutant can grow. While negative selections are often the only option for many types of mutants, screening for mutants is much easier with positive selections, and geneticists expend much effort trying to design positive selections. In a sense, negative selections are not really selections at all; they are screens (see below), because the selective conditions are being used to screen for the mutants rather than to eliminate all other organisms that are not mutated in the same way. Nevertheless, the common terms are used in this discussion.

ISOLATING MUTANTS BY NEGATIVE SELECTIONS

Most mutants, such as those that are auxotrophic or temperature sensitive, can be isolated only by using negative selections. Most of an organism's gene products help it to multiply. Therefore, mutations that inactivate a gene product are more likely to make the organism unable to multiply under a given set of conditions rather than able to multiply when the wild type cannot. To isolate mutants by negative selection, the bacteria are first plated on a nonselective plate, on which both the mutant and the wild type can multiply. When the colonies have developed, some of the bacteria in each colony are transferred to a selective plate to determine which colonies contain mutant

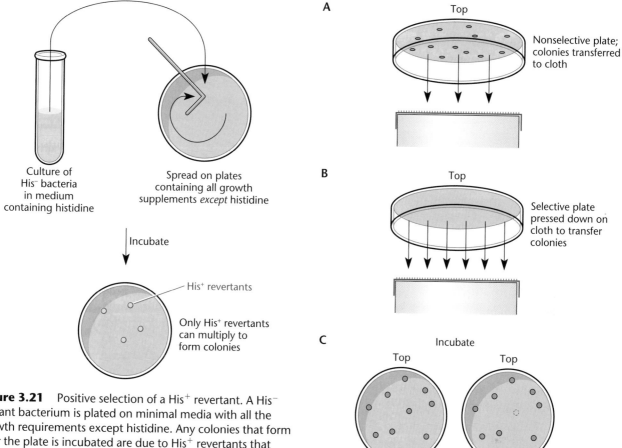

Figure 3.21 Positive selection of a His⁺ revertant. A His⁻ mutant bacterium is plated on minimal media with all the growth requirements except histidine. Any colonies that form after the plate is incubated are due to His⁺ revertants that can multiply without histidine in the medium.

Figure 3.22 Replica plating. (A) A few hundred bacteria are spread on a nonselective plate, and the plates are incubated to allow colonies to form. The plate is then inverted over velveteen cloth to transfer the colonies to the cloth. (B) A second plate is then inverted and pressed down over the same cloth and then incubated. (C) Both plates after incubation. The dotted circle indicates the position of a colony missing from the selective plate. See the text for details.

bacteria that cannot multiply to form a colony under those conditions. Once such a colony has been identified, the mutant bacterial strain can be retrieved from the corresponding colony on the original nonselective plate. Figure 3.1 showed the detection of two types of auxotrophic mutants, His⁻ (unable to make histidine) and Bio⁻ (unable to make biotin), by negative selections.

Replica Plating

Because of the general rarity of mutants, many colonies usually have to be screened to find a mutant when using negative selection. **Replica plating** can be used to streamline this process and is illustrated in Figure 3.22. A few hundred bacteria are spread on a nonselective plate, and the plate is incubated to allow colonies to form. A replica is then made of this plate by inverting the plate and pressing it down over a piece of fuzzy cloth, such as velveteen. Then a selective plate is inverted and pressed down over the same cloth so that the colonies are transferred from the cloth to the selective plate. After the selective plate has been incubated, it can be held in front of the original nonselective plate to identify colonies

that did not reappear on the selective plate. The missing colonies presumably contain descendants of a mutant bacterium that are unable to multiply on the selective plate. The mutant bacteria can then be taken from the colony on the original, nonselective plate. Replica plating was used by the Lederbergs to demonstrate that bacteria behave by the principles of Darwinian inheritance, as discussed earlier in the chapter.

Enrichment

If a type of bacterial mutant being sought is rare, finding it by negative selection can be very laborious, even with replica plating. No more than about 500 bacteria can be spread on a plate and still give discrete colonies. So, for example, if the mutant occurs at a frequency of 1 in 10^6, more than 2,000 plates might have to be replicated to find a mutant!

Many fewer colonies need to be screened if the frequency of mutants is first increased through mutant **enrichment**. This method depends on the use of antibiotics such as ampicillin and 5-bromouracil (5-BU) that kill growing but not nongrowing cells. Ampicillin inhibits cell wall synthesis and causes a growing bacterial cell literally to grow out of its skin and lyse. A mutant cell that was not growing while the ampicillin was present does not grow out of its skin and so is not killed. 5-BU also kills only growing cells but by a very different mechanism. DNA containing 5-BU (an analog of thymine) is much more sensitive to UV light than is normal DNA containing only thymine. Cells replicate their DNA only while they are growing, so they take 5-BU into their DNA and become more UV sensitive only if they were growing while 5-BU was present in the medium.

To enrich for mutants that cannot grow under a particular set of selective conditions, the population of mutagenized cells is placed under the selective conditions in which the desired mutants stop growing. Meanwhile, the nonmutant wild-type cells continue to multiply. The antibiotic—either ampicillin or 5-BU—is then added to kill any multiplying cells. The cells are then filtered or centrifuged to remove the antibiotic and transferred to nonselective conditions. The mutant cells will have survived preferentially because they were not growing in the presence of the antibiotic; therefore, they will have become a higher percentage of the population. No enrichment is 100% effective; however, even if the enrichment makes the mutant only 100 times more frequent, only 1/100 as many colonies and therefore 1/100 as many plates must be replicated to find a mutant after an enrichment. In the example given above, after an enrichment we would need to replicate only 20 plates instead of 2,000 to find a mutant.

Unfortunately, enrichments cannot be applied to all types of mutants. Some mutants are killed by the selective conditions and so cannot be enriched by these procedures. To be enriched, the mutant must still be alive and resume multiplying after it is removed from the selective conditions.

Genetic Mapping by Recombination in Bacteria

RECOMBINATION TESTS

Once we have our collection of mutants, we wish to further characterize the responsible mutations. One thing we can do is locate them in the DNA of the bacterium. If a genetic map of the organism is available, we can locate mutations through genetic recombination; this is called genetic mapping.

Recombination is defined as the breakage and rejoining of two DNA molecules in new combinations. The site of breakage and rejoining is called a **crossover**. Recombination can be either **site specific** or **general**. Site-specific recombination uses specialized enzymes that cut and religate DNA but only at unique sequences, so it is not useful for genetic mapping. Site-specific recombination is discussed in detail in chapter 9. Genetic mapping requires **generalized recombination**, sometimes called **homologous recombination** because it can occur anywhere but occurs only between two DNA regions that have the same or homologous sequences ("homo-logos" means "same-word" in Greek). Homologous recombination probably occurs naturally in all organisms and serves the purpose of increasing genetic diversity within a species and/or repairing damage to DNA by restarting replication forks and making one good DNA molecule out of two damaged ones.

Generalized recombination is quite complex and uses a number of different enzymes and pathways depending on the situation. These details are discussed further in chapter 10. However, for now, the simplified model of recombination shown in Figure 3.23 is sufficient. After all, people used recombination to breed new strains and for genetic mapping for decades before it was understood in any depth, in fact before it was even known that DNA is the hereditary material. According to the simplified model, the two DNAs that are going to recombine first pair in a region where their nucleotide sequences are homologous. This allows the two opposing strands to base pair (see below). This requirement for base pairing ensures that recombination occurs in the same place in the two DNAs since only in homologous regions are the opposing strands of the two DNAs complementary and able to base pair. Restricting recombination to regions of homology between two DNAs helps ensure that the order of genes in the DNA of a species is not scrambled each time recombination occurs. In fact, we saw earlier in the chapter how recombination between the wrong places in DNA (ectopic recombination) can cause deletion, duplication, and inversion mutations. In the next step, staggered breaks are made in the two DNA molecules at the same position in both molecules. This allows the strands of the two DNAs to cross over and pair with each other by complementary base pairing, as shown. Then the broken ends of one DNA are joined to the broken ends of the other DNA to make two new DNA molecules. The site where the breakages occurred is the crossover point.

GENETIC MARKERS

If the breaks and rejoining occur in the same place on two identical DNAs, the two new DNA molecules created by the crossover will have the same sequences as the

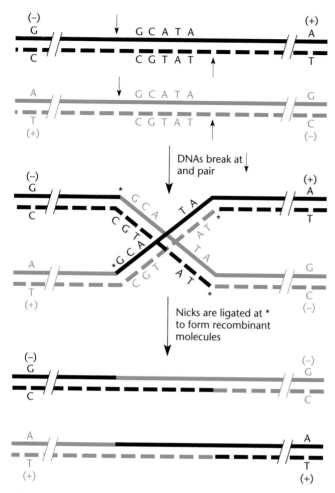

Figure 3.23 A simplified diagram of recombination. Staggered breaks are made in the two DNAs at the same sites on both DNAs. The strands then cross over, and the ends are joined to form two new DNA molecules. The positions of bases are shown only where the mutations have occurred and where the crossing over occurs. (−), mutation; (+), wild type.

original two DNA molecules and there would be no way of knowing that a crossover had occurred between them. However, if, as in the example shown in Figure 3.23, the two DNAs have some difference in their sequences on either side of where the recombination occurred, a crossover between the regions of the differences will then yield two DNAs with different sequences from the original DNAs. Two bacteria of the same species that have some sequence differences in their DNAs are different **strains** of the same species. If the mutations that create the sequence differences in the two strains cause observable phenotypes to the organism, we can use these phenotypes to determine that a crossover has occurred between the two sites. In so doing, we are using the sequence differences as **genetic markers**. When the two

DNAs that have formed as a result of a crossover between the two genetic markers **segregate** into progeny, these progeny are genetically different and correspondingly phenotypically different from either parent since they have phenotypes of both parents. Progeny that are genetically different from either parent as a result of a recombination are called **recombinant types**. Progeny that are genetically identical to one or the other of the two parents are **parental types**. The frequency of recombination types is a measure of how often recombination has occurred, which gives us information about where the genetic markers are in the DNA. However, to measure recombination between genetic markers, we need a way of exchanging DNA between different strains; in other words, a way of doing genetic crosses. These methods are discussed later in the chapter.

Complementation Tests

The other general method in genetic analysis is the complementation test. Rather than depending on breaking and joining DNA in new combinations, complementation depends on the functional interaction of gene products made from different DNAs. Complementation allows us to determine how many gene products are represented by a collection of mutations and allows us to obtain preliminary information about the functions affected by the mutations. To perform a complementation test, we must put two copies of the regions of DNA from different strains containing two different mutations into the same cell and see what effect this has on the phenotypes of the mutations. With a diploid organism, which contains two homologous chromosomes of each type, this is no problem since they normally have two copies of each gene. With phages and other viruses, it is also no problem because we can infect cells with two different mutant viruses simultaneously. However with bacteria, which are naturally haploid, complementation tests are more difficult. Rather than being made diploids, bacteria can be made **partial diploids** by stably introducing a small region of the chromosome of one strain into another strain, using plasmids or prophages that can coexist with the chromosome. Production of such partial diploids is described in chapters 5 and 8. We can also sometimes make transient partial diploids, for example on tandem duplications and after some types of matings. These are, by definition, not stable but can last long enough for complementation tests.

ALLELISM TESTS

One application of complementation is its use in determining how many different genes (or regions encoding a particular gene product) can be mutated to give a particular phenotype. Another name for this is an **allelism test**, because we are asking whether any two mutations are allelic, i.e., whether they affect the same gene (see above).

Returning to our example of histidine biosynthesis, assume that we have isolated a collection of mutants, all of which exhibit the His⁻ phenotype, and want to know how many genes they represent. This should tell us how many enzymes (or, more accurately, separate polypeptides, since some enzymes are composed of more than one different polypeptide [see chapter 2]) are required to make the amino acid histidine and allow the cell to multiply in the absence of histidine in the medium. Each of these polypeptides should be encoded by a different gene, and if our collection of mutations is large and varied enough, each of these genes should be inactivated by at least one of our mutations. The allelism test is performed on the mutations two at a time, as illustrated in Figure 3.24. If the two mutations are in different genes, each DNA can furnish the polypeptide that cannot be furnished by the other, so that all the polypeptides are present and the diploid cell is phenotypically His⁺. If, however, the two mutations are allelic (see above), neither DNA can make that gene product, the two mutations cannot complement each other, and the cells remain phenotypically His⁻. We can then extend this analysis to include the other mutations in our collection, two at a time, to place them in complementation groups and determine how many total genes or complementation groups are represented in the collection of mutations.

Usually these rules apply, and complementation between two mutations indicates that the mutations are in different genes while lack of complementation indicates they are in the same gene. However, in some cases, complementation can occur between two mutations even if they are in the same gene. Complementation between two mutations in the same gene is called **intragenic complementation** and usually occurs only if the protein product of the gene is a multimer that contains more than one polypeptide product of the gene (see chapter 2 for a definition of "multimer"). Also, the polypeptide usually has more than one functional domain, with one domain of the polypeptide having one activity and the other domain being responsible for the other activity. For example, DNA polymerase I has more than one domain, one with the polymerizing activity and the other with the 5′ exonuclease activity. When the organism only has one copy of a defective gene, all the polypeptides of that gene in the protein have the same defect and the protein will be nonfunctional. However, if the organism is diploid for the gene, one copy can have a defect in one domain and the other can have the defect in the other domain. When the protein is assembled, some of the proteins have polypeptides encoded by both genes. The protein then has good copies of both the domains and could be active.

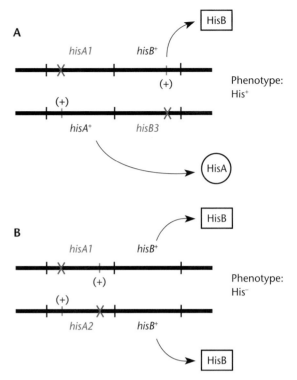

Figure 3.24   Complementation tests for allelism. Three mutations, *hisA1*, *hisA2*, and *hisB3*, are being tested to determine which of them are allelic. (A) The *hisA1* and *hisB3* mutations are in different DNAs in the same cell, and the two mutations are in different genes whose products are required for synthesis of histidine. The DNA with the *hisA1* mutation can make HisB, and the DNA with the *hisB3* mutation can make HisA; hence, the cell is His⁺. (B) The *hisA1* and *hisA2* mutations are in the same gene. Neither DNA can make HisA; hence, the cell is His⁻.

However, intragenic complementation is rare and occurs only between certain mutations in the gene; it is usually interpreted to mean that the product of the gene has more than one domain and is part of a homomultimer, made up of more than one copy of the polypeptide product of the gene.

Sometimes complementation does not occur between two mutations even though they are in different genes. This can happen if the mutation in one of the two genes is polar on the other gene or if the two genes are translationally coupled. Then a mutation that terminates translation in one gene can prevent the transcription or translation of another gene downstream of it and transcribed into the same mRNA (see chapter 2 for explanations of polarity and translational coupling). Table 3.5 outlines the interpretation of complementation experiments and their possible complications.

TABLE 3.5	Interpretation of complementation tests
Test result	**Possible explanations**
x and y complement	Mutations are in different genes Intragenic complementation has occurred[a]
x and y do not complement	Mutations are in the same gene One of the mutations is dominant One of the mutations affects a regulatory site or is polar

[a] See the text for an explanation of intragenic complementation. This is a less likely explanation than the mutations being in different genes.

RECESSIVE OR DOMINANT

Complementation can also be used to tell if a mutation is **recessive** or **dominant** to the wild-type allele. A recessive mutation is subordinate to the wild type, in the sense that an allele with the mutation does not exert its phenotype if the wild-type allele of the gene is present in the same cell. A dominant mutation exerts its phenotype even if the wild-type allele is present. Recessive mutations have generally inactivated the gene product, while dominant mutations often have subtly changed the gene product so that it can function in a situation where the wild type cannot function or that it can perform a function that the wild type cannot perform. Recessive mutations are much more common than dominant mutations because many more types of changes in the DNA inactivate the gene product than change it in some subtle way. Whether a particular type of mutation in a gene that gives a particular phenotype is dominant or recessive can tell us something about the normal functioning of the gene product.

To determine whether a mutation is recessive or dominant, we need to make a partially diploid cell which has both the wild-type allele and the mutant allele and ask whether the wild-type phenotype or the mutant phenotype prevails. We return again to the example of the *his* pathway to illustrate the difference between recessive and dominant mutations. Most mutations which make the cell His⁻ have inactivated one of the enzymes required to make histidine. These mutations are all recessive to the wild type because, in the presence of the wild-type allele for each of the genes, all the enzymes required to make histidine will be made and the cell will be His⁺, the phenotype of the wild-type alleles. However, assume that there is an inhibitor of the pathway, such as an analogue of histidine that binds to the first enzyme of the pathway and inactivates it. In the presence of this inhibitor, the cell is also unable to make histidine and will be His⁻, so the phenotype caused by the wild-type allele in the presence of the inhibitor is His⁻. However, a mutation in the gene for the enzyme can make the enzyme insensitive to the inhibitor, so that the mutant

cell can make histidine even in the presence of the inhibitor. The phenotype of the cell containing this mutation is His⁺ even in the presence of the inhibitor. The mutant enzyme might continue to function to make histidine in the presence of the inhibitor even if the sensitive wild-type enzyme is present. If so, in a diploid containing both the mutant and wild-type alleles, the phenotype would be His⁺ in the presence of the inhibitor, the phenotype of the mutant allele, and the resistance mutation is dominant.

CIS-TRANS TESTS

Another use of complementation is to determine whether a mutation is **trans** acting or **cis** acting. These prefixes come from Latin and mean "on the other side" and "on this side," respectively. A *trans*-acting mutation usually affects a diffusible gene product, either a protein or an RNA. If the mutation affects a protein or RNA product, it can be complemented, and it does not matter which DNA has the mutation in a complementation test because the gene product is free to diffuse around in the cell (i.e., the mutation acts in *trans*). A *cis*-acting mutation usually has changed a site on the DNA such as a promoter or an origin of replication. If the mutation affects a site on the DNA, it affects only that DNA and cannot be complemented (i.e., it acts in *cis*). In our example of the histidine synthesis genes, mutations, either recessive or dominant, that affect the enzymes which make histidine would be *trans* acting while a promoter mutation that prevents transcription of the genes for histidine synthesis would be *cis* acting. In subsequent chapters we discuss how *cis-trans* tests have been used to analyze regulation of gene expression and other cellular functions.

CLONING BY COMPLEMENTATION

Another very useful application of complementation in bacteria is in cloning. Complementation can be used to identify clones carrying a particular gene by their ability to complement a mutation in the chromosome and restore the normal or wild-type phenotype. Figure 3.25 illustrates the use of complementation to identify clones of the gene for thymidylate synthetase (*thyA*) in *E. coli*. This enzyme is needed to synthesize dTMP from dUMP, and so a *thyA* mutant will not be able to replicate its DNA and multiply to form a colony unless thymine is provided in the medium (see chapter 1). A library of the chromosome of wild-type (*thyA⁺*) *E. coli* is introduced into a strain of *E. coli* with a *thyA* mutation in its chromosome. The bacteria are then plated on selective plates containing all the necessary growth supplements but without thymine. Any colonies that appear may be due to bacteria containing a clone expressing the *thyA* gene

A

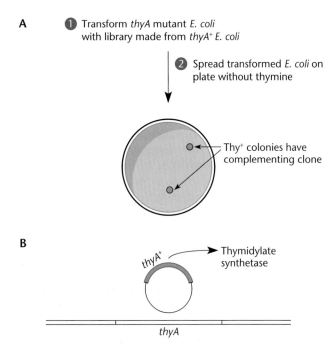

❶ Transform *thyA* mutant *E. coli* with library made from *thyA⁺ E. coli*

❷ Spread transformed *E. coli* on plate without thymine

Thy⁺ colonies have complementing clone

B

thyA⁺

Thymidylate synthetase

thyA

Figure 3.25   Identification of clones of the *thyA* gene of *E. coli* by complementation. (A) A *thyA* mutant of *E. coli* is transformed by a library of DNA from *thyA⁺ E. coli,* and the transformants are selected directly on plates lacking thymidine but containing the antibiotic to which the cloning vector confers resistance. (B) Thy⁺ transformants that contain a clone of the *thyA* gene synthesize the thymidylate synthetase, thus complementing the mutation in the chromosome.

from the clone, which is complementing the *thyA* mutation in the chromosome. The clone containing the wild-type copy of the *thyA* gene can then be recovered from the cells in these colonies. For the clone to complement, it usually must contain the entire gene and the gene must be expressed from the cloning vector. Genes from distantly related organisms generally cannot be expressed in a particular host, so this method is generally useful only for identifying clones in the "host of origin," for example a bacterial gene in the bacterium it came from.

Genetic Crosses in Bacteria

We are all the result of a genetic cross. This is because humans and most other higher eukaryotes have an obligate sexual cycle and progeny can be produced only by mating. Two haploid germ cells fuse to form a diploid **zygote** from which the organism develops. In contrast, bacteria do not need to mate to produce progeny, they just divide. However, even though they do not have an obligate sexual cycle, they do have ways of exchanging DNA between individual bacteria. These are called

conjugation, transformation, and **transduction.** The discovery of these mechanisms of gene exchange was an important chapter in the history of molecular genetics and was discussed in the Introductory chapter. Genetic data are interpreted differently by these different methods, so we discuss them separately. However, they do have some features in common, which are reviewed here.

DONOR VERSUS RECIPIENT

In all three of these methods, a small region of DNA from one strain is transferred into the other strain. The strain from which the small piece of DNA comes is the **donor strain,** and the strain that receives the piece of DNA is called the **recipient strain.** Recombination can occur only in the recipient strain since, in analogy to sexual mating in higher organisms, this is the strain that forms the zygote which contains DNA from both strains.

SELECTED AND UNSELECTED MARKERS

Unlike higher organisms with an obligate sexual cycle, transfer of DNA generally occurs rather rarely in bacteria, and so we need a way of selecting those few bacteria that have participated in a mating event. Fortunately, bacteria that have participated in a mating event can often be detected even if they are very rare. The process is much like detecting mutants with a positive selection. Conditions are established where only a particular recombinant type can multiply, so that the recombinants can form colonies when the other bacteria cannot. To illustrate the selection of recombinant types, refer to the example in Figure 3.26. In this example, a piece of DNA from one strain that has one of the mutations has been transferred into another strain that has another mutation. Both mutations inactivate a gene whose product is required for growth under some conditions. Therefore, the recipient strain is not able to grow under the selective conditions, because it has the m_y mutation. However, if crossovers occur between the incoming DNA and the homologous region in the recipient DNA, recombinant-type progeny can arise in which the sequence of the donor DNA, which does not have the mutation (indicated by "+" in Figure 3.26) has replaced the sequence of the recipient strain, which did have the mutation, allowing the recipient strain to grow under the selective conditions. We decided to use the m_y marker as the **selected marker** to select recipient bacteria in which crossovers have occurred that make the strain recombinant for the selected marker, because these can be selected by a positive selection. Bacteria selected for being recombinant for this marker must have participated in a mating event, no matter how rare these mating events may be.

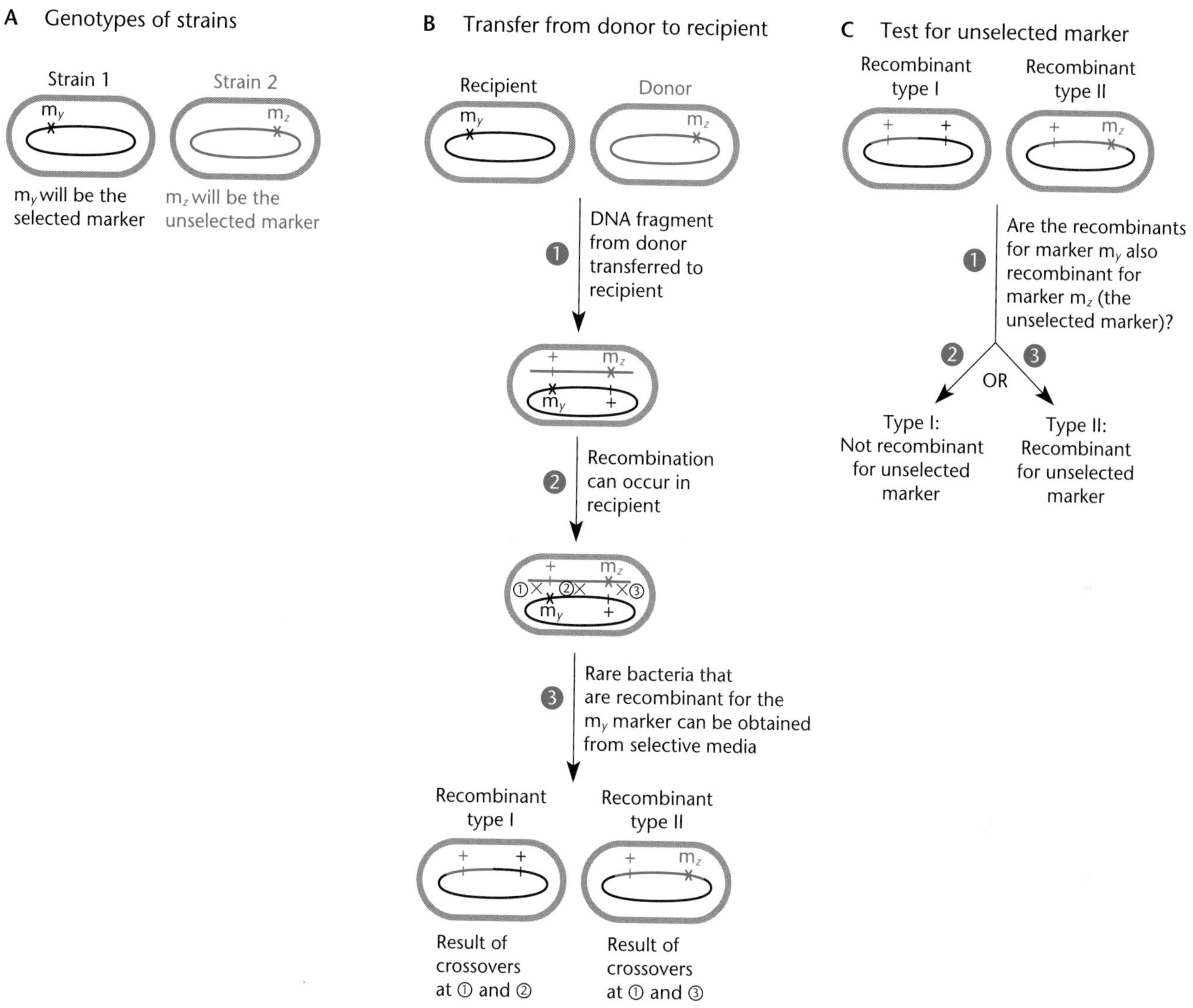

A   Genotypes of strains

Strain 1   Strain 2

m_y will be the selected marker m_z will be the unselected marker

B   Transfer from donor to recipient

Recipient   Donor

❶ DNA fragment from donor transferred to recipient

❷ Recombination can occur in recipient

❸ Rare bacteria that are recombinant for the m_y marker can be obtained from selective media

Recombinant type I   Recombinant type II

Result of crossovers at ① and ②   Result of crossovers at ① and ③

C   Test for unselected marker

Recombinant type I   Recombinant type II

❶ Are the recombinants for marker m_y also recombinant for marker m_z (the unselected marker)?

❷   OR   ❸

Type I: Not recombinant for unselected marker   Type II: Recombinant for unselected marker

Figure 3.26   Selected versus unselected markers in a bacterial cross. Replacement by recombination of the sequence of the recipient DNA by the sequence of the donor DNA creates a recombinant type. Recipient bacteria, recombinant for one marker, the selected marker, are tested to see how many are also recombinant for a second marker. In the example, m_y is the selected marker and m_z is the unselected marker, so chosen because it is easier to select recombinants for the m_y marker. (A) The genotypes of the strains. (B) A fragment of the donor DNA is transferred into the recipient cell. Recombination occurs between the incoming donor DNA and the recipient DNA, replacing regions of the recipient DNA with donor DNA. The recipients in which the m_y region of the donor has replaced that of the recipient are selected and purified on selective plates. (C) The recipients that have been selected for being recombinant for the m_y marker are tested to see if they are also recombinant for the m_z marker. For details, see the text.

Once we have selected for recipient bacteria that have participated in a mating event because they have become recombinant for one of the markers, we can test them to see if they have also become recombinant for any other markers in the recipient strain. The other markers then become the **unselected markers.** In our example, the recipient will have become recombinant for the other marker if it now has the m_z mutation and is unable to grow without the growth supplement, since that is the sequence of the donor DNA, which had the mutation. In other words, the recipient is recombinant for the region of the m_z mutation when the sequence of the donor, which had the mutational change, has replaced the sequence of the recipient, which did not have the mutation. If any of the recipients that became recombinant for the selected marker have also become recombinant for the unselected marker, the donor DNA from this region must also have entered the same recipient cell and recombined with the recipient chromosome. If the mating events are rare enough, it is highly unlikely that two DNAs transferred in separately, and the two regions must have come in on the same DNA molecule. Therefore, the frequency with which bacteria selected for one marker have become recombinant for other markers gives us information about where the mutations that gave rise to the phenotypic differences were on the chromosome. Note that once we have selected for one marker, we do not need to screen nearly as many bacteria to get meaningful data about the frequency of recombinants for the other markers. Our screen is restricted to bacteria that have participated in a mating event. Specific examples of this process, using the various methods of gene exchange which differ in how data are interpreted, are given below.

Genetic Mapping by Hfr Crosses

One way bacteria can exchange DNA between strains is by conjugation. **Conjugation** is the transfer of DNA from one bacterium to another by a **self-transmissible plasmid.** Plasmids are small, usually circular, DNA elements that have the capacity to replicate independently of the chromosome (see chapter 4). Some of these plasmids have the capacity to transfer themselves from one bacterium to another. If so, they are called self-transmissible plasmids (see chapter 5). If such a self-transmissible plasmid integrates into the chromosome of the bacterium that carries it, the chromosome can also be transferred when the plasmid transfers itself. A bacterial strain with a self-transmissible plasmid integrated into its chromosome is called an **Hfr strain.** This phenomenon was first detected in 1947 by Joshua Lederberg and Edward Tatum when they observed recombinant types after mixing some strains of *E. coli* with other strains. We now know that one of the strains contained a self-transmissible plasmid called the F plasmid. In a few bacteria in the population, the F plasmid had integrated into the chromosome, and these bacteria were transferring chromosomal DNA into the other strain, leading to the formation of recombinant types. In retrospect, it was fortuitous that some of the strains used by Lederberg and Tatum contained the F plasmid. Their experiment would not have succeeded if none of the strains they used had contained a self-transmissible plasmid. Also, any plasmid other than the

F plasmid would not have worked as well, since, as discussed in chapter 5, the F plasmid is a mutant that is always ready to transfer.

Figure 3.27 illustrates the process by which chromosomal DNA is transferred in a mating between a donor Hfr strain containing an integrated F plasmid and a recipient

Figure 3.27 Transfer of chromosomal DNA by an integrated plasmid. Formation of mating pairs, nicking of the F *oriT* sequence, and transfer of the 5′ end of a single strand of DNA proceed as in transfer of the plasmid. Transfer of the covalently linked chromosomal DNA also occurs as long as the mating pair is stable. Complete chromosome transfer rarely occurs, and so the recipient cell remains F⁻, even after mating. Replication in the donor usually accompanies DNA transfer. Some replication of the transferred single strand may also occur. Once in the recipient cell, the transferred DNA may recombine with homologous sequences in the recipient chromosome.

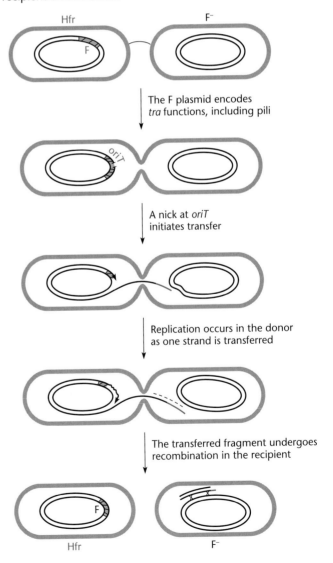

strain that does not contain the F plasmid (F$^-$). On contact with a recipient cell, the DNA in the donor is nicked at a site in the integrated plasmid, and one strand is displaced into the recipient cell. Normally the displaced plasmid DNA alone would be transferred into the recipient cell, but because the plasmid is integrated into the chromosome, the chromosomal DNA is also transferred, beginning at one side of the location where the plasmid has integrated and proceeding around the entire chromosome. If the transfer continues long enough (in *E. coli* approximately 100 min at 37°C to transfer the entire 1-mm-long chromosome), it will come full circle, ending with the chromosomal sequences on the other side of the location where the plasmid has integrated. However, transfer of the entire chromosome is rare, perhaps because the union between the cells is frequently broken or because the DNA is often broken during conjugation.

FORMATION OF RECOMBINANT TYPES

Because the entire chromosome is seldom transferred, the DNA is unable to replicate after it enters the recipient cell. Hence, the transferred DNA is lost unless it recombines with the chromosome in the recipient cell. If this happens and if the donor and recipient strains are different in some genetic markers, recombinant types might arise that can be identified because they are different from both the Hfr donor strain and the recipient F$^-$ strain.

In the example shown in Figure 3.28, the donor Hfr strain has an *arg* mutation and therefore does not form colonies on agar plates containing all the growth supplements except arginine, while the recipient strain has a *trp* mutation and therefore does not form colonies on agar plates lacking tryptophan. When the two are mixed, the Hfr strain can transfer DNA into the recipient cell, and sometimes this DNA replaces the recipient DNA at the *trp* allele, replacing it with the wild-type *trp$^+$* allele, as shown. The recipient bacteria then require neither arginine nor tryptophan and multiply to form colonies on minimal plates containing neither of these growth supplements. These Trp$^+$ Arg$^+$ bacteria are recombinant types because they are genetically unlike either parent. This is basically the experiment that allowed Lederberg and Tatum to discover conjugation. They mixed different strains of *E. coli* that required different growth supplements and showed that some mixtures gave rise to recombinant types with neither growth requirement.

MAPPING BY GRADIENT OF TRANSFER

Hfr crosses offer one of the most convenient methods of genetic mapping in bacteria. However, because of the unusual nature of the genetic exchange, data are interpreted differently from the data for other types of genetic

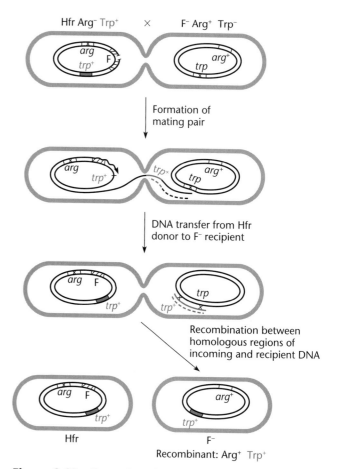

Figure 3.28 Formation of recombinant types after DNA transfer by an Hfr strain. If the *trp* region is transferred from the donor to the recipient, it can recombine to replace the homologous region in the recipient, giving rise to a Trp$^+$ Arg$^+$ recombinant.

crosses. Most methods of mapping by Hfr crosses depend on the fact that the chromosomal DNA is transferred from the donor to the recipient in order, beginning at the location where the plasmid has integrated into the chromosome and proceeding around the chromosome until it arrives back at the site of integration. The chromosome is very long, and this process takes more than an hour to complete. Therefore, it is possible to tell the order of markers in the chromosome by determining when the DNA regions containing the markers enter the recipient cell during conjugation. This can be done either directly, by periodically disrupting the mating and seeing which markers have entered by that time (interrupted mating), or indirectly, by taking advantage of the fact that the mating bridges between the cells are often disrupted during the long period of chromosome transfer. Therefore, the frequency of transfer of a marker decreases the farther the marker is from the origin of transfer in the integrated self-transmissible plasmid (the **gradient of transfer**). We discuss only the gradient-of-transfer method here.

As mentioned, genetic mapping by the gradient of transfer depends on the fact that the entire donor chromosome is seldom transferred during conjugation. Because the chromosome is enormously long and the entire transfer takes more than an hour, the mating bridge usually breaks long before the transfer is complete. These frequent interruptions of mating lead to an approximately exponential decay in the transfer frequency of markers as their distance from the origin of transfer increases.

Table 3.6 lists some data for a typical Hfr cross, and Figures 3.29 through 3.31 illustrate how these data are analyzed. In this example, we are trying to map a genetic marker on the *E. coli* chromosome defined by the mutation, *rif-8*, that confers resistance to the antibiotic rifampin (Rif^r). A partial *E. coli* genetic map is shown in Figure 3.30, which shows the positions on the chromosome of all of the markers we are using except the *rif-8* marker, which we are trying to map. Also shown on the map is the site of integration of the F plasmid in some Hfr strains and the direction in which they transfer, drawn as an arrow. According to convention, they transfer as though the arrow was being shot into the recipient cell. The donor Hfr strain we are using is PK191, which can be seen to have the F plasmid integrated at 42 min and oriented so that it transfers the chromosome in the clockwise direction. This strain also requires proline because of a small deletion that removes a few genes around the *lac* region including the *proC* gene Δ(*lac-pro*) but is wild type for all of the other genetic markers being used. Normally, if the alleles are not mentioned, they are assumed to be wild type; therefore, the donor can be assumed to be wild type for the other genetic markers. The recipient strain has mutations that make it require the amino acids histidine (*hisG1*), arginine (*argH5*), and tryptophan (*trpA3*). It also has the *rif-8* mutation being mapped, so it is resistant to rifampin (Rif^r). *E. coli* and most other bacteria are sensitive to this antibiotic, which binds to RNA polymerase, unless they have a resistance mutation, which changes the RNA polymerase β subunit (see chapter 2).

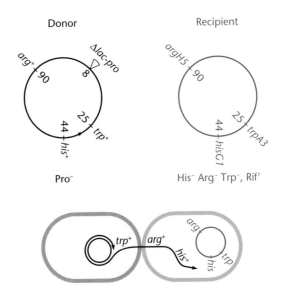

Figure 3.29 Mapping by Hfr crosses. The phenotypes and positions of the mutations in the genetic maps of the donor and recipient bacteria are shown. The chromosome is transferred from the donor to the recipient, starting at the position of the integrated self-transmissible plasmid (arrowhead). The plating media used to select the markers are also shown.

Selecting Recombinants for One of the Markers

First we need to cross the donor and recipient and select recombinants for one of the markers. The Hfr donor strain is mixed with the recipient strain, and the mixture is incubated for a sufficient time to permit transfer of the entire chromosome (more than 100 min for *E. coli* at 37°C). The mating mixture is then plated under conditions in which *neither* the donor *nor* the recipient can grow, only the recombinants being selected; otherwise one or both of the parent strains would grow up and cover the plates, making the detection of recombinants impossible. Plating under conditions where the donor cannot multiply is known as **counterselecting** the donor. In this case, the donor can be counterselected by omitting proline from the plates, since the donor requires proline for growth. To plate under conditions where only recipients that are recombinant for a particular marker can multiply, the selective plates should contain two of the three amino acids required by the recipient but lack the one that corresponds to the selected marker (e.g., if the selected marker is the region of the *hisG* mutation, the medium should lack histidine). Then only recipient bacteria that are recombinant for the *hisG* marker or

TABLE 3.6	Typical results of an Hfr cross			
Selected marker	Percent recombinant for unselected markers			
	hisG	*trpA*	*argH*	*rif*
hisG		1	7	6
trpA	33		29	31
argH	28	12		89

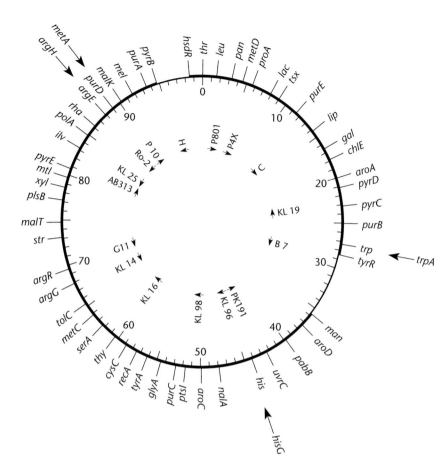

Figure 3.30 Partial genetic linkage map of *E. coli* showing the positions (large arrows) of the known markers used for the Hfr gradient of transfer in Figure 3.31. The small arrows indicate the position of integration of the F plasmid in some Hfr strains, including PK191 (located near the position of *hisG* at 44 min). In each of these Hfr strains, the chromosomal DNA is transferred like an arrow being shot into the recipient cell, beginning from the tip of the arrow.

His+ will be able to multiply and form a colony on the plates. The plates should also contain arginine and tryptophan and lack rifampin so that the His+ recombinants will form colonies whether or not they are also recombinant for any of the other markers, in other words whether they are Arg+ or Arg−, Trp+ or Trp−, or Rifʳ or Rifˢ. Similarly, to select the region of the *argH* marker, the mating mixture is plated on minimal plates plus tryptophan and histidine; to select the *trpA* marker, the mixture is plated on minimal plates plus arginine plus histidine. Bacteria that have been selected for one of the markers, and must therefore have received DNA from this region of the donor DNA, are called **transconjugants.**

Testing for Recombinants for Unselected Markers

After transconjugants recombinant for one of the markers have been selected, they are further purified on the selective plates to eliminate any contaminating parental bacteria. They are then tested to determine if they are also recombinant for one or more of the unselected markers by patching them on plates selective for the other markers. For example, if we are testing the region of the *arg* mutation as the unselected marker, we test the

transconjugants on plates lacking arginine but containing the other growth supplements. If they grow on these plates, they are also recombinant for the *arg* marker since that is the genotype of the donor. If the *rif* mutation is being tested, we patch on plates containing all the growth supplements and rifampin. If they do not grow, they are recombinant for the *rif* marker since that is the genotype of the donor. Table 3.6 shows some representative data in which one of the markers has been selected, and the percentage of transconjugants recombinant for this marker that are also recombinant for each of the unselected markers is given. Remember that the recipient is recombinant for a marker when it has the allele of the donor.

It is apparent from the data that the frequency of recombinants for unselected markers depends greatly on whether the region of the marker is transferred in before or after the region of the selected marker. Any marker that transfers before the selected marker must have already entered the recipient cell in order for the cell to become recombinant for the selected marker. Once it enters the recipient cell, a marker will show about the same frequency of recombination into the chromosome

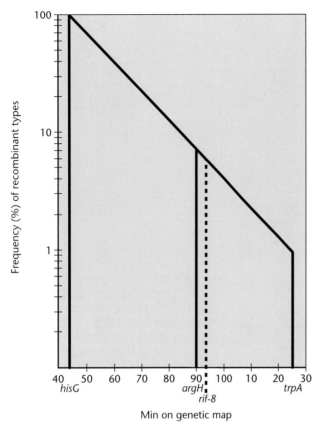

Figure 3.31 Mapping by gradient of transfer during an Hfr cross. The ordinate shows the frequency of each unselected marker with *his* as the selected marker. The abscissa is the distance in minutes from the selected marker. The dashed line shows an estimate of the position of *rif-8* based on the percentage of Rif^s recombinants from the data in Table 3.6.

(about 30% in *E. coli* K-12), so that all the unselected markers that come in before the selected marker show a recombination frequency of about 30% and are of little use in mapping. This is why, to map by gradient of transfer, we need to select a marker that comes in before all of the other markers. To determine which marker will come in first in our mating, refer to the *E. coli* genetic map shown in Figure 3.30. From the picture, we can see that PK191, in which the F plasmid is integrated at about 42 min and transfers counterclockwise, should transfer the *hisG* marker at about 44 min very early, probably before the unknown *rif* marker, and so we use the data in which *hisG* was the selected marker. We then construct a standard curve on semilog paper on which we plot the frequency of recombinants for each known unselected marker versus its distance in minutes from the *hisG* marker on the genetic map. Such a plot is shown in Figure 3.31. This standard curve should appear as a more or

less straight line on semilog paper since, as mentioned, the frequency of transmission of markers falls off approximately exponentially the farther they are from the selected marker. The frequency of recombinants of the unknown *rif* marker (those that are Rif^s) is then placed on this line, reading down to determine what map position would give rise to this recombination frequency. These data place the site of the *rif-8* mutation at approximately 90 min, close to the *argH* marker.

The placement of the site of the *rif-8* mutation close to *argH5* is also supported by the results obtained when *argH5* was the selected marker. A very high percentage (89%) of the recombinants selected for being Arg$^+$ are also Rif^s and so are recombinant for the *rif* marker. Apparently, few crossovers occurred between the regions of the *argH* and *rif* markers when the *argH* region of the donor replaced the *argH* region of the recipient, indicating that the two markers are very closely linked. If markers are much farther apart than this, so many crossovers occur between the two markers that such genetic linkage is not apparent and only the time they enter the cell determines their recombination frequency.

A CAVEAT

The interpretation of mapping data from Hfr crosses is fairly straightforward but can be complicated if the marker being mapped is too close to the marker used to counterselect the donor. In the example, if the *rif-8* mutation were very close to the *proC* mutation of the Hfr donor, there would have been very few crossovers between the two mutations, so that most of the transconjugants recombinant for the *rif* marker would also have been Pro$^-$ and not able to grow on the selective plates lacking proline. Accordingly, to get a reliable map position for an unknown marker, it is best to counterselect the donor with a marker that comes in very late, so that it does not interfere with the frequency of recombinants for any of the other markers. Detailed protocols for use of Hfr strains for mapping are given in Low, Suggested Reading.

Hfr mapping is a powerful technique to locate genetic markers on the entire chromosome, but it can be used only if some previous mapping information is available for the bacterium so that the positions of some markers on the chromosome are known. Strains must also be available that have an integrated self-transmissible plasmid which transfers the chromosome at a reasonably high frequency. Therefore, this method has now been largely supplanted by genomic sequencing. The chromosomal DNA of many types of bacteria has now been sequenced, and this sequence information is often used to locate genetic markers on the chromosome. A piece of DNA containing the region of the genetic marker is somehow cloned, for example by complementation or marker rescue (see below), and partially sequenced. Using readily available software and databases, the sequence can then quickly be located in the entire DNA of the bacterium.

Mapping of Bacterial Markers by Transduction and Transformation

Transformation and transduction are the two other ways that DNA can be transferred from one bacterium to another. In transformation, the DNA from one bacterium can enter another bacterium. In transduction, DNA from one bacterium is transferred to another bacterium in a bacteriophage head, so that when the bacteriophage infects another bacterium the bacterial DNA is injected. In both cases, pieces of chromosomal DNA from one strain, the donor strain, enter another strain, the recipient strain. If the two strains have different genetic markers, recombination can occur producing recombinant types and allowing genetic mapping experiments. An illustration of transduction creating a recombinant is shown in Figure 3.32. Recipient bacteria that have received DNA from the donor by transformation or transduction are called **transformants** or **transductants,** respectively. The major difference is that single-stranded DNA enters the cell during transformation whereas double-stranded DNA enters during transduction. However, the double-stranded DNA introduced by transduction is probably quickly converted into single-stranded DNA by an enzyme called the RecBCD nuclease, as discussed in chapter 10. The interpretation of mapping data is similar for these two methods, so they are treated together in this section. Details of how each of these occurs is given in chapters 6 and 7, respectively.

Like Hfr mapping, mapping by transformation or transduction is based on genetic markers, with one marker being the selected marker and the other markers being the unselected markers. However, rather than being based on gradient of transfer as in Hfr mapping, mapping by transformation or transduction is based on whether the regions of markers can be carried in on the same piece of DNA, i.e., can be **cotransformed** or **cotransduced.** If a strain which has become recombinant for the selected marker sometimes also becomes recombinant for another unselected marker, the regions of the two markers are **cotransformable** or **cotransducible,** respectively. Only fairly small pieces of DNA can enter the cell in transformation, and a phage head will hold only a small piece of DNA in transduction; therefore, markers that are cotransformable or cotransducible must be very close to each other. Furthermore, the regions of the two markers are usually on the same piece of DNA and not on different pieces that came in separately. Both

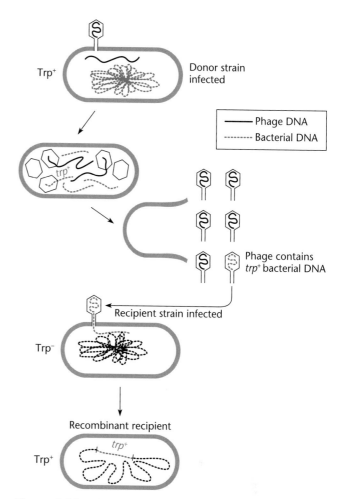

Figure 3.32 An example of generalized transduction. A phage infects a Trp⁺ bacterium, and in the course of packaging DNA into heads, the phage mistakenly packages some bacterial DNA containing the *trp* region instead of its own DNA into a head. In the next infection, this transducing phage injects the Trp⁺ bacterial DNA instead of phage DNA into the Trp⁻ bacterium. If the incoming DNA recombines with the chromosome, a Trp⁺ recombinant transductant may arise. Only one strand of the DNA is shown.

methods are inefficient enough that the chances of more than one piece of DNA entering the same cell and producing recombinants can be considered negligible.

Not only does the appearance of cotransformants or cotransductants which are recombinant for two markers signify that the markers are close to each other in the DNA, but also the higher the percentage of recombinants for the selected marker that are also recombinant for the unselected marker, the closer together the two markers are likely to be. The percentage of the total transformants or transductants selected for one marker that are also recombinant for the other marker is called the cotransformation frequency or cotransduction frequency of the two markers, respectively. In principle, this frequency between two markers should be a constant for any two markers and should be independent of which of the two markers is the selected marker and which is the unselected marker. A cross with the selected and unselected markers reversed is called a **reciprocal cross.** Note that, as with Hfr crosses, recombinants form only in the recipient cell and a recombinant for a marker has the allele of the donor. In the next section, we illustrate how mapping data from transductional crosses is interpreted by using an actual example. Similar reasoning would apply to transformational crosses.

MAPPING BY COTRANSDUCTION FREQUENCIES
By the Hfr crosses above, we had determined that the *rif-8* mutation lies somewhere in the vicinity of the *argH* gene in the chromosome, and we wish to further localize it with respect to the other genes in this region. Transduction or transformation can be used to further define the map position of a genetic marker.

In the example shown in Figure 3.33, phage P1 is being used for transduction to further refine the mapping of the *rif-8* mutation to resistance to the antibiotic rifampin. The first step is to determine if the regions of the *argH* and *rif* markers are close enough to each other on the DNA to be cotransducible. If they are cotransducible, transductants selected for one of the markers are sometimes also recombinant for the other marker. For practical reasons, it is easier to use the *argH* marker than the *rif* marker as the selected marker. Rifampin resistance is recessive and takes many generations to be expressed because the rifampin-sensitive RNA polymerase molecules bind to promoters and block the resistant ones until the sensitive ones are diluted out by many cell divisions. Therefore, to select rifampin-resistant transductants, it is necessary to express the transductants for many generations in the absence of rifampin before they are plated in the presence of the antibiotic. It is also much easier to select Arg⁺ transductants than Arg⁻ transductants, since the former can be selected on minimal plates without arginine and any transductants that form colonies are Arg⁺ and therefore recombinant for the *arg* marker. Thus, we shall select the *argH* marker by using a donor bacterium that is Arg⁺ and a recipient that has the *argH5* mutation. The *rif-8* mutation, the unselected marker, can be in either the donor or the recipient; we will have it be in the recipient so that we can use the same recipient strain we used for the Hfr mapping.

In the experiment illustrated in Figure 3.33, the transducing phage are grown on donor cells that are wild type for the *argH* and *rif* genes and so are phenotypically Arg⁺ Rifˢ. A few of these phage will pick up chromosomal

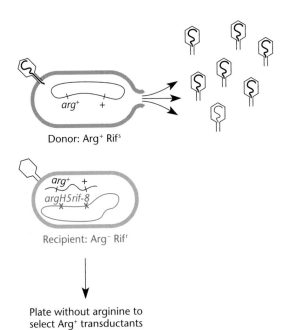

Figure 3.33 Cotransduction of two bacterial genetic markers. The regions of the *argH* and *rif-8* mutations are in close enough proximity that both regions can be carried on a piece of DNA fitting into a phage head (not shown to scale). After transduction, some of the Arg⁺ transductants are also rifampin sensitive. Thick lines indicate phage genomes in progeny phage; thin lines indicate bacterial DNA in transducing particles. See the text for details.

DNA by mistake, and some of these will contain the *arg* region. When this mixture of phage is then used to infect recipient cells, the phage that contain chromosomal DNA from the *arg* region will inject this DNA, which can then recombine with the chromosome to make an Arg⁺ recombinant as shown. Even though rare, these few Arg⁺ transductants can be selected by plating the infected cells on minimal plates without arginine. If the region of the *rif-8* mutation is close enough to the region of the *argH5* mutation, sometimes the same piece of DNA includes this *rif* region. If so, then some of the Arg⁺ transductants will also become recombinant for the rif marker. This becomes apparent when the Arg⁺ transductants are purified and tested on plates containing rifampin and some of them are found to be rifampin sensitive and do not grow. As before, the transductants that are recombinant for the *rif-8* marker are rifampin sensitive since that is the allele of the donor. The data in Table 3.7 show that 22 + 11 = 33 of 96, or ~34%, of the Arg⁺ transductants are also Rifˢ. Thus, the two markers are cotransducible, and the cotransduction frequency is about 34%. Note that if we did the reciprocal cross,

TABLE 3.7	Typical transductional data from a three-factor cross
Recombinant phenotype	No. of recombinants
Arg⁺ Met⁺ Rifʳ	61
Arg⁺ Met⁺ Rifˢ	22
Arg⁺ Met⁻ Rifʳ	2
Arg⁺ Met⁻ Rifˢ	11

grew the phage on a donor with the *rif-8* mutation and used it to transduce an *argH5* mutant, selecting for rifampin-resistant transductants on plates containing rifampin, and then tested for the *argH* marker, the result should be about the same. About 34% of the Rifʳ transductants should be Arg⁻, again indicating that the regions of the *rif* and *argH* markers are 33% cotransducible. However, as mentioned, it is difficult to select Rifʳ transductants.

We can estimate how close together on the *E. coli* DNA the *argH* and *rif* markers would have to be so that they could be cotransducible by P1 phage. The chromosome is 100 min long, and the P1 phage head holds only about 2% of that length (see Table 7.2); therefore, the two markers must be less than 2 min apart. Translating this distance into base pairs of DNA, the *E. coli* chromosome is about 4.5×10^6 bp long, and the P1 phage head holds only $0.02 \times 4.5 \times 10^6$ or 90,000 bp. Thus, to be cotransducible, two markers in the DNA must be less than 90,000 bp or 2 min apart on the 100-min map shown in Figure 3.30.

ORDERING THREE MARKERS BY COTRANSDUCTION FREQUENCIES

As mentioned, the closer together two markers are in the DNA, the more likely they are to be carried in the same phage head and the higher their cotransduction frequency. Therefore, cotransduction frequencies can also be used to determine which markers are closest to each other on the DNA and therefore to determine the order of markers. To illustrate, we shall use cotransduction to order the *argH5* and *rif-8* markers with respect to another marker in this region, due to a mutation in the *metA* gene, whose product is required to make methionine. In the transductional cross shown in Table 3.7, the donor also had the *metA15* marker and so it required methionine (Met⁻). The transductants recombinant for the *metA* marker are therefore Met⁻, since this is the allele of the donor, and do not grow on plates lacking methionine. In the example, $2 + 11 = 13$ of the Arg⁺ transductants are Met⁻, so the cotransduction frequency of the *argH* and *metA* markers is ~14%. We have already determined that the *argH* and *rif* markers are

about 34% cotransducible. Hence, the order seems to be *argH-rif-metA*, with the *arg* marker closer to the *rif* marker than it is to the *met* marker. It also seems possible that the order is *rif-argH-metA*, with *argH* in the middle. However, if the *rif* marker and the *metA* marker are on opposite sides of the *argH* marker, the Arg⁺ transductants that were recombinant for the *metA* marker should be less apt to be recombinant for the *rif* marker on the other side. This does not seem to be the case since 11 of 13 of the transductants that were Met⁻ were Rifˢ and only 2 were Rifʳ. However, to be sure of this order and to determine that *argH* is not in the middle, we might want to do the reciprocal cross, growing the phage on the Met⁺ Rifʳ Arg⁻ strain and using them to transduce the strain that is Met⁻ due to the *metA15* mutation, selecting Met⁺ transductants. If *argH* is not in the middle, the cotransduction frequency of the *rif* marker with the *metA* marker should be higher than that of the *argH* marker with the *metA* marker.

ORDERING MUTATIONS BY THREE-FACTOR CROSSES

A careful determination of cotransduction frequencies can reveal the order of markers in the DNA. However, three-factor crosses offer a less ambiguous way to determine marker order. This technique can be used in the genetic mapping of any organism in which genetic crosses are possible and is mentioned in chapter 7 in connection with ordering mutations in phage. This analysis is based on how many crossovers are required to make a certain recombinant type with a given order of markers. We have already pointed out that a single crossover between a short linear piece of the chromosome and the entire chromosome will break the chromosome and be lethal. Therefore, in bacterial crosses, a minimum of two crossovers are required to replace the chromosome sequence with the sequence on the incoming donor DNA and form a recombinant type. In general, in such a cross, odd numbers of crossovers (one, three, five, etc.) will break the chromosome and be lethal; therefore, any viable recombinant types must have originated from an even number of crossovers (two, four, six, etc.).

To illustrate the ordering of bacterial markers by transductional three-factor crosses, we again use the example of the *argH*, *metA*, and *rif* markers and the data in Table 3.7 in which the donor has the *metA15* mutation and the recipient has the *argH5* and *rif-8* mutations. To obtain these data, the *argH* marker was selected and the Arg⁺ transductants were tested for the unselected *metA* and *rif* markers. There are four recombinant types possible in a cross of this type, and they are listed in Table 3.7. With any particular order of the three markers, three of the four recombinant types listed require only two crossovers and the fourth requires four crossovers. The recombinant type that requires four crossovers should be rarer than the others. Therefore, a determination of which recombinant type is rarest should reveal the order of the three markers. In Table 3.7, the rarest recombinant type is clearly Arg$^+$ Met$^-$ Rif^r, since only 2 of the 100 Arg$^+$ transductants were of this type. Figure 3.34 shows how many crossovers should be required to make this recombinant type if the order is *argH-metA-rif* or if the order is, as we suspect, *argH-rif-metA*. Only two crossovers are required to make this recombinant type if the order is *argH–metA–rif* (with the *metA* marker in the middle), while four crossovers are required if the order is *argH-rif-metA*. It seems clear that the order is *argH–rif–metA*, which is also consistent with the order we obtained based on cotransduction frequencies alone.

Incidentally, since resistance to rifampin is due to a mutation in the gene for the β subunit of RNA polymerase (see chapter 2), the gene for the β subunit of RNA polymerase must lie between the *argH* and *metA* genes in the chromosome of *E. coli*.

Other Uses of Transformation and Transduction: Strain Construction

Unlike Hfr crosses, which have been largely supplanted by genomics, transformation and transduction continue to be very useful in modern bacterial molecular genetics. For many strains of bacteria, there still is no better way to do strain construction and introduce known mutations into the chromosome. Not all strains of bacteria can be transformed efficiently by chromosomal DNA, and this process is limited largely to bacteria that exhibit natural transformation. However, electroporation, a way of forcing DNA into cells via an electric field, has expanded this technique to many other species (see chapter 6). Transduction requires finding a transducing phage strain for the bacterium, which is not always easy (see chapter 7). However, transduction is so useful in molecu-

lar genetic studies that the search continues for transducing phages for many types of bacteria.

USING TRANSFORMATION AND TRANSDUCTION FOR STRAIN CONSTRUCTION

One of the major uses of transformation and transduction is in constructing isogenic bacterial strains. Different strains of the same species can differ at a number of genetic loci. To be certain that a phenotypic difference between two strains is due to a particular genetic difference, it is essential to eliminate other genetic differences as being at least partially responsible. Therefore, meaningful experiments often require comparing strains that differ by only one small genetic difference but are otherwise identical. Such strains are said to be **isogenic.** Transformation and transduction introduce only a small region of the chromosome, and so any differences between the original recipient strain and a transductant must have been carried in on the same piece of DNA. If other genetic differences are contributing to phenotypic differences between the two strains, they must be very closely linked to the mutation being introduced.

Sometimes we can use transformation or transduction to move mutations into a strain even if the mutation has no easily selectable phenotype. For example, we might use a closely linked transposon carrying an antibiotic resistance gene to move such a mutation. For this purpose, collections of *E. coli* strains have been assembled which have transposon insertions around the genome that are all cotransducible with at least one other transposon insertion in the collection. Therefore, the site of any mutation is cotransducible with at least one of the transposon insertions. We can use the collection as a donor for transduction, selecting the antibiotic resistance gene on the transposon and testing a number of the transductants for the mutation. A recipient strain that has lost the mutation will probably have the transposon integrated close to the site of the mutation. We can then repeat the transduction with this isolated strain as a donor but this time save a transductant that has retained the mutation, discarding the ones which have lost it due to cotransduction. These transductants have the transposon inserted close to the mutation, and so they can be used to move the mutation into other strains, selecting the antibiotic resistance gene on the closely linked transposon and thereby easily constructing many isogenic strains with the mutation, even if it has no easily selectable phenotype.

REVERSION VERSUS SUPPRESSION

Another use of transformation or transduction is to distinguish revertants from strains with suppressor mutations. As mentioned earlier in this chapter, the phenotypic

Figure 3.34 The number of crossovers required to make the rarest recombinant type, Arg$^+$ Rif^r Met$^-$, with two different orders of the three markers. Since order II, *argH–rif-8–metA*, requires four crossovers, this is probably the order of the three markers. See the text for details.

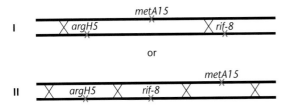

change due to these two types of mutations can be very similar even though their molecular basis is very different. Figure 3.35 shows the generic recombination test for suppression, applied to apparent revertants of a *his* mutation obtained as in Figure 3.21. If the His⁺ apparent revertants are true revertants, the original *his* mutation no longer exists and the recombinants are all still His⁺. However, if the original mutation has been suppressed, recombination can sometimes separate the original mutation from the suppressing mutation, called *supX* in the figure, and some of the recombinants will be phenotypically His⁻, because the original *his* mutation is no longer suppressed in these recombinants.

Any of the means of genetic exchange in bacteria can be used to distinguish reversion from suppression. However, transduction and transformation are particularly useful in this regard. Returning to the transduction example in Figure 3.34, suppose we have plated large numbers of our donor bacterium with the *metA15* mutation on plates without methionine and get a few apparent Met⁺ revertants that can now grow without methionine in the medium. We want to know whether the original *metA15* mutation has reverted or whether it has been suppressed in some cases by a mutation else-

where in the chromosome. For example, what if the original *metA15* mutation were a nonsense mutation? Then a mutation in a tRNA gene elsewhere in the chromosome could create a nonsense suppressor and suppress the mutation, leading to the Met⁺ phenotype even though the original *metA15* mutation is still present.

Figure 3.36 illustrates how transduction could be used to distinguish reversion from suppression of the *metA15* mutation in *E. coli*. In the example, a Met⁺ apparent revertant is used as a donor to transduce the Arg⁻ recipient, selecting for Arg⁺ transductants. If the *metA15* mutation has reverted, none of the Arg⁺ transductants will be Met⁻ because the *metA15* mutation no longer exists in this strain since it has reverted. However, if the *metA15* mutation has not reverted but has been suppressed by a mutation elsewhere in the chromosome, about 14% of the Arg⁺ transductants should also be Met⁻, as before, since the *metA15* mutation is still there and the recipient strain presumably does not have the suppressor mutation.

MARKER RESCUE

A very useful application of transformation is in identifying clones of chromosomal genes by **marker rescue**. Recombination between a piece of DNA introduced into

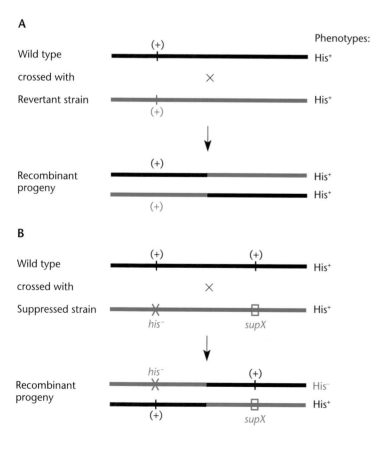

A

Phenotypes:

Wild type — (+) — His⁺

crossed with ✕

Revertant strain — (+) — His⁺

Recombinant progeny — (+) — His⁺
(+) — His⁺

B

Wild type — (+) (+) — His⁺

crossed with ✕

Suppressed strain — ✕ □ — His⁺
his⁻ *supX*

Recombinant progeny — *his⁻* (+) — His⁻
✕
(+) □ — His⁺
supX

Figure 3.35 Test for reversion versus suppression. (A) The mutation had reverted, giving the His⁺ phenotype. Purple (+) shows the site of the reversion mutation. When the revertant strain is crossed with the wild type (in black), no His⁻ recombinants appear in the progeny. (B) A suppressor mutation, *supX*, has suppressed the mutation, giving the His⁺ phenotype. When the suppressed strain is crossed with the wild type (in black), it gives some His⁻ recombinants. The site on the DNA with the *his* mutation is shown as a purple ✕, and the site of the suppressor mutation is shown as the purple box.

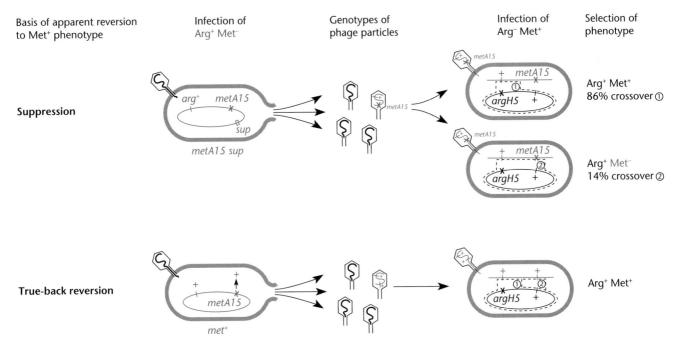

| Basis of apparent reversion to Met+ phenotype | Infection of Arg+ Met− | Genotypes of phage particles | Infection of Arg− Met+ | Selection of phenotype |

Figure 3.36 Using transduction to distinguish reversion from suppression. If the *metA15* mutation has been suppressed, about 14% of the Arg+ transductants will be Met−. If the mutation has reverted, none of the Arg+ transductants will be Met− because the *metA15* mutation no longer exists.

the cell by transformation and the corresponding region in the chromosome containing a mutation can "rescue" the mutation in the chromosome, restoring the wild-type phenotype. Figure 3.37 illustrates the use of marker rescue to identify a clone containing at least part of the *thyA* gene of *E. coli*. As in complementation cloning of the *thyA* gene illustrated in Figure 3.25, a library of wild-type *E. coli* DNA is introduced into a *thyA* mutant strain of *E. coli*. However, the cells containing the various clones are now plated on nonselective plates containing thymine and the plates are incubated to allow the colonies to develop. Each plate is then replicated onto a selective plate containing all the necessary growth supplements but lacking thymine. If a particular clone includes the part of the *thyA* gene containing the site of the mutation, recombination between the clone and the chromosome can give rise to some Thy+ recombinants within the original colony. These Thy+ recombinants can grow to produce small colonies on the selective plate, as shown. Cloning by marker rescue has advantages over cloning by complementation in that the clone need not contain the entire gene and the gene does not need to be expressed. However, it has the disadvantage that the clone is irretrievably altered by the marker rescue recombination, so that clones carrying the gene cannot be selected directly and there must be some way of return-

ing to the original unaltered clone that showed marker rescue. That is why we replica plated the original colonies onto the selective plates, so that we could go back to the original colony on the permissive plate to obtain the clone once it had been identified.

Another use of marker rescue is in mapping the sites of mutations within a gene. Figure 3.38 shows an example of using this method to map mutations within the *thyA* gene of *E. coli*. Clones of the *thyA* gene were constructed with deletions extending different distances into the gene from one side (i.e., nested deletions). If a deleted clone can give Thy+ recombinants when introduced into a cell containing a particular *thyA* mutation in the chromosome, then the deleted clone must retain the region of the gene containing the mutation. The pattern of deletions that still show marker rescue localizes the mutation to a particular site in the gene.

Gene Replacements and Reverse Genetics

One of the most useful current technologies involving recombination following transformation is its use in introducing foreign DNA into the chromosome of an organism. In some applications, recombination is used to replace the normal gene of an organism with a particular mutated

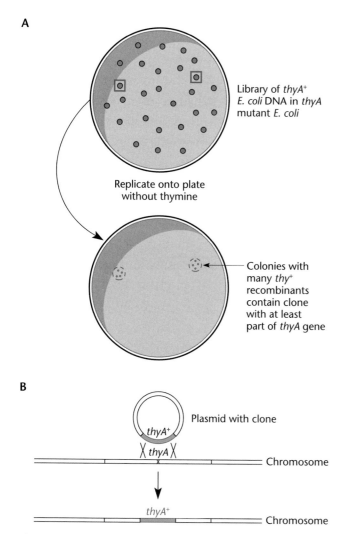

Figure 3.37 Use of marker rescue to identify a clone containing at least part of the *thyA* gene of *E. coli*. (A) The boxed colonies correspond to Thy⁺ recombinants grown on a replicated plate. See the text for details. (B) The *thyA*⁺ gene on the cloning vector recombines with the *thyA* mutant gene on the chromosome to produce Thy⁺ recombinants.

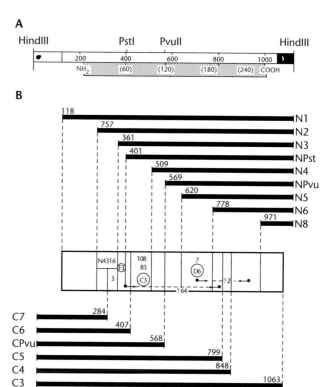

Figure 3.38 (A) Map of the *thyA* gene of *E. coli*. (B) Mutations in the chromosome were mapped by being crossed with deletions extending various distances into the cloned *thyA* gene of *E. coli*. Solid bars show the regions deleted in each of the constructs. N is the amino terminus of the gene, and C is the carboxyl terminus.

allele of the same gene, often containing a mutation that we have made by site-specific mutagenesis (see chapter 1). Once the mutated DNA sequence has replaced the normal sequence in the chromosome, we can determine the effect of the specific mutation on the phenotypes of the organism. This process of introducing a predetermined mutation into the DNA of an organism is sometimes called **reverse genetics** because it is essentially the reverse of normal genetic analysis. In reverse genetics, first we make the mutation and only afterward do we see the effect of the mutation on the organism. In classical genetics, we know that the mutation has occurred because of its effect on the organism, and only afterward do we clone and sequence the DNA to determine what kind of mutation

caused the phenotype. In some cases we wish to use reverse genetics to inactivate a specific gene and see what effect this has on the organism. This is often called a knockout, and it can be achieved by introducing an **antibiotic resistance gene cassette** into the gene or deleting part or all of the gene in the clone before it is reintroduced. Recombination can also be used to introduce new genes into the chromosome of an organism, for example a gene for antibiotic resistance. An organism with foreign DNA in its chromosome is sometimes called a **transgenic organism,** and this process is called **transgenics.**

An overview of gene replacement is provided in Figure 3.39. In the illustration, a short piece of DNA that is homologous to part of the chromosome is introduced into a cell by transformation. This is a similar situation to the recombination following bacterial crosses, where generally only a small part of the DNA of one parent is introduced into the other parent. Gene replacement occurs through homologous recombination between the introduced DNA and homologous sequences in the chromosome. The number of crossovers required to introduce foreign DNA into the chromosome depends on

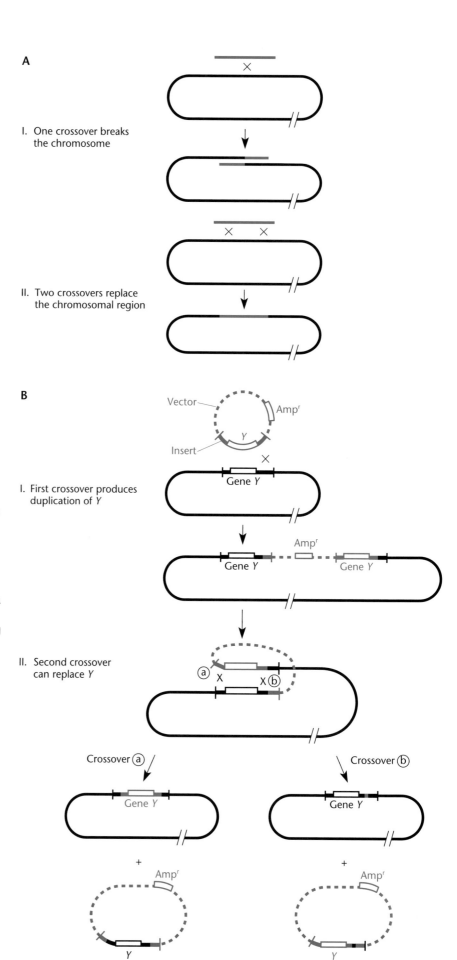

Figure 3.39   Gene replacements.
(A) The introduced DNA is a linear
piece of the chromosome with a
slightly altered sequence. (I) A single
crossover between the short linear
piece of DNA (shown in purple) and
the corresponding homologous region
in the chromosome (shown in black)
will break the chromosome and be
potentially lethal. (II) Two crossovers
are required to replace the sequence
in the chromosome with the altered
sequence in the introduced linear DNA
(shown in purple). (B) The introduced
piece of chromosomal DNA containing
gene *Y* with a specific mutation is
cloned in a circular plasmid carrying
a gene for resistance to ampicillin
(shown in purple). (I) A single
crossover between the cloned DNA
and the corresponding homologous
region in the chromosome will insert
the plasmid, bracketing it with the
chromosomal region containing a
normal gene *Y* (shown in black) and
the plasmid clone containing the
mutated copy of gene *Y* (shown in
purple). (II) A second crossover can
loop out the plasmid, leaving only
one copy of gene *Y* in the chromosome. Depending on where this
second crossover occurs, the copy
of gene *Y* left in the chromosome
can be either the mutant copy (shown
in purple; crossover a) or the original,
wild-type copy (shown in black;
crossover b).

whether the introduced DNA is linear or circular. As illustrated in Figure 3.39A and as mentioned earlier in the chapter, a single crossover between a short linear DNA and the much longer circular chromosome leads to breakage of the DNA, usually a lethal event. Two crossovers are required to replace the chromosomal sequence between the regions of the two crossovers with the sequence of the introduced DNA.

A very different situation prevails if the introduced DNA is circular (Figure 3.39B). This situation is analogous to a region of the chromosome cloned into a circular plasmid cloning vector and then altered by site-specific mutagenesis in vitro, which is often the starting point for a gene replacement. When the plasmid containing the cloned DNA is then introduced into the cell by transformation or any other means, a single crossover between the mutated cloned DNA and the corresponding homologous sequence in the chromosome is not lethal and will integrate the circular plasmid into the chromosome, as shown. However, the altered cloned DNA sequence does not replace the corresponding sequence in the chromosome. Rather, the homologous sequences on the cloned DNA and the chromosome where the crossover occurred now bracket the integrated plasmid vector as shown, leading to a duplication of these sequences. A second crossover between the original sequence in the chromosome and the same chromosomal sequence duplicated in the plasmid, which are now flanking the plasmid sequences, can excise the circular plasmid DNA. Depending on where it occurs, this second crossover can either restore the original sequence in the chromosome or replace it with the mutated sequence that had been in the plasmid.

Figure 3.40 shows a similar situation, but now we want to replace a gene in the chromosome with the same gene into which we have introduced a cassette that confers resistance to the antibiotic kanamycin (Kan^r). This inactivates the gene product, allowing us to determine what role the gene product plays in the cell. This shows that the sequences of the introduced DNA and the chromosome need not be homologous over their entire lengths but only over the regions where the two crossovers occur. If two homologous sequences are on either side of a foreign DNA sequence, in the example the gene for kanamycin resistance, recombination between the two homologous sequences called the **flanking sequences** and their corresponding sequences in the chromosome can insert the foreign DNA into the chromosome and make a transgenic organism containing the newly introduced Kan^r gene.

In practice, gene replacements are not as straightforward as presented above and there are technical difficulties that must be overcome. First, there must be some way of selecting the cells in which the gene replacement has

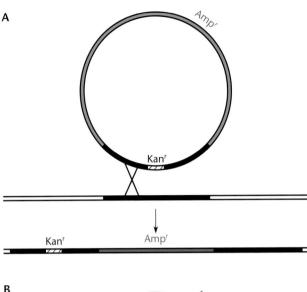

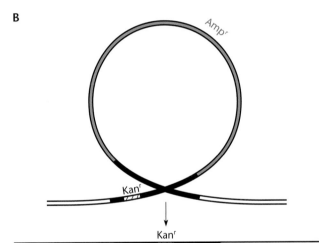

Figure 3.40   Gene replacement. (A) With a single crossover, the cloning vector integrates into the chromosome and the cells become ampicillin resistant (Amp^r) and kanamycin resistant (Kan^r). (B) With a second crossover, the cells lose the cloning vector. Depending on where this second crossover occurs, the cells may be left with only the cloned sequence with the kanamycin resistance cassette and be resistant to kanamycin alone. See the text for details. Plasmid cloning vector sequences are in purple. The cloned region of the chromosome is in black.

occurred. The recombination events that lead to gene replacement are relatively rare, and only a few cells in a population ever have their normal gene replaced by the mutated copy. Gene replacements are relatively easy to select if the mutation that alters the cloned gene is an insertion of a selectable gene such as a gene for antibiotic resistance. Not only does introducing an antibiotic resistance cassette into the gene both disrupt the gene in the clone and almost certainly inactivate the gene product, but also it introduces a selectable marker, which can be used later to select the few cells in which the replacement has occurred. To illustrate how these selections work, return to the example shown in Figure 3.40, in which a kanamycin resistance cassette (Kan^r) has been introduced into a cloned gene. To select cells in which the Kan^r cassette has been introduced, the cells are plated on medium containing kanamycin. This allows the growth only of cells in which the clone containing the kanaymcin resistance cassette has recombined into the chromosome. In the example, the plasmid cloning vector has a gene for ampicillin resistance (Amp^r). If the plasmid has integrated into the chromosome by a single crossover, the cells will become ampicillin resistant in addition to kanamycin resistant. However, if a second crossover excises the plasmid from the chromosome and replaces the normal gene in the chromosome with the gene disrupted by the kanamycin resistance cassette, the cell will be kanamycin resistant but ampicillin sensitive, allowing detection of the cells in which the second crossover has occurred.

Another problem arises if the DNA is introduced into the cell in a vector that can replicate autonomously in the cell, i.e., is a replicon. This is often the case if the DNA is introduced in a plasmid cloning vector which has its own origin of replication and so can replicate autonomously. Then the cells retain the antibiotic resistance gene carried on the plasmid, even if the introduced DNA does not recombine with the chromosome since the plasmid can maintain itself independently of the chromosome. This is not usually a problem if the DNA that is introduced into the cell is linear, because most linear DNAs do not replicate in bacteria even if they contain an origin of replication. However, using linear DNA for gene replacements in bacteria has its own limitations since linear DNA is often degraded in the cell by the RecBCD nuclease (see chapter 10). Methods have been devised to allow gene replacements with linear DNA in *E. coli* in which the RecBCD nuclease is partially or completely inactivated; some of these methods are discussed in subsequent chapters. Alternatively, if the cloned DNA for the gene replacement is introduced in a circular plasmid, it is necessary that the cloning vector be somehow converted into a form in which it cannot replicate, i.e., a **suicide vector**. A protein required for replication of the plasmid may have been inactivated by a mutation, or the plasmid might be from an unrelated host and may be unable to replicate in the cells. Then the plasmid containing the clone will be lost from the cell unless the cloned gene recombines with the chromosome. Suicide vectors are also often used for transposon mutagenesis, as discussed in chapter 9.

In many places throughout the book, we introduce methods for gene replacement and site-specific mutagen-

esis in a variety of applications and organisms. Some of these require specialized knowledge, but they are the foundation of **functional genomics**.

Isolation of Tandem Duplications of the *his* Operon in *Salmonella enterica* Serovar Typhimurium

We finish this chapter with an example of genetic analysis in bacteria, the selecting and analysis of tandem duplication mutations of the *his* operon of *Salmonella*. This specific example illustrates the properties of mutations and how genetic data are interpreted. As discussed in a general way earlier in this chapter, tandem-duplication mutations can occur by recombination between directly repeated sequences, causing the duplication of the DNA between the repeated sequences. Such ectopic recombination is fairly rare because repeated sequences are not very common in bacteria. However, once they form, tandem-duplication mutations are usually very unstable because recombination anywhere within the duplicated segments can destroy the duplication. Also, as mentioned, most long tandem-duplication mutations do not cause easily detectable phenotypes because they do not inactivate any genes. Even the genes in which the mistaken recombination occurred to create the duplication exist in a functional copy at the other end of the duplication. Therefore, special methods must be used to select bacteria with duplication mutations.

In our chosen example, transduction was used to select tandem duplications of the *his* region of *Salmonella enterica* serovar Typhimurium (see Anderson et al., Suggested Reading). Their selection depends on the properties of two deletion mutations in the *his* region, Δ*his2236* and Δ*his2527* (Figure 3.41). These deletion mutations complement each other because one ends in *hisC* and the other ends in *hisB*, so they inactivate different genes. However, because the endpoints of the two deletion mutations are very close to each other, crossovers between them occur very infrequently.

The process of using these deletions and transduction to select duplications of the *his* region is illustrated in Figure 3.42. P22 transducing phage were propagated on a strain with one of the deletion mutations and used to transduce a strain with the other deletion mutation. The His$^+$ tranductants were then selected by plating on minimal plates without histidine. A few His$^+$ transductants arose, even though there should be very little recombination between the deletions. Moreover, many of the His$^+$ transductants that arose were unusual. Most were very unstable, spontaneously giving off His$^-$ segregants at a high frequency when they multiplied in the presence of histidine, when they were not exposed to selective

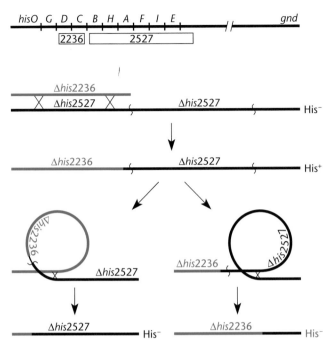

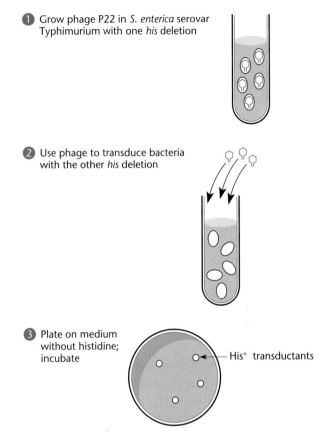

Figure 3.41 Mechanism for generation and destruction of *his* duplications. One of the two copies of the *his* region of bacteria with a preexisting duplication (in the example, Δ*his2527*) recombines with incoming transducing DNA carrying the other deletion (in the example, Δ*his2236*) to replace one copy of the duplicated region with the corresponding region of the donor. The two deletions complement each other to make the bacteria His⁺. The duplication can be destroyed by looping out one of the two duplicated segments, giving rise to His⁻ haploid segregants with one or the other of the two deletions.

Figure 3.42 Using transduction to select bacteria with duplication mutations of the *his* region of *Salmonella enterica* serovar Typhimurium. The P22 transducing phage are grown on bacteria with one deletion and used to transduce bacteria with the other deletion. The transductants are plated on medium without histidine to select for the His⁺ phenotype. See the text for details and conclusions.

pressure. Also, some of the His⁺ transductants were slimy (mucoid) in appearance but the His⁻ segregants had lost this mucoidy. On the basis of these observations, these investigators concluded that unstable His⁺ transductants had tandem duplications of the *his* region (Figure 3.41), in which one copy has the Δ*his2236* deletion whereas the other has the Δ*his2527* deletion. The two deletion mutations are complementing each other, making the transductants His⁺.

Figure 3.41 also shows how these duplications might have arisen. While the recipient bacteria were multiplying prior to the transduction, recombination might have occurred between directly repeated sequences in the DNA of some of the recipient bacteria, creating a tandem duplication of the *his* region as shown. These bacteria are still His⁻ because both copies of the duplication have the same *his* deletion. However, when a bacterium that contains such a duplicated *his* operon is transduced with the *his* region of the donor having the other deletion, the *his* region of the donor DNA can replace one of the dupli-

cated *his* regions of the recipient cell, giving rise to the duplication in which one copy has one deletion and the other copy has the other deletion. The two deletion mutations can then complement each other, giving rise to a His⁺ cell. This type of His⁺ transductant, although rare, might be more frequent than His⁺ transductants produced from recipients that do not have a duplicated *his* operon, since the latter would require recombination in the small region between the two deletions to give a His⁺ recombinant.

The properties of the *his* duplications illustrate many of the characteristics of tandem duplications. One property is their instability. Growing the duplications in the absence of selective pressure (i.e., in the presence of histidine), gives rise to His⁻ segregants at a high frequency. Recombination anywhere in the duplicated regions leaves only one copy of the *his* region with one or the other of the deletions, as is also illustrated in Figure 3.41. These cells are called haploid segregants because they have only one copy of the duplicated region and so they are no longer diploid but rather haploid for this region. They also occur spontaneously without the need for genetic crosses, which is the definition of a segregant (see earlier in this chapter for definitions). The mucoidy (see above) of some of the His⁺ transductants can also be explained. The duplicated region in some of the His⁺ transductants may contain another gene that is now also duplicated and so makes twice as much of its gene product. The colonies appear mucoid when twice as much of this gene product is synthesized. However, in the haploid segregants, there is only one copy of this "mucoidy" gene and the colonies appear normal.

If the instability of the putative duplications results from recombination between the repeated segments, then *recA* mutations, which prevent homologous recombination, should stabilize the duplications (RecA and its role in recombination are discussed in detail in chapter 10). To test this hypothesis, the investigators introduced a *recA* mutation by Hfr crosses into cells containing some of the putative duplications. They introduced the *recA* mutation by crossing with an Hfr strain carrying a *recA* mutation, selecting for a *serA* marker closely linked to *recA*. Ser⁺ transconjugants that had the unselected *recA* mutation were identified by their sensitivity to UV light (see chapter 11). The tandem duplication in these *recA* recombinants was now stable and did not give His⁻ haploid segregants even when grown under nonselective conditions in medium with histidine.

Length of Tandem Duplications

The investigators also used genetic experiments to determine the length of the segment duplicated in some of the strains. In particular, they wanted to know if the duplicated regions in some of the strains ever extended as far as the *metG* gene, about 2 min (~100,000 bp) away from the *his* region in the *E. coli* chromosome (Figure 3.43). To test this, they selected *his* duplications in a strain that also had a *metG* mutation. Strains containing these duplications were propagated without histidine in the medium to maintain selective pressure for the duplication and eliminate haploid segregants. The investigators then transduced the strains a second time with phage propagated on a *metG⁺* donor, selecting for Met⁺ transductants. The Met⁺ transductants were then allowed to segregate into haploids by being grown with histidine and methionine in the medium. Isolated His⁻ segregants were then tested to determine if any were also Met⁻. The reasoning was that if the *metG* gene is included in the duplicated region in a particular duplication, there should be two copies of the *metG* gene, and only one of these will have been transduced to *metG⁺*, as shown in the figure. If so, then some of the His⁻ haploid segregants should also be Met⁻, depending on whether

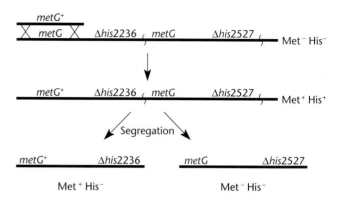

Figure 3.43 Determining if the duplicated segment in a *his* duplication extends as far as the *metG* gene. A duplication is made as before but in a strain with a *metG* mutation. A strain with the duplication to be tested is transduced with phage grown on *metG⁺* bacteria. If the *metG* gene is in fact duplicated, only one of the two copies of the *metG* gene will be transduced to *metG⁺* and when the strain gives rise to His⁻ haploid segregants, some of them will be Met⁻.

they lose the copy which has been transduced to *metG⁺* or the copy which retains the *metG* mutation. The results of this work revealed that many duplications of the *his* operon did include the *metG* gene. In fact, some duplications extended much farther, even as far as the *aroD* gene, which is 10 min (or 10% of the entire genome) away from *his*. The original duplications must have been very long and included many hundreds of genes.

Frequency of Spontaneous Duplications

Lastly, the investigators attempted to estimate the frequency of spontaneous tandem-duplication mutations in a growing population of bacteria. They did this by comparing transduction frequencies when the donor and recipient bacteria had the different deletion mutations used above with the normal transduction frequencies, when only the recipient had one of the deletions. In the first case, most of the recipient cells that were transduced must have had a tandem duplication of the *his* region, whereas in the second case, most of the recipients will have had only one copy of the *his* region. The investigators estimated that duplications occur as frequently as once every 10^4 times a cell divides, which is hundreds of times greater than normal mutation frequencies. Apparently, spontaneous duplications in bacteria occur quite often during cell multiplication and can occur over very large regions of the chromosome. However, because they cause few phenotypic changes, most of these large duplications have little effect on the organism. Nevertheless, they are probably important in evolution, as mentioned earlier in this chapter.

SUMMARY

1. A mutation is any heritable change in the sequence of DNA of an organism. The organism with a mutation is called a mutant, and that organism's mutant phenotype includes all of the characteristics of the mutant organism that are different from the wild-type, or normal, organism.

2. The mutation rate is the chance of occurrence of a mutation to a particular phenotype each time a cell divides. The mutation rate offers clues to the molecular basis of the phenotype. Quantitative determination of mutation rates is often difficult because it is necessary to determine how many mutations have occurred.

3. One important conclusion from an analysis of population genetics is that the fraction of mutants increases as the population grows. This causes practical problems in genetics, which can be partially overcome by storing organisms without growth or by periodically colony purifying one or very few organisms.

4. The type of mutation causing a phenotype can often be ascertained from the properties of the mutation. Base pair changes revert and are often leaky. Frameshift mutations also revert but are seldom leaky. Deletion mutations do not revert and are seldom leaky. They also often inactivate more than one gene simultaneously and can fuse one gene to another. Tandem duplication mutations revert at a high frequency and often have no observable phenotypes, except that they can fuse one gene to another. Inversion mutations also often have no observable phenotype and often revert. Insertion mutations seldom revert and are usually not leaky.

5. Deletions, inversions, and tandem duplication mutations can be caused by recombination between different regions in the DNA. Deletion and tandem duplication mutations are caused by recombination between directly repeated sequences in the DNA and inversions by recombination between inverted repeats. Deletions and tandem duplications arise as reciprocal recombination products between directly repeated sequences in different DNA molecules.

6. If a secondary mutation returns the DNA to its original sequence, the mutation is said to have reverted. If a mutation somewhere else in the DNA restores function, the mutation is said to have been suppressed. Suppressors can be either intragenic or intergenic depending on whether they occur in the same gene or in a different gene from the original mutation, respectively. An example of an intragenic suppressor is a frameshift mutation within the same gene that restores the reading frame shifted by another frameshift mutation. An example of an intergenic suppressor is a mutation in a tRNA gene that changes the tRNA so that it recognizes one or more of the nonsense codons, allowing translation of nonsense mutations in other genes.

7. To isolate a mutant means to separate the mutant strain from the many other members of the population that are normal for the phenotype. Bacteria have advantages in genetic analysis because of the ease of isolating mutants. They multiply asexually, they are usually haploid, and large numbers can multiply on a single petri plate.

8. Mutations can be either spontaneous or induced. Spontaneous mutations often occur as mistakes while the DNA is replicating, while induced mutations are deliberately caused by using mutagenic chemicals or irradiation. Induction of mutations with mutagens has the advantage that mutations are more frequent and that a specific mutagen often causes only a specific type of mutation. To ensure that mutations are as representative as possible, different mutagens should be used and the isolation of siblings should be avoided by mutagenizing separate cultures and only using one mutant from each culture.

9. Screening for a mutant means devising a way to distinguish the mutant from the normal or wild type. Selecting a mutant means devising conditions under which either only the mutant or the wild type can multiply. Selections can be either positive or negative. In a positive selection, conditions are devised under which the mutant but not the wild type can multiply; in a negative selection, the wild type but not the mutant can multiply under the selective conditions. Most types of mutants can be selected only by negative selection. However, enrichment can sometimes be used to increase the frequency of the mutant in a population by killing the normal multiplying cells under the negative-selection conditions.

10. In recombination, two DNAs are broken and rejoined in new combinations. Generalized or homologous recombination occurs only between two DNAs with the same sequence. Progeny organisms that are different genetically from either parent as a result of recombination are called recombinant types, while progeny that are the same as one of the parents are parental types.

11. Complementation tests reveal whether different mutations inactivate different gene products. To do complementation tests, we introduce two copies of the region of DNA containing the two different mutations into a cell and observe whether the wild-type phenotype is restored. Bacteria can generally be made partially diploid for only a small region of the chromosome by introducing a region of the chromosome on another DNA element such as a plasmid or prophage. Complementation or allelism tests can be used to determine how many separate genes or, more precisely, regions encoding different gene products are represented in a collection of mutations exhibiting the same phenotype. It can also be used to determine if a mutation is dominant or recessive or is *cis* acting or *trans* acting and for cloning.

(continued)

QUESTIONS FOR THOUGHT

1. A single inversion mutation greatly alters a genetic map, or the order of genes in the DNA. Why, then, are the genetic maps of *S. enterica* serovar Typhimurium and *E. coli* so similar?

2. Do you suppose that duplication mutations play a role in evolution? If so, why are they not always destroyed by recombination as quickly as they form?

3. Why do you suppose that nonsense suppressors are possible in lower organisms but not higher organisms?

4. Can you propose a mechanism by which "directed" mutations might occur?

PROBLEMS

1. In a collection of cultures, all started from a few wild-type bacteria, which cultures are most likely to have had the earliest mutation to a particular mutant phenotype: those with a few mutant bacteria or those with many mutant bacteria?

2. Which phenotype would you expect to have the higher mutation rate, rifampin resistance or Arg⁻ (arginine auxotrophy)? Rifampin inhibits transcription by binding to RNA polymerase, and Rif^r mutations change the RNA polymerase so that it no longer binds rifampin but still functions to make RNA.

3. Luria and Delbrück grew 100 cultures of 1 ml each to 2×10^9 bacteria per ml. They then measured the number of bacteria resistant to T1 phage in each culture: 20 cultures had no resistant bacteria, 35 had one resistant mutant, 20 had two resistant mutants, and 25 had three or more resistant mutants. Calculate the mutation rate to T1 resistance by using the Poisson distribution.

4. Newcombe spread an equal number of bacteria on each of four plates. After 4 h of incubation, he sprayed plate 1 with T1 and put it back in the incubator. At the same time, he washed the bacteria off of plate 2, diluted them 10^7-fold, and replated them to determine the total number of bacteria. After a further 2 h of incubation, he sprayed plate 3 and washed the bacteria off of plate 4 and diluted them 10^8-fold before replating them. The next morning, he counted the colonies on each plate. He found that plate 1 had 10 colonies; plate 2 had 20 colonies; plate 3 had 120 colonies, and plate 4 had 22 colonies. Calculate the rate of mutation to T1 resistance.

5. You have isolated Arg⁻ auxotrophs of *Klebsiella pneumoniae*.

a. If you plate a few cells with a mutation, *arg-1*, on plates without arginine, they multiply to make tiny colonies. If you plate 10^8 cells, you get some large, rapidly growing colonies. What kind of mutation is *arg-1* likely to be?

b. If you plate another mutant with a different mutation, *arg-2*, you get no growth on plates without arginine. Even if you plate large numbers of mutant bacteria (>10^8), you get no colonies. What type of mutation is *arg-2* likely to be?

c. What are some other possible explanations for (a) and (b)?

6. Design an experiment to show that *dam* mutations of *E. coli* are mutagenic, i.e., that they show higher than normal rates of spontaneous mutations. Assume that you can get a Dam⁻

mutant and its isogenic parent in the mail and do not have to isolate the mutant yourself.

7. Why is it necessary to isolate mutants from different cultures to be certain of getting independent mutations, i.e. ones that are not siblings?

8. Design a positive selection for each of the following types of mutants. Discuss what kind of selective plates and/or conditions you would use.

a. Mutants resistant to the antibiotic coumermycin.

b. Revertants of a *trp* mutation that makes cells require tryptophan.

c. Double mutants with a suppressor mutation that relieves the temperature sensitivity due to another mutation in *dnaA*.

d. Mutants with a suppressor in *araA* that relieves the sensitivity to arabinose due to a mutation in *araD*.

e. Mutants with a mutation in *suA* (the gene for the transcription terminator protein Rho) that relieves the polarity of a *hisC* mutation on the *hisB* gene. Hint: make a partial diploid to perform complementation tests.

9. Design an enrichment procedure for reversible, temperature-sensitive mutants with mutations in genes whose products are required for cell growth.

10. Are nonsense suppressor mutations dominant or recessive? Why?

11. You have a strain of *Pseudomonas* that requires arginine, histidine, and serine as a result of nonsense mutations in genes encoding enzymes to make these amino acids. When you plate large amounts of this strain on media lacking all three of these amino acids, a few colonies arise that are Arg⁺ His⁺ Ser⁺. These mutations are almost as frequent as mutations which revert each of the mutations separately. What kind of mutation do you think caused the apparent reversion of all three of these mutations? How would you test your hypothesis?

12. An Hfr strain that is His⁺ Trp⁺ but has an *argH* mutation is crossed with a recipient that is Arg⁺ but has *hisG* and *trpA* mutations, and the cross is plated on minimal plates containing histidine and arginine but no tryptophan. Which is the selected marker, and which are the unselected markers?

13. You have crossed an *E. coli* Hfr strain that has *hisB4* and *recA1* mutations with a strain with the *thyA8* mutation and plated the cross on minimal plates with no growth supplements. Almost 80% of the Thy⁺ recombinants are very UV sensitive. Where is the *recA1* marker located in the chromosome? Note that a *recA* mutation makes the cell very sensitive to UV light (see chapter 11).

14. An *E. coli* strain has *metB1* (90 min) and *leuA5* (2 min) mutations that make it require methionine and leucine, respectively. It also has an *strA7* (73 min) mutation, which makes it resistant to streptomycin, and a Tn*5* transposon, which confers kanamycin resistance, inserted somewhere in its chromosome.

You want to know where the transposon is inserted. You cross the mutant strain with an Hfr strain that is streptomycin sensitive and that transfers counterclockwise from 0 min (Figure 3.30) and has a *hisG2* mutation (44 min) that makes it require histidine. After incubation for 100 min, you plate the cells on minimal plates plus leucine and histidine to select the *metB* marker. The plates also contain streptomycin to counterselect the donor. After purifying 100 of the Met⁺ transconjugants, you test them for the other markers. You find that 15 of them are His⁻, only 2 are Leu⁺, and 12 are kanamycin sensitive. Which is the selected and which are the unselected markers? Where is the transposon probably inserted?

15. You have isolated a His⁻ mutant of *E. coli* that you suspect has a nonsense mutation because it is suppressed by intergenic suppressors. You want to map one of the suppressors. To do this, you use an Hfr strain that has the original mutation and transfers clockwise from 30 min on the *E. coli* map. As recipient, you use a strain that has the suppressor as well as the *his* mutation and an *argG* mutation. You find that 80% of the Arg⁺ recombinants are His⁻. Where is the suppressor mutation located?

16. You wish to use transduction to determine if the order of three *E. coli* markers is *metB1–argH5–rif-8* or *argH5–metB1–rif-8*, so you do a three-factor cross. The donor for the transduction has the *rif-8* mutation, and the recipient has the *metB1* and *argH5* mutations. You select the *argH5* marker by plating on minimal plates plus methionine, purify 100 of the Arg⁺ transductants, and test for the other markers. You find that 17 are Arg⁺ Met⁺ Rif^r, 20 are Arg⁺ Met⁺ Rif^s, 60 are Arg⁺ Met⁻ Rif^s, and 3 are Arg⁺ Met⁻ Rif^r. What is the cotransduction frequency of the *argH* and *metB* markers? What is the cotransduction frequency of the *argH* and *rif-8* markers? What is the order of the three markers deduced from the three-factor cross data? Are the results consistent?

17. In the test for suppression versus reversion in Figure 3.36, what would you expect in the two cases if you used the Met⁺ apparent revertant as a donor and a strain with the *metA15* and *argH6* mutations as a recipient? Would you expect any Met⁺ transductants if the mutation had been suppressed? If you did get Met⁺ transductants, what percentage of them would you expect to be Arg⁻? What percentage would be Arg⁺?

18. You have isolated a *hemA* mutant of *E. coli* that requires δ-aminolevulinic acid for growth. You wish to move this mutation into other genetic backgrounds. You obtain a strain of *E. coli* from the Yale Stock Collection that has a Tn*10* transposon conferring resistance to tetracycline inserted only 0.5 min away from the *hemA* gene. Explain the steps involved in using transduction to move your *hemA* mutation into other *E. coli* strains.

19. You have isolated a nonrevertible *hemA* mutant of *E. coli* that requires δ-aminolevulinic acid to make hemes required for growth on succinate. You wish to use your mutant to clone the *hemA* gene. You do a partial digest of *E. coli* DNA with Sau3A, in which the average-sized piece is about 2 kbp, and clone the pieces into plasmid pBR322 cut with BamHI and treated with phosphatase. You than use the ligation mix to transform the *hemA* mutant, selecting the ampicillin resistance gene on the plasmid. You test the colonies for growth on plates lacking δ-aminolevulinic acid but containing succinate as the sole carbon source. How many Amp^r transformants should you have to test to have a reasonable chance of finding a colony that contains bacteria no longer requiring δ-aminolevulinic acid for growth on succinate? There are about 4,500 kbp of DNA in the *E. coli* genome.

20. One of the transformant colonies you test contains a few bacteria that no longer require δ-aminolevulinic acid, but most of the bacteria in the colony still require it. Do you think the clone has all of the *hemA* gene on it? Why or why not? Is the HemA$^+$ phenotype in these few transformants due to recombination or complementation?

21. Outline how you would replace the *argH* gene, responsible for making an enzyme for arginine synthesis, with the corresponding gene into which you have inserted a gene for chloramphenicol resistance (Cm^r) in *E. coli*. What would you expect the phenotypes of your mutation to be?

SUGGESTED READING

Anderson, R. P., C. G. Miller, and J. R. Roth. 1976. Tandem duplications of the histidine operon observed following generalized transduction in *Salmonella typhimurium*. *J. Mol. Biol.* **105**:201–218.

Hill, C. W., J. Foulds, L. Soll, and P. Berg. 1969. Instability of a nonsense suppressor resulting from a duplication of genes. *J. Mol. Biol.* **39**:563–581.

Jones, M. E., S. M. Thomas, and K. Clarke. 1999. The application of linear algebra to the analysis of mutation rates. *J. Theor. Biol.* **199**:11–23.

Lea, D. E., and C. A. Coulson. 1949. The distribution of numbers of mutants in bacterial populations. *J. Genet.* **49**:264–285.

Lederberg, J., and E. M. Lederberg. 1952. Replica plating and the indirect selection of bacterial mutants. *J. Bacteriol.* **63**:399.

Lederberg, J., and E. L. Tatum. 1946. Gene recombination in *E. coli. Nature* (London) **158**:558.

Low, K. B. 1996. Genetic mapping, p. 2511–2517. *In* F. C. Neidhardt, R. Curtiss III, J. L. Ingraham, E. C. C. Lin, K. B. Low, B. Magasanik, W. S. Reznikoff, M. Riley, M. Schaechter, and H. E. Umbarger (ed.), Escherichia coli *and* Salmonella: *Cellular and Molecular Biology*, 2nd ed. ASM Press, Washington, D.C.

Luria, S., and M. Delbrück. 1943. Mutations of bacteria from virus sensitivity to virus resistance. *Genetics* **28**:491–511.

Masters, M. 1996. Generalized transduction, p. 2421–2441. *In* F. C. Neidhardt, R. Curtiss III, J. L. Ingraham, E. C. C. Lin, K. B. Low, B. Magasanik, W. S. Reznikoff, M. Riley, M. Schaechter, and H. E. Umbarger (ed.), Escherichia coli *and* Salmonella: *Cellular and Molecular Biology*, 2nd ed. ASM Press, Washington, D.C.

Newcombe, H. 1949. Origin of bacterial variants. *Nature* (London) **164**:150–151.

Yanofsky, C., B. C. Carlton, J. R. Guest, D. R. Helinski, and U. Henning. 1964. On the colinearity of gene structure and protein structure. *Proc. Natl. Acad. Sci. USA* **51**:266–272.

Zinder, N. D., and J. Lederberg. 1952. Genetic exchange in *Salmonella. J. Bacteriol.* **64**:679–699.

CHAPTER 4

Plasmids

What Is a Plasmid?

In addition to the chromosome, bacterial cells often contain **plasmids.** These DNA molecules are found in essentially all types of bacteria and, as discussed below, play a significant role in bacterial adaptation and evolution. They also serve as important tools in studies of molecular biology. We address such uses later in the chapter.

Plasmids, which vary widely in size from a few thousand to hundreds of thousands of base pairs (a size comparable to that of the bacterial chromosome), are most often circular molecules of double-stranded DNA. However, some bacteria have linear plasmids, and some plasmids, most often those from gram-positive bacteria, can accumulate single-stranded DNA owing to aberrant rolling-circle replication (discussed below). The number of copies also varies among plasmids, and bacterial cells can harbor more than one type. Thus, a cell can harbor two or more different types of plasmids, with hundreds of copies of some plasmid types and only one or a few copies of other types.

Like chromosomes, plasmids encode proteins and RNA molecules and replicate as the cell grows, and the replicated copies are usually distributed into each daughter cell when the cell divides. They even share some of the same types of Par functions and site-specific recombinases with the host chromosome (see below). However, unlike chromosomes, plasmids generally do not encode functions essential to bacterial growth. Instead, they provide gene products that can benefit the bacterium under certain circumstances but are not always essential. However, there are even exceptions to these rules. For example, the pSymB plasmid of some *Rhizobium* species is about half as big as the chromosome and carries essential genes, including a gene for an arginine transfer RNA (tRNA) and the *minCDE* genes involved

197

in division site selection. Also, *Vibrio cholerae* has two large DNA molecules, both of which carry essential genes. *Agrobacterium tumefaciens* has two large DNAs, one circular and the other linear, both of which carry essential genes. In cases where a plasmid is almost as big and carries essential genes, which one is the chromosome and which is a plasmid? Probably a better criterion for whether a DNA is a plasmid or the chromosome is the nature of its origin of replication. In all known cases, one of the large DNAs has a typical bacterial origin of replication with an *oriC* site and closely linked *dnaA*, *dnaN*, and *gyrA* genes, among others, while the other DNA has a typical plasmid origin with *repABC*-like genes more characteristic of plasmids.

Naming Plasmids

Before methods for physical detection of plasmids became available, plasmids made their presence known by conferring phenotypes on the cells harboring them. Consequently, many plasmids were named after the genes they carry. For example, R-factor plasmids contain genes for resistance to several antibiotics (hence the name R for resistance). These were the first plasmids discovered, when *Shigella* and *Escherichia coli* strains resistant to a number of antibiotics were isolated from the fecal flora of patients in Japan in the late 1950s. The ColE1 plasmid, from which many of the cloning vectors were derived, carries a gene for the protein colicin E1, a bacteriocin that kills bacteria that do not carry this plasmid. The Tol plasmid contains genes for the degradation of toluene, and the Ti plasmid of *A. tumefaciens* carries genes for *tumor initiation* in plants. This system of nomenclature has led to some confusion, because plasmids carry various genes besides the ones for which they were originally named. Also, many of these plasmids have been altered beyond recognition to make plasmid cloning vectors (see below) and for other purposes.

To avoid further confusion, the naming of plasmids is now standarized. Plasmids are given number and letter names much like bacterial strains. A small "p," for *plasmid*, precedes capital letters that describe the plasmid

or sometimes give the initials of the person or persons who isolated or constructed it. These letters are often followed by numbers to identify the particular construct. When the plasmid is further altered, a different number is assigned to indicate the change. For example, plasmid pBR322 was constructed by *Bolivar* and *Rodriguez* from the ColE1 plasmid and is derivative number 322 of the plasmids they constructed. pBR325 is pBR322 with a chloramphenicol resistance gene inserted. The new number 325 distinguishes this plasmid from pBR322.

Functions Encoded by Plasmids

Depending on their size, plasmids can encode a few or hundreds of different proteins. However, as mentioned above, plasmids rarely encode gene products that are always essential for growth, such as RNA polymerase, ribosomal subunits, or enzymes of the tricarboxylic acid cycle. Instead, plasmid genes usually give bacteria a selective advantage under only some conditions.

Table 4.1 lists a few naturally occurring plasmids and some traits they encode, as well as the host in which they were originally found. Gene products encoded by plasmids include enzymes for the utilization of unusual carbon sources such as toluene, resistance to substances such as heavy metals and antibiotics, synthesis of antibiotics, and synthesis of toxins and proteins that allow the successful infection of higher organisms (Box 4.1).

It is interesting to speculate about why so many nonessential functions are encoded on plasmids and not on the chromosome. If plasmid genes, such as those for antibiotic resistance and toxin synthesis, were part of the chromosome, all bacteria of the species, not just the ones with the plasmid, would have the benefits of those genes. Consequently, all the members of that species would be more competitive in environments where these traits were desirable. Maybe having some genes on plasmids makes the host species able to survive in more environments without the burden of a larger chromosome. Bacteria must be able to multiply very quickly under some conditions to obtain a selective advantage, and smaller bacterial chromosomes can replicate faster than

TABLE 4.1	Some naturally occurring plasmids and the traits they carry	
Plasmid	**Trait**	**Original source**
ColE1	Bacteriocin which kills *E. coli*	*E. coli*
Tol	Degradation of toluene and benzoic acid	*Pseudomonas putida*
Ti	Tumor initiation in plants	*Agrobacterium tumefaciens*
pJP4	2,4-D (dichlorophenoxyacetic acid) degradation	*Alcaligenes eutrophus*
pSym	Nodulation on roots of legume plants	*Rhizobium meliloti*
SCP1	Antibiotic methylenomycin biosynthesis	*Streptomyces coelicolor*
RK2	Resistance to ampicillin, tetracycline, and kanamycin	*Klebsiella aerogenes*

BOX 4.1

Plasmids and Bacterial Pathogenesis

Plasmids often carry virulence genes required for bacterial pathogenicity. For example, while many strains of *Escherichia coli* are nonpathogenic denizens of the human intestine, others are pathogenic. The pathogenic strains, including *Shigella*, which is now considered a type of pathogenic *E. coli*, contain large plasmids that carry many of the virulence genes, while others are carried on prophages and DNA elements called genetic islands (see Box 2.7 and chapters 8 and 9). Not only must *E. coli* harbor these plasmids to be pathogenic, but also the nature of the plasmids harbored by these bacteria determines the characteristics of the diseases they cause. While they all cause diarrhea, they do so by different mechanisms. For example, the enterohemorrhagic *E. coli* strain O157:H7, which has been responsible for many of the most serious bacterial dysentery outbreaks worldwide, harbors a large plasmid, pO157, which carries many virulence genes including a toxin which may affect GTPases that regulate actin structures in eukaryotic cells and a specific protease that cleaves the human C1 esterase inhibitor and may thereby enhance the inflammation and tissue damage characteristic of the disease (see Lathem et al., below).

Not only is the presence of plasmids and other moveable elements required for pathogenicity, but it is often the major difference between a pathogen and its free-living cousins. Almost the only difference between the dreaded *Bacillus anthracis* (the agent of anthrax) and its close relative the soil bacterium *Bacillus cereus,* which only causes food poisoning, is the presence of two plasmids encoding virulence genes in *B. anthracis*. Another close relative, *Bacillus thuringiensis,* differs from *B. cereus* only because it has a large plasmid that encodes insect toxins, which are widely used in mosquito and other insect control (see Helgason et al., below).

An especially striking example of the importance of plasmids in bacterial virulence is in the genus *Yersinia*. The three species of *Yersinia, Y. enterocolitica, Y. pseudotuberculosis,* and *Y. pestis,* all cause disease, ranging in severity from mild enteritis in the case of *Y. enterocolitica* and *Y. pseudotuberculosis* to

the devastating bubonic plague in the case of *Y. pestis.* To be pathogenic, all three species must harbor the Lcr plasmid, which is about 70 kb long. This plasmid encodes a type III secretion system (see chapter 14) and effector proteins called Yops, which it injects directly into white blood cells called phagocytes. These white blood cells normally defend against bacterial invaders by ingesting and destroying them. However, once in the phagocytic cell, these effectors disrupt intracellular signaling and cause cytoskeletal changes that prevent the phagocytosis and allow the bacteria to persist in the phagocytic cells. The Yops are synthesized only under conditions of limiting calcium ions and high concentrations of sodium ions—conditions that may mimic the environment inside eukaryotic cells, hence the name Lcr plasmid (for *l*ow *c*alcium *r*esponse plasmid). What distinguishes the dreaded "plague bacillus," *Y. pestis,* from the other two relatively innocuous species of *Yersinia* is the presence of two other plasmids. These plasmids encode many proteins, but one is known to encode a toxin and an antiphagocytic protein similar to Yops and the other is known to encode a protease that increases invasiveness. These other plasmids may help *Y. pestis* survive inside fleas and infect mammals through the bite of the infected fleas, a life-style not available to the less pathogenic species of *Yersinia* (see Hinnebusch et al., below).

References

Helgason, E., O. A. Økstad, D. A. Caugant, H. A. Johansen, A. Fouet, M. Mock, I. Hegna, and A.-B. Kolstø. 2000. *Bacillus anthracis, Bacillus cereus,* and *Bacillus thuringiensis*—one species on the basis of genetic evidence. *Appl. Environ. Microbiol.* **66:**2627–2630.

Hinnebusch, B. J., A. E. Rudolph, P. Cherepanov, J. E. Dixon, T. G. Schwan, and A. Forsberg. 2002. Role of *Yersinia* murine toxin in survival of *Yersinia pestis* in the midgut of the flea vector. *Science* **296:**733–735.

Lathem, W. W., T. E. Grys, S. E. Witowski, A. G. Torres, J. B. Kaper, P. I. Tarr, and R. A. Welch. 2002. StcE, a metalloprotease secreted by *Escherichia coli* O157:H7, specifically cleaves C1 esterase inhibitor. *Mol. Microbiol.* **45:**277–288.

larger ones. Plasmids encoding different traits can then be distributed among different members of the population, where they do not burden any single bacterium too heavily. However, if the environment abruptly changes so that the genes carried on one of the plasmids become essential, the bacteria that carry the plasmid suddenly have a selective advantage and will survive, thereby ensuring survival of the species. In this way, plasmids allow bacteria to occupy a larger variety of ecological niches and contribute to the evolutionary success of not

only the bacterial species but also the plasmids found in that species.

Plasmid Structure

Most plasmids are circular with no free ends, although a few known plasmids are linear. In a circular plasmid, all of the nucleotides in each strand are joined to another nucleotide on each side by covalent bonds to form continuous strands that are wrapped around each other. Such DNAs are said to be **covalently closed circular.** This structure prevents the strands from separating, and there are no ends to rotate, so that the plasmid can be supercoiled. As discussed in chapter 1, in a DNA that is supercoiled, the two strands are wrapped around each other more or less often than once in about 10.5 bp, as predicted from the Watson-Crick double-helical structure of DNA. If they are wrapped around each other more often than once every 10.5 bp, the DNA is positively supercoiled; if they are wrapped around each other less often, the DNA is negatively supercoiled. Like the chromosome, covalently closed circular plasmid DNAs are usually negatively supercoiled (see chapter 1). Because DNA is stiff, the negative supercoiling introduces stress, and this stress is partially relieved by the plasmid wrapping up on itself, as illustrated in Figure 4.1A. This makes the plasmid more compact, so that it runs more quickly in an agarose gel (Figure 4.1B). In the cell, the DNA wraps around proteins, which relieves some of the stress. The remaining stress facilitates some reactions involving the plasmid, such as separation of the two DNA strands for replication or transcription.

PURIFYING PLASMIDS

The structure of plasmids can be used to purify them away from the chromosomal and other DNA in the cell. Cloning manuals usually give detailed protocols for these methods (see chapter 1, Suggested Reading), but we review them briefly in this section. Many purification procedures are based on the relatively small size of most plasmids. The plasmids often do not precipitate at the salt concentrations at which the chromosome precipitates. Therefore, if extracts of cells are treated at high salt concentrations, the chromosome often precipitates and can be removed by centrifugation while the much smaller plasmids stay in the supernatant.

Other purification steps take advantage of the fact that most plasmids are covalently closed circular and supercoiled. One such purification involving the acridine dye ethidium bromide (EtBr) is illustrated in Figure 4.2. This procedure is based on the fact that covalently closed circular DNAs bind less EtBr than do linear or nicked circular DNAs. EtBr intercalates (inserts itself) between DNA bases, pushing the bases apart and rotating the two strands of the DNA around each other. If the two strands of the DNA are not free to rotate, as in a covalently closed circular plasmid, the binding of EtBr eventually introduces positive supercoils and increases the stress on the DNA until no more EtBr can bind. EtBr bound to DNA makes it *less* dense in salt solutions made with heavy atoms such as cesium chloride (CsCl). As a consequence, if the DNA is mixed with a solution of CsCl and EtBr and centrifuged to establish a gradient of CsCl concentration, the covalently closed circular plasmid DNAs will band lower, at a position where the solution is more dense (Figure 4.3).

The methods discussed above work well with plasmids that have many copies per cell and are not too large. However, large, low-copy-number plasmids are much more difficult to detect. Most methods for detecting large plasmids involve separating them from the chromosome directly by electrophoresis on agarose gels (see chapter 1). The cells are often broken open directly on the agarose gel to avoid breaking the large plasmid DNA. The plasmid, because of its unique size, makes a sharp band on the gel, distinct from that due to chromosomal DNA, which is usually broken and so gives a more diffuse band. Also, methods such as pulsed-field gel electrophoresis have been devised to allow the separation of long pieces of DNA based on size. These methods depend on periodic changes in the direction of

Figure 4.1 Supercoiling of a covalently closed circular plasmid. (A) A break in one strand relaxes the DNA, eliminating the supercoiling and making the DNA less compact. (B) A schematic diagram of an agarose gel showing that the covalently closed supercoiled circles run faster on a gel than the nicked relaxed circles. Depending on the conditions, linear DNA and covalently closed circular DNA run in approximately the same position as nicked relaxed circles of the same length. The arrow shows the direction of migration.

A

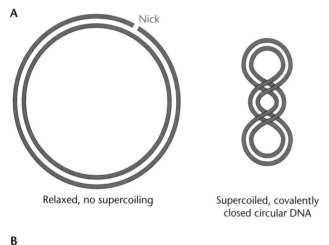

Relaxed, no supercoiling

Supercoiled, covalently closed circular DNA

B

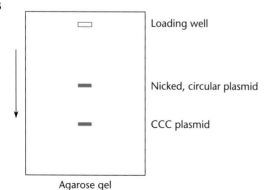

Loading well

Nicked, circular plasmid

CCC plasmid

Agarose gel

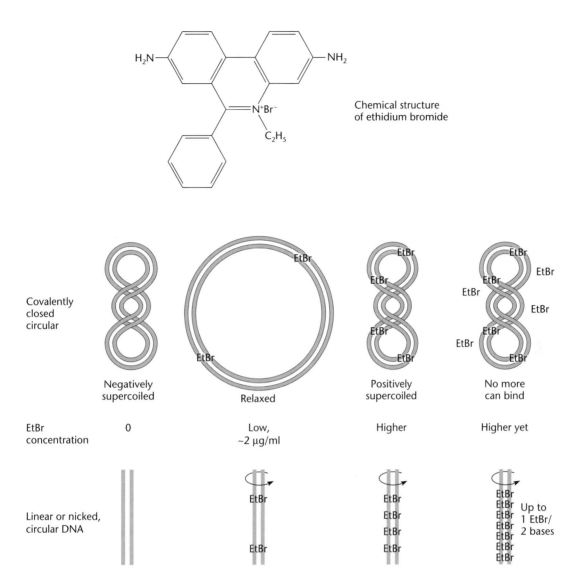

Figure 4.2 Less EtBr can bind to a covalently closed circular DNA than to a linear or nicked circular DNA. Also, progressively higher EtBr concentrations shift DNA supercoiling from negative to positive. At about 2 µg of EtBr per ml, most DNAs are completely relaxed. The arrow indicates the free rotation of linear DNA.

the electric field. The molecules attempt to reorient themselves each time the field shifts, and the longer molecules will take longer to reorient than the shorter ones and so move more slowly on the gel. Such methods have allowed the separation of DNA molecules hundreds of thousands of base pairs long and the detection of very large plasmids.

Properties of Plasmids

Replication

To exist free of the chromosome, plasmids must have the ability to replicate independently. DNA molecules that can replicate autonomously in the cell are called **replicons**.

Plasmids, phage DNA, and the chromosomes are all replicons, at least in some types of cells.

To be a replicon in a particular type of cell, a DNA molecule must have at least one origin of replication, or *ori* site, where replication begins (see chapter 1). In addition, the cell must contain the proteins that enable replication to initiate at this site. Plasmids encode only a few of the proteins required for their own replication. In fact, many encode only one of the proteins needed for initiation at the *ori* site. All of the other required proteins, including DNA polymerases, ligases, primases, helicases, and so on, are borrowed from the host.

Each type of plasmid replicates by one of two general mechanisms, which is determined along with other

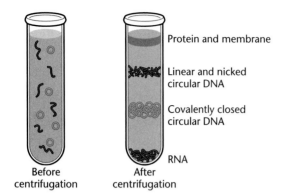

Protein and membrane

Linear and nicked circular DNA

Covalently closed circular DNA

RNA

Before centrifugation

After centrifugation

Figure 4.3   Separation of covalently closed circular plasmid DNA from linear and nicked circular DNAs on EtBr-CsCl gradients. After centrifugation, the plasmid DNA bands below the other DNAs because it has a higher buoyant density as a result of less binding of EtBr.

properties by its *ori* region (see the section on the *ori* region, below). The plasmid replication origin is often named *oriV* for *ori vegetative*, to distinguish it from *oriT*, which is the site at which DNA transfer initiates in plasmid conjugation (see chapter 5). Most of the evidence for the mechanisms described below came from observations of replicating plasmid DNA with the electron microscope.

THETA REPLICATION
Some plasmids begin replication by opening the two strands of DNA at the *ori* region, creating a structure that looks like the Greek letter θ—hence the name **theta replication** (Figure 4.4A and B). In this process, an RNA primer begins replication, which can proceed in one or both directions around the plasmid. In the first case, a single replication fork moves around the molecule until it returns to the origin and then the two daughter DNAs

separate. In the other case (bidirectional replication), two replication forks move out from the *ori* region, one in either direction, and replication is complete (and the two daughter DNAs separate) when the two forks meet somewhere on the other side of the molecule.

The theta mechanism is the most common form of DNA replication, especially in gram-negative bacteria. It is used not only by most plasmids, including ColE1, RK2, F, and P1, but also by the chromosome in most bacteria (see chapter 1).

ROLLING-CIRCLE REPLICATION
Other types of plasmids replicate by very different mechanisms. One type of replication is called **rolling-circle replication** because it was first discovered in a type of phage where the template circle seems to roll like a signet ring which has been dipped in ink and rolled on paper, making a copy of its design on the paper. Plasmids which replicate by this mechanism are called **RC plasmids**. This type of plasmid is widespread, being found in both gram-negative and gram-positive eubacteria as well as archaea.

In an RC plasmid, the replication occurs in two stages. In the first stage, the double-stranded circular plasmid DNA replicates to form another double-stranded circular DNA and a single-stranded circular DNA. This stage is analogous to the replication of the DNA of some single-stranded DNA phages (see chapter 7) and to DNA transfer during plasmid conjugation (see chapter 5). In the second stage, the complementary strand is synthesized on the single-stranded DNA to make another double-stranded DNA.

The details of the rolling-circle mechanism of plasmid replication are shown in Figure 4.4C. First the Rep protein recognizes and binds to a palindromic sequence which contains the double-strand origin (DSO) on the DNA. Binding of the Rep protein to this sequence might

Figure 4.4   Some common schemes of plasmid replication. (A) Unidirectional replication. The origin region is designated *oriV*. Replication terminates when the replication fork gets back to the origin. (B) Bidirectional replication. Replication terminates when the replication forks meet somewhere on the DNA molecule opposite the origin. (C) Rolling-circle replication. A nick is made at the double-strand origin (DSO) by the plasmid-encoded Rep protein, which remains bound to the 5′ phosphate end at the nick. The free 3′ OH end then serves as a primer for the DNA polymerase III (Pol III) that replicates around the circle, displacing one of the old strands as a single-stranded DNA. The Rep protein then makes another nick, releasing the single-stranded circle, and also joins the ends to form a circle by a phosphotransferase reaction (see the text). The DNA ligase then joins the ends of the new DNA to form a double-stranded circle. The host RNA polymerase makes a primer on the single-stranded DNA origin (SSO), and Pol III replicates the single-stranded (SS) DNA to make another double-stranded circle. DNA Pol I removes the primer, replacing it with DNA, and ligase joins the ends to make another double-stranded circular DNA. CCC, covalently closed circular; SSB, single-strand-DNA-binding protein.

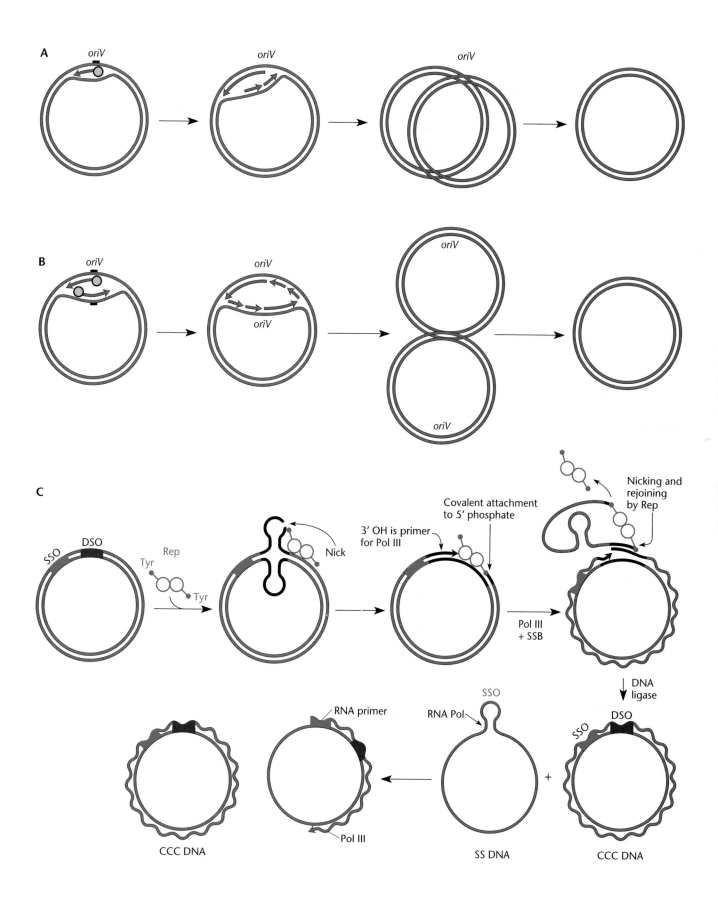

allow the formation of a cruciform structure by base pairing between the inverted repeated sequences in the cruciform as shown in the figure. Once the cruciform forms, the Rep protein can make a nick in the sequence as shown in the figure. It is important for the models that the Rep protein is also known to function as a dimer, at least in some plasmids, as shown in the figure. After the Rep protein has made a break in the DSO sequence, it remains covalently attached to the phosphate at the 5′ end of the DNA at the nick through a tyrosine in one copy of the Rep protein in the dimer, as shown. The DNA polymerase III (the replicative polymerase [see chapter 1]) uses the free 3′ hydroxyl end at the break as a primer to replicate around the circle, displacing one of the strands. It may use a host helicase to help separate the strands, or the Rep protein itself may have the helicase activity, depending on the plasmid. Once the circle is complete, the 5′ phosphate is transferred from the tyrosine on the Rep protein to the 3′ hydroxyl on the other end of the displaced strand, producing a single-stranded circular DNA. This process is called a phosphotransferase reaction and requires little energy. The same reaction is used to re-form a circular plasmid after conjugational transfer (see chapter 5).

It is less certain what happens to the newly formed double-stranded DNA when the DNA polymerase III has made its way all around the circle and gets back to the site of the DSO. Why does it not just keep going, making a longer molecule with individual genomes linked head to tail, in a structure called a concatemer? Such structures are created when some phage DNAs replicate by a rolling-circle mechanism (see chapters 7 and 8). One idea is that the DNA polymerase III does keep going past the DSO for a short distance, creating another double-stranded DSO. The other copy of Rep protein in the dimer may then nick the newly created DSO, transferring the 5′ end to itself as above. This might inactivate the Rep protein, releasing it with a short oligonucleotide attached. Other reactions, probably involving host DNA ligase, then cause the nick to be resealed, resulting in a circular double-stranded DNA molecule.

The displaced circular single-stranded DNA now replicates by a completely different mechanism using only host-encoded proteins. The RNA polymerase first makes a primer at a different origin, the single-strand origin (SSO), and this RNA then primes replication around the circle by DNA polymerase III. However, the RNA polymerase does not make this primer until the single-stranded DNA is completely displaced during the first stage of replication. This delay is accomplished by locating the SSO immediately counterclockwise of the DSO (Figure 4.4C), so that the SSO does not appear in the displaced DNA until the displacement of the single-stranded DNA is almost complete. After the entire complementary strand has been synthesized, the 5′ exonuclease activity of DNA polymerase I removes the RNA primer, replacing it with DNA, and host DNA ligase joins the ends to make another double-stranded plasmid. The net result is two new double-stranded plasmids synthesized from the original double-stranded plasmid.

In order for the complementary strand of the displaced single-stranded DNA to be synthesized, the RNA polymerase of the host cell must recognize the SSO on the DNA. In some hosts the SSO is not well recognized and single-stranded DNA accumulates. For this reason, some RC plasmids were originally called single-stranded DNA plasmids, although we now know that this is not their normal state. Broad-host-range RC plasmids presumably have an SSO that is recognized by the RNA polymerases of a wide variety of hosts, which allows them to make the complementary strand of the displaced single-stranded DNA in a variety of hosts.

The Rep protein is used only once for every round of plasmid DNA replication and is destroyed after the round is completed. This allows the replication of the plasmid to be controlled by the amount of Rep protein in the cell and keeps the total number of plasmid molecules in the cell within narrow limits dictated by the copy number. A little later in this chapter, we discuss how the copy number of other types of plasmids is controlled.

REPLICATION OF LINEAR PLASMIDS

As already mentioned, some plasmids are linear rather than circular (Box 4.2). In general, linear DNAs face a problem with replicating the lagging strand, the strand that ends with a 5′ phosphate, all the way to the end of the DNA. This has been called the "primer problem" because DNA polymerases cannot initiate the synthesis of a new strand of DNA. They can only add nucleotides to a preexisting primer, and, in a linear DNA, there is no upstream primer on this strand from which to grow. Different linear DNAs solve the primer problem in different ways. Some linear plasmids have hairpin ends, with the 5′ and 3′ ends joined to each other. These plasmids replicate from an internal origin of replication to form dimeric circles, composed of two plasmids joined head to tail to form a circle. These dimeric circles are then resolved into individual linear plasmid DNAs by enzymes called prototelomerases, which cut the dimeric circles into the two linear plasmid DNAs and join the ends to each other to form the closed hairpins. Other linear plasmids seem to replicate their ends by a completely different mechanism. They have extensive inverted repeated sequences at their ends and a terminal protein attached to their 5′ ends. It is not clear how these plasmids replicate, but they might use some sort of slippage mechanism,

BOX 4.2

Linear Plasmids

Not all plasmids are circular. Linear plasmids have been found in many bacteria, including members of the genus *Streptomyces,* which are soil bacteria responsible for making most useful antibiotics, and *Borrelia burgdorferi,* the causative agent of Lyme disease. Also in some phages, for example the N15 phage of *E. coli,* the prophage is a linear plasmid with different ends from the linear DNA in the phage head. Linear plasmids, like linear chromosomes and phages, face a "primer problem" because of the properties of DNA polymerases (see Box 1.1). No known DNA polymerase can initiate the synthesis of a new DNA strand; they can only add deoxynucleotides to a preexisting primer. This means that when replication proceeds to the ends of a linear DNA, the polymerase cannot replicate to the extreme 5′ end of the lagging strand because there is no upstream primer. This is not a problem for circular plasmids because there is always DNA upstream to provide a primer.

Different types of linear plasmids and linear chromosomes solve the primer problem in different ways. In fact, quite often linear plasmids are found in cells with linear chromosomes for unknown reasons, and they solve the "primer problem" in the same way. Some linear plasmids, for example the 16-kb plasmid from *B. burgdorferi,* use the same mechanism to replicate their ends as does the chromosome, which is also linear (see Box 1.1). This plasmid and other linear plasmids in the cell have hairpin ends, which means that the 3′ end is attached to the 5′ end on the other antiparallel strand (see the

figure in Box 1.1). The plasmid replicates from an internal origin of replication and replicates right around the ends to form a large circular DNA composed of two genomes linked head to tail (a dimerized circle). An enzyme called a prototelomerase then cuts at what were the ends of the linear DNAs and rejoins the ends to reform the hairpins, in a recombinase–like reaction. A similar reaction occurs for the *E. coli* plasmid prophage N15 (mentioned above).

Linear plasmids of *Borrelia* carry genes for the major surface proteins of the bacteria, again breaking the rule that essential genes are never found on plasmids. It is also possible that the chromosomes from these bacteria are linear and the plasmids are directly derived from the chromosome, perhaps through the deletion of large chromosomal segments.

Another interesting feature of the hairpin ends of these linear plasmids is their similarity to the ends of the linear DNAs of poxviruses and African swine fever virus, as shown in the figure (see Hinnebusch and Barbour, below). These virus DNAs also have hairpin ends and may replicate by a similar mechanism. Whether this similarity is significant in terms of the origins of these bacteria and viruses remains to be seen.

Another solution to the primer problem is suggested by the ends of the linear plasmid pSLA2 of *Streptomyces.* This plasmid has a protein covalently attached to its ends, as does the chromosome in some strains. However, this protein does not simply serve as a primer for lagging-strand synthesis as proteins do for some phages (see Box 7.3), because

Comparison of the telomeres from a linear plasmid of *B. burgdorferi* (A) and from African swine fever virus and vaccinia virus (B). The ends of the DNAs are covalently joined as shown.

(continued)

Linear Plasmids

the replication origin for this plasmid is known not to be at the ends but, rather, internal to the plasmid, and replication proceeds bidirectionally toward both ends. This plasmid contains a number of tandem inverted repeated sequences or palindromes, which may allow slippage by pairing between the inverted repeated sequences. Models have been proposed to explain how these palindromic sequences may allow the leading strand to fold back on itself and form hairpins with the lagging strand to provide a primer for lagging-strand synthesis, but it is not clear which mechanism is actually used. The protein attached to the 5′ end of the leading strand may also play a role by acting as a recombinase or as a Rep protein, as in rolling-circle replication.

References

Casjens, S., N. Palmer, R. van Vugt, W. M. Huang, B. Stevenson, P. Rosa, R. Lathigra, G. Sutton, J. Peterson, R. J. Dodson, D. Haft, E. Hickey, M. Gwinn, O. White, and C. M. Fraser. 2000. A bacterial genome in flux: the twelve linear and nine circular extrachromosomal DNAs in an infectious isolate of the Lyme disease spirochete *Borrelia burgdorferi. Mol. Microbiol.* **35:**490–516.

Hinnebusch, J., and A. G. Barbour. 1991. Linear plasmids of *Borrelia burgdorferi* have a telomeric structure and sequence similar to those of a eukaryotic virus. *J. Bacteriol.* **173:**7233–7239.

Huang, W. M., L. Joss, T. T. Hsieh, and S. Casjens. 2004. Protelomerase uses a topoisomerase 1β/γ recombinase type mechanism to generate hairpin ends. *J. Mol. Biol.* **337:**77–92.

Yang, C.-C., C.-H. Huang, C.-Y. Li, Y.-G. Tsay, S.-C. Lee, and C. W. Chen. 2002. The terminal proteins of linear *Streptomyces* chromosomes and plasmids: a novel class of replication priming proteins. *Mol. Microbiol.* **43:**297–305.

using the terminal protein as either a recombinase or a primer or both. It is interesting that bacteria with linear plasmids also often have linear chromosomes and the two DNAs replicate by similar mechanisms. The way in which linear chromosomes replicate to their ends is discussed in some detail in Box 1.1.

Functions of the *ori* Region

In most plasmids, the genes for proteins required for replication are located very close to the *ori* sequences at which they act. Thus, only a very small region surrounding the plasmid *ori* site is required for replication. As a consequence, the plasmid still replicates if most of its DNA is removed, provided that the *ori* region remains and the plasmid DNA is still circular. Smaller plasmids are easier to use as cloning vectors, as discussed later in this chapter, so often the only part of the orginal plasmid that remains in a cloning vector is the *ori* region.

In addition, the genes in the *ori* region often determine many other properties of the plasmid. Therefore, any DNA molecule with the *ori* region of a particular plasmid will have most of the characteristics of that plasmid. The following sections describe the major plasmid properties determined by the *ori* region.

HOST RANGE

The **host range** of a plasmid includes all the types of bacteria in which the plasmid can replicate; it is usually determined by the *ori* region. Some plasmids, such as those with *ori* regions of the ColE1 plasmid type, including pBR322, pET, and pUC, have a **narrow host range.** These plasmids replicate only in *E. coli* and some other closely related bacteria such as *Salmonella* and *Klebsiella* species. In contrast, plasmids with a **broad host range** include the RK2 and RSF1010 plasmids, as well as the rolling-circle plasmids from gram-positive bacteria mentioned above. The host range of these plasmids is truly remarkable. Plasmids with the *ori* region of RK2 can replicate in most types of gram-negative bacteria, and RSF1010-derived plasmids even replicate in some types of gram-positive bacteria. Many of the plasmids isolated from gram-positive bacteria also have quite broad host ranges. For example, pUB110, which was first isolated from the gram-positive *Staphylococcus aureus*, also replicates in many other gram-positive bacteria, including *Bacillus subtilis*. However, most plasmids isolated from gram-negative bacteria do not replicate in gram-positive bacteria and vice versa, which reflects the evolutionary divergence of these groups (see the introductory chapter).

It is perhaps surprising that the same plasmid can replicate in bacteria that are so distantly related to each other. Broad-host-range plasmids must encode all of their own proteins required for initiation of replication, and so they do not have to depend on the host cell for any of these functions. They also must be able to express these genes in many types of bacteria. Apparently, the promoters and ribosome initiation sites for the replication genes of broad-host-range plasmids have evolved so that they can be recognized in a wide variety of bacteria.

Determining the Host Range

The actual host range of most plasmids is unknown because it is sometimes difficult to determine if a plasmid can replicate in other hosts. First, we must have a way of introducing the plasmid into other bacteria. Transformation systems (see chapter 6) have been developed for some but not all types of bacteria, and if one is available, it can be used to introduce plasmids into the bacterium. Electroporation can often be used. Plasmids that are self-transmissible or mobilizable (see chapter 5) can sometimes be introduced into other types of bacteria by conjugation, a process in which DNA is transferred from one cell to another.

Even if we can introduce the plasmid into other types of bacteria, we still must be able to select cells that have received the plasmid. Most plasmids, as isolated from nature, are often not known to carry a convenient selectable gene such as one for resistance to an antibiotic, and even if they do, the selectable gene may not be expressed in the other bacterium since most genes are not expressed in bacteria distantly related to those in which they were originally found. Sometimes we can introduce a selectable gene, chosen because it is expressed in many hosts, into the plasmid. For example, the kanamycin resistance gene, first found in the Tn5 transposon, is expressed in most gram-negative bacteria, making them resistant to the antibiotic kanamycin. We can either clone a marker gene into the plasmid or introduce a transposon carrying a selectable marker into the plasmid by methods discussed in chapter 9.

If all goes well and we have a way to introduce the plasmid into other bacteria and the plasmid carries a marker that is likely to be expressed in other bacteria, we can see if the plasmid can replicate in bacteria other than its original host. Clearly, this could be a laborious process since the mechanisms for introducing DNA into different types of bacteria differ and there are many barriers to plasmid transfer between species. Therefore, the host range of plasmids is often extrapolated from only a few examples.

REGULATION OF COPY NUMBER

Another characteristic of plasmids that is determined mostly by their *ori* region is their **copy number,** or the average number of that particular plasmid per cell. More precisely, we define the copy number as the number of copies of the plasmid in a newborn cell immediately after cell division. All plasmids must regulate their replication; otherwise they would fill up the cell and become too great a burden for the host, or their replication would not keep up with the cell replication and they would be progressively lost during cell division. Some plasmids, such as pIJ101 of *Streptomyces coelicolor*, replicate

enough times to populate the cell with hundreds of copies. However, others, such as the F plasmid of *E. coli*, replicate only once or a few times during the cell cycle. Table 4.2 lists the copy numbers of these and other plasmids.

The regulation mechanisms used by plasmids with higher copy numbers often differ greatly from those used by plasmids with lower copy numbers. Plasmids that have high copy numbers, such as ColE1 plasmids, need only have a mechanism that inhibits the initiation of plasmid replication when the number of plasmids in the cell reaches a certain level. Consequently, these molecules are called **relaxed plasmids.** By contrast, low-copy-number plasmids such as F must replicate only once or very few times during each cell cycle and so must have a tighter mechanism for regulating their replication. Hence, these are called **stringent plasmids.** Much more is understood about the regulation of replication of relaxed plasmids than about the regulation of replication of stringent plasmids.

The regulation of relaxed plasmids falls into three general categories. Some plasmids are regulated by an antisense RNA, sometimes called a countertranscribed (ctRNA) because it is transcribed from the same region of the plasmid but from the opposite strand of an RNA essential for plasmid replication. Because the ctRNA is transcribed from the opposite strand from the essential RNA, it is able to hybridize to the essential RNA and inhibit its function. The ctRNA of these plasmids is often assisted in its inhibitory role by a protein. Other plasmids are regulated by a ctRNA alone, which inhibits the translation of a protein essential for replication. Yet others are regulated by a protein alone, which binds to repeated sequences in the plasmid DNA called iterons, thereby inhibiting plasmid replication. Examples of these three types of regulation are discussed in later sections.

INCOMPATIBILITY

Another function of plasmids that is controlled by the *ori* region is incompatibility. Incompatibility refers to the ability of two plasmids to coexist stably in the same cell.

TABLE 4.2	Copy numbers of some plasmids
Plasmid	**Approximate copy number**
F	1
P1 prophage	1
RK2	4–7 (in *E. coli*)
pBR322	16
pUC18	~30–50
pIJ101	40–300

Many bacteria, as they are isolated from nature, contain more than one type of plasmid. These plasmid types coexist stably in the bacterial cell and remain there even after many cell generations. In fact, bacterial cells containing multiple types of plasmids are not cured of each plasmid any more frequently than if the other plasmids were not there.

However, sometimes two different types of plasmid cannot coexist stably in the same cell. In this case, one or the other plasmid is lost as the cells multiply; this loss is more frequent than would occur if the plasmids were not occupying the cells along with the other plasmid. If two plasmids cannot coexist stably, they are said to be members of the same **incompatibility (Inc) group.** If two plasmids can coexist stably, they belong to different Inc groups. There are a number of ways in which plasmids can be incompatible. One way is if they can each regulate the other's replication. Another way is if they share the same partitioning (*par*) functions, which are often closely associated with replication control in the *ori* region. There may be hundreds of different Inc groups, and plasmids are usually classified by the Inc group to which they belong. For example, RP4, also called RK2, is an IncP (incompatibility group P) plasmid. In contrast, RSF1010 is an IncQ plasmid: it can therefore be stably maintained with RP4 because it belongs to a different Inc group but cannot be stably maintained with another IncQ plasmid.

Determining the Incompatibility Group

To classify a plasmid by its incompatibility group, we must determine if it can coexist with other plasmids of known incompatibility groups. In other words, we must measure how frequently cells are cured of the plasmid when it is introduced into cells carrying another plasmid of a known incompatibility group. However, we can know that cells have been cured of a plasmid only when it encodes a selectable trait, such as resistance to an antibiotic. Then the cells become sensitive to the antibiotic if the plasmid is lost.

The experiment shown in Figure 4.5 is designed to measure the curing rate of a plasmid that contains the Cam^r gene, which makes cells resistant to the antibiotic chloramphenicol. To measure the frequency of plasmid curing, we grow the plasmid-containing cells in medium with all the growth supplements and *no* chloramphenicol. At different times, we take a sample of the cells, dilute it, and plate the dilutions on agar containing the same growth supplements but again no chloramphenicol. After incubation of the plates, we replicate the plate onto another plate containing chloramphenicol (see chapter 3). If we do not observe any growth of a colony, the bacteria in that colony must all have been sensitive to the antibiotic and hence the original bacterium that had multiplied to form the colony must have been cured of the plasmid. The percentage of colonies that contain no

Figure 4.5 Measurement of the curing of a plasmid carrying resistance to chloramphenicol. See the text for details.

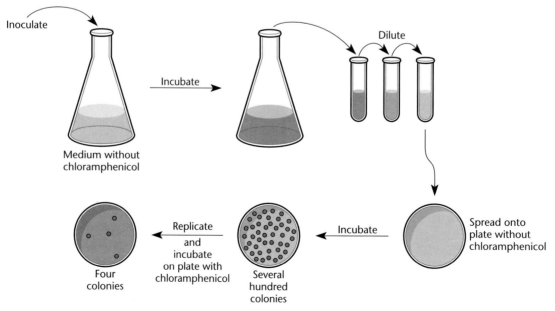

resistant bacteria is the percentage of bacteria that were cured of the plasmid at the time of plating.

To apply this test to determine if two plasmids are members of the same Inc group, both plasmids must contain different selectable genes, for example, genes encoding resistance to different antibiotics. Then one plasmid is introduced into cells containing the other plasmid. Resistance to both antibiotics is selected for. Then cells containing both plasmids are incubated without either antibiotic and finally grown on antibiotic-containing plates, as above. The only difference is that the colonies are transferred onto two plates, each containing one or the other antibiotic. If the percentage of cells cured of one or the other plasmid is no higher than the percentage cured of either plasmid when it was alone, the plasmids are members of different Inc groups. We continue to apply this test until we find a known plasmid, if any, which is a member of the same Inc group as our unknown plasmid.

Maintaining Plasmids Belonging to the Same Incompatibility Group

Two plasmids of the same Inc group can usually be maintained in the same population of cells, provided that they are high-copy-number relaxed plasmids and both plasmids are continuously selected for. This requires that the two plasmids carry different antibiotic resistance genes or other selectable markers. One or the other will be cured with a higher than normal frequency (see above), but by growing the cells in medium containing both antibiotics, a cell that has lost one or the other plasmid has a limited life span, and it, or its immediate progeny, will die. Therefore, most of the surviving cells at any time have both plasmids.

Incompatibility Due to Shared Replication Control

One way in which two plasmids can be incompatible is if they share the same mechanism of replication control. The replication control system does not recognize the two as different, and so either plasmid may be randomly selected for replication. At the time of cell division, the total copy number of the two plasmids will be the same, but one may be represented much less than the other. Figure 4.6 illustrates this by contrasting the distribution at cell division of plasmids of the same Inc group with plasmids of different Inc groups. Figure 4.6A shows a cell containing two types of plasmids that belong to different Inc groups and use different replication control systems. In the illustration, the two plasmids exist in equal numbers before cell division, but after division, the two daughter cells are not likely to get the same number of each plasmid. However, in the new cells, each plasmid replicates to reach its copy number, so that at the time of

the next division, both cells again have the same numbers of the plasmids. This process is repeated each generation, so very few cells will be cured of either plasmid.

Now consider the situation illustrated in Figure 4.6B, in which the cell has two plasmids that belong to the same Inc group and therefore share the same replication control system. As in the first example, both plasmids originally exist in equal numbers, and when the cell divides, it is still likely that the two daughter cells will not receive the same number of the two plasmids. Note that in the original cell, the copy number of each plasmid is only half its normal number; both plasmids contribute to the total copy number since they both have the same *ori* region and inhibit each other's replication. After cell division, the two plasmids replicate until the *total* number of plasmids in each cell equals the copy number. The underrepresented plasmid (recall that the daughters may not receive the same number of plasmids if the plasmid is high copy number) does not necessarily replicate more than the other plasmid, so that the imbalance of plasmid number might remain or become even worse. At the next cell division, the underrepresented plasmid has less chance of being distributed to both daughter cells since there are fewer copies of it. Consequently, in subsequent cell divisions, the daughter cells are much more likely to be cured of one of the two plasmid types by chance alone.

Incompatibility due to copy number control is probably more detrimental to low-copy-number plasmids than to high-copy-number plasmids. If the copy number is only 1, then only one of the two plasmids can replicate; each time the cell divides, a daughter is cured of one of the two types of plasmids.

Incompatibility Due to Partitioning

Two plasmids can also be incompatible if they share the same Par (partitioning) system. Par systems help segregate plasmids or chromosomes into daughter cells upon cell division (see below). Normally this helps ensure that both daughter cells get at least one copy of the plasmid and neither daughter cell is cured of the plasmid. If coexisting plasmids share the same Par system, one or the other is always distributed into the daughter cells during division. However, sometimes one daughter cell receives one plasmid type and the other cell gets the other plasmid type, producing cells cured of one or the other plasmid. We discuss what is understood about the mechanisms of partitioning later in the chapter.

Plasmid Replication Control Mechanisms

The mechanisms used by some plasmids to regulate their copy number have been studied in detail. Some of the better understood mechanisms are reviewed in this section.

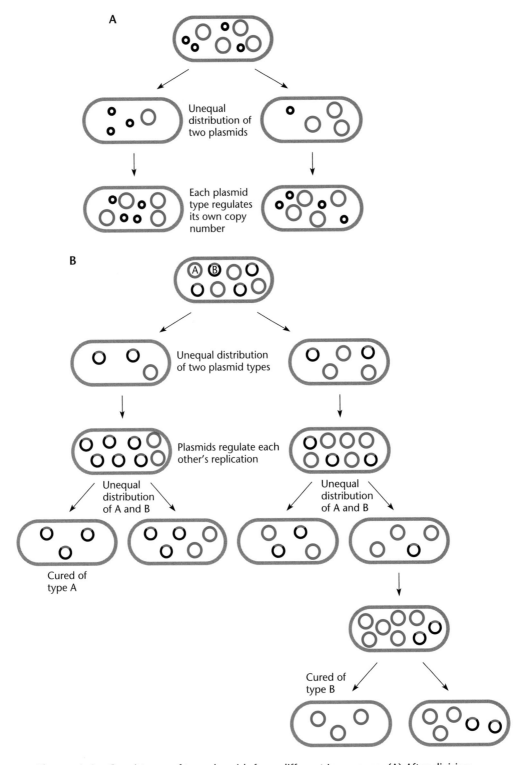

Figure 4.6   Coexistence of two plasmids from different Inc groups. (A) After division, both plasmids replicate to reach their copy number. (B) Curing of cells of one of two plasmids when they are members of the same Inc group. The sum of the two plasmids is equal to the copy number, but one may be underrepresented and lost in subsequent divisions. Eventually, most of the cells contain only one or the other plasmid. The light purple region in the smaller plasmid is shown to indicate that it shares the *ori* region with the larger plasmid.

ColE1-DERIVED PLASMIDS: REGULATION OF PROCESSING OF PRIMER BY COMPLEMENTARY RNA

The mechanism of copy number regulation of the plasmid ColE1 was one of the first to be studied. Figure 4.7 shows a partial genetic map of the original ColE1 plasmid. This plasmid has been put to use in numerous molecular biology studies, and many vectors have been derived from it or its close relative, pMB1. These vectors include the commonly used pBR322, the pUC plasmids, the pBAD plasmids, pACYC184, and the pET series of plasmids discussed below and in chapters 2 and 7. Although the genetic maps of these cloning vectors have been changed beyond recognition, they all retain the original ColE1 *ori* region and hence they share many of its properties, including the mechanism of replication regulation.

The mechanism of regulation of ColE1-derived plasmids is shown in Figure 4.8. Replication is regulated mostly through the effects of a small plasmid-encoded RNA called RNA I. This small RNA inhibits plasmid replication by interfering with the processing of another RNA called RNA II, which forms the primer for plasmid DNA replication. In the absence of RNA I, RNA II forms an RNA-DNA hybrid at the replication origin. RNA II

is then cleaved by the RNA endonuclease RNase H, releasing a 3′ hydroxyl group that serves as the primer for replication first catalyzed by DNA polymerase I. Unless RNA II is processed properly, it does not function as a primer and replication does not ensue.

RNA I inhibits DNA replication by interfering with RNA II primer formation by forming a double-stranded RNA with it, as illustrated in Figure 4.8. It can do this because the two RNAs are transcribed from opposite strands in the same region of DNA. Figure 4.9 illustrates how any two RNAs transcribed from the same region of DNA but from opposite strands are complementary. Initially, the pairing between RNA I and RNA II occurs through short exposed regions on the two RNAs that are not occluded by being part of secondary structures. This initial pairing is very weak and therefore has been called a "kissing complex." The protein named Rop (Fig. 4.7) helps stabilize the kissing complex, although it is not essential. The kissing complex can then extend into a "hug," with the formation of the double-stranded RNA as shown. Formation of the double-stranded RNA prevents the RNA II from forming the secondary structure required for it to hybridize to the DNA before being processed by RNase H to form the mature primer.

Even though Rop (sometimes called Rom) is known to help RNA I to pair with RNA II and therefore help inhibit plasmid replication, it is not clear how Rop works, nor is this protein essential to maintain the copy number. Mutations that inactivate Rop cause only a moderate increase in plasmid copy number.

This mechanism provides an explanation for how the copy number of ColE1 plasmids is maintained. Since RNA I is synthesized from the plasmid, more RNA I is made when the concentration of the plasmid is high. A high concentration of RNA I interferes with the processing of most of the RNA II, and replication is inhibited. The inhibition of replication is almost complete when the concentration of the plasmid reaches about 16 copies per cell, the copy number of the ColE1 plasmid.

We can predict from the model what the effect of mutations in RNA I should be. Formation of the kissing complex involves pairing between very small regions of RNA I and RNA II. However, these regions must be completely complementary for this pairing to occur and plasmid replication to be inhibited. Changing even a single base pair in this short sequence makes the mutated RNA I no longer complementary to the RNA II of the original nonmutant ColE1 plasmid, so it is no longer able to "kiss" it and regulate its replication. However, a mutation in the region of the plasmid DNA encoding RNA I also changes the sequence of RNA II made by the same plasmid in a complementary way, since they are encoded in the same region of the DNA but from the opposite

Figure 4.7 Genetic map of plasmid ColE1. The plasmid is 6,646 bp long. On the map, *oriV* is the origin of replication; p_{RNAII} is the promoter for the primer RNA II, *inc* encodes RNA I, *rop* encodes a protein that helps regulate the copy number, *bom* is a site that is nicked at *nic*, *cea* encodes colicin ColE1, and *mob* encodes functions required for mobilization (discussed in chapter 5).

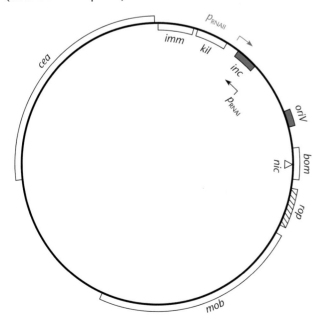

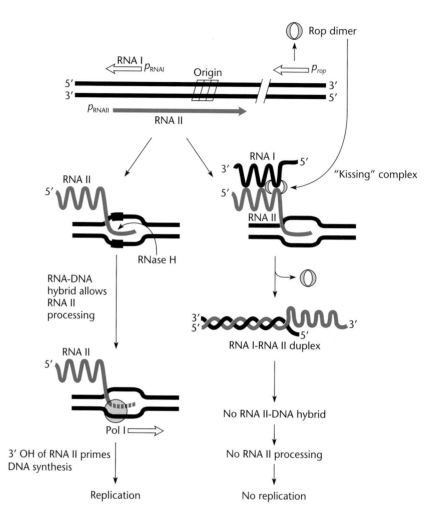

Figure 4.8 Regulation of the replication of ColE1-derived plasmids. RNA II must be processed by RNase H before it can prime replication. "Origin" indicates the transition point between the RNA primer and DNA. RNA I binds to RNA II and inhibits the processing, thereby regulating the copy number. p_{RNAI} and p_{RNAII} are the promoters for RNA I and RNA II transcription, respectively. RNA II is shown in purple. The Rop protein dimer enhances the initial pairing of RNA I and RNA II.

Figure 4.9 Pairing between an RNA and its antisense RNA. (A) An antisense RNA is made from the opposite strand of DNA in the same region. (B) The two RNAs are complementary and can base pair with each other to make a double-stranded RNA.

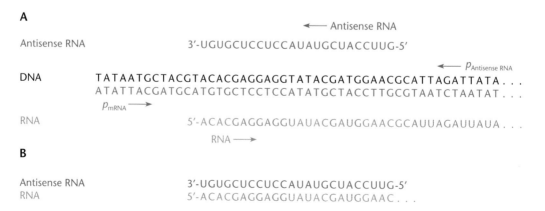

strands. Therefore, the mutated RNA I should still form a complex with the mutated RNA II made from the same mutated plasmid and prevent its processing; it just cannot interfere with the processing of RNA II from the original nonmutant plasmid. Therefore, a single-base-pair mutation in the RNA I coding region of the plasmid should effectively change the Inc group of the plasmid to form a new Inc group, of which the mutated plasmid is conceivably the sole member! In fact, the naturally occurring plasmids ColE1 and its close relative p15A, from which other cloning vectors such as pACYC184 have been derived, are members of different Inc groups even though they differ by only 1 base in the kissing regions of their RNA I and RNA II. The genetic experiments which support this aspect of the model are reviewed in Box 4.3.

R1 AND ColIb-P9 PLASMIDS: REGULATION OF TRANSLATION OF Rep PROTEIN BY ANTISENSE RNA

The ColE1-derived plasmids are unusual in that they do not require a plasmid-encoded protein to initiate DNA replication at their *oriV* region, only an RNA primer synthesized from the plasmid. Most plasmids require a plasmid-encoded protein, often called Rep, to initiate replication. The Rep protein is required to separate the strands of DNA at the *oriV* region, often with the help of host proteins including DnaA (see chapter 1). Opening the strands is a necessary first step that allows the replication apparatus to assemble at the origin. The Rep proteins are very specific in that they bind only to the *oriV* of the same type of plasmid because they bind to certain specific DNA sequences within *oriV*. The amount of Rep protein is usually limiting for replication, meaning that there is never more than is needed to initiate replication. Therefore, the copy number of the plasmid can be controlled, at least partially, by controlling the synthesis of the Rep protein.

The R1 Plasmid

One type of plasmid which regulates its copy number by regulating the amount of a Rep protein is the R1 plasmid, a member of the IncFII family of plasmids. Like ColE1 plasmids, this plasmid uses a small antisense RNA to regulate its copy number and this small RNA forms a kissing complex with its target RNA (see Kolb et al., Suggested Reading). Also like ColE1 plasmids, the more copies of the plasmid in the cell, the more of this antisense RNA is made and the more plasmid replication is inhibited. However, rather than inhibiting primer processing, the R1 plasmid uses its antisense RNA to inhibit the translation of Rep protein and thereby inhibit the replication of the plasmid DNA.

Figure 4.10 illustrates the regulation of R1 plasmid replication in more detail. The plasmid-encoded protein, RepA, is the only plasmid-encoded protein that is required for the initiation of replication. The *repA* gene can be transcribed from two promoters. One of these promoters, called p_{copB}, transcribes both the *repA* and *copB* genes, making an mRNA that can be translated into the proteins RepA and CopB. The second promoter, p_{repA}, is in the *copB* gene and so makes an RNA that can encode only the RepA protein. Because the p_{repA} promoter is repressed by the CopB protein, it is turned on only immediately after the plasmid enters a cell and before any CopB protein is made. The short burst of synthesis of RepA from p_{repA} after the plasmid enters a cell causes the plasmid to replicate until it attains its copy number. Then the p_{repA} promoter is repressed by CopB protein, and the *repA* gene can be transcribed only from the p_{copB} promoter.

Once the plasmid has attained its copy number, the regulation of synthesis of RepA, and therefore the replication of the plasmid, is regulated by the antisense RNA, CopA. The *copA* gene is transcribed from its own promoter, and the RNA product affects the stability of the messenger RNA (mRNA) made from the p_{copB} promoter. Because the CopA RNA is made from the same region encoding the translation initiation region for the *repA* gene, but from the other strand of the DNA, the two RNAs are complementary and can pair to make double-stranded RNA. Then a ribonuclease (RNase) called RNase III, a chromosomally encoded enzyme that cleaves some double-stranded RNAs (see chapter 2), cleaves the CopA-RepA duplex RNA.

The reasons why cleavage of this RNA prevents the synthesis of RepA are a little complicated. The 5′ leader region of the mRNA, upstream of where the RepA protein is encoded, encodes a short leader polypeptide, which has no function of its own but just exists to be translated. The translation of RepA is coupled to the translation of this leader polypeptide (see chapter 2 for an explanation of translational coupling). Cleavage of the mRNA by RNase III in the leader region interferes with the translation of this leader polypeptide and, by blocking its translation, also blocks translation of the downstream RepA. Therefore, by having the CopA RNA activate cleavage of the mRNA for the RepA protein upstream of the RepA coding sequence, the plasmid copy number is controlled by the amount of CopA RNA in the cell, which in turn depends on the concentration of the plasmid. The higher the concentration of the plasmid, the more CopA RNA is made and the less RepA protein is synthesized, maintaining the concentration of plasmid around the plasmid copy number.

BOX 4.3

An Incompatibility Group of One's Own

Some of the evidence for the model of replication control of ColE1-type plasmids presented in the text came from genetic experiments (see Lacatena and Cesareni, below). Genetic studies of plasmid replication control are complicated by the fact that mutations that inactivate the plasmid origin of replication lead to loss of the plasmid. On the other hand, mutations that remove the regulation and allow plasmid replication to run amok cannot be isolated since they kill the host cell. Therefore, these experiments required a different approach. This involved using λ phage, which is not discussed in detail until chapter 8; however, the experiments are reviewed here.

To isolate mutations in the ColE1 plasmid origin of replication, the investigators constructed a **phasmid** (part phage and part plasmid) by inserting the ColE1 plasmid into a λ cloning vector phage (see panel A of the figure). The ColE1 plasmid in the phasmid was bracketed by *attB* and *attP* sites on either side. The plasmid could then be excised from the phasmid merely by introducing the phasmid into cells containing excess λ integrase enzyme, which promotes recombination between the *attB* and *attP* sites, excising the plasmid, as shown in the figure (also see chapter 8). Once out of the phasmid, the mutant plasmids could be studied to determine whether they could regulate their own replication and that of other plasmids and phasmids.

The most important feature of the phasmid is that it has two origins of replication, one from the ColE1 plasmid and the other from λ phage. This allows the isolation of mutants with mutations in the *ori* region of the plasmid. With two origins of replication, the phasmid can replicate and form plaques under conditions that inactivate either one of its origins of replication, as long as the other origin remains functional (see panel A of the figure). The plasmid origin of replication is inactive if the cells contain another ColE1-derived plasmid of the same Inc group, and the λ phage origin of replication is inactive if the phasmid is plated on a λ lysogen. In the former case, the RNA I from the resident ColE1 plasmid inhibits replication from the ColE1 *ori* site of the phasmid but the λ *ori* region is active. In the latter case, the repressor made by the λ prophage prevents the synthesis of the λ O and P proteins as well as the transcription of the λ *ori* region, all of which are required for replication from the λ origin (see chapter 8). However, in this case the ColE1 *ori* region is active. Only if the cells harbor both a λ prophage and a ColE1-derived plasmid of the same Inc group do plaques not form.

The investigators reasoned that it might be possible for the phasmid to replicate on a λ lysogen carrying a ColE1-type plasmid if the ColE1 replication origin were mutated so that its replication was no longer inhibited by the resident plasmid. Accordingly, they plated millions of mutagenized phasmids on a λ lysogen carrying a ColE1 plasmid. A few plaques formed, presumably from phasmids with mutations in either the plasmid or the λ origin of replication. Phasmids picked from these plaques were then used to infect cells with excess λ integrase in order to excise the plasmid, as described above, and determine whether its control of replication had been altered. Two types of mutations in the ColE1 *ori* region were observed and are discussed below.

1. Mutations that prevent interaction of RNA I and RNA II. No transformants appeared when cells were transformed with plasmids with this type of mutation. Presumably, these mutations cause the transformants to be killed, apparently by uncontrolled runaway replication of the plasmid. Under these conditions, the RNA I is no longer able to interact with RNA II to inhibit primer formation and the plasmid continues to replicate until the cells are killed. Note that such a mutation does not prevent plaque formation by the phasmid, since the phasmid replication does not need to be controlled for a plaque to form.

2. Mutations that change the Inc group. The other type of mutation, illustrated in the figure, parts B and C, is more intriguing. The plasmids in these mutant phasmids could be excised and used to transform bacteria that do not contain a plasmid, where they were able to maintain their copy number and were therefore not lethal. However, the mutant phasmids from which these plasmids were excised did not form plaques on a λ lysogen containing their own excised plasmid, even though they did form plaques on cells containing the original plasmid or any other excised plasmid with a different *ori* mutation. In the example shown in the figure, a phasmid containing the *svir2* mutation in its ColE1 *ori* region does not form plaques on a λ lysogen containing the plasmid with the *svir2* mutation but does form plaques on cells containing the plasmid with the *svir11* mutation and vice versa.

We can deduce what is happening by examining the model for copy number control of ColE1 plasmids shown in Figure 4.8 in the text and panel C of the figure. According to the model, RNA I must bind to RNA II through complementary base pairing to inhibit plasmid replication by interfering with the processing of the primer RNA II. Because the two RNAs are made from opposite strands of the DNA in the same region, a mutation which changes RNA I also makes a complementary change in RNA II, and so the two still pair

BOX 4.3 (continued)

An Incompatibility Group of One's Own

with each other and inhibit plasmid replication. Therefore, the mutant phasmid, which depends on its ColE1 replication origin to replicate and form a plaque in a λ lysogen, is not able to do so in the presence of the mutant plasmid excised from it, which has the same mutation and therefore the same mutant RNA I. However, it is able to replicate and form a plaque in any cell containing a different mutant plasmid which makes a different mutant RNA I. Apparently, the origin of replication of the mutant plasmid now belongs to a different Inc group from other ColE1-derived plasmids. Indeed, it now defines its own Inc group and has an Inc group of its own!

The positions of some of the changes that alter the incompatibility group are also intriguing. Panel C of the figure shows some of these changes. RNA I can be drawn as a

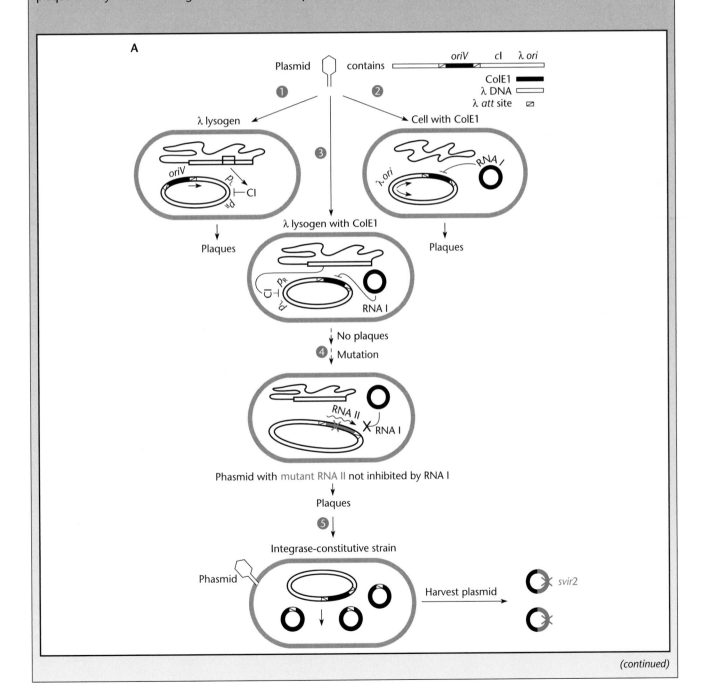

(continued)

BOX 4.3 (continued)

An Incompatibility Group of One's Own

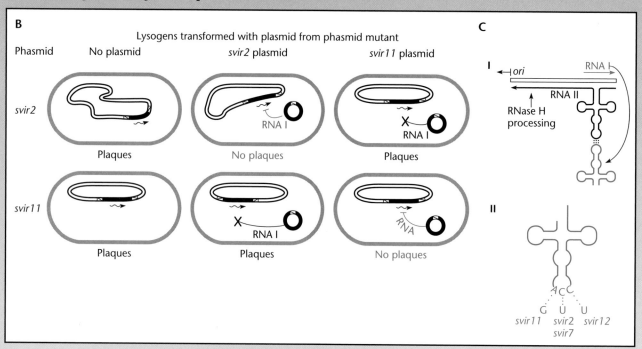

A phasmid method for isolating copy number mutants of the ColE1 plasmid and changing its Inc group. (A) The phasmid has two origins of replication and so can replicate in either a λ lysogen or cells harboring a ColE1-derived plasmid (steps 1 and 2). Only if the cells are λ lysogens and also harbor a resident ColE1 plasmid of the same Inc group do plaques not form (step 3). Mutations in the phasmid that allow plaque formation prevent copy number control or have changed the Inc group (step 4). The plasmid can be excised from the λ phage-cloning vector in cells containing λ integrase enzyme (step 5). (B) Specificity tests for incompatibility groupings. Some RNA I mutants of ColE1-derived plasmids form their own Inc group. The mutant phasmid does not form plaques when plated on cells containing its own excised mutant plasmid but does form plaques on cells containing the original ColE1 plasmid or any other mutant plasmid. (C) Location of *svir* copy number control mutations in RNA I of the ColE1-type plasmid. The mutations lie in what would be the anticodon region of tRNA-like structures as they are commonly drawn.

cloverleaf structure reminiscent of how tRNAs are often drawn, and the changes occur in what would correspond to the anticodon loop. However, it is not clear whether this cloverleaf structure ever forms and what relationship, if any, this structure (if it does form) has to the regulation. In any case, this loop in RNA I must be important for the pairing of RNA I with RNA II and forms the first "kissing complex" with RNA II. The finding that changing RNA I and RNA II in com-

plementary ways in this region can create new Inc groups offers a dramatic confirmation of the model for ColE1 regulation presented in the text.

Reference
Lacatena, R. M., and G. Cesareni. 1981. Base pairing of RNA I with its complementary sequence in the primer precursor inhibits ColE1 replication. *Nature* (London) **294:**623–626.

The ColIb-P9 Plasmid

Yet another level of complexity of the regulation of the copy number by an antisense RNA is provided by the ColIb-P9 plasmid (Figure 4.11) (see Azano and Mizobuchi, Suggested Reading). As in the R1 plasmid,

the Rep protein-encoding gene (called *repZ* in this case) is translated downstream of the leader peptide open reading frame, called *repY*, and the two are also translationally coupled. The translation of *repY* opens an RNA secondary structure which normally occludes the

Shine-Dalgarno (S-D) sequence of the TIR of *repZ*. A sequence in the secondary structure then can pair with the loop of a hairpin upstream of *repY*, forming a pseudoknot (see Figure 2.2 for the definition of a pseudoknot), thus permanently disrupting the secondary structure and leaving the S-D sequence for *repZ* exposed. A ribosome can then bind to the translational initiation region TIR for *repZ* and translate the initiator protein. The antisense Inc RNA pairs with the loop of the upstream hairpin and prevents hairpin formation, leaving

the S-D sequence of the *repZ* coding sequence blocked and preventing translation of the leader polypeptide.

THE pT181 PLASMID: REGULATION OF TRANSCRIPTION OF THE *rep* GENE BY ANTISENSE RNA

Not all plasmids that have an antisense RNA to regulate their copy number use it to inhibit translation or primer processing. Some plasmids of gram-positive bacteria, including the *Staphylococcus* plasmid pT181, use antisense

A Plasmid genetic organization

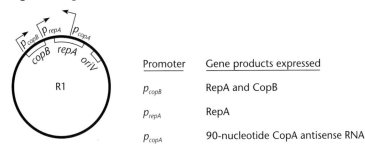

Promoter	Gene products expressed
p_{copB}	RepA and CopB
p_{repA}	RepA
p_{copA}	90-nucleotide CopA antisense RNA

B Replication occurs after plasmid enters cells

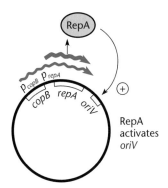

Figure 4.10 Regulation of replication of the IncFII plasmid R1. (A) The locations of promoters, genes, and gene products involved in the regulation. (B) Immediately after the plasmid enters the cell, most of the *repA* mRNA is made from promoter p_{repA} until the plasmid reaches its copy number. (C) Once the plasmid reaches its copy number, CopB protein represses transcription from p_{repA}. Now *repA* is transcribed only from p_{copB}. (C') The antisense RNA CopA hybridizes to the leader peptide coding sequence in the *repA* mRNA, and the double-stranded RNA is cleaved by RNase III. This prevents translation of RepA, which is translationally coupled to the translation of the leader peptide.

C Replication shutdown

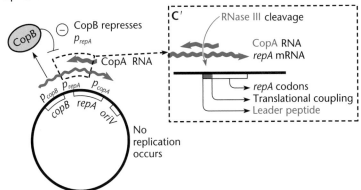

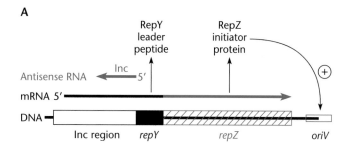

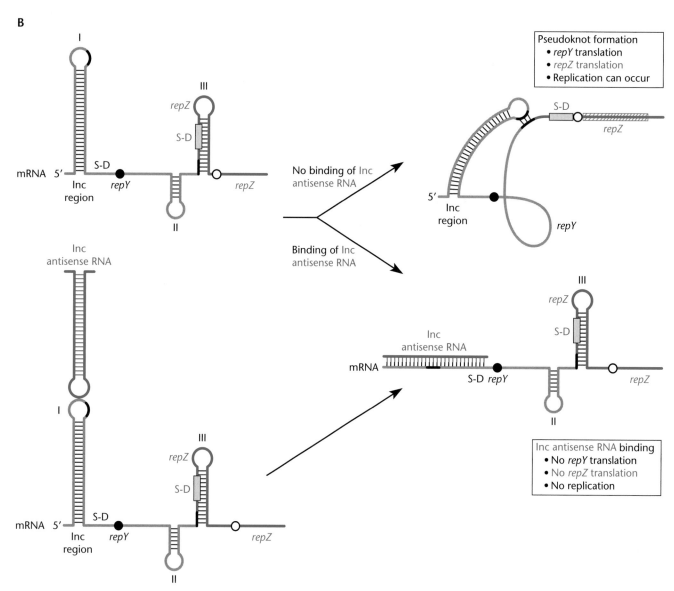

Figure 4.11 Regulation of plasmid ColIb-P9 copy number by antisense RNA inhibition of pseudoknot formation. (A) The minimal replicon with the *repY* (leader peptide) (black box) and *repZ* genes is shown. The Inc region encodes both the 5′ end of the *repYZ* mRNA and the antisense RNA. (B) The *repY* and *repZ* genes are translationally coupled. On the mRNA, the *repY* S-D sequence is exposed, whereas structure III sequesters the *repZ* S-D sequence (purple rectangle) and thereby prevents *repZ* translation. Also shown by thick black bars are regions in structures I and III that are complementary and so can pair, resulting in pseudoknot formation. The closed circle indicates the *repY* start codon; the

RNAs to regulate transcription of the *rep* gene, in this case called *repC*, through a process called attenuation (Figure 4.12) (see Novick et al., Suggested Reading). The pT181 plasmid replicates by a rolling-circle mechanism, and the RepC protein is required to initiate replication of the leading strand at *oriV*. Also, the RepC protein is inactivated each time the DNA replicates (see above). This makes the RepC protein rate-limiting for replication; i.e., the more RepC protein there is, the more plasmid is made. The antisense RNA binds to the mRNA for the RepC protein as the mRNA is being made and prevents formation of a secondary structure. This secondary structure would normally prevent the formation of a hairpin that is part of a factor-independent transcriptional terminator (see chapter 2). Therefore, if the secondary structure does not form, the hairpin forms and transcription terminates (i.e., is attenuated). Transcriptional regulation by attenuation is discussed in more detail in chapter 12.

This regulation works well only because the antisense RNA is so unstable that its concentration drops quickly if the copy number of the plasmid decreases, allowing fine-tuning of the replication of the plasmid with copy number. In other plasmids of gram-positive bacteria, the antisense RNA is much more stable. These plasmids also use a transcription repressor to regulate transcription of the *rep* gene.

THE ITERON PLASMIDS: REGULATION BY COUPLING

Many commonly studied plasmids use a very different mechanism to regulate their replication. These plasmids are called iteron plasmids because their *oriV* region contains several repeats of a certain set of DNA bases called an **iteron sequence**. The iteron plasmids include pSC101, F, R6K, P1, and the RK2-related plasmids. The iteron sequences of these plasmids are typically 17 to 22 bp long and exist in about three to seven copies in the *ori* region. In addition, there are usually additional copies of these repeated sequences a short distance away.

One of the simplest of the iteron plasmids is pSC101. For our purposes, the essential features of the *ori* region of this plasmid (Figure 4.13) are the gene *repA*, which encodes the RepA protein required for initiation of replication, and three repeated iteron sequences, R1, R2, and R3, through which RepA regulates the copy number. The RepA protein is the only plasmid-encoded protein required for the replication of the pSC101 plasmid and many other iteron plasmids. It serves as a positive activator of replication, much like the RepA protein of the R1 plasmid. The host chromosome encodes the other proteins that bind to this region to allow initiation of replication; these include DnaA, DnaB, DnaC, and DnaG (see chapter 1).

Iteron plasmid replication is regulated by two superimposed mechanisms. First, the RepA protein represses its own synthesis by binding to its own promoter region and blocking transcription of its own gene. Therefore, the higher the concentration of plasmid, the more RepA protein is made and the more it represses its own synthesis. Thus, the concentration of RepA protein is maintained within narrow limits and the initiation of replication is strictly regulated. This type of regulation, known as **transcriptional autoregulation,** is discussed in chapter 2.

However, this mechanism of regulation by itself is not enough to regulate the copy number of the plasmid, especially low-copy-number stringent plasmids such as F and P1. Iteron plasmids must have another mechanism to regulate their copy number within narrow limits. This other form of regulation has been hypothesized to be due to the coupling of plasmids through the Rep protein and their iteron sequences (see McEachern et al., Suggested Reading). The **coupling hypothesis** for regulation of plasmid replication is illustrated in Figure 4.14. When the concentration of plasmids is high enough, the concentration of RepA protein also becomes very high. Two copies of the RepA protein may then bind to each other to form dimers, as shown in the figure. These dimers can then bind to iteron sequences on two different plasmids, coupling them to each other. This inhibits the replication of both coupled plasmids. The coupling mechanism allows plasmid replication to be controlled not only by how much RepA protein is present in the cell but also by the concentration of the plasmid itself or, more precisely, the concentration of the iteron sequences on the plasmid.

open circle indicates the *repY* stop codon. Unfolding of structure II by the ribosome stalling at the *repY* stop codon results in the formation of a pseudoknot by base pairing between the complementary sequences and allows the ribosome to access the *repZ* S-D sequence. Binding of Inc antisense RNA to the loop of structure I directly inhibits formation of the pseudoknot and the subsequent IncRNA-mRNA duplex inhibits RepY translation, and consequently RepZ translation, since the two are translationally coupled.

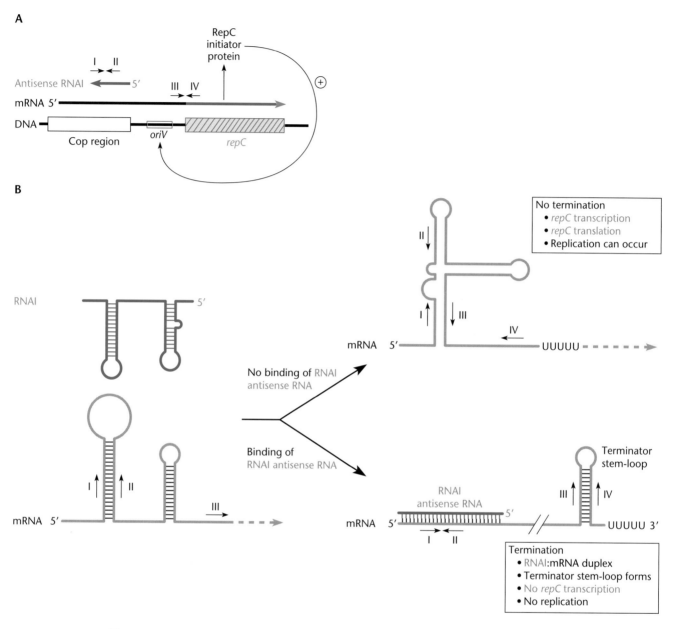

Figure 4.12   Regulation of plasmid pT181 copy number by antisense RNA regulation of transcription of the *repC* gene. (A) The genetic structure of the minimal replicon of pT181. Shown are the mRNA that encodes the RepC protein that initiates leading-strand replication at *oriV*, the Cop region that encodes the antisense RNA (RNA I) that regulates copy number, and regions in the mRNA and antisense RNA, indicated by arrows labeled I, II, III, and IV, that can pair to form alternative secondary structures. (B) Formation of an antisense RNA-mRNA duplex regulates RepC expression by a transcriptional attenuation mechanism. The antisense RNA I can form a duplex with the 5′ end of the mRNA that encodes RepC and disrupt a secondary structure, allowing instead the formation of a terminator loop that causes transcription termination.

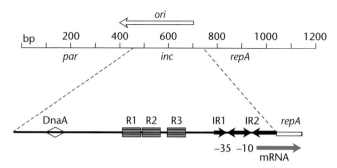

Figure 4.13 The *ori* region of pSC101. R1, R2, and R3 are the three iteron sequences (CAAAGGTCTAGCAGCAGAATT-TACAGA for R3) to which RepA binds to handcuff two plasmids. RepA autoregulates its own synthesis by binding to the inverted repeats IR1 and IR2. The location of the partitioning site *par* (see "Partitioning") and the binding sites for the host protein DnaA are also shown.

The observations that inspired the coupling model are illustrated in Figure 4.15. One puzzling observation came from experiments to determine the effect on the copy number of the plasmid if the concentration of RepA protein is vastly increased (Figure 4.15A). If the function of the iteron sequences is simply to bind excess RepA protein to limit replication, increasing the concentration of RepA protein should increase the copy number of the plasmid once the capacity of the iteron sequences to bind excess RepA protein has been exceeded. To make the experiment easier to interpret, the *repA* gene was expressed from a very strong promoter on a plasmid that also lacked the iteron sequences. The copy number of a plasmid that did contain the iteron sequences was then measured. This allowed the effects of increasing the RepA concentration on the plasmid copy number to be determined without the added complications of also increasing the number of copies of the *repA* gene. The researchers observed that increasing the concentration of the RepA protein even by factors of hundreds led to only a modest increase in the copy number of the iteron-containing plasmid, consistent with the coupling hypothesis.

Other experiments studied the effects of increasing the number of copies of the iteron sequence in the cell, as illustrated in Figure 4.15B. In these experiments, another plasmid that contains a few copies of the iteron sequence but does not depend on the RepA protein for its own replication was introduced into the same cell with the original plasmid. The extra copies of the iteron sequence caused the copy number of the original plasmid to decrease. This effect is not merely due to the iteron sequences on the second plasmid "soaking up" RepA protein, since overproduction of the RepA protein did not overcome the negative effect on the plasmid copy

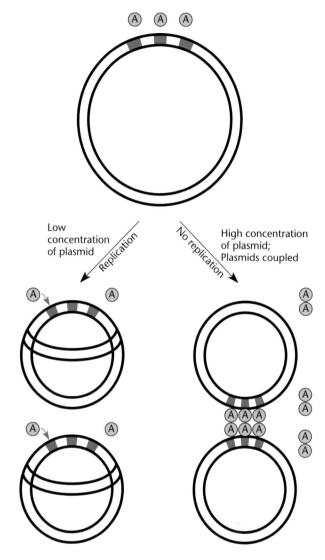

Figure 4.14 The "handcuffing" or "coupling" model for regulation of iteron plasmids. At low concentrations of plasmids, the RepA protein binds to only one plasmid at a time, initiating replication. At high plasmid and RepA concentrations, the RepA protein may dimerize and bind to two plasmids simultaneously, handcuffing them and inhibiting replication.

number of the extra iteron sequences (not shown in the figure). Apparently, the second plasmid is inhibiting the replication of the first plasmid by being coupled to it through its iteron sequences, as shown.

The coupling model also explains the existence of mutations in the *repA* gene called "copy-up" mutations. These mutations are thought to increase the copy number of the plasmid by weakening the binding of RepA protein to the iteron sequences, thereby preventing coupling,

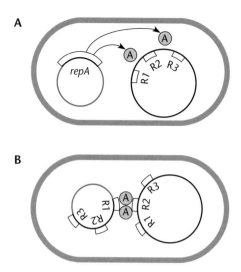

Figure 4.15 Molecular genetic analysis of the regulation of iteron plasmids. (A) The RepA protein is expressed from a clone of the *repA* gene in an unrelated plasmid cloning vector. (B) Additional iteron sequences in an unrelated plasmid can lower the copy number of an iteron plasmid. R1, R2, and R3 are iteron sequences.

without affecting the function of RepA protein in initiation of replication. As predicted by the model, increasing the concentration of copy-up mutant RepA protein does overcome the negative effect of extra iterons on the plasmid copy number, unlike increasing the concentration of wild-type RepA protein. Apparently, the extra copies of the mutant RepA protein do not couple the plasmids and inhibit replication, they only cause new initiation of replication and an increase in the copy number.

Direct support for the coupling model in the replication control of iteron plasmids has come from electron micrographs of purified iteron plasmids mixed with the purified RepA protein for that plasmid. In these pictures, two plasmid molecules can often be seen coupled by RepA protein. However, it still seems possible that this is an in vitro artifact and the plasmid coupling is not the real explanation for the second level of copy number control in iteron plasmids.

HOST FUNCTIONS INVOLVED IN REGULATING PLASMID REPLICATION

As mentioned above, in addition to Rep, many plasmids require host proteins to initiate replication. For example, some plasmids require the DnaA protein, which is normally involved in initiating replication of the chromosome, and have *dnaA* boxes in their *oriV* region to which DnaA binds (Figure 4.13). The DnaA protein may also directly interact with Rep proteins of some plasmids.

This may explain why some broad-host-range plasmids such as RP4 make two Rep proteins, one of which is the runoff product of the other. The different forms of Rep protein might better interact with the DnaA protein of different species of bacteria (see Caspi et al., Suggested Reading). The DnaA protein is involved in coordinating replication of the chromosome with cell division (see chapter 1); making their own replication dependent on DnaA may allow plasmids to better coordinate their own replication with cell division. Like the chromosome origin (*oriC*), some *E. coli* plasmids also have Dam methylation sites close to their *oriV*. These methylation sites presumably help to further coordinate their replication with cell division. As with the chromosomal origin of replication, both strands of DNA at these sites must be fully methylated for initiation to occur. Immediately after initiation, only one strand of these sites is methylated (hemimethylation), delaying new initiations at these sites (see the discussion of sequestration of chromosome origins in chapter 1). Despite substantial progress, however, the method by which the replication of very stringent plasmids, such as P1 and F (with a copy number of only 1), is controlled to within such narrow limits is still something of a mystery and the object of current research.

Mechanisms To Prevent Curing of Plasmids

Cells that have lost a plasmid during cell division are said to be **cured** of the plasmid. Several mechanisms prevent curing, including plasmid addiction systems (Box 4.4), site-specific recombinases that resolve multimers, and partitioning systems. The last two are reviewed below.

RESOLUTION OF MULTIMERIC PLASMIDS

The possibility that a cell will lose a plasmid during cell division is increased if the plasmids form dimers or higher multimers during replication. A dimer consists of two individual copies of the plasmid molecules linked head to tail to form a larger circle, and a multimer links more than two such monomers. Such dimers and multimers probably occur as a result of recombination between monomers. Recombination between two monomers forms a dimer, and subsequent recombination can form higher and higher multimers. Also, rolling-circle replication of RC plasmids can form multimers if termination after each round of replication is not efficient. Multimers may replicate more efficiently than monomers, perhaps because they have more than one origin of replication, so they tend to accumulate if the plasmid has no effective way to remove them. The formation of multimers creates a particular problem when the plasmid attempts to segregate into the daughter cells on cell division. One reason is that multimers lower the effective copy number. Each

BOX 4.4

Plasmid Addiction

Even with their partitioning functions, plasmids are often lost from multiplying cells. In what seems like revenge, some plasmids encode proteins that will kill a cell if it is cured of the plasmid. Such functions have been found in many plasmids including the F plasmid, the R1 plasmid, and the P1 prophage, which replicates as a plasmid (see chapter 8).

These systems have been called plasmid addiction systems because they cause the cell to undergo severe withdrawal symptoms and die if they are cured of the plasmid to which they are addicted.

Plasmid addiction systems all use basically the same strategy. They consist of two components which can be either proteins or RNA. One component functions as a toxin, and the other functions as an antitoxin or antidote. While the cell contains the plasmid, both the toxin and the antitoxin are made and the antitoxin somehow inactivates the effect of the toxin, either by binding to it and inactivating it directly or by somehow indirectly alleviating its effect. The PhD-Doc system is an example of the former type (see the figure). The PhD protein is the antitoxin that binds to the Doc toxin and inactivates it. Restriction modification systems are an example of the latter type. The restriction endonuclease component of the restriction system is the toxin that cuts the chromosome and kills the cell, but not if the modification component, the antitoxin, has methylated a base in its recognition sequence (see chapter 1). Once the cell is cured of the plasmid, however, neither the toxin nor the antitoxin continues to be made. The toxin, however, is longer lived than the antitoxin, so eventually the antitoxin is degraded and the cured cells contain only the toxin. Without the antitoxin to counteract it, the toxin kills the cell.

The toxic protein kills the cell by various mechanisms, depending on the source. For example, the toxic protein of the F plasmid, Ccd, kills the cells by altering DNA gyrase, so it causes double-strand breaks in the DNA. Killer protein Hok of plasmid R1 destroys the cellular membrane potential, causing loss of cellular energy. The mechanism of killing by the P1-encoded killer protein, Doc, is not known, but it blocks translation and seems to work indirectly through another addiction system, MazEF, encoded in the chromosome, that kills cells in which translation has been blocked (see below).

There is a good rationale for plasmid addiction systems in that they prevent cells cured of the plasmid from accumulating and thus help ensure survival of the plasmid. Therefore, it was a surprise to discover that similar toxin-antitoxin systems also occur in the chromosome. Some of these are on exchangeable DNA elements such as genetic islands and superintegrons (see chapter 9) and may play a similar role to

the addiction systems of plasmids, preventing loss of the DNA element. They could be considered selfish genes that prevent themselves, and therefore the DNA element in which they reside, from being lost from the cell. However, other toxin-antitoxin modules seem to be encoded by normal genes in the chromosome. Two examples of these are the MazEF and RelBE systems, both found in *E. coli* K-12. These two systems work in remarkably similar ways. MazF and RelE are the toxins that are ribonucleases which cleave mRNA in the ribosome and block translation, killing the cell. MazF and RelE are the antitoxins that bind to the toxins MazF and RelE, respectively, and inactivate them. The toxins and antitoxins are not made if translation is inhibited, but the toxin is more stable and longer-lived than the antitoxin. These systems could therefore be considered suicide modules in that they cause the cell to kill itself if translation is inhibited, for example by antibiotics or a plasmid addiction module such as PhD-Doc if the cell is cured of the P1 plasmid. A number of hypotheses have been

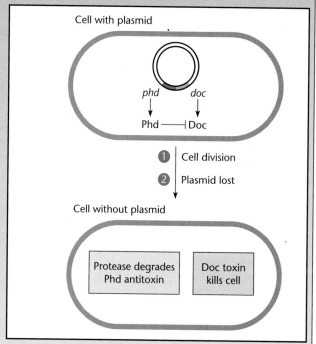

The P1 phage-encoded plasmid addiction system Phd-Doc. Cells containing the plasmid contain both Phd and Doc; Phd is the antidote to Doc, binding to it and inactivating it. If the cell is cured of the plasmid, neither Phd nor Doc is made, but Doc is more stable than Phd and outlives it. Once Phd is degraded by a protease, Doc kills the cell (actually Doc only inhibits translation and MazE kills the cell in response).

(continued)

BOX 4.4 (continued)

Plasmid Addiction

proposed to explain the existence of these suicide modules. One is that they help prevent the spread of phages that inhibit host translation. Another is that they help shut down cellular metabolism in response to starvation and help ensure the long-term survival of some of the cells. It might be relevant that bacteria with a free-living life-style in the environment tend to have many more such suicide modules than do obligate parasitic bacteria, which can live only in the more stable nurturing environment of a eukaryotic host.

Another apparent suicide system in *Bacillus subtilis* lends credence to the idea that the purpose of suicide systems might be to kill some bacteria so that others may live (see Ellermeier et al., below). This system is much more complex than the others we have mentioned and consists of many genes in two operons, *skf* and *sdp*. To summarize, when a population of *B. subtilis* cells are starved for nutrients, some of them begin to sporulate. These cells produce a toxin that kills other cells that were slow to start sporulating. The killed cells

are then devoured by the cells producing the toxin, which can then reverse their sporulation process. This buys the sporulating cells more time, in case the situation changes and they do not really need to sporulate, allowing them to avoid a need to undergo a drastic measure to ensure survival.

References

Ellermeier, C. D., E. C. Hobbs, J. E. Gonzales-Pastor, and R. Losick. 2006. A three-protein signaling pathway governing immunity to a bacterial cannibalism toxin. *Cell* **124**:549–559.

Engelberg-Kulka, H., R. Hazan, and S. Amitai. 2005. *mazEF*: a chromosomal toxin-antitoxin module that triggers programmed cell death in bacteria. *J. Cell Sci.* **118**:4327–4332.

Greenfield, T. J., E. Ehli, T. Kirshenmann, T. Franch, K. Gerdes, and K. E. Weaver. 2000. The antisense RNA of the *par* locus of pAD1 regulates the expression of a 33-amino-acid toxic peptide by an unusual mechanism. *Mol. Microbiol.* **37**:652–660.

Hazan, R., B. Sat, M. Reches, and H. Engelberg-Kulka. 2001. Postsegregational killing by the P1 phage addiction module *phd-doc* requires the *Escherichia coli* programmed cell death system *mazEF*. *J. Bacteriol.* **183**:2046–2050.

multimer segregates into the daughter cells as a single plasmid, and if all of the plasmid is taken up in one large multimer, it can segregate into only one daughter cell. Also, the presence of more than one *par* site on the multimer may cause it to be pulled to both ends of the cell at once, much like a dicentric chromosome can lead to nondisjunction in higher organisms. Therefore, multimers greatly increase the chance of a plasmid being lost during cell division.

To avoid this problem, many plasmids have site-specific recombination systems that resolve multimers as soon as they form. These systems can be either chromosomally encoded or encoded by the plasmid itself. A site-specific recombination system promotes recombination between specific sites on the plasmid if the same site occurs more than once in the molecule, as it would in a dimer or multimer. This recombination has the effect of resolving multimers into separate monomeric plasmid molecules.

A well-studied example of a plasmid-encoded site-specific recombination system is the Cre/*loxP* system encoded by P1 phage. This phage is capable of lysogeny, and its prophage form is a plasmid, subject to all the problems faced by other plasmids, including multimerization due to recombination. The Cre protein, a tyrosine (Y) recombinase, promotes recombination between two *loxP* sites on a dimerized plasmid, resolving the dimer into two monomers. This system is very efficient and relatively simple and has been useful in a number of

studies, including demonstrations of the interspecies transfer of proteins (see chapter 5). It has also been used as a model for Y recombinases, since the recombinase has been crystallized with its *loxP* DNA substrate. Y recombinases and their mechanism of action are discussed in chapter 9.

The best known examples of host-encoded site-specific recombination systems used to resolve plasmid dimers are the *cer*-XerC,D and the *psi*-XerC,D site-specific recombination systems used by the ColE1 plasmid and the pSC101 plasmid, respectively. The XerC,D system is mentioned in chapter 1 in connection with segregation of the chromosome of *E. coli*. The XerC and XerD proteins are part of a site-specific recombinase that acts on a site, *dif*, close to the terminus of replication of the chromosome to resolve chromosome dimers created during chromosome replication. The plasmids have commandeered this site-specific system of the host to resolve their own dimers by having sites at which the recombinase can act. The site on the ColE1 plasmid is called *cer*, and the site on pSC101 is called *psi*. No accessory proteins are required for Xer recombination at *dif* sites to resolve chromosome dimers. In contrast, the *cer* site on the ColE1 plasmid is not recognized as such but is recognized only if two other host proteins called PepA and ArgR are bound close by in the DNA, as shown in Figure 4.16. A similar situation occurs in the pSC101 plasmid, but now the auxiliary host proteins are PepA

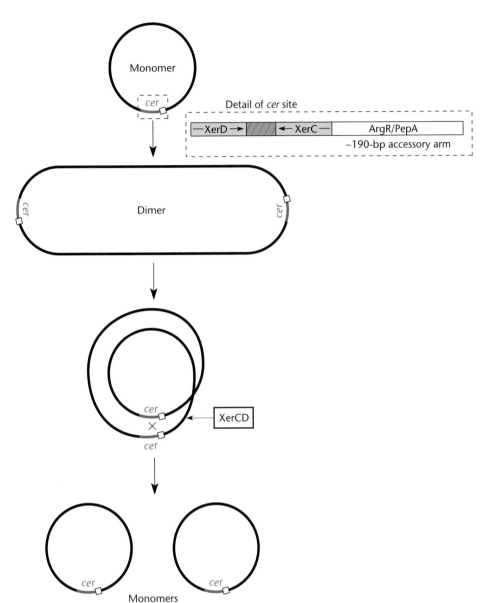

Figure 4.16 The Xer functions of *E. coli* catalyze site-specific recombination at the ColE1 plasmid *cer* site to resolve plasmid dimers. The sites of binding of the host proteins ArgR and PepA are shown.

and ArcA~P (phosphorylated ArcA), which binds close to where ArgR binds in *cer*. Apparently, these other proteins bind to XerC,D recombinase at the plasmid *cer* or *psi* sites and help orient it for the recombination process. However, it is not clear how these particular host proteins came to play this role. The only thing these accessory proteins have in common is that they all normally bind to DNA because they are all transcription factors. This is yet another example of a case where plasmids commandeer host functions for their own purposes, in this case a site-specific recombination system normally used for resolving chromosome dimers. XerC,D is also a Y recombinase, and its mechanism of action is discussed

in more detail in the section on tyrosine recombinases in chapter 9.

PARTITIONING

The most effective mechanism that plasmids have to avoid being lost from dividing cells is their set of **partitioning** systems. These systems ensure that at least one copy of the plasmid segregates into each daughter cell during cell division. The functions involved in these systems are called **Par functions,** and in many ways they are analogous to the Par functions involved in chromosome segregation. In fact, the discovery of Par systems in plasmids preceded their discovery in chromosome segregation.

The first evidence for plasmid Par systems came from calculations, using combinatorial probability, of how often cells would be cured of a plasmid if the plasmid had no active way of segregating plasmids into daughter cells. In the simple example shown in Figure 4.17, the copy number of the plasmid is 4. Immediately after cell division, a cell contains four copies, and immediately before the cell divides, it contains eight copies. If the plasmids are equally divided into the two daughter cells, each daughter cell gets four plasmids. However, chance dictates that the plasmids usually are not equally distributed between the two daughter cells: one daughter cell gets more than four, and the other gets less than four. In fact, there is a certain probability that one cell will get all of the plasmids and the other cell will be cured. Since each plasmid can go into either one cell or the other, the probability that one daughter cell will be cured of the plasmid is the same as the probability of tossing eight heads or tails of a coin in a row. Thus, the probability that the first plasmid will go into one cell is 1/2, and the probability that the first two plasmids will go into the same cell is 1/2 times 1/2 = 1/4, and so on. The probability that all eight will go into one cell is therefore $(1/2)^8$, or 1/256. Since it is irrelevant which of the two cells is cured, the frequency of curing is twice this value, so that $2(1/2)^8$, or 1/128, of the cells are cured each time the cells divide. In general, for a plasmid with a copy number n, the frequency of curing is $2(1/2)^{2n}$, since the number of plasmids at the time of division is twice the copy number. Also, as the cells divide once every generation time, the frequency of cured cells in the population is roughly equal to the number of generation times that have elapsed times this number. This is the frequency of curing if the sorting of the plasmids into daughter cells is completely random. Therefore, if the fraction of the cells that are cured of the plasmid is less than $2(1/2)^{2n}$ times the number of generation times that have elapsed, the plasmid must have some sort of partitioning function.

This calculation indicates that few cells would be cured each generation if the plasmid has a high enough copy number, even without a partitioning mechanism. Nevertheless, a significant fraction of cells would be cured if the plasmid has a low copy number. In fact, with

Figure 4.17   Random curing of a plasmid with no Par system. Each plasmid has an equal chance of going to one of the daughter cells when the cell divides, and the occasional cell inherits no plasmids.

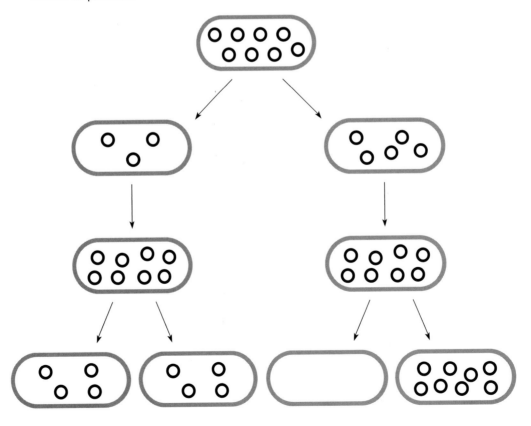

a plasmid with a copy number of only 1, such as F or P1, $2(1/2)^2$ or 1/2 of the cells would be cured each generation. Since cells are seldom cured of even low-copy-number plasmids, some mechanism must ensure that plasmids, especially those with low copy numbers, are partitioned faithfully into the daughter cells each time the cell divides.

The Par Systems of Plasmids

Because of their relevance to chromosome segregation and to bacterial cell biology, the Par systems of plasmids have been studied extensively. While these systems are still not completely understood, the results have been very interesting and have helped create a paradigm shift in how we view the bacterial cell (see "Chromosome Segregation" in Chapter 1).

The Par systems of low-copy-number plasmids fall into at least two groups whose members are related to each other by sequence and function (Table 4.3). At least one of these groups of plasmid Par functions is also related to the putative chromosomal Par systems of some bacteria, as also shown in Table 4.3. One group is represented by the Par system of the R1 plasmid of *E. coli*, the other, much larger, group is represented by the Par systems of the F, P1, and broad-host-range RK2 plasmids, among others. This is also the group to which the chromosomal partitioning systems from the bacteria *B. subtilis* and *Caulobacter crescentus* belong. The two groups of partitioning systems differ in the details of how they achieve the feat of plasmid partitioning, but they seem to share an overall strategy. They both help localize the plasmid to the site of replication, which is usually the center of the cell, where the chromosome also replicates, and promote the pairing of the daughter plasmids after they replicate. They then somehow push (or pull) the daughter plasmids apart to new positions prior to cell division, in what will become the replication site in the daughter cells. Evidence is accumulating that they do this by forming dynamic filaments in the cell, which are in some ways reminiscent of the mitotic spindle of eukaryotes that perform similar functions. The R1 plasmid partitioning system is addressed first, since that seems to be the better understood of the two systems.

THE R1 PLASMID PAR SYSTEM

The mechanism of partitioning by the R1 plasmid is illustrated in Figure 4.18. The partitioning system of the R1 plasmid consists of two protein-coding genes, *parM* and *parR*, as well as a centromere-like *cis*-acting site, *parC* (see Moller-Jensen et al., Suggested Reading). First, a few dimers of the ParR protein bind to the *parC* site. The ParM protein can then bind to ParR, but only if the latter is already bound to *parC*. In this way, the ParR-*parC* complex serves as a sort of nucleation site for the assembly of ParM. While the plasmid is replicating, this complex of the two ParR and ParM proteins is localized to the midpoint of the cell and thereby localizes the plasmid to this point. As evidence, in the absence of any one of the components of the Par system, the plasmid is more or less evenly distributed around the cell.

Dramatic changes occur, however, when replication is completed and it is time to segregate the two daughter plasmids into the daughter cells. The ParM protein begins to polymerize into helical filaments that extend from the center of the cell toward the cell poles. This polymerization might require pairing of the newly synthesized plasmids because it occurs only after plasmid replication is completed. Growth of the filaments seems to occur by the addition of ParM subunits to the end to which the plasmid and ParR are attached. The ParM

TABLE 4.3	Partitioning systems	
DNA	**Par system**	**Centromere-like site**[a]
ParA-ParB homologs		
P1 prophage	ParA-ParB	*parS*
F plasmid	SopA-SopB	*sopC*
RK2	IncC2-KorB	O_B3?
Ti	RepA-RepB	*inc2*?
B. subtilis	Soj-Spo0J	*parS*
C. crescentus	ParA-ParB	*parS*
ParM-ParR homologs		
R1	ParM-ParR	*parC*
Collb-P9	ParA-ParB	Unknown
C. crescentus	MreB	Unknown

[a] See Yao et al., Suggested Reading, and references therein.

A *parCMR* locus

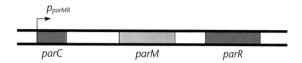

B   Plasmid R1 partitioning

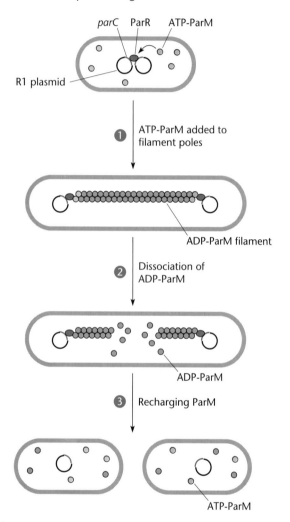

Figure 4.18   Partitioning of the R1 plasmid. (A) Structure of the *par* locus of R1, showing the positions of the *parM* and *parR* genes as well as the *cis*-acting *parC* site. The transcription start site is in the *parC* site. (B) While the plasmid is replicating, ParR bound to *parC* pairs the replicating plasmids in the cell center and provides a nucleation site for ParM. (1) After replication is complete, the filaments grow by adding successive ATP~ParM subunits to the ends where *parC* and ParR are bound while the ATP on the ParM subunit added before it is cleaved to ADP to provide energy. (2) After the plasmid copies have been pushed to the ends (poles) of the cell, the ADP–ParM subunits dissociate from one or both ends and the filaments disappear. (3) ParM can then be recharged with ATP before the plasmids are partitioned again before the next cell division. The *parC* site on the plasmid is shown in purple.

subunit is in the ATP-bound form when it first attaches. Then its ATP is cleaved to ADP when the next subunit attaches. In this way, cleavage of ATP on the penultimate ParM subunit provides energy for the attachment of the next subunit and for growth of the filament. As they grow, the helical ParM filaments may push the daughter copies of the plasmid to opposite cell poles, as suggested by the fact that the plasmids seem to be associated with the tips of the growing filaments. The filamentation occurs only in the presence of ParR and the *parC* site; therefore, it presumably requires nucleation by the ParR-*parC* complex in the cell center. The ability of ParM to polymerize into filaments was suspected because the ParM protein is an ATPase with a similar structure to eukaryotic actin and the actin-like MreB protein, required for determining cell shape and perhaps chromosomal segregation in some bacteria (see chapter 1). Like actin, the ParM protein can polymerize in vitro into dynamic double-stranded filaments, which can get shorter and longer with the cleavage of ATP to ADP, and this polymerization requires nucleation by ParR and *parC*.

The relatedness of the ParM protein to actin and the actin-like bacterial protein MreB has attracted much attention (see Box 1.3). The role of actin filaments in eukaryotic cells is to move things around in the cell. In fact, some intracellular pathogenic bacteria, including *Shigella* and *Listeria*, move within eukaryotic cells by attaching themselves to actin filaments and promoting the elongation of the filaments. The filaments formed by MreB may also promote the migration of the bacterial chromosome, at least in some species. Therefore, contrary to what was previously thought to be the case, bacteria do have a cytoskeleton that moves cellular constituents around, including plasmids and the chromosome. It is curious that while ParM, MreB, and other actin-like proteins play a role in chromosome and plasmid segregation in bacteria, this role in eukaryotes is played by microtubules, which are composed of tubulin, not actin. In bacteria, the FtsZ protein is the homolog of tubulin, a protein that plays a role in cell division, which is more like the role of actin in eukaryotes. Thus, although the cytoskeleton apparently existed in the earliest organisms, before the separation of bacteria and eukaryotes, the roles of actin and tubulin filaments seem to be reversed in the two kingdoms.

THE P1 AND F PLASMID Par SYSTEMS

The larger group of plasmid-partitioning systems are related to the systems of the P1 plasmid prophage and the F plasmid. Also, some bacteria, including *B. subtilis* and *C. crescentus*, seem to use a similar Par system to partition their chromosomes. These Par systems are similar to the Par system of the R1 plasmid in that they usually consist of two proteins and an adjacent *cis*-acting site on the DNA to which one of these proteins, often called ParB, binds. Also, the other protein, often called ParA, has ATP binding motifs. However, the ATP binding protein has no homology to actin or any other cytoskeletal protein. Rather, it is distantly related to MinD, a protein that helps select the division site in bacteria and, in *E. coli* at least, oscillates in a helical path from one end of the cell to the other during the cell cycle. In the F plasmid, the corresponding proteins are called SopA and SopB and the site is called *sopS* (for *s*tability of *p*lasmid). For simplicity, we refer to them as ParA, ParB, and *parS*. In some plasmids, the Par systems are more closely interspersed with the replication proteins; some of them may lack an autonomous ParB protein, and perhaps a larger ParA-related protein may play both roles.

The mechanism of action of these Par systems may be superficially similar to that of the R1 plasmid discussed above. First, a dimer of ParB binds to the *parS* site. Then many more dimers of ParB bind cooperatively nearby, coating the DNA with ParB protein for some distance around the *parS* site. Cooperative binding means that the second copy binds much more tightly if the first copy is already bound. Cooperative binding is discussed in chapter 8, in connection with the binding of λ repressor to the operators. The function, if any, of this coating of the DNA around the *parS* site is not known.

Once the ParB protein has bound and coated the DNA around the *parS* site, the ParA protein can bind to the complex. Like ParM, the ParA protein is also an ATPase that can polymerize in the presence of ATP; it also requires the ParB-*parS* complex to nucleate the polymerization. However, ParA proteins polymerize into shorter and more dynamic filaments than the ParM protein of the R1 plasmid (see Lim et al., Suggested Reading). Rather than forming a single filament, a number of ParA filaments seem to radiate out from the ParB-*parS* complex, in what have been called radial asters, like the petals on a daisy flower. These filaments may then push the plasmids to the one-quarter and three-quarter positions of the cell, which will be the new centers of the daughter cells after the cell divides. It is tempting to speculate that the ParA filaments might be shorter than the ParM filaments because they do not have to push the plasmids as far, only to the one-quarter and three-quarter positions rather than all the way to the poles of the cell. The radiation of the ParA filaments from the ParB-*parS* complex is more reminiscent of the eukaryotic spindle apparatus than the single filaments formed by the ParM protein of the R1 plasmid. The subject of plasmid partitioning is closely related to the subject of bacterial cell biology in general and promises to be an active research area for years to come.

INCOMPATIBILITY DUE TO PLASMID PARTITIONING

If two plasmids share the same partitioning system, they will be incompatible, even if their replication control systems are different. Incompatibility due to shared partitioning systems makes sense, considering the models presented above. If two plasmids that are otherwise different share the same Par system, they can pair with each other after they replicate. Then one plasmid of each type is directed to opposite ends of the cell before the cell divides. In this way, one daughter cell can get plasmids of one type while the other daughter cell gets the plasmid of the other type, and cells are cured of one or the other plasmid. However, even though shared partitioning systems can cause incompatibility, this is usually not the sole cause of their incompatibility. Usually cells with the same partitioning system also share the same replication control system, since the two are often closely associated on the plasmid; therefore, the incompatibility is due to both systems. In fact, in some cases, the replication control genes and the partitioning genes are intermingled around the origin of plasmid replication.

Constructing a Plasmid Cloning Vector

As discussed in chapter 1, a cloning vector is an autonomously replicating DNA (replicon) into which other DNAs can be inserted. Any DNA inserted into the cloning vector replicates passively with the vector so that many copies (clones) of the original piece of DNA can be obtained.

Plasmids offer many advantages as cloning vectors, and many plasmids have been engineered to serve as plasmid cloning vectors. They generally do not kill the host cell and are relatively easy to purify to obtain the cloned DNA. They also can be made relatively small because few plasmid-encoded functions are required for their replication. In fact, in one of the first cloning experiments, a frog gene was cloned into plasmid pSC101 (see Cohen et al., Suggested Reading).

Most plasmids, as they are isolated from nature, are too large to be convenient as cloning vectors and/or often do not contain easily selectable genes that can be used to move them from one host to another. In this section, we outline some of the steps in making a plasmid cloning vector from a wild-type plasmid and some of the desirable features that have been engineered into plasmid cloning vectors.

Finding the Plasmid *ori* Region

The first step in making a plasmid cloning vector from a wild-type plasmid we have isolated is to find the *ori* region of the plasmid. As discussed above, the *ori* region is usually responsible for most of the properties of the plasmid, including replication, copy number control, and partitioning. The origins of some commonly used cloning vectors are listed in Table 4.4.

Recombinant DNA techniques are particularly suited for locating and studying these regions. As shown in Figure 4.19, the plasmid is cut into several pieces with a restriction endonuclease (arrows in the figure) and the pieces are ligated (joined) to another piece of DNA that has a selectable marker, such as resistance to ampicillin (Amp^r). For the experiment to work, the second piece of DNA cannot have a functional origin of replication. The ligated mixture is then used to transform bacteria, and the antibiotic-resistant transformants are selected by plating the mixture on agar plates containing growth medium and the antibiotic. The only DNA molecules able to replicate and also confer antibiotic resistance on the cells are hybrids with both the *ori* region of the plasmid and the piece of DNA with the antibiotic resistance gene. Therefore, only cells harboring these hybrid molecules can grow on the antibiotic-containing medium. We can determine which plasmid fragment these transformants have by using methods described in chapter 1, and this will tell us approximately where the *ori* region is located on the plasmid.

We can further localize the *ori* region by cutting the fragment known to contain it into even smaller pieces and repeating the process above until the smallest piece from the plasmid that can function as an origin has been identified.

This same type of analysis also can be used to identify the partitioning region and the region responsible for incompatibility. To locate the *par* sequences, smaller and smaller pieces of DNA that retain origin function are tested to see if they also confer the ability to partition properly. If the plasmid containing the smaller origin region cures at a higher frequency, it has lost its *par* region. Once we identify a plasmid that lacks a *par* region, we can clone

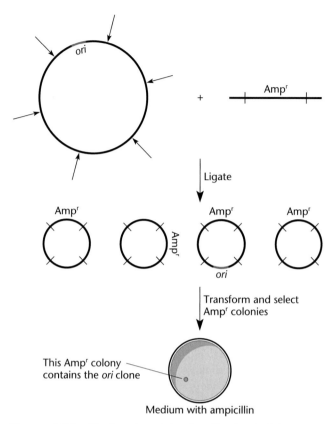

Figure 4.19   Finding the origin of replication (*ori*) in a plasmid. Random pieces of the plasmid are ligated to a piece of DNA containing a selectable gene but no origin of replication and introduced into cells. Cells that can form a colony on the selective plates contain the selectable gene ligated to the piece of DNA containing the origin.

pieces of the same type of plasmid back into it to find a piece that restores proper partitioning. In fact, the *par* regions of plasmids were discovered by this means.

INTRODUCING SELECTABLE GENES

Once we have identified the *oriV* region of the plasmid, the next step is to introduce genes into the cloning vector that confer selectable properties on cells containing the plasmid. This allows us to select cells which contain the plasmid vector and to move it from one host cell to another. Genes that confer resistance to an antibiotic make convenient selectable genes. Cells containing the plasmid can then be selected by plating on medium containing the antibiotic, and only cells containing the plasmid will multiply to form a colony. Antibiotic resistance genes are often taken from transposons and other plasmids. Some antibiotic resistance genes which have been introduced into cloning vectors are the chloramphenicol

TABLE 4.4	Replication origins of several *E. coli* plasmid vectors	
Plasmid	*ori*	Copy number
pBR322	pMB1 (ColE1)	15–20
pUC vectors	pMB1 mutant	Hundreds
pET vectors	pMB1	15–20
pBluescript	pMB1 mutant	Hundreds
pACYC184	p15A	10–12
pSC101	pSC101	5
pBAC	F	1

resistance gene, Cam^r, of transposon Tn9; the tetracycline resistance gene, Tet^r, of plasmid pSC101; the ampicillin resistance gene, Amp^r, of transposon Tn3; and the kanamycin resistance gene, Kan^r, of transposon Tn5. The antibiotic resistance gene that is chosen depends on the uses to which the cloning vector will be put. Some antibiotic resistance genes such as the Tet^r gene from pSC101 are expressed only in some types of bacteria closely related to *E. coli*, while others such as Kan^r from Tn5 are expressed in most gram-negative bacteria.

INTRODUCING UNIQUE RESTRICTION SITES

Since many applications of plasmid cloning vectors require that clones be introduced into restriction sites, it is necessary that our cloning vector have some restriction sites that are unique. If a site is unique, the cognate restriction endonuclease cuts the vector at only that one site when it is used to cut the plasmid. We can then clone pieces of foreign DNA into the unique site, and the cloning vector will remain intact. Restriction sites for six-hitters occur on average about once every 1,000 bp (see chapter 1), so that a cloning vector of 3,000 to 4,000 bp is apt to have more than one site for any particular six-hitter restriction endonuclease. There are many tricks for removing the extra sites. We can change the sequence at some of the sites by site-specific mutagenesis so that they are no longer recognized by the restriction endonuclease. Alternatively, we can make a partial digest of the plasmid so that it is cut at only one of the sites. The site at which it was cut can then be removed by filling in the overhang with DNA polymerase, if the restriction endonuclease leaves a 5′ overhang, or by removing the overhang with a single-strand-specific DNase before religating.

Not only do we want to have unique sites for some restriction nucleases in our cloning vector, but also these sites should be located in the plasmid in a way which makes them most useful for cloning. In many plasmid cloning vectors, the unique sites are located in a selectable gene, so that insertion of a foreign piece of DNA in the site inactivates the selectable gene. This is called **insertional inactivation** and is discussed below. Normally, during a cloning operation, only a small percentage of the cloning vectors pick up a foreign DNA insert. If those that have picked up an insert no longer confer the selectable trait, for example resistance to an antibiotic, they can be more easily identified. Many cloning vectors also have the unique restriction sites for many different restriction endonucleases all grouped in one small region on the plasmid called a polyclonal or multiple-restriction site. This offers the convenience of choosing among a variety of restriction endonucleases for cloning, and the cloned DNA is always inserted at the same site, independent of the restriction endonuclease used. Polyclonal sites

can also be used for **directional cloning.** If the cloning vector is cut by two different incompatible restriction endonucleases with unique sites within the polyclonal site, the resulting overhangs cannot pair to recyclize the plasmid. The plasmid can recyclize only if it picks up a piece of foreign DNA. If the piece of DNA to be cloned has overhangs for the two different sites at its ends, it is usually cloned in only one orientation into the polyclonal site.

The unique restriction sites can also be placed so that genes cloned into them will be expressed from promoters and translation initiation regions (TIRs) on the plasmid. These are called expression vectors and can be used to express foreign genes in *E. coli* and other convenient hosts. Such vectors can also be used to attach affinity tags to proteins to aid in their purification. Expression vectors and affinity tags are discussed in connection with translational and transcriptional fusions in chapter 2.

Examples of Plasmid Cloning Vectors

A number of plasmid cloning vectors have been engineered for special purposes. Almost all of these plasmids have at least some of the features mentioned above for a desirable cloning vector. To reiterate:

1. They are small, so that the plasmid can be easily isolated and introduced into various bacteria.
2. They have a relatively high copy number, so that the plasmid can be easily purified in sufficient quantities.
3. They carry easily selectable traits, such as a gene conferring resistance to an antibiotic, which can be used to select cells that contain the plasmid.
4. They have one or a few sites for restriction endonucleases, which cut DNA and allow the insertion of foreign DNAs. Also, these sites usually occur in selectable genes to facilitate the detection of plasmids with foreign DNA inserts by insertional inactivation.

Many plasmid cloning vectors have other special properties that aid in particular experiments. For example, some contain the sequences recognized by phage packaging systems (*pac* or *cos* sites), so that they can be packaged into phage heads (see chapters 7 and 8). Expression vectors can be used to make foreign proteins in bacteria. Mobilizable plasmids have mobilization (*mob*) sites and so can be transferred by conjugation to other cells (see chapter 5). Some broad-host-range vectors have *ori* regions that allow them to replicate in many types of bacteria or even in organisms from different kingdoms. Shuttle vectors contain more than one type of replication origin and so can replicate in unrelated organisms. These and some other types of specialty plasmid cloning vectors are discussed in more detail below and in later chapters.

CLONING VECTOR PBR322

Figure 4.20 shows a map of pBR322, which embodies many of the desirable traits of a cloning vector. This plasmid is fairly small (only 4,360 bp) and has a relatively high copy number (~16 copies per cell), making it easy to isolate. The vector was constructed by removing all but the essential *ori* region from pMB1, a ColE1-like plasmid, and adding two resistance genes for the antibiotics tetracycline and ampicillin, which were taken from plasmid pSC101 and transposon Tn*3*, respectively. The plasmid also has several unique sites for restriction endonucleases, including BamHI, EcoRI, and PstI. These enzymes cut DNA at specific sites, allowing a piece of foreign DNA to be inserted into the plasmid and then studied.

Insertional inactivation offers a simple genetic test for determining whether a cell contains a pBR322 plasmid with a foreign DNA insert. To illustrate, Figure 4.21 shows a piece of foreign DNA inserted into the BamHI site in the tetracycline resistance (Tet^r) gene in pBR322. The piece of foreign DNA in the BamHI site disrupts the Tet^r gene and causes the plasmid to lose the ability to confer tetracycline resistance on a bacterium that carries it. The plasmid still confers ampicillin resistance, however, since the Amp^r gene remains intact. Therefore, cells containing a plasmid with a foreign DNA insert are ampicillin resistant but tetracycline sensitive, which is easy to test on agar plates containing one or the other antibiotic.

pUC PLASMIDS

Some of the most commonly used plasmid cloning vectors are the pUC vectors and vectors derived from them. One pUC vector, pUC18, is shown in Figure 4.22. The

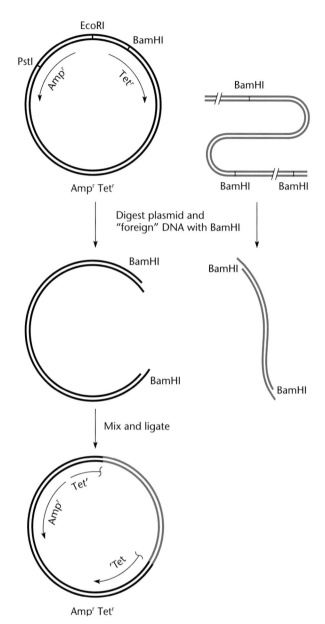

Figure 4.21   Insertional inactivation of the tetracycline resistance (Tet^r) gene of pBR322 by insertion of a foreign DNA into the BamHI site. The Tet^r gene is disrupted, but the plasmid still confers ampicillin resistance because the Amp^r gene is still active.

Figure 4.20   The plasmid cloning vector pBR322. Only unique restriction sites are labeled. Amp^r, ampicillin resistance; Tet^r, tetracycline resistance.

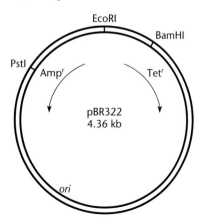

pUC plasmids are very small (with only 2,700 bp of DNA) and have a very high copy number of 30 to 50, making them relatively easy to purify. They also have the easily selectable ampicillin resistance (Amp^r) gene. One of the most useful features of these plasmids is the ease with which they can be used for insertional inactivation. They encode the N-terminal region of the *lacZ* gene product, called the α-peptide, which is not active in the

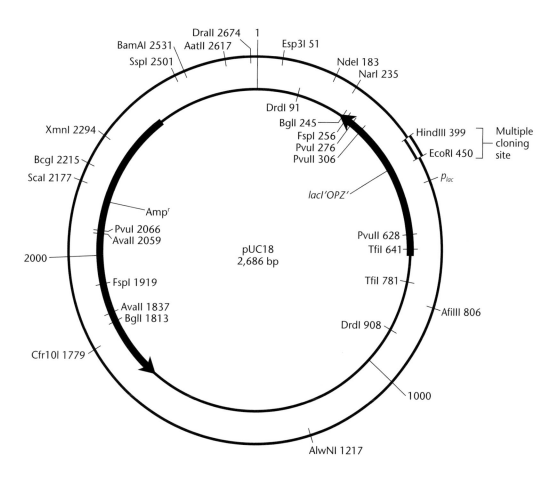

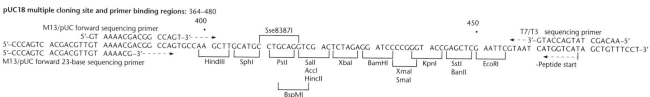

Figure 4.22 A pUC expression vector. A gene cloned into one of the restriction sites in the multiple-cloning site almost invariably disrupts the coding sequence for the *lacZ* α-peptide. If it is inserted in the correct orientation, the gene is transcribed from the *lac* promoter called p_{lac} in the figure. If the open reading frame for the gene is in the same reading frame as that for the *lacZ* coding sequence, the gene is also translated from the *lacZ* TIR, and the N-terminal amino acids of *lacZ* become fused to the polypeptide product of the gene.

cell by itself but complements the C-terminal portion of the protein called the *lacZ* β-polypeptide, to make active *lacZ* polypeptide, which turns colonies blue on 5-bromo-4-chloro-3-indolyl-β-D-galactopyranoside (X-Gal) plates. Some host strains such as *E. coli* JM109 have been engineered to make the β-polypeptide of *lacZ*. As a consequence, *E. coli* JM109 containing a pUC plasmid forms blue colonies on X-Gal plates. The pUC plasmids have a

multicloning site containing the recognition sequences for many different restriction endonucleases in the coding region for the α-peptide (Figure 4.22). If a foreign DNA is cloned into any one of these sites, the bacterium does not make the α-peptide and the colonies are colorless on X-Gal plates; bacteria containing plasmids with inserts are therefore easy to identify. These plasmids are also transcription vectors (see chapter 2) because a gene in a piece of DNA directionally cloned into the multi-cloning site on the plasmid is immediately downstream of the strong *lac* promoter on the plasmid, called p_{lac} in the figure, and so it is transcribed from the *lac* promoter. The *lac* promoter is also inducible and is turned on only if an inducer, such as isopropyl-β-D-thiogalactopyranoside (IPTG) or lactose, is added. Thus, the cells can be propagated before the synthesis of the gene product is induced, a feature that is particularly desirable if the gene product is toxic to the cell. Genes cloned into one of the multi-cloning sites in the *lacZ* gene can also be translated from the *lacZ* TIR on the plasmid, provided that there are no intervening nonsense codons and the gene is cloned in the same reading frame as the upstream *lacZ* sequences.

BAC VECTORS

One problem with using high-copy-number cloning vectors such as pUC vectors is that the clones are very unstable, particularly if they are large. If the clone exists in many copies, recombination between repeated sequences in the clones can rearrange the sequences in the clone. This is a particular problem in some applications, for example the sequencing of large genomes such as the human genome, where it is necessary to obtain plasmid libraries containing very large clones. The DNA of higher eukaryotes including humans contains many repeated sequences. For this reason, **bacterial artificial chromosome (BAC) cloning vectors** have been designed (Figure 4.23). These plasmid vectors are based on the F plasmid origin of replication, so they have a copy number of only 1 in *E. coli*. They can also accommodate very large inserts, on the order of 300,000 bp of DNA. This was expected, since it was known that F' factors can be very large and are quite stable, especially in a RecA⁻ host. An F' factor is a naturally occurring plasmid in which the F plasmid has incorporated a large region of the *E. coli* chromosome (see chapter 5).

The original pBAC vector shown in Figure 4.23 (see Shizuya et al., Suggested Reading) contains the F plasmid origin of replication and partitioning functions, and a selectable chloramphenicol resistance gene. It also contains unique HindIII and BamHI cloning sites into which large DNA fragments can be introduced, as well as a number of other features that are helpful in the cloning and sequencing of large fragments. Surrounding the cloning sites are the sites for other restriction endonucleases, chosen to be very GC rich so that they are not likely to exist in human DNA, which is relatively low in GC base pairs. This allows the DNA inserts to be excised from the cloning vector without (usually) cutting the DNA insert as well. The cloning sites are also flanked by the sequences of some phage RNA polymerase promoters, so that the phage RNA polymerases can be added

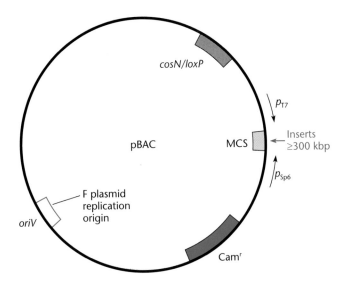

Figure 4.23 A pBAC cloning vector for cloning large pieces of DNA for genome sequencing. Shown is the multiple cloning site (MCS), where clones are inserted, and promoters for phages T7 and SP6 that bracket the sites. Also shown are the sites *loxP* and *cosN*, where the plasmid can be cut by the Cre recombinase or λ terminase, respectively, for restriction mapping of the insert. These recognition sites are long enough that they almost never occur by chance in the insert.

to the DNA in vitro and the resulting RNA will be complementary to the ends of the cloned DNA fragment. Hybridization of these RNAs to other clones in the library allows the identification of overlapping clones for the DNA sequencing. Two other sites, *loxP* and *cosN*, allow the plasmid to be cut at unique sites for restriction site mapping of the clone.

Broad-Host-Range Cloning Vectors

Many of the common *E. coli* cloning vectors such as pBR322, the pUC plasmids, and the pET plasmids have been constructed with the pMB1 *ori* region related to ColE1 and thus are very narrow in their host range. They replicate only in *E. coli* and a few of its close relatives. However, some cloning applications require a plasmid cloning vector that replicates in other gram-negative bacteria, and so cloning vectors have been derived from the broad-host-range plasmids RSF1010 and RK2, which replicate in most gram-negative bacteria. In addition to the broad-host-range *ori* region, these cloning vectors sometimes contain a *mob* site, which can allow them to be mobilized into other bacteria (see chapter 5). This trait is very useful, because ways of introducing DNA other than conjugation have not been developed for many types of bacteria, although electroporation works for most (see chapter 6).

SHUTTLE VECTORS

Sometimes, an experiment requires that a plasmid cloning vector be transferred from one organism into another. If the two organisms are not related, the same plasmid *ori* region is not likely to function in both organisms. Such applications require the use of **shuttle vectors,** so named because they can be used to "shuttle" genes between the two organisms. A shuttle vector has two origins of replication, one that functions in each organism. Shuttle vectors also must contain selectable genes that can be expressed in both organisms.

In most cases, one of the organisms in which the shuttle vector can replicate is *E. coli.* The genetic tests can be performed with the other organism, but the plasmid can be purified and otherwise manipulated by the refined methods developed for *E. coli.*

Some shuttle vectors can replicate in gram-positive bacteria and *E. coli,* whereas others can be used in lower or even higher eukaryotes. For example, plasmid YEp13 (Figure 4.24) has the replication origin of the 2μm circle, a plasmid found in the yeast *Saccharomyces cerevisiae,* and so it can replicate in *S. cerevisiae.* It also has the pBR322 *ori* region and thus can replicate in *E. coli.* In addition, the plasmid contains the yeast gene *LEU2,* which can be selected in yeast, as well as Amp^r, which confers ampicillin resistance in *E. coli.* Similar shuttle vectors that can replicate in mammalian or insect cells and *E. coli* have been constructed. Some of these plasmids have the replication origin of the animal virus simian virus 40 and the ColE1 origin of replication.

Figure 4.24   Shuttle plasmid YEp13. The plasmid contains origins of replication that function in the yeast *Saccharomyces cerevisiae* and the bacterium *E. coli.* It also contains genes that can be selected in *S. cerevisiae* and *E. coli.*

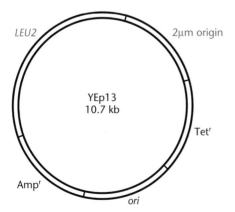

Using Plasmid Vectors for Gene Replacement and Functional Genomics

Some plasmid vectors have been designed to use to replace genes in the chromosome with genes that have been manipulated in the cloning vector (see chapter 3) and for systematic gene disruptions in functional genomics. In some replacement vectors, the replication of the plasmid cloning vector is made conditional, so that the plasmid can replicate only under certain conditions. Under other conditions, the only way the cloned gene can be maintained is if it recombines with the chromosome, replacing the sequence in the chromosome with the sequence of the gene in the plasmid. Recombination is always rare, so there must be some way of selecting cells in which the replacement recombination has occurred. One such procedure is addressed in this section.

SELECTING GENE REPLACEMENTS IN *E. COLI* BY USING THE LETHAL EFFECT OF A SECOND REPLICATION ORIGIN

As discussed in chapter 3, most methods for gene replacement require that a selectable gene cassette, such as for antibiotic resistance, be introduced into the cloned gene. Otherwise, it is difficult to select the few cells in which a second crossover has replaced the chromosomal gene with the mutant cloned copy. However, sometimes we want to introduce another, more subtle, type of mutation into the chromosome, for example one which changes a single amino acid in a protein.

A gene replacement method has been developed which makes it easier to introduce such minor sequence changes (see Hamilton et al., Suggested Reading). This method is based on the fact that *E. coli* cells are killed if their chromosome contains two active replication origins. This method is illustrated in Figure 4.25. The mutant gene is first cloned into a derivative of the plasmid vector pSC101 that confers tetracycline resistance (Tet^r) and has a temperature-sensitive mutation in its *repA* gene [*repA*(Ts)]. Because of the *repA*(Ts) mutation, the pSC101 *ori* region is not active at higher temperatures (around 42°C), so that at these temperatures the plasmid cannot replicate and becomes lost from the cells. Only at lower temperatures (around 30°C) is the plasmid origin active so the plasmid can replicate itself.

To use this method, the plasmid with the *repA*(Ts) mutation and the cloned mutant gene is transformed into *E. coli* and transformants are selected on tetracycline-containing plates at the high temperature of 42°C. Since the plasmid cannot replicate at this temperature, the only cells that become Tet^r are those in which the plasmid has integrated into the chromosome by a single crossover between the cloned mutant gene and the normal gene in the chromosome as shown in Figure 4.25. These cells can

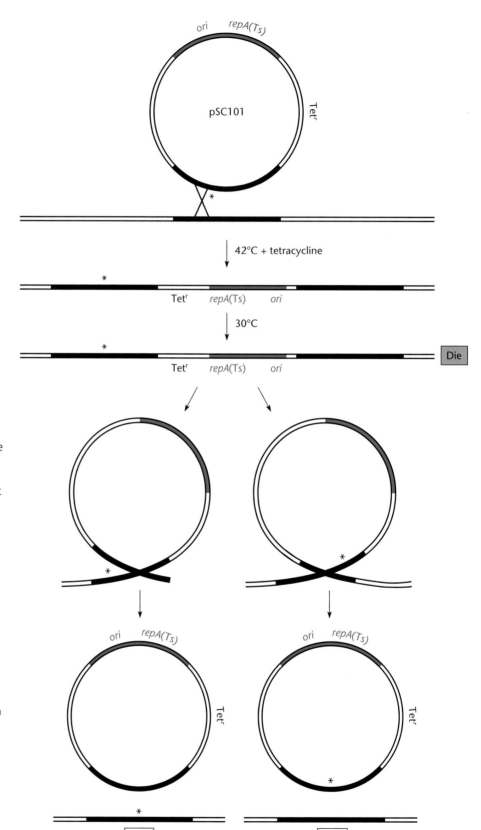

Figure 4.25   A method of selecting cells in which a modified sequence (shown by an asterisk) cloned in a plasmid has replaced the normal sequence in the chromosome that depends on the lethal effect of a second origin of replication. The RepA protein of the pSC101 vector is temperature sensitive [RepA(Ts)], and so the pSC101 origin does not function at 42°C. If cells are transformed with the plasmid at 42°C and plated on tetracycline-containing medium, only cells in which the plasmid has integrated by a single crossover between the cloned sequence and the chromosomal sequence will be tetracycline resistant (Tet^r) and form colonies. If the temperature is then lowered to 30°C, the pSC101 origin becomes active, killing most of the cells because their chromosome now has two functional origins of replication. The only cells that survive are those in which a second crossover has excised the plasmid. Depending on where the second crossover occurs, some of these cells retain the altered sequence in the chromosome while the plasmid has the original chromosomal sequence. *ori* denotes the plasmid origin of replication.

be selected on tetracycline-containing plates. The chromosome in these transformants has two potential origins of replication, its own *oriC* and the origin on the integrated plasmid, but the cells survive because the plasmid *ori* is inactive owing to the high temperature. However, when these Tet[r] cells are shifted to 30°C, the temperature at which the pSC101 origin is active, the only cells that survive are those in which a second crossover has occurred, excising the plasmid from the chromosome (as shown in Figure 4.25) and leaving only the chromosomal origin of replication. Therefore, to select the few cells in which a second crossover has excised the plasmid, it is necessary only to plate the cells at 30°C. Depending on where this second crossover occurred, some of these cells have the normal gene restored in the chromosome while some now have the mutant copy instead of the normal gene.

This method has an advantage over most methods in that the excised plasmid remains in the cells because it is a replicon at 30°C. This feature helps identify the cells in which the normal gene has been replaced with the mutant copy because these are often the cells in which the excised plasmid contains the normal copy which it picked up when it left the mutant copy behind in the chromosome by a reciprocal crossover. Plasmids are purified from the bacteria in a few of the surviving colonies, and the inserted DNA is sequenced to find one which contains the normal sequence from the chromosome. As with all gene replacement techniques, the presence of the gene replacement in the chromosome should still be confirmed by some method, for example by direct PCR sequencing of the gene in the chromosome (see chapter 1).

A VECTOR FOR GENOME-WIDE GENE DISRUPTION IN *B. SUBTILIS*

The advent of whole-genome sequencing has spurred the development of methods to determine systematically the function of each of the thousands of open reading frames of an organism. This requires methods to inactivate each of the open reading frames to begin to determine the function of the product of each of them. A plasmid vector, pMUTIN, that was used for such a systematic analysis has been developed for *B. subtilis* (see Vagner et al., Suggested Reading). Besides allowing gene-by-gene disruption, this vector allows measurements of the expression of the gene as well as allowing the expression of downstream genes, thereby preventing polarity effects on downstream genes (see the section on polarity in chapter 2). This latter feature is important since many *B. subtilis* genes lie in operons. This analysis was facilitated by the highly efficient natural transformation system of *B. subtilis* (see chapter 6), which made it easier to disrupt all the genes.

A picture of pMUTIN is shown in Figure 4.26A. The plasmid can be grown in *E. coli*, selecting the Amp[r] gene on the plasmid. However, if the plasmid is transformed into *B. subtilis,* the plasmid is unable to replicate because it has the narrow-host-range ColE1 origin of replication. The only way it can be maintained in *B. subtilis* is if it recombines with the chromosome. A single crossover between the gene in the chromosome and homologous sequences in the plasmid will integrate the plasmid. These integration events can be selected because the plasmid also has a gene for resistance to the antibiotic erythromycin, Erm[r], which is expressed in *B. subtilis*. Once integrated, a *lacZ* reporter gene on the plasmid, which has a TIR from *B. subtilis*, is transcriptionally fused to the promoter for the gene into which it has integrated and therefore makes β-galactosidase only under conditions where the target gene is normally transcribed. An inducible p_{spac} promoter on the integrated plasmid also allows transcription of downstream genes in the operon. This promoter, which contains the *lac* operators, is active only in the presence of the inducer IPTG, because the *E. coli lacI* gene for the Lac repressor has also been introduced into the plasmid (see chapter 12). Of course, being an *E. coli* gene, it first had to be modified so that it can be expressed in *B. subtilis*. A λ phage transcription terminator upstream of the p_{spac} promoter blocks any transcription from other promoters.

Figure 4.26B illustrates the cloning steps needed to use the pMUTIN vector to disrupt the middle gene, *orf2*, of a three-gene operon. A fragment internal to *orf2* is PCR amplified and cloned into the multiple cloning site just downstream of the p_{spac} promoter. Figure 4.26C illustrates what happens when the plasmid containing this clone is transformed into *B. subtilis* and Erm[r] transformants are selected. Recombination between the cloned sequences on the plasmid and *orf2* will have integrated the plasmid by a single crossover, making the cells Erm[r]. Because the fragment was internal to *orf2*, both the upstream and downstream copies of *orf2* are incomplete and presumably inactive. If β-galactosidase is expressed, the promoter for the operon, called p_{orf123} in the figure, must be active under the given conditions. This allows conclusions to be drawn about the conditions under which the operon is normally expressed. If IPTG is added, *orf3* is also expressed, preventing polarity. Therefore, the only gene that is disrupted in the presence of IPTG is *orf2*, so that any phenotypes in the presence of IPTG are due to the disruption of *orf2*, allowing the function of the *orf2* gene to be deduced. An important caveat to this approach is that *orf2* is disrupted but not totally deleted and the N terminus could still retain some activity. A second caveat is that the downstream *orf3* is being expressed from a different promoter, so that its

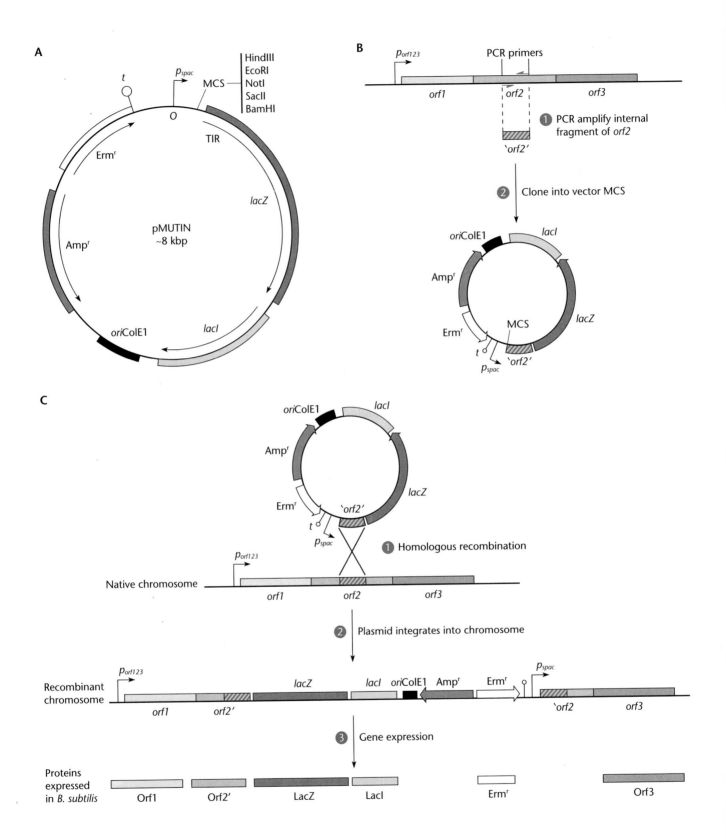

product could be made in greater or lesser amounts than normal; this also has the potential to cause phenotypes.

This vector has been used to disrupt more than 4,100 annotated open reading frames in *B. subtilis*. This effort involved a consortium of laboratories worldwide, especially in Europe and Japan. One outcome was to define a set of approximately 270 essential genes (see Kobayashi et al., Suggested Reading). To identify which open reading frames encode proteins essential for growth of *B. subtilis*, it was necessary to adapt somewhat the approach shown in Figure 4.26. In the example, the PCR fragment that was cloned was internal to the *orf2* coding region. When the plasmid containing this clone then inserts by a single crossover, both flanking *orf2* sequences are incomplete, there is no good copy of *orf2* in the cell, and the transformants will die. Therefore, if the product of *orf2* is essential, no Erm^r transformants will be observed, providing evidence that the product of *orf2* is essential. However, this is negative evidence, and it is always better not to have to rely solely on negative evidence. To obtain positive evidence that the product of the *orf* is essential, the PCR fragment that is cloned should include the coding sequence for either the N terminus or the C terminus of

the product of the *orf*. When such a clone is then used to transform the cells, one or the other duplicated copy of *orf2* is complete (see chapter 3), the complete product of *orf2* is made, and the cell will be viable, even if the product of *orf2* is essential. In general, if Erm^r transformants were obtained when the cloned PCR fragment included one terminus of the *orf* or the other, but not when the cloned fragment was completely internal to the *orf*, the product of the *orf* was deemed to be essential. Approximately 70% of the essential genes were found to have homologues in eukaryotes and archaea. This analysis would have missed essential genes that are redundant or expressed in media or growth conditions different from those that were used, but more recent analyses have specifically studied duplicated genes (see Thomaides et al., Suggested Reading).

VECTORS FOR GENOME-WIDE GENE INACTIVATION

Later chapters describe methods for construction of plasmids that can create in-frame deletions in chromosomal genes, thus producing true null mutants.

Figure 4.26   A plasmid-based method for genome-wide gene disruption in *B. subtilis*. (A) Map of the pMUTIN vector showing the Amp^r gene for selection in *E. coli* and the Erm^r gene for selection in *B. subtilis*. Also shown is the ColE1 origin of replication, which allows replication in *E. coli* but not *B. subtilis*. The *lacZ* reporter gene includes a translation initiation region (TIR) of a *B. subtilis* gene and the multiple cloning site (MCS) into which PCR fragments can be cloned. The LacI repressor is made from the *lacI* gene of *E. coli,* modified so that it can be expressed in *B. subtilis*. The p_{spac} promoter is an inducible hybrid promoter that contains sequences of a promoter from the *B. subtilis* phage SP01 and three *lac* operators (*o*) to which the LacI repressor binds to make it inducible by IPTG. *t* is a strong hybrid transcriptional terminator from λ phage and an rRNA operon. (B) Cloning into pMUTIN. (1) A fragment internal to the target gene is PCR amplified with primers that add restriction sites compatible with those in the MCS for directional cloning (see chapter 1). (2) The fragment is cloned into the MCS on the plasmid. (C) Integration of recombinant vector into the *B. subtilis* chromosome. (1) Homologous recombination into the native chromosomal locus. (2) Structure of the recombinant chromosome after plasmid integration. (3) The products of gene expression. The prime before or after an *orf* or protein indicates that only part of the *orf* or protein remains.

SUMMARY

1. Plasmids are DNA molecules that exist free of the chromosome in the cell. Most plasmids are circular, but some are linear. The sizes of plasmids range from a few thousand base pairs to almost the length of the chromosome itself. Probably the best distinguishing characteristic of a plasmid is that it has a more typically plasmid origin of replication with an adjacent gene for a Rep protein rather than a typical chromosome origin with an *oriC*, along with a *dnaA* gene and other genes typical of the chromosomal origin of replication.

2. Plasmids usually carry genes for proteins that are necessary or beneficial to the host under some situations but are not essential under all conditions. By carrying nonessential genes on plasmids, bacteria are able to keep their chromosome small but still respond quickly as a population to changes in the environment.

3. Plasmids replicate from a unique origin of replication, or *oriV* region. Many of the characteristics of a given plasmid derive from this *ori* region. These include the mechanism of replication, copy number control, partitioning, and incompatibility. If other genes are added to or deleted from the plasmid, it will retain most of its original characteristics, provided that the *ori* region remains.

4. Many plasmids replicate by a theta mechanism, with replication forks moving from a unique origin with leading and lagging strands much like circular bacterial chromosomes. Others use a rolling-circle mechanism, similar to that used to replicate some phage DNAs and during bacterial conjugation. In rolling-circle replication, the plasmid is cut at a unique site, and the Rep protein remains attached to the 5′ end at the cut through one of its tyrosines. The free 3′ end is used as a primer to replicate around the circle, displacing one of the strands. When the circle is complete, the 5′ phosphate is transferred from the Rep protein to the 3′ hydroxyl to form a single-stranded circle. The host ligase rejoins the ends to form a double-stranded circular DNA. A complementary strand is made to the single-stranded circle, using a different origin, to form two double-stranded circular DNAs. Linear plasmids replicate by more than one mechanism. Some have hairpin ends and replicate from an internal origin around the ends to form dimeric circles that are then processed by protelomerases to form two linear plasmids. Others have a terminal protein at both 5′ ends and extensive inverted repeated sequences at their ends. They may replicate the ends by some sort of slippage mechanism.

5. The copy number of a plasmid is the number of copies of the plasmid per cell immediately after cell division.

6. Different types of plasmids use different mechanisms to regulate their initiation of replication and therefore their copy number. Some plasmids use antisense RNA (ctRNA) transcribed from the other strand in the same region (countertranscribed) to regulate their copy number. In ColE1-derived plasmids, the ctRNA, called RNA I, interferes with the processing of the primer for leading-strand replication, called RNA II. In other cases, including the R1, ColIB-P9 and pT181 plasmids, the ctRNA interferes with the expression of the Rep protein required to initiate plasmid DNA replication.

7. Other plasmids called iteron plasmids regulate their copy number by two interacting mechanisms. They control the amount of the Rep protein required to initiate plasmid replication, and the Rep protein also couples plasmids through their iteron sequences.

8. Some plasmids have a special partitioning mechanism to ensure that each daughter cell gets one copy of the plasmid as the cells divide. These partitioning systems usually consist of two genes for proteins and a *cis*-acting centromere-like site. One of the proteins binds to the *cis*-acting site and serves as a nucleation point for the attachment of replicating plasmid to the midpoint of the cell. The other protein is an ATPase, which is capable of polymerizing into filaments in the presence of ATP. The ATPase protein of the R1 plasmid is analogous to MreB, a bacterial actin homolog, and forms actin-like filaments in the presence of ATP that may push the daughter plasmids to opposite ends of the cell before cell division. The ATPase protein of most other plasmids is more analogous to the corresponding protein from the partitioning system of the chromosome of some bacteria and forms a number of shorter filaments that radiate out from the cell center, perhaps pushing the plasmid to the quarter and three-quarter positions in the cell before cell division.

9. If two plasmids cannot stably coexist in the cells of a culture, they are said to be incompatible or to be members of the same Inc group. They can be incompatible if they have the same copy number control system or the same partitioning functions.

10. The host range for replication of a plasmid is defined as all the different organisms in which the plasmid can replicate. Some plasmids are very broad in their host range and can replicate in a wide variety of bacteria. Others are very narrow in their host range and can replicate in only very closely related bacteria.

11. Many plasmids have been engineered for use as cloning vectors. They make particularly desirable cloning vectors for some applications because they do not kill the host, can be small, and are easy to isolate. Some plasmids can carry large amounts of DNA and are used to make bacterial artificial chromosomes for eukaryotic genome sequencing. Plasmids have been adapted to be used to express cloned genes in bacteria, to do gene replacements in the chromosome, and to perform systematic gene inactivation for functional genomics.

QUESTIONS FOR THOUGHT

1. Why are genes whose products are required for normal growth not carried on plasmids? List some genes which you would not expect to find on a plasmid and some genes you might expect to find on a plasmid.

2. Why do you suppose some plasmids are broad host range for replication? Why are not all plasmids broad host range?

3. How do you imagine a partitioning system for a single-copy plasmid such as F could work? How might a copy number control mechanism work?

4. How would you find the genes required for replication of the plasmid if they are not all closely linked to the *ori* site?

5. How would you determine which of the replication genes of the host *E. coli* (e.g., *dnaA* and *dnaC*) are required for replication of a plasmid you have discovered?

6. The R1 plasmid has a leader polypeptide translated upstream of the gene for RepA, and cleavage of the mRNA by RNase III occurs in the coding sequence for this leader polypeptide. This blocks the translation of the leader polypeptide and also the translation of the downstream *repA* gene to which it is translationally coupled. Do you think it would have been easier just to have the cleavage occur in the coding sequence for the RepA protein itself? Why or why not?

7. Recent evidence suggests that many stringent plasmids replicate in the center of the cell and then quickly move to the quarter positions of the cell before it divides. What do you think might define the one-quarter and three-quarter positions of the cell at this time?

8. Try to design a mechanism that uses inverted repeated sequences at the ends to replicate to the ends of a linear plasmid without the DNA getting shorter each time it replicates.

PROBLEMS

1. The IncQ plasmid RSF1010 carries resistance to the antibiotics streptomycin and sulfonamide. Suppose you have isolated a plasmid that carries resistance to kanamycin. Outline how you would determine whether your new plasmid is an IncQ plasmid.

2. A plasmid has a copy number of 6. What fraction of the cells are cured of the plasmid each time the cells divide if the plasmid has no partitioning mechanism?

3. You wish to clone a fragment of human DNA cut with the restriction endonuclease BamHI into the BamHI site of pBR322 (Figure 4.20). You cut both human DNA and pBR322 with BamHI and ligate them. You transform the ligation mix into *E. coli*, selecting for Amp^r. Outline how you would determine which of the transformants probably contain a plasmid with a human DNA insert.

4. The ampicillin resistance gene of plasmid RK2 is unregulated. The more copies of the gene a bacterium has, the more gene product is made. In this case, the resistance of the cell to ampicillin is higher the more of these genes it has. Use this fact to devise a method to isolate mutants of RK2 that have a higher than normal copy number (copy-up mutations). Determine whether your mutants have mutations in the Rep-encoding gene.

5. Outline how you would determine whether a plasmid has a partitioning system.

6. Outline how you would use the phage promoters bracketing the cloning site of a pBAC vector to synthesize RNA on the ends of the clone. How would you use these RNAs to identify overlapping clones in your library?

7. What would be the effect of mutating one of the two complementary sequences in structure I and III of the ColIb-P9 plasmid origin region on the copy number of the plasmid? In the presence and absence of the Inc antisense RNA?

SUGGESTED READING

Azano, K., and K. Mizobuchi. 2000. Structural analysis of late intermediate complex formed between plasmid ColIb-P9 Inc RNA and its target RNA. How does a single antisense RNA repress translation of two genes at different rates? *J. Biol. Chem.* **275**:1269–1274.

Bagdasarian, M., R. Lurz, B. Ruckert, F. C. H. Franklin, M. M. Bagdasarian, J. Frey, and K. N. Timmis. 1981. Specific purpose plasmid cloning vectors. II. Broad host, high copy number, RSF1010-derived vectors, and a host vector system for cloning in *Pseudomonas*. *Gene* **16**:237–247.

Caspi, R., M. Pacek, G. Consiglieri, D. R. Helinski, A. Toukdarian, and I. Konieczny. 2001. The broad host range replicon with different requirements for replication initiation in three bacterial species. *EMBO J.* **20**:3262–3271.

Cohen, S. N., A. C. Y. Chang, H. W. Boyer, and R. B. Helling. 1973. Construction of biologically functional bacterial plasmids *in vitro*. *Proc. Natl. Acad. Sci. USA* **70**:3240–3244.

Funnell, B. E., and G. J. Phillips (ed.). 2004. *Plasmid Biology*. ASM Press, Washington, D.C.

Hamilton, C. M., M. Aldea, B. K. Washburn, P. Babitzke, and S. R. Kushner. 1989. A new method for generating deletions and gene replacements in *Escherichia coli*. *J. Bacteriol.* **171**: 4617–4622.

Khan, S. A. 2000. Plasmid rolling circle replication: recent developments. *Mol. Microbiol.* **37**:477–484.

Kim, G. E., A. I. Derman, and J. Pogliano. 2005. Bacterial DNA segregation by dynamic SopA polymers. *Proc. Natl. Acad. Sci. USA* **102**:17658–17663.

Kobayashi, K., S. D. Ehrlich, A. Albertini, G. Amati, K. K. Andersen, M. Arnaud, et al. 2003. Essential *Bacillus subtilis* genes. *Proc. Natl. Acad. Sci. USA* **100**:4678–4683.

Kolb, F. A., C. Malmgren, E. Westof, C. Ehresmann, B. Ehresmann, E. G. H. Wagner, and P. Romby. 2000. An unusual structure formed by anti-sense target binding involves an extended kissing complex and a four-way junction and side-by-side helical alignment. *RNA* **6**:311–324.

McEachern, M. J., M. A. Bott, P. A. Tooker, and D. R. Helinski. 1989. Negative control of plasmid R6K replication: possible role of intermolecular coupling of replication origins. *Proc. Natl. Acad. Sci. USA* **86**:7942–7946.

Meacock, P. A., and S. N. Cohen. 1980. Partitioning of bacterial plasmids during cell division: a *cis*-acting locus that accomplishes stable plasmid maintenance. *Cell* **20**:529–542.

Moller-Jensen, J., R. B. Jensen, J. Lowe, and K. Gerdes. 2002. Prokaryotic DNA segregation by an actin-like filament. *EMBO J.* **21**:3119–3127.

Novick, R. P., and F. C. Hoppensteadt. 1978. On plasmid incompatibility. *Plasmid* **1**:421–434.

Novick, R. P., S. Iordanescu, S. J. Projan, J. Kornblum, and I. Edelman. 1989. pT181 plasmid regulation is regulated by a countertranscript-driven transcriptional attenuator. *Cell* **59**:395–404.

Radloff, R., W. Bauer, and J. Vinograd. 1967. A dye-buoyant-density method for the detection and isolation of closed circular duplex DNA: the closed circular DNA in HeLa cells. *Proc. Natl. Acad. Sci. USA* **57**:1514–1521.

Shizuya, H., B. Birren, U.-J. Kim, V. Mancino, T. Slepak, Y. Tachiiri, and M. Simon. 1992. Cloning and stable maintenance of 300-kilobase-pair fragments of human DNA in *Escherichia coli* using a F-factor-based vector. *Proc. Natl. Acad. Sci. USA* **89**:8794–8797.

Thomaides, H. B., E. J. Davison, L. Burston, H. Johnson, D. R. Brown, A. C. Hunt, J. Errington, and L. Czaplewski. 2007. Essential bacterial functions encoded by gene pairs. *J. Bacteriol.* **189**:591–602.

Vagner, V., E. Dervyn, and S. D. Ehrlich. 1998. A vector for systematic gene inactivation in *Bacillus subtilis*. *Microbiology* **144**:3097–3104.

Yao, S., D. R. Helinski, and A. Toukdarian. 2007. Localization of the naturally occurring plasmid ColE1 at the cell pole. *J. Bacteriol.* **189**:1946–1953.

CHAPTER 5

Conjugation

A remarkable feature of many plasmids is the ability to transfer themselves and other DNA elements from one cell to another in a process called **conjugation.** Joshua Lederberg and Edward Tatum first observed this process in 1947, when they found that mixing some strains of *Escherichia coli* with others resulted in strains that were genetically unlike either of the originals. As discussed later in this chapter, Lederberg and Tatum suspected that bacteria of the two strains exchanged DNA—that is, two parental strains mated to produce progeny unlike themselves but with characteristics of both parents. At that time, however, plasmids were unknown, and it was not until later that the basis for the mating was understood.

Overview

During conjugation, the two strands of a plasmid separate in a process resembling rolling-circle replication (see "Mechanism of DNA Transfer during Conjugation in Gram-Negative Bacteria" below), and one strand moves from the bacterium originally containing the plasmid—the **donor**—into a **recipient** bacterium. Then the two single strands serve as templates for the replication of complete double-stranded DNA molecules in both the donor cell and the recipient cell. A recipient cell that has received DNA as a result of conjugation is called a **transconjugant.** A simplified view of conjugation is shown in Figure 5.1.

Many naturally occurring plasmids can transfer themselves. If so, they are said to be **self-transmissible.** The prevalence of conjugation systems suggests that plasmid conjugation is advantageous for plasmids and their hosts. Self-transmissible plasmids encode all the functions they need to move among cells, and sometimes they also aid in the transfer of mobilizable plasmids.

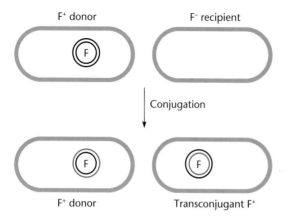

F⁺ donor   F⁻ recipient

Conjugation

F⁺ donor   Transconjugant F⁺

Figure 5.1 A simplified view of conjugation by a self-transmissible plasmid, the F plasmid. A replica of the plasmid is transferred from the donor to the recipient cell so that both the donor and recipient cells have the plasmid. A cell that has received the plasmid by conjugation is a transconjugant.

Mobilizable plasmids encode some but not all of the proteins required for transfer and consequently need the help of self-transmissible plasmids to move.

Any bacterium harboring a self-transmissible plasmid is a potential donor, because it can transfer DNA to other bacteria. In gram-negative bacteria, such cells produce a structure, called a **sex pilus,** which facilitates conjugation (discussed in a later section). Bacteria that lack the self-transmissible plasmid are potential **recipients,** and conjugating bacteria are known as **parents.** Potential donor strains are sometimes referred to as **male strains.**

Self-transmissible plasmids probably exist in all types of bacteria, but those that have been studied most extensively are from the gram-negative genera *Escherichia* and *Pseudomonas* and the gram-positive genera *Enterococcus, Streptococcus, Bacillus, Staphylococcus,* and *Streptomyces.* The best-known transfer systems are those of plasmids isolated from *Escherichia* and *Pseudomonas* species, and so we focus our attention on these gram-negative systems and do not address conjugation in gram-positive bacteria until the end of this chapter.

Classification of Self-Transmissible Plasmids

Bacterial plasmids have many different types of transfer systems, which are encoded by the plasmid *tra* genes [see "Transfer (*tra*) Genes" below]. However, as discussed in chapter 4, plasmids are usually classified by their incompatibility (Inc) group. Accordingly, the F-type plasmids use a transfer system known as the Tra system of IncF plasmids and the RP4 plasmid uses the Tra system of IncP plasmids.

Despite this nomenclatural link, transfer systems have no direct relationship to the replication and partitioning functions of a plasmid, the characteristics that determine its Inc group. In fact, the genes for these functions and the transfer genes are located in different regions of the plasmid, and there is no a priori reason for any correlation between them. Nevertheless, there is a high degree of correlation between the type of transfer system and the Inc group. There may be a good reason for this. Some products of plasmid transfer genes inhibit the entry of plasmids with the same Tra functions (see below). If the *tra* genes did not correlate with the Inc group, a plasmid would sometimes transfer into a cell that already had a plasmid of the same Inc group, and one of the two plasmids would subsequently be lost.

Mechanism of DNA Transfer during Conjugation in Gram-Negative Bacteria

Much has been learned in the last few years about the detailed mechanism of plasmid conjugation, especially in gram-negative bacteria. Some of this progress has come from the convergence of two seemingly different fields: conjugation and some types of protein secretion. The practical applications of plasmid conjugation systems, especially in plant biotechnology, have also inspired some of this work. The process of conjugation is outlined in some detail in this section.

Transfer (*tra*) Genes

Conjugation is a complicated process that requires the products of many genes. As mentioned above, the genes required for transfer are called the ***tra* genes.** The products of the *tra* genes are *trans*-acting and can act on another plasmid in the same cell. The map of the F plasmid (Figure 5.2) shows that a large region of a self-transmissible plasmid is devoted to encoding plasmid transfer functions. This plasmid contains at least 20 of these genes (Table 5.1), as well as genes such as *traST* (entry exclusion) that are not required for transfer but play related roles, in this case preventing the entry of other plasmids with the same Tra functions. In addition to the *tra* genes, a *cis*-acting site called *oriT* is required for transfer. With so many genes required for transfer-related functions, self-transmissible plasmids are quite large, by necessity.

The *tra* genes of a self-transmissible plasmid required for plasmid transfer can be divided into two components. Some of the *tra* genes encode proteins involved in the processing of the plasmid DNA to prepare it for transfer. This is called the **Dtr component** (for DNA Transfer and Replication). The *tra* genes encoding the Dtr component tend to cluster around the *oriT* site. The bulk of the *tra* genes encode proteins of the **Mpf component** (for

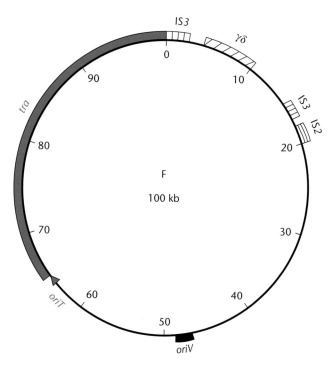

Figure 5.2 Partial genetic and physical map of the 100-kbp self-transmissible plasmid F. The regions IS3 and IS2 are insertion sequences; γδ is also known as transposon Tn*1000*. *oriV* is the origin of replication; *oriT* is the origin of conjugal transfer; the *tra* region encodes numerous *tra* functions.

*m*ating-*p*air *f*ormation). This large membrane-associated structure includes the pilus that is responsible for holding the mating cells together and the channel between the mating cells through which the plasmid is transferred. Below, we outline what is known of these two components.

TABLE 5.1	Some F-plasmid genes and sites
Symbol	**Function**
ccdAB	Inhibition of host cell division
incBCE	Incompatibility
oriT	Site of initiation of conjugal DNA transfer
oriV	Origin of bidirectional replication
sopAB	Partitioning
traABCEFGHKLQUVWX	Pilus biosynthesis, assembly
traGN	Mating-pair stabilization
traD	Coupling protein
traI	Relaxase
traYM	Accessories for relaxosome
traJ, finOP	Regulation of transfer
traST	Entry exclusion

THE Mpf SYSTEM

The function of the Mpf system is to hold a donor cell and a recipient cell together during the mating process and to form a channel through which proteins and DNA are transferred during the mating. It also includes the protein that communicates news of mating-pair formation to the Dtr system, beginning the transfer of plasmid DNA. A representation of the entire Mpf system of a self-transmissible plasmid, the F plasmid, is shown in Figure 5.3.

The Pilus

The most dramatic feature of the Mpf structure is the **pilus,** a tube-like structure that sticks out of the cell surface (Figure 5.3). These pili are 10 nm or more in diameter with a central channel. Each pilus is constructed of many copies of a single protein called the **pilin** protein. The assembly of a pilus is shown in Figure 5.4. The pilin protein is synthesized with a long signal sequence that is removed as it passes through the membrane to assemble on the cell surface. The pilin protein is also cyclized, with its head attached to its tail, which is unusual among proteins (see Eisenbrandt et al., Suggested Reading).

The structure of the pilus differs markedly among plasmid transfer systems. For example, the F plasmid encodes a long, thin, flexible pilus; the pKM101 plasmid (see below) makes a long, rigid pilus; and IncP plasmids such as RP4 make a short, thick, rigid pilus. The structure of the pilus of an Mpf system can determine the efficiency of transfer under various conditions. For example, the long, flexible pilus of the F plasmid allows transfer in liquid medium while the cells are suspended in broth, while the short, thick, rigid pilus of RP4 allows efficient transfer only when the mating cells are fixed to a solid surface such as a membrane, where they are less free to move. A long, flexible pilus may facilitate mating in liquid by helping bring more widely dispersed cells together, while a short, rigid pilus may be able to hold mating cells together only when they are concentrated on a solid surface. To make them more versatile, IncI plasmids such as Col1b-P9 make two pili, a long thin one and a short rigid one, the former increasing the frequency of mating in liquid medium and the latter increasing the frequency of mating on a solid surface.

Even though the male-cell pilus was observed a long time ago, the function of pili in conjugation is still unclear. They may only hold cells together during mating, or they may play a more direct role in the DNA transfer. What is clear, however, is that the early assumption, doubtless inspired by anthropomorphic considerations, that DNA passes through the pilus during the mating seems not to be true.

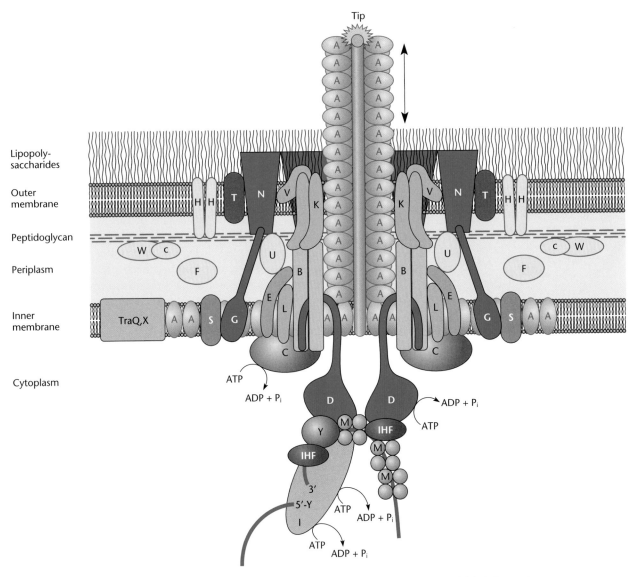

Figure 5.3 A representation of the F transfer apparatus drawn from available information. The pilus is assembled with five TraA (pilin) subunits per turn that are inserted into the inner membrane via TraQ and acetylated by TraX. The pilus is shown extending through a pore constructed of TraB and TraK, a secretion-like protein anchored to the outer membrane by the lipoprotein TraV. TraB is an inner membrane protein that extends into the periplasm and contacts TraK. TraL seeds the site of pilus assembly and attracts TraC to the pilus base, where it acts to drive assembly in an energy-dependent manner. A channel formed by the lumen is indicated, as is a specialized structure at the pilus tip that remains uncharacterized. A two-way arrow indicates the opposing processes of pilus assembly and retraction. The mating-pair formation (Mpf) proteins include TraG and TraN, which aid in mating-pair stabilization (Mps), and TraS and TraT, which disrupt mating-pair formation through entry and surface exclusion, respectively. TraF, TraH, TraU, TraW, and TrbC, which together with TraN are specific to F-like systems, appear to play a role in pilus retraction, pore formation and mating-pair stabilization. The relaxosome, consisting of TraY, TraM, TraI, and host-encoded IHF bound to the nicked DNA in *oriT,* is shown interacting with the coupling protein, TraD, which in turn interacts with TraB. The 5' end of the nicked strand is shown bound to a tyrosine in TraI, and the 3' end is shown as being associated with TraI in an unspecified way. The retained, unnicked strand is not shown. TraC, TraD, and TraI (two sites for both relaxase and helicase activity) have ATP utilization motifs represented by curved arrows, with ATP being split into ADP and inorganic phosphate (P$_i$).

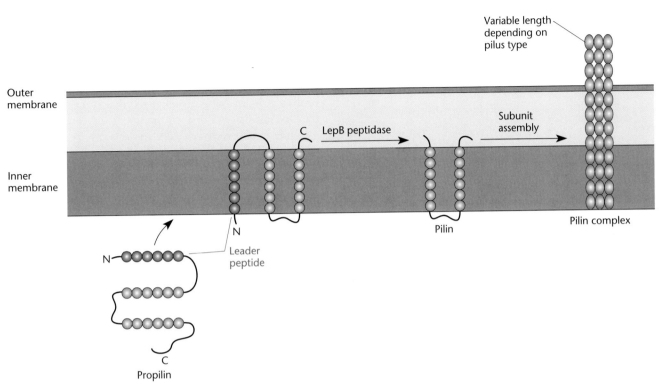

Figure 5.4   Assembly of the pilus on the cell surface. The propilin is processed by LepB peptidase as it passes through the membrane, and then it assembles between the inner and outer membranes (details are given in text).

The Channel

In addition to a pilus, the Mpf system encodes a channel or pore through which DNA passes during conjugation. Some of the *tra* gene products making up this pore are known, but the pore itself has so far escaped detection, and so little is known of its exact structure (see Samuels et al., Suggested Reading).

Coupling Proteins

The Mpf component is the first to make contact with a recipient cell. Then the information that it has contacted another cell is communicated to the Dtr component before DNA transfer occurs. The communication between the Dtr and Mpf systems is provided by proteins called **coupling proteins,** which are part of the Mpf system. These coupling proteins provide the specificity for the transport process, so that only some plasmids are transferred. The coupling protein is bound to the membrane channel (Figure 5.5). Information that the Mpf system has encountered a recipient cell is somehow communicated to the coupling protein, which in turn activates the relaxase to nick the plasmid DNA to initiate the transport process. Coupling proteins are sometimes called docking proteins because they bind to or "dock" proteins on the membrane channel that are to be transported. The coupling protein specifically recognizes the relaxase of the Dtr components of certain plasmids as well as any other proteins to be transferred. In order to be "docked," a protein must contain certain amino acid sequences that identify it as a protein to be transported by the system (see below).

THE Dtr COMPONENT

The Dtr (or DNA-processing) component of a self-transmissible plasmid is involved in preparing the plasmid DNA for transfer. A number of proteins make up this component, and the functions of many of these are known.

Relaxase

A central part of the Dtr component of plasmids is the **relaxase.** This is a specific DNA endonuclease which makes a single-strand break or "nick" at the specific site called the *nic* site in the *oriT* sequence (see below) to initiate the transfer process. It also recyclizes the plasmid after transfer. The way in which the relaxase works is similar to the action of Rep protein in rolling-circle plasmid replication and is illustrated in Figure 5.6. The relaxase breaks a phosphodiester bond at the *nic* site by transferring the bond from a deoxynucleotide to one of its own tyrosines. Such a reaction is called a transesterification reaction and

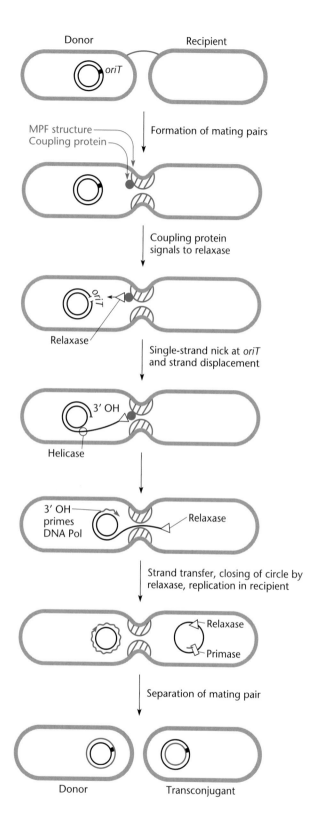

requires very little energy because there is no net breakage or formation of new chemical bonds. This transfer leaves the relaxase protein bound to the 5′ end of the nick through its tyrosine, and the relaxase protein is transferred into the recipient cell along with the DNA. In fact, it is probably the relaxase protein itself that is transferred and the DNA just goes along for the ride (see below).

Once in the recipient cell, the relaxase recyclizes the plasmid by doing essentially the reverse of what it had done in the donor cell. It binds to the two halves of the cleaved *oriT* sequence, holding them together while it transfers the phosphate bond from its tyrosine back to the 3′ hydroxyl deoxynucleotide in the DNA (Figure 5.6). This transesterification reaction reseals the nick in the DNA and releases the relaxase, which has done its job and is degraded. The transferred plasmid is now a single-stranded circle in the recipient cell.

Relaxosome

The relaxase protein in the donor cell is part of a larger structure called the **relaxosome,** which is made up of a number of proteins that are normally bound to the *oriT* sequence of the plasmid. The function of most of the proteins of the relaxosome is unclear. They might help the relaxase bind to the *oriT* sequence or help separate the strands at the *oriT* sequence to initiate transfer. They might also help in the communication with the coupling

---

Figure 5.5 Mechanism of DNA transfer during conjugation, showing the mating-pair formation (Mpf functions) in purple. The donor cell produces a pilus, which forms on the cell surface and which may contact a potential recipient cell and bring it into close contact or may help hold the cells in close proximity after contact has been made, depending on the type of pilus. A pore then forms in the adjoining cell membranes. On receiving a signal from the coupling protein that contact with a recipient has been made, the relaxase protein makes a single-stranded cut at the *oriT* site in the plasmid. A plasmid-encoded helicase then separates the strands of the plasmid DNA. The relaxase protein, which has remained attached to the 5′ end of the single-stranded DNA, is then transported out of the donor cell through the channel directly into the recipient cell, dragging the single-stranded attached DNA along with it. Once in the recipient, the relaxase protein helps recyclize the single-stranded DNA. A primase, encoded either by the host or by the plasmid and injected with the DNA, then primes replication of the complementary strand to make the double-stranded circular plasmid DNA in the recipient. The 3′ end at the nick made by the relaxase in the donor can also serve as a primer, making a complementary copy of the single-stranded plasmid DNA remaining in the donor. Therefore, after transfer, both the donor and the recipient bacterium end up with a double-stranded circular copy of the plasmid. Details are given in the text.

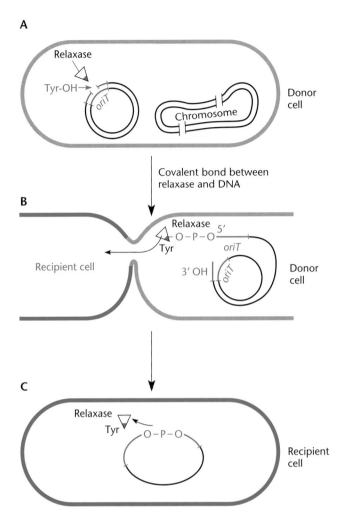

Figure 5.6 Reactions performed by the relaxase.
(A) The relaxase nicks the DNA at a specific site in *oriT*, and the 5′ phosphate is transferred to one of its tyrosines in a transesterification reaction. (B) The relaxase is transferred to the recipient cell, dragging the DNA along with it.
(C) In a reversal of the original transesterification reaction, the phosphate is transferred back to the 3′ hydroxyl of the other end of the DNA, recyclizing the DNA and releasing the relaxase.

protein of the Mpf system, which tells the relaxase when to cut the plamid DNA at the *oriT* site (see above). In some plasmids, one of the other proteins of the relaxosome may be the helicase, which helps separate the strands of DNA beginning at the *oriT* sequence, while in others, the helicase seems to be part of the relaxase protein itself. Whatever their function, the other proteins of the relaxosome are not transferred to the recipient, perhaps because the transferred DNA is already single stranded and so their function is not needed.

Primase

Another component of the Dtr system made in the donor is the **primase.** Primases are needed for chromosomal DNA replication to make an RNA primer to prime the synthesis of the lagging strand of DNA replication (see chapter 1) and to prime plasmid replication (see chapter 4). However, at first the role that a primase would play in the donor was not clear. A primase should not be necessary to prime replication in the donor cell since the free 3′ hydroxyl end of DNA created at the nick in *oriT* is the primer for replication during transfer, similar to the priming of the first stage of replication of rolling-circle plasmids (see chapter 4). Synthesis of the complementary strand of DNA in the recipient after DNA transfer should require synthesis of RNA primers, but the primase is made in the donor cell, not in the recipient cell.

One way the plasmid could escape the dilemma of the misplaced primase would be to transfer its own primase into the recipient cell along with the DNA. Clever experiments showed that at least some types of plasmids do just that (see Wilkins and Thomas, Suggested Reading). The researchers reasoned that transfer of the primase would have been difficult to detect biochemically since only a few molecules need be transferred and not all of the cells are involved in the mating and therefore in the transfer. However, if the plasmid primase is transferred to the recipient cell, it might substitute for the host primase in replication in these cells. Therefore they used a recipient cell in which the primase gene has a temperature-sensitive (Ts) mutation (see chapter 3). When the mutant strain is raised to its nonpermissive temperature, chromosomal DNA replication stops for want of a primase. However, if the bacterium has just received a self-transmissible plasmid, replication continues in some of the cells, presumably using the newly transferred plasmid primase in lieu of its own inactive primase.

But why would a plasmid bother to make its own primase and transfer it into the recipient cell if it can use the host cell primase instead? The answer may be that it does this to make itself more promiscuous and able to transfer into a wider variety of bacterial species. Sometimes a promiscuous plasmid may find it has transferred itself into a type of bacterium which is so distantly related to its original host that the primase in this bacterium does not recognize the sequences on the plasmid DNA necessary to prime the replication of the complementary strand. The plasmid faces a "Catch-22" because it cannot make its own primase in this cell (since the plasmid DNA that has been transferred is single stranded and unable to serve as a template for transcription) and it also cannot make the double-stranded DNA to use as template without its own primase. By transferring its own primase with the DNA, it avoids this problem. However, some plasmids do have primase genes that are known to be transcribed from single-stranded DNA, so it is not necessarily a problem for all plasmids.

In addition to primases, conjugation systems may secrete other proteins including the proteins that form a channel in the recipient cell membrane. This could explain the extreme promiscuity of some transmissible plasmids that can even transfer themselves into eukaryotic cells because the plasmid proteins make their own channel in the recipient cell. Host proteins including RecA may also be transferred by some plasmids. However, it is important to realize that only certain proteins are transferred by a particular transfer system, and these are proteins that are recognized by the coupling protein so that they can "dock" on the channel and be transported (see above).

The *oriT* Sequence

The *oriT* site is not only the site at which plasmid transfer initiates but also the site at which the DNA ends rejoin to recyclize the plasmid after transfer. Plasmid transfer initiates specifically at the *oriT* site because the specific relaxase encoded by one of the *tra* genes cuts DNA only at this sequence. Also, presumably, the plasmid-encoded helicase enters DNA only at this sequence to separate the strands. Moreover, after transfer, the two ends of the DNA are probably held together at the *oriT* sequence so that they can be religated. Therefore, to be transferred, the plasmid must have this specific *oriT* sequence. In fact, a self-transmissible plasmid mobilizes any DNA that contains its *oriT* sequence, as discussed below.

The essential features of *oriT* sequences are currently being investigated. The *oriT* sequence of the F plasmid is known to be shorter than 300 bp and contains inverted repeated sequences and a region rich in AT base pairs. The importance of these sequences for *oriT* function is under investigation.

Male-Specific Phages

Some types of phages can infect only cells that express a certain type of pilus on their cell surface. All phages adsorb to specific sites on the cell surface to initiate infection (see chapter 7), and some phages use the pilus of a self-transmissible plasmid as their adsorption site. Phages that adsorb to the sex pilus of a self-transmissible plasmid are called **male-specific phages** because they infect only donor or "male" cells capable of DNA transfer. Examples of male-specific phages are M13 and R17, which infect only cells carrying the F plasmid, and Pf3 and PRR1, which infect only cells containing an IncP plasmid such as RP4.

Because male-specific phages infect only cells expressing a pilus, mutations in any *tra* gene required for pilus

assembly prevent infection by the phage. This offers a convenient way to determine which of the *tra* genes of a plasmid encode proteins required to express a pilus on the cell surface and which *tra* genes encode other functions required for DNA transfer. To apply this test to a particular *tra* gene, the phages are used to infect cells containing the plasmid with a mutation in the *tra* gene. If the phage can multiply in the host cell, the *tra* gene which has been mutated must not be one of those that encode a protein required for pilus expression.

Incidentally, the susceptibility of pilus-expressing cells to some phages may explain why the *tra* genes of plasmids are usually tightly regulated. Most self-transmissible plasmids express a pilus only immediately after entering a cell and then only intermittently thereafter (see "An Example: Regulation of *tra* Genes in IncF Plasmids" below). If cells containing the plasmid always expressed the pilus, a male-specific phage could spread quickly through the population, destroying many of the cells and, with them, the plasmid they contain. By only intermittently expressing a pilus, cells containing a self-transmissible plasmid limit their susceptibility to phages that use their pilus as an adsorption site.

Efficiency of Transfer

One of the striking features of many transfer systems is their efficiency. Under optimal conditions, some plasmids can transfer themselves into other cells in almost 100% of cell contacts. This high efficiency has been exploited in the development of methods for transferring cloned genes between bacteria and in transposon mutagenesis, both of which require highly efficient transfer of DNA. Such methods are discussed in subsequent chapters.

REGULATION OF THE *tra* GENES

Many naturally occurring plasmids transfer with a high efficiency for only a short time after they are introduced into cells and then transfer only sporadically thereafter. Most of the time the *tra* genes are repressed, and without the synthesis of pilin and other *tra* gene functions, the pilus is lost. For unknown reasons, the repression is relieved occasionally in some of the cells, allowing this small percentage of cells to transfer their plasmid at a given time.

As mentioned, plasmids may normally repress their *tra* genes to prevent infection by some types of phages. The pilus serves as the adsorption site for some phages. If all the cells in a population had a pilus all the time, such a phage could multiply quickly, infecting and killing all the bacteria carrying the plasmid.

This property of only periodically expressing their *tra* genes probably does not prevent the plasmids from spreading quickly through a population of bacteria that does not contain them. When a plasmid-containing population of cells encounters a population that does not contain the plasmid, the plasmid *tra* genes in one of the plasmid-containing cells are eventually expressed and the plasmid transfers to another cell. Then when the plasmid first enters a new cell, efficient expression of the *tra* genes leads to a cascade of plasmid transfer from one cell to another. As a result, the plasmid soon occupies most of the cells in the population. This is the rationale behind triparental matings (discussed below).

AN EXAMPLE: REGULATION OF *Tra* GENES IN IncF PLASMIDS

Regulation of the *tra* genes in IncF plasmids has been studied more extensively than that of other types. This regulation is illustrated in Figure 5.7. Transfer of these plasmids depends on TraJ, a transcriptional activator. A **transcriptional activator** is a protein required for initiation of RNA synthesis at a particular promoter (see chapter 2). If TraJ were always made, the other *tra* gene products would always be made and the cell would

always have a pilus. However, the translation of TraJ is normally blocked by the concerted action of the products of two plasmid genes, *finP* and *finO*, which encode an RNA and a protein, respectively. The FinP RNA is an antisense RNA that is transcribed constitutively from a promoter within and in opposite orientation to the *traJ* gene. Complementary pairing of the FinP RNA and the *traJ* transcript prevents translation of TraJ. The FinO protein stabilizes the FinP antisense RNA. When the plasmid first enters a cell, neither FinP RNA nor FinO protein is present, and so TraJ and the other *tra* gene products are made. Consequently, a pilus appears on the cell, and the plasmid can be transferred. Initially the transferred plasmid is in a single-stranded state. However, primases in the recipient cell synthesize the complementary strain to make the double-stranded form. After the plasmid has become established in the double-stranded state, the FinP RNA and FinO protein can be synthesized, the *tra* genes are repressed, and the plasmid can no longer transfer. Later, the *tra* genes are expressed only intermittently.

Figure 5.7 Fertility inhibition of the F plasmid. Only the relevant *tra* genes discussed in the text are shown. (A) Genetic organization of the *tra* region. (B) The *traJ* gene product is a transcriptional activator that is required for transcription of the other *tra* genes, Y-X, and *finO* from promoter p_{traY}. (C) Translation of the *traJ* mRNA is blocked by hybridization of an antisense RNA, FinP, which is transcribed in the same region from the complementary strand. A protein, FinO, stabilizes the FinP RNA. Details are given in the text.

A   Genetic organization of *tra* region

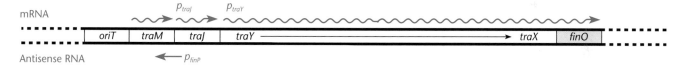

B   Immediately after entry into cell

C   After plasmid establishment

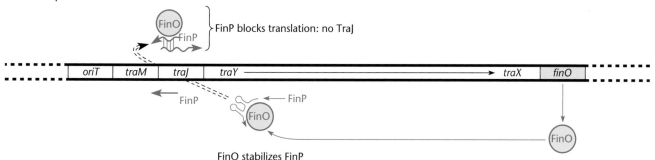

The F plasmid was the first transmissible plasmid discovered (Figure 5.2; Table 5.1), and its discovery may have resulted from a happy coincidence involving its *finO* gene. Because of an insertion mutation in this gene (IS*3* in Figure 5.2), the F plasmid is itself a mutant that always expresses the *tra* genes. Consequently, a sex pilus almost always extends from the surface of cells harboring this F plasmid, and the F plasmid can always transfer, provided that recipient cells are available, increasing the efficiency of transfer and facilitating the discovery of conjugation. Mutations that increase the efficiency of plasmid transfer, thereby increasing their usefulness in gene cloning and other applications, have been isolated in other commonly used transfer systems.

Interspecies Transfer of Plasmids

Many plasmids have transfer systems that enable them to transfer DNA between unrelated species. These are known as **promiscuous plasmids** and include the IncW plasmids, represented by R388; the IncP plasmids, represented by RP4; and the IncN plasmids, represented by pKM101 (see below). The IncP plasmids can transfer themselves or mobilize other plasmids from *E. coli* into essentially any gram-negative bacterium. Recent studies showed that plasmids of this group transfer at a low frequency into cyanobacteria, gram-positive bacteria such as *Streptomyces* species, and even plant cells. The F plasmid, which was not known to be particularly promiscuous, can transfer itself from *E. coli* into yeast cells (Box 5.1).

Transfer of DNA by promiscuous plasmids probably plays an important role in evolution. Such transfer could explain why genes with related functions are often very similar to each other regardless of the organism that harbors them. These genes could have been transferred by promiscuous plasmids fairly recently in evolution, which would account for their similarity relative to the other genes of the two organisms.

The interspecies transfer of plasmids also has important consequences for the use of antibiotics in treating human and animal diseases. Many of the most promiscuous plasmids, including those of the IncP group, such as RP4, and the IncW plasmid R388 were isolated in hospital settings. These large plasmids (commonly called R-plasmids because they carry genes for antibiotic resistance) presumably have become prevalent in recent years in response to the indiscriminate use of antibiotics in medicine and agriculture. The source of the resistance genes may be soil bacteria, such as actinomycetes, that are the producers of antibiotics. In chapter 9, we discuss how transposons might have helped assemble antibiotic resistance genes in these promiscuous R-plasmids.

Whatever their source, the emergence of R-plasmids indicates why antibiotics should be used only when they are absolutely necessary. In humans or animals treated indiscriminately with antibiotics, bacteria that carry R-plasmids are selected from the normal flora. R-plasmids can be quickly transferred into an invading pathogenic bacterium, making it antibiotic resistant. Consequently the infection will be difficult to treat.

CONJUGATION SYSTEMS AND TYPE IV SECRETION SYSTEMS

The Mpf structures involved in conjugal DNA transfer are remarkably similar to type IV protein secretion systems that transfer virulence proteins from pathogenic bacteria directly into eukaryotic cells (Box 5.2). They are also related to some DNA uptake systems in naturally transformable bacteria (see chapter 6). Type IV protein secretion systems are described in chapter 14.

Mobilizable Plasmids

Some plasmids are not self-transmissible but can be transferred by another self-transmissible plasmid sharing the same cell. Plasmids that cannot transfer themselves but can be transferred by other plasmids are said to be mobilizable, and the process by which they are transferred is called **mobilization.** The simplest mobilizable plasmids merely contain the *oriT* sequence of a self-transmissible plasmid, since any plasmid that contains the *oriT* sequence of a self-transmissible plasmid can be mobilized by that plasmid. Expressed in genetic terminology, the Mpf and Dtr systems of the self-transmissible plasmid can act in *trans* on the *cis*-acting *oriT* site of the plasmid and mobilize it.

The ability to allow mobilization can be used to locate the *oriT* sequence in a plasmid (Figure 5.8). Random clones of the DNA of the self-transmissible plasmid are introduced into a nonmobilizable cloning vector, and the mixture is introduced into a cell containing the self-transmissible plasmid. Any vector plasmids that are mobilized into recipient cells probably contain a DNA insert including the *oriT* sequence. Transposons and plasmid cloning vectors containing the *oriT* site of a self-transmissible plasmid have many applications in molecular genetics because they can be mobilized into other cells.

While we can construct such a plasmid and they are mobilizable, minimal mobilizable plasmids containing only the *oriT* site of a self-transmissible plasmid do not seem to occur naturally. All mobilizable plasmids isolated so far encode their own Dtr systems, including their own relaxase and helicase. For historical reasons, the *tra* genes of the Dtr system of mobilizable plasmids are called the ***mob* genes** and the region required for mobilization is called the ***mob* region.** The function of the *mob* gene products of mobilizable plasmids seems to be to expand the range of self-transmissible plasmids by

Gene Exchange between Kingdoms

Not only can some plasmids transfer themselves into other types of bacteria, but also they can sometimes transfer themselves into eukaryotes, that is, into organisms of a different kingdom.

A. tumefaciens and Crown Gall Tumors in Plants

The first discovery of transfer of bacterial plasmids into eukaryotes occurred in the plant disease crown gall. Crown gall disease is caused by *Agrobacterium tumefaciens*; it is identified by a tumor that appears on the plant, usually where the roots join the stem (the crown). Virulent strains of *A. tumefaciens* contain a plasmid called the Ti plasmid, for *tumor initiation*. The Ti plasmids of *A. tumefaciens* are in most respects normal bacterial self-transmissible plasmids. A typical Ti plasmid is shown in panel A of the figure. Like other self-transmissible plasmids, Ti plasmids encode Tra functions that enable them to transfer themselves into other bacteria. What makes these plasmids remarkable is that they can also transfer part of themselves, called the T-DNA region, into plants. This discovery, made in the 1970s, has allowed the construction of transgenic plants because any foreign genes cloned into one of the T-DNA regions of the Ti plasmid will be transferred into the plant along with the T-DNA and integrated into the plant DNA. The integrated foreign genes can alter the plant, provided that they are transcribed and translated in the plant. Panel B of the figure shows the general procedure which is followed. A gene for kanamycin resistance has been inserted into the T-DNA of a Ti plasmid in such a way that it will be expressed in the plant. A piece of the plant leaf is floated in a bath of the bacterium containing the engineered Ti plasmid. Transgenic plants with the T-DNA inserted in their chromosome can be regenerated and selected on plates containing kanamycin. A whole industry has developed around this technology, and agrobacteria have been used to genetically engineer plants to make their own insecticides, to be more nutritious, and to survive more severe growing conditions.

The functions required for transfer of the T-DNA into plants are encoded by a region called the *vir* region (see panel A of the figure). This region is distinct from the *tra* region, which is required for the transfer of the plasmid into other bacteria, but its functions are remarkably similar to those of both other Tra functions and other type IV secretion systems. Panel C of the figure shows the structure of the type IV secretion system encoded by the *vir* region of the Ti plasmid. Like the *tra* region, the *vir* region encodes both a mating-pair formation (Mpf) system, which elaborates a pilus, and a Dtr system, which processes the DNA for transfer. The pilus is composed of the pilin protein, which is the product of the *virB2* gene and is cyclized, like the pilins of the pili of other type IV secretion systems. The Mpf system also includes a coupling protein, the product of the *virD4* gene, which communicates with the relaxosome, which includes the specific relaxase. The relaxase, the product of the *virD2* gene, cleaves the sequences bordering the T-DNA in the plasmid and remains covalently attached to the 5' ends of the single-stranded T-DNAs during transfer into the plant. The sequences at which the relaxase cuts these border sequences are similar to the *oriT* sequences of IncP plasmids, and the relaxase makes a cut in exactly the same place in the sequences where the relaxases of IncP plasmids cut their *oriT*.

This is where the T-DNA transfer system begins to differ from normal *tra* DNA transfer systems. In addition to its role as a relaxase, the VirD2 protein contains amino acid sequences that target it to the plant cell nucleus once it is in the plant. These sequences, called nuclear localization signals, are

(continued)

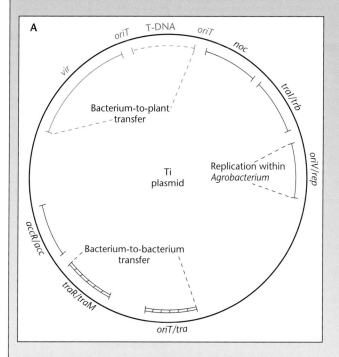

The Ti plasmid. **(A) The structure of a Ti plasmid, showing the various regions discussed clockwise from the top. The T-DNA is bordered by the *oriT* sequence, which is transferred into plants. The T-DNA contains the genes that are expressed in the plant to make opines and plant hormones (not shown); the *noc* genes, encoding enzymes for the catabolism of the opine nopaline in the bacterium; some *tra* genes, for transfer into other bacteria; the *oriV* region, for replication of the plasmid in the bacterium; *oriT* and *tra* function genes, for transfer into other bacteria; *acc* genes, for catabolism of another opine; and *vir* genes, for transfer into plants.**

BOX 5.1 (continued)

Gene Exchange between Kingdoms

essentially "passwords" that tell the plant that a particular plant nuclear protein should be transported into the nucleus after it has been translated in the cytoplasm. By imitating the password, the VirD2 protein tricks the plant into transporting it into the nucleus, dragging the attached T-DNA with it. Once in the nucleus, the T-DNA can enter the plant DNA by recombination. Once integrated into the plant DNA, the T-DNA of the plasmid encodes the synthesis of plant hormones which induce the plant cells to multiply and form tumors (galls) on the plant, hence the name "crown gall tumors." The T-DNA also encodes enzymes which synthesize unusual small molecules composed of an amino acid such as

arginine joined to a carbohydrate such as pyruvate. These compounds, called opines, are excreted from the tumor. The plant is able to express the genes on the T-DNA and make these compounds because the genes on the T-DNA are essentially plant genes with plant promoters and plant translational initiation regions, so they can be expressed once they are in the plant. Meanwhile, back in the bacterium, the Ti plasmid carries genes that allow it to use the particular opine made by that strain as a carbon and nitrogen source. Very few types of bacteria can degrade opines, which gives the *Agrobacterium* species containing this particular Ti plasmid an advantage. In this way, the bacterium creates its own special "ecological niche" at the expense of the plant.

The interaction of *A. tumefaciens* with the host plant presents some other interesting points of ecology. When a bacterium containing a Ti plasmid encounters a plant, phenolic residues and monosaccharides given off by the plant activate a two-component regulatory system, VirA-VirG, which in turn activates the transcription of the *vir* genes, which transport the T-DNA into the plant. As if to share its good fortune in finding a susceptible plant, the plasmid also induces the VirA-VirB system to activate the transcription of the *tra* genes, which transfer the Ti plasmid into any other surrounding agrobacteria which may lack it and so cannot use that particular opine

B

Cut out a piece of the leaf

Float in plate containing an *Agrobacterium* strain with engineered Ti plasmid

Incubate on plate containing plant regeneration medium

— Regenerating plants

Excise germinating shoots and transplant to plates containing kanamycin

— Kanamycin-resistant shoots

— Transgenic plant

(B) A procedure for making a transgenic plant (for details, see the text).

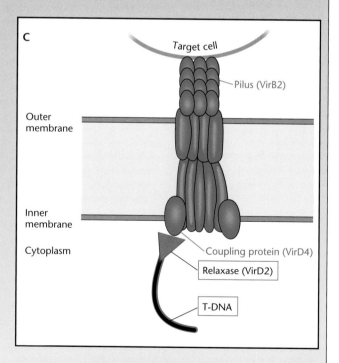

(C) Structure of the type IV secretion system for transfer of T-DNA into plant (for details, see the text).

BOX 5.1 (continued)

Gene Exchange between Kingdoms

as a carbon, nitrogen, and energy source. These other surrounding bacteria can then be recruited to aid in the infection process. A more detailed discussion of two-component systems and how they activate genes in response to extracellular signals is presented in chapter 13.

Another plasmid-encoded protein, VirE2, is also transported into the plant cytoplasm. This protein might not enter the nucleus but, rather, might form the channel in the plant membrane through which the T-DNA enters. If other *tra* systems encode such proteins, it would help explain the extreme promiscuity of some conjugation systems (see below). If other transport systems also secrete proteins to form their own channel in the recipient cell membrane through which the DNA must pass, the *tra* system becomes less restricted in its choice of recipient cell. The membrane channel may be able to assemble in almost any type of cell, since bipolar lipid membranes are similar in all organisms.

Transfer of Broad-Host-Range Plasmids into Eukaryotes

The Ti plasmid is obviously designed to transfer part of itself into plant cells. The surprising result of recent studies is that other bacterial plasmids can also transfer themselves or mobilize other plasmids into eukaryotic cells. One striking example is the mobilization of other plasmids into plant cells by the Ti plasmid. As mentioned, the sequences bracketing the T-DNA

in the Ti plasmid can be thought of as *oriT* sites, and the Tra functions of the Ti plasmid can be thought of as mobilizing the T-DNA into plant cells. Plasmid RSF1010, and plasmids derived from it, can also be mobilized into plant cells by the Ti plasmid, provided that they contain the correct *mob* sequence.

Not only do plasmids transfer into plants, but also they can sometimes transfer into lower fungi. This observation is very surprising because the cell surfaces of bacteria and eukaryotes are very different. So are the surfaces of plant cells, but in the case of the Ti plasmid, we can assume that the transfer functions have evolved to recognize plant cells. There is no apparent reason why bacterial plasmids should have evolved to transfer into other kingdoms. Whatever the reason, the transfer of genes between eukaryotes and bacteria may play an important role in evolution.

References

Bates, S., A. M. Cashmore, and B. M. Wilkins. 1998. IncP plasmids are unusually effective in mediating conjugation of *Escherichia coli* and *Saccharomyces cerevisiae*: involvement of the Tra2 mating system. *J. Bacteriol.* **180:**6538–6543.

Brencic, A., and S. C. Winans. 2005. Detection of and response to signals involved in host-microbe interactions by plant-associated bacteria. *Microbiol. Mol. Biol. Rev.* **69:**155–194.

Buchanan-Wollasten, U., J. E. Passiatore, and F. Cannon. 1987. The *mob* and *oriT* mobilization functions of a bacterial plasmid promote its transfer to plants. *Nature* (London) **328:**172–175.

which they can be mobilized (see below). A plasmid containing only the *oriT* sequence of a self-transmissible plasmid can be mobilized only by the *tra* system of that self-transmissible plasmid and not by that of other self-transmissible plasmids which do not share the same *oriT* site, while naturally occurring mobilizable plasmids can often be mobilized by a number of *tra* systems.

The process of mobilization of a plasmid by a self-transmissible plasmid is illustrated in Figure 5.9. The process is identical to the transfer of a self-transmissible plasmid, except that the Mpf system of the self-transmissible plasmid is acting not only on its own Dtr system but also on the Dtr system of the mobilizable plasmid. The self-transmissible plasmid forms a mating bridge with a recipient cell and communicates this information via its coupling protein not only to its own relaxase but also to the relaxase of a mobilizable plasmid that happens to be in the same cell. The relaxase of the mobilizable plasmid then makes a single-stranded break at its *oriT* site, and the helicase separates the strands. The relaxase remains bound to

the 5′ end of the single-stranded DNA and is transferred into the recipient cell, dragging the single-stranded DNA of the mobilizable plasmid with it. The self-transmissible plasmid is also often transferred at the same time that it mobilizes other plasmids. However, generally either one plasmid or the other is transferred into a particular recipient cell, due to competition between the two plasmids for coupling protein (see below).

The secret of being mobilized by another plasmid is to be recognized by the coupling protein of the other plasmid. Any plasmid encoding its own Dtr system can be mobilized by a coresident self-transmissible plasmid, provided that its relaxase can communicate with the coupling protein of the Mpf system of the coresident plasmid. Accordingly, the relaxases of mobilizable plasmids are designed to communicate with a broader range of coupling proteins of self-transmissible plasmids so that they can take advantage of a number of different Mpf systems, unlike the relaxases of self-transmissible plasmids, which seem to be more specific. This suggests that mobilizable

BOX 5.2

Conjugation and Type IV Protein Secretion Systems

One of the more intriguing discoveries in modern cell biology is the extent to which systems developed for one purpose have been reassembled and adapted to serve other purposes. We have already seen examples of such molecular "battlebots" that are assembled from the parts of other molecular machines. One example is the relatedness between the syringe-like type III secretion systems, which secrete proteins directly into eukaryotic cells as part of the disease-causing process, and the bacterial flagellum systems, which help bacteria to swim. Such discoveries have become almost routine since the development of computer technology for searching databases for related sequences. Databases, such as GenBank, that contain the sequences of hundreds of thousands of genes have been assembled over the years. Once you have sequenced a gene, you can search these databases to determine whether other genes in these databases have related sequences (see Box 2.7). When the sequences of the *tra* genes of many conjugation systems are compared to the sequences of the *vir* genes of type IV secretion systems, it is obvious that many of them have a common ancestry.

Apparently, the basic machinery developed to transfer macromolecules from one cell to another has been adapted to many different specialized functions.

Conjugation and type IV protein secretion do have much in common. In conjugation, the Tra functions transfer DNA as well as some accompanying proteins from one bacterial cell to another cell. This other cell can be either bacterial or, in some cases such as the Ti plasmid, a eukaryotic plant cell. Type IV protein secretion systems do something similar. They transfer proteins from a bacterial cell directly into a eukaryotic cell as part of the disease-causing mechanism. Both types of systems require pili or another adhesin to hold the cells together during the transfer and special membrane structures through which the macromolecules must pass. Both processes are very specific, and only some types of proteins or plasmid DNAs can be transferred. Nevertheless, it came as a surprise how closely related these two types of systems can be. In fact, they may simply be different manifestations of the same process.

The relatedness of type IV secretion to conjugation is dramatically illustrated by comparisons of the T-DNA transfer

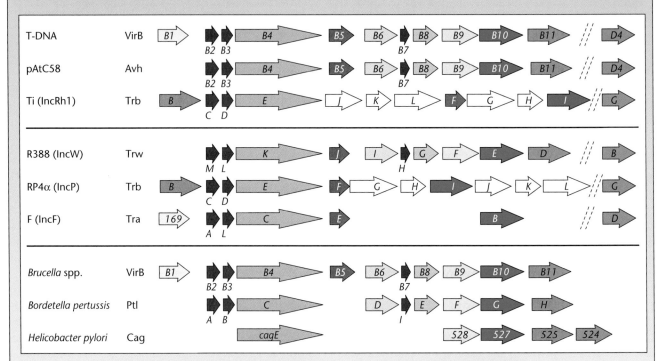

Gene arrangements of type IV secretion loci. Genes encoding VirB homologues are similarly shaded, and those encoding proteins unrelated to VirB are not shaded. (Top) *A. tumefaciens* is the first species shown to carry three distinct type IV secretion systems whose substrates are defined; e.g., the VirB system transfers T-DNA and protein effectors to plants, whereas Avh transfers pATC58 and Trb transfers the Ti plasmid, respectively, to other bacteria. (Middle) Representative type IV secretion systems of other species that direct conjugal DNA transfer. (Bottom) Representative type IV secretion systems that direct protein transfer during the course of infections.

BOX 5.2 (continued)

Conjugation and Type IV Protein Secretion Systems

system of *Agrobacterium tumefaciens* called VirB to some type IV secretion systems. A diagram of Mpf of the VirB transfer system is shown in the figure in Box 5.1. As discussed in Box 5.1, the T-DNA transfer system transfers the T-DNA part of a plasmid from bacterial to plant cells. It also transfers a protein that forms a channel in the plant membrane and the relaxase protein, which doubles as a protein that can target the T-DNA to the plant nucleus, where it can enter the plant DNA. This transfer system shares features with other plasmid conjugation systems in that it encodes a pilus, relaxase, coupling proteins, and chaperones, as well as many other proteins involved in the transfer process. In fact, most of the proteins of the Tra systems of the F and R388 plasmids can be assigned homologues in the T-DNA transfer system (see the figure). Moreover, some pathogenic bacteria that transfer proteins into eukaryotic cells as part of the disease-causing process also have analogous functions. One of the most striking similarities is to the CagA toxin-secreting system of *Helicobacter pylori*, implicated in some types of gastric ulcers. This type IV toxin-secreting system has at least five protein homologues to the T-DNA system of the Ti plasmid of *Agrobacterium*. The system delivers a toxin directly through the bacterial membranes and into the eukaryotic cell, where the toxin is phosphorylated on one of its tyrosines. In the phosphorylated state, the toxin causes many changes in the cell including alterations in its actin cytoskeleton. Another pathogenic bacterium, *Bordetella pertussis*, the causative agent of whooping cough, also has a type IV secretion system which has nine homologous proteins

to the T-DNA system. This system secretes the pertussis toxin through the outer membrane of the bacterial cell. Once outside the cell, the pertussis toxin assembles into a form that can then enter the eukaryotic cell, where it can ADP-ribosylate G proteins, thereby interfering with signaling pathways and causing disease symptoms.

However, the most striking evidence that conjugation is related to type IV secretion has come from the virulence system of *Legionella pneumophila*, which causes Legionnaires' disease. Like many pathogenic bacteria, this bacterium can multiply in macrophages, specialized white blood cells that are designed to kill them (see Vogel et al., below). These bacteria are taken up by the macrophage but then secrete proteins which disarm the phagosomes that have engulfed them. The components of this type IV secretion system are analogous to the Tra functions of some self-transmissible plasmids, and this type IV system can even mobilize the plasmid RSF1010 at a low frequency!

References

Christie, P. J., K. Atmakuri, V. Krishnamoorthy, S. Jakubowski, and E. Cascales. 2005. Biogenesis, architecture, and function of bacterial type IV secretion systems. *Annu. Rev. Microbiol.* **59**:451–485.

Covacci, A., J. L. Telford, G. Del Giudice, J. Parsonnet, and R. Rappuoli. 1999. *Helicobacter pylori* virulence and genetic geography. *Science* **284**:1328–1333.

Vogel, J. P., H. L. Andrews, S. K. Wong, and R. R. Isberg. 1998. Conjugative transfer by the virulence system of *Legionella pneumophila. Science* **279**:873–876.

plasmids are designed to be parasitic for transfer on self-transmissible plasmids rather than just being erstwhile self-transmissible plasmids that have lost their own Mpf system. Moreover, some mobilizable plasmids overlap the functions of replication and mobilization so that they can use the same helicase and primase for both processes and their *oriV* site is often placed close to their *oriT* site, again unlike self-transmissible plasmids.

In spite of their versatility, however, not all mobilizable plasmids can be mobilized by all self-transmissible plasmids. For example, the IncQ plasmid RSF1010 can be mobilized by the IncP plasmid RP4 but not by the IncF plasmid F. Apparently, the coupling protein of RP4 can communicate with the relaxase of RSF1010 while the coupling protein of the F plasmid cannot. However, the F plasmid can mobilize the ColE1 plasmid, and so its coupling protein can communicate with the relaxase of this mobilizable plasmid. This complicated interplay between the Mpf and Dtr systems is particularly dramatic

for the Ti plasmid and RSF1010. The *tra* system of the Ti plasmid does not mobilize the RSF1010 plasmid into other bacteria, but the *vir* system of the Ti plasmid mobilizes RSF1010 into plants (see Buchanan-Wollasten et al., Box 5.1).

PLASMID MOBILIZATION IN BIOTECHNOLOGY
Mobilization plays an important role in biotechnology because it can be used to efficiently introduce foreign DNA into bacteria. A *mob* site is often introduced into cloning vectors so that they can be efficiently transferred into cells (see Bagdasarian et al., Suggested Reading). Smaller is better in cloning vectors (see chapter 4), and a mobilizable plasmid can be much smaller than a self-transmissible plasmid because it does not need the 15 or so Mpf genes required to assemble the mating bridge, only the 4 or so Dtr genes required for DNA processing. Once foreign DNA has been cloned into such a cloning vector, it can be introduced into even distantly related

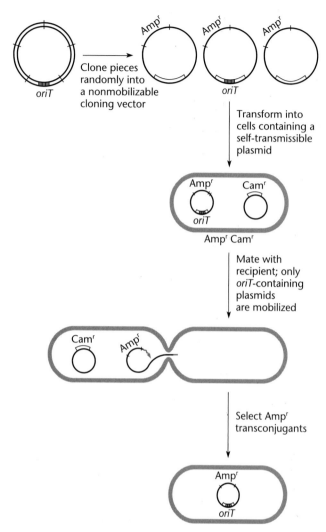

Figure 5.8 Identifying the *oriT* site on a plasmid. Pieces of the plasmid are cloned randomly into a nonmobilizable cloning vector. The mixture is transformed into cells containing the self-transmissible plasmid and mixed with a proper recipient. Pieces of DNA that allow the cloning vector to be mobilized contain the *oriT* site of the plasmid.

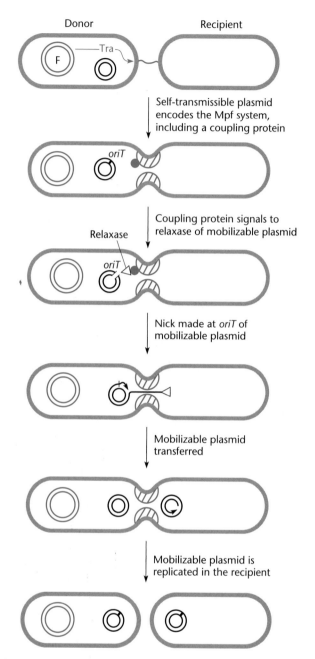

Figure 5.9 Mechanism of plasmid mobilization. The donor cell carries two plasmids, a self-transmissible plasmid, F, which encodes the *tra* functions that promote cell contact and plasmid transfer, and a mobilizable plasmid (purple). The *mob* functions encoded by the mobilizable plasmid make a single-stranded nick at *oriT* in the *mob* region. Transfer and replication of the mobilizable plasmid then occur. The self-transmissible plasmid may also transfer. Details are given in the text.

bacteria by the Mpf system of a larger, promiscuous, self-transmissible plasmid. In addition, some self-transmissible plasmids have been crippled so that they cannot transfer themselves but can transfer only mobilizable plasmids. Then the recipient cell receives only the mobilized cloning vector in such a transfer and not the self-transmissible plasmid which mobilized it.

A common application of plasmid mobilization technology is in transposon mutagenesis. These methods are most highly developed for gram-negative bacteria. Some plasmids, such as RP4, are so promiscuous that they can transfer themselves into essentially any gram-negative

bacterium. If such a plasmid is used to introduce a small mobilizable plasmid which has a narrow-host-range origin of replication like the ColE1 origin (see chapter 4), the smaller plasmid is mobilized into the bacterium but probably cannot replicate there and is eventually lost (e.g., is a suicide vector). If the smaller plasmid also contains a transposon such as Tn5, containing the selectable kanamycin resistance gene and with a broad host range for transposition, the only way the recipient cell can become resistant to kanamycin is if the transposon hops into the chromosome of the recipient strain, causing random insertion mutations. The transposon insertion mutants facilitate cloning and can even be used for Hfr mapping if an *oriT* sequence has been introduced on the transposon. We return to such methods of transposon mutagenesis in chapter 9. Mobilizable plasmids can also be used to detect the transferability of naturally occurring plasmids for which no selection is available. A resident plasmid may have transfer functions and be self-transmissible if it can mobilize another plasmid carrying an easily selectable marker.

However, mobilizable plasmids also present regulatory complications. To meet regulatory requirements or for other reasons, genetic engineers often have to prove that a plasmid containing recombinant genes that confer desirable properties on one bacterium will not be mobilized into another, unknown, bacterium, where the genes might be harmful. But how do we really know that a plasmid does not contain an *oriT* site that will be recognized by some set of Tra functions? Unfortunately, negative evidence of mobilization by all the known self-transmissible plasmids does not mean that a given plasmid cannot be mobilized by *some* plasmid.

TRIPARENTAL MATINGS

Mobilization of a plasmid into a recipient cell is often used for cloning, transposon mutagenesis, or other procedures. As mentioned, mobilizable plasmids have an advantage over self-transmissible plasmids in being smaller. Nevertheless, difficulties can be encountered before these plasmids can be mobilized. For example, the self-transmissible plasmid and the plasmid to be mobilized may be members of the same Inc group and so do not stably coexist in the same cell. Also, the self-transmissible plasmid may express its *tra* genes only for a short time after entering a recipient cell so that transfer is inefficient.

Triparental matings help overcome some of the barriers to efficient plasmid mobilization. Figure 5.10 illustrates the general method. As the name implies, three bacterial strains participate in the mating mixture. The first strain contains a self-transmissible plasmid, the second contains the plasmid to be mobilized, and the third is the eventual recipient. After the cells are mixed, some of the self-transmissible plasmids in the first strain are transferred into the strain carrying the plasmid to be mobilized. Because it is fertile when it first enters the cell, the

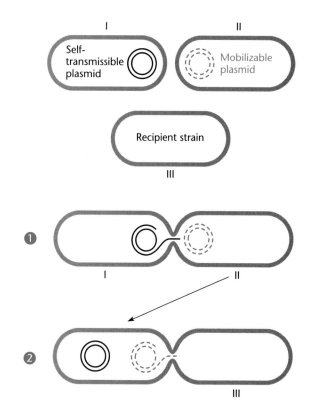

Figure 5.10 Triparental matings. In step 1, a self-transmissible plasmid from parent I transfers into parent II. In step 2, the self-transmissible plasmid transfers the mobilizable plasmid into parent III. This method works even if the self-transmissible plasmid and the mobilizable plasmid are members of the same Inc group (see the text) and if the self-transmissible plasmid cannot replicate in parent II.

self-transmissible plasmid quickly spreads through the population of cells containing the mobilizable plasmid. It is then able to mobilize the mobilizable plasmid into the third strain with a high efficency because new transconjugants retain their ability to transfer for at least six generations. Contrast this to a mating involving only two strains, one of which contains both the self-transmissible plasmid and the mobilizable plasmid. Only a small fraction of the cells are fertile and can mobilize the mobilizable plasmid into the recipient strain. Also in a triparental mating, even if the two plasmids are members of the same Inc group, they coexist long enough for the mobilization to occur.

Genetic Analysis of Tra Systems in Gram-Negative Bacteria

Genetic analysis allowed the detection and mapping of the *tra* genes on plasmids and preliminary studies on their functions in conjugation. In this section we describe experiments for the identification and mapping of *tra* genes on plasmid pKM101 (see Winans and Walker, Suggested Reading). Similar experiments were done with other plasmids, but these were reported in a single paper, which makes them more convenient to discuss.

Isolation of *tra* Mutant Plasmids

The first step in identifying the *tra* genes of plasmid pKM101 was to mutagenize the plasmid with the Tn*5* transposon to obtain a large number of transferless (Tra⁻) mutants. One of the two methods used is illustrated in Figure 9.21B in the discussion of random transposon mutagenesis of a plasmid. The experiments used a mutant form of λ as the suicide vector. It has nonsense mutations in its replication genes, and so it cannot replicate in cells that lack a nonsense suppressor. It also has a deletion of its *attP* site so that it cannot integrate into the chromosomes (see chapter 8). Cells containing plasmid pKM101 were infected with the λ suicide vector containing transposon Tn*5*, and plasmids with transposon insertion mutations were isolated by preparing the plasmids and using them for transformation. Kanamycin-resistant transformants, each of which contained the plasmid with a transposon insertion in a different region of the plasmid, were isolated. The second method was the same initially, in that the λ suicide vector carrying the transposon was used to infect cells containing the plasmid. However, after being infected, the cells were incubated only long enough to give the transposon time to hop. The cells were then mixed with recipient bacteria, and kanamycin-resistant transconjugants were isolated. In general, only cells in which the transposon had hopped into the plasmid would be able to transfer kanamycin resistance to the recipient, since usually only the plasmid would be transferred. Even if the transposon had happened to hop into a *tra* gene in the plasmid, the plasmid would still transfer, because after such a short incubation, the cells would still contain some of the Tra function owing to phenotypic lag (see chapter 3). Using conjugation rather than transformation to isolate plasmids with transposon insertions has the advantage that conjugation is often more efficient than transformation, making it easier to isolate many mutant plasmids.

Once the investigators had isolated several transposon insertions in the plasmid, they needed to determine which of the transposon insertions had inactivated a *tra* gene. If the transposon had hopped into, and therefore inactivated, a *tra* gene, the plasmid would no longer be able to transfer itself. Therefore, each of the mutagenized plasmids was tested individually for its ability to transfer.

To streamline the process of screening the mutants, the investigators devised the replica-plating test for transfer illustrated in Figure 5.11. In the example, a number of plasmid-containing bacteria were patched

onto a plate so that each patch would have bacteria containing a different mutant plasmid. After incubation to allow multiplication of the bacteria in the patches, this plate was replicated onto another plate containing rifampin on which rifampin-resistant recipient bacteria had been spread. The antibiotic rifampin counterselected the donor (see the discussion of Hfr mapping in chapter 3). After incubation to allow any potential transconjugants to form and multiply, the second plate was replicated onto a third plate containing medium with both kanamycin and rifampin. If the Tn*5* insertion in the plasmid did not inactivate a *tra* gene, the plasmid would have transferred into some of the recipient bacteria on the second plate, yielding transconjugants that are both kanamycin and rifampin resistant. However, if the transposon had inserted into a *tra* gene, no colonies should appear in a patch on the third plate since there would be no transconjugants in the patch that are both kanamycin and rifampin resistant. In this way, several mutant plasmids were identified that had a transposon insertion in one of their *tra* genes. The site of the transposon insertions in these *tra* plasmids could then be physically mapped by methods such as those described in chapter 9 (Figures 9.22 to 9.24) in the discussion of mapping the site of transposon insertions in a plasmid.

Complementation Tests To Determine the Number of *tra* Genes

Mapping the site of insertion of the transposon in a *tra* mutant plasmid reveals where a *tra* gene is located on the plasmid but does not, by itself, reveal the number of *tra* genes. Only complementation tests can determine how many genes are represented in the collection of *tra* mutants. To do complementation tests, plasmids containing two different *tra* mutations must be introduced into the same cell. If the mutations inactivate different genes, each mutant plasmid furnishes the Tra function that the other one lacks, and one or both plasmids are able to transfer. If, however, the two mutations inactivate the same *tra* gene, neither plasmid is able to make the product of that gene and neither plasmid is able to transfer. To save time, these complementation tests need to be done only between *tra* mutants with insertion mutations close to each other on the plasmid. Any two transposon insertion mutations which are not closely linked or which have transposon insertions between them that do not inactivate a *tra* gene are almost certainly not in the same *tra* gene, and so they do not need to be tested for complementation.

Complementation tests between different *tra* mutations in the same plasmid, in this case pKM101, are complicated by the fact that they are in the same Inc group and so cannot coexist for long. Figure 5.12 shows one

systems which are similar to the gram-negative transfer systems discussed above. They are transferred complete with their own relaxases and *oriT* sequences. In fact, the *oriT* sequences of gram-positive plasmids are sometimes closely related to those of gram-negative bacteria (see Kurenbach et al., Suggested Reading). Other plasmids found in the gram-positive *Streptomyces* and presumably other related bacteria are very different and are discussed in Box 5.3.

The major differences between plasmids from the two bacterial groups come in the mating-pair formation (Mpf) systems, which can be simpler in gram-positive bacteria because of the lack of an outer membrane. We discuss only plasmids from *Enterococcus* in this section because of their interesting method of attracting mating cells and their medical importance (for a review, see Clewell and Francia, Suggested Reading).

Plasmid-Attracting Pheromones

Some strains of *Enterococcus faecalis* excrete pheromone-like compounds that can stimulate mating with donor cells. These pheromones are small peptides, each of which stimulates mating with cells containing a particular plasmid. The pheromone-like peptides act by specifically stimulating the expression of *tra* genes in the plasmids of neighboring bacteria, thereby inducing aggregation and mating. Once a cell has acquired a plasmid, it no longer excretes the specific peptide, but it continues to excrete other peptides that stimulate mating with cells containing other plasmid types. By inducing its Tra functions

---

BOX 5.3

Conjugation in Streptomycetes

The streptomycetes are important for their production of many of our most useful antibiotics. They are also capable of conjugation, which was, in fact, discovered about the same time as conjugation in *E. coli*. The mechanism for conjugation in streptomycetes, which are gram-positive bacteria, differs greatly from conjugation mechanisms that we have discussed so far. For example, stable mating-pair formation requires numerous *tra* gene products in most bacteria. However, the streptomycete morphology and life cycle facilitate cell contact without plasmid-encoded genes. Commonly found in soil, streptomycetes grow from spores into branching, filamentous hyphae that form an intertwined network. Specialized hyphae within this network eventually differentiate to form haploid spores (see the figure).

As shown in the figure, the growth of hyphae toward and around each other evidently provides stable contact between different parent strains, since no plasmid-encoded genes involved in establishing or maintaining cell contact have been discovered. Streptomycete transmissible plasmids also lack genes encoding the elaborate proteinaceous structures involved in DNA transfer across cell membranes and walls. Partial fusion of the hyphae may occur, creating an opportunity for DNA transfer.

The single plasmid-encoded *tra* gene product required for conjugation of the well-studied plasmid pIJ101 resembles gene products known to function in DNA translocation in *E. coli* and *Bacillus subtilis* (see the discussion of *ftsK*/*spoIIIE* in chapter 1). The precise role of this protein, TraB, is unknown, but it localizes to hyphal tips. It could be speculated that it functions to move DNA between partially fused hyphae.

The efficiency of transfer of plasmid pIJ101 is extremely high—even 100% following the plating of a mixture of donor and recipient spores on agar plates. Moreover, dissemination of the plasmid throughout the recipient mycelium is also highly efficient. The streptomycete hyphae within a mycelial network have few cross walls, and cell compartments are multinucleoid. Both high- and low-copy-number transmissible plasmids are able to spread throughout mycelia and become incorporated in spores. A plasmid genetic locus named *spd* is responsible for efficient spread through hyphae, but its mechanism of action is poorly understood.

Another major difference between gene transfer in streptomycetes and most other bacteria is that chromosome transfer occurs at the same time as plasmid transfer even if the plasmid is not integrated. Also, recombination (and apparently chromosome transfer) is not limited to one parent. Rather, the genotypes of both parents can be changed through recombination, so that as many as 1% of the progeny genomes found in haploid spores can be recombinant. Usually, there is no covalent association of the plasmid with the chromosome; therefore, no gradient of transfer is created. However, if the conjugative plasmid contains chromosomal sequences, it can integrate through homologous recombination, and then a gradient of transfer is apparent.

The mechanisms of generation of recombinant chromosomes are clearly different in many ways from those in Hfr strains of gram-negative bacteria. It is not known whether transfer of either plasmid or chromosomal DNA requires nicking at an *oriT* sequence and subsequent transfer of a single

(continued)

BOX 5.3 (continued)

Conjugation in Streptomycetes

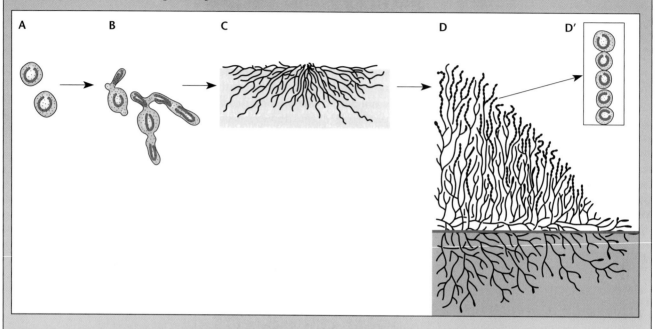

Development of a streptomycete colony including plasmid-mediated conjugation. (A) Streptomycete spores. The chromosome is shown by purple lines. The chromosomes are linear, but the ends may be held together (see Box 1.1). The small dots indicate a conjugative plasmid such as the multicopy pIJ101. (B) Germinating spore. Hyphae begin to form when spores germinate. (C) A growing colony. Growing branching hyphae form so-called "substrate" mycelia (agar shown in gray). Growing hyphae contain very few cross walls, and so chromosomes (not shown) are not separated into individual compartments. During the transition from panel B to panel C, hyphal fusions may have occurred, allowing an opportunity for recombination between chromosomes from different parents. (D) Sporulating mycelia. The vertically directed "aerial" hyphae have differentiated, forming chains of spores. As shown in panel D′, the spores are haploid. A small percentage of the genomes are recombinant, composed of segments of the different parental chromosomes (recombinants not shown). In the mature colony, the substrate mycelium produces antibiotics; in this example the antibiotic has a purple color.

strand from the donor parent. The *Streptomyces* plasmid pIJ101 does contain a *cis*-acting *oriT*-like site, which is required for plasmid transfer, but there is no evidence for nicking at this site. It is notable that this site is not required for generation of recombinant chromosomes in a mating culture.

References
Kieser, T., M. J. Bibb, M. J. Buttner, K. F. Chater, and D. A. Hopwood. 2000. *Practical* Streptomyces *Genetics.* The John Innes Foundation, Norwich, United Kingdom.

Pettis, G. S., and S. N. Cohen. 2000. Mutational analysis of the *tra* locus of the broad-host-range *Streptomyces* plasmid pIJ101. *J. Bacteriol.* **182:**4500–4504.

Possoz, C., C. Ribard, J. Cagnat, J. L. Pernodet, and M. Guerineau. 2001. The integrative element pSAM2 from *Streptomyces:* kinetics and mode of conjugal transfer. *Mol. Microbiol.* **42:**159–166.

Reuther, J., C. Gekeler, Y. Tiffert, W. Wohlleben, and G. Muth. 2006. Unique conjugation mechanism in mycelial streptomycetes: a DNA-binding ATPase translocates unprocessed plasmid DNA at the hyphal tip. *Mol. Microbiol.* **61:**436–446.

only when a potential recipient is nearby, the plasmid saves energy and limits the expression of surface antigens that may provide targets for the host immune system and receptors for male-specific phages. There are many families of transmissible plasmids in the enterococci. All of these plasmids share certain properties and the essential mechanisms for conjugation.

The mechanism of pheromone sensing of recipient cells is outlined in Figure 5.18. All of the genes that encode the peptide pheromones are located on the *Enterococcus* chromosome and are cut from the signal peptides of normal cell lipoproteins (see De Boever et al., Suggested Reading). After the signal sequences are cut from the lipoprotein being transported, the active pheromones are produced by proteolysis of the C-terminal 7 or 8 amino acids of the signal sequence. Processing occurs as the pheromone is excreted from the cell. By cutting the pheromones from the signal sequences of many different lipoproteins, each cell is able to excrete a number of different pheromones and attract a number of different plasmids.

The pheromone-sensing mechanism is also conserved in the various plasmid families. Sensing of the pheromones by plasmid-containing potential donor cells requires specific proteins located on the cell surface of a plasmid-containing donor cell. The sensing proteins are encoded by the transmissible plasmids. Each type of transmissible plasmid expresses a protein specific for one type of pheromone, and the pheromone is named after the plasmid it attracts. For example, in the well-studied plasmid pAD1, the plasmid-encoded pheromone-binding protein is called TraC and the pheromone is called cAD1. A given pheromone-binding protein that has bound a specific pheromone signals the cell's peptide uptake system, called the oligopeptide permease, to take up the peptide. Once inside the cell, the pheromone can induce the expression of plasmid genes involved in plasmid transfer, including aggregation substance. This protein coats the donor cell surface and initiates contact with the recipient cell. Following cell-cell contact, the plasmid transfers through a mating channel, much like plasmid transfer in gram-negative bacteria.

Once the plasmid has entered the recipient, the new transconjugant no longer functions as a recipient and is unable to acquire additional plasmids of the same family. The limitation of plasmid uptake by a donor cell involves three mechanisms. One mechanism involves the expression of surface exclusion proteins which function much like the entry exclusion *eex* systems of gram-negative bacteria. The second mechanism involves the shutdown of pheromone sensing due to synthesis of an inhibitor encoded by a gene of the plasmid. The inhibitor gene

product is a peptide of 7 or 8 amino acids, much like the pheromone itself, but it differs from the pheromone by only 1 or a few amino acids, which allows it to bind to the specific pheromone-binding protein. This allows it to competitively inhibit the secretion of the pheromone by the donor cell and prevent autoinduction (i.e., self-induction) of its own mating system and that of other potential donors. The third system involves a shutdown protein called TraB in pAD1 and PrgY in a related plasmid, pCF10, which has a similar method of pheromone sensing (Figure 5.19) (see Chandler et al., Suggested Reading). The function of these inhibitor proteins is unclear, but they are membrane proteins that somehow interfere with the processing or distribution of the pheromone to further prevent autoinduction and to avoid attracting other donors (see Buttaro et al., Suggested Reading).

In contrast to the similarities in pheromone production and sensing, the mechanisms for pheromone induction of *tra* gene expression differ from one type of plasmid to another. For example, in pAD1, the *traA* gene encodes a negative regulator (repressor) (see chapter 2) that represses the transcription of genes encoding aggregation substance and other transfer gene products. The pheromone cAD1 binds to TraA, releasing it from DNA to allow transcription. It is clear that other plasmids use different mechanisms, but these mechanisms are less well understood.

The enterococci and their plasmids are of especial importance because they are important hospital-acquired (nosocomial) pathogens. Their diverse plasmids carry both genes that enhance virulence and genes that confer resistance to multiple antibiotics. The enterococcal plasmids can also transfer their genes to other gram-positive bacteria including the very dangerous pathogen *Staphylococcus aureus*. This is because *S. aureus* can produce pheromones to attract enterococcus plasmids (see De Boever et al., Suggested Reading). This type of transfer is of particular medical importance because enterococcal plasmids can confer resistance to vancomycin, which is often a "last-resort" antibiotic in the treatment of *S. aureus* infections.

Other Types of Transmissible Elements

Plasmids are not the only DNA elements in bacteria that are capable of transferring themselves by conjugation. There are numerous examples of other elements, often called ICEs (integrating conjugative elements) that are normally integrated into the chromosome and do not exist autonomously like plasmids. Nevertheless, they often encode Tra functions and can transfer themselves

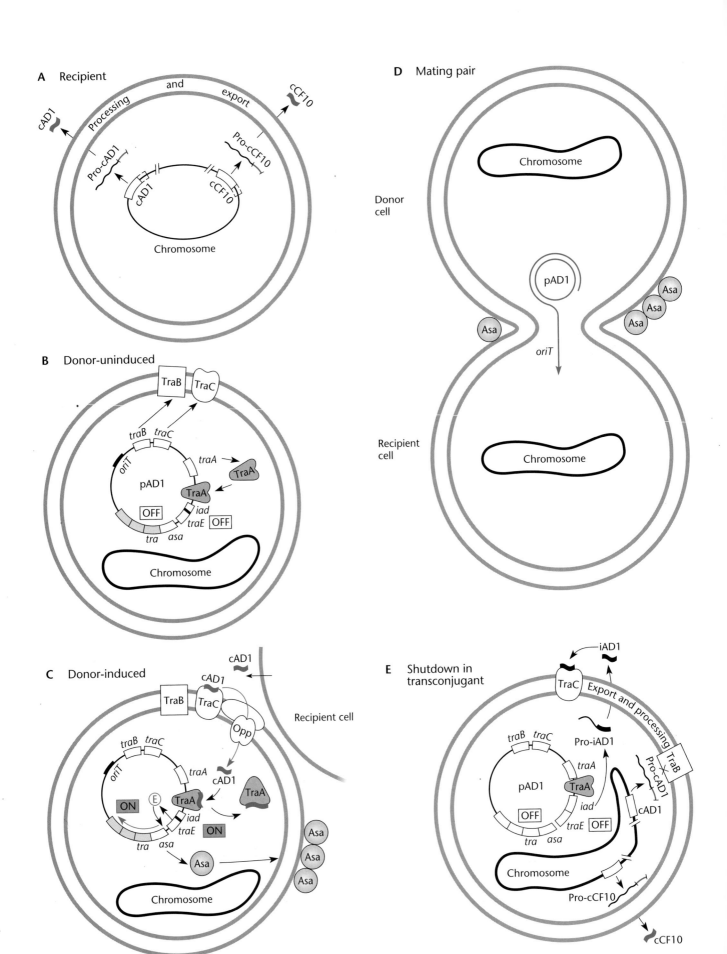

or mobilize other elements into recipient cells. Some of these elements are discussed in this section and in Box 5.4.

Conjugative Transposons

Some integrating elements, called **conjugative transposons,** also encode Tra functions to promote their own transfer. An example, Tn*916* from *E. faecalis* is not a replicon capable of autonomous replication, but it can transfer itself from the chromosome of one bacterium to the chromosome of another bacterium without transferring chromosomal genes. It transiently excises from one DNA, transfers itself into another cell, and then integrates into the DNA of the recipient bacterium (see Marra et al., Suggested Reading).

The Tn*916* conjugative transposon and its relatives are known to be promiscuous, and they transfer into many types of gram-positive bacteria and even into some gram-negative bacteria. The antibiotic resistance gene they carry, *tetM*, has also been found in many types of gram-positive and gram-negative bacteria. It is tempting to speculate that conjugative transposons such as Tn*916* are responsible for the widespread dissemination of the *tetM* gene. The integration of conjugative transposons is discussed in more detail in chapter 9.

Other elements similar to conjugative transposons have been found in the genus *Bacteroides*. These elements not only transfer themselves but also mobilize other small DNA elements in the chromosome (see Schmidt et al., Suggested Reading).

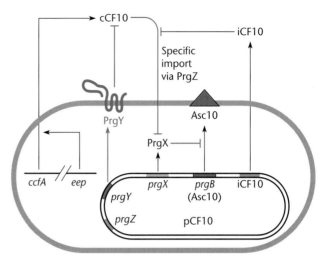

Figure 5.19 Model of pCF10 pheromone production, control, and response. *ccfA*, which encodes the pheromone of cCF10, and *eep* are both expressed from the chromosome. Mature cCF10 is processed from the signal sequence of the lipoprotein CcfA by the membrane protease Eep as the lipoprotein exits the cell. PrgY, PrgX, PrgZ, Asc10 (aggregation substance, expressed from *prgB*), and iCF10 are all encoded within pCF10. In the absence of PrgY or the inhibitor peptide, iCF10, endogenous mature cCF10 can internalize via PrgZ into donor cells to continually induce pCF10 transfer proteins. PrgY (in purple) and the inhibitor peptide iCF10 prevent this self-induction by endogenous cCF10. iCF10 neutralizes endogenous cCF10 in the medium, and PrgY sequesters or blocks the activity of endogenous cCF10. PrgX functions as the on-off switch for induction and, on interaction with pheromone, releases the repression of Asc10 expression. Asc10 mediates aggregation of the cells, which is necessary for efficient transfer of pCF10 to recipient cells.

Figure 5.18 Role of pheromones in plasmid transfer. (A) The recipient cell. The pheromone genes are located on the enterococcal chromosome; several examples are shown. The propheromone peptides cAD1 and cCF10 are processed from the signal sequences cut off of normal cellular proteins. The pheromones are processed from the propheromones Pro-cCF10 and Pro-cAD1 when exported. (B) The donor cell. The plasmid-carrying cell expresses TraA, which represses transcription of the other *tra* genes except *traC*, which encodes a cell surface protein that can sense the pheromone. Also shown is TraB, which is discussed in panel E. (C) Mating induction. The pheromone corresponding to the plasmid, in this case cAD1, binds TraC on the cell surface of a donor cell in close proximity and enters the cell via the oligopeptide permease system (Opp). The pheromone binds to the repressor TraA, releasing it from the DNA and derepressing the synthesis of TraE, which activates the expression of the *tra* genes including the gene encoding the aggregation substance (Asa). (D) Plasmid transfer. The donor cell establishes contact with the recipient cell, and the plasmid transfers, producing a transconjugant. (E) Pheromone shutdown in the transconjugant. Once the cell has become a transconjugant, the inhibitor peptide iAD1 binds to TraC and prevents autoinduction or pheromone stimulation of mating with other donor cells. Also, TraB is an inhibitor protein that somehow functions to shut down induction by preventing the excretion of pheromone cAD1, but pheromone cCF10 continued to be expressed.

BOX 5.4

Conjugative Transposons (Integrating Conjugative Elements)

The conjugative transposons are essentially transposons combined with transfer functions, such as those of a self-transmissible plasmid (see the text). A better name is probably integrating conjugative elements (ICEs) because the mechanism they use to excise from the donor cell DNA and to integrate into the recipient cell DNA after transfer is more analogous to the integration of phages than it is to true transposition (see chapter 9).

ICEs are very large because they must carry *tra* genes as well as integration functions. The first ICE to be discovered, Tn*916*, was found in *Enterococcus faecalis* and carries tetracycline resistance (see Senghas et al., below, and the figure). Later, other ICEs such as CTnDOT, which also carries tetracycline resistance, were found in *Bacteroides* species. These *Bacteroides* ICEs also mobilize other smaller elements in the chromosome of *Bacteroides*, called **nonreplicating Bacteroides units (NBUs),** much as self-transmissible plasmids mobilize other smaller plasmids. In fact, NBUs can be mobilized by IncQ plasmids.

To move from the DNA of one cell to the DNA of another cell, an ICE must first excise from the DNA of the cell in which it resides, be transferred into the other cell, and then integrate into the DNA of the second cell. This process has been studied extensively for Tn*916*. Like phage λ (see chapter 8), Tn*916* requires two proteins, Int and Xis, to excise. Excision of the element requires cutting of the DNA next to the ends of the element. The integrase first makes a staggered break in the donor host DNA near the ends of the transposon, to leave single-stranded ends 6 nucleotides long. The flanking sequences shown in the figure are arbitrary because they differ depending on where the transposon was inserted, although the element has a tendency to integrate next to a sequence similar to the sequence on one side of where it integrated previously. To the extent that they are random, these single-stranded ends (called the coupling sequences) are not complementary to each other. Nevertheless, these ends seem to pair to form a circular intermediate of the element, including the unpaired region formed by the ends of the element, which are thought to form a heteroduplex sequence called the coupling sequence. It has proven difficult to confirm various aspects of this model because of the possibility of directed mismatch repair in the coupling sequence before or after transfer.

The circular intermediate cannot replicate but can be transferred into another cell by a process that is much like plasmid transfer during conjugation. The element has its own *oriT* sequence and *tra* genes (see the figure). In fact, this *oriT* sequence is related to those of the self-transmissible plasmids RP4 and F. The *tra* genes are *orf23-orf13*, and the *oriT* sequence lies between *orf20* and *orf21*. Orf20 functions as the Tn*916* relaxase (see Rocco and Churchward, below). To initiate the transfer, a single-stranded break is made in the *oriT* sequence of the excised circular element and a single strand of the element is transferred to the recipient cell by the element-encoded Tra functions. Once in the recipient cell, the ends rejoin and a complementary strand is synthesized to make another double-stranded circular element. The Int protein of the element alone then integrates the element into the DNA of the recipient cell; this is similar to the way the Int protein but not Xis is required to integrate the λ phage genome into the chromosome (see chapter 8).

It would seem that a bacterial strain harboring an ICE in its chromosome would be an Hfr strain, capable of transferring chromosomal DNA into recipient cells. However, this does not happen, because known ICEs, including Tn*916*, can transfer only after they have excised, not while they are still integrated in the chromosome. Tn*916* uses a clever mechanism to ensure that it will be transferred only after it is excised. The *tra* genes are arranged so that they can be transcribed only after the element has excised and has formed a circle. It accomplishes this by positioning the promoter, called p_{orf7} (see the figure), such that the *tra* genes (*orf23-orf13*) are transcribed only when the element has excised and formed a circle. The Int protein may also bind to the *oriT* sequence while the ICE is inserted in the chromosome, blocking access to the *oriT* sequence by the Tra functions and precluding transfer. The advantage to the element of thus regulating its transfer seems obvious. If they could transfer themselves while still integrated in the chromosome, they would transfer the chromosome, much as self-transmissible plasmids transfer the chromosome in Hfr strains. However, then, as occurs during *Hfr* transfer, only the part of the element on one side of the *oriT* site would enter the recipient, and the element would essentially decapitate itself.

The process of transfer of the *Bacteroides* conjugative elements such as CTnDOT may be similar, but there are important differences (see Cheng et al., below). The CTnDoT element seems to be more restricted in its target site selection than Tn*916*, preferring sites with a 10-bp sequence that is similar to a sequence on the element. The transfer of the CTnDOT element is also induced by tetracycline in the medium, in much the same way that opines induce transfer of the Ti plasmid and unlike the transfer of Tn*916*, which is not induced by tetracycline. Interestingly, tetracycline also induces CTnDoT to mobilize the smaller NBU elements. This seems to be a clear case of cooperation between DNA elements and their hosts.

BOX 5.4 (continued)

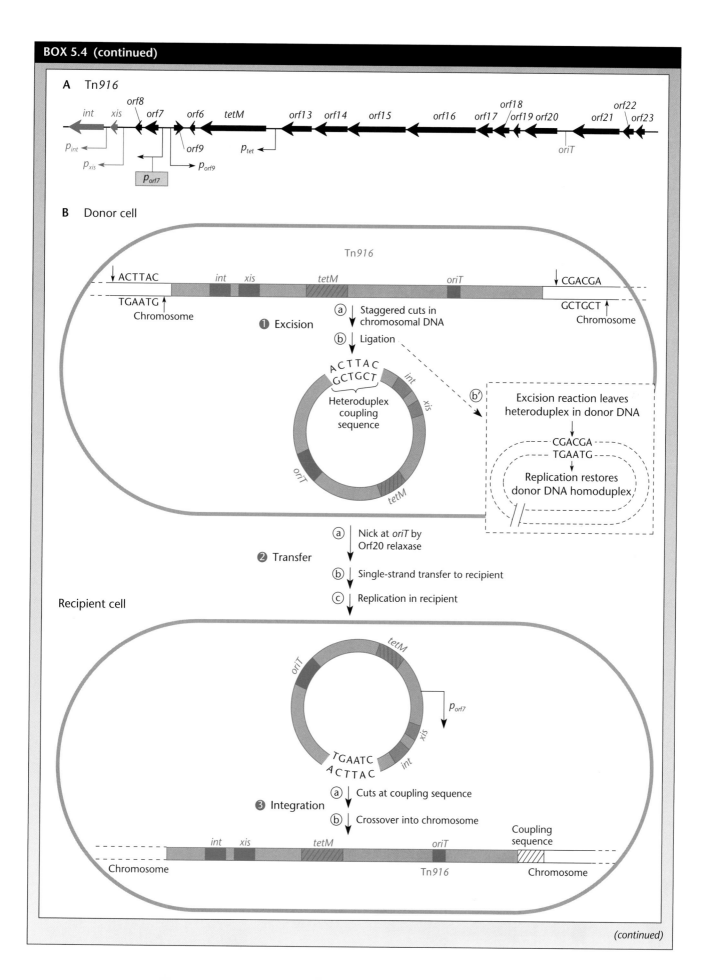

BOX 5.4 (continued)

Conjugative Transposons (Integrating Conjugative Elements)

References

Cheng, Q., B. J. Paszkiet, N. B. Shoemaker, J. F. Gardner, and A. A. Salyers. 2000. Integration and excisions of a *Bacteroides* conjugative transposon, CTnDOT. *J. Bacteriol.* **182:**4035–4043.

Hinerfeld, D., and G. Churchward. 2001. Specific binding of integrase to the origin of transfer (*oriT*) of the conjugative transposon Tn*916*. *J. Bacteriol.* **183:**2947–2951.

Marra, D., and J. R. Scott. 1999. Regulation of excision of the conjugative transposon Tn*916*. *Mol. Microbiol.* **31:**609–621.

Rocco, J. M., and G. Churchward. 2006. The integrase of the conjugative transposon Tn*916* directs strand- and sequence-specific cleavage of the origin of conjugal transfer, *oriT*, by the endonuclease Orf20. *J. Bacteriol.* **188:**2207–2213.

Senghas, E., J. M. Jones, M. Yamamoto, C. Gawron-Burke, and D. B. Clewell. 1988. Genetic organization of the bacterial conjugative transposon, Tn*916*. *J. Bacteriol.* **170:**245–249.

SUMMARY

1. Self-transmissible plasmids can transfer themselves to other bacterial cells, a process called conjugation. Some plasmids can transfer themselves into a wide variety of bacteria from different genera. Such plasmids are said to be promiscuous.

2. The plasmid genes whose products are involved in transfer are called the *tra* genes. The site on the plasmid DNA at which transfer initiates is called the origin of transfer (*oriT*). The *tra* genes can be divided into two groups: those whose products are involved in mating-pair formation (Mpf) and those whose products are involved in processing the plasmid DNA for transfer (Dtr).

3. The Mpf component includes a sex pilus that extrudes from the cell and holds mating cells together. The pilus is the site to which male-specific phages adsorb. The Mpf system also includes the channel in the membrane through which DNA and proteins pass, as well as a coupling protein that lies on the channel and docks the relaxase of the Dtr component and other proteins to be transferred through the channel.

4. The Dtr component includes the relaxase, which makes a cut at the *nic* site within the *oriT* sequence and rejoins the ends of the plasmid in the recipient cell. The relaxase also often contains a helicase activity, which separates the strands of DNA during transfer. The Dtr component also includes proteins that bind to the *oriT* sequence to form the multiprotein complex called the relaxosome and a primase which primes replication in the recipient cell and is sometimes transferred along with the DNA.

5. Most plasmids transiently express their Mpf *tra* genes immediately after transfer to a recipient cell and then only intermittently thereafter. Presumably, this helps them avoid infection by male-specific phages that use the sex pilus as their adsorption site.

6. Mobilizable plasmids cannot transfer themselves but can be transferred by other plasmids. Mobilizable plasmids encode only a Dtr component; they lack genes to encode an Mpf component. The *tra* genes of a mobilizable plasmid encoding its Dtr component are called the *mob* genes. A mobilizable plasmid can be mobilized by a self-transmissible plasmid only if the coupling protein of the self-transmissible plasmid can dock the relaxase of the mobilizable plasmid. Because they lack an Mpf component, mobilizable plasmids can be much smaller than self-transmissible plasmids, which makes them very useful in molecular genetics and biotechnology.

7. Hfr strains of bacteria have a self-transmissible plasmid integrated into their chromosome. Hfr strains are useful for genetic mapping in bacteria because they transfer chromosomal DNA in a gradient, beginning at the site of integration of the plasmid. Hfr crosses are particularly useful for locating genetic markers on the entire genome.

8. Prime factors are self-transmissible plasmids that have picked up part of the bacterial chromosome. They can be used to make partial diploids for complementation tests. If a prime factor is transferred into a cell, the cell will be a partial diploid for the region of the chromosome carried on the prime factor, making it useful for complementation experiments.

9. Self-transmissible plasmids also exist in gram-positive bacteria. However, these plasmids do not encode a sex pilus. Some gram-positive bacteria have the interesting property of excreting small pheromone-like compounds that stimulate mating with certain plasmids. The existence of such systems emphasizes the importance of plasmid exchange to bacteria.

10. Some integrating elements of gram-positive bacteria are also self-transmissible. These so-called conjugative transposons can excise and transfer themselves into other cells even though they cannot replicate and are therefore not replicons.

QUESTIONS FOR THOUGHT

1. Why do you suppose the Tra functions, including the *eex* genes, which exclude other plasmids of the same type, are usually the same for all the members of the same Inc group?

2. Why are the *tra* genes whose products are directly involved in DNA transfer usually adjacent to the *oriT* site?

3. Why do you suppose plasmids with a certain *mob* site are mobilized by only certain types of self-transmissible plasmids?

4. Why do self-transmissible plasmids usually encode their own primase function?

5. What do you think is different about the cell surfaces of gram-positive and gram-negative bacteria that causes only the self-transmissible plasmids of gram-negative bacteria to encode a pilus?

6. Why do so many types of phages use the sex pilus of plasmids as their adsorption site?

7. Why are so many plasmids either self-transmissible or mobilizable? Why are so many promiscuous?

8. What is the evidence that mobilizable plasmids are not just self-transmissible plasmids that have lost their Mpf?

PROBLEMS

1. After mixing two strains of a bacterial species, you observe some recombinant types that are unlike either parent. These recombinant types seem to be the result of conjugation, because they appear only if the cells are in contact with each other. How would you determine which is the donor strain and which is the recipient? Whether the transfer is due to an Hfr strain or to a prime factor?

2. How would you determine which of the *tra* genes of a self-transmissible plasmid encodes the pilin protein? The site-specific DNA endonuclease that cuts at *oriT*? The helicase?

3. How would you show that only one strand of the plasmid DNA enters the recipient cell during plasmid transfer?

4. You have discovered that tetracycline resistance can be transferred from one strain of a bacterial species to another. How would you determine whether the tetracycline resistance gene being transferred is on a self-transmissible plasmid or on a conjugative transposon?

5. Can a male-specific phage infect bacteria containing only a mobilizable plasmid? Why or why not?

6. Outline how you would determine which of the open reading frames in the *tra* region of a plasmid encodes the Eex (entry exclusion) protein.

7. Outline how you would use mobilizable plasmids to determine if a plasmid indigenous to a wild-type bacterium you have isolated is self-transmissible if the plasmid does not have any selectable gene you know of.

8. Why do you suppose a pheromone-responsive plasmid such as pAD1 encodes both an inhibitory peptide iAD1 as well as an inner membrane protein that interferes with pheromone processing?

SUGGESTED READING

Bagdasarian, M., R. Lurz, B. Rukert, F. C. H. Franklin, M. M. Bagdasarian, J. Frey, and K. N. Timmis. 1981. Specific-purpose plasmid cloning vectors. II. Broad host range, high copy number, RSF1010-derived vectors, and a host-vector system for gene cloning in *Pseudomonas. Gene* **16:**237–247.

Buttaro, B. A., M. H. Antiporta, and G. M. Dunny. 2000. Cell associated pheromone peptide (cCF10) production and pheromone inhibition in *Enterococcus faecalis. J. Bacteriol.* **182:**4926–4933.

Chandler, J. R., A. R. Flynn, E. M. Bryan, and G. M. Dunny. 2005. Specific control of endogenous cCF10 pheromone by a conserved domain of the pCF10-encoded regulatory protein PrgY in *Entercoccus faecalis. J. Bacteriol.* **187:**4830–4843.

Clewell, D. B., and M. V. Francia. 2004. Conjugation in gram-positive bacteria, p. 227–256. *In* B. E. Funnell and G. J. Phillips (ed.), *Plasmid Biology.* ASM Press, Washington, D.C.

De Boever, E. H., D. B. Clewell, and C. M. Fraser. 2000. *Enterococcus faecalis* conjugative plasmid pAM373: complete nucleotide sequence and genetic analyses of sex pheromone response. *Mol. Microbiol.* **37:**1327–1341.

Derbyshire, K. M., G. Hatfull, and N. Willets. 1987. Mobilization of the nonconjugative plasmid RSF1010: a genetic and DNA sequence analysis of the mobilization region. *Mol. Gen. Genet.* **206:**161–168.

Eisenbrandt, R., M. Kalkum, R. Lurz, and E. Lanka. 2000. Maturation of IncP pilin precursors resembles the catalytic dyad-like mechanism of leader peptidases. *J. Bacteriol.* **182:**6751–6761.

Firth, N., K. Ippen-Ihler, and R. A. Skurray. 1996. Structure and function of the F factor and mechanism of conjugation, p. 2377–2401. *In* F. C. Neidhardt, R. Curtiss III, J. L. Ingraham, E. C. C. Lin, K. B. Low, B. Magasanik, W. S. Reznikoff,

M. Riley, M. Schaechter, and H. E. Umbarger (ed.), Escherichia coli *and* Salmonella: *Cellular and Molecular Biology*, 2nd ed. ASM Press, Washington, D.C.

Gilmour, M. W., T. D. Lawley, and D. E. Taylor. 15 November 2004, posting date. The cytology of bacterial conjugation. *In* A. Böck et al. (ed.), *EcoSal*—Escherichia coli *and* Salmonella: *Cellular and Molecular Biology.* www.ecosal.org. ASM Press, Washington, D.C.

Grahn, A. M., J. Haase, D. H. Bamford, and E. Lanka. 2000. Components of the RP4 conjugative transfer apparatus form an envelope structure bridging inner and outer membranes of donor cells: implications for related macromolecule transport systems. *J. Bacteriol.* **182:**1564–1574.

Kurenbach, B., D. Grothe, M. E. Farias, U. Szewzyk, and E. Grohmann. 2002. The *tra* region of the conjugative plasmid pIP501 is organized in an operon with the first gene encoding the relaxase. *J. Bacteriol.* **184:**1801–1805.

Lawley, T., B. M. Wilkins, and L. S. Frost. 2004. Bacterial conjugation in gram-negative bacteria, p. 203–226. *In* B. E. Funnell and G. J. Phillips (ed.), *Plasmid Biology.* ASM Press, Washington, D.C.

Lederberg, J., and E. L. Tatum. 1946. Gene recombination in *E. coli. Nature* (London) **158:**58.

Low, K. B. 1968. Formation of merodiploids in matings with a class of Rec⁻ recipient strains of *E. coli* K12. *Proc. Natl. Acad. Sci. USA* **60:**160–167.

Manchak, J., K. G. Anthony, and L. S. Frost. 2002. Mutational analysis of F-pilin reveals domains for pilus assembly, phage infection, and DNA transfer. *Mol. Microbiol.* **43:**195–205.

Marra, D., B. Pethel, G. G. Churchward, and J. R. Scott. 1999. The frequency of conjugative transposition of Tn*916* is not determined by the frequency of excision. *J. Bacteriol.* **181:**5414–5418.

Matson, S. W., J. K. Sampson, and D. R. N. Byrd. 2001. F plasmid conjugative DNA transfer: the TraI helicase activity is essential for DNA strand transfer. *J. Biol. Chem.* **276:**2372–2379.

Samuels, A. L., E. Lanka, and J. E. Davies. 2000. Conjugative junctions in RP4-mediated mating of *Escherichia coli. J. Bacteriol.* **182:**2709–2715.

Schmidt, J. W., L. Rajeev, A. A. Salyers, and J. F. Gardner. 2006. NBU1 integrase: evidence for an altered recombination mechanism. *Mol. Microbiol.* **60:**152–164.

Schroder, G., and E. Lanka. 2005. The mating pair formation system of conjugative plasmids—a versatile secretion machinery for transfer of proteins and DNA. *Plasmid* **54:**1–25.

Watanabe, T., and T. Fukasawa. 1961. Episome-mediated transfer of drug resistance in *Enterobacteriaceae.* 1. Transfer of resistance factors by conjugation. *J. Bacteriol.* **81:**669–678.

Wilkins, B. M., and A. T. Thomas. 2000. DNA transfer independent transport of plasmid primase protein between bacteria by the I1 conjugation system. *Mol. Microbiol.* **38:**650–657.

Winans, S. C., and G. C. Walker. 1985. Conjugal transfer system of the IncN plasmid pKM101. *J. Bacteriol.* **161:**402–410.

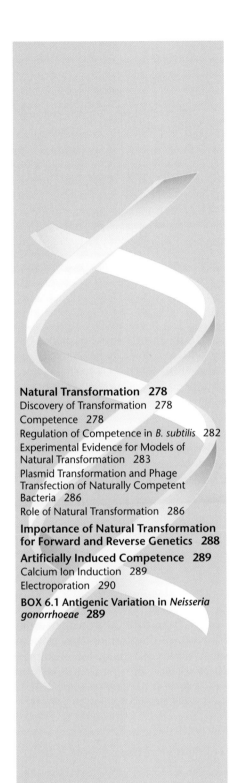

CHAPTER 6

Transformation

DNA can be exchanged among bacteria in three ways: conjugation, transduction, and transformation. Chapter 5 covers the mechanism of conjugation, in which a plasmid or other self-transmissible DNA element transfers itself and sometimes other DNA into another bacterial cell. In transduction, whose mechanism is discussed in detail in chapter 7, a phage carries DNA from one bacterium to another. In this chapter, we discuss transformation, in which cells take up free DNA directly from their environment.

Transformation is one of the cornerstones of molecular genetics because it is often the best way to reintroduce experimentally altered DNA into cells. Transformation was first discovered in bacteria, but ways have been devised to transform many types of animal and plant cells as well.

The terminology of genetic analysis by transformation is similar to that of conjugation and transduction (see chapter 3). DNA is derived from a **donor bacterium** and taken up by a **recipient bacterium,** which is then called a **transformant.** If the incoming DNA recombines with an indigenous DNA in the cell, such as the chromosome, recombinant types can form. The frequency of recombinant types for various genetic markers can be used for genetic analysis. Such genetic data obtained by transformation is analyzed similarly to that obtained by transduction. If the regions of two markers can be carried on the same piece of transforming DNA, the two markers are said to be **cotransformable.** The higher the **cotransformation frequency,** the more closely linked are the two markers on the DNA. The principles of using transduction and transformation for genetic mapping are outlined in chapter 3. In this chapter we concentrate on the mechanism of transformation in various bacteria and its relationship to other biological phenomena.

Natural Transformation

Most types of cells cannot take up DNA efficiently unless they have been exposed to special chemical or electrical treatments to make them more permeable. However, some types of bacteria are **naturally transformable,** which means they can take up DNA from their environment without requiring special treatment. Even naturally transformable bacteria are not always capable of taking up DNA but do so only at certain stages in their life cycle. Bacteria at the stage in which they can take up DNA are said to be **competent,** and bacteria that are naturally capable of reaching this state are said to be **naturally competent.** At latest count, about 40 species have been found to be naturally competent and transformable. Naturally competent transformable bacteria are found in several genera, including both gram-positive bacteria such as *Bacillus subtilis*, a soil bacterium, and *Streptococcus pneumoniae*, which causes throat infections; and gram-negative bacteria such as *Haemophilus influenzae*, a causative agent of spinal meningitis, *Neisseria gonorrhoeae*, which causes gonorrhea, and *Helicobacter pylori*, a stomach pathogen. *Acinetobacter baylyi*, a soil bacterium, is very highly transformable, as are some species of marine cyanobacteria.

Discovery of Transformation

Transformation was the first mechanism of bacterial gene exchange to be discovered. In 1928, Fred Griffith found that one form of the pathogenic pneumococci (now called *Streptococcus pneumoniae*) could be mysteriously "transformed" into another form. Griffith's experiments were based on the fact that *S. pneumoniae* makes two types of different-appearing colonies, one type made by pathogenic bacteria and the other type made by bacteria that are incapable of causing infections (i.e., are nonpathogenic). The colonies made by the pathogenic strains appear smooth on agar plates, because the bacteria excrete a polysaccharide capsule. The capsule apparently protects them and allows them to survive in vertebrate hosts, including mice, that they can infect and kill. However, rough-colony-forming mutants that cannot make the capsule sometimes arise. These are nonpathogenic in mice and sometimes arise from the smooth-colony formers.

In his experiment, Griffith mixed dead *S. pneumoniae* cells that made smooth colonies with live nonpathogenic cells that made only rough colonies and injected the mixture into mice (Figure 6.1). Mice given injections of only the rough-colony-forming bacteria survived, but mice that received a mixture of dead smooth-colony formers and live rough-colony formers died. Furthermore, Griffith isolated live smooth-colony-forming bacteria

	Bacterial type	Effect in mouse	Bacteria recovered
A	Live type R	Nonpathogenic	None
B	Live type S	Pathogenic	Live type S
C	Heat-killed type S	Nonpathogenic	None
D	Mixture of live type R and heat-killed type S	Pathogenic	Live type S

Figure 6.1 The Griffith experiment. Heat-killed pathogenic encapsulated bacteria can convert nonpathogenic noncapsulated bacteria to the pathogenic capsulated form. Type R indicates rough-colony formers; type S indicates smooth-colony formers.

from the blood of the dead mice. Concluding that the dead pathogenic bacteria gave off a "transforming principle" that changed the live nonpathogenic rough-colony-forming bacteria into the pathogenic smooth-colony form, he speculated that this transforming principle was the polysaccharide itself. Later, other researchers did an experiment in which they transformed rough-colony formers into smooth-colony formers by mixing the rough forms with extracts of the smooth forms in a test tube. Then, about 16 years after Griffith did his experiments with mice, Oswald Avery and his collaborators purified the "transforming principle" from the extracts of the smooth-colony formers and showed that it is DNA (see Avery et al., Suggested Reading). Thus, Avery and colleagues were the first to demonstrate that DNA, and not protein or other factors in the cell, is the hereditary material (see the introductory chapter).

Competence

As mentioned above, the term "competence" refers to the state that some bacteria can enter, in which they can take up naked DNA from their environment. This capability is genetically programmed, and the process of DNA uptake is often called "natural transformation," to distinguish it from transformation induced by electroporation, heat shock, Ca^{2+} treatment of cells, or protoplast uptake of DNA. The genetic programming of competence is widespread but not universal. Generally, more than a dozen genes are involved, encoding both regulatory and structural components of the transformation process.

The general steps that occur in natural transformation differ somewhat depending on whether the bacteria are gram negative or gram positive. If they are gram negative, the steps are (i) binding of double-stranded DNA to the outer cell surface of the bacterium, (ii) movement of the DNA across the cell wall and outer membrane, (iii) degradation of one of the DNA strands, and (iv) translocation of the remaining single strand of DNA into the cytoplasm of the cell across the inner membrane. Once in the cell, the single-stranded transforming DNA might synthesize the complementary strand and reestablish itself as a plasmid, stably integrate into the chromosome by homologous recombination of the translocated single strand into the chromosome or other recipient DNA, or be degraded. In a gram-positive organism that lacks an outer membrane, the process is similar except that movement through the outer membrane can be dispensed with, and it is necessary only to transport the DNA through the cell wall and one membrane. The uptake of DNA in both gram-positive and gram-negative bacteria is discussed in more detail below. While the DNA uptake systems of gram-positive and gram-negative bacteria have features in common, they do seem to differ in certain important respects; therefore, they are discussed separately.

COMPETENCE IN GRAM-POSITIVE BACTERIA

Two gram-positive species that have been particularly well studied are *B. subtilis* and *S. pneumoniae*. The proteins involved in transformation in these bacteria were discovered on the basis of isolation of mutants that are completely lacking in the ability to take up DNA. The genes affected in the mutants were named *com* (for *com*petence defective). In *B. subtilis*, the *com* genes are organized into several operons. The products of several of these, including the *comA* and *comK* operons, are involved in regulation of competence (see below). Others, including the products of genes in the *comE*, *comF*, and *comG* operons, become part of the competence machinery in the membrane that takes DNA up into the bacterium. The genes in these operons are given two letters, the first for the operon and the second for the position of the gene in the operon. For example, *comFA* is the first gene of the *comF* operon, while *comEC* is the third gene of the *comE* operon. The corresponding protein products of the genes have the same name with the first letter capitalized, e.g., ComFA and ComEC, respectively.

The role played by some of the Com proteins in the competence machinery of *B. subtilis* is diagrammed in Figure 6.2A. The first gene of the *comE* operon, *comEA*, encodes the protein that directly binds extracellular double-stranded DNA. The *comF* genes encode proteins that translocate the DNA into the cell. For example, ComFA is an ATPase that may provide the energy for translocation of DNA through the membrane (not

shown). It has been proposed that ComEA, ComEC, and ComFA form a sort of ATP-binding cassette (ABC) transporter, which transports DNA into the cell. A large number of different ABC transporters are known that transport molecules into and out of cells. The genes in the *comG* operon encode proteins that might form a "pseudopilus," which helps move DNA through the ComEC channel. They might bind to extracellular DNA, perhaps acting through the ComEA DNA-binding protein, and then retract, drawing the DNA into the cell. Such speculation is inspired by their similarity to type IV pilin proteins in other systems (see below).

The *comE*, *comF*, and *comG* operons are all under the transcriptional control of ComK, a transcription factor that is itself regulated by ComA, as discussed below.

Some of the genes involved in the transformation process are not designated *com*, because such genes were first discovered on the basis of their involvement in another process. For example, one of the nuclease activities which makes double-strand breaks in extracellular DNA is the *nucA* gene product (see Provedi et al., Suggested Reading). These free DNA ends become the substrates for the competence proteins. Other examples of proteins with multiple roles include single-stranded-DNA-binding protein (SSB), and RecA, which functions in the recombination of transforming DNA with the chromosome as well as generally in recombination (see chapter 10).

The lengths of single-stranded DNA incorporated into the recipient chromosome are about 8.5 to 12 kb, as shown by cotransformation of genetic markers; the incorporation takes only a few minutes to be completed.

Transformation in *S. pneumoniae* utilizes similar proteins and mechanisms to those of transformation in *B. subtilis*, although the names of the *com* genes are often different (see Berge et al., Suggested Reading).

COMPETENCE IN GRAM-NEGATIVE BACTERIA

As mentioned above, a variety of gram-negative bacteria are also capable of acquiring natural competence. Some examples are the bacterium *Acinetobacter calcoaceticus* as well as the pathogens *H. pylori, Neisseria* spp., and *Haemophilus* spp. In the last two, specific uptake sequences are required for the binding of DNA, so that these species usually take up DNA only of the same species (see below). This differs from the gram-positive bacteria and also from many other gram-negative bacteria, which do not have specific uptake sequences (discussed below).

Transformation Systems Based on Type II Secretion

Gram-negative bacteria utilize one of two fundamentally different types of DNA uptake systems. Most of them

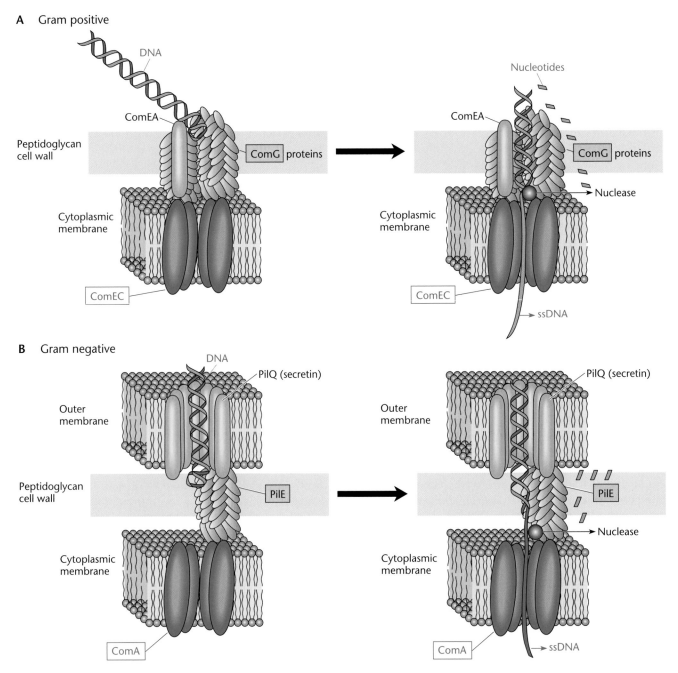

A Gram positive

DNA

ComEA

Peptidoglycan cell wall

ComG proteins

Cytoplasmic membrane

ComEC

Nucleotides

ComEA

ComG proteins

Nuclease

Cytoplasmic membrane

ComEC

ssDNA

B Gram negative

DNA

PilQ (secretin)

Outer membrane

Peptidoglycan cell wall

PilE

Cytoplasmic membrane

ComA

PilQ (secretin)

Outer membrane

PilE

Nuclease

Cytoplasmic membrane

ComA

ssDNA

Figure 6.2 Structure of DNA uptake competence systems in gram-positive (A) and gram-negative (B) bacteria. Shown are the some of the proteins involved and the channels they form. The nomenclature in panel A is based on *Bacillus subtilis*, and that in panel B is based on *Neisseria gonorrhoeae*. Some of the *B. subtilis* ComG proteins are analogous to the *Neisseria* PilE protein (shaded boxes). The *B. subtilis* ComEC protein is an ortholog of the *Neisseria* ComA protein (unshaded boxes). ss, single stranded. The DNA is shown running through the cell wall alongside the pseudopilus (ComG proteins in *B. subtilis*; PilE in *Neisseria*), although the actual mechanism is unknown. Both gram-positive and gram-negative organisms' competence systems are related to type II protein secretion systems, which are discussed in chapter 14.

use a system related to type II secretion systems, which are used to assemble type IV pili on the cell surface and are similar to the secretion system used in the gram-positive bacteria discussed above (Figure 6.2B). The major difference between the gram-negative bacterial competence systems and the gram-positive systems is necessitated by the presence of an outer membrane in the gram-negative bacteria. In gram-negative bacteria, the water-soluble (hydrophilic) DNA must first pass through this hydrophobic outer membrane before it can pass through the cytoplasmic membrane into the cytoplasm of the cell. To facilitate DNA transfer through the outer membrane, the competence systems of gram-negative bacteria also have a pore through the outer membrane, made up of 12 to 14 copies of a secretin protein (called PilQ in *Neisseria*). This pore has a hydrophilic aqueous channel through which the double-stranded DNA can pass. One strand of the DNA is then degraded as it passes through a channel in the inner membrane; this channel is formed by a protein called ComA in *Neisseria*, which has sequences in common with the ComEC protein that forms a similar channel in *B. subtilis* (Figure 6.2B).

While these competence systems are very similar to type II secretion systems that assemble type IV pili, to the extent of often elaborating type IV pili on the cell surface, the role of pili in this process is somewhat obscure. Type IV pili are long, thin hairlike appendages that stick out from the cell and are used to attach cells to solid surfaces such as the surfaces of eukaryotic cells; they are often involved in pathogencity. They are also required for a type of cell movement called "twitching" motility in some types of bacteria such as *Myxococcus xanthus*. In these bacteria, the pili are located on the leading edge of the cell and move the cell by reaching out and attaching to the solid surface and then shortening or retracting, pulling the cell toward themselves, much like a mountain climber would climb a cliff by inserting a piton ahead of herself and pulling herself toward it. Speculation is that type IV pili systems involved in DNA transformation might work by a similar mechanism, binding to DNA on the cell surface and then retracting, pulling the DNA into the cell. However, the actual mechanism does not seem to be quite this simple. While it is true that bacteria exhibiting type IV pili on their cell surface seem to need this system for transformation, and most bacteria that are competent for transformation do seem to require systems related to type II secretion systems, not all bacteria capable of natural transformation actually exhibit type IV pili on their cell surface. Furthermore, while competence requires the protein that makes up most of the pilus, i.e., the major pilin protein (called PilE in *Neisseria* [Figure 6.2B]), some other minor pilin proteins are required for competence but not for pilus formation.

This suggests that when the pilin protein is associated with these minor pilus proteins, it is in a different state from when it is in a type IV pilus. One way out of this dilemma is to propose that competence systems and type II secretion systems elaborate a modified pilus, sometimes called a pseudopilus, to play these other roles. The pseudopilus is composed of the major pilin protein, but now it is associated with different minor pilus proteins so that it does not actually extrude from the cell but, rather, remains closely associated with the cell surface. In type II secretion systems, this pseudopilus may grow and push proteins out of the cell through the secretin channel in the outer membrane. In the case of transformation, the pseudopilus may do the reverse; it may bind DNA and retract, pulling the DNA into the cell through the secretin channel in the outer membrane. There is no evidence that any of the pilus proteins actually bind DNA, although they could of course work through other DNA-binding proteins, analogous to ComEA in *B. subtilis* (see above).

Competence Systems Based on Type IV Secretion Systems

As mentioned above, most competent bacteria have competence systems based on type II secretion systems. To date, the only known exception is *H. pylori*, which has a system based on type IV secretion-conjugation systems, discussed in chapter 5. *H. pylori* is an opportunistic pathogen involved in gastrointestinal diseases. In fact, two Australian scientists were awarded the Nobel Prize in physiology or medicine in 2005 for showing that it is the major cause of gastric ulcers, which were long thought to be due to stress. The similarity between the competence system of *H. pylori* and type IV secretion-conjugation was discovered because of the similarity between the proteins in this system and the VirB conjugation proteins in *Agrobacterium tumefaciens* that transfer T-DNA from the Ti plasmid into plants (see Boxes 5.1 and 5.2). The Com proteins of *H. pylori*, a human pathogen, were therefore given letters and numbers corresponding to their orthologs (see Box 2.7 for definitions) in the T-DNA transfer system of the Ti plasmid in *A. tumefaciens*, a plant pathogen. Table 6.1 lists these Com proteins and their orthologs in the *Agrobacterium* Ti plasmid (see Karnholz et al., Suggested Reading). Apparently, type IV secretion pathway systems can function as two-way DNA transfer systems, capable of moving DNA both into and out of the cell. Interestingly, in addition to its transformation system, *H. pylori* has a bona fide type IV secretion system that secretes proteins directly into eukaryotic cells (see Box 5.2). However, even though these two systems are related, they function independently of each other and have no proteins in common.

TABLE 6.1	Orthologous Com proteins of *H. pylori* and Vir proteins in *A. tumefaciens*	
Helicobacter protein	**Function**	*Agrobacterium* ortholog
ComB2	Pseudopilus?	VirB2
ComB3	Unknown	VirB3
ComB4	ATPase	VirB4
ComB6	DNA binding?	VirB6
ComB7	Channel	VirB7
ComB8	Channel	VirB8
ComB9	Channel	VirB9
ComB10	Channel	VirB10

The nomenclature of competence is admittedly very confusing. To reiterate, type IV pili and type IV secretion systems are not related to each other. Type IV secretion systems actually have type II pili, and type IV pili are assembled on the cell surface by systems related to type II secretion systems! Some conjugation systems have type IV pili in addition to their type II pili. The type IV pili help to hold the cells together during DNA transfer. There is also evidence that a type IV secretion system in *N. gonorrhoeae* is involved in releasing DNA into the environment, where it can be taken up by a transformation system related to type II secretion systems (see above). When the secretion systems, transformation systems, and pili were being named, no one could have predicted their relationships to each other; this confusion is the result.

Regulation of Competence in *B. subtilis*

The regulation of competence in *B. subtilis* is achieved through a **two-component regulatory system** analogous to those used to regulate many other systems in bacteria (see chapter 13). First, information that the cell is running out of nutrients and the population is reaching a high density is registered by ComP, a **sensor protein** in the membrane (Figure 6.3). The high cell density causes this sensor-kinase protein to transfer a phosphate from ATP to itself, in other words to phosphorylate itself. The phosphate is then transferred from ComP to ComA, a **response regulator protein**. In the phosphorylated state, the ComA protein is a transcriptional activator (see chapter 2) for several genes, including some required for competence. Eventually another transcriptional activator, ComK, is made; this activator is directly responsible for activating the transcription of other *com* genes, including those that form the transformation machinery illustrated in Figure 6.2A.

COMPETENCE PHEROMONES

How does the cell know that other *B. subtilis* cells are nearby and that it should induce competence? High cell

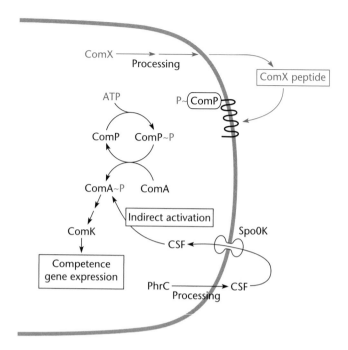

Figure 6.3 Regulation of competence development in *Bacillus subtilis* by quorum sensing. The ComP protein in the membrane senses a high concentration of the ComX peptide (shown in purple) and phosphorylates itself by transferring a phosphate from ATP. The phosphate is then transferred to ComA, which allows the transcription of many genes including ComK, the activator of the *com* genes. In a separate pathway, a peptide sometimes called CSF (competence-stimulating factor), processed from the signal sequence of another protein (PhrC), is imported into the cell by the SpoOK oligopeptide permease and indirectly activates ComA~P by inactivating another protein, RapC (not shown). There is at least one other such pathway; these pathways' purpose may be to coordinate competence with sporulation and other functions when the cell runs out of nutrients and enters stationary phase.

density is signaled through small peptides called **competence pheromones** that are excreted by the bacteria as they multiply (see Lazazzera et al., Suggested Reading). Cells become competent only in the presence of high concentrations of these peptides, and the concentration of these peptides in the medium is high only when the concentration of cells giving them off is high. The requirement for competence pheromones ensures that cells are able to take up DNA only when other *B. subtilis* cells are nearby and giving off DNA to be taken up. This is one example of a phenomenon called quorum sensing, by which small molecules given off by cells send signals to other cells in the population that the cell concentration is high. Many such small molecules are known, including homoserine lactones that signal cell density in some gram-negative bacteria. Other examples of quorum sensing are discussed in later chapters.

A second question is that of how the cell knows that the competence pheromone peptide came from other cells and was not produced internally. It does this by cutting the peptide out of a larger protein as the larger protein passes through the membrane, as shown in Figure 6.3. Once outside the cell, the peptide is diluted by the surrounding medium, and it can achieve high enough concentrations to induce competence only if the cell density is high and many surrounding cells are also giving off the peptide. In *B. subtilis*, the major competence pheromone peptide is called ComX and is cut out of a longer polypeptide, the product of the *comX* gene. Another gene, *comQ*, which is immediately upstream of *comX*, is also required for synthesis of the competence pheromone because its product is the protease enzyme that cuts the competence pheromone out of the longer polypeptide. Once the peptide has been cut out of the longer molecule, it can bind to the ComP protein in the membrane and trigger its autophosphorylation, although the mechanism remains unknown.

At best, only about 10% of *B. subtilis* cells ever become competent, no matter how favorable the conditions or how high the cell density. The advantages of this to the bacteria are obvious: if all the cells were competent, which cells would be giving off DNA to be taken up by the competent cells? This has been called a bistable state and seems to be determined somehow by autoregulation of the ComK activator protein. Bistable states are common in biological phenomena, and competence in *B. subtilis* has been used as an experimental model for such phenomena (see Maamer and Dubnau, Suggested Reading).

RELATIONSHIP BETWEEN COMPETENCE, SPORULATION, AND OTHER CELLULAR STATES

At about the same time as *B. subtilis* reaches the stationary phase, some cells acquire competence and some cells sporulate (see chapter 14). Sporulation, a developmental process common to many bacteria, allows a bacterium to enter a dormant state and survive adverse conditions such as starvation, irradiation, and heat. During sporulation, the bacterial chromosome is packaged into a resistant spore, where it remains viable until conditions improve and the spore can germinate into an actively growing bacterium. To coordinate sporulation and competence, *B. subtilis* cells may produce other competence peptides (see Bongiorni et al., Suggested Reading). There are at least two such peptides that regulate ComA indirectly by inhibiting proteins called Rap proteins, which bind to the C-terminal DNA-binding domain of phosphorylated ComA and prevent it from binding to DNA and activating transcription. These peptides are processed from the signal sequences of longer polypeptides, the

products of the *phr* genes, and are transported into the cell by the oligopeptide permease, Spo0K (see below and Figure 6.3).

The *spo0K* gene is an example of a regulatory gene that is required for sporulation and also for the development of competence. This gene was first discovered because of its role in sporulation. A *spo0K* mutant is blocked in the first stage, the "0" stage, of sporulation. The *K* means that it was the 11th gene (as K is the 11th letter in the alphabet) involved in sporulation to be discovered in that collection.

Experimental Evidence for Models of Natural Transformation

The above models for natural transformation are based on experiments with a number of different systems. Experiments directed toward an understanding of DNA uptake during natural transformation have sought to answer three obvious questions. (i) How efficient is DNA uptake? (ii) Can only DNA of the same species enter a given cell? (iii) Are both of the complementary DNA (cDNA) strands taken up and incorporated into the cellular DNA?

EFFICIENCY OF DNA UPTAKE

The efficiency of uptake is fairly easy to measure biochemically. Figure 6.4 shows an experiment based on the fact that transport of free DNA into the cell makes the DNA insensitive to DNases, which cannot enter the cell because competent cells permit only DNA, not proteins,

Figure 6.4 Determining the efficiency of DNA uptake during transformation. DNA in the cell is insensitive to DNase. Degraded DNA passes through a filter. The asterisk refers to radioactively labeled DNA.

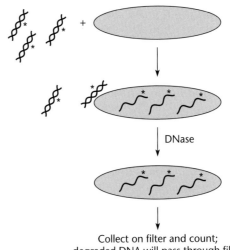

Collect on filter and count;
degraded DNA will pass through filter

to enter. In the experiment, donor DNA is radioactively labeled by growing the cells in medium in which the phosphorus has been substituted with phosphorus-32, the radioisotope of phosphorus. The radioactive DNA is then extracted and mixed with competent cells, and the mixture is treated with DNase at various times. Any DNA that is not degraded and survives intact must have been taken up by the cells, where it is protected from the DNase. Undegraded DNA can be distinguished from degraded DNA because the former DNA can be precipitated with acid and collected on a filter whereas degraded DNA does not precipitate and passes through the filter. Therefore, if the medium containing the cells is precipitated with acid and collected on a filter, the radioactivity on the filter is due to undegraded DNA that must have been taken up by the cells. If the radioactivity on the filter is counted and compared with the total radioactivity of the DNA that was added to the cells, the percentage of DNA that is taken up, or the efficiency of DNA uptake, can be calculated. Experiments such as these have shown that some competent bacteria take up DNA very efficiently.

SPECIFICITY OF DNA UPTAKE

The second question, i.e., whether DNA from only the same species is taken up, is also fairly easy to answer. By using the same assay of resistance to DNases, it has been determined that some types of bacteria take up DNA from only their own species whereas others can take up DNA from any source. The first group includes *N. gonorrhoeae* and *Haemophilus influenzae*.

Bacteria that preferentially take up the DNA of their own species do so because their DNA contains specific **uptake sequences**. Figure 6.5 shows the minimal uptake sequences for *H. influenzae* and *N. gonorrhoeae*. Uptake sequences are long enough that they almost never occur by chance in other DNAs. In contrast, bacteria such as *B. subtilis* seem to take up any DNA. Possible reasons why some bacteria should preferentially take up DNA from their own species while others take up any DNA are subjects of speculation and are discussed later.

FATE OF DNA TAKEN UP IN *S. PNEUMONIAE*

Although the genetic requirements for transformation are best known for *B. subtilis*, the more efficient uptake

of DNA by some other naturally transformable bacteria has allowed investigators to perform biochemical experiments on the uptake of DNA by these species, which has led to the models described above. The general pathway for DNA uptake was first worked out for *S. pneumoniae* but is probably similar in most other naturally transformable bacteria.

Figure 6.6 shows a general scheme for DNA uptake during the transformation of *S. pneumoniae*. In the first step, double-stranded DNA released by lysis of the donor bacteria is bound to specific receptors on the cell surface of the recipient bacterium. The bound DNA is then broken into smaller pieces by endonucleases; one of the two complementary strands is degraded by an exonuclease; and the remaining strand is transported into the

Figure 6.6 Transformation in *Streptococcus pneumoniae*. Competence-stimulating peptide accumulates as the cells reach a high density. Double-stranded DNA binds to the cell, and one strand is degraded. The remaining single strand protected by a DNA-binding protein replaces the strand of the same sequence in the chromosome, creating a "heteroduplex" in which one strand comes from the donor and one comes from the recipient.

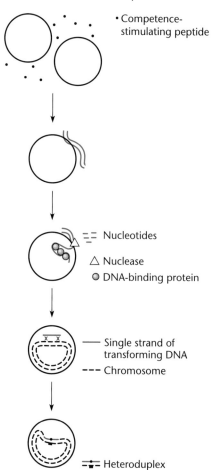

Figure 6.5 The uptake sequences on DNA for some types of bacteria. Only DNA with these sequences is taken up by the bacteria indicated. Only one strand of the DNA is shown.

Haemophilus influenzae	5′ AAGTGCGGT 3′
Neisseria gonorrhoeae	5′ GCCGTCTGAA 3′

cell, being protected by a DNA-binding protein. The transforming DNA integrates into the cellular DNA in a homologous region by means of strand displacement, a mechanism in which the new strand invades the double helix and displaces an old strand with the same sequence. The old strand is then degraded. If the donor DNA and recipient DNA sequences differ slightly in this region, recombinant types can appear. Evidence for this model comes from several different experiments, some of which are discussed below. Also, the gene for a membrane-bound DNase that may be involved in degrading one of the two strands of the incoming DNA has been found in *S. pneumoniae*.

TRANSFORMASOMES

The basic scheme described above probably differs among different types of naturally competent bacteria. For example, *H. influenzae* may first take up double-stranded DNA in subcellular compartments called **transformasomes** (Figure 6.7). The new DNA may not become single stranded until it enters the cytoplasm. However, the basic process of all natural transformation is the same. Only one strand of the DNA enters the

Figure 6.7   Transformation in *Haemophilus influenzae*. Double-stranded DNA is first taken up in transformasomes. One strand is degraded, and the other strand invades the chromosome, displacing one chromosome strand.

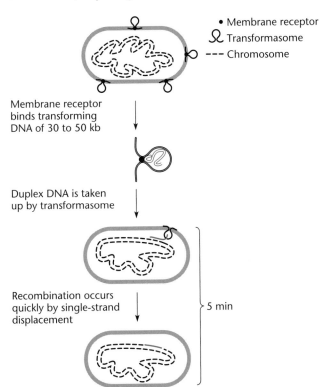

Membrane receptor binds transforming DNA of 30 to 50 kb

Duplex DNA is taken up by transformasome

Recombination occurs quickly by single-strand displacement

5 min

interior of the cell and integrates with the cellular DNA to produce recombinant types.

GENETIC EVIDENCE FOR SINGLE-STRANDED DNA UPTAKE

Genetic experiments taking advantage of the molecular requirements for transformation can be used to study the molecular basis for transformation; in other words, transformation can be used to study itself. Evidence that DNA has transformed cells is usually based on the appearance of recombinant types after transformation. A recombinant type can form only if the donor and recipient bacteria differ in their genotypes and if the incoming DNA from the donor bacterium changes the genetic composition of the recipient bacterium. The chromosome of a recombinant type has the DNA sequence of the donor bacterium in the region of the transforming DNA.

Experiments have shown that only double-stranded DNA can bind to the specific receptors on the cell surface, so that double-stranded but not single-stranded DNA can transform cells and yield recombinant types. However, we can also conclude from these experiments that the cells actually take up only *single*-stranded DNA, because the DNA enters an "eclipse" phase in which it cannot transform. For example, in the experiment shown in Figure 6.8, an Arg⁻ mutant requiring arginine for growth is used as the recipient strain and the corresponding Arg⁺ prototroph is the source of donor DNA. At various times after the donor DNA has been mixed with the recipient cells, the recipients are treated with DNase, which cannot enter cells but destroys any DNA remaining in the medium. The surviving DNA in the recipient cells is then extracted and used for retransformation of more auxotrophic recipients, and Arg⁺ transformants are selected on agar plates without the growth supplement arginine. Any Arg⁺ transformants must have been due to double-stranded donor DNA in the recipient cells.

Whether transformants were observed depends on the time the DNA was extracted from the cells. When the DNA is extracted at time 1 in Figure 6.8, while it is still outside the cells and accessible to the DNase, no Arg⁺ transformants are observed because the Arg⁺ donor DNA is all destroyed by the DNase. At time 2, some of the DNA is now inside the cells, where it cannot be degraded by the DNase, but this DNA is single stranded. It has not yet recombined with the chromosome, and so Arg⁺ transformants are still not observed in step 4. Only at time 3, when some of the DNA has recombined with the chromosomal DNA and so is again double stranded, do Arg⁺ transformants appear in step 4. Thus, the transforming DNA enters the eclipse period for a short time after it is added to competent cells, as expected if it enters the cell in a single-stranded state.

• Membrane receptor

Ω Transformasome

--- Chromosome

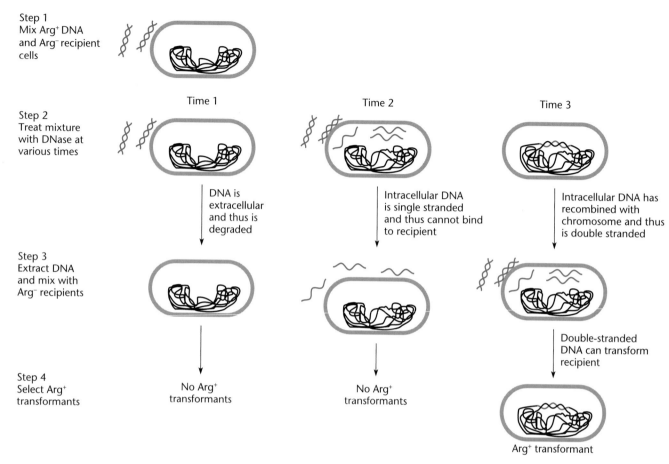

Step 1
Mix Arg⁺ DNA
and Arg⁻ recipient
cells

Step 2
Treat mixture
with DNase at
various times

Time 1

Time 2

Time 3

DNA is
extracellular
and thus is
degraded

Intracellular DNA
is single stranded
and thus cannot bind
to recipient

Intracellular DNA has
recombined with
chromosome and thus
is double stranded

Step 3
Extract DNA
and mix with
Arg⁻ recipients

Double-stranded
DNA can transform
recipient

Step 4
Select Arg⁺
transformants

No Arg⁺
transformants

No Arg⁺
transformants

Arg⁺ transformant

Figure 6.8 Genetic assay for the state of DNA during transformation. Only double-stranded DNA binds to the cell to initiate transformation. The appearance of transformants in step 4 indicates that the transforming DNA was double stranded at the time of DNase treatment.

Plasmid Transformation and Phage Transfection of Naturally Competent Bacteria

Chromosomal DNA can efficiently transform any bacterial cells from the same species that are naturally competent. However, neither plasmids nor phage DNAs can be efficiently introduced into naturally competent cells for two reasons. First, they must be double stranded to replicate. Natural transformation requires breakage of the double-stranded DNA and degradation of one of the two strands so that a linear single strand can enter the cell. Second, they must recyclize. However, pieces of plasmid or phage DNAs cannot recyclize if there are no repeated or complementary sequences at their ends.

Transformation of naturally competent bacteria with plasmid or phage DNA usually occurs only with DNAs that are **dimerized** or **multimerized** into long **concatemers** (see chapters 4 and 7). A dimerized or multimerized DNA is one in which two or more copies of the molecule are linked head to tail, as illustrated in Figure 6.9. If a dimerized plasmid or phage DNA is cut only once, it still has complementary sequences at its ends that can recombine to recyclize the plasmid, as illustrated in the figure. Such dimers and higher multimers often form naturally while plasmid or phage DNA is replicating, so that most preparations of plasmid or phage DNAs contain some dimers. The fact that only dimerized plasmid or phage DNAs can transform naturally competent bacteria supports the model of uptake of single-stranded DNA during transformation described earlier in this chapter.

Role of Natural Transformation

The fact that so many gene products play a direct role in competence indicates that the ability to take up DNA from the environment is advantageous. Below, we discuss three possible advantages and the arguments for and against them.

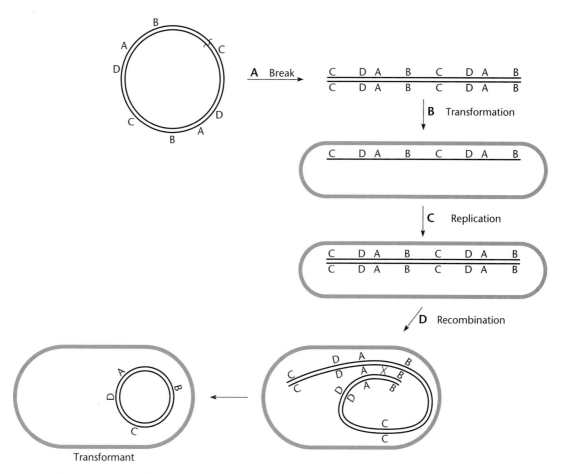

Figure 6.9 Transformation by dimeric plasmids. After the single-stranded dimeric plasmid DNA is taken up, it can serve as a template to make the double-stranded DNA. The repeated ends can recombine with each other to re-form the circular plasmid.

NUTRITION

Organisms may take up DNA for use as a carbon and nitrogen source (see Redfield, Suggested Reading). One argument against this hypothesis is that taking up whole DNA strands for degradation inside the cell may be more difficult than degrading the DNA outside the cell and then taking up the nucleotides. In fact, *B. subtilis* excretes a deoxyribonuclease (DNase) which degrades DNA so that it can be taken up more easily. The major argument against this hypothesis as a general explanation for transformation in all bacteria is that some bacteria take up only DNA of their own species, since DNA from other organisms should offer the same nutritional benefits. Moreover, the fact that competence develops only in a minority of the population, at least in *B. subtilis*, argues against the nutrition hypothesis, since all the bacteria in the population would presumably need the nutrients.

These arguments are attractive but do not disprove the nutrition hypothesis. The bacteria may consume DNA of only their own species because of the danger inherent in taking up foreign DNAs, which might contain prophages, transposons, or other elements that could become parasites of the organism. Furthermore, consumption of DNA from the same species may be a normal part of colony development; cell death and cannibalism are thought to be part of some prokaryotic developmental processes. These processes would require that only some of the cells in the population become DNA consumers while the others become the "sacrificial lambs." The existence of specific cell-killing mechanisms that kill some cells in the population as *B. subtilis* enters the stationary phase lends support to such interpretations (see Box 4.3 in chapter 4).

REPAIR

Cells may take up DNA from other cells to repair damage to their own DNA (see Mongold, Suggested Reading). Figure 6.10 illustrates this hypothesis, in which a population of cells is exposed to UV irradiation. The radiation damages the DNA, causing pyrimidine dimers

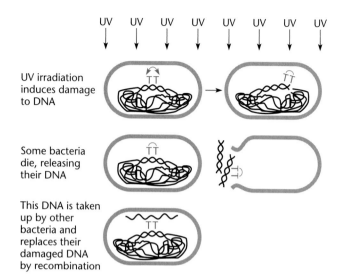

Figure 6.10 Repair of DNA damage by transforming DNA. A region containing a thymine dimer (TT) induced by UV irradiation is replaced by the same, but undamaged, sequence from the DNA of a neighbor killed by the radiation. This mechanism could allow survival of the species.

and other lesions to form (see chapter 11). DNA leaks out of some of the dead cells and enters other bacteria. Because the damage to the DNA has not occurred at exactly the same places, undamaged incoming DNA sequences can replace the damaged regions in the recipient, allowing at least some of the bacteria to survive. This scenario explains why some bacteria take up DNA of only the same species, since, in general, this is the only DNA that can recombine and thereby participate in the repair.

If natural transformation helps in DNA repair, we might expect that repair genes would be induced in response to developing competence and that competence would develop in response to UV irradiation or other types of DNA damage. In fact, in some bacteria, including *B. subtilis* and *S. pneumoniae*, the *recA* gene required for recombination repair is induced in response to the development of competence (see Haijema et al. and Raymond-Denise and Guillen, Suggested Reading). However, in other bacteria, such as *H. influenzae,* the *recA* gene is not induced in response to competence. There is also no evidence that competence genes are induced in response to DNA damage. Nevertheless, the need for DNA repair is an attractive explanation for why at least some types of bacteria develop competence.

RECOMBINATION

The possibility that transformation allows recombination between individual members of the species is also an attactive hypothesis but is difficult to prove. According to this hypothesis, transformation serves the same function that sex serves in higher organisms: it allows the assembly of new combinations of genes and thereby increases diversity and speeds up evolution. Bacteria do not have an obligatory sexual cycle; therefore, without some means of genetic exchange, any genetic changes that a bacterium accumulates during its lifetime are not necessarily exchanged with other members of the species.

The gene exchange function of transformation is supported by the fact that cells of some naturally transformable bacteria leak DNA as they grow. It is hard to imagine what function this leakage could perform unless the leaked DNA is supposed to be taken up by other bacteria.

In several *Neisseria* species, including *N. gonorrhoeae*, transformation may enhance antigenic variability, allowing the organism to avoid the host immune system (Box 6.1). In mixed laboratory cultures, transformation does contribute substantially to the antigenic diversity in this species. However, under natural conditions, it is debatable whether most of this antigenic diversity results from recombination between DNAs brought together by transformation or simply from recombination between sequences within the chromosomal DNA of the bacterium itself.

We still do not know why some types of bacteria are naturally transformable and others are not. It seems possible that most types of bacteria are naturally transformable at low levels. Transformation may serve different purposes in different organisms. Perhaps transformation is used for DNA repair in soil bacteria such as *B. subtilis* but is used to increase genetic variability in obligate parasites such as *N. gonorrhoeae*.

Importance of Natural Transformation for Forward and Reverse Genetics

Whatever its purpose for individual bacterial species, natural transformation has many uses in molecular genetics. Transformation has been used in many bacteria to map genetic markers in chromosomes in many bacteria and to reintroduce DNA into cells after the DNA has been manipulated in the test tube, which has made them ideal model systems for molecular genetic studies. As mentioned, the interpretation of genetic data obtained by transformation is similar to the interpretation of data obtained by transduction. The interpretation of genetic data obtained by transduction and transformation is discussed in chapter 3. More recently, the presence of efficient natural transformation systems in some bacteria, for example, species of *Neisseria, Acinetobacter,* and cyanobacteria, has made them ideal subjects for functional genomics, i.e., attempts to determine the function

Antigenic Variation in *Neisseria gonorrhoeae*

Many types of pathogenic microorganisms avoid the host immune system by changing the antigens on their cell surface. Well-studied examples include trypanosomes, which cause sleeping sickness, and *Neisseria gonorrhoeae*, which causes a sexually transmitted disease.

The pili of *N. gonorrhoeae* are involved in attaching the bacteria to the host epithelial cells. These pili can undergo spontaneous alterations that can change the specificity of binding and confound the host immune system. *N. gonorrhoeae* appears to be capable of making millions of different pili.

The mechanism of pilin variation in *N. gonorrhoeae* is understood in some detail. The major protein subunit of the pilus is encoded by the *pilE* gene. In addition, silent copies of *pilE*, called *pilS*, lack promoters or have various parts deleted. These silent copies share some conserved sequences with each other and with *pilE* but differ in the so-called variable regions. Pilin protein is usually not expressed from these silent copies. However, recombination between a *pilS* gene and *pilE* can change the *pilE* gene and result in a somewhat different pilin protein. This recombination is a type of gene conversion because reciprocal recombinants are not formed (see chapter 10). Interestingly, the availability of iron affects the frequency of antigenic variation. Many bacteria use the availability of iron as an indicator that they are in a eukaryotic host, or in a particular tissue of that host, suggesting that the variation becomes activated in certain tissues.

Because *N. gonorrhoeae* is naturally transformable, not only could recombination occur between a *pilS* gene and the *pilE* gene in the same organism, but also transformation could allow even more variation through the exchange of *pilS* genes with other strains. Experiments indicate that pilin variation is affected by the presence of DNase, suggesting that transformation between individuals is contributing to the variation. Also, the recombination seems to utilize mostly the RecFOR pathway rather than the RecBCD pathway, which is expected since only a single strand of DNA enters during transformation (see the text) and single-stranded DNA might be a better substrate for the RecFOR pathway. Also, experiments with marked *pil* genes indicate that transformation can result in the exchange of *pil* genes between bacteria. These experiments suggest but do not prove that transformation plays an important role in pilin variation during infection. To prove this would require experiments in the infected host, and humans are the only known host for *Neisseria gonorrhoeae*.

References

Sechman, E. V., M. S. Rohrer, and H. S. Seifert. 2005. A genetic screen identifies genes and sites involved in pilin antigenic variation in *Neisseria gonorrhoeae*. *Mol. Microbiol.* **57:**468–483.

Seifert, H. S., R. S. Ajioka, C. Marchal, P. F. Sparling, and M. So. 1988. DNA transformation leads to pilin antigenic variation in *Neisseria gonorrhoeae*. *Nature* (London) **336:**392–395.

of every gene of an organism. One example has been the use of pMUTIN in *B. subtilis* (see chapter 4). In the bacterium *Acinetobacter baylyi*, transformation is so efficient that it can be done merely by spotting DNA restriction fragments or PCR-amplified fragments on streaks of recipient bacteria on plates. This offers opportunities to construct many different types of mutations in genes, including **loss-, gain-,** or **change-of-function mutations** (see Young et al., Suggested Reading). Such manipulations are more difficult in bacteria that do not have efficient natural competence systems.

Artificially Induced Competence

Most types of bacteria are not naturally transformable, at least not at easily detectable levels. Left to their own devices, these bacteria do not take up DNA from the environment. However, even these bacteria can sometimes be made competent by certain chemical treatments, or DNA can be forced into them by a strong electric field in a process called electroporation.

Calcium Ion Induction

Treatment with calcium ions (see Cohen et al., Suggested Reading) can make some bacteria competent, including *Escherichia coli* and *Salmonella* spp. as well as some *Pseudomonas* spp., although the reason is not understood.

Chemically induced transformation is usually inefficient, and only a small percentage of the cells are ever transformed. Accordingly, the cells must be plated under conditions selective for the transformed cells. Therefore, the DNA used for the transformation should contain a selectable gene such as one encoding resistance to an antibiotic.

TRANSFORMATION BY PLASMIDS

In contrast to naturally competent cells, cells made permeable to DNA by calcium ion treatment will take up both single-stranded and double-stranded DNA. Therefore, both linear and double-stranded circular plasmid DNAs can be efficiently introduced into chemically treated cells. This fact has made calcium ion-induced competence very useful for cloning and other applications that require the introduction of plasmid and phage DNAs into cells.

TRANSFECTION BY PHAGE DNA

In addition to plasmid DNAs, viral genomic DNAs or RNAs can often be introduced into cells by transformation, thereby initiating a viral infection. This process is called **transfection** rather than transformation, although the principle is the same. To detect transfection, the potentially transfected cells are usually mixed with indicator bacteria and plated using top agar (see the introductory chapter). If the transfection is successful, a plaque forms where the transfected cells had produced phage which then infected the indicator bacteria.

Some viral infections cannot be initiated merely by transfection with the viral DNA. These viruses cannot transfect cells, because in a natural infection, proteins in the viral head are normally injected along with the DNA, and these proteins are required to initiate the infection. For example, the *E. coli* phage N4 carries a phage-specific RNA polymerase in its head that is injected with the DNA and used to transcribe the early genes (see chapter 7). Transfection with the purified phage DNA does not initiate an infection, because the early genes are not transcribed without this phage-encoded RNA polymerase. Another example of a phage in which the infection cannot be initiated by the nucleic acid alone is phage φ6 (see Box 7.1 in chapter 7). This phage has RNA instead of DNA in the phage head and must inject an RNA replicase to initiate the infection, so that the cells cannot be transfected by the RNA alone. Such examples of phages that inject required proteins are rather rare; for most phages, the infection can be initiated by transfection.

As an aside, many animal viruses do inject proteins required for multiplication, and these proteins cannot be made after injection of the naked DNA or RNA. For example, a retrovirus such as human immunodeficiency virus, which causes AIDS, injects a reverse transcriptase required to make a DNA copy of the incoming RNA before it can be transcribed to make viral proteins. Therefore, human cells cannot be transfected with human immunodeficiency virus RNA alone.

TRANSFORMATION OF CELLS WITH CHROMOSOMAL GENES

Transformation with linear DNA is one method used to replace endogenous genes with genes altered in vitro. However, most types of bacteria made competent by calcium ion treatment are transformed poorly by chromosomal DNA because the linear pieces of double-stranded DNA entering the cell are degraded by an enzyme called the RecBCD nuclease. This nuclease degrades DNA from the ends; therefore, it does not degrade circular plasmid and phage DNAs. However, inactivating the RecBCD nuclease by a mutation would preclude recombination between the incoming DNA and the chromosome, because the enzyme is required for normal recombination in *E. coli* and other bacteria (see chapter 10).

Nevertheless, methods have been devised to transform competent *E. coli* with linear DNA. One way is to use a mutant *E. coli* lacking the D subunit of the RecBCD nuclease. These *recD* mutants are still capable of recombination, but because they lack the nuclease activity that degrades linear double-stranded DNA, they can be transformed with linear double-stranded DNAs. Other methods, sometimes called recombineering, use the recombination systems of phages such as λ phage. Double-stranded DNA can be introduced into cells expressing the λ recombination functions because of the ability of a λ protein named γ (gamma) to inhibit RecBCD. The method can also be used with single-stranded DNAs, such as PCR primers, which are not degraded by RecBCD. Such procedures are discussed further in chapter 10.

Electroporation

Another way in which DNA can be introduced into bacterial cells is by **electroporation**. In the electroporation process, the bacteria are mixed with DNA and briefly exposed to a strong electric field. It is important that the recipient cells first be washed extensively in buffer with very low ionic strength such as distilled water. The buffer usually also contains a nonionic solute such as glycerol to prevent osmotic shock. The brief electric fields across the cellular membranes might create artificial pores of H_2O lined by phospholipid head groups. DNA can pass through these temporary hydrophilic pores (see Tieleman, Suggested Reading). Electroporation works with most types of cells, including most bacteria, unlike the methods mentioned above, which are very specific for certain species. Also, electroporation can be used to introduce linear chromosomal and circular plasmid DNAs into cells. However, electroporation requires specialized equipment.

SUMMARY

1. In transformation, DNA is taken up directly by cells. Transformation was the first form of genetic exchange to be discovered in bacteria, and the demonstration that DNA is the transforming principle was the first direct evidence that DNA is the hereditary material. The bacteria from which the DNA was taken are called the donors, and the bacteria to which the DNA has been added are called the recipients. Bacteria that have taken up DNA are called transformants.

2. Bacteria that are capable of taking up DNA are said to be competent.

3. Some types of bacteria can take up DNA naturally during part of their life cycle. A number of genes whose products form the competence machinery have been identified. Some of these encode proteins related to type II secretion systems, which form type IV pili, or to type IV secretion-conjugation systems.

4. The fate of the DNA during natural transformation is fairly well understood. The double-stranded DNA first binds to the cell surface and then is broken into smaller pieces by endonucleases. Then one strand of the DNA is degraded by an exonuclease. The single-stranded pieces of DNA then invade the chromosome in homologous regions, displacing one strand of the chromosome at these sites. Through repair or subsequent replication, the sequence of the incoming DNA may replace the original chromosomal sequence in these regions.

5. Naturally competent cells can be transformed with linear chromosomal DNA but usually not with monomeric circular plasmid or circular phage DNAs. Transformation by plasmid DNA usually occurs with dimers or higher multimers of the plasmid that can recyclize by recombination between the repeated sequences at the ends.

6. Some types of bacteria, including *Haemophilus influenzae* and *Neisseria gonorrhoeae*, take up DNA of only the same species. Their DNA contains short uptake sequences that are required for uptake of DNA into the cells. Other types of bacteria, including *Bacillus subtilis* and *Streptococcus pneumoniae*, seem to be capable of taking up any DNA.

7. There are three possible roles for natural competence: a nutritional function allowing competent cells to use DNA as a carbon, energy, and nitrogen source; a repair function in which cells use DNA from neighboring bacteria to repair damage to their own chromosomes, thus ensuring survival of the species; and a recombination function in which bacteria exchange genetic material among members of their species, increasing diversity and accelerating evolution. Different types of bacteria may use transformation for different purposes.

8. Some types of bacteria that do not show natural competence can nevertheless be transformed after some types of chemical treatment or by electroporation. The standard method for making *E. coli* permeable to DNA involves treatment with calcium ions. Cells made competent by calcium treatment can be transformed with plasmid and phage DNAs, making this method one of the cornerstones of molecular genetics.

9. If the cell is transformed with viral DNA to initiate an infection, the process is called transfection.

10. Brief exposure of cells to an electric field also allows them to take up DNA, a process called electroporation.

QUESTIONS FOR THOUGHT

1. Why do you think some types of bacteria are capable of developing competence? What is the real function of competence?

2. How would you determine if the competence genes of *Bacillus subtilis* are turned on by UV irradiation and other types of DNA damage?

3. How would you determine whether antigenic variation in *Neisseria gonorrhoeae* is due to transformation between different bacteria or to recombination within the same bacterium?

PROBLEMS

1. How would you determine if a type of bacterium you have isolated is naturally competent? Outline the steps you would use.

2. You have isolated a naturally competent bacterium from the soil. Outline how you would isolate mutants of your bacterium that are defective in transformation. Distinguish those that are defective in recombination from those that are defective in the uptake of DNA.

3. How would you determine if a naturally transformable bacterium can take up DNA of only its own species or can take up any DNA?

4. How would you determine if a piece of DNA contains the uptake sequence for that species?

5. How would you determine if the DNA of a phage can be used to transfect *E. coli*?

SUGGESTED READING

Avery, O. T., C. M. MacLeod, and M. McCarty. 1944. Studies on the chemical nature of the substance inducing transformation of pneumococcal types. Induction of transformation by a deoxyribonucleic acid fraction isolated from pneumococcus type III. *J. Exp. Med.* **79:**137–159.

Berge, M., M. Moscoso, M. Prudhomme, B. Martin, and J. P. Claverys. 2002. Uptake of transforming DNA in Gram-positive bacteria: a view from *Streptococcus pneumoniae. Mol. Microbiol.* **45:**411–421.

Bongiorni, C., S. Ishikawa, S. Stephenson, N. Ogasawara, and M. Perego. 2005. Synergistic regulation of competence development in *Bacillus subtilis* by two Rap-Phr systems. *J. Bacteriol.* **187:**4353–4361.

Chen, I., and D. Dubnau. 2004. DNA uptake during bacterial transformation. *Nat. Rev. Microbiol.* **2:**241–249.

Clerico, E. M., J. L. Ditty, and S. F. Golden. 2006. Specialized techniques for site directed mutagenesis in cyanobacteria, p. 155–171. *In* E. Rosato (ed.), *Methods in Molecular Biology.* Humana Press, Totowa, N.J.

Cohen, S. N., A. C. Y. Chang, and L. Hsu. 1972. Nonchromosomal antibiotic resistance in bacteria: genetic transformation of *Escherichia coli* by R-factor DNA. *Proc. Natl. Acad. Sci. USA* **69:**2110.

Cornella, N., and A. D. Grossman. 2005. Conservation of genes and processes controlled by the quorum response in bacteria: characterization of genes controlled by the quorum-sensing transcription factor ComA in *Bacillus subtilis. Mol. Microbiol.* **57:**1159–1174.

Haijema, B. J., D. van Sinderen, K. Winterling, J. Kooistra, G. Venema, and L. W. Hamoen. 1996. Regulated expression of the *dinR* and *recA* genes during competence development and SOS induction in *Bacillus subtilis. Mol. Microbiol.* **22:**75–86.

Karnholz, A., C. Hoefler, S. Odenbreit, W. Fischer, D. Hofreuter, and R. Haas. 2006. Functional and topological characterization of novel components of the *comB* DNA transformation system in *Helicobacter pylori. J. Bacteriol.* **188:** 882–893.

Lazazzera, B. A., T. Palmer, J. Quisel, and A. D. Grossman. 1999. Cell density control of gene expression and development in *Bacillus subtilis*, p. 27–46. *In* G. M. Dunny and S. C. Winans (ed.), *Cell-Cell Signaling in Bacteria.* ASM Press, Washington, D.C.

Maamer, H., and D. Dubnau. 2005. Bistability in the *Bacillus subtilis* K-state (competence) system requires a positive feedback loop. *Mol. Microbiol.* **56:**615–624.

Mongold, J. A. 1992. DNA repair and the evolution of competence in *Haemophilus influenzae. Genetics* **132:**893–898.

Peterson, S. N., C. K. Sung, R. Cline, B. V. Desai, E. C. Snesrud, P. Luo, J. Walling, H. Li, M. Mintz, G. Tsegaye, P. C. Burr, Y. Do, S. Ahn, J. Gilbert, R. D. Fleischmann, and D. A. Morrison. 2004. Identification of competence pheromone responsive genes in *Streptococcus pneumoniae* by use of DNA microarrays. *Mol. Microbiol.* **51:**1051–1070.

Provedi, R., I. Chen, and D. Dubnau. 2001. NucA is required for DNA cleavage during transformation of *Bacillus subtilis. Mol. Microbiol.* **40:**634–644.

Raymond-Denise, A., and N. Guillen. 1992. Expression of the *Bacillus subtilis dinR* and *recA* genes after DNA damage and during competence. *J. Bacteriol.* **174:**3171–3176.

Redfield, R. J. 1993. Genes for breakfast: the have your cake and eat it too of bacterial transformation. *J. Hered.* **84:**400–404.

Tieleman, D. P. 2004. The molecular basis of electroporation. *BMC Biochem.* **5:**10.

Young, D. M., D. Parke, and L. N. Ornston. 2005. Opportunities for genetic investigation afforded by *Acinetobacter baylyi*, a nutritionally versatile bacterial species that is highly competent for natural transformation. *Annu. Rev. Microbiol.* **59:**519–551.

CHAPTER **7**

Lytic Bacteriophages: Development, Genetics, and Generalized Transduction

Probably all organisms on Earth are parasitized by viruses, and bacteria are no exception. For purely historical reasons, viruses that infect bacteria are usually not called viruses but are called **bacteriophages** (**phages** for short), even though they have lifestyles similar to those of plant and animal viruses. Bacteriophages are probably the most abundant biological entities. It has been estimated that there are on the order of 10^{31} phages on Earth. As mentioned in the introduction, the name *phage* derives from the Greek verb "to eat," and it describes the eaten-out places, or **plaques,** that are formed on bacterial lawns. The plural of phage is phage when a specific quantity is discussed, but we add an "s" (phages) when we are discussing more than one type of phage.

Like all viruses, phages are so small that they can be seen only under the electron microscope. As shown in Figure 7.1A, phages are often spectacular in appearance, with **capsids,** or icosahedral heads, and elaborate tail structures that make them resemble lunar landing modules. The tail structures allow them to penetrate bacterial membranes and cell walls to inject their DNA into the cell. Animal and plant viruses have much simpler shapes because they do not need such elaborate tail structures. They either are engulfed by the cell, in the case of animal viruses, or enter through wounds, in the case of plant viruses.

Phages differ greatly in their complexity. Smaller phages, such as MS2, usually have no tail, and their heads may consist of as few as two different types of proteins. The heads and tails of some of the larger phages such as T4 have up to 20 different proteins, each of which can exist in as few as 1 copy to as many as 1,000 copies depending on the structural role they play in the phage particle. Phages also infect specific bacterial hosts, and different phages have very different host ranges, as discussed below. Phages fall into

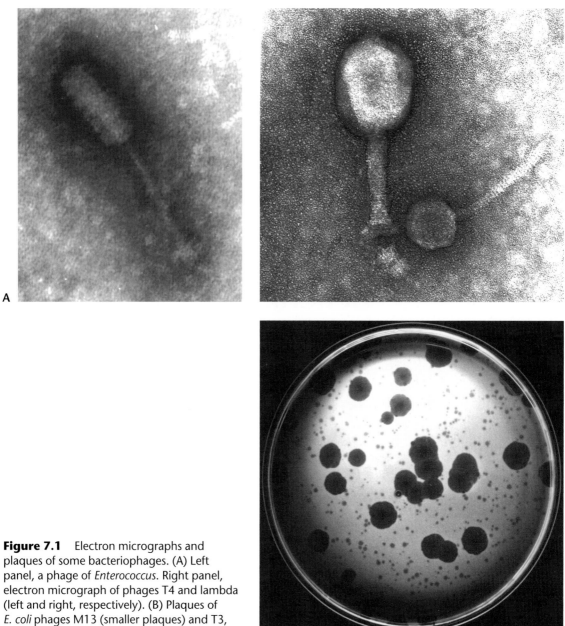

Figure 7.1 Electron micrographs and plaques of some bacteriophages. (A) Left panel, a phage of *Enterococcus*. Right panel, electron micrograph of phages T4 and lambda (left and right, respectively). (B) Plaques of *E. coli* phages M13 (smaller plaques) and T3, a relative of phage T7 (larger plaques). **B**

one of a relatively small number of families that can be composed of a large number of types that infect different bacterial hosts. Also, recent work comparing the genomic sequences of phages from within these families has revealed their "mosaic" nature, in which different phages of the same family seem to be assembled from "tiles" composed of groups of genes for the same function, for example genes for DNA replication or for formation of the phage head, rather than from individual genes (see Hendrix, Suggested Reading). The assumption is that these genes have been exchanged as an interacting

group between different phages from different species of bacteria fairly recently in evolutionary times. This suggests that phages do not just evolve with their host but evolve as a family, by somehow exchanging building tiles, even though they infect different hosts.

Also like all viruses, phages are not live organisms but merely a nucleic acid—either DNA or RNA depending on the type of phage—wrapped in a protein and/or membrane coat for protection. This nucleic acid carries genes that direct the synthesis of more phage. In phages, either type of nucleic acid carried in the head is called the **phage** genome. These molecules can be very long, because the genome must be long enough to have at least one copy of each of the phage genes. The length of the DNA or RNA genome therefore reflects the size and complexity of the phage. For instance, the small phage MS2 has only four genes and a rather small RNA genome whereas phage T4 has more than 200 genes and a DNA genome that is almost 10 μm long. Long genomes, which can be as much as 1,000 times longer than the head, must be very tightly packed into the head of the phage.

Because phage are so small, they are usually detected only by the plaques (Figure 7.1B) they form on **lawns** of susceptible host bacteria (see the introductory chapter). Each type of phage makes plaques on only certain host bacteria, which define its **host range.** Mutations in the phage DNA can alter the host range of a phage or the conditions under which the phage can form a plaque, which is usually how mutations are detected. In this chapter, we discuss what is known about how some representative phages multiply and some of the genetic experiments that have contributed to this knowledge. We also discuss why some phages, but not others, can be used for genetic mapping and strain construction in bacteria in a process called transduction. First, however, we review some general features of phage development.

The Bacteriophage Lytic Development Cycle

Because phages, like all viruses, are essentially genes wrapped in a protein or membrane coat, they cannot multiply without benefit of a host cell. The virus injects its genes into a cell, and the cell furnishes some or all of the means to express those genes and make more viruses.

Figure 7.2 illustrates the multiplication process for a typical large DNA phage. To start the infection, a phage adsorbs to an actively growing bacterial cell by binding to a specific receptor on the cell surface. In the next step, the phage injects its entire DNA into the cell, where transcription of RNA, usually by the host RNA polymerase, begins almost immediately. However, not all the genes of a phage are transcribed into mRNA when the DNA first enters the cell. Only some of the genes of the phage have promoters that mimic those of the host cell DNA and so are recognized by the host RNA polymerase. Those transcribed soon after infection are called the **early genes** of the phage and encode mostly enzymes involved in DNA synthesis such as DNA polymerase, primase, DNA ligase, and helicase. With the help of these enzymes, the phage DNA begins to replicate and many copies accumulate in the cell.

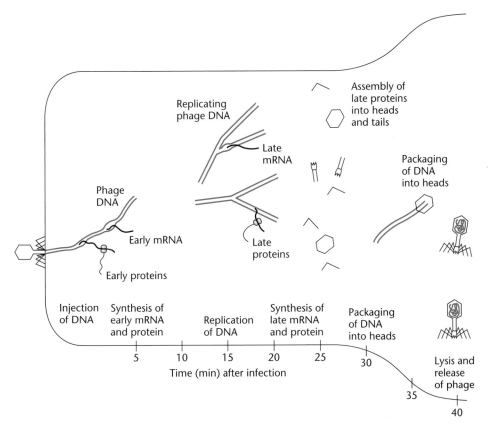

Figure 7.2 A typical bacteriophage multiplication cycle. After the phage injects its DNA, the early genes, most of which encode products involved in DNA replication, are transcribed and translated. Then DNA replication begins, and the late genes are transcribed and translated to form the head and tail of the phage. The DNA is packaged into the heads; the tails are attached; and the cells lyse, releasing the phage to infect other cells.

Replicating phage DNA

Assembly of late proteins into heads and tails

Late mRNA

Packaging of DNA into heads

Phage DNA

Early mRNA

Early proteins

Late proteins

Injection of DNA

Synthesis of early mRNA and protein

Replication of DNA

Synthesis of late mRNA and protein

Packaging of DNA into heads

Lysis and release of phage

5   10   15   20   25   30   35   40

Time (min) after infection

Next, mRNA is transcribed from the rest of the phage genes, the **late genes,** which may or may not be intermingled with the early genes in the phage DNA, depending on the phage. These genes have promoters that are unlike those of the host cell and so are not recognized by the host RNA polymerase alone. Most of these genes encode proteins involved in assembly of the head and tail. After the phage particle is completed, the DNA is taken up by the heads and the tails are attached. Finally, the cells break open, or **lyse,** and the new phage are released to infect other sensitive cells. This whole process, known as the **lytic cycle,** takes less than 1 h for many phages, and hundreds of progeny phage can be produced from a single infecting phage.

Actual phage development is usually much more complex than this basic process, proceeding through several intermediate stages in which the expression of different genes is regulated by specific mechanisms. Most of the regulation is achieved by having genes be transcribed into mRNA only at certain times; this type of regulation is called **transcriptional regulation** (see chapter 2). However, some genes undergo **posttranscriptional regulation,** which occurs after the mRNA has been made. For example, regulation may operate at the level of whether the mRNAs are translated; this is known as **translational regulation.** Other types of posttranscriptional regulation involve the stability of certain RNAs that quickly degrade unless they are synthesized at the right stage of development.

Figure 7.3 shows the basic process of phage gene transcriptional regulation, in which one or more of the gene products synthesized during each stage of development turns on the transcription of the genes in the next stage of development. The gene products synthesized during each stage can also be responsible for turning off the transcription of genes expressed in the preceding stage. Genes whose products are responsible for regulating the transcription of other genes are called **regulatory genes,** and this type of regulation is called a **regulatory cascade** because each step triggers the next step and stops the preceding step. By having such a cascade of gene expression, all the information for the step-by-step development of the phage can be preprogrammed into the DNA of the phage.

Regulatory genes can usually be easily identified by mutations. Mutations in most genes affect only the product of the mutated gene. However, mutations in regulatory genes can affect the expression of many other genes. This fact has been used to identify the regulatory genes of many phages. Below, we discuss some of these genes and their functions, selecting our examples either because they are the basis for cloning technologies or because of the impact they have had on our understanding of

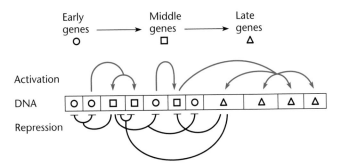

Figure 7.3 Transcriptional regulation during development of a typical large DNA phage. The purple arrows indicate activation of gene expression; the black bars indicate repression of gene expression.

regulatory mechanisms in general. All the phages presented in this chapter contain DNA; Box 7.1 briefly describes the properties of some RNA phages.

Phage T7: a Phage-Encoded RNA Polymerase

Compared with some of the larger phages, phage T7 has a relatively simple program of gene expression after infection, with only two major classes of genes, the early and late genes. The phage has about 50 genes, many of which are shown on the genome map in Figure 7.4. After infection, expression of the T7 genes proceeds from left to right, with the genes on the extreme left of the genetic map, up to and including gene *1.3*, expressed first. These are the early genes. The genes to the right of *1.3*—the middle DNA metabolism and late phage assembly genes—are transcribed after a few minutes' delay.

Nonsense and temperature-sensitive mutations were used to identify which of the early-gene products is responsible for turning on the late genes. Under nonpermissive conditions, in which the mutated genes were inoperable, amber and temperature-sensitive mutations in gene *1* prevented transcription of the late genes, and so gene *1* was a candidate for the regulatory gene. Later work showed that the product of gene *1* is an RNA polymerase that recognizes the promoters used to transcribe the late genes. In fact, transcription of these late genes by the gene *1* product may help pull the DNA into the cell, causing sequential gene expression. The sequence of these promoters differs greatly from those recognized by bacterial RNA polymerases, so that these phage promoters are recognized only by this T7-specific RNA polymerase. Other phages, including T7's close relative T3 and phage φ29 *of Bacillus subtilis* (see below), also synthesize their own RNA polymerase, which exclusively recognizes their own promoters. The specificity of phage RNA polymerases for their own promoters has been exploited in many applications in molecular genetics, some of which are discussed below and in chapters 2 and 4.

BOX 7.1

RNA Phages

The capsids (i.e., heads) of many animal and plant viruses contain RNA instead of DNA. Some of these viruses, the so-called retroviruses, use enzymes called reverse transcriptases to transcribe the RNA into DNA, and these enzymes, because they are essentially DNA polymerases, need primers. In contrast, other RNA-containing animal viruses, such as the influenza viruses, which cause flu, and the reoviruses, which cause colds, replicate their RNA by using RNA replicases and need no DNA intermediate. Because these RNA replicases have no need for primers, the genomes of RNA viruses can be linear without repeated ends. As we might expect, RNA viruses seem to have higher spontaneous mutation rates during replication, probably because their RNA replicases have no editing functions.

Some phages also have RNA as their genome. Examples include Qβ, MS2, R17, f2, and φ6. The *E. coli* RNA phages Qβ, MS2, R17, and f2 are similar to each other. All have a single-stranded RNA genome that encodes only four proteins: a replicase, two head proteins, and a lysin. Immediately after the RNA enters the cell, it serves as an mRNA and is translated into the replicase. This enzyme replicates the RNA, first by making complementary minus strands and then by using these as a template to synthesize more plus strands. The phage genomic RNA must serve as an mRNA to synthesize the replicase, because no such replicase enzyme exists in *E. coli* to synthesize RNA from an RNA template. Interestingly, the phage Qβ replicase has four subunits, only one of which is encoded by the phage. The other three are components of the host translational machinery: two of the elongation factors for translation, EF-Tu and EF-Ts, and a ribosomal protein, S1. They also need another protein Hfq (Host Factor for Qβ) that is required to replicate their genome. This latter protein is used for regulation by small noncoding RNAs and is discussed extensively in chapter 13. Because the genomes of these RNA phages also function as an mRNA, they have served as a convenient source of a single species of mRNA in studies of translation.

Another RNA phage, φ6, was isolated from the bean pathogen *Pseudomonas syringae* subsp. *phaseolicola*. The RNA genome of this phage is double stranded and exists in three segments in the phage capsid, much like the reoviruses of mammals. These three segments are called the S, M, and L segments, for Small, Medium, and Large. Also like animal viruses, this phage is surrounded by membrane material derived from the host cell, i.e., an envelope, and the phage enters its host cells in much the same way that animal viruses enter their hosts. However, unlike most animal viruses, φ6 is released by lysis.

The replication, transcription, and translation of the double-stranded RNA of a virus such as φ6 present special problems. Not only must the phage replicate its double-stranded RNA, but also it must transcribe it into single-stranded mRNA since double-stranded RNA cannot be translated. The uninfected host cell contains neither of the enzymes required for these functions, which therefore must be virus encoded and packaged into the phage head so that they enter the cell with the RNA. Otherwise, neither the transcriptase nor the replicase could be made. Another interesting question with this phage is how three separate RNAs are encapsidated in the phage head. The three segments of the genome are transcribed into + strand transcripts that are then packaged into a preformed head in sequential order with the S segment first, the M segment second, and the L segment last. Packaging is initiated at 200-base *pac* sequences that seem to share little sequence or structural similarity among the three RNAs. Only after the single + strand enters the head is its − strand complement synthesized, to make the double-stranded genomic RNA.

References

Blumenthal, T., and G. G. Carmichael. 1979. RNA replication: function and structure of Qβ replicase. *Annu. Rev. Biochem.* **48:**525–548.

Qiao, X., J. Qiao, and L. Mindich. 2003. Analysis of specific binding involved in genomic packaging of the double-stranded RNA bacteriophage φ6. *J. Bacteriol.* **185:**6409–6414.

Figure 7.4 Genetic map of phage T7. The genes for the RNA polymerase used for expression vectors and the major capsid protein used for phage display are indicated.

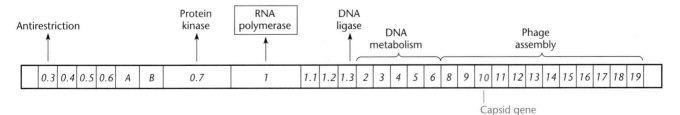

T7 PHAGE-BASED EXPRESSION VECTORS

Some of the most useful expression vectors use the T7 RNA polymerase to express foreign genes in *Escherichia coli*. The pET vectors (for *p*lasmid *e*xpression *T7*) are a family of plasmid expression vectors that use the T7 phage RNA polymerase and T7 gene 10 promoter to express foreign genes in *E. coli* (see Figure 2.45 for an illustration of one such vector, pET15b). The pET expression vectors use the promoter of the head protein gene (gene *10*) of T7, which is a very strong promoter. Hundreds of thousands of copies of T7 head protein must be synthesized in a few minutes after infection, making this one of the strongest known promoters. Downstream of the T7 promoter are a number of restriction sites into which foreign genes can be cloned. Any foreign gene cloned downstream of this T7 promoter is transcribed at very high rates by the T7 RNA polymerase. A number of variations of the vector shown in Figure 2.45 have been designed. Some of the pET vectors have strong translational initiation regions (see chapter 2) for making translational fusions to affinity tags such as a His tag, which makes the protein easy to detect and purify on nickel columns (see chapter 2 for a discussion of translational fusions and affinity tags). The T7 promoter can also be made inducible by providing the T7 RNA polymerase only when the foreign gene is to be expressed. This is important in cases where the fusion protein is toxic to the cell.

One general strategy for using a pET vector is illustrated in Figure 7.5. To provide a source of inducible phage T7 RNA polymerase, *E. coli* strains have been constructed in which phage gene *1* for RNA polymerase is cloned downstream of the inducible *lac* promoter and integrated into their chromosome. In these strains, which often have the DE3 suffix (as in *E. coli* JM109DE3), the phage polymerase gene is transcribed only from the *lac* promoter, so that the T7 RNA polymerase is synthesized only if an inducer of the *lac* promoter, such as isopropyl-β-D-thiogalactopyranoside (IPTG), is added. The newly synthesized T7 RNA polymerase then makes large amounts of mRNA on the foreign gene cloned into the pET plasmid and large amounts of its protein product.

Figure 7.5 Strategy for regulating the expression of genes cloned into a pET vector. The gene for T7 RNA polymerase (gene 1) is inserted into the chromosome of *E. coli* and transcribed from the *lac* promoter; therefore, it is expressed only if the inducer IPTG is added. The T7 RNA polymerase then transcribes the gene cloned into the pET vector downstream of the T7 late promoter on the pET cloning vector. If the protein product of the cloned gene is toxic, it may be necessary to further reduce the transcription of the cloned gene before induction. The T7 lysozyme encoded by a compatible plasmid, pLysS, binds to any residual T7 RNA polymerase made in the absence of induction and inactivates it. Also, the presence of *lac* operators between the T7 promoter and the cloned gene further reduces transcription of the cloned gene in the absence of IPTG.

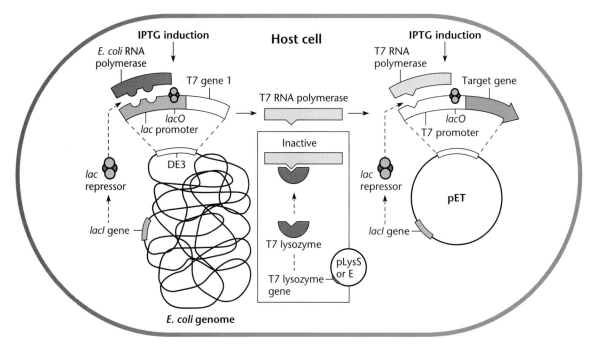

RIBOPROBES AND PROCESSING SUBSTRATES

Another application of phage T7 and RNA polymerases from related phages is in making specific RNAs in vitro. Specific RNAs made from a single gene are useful as probes for hybridization experiments (**riboprobes**) or as RNA substrates for processing reactions, such as splicing. This technology is also based on the fact that the phage RNA polymerases transcribe RNA only from their own promoters. pBAC vectors, which are used in the Human Genome Project and other large genome projects, use phage promoters to make RNAs as hybridization probes to identify clones of neighboring sequences (Figure 4.23). In this and other applications, the vectors are designed with multiple restriction sites bracketed by promoters for phage RNA polymerases. The gene on which RNA is to be made is cloned into one of the restriction sites on the vector, and the vector DNA is purified and cut with a restriction endonuclease on the other side of the cloned gene from the phage promoter. When purified phage RNA polymerase is added, along with the other ingredients including the ribonucleoside triphosphates needed for RNA synthesis, the only RNA that is made is complementary to the transcribed strand of the cloned gene. To make RNA complementary to the other strand, the gene can be cloned in the opposite orientation in the cloning vector or a special vector can be used that has the promoter for another phage on the other side of the cloning site. For example, in one such vector, the cloning site is bracketed by a T7 promoter on one side and a Sp6 promoter on the other. One strand of the cloned gene DNA is transcribed into RNA if purified T7 RNA polymerase is added, but the other strand is transcribed into RNA if Sp6 RNA polymerase is added. Purified RNA polymerase of T7 and other phages are available from biochemical supply companies.

PHAGE DISPLAY

Another application of phage T7 is in phage display (Box 7.2). This is a way of detecting peptides that bind to another molecule such as a hormone or a specific antibody. To use this technology, a randomized peptide-coding sequence is fused to the T7 head protein so that different phages display different versions of the peptide on their surface. The phage that display a version of the peptide which binds to the other molecule can then be "panned for" and isolated. The cloned DNA can then be

---

BOX 7.2

Phage Display

One of the most powerful current applications of phages is in phage display. This remarkable technology allows the identification and synthesis of peptides and proteins that bind tightly to ligands including other proteins. This technology is currently being used to identify antigens, for example, in autoimmune diseases such as multiple sclerosis; to purify human antibodies against specific antigens; to identify drug targets; and to develop peptides that target chemotherapeutics to specific tissues, just to list a few examples.

Phage display technology depends on the fact that phage are made up of DNA coated with protein encoded by that DNA. If a particular peptide-coding sequence is translationally fused to the coding sequence for one of the head proteins of the phage, all the progeny of that phage will display the particular peptide on their surface. The phage expressing that particular peptide can then be purified away from the other phage in the population by virtue of the ability of the peptide to bind to a particular target. Once the phage is isolated, it can be grown in large amounts and the gene encoding the fusion protein can be sequenced to reveal its DNA sequence and hence the sequence of the peptide that binds to the target. The sequence of the peptide might then be entered into a database to identify a protein that might contain this peptide sequence.

The figure outlines in more detail the steps in phage display with T7, one of the phages that has been adapted to use for phage display. For purposes of illustration, say we want to use this modified T7 phage to determine the peptide antigen to which a particular antibody binds. In panel A, a randomized protein-coding sequence (shown in purple) is cloned into the polycloning site in the phage DNA to make translational fusions in which random peptides have been fused to the head protein-coding region, gene *10*, of the phage. The DNA is then introduced into cells by transfection or by packaging the DNA into a phage head and using the phage to infect cells. When the phages multiply, the progeny of each phage all display the same peptide sequence on their surface fused to their head proteins (panel B). However, since a random collection of peptide-coding sequences has been fused to the gene *10* protein-coding sequence, the descendants of each phage in the population express different fusion peptides on their surfaces. To identify the phage expressing a particular peptide, we do something called panning, so named because of its analogy to panning for gold and other minerals. Panel C illustrates the principle behind panning. The ligand molecule to which the desired protein sequence binds is fixed to a solid surface such as a well in a microtiter plate, cellulose, or Sepharose. Ligands are molecules which

(continued)

BOX 7.2 (continued)

Phage Display

bind to proteins and can be RNA, DNA, another protein, or even a small molecule such as a hormone. The ligand can be fixed to the solid surface in a number of ways, depending on the nature of the ligand. In our case the antibody, being a protein, might be chemically cross-linked to the solid surface through one or more of its amino acids. In the next step, the mixture of phages, some of which are presumably displaying

the peptide sequence to which the antibody binds, are put in the well of the plate, and the excess phages are washed out or eluted. Any phages that bind to the ligand in the well are preferentially retained. The retained phages can then be separated from whatever they are bound to and eluted. The methods used to separate them from the ligand depend on what the phage can tolerate and how tightly the peptide

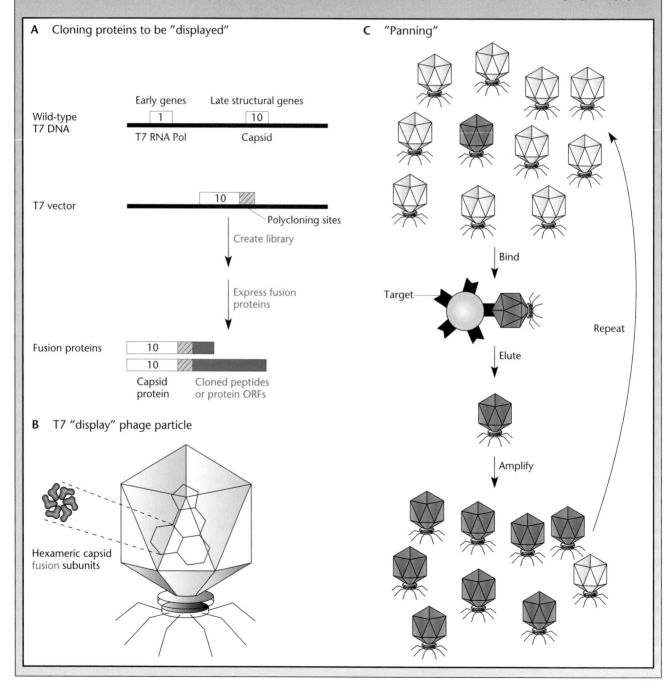

A Cloning proteins to be "displayed"

Early genes   Late structural genes

Wild-type T7 DNA

1   10

T7 RNA Pol   Capsid

T7 vector

10

Polycloning sites

Create library

Express fusion proteins

Fusion proteins

10

10

Capsid protein   Cloned peptides or protein ORFs

B T7 "display" phage particle

Hexameric capsid fusion subunits

C "Panning"

Bind

Target

Elute

Amplify

Repeat

BOX 7.2 (continued)

Phage Display

binds to the ligand. The T7 phage used in our example is very rugged and can be eluted by adding a detergent such as sodium dodecyl sulfate without inactivating it. Others, such as M13 and T4, must be eluted more gently. Once the bound phage has been eluted, it is added to its host bacterium, in this case *E. coli*, and propagated. One such purification step is usually not sufficient to purify the phage, and the process must be repeated several times to sufficiently enrich for phage displaying the desired ligand-binding peptide sequence. The DNA of the phage displaying the desired protein sequence can then be sequenced to determine the sequence of the peptide that binds to the ligand or target molecule, in our example the peptide antigen that binds to the antibody.

The first phage vectors to be developed for phage display were based on M13, a single-stranded DNA phage. Any head protein of the phage could be used to make the translational fusion, but in general, it was best to use the gpVIII protein because the head contains many copies of this protein. The M13 vectors have been used successfully for a number of applications. However, the disadvantages of M13-based systems are that only very short peptides can be fused to the head proteins, and the fused peptide must be secreted from the cell. These phages are assembled in the periplasm and secreted out of the cell (see the text), so the fusion peptides must also be secreted along with the head protein to which they are fused. Some highly charged peptides containing many acidic or basic amino acids might not be secreted easily.

Phage display with T7 or T4 has the advantage that these phages lyse the cell, so that the fused peptide need not be secreted. The phages can also accommodate longer polypeptides fused to their head proteins. The T4-based systems are particularly promising in this regard. In this system, the polypeptide to be panned for is fused to the HOC and/or SOC head proteins of the phage (Figure 7.8). These proteins are not required for head formation, but between them about 10^3 copies bind tightly to the outside of the head after it is assembled (Figure 7.8). The coding sequence for the polypeptide to be panned for is not cloned directly into the phage, as it is with M13 or T7, but is cloned into a plasmid that contains the HOC or SOC coding sequence to make the translational fusion. Cells containing the plasmid with the fusion are then infected by T4, and the fusion is crossed into the phage by recombination. The phage they are crossed into has a deletion mutation in its *e* gene, which encodes the lysozyme. This mutation prevents release of the phage unless external egg white lysozyme is added. Recombination with the cloned fusion gene restores the *e* gene and allows phage release in the absence of added lysozyme, allowing the selection of phage that have integrated the fusion gene expressing the peptide so that they will express the fusion protein on their head surface.

References

Gupta, A., A. B. Oppenheim, and V. K. Chaudhary. 2005. Phage display: a molecular fashion show, p. 415–429. *In* M. K. Waldor, D. I. Friedman, and S. L. Adhya (ed.), *Phages: Their Role in Bacterial Pathogenesis and Biotechnology.* ASM Press, Washington, D.C.

Harrison, J. L., S. C. Williams, G. Winter, and A. Nissim. 1996. Screening phage antibody libraries. *Methods Enzymol.* **267:**83–109.

Malys, N., D.-Y. Chang, R. G. Baumann, D. Xie, and L. W. Black. 2002. A bipartite bacteriophage T4 SOC and HOC randomized peptide display library: detection and analysis of phage T4 terminase and late σ factor (gp55) interaction. *J. Mol. Biol.* **319:**289–304.

sequenced to determine the particular peptide sequence that binds to the molecule.

Phage T4: Transcriptional Activators, Antitermination, a New Sigma Factor, and Replication-Coupled Transcription

Bacteriophage T4 is one of the largest known viruses, with a complex structure, making it a popular cover for biology textbooks (Figure 7.1A). Experiments with this phage have been so important in the development of molecular genetics that this phage deserves equal status with Mendel's peas (see the introductory chapter). The function of ribosomes, the existence of mRNA, the nature of the genetic code, the confirmation of codon assignments, and many other basic insights originally came from studies with this phage.

Phage T4 is much larger than T7, with over 200 genes (the T4 genome map is shown in Figure 7.6), and the regulation of its gene expression is predictably more complex. In fact, T4 uses many of the known mechanisms of regulation of gene expression at some stage of its life cycle.

Figure 7.7 shows the time course of T4 protein synthesis after infection. Each band in the figure is the polypeptide product of a single phage T4 gene, and some of the gene products are identified by the gene that encodes them (details on how the bands were obtained are given in the legend to Figure 7.7). For example, p37

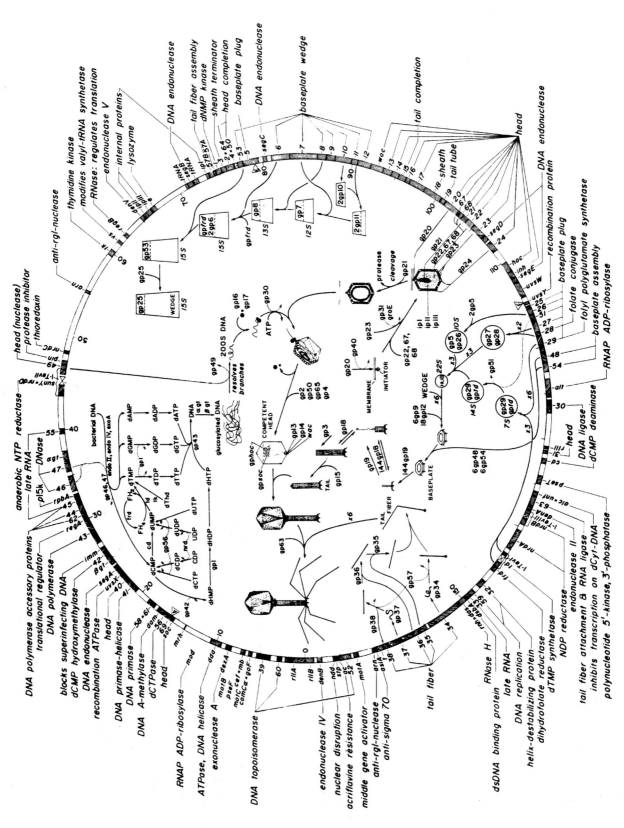

Figure 7.6 Genomic map of phage T4.

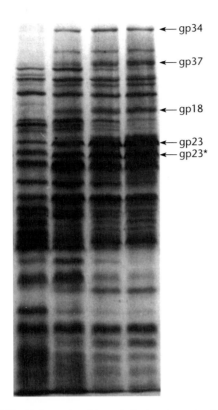

Figure 7.7 Polyacrylamide gel electrophoresis of proteins synthesized during the development of phage T4. The proteins were labeled by adding amino acids containing the ¹⁴C radioisotope of carbon at various times after the phage were added to the bacteria. Because phage T4 stops host protein synthesis after infection, radioactive amino acids are incorporated only into phage proteins, and therefore only the phage proteins become radioactive. Moreover, only the phage proteins made when the radioactive amino acids were added are labeled. Hence, we can tell which phage proteins are made at any given time. To separate the proteins, the cells were broken open with the detergent sodium dodecyl sulfate, which also separates the polypeptides that are part of the multimeric proteins. Electrophoresis caused the polypeptides to migrate on an acrylamide gel, forming bands in columns. The smaller polypeptides move faster on these gels, so that the polypeptides are arranged by size from the smallest at the bottom to the largest at the top. The gel was subsequently dried on a piece of filter paper and then laid next to a photographic film, which was exposed to the high-energy light waves given off from the radioactive disintegrations. Each band represents the polypeptide product of a single T4 gene. Lane 1 (from the left), proteins synthesized between 5 and 10 min after infection; lane 2, 10 to 15 min; lane 3, 15 to 20 min; lane 4, 30 to 35 min.

is the *p*roduct of gene *37* (Figure 7.6). Because of the way the polypeptides were labeled, the time at which a band first appears is the time at which that gene begins to be expressed, and the time a band disappears is the time that gene is shut off. Clearly, some genes of T4 are expressed immediately after infection. These are called the immediate-early genes. Other genes, called the delayed-early and middle genes, are expressed only a few minutes after infection. This is followed by expression of the true-late genes, so called to distinguish them from some of the delayed-early and middle genes that continue to be expressed throughout infection. Overall, the regulation of protein synthesis during T4 phage development is very complex, as might be expected from such a large virus.

The assigments of the polypeptide products to genes were made by using amber mutations in the genes. If nonsuppressor cells (i.e., cells that lack a nonsense suppressor [see chapter 3]) are infected by a phage with an amber mutation in a gene, the band corresponding to the product of that gene will be missing. Translation of the gene stops at the amber mutation in a nonsuppressor host, leading to the synthesis of a shorter polypeptide, which is sometimes detected elsewhere on the gel. Sometimes two or more bands are missing as a result of a single amber mutation, such as the two bands missing as a result of an amber mutation in gene *23*, identified as p23 and p23* in Figure 7.7. Several factors could account for the absence of multiple bands as a result of a single mutation. In this case, the polypeptide product of gene *23*, which makes up the phage head, is cleaved after it is synthesized. Normally, approximately 1,000 copies of this polypeptide are used to build every phage head. First, the head is assembled with the p23 polypeptide, and then part of the N terminus of p23 is cut off to form the shorter polypeptide p23* as the head matures into its final form before DNA is encapsidated (i.e., put inside the head). Thus, by disrupting the synthesis of p23, the mutation also prevents the appearance of p23*. The T4 gene products are often referred to as gp23, etc., for *g*ene *p*roduct of *23*. The locations where each of these gene products fit into the T4 particle are shown in Figure 7.8.

REGULATION OF GENE EXPRESSION DURING T4 DEVELOPMENT

Experiments like those described in the legend to Figure 7.7 were also used to identify the regulatory genes of a phage. Mutations in these genes prevent the expression of many other genes and thus cause the disappearance of many bands from the gel. In T4, mutations in genes named *motA* and *asiA* prevent the appearance of the middle-gene products. Mutations in genes *33* and *55*, as well as mutations in many of the genes whose products are required for T4 DNA replication, prevent the appearance of the true-late-gene products. Therefore, *motA*, *asiA*, *33*, and *55*, as well as some genes whose products are required for DNA replication, are predicted

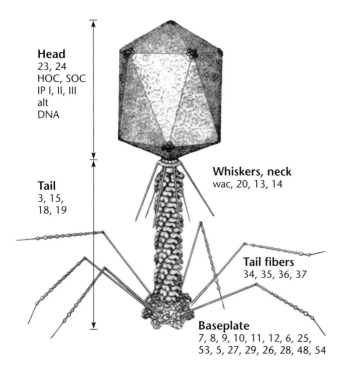

Figure 7.8 Structural components of the T4 particle. Features of the particle have been resolved to about 3 nm. The positions of the baseplate and tail fiber proteins are indicated. The HOC and SOC proteins used for phage display coat the head after it has assembled (see the text).

Head
23, 24
HOC, SOC
IP I, II, III
alt
DNA

Tail
3, 15,
18, 19

Whiskers, neck
wac, 20, 13, 14

Tail fibers
34, 35, 36, 37

Baseplate
7, 8, 9, 10, 11, 12, 6, 25,
53, 5, 27, 29, 26, 28, 48, 54

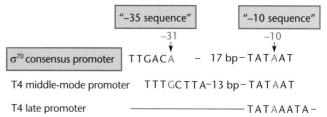

Figure 7.9 Sequence of T4 middle-mode and late promoters. Only the sequences important for recognition by RNA polymerase are shown.

to be regulatory genes. Many genetic and biochemical experiments have been directed toward understanding how these T4 regulatory gene products turn on the synthesis of other proteins.

MIDDLE-MODE TRANSCRIPTION

The genes that are turned on a few minutes after infection can be divided into two classes, the delayed-early genes and the middle genes. The delayed-early genes of T4 are transcribed from the same normal σ70 promoters as the immediate-early genes but are regulated by an antitermination mechanism. Without some T4 regulatory gene products, the RNA polymerase which has initiated at an immediate-early promoter will stop before it gets to the delayed-early genes. Therefore, the transcription of these genes must await the synthesis of antitermination factors encoded by the phage. Regulation by antitermination of transcription is discussed in the case of phage λ in chapter 8, where the mechanism is explained. The middle-gene products, however, are transcribed from their own promoters: the so-called "middle-mode" promoters, which look somewhat different from normal σ70 promoters in that their −35 sequence is replaced by a sequence centered at −30 called a Mot box (Figure 7.9). Because they are somewhat different, transcription from

these promoters requires the phage-encoded MotA and AsiA proteins, themselves the products of delayed-early genes. Current thinking is that the AsiA protein binds to region 4 of σ70 and inhibits its binding to the −35 sequence (see chapter 2). In fact, this protein was first identified as a σ70 inhibitor. Once bound to σ70 region 4, however, the AsiA protein allows MotA to bind. With MotA bound to region 4, it can now recognize the −30 sequence of the middle T4 promoters rather than the −35 sequence normally recognized by σ70 region 4 in uninfected cells.

TRUE-LATE TRANSCRIPTION

The latest genes of T4 to be transcribed are the true-late genes. The products of these genes are mostly the head, tail, and tail fiber components of the phage and enzymes and proteins needed to lyse the cell and release the phage. The initiation of transcription of the true-late genes of T4 is of particular interest because it is coupled to the replication of the DNA. Other viruses, including some human viruses, are known also to couple their late transcription to the replication of their DNA, although it is not known if they use a similar mechanism. This type of transcription regulation also has obvious advantages in coordinating gene expression and cell division with replication during the normal cell cycle, since genes are then transcribed only when they are replicated. Thus, a detailed understanding of the mechanism of true-late transcription in T4 phage may point to similar mechanisms in other systems, contributing yet another universal basic cellular regulatory mechanism to the list of those first discovered with this phage.

Like the middle genes, the true-late genes of T4 are transcribed from promoters that are different from those of its host (Figure 7.9). They have the sequence TATAAATA rather than the −10 sequence of a typical bacterial σ70 promoter, and they lack a −35 sequence (see chapter 2). Because of this difference, the host RNA polymerase does not normally recognize the T4 promoters. However, the product of the T4 regulatory gene *55* is an alternate sigma factor that binds to the host RNA polymerase, changing its specificity so that it recognizes only the promoters for the T4 true-late genes. This sigma factor has little sequence homology to other sigma factors, but it has an identifiable region 2 that recognizes the altered −10 sequence of the T4 true-late promoters. It seems to lack a region 4 to recognize −35 sequences, however, and this makes it unable to form open complexes and initiate transcription efficiently. Another protein, gp33, which is required for T4 late transcription, binds to the RNA polymerase β flap (see Nechaev et al., Suggested Reading), where region 4 of σ^{70} normally binds. In this sense, the gp33 substitutes for the region 4 of the σ^{70}. However, the gp33 protein cannot activate transcription by itself but can do so only if it binds to the sliding clamp composed of gp45 (see below), which allows open-complex formation and the initiation of transcription.

Phage T4 and its close relatives are not the only phages to use alternate sigma factors to activate the transcription of their late genes. For example, the *B. subtilis* phage SPO1 also uses this regulatory mechanism. The SPO1 late promoters are very unlike the normal bacterial promoters, but they are also quite unlike the T4 late promoters. In fact, host RNA polymerases can be adapted to recognize a wide variety of promoter sequences merely through the attachment of an alternate sigma factor. This general strategy is also used during many bacterial developmental processes such as sporulation, as discussed in chapter 14.

REPLICATION-COUPLED TRANSCRIPTION

In addition to the alternate sigma factor, gp55, and the accessory protein, gp33, the products of other T4 genes are required to turn on the transcription of the late genes, most notably those of genes *44*, *62*, and *45*. The products of these genes are also required for replication of the phage DNA. The gp45 protein is a DNA polymerase accessory protein that acts like the β clamp in *E. coli* DNA replication and wraps around the DNA to form a "sliding clamp," which moves with the DNA polymerase and helps prevent it from falling off the DNA during replication (Table 7.1) (see chapter 1). The gp44 and gp62 proteins are "clamp loaders," analogous to the γ complex of *E. coli*, and load the gp45 clamp onto the DNA. These observations led to the hypothesis that T4 DNA replication is also required for expression of the late genes and that the complex of the host RNA polymerase and the gene *55* protein initiates RNA synthesis efficiently only if the T4 DNA is replicating. Coupling the transcription of the true-late genes to the replication of the phage DNA makes sense from a strategic standpoint. Many of the true-late genes encode parts of the phage particle including the head, and phage heads are not needed until phage DNA is available to be packaged inside them. However, exactly how replication could activate transcription was not clear. One line of evidence was that the T4 DNA does not need to actually replicate in order to

TABLE 7.1	**T4 gene products involved in replication and their homologs in *E. coli* and eukaryotes**		
T4 gene product	***E. coli* function**	**Eukaryotic function**[a]	
Origin-specific replication			
gp43	Pol III α and ε	DNA pol α, β, γ	
gp45	β sliding clamp	PCNA	
gp44, gp62	Clamp loader (γ complex)	RFC	
gp41	Replicative helicase (Dna B)	Mcm complex	
gp61	Primase (Dna G)	Pol α	
gp39, gp52, gp60	GyrAB, topoisomerase IV	Topoisomerase II	
gp30	DNA ligase	—[b]	
Rnh	RNase H	—	
Recombination-dependent replication			
UvsW	RecG, RuvAB	—	
UvsX	RecA	Rad51 (yeast)	
UvsY	RecFOR	Rad52 (yeast)	
gp46, gp47	RecBCD, SbcCD	Rad50, MreII (yeast)	
gp32	SSB	RPA	
gp59	PriABC, DnaT, DnaC	—	
gp49	RuvC	—	

[a] PCNA, proliferating-cell nuclear antigen; RPA, replication protein A; RFC, replication factor C; MreII, double-strand break processing.

[b] —, not identified.

activate late transcription since the DNA polymerase and other replication proteins are not needed if the T4 DNA ligase was inactivated and the T4 DNA was accumulating nicks and ends. However, even under these conditions, the gp45 clamp was still required.

Figure 7.10 shows a model for how the gp33 protein and T4 DNA replication proteins activate true-late-gene transcription. According to the model, the promoters for the true-late genes are activated when the gp33 protein, which is bound to the β flap of RNA polymerase (see above), makes contact with the gp45 clamp, which is normally part of the replication apparatus. The gp45 protein sliding clamp can load on the DNA at a nick or an end or can be loaded on the double-stranded DNA by the gp44 and gp62 clamp loaders. Once it is loaded on the DNA, it remains loaded as the DNA polymerase replicating the lagging strand comes on and off, synthesizing Okazaki fragments (see the description of the trombone model in chapter 1). In fact, the trombone model, in which the DNA polymerase cycles on and off as it replicates the lagging strand, has been most clearly demonstrated with the T4 DNA replication apparatus (see below and Chastian et al., Suggested Reading). Presumably the gp45 clamp stays on the DNA when the DNA polymerase comes off, and the gp45 clamps that are no longer associated with the replication apparatus can slide along the DNA in either direction. If such a wayward clamp makes contact with a gp33 protein bound to the RNA polymerase at a true-late promoter, it allows the RNA polymerase to initiate transcription from the true-late promoter. This explains why the gp33 and gp45 proteins are required for optimal true-late gene transcription, as well as gp44 and gp62, since the latter are required under normal conditions for gp45 protein to load on the DNA to form the sliding clamp. It also explains why the clamp loaders and DNA replication are no longer needed in the absence of DNA ligase. The gp45 clamp can load on DNA at the nicks and ends that accumulate in the absence of DNA ligase, and it no longer needs the clamp loaders or other components of the replication apparatus.

Phage DNA Replication

Unlike the replication of chromosomal or plasmid DNAs, which must be coordinated with cell division, phage DNA replication is governed by only one purpose: to make the greatest number of copies of the phage genome in the shortest possible time. Phage DNA replication can be truly impressive. A single phage DNA molecule initially entering the cell can replicate to make hundreds or even thousands of copies to be packaged into phage heads in as little as 10 or 20 min. This unchecked replication often makes phage DNA replication easier to study than replication in other systems. Nevertheless, phage replication shares many of the features of cellular replication in all types of living organisms, and phage DNA replication has served as a model system to understand DNA replication in bacteria and even in humans.

As with linear chromosomes and plasmids, the linear structure of many phage genomes presents special problems for replication to the ends. This is sometimes called the "primer problem" because DNA polymerases cannot initiate the synthesis of a new DNA but can only add to a prexisiting primer (see chapters 1 and 4). When lagging-strand replication gets to the end of a linear template molecule, there is no DNA upstream on which to synthesize a primer. RNA polymerases can initiate the synthesis of a new strand of RNA, but, even if the primers for the 5′ ends are synthesized as RNA, once the RNA primer is removed there is no DNA upstream to serve as a primer to replace the RNA primer. Because of this priming problem, a linear DNA molecule would get smaller each time it replicated until essential genes were lost. Some phage genomes, for example that of phage M13, are single-stranded circles, which solves the problem. Eukaryotic chromosomes are linear but have throwaway lengths of DNA at their ends, called telomeres, which are enzymatically synthesized from the shortened end after each replication without the need for a DNA template, and are dispensable. However, phages with linear genomes do not have telomeres and solve this primer problem in various ways. Some use protein primers (Box 7.3). Other phages have repeated sequences at the ends of their genomic DNA called terminal redundancies, as discussed below. Still others have hairpin ends like some linear plasmids, which allow them to replicate around the ends and form dimeric circles that can then be resolved by protelomerases (see chapter 4). Therefore, phages use a surprising variety of mechanisms to solve their replication problems; some of these are described in this section.

Phages with Single-Stranded Circular DNA

The genome of some small phages consists of circular single-stranded DNA. The small *E. coli* phages that fit into this category can be separated into two groups. The representative phage of one group, ϕX174, has a spherical capsid with spikes sticking out, resembling the ball portion of the medieval weapon known as a morning star. These are called icosahedral phages because their capsid is an icosahedron like a geodesic dome with mostly six-sided building blocks and an occasional five-sided block. In the other group, represented by M13 and f1 and often referred to as **filamentous phages,** the phages have a single layer of protein covering the extended DNA molecule, making the phages filamentous in appearance.

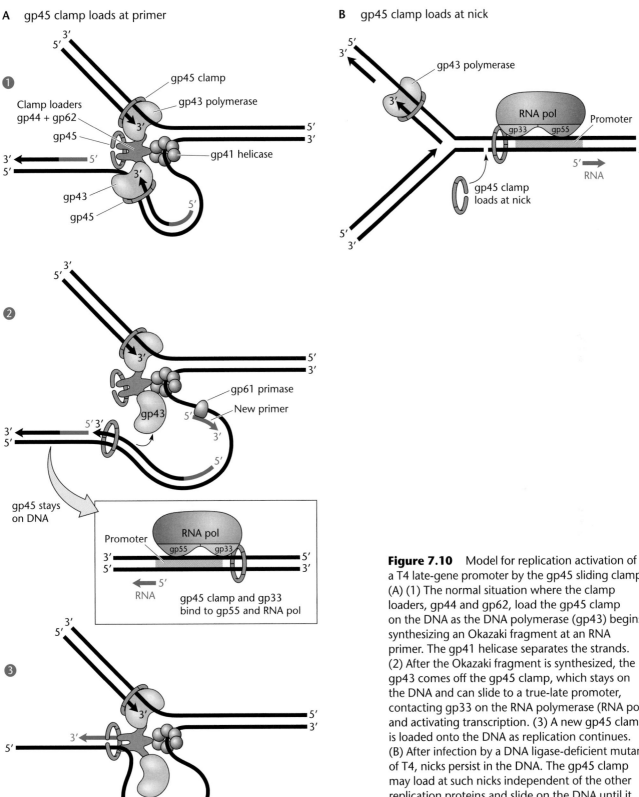

A gp45 clamp loads at primer

1

Clamp loaders
gp44 + gp62

gp45

gp45 clamp

gp43 polymerase

gp41 helicase

gp43

gp45

2

gp61 primase

New primer

gp43

gp45 stays
on DNA

Promoter

RNA pol

gp55   gp33

RNA

gp45 clamp and gp33
bind to gp55 and RNA pol

3

B gp45 clamp loads at nick

gp43 polymerase

RNA pol

gp33   gp55

Promoter

RNA

gp45 clamp
loads at nick

Figure 7.10 Model for replication activation of a T4 late-gene promoter by the gp45 sliding clamp. (A) (1) The normal situation where the clamp loaders, gp44 and gp62, load the gp45 clamp on the DNA as the DNA polymerase (gp43) begins synthesizing an Okazaki fragment at an RNA primer. The gp41 helicase separates the strands. (2) After the Okazaki fragment is synthesized, the gp43 comes off the gp45 clamp, which stays on the DNA and can slide to a true-late promoter, contacting gp33 on the RNA polymerase (RNA pol) and activating transcription. (3) A new gp45 clamp is loaded onto the DNA as replication continues. (B) After infection by a DNA ligase-deficient mutant of T4, nicks persist in the DNA. The gp45 clamp may load at such nicks independent of the other replication proteins and slide on the DNA until it contacts gp33 on RNA polymerase at a late promoter.

BOX 7.3

Protein Priming

Some viruses, including the adenoviruses and the *Bacillus subtilis* phage ф29, have solved the primer problem by using proteins, rather than RNA, to prime their DNA replication. In the virus head, a protein is covalently attached to the 5′ end of the virus DNA. After infection, the DNA grows from this protein, with the first nucleotide attached to a specific serine on the protein. Thus, the virus DNA does not need to form circles or concatemers. The phage DNA polymerase uses this protein to prime its replication by an unusual "sliding-back" mechanism. First, the DNA polymerase adds a dAMP to the hydroxyl group of a specific serine on the protein. The incorporation of this dAMP is directed by a T in the template DNA. However, the T is the second nucleotide from the 3′ end of the template, not the first deoxynucleotide. The DNA polymerase then backs up to recapture the information in the 3′ deoxynucleotide before replication continues. After replication, the extra dAMP at the 5′ end of the newly synthesized strand is removed, the protein is transferred to the 5′ end of

the newly replicated strand, and replication continues. In this way, no information is lost during replication.

Phage ф29 has also been an important model system with which to study phage maturation because the phage DNA can be packaged very efficiently into phage heads in a test tube. Interestingly, its packaging motor contains an RNA, six copies of which are joined to form a ring around the entering DNA. This RNA ring binds ATP and might somehow rotate to help pump the DNA into the phage head. RNA motors to pump DNA have not yet been identified in other systems.

References

Escarmis, C., D. Guirao, and M. Salas. 1989. Replication of recombinant ф29 DNA molecules in *Bacillus subtilis* protoplasts. *Virology* **169:**152–160.

Lee, T.-J., and P. Guo. 2006. Interaction of gp16 with pRNA and DNA for genome packaging by motor of bacterial virus ф29. *J. Mol. Biol.* **356:**589–599.

Meijer, W. J., J. A. Horcajadas, and M. Salas. 2001. ф29 family of phages. *Microbiol. Mol. Biol. Rev.* **65:**261–287.

The different shapes of phages of these two families determine how they enter and infect cells and then leave the infected cells. The icosahedral phages enter and leave cells much like other phages. They bind to the cell surface, and only the DNA enters the cell; they then lyse or break open the cell to exit after they have developed. In contrast, M13 and other filamentous phages are male-specific phages because they specifically adsorb to the sex pilus encoded by certain plasmids and so infect only "male" strains of bacteria (see chapter 5). Unlike most other phages, filamentous phages do not inject their DNA. Instead, the entire phage is ingested by the cell, and the protein coat is removed from the DNA as the phage passes through the inner cytoplasmic membrane of the bacterium. After the phage DNA has replicated, it is again coated with protein as it leaks back out through the cytoplasmic membrane. These phage do not lyse infected cells and leak out only slowly. Consequently, cells infected with M13 or other filamentous phages are "chronically" rather than "acutely" infected. Nevertheless, the filamentous phages form visible plaques, because the chronically infected cells grow more slowly than uninfected cells do.

The process of infection of the cell by a filamentous single-stranded DNA phage and its release from the cell has been studied in some detail because it serves as a model system for the ability of a large particle like a virus to get through a membrane (see Rakonjac et al., Suggested Reading). Most of these studies have been per-

formed with phage f1, but related phages probably use a similar mechanism.

Phage f1 particle has only five proteins, one of which, the major head protein, pVIII, exists in about 2,700 copies and coats the DNA. The other four proteins are on the ends of the phage and exist in only four or five copies per phage particle. Two of these proteins, pVII and pIX, are located on one end of the phage, and the other two, pVI and pIII, are on the other end. To start the infection, the pIII protein on one end of the phage first makes contact with the end of the sex pilus. The sex pilus makes a good first contact point because it sticks out of the cell and hence is very accessible. The pilus retracts when the phage binds to it, drawing the phage to the cell surface. A different region of the same pIII protein then contacts a host inner membrane protein called TolA. How this contact is made is somewhat unclear. The TolA protein sticks into the periplasmic space and might make contact with the outer membrane since it is part of a larger structure whose role has something to do with keeping the outer membrane intact. The phage DNA then enters the cytoplasm, while the major coat protein is stripped off into the host inner membrane.

Release of the phage from the cell uses a different process. This process is illustrated in Figure 7.11. Unlike injection of the phage DNA, which must rely exclusively on host proteins, secretion of the phage from infected cells can use newly synthesized phage proteins synthesized during the course of the infection. After the

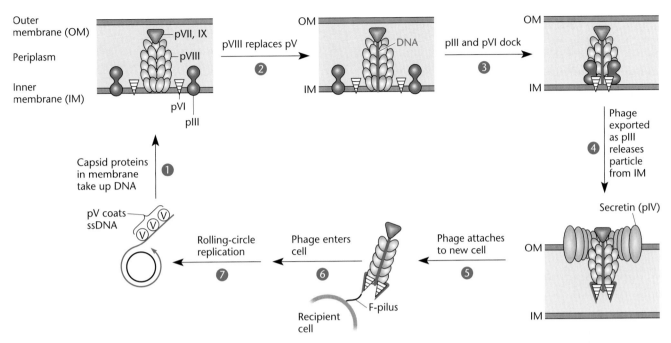

Figure 7.11   Infection cycle of the single-stranded DNA phage f1. Steps 1 through 7 show the encapsidation of phage DNA as it is secreted through the membrane pore formed by the pIV secretin to release the phage and infect a new cell. Details are given in the text. ssDNA, single-stranded DNA.

phage DNA has replicated a few times to produce the replicative-form DNA (see below), it enters the rolling-circle stage of replication. As it rolls off the circle, the newly synthesized single-stranded DNA is coated by another protein, pV. The proteins which make up the phage coat are waiting in the membrane, and the major head protein, pVIII, replaces the pV protein on the DNA as it enters the membrane. Only DNA containing the sequence of deoxynucleotides of the *pac* site of the phage is packaged. The other phage proteins are then added to the particle. Meanwhile, the phage-encoded secretin protein, pIV, has formed a channel in the outer membrane, through which the assembled phage can pass. This channel is related to the channels formed by type II secretion systems to assemble pili on the cell surface and by competence systems to allow DNA into the cell (see chapters 2, 6, and 14).

REPLICATION OF SINGLE-STRANDED PHAGE DNA
Studies of the replication of single-stranded phage DNA have contributed much to our understanding of replication in general. It was with these phages that rolling-circle replication was discovered, as well as many proteins required for host DNA replication including the proteins PriA, PriB, PriC, and DnaT, which are now known to be involved in restarting chromosomal

replication forks after they have dissociated upon encountering DNA damage (see Box 1.2). Many of the genes for these host proteins were found in searches for host mutants that cannot support the development of these phages and by reconstituting replication systems in vitro by adding host DNA replication proteins to phage DNA until replication was achieved.

The groups working on the secretion of single-stranded DNA phage from the cell have been different from those working on phage DNA replication, and these groups have used different phages. Much of the work on phage DNA replication has been done with M13 and φX174, and the genes for these phages are named somewhat differently from those for phage f1.

First we talk about the replication of the DNA of the filamentous phage M13, and then we compare it to the replication of the icosahedral phage φX174, which turns out to be surprisingly different.

Synthesis of the Complementary Strand To Form the First RF
The steps in the replication of phage M13 DNA are outlined in Figure 7.12. The DNA strand of the phage encapsidated in the phage head is called the + strand. Immediately after the single-stranded + strand DNA enters the cell, a complementary − strand is synthesized

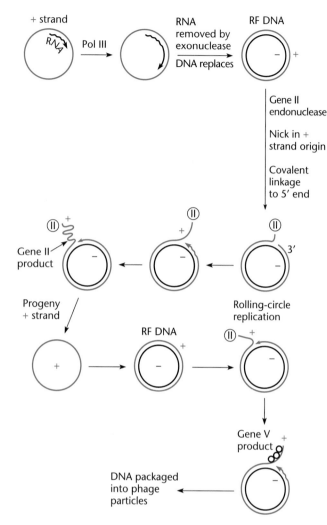

Figure 7.12 Replication of the circular single-stranded DNA phage M13. First, an RNA primer is used to synthesize the complementary minus strand (in black) to form double-stranded replicative-form (RF) DNA. The product of gene II, an endonuclease, nicks the plus strand of the RF and remains attached to the 5′ phosphate at the nick. Then more + strands are synthesized via rolling-circle replication, and their − strands are synthesized to make more RFs. Later, the gene V product binds to the plus strands as they are synthesized, preventing them from being used as templates for more RF synthesis and helping package them into phage heads.

on the + strand to form a double-stranded DNA called the **replicative form (RF)**. The formation of this first RF is dependent entirely on host functions, as it must be since no phage proteins enter the cytoplasm with the phage DNA and phage proteins cannot be synthesized from single-stranded DNA. The synthesis of the complementary strand is primed by an RNA made by the normal host RNA polymerase. Normally the host RNA

polymerase recognizes only double-stranded DNA, but the single-stranded phage DNA forms a hairpin at the origin of replication, making it double-stranded in this region. Once the RNA primer is synthesized, the DNA polymerase III and accessory proteins load on the DNA and synthesize the complementary strand until they have come full circle on the DNA and encounter the RNA primer. The 5′ exonuclease activity of DNA polymerase I then removes the RNA primer, and the nick is sealed by DNA ligase to leave a double-stranded covalently closed RF, which can then be supercoiled (see chapter 1). The replication also occurs on a specific site on the membrane of the bacterium, sometimes called the reduction sequence. This binding may direct the phage DNA to sites where the replication machinery of the host is located.

The icosahedral phages such as φX174 use a much more complicated mechanism to initiate synthesis of the first RF. Rather than using just the host RNA polymerase, they assemble a large primosome at a unique site on the single-stranded phage DNA. This primosome is composed of many copies of seven different proteins including DnaB (the replicative helicase), DnaC (which loads the helicase on the DNA), and DnaG (the primase). Most of these proteins are required to initiate replication at the chromosomal origin *oriC*, and they have been enlisted by the phage for the same purpose. However, the primosome also includes other proteins, PriA, PriB, PriC, and DnaT, which are not required for initiation of chromosomal replication at *oriC* but, rather, are required to restart chromosomal replication after it has been blocked due to encountering damage in the DNA template (see below). The role of the PriA, PriB, PriC, and DnaT proteins in the initiation of replication of φX174 DNA is not yet clear. Some of these proteins are helicases and may be required to open up a hairpin at the unique origin of replication of the single-stranded DNA and to allow the replication apparatus to load on the DNA and the primosome to move on the DNA.

The discovery of the function of Pri proteins is an interesting lesson in history with important general ramifications in science and medicine. Studies of the replication of φX174 DNA originally led to the discovery of the Pri proteins (see Box 1.2). At first, it was puzzling why these proteins would be required for the initiation of replication of phage φX174 DNA since they were not known to be required for replication of the host DNA. At the *oriC* site, where chromosomal replication normally initiates, only DnaC is required to load the replicative helicase, DnaB, and initiate replication. Later it was shown that the Pri proteins and DnaT protein were required to restart chromosomal replication forks, after they had collapsed at damage to the DNA. If the replication fork encounters damage in the template DNA and collapses, the recombination functions can promote the formation of a recombinational intermediate, and the Pri proteins and DnaT are required to load DnaB helicase at such a structure and reinitiate replication. This is analogous to the recombination-dependent replication (RDR) of phage T4 (see below). The φX174 phage enlists the host functions to initiate its own RF replication, presumably by having an origin of replication that mimics the recombinational intermediate. Presumably, eukaryotes, including humans, use a similar mechanism to restart replication forks, and this plays an important role in preventing DNA damage and therefore cancer. This is yet another case where studies of phage have led to the discovery of universal phenomena applicable to all organisms.

Synthesis of More RFs and Phage DNA

The subsequent steps in replication are probably similar in all single-stranded DNA phage but are best understood in M13. Once the first RF of M13 is synthesized, more RFs are made by semiconservative replication. This process requires phage proteins that are synthesized from the first RF. The two strands of the RF are replicated separately and by very different mechanisms. The + strand is replicated by rolling-circle replication from a different origin by a process similar to the replication of DNA during transfer of a plasmid by conjugation (see chapter 5). First, a nick is made in the RF, at the origin of + strand synthesis, by a specific endonuclease, the product of gene II in M13. A host protein called Rep, a helicase, helps unwind the DNA at the nick. The gene II protein remains attached through one of its tyrosines to the 5′ end of the DNA at the nick, and the DNA polymerase III, with its accessory proteins, extends the 3′ end to synthesize more + strand, displacing the old + strand. The gene II protein bound to the 5′ end of the old displaced strand then reseals the ends of the displaced strand by a transesterification reaction in which the phosphate attached to its tyrosine is passed back to the free 3′ end of the old strand. This recyclizes the old + strand, which can then serve again as the template for − strand synthesis to create another RF. Such transesterification reactions use little energy and are also used to recyclize plasmids after conjugation (see chapter 5) and in site-specific recombinases and some transposases (see chapter 9).

This process of accumulating RFs continues until the product of phage gene V begins to accumulate. This protein coats the single-stranded + strand of DNA, probably with the help of the attached gene II product, and prevents the synthesis of more RF by a complicated process that is only incompletely understood. The single-stranded viral DNA is then encapsulated in the head, and the cell is lysed (for icosahedral phages such as φX174) or transferred to the assembling viral particle in the membrane and leaked out of the cell (for filamentous phages such as f1 and M13) as described above.

M13 CLONING VECTORS

Because M13 and related phages encapsidate only one of the two DNA strands in their head, these phages provide a convenient vehicle for cloned DNA that we might want to sequence, use as a probe, or use in other applications that involve single-stranded DNA. Also, because filamentous phages such as M13 have no fixed length and the phage particle is as long as its DNA, foreign DNA of different lengths can be cloned into the phage DNA, producing a molecule longer than normal without disrupting the functionality of the phage.

Figure 7.13 shows one such phage-derived cloning vector, the M13 cloning vector M13mp18 (see Yanisch-Perron et al., Suggested Reading). Like pUC plasmid vectors (Figure 4.22), the mp series of M13 phage vectors contain the α-fragment-coding portion of the *E. coli lacZ* gene, into which has been introduced some convenient restriction sites (shown as the polylinker cloning site in Figure 7.13, with the multiple restriction sites shown at the bottom of the figure). Phage with a foreign DNA insert in one of these sites can be identified easily by insertional inactivation (see chapter 4) because they make colorless instead of blue plaques on plates containing X-Gal (5-bromo-4-chloro-3-indolyl-β-D-galactopyranoside).

To use a single-stranded DNA phage vector, the double-stranded RF must be isolated from infected cells, since most restriction endonucleases and DNA ligase require double-stranded DNA. A piece of foreign DNA is cloned into the RF by using restriction endonucleases, and the recombinant DNA is used to **transfect** competent bacterial cells. The term "transfection" refers to the artificial initiation of a viral infection by viral DNA (see chapter 6). When the RF containing the clone replicates to form single-stranded progeny DNA, a single strand of the cloned DNA is packaged into the phage head. The phage plaques obtained when these phage are plated are a convenient source of one strand of the cloned DNA. Which strand of the cloned DNA is represented in the single-stranded DNA will depend on the orientation of the cloned DNA in the cloning vector. If it is cloned in one orientation, one of the strands is obtained; if it is cloned in the other orientation, the other strand is obtained.

Some plasmid cloning vectors have also been engineered to contain the *pac* site of a single-stranded DNA phage. If cells containing such a plasmid are infected with the phage, the plasmid is packaged into the phage head in a single-stranded form. Such phages are called phasmids, indicating that they are a cross between a plasmid and a phage. The use of phasmids to study ColE1 plasmid incompatibility is discussed in chapter 4.

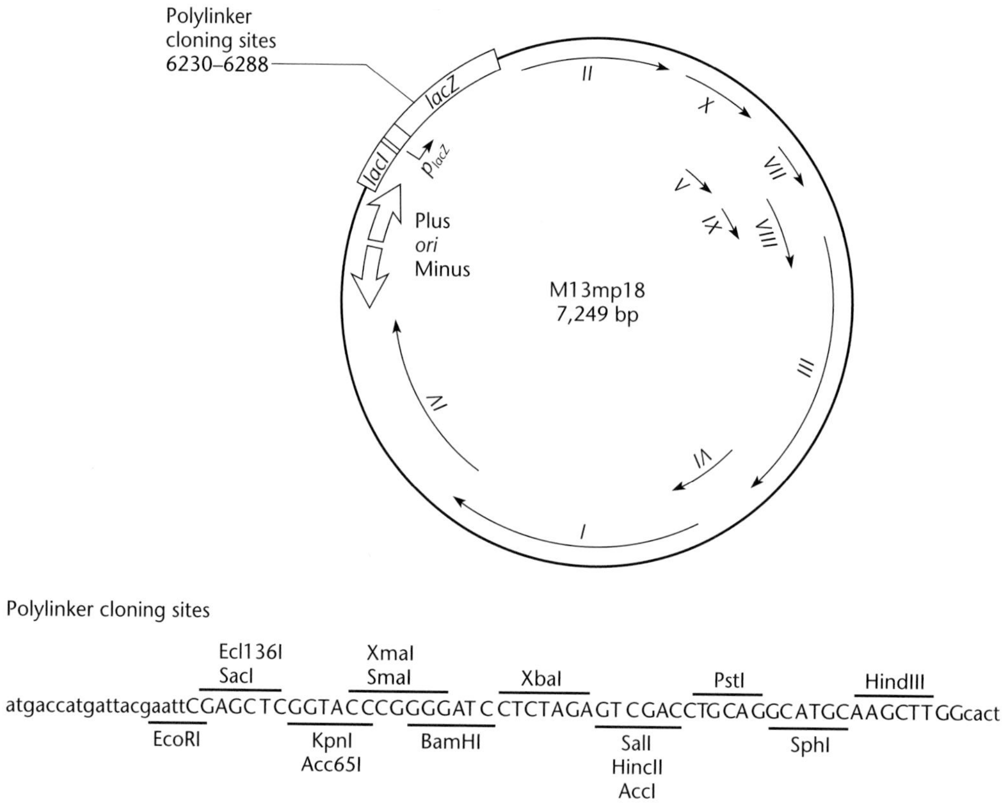

Polylinker
cloning sites
6230–6288

lacZ

II

X

VII

V

IX

VIII

*p*lacZ

Plus
ori
Minus

lacI

M13mp18
7,249 bp

III

IV

VI

I

Figure 7.13 Map of the M13mp18 cloning vector. The positions of the genes of M13 and the polylinker cloning site containing multiple restriction sites are shown below the map. Cloning into one of these sites inactivates the portion of the *lacZ* gene on the cloning vector, a process called insertional inactivation. The cloning vector also contains the *lacI* gene, whose product represses transcription from the p_{lac} promoter (see chapter 11).

Polylinker cloning sites

Ecl136I
SacI
XmaI
SmaI
XbaI
PstI
HindIII

atgaccatgattacgaattCGAGCTCGGTACCCGGGGATCCTCTAGAGTCGACCTGCAGGCATGCAAGCTTGGcact

EcoRI
KpnI
Acc65I
BamHI
SalI
HincII
AccI
SphI

Phage Display with Filamentous Phages

Single-stranded filamentous phages are also useful for phage display (Box 7.2). In fact, these were the first phages to be adapted for such uses. Phage display uses different vectors from those used for cloning and sequencing. In phage display, the polylinker cloning site is in one of the regions encoding a head protein, so that the coding sequence for a peptide cloned into the site is fused to the head protein sequence in such a way that the peptide is exposed or "displayed" on the surface of the phage particle. Figure 7.14 shows the head proteins that can be used for display. Phage displaying a desired peptide can be isolated by "panning" with another molecule that binds to the peptide. These phage have the advantage for phage display that they are small, making panning of large numbers of phage easier. However, only short peptides and those which can be transported through the membrane channels with the head protein can be displayed on their surface, since the phage must be secreted through the cell membranes (Fig. 7.11).

SITE-SPECIFIC MUTAGENESIS OF M13 CLONES

Because single-stranded DNA phages such as M13 offer a convenient source of only one of the two strands of a cloned DNA, they were used in the first applications of site-specific mutagenesis. As discussed in chapter 1, site-specific mutagenesis involves making a predetermined change in the DNA sequence, unlike random mutagenesis, where the change is made by chance. The standard method of using M13 for site-specific mutagenesis is diagrammed in Figure 7.15. The gene to be mutated has been cloned into a cloning vector such as M13mp18. An oligonucleotide complementary to the region to be

Figure 7.14 Schematic representation of the filamentous bacteriophage M13. The single-stranded circular DNA is coated with five viral proteins. The schematic locations of the different proteins are shown. The gpVIII protein is present at about 2,700 copies, while gpIII, gpVI, gpVII, and gpIX are present at about 5 copies each. All of the coat proteins can be used as platforms for protein display. With the exception of gpIII, the capsid proteins are small, with 33 to 112 amino acids.

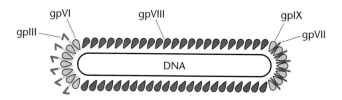

gpVI
gpVIII
gpIX
gpIII
gpVII
DNA

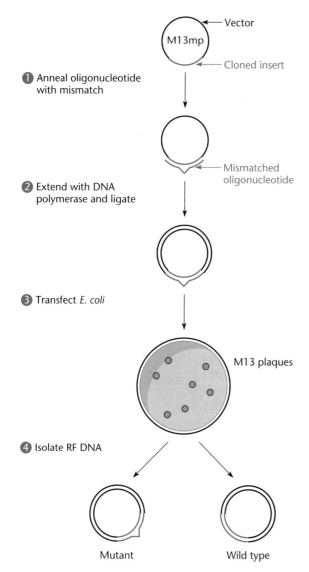

Figure 7.15 Site-specific mutagenesis with M13 and a mismatched oligonucleotide primer. See the text for details. RF, replicative form.

mutagenized, except for the change to be made, is synthesized and then hybridized to the single-stranded DNA. DNA polymerase is added, which uses the oligonucleotide as a primer to synthesize the complementary (or −) strand of the M13 DNA, including the clone. The double-stranded RF DNA is then ligated to give the covalently closed RF. When this DNA is transfected into cells, some of the plaques contain progeny phage which have the altered sequence of the oligonucleotide primer rather than the original sequence in the clone, depending on whether they are descended from the original template strand or the mutagenized complementary strand in the RF.

This method can also be adapted to allow random mutagenesis of a DNA region so that every possible base pair change will be represented in the population of molecules. Instead of a well-defined primer, a mixture of "spiked," or contaminated, oligonucleotide primers is used to mutagenize the DNA. These oligonucleotide primers are synthesized with the deliberate intention of making mistakes. The nucleotide added at each step of the synthesis is deliberately contaminated with a low concentration of the other three nucleotides; the concentration of the contaminating nucleotides is adjusted to make one mistake, on average, in each of the oligonucleotides. When these misfits are used as primers for the synthesis of complementary strands, as above, the DNA synthesized has a random collection of changes in the region being mutagenized. This method has the advantage that it can be used to make all of the possible base pair changes in a region without preferentially mutagenizing hot spots as chemical mutagens do.

The major difficulty with most methods of site-specific mutagenesis based on M13 lies in finding which of the plaques contain the few mutated M13 clones among the majority of plaques containing phage with the original wild-type sequence. A number of methods have been devised to eliminate phages with the original sequence. One of these is illustrated in Figure 7.16. In this method, the thymines in the M13 phage cloning vector are replaced with uracils by propagating the phage in a dUTPase- and uracil-N-glycosylase-deficient (Dut⁻ Ung⁻) host (see chapter 1). When the double-stranded RF molecules are synthesized from such a template, the newly synthesized strand contains thymines while the template strand still contains uracils. When these RFs are transfected into Ung⁺ cells, the uracil-containing template strands are preferentially degraded by the uracil-N-glycosylase, so that most of the phage that survive will be descended from the mutated complementary strands. In chapter 1, we describe another method for eliminating the nonmutated parental DNA after site-specific mutagenesis. This method uses two primers, one that alters a restriction site in the cloning vector and one that makes the desired change in the sequence of the cloned gene (see Figure 1.32). Most of the complementary strands are made using both primers so that they do not contain the restriction site. The RFs are transfected into cells, and the phage are allowed to multiply. The double-stranded RF phage DNAs are isolated from the infected cells and cut with the restriction endonuclease. The RF molecules with the restriction site intact are all cut, leaving mostly RFs that are descended from the newly synthesized strand with the desired mutation.

Methods for site-specific mutagenesis of M13 clones have been largely replaced by methods using PCR or

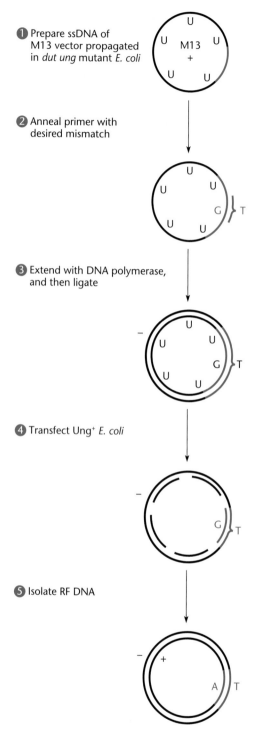

① Prepare ssDNA of M13 vector propagated in *dut ung* mutant *E. coli*

② Anneal primer with desired mismatch

③ Extend with DNA polymerase, and then ligate

④ Transfect Ung⁺ *E. coli*

⑤ Isolate RF DNA

Figure 7.16 Use of uracil-*N*-glycosylase to eliminate the wild-type sequence after site-specific mutagenesis. See the text for details. ssDNA, single-stranded DNA; RF, replicative form.

recombineering, which do not require additional cloning steps. However, the principles are similar, at least for PCR mutagenesis.

Phage T7: Linear DNA That Forms Concatemers

Phages such as M13 and φX174 solve the primer problem by having circular DNA, so that there is always DNA upstream to prime the synthesis of new DNA. Other phages, such as λ and P22, have cohesive ends at the ends of their DNA that can pair to form circles after infection (see chapter 8). However, some phages, such as T7 and T4, never cyclize their DNA but form **concatemers** composed of individual genome-length DNAs linked end to end. The phage DNA can then be cut out of these concatemers so that no information is lost when the phage DNA is packaged.

As shown in Figure 7.17, T7 DNA replication begins at a unique *ori* site and proceeds toward both ends of the molecule, leaving the 3′ ends single stranded because there is no way to prime replication at these ends. However, because T7 has the same sequence at both ends, these single strands are complementary to each other and so can pair, forming a concatemer with the genomes linked end to end. Consequently, the information missing as a result of incomplete replication of the 3′ ends is provided by the complete information at the 5′ end of the other daughter DNA molecule. Individual molecules are then cut out of the concatemers at the unique *pac* sites at the ends of the T7 DNA and packaged into phage heads. It is not clear how the terminal redundancies are re-formed in the mature phage DNAs, but it might be done by making staggered breaks and them filling them in with DNA polymerase or merely by discarding every other genome in the concatemer, which seems wasteful.

GENETIC REQUIREMENTS FOR T7 DNA REPLICATION

In contrast to single-strand DNA phages, which encode only two of their own replication proteins (the products of genes II and V in M13) and otherwise depend on the host replication machinery, T7 encodes many of its own replication functions, including DNA polymerase, DNA ligase, DNA helicase, and primase. The phage T7 RNA polymerase is also required to synthesize the initial primer for phage T7 DNA synthesis. In addition to these proteins, the phage encodes a DNA endonuclease and exonuclease that degrade host DNA to mononucleotides, thereby providing a source of deoxynucleotides for phage DNA replication. Analogous host enzymes can substitute for some of these T7-encoded gene products, so that they are not absolutely required for T7 DNA replication. For example, the T7 DNA ligase is not required because the host ligase can act in its stead. T7 DNA replication is a remarkably simple process that requires fewer gene products overall than the replication

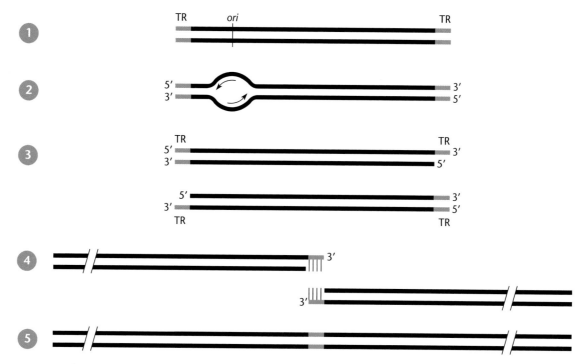

Figure 7.17   Replication of phage T7 DNA. Replication is initiated bidirectionally at the origin (*ori*). The replicated DNAs could pair at their terminally repeated ends (TR) to give long concatemers as shown.

of bacterial chromosomes and of many other large DNA phages.

Phage T4: Another Linear DNA That Forms Concatemers

Phage T4 also has linear DNA in its head that never cyclizes. It forms concatemers like T7, except that it forms them by recombination rather than by pairing between complementary single-stranded ends. However, T4 and T7 differ greatly in how the DNA replicates and is packaged. Also, befitting its larger size, T4 has many more gene products involved in replication than T7 does. As many as 30 T4 gene products participate in replication (Figure 7.6; Table 7.1). In fact, one of the advantages of studying replication with T4 is that it encodes many of its own replication proteins rather than just using those of its host. It encodes its own DNA polymerase, sliding clamp, clamp-loading proteins, primase, replicative helicase, DNA ligase, etc. These proteins are analogous to the replication proteins of bacteria and even eukaryotes (Table 7.1), and it was often in T4 phage where these functions were first discovered. Only later were their analogous functions found in uninfected bacteria and eukaryotes. Interestingly, the replication functions of T4 phage are often more similar to those of eukaryotes than to those of the bacteria they infect. For

example, the sliding clamp of T4 (the product of gene *45*) is more structurally similar to the sliding clamp of eukaryotes, called proliferating-cell nuclear antigen, then it is to the sliding clamp of *E. coli*.

OVERVIEW OF T4 PHAGE DNA REPLICATION AND PACKAGING
Phage T4 replication occurs in two stages, which are illustrated in Figure 7.18. In the first stage, T4 DNA replicates from a number of well-defined origins around the DNA. This type of replication is analogous to the replication of bacterial chromosomes and leads to the accumulation of single-genome-length molecules. However, these two daughter molecules have single-stranded 3' ends because of the inability of DNA polymerase to completely replicate the ends. They lose no information, however, because the sequences at the ends of T4 DNA are repeated; i.e., the DNAs are terminally redundant (see below). Somewhat later, this type of replication ceases and an entirely new type of replication ensues. The single-stranded repeated sequences at the ends of the genome-length molecules (called **terminal redundancies**) can invade the same sequence in other daughter DNAs, forming D-loops, which prime replication to form large branched concatemers. This replication, RDR (see "Discovery of Recombination-Dependent Replication" below)

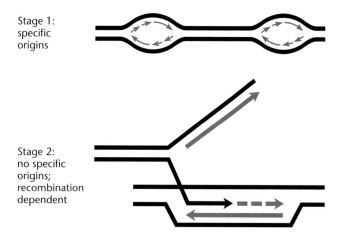

Stage 1:
specific
origins

Stage 2:
no specific
origins;
recombination
dependent

Figure 7.18 Initiation of replication of phage T4 DNA. In stage 1, replication initiates at specific origins, using RNA primers. In stage 2, recombinational intermediates furnish the primers for initiation. See also Figure 7.10.

is analogous to the "replication restarts" discussed in chapter 1 (see Box 1.2) and is now known to occur in all organisms. However it was in T4 that this type of replication was first discovered. The two stages of T4 DNA replication are discussed in more detail below.

Like T7, T4 DNA is packaged into the phage head from concatemers. Periodic cycles of RDR lead to the synthesis of very large, branched concatemers from which individual genome-length DNAs are cut out and packaged into phage heads. However, unlike phage T7, where the DNA is cut at unique *pac* sites, T4 DNA is cut out of the concatemers in phage "headfuls," as illustrated in Figure 7.19. It is like sucking a very long strand of spaghetti into your mouth until your mouth is full and then biting it off—not very polite, but effective. This is also how the terminal redundancies or repeated

sequences at the ends of T4 DNA are created. The head of phage T4 holds about 3% more DNA than a single length of the T4 genome, so that each molecule that is cut off includes some sequences from the next genome sequence in the concatemer. These sequences are then repeated at both ends of the DNA molecule in the head. Also, because it does not cut at unique *pac* sites, each T4 DNA that is packaged will have different sequences at its ends. In other words the genomes of the phages that come out of the infection are **cyclic permutations** of each other. The mathematical definition of a cyclic permutation is a permutation which shifts all elements of a set by a fixed offset, with the elements shifted off the end inserted back at the beginning. This explains why the genetic map of T4 is circular (Figure 7.6) even though T4 DNA itself never forms a circle. The way in which different modes of replication and packaging give rise to the various different genetic maps of phages is discussed later in this chapter (see "Genetic Analysis of Phages" below).

DETAILS OF T4 STAGE 1 REPLICATION: REPLICATION FROM DEFINED ORIGINS

As mentioned, the first stage of T4 DNA replication, from defined origins, is analogous to chromosome replication from unique origins in other organisms including bacteria and eukaryotes. Consistent with its large size, T4 has a number of defined origins around the chromosome, which can be used to initiate replication. This is unlike most bacteria and other phages including T7, which usually use only one unique origin to initiate replication. However, T4 most often uses only one of these origins, *oriE*, to replicate each chromosome.

The first step in initiating replication from a T4 origin is to synthesize RNA primers on the origin, using the host RNA polymerase. These primer RNAs also sometimes double as mRNAs for the synthesis of middle-mode

Figure 7.19 T4 DNA headful packaging. Packaging of DNA longer than a single genome equivalent gives rise to repeated terminally redundant ends and cyclically permuted genomes. (A) Headfuls of DNA are packaged sequentially from concatemers. Vertical arrows indicate the site of cleavage during packaging. (B) Each packaged genome is a different cyclic permutation with different terminal redundancies.

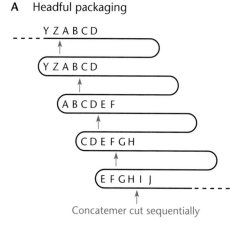

A Headful packaging

Y Z A B C D

Y Z A B C D

A B C D E F

C D E F G H

E F G H I J

Concatemer cut sequentially

B Cyclically permuted genomes

A B C D	Y Z A B
C D E F	A B C D
E F G H	C D E F
G H I J	E F G H

Terminal redundancies of packaged genomes

proteins and are made from middle-mode-type promoters. These promoters are first turned on a few minutes after infection and require RNA polymerase whose σ^{70} has been remodeled by binding the MotA and AsiA proteins (see above). In their role as primers, these short RNAs invade the double-stranded DNA at the origin and hybridize to the strand of the DNA to which they are complementary, displacing the other strand to create a structure called an R-loop. The invading RNA can then prime the leading strand of DNA replication from the origin. The gp41 replicative helicase, which plays the role of DnaB in uninfected *E. coli*, is then loaded on the DNA. The gp59 helicase-loading protein seems to assist in this but is not absolutely required. Other helicase-loading proteins may assist at particular promoters. Once the gp41 helicase is loaded, replication is under way.

Many of the proteins and functions of the replication apparatus of T4 have their counterparts in the replication apparatus of *E. coli* and eukaryotes (see chapter 1), with a few minor differences. A comparison of these functions in T4, *E. coli*, and eukaryotes is shown in Table 7.1. The gp41 helicase is associated with the lagging-strand primase (gp61), which primes replication of the lagging strand, similar to the role of the primase DnaG in *E. coli* DNA replication. Once replication is under way, the DNA polymerase (gp43) is held on the DNA by a sliding clamp (gp45), which has been loaded on the DNA by the clamp-loading proteins (gp44 and gp62) (a schematic of the T4 replication machinery is shown in Figure 7.10). The T4 sliding clamp is much like the β clamp of *E. coli* in that it forms a ring around the DNA. One difference is that each clamp is composed of three subunits, rather than two like the β clamp of *E. coli*, making the T4 sliding clamp more like the corresponding sliding clamp of eukaryotes, proliferating-cell nuclear antigen, than it is like the bacterial sliding clamp. Once synthesis of Okazaki fragments is complete, a T4-encoded DNA ligase (gp30) joins the pieces together, although the host DNA ligase can substitute to some extent for this function. After one or very few copies of the DNA have been made from defined origins, a helicase called UvsW can displace these R-loops, suppressing origin-specific replication in favor of recombination-dependent replication (see below).

STAGE 2: RECOMBINATION-DEPENDENT REPLICATION (RDR)

In the second stage of replication, the leading strand of T4 replication is primed by recombination intermediates rather than by primer RNAs synthesized by RNA polymerase (see Mosig, Suggested Reading). This T4 recombination-dependent replication is similar to replication restarts in uninfected cells (see chapter 1),

and the similarities and related proteins are mentioned in the following discussion (Table 7.1).

The first step in recombination-dependent replication is the invasion of a complementary double-stranded DNA by a single-stranded 3' end to form a three-stranded **D-loop** (Figure 7.18). This invading single-stranded 3' end is created during an earlier round of origin-specific replication by the inability of the DNA polymerase to replicate to the end of the molecule and could be extended by the action of exonucleases such as gp46 and gp47 on the ends of the molecule. In this respect, gp46 and gp47 are analogous to the RecBCD protein of *E. coli* (Table 7.1) (see chapter 10). If the cell has been infected by more than one T4 phage particle, the complementary sequence that the free 3' end invades could be anywhere in the DNA of a coinfecting phage, since T4 DNAs are cyclically permuted (see above). However, if the cell has been infected by a single phage, the newly replicated phage DNAs have the same sequences at their ends, and the single-strand invasion would be into the terminal redundancy of the other daughter DNA. This pairing of the invading strand with the complementary strand in the invaded DNA is promoted by the T4 *uvsX* gene product (Table 7.1), which is analogous in function to the RecA protein, the *E. coli* function that promotes single-strand invasion in uninfected cells (see chapter 10). Normally, single-stranded T4 DNA is coated with the T4-coded single-stranded-DNA-binding protein gp32, and the UvsX protein might need the help of another T4 protein, UvsY, to displace the gp32 protein, much as RecFOR proteins displace the *E. coli* single-stranded-DNA-binding protein SSB in uninfected cells. Once the D-loop has formed, the replicative helicase gp41 is loaded on the DNA by gp59, in a process that seems similar to the loading on of the DnaB helicase by the Pri proteins in uninfected cells during replication restarts. The invading 3' end can then serve as a primer for new leading-strand DNA replication, and the primase gp62 can be loaded on the displaced strand for lagging-strand replication. Later, when replication from defined origins has ceased and there are no double-stranded ends, recombination is initiated by other proteins, gp17 and gp49, that break the DNA and create ends for single-strand invasion (as above), and the process continues. Repeated rounds of strand invasion and replication lead to very long branched concatemers which can then be packaged into phage heads.

DNA PACKAGING FROM CONCATEMERS

Packaging of DNA into T4 phage heads is initiated by cutting the DNA by a terminase complex which remains attached to the end. This complex then binds to the head at an opening called the portal, and the DNA begins to be sucked into the head. However, the concatemers are branched as a result of recombination, and packaging from branched concatemers presents a potential problem for the phage. What keeps the phage head from "choking" on the branch when it tries to package a headful past a branch? This is another place where the gp49 protein comes into play. The gp49 protein is an X-phile (see chapter 10) capable of cutting Holliday junctions and DNA with branches. The gp49 protein cuts the branches, allowing the phage to package past branches to fill its head with DNA.

This is a simplified version of what must be a much more complicated mechanism of recombination-dependent replication. It ignores some known features of RDR such as its bidirectionality, as well as details about the roles of some of the helicases and exonucleases, among other enzymes, known to be required for this process. The details of RDR in this and other systems are still being uncovered.

DISCOVERY OF RECOMBINATION-DEPENDENT REPLICATION

The discovery of RDR is an interesting example of how progress in basic cell biology often occurs. First, a basic cellular mechanism is discovered and characterized in a relatively simple organism or a phage, which are more accessible to experimentation than higher organisms. Once the basic cellular function is characterized in a more malleable system, it is found to exist in all organisms. The number of developments that have come this way are too numerous to mention but include essential features of the mechanisms of protein synthesis and folding and of DNA replication.

The discovery of RDR followed the same pattern. It had been known since the 1960s that T4 DNA replication ceases prematurely in the absence of recombination. As evidence, when cells are infected by T4 with mutations in any of the recombination function genes, genes *46, 47, 49, 59, uvsX*, etc. (see chapter 10), replication begins normally but soon ceases. This was named the DNA-arrested phenotype. In the early 1980s, Gisela Mosig proposed that recombination functions were required for T4 DNA replication later in infection because replication at later times is dependent on recombination intermediates. She proposed a model whereby D-loops formed by strand invasion could prime the leading strand of DNA replication. It was some time before this model was generally accepted and even longer before it was thought to be anything but an arcane mechanism confined to this type of phage. Only relatively recently has it come to be recognized that ubiquitous phenomena such as double-strand break repair, replication restart, and intron and intein mobility—processes common to all

organisms—use basically the same mechanism as RDR (see Kreuzer, Suggested Reading). Developments in these fields are covered in chapters 1, 10, and 11.

Phage Lysis

Once the phage DNA has replicated and been packaged into heads, the remainder of the phage assembles and the phage are released from the infected cell to infect other cells. Some phages, such as the small filamentous DNA phages, including M13 and F1, assemble in the membrane and then leak out of the cell, using a modified type II secretion system. Type II systems have already been mentioned in chapter 6 because of their relationship to some competence systems for DNA transformation and because they are used to secrete the pilin proteins of type IV pili, but these phages have adapted them to their own use by encoding their own secretin proteins that form a channel through the outer membrane through which the assembled phage can pass.

As mentioned, phages that leak out of the infected cell without killing them could be said to cause a chronic infection because the host is not killed and continues to produce phage. However, most phages cause an acute infection, in that a cell infected by one of these phages begins to round up later in infection and then suddenly explodes, releasing the phage. Phages that cause the cell to lyse (explode) encode two proteins: one is a lysozyme that cleaves bonds in the peptidylglycan cell wall, breaking it open, and the other is a holin protein that resides in the inner membrane and somehow activates the lysozyme at the appropriate time for cell lysis. The name *holin* implies that these proteins form holes or pores in the membrane through which the lysozyme can pass. Their role in making pores is suggested by the fact that they have one or more transmembrane domains which are strings of mostly hydrophobic amino acids that can traverse the membrane (see chapter 2). However, some lysozymes, particularly in phages of gram-positive bacteria, have signal sequences and seem to be transported through the membrane by the *sec* system (see chapter 2), so they have no need for pores in the membrane. Nevertheless, these phages also encode holins that are required for activation of the lysozyme. In these cases, at least, the holin apparently works by a different mechanism to time the lysis.

Timing of lysis is obviously very important for the phage. If the cell lyses too soon, no or very few phage will have been produced; if it lyses too late, time will be lost and the phage will take too long to spread through a population of bacteria to compete effectively. To ensure that the timing of lysis is exquisitely regulated, many phages also make antiholins that inhibit the holins until it is time to activate the lysozyme and lyse the cell. The antiholin-holin pairs of some phages are shown in Figure 7.20. The antiholin is often a shorter or somehow altered form of the holin and presumably binds to it, inactivating it. For example, the antiholin of phage λ (see chapter 8) is a slightly longer version of its holin. It is translated from the same open reading frame, but its translation starts somewhat upstream from a different

Figure 7.20   Timing of phage lysis by activation of holins. The antiholin keeps the holin inactive until the time of lysis. The holin then becomes active, forming a pore that allows the lysozyme (in purple) to traverse the membrane and then to degrade the cell wall and lyse the cell. (A) λ phage. The antiholin (S107) and holin (S105) differ only in that the antiholin has an extra two amino acids at its N terminus. This makes the antiholin inactive as a holin, but it still binds to the holin, inactivating it. At the time of lysis, the extra two amino acids might be removed from the antiholin, converting it into active holin and allowing it to participate in the formation of pores through the membrane. (B) T4 phage. The antiholin (gpr1) binds to the holin (gpt) in the periplasm, inactivating it. At the time of lysis, the antiholin somehow becomes inactive, allowing the holin to form a pore in the membrane.

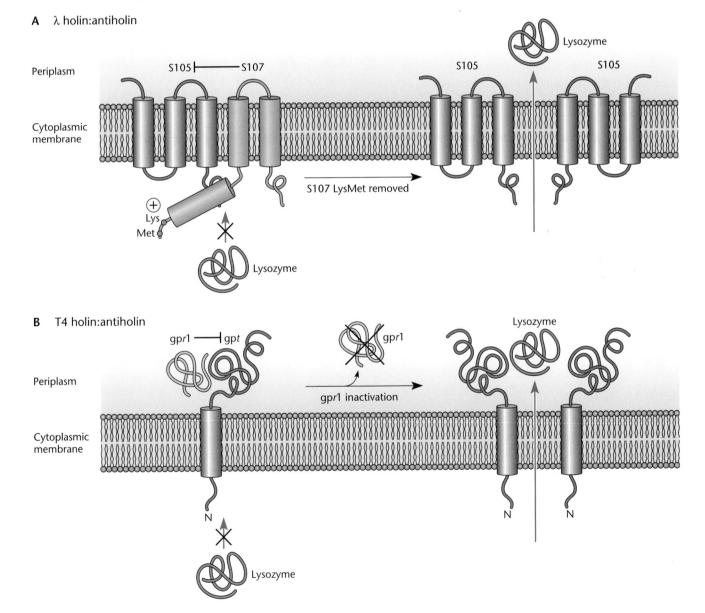

A λ holin:antiholin

B T4 holin:antiholin

AUG codon, adding an extra methionine and lysine at its N terminus. For some reason these two extra amino acids at its N terminus make the antiholin inactive as a holin, perhaps because the positively charged lysine cannot enter the membrane. However, it can still bind to the holin and inactivate it, perhaps by contributing to the formation of channels that are then inactive as shown. At the appropriate time for lysis, the antiholin might lose its extra amino acids and become an active holin, thereby releasing its inhibition of the holin, and itself contributing to lysis.

By contrast, the T4 antiholin, the product of gene *r1*, gp*r1*, is unrelated to the holin, which is encoded by a different gene, gene *t*. Nevertheless, it can inhibit the holin until time for lysis by an unknown mechanism. The *r1* and *t* genes have been known for a long time, because mutations in gene *r1* cause premature lysis and mutations in gene *t* delay lysis. However, only recently have their roles as antiholins and holins begun to be appreciated.

The players in the timing of lysis are now known for a number of phages, and mutational studies have been done on the role of specific amino acids and regions of the proteins on the timing. However, in no case do we know for certain the cause of the timing of inactivation of the antiholin and activation of the holin and cell lysis. In some cases at least, it seems to be linked, somehow, to loss of the membrane potential called the proton motive force (PMF). The membrane potential drives ATP synthesis by the membrane ATPase, among other cellular functions. The PMF is abruptly lost at about the same time as the antiholin becomes inactive, the holin becomes active, and the cell lyses. However, which causes which is not clear. Either activation of the holin is causing loss of the PMF, perhaps by making the membrane permeable to protons, or loss of the PMF due to some other change in the cell, perhaps accumulated damage to the membrane, is causing inactivation of the antiholin, activation of the holin, and cell lysis.

Genetic Analysis of Phages

Phages are ideal for genetic analysis (see the introductory chapter). They have short generation times and are haploid. Mutant strains can be stored for long periods and resurrected only when needed. Also, phages multiply as clones in plaques, and large numbers can be propagated on plates or in small volumes of liquid media. Different phage mutants can be easily crossed with each other, and the progeny can be readily analyzed. Because of these advantages, phages were central to the development of molecular genetics, and important genetic principles such as recombination, complementation, suppression, and *cis*- and *trans*-acting mutations are most easily demonstrated with phages. In this section, we discuss the general principles of genetic analysis of phages. However, most of the genetic principles presented here are the same for all organisms, including humans. Only the details of how genetic experiments are performed differ from organism to organism.

Infection of Cells

The first step in doing a genetic analysis of phages, or any other virus for that matter, is to infect cells with the phage. Phages can infect only cells that are sensitive or susceptible to them, and they can multiply only in cells which are permissive for their development. To multiply in a cell, not only must the phage adsorb to the cell surface of the bacterium and inject its nucleic acid, either DNA or RNA, but also a permissive host cell must provide all of the functions needed for multiplication of the phage. Therefore, most phages can infect and multiply in only a very limited number of types of bacteria. The types of cells which a phage can multiply are called its host range. Sometimes a normally permissive type of cell can become a **nonpermissive host** for the phage as a result of a single mutation or other genetic change. Alternatively, a mutant virus or phage may be able to multiply in a particular type of host cell under one set of conditions, for example at lower temperatures, but not under a different set of conditions, for example at higher temperatures. The conditions under which it can multiply are **permissive conditions**, while the conditions under which it cannot multiply are **nonpermissive conditions**.

MULTIPLICITY OF INFECTION

Infecting permissive cells with a phage is simple enough in principle. The phage and potential bacterial host need only be mixed with each other, and some bacteria and phage will collide at random, leading to phage infection. However, the percentage of the cells that are infected depends on the concentration of phage and bacteria. If the phage and bacteria are very concentrated, they collide with each other to initiate an infection more often than if they are more dilute.

The efficiency of infection is affected not only by the concentration of phage and bacteria but also by the ratio of phage to bacteria, the **multiplicity of infection (MOI)**. For example, if 2.5×10^9 phage are added to 5×10^8 bacteria, there are $2.5 \times 10^9 / 5 \times 10^8 = 5$ phage for every cell, and the MOI is 5. If only 2.5×10^8 phage had been added to the same number of bacteria, the MOI would have been 0.5.

The MOI can be either high or low. If the number of phage greatly exceeds the number of cells to infect, the cells are infected at a **high MOI**. Conversely, a **low MOI** indicates that the cells outnumber the phage. To illustrate, an MOI of 5 is considered high; there are five times as many phage as bacteria. An MOI of 0.5 is low; there is only one phage for every two bacteria. Whether a high or low MOI is used depends on the nature of the experiment. At a high enough MOI, most of the cells are infected by at least one phage; at a low MOI, many of the cells remain uninfected but each infected cell is usually infected by only one phage.

Even at very high MOI, not all the cells are infected. There are two reasons for this. First, infection by phage is never 100% efficient. The surface of each cell may have only one or a very few receptors for the phage, and a phage can infect a cell only if it happens to bind to one of these receptors. There is also the statistical variation in the number of phages which bind to each cell. Because the chance of each phage binding to a cell is random, the number of phages infecting each cell follows a normal distribution. Some cells are infected by five phages—the average MOI—but some are infected by six phages, some by four, some by three, and so on. Even at the highest MOIs, some cells by chance receive no phage and so remain uninfected.

The minimum fraction of cells that escape infection due to statistical variation can be calculated by using the Poisson distribution, which can be used to approximate the normal distribution in such situations. In chapter 3 we discuss how Luria and Delbrück used the Poisson distribution to estimate mutation rates. According to the Poisson distribution, the probability of a cell receiving no phages and remaining uninfected (P_0) is at least e^{-MOI}, since the MOI is the average number of phage per cell. If the MOI is 5, then $P_0 = e^{-5} = $ ~0.0067; i.e., at least 0.67% of the cells remain uninfected. At an MOI of only 1, e^{-1}, or at least ~37%, of the cells remain uninfected. In other words, at most ~63% of the cells are infected at an MOI of 1. Even this is an overestimation of the fraction of cells infected, since, as mentioned, some of the viruses never actually infect a cell.

Phage Crosses

As with any genetic analysis, the first step in a phage genetic analysis is to isolate mutant phage with the desired mutations. Once the mutations to be tested are chosen, the mutated DNAs of two members of the same species must be put together into the same cell. This is called **crossing**. If the DNAs of the two different organisms are in a cell at the same time, the genes of both mutant strains can be expressed and the two DNAs can recombine with each other. Crosses in cellular organisms are usually performed by mating the two organisms to form zygotes that can develop into the mature organism. Crosses in phage and other viruses are performed by

infecting the same cell with different strains of the virus at the same time.

To be certain that many of cells in a culture are simultaneously infected by both strains of a phage, we must use a high MOI of both phages. The Poisson distribution can again be used to calculate the maximum fraction of cells that will be infected by both mutant phages at a given MOI of each. If an MOI of 1 for each mutant phage is used for the infection, then at least e^{-1} or ~0.37 (37%) of the cells will be uninfected by each phage strain and at most $1 - 0.37 = 0.63$ (63%) of the cells will be infected with each mutant strain of the phage. Since the chance of being infected with one strain is independent of the chance of being infected by the other strain, at most $0.63 \times 0.63 \approx 0.40$ (40%) of the cells will be infected by *both* phage strains at an MOI of 1. This shows that only when both phage strains have a high MOI will most of the bacteria be infected by both strains.

Recombination and Complementation Tests with Phages

As discussed in chapter 3, two basic concepts in classical genetic analysis are recombination and complementation. The types of information derived from these tests are completely different. In recombination, the DNA of the two parent organisms is assembled in new combinations, so that the progeny have DNA sequences from both parents. In complementation, the gene products synthesized from two different DNAs interact in the same cell to produce a phenotype.

RECOMBINATION TESTS

The principles of recombination are the same for all organisms, but they are most easily illustrated with phage. Figure 7.21 gives a simplified view of what happens when two DNA molecules from different strains of the same phage recombine. The two mutant phage strains infecting the cell are almost identical, except that one has a mutation at one end of the DNA and the other has a mutation at the other end. The sequences of the two DNAs therefore differ only at the ends, the sites of the two mutations. Recombination occurs by means of a crossover between the two DNA molecules, where the two DNAs are broken at the same place. The ends of one DNA created by the break are joined to the ends of the other DNA to create two new molecules which are identical in sequence to the original molecules except that one part now comes from one of the original DNA molecules and the other part comes from the other. The effect, if any, of the crossover depends on where it occurs. If the crossover occurs between the sites of the two mutations, two new types of recombinant DNA molecules

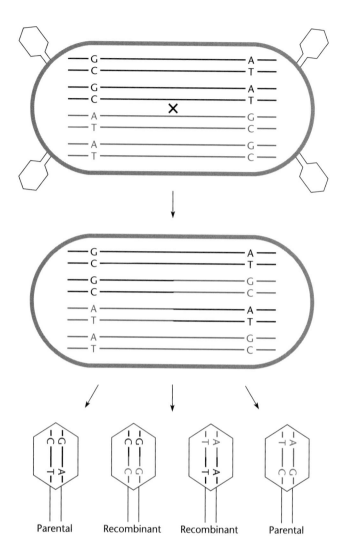

Figure 7.21 Recombination between two phage mutations. The two different mutant parent phages infect the same permissive host cell, and their DNA replicates. Crossovers occur in the region between the two mutations, giving rise to recombinant types that are unlike either parent phage. Only the positions of the mutated base pairs are shown. The DNA of one parent phage is shown in black, and that of the other is shown in purple.

appear: one has neither mutation, and the other has both mutations. Progeny phage that have packaged the DNAs with these new DNA sequences are **recombinant types** because they are unlike either parent (see chapter 3). Progeny phage that have packaged a DNA molecule with only one of the mutations are called **parental types** because they are like the original phages that infected the cell. The appearance of recombinant types tells us that recombination has occurred. Note that the decision about what will be a recombinant type and what will be a parental type depends on how the mutations were

distributed between the parent phages. In Figure 7.21, one parent strain had one mutation and the other parent strain had the other mutation. However, one strain could have had both mutations and the other strain could have had neither. In that case, the recombinant types would have had only one or the other of the two mutations and the parental types would have had either both mutations or neither.

RECOMBINATION FREQUENCY

The closer together the regions of sequence difference are to each other in the DNA, the less room there is between them for a crossover to occur. Therefore, the frequency of recombinant-type progeny is a measure of how far apart the mutations are in the DNA of the phage. This number is usually expressed as the recombination frequency. In general, independent of the type of organism involved in the cross, the recombination frequency is defined as the number of recombinant progeny divided by the total number of progeny produced in the cross (see chapter 3). When the recombination frequency is expressed as a percentage, it is called the map unit. For example, the regions of two mutations in the DNA give a recombination frequency of 0.01 if 1 in 100 of the progeny are recombinant types. The regions of the two mutations are then $0.01 \times 100 = 1$ map unit apart.

Different organisms differ greatly in their recombination activity; therefore, map distance is only a relative measure and a map unit represents a different physical length of DNA for different organisms. Also, the recombination frequency can indicate the proximity of two mutated regions only when the mutations are not too far apart. If they are far apart, two crossovers often occur between them, reducing the apparent recombination frequency. Note that while one crossover between the regions of two mutations will create recombinant types, two crossovers will recreate the parental types. In general, odd numbers of crossovers produce recombinant types and even numbers recreate parental types.

COMPLEMENTATION TESTS

Complementation is also most easy to demonstrate with phages and other viruses, although the nomenclature and concepts are the same for all organisms. As with recombination tests, to perform a complementation test with phages, cells are infected simultaneously by different strains of a particular phage. However, rather than measure the frequency with which recombination occurs between the regions of mutations, complementation measures the interaction between gene products synthesized from different DNAs in the same cell (see Table 3.5). Usually, with phage, we are asking whether the two mutations complement each other to allow the phage to multiply under conditions that are nonpermissive for either of them alone. If they complement each other, both mutant phage multiply; if they do not complement each other, neither will multiply. If the two mutations complement each other, they are probably in different genes.

Figure 7.22 illustrates complementation tests with phages. In the example, two different mutant strains of a phage infect the same host cell. This host cell is normally a permissive host for the wild-type phage but cannot propagate either of the mutant phages by itself because they each cannot make a gene product required for multiplication on that host. The outcome of the infection depends on whether the mutations are in the same or different genes. If they are in different genes (left side of the figure), each DNA furnishes one of the needed gene products and so the two mutations complement each other and both mutant viruses can multiply. If, however, the two mutations are in the same gene (right side of figure), neither DNA furnishes that gene product and so the

mutations do not complement each other and neither mutant strain multiplies. Usually the interpretation of complementation tests follows this simple rule. However, as discussed in chapter 3, the interpretation of complementation experiments can be complicated by intragenic complementation and by polarity and translational coupling. Complementation tests can also be used to determine whether a mutation affects a gene product (i.e., is *trans* acting) or whether it affects a site on the DNA such as a promoter or origin of replication (i.e., is *cis* acting). If the mutation affects a *trans*-acting gene product, it can be complemented; if it affects a *cis*-acting site, it cannot.

Note the difference in the hosts used for recombination and complementation tests with phage or other viruses. In a recombination test, the permissive host cells are infected with the two strains and the phage are allowed to multiply before the genotypes of the progeny phage are tested for recombinant types. However, in the complementation test, the host cells are infected with both mutant strains under nonpermissive conditions, and only if the mutations complement each other do the phage multiply. If complementation occurs, most of these progeny phage will still be the parental types, unable to multiply alone in subsequent infections under nonpermissive conditions.

Genetic Experiments with the *r*II Genes of Phage T4

We illustrate the basic principles of phage genetics by using the *r*II genes of phage T4. Experiments with these genes were responsible for many early developments in molecular genetics, including the discovery of nonsense codons, the definition of the nature of the genetic code, and the discovery of gene divisibility. Considering their historical importance, it is ironic that we still do not know what these gene products do for the phage. However, as discussed in chapter 3, one of the advantages of genetic analysis is that one can perform a genetic analysis without knowing the functions of the genes involved.

The name *r*II means "rapid-lysis mutants type II." Many genes of this phage were named before the current three-letter names for genes became conventional. Phage with an *r*-type mutation cause the infected cells to lyse more quickly than the normal (*r*+) phage, a property that *r*II mutant phage share with the other rapid-lysis mutants types *r*I and *r*III. Recall from above that the product of *r*I is an antiholin and the function of the *r*III product is unknown. Phage with a rapid-lysis mutation can be distinguished by the appearance of their plaques on *E. coli* B indicator bacteria. The plaques formed by wild-type *r*+ phage have fuzzy edges because of a phenomenon called lysis inhibition, which delays lysis of the infected cells. However *r*-type mutants do not show lysis

Figure 7.22 Tests of complementation between phage mutations. Phages with different mutations infect the same host cell, in which neither mutant phage can multiply. (Left) The mutations, represented by the minus signs, in different genes (*M* and *N*). Each mutant phage synthesizes the gene product that the other one cannot make; complementation occurs, and new phage are produced. (Right) Both mutations (minus signs) prevent the synthesis of the *M* gene product. There is no complementation, the mutants cannot help each other multiply, and no phages are produced.

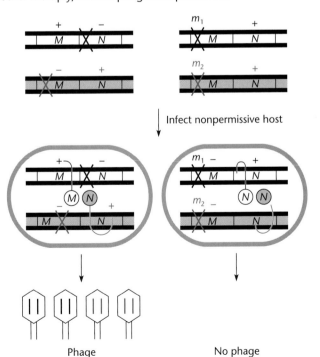

inhibition, which causes them to form hard-edged, clear plaques. The hard-edge, clear-plaque phenotype makes it easy to distinguish rapid-lysis mutants from the wild type (Figure 7.23).

The property of rII mutants that distinguishes them from the other types of rapid-lysis mutants is that they cannot multiply in strains of *E. coli* that are lysogenic for the λ prophage; these strains are designated *E. coli* K-12λ or Kλ to indicate that they harbor the λ prophage in their chromosome. Lysogeny is discussed in chapter 8. As mentioned below, the inability of rII mutants to multiply in λ lysogens greatly facilitates complementation tests with rII mutants and makes possible the detection of even very rare recombinant types.

COMPLEMENTATION TESTS WITH rII MUTANTS

In about 1950, Seymour Benzer and others realized the potential of using the rII genes of T4 to determine the detailed structure of genes. The first question asked was that of how many genes, or **complementation groups,** are represented by rII mutations. To obtain an answer, numerous phage with rII mutations were isolated and pairwise complementation tests were performed in which two different rII mutants infected cells of the nonpermissive host, *E. coli* Kλ, at the same time. When the two rII mutations complemented each other, phage were produced. These complementation tests revealed that all rII mutations could be sorted into two complementation groups, or genes, which were named rIIA and rIIB. The

Figure 7.23 Plaques of phage T4. Most plaques are fuzzy edged, but some due to rII or other r-type mutants have hard edges because of rapid lysis of the host cells.

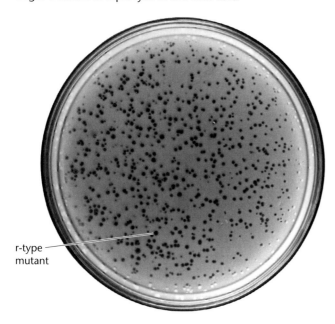

r-type mutant

investigators concluded that the rIIA and rIIB genes encode different polypeptides, both of which are required to make normal-appearing plaques and for multiplication in *E. coli* Kλ.

RECOMBINATION TESTS WITH rII MUTANTS

The next step in the genetic analysis was to perform recombination tests between rII mutations to determine the location of the rIIA and rIIB genes with respect to each other and to order the mutations within these genes. Recombination between rII mutations was measured by infecting permissive *E. coli* B with two different rII mutants and allowing the phage to multiply. The progeny phage were then plated to measure the frequency of recombinant types.

If recombination can occur between the two rII mutations, two different recombinant-type progeny would appear: double mutants with both rII mutations, and wild-type, or r^+, recombinants with neither mutation. The recombinant types with both mutations are difficult to distinguish from the parental types. However, the r^+ recombinants with neither rII mutation are easy to detect because they can multiply and form plaques on *E. coli* Kλ. Therefore, when the progeny of the cross are plated on *E. coli* Kλ, any plaques that appear are due to r^+ recombinants. As discussed above, the recombination frequency equals the *total* number of recombinant types divided by the total progeny of the cross. We can assume that about half of the recombinants are not being detected because they are double-mutant recombinants, and so the total number of recombinant types is the number of r^+ recombinants multiplied by 2. All the progeny should form plaques on *E. coli* B, so this is a measure of the total progeny. Hence, the recombination frequency between the two rII mutations is twice the number of phage that form plaques on *E. coli* Kλ divided by the number of phage that form plaques on *E. coli* B. However, for practical reasons, we cannot merely plate the cross on the two types of bacteria and count the plaques. There are many fewer r^+ recombinant progeny than total progeny, so that if the cross is plated directly on the two types of bacteria, there may be only a few plaques on *E. coli* Kλ and millions of plaques on *E. coli* B—too many to count. Therefore, the phage progeny of the cross must be serially diluted by different amounts before they are plated on the two types of bacteria, and the final recombination frequency must take these differences in dilution into account.

To illustrate, let us cross an rII mutant that has the mutation *r168* with another rII mutant that has the mutation *r131*. We infect *E. coli* B with the two mutants and incubate the infected cells to allow the phage to multiply. To determine the number of r^+ recombinants, we dilute the phage by a factor of 10^5 and plate on the *E. coli* Kλ indicator bacteria. To determine the total number of progeny, we dilute the phage by a factor of 10^7 and plate on the *E. coli* B indicator bacteria. After incubating the plates overnight, we observe 108 plaques on the *E. coli* Kλ plate and 144 plaques on the *E. coli* B plate. The total number of recombinant-type plaques is twice the number of plaques on *E. coli* K-12λ or 2 × 108 = 216, since we are counting only the r^+ recombinants and not the double-mutant recombinants, which are produced in equal numbers. This number must be multiplied by the dilution 10^5 to get the total number of recombinant-type phage, which is $2.16 × 10^7$. The total number of progeny phage is the number of plaques on the permissive bacteria *E. coli* B times the dilution factor or $144 × 10^7 = 1.44 × 10^9$.

From the equation for recombination frequency (RF)

$$RF = \text{total recombinant progeny/total progeny}$$

the recombination frequency between the two mutations is 0.015. If we want to express this in map units, we multiply the recombination frequency by 100, which gives 1.5. The two mutations *r168* and *r131* are therefore only 1.5 map units apart, which is very close in T4 DNA. Therefore, crosses such as these revealed that *r*IIA mutations are close to *r*IIB mutations, suggesting that the *r*IIA and *r*IIB genes are adjacent in the T4 DNA.

ORDERING *r*II MUTATIONS BY THREE-FACTOR CROSSES

Measuring the recombination frequency between two mutations can give an estimate of how close together the two mutations are in the DNA. However, to determine the relative order of mutations from such crosses, it is necessary to measure recombination frequencies very accurately. For example, let us say we have three *r*IIA mutations, *r21*, *r3*, and *r12*, and we want to use recombination frequencies to determine their order in the *r*IIA gene. When we cross *r12* with *r3*, we obtain a recombination frequency of approximately 0.01. When we cross *r21* with *r12*, we also obtain a recombination frequency of approximately 0.01. When we cross *r3* with *r21*, we get a recombination frequency of approximately 0.02. From these data alone, we suspect that *r12* is between *r3* and *r21* and that the order of the three mutations is *r3*-*r12*-*r21*. However, it is difficult to measure recombination frequencies accurately enough to be certain.

Three-factor crosses offer a less ambiguous method for ordering mutations. The principle behind a three-factor cross is illustrated in Figure 7.24. In this method, a mutant strain that has two mutations is crossed with another strain that has the third mutation. The number of wild-type recombinants is then determined. In such a

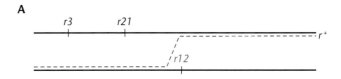

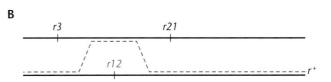

Figure 7.24   A three-factor cross to map *r*II mutations. See the text for details.

cross, the number of crossovers required to make a wild-type recombinant depends on the order of the three mutations; the more crossovers required to make a wild-type recombinant, the less frequent that recombinant type will be.

In Figure 7.24, a three-factor cross is being used to order the three *r*II mutations. First, a double mutant is constructed with the *r21* and *r3* mutations. Then this double mutant is crossed with a mutant that had only the *r12* mutation. If the order is *r3*-*r21*-*r12*, only one crossover between *r21* and *r12* is required to make an r^+ recombinant, and the frequency of r^+ recombinants should be about 0.01, the frequency of recombination between the two single mutations *r21* and *r12*. The same is true if the order is *r21*-*r3*-*r12*, where, again, only one crossover is required to make r^+ recombinants (data not shown). If, however, the order is *r3*-*r12*-*r21*, with the *r12* mutation in the middle, as we suspected from the earlier two-factor crosses, two crossovers are required and the frequency of r^+ recombinants is much lower than 0.01.

Theoretically, if the two crossovers were independent, the frequency of double crossovers should be the product of the frequencies of each of the single crossovers, or about 0.01 × 0.01 = 0.0001, which is only 1/100 of the frequency of the single crossover. However, because of high negative interference, also called gene conversion (see chapter 10), crossovers close to each other in the DNA are not truly independent, and one crossover greatly increases the likelihood of what appears to be a second crossover nearby, making the frequency of apparent double crossovers much higher than predicted. Nevertheless, the frequency of double crossovers is generally much lower than the frequencies of the individual crossovers, permitting the use of three-factor crosses to unambiguously order mutations.

ORDERING LARGE NUMBERS OF *r*II MUTATIONS BY DELETION MAPPING

One of the early contributions of *r*II genetics was the ordering by Benzer of large numbers of mutations in the *r*II genes, both spontaneous mutations and mutations induced by mutagens (see Benzer, Suggested Reading). As part of his genetic analysis of the structure of the *r*II genes, Benzer wanted to determine how many sites there are for mutations in *r*IIA and *r*IIB and to determine whether all sites within these genes are equally mutable or whether some are preferred. To find these answers, he turned to **deletion mapping**. The principle behind this method is that if a phage with a point mutation is crossed with another phage with a deletion mutation, no wild-type recombinants appear when the point mutation lies within the deleted region. It is much easier to determine whether there are any r^+ recombinants at all than it is to carefully measure recombination frequencies. Therefore, deletion mapping offers a convenient way to map large numbers of mutations.

For this approach, Benzer needed deletion mutations extending for known distances into the *r*II genes. Some of the *r*II mutants he had already isolated had the properties of deletion mutations (see chapter 3). First, these *r*II mutations did not revert, as indicated by the lack of plaques due to r^+ revertant phage, even when very large numbers of *r*II mutant phage were plated on *E. coli* Kλ. Second, these mutations did not map at a single position, or point, as would base pair changes or frameshift mutations. They did not give r^+ recombinants when crossed with many different *r*II mutations, at least some of which gave r^+ recombinants when crossed with each other and so must have been at different positions in the *r*II genes.

Figure 7.25 shows a set of Benzer deletions that are particularly useful for mapping *r*IIA mutations. These lengthy deletions begin somewhere outside of *r*IIB and remove all of that gene, extending for various distances into *r*IIA. One deletion, *r1272*, extends through the entire *r*II region, completely removing both *r*IIA and *r*IIB.

Armed with such deletions with known endpoints, Benzer was able to quickly localize the position of any new *r*II point mutation by crossing the mutant phage separately with phage containing each of these deletions. For example, if a point mutation gives r^+ recombinants when crossed with the deletion *rA105* but not when crossed with *rpB242*, this point mutation must lie in the short region between the end of *rA105* and the end of *rpB242*. Therefore, with no more than seven crosses, a mutation could be localized to one of seven segments of the *r*IIA gene. The position of the mutation could be located more precisely through additional crosses with other mutations located within this smaller segment.

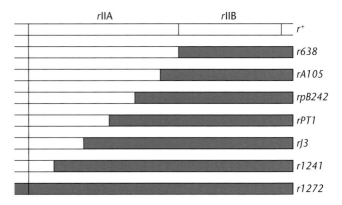

Figure 7.25 Some of Benzer's *r*II deletions in phage T4. The deletions remove all of *r*IIB and extend various distances into *r*IIA. The shaded purple bars show the region deleted in each of the mutations.

MUTATIONAL SPECTRA

The numerous point mutations within the *r*IIA and *r*IIB genes that Benzer found by deletion mapping included spontaneous as well as induced mutations. Figure 7.26 illustrates the map locations of some of the spontaneous mutations. Spontaneous mutations can occur everywhere in the *r*II genes, but Benzer noted that some sites are "hot spots," in which many more mutations occur than at other sites. Mutagen-induced mutations also have hot spots, which differ depending on the mutagen and from those of spontaneous mutations (data not shown).

The tendency of different mutagens to mutate some sites much more frequently than others has practical consequences in genetic analysis (see chapter 3). It is apparent from Figure 7.26 that if Benzer had studied only spontaneous mutations in the *r*II genes, almost 30% of these would have been at one site in A6c, a major hot spot for spontaneous mutations. Therefore, to obtain a random collection of mutations in a gene requires isolating not only spontaneous mutations but also ones induced with different mutagens.

Methods are now available that allow essentially random mutagenesis of selected regions of DNA. These methods involve the use of special oligonucleotide primers for site-specific mutagenesis and PCR mutagenesis. Some of these methods are discussed in chapter 1.

THE *r*II GENES AND THE NATURE OF THE GENETIC CODE

Of all the early experiments with the T4 *r*II genes, some of the most elegant were those that revealed the nature of the genetic code. These were conducted by Francis Crick and his collaborators (see Crick et al., Suggested Reading). These experiments not only have great historical

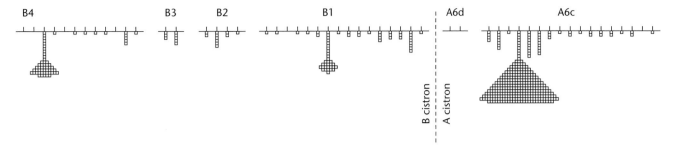

Figure 7.26 Mutational spectrum for spontaneous mutations in a short region of the *r*II genes. Each small box indicates one mutation observed at that site. Large numbers of boxes at a site indicate hot spots, where spontaneous mutations often occur.

importance but also are a good illustration of classical genetic principles and analysis.

At the time Crick and his collaborators began these experiments, he and James Watson had used the X-ray diffraction data of Rosalind Franklin and Maurice Wilkins and the biochemical data of Erwin Chargaff and others to solve the structure of DNA (see the introductory chapter). This structure indicated that the sequence of bases in DNA determines the sequence of amino acids in protein. However, the question remained of how the sequence of bases is read. For example, how many bases in DNA encode each amino acid? Does every possible sequence of bases encode an amino acid? Is the code "punctuated," with each code word demarcated, or does the cell merely begin reading at the beginning of a gene and continue to the end, reading a certain number of bases each time? The ease with which the *r*II genes of T4 could be manipulated made this system the obvious choice for use in experiments to answer these questions.

The experiments of Crick et al. were successful for two reasons. First, the extreme N-terminal region of the *r*IIB polypeptide, the so-called B1 region, is nonessential for activity of the *r*IIB protein. The B1 region can be deleted or all the amino acids it encodes can be changed without affecting the activity of the polypeptide. Note that this is not normal for proteins. Most proteins cannot tolerate such extensive amino acid changes in any region.

The second reason for the success of these experiments is that acridine dyes specifically induce frameshift mutations by causing the removal or addition of a base pair in DNA (see chapter 11). This conclusion required a leap of faith at the time. The mutations caused by acridine dyes usually are not leaky but are obviously not deletions because they map as point mutations and revert. Also, the frequency of revertants of mutations due to acridine dyes increases when the mutant phage are again propagated in the presence of acridine dyes but not when they are propagated in the presence of base analogs, which at the time were suspected to cause only

base pair changes. It was reasoned that if acridine dye-induced mutations could not be reverted by base pair changes, the mutations induced by acridine dyes themselves could not be base pair changes. This evidence that acridine dyes cause frameshift mutations may seem flimsy in retrospect, yet it was convincing enough to Crick et al. that they proceeded with their experiments on the nature of the genetic code.

Intragenic Suppressors of a Frameshift Mutation in *r*II B1

The first step in their analysis was to induce a frameshift mutation in the *r*II B1 region by propagating cells infected with the phage in the presence of the acridine dye proflavin. Crick and his colleagues named their first *r*II mutation FC0 for "Francis Crick Zero." The FC0 mutation prevents T4 multiplication in *E. coli* Kλ because it inactivates the *r*IIB polypeptide. That an acridine-induced mutation in the region encoding the nonessential B1 portion of the B polypeptide can inactivate the *r*IIB polypeptide in itself suggests that the mutations are frameshifts, since, as mentioned above, merely changing an amino acid in the B1 region should not inactivate the gene.

Selecting Suppressor Mutations of FC0

The next step in the Crick et al. analysis was to select suppressor mutations of FC0. As discussed in chapter 3, a suppressor mutation restores the wild-type phenotype by altering the DNA sequence somewhere other than the site of the original mutation. To select suppressors of FC0, Crick et al. merely needed to plate large numbers of FC0 mutant phage on *E. coli* Kλ. A few plaques due to phenotypically r^+ phage appeared. These phage could either have been revertants of the original FC0 mutation or have had two mutations, the original FC0 mutation plus a suppressor.

To determine which of the r^+ phage had suppressor mutations, Crick et al. applied the classic genetic test for suppression (see chapter 3). In such a test, an apparent revertant is crossed with the wild type. If any of the progeny are recombinant types with the mutant phenotype, the interpretation is that the mutation had not reverted but had been suppressed by a second-site mutation that restored the wild-type phenotype.

Figure 7.27 illustrates the principle behind this test as applied to the apparent revertants of the FC0 rII mutation of T4. The apparent wild-type revertant is crossed with the wild-type phage. If the mutation has reverted, all of the progeny are r^+ and there are no rII mutant recombinants. In contrast, if the mutation has been suppressed, the suppressing mutation can be crossed away from the FC0 mutation, and rII mutant recombinant types appear that cannot multiply in E. coli Kλ. In the test used by Crick et al., most of the apparent r^+ revertants of FC0 gave some rII mutant recombinants when crossed with the wild type; therefore, the FC0 mutation in these apparent revertants was being suppressed rather than reverted. Moreover, there were very few rII mutant recombinants, and so the suppressing mutations must have been very close to the original FC0 mutation, presumably also in the B1 region of the rIIB gene, since only crossovers between the regions of the FC0 mutation and the suppressing mutation give rise to rII mutant recombinants.

Isolating the Suppressor Mutations

A double mutant with both the FC0 mutation and the suppressor mutation is r^+ and multiplies in E. coli Kλ.

Figure 7.27 Classical genetic test for suppression. Phages that have apparently reverted to wild type (r^+) are crossed with wild-type r^+ phage. (A) If the mutation has reverted, all of the progeny will be r^+. (B) If the mutation has been suppressed by another mutation, x, there will be some rII mutant recombinants among the progeny.

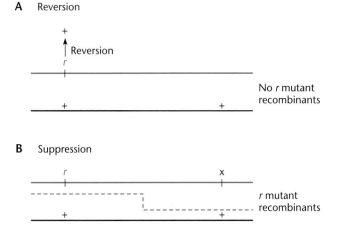

But would a mutant with a suppressor mutation alone be phenotypically rII or r^+? If the suppressor mutations by themselves produce a phenotypically rII mutant, presumably some of the rII mutant recombinants obtained by crossing the suppressed FC0 mutant with wild-type T4 would be single-mutant recombinants with only the suppressor mutation. This can be easily tested. If a recombinant is phenotypically rII^- because it has the suppressor mutation rather than the FC0 mutation, it should give some r^+ recombinants when crossed with the FC0 single mutant. Some of the rII mutant recombinant phages did give some r^+ recombinants when crossed with FC0 mutants, indicating that these phages have an rII mutation different from the FC0 mutation, presumably the suppressor mutation by itself.

Selecting Suppressor-of-Suppressor Mutations

Because the suppressor mutations of FC0 by themselves make the phage $rIIB^-$ and prevent multiplication on E. coli Kλ, the next question was whether the suppressor mutations of FC0 could be suppressed by "suppressor-of-suppressor" mutations. As with the original FC0 mutation, when Crick et al. plated large numbers of T4 with a suppressor mutation on Kλ, they observed a few plaques. Most of these resulted from second-site suppressors. Moreover, these suppressor-of-suppressor mutations were $rIIB^-$ when isolated by themselves. This process could be continued indefinitely.

Frameshift Mutations and Implications for the Genetic Code

To explain these results, Crick and his collaborators proposed the model shown in Figure 7.28. They proposed that FC0 is a frameshift mutation that alters the reading frame of the rIIB gene by adding or removing a base pair so that all the amino acids inserted in the protein from that point on are wrong. This explains how the FC0 mutation can inactivate the rIIB polypeptide, even though it occurs in the nonessential N-terminal-encoding B1 region of the gene.

The suppressors of FC0 are also frameshift mutations in the rIIB1 region. The suppressors either remove or add a base pair, depending on whether FC0 adds or removes a base pair, respectively. As long as the other mutation has the opposite effect to FC0, an active rIIB polypeptide is often synthesized. The results of these experiments had several implications for the genetic code.

The code is unpunctuated. At the time, it was not known if something demarcated the point where a code word in the DNA begins and ends. Consider a language in which all the words have the same number of letters. If there were spaces between the words, we could always

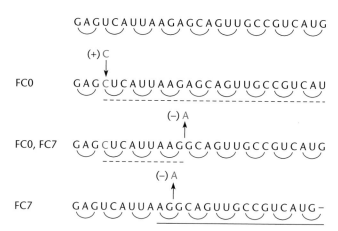

Figure 7.28 Frameshift mutations and suppression. The FC0 frameshift is caused by the addition of 1 bp, which alters the reading frame and makes an *rII* mutant phenotype. FC0 can be suppressed by another mutation, FC7, which deletes 1 bp and restores the proper reading frame and the *r*⁺ phenotype. The FC7 mutation by itself confers an *rII* mutant phenotype. Regions translated in the wrong frame are underlined.

read the words of a sentence correctly because we would know where one word ended and the next word began. However, if the words did not have spaces between them, the only way we would know where the words began and ended would be to count the letters. If a letter were left out or added to a word, we would read all the following words wrong. This is what happens when a base pair is added to or deleted from a gene. The remainder of the gene is read wrong; therefore, the code must be unpunctuated.

The code is three lettered.   The experiments of Crick et al. also answered the question of how many letters are in each word of this language; i.e., how many bases in DNA are being read for each amino acid inserted in the protein. At the time, there were theoretical reasons to believe that the number is larger than 2. Since DNA contains four "letters" or bases (A, G, T, and C), only $4 \times 4 = 16$ possible amino acid code words could be made out of only two of these letters. However, at least 20 amino acids were known to be inserted into proteins (the known number is now 22 if you do not count selenocysteine, which is made on the tRNA [see Box 2.3]), and so a two-letter code would not yield enough code words for all the amino acids. However, three bases per code word results in $4 \times 4 \times 4 = 64$ possible code words, plenty to encode all 20 amino acids.

They could test the assumption that the code is three lettered. The reading frame of a three-letter code would not be altered if 3 bp was added to or removed from the

*r*IIB1 region. Continuing with the letter analogy, an extra word would then be put in or left out but all the other words would be read correctly. Therefore, in the B1 region, if three suppressors of FC0 or three suppressors of suppressors were combined in the same phage DNA, a complete new code word would be added to or subtracted from the molecule and the correct reading frame would be restored. Thus, the phage should be *r*⁺ and should multiply in *E. coli* Kλ. Experimental results were consistent with this hypothesis, indicating that 3 bp in DNA encodes each amino acid inserted into a protein.

The code is redundant.   The results of these experiments also indicated that the code is redundant; that is, more than one word codes for each amino acid. Crick et al. reasoned that if the code were not redundant, most of the code words, i.e., $64 - 20 = 44$, would not encode an amino acid; then a ribosome translating in the wrong frame would almost immediately encounter a code word that does not encode an amino acid, and translation would cease. The fact that most combinations of suppressors with FC0 and with suppressors of suppressors restored the *r*⁺ phenotype indicated that most of the possible code words do encode an amino acid.

Some code words are nonsense and terminate translation.   Although their evidence indicated that most code words encode an amino acid, it also indicated that not all of them do. If all possible words signified an amino acid, the entire *r*IIB1 region should be translatable in any frame and a functional polypeptide would result, provided that the correct frame was restored before the translation mechanism entered the remainder of the *r*IIB gene. However, if not all the words encode an amino acid, a "forbidden" code word that does not encode an amino acid might be encountered during translation in a wrong frame. In this situation, not all combinations of suppressors and suppressors of suppressors would restore the *r*⁺ phenotype. Crick et al. observed that some combinations of suppressors and suppressors of suppressors did cause "forbidden" code words to be encountered in the *r*IIB region. However, other combinations in the same region resulted in a functional *r*IIB polypeptide. For example, in Figure 7.29, a nonsense codon (UAA) is encountered when a region is translated in the +1 frame because 1 bp was removed. However, no nonsense codons are encountered when 1 bp is added and the same region is translated in the −1 frame.

Even more convincing evidence that some code words do not encode an amino acid came from experiments with the deletion *r1589* (Figure 7.30) (see Benzer and Champe, Suggested Reading). The deletion *r1589* removes much of *r*IIA and the nonessential *r*IIB1 region, eliminating

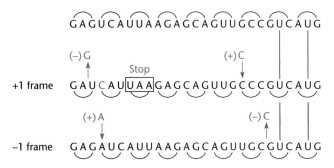

Figure 7.29 Frameshift suppression and nonsense codons. The nonsense codon UAA is encountered in the +1 frame because of a deletion of 1 bp in the DNA. Translation terminates even if the correct translational frame is restored farther downstream. In the −1 frame, due to addition of 1 bp, no nonsense codons are encountered. A downstream deletion restores the correct reading frame, and the active polypeptide is translated.

the nonsense codon or codons that are normally at the end of the *r*IIA gene and the translation initiation region of *r*IIB. This deletion mutation thereby causes translation initiated at *r*IIA to proceed into *r*IIB, resulting in a fusion protein in which the N terminus comes from *r*IIA and the rest of the protein comes from *r*IIB. Since most of *r*IIA but only the nonessential B1 region of *r*IIB is deleted, this fusion protein has *r*IIB activity but not *r*IIA activity, as can be demonstrated by complementation tests.

Although the fusion protein does not require the *r*IIA portion for *r*IIB activity, Benzer and Champe found that some base pair change mutations in the *r*IIA region prevented *r*IIB activity. These base pair changes presumably caused nonsense codons that stopped translation in the *r*IIA region. Other base pair change mutations that did not disrupt *r*IIB activity were presumably missense mutations, which resulted in insertion of the wrong amino

Figure 7.30 The *r*II deletions *r638* and *r1589*. An *r*IIAB fusion protein is made in a strain with *r1589*. See the text for details.

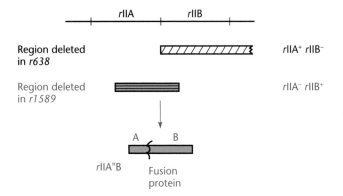

acid in the *r*IIA portion of the fusion protein but did not stop translation.

Benzer and Champe also found that even the presumed nonsense mutations did not prevent *r*IIB activity in some strains of *E. coli* and so were in a sense "ambivalent." We now know that these "permissive" strains of *E. coli* are nonsense suppressor strains with mutations in tRNA genes that allow readthrough of one or more of the nonsense codons (see chapter 3).

Postscript on the Crick et al. Experiments

The experiments of Crick et al. laid the groundwork for the subsequent deciphering of the genetic code by Marshall Nirenberg and his colleagues, who assigned an amino acid to each of the 61 3-base sense codons. Other researchers later used reversion and suppression studies to determine that the nonsense codons are UAG, UAA, and UGA.

ISOLATING DUPLICATION MUTATIONS OF THE *r*II REGION

Our final example of the genetic manipulation of the *r*II genes of T4 is the isolation of tandem duplication mutations (see, for example, Symonds et al., Suggested Reading). These experiments help contrast the differences between complementation and recombination and also illustrate some of the genetic properties of tandem duplication mutations. In some ways, this analysis is similar to the analysis of *his* duplications presented in chapter 3, but it helps distinguish the manipulations in bacterial and phage genetics.

The isolation of tandem duplication mutations of the *r*II region depended on the properties of two deletions in the *r*II region, the aforementioned *r1589* and *r638* deletions (Figure 7.30). As mentioned, the *r1589* deletion removes the N-terminal-coding B1 region of the *r*IIB gene as well as the C-terminal-coding part of *r*IIA. Phages with this deletion make a fusion protein with the N terminus of the *r*IIA protein fused to most of *r*IIB and are phenotypically *r*IIA⁻ *r*IIB⁺. The deletion mutation *r638* deletes all of *r*IIB but does not enter *r*IIA, so that phages with this deletion are *r*IIA⁺ *r*IIB⁻. Because one deleted DNA makes the product of the *r*IIA gene and the other makes the product of the *r*IIB gene, the two deletion mutations can complement each other. However, they cannot recombine to give *r*⁺ recombinants, because they overlap, both deleting the B1 region of the *r*IIB gene.

Even though recombination should not occur between the two deletions to give *r*⁺ recombinants, when *E. coli* B is infected simultaneously with the two deletion mutants and the progeny are plated on *E. coli* Kλ, a few rare plaques due to *r*⁺ phages arise. These phenotypically *r*⁺ phages have tandem duplications of the *r*II region

(Figure 7.31A). Each copy of the *r*II region has a different deletion mutation, and these mutations complement each other to give the r^+ phenotype.

Figure 7.31A also illustrates how these tandem duplication mutations might arise. Sometimes, while the DNA is replicating, recombination mistakenly occurs between two short directly repeated regions on either side of the *r*II region (ectopic recombination). Such mistaken crossovers are rare because repeated sequences in DNA, when they exist at all, are usually very short. However, once

such a crossover occurs, one of the recombinant-type phage will have a duplicate of the *r*II region, both copies of which, in the example, have the *r1589* mutation. If a phage with such a duplication then infects the same cell as a phage with the *r638* deletion, one of the copies of the *r*II region can recombine with the DNA of the other parent and the *r638* deletion replaces the *r1589* deletion in one of the copies. This second recombination occurs very frequently, because of the extensive homology between the duplicated regions. The phages that package

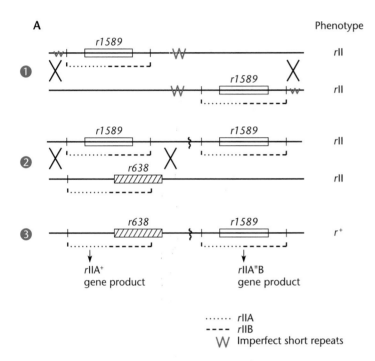

Figure 7.31 (A) Model for how tandem duplications in the *r*II region form. Recombination between short repeated sequences flanking the *r*II region (W) may occur at a low frequency, giving rise to a duplication of one or more genes. This duplication can recombine with the other deletion strain in the region of one of the repeated sequences, giving rise to a duplication in which one copy of *r*IIB has the *r1589* deletion and the other copy has the *r638* deletion. The two deletions complement each other, so that the phage is phenotypically r^+. (B) Tandem duplications are unstable because recombination between the duplicated regions can destroy the duplication. Which deletion mutation remains in the haploid segegant depends on where the recombination occurs. (1) Recombination between the y duplicated segments gives rise to a haploid segregant with only the *r638* deletion. (2) Recombination between the duplicated segments designated x gives rise to a haploid segregant with only the *r1589* deletion. In this example, *r1589* haploid segegants are about three times more frequent than *r638* segregants because x is approximately three times as long as y. A circled x indicates a crossover.

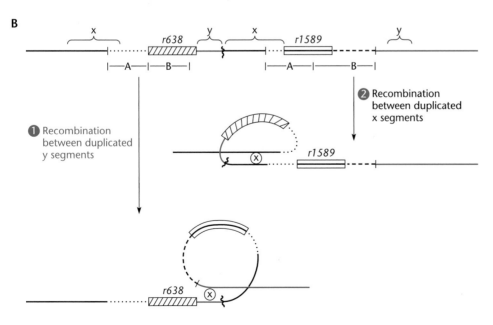

this DNA then have two copies of the *r*II region, one copy with the *r1589* deletion and the other copy with the *r638* deletion. In subsequent infections, the two deletion mutations can complement each other to make the phages phenotypically *r*⁺. These phages are diploid for the *r*II regions and surrounding genes because they have two copies of these genes.

The salient property of tandem duplication mutations is that they are very unstable, because recombination anywhere in the long duplicated region destroys the duplication (see chapter 3). These *r*⁺ phages exhibit this instability. If the *r*⁺ phages with the putative duplication are propagated in *E. coli* B, where there is no selection for phage with the duplication, a very high percentage of the progeny phages will be unable to multiply in *E. coli* Kλ.

Figure 7.31B illustrates why the duplications are unstable. A crossover between either of the duplicated segments, x or y, causes the intervening sequences to be deleted and one copy of the duplication to be lost. The resulting phages, some of which have the *r1589* deletion whereas others have the *r638* deletion, are haploid segregants that now have only one copy of the duplicated region. The term "segregants" is used rather than "recombinants" because the recombination that destroys the duplication occurs spontaneously while the phage is multiplying and does not require crosses. The x and y regions are usually quite long and identical in sequence, so that haploid segregants appear quite frequently.

The two haploid types segregate at a characteristic frequency for each duplication mutation. As we can see from the duplication shown in Figure 7.31B, a crossover in region x yields the *r1589* haploid whereas a crossover in region y yields the *r638* haploid. Therefore, the haploid type that segregates at the highest frequency depends on which region, x or y, is longer. If x is longer than y, the *r1589* haploid segregates more frequently. However, if x is shorter than y, *r638* segregates more frequently.

Constructing the Genetic Linkage Map of a Phage

The *r*II genes are only two genes out of the hundreds of genes of phage T4 (Figure 7.6), and these genes do not even exist in other types of phage. To obtain the genetic map of a phage, we need a way to identify many more genes of the phage. A picture that shows many of the genes of an organism and how they are ordered with respect to each other is known as the **genetic linkage map** of the organism, so named because it shows the proximity or linkage of the genes to each other. This linkage is determined by genetic crosses. Physical methods for mapping DNA, discussed in chapter 1, give rise to a physical map, which can often be correlated with the genetic map.

Conditional-lethal mutations, including temperature-sensitive and nonsense mutations, are very useful types of mutations in phages and can be used to identify any gene that is essential for multiplication of the phage. If a phage has a mutation to a nonsense codon (UAA, UAG, or UGA) in an essential gene, it can multiply and form a plaque only on a permissive host with a nonsense suppressor tRNA (see chapter 2). Phages with a temperature-sensitive mutation in an essential gene multiply and form plaques at a lower (permissive) temperature but not at a higher (nonpermissive) temperature. Because such mutations can be isolated in any essential gene, nonsense and temperature-sensitive mutations can be used to identify many of these genes and to construct a nearly complete genetic linkage map of a phage. Not all genes can be identified this way, however, since the products of some genes of the phage may be nonessential on a particular host; these genes cannot be identified by conditional-lethal mutations, and other methods must be used. For example, the *r*II genes could not be found in this way, unless the host that was used was a λ lysogen.

The first step in constructing the conditional-lethal map of a phage is to isolate a large number of temperature-sensitive and nonsense mutations of the phage by mutagenizing the phage with various mutagens and then plating the surviving phage on suppressing bacteria at the permissive temperature. Then plaques are picked, and the phage is tested for multiplication on nonsuppressing bacteria and at the nonpermissive temperature. Phages that cannot form plaques on the nonsuppressing bacteria or at the nonpermissive temperature have nonsense mutations or temperature-sensitive mutations, respectively, in essential genes.

IDENTIFYING PHAGE GENES BY COMPLEMENTATION TESTS

Once a large collection of mutations of the phage have been assembled, the mutations can be placed into complementation groups or genes. As discussed above, two mutations that do not complement each other are probably in the same complementation group or gene. Complementation tests are done under conditions where neither mutant can multiply. For example, to test for complementation between two different amber (UAG) nonsense mutations, nonsuppressing bacteria are infected with the two mutants simultaneously and the progeny are plated on amber-suppressing bacteria to determine how many phage progeny are produced. To test for complementation between a temperature-sensitive mutation and a nonsense mutation, nonsuppressing bacteria are infected at the high (nonpermissive) temperature. The progeny phage is then plated on amber-suppressing bacteria at the permissive temperature to determine the quantity of phage produced. Note that a temperature-sensitive mutation and an amber mutation could be in the same gene; in other words, they could be allelic. Even though they are different types of mutations, they do not complement each other if they are allelic.

Each time we find a mutation that complements all the other mutations in the collection, we have found a new gene. Once two mutations have been found that do not complement each other, only one of the two mutations need be used for further complementation tests. If many of the complementation groups are represented by only a single mutation, many other essential genes are probably not yet represented by any mutations. More mutants must then be isolated to identify more of the essential genes. Eventually, more and more of the new mutations will sort into one of the previously identified complementation groups and the collection of mutations in essential genes will be almost complete.

MAPPING PHAGE GENES

Once most of the genes of the phage have been identified, representative mutations in each of the genes can be mapped with respect to each other. To measure the frequency of recombination between two mutations, cells are infected under conditions that are permissive for both mutations. For example, to cross a temperature-sensitive mutation with an amber mutation, amber suppressor cells are infected at the low (permissive) temperature. After phage is produced, the progeny phage is plated under conditions permissive for both mutations, to measure the total progeny, and under conditions that are nonpermissive for both mutations, to measure the number of wild-type recombinants. Hence, if a mutant strain with a nonsense mutation in one gene is crossed with another mutant with a temperature-sensitive mutation in a different gene, the progeny should be plated on nonsuppressing bacteria at the high (nonpermissive) temperature to measure the number of wild-type recombinants and at the low (permissive) temperature on nonsense-suppressing bacteria to measure the total progeny. From the frequency of wild-type recombinants, the map distance between the mutations and therefore the genes they are in can be calculated by using the equation for recombination frequency given above. Genes are said to be **linked** when the recombination frequency between mutations in the genes is lower than random, indicating that fewer than a random number of crossovers are occurring between them.

ORDERING MUTATIONS BY THREE-FACTOR CROSSES

It is often difficult to order mutations in closely linked genes by measuring recombination frequencies alone.

The recombination frequencies must be measured very carefully to be certain of their order. In such cases, the use of three-factor crosses may be a less ambiguous way to order the mutations.

Figure 7.32 illustrates how a three-factor cross can be used to order two amber mutations with respect to a temperature-sensitive mutation, although the same reasoning could apply to any combination of mutations. In particular, we want to know whether the site of the temperature-sensitive mutation lies between the sites of the two amber mutations or outside them. In method I, we cross a double mutant containing both amber mutations with a single mutant containing only the temperature-sensitive mutation. We then plate the progeny on nonsuppressing bacteria at the nonpermissive temperature for the temperature-sensitive mutation, so that only recombinant phage that lack all three mutations can form phage. If the temperature-sensitive mutation lies outside the site of the two amber mutations, only one crossover between the site of the *am2* mutation and the site of the *ts3* mutation will create a wild-type recombinant and the frequency of wild-type recombinants should be quite high. In fact, it should be about as high as if a phage with just the *am2* mutation were crossed with a phage with just the *ts3* mutation. However, if the site of the temperature-sensitive mutation lies between the sites of the two amber mutations, two crossovers are required to make the wild-type recombinant and the frequency of wild-type recombinants should be much lower. As discussed above, the frequency of the double crossover between closely linked mutations is often much higher than the product of the frequencies of the single crossovers due to high negative interference. Phenomenona associated with recombination such as high negative interference and gene conversion are discussed in more detail in chapter 10.

A second method for doing the three-factor cross to order the three mutations that is even less ambiguous is illustrated in Figure 7.32 as method II. In this method, a double-mutant phage having one of the amber mutations, *am1*, and the temperature-sensitive mutation, *ts3*, is crossed with a phage having the other amber mutation, *am2*. The cross is plated on nonsuppressing bacteria at the low, permissive temperature for the temperature-sensitive mutation. After the plaques develop, they are picked and spotted on two plates, one incubated at the permissive temperature and the other incubated at the nonpermissive temperature, to determine what percentage of the progeny have the temperature-sensitive mutation. If the *ts3* mutation lies outside the region of the *am1* and *am2* mutations, most of the progeny will be temperature sensitive and plate at the permissive temperature but not the nonpermissive temperature. However, if it lies between them, some progeny will be

Method 1

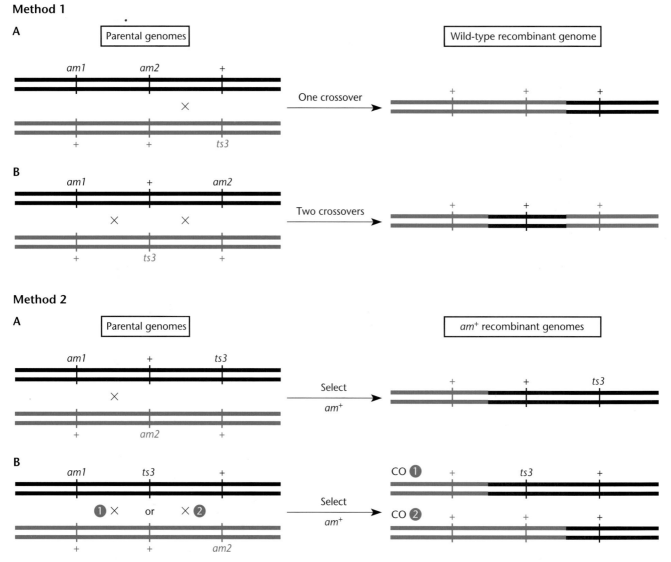

Figure 7.32 Three-factor cross for ordering conditional-lethal mutations. Method 1: In a three-factor cross, the frequency of wild-type recombinants depends on the order of the three mutations. (A) The temperature-sensitive mutation, *ts3,* lies outside the region of the two amber mutations, *am1* and *am2.* A single crossover gives a wild-type recombinant that has none of the mutations. (B) The *ts3* mutation lies between the two amber mutations. Now two crossovers are required to give the wild-type recombinant, making it much rarer. Method 2: One of the parental phages has the *am1* mutation and the *ts3* mutation, and the other has the *am2* mutation. Am⁺ recombinants are tested for the Ts phenotype. (A) If the *ts3* mutation lies outside the two amber mutations, most will be Ts. (B) If it lies between them, some will be Ts and some will not. CO 1: crossover in region 1. CO 2: crossover in region 2. Details are given in the text. See also problem 4 at end of this chapter.

temperature sensitive and some will not, depending on where the crossover occurred.

GENETIC LINKAGE MAPS OF SOME PHAGES
Using methods such as those described above, genetic linkage maps have been determined for several commonly used phages. Some of these maps appear throughout this chapter and the next, along with the functions of some of the gene products where known (examples are given in Figures 7.4, 7.6, and 7.13).

One noticeable feature of most phage genetic maps is that genes whose products must physically interact, such as the products involved in head or tail formation, tend to be clustered. The argument is that this clustering may allow recombination between closely related phages without disruption of gene function. If the genes whose products must physically interact were not close to each other, recombination between two closely related phages could separate the genes and give rise to inviable phage. For example, if the head genes were not clustered, recombination with the DNA of another related phage would often replace some of the head genes of the first phage with the corresponding head genes from the other phage. If the head of the phage could not be assembled from this mixture of head proteins, the phage would be inviable. Therefore, the potential advantages of the new recombinant phage would be lost. This hypothesis has received support from the structure of the family of phages related to λ. These phages seem to be modular in construction, being made up of regions, or cassettes, assembled from different regions drawn from this large family of phages.

FACTORS THAT DETERMINE THE FORM OF THE LINKAGE MAP

Another striking feature of phage genetic maps is that some are linear while others are circular (compare the T7 map in Figure 7.4 with the T4 map in Figure 7.6). A genetic map is found to be circular when, as the genes are ordered from left to right by genetic crosses, the last gene to be ordered is linked to the first gene, so that you have come full circle. The form of the genetic map does not necessarily correlate with the linearity or circularity of the phage DNA itself, however. Some phages with circular DNA in the cell, such as λ (see chapter 8), have a linear linkage map, while some with linear DNA in the cell, such as T4, have a circular map. To understand how the genetic maps arise, we need to review how some phage replicate their DNA and how it is packaged into phage heads.

Phage λ

Phage λ has a linear genetic map, even though the DNA forms a circle after it enters the cell. As discussed in chapter 8, phage λ has a linear map because its concatemers are cleaved at unique *cos* sites before being packaged into the phage head. The position of these *cos* sites determines the end of the linear genetic map. As an illustration, consider a cross between two phages with mutations in the *A* and *R* genes at opposite ends of the phage DNA (see the λ map in chapter 8). Even though different parental alleles of the *A* and *R* genes can be next to each other in the concatemers prior to packaging, these alleles are separated when the DNA is cut at the *cos* site during packaging of the DNA. Therefore, the *A* and *R* genes appear to be far apart and essentially unlinked when one

measures recombination frequencies in genetic crosses, and the genetic map is linear with the *A* and *R* genes at its ends. For this reason, all types of phages which package DNA from unique *pac* or *cos* sites have linear genetic linkage maps with the ends defined by the position of the *pac* or *cos* sites.

Phage T4

Phage T4 has a circular genetic map even though its DNA never forms a circle. The reason for its circular map is that T4 has no unique *pac* site and the DNA is packaged by a headful mechanism from long concatemers (Figure 7.19). Consequently, any two T4 phage DNAs in different phage heads from the same infection do not have the same ends but are cyclic permutations of each other. Therefore, genes that are next to each other in the concatemers will still be together in most of the phage heads unless they happen to be on the terminal redundancy, which is only 3% of the genome, and so will appear linked in crosses, producing a circular map.

Phage P22

Phage P22 is a phage of *Salmonella*. It is closely related to λ and replicates by a similar mechanism. However, unlike λ, it has a circular linkage map. The difference is that P22 begins packaging at a unique *pac* site or *cos* site, like λ, but then packages a few genomes by a processive headful mechanism, like T4, giving rise to a circular genetic map.

Phage P1

Phage P1 has linear DNA in the head, which forms a circle by recombination between terminally repeated sequences at its ends after infection. The DNA then replicates as a circle and forms concatemers from which the DNA is packaged by a headful mechanism. However, unlike most phages that package DNA by a headful mechanism, the genetic map of P1 is linear. It is linear because it has a very active site-specific recombination system called *cre-lox*, which promotes recombination at a particular site in the DNA. Because recombination at this site is so frequent, genetic markers on either side of the site appear to be unlinked, giving rise to a linear map terminating at the *cre-lox* site. The function of this site-specific recombination system in phage development is unknown, but it may resolve dimeric circles in the prophage state. Because of its sensitivity and specificity, the *cre-lox* recombination system of P1 has many uses. One of these uses is mentioned in chapter 5, where the system is used to show that some proteins encoded by the Ti plasmid enter the plant cell nucleus along with the T-DNA during the formation of crown gall tumors. It also has served as a model system for site-specific recombinases, as discussed in chapter 9.

Generalized Transduction

Bacteriophages not only infect and kill cells but also sometimes transfer bacterial DNA from one cell to another in a process called **transduction**. There are two types of transduction in bacteria: **generalized transduction,** in which essentially any region of the bacterial DNA can be transferred from one bacterium to another, and **specialized transduction,** in which only certain genes close to the attachment site of a lysogenic phage in the chromosome can be transferred. These two types of transduction have fundamentally different mechanisms and are considered separately. In this section, we discuss only the mechanism of generalized transduction; coverage of specialized transduction is deferred until it can be addressed in the context of lysogeny in chapter 8. Also, since the analysis of genetic data obtained by generalized transduction has been discussed in chapter 3, we restrict ourselves in this section to the mechanism of transduction and what constitutes a transducing phage.

Figure 7.33 gives an overview of the process of generalized transduction. While phages are packaging their own DNA, they sometimes mistakenly package the DNA of the bacterial host instead. These phages are still capable of infecting other cells, but progeny phage are not produced. What happens to the DNA after it enters the cell depends on the source of the bacterial DNA. If the bacterial DNA is a piece of the bacterial chromosome of the same species, it usually has extensive sequence homology to the chromosome and may recombine with the host chromosome to form recombinants. If the piece of DNA that was picked up and injected is a plasmid, it may replicate after it enters the cell and thus be maintained. If the incoming DNA contains a transposon, the transposon may hop, or insert itself, into a host plasmid or chromosome, even if the remainder of the DNA contains no sequences in common with the DNA of the bacterium it entered (see chapter 9).

The nomenclature of transduction is much like that of transformation and conjugation. Phages capable of transduction are called **transducing phages**. A phage that has picked up bacterial DNA is called a **transducing particle.** The original bacterial strain in which the transducing particle had multiplied and picked up host DNA is called the **donor strain.** The bacterial strain it infects is called the **recipient strain.** Cells that have received DNA from another bacterium by transduction are called **transductants.**

Transduction occurs very rarely for a number of reasons. First, mistaken packaging of host DNA is itself

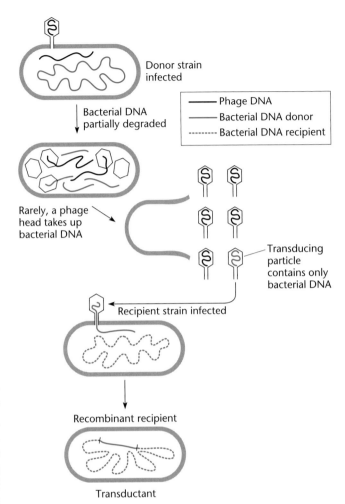

Figure 7.33 Generalized transduction. A phage infects one bacterium, and in the course of packaging DNA into heads, the phage mistakenly packages some bacterial DNA instead of its own DNA into a head. In the next infection, this transducing particle is different from most of the phage in the population in that it injects the bacterial DNA instead of phage DNA into the recipient bacterium. If the bacterial DNA that was picked up is chromosomal DNA, it might recombine with the chromosome and form a recombinant type.

rare, and transduced DNA must survive in the recipient cell to form a stable transductant. Since each of these steps has a limited probability of success, transduction can be detected only by powerful selection techniques.

What Makes a Transducing Phage?

Not all phages can transduce. To be a generalized transducing phage, the phage must have a number of characteristics. It must not degrade the host DNA completely after infection, or no host DNA will be available to be packaged into phage heads when packaging begins. As already discussed, most phages package DNA from sites on the DNA called *pac* or *cos* sites. If a DNA lacks such specific sites, it is usually not packaged. The packaging sites, or *pac* sites, of the phage must not be so specific that such sequences do not occur by chance in host DNA. Also, if the phage has a broad host range for absorption, it might be possible to use it to introduce DNA into a broad variety of bacteria. It is important to realize that it does not actually have to be capable of multiplying in the recipient host, only of absorbing to it and injecting its DNA.

Table 7.2 compares two good transducing phages, P1 and P22. Phage P1, which infects gram-negative bacteria, is a good transducer because it has less *pac* site specificity than most phages and packages DNA via a headful mechanism; therefore, it efficiently packages host DNA. About 1 in 10^6 phage P1 particles transduce a particular marker. It also has a very broad host range for adsorption and can transduce DNA from *E. coli* into a wide variety of other gram-negative bacteria including members of the genera *Klebsiella* and *Myxococcus*. It cannot multiply in hosts other than *E. coli*, but it can transfer plasmids, etc., from *E. coli*, on which it can be propagated, into these other hosts.

The *Salmonella enterica* serovar Typhimurium phage P22 is also a very good transducer and, in fact, was the first transducing phage to be discovered (see Zinder and Lederberg, Suggested Reading). Like P1, P22 has *pac* sites that are not too specific and packages DNA by a headful mechanism. From a single *pac*-like site, about 10 headsful of DNA can be packaged before the mechanism requires another *pac* site. Because of even this limited *pac* site specificity, however, some regions of *Salmonella* DNA are transduced by P22 at a much higher frequency than others.

Other phages are not normally transducing phages but can be converted into them by special treatments. For example, T4 normally degrades the host DNA after infection but works extremely well as a transducing phage if its genes for the degradation of host DNA have been inactivated. Because phage T4 DNA packaging does not require *pac* sites, it packages any DNA, including the host DNA, with equal efficiency.

In contrast, phage λ does not work well for generalized transduction, because it normally packages DNA between two *cos* sites rather than by a headful mechanism. It very infrequently picks up host DNA by mistake, but then it does not cut the DNA properly when the head is filled unless another *cos*-like sequence happens to lie a genome length distance along the DNA. Thus, potential transducing particles usually have DNA hanging out of them that must be removed with DNase before the tails can be added. Even with these and other manipulations, λ works poorly as a generalized transducer.

Transducing phages have been isolated for a wide variety of bacteria and have greatly aided genetic analysis of these bacteria. Transduction is particularly useful for moving alleles into different strains of bacteria and making isogenic strains that differ only in a small region of their chromosomes so that they are very useful for strain construction and gene knockouts for functional genomics (see chapter 3). However, if no transducing phage is known for a particular strain of bacterium, finding one can be very time-consuming. Therefore, for such bacteria, generalized transduction is often replaced by other methods.

Shuttle Phasmids

One variation on transduction that is sometimes used in a bacterium for which no transducing phage is available involves **shuttle phasmids**, which have been constructed and used effectively in strains of *Mycobacterium* and *Streptomyces*, both gram-positive bacteria for which no transducing phages are available (see, for example, Bardarov et al., Suggested Reading). A phasmid is part

TABLE 7.2	Characteristics of generalized transducing phages P1 and P22	
	Phage	
Characteristic	**P1**	**P22**
Length (kb) of DNA packaged	100	44
Length (%) of chromosome transduced	2	1
Packaging mechanism	Sequential headful	Sequential headful
Specificity of markers transduced	Almost none	Some markers transduced at low frequency
Packaging of host DNA	Packaged from ends	Packaged from *pac*-like sequences
% of transducing particles in lysate	1	2
% of transduced DNA recombined into chromosome	1–2	1–2

phage and part plasmid. The use of a phasmid to study incompatibility in plasmids is discussed in chapter 5. In that case the phasmid was constructed from a ColE1 plasmid and λ phage, both of which function only in *E. coli*. However, a shuttle phasmid is constructed by combining parts of a phage of the bacterium being studied with a plasmid from *E. coli*, where the bacterium being studied is not related to *E. coli*. In fact, problems can arise if attempts are made to construct a shuttle phasmid for a bacterium even distantly related to *E. coli* so that some of its genes are expressed in *E. coli*. Many phage gene products are toxic to bacteria. Because the DNA is introduced into the bacterium by infection with one of its own phages, shuttle phasmids allow the very efficient transfer of DNA into bacteria for which no other efficient DNA transfer system is available. Many applications in molecular genetics require efficient DNA transfer including transposon mutagenesis and allele replacement.

To construct a shuttle phasmid for a particular bacterium to be studied, a phage is isolated that can form plaques on a lawn of the bacterium. Nothing need be known about this phage except that it should have a genome size similar to that of λ (about 40 to 50 kb), which is a common size for phages. Also, like λ, it should have sticky cohesive ends that pair with each other; this is also a common characteristic of many phages. The phage are purified, and the DNA is extracted. The DNAs are then ligated to each other to form a large concatemer. The concatemers are partially digested with a four-hitter restriction endonuclease such as Sau3A to leave pieces of about 40 to 50 kb, about the genome length of the phage but with random ends. The pieces are then ligated to the halves of a cosmid. A cosmid is a plasmid that contains the *pac* (*cos*) site of λ phage, a plasmid origin of replication that functions in *E. coli*, and a selectable antibiotic resistance gene. Cosmids are discussed in chapter 8. After ligation, long concatemers form in which phage genomes are sometimes bracketed by the halves of the cosmid. The cosmids can be packaged in λ heads by a process called in vitro packaging, and the λ phage particles are used to infect *E. coli*, selecting for the antibiotic resistance gene on the cosmid. Many of these cells contain the cosmid inserted into the phage genome at various places, rather than just the cosmid (which is not long enough to be packaged into a λ phage head) or just the DNA of the phage of the bacterium being studied (which cannot replicate in *E. coli* and, even if it could, could not confer antibiotic resistance on *E. coli*).

To determine which of these antibiotic-resistant *E. coli* cells contain a usable phasmid, they are pooled and the plasmid DNA is isolated and transformed or electroporated into the bacterium being studied. These cells are then mixed with indicator bacteria of the same type and incubated to allow plaques to form. Any plaques may contain phage which have packaged a useful phasmid with the cosmid inserted into a nonessential region of the phage DNA. However, they should be tested by repeating the process of using their DNA to form concatemers, packaging them into λ phage heads, and reinfecting *E. coli*, selecting the antibiotic resistance on the cosmid.

Once a useful phasmid has been obtained for the bacterium being studied, any DNA can be introduced into it by replacing the cosmid in the phasmid with a cosmid containing a cloned piece of DNA from the bacterium being studied. Then the DNA can be reintroduced into the bacterium with high efficiency merely by infecting the bacterium with the phasmid. In some of these applications, such as transposon mutagenesis or allele replacement, it may be necessary to make a suicide vector out of the phasmid. This can be done by isolating mutants of the phasmid that cannot replicate under some conditions, for example at higher temperatures. The use of phage suicide vectors for transposon mutagenesis is discussed in chapter 9, and the process is similar for allele replacement.

One promising recent application of shuttle phasmids is in phage typing and testing for antibiotic susceptibility (see, for example, Banaiee et al., Suggested Reading). Standard techniques for testing bacteria responsible for an infection involve culturing the bacteria on plates to identify their serotype and to test their susceptibility to various antibiotics. Culturing bacteria takes time, especially since some types of pathogenic bacteria grow slowly. Often, an antibiotic is administered to the patient before susceptibility tests are completed; if the bacterium turns out to be insensitive to the antibiotic, the patient must be given a different antibiotic and valuable time is lost. Shuttle phasmids containing the luciferase reporter gene offer the opportunity to make diagnostic tests more quickly. The luciferase gene is cloned into the cosmid part of a series of shuttle phasmids that infect different serotypes of the bacterium. The luciferase must be cloned in such a way that it is expressed in the bacteria being tested. If the phage can infect the bacterium, the luciferase is expressed and the bacterial cells in a culture "light up"; this can be detected with a luminometer. Also, if the bacterium has been pretreated with an antibiotic to which it is susceptible, particularly one that blocks translation, as most of them do (see chapter 2), the luciferase is not expressed after infection and no light is given off. This antibiotic can then be used to treat the infection without the need to wait for the results of other antibiotic sensitivity tests, which generally take longer.

Role of Transduction in Bacterial Evolution

Phages may play an important role in evolution by promoting the horizontal transfer of genes between individual members of a species, as well as between distantly related bacteria. The DNA in phage heads is usually more stable than naked DNA and so may persist longer in the environment. Also, many phages have a broad host range for adsorption. We have given the example of phage P1, which infects and multiplies in *E. coli* but also injects its DNA into a number of other gram-negative bacterial species including *Myxococcus xanthus*. The host range of P1 is partially affected by the orientation of an invertible DNA segment encoding the tail fibers (see chapter 9). Although incoming DNA from one species does not recombine with the chromosome of a different species if they share no sequences, stable transduction of genes between distantly related bacteria becomes possible when the transduced DNA is a broad-host-range plasmid that can replicate in the recipient strain or contains a broad-host-range transposon that can hop into the DNA of the recipient cell.

SUMMARY

1. Viruses that infect bacteria are called bacteriophages, or phages for short.

2. The productive developmental cycle of a phage is called the lytic cycle. The larger DNA phages undergo a complex program of gene expression during development.

3. The products of phage regulatory genes regulate the expression of other phage genes during development. One or more regulatory genes in each stage of development turn on the genes in the stage to follow and turn off the genes in the preceding stage, creating a regulatory cascade. In this way, all the information for the stepwise development of the phage can be preprogrammed into the phage DNA.

4. Phage T7 encodes an RNA polymerase that specifically recognizes the promoters for the late genes of the phage. Because of its specificity and because transcription from the T7 promoters is so strong, this system has been used to make cloning vectors for expressing large amounts of protein from cloned genes.

5. All the genes of phage T4 are transcribed by the host RNA polymerase, which undergoes many changes in the course of infection. Phage T4 goes through a number of steps of gene regulation in its development, and the genes are named based on when their products appear in the infected cell. The immediate-early genes are transcribed immediately after infection from σ^{70}-type promoters. The delayed-early genes are transcribed through antitermination of transcription from immediate-early genes. The transcription of the middle genes is from phage-specific middle-mode promoters and requires a transcriptional activator, MotA, as well as a polypeptide, AsiA, which binds tightly to the RNA polymerase. The true-late genes are also transcribed from phage-specific promoters and require a phage-encoded sigma factor, gp55. In addition, T4 couples the transcription of its late genes to the replication of the phage DNA, so that the late genes are not transcribed until the phage DNA begins to replicate.

6. All phages encode at least some of the proteins required to replicate their nucleic acids. They borrow others from their hosts.

7. The requirement of DNA polymerases for a primer prevents the replication of the extreme 5′ ends of linear DNAs. Phages solve this replication primer problem in different ways. Some replicate as circular DNA. Others form long concatemers by rolling-circle replication, by linking single DNA molecules end to end by recombination, or by pairing through complementary ends.

8. Bacteriophages are ideal for illustrating the basic principles of classical genetic analysis, including recombination and complementation. To perform recombination tests with phages, cells are infected by two different mutants or strains of phage and the progeny are allowed to develop. The progeny are then tested for recombinant types. To perform complementation tests with phages, the cells are infected by two different strains under conditions which are nonpermissive for both mutants. If the mutations complement each other, the phage will multiply. Recombination tests can be used to order mutations with respect to each other. Complementation tests can be used to determine whether two mutations are in the same functional unit or gene.

9. Nonsense and temperature-sensitive mutations are very useful for identifying the genes of a phage whose products are essential for multiplication.

10. The linkage map of a phage shows the relative positions of all of known genes of the phage with respect to each other. Whether the linkage map of a phage is circular or linear depends on how the DNA of the phage replicates and how it is packaged into phage heads.

11. Transduction occurs when a phage accidentally packages bacterial DNA into a head and carries it from one host to another. In generalized transduction, essentially any region of bacterial DNA can be carried.

(continued)

12. Not all types of phages make good transducing phages. To be a good transducing phage, the phage must not degrade the host DNA after infection, must not have very specific *pac* sites, and must package DNA by a headful mechanism.

13. Transduction is very useful for genetic mapping in bacteria and for strain construction. Using transduction to map bacterial genetic markers is based on the fact that only a small part of the bacterial genome can be packaged in a phage head. If two markers are close enough together to be packaged in a phage head, they are cotransducible. Since transduction is so infrequent, one marker is selected and the transductants are tested for the cotransduction of the region of the other marker. The frequency of cotransduction increases the closer two markers are to each other.

QUESTIONS FOR THOUGHT

1. Why do you suppose phages regulate their gene expression during development so that genes whose products are involved in DNA replication are transcribed before genes whose products become part of the phage particle? What do you suppose would happen if they did not do this?

2. What do you suppose the advantages are to a phage encoding its own RNA polymerase instead of merely changing the host RNA polymerase? What are the advantages of using the host RNA polymerase?

3. Why do phages often encode their own replicative machinery rather than depending on that of the host?

4. Why do some single-stranded DNA phages such as φX174 use a complicated mechanism to initiate replication that uses the cellular proteins PriA, PriB, PriC, and DnaT, which are normally used by the cell to reinitiate replication at blocked forks, while others, such as f1, use a much simpler process, involving only the host RNA polymerase, to prime replication?

5. Phages often change the cell to prevent subsequent infection by other phage of the same type once the infection is under way. This is called superinfection exclusion. What do you suppose would be the consequences of another phage of the same type superinfecting a cell that is already in the late state of development of the phage if this mechanism did not exist?

PROBLEMS

1. We have precisely determined the titer of a stock of virus by counting the viruses under the electron microscope. How would you determine the effective MOI of the virus (i.e., the fraction of viruses that actually infect a cell) under a given set of conditions?

2. Phage components (e.g., heads and tails) can often be seen in lysates by electron microscopy, even before they are assembled into phage particles. In studying a newly discovered phage of *Pseudomonas putida*, you have observed that amber mutations in gene *C* of the phage prevent the appearance of phage tails in a nonsuppressor host. Similarly, mutations in gene *T* prevent the appearance of heads. However, mutations in gene *M* prevent the appearance of either heads or tails. Which gene, *C*, *T*, or *M*, is most likely to be a regulatory gene? Why?

3. Would you expect to be able to isolate amber mutations in the *ori* sequence of a phage? Why or why not?

4. To order the three genes *A*, *M*, and *Q* in a previously uncharacterized phage you have isolated, you cross a double mutant having an amber mutation in gene *A* and a temperature-sensitive mutation in gene *M* with a single mutant having an amber mutation in gene *Q*. About 90% of the Am⁺ recombinants that can form plaques on the nonsuppressor host are temperature sensitive. Is the order *Q-A-M* or *A-Q-M*? Why?

5. Phage T1 packages DNA from concatemers beginning at a unique *pac* site and then packaging by a processive headful mechanism, cutting about 6% longer than a genome length each time. However, it packages a maximum of only three headfuls from each concatemer. Would you expect T1 to have a linear or a circular genetic map? Draw a hypothetical T1 map.

6. Most types of phage encode lysozyme enzymes which break the cell wall of the infected cell late in infection to release the phage. How would you determine which of the genes in the linkage map of a phage you have isolated encodes the lysozyme? Hint: You can purchase egg white lysozyme from biochemical supply companies.

7. Chloroform (CHCl₃) dissolves the membrane of bacteria but does not affect the cell wall. What would be the effect of adding CHCl₃ to an *rI* (antiholin) mutant of T4? To a *t* (holin) mutant? To an *e* (lysozyme) mutant?

8. Why is it important to propagate phage for phage display at a low MOI?

SUGGESTED READING

Banaiee, N., M. Bobadilla-del-Valle, P. F. Riska, S. Bardarov, Jr. P. M. Small, A. Ponce-de-Leon, W. R. Jacobs, Jr., G. F. Hatfull, and J. Sifuentes-Osornio. 2003. Rapid identification and susceptibility testing of *Mycobacterium tuberculosis* from MGIT cultures with luciferase reporter mycobacteriophages. *J. Med. Microbiol.* **52:**557–561.

Bardarov, S., S. Bardarov, Jr., M. S. Pavelka, Jr., V. Sambandamurthy, M. Larsen, J. Tufariello, J. Chan, G. Hatfull, and W. R. Jacobs, Jr. 2002. Specialized transduction: an efficient method for generating marked and unmarked targeted gene disruptions in *Mycobacterium tuberculosis*, *M. bovis* and *M. smegmatis*. *Microbiology* **148:**3007–3017.

Benzer, S. 1961. On the topography of genetic fine structure. *Proc. Natl. Acad. Sci. USA* **47:**403–415.

Benzer, S., and S. P. Champe. 1962. A change from nonsense to sense in the genetic code. *Proc. Natl. Acad. Sci. USA* **48:**1114–1121.

Brenner, S., A. O. W. Stretton, and S. Kaplan. 1965. Genetic code: the nonsense triplets for chain termination and their suppression. *Nature* (London) **206:**994–998.

Calender, R. (ed.). 2005. *The Bacteriophages*, 2nd ed. Oxford University Press, New York, N.Y.

Chastian, P. D., Jr., A. M. Makhov, N. G. Nossal, and J. Griffith. 2003. Architecture of the replication complex and DNA loops at the fork generated by the bacteriophage T4 proteins. *J. Biol. Chem.* **278:**21276–21285.

Colland, F., G. Orsini, E. N. Brody, H. Buc, and A. Kolb. 1998. The bacteriophage T4 AsiA protein: a molecular switch for σ70 dependent promoters. *Mol. Microbiol.* **27:**819–829.

Crick, F. H. C., L. Barnett, S. Brenner, and R. J. Watts-Tobin. 1961. General nature of the genetic code for proteins. *Nature* (London) **192:**1227–1232.

Fluck, M. M., and R. H. Epstein. 1980. Isolation and characterization of context mutations affecting the suppressibility of nonsense mutations. *Mol. Gen. Genet.* **177:**615–627.

Hendrix, R. W. 2005. Bacteriophage evolution and the role of phages in host evolution, p. 55–65. *In* M. K. Waldor, D. I. Friedman, and S. L. Adhya (ed.), *Phages: Their Role in Bacterial Pathogenesis and Biotechnology*. ASM Press, Washington, D.C.

Herendeen, D. R., G. A. Kassavetis, and E. P. Geiduschek. 1992. A transcriptional enhancer whose function imposes a requirement that proteins track along the DNA. *Science* **256:**1298–1303.

Kreuzer, K. N. 2005. Interplay between replication and recombination in prokaryotes. *Annu. Rev. Microbiol.* **59:**43–67.

Marciano, D. K., M. Russel, and S. M. Simon. 2001. Assembling filamentous phage occlude pIV channels. *Proc. Natl. Acad. Sci. USA* **98:**9359–9364.

Miller, E. S., E. Kutter, G. Mosig, F. Arisaka, T. Kunisawa, and W. Ruger. 2003. Bacteriophage T4 genome. *Microbiol Mol. Biol. Rev.* **67:**86–156.

Molineux, I. J. 2001. No syringes please, ejection of phage T7 DNA from the virion is enzyme driven. *Mol. Microbiol.* **40:**1–8.

Mosig, G. 1998. Recombination and recombination-dependent replication in bacteriophage T4. *Annu. Rev. Genet.* **32:**379–413.

Nechaev, S., M. Kamali-Moghaddam, E. Andre, J. P. Leonetti, and E. P. Geiduschek. 2004. The bacteriophage T4 late-transcription coactivator gp33 binds the flap domain of *Escherichia coli* RNA polymerase. *Proc. Natl. Acad. Sci USA* **101:**17365–17370.

Nelson, S. W., J. Yang, and S. J. Benkovic. 2006. Site-directed mutations of T4 helicase loading protein (gp59) reveal multiple modes of DNA polymerase inhibition and the mechanism of unlocking by gp41 helicase. *J. Biol. Chem.* **281:**8697–8706.

Nomura, M., and S. Benzer. 1961. The nature of deletion mutants in the *r*II region of phage T4. *J. Mol. Biol.* **3:**684–692.

Rakonjac, J., J. Feng, and P. Model. 1999. Filamentous phage are released from the bacterial membrane by a two-step mechanism involving a short C-terminal fragment of pIII. *J. Mol. Biol.* **289:**1253–1265.

Shutt, T. E., and M. W. Gray. 2006. Bacteriophage origins of mitochondrial replication and transcription protein. *Trends Genet.* **22:**90–95.

Studier, F. W. 1969. The genetics and physiology of bacteriophage T7. *Virology* **39:**562–574.

Studier, F. W., A. H. Rosenberg, J. J. Dunn, and J. W. Dubendoff. 1990. Use of T7 RNA polymerase to direct expression of cloned genes. *Methods Enzymol.* **185:**60–89.

Symonds, N., P. Vander Ende, A. Dunston, and P. White. 1972. The structure of *r*II diploids of phage T4. *Mol. Gen. Genet.* **116:**223–238.

Yanisch-Perron, C., J. Vieira, and J. Messing. 1985. Improved M13 phage cloning vectors and host strains: nucleotide sequences of the M13mp18 and pUC19 vectors. *Gene* **33:**103–119.

Zinder, N. D., and J. Lederberg. 1952. Genetic exchange in *Salmonella*. *J. Bacteriol.* **64:**679–699.

CHAPTER 8

Lysogeny: the λ Paradigm and the Role of Lysogenic Conversion in Bacterial Pathogenesis

In chapter 7, we reviewed the lytic development of some representative phages. During lytic development, the phage infects a cell and multiplies, producing more phage that can then infect other cells. However, this is not the only lifestyle of which phages are capable. Some phages are able to maintain a stable relationship with the host cell in which they neither multiply nor are lost from the cell. Such a phage is called a lysogen-forming or **temperate phage.** In the lysogenic state, the phage DNA either is integrated into the host chromosome or replicates as a plasmid. The phage DNA in the lysogenic state is called a **prophage,** and the bacterium harboring a prophage is a **lysogen** for that phage. Thus, a bacterium harboring the prophage P2 would be a P2 lysogen.

In a lysogen, the prophage acts like any good parasite and does not place too great a burden on its host. The prophage DNA is mostly quiescent; most of the genes expressed are those required to maintain the lysogenic state, and most of the others are turned off. Often the only indication that the host cell carries a prophage is that the cell is immune to superinfection by another phage of the same type. The prophage state can continue almost indefinitely unless the host cell suffers potentially lethal damage to its chromosomal DNA or is infected by another phage of the same or a related type. Then, like a rat leaving a sinking ship, the phage can be **induced** and enter lytic development, producing more phage. The released phage can then infect other cells and develop lytically to produce more phage or lysogenize the new bacterial cell, hopefully one with a brighter future than its original host.

Studies with phage lysogeny were important in forming concepts of how viruses can remain dormant in their hosts and how they can convert cells into cancer cells. Also, many genes of benefit to the host bacterium, including virulence genes, whose products are required to make bacteria pathogenic, are carried on prophages, a topic covered later in this chapter. In addition to prophages that can be induced under some circumstances, bacteria carry many **defective prophages** that no longer can form infective phage due to deletion of some of their essential genes. These DNA elements are suspected of being defective prophages rather than normal parts of the chromosome because they are not common to all the strains of a species of bacterium and they carry genes related to the genes of other known phages. Defective prophages may be important in evolution because they eventually lose their identity and contribute useful genes to the chromosome of their host.

In this chapter, we discuss some examples of lysogen-forming phages and how the lysogenic state is achieved and maintained. One striking insight to come from these studies is the extent to which bacteria and their prophages depend on each other, to the point where the distinction between the bacterial host and the parasitic virus begins to blur.

Phage λ

Phage λ is the lysogen-forming phage which has been studied most extensively and the one to which all others are compared. Although lysogeny was suspected as early as the 1920s, the first convincing demonstration that bacterial cells could carry phages in a quiescent state was made with λ in about 1950 (see Lwoff, Suggested Reading). In this experiment, apparently uninfected *Escherichia coli* cells could be made to produce λ by being irradiated with UV light. Since then, phage λ has played a central role in the development of the science of molecular genetics. A large number of very clever and industrious researchers have collaborated to make the interaction of λ with its host, *E. coli*, our best understood biological system (see Gottesman, Suggested Reading). This research has revealed not only the complexity and subtlety of biological systems but also their utility, robustness, and beauty.

Figure 8.1 gives an overview of the two life cycles of which λ is capable and the fate of the DNA in each cycle, while Figure 8.2 gives more detailed maps of the phage genome for reference. The phage DNA is linear in the phage head. Immediately after the DNA is injected into the cell to initiate the infection, the DNA cyclizes by pairing between the *cos* sites at the ends (Figure 8.3). This brings the lysis genes (*S* and *R*) and the head and tail genes (*A* to *J*) of the phage together and allows them to be transcribed from the late promoter p'_R, as discussed later. This circular DNA can then either integrate into the host chromosome (lysogenic cycle) or replicate and be packaged into phage heads to form more phage (lytic

cycle). Which decision is made depends on the physiological state of the cell. Later, the integrated DNA can also excise and replicate and form more phage (induction). We first discuss in detail how λ replicates lytically and then describe how it can form a lysogen and finally how these two states are coordinated.

Lytic Development

Phage λ is fairly large, with a genome intermediate in size between those of T7 and T4. Some of the gene products and sites encoded by λ that are required for transcriptional regulation are listed in Tables 8.1 and 8.2, respectively. Phage λ goes through three major stages of gene expression during development. The first λ genes to be expressed after infection are *N* and *cro*. Most of the genes expressed next play a role in replication and recombination. Finally, the late genes of the phage are expressed, encoding the head and tail proteins of the phage particle and enzymes involved in cell lysis.

TRANSCRIPTION ANTITERMINATION

Many mechanisms of gene regulation now known to be universal were first discovered in phage λ. One of these is **transcription antitermination** (see chapter 2). In regulation by transcription antitermination, transcription begins at the promoter but then soon terminates until certain conditions are met. Then the transcription continues into other downstream genes, hence the name "antitermination" because the mechanism works against, or "anti to," termination. After being discovered in λ, this type of regulation was found in many other systems including in the regulation of transcription of human immunodeficiency virus (see below).

Phage λ uses antitermination of regulation at two stages in its development. At an early stage, it uses antitermination protein N to regulate the synthesis of its recombination and replication functions. At a later stage, it uses antitermination protein Q to regulate the synthesis of its late proteins including the head and tail proteins.

The N Protein

The N protein is responsible for the first stage of λ antitermination, as illustrated in Figure 8.4. When the λ DNA first enters the cell, transcription immediately begins from two promoters, p_L and p_R, that face outward from the immunity region. This leads to the synthesis of two RNAs, one leftward on the genome and the other rightward. However, the synthesis of both RNAs terminates at transcription termination sites, t_L^1 and t_R^1, after only short RNAs are synthesized (Figure 8.4A). One of these short RNAs encodes the Cro protein, which is an inhibitor of repressor synthesis, as discussed in the section on lysogeny (see below). The other encodes the

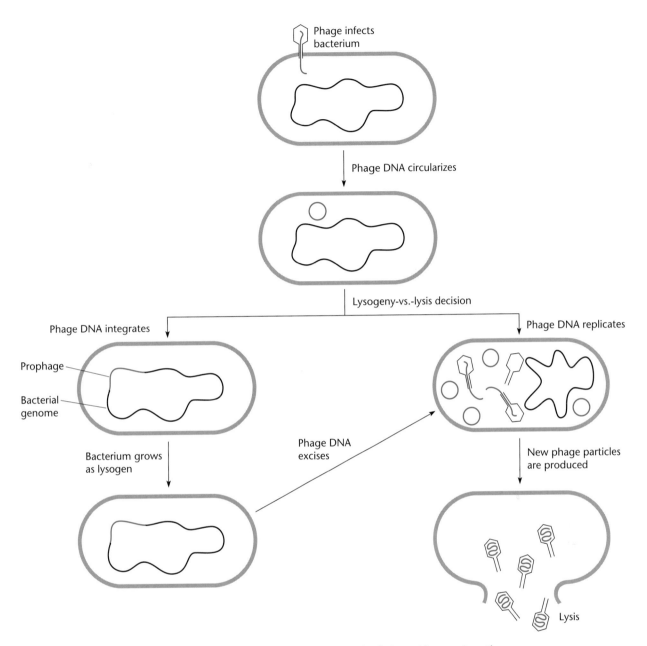

Figure 8.1 Overview of the fate of λ DNA during the lytic and lysogenic pathways.

N protein, the antitermination factor that permits the RNA polymerase to bypass the transcription termination sites and continue along the DNA, as shown in Figure 8.4B.

Figure 8.4C and D outline the current picture for how N protein antiterminates, showing only rightward transcription. Initially, transcription initiated at the rightward p_R promoter terminates at the transcription terminator designated $t_R{}^1$. One of the sequences transcribed into RNA is *nutR* (for N *utilization rightward*). In the meantime, the N protein is being translated from the leftward

RNA. It binds to the RNA polymerase but only after the *nutR* region on DNA has been transcribed (Figure 8.4D). Presumably, the N protein can form a complex with RNA polymerase only if it is also bound to the *nutR* RNA, which might also remain bound to the complex. Other host proteins, called the Nus proteins, help it bind (see below), and together they all form an antitermination complex that is resistant to almost all types of transcription termination signals, including $t_R{}^1$. When this antitermination complex transcribes through $t_R{}^1$, it reaches the genes *O* and *P*, which encode the replication

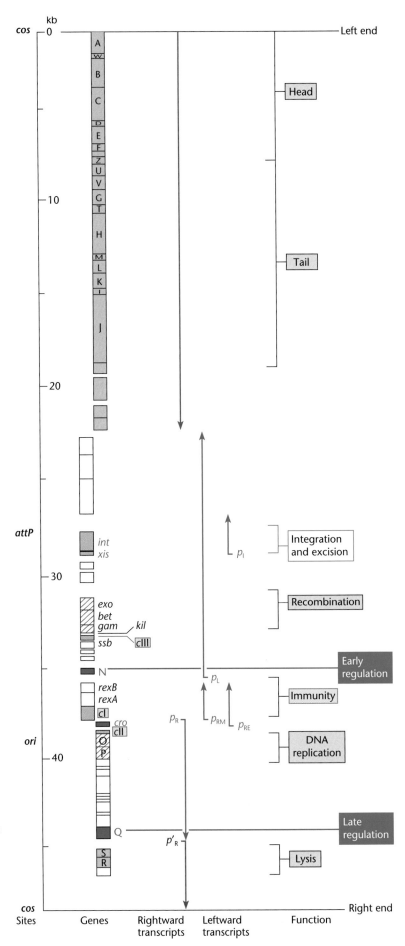

Figure 8.2 Genetic map of phage λ. Regulatory genes for the lytic pathway are shown in dark purple. Additional genes emphasized in text are shown in light purple. Recombination and replication genes are indicated by hatched boxes. The structural genes for the phage particle and lysis genes are in gray. Promoters and transcripts discussed in text are also indicated. The GenBank accession number is NC_001416.

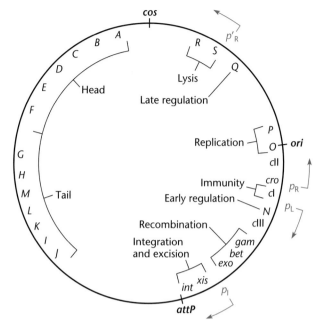

Figure 8.3 Genetic map of λ cyclized by pairing at the *cos* sites, shown at the top.

This model for how N regulates the expression of the early genes by antiterminating transcription was first proposed on the basis of indirect genetic experiments. First it was shown that transcription initiated at the λ promoters p_R and p_L (see the genetic map in Figure 8.2) soon terminates unless N protein is present. This led to the conclusion that N was acting as an antiterminator by allowing transcription through downstream transcription terminators. However, surprisingly, N could antiterminate only if transcription had initiated at the p_R and p_L promoters and not if transcription had initiated from other promoters closer to the terminators. This led to the conclusion that N does not act only at the terminators to prevent termination but that some sites upstream close to the P_R and P_L promoters are required for N action, possibly the p_R and p_L promoters themselves. Then it was shown that the sites required for N antitermination were actually not the p_R and p_L promoters themselves but nearby sites somewhat downstream of the promoters. It was hypothesized that N must bind to these sites to allow transcription through the downstream termination sites, and these were named the *N utilization* sites (or *nut* sites).

Identifying the hypothetical *nut* sites on mRNA involved some clever selectional genetics (see Salstrom and Szybalski, Suggested Reading). This quest must have required a leap of faith, since all the evidence for the existence of the *nut* sites up to this point was indirect. However, the investigators were able to isolate *nutL* mutations that had all the predicted characteristics of *nut* mutations. These mutations prevent the expression of genes downstream of the terminator site; hence, they no longer allow antitermination, and they are *cis* acting and affect only mRNA from the DNA which has the mutation. They also map in approximately the right place for a predicted *nutL* mutation, just downstream of the p_L

proteins, and replication begins. The product of another gene, Q, is also made, and this protein turns on the transcription of the late genes (see below).

Similar events are occurring during leftward transcription. When the *nut* site on the other side, called *nutL* (for *N u*tilization *l*eftward), is transcribed, the N protein binds to the RNA polymerase and prevents any further termination, allowing transcription to continue into other genes on the left side, including *gam* and *red*, which are λ recombination functions. The important sequences in the *nutL* site, including the BoxA and BoxB sequences, are identical to those in the *nutR* site (see below).

TABLE 8.1	Some λ gene products and their function
Gene product	**Function**
N	Antitermination protein acting at t_L^1, t_R^1 and t_R^2
O, P	Initiation of λ DNA replication
Q	Antitermination protein acting at t_R'
CI	Repressor; protein inhibitor of transcription from p_L and p_R
CII	Activator of transcription of *cI* and *int*
CIII	Stabilizer of CII
Cro	Protein inhibitor of CI synthesis
Gam	Protein required for rolling-circle replication
Red	Proteins involved in λ recombination
Int	Integrase; protein required for site-specific recombination with chromosome
Xis	Excisase; protein forms complex with Int and functions in excision of prophage

TABLE 8.2	Some sites involved in phage λ transcription and replication[a]
Site(s)	**Function(s)**
P_L	Left promoter
p_R, p_R'	Right promoters
Q_L	Operator for leftward transcription; binding sites for CI and Cro repressors
Q_R	Operator for rightward transcription; binding sites for CI and Cro repressors
t_L^1, t_L^2	Termination sites of leftward transcription
t_R^1, t_R^{234}, t_R'	Termination sites of rightward transcription
nutL	N utilization site for leftward transcribing RNA Pol (i.e., the site at which N binds to RNA Pol)
nutR	N utilization site for rightward transcribing RNA Pol
qut	Q utilization site for antitermination at p_R'
p_{RE}	Promoter for repressor establishment; activated by CII
p_{RM}	Promoter for repressor maintenance; activated by CI
p_I	Promoter for *int* transcription; activated by CII
POP′	Attachment site (*attP*)
cos	Cohesive ends of λ genome (12-bp single-stranded ends in linear genome anneal to form circular genome after infection)

[a] In λ, essential genes have single-letter names while nonessential genes have more conventional three-letter names.

promoter and just upstream of the *N* gene. Once such mutations had been isolated, the exact base pair change in the mutations was determined by comparing the DNA sequence of the *nutL* mutants to the DNA sequence of wild-type λ in this region. Once the site of *nutL* mutations was identified by DNA sequencing, a region with similar sequences was found just to the right of the *cro* gene and was assumed to be *nutR*. The experiments of Salstrom and Szybalski are discussed in detail later in this chapter (see "Genetic Experiments with Phage λ").

The location of the *nut* sites supported another element of the model, i.e., that the N protein binds to the *nut* site sequence in the mRNA rather than in the DNA. Because of their location between genes, the *nut* sequences are not normally translated into protein. In fact, translating the *nut* sites interferes with antitermination. Apparently, ribosomes translating the mRNA can interfere with the binding of N to a *nut* site and thereby interfere with antitermination.

Figure 8.5 shows the sequence of the *nut* sites of λ. The *nut* sites consist of a sequence, called BoxB, that forms a "hairpin" secondary structure in the mRNA because it is encoded by a region of the DNA with a twofold rotational symmetry (see chapter 2). The original *nutL* mutations all change bases in BoxB and disrupt the twofold symmetry of the sequence, preventing formation of the hairpin in the mRNA. Thus, apparently the formation of the BoxB hairpin in the mRNA is important for the binding of the N protein. In fact, structural studies performed more recently have indicated that the N protein can bind to RNA with just the BoxB

hairpin and that both the N protein and the BoxB secondary structure change as a result of this binding. This supports the idea that the N protein changes its conformation on binding to the BoxB sequence and that only N in the changed conformation can bind to RNA polymerase and prevent termination.

The function of the BoxA and BoxC sequences in *nut* sites is more obscure. These sequences are common to the *nut* sites of all λ-related phages, and BoxA-like sequences occur in some bacterial genes including the genes for rRNA, where they play an important role in preventing premature termination of transcription (see chapter 13). Adding to the mystery is the fact that point mutations in BoxA of a phage λ *nut* site can prevent antitermination but deletion of the entire BoxA sequence does not. It is an attractive idea that the BoxA sequences in the *nut* sites and other antitermination sequences help bind host Nus proteins to promote antitermination (see below). No one has found a function for BoxC. It could be that these sites are required to regulate antitermination in subtle ways not easily detectable in a laboratory situation.

Host Nus Proteins

As mentioned above, the N protein does not act alone to cause antitermination. A number of *E. coli* proteins, some of which are involved in transcription termination and antitermination in the uninfected *E. coli* cell, are also involved. Host proteins which collaborate with N to cause antitermination are called Nus proteins (for *N* utilization *s*ubstance). Many of the *nus* genes were first

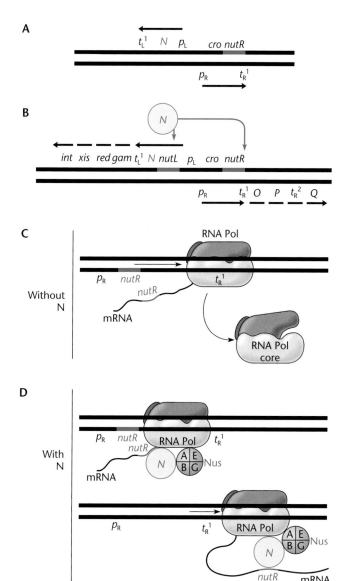

Figure 8.4 Antitermination of transcription in phage λ.
(A) Before the N protein is synthesized, transcription starts at promoters p_L and p_R and stops at transcription terminators t_L^1 and t_R^1. (B) The N protein causes transcription to continue past t_L^1 and t_R^1 into *gam-red-xis-int* and *O-P-Q*, respectively. (C and D) Mechanism of antitermination by N, showing rightward transcription only. (C) In the absence of N, transcription initiated at p_R terminates at the terminator t_R^1. (D) If N has been made, it binds to RNA polymerase (RNA pol) as the polymerase transcribes the *nutR* site, possibly because N undergoes a conformational change when it binds to the *nutR* sequence in the RNA. This change is required before N can bind to the RNA polymerase. The binding of N to RNA polymerase is facilitated by the host Nus proteins A, B, E, and G. The antitermination complex composed of RNA polymerase, N, *nutR*, and Nus ABEG then transcribes past transcriptional terminator t_R^1 plus any other transcriptional terminators downstream. The sites and λ gene products shown here are defined in Tables 8.1 and 8.2.

identified by *E. coli* mutations which prevent killing by phage λ after induction (see Friedman et al., Suggested Reading, and "Genetic Experiments with Phage λ" below). Six *nus* genes, *nusA* to *nusG*, have been identified thus far by using this and other types of selections. Four of these, *nusA*, *nusB*, *nusE*, and *nusG*, encode proteins involved in transcription termination and/or antitermination in the uninfected host. The products of these genes travel with the N-*nut*-RNA polymerase complex and may help hold the complex together (Figure 8.4).

Surprisingly, *nusE* mutations are in a gene for a ribosomal protein, S10. This is surprising because translation is not thought to be required for antitermination and, in fact, inhibits it (see above). Perhaps the S10 protein plays a dual role in the cell, one in translation and another in transcription antitermination. Other *nus* mutations have a less direct effect on antitermination. For example, *nusD* mutations affect the host ρ factor, which is required for transcription termination at ρ-dependent termination sites (see chapter 2). These mutations may affect N antitermination by causing stronger termination that cannot be overcome by N. Other *nus* mutations called *nusC* are present in the RNA polymerase β subunit. They may alter the binding of the antitermination complex to the RNA polymerase and thereby reduce antitermination.

The Q Protein

One of the genes under the control of the N antiterminator is gene *Q*, whose product is responsible for the transcription of the late genes of λ including the head and tail genes. Thus, λ marches through a regulatory cascade, with one of the earliest gene products (N) directing the synthesis of another gene product (Q), which in turn directs the transcription of the late genes. Like N, the Q protein of λ is an antiterminator, which allows transcription from the late promoter p_R' to proceed through terminators into downstream genes. The mechanism of antitermination by Q is very different from that of antitermination by N. Like N, the Q protein loads on RNA polymerase in response to a sequence located close to the promoter, called *qut* (for *Q* utilization site) (Table 8.2). However, the similarity ends here. The *qut* site is not in the mRNA; but, rather, in the DNA. Some of the required *qut* sequence is not even transcribed into mRNA, being upstream of the start site of transcription at p_R'.

The details of Q protein antitermination have been studied in some detail (see Nickels et al., Suggested Reading). As a prelude to antitermination, the RNA polymerase transcribes a short RNA of only 16 to 17 nucleotides from the late promoter p_R' before it pauses. The RNA polymerase pauses because it confuses the

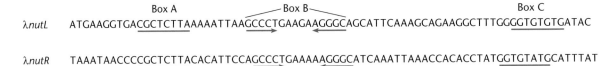

Figure 8.5 The sequences of the *nutL* and *nutR* regions of bacteriophage λ. Box A, Box B, and Box C are underlined. The twofold rotational symmetry in Box B that causes a hairpin to form in the mRNA is shown in purple.

pause site for a promoter. The σ factor has not yet cycled off the RNA polymerase, and the pause site has a −10 sequence similar to the −10 sequence at promoters to which the region 2 of the σ^{70} factor binds (see chapter 2). The Q protein then loads on the paused RNA polymerase, displacing region 4 of σ^{70} and allowing the transcript to exit through the exit pore and the RNA polymerase to escape from the pause site. Once bound, Q then travels with the RNA polymerase, making it oblivious to further transcription termination and pause sites and allowing it to proceed untrammeled through the late genes of the phage, much as the N protein prevents transcription termination by remaining bound to the RNA polymerase. Apparently, the Q protein can load on RNA polymerase only if the RNA polymerase is positioned correctly at the pause site. Sometimes the RNA polymerase overshoots the pause site and makes an RNA 17 nucleotides or more in length. Then the RNA polymerase must backtrack to 16 nucleotides while the GreA and GreB proteins cleave the extra RNA that extrudes from the RNA polymerase as a result of the backtracking (see chapter 2). The Q protein can then load on the RNA polymerase and send it on its way at last.

ANTITERMINATION IN OTHER SYSTEMS

Like many regulatory systems, antitermination was first discovered in phage, but it is used in many other systems. This mechanism operates not only in related phages such as P22 but also for many bacterial genes. As mentioned, antitermination regulates transcription of the rRNA genes of all bacteria. It is also known to regulate the transcription of bacterial operons including the *bgl* operon of *E. coli* and the aminoacyl-tRNA synthetase genes of *Bacillus subtilis*, although in these cases it involves a specific action on the terminator itself (see chapter 12). Often, where regulation by antiterminator proteins bound to RNA polymerase occurs, sequences similar to λ *nut* sequences also are present in the RNA. Antitermination may also be used to regulate eukaryotic genes. For example, the *myc* oncogene of mammals and, as mentioned above, the transcription of human immunodeficiency virus, which causes AIDS, are also regulated through antitermination, although the mechanisms used differ. However, the conceptualization of antitermination as a way of regulating transcription first came from λ.

Replication of λ DNA

The replication of λ DNA has also been studied extensively and has been one of the major model systems for understanding replication in general. The λ DNA is linear in the phage head but cyclizes, that is, forms a circular molecule, after it enters the cell through pairing between its cohesive ends, or *cos* sites (Figure 8.3). These sites are single stranded and complementary to each other for 12 bases and so can join by complementary base pairing, which makes them cohesive or "sticky." Once the cohesive ends are paired, DNA ligase can join the two ends to form covalently closed circular λ DNA molecules. These circular DNA molecules can replicate in their entirety because there is always DNA upstream to serve as a primer (see chapter 7).

CIRCLE-TO-CIRCLE, OR θ, REPLICATION OF λ DNA

Once circular λ DNA molecules have formed in the cell, they replicate by a mechanism similar to the θ replication described for the chromosome in chapter 1 and for plasmids in chapter 4. Replication initiates at the *ori* site in gene O (see the map in Figure 8.2) and proceeds in both directions, with both leading- and lagging-strand synthesis in the replication fork (Figure 8.6). When the two replication forks meet somewhere on the other side of the circle, the two daughter molecules separate and each can serve as the template to make another circular DNA molecule.

ROLLING-CIRCLE REPLICATION OF λ DNA

After a few circular λ DNA molecules have accumulated in the cell by θ replication, the rolling-circle type of replication ensues. The initiation of λ rolling-circle replication is similar to that of M13 in that one strand of the circular DNA is cut and the free 3′ end serves as a primer to initiate the synthesis of a new strand of DNA that displaces the old strand. DNA complementary to the displaced strand is also synthesized to make a new double-stranded DNA. The λ process differs in that the

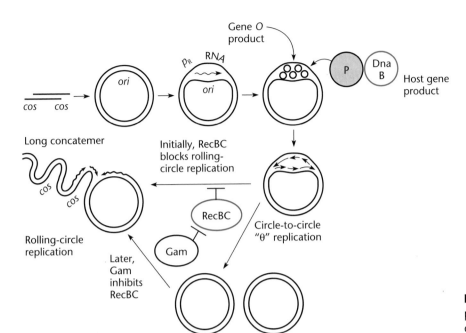

Figure 8.6 Overview of replication of phage λ. The *E. coli* gene products are outlined in purple.

displaced individual single-stranded molecules are not released when replication around the circle is completed. Rather, the circle keeps rolling, giving rise to long tandem repeats of individual λ DNA molecules linked end to end, called **concatemers** (Figure 8.6). The formation of end-to-end concatemers by rolling-circle replication can be compared to what happens when an engraved ring, dipped in ink, is rolled across a piece of paper: the pattern on the ring is repeated over and over again on the paper.

In the final step, the long concatemers are cut at the *cos* sites into λ-genome-length pieces as they are packaged into phage heads. Phage λ can package only DNA from concatemers, and at least two λ genomes must be linked end to end in a concatemer, because the packaging system in the λ head recognizes one *cos* site on the concatemeric DNA and takes up DNA until it arrives at the next *cos* site, which it cleaves to complete the packaging. The dependence on concatemeric DNA for λ DNA packaging was important in the discovery of *chi* sites and their role in RecBCD recombination (see chapter 10).

GENETIC REQUIREMENTS FOR λ DNA REPLICATION

Unlike T4 and T7, which encode many proteins for replication, the products of only two λ genes, *O* and *P*, are required for λ DNA replication. As Figure 8.6 illustrates, both of these proteins are required for priming DNA replication at the *ori* site. The O protein is thought to bend the DNA at this site by binding to repeated sequences, similar to the mechanism by which DnaA protein initiates chromosome replication (see chapter 1),

and the P protein binds to the O protein and to the replicative helicase DnaB of the host replication machinery, thus comandeering it for λ DNA replication; it therefore acts like DnaC. Appropriately, the P of the P gene product stands for "pirate."

RNA synthesis must also occur in the *ori* region for λ replication to initiate. It normally initiates at the p_R promoter and may be required to separate the DNA strands at the origin and/or serve as a primer for rightward replication.

A third λ protein, the product of the *gam* gene, is required for the shift to the rolling-circle type of replication, albeit indirectly. The RecBCD nuclease (an enzyme that facilitates recombination of *E. coli*) somehow inhibits the switch to rolling-circle replication, perhaps by degrading the free 3′ end that forms, but Gam inhibits the RecBCD nuclease. Therefore, a *gam* mutant of λ is restricted to the θ mode of replication, and concatemers can form from individual circular λ DNA molecules only by recombination. Because λ requires concatemers for packaging, a *gam* mutant of λ cannot multiply in the absence of a functional recombination system, either its own or the RecBCD pathway of its host. This fact was also important in the detection of *chi* sequences, as discussed further in chapter 10.

PHAGE λ CLONING VECTORS

The many cloning vectors derived from phage λ offer numerous advantages. The phage multiply to a high copy number, allowing the synthesis of large amounts of DNA and protein. Also, it matters less if the cloned gene encodes a toxic protein than with plasmid cloning vectors. The toxic protein is not synthesized until the phage infects the cell, and the infected cell is destined to die anyway. It is also relatively easy to store libraries in the relatively stable λ phage head.

COSMIDS

As mentioned, λ packages DNA into its head by recognizing *cos* sites in concatemeric DNA; therefore, any DNA containing two *cos* sequences is packaged into phage heads if the two *cos* sequences are about 50 kb apart. In particular, plasmids containing *cos* sites can be packaged into λ phage heads. Such plasmid cloning vectors, called **cosmids,** also offer many advantages for genetic engineering, including **in vitro packaging.** In this procedure, plasmid DNA is mixed with extracts of λ-infected cells containing heads and tails of the phage. The DNA is taken up by the heads, and because λ particles self-assemble in the test tube, the tails are attached to the heads to make infectious λ particles, which can then be used to introduce the cosmid into cells by infection. Any λ cloning vector can serve in this method, and infection is a more efficient way of introducing DNA into bacteria than is transfection or transformation. We discussed phasmids based on cosmids in chapter 7.

Another major advantage of cosmids is that the size of the cloned DNA is limited by the size of the phage head. If the piece of DNA cloned into a cosmid is too large, the *cos* sites will be too far apart and the DNA will be too long to fit into a phage head. However, if the cloned DNA is too small, the *cos* sites will be too close to each other and the phage heads will have too little DNA and be unstable. Therefore, the use of cosmids ensures that the pieces of DNA cloned into a vector will all be approximately the same size, which is sometimes important for making DNA libraries (see chapter 1).

Lysogeny

Phage λ is the classical example of a phage that can form lysogens. In the lysogenic state, very few λ genes are expressed, and essentially the only evidence that the cell harbors a prophage is that the lysogenic cells are immune to superinfection by more λ. The growth of the immune lysogens in the plaque is what gives the λ plaque its characteristic "fried-egg" appearance, with the lysogens forming the "yolk" in the middle of the plaque (Figure 8.7).

Some phage λ mutants form plaques that are clear because they do not contain immune lysogens. These phage have mutations in either the *cI*, *cII*, or *cIII* gene, where the "c" stands for *c*lear plaque. These mutations prevent the formation of lysogens. Understanding the

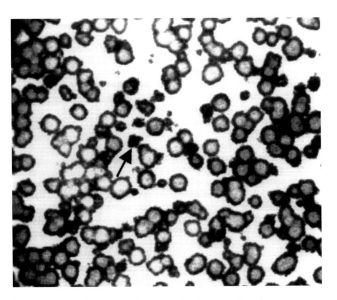

Figure 8.7   Phage λ plaques with typical cloudy centers, giving them a fried-egg appearance. The arrow points to a plaque formed by a clear-plaque mutant.

regulation of the λ lysogenic pathway and the function of the *cI*, *cII*, and *cIII* gene products in forming lysogens required the concerted effort of many people. Their findings illustrate the complexity and subtlety of biological regulatory pathways and serve as a model for other systems (see Gottesman, Suggested Reading).

The cII Gene Product

Figure 8.8 illustrates the process of forming a lysogen after λ infection, how the *cI*, *cII*, and *cIII* gene products are involved, and the central role of CII protein in this decision. After λ infects a cell, the decision whether the phage enters the lytic cycle and makes more phage or forms a lysogen depends on the outcome of a competition between the product of the *cII* gene, which acts to form lysogens, and the products of genes in the lytic cycle that replicate the DNA and make more phage particles. Most of the time, the lytic cycle wins, the λ DNA replicates, and more phage are produced. However, about 1% of the time, depending on environmental factors such as the richness of the medium, the *cII* gene product wins the race and a lysogen is formed.

The CII protein promotes lysogeny by activating the RNA polymerase to begin transcribing at two promoters, which are otherwise inactive (Figure 8.8). Proteins that enable RNA polymerase to begin transcription at certain promoters are called **transcriptional activators** (see chapter 2). One of the promoters activated by CII is p_{RE}, which allows transcription of the *cI* gene. The product of this gene, the **CI repressor**, prevents transcription from the promoters p_R and p_L, which service many of the

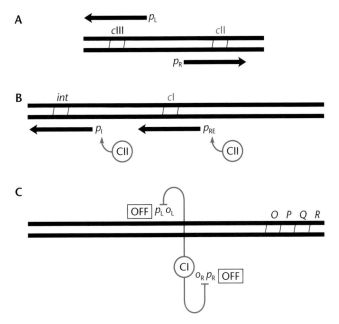

Figure 8.8 Formation of lysogens after λ infection. (A) The *c*II and *c*III genes are transcribed from promoters p_R and p_L, respectively. (B) CII activates transcription from promoters p_{RE} and p_I, leading to the synthesis of CI repressor and the integrase Int, respectively. (C) The repressor shuts off transcription from p_L and p_R by binding to o_R and o_L. Finally, the Int protein integrates the λ DNA into the chromosome (see Figure 8.9).

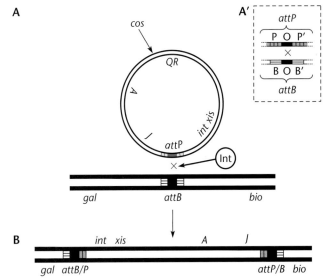

Figure 8.9 Integration of λ DNA into the chromosome of *E. coli*. (A) The Int protein promotes recombination between the *attP* sequence in the λ DNA and the *attB* sequence in the chromosome. The inset (A′) shows the region in more detail, with sequences POP′ and BOB′. The common core sequence of the two sites is shown in black. (B) The gene order in the prophage. The *cos* site is the location where the λ DNA is cut for packaging and recircularization after infection. The location of the *int*, *xis*, *A*, and *J* genes in the prophage is shown (refer to the λ map in Figure 8.2). The *E. coli gal* and *bio* operons are on either side of the prophage DNA in the chromosome.

remaining genes of λ. The CI repressor is discussed in more detail below. The other promoter activated by the CII protein, p_I, allows transcription of the integrase (*int*) gene. The Int enzyme integrates the λ DNA into the bacterial DNA to form the lysogen.

The role of the *c*III gene product in lysogeny is less direct. CIII inhibits a cellular protease that degrades CII. Therefore, in the absence of CIII, the CII protein is rapidly degraded and no lysogens form.

Phage λ Integration

As discussed above, the λ DNA forms a circle immediately after infection by pairing between the *cos* sequences at its ends. The Int protein can then promote the integration of the circular λ DNA into the chromosome, as illustrated in Figure 8.9. Int is a site-specific recombinase that specifically promotes recombination between the attachment sequence (called *attP* for *att*achment *p*hage) on the phage DNA and a site on the bacterial DNA (called *attB* for *att*achment *b*acteria) that lies between the galactose (*gal*) and biotin (*bio*) operons in the chromosome of *E. coli*. This is a nonessential region of the *E. coli* chromosome, so integration of λ at this site causes no observed phenotypes. Some phages do integrate their

DNA into essential genes of the bacterium, which requires special adaptations (Box 8.1). Because the Int-promoted recombination does not occur at the ends of λ DNA but, rather, at the internal *attP* site, the prophage map is different from the map of DNA found in the phage head. In the phage head, the λ DNA has the *A* gene at one end and the *R* gene at the other end (see the λ map in Figure 8.2). In contrast, in the prophage, the *int* gene is at one end and the *J* gene is at the other (Figure 8.9). The relative order of the genes in the maps is still the same, but the prophage and phage maps are cyclic permutations of each other. It was the difference between the phage genetic map and the prophage map that led to this model of integration, which is sometimes called the **Campbell model** after the person who first proposed it.

The recombination promoted by Int is called **site-specific recombination** because it occurs between specific sites, one on the bacterial DNA and another on the phage DNA. This site-specific recombination is not normal homologous recombination but, rather, a type of nonhomologous recombination because the sequences of the phage and bacterial *att* sites are mostly dissimilar. They have a common core sequence, O, of only

BOX 8.1

Effects of Prophage Insertion on the Host

The insertion of the prophage into the host cell DNA can have many effects on the host. For example, the prophage may encode virulence proteins that increase the pathogenicity of the host, in a process called lysogenic conversion. However, sometimes the insertion of the prophage causes phenotypes by itself, by disrupting genes of the host. Surprisingly, this does not usually happen, for a number of reasons. The archetypal phage λ avoids causing phenotypes by integrating into a nonessential region between the *gal* and *bio* operons of *E. coli* (see the text). However, we now know that some phages and other DNA elements such as pathogenicity islands (see chapter 9) integrate directly into genes, often into genes for tRNAs. Examples include the *Salmonella* phage P22, which integrates into a threonine tRNA gene; the *E. coli* phage P4, which integrates into a leucine tRNA gene; the *Haemophilus influenzae* phage HPc1, which also integrates into a leucine tRNA gene; and the virus-like element SSV1 of the archaeon *Sulfolobus* sp., which integrates into an arginine-tRNA gene. It is not known why so many phages use tRNA genes as their attachment sites. Perhaps it is because tRNA genes are relatively highly conserved in evolution. A phage could lysogenize a different species of bacterium if the sequence of its attachment site were highly conserved. If its *attB* site is in the highly conserved region of a tRNA gene, it could be found, virtually unchanged, in the chromosome of another host. Another possible explanation is that phages seem to prefer sequences with twofold rotational symmetry for their attachment sites. The sequences of tRNA genes have such symmetry, since the tRNA products of the genes can form hairpin loops, and in fact, most phage seem to integrate into the region of the tRNA gene that encodes the anticodon loop.

Not all phages that integrate into genes use tRNA genes for their attachment sites, however. For example, phage φ21, a close relative of λ, and phage e14, a defective prophage, both integrate into the isocitrate dehydrogenase (*icd*) gene of *E. coli*. The product of the *icd* gene is an enzyme of the tricarboxylic acid cycle and is required for optimal utilization of most energy sources, as well as for the production of precursors for some biosynthetic reactions. Inactivation of the *icd* gene would cause the cell to grow poorly on most carbon sources.

How can phages integrate into essential genes and not inactivate the gene, thereby compromising the host? Sometimes the answer is that they duplicate part of the gene in their *attP* site. The 3' end of the gene is repeated in the phage *attP* site with very few changes, so that when the phage integrates, the normal 3' end of the gene is replaced by the very similar phage-encoded sequence. This is true both of phages like φ21 that integrate into a protein-coding sequence and of phages that integrate into an essential tRNA gene. It is an interesting question in evolution how the bacterial sequence could have arisen in the phage. Perhaps the phage first arose as a specialized transducing particle, which then adapted to using the substituted bacterial genes as their normal attachment site in the chromosome.

Other phages may disrupt the gene into which they integrate, for example a tRNA gene, but they carry genes for tRNAs that may substitute for the tRNA gene they disrupt (see Ventura et al., below). However, there are known cases where insertion of the prophage does cause phenotypes. In some bacterial pathogens, disruption of a gene by insertion of a prophage may actually contribute to the pathogenicity of the bacterium in a case where the product of the disrupted gene interferes with pathogenicity (see Al Mamun et al., below).

References

Al Mamun, A., A. Tominaga, and M. Enomoto. 1997. Cloning and characterization of the region III flagellar operons of the four *Shigella* subgroups: genetic defects that cause loss of flagella of *Shigella boydii* and *Shigella sonnei*. *J. Bacteriol.* **179**:4493–4500.

Hill, C. W., J. A. Gray, and H. Brody. 1989. Use of the isocitrate dehydrogenase structural gene for attachment of e14 in *Escherichia coli* K-12. *J. Bacteriol.* **171**:4083–4084.

Ventura, M., C. Canchaya, D. Pridmore, B. Berger, and H. Brussow. 2003. Integration and distribution of *Lactobacillus johnsonii* prophages. *J. Bacteriol.* **185**:4603–4608.

15 bp—GCTT T(TTTATAC)TAA—flanked by two dissimilar sequences, B and B′ in *attB*, and P and P′ in *attP* (Figure 8.9, inset). The recombination always occurs within the bracketed 7-bp sequence. Because the region of homology is so short, this recombination would not occur without the Int protein, which recognizes both *attP* and *attB* and promotes recombination between them. The λ integrase is a member of the Y recombinase family because it has an active-site tyrosine (Y) to which the 3′ phosphate end of the DNA is covalently attached

after the DNA is cut and because it forms a Holliday junction as an intermediate in the recombination process. The mechanism of action of Y recombinases and other types of site-specific recombinases is discussed in chapter 9.

Maintenance of Lysogeny

After the lysogen has formed, the *cI* repressor gene is one of the few λ genes to be transcribed. The CI repressor binds to two regions, called **operators**. These operators, o_R and o_L, are close to promoters p_R and p_L, respectively, and, by binding the repressor, prevent transcription of most of the other genes of λ. Repressors and operators are discussed further in chapter 12. In the prophage state, the *cI* gene is transcribed from the p_{RM} promoter (for *repression maintenance*), which is immediately upstream of the *cI* gene, rather than from the p_{RE} promoter used immediately after infection (see above). The p_{RM} promoter is not used immediately after infection because its activation requires the CI repressor. The regulation of CI synthesis is discussed in more detail below.

REGULATION OF REPRESSOR SYNTHESIS

The CI repressor is the major protein required to maintain the lysogenic state; therefore, its synthesis must be regulated even after a lysogen has formed. If the amount of repressor drops below a certain level, transcription of the lytic genes begins and the prophage is induced to produce phage. However, if the amount of repressor increases beyond optimal levels, cellular energy is wasted in making excess repressor and it might be too difficult to induce the prophage should the need arise. The mechanism of regulation of repressor synthesis in lysogenic cells is well understood and has served as a model for gene regulation in other systems.

Figure 8.10 illustrates the regulation of CI repressor synthesis. Each CI polypeptide consists of two parts, or **domains**. In fact, the λ repressor was probably the first protein shown to have separable domains (see the discussion of the Lieb experiments on the λ repressor in "Genetic Experiments with Phage λ" below). We now know that many proteins have a similar modular construction, with different functions of the protein separated into different domains. One of the domains of the CI polypeptide promotes the formation of dimers and tetramers by binding to the corresponding sites on other CI polypeptides. The other domain on each polypeptide binds to an operator sequence on the DNA. To illustrate the two-domain structure of the CI polypeptide, it is traditionally drawn as a dumbbell, with the weights at the ends of the dumbbell indicating the two domains (Figure 8.10). For the CI repressor to function, two of these dumbbells must bind to each other through their dimerization domains to form a dimer made up of two copies of the polypeptide. In turn, a tetramer forms when two of these dimer dumbbells bind to each other through their tetramerization regions in the same domain. At very low concentrations of CI polypeptide, the dimers do not form and the repressor is not active. At very high concentrations, the dimers form and the repressor is active. It is now able to regulate its own synthesis, whose mechanism we describe next.

The repressor regulates its own synthesis as well as that of other λ gene products by binding to the operator

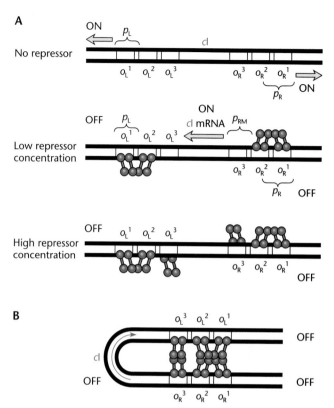

Figure 8.10 Regulation of repressor synthesis in the lysogenic state. The dumbbell shape represents the two domains of the repressor. (A) The dimeric repressor, shown as two dumbbells, binds cooperatively to o_R^1 and o_R^2 (and o_L^1 and o_L^2), repressing transcription from p_R (and p_L) and activating transcription from p_{RM}. (B) At higher repressor concentrations, it also binds to o_R^3 and o_L^3. Formation of tetramers bends the DNA, repressing transcription from p_{RM}. The relative affinity of the repressor for the sites is $o_R^1 > o_R^2 > o_R^3$ and $o_L^1 > o_L^2 > o_L^3$, respectively.

sequences, the one to the right of the CI gene called o_R and the other to the left of CI, called o_L. CI can be either a repressor or an activator of transcription, depending on where it is bound to these operators, as discussed later. The most important operator for regulating repressor synthesis is o_R, to the right of the repressor gene, although both operators have the same structure. The operator o_R can be divided into three repressor-binding sites, o_R^1, o_R^2, and o_R^3. If the concentration of repressor is low, only the o_R^1 site is occupied, which is sufficient to repress transcription from promoter p_R, which overlaps the operator site (Figure 8.10). This prevents transcription of the replication genes *O* and *P*. However, as the repressor concentration increases, eventually o_R^2 also becomes occupied, because tetramers can form between the dimers bound at o_R^1 and o_R^2 (Figure 8.10). The formation of a tetramer stabilizes the binding of a CI dimer to o_R^2. Only repressor bound at o_R^2 can activate transcription from the promoter p_{RM}, which is why p_{RM} is used to transcribe the repressor gene in the lysogen only when there is some repressor in the cell; immediately after infection, another promoter, p_{RE} is used to make repressor instead (see above). With very high concentrations of repressor, o_R^3 is also occupied and transcription from p_{RM} is blocked, blocking synthesis of more repressor.

How binding to o_R^3 blocks repressor synthesis is a little more complicated and involves o_L^3 and the bending of the DNA (Figure 8.10B). Since o_R and o_L have the same structure, o_L^1, o_L^2 and o_L^3 are occupied by repressor at the same repressor concentrations as the corresponding sites in o_R. At very high concentrations, CI repressor is weakly bound at o_R^3, but it cannot form a tetramer by binding to the dimer at o_R^2, which already has its partner, the dimer bound at o_R^1. However, it can stabilize its binding by forming a tetramer with the dimer bound at o_L^3, which is also without a partner. This stabilizes the binding of the CI dimer to o_R^3 and also causes the DNA to bend between the two operators as shown, further interfering with the functioning of the p_{RM} promoter, which lies between them. It is also possible that the dimers bound at o_R^1 and o_R^2 form octamers with dimers bound at o_L^1 and o_L^2, which helps stabilize the DNA loop as shown in the figure. This complex regulation allows the prophage to synthesize more repressor when there is less in the cell and vice versa, so that the cell maintains the levels of repressor within narrow limits. The synthesis of repressor also responds quickly to perturbations in the cell and explains why λ lysogens are very stable and usually release phage only under unusual circumstances. The term **robust regulation** has been used to describe this interactive system that maintains the CI repressor concentrations within narrow limits, thereby preventing spontaneous induction of the phage under varying conditions.

The regulation of λ repressor protein synthesis in the prophage state illustrates many important features of biological regulatory systems, some of which were first conceptualized in λ phage. One, which has been mentioned already, is how regulatory proteins often have separate domains that perform different functions of the protein. In the case of λ CI repressor, one domain is involved in binding of the individual CI repressor polypeptides to each other to form dimers and tetramers. The other domain binds to the DNA at operator sequences. Many regulatory proteins have such structures, and many examples are discussed in later chapters. Another feature of many regulatory systems that was first described in studies of λ repressor regulation is the concept of **cooperative binding.** A dimer of repressor

polypeptide bound at one subsite in the operator makes contact with another dimer through its tetramerization domain, allowing the other dimer to bind at the adjacent subsite more stably. This is called cooperative binding because protein bound at one subsite cooperates in the binding of a protein to an adjacent subsite, ensuring that the binding occurs in the correct order. Cooperative binding is also used in many regulatory systems, including the cooperative binding of RNA polymerase and activator proteins to promoters, as discussed in later chapters. The inhibition of transcription by binding of repressor to more than one operator simultaneously and bending the DNA at the promoter was not discovered in λ, and it is also known to occur in a number of other regulatory systems, some of which are discussed in chapter 12.

Immunity to Superinfection

The CI repressor in the cell of a lysogen prevents not only the transcription of the other prophage genes by binding to operators o_L and o_R but also the transcription of the genes of any other λ phage infecting the lysogenic cell by binding to the operators of that phage. Thus, bacteria lysogenic for λ are immune to λ superinfection. However, λ lysogens can still be infected by any relative of λ phage that has different operator sequences to which the λ CI repressor cannot bind. Any two phages that differ in their operator sequences are said to be **heteroimmune.** If they have the same operator sequences, they can inhibit each other's transcription and are said to be **homoimmune,** no matter how different they are in their other genes.

Induction of λ

Phage λ remains in the prophage state until the host cell DNA is severely damaged by irradiation or some types of chemicals. The prophage is then induced to go through its lytic cycle. Figure 8.11 outlines the process of induction of λ. When the cell attempts to repair the damage to its DNA, short pieces of single-stranded DNA accumulate and bind to the RecA protein of the host. The RecA protein with single-stranded DNA attached then binds to the λ CI repressor, causing it to cleave itself. The cleavage separates the DNA-binding domain in the polypeptide from the domain involved in dimer formation. Without the dimerization domain, the CI repressor can no longer form dimers and the DNA-binding domains can no longer bind to the operators. As the repressors drop off the operators, transcription initiates from the promoters p_R and p_L and the lytic cycle begins.

THE Cro PROTEIN

Early during induction, more repressor could be made to interfere with later stages in lytic development or even reestablish lysogeny. However, the *cro* gene product,

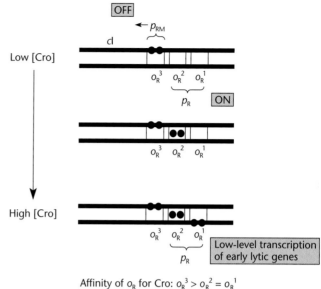

Affinity of o_R for Cro: $o_R^3 > o_R^2 = o_R^1$

Figure 8.12 Cro prevents repressor binding and synthesis by binding to the operator sites in reverse order from the repressor. By binding to o_R^3, Cro prevents repressor activation of transcription from p_{RM} while allowing transcription from p_R. Eventually Cro accumulates to the point where it binds to o_R^1 and o_R^2 and blocks transcription of early RNA.

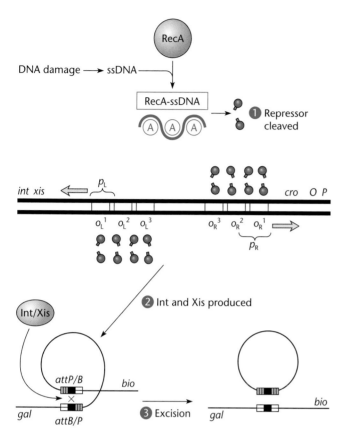

Figure 8.11 Induction of λ. Accumulation of single-stranded DNA (ssDNA) due to damage to the DNA results in activation of the RecA protein, which promotes the autocleavage of the CI repressor protein, separating the dimerization domain of the protein from the DNA-binding domain so that the repressor can no longer form dimers and bind to DNA. Transcription of *int-xis* and *cro*, *O*, and *P* ensues, and the phage DNA (in purple) excises from the chromosome and replicates.

which is one of the first λ proteins to be made after induction, prevents the synthesis of more repressor. Cro does this by binding to the operator sequences, although in reverse order of repressor binding. This binding is illustrated in Figure 8.12. Cro binds first to the o_R^3 site and then to the o_R^2 site, thereby preventing the CI repressor from binding to the o_R^2 site and activating its own synthesis from the p_{RM} promoter. Cro also binds to o_L^3, thereby preventing CI repressor binding to o_L^2 and o_L^1. Thus, p_L is also no longer repressed, leading to synthesis of Int and Xis.

EXCISION
Once the repressor is out of the way, transcription from p_L and p_R can begin in earnest. Some of the genes transcribed from p_L are required to excise the λ DNA from the chromosome. Excision requires site-specific

recombination between hybrid *attP-attB* sequences that exist at the junctions between the prophage DNA and the chromosomal DNA. These hybrid sequences are different from either *attB* or *attP* and contain sequences from both; therefore, Int alone is not capable of recognizing them and promoting recombination between them to excise the prophage. Another protein called excisase (Xis) is also required to allow the Int protein to recognize these hybrid sequences. Accordingly, unlike after infection, when only Int is synthesized, after induction both the Int and Xis proteins are synthesized (see below). In fact, it is necessary that only Int be made after infection, because if both Int and Xis were synthesized after infection, the λ prophage would be excised as soon as it integrates and lysogens could not form. However, this created a puzzle (Box 8.2). How could only Int, and not Xis, be synthesized after infection when both the *int* and *xis* genes are transcribed from p_L into the same messenger RNA (mRNA)? Not only are they transcribed into the same mRNA, but *xis* is transcribed before *int*, so it could not be that transcription merely stops before *xis* (figure in Box 8.2). To achieve this differential gene expression, λ takes advantage of the fact that the ends of the λ prophage are different from the ends of the DNA in the phage head. Briefly, after infection, both the *xis* and *int* genes are transcribed from the p_L promoter, but because of antitermination, the transcription proceeds

BOX 8.2

Retroregulation

The term **retroregulation** means that the expression of a gene is regulated by sequences downstream of it rather than upstream, such as at the promoter, where most genes are regulated. The way in which phage λ ensures that only Int is made after infection but that both Int and Xis are made after prophage induction is an example of retroregulation. Initially, after infection, both the *int* and *xis* genes are transcribed from the promoter p_L. However, the *int* and *xis* coding parts of this RNA do not survive long enough to be translated, because the RNA contains an RNase III cleavage site downstream of the *int* and *xis* coding sequences. The RNA is cleaved at this site and degraded past the *int* and *xis* coding sequences by a 3′ exonuclease, probably RNase II (see the table in Box 2.5). This regulation was named retroregulation because mutations downstream of the gene could change the expression of the gene. Later it was determined that the mutations changed the RNase III cleavage site, preventing enzyme recognition, and thus stabilized the RNA and allowed both *int* and *xis* to be translated from the transcript initiated at p_L.

As discussed in the text, the *int* gene is also transcribed from the promoter p_I, which is actually located in the *xis* gene (Figure 8.2). This resulting RNA does not contain a *nut* site, and so the N protein cannot bind to the RNA polymerase and allow it to proceed past termination signals as far as the coding sequence for the RNase III cleavage site; therefore, the RNA is stable. Moreover, the *int* RNA contains all of the *int* sequence but only part of the *xis* sequence, so that only Int can be made from this RNA.

Immediately after induction, however, both Int and Xis can be made from the RNA produced from the p_L promoter, because *xis-int* RNA produced from this promoter are now stable. As shown in the figure, during integration of the phage, the coding region for the RNase III cleavage site has been separated from the *xis-int* coding region, since this region is on the other side of the *attP* site, which is split during integration of the phage DNA. Therefore, the long RNA transcript initiated at p_L no longer contains the RNase III cleavage site at its 3′ end and so is stable, and both Int and Xis can be translated from this RNA after induction. This is an example of posttranscriptional regulation, because it occurs after the RNA synthesis has occurred on the gene (see chapter 12). There are other known examples where the expression of genes is prevented by separating them from their promoter during insertion of the element. For example, the *tra* genes of the conjugative insertion element Tn*916* are not expressed until the element is excised and forms a circle (see Chapter 5).

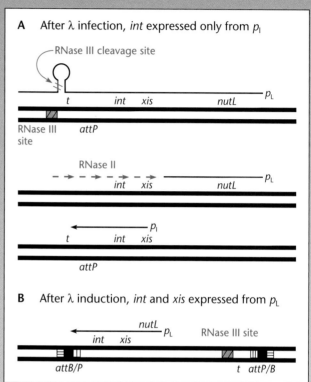

A After λ infection, *int* expressed only from p_I

B After λ induction, *int* and *xis* expressed from p_L

(A) After infection, the *xis* and *int* genes cannot be expressed from the p_L promoter. Because of N, transcription from p_L continues past the terminator *t* into an RNase III cleavage site on the other side of *attP*. The RNA is cleaved and digested back into *xis* and *int*, removing them from the RNA. Xis also cannot be expressed from p_I because the p_I promoter is in the *xis* gene. The RNA from p_I is stable, however, because this transcript does not contain a *nut* site and so does not continue through *t* to the RNase III cleavage site beyond. The purple region indicates the location of the coding information for the RNase III site, but RNase III cleaves only the mRNA transcript. **(B)** When the prophage is first induced, however, and before it excises, the sequence encoding the RNase III cleavage site is separated from the *xis-int* coding sequence, so that the RNA made from p_L is stable and both Int and Xis are made. **(A)** Early after infection; **(B)** early after induction.

References

Guarneros, G., C. Montanez, T. Hernandez, and D. Court. 1982. Posttranscriptional control of bacteriophage λ *int* gene expression from a site distal to the gene. *Proc. Natl. Acad. Sci. USA* **79:**238–242.

Schmeissner, U., K. M. McKenny, M. Rosenberg, and D. Court. 1984. Removal of a terminator structure by RNA processing regulates *int* gene expression. *J. Mol. Biol.* **176:**39–53.

past *xis* and *int* into sequences on the other side of *attP*, as shown in the figure in Box 8.2. One of these sequences on the other side of *attP* is cleaved by ribonuclease III (RNase III) if it shows up in an RNA (see the table in Box 2.5 for a listing of *E. coli* RNases). Cleavage by RNase III at this site sets up the RNA to be degraded by the 3′-5′ exonuclease, RNase II, before either Int or Xis can be translated from this RNA. However, when CII is made, it activates transcription from another promoter, p_I, which lies within the *xis* gene. This mRNA does not contain the entire coding sequence for Xis, only for Int, so that only Int can be translated from this mRNA. Also, this mRNA is stable for two reasons. Transcription from the p_I promoter does not include *nutL*, so that it is not antiterminated and stops before it transcribes the RNase III cleavage site. Also, initially at least, the λ DNA is still integrated in the chromosome; therefore, the DNA on the other side of *attP* is *E. coli* chromosomal DNA, which does not contain the sequences for an RNase III cleavage site, and so the mRNA is not degraded even if it does extend past *attP*.

While the Int and Xis proteins are excising λ DNA from the chromosome, the *O* and *P* genes are being transcribed from p_R. These proteins promote replication of the excised λ DNA. Therefore, a few minutes after the cellular DNA is damaged, the phage DNA is replicating, repressor levels are dropping, and the phage is irreversibly committed to lytic development. In about 1 h, depending on the medium and the temperature, the cell lyses, spilling about 100 phage into the medium from a cell that, an hour before, showed few signs of harboring the phage.

Competition between the Lytic and Lysogenic Cycles

As mentioned above, some cells infected by λ follow the lytic pathway while others become lysogens. Figure 8.13 and Table 8.3 review the competition for entry into the lysogenic cycle versus the lytic cycle. After infection, when there is no CI repressor in the cell, the *N* and *cro* genes are transcribed. As discussed earlier in this chapter, the *N* gene product acts as an antiterminator and allows the transcription of many genes, including *cII* and *cIII*, as well as the genes encoding the replication proteins O and P.

Whether the phage enters the lytic or the lysogenic cycle depends on the outcome of a race between the CII activator protein and the Cro protein, which is determined by the multiplicity of infection but is influenced by the metabolic state of the host cell. If the CII protein wins, it activates the synthesis of the CI repressor from the p_{RE} promoter and the integrase from the p_I promoter. The CI repressor binds to the operators o_L and o_R and represses the synthesis of more Cro as well as O and P, the DNA

integrates, and the lysogen forms. However, if the Cro protein wins, it prevents the synthesis of more CI repressor. Then, without more CI repressor, some transcription occurs from genes *O* and *P* and replication of the λ DNA begins. Eventually, there is too much DNA for the repressor to bind to all of it, and transcription of *O* and *P* increases further, followed by yet more DNA replication. The Q protein is synthesized next, allowing transcription of the head, tail, and lysis genes. However, it is inhibited by an antisense RNA activated by CII so it is not activated too soon (see Kobiler et al., Suggested Reading).

Specialized Transduction

As discussed in chapter 7, some lytic phages are capable of transduction, in which they carry or transduce host DNA, instead of their own DNA, from one cell to another. This is often called **generalized transduction** because any region of the chromosome can be transduced. Some lysogenic phages are also capable of another type of transduction, called **specialized transduction** because in this type of transduction only bacterial genes close to the attachment site of the prophage can be transduced. Also, the specialized transducing phage particle carries *both* bacterial genes and phage genes instead of only bacterial genes, like a generalized transducing particle.

Figure 8.14 illustrates how a phage particle capable of specialized transduction arises in phage λ. In a λ lysogen of *E. coli*, the λ prophage is integrated between the closely linked *gal* and *bio* genes in the chromosome. The *gal* gene products degrade galactose for use as a carbon and energy source, while the *bio* gene products make the vitamin biotin.

Specialized transduction can occur when a phage picks up neighboring bacterial genes during induction of the prophage. As shown in Figure 8.14, a specialized transducing phage carrying the *gal* genes, called λd*gal*, forms as the result of a mistake during the excision recombination. As mentioned in the preceding section, when the phage DNA is excised from the bacterial DNA, recombination occurs between the hybrid *attP*-*attB* sites at the junction between the prophage and host DNA. However, recombination sometimes occurs by mistake between the prophage DNA and a neighboring site in the bacterial DNA. The DNA later packaged into the head includes some bacterial sequences, as shown. Such transducing phage are very rare because the erroneous recombination that gives rise to them is extremely infrequent, occurring at one-millionth the frequency of normal excision. Furthermore, the recombination must, by chance, occur between two sites that are approximately a λ genome length apart, or the DNA would not fit into a phage head. The DNA

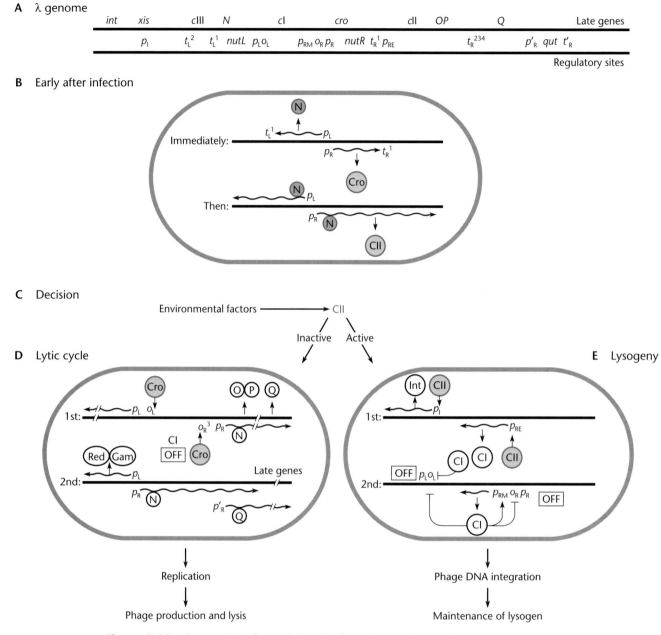

Figure 8.13 Competition determining whether phage will enter the lytic or lysogenic cycle. (A) Key genes (top line) and sites (bottom line). (B) Gene expression early after infection. (C) The abundance of active CII protein determines whether the phage enters the lytic or lysogenic cycle. (D) The synthesis of Cro promotes lytic development by repressing the synthesis of CI repressor. Once O and P are synthesized, the replication of λ DNA dilutes out the CI repressor. (E) The synthesis of CII promotes lysogeny.

that is excised must also contain at least some phage genes to be recognizable as the phage.

Because of the rarity of these transducing phage, powerful selection techniques are required to detect them. To select λ phage carrying *gal* genes of the host, induced phage are used to infect Gal⁻ recipient bacteria, and Gal⁺ transductants are selected on plates with galactose as the sole carbon source. In the rare Gal⁺ transductants, a λ phage carrying *gal* genes may have integrated into the chromosome, providing, by complementation, the *gal* gene product that the mutant lacks. If such a Gal⁺ lysogen is colony purified and the prophage is induced from

TABLE 8.3	Steps leading to lytic growth and lysogeny	
Steps leading to lytic growth	**Steps leading to lysogeny**	
1. Transcription from p_L and p_R	1. Same as for lytic growth	
2. N and Cro are made	2. Same as for lytic growth	
3. N allows CII expression	3. Same as for lytic growth	
4. **CII degraded**	4. **CII stable**	
5. Low CII concentration means that little CI is made	5a. High CII concentration activates p_I, and so Int is made and λ DNA integrates	
	5b. High CII concentration activates p_{RE}, and so CI is made	
6. Cro binds at $O_R{}^3$ and $O_L{}^3$, blocking binding by any low level of CI that is made	6. CI outcompetes Cro, and so CI binding at o_R and o_L both represses p_L and p_R and positively autoregulates at p_{RM}, maintaining lysogeny	
7. Meanwhile, N allows O and P replication gene transcription		
8. A second antiterminator, Q, allows late-gene transcription, and so λ phage particles are made		

it, a high percentage of the resultant phage progeny will carry the *gal* genes. Such a lysogenic strain produces an **HFT lysate** (for *high-frequency transduction*) because it produces phage that can transduce bacterial genes at a very high frequency.

Normally, not all of the induced phage particles in a *gal* transducing HFT lysate produced in this way carry

Figure 8.14 Formation of a λd*gal* transducing particle. A rare mistake in recombination between a site in the prophage DNA (in this case between *A* and *J*) and a bacterial site to the left of the prophage in the *gal* operon results in excision of a DNA particle in which some bacterial DNA including *gal* has replaced phage DNA.

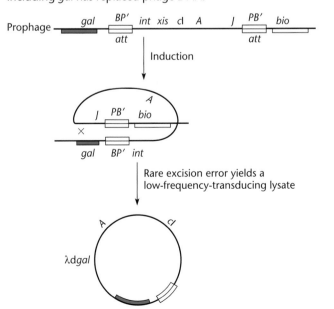

the *gal* genes, and the lysate contains a mixture of transducing phage particles and wild-type λ. The reason for this is also apparent from Figure 8.14. Because the λ phage head can hold DNA of only a certain length, the transducing particles have of necessity lost some phage genes to make room for the bacterial genes. Which phage genes are lost depends on where the mistaken recombination occurs. If the recombination occurs to the left of the prophage, some of the head and tail genes are replaced by *gal* genes of the host (Figure 8.14). However, if the mistake in recombination occurs to the right, the *int* and *xis* genes are replaced by the *bio* genes of the host (not shown).

Clearly, the properties of the transducing particle are determined by which phage genes are lost. For example, the λd*gal* phage shown in Figure 8.14 lacks essential head and tail genes, beginning with the *J* gene, and so cannot multiply without a wild-type λ **helper phage** to provide the missing head and tail proteins. These phage particles are thus called λd*gal*, where the "d" stands for *defective*. Usually they can be produced only by induction from dilysogens that contain the λd*gal* prophage and a wild-type phage integrated next to each other. After induction, the wild-type phage DNA provides the head and tail genes which the λd*gal* DNA has lost, and the two can multiply together. Roughly half the phage produced are λd*gal*; the other half are wild-type λ (Figure 8.15).

However, this simple picture hides some of the complexity that depends on how the λd*gal* DNA integrated originally. When the λd*gal* integrated next to a preexisting wild-type prophage in the chromosome, it could have involved recombination between the hybrid BP′ site on the λd*gal* DNA and the PB′ site on one side of the wild-type prophage. If this occurred, this recombination

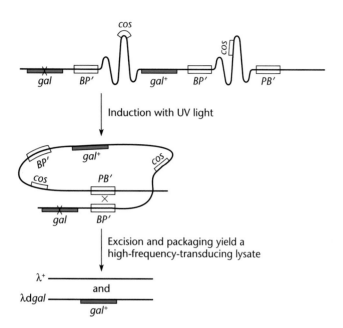

Figure 8.15 Induction of the λd*gal* phage from a dilysogen containing both λd*gal* and a wild-type λ in tandem. Recombination between the hybrid *PB′* and *BP′* sites at the ends excises both phages. The wild-type "helper" phage helps the λd*gal* phage to form phage particles, and both are packaged from repeated *cos* sites in long concatemers. See the text for details.

should require both Int and Xis since it occurred between the two hybrid sites. But then the question arises of how both phages excise, since the wild-type phage is now bracketed by a hybrid BP′ site and a nonhybrid *attP* site, and the λd*gal* prophage is bracketed by nonhybrid *attP* and *attB* sites. One possibility is that they excise one after the other, with the λd*gal* prophage excising first, using only Int, and creating a hybrid P′ B site. This leaves the wild-type prophage behind, bracketed by hybrid BP′ and BP′ sites, to be excised later by using Int and Xis. Alternatively, when the λd*gal* integrated, it could have done so by homologous recombination between λ phage DNA sequences carried by both the λd*gal* DNA and the wild-type λ DNA, creating a structure like that shown in Figure 8.15. Then the two could be excised together by Int and Xis as shown. They could then be packaged from *cos* sites on the concatemers that form. We leave these questions to the end of the chapter in the Questions for Thought.

The situation is very different if the HFT transducing phage are created by a mistaken recombination on the other side, replacing the *int* and *xis* genes with *bio* genes. These phage are able to multiply, since the genes on this side of *attP* are not required for multiplication. However, they cannot form a lysogen or be induced without the

help of a wild-type phage because they lack an *attP* site and *int* and *xis* genes. Because they can multiply and form plaques, *bio*-transducing phages are called λp*bio*, in which the "p" stands for *p*laque forming.

Specialized transducing phage particles played a major role in the development of microbial molecular genetics, including the first isolation of genes and the discovery of IS elements in bacteria. They can also be used to map phage genes and sites (see below). Although their general use has been largely supplanted by recombinant DNA techniques, they continue to have special applications.

Other Lysogen-Forming Phages

Phage λ was the first lysogen-forming phage to be extensively studied and thus serves as the archetypal temperate phage. However, many other types of lysogen-forming phages are known. Many use somewhat different strategies to achieve and maintain the prophage state. Some of them are described briefly here.

Phage P2

Phage P2 is another lysogen-forming phage of *E. coli*. The phage DNA is linear in the phage head but has cohesive ends like λ, which cause the DNA to cyclize immediately after infection. The phage replicates as a circle, and the DNA is packaged from these circles instead of from concatemers as λ does normally. Also, like λ, the genetic map of P2 phage is linear because it has a unique *cos* site at which the circles are cut during packaging.

One way in which P2 differs significantly from λ, which almost always integrates into a single site in the *E. coli* chromosome, is that P2 can integrate into many sites in the bacterial DNA, although it uses some sites more than others. Like λ, P2 requires one gene product to integrate and two gene products to excise. P2 prophage is much more difficult to induce than λ, however. It is not inducible by UV light, and even temperature-sensitive repressor mutations cannot efficiently induce it. The only known ways to induce it are to infect with another P2 or P4 (see below).

Phage P4

Even viruses can have parasites! Phage P4 is a parasite that depends on phage P2 for its lytic development (see Kahn et al., Suggested Reading). Thus, it is a representative of a group called **satellite viruses**, which need other viruses to multiply. Phage P4 does not encode its own head and tail proteins but, rather, uses those of P2. Thus, P4 can multiply only in a cell that is lysogenic for P2 or that has been simultaneously infected with a P2 phage. When P4 multiplies in bacteria lysogenic for P2, it induces transcription of the head and tail genes of the P2 prophage, which are normally not transcribed in the P2 lysogen. This is illustrated in Figure 8.16A. P4 uses two mechanisms to induce transcription of the late genes of P2. It induces the P2 lysogen because it makes an inhibitor of the P2 repressor protein, which binds to the P2 repressor, inactivating it and inducing P2 to enter the

lytic cycle. However, even though the P2 DNA replicates after induction by P4 and all the P2 proteins are made, the phage that is made contains mostly P4 DNA. This is because P4 makes a protein called Sid, which causes the P2 proteins to assemble into heads that are smaller than normal, with only one-third the volume of a normal P2

Figure 8.16 P2 can't win for losing. (A) A P2 lysogen is infected with P4. P4 makes a protein which binds to and inhibits the P2 repressor inducing the P2 prophage. The P4 protein Sid makes the P2 head proteins form a smaller head, which packages the shorter P4 DNA rather than P2 DNA. Therefore, P4 phage particles are released preferentially when the cells lyse. (B) A P4 lysogen is infected with P2. Now the P4 prophage is induced, and its replicating DNA is packaged by head proteins made by the infecting P2. Again, some P4 phage particles are released from the lysed cell even though it was a P2 phage that infected the cell. Details are given in the text.

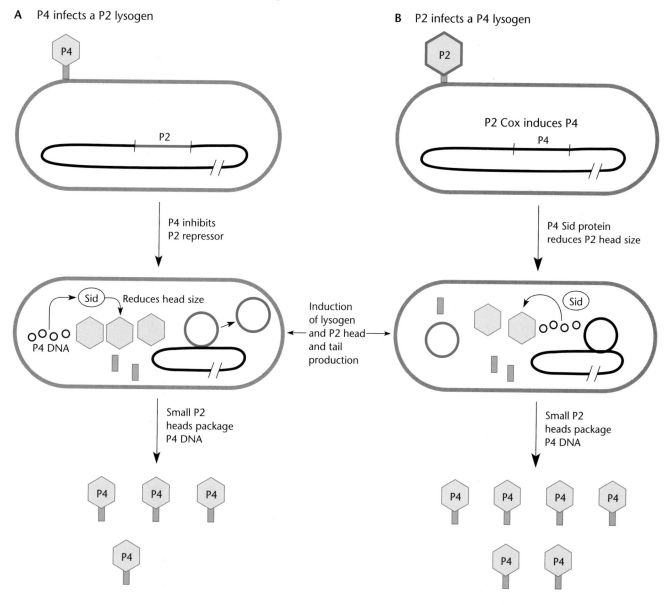

head. These heads are too small to hold P2 DNA but large enough to hold P4 DNA, which is only about one-third the length of P2 DNA, so that the heads are filled with P4 DNA instead. Nevertheless, a P4 *sid* mutant, which cannot make Sid protein, can still multiply in a P2 lysogen. The heads in the lysate, which are now the larger P2 size, contain either P2 DNA or P4 DNA. However, those which contain P4 DNA have two or three copies of the P4 DNA to fill the larger heads.

P4 can still multiply in P2 lysogens even if it cannot induce the P2 prophage, which remains in the chromosome. At first it was not obvious how P4 could induce the transcription of the late genes of P2, since, like T4 phage, the transcription of the late genes of P2 is normally coupled to replication and the P2 DNA does not replicate if the prophage is not induced. P4 accomplishes this by *trans* activating the transcription of the head and tail genes of P2 via synthesis of a protein called δ, which activates the transcription of the P2 late genes without P2 replication, even though the transcription seems to occur from the same promoters as the normal P2 replication-dependent transcription.

Because it wears the protein coat of P2, the phage P4 particle looks similar to P2. Only the head of P4 is smaller, to accommodate the shorter DNA. While the DNAs of P2 and P4 have otherwise very different sequences, the *cos* sites at the ends of the DNA are the same, so that the head proteins of P2 can package either DNA.

Phage P4 can also form a lysogen; when it does so, it usually integrates into a unique site on the chromosome. Not only can P4 infection induce a P2 prophage, but also P2 infection can induce a P4 prophage. It does this inadvertently, by making a protein called Cox, which induces the P4 prophage (Figure 8.16B). Apparently P4, which cannot multiply by itself, does not want to be caught sleeping as a prophage if the cell happens to be infected by P2. Not only would it die along with its host, but also it would miss the opportunity to multiply and infect new hosts. Again, at least some of the phage which emerge from the infection after P4 is induced have P4 DNA wrapped in a smaller-than-normal P2 coat, even though it was a P2 phage that infected the cell. One phage enters the cell and emerges as a different phage. No matter who starts the infection, P2 comes out the loser.

Not only can phage P4 integrate into the chromosome, but also it can replicate autonomously as a circle in the prophage state, as does P1 (see below). Because of this ability to maintain itself as a circle, phage P4 has been engineered for use as a cloning vector.

Phages P2 and P4, as well as their many relatives, have a very broad host range and infect many members of the *Enterobacteriaceae* including *Salmonella* and *Klebsiella*

spp., as well as some *Pseudomonas* spp. They are also related to phage P1, although their lifestyles and strategies for lytic development and lysogeny are very different.

As is often the case, once the interaction of P2 and P4 had been discovered and characterized, other examples of DNAs that parasitize phages were discovered. One of the more intriguing is the parasitization of a *Staphylococcus aureus* phage by a pathogenicity island carrying the toxin gene for toxic shock syndrome. The phage gives the pathogenicity island its mobility, allowing it to move between strains of *S. aureus* (see Box 8.3).

Phages P1 and N15: Plasmid Prophages

Not all prophages integrate into the chromosome of the host to form a lysogen. Some, represented by P1, form a prophage that replicates autonomously as a plasmid. Other phages are known to sometimes exist as plasmids in the prophage state, including P4 (see above) and some mutants of λ. In these cases, partial repression of gene expression of the phage limits replication and keeps the copy number of the plasmid low but very variable, suggesting that this is not their normal state. However, P1 in the prophage state is a bona fide plasmid. The P1 plasmid prophage maintains a copy number of 1 and combines many of the other features of true plasmids including a partitioning system and plasmid addiction system. Because it is a true plasmid, combined with the convenience of having a phage cycle which facilitates the isolation of DNA, etc., plasmid P1 is one of the major model systems for studying plasmid copy number control, segregation, and partitioning (see chapter 4). Another interesting aspect of this phage is that it has an invertible segment (see chapters 7 and 9). A region of the phage DNA encoding the tail fibers frequently inverts, and the host range of the phage depends on the orientation of this invertible segment. This invertible segment thereby contributes to the very broad host range of P1. It also has a site-specific Y recombinase, Cre, that acts on the *lox* site to resolve plasmid dimers and prevent curing of the prophage. The Cre-*lox* system is the best understood of the Y recombinases and has been put to many uses in molecular genetics.

Another *E. coli* phage, N15, has a plasmid as its prophage, but this plasmid is linear rather than circular. It has served as a model system for how some types of linear plasmids replicate. It has hairpin ends with the 3′ and 5′ ends joined to each other, and replicates around the ends from an internal origin to yield a dimeric circle. The dimeric circles are then resolved by a prototelomerase. The replication of N15 prophage and other linear plasmids is discussed in chapter 4.

BOX 8.3

How a Pathogenicity Island Gets Around

Many bacteria have large DNA elements integrated in their chromosome; these elements are called genetic islands, and they contain genes which confer special properties on the bacteria that carry them (see chapter 9). Like prophages and integrated plasmids, genetic islands are not normal regions of the chromosome. They are not carried by all the strains of a particular species and also often precisely excise from the chromosome, deleting no chromosomal sequences and leaving no part of themselves behind. They also carry genes which allow a bacterium carrying them to occupy special ecological niches. Pathogenicity islands (PIs) are a type of genetic island, which carry genes required for pathogenicity. For example, *Yersinia* species have a pathogenicity island which carries genes for iron scavenging in the animal host, and the *cag* pathogenicity island of *Helicobacter pylori* encodes a type IV protein secretion system to secrete a toxin required for pathogenicity (see Box 5.2). However, PIs are neither integrated plasmids nor prophages, even defective ones. They are not capable of autonomous replication, nor do they encode any gene products required to make a phage on induction. Nevertheless, they seem to be mobile and able to move from one strain of bacterium to another, because identical genetic islands are sometimes found in otherwise less closely related bacterial strains. They carry an integrase (*int*) gene, whose product allows them to integrate specifically into a region of the host DNA, often a tRNA gene. They also carry inverted repeated sequences at their ends, which are presumably involved in their integration. Because they are not found in all strains of a species, pathogenicity islands are assumed to be able to move from one host to another. However, very few genetic islands have actually been demonstrated to move in a laboratory situation.

The first pathogenicity island whose movement was demonstrated is SaPI1, which is found in some strains of *Staphylococcus aureus*. SaPI1 is the prototype of a family of pathogenicity islands of *S. aureus*, one member of which carries the gene for the toxin that causes toxic shock syndrome. The SaPI1 pathogenicity island moves by specifically parasitizing an *S. aureus* phage called 80α in a process strikingly similar to the way in which P4 parasitizes P2 phage (see the text). When phage 80α infects an *S. aureus* bacterium carrying SaPI1, the pathogenicity island excises from the chromosome and replicates, apparently with the help of phage replication proteins. Like P4, the pathogenicity island directs the phage to make smaller heads, which then package the pathogenicity island rather than the phage DNA. When such a phage infects another cell, the pathogenicity island is injected and can integrate into the chromosome of its new host, using its Int protein. Thus, the major difference between phage P4 and the SaPI1 pathogenicity island is that P4 encodes all the proteins to replicate its own DNA while SaPI1 depends on the infecting phage for its replication proteins. Maybe P4 should be called a genetic island rather than a phage and you should never judge a DNA element by its coat alone.

The mobilization of pathogenicity islands by phages raises special questions concerning the use of antibiotics. Some antibiotics, particularly those that inhibit the DNA gyrase such as the fluoroquinolones including ciprofloxacin (see chapter 1), cause DNA damage that can induce prophages. If the chromosomes of the cells in which the prophages exist also contain pathogenicity islands, these pathogenicity islands may be moved into other harmless bacteria, making them pathogenic. Therefore, sublethal doses of these antibiotics could do more harm than good in some situations (see Ubeda et al., below).

References

Ruzin, A., J. Lindsay, and R. P. Novick. 2001. Molecular genetics of SaPI1, a mobile pathogenicity island in *Staphylococcus aureus*. *Mol. Microbiol.* **41:**365–377.

Ubeda, C., E. Maiquez, E. Knecht, I. Lasa, R. P. Novick, and J. R. Penades. 2005. Antibiotic-induced SOS response promotes horizontal dissemination of pathogenicity island-encoded virulence factors in staphylococci. *Mol. Microbiol.* **56:**836–844.

Phage Mu

Another phage that forms lysogens is Mu, which integrates randomly into the chromosome. Because it integrates randomly, it often integrates into genes and causes random insertion mutations, hence its name Mu (for *mu*tator phage). This phage is essentially a transposon wrapped in a phage coat, and it integrates and replicates by transposition. For this reason, the discussion of phage Mu is deferred to chapter 9.

Use of Lysogen-Forming Phages as Cloning Vectors

Temperate phages offer some distinct advantages as cloning vectors. Because they can multiply as a phage, obtaining large amounts of the cloned DNA is relatively easy. Since they can also integrate into the DNA of the host, a cloned gene exists in only two copies, one in the prophage and one at the normal site, which is important in complementation studies with bacteria.

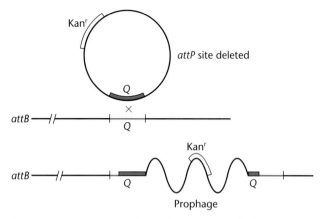

Figure 8.17 Use of a phage cloning vector lacking its *attP* site to mark a cloned chromosomal gene for mapping or for gene replacement. In the example, a phage lysate containing a DNA clone of the bacterial gene *Q* is used to infect the bacterium. Because the phage has its *attP* site deleted, it can integrate only by recombination between the *Q* gene in the phage and in the chromosome. The location of the *Q* gene in the chromosome can then be ascertained by mapping the kanamycin resistance (Kan^r) gene on the prophage.

Cloning into lysogen-forming phage can also facilitate genetic mapping or gene replacements. Such experiments use a phage that lacks its *att* site but has an easily selectable marker (such as resistance to an antibiotic) introduced into it. The gene of interest is cloned into the phage DNA, which is introduced into the cell either by in vitro packaging and infection or by transfection. Lysogens form by recombination between the cloned gene in the phage and its counterpart in the chromosome (Figure 8.17). The phage is integrated at this site in the chromosome rather than at the *attB* site because the phage *attP* site is deleted. When the antibiotic resistance gene on the phage DNA is mapped genetically by methods discussed in chapter 3, the original location in the chromosome of the gene will be known. This may be the preferred way to map a cloned gene for which no convenient phenotype is available, since antibiotic resistance markers are relatively easy to map genetically. Gene replacement is also discussed in chapter 3.

Lysogenic Conversion and Bacterial Pathogenesis

In a surprising number of instances, prophages carry genes for virulence factors or toxins required for virulence by the pathogenic bacteria they lysogenize. These genes are sometimes called **morons** (for "more DNA") and are not found in all the phages of that type, suggesting that they were recently acquired. They also often are expressed from their own promoter, so they are expressed in the lysogen where other prophage genes are usually repressed. Some examples of bacteria carrying prophages with morons that contribute to the diseases they cause are the bacteria that cause diphtheria, scarlet fever, botulism, tetanus, and cholera. Even λ phage carries genes that confer on its *E. coli* host serum resistance and the ability to survive in macrophages. As mentioned earlier, the process by which a prophage converts a nonpathogenic bacterium to a pathogen is an example of **lysogenic conversion.**

E. coli and Dysentery: Shiga Toxins

Pathogenic strains of *E. coli* are prime examples of bacteria that are not pathogenic unless they harbor prophages or other DNA elements carrying virulence genes. These bacteria are part of the normal intestinal flora unless they carry certain DNA elements. Then they can cause severe diseases, including bacterial dysentery with symptoms such as bloody diarrhea. The infamous *E. coli* strain O157:H7, which has caused many outbreaks of bacterial dysentery worldwide, is one example of such a lysogenic *E. coli*. In fact, bacterial dysentery due to these bacteria is the major cause of infant mortality worldwide.

In one particularly clear example, a group of prophages very closely related to λ can make *E. coli* pathogenic by encoding toxins called Shiga toxins, so named because they were first discovered in *Shigella dysenteriae*, which is so closely related to *E. coli* that it has recently been moved into the same genus. Like cholera toxin and many other toxins, the Shiga toxin is composed of two subunits, A and B. The B subunit helps the A subunit enter an endothelial cell of the host by binding to a specific receptor on the cell surface of some tissues. The A subunit is an *N*-glycosylase, a type of enzyme that cleaves the bond between the base and the sugar in nucleotides, removing the base from RNA or DNA. There are many *N*-glycosylases, including uracil-*N*-glycosylase, which removes uracil from DNA (see chapter 1), and the *N*-glycosylases that remove other damaged bases from DNA to avoid mutagenesis (see chapters 1 and 11). However, the Shiga toxin A subunit is very specific in that it removes only a certain adenine base from the 28S rRNA. Removal of this adenine from the 28S rRNA in a ribosome blocks translation by interfering with binding of the translation factor EF-1α, the eukaryotic equivalent of EF-Tu, to the ribosome. Interestingly, this adenine in the 28S rRNA seems to be the "Achilles heel" of the ribosome and is a popular target of translation-blocking systems. The rRNAs are highly conserved, and an adenine occurs in this position in the rRNAs of all eukaryotes. Plant enzymes such as ricin that protect the plant by blocking translation in cells infected by virus are also *N*-glycosylases that remove this same adenine from their own 28S rRNA, killing the cell and preventing multiplication of the virus. Yeast also make an enzyme called saracin, which has the same target, although the function of this enzyme is unknown.

Shiga toxins can be divided into two groups based on their amino acid sequence: the Stx1 group, encoded by *E. coli* prophages, which also includes the toxin encoded in the chromosome of *S. dysenteriae* and the ricin toxin in plants, and the Stx2 group, which so far has been found only in prophages carried by *E. coli*. Usually, the bacteria responsible for the most serious human diseases carry the Stx1 type. Apparently, expression of the toxin is required to convert the disease from just watery diarrhea to hemolytic-uremic syndrome (HUS), which is the leading cause of kidney failure in children.

The *stx* genes in prophages are usually in one of two places in the phage genome. A prophage genetic map of phage φ361 encoding Shiga toxin 2 (Stx2) is shown in Figure 8.18. Note the remarkable similarity between the genetic map of this prophage and the genetic map of the λ prophage shown in Figure 8.2. Obviously, they are very close relatives. The toxin genes *stx2A* and *stx2B* lie just downstream of the *Q* gene and upstream of the lysis genes *S* and *R*. The genes in this region of phage λ are late genes, transcribed with the other late genes from the p_R' promoter, and so they would be transcribed only if the lysogen is induced. However, the toxin genes in φ361 have their own weak promoter, p_{stx}, and so they are weakly expressed, even in the lysogen (see Wagner et al., Suggested Reading).

The regulation and secretion of the Shiga toxin present intriguing clues about the etiology of the diseases caused by these bacteria. Because many of the *stx* genes

are inserted downstream of the *Q* gene in the λ-like phage, they are expressed from the phage p_R' promoter only late in induction. Some of them have their own promoter, which does allow them to be expressed in the lysogenic state. Interestingly, this promoter is sometimes regulated by the presence of iron, as is the *stx1* gene found in the chromosome of *Shigella*. Iron deficiency is often used as a sensor of the eukaryotic environment, so that the gene is turned on only in the eukaryotic intestine (see chapter 13 for a discussion of iron regulation). However, even if the toxin protein is made, it apparently cannot be secreted from the bacterial cell because it does not contain a signal sequence, nor does the bacterial host harbor another type of secretion system capable of secreting it into the extracellular environment of the intestine. Therefore, we surmise that the only way it can get out of the bacterial cell is if the prophage is induced and lyses the cell. This may happen, but it kills the pathogenic bacterium and thus is counterproductive. Another possibility is that some of the bacteria lyse and release the phage, which then infect and lysogenize nonpathogenic *E. coli* strains that are part of the normal bacterial flora. If the phage in these normally nonpathogenic bacteria are then induced, they kill the nonpathogenic bacteria that harbored the prophage but the released Shiga toxin can participate in pathogenesis by the pathogenic strain and perhaps lead to HUS. This raises questions about the effect of some types of antibiotics that damage DNA and therefore induce phages. They may help to spread these phages and increase, rather than decrease, the severity of the disease. In fact, there is some evidence that the use of ciprofloxacin and similar antibiotics to treat people with bacterial dysentery due to *E. coli* can increase the chance of the disease developing into the more serious HUS.

Figure 8.18 Close relatives of λ encode Shiga toxins. Shown is the prophage genetic map of phage φ361 indicating the positions of the toxin genes. The purple shading indicates that the repressor and toxin genes are expressed in the lysogen. Details are given in the text.

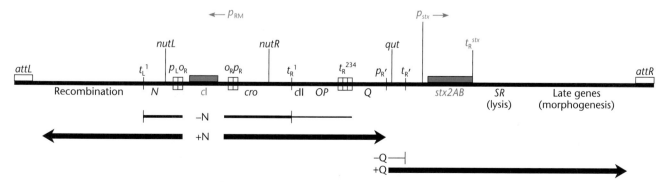

Diphtheria

Diphtheria is the classic example of a disease caused in part by the product of a gene carried by a phage. Pathogenic strains of *Corynebacterium diphtheriae* differ from nonpathogenic strains in that they are lysogenic for phage β, or a closely related phage (see Freeman, Suggested Reading). These prophages carry a gene, *tox*, for the diphtheria toxin (Figure 8.19). The diphtheria toxin is an enzyme that kills eukaryotic cells by ADP-ribosylating (attaching adenosine diphosphate to) the EF-2 translation factor, thereby inactivating it and blocking translation in the cell. The *tox* gene of the β prophage is transcribed only when *C. diphtheriae* infects its eukaryotic host or under conditions that mimic this environment. Even though the *tox* gene is on the prophage, it is regulated by the products of chromosomal genes, illustrating the close relationship between these bacteria and their phages. The mechanism of action of the diphtheria toxin and the regulation of the *tox* gene are discussed in chapter 13 under "Global Regulation of Virulence Genes."

Cholera

Another more recently discovered example of toxin genes carried by a phage involves the bacterium that causes cholera, *Vibrio cholerae* (see Waldor and Mekalanos, Suggested Reading). In this case, the toxin genes are carried on a single-stranded filamentous phage called CTXφ (for *c*holera *tox*in phage). In many ways this phage is like any other filamentous single-stranded DNA phage such as fd (see chapter 7). It infects the cell by attaching to a pilus and then enters the cell through the TolA, TolQ, and TolR channel. The similarity between this phage and the *E. coli* phages was dramatically confirmed by the discovery that exchanging the region of the pIII head protein of coliphage fd with the corresponding region from CTXφ allowed the coliphage to infect *V. cholerae* (see Heilpern and Waldor, Suggested Reading).

The CTXφ phage differs from what we know about the coliphages in that it can also form a lysogen and exist as a prophage. The prophage integrates at a specific bacterial attachment site in the *V. cholerae* chromosome, using the XerCD recombinase of the host, which is used to resolve chromosome and plasmid dimers (see chapters 1, 4, and 9). If the host lacks this site in its chromosome, the prophage can maintain itself as a double-stranded plasmid which replicates by a rolling-circle mechanism analogous to the rolling-circle mechanism by which other single-stranded phages replicate their replicative form (see chapter 7). When it is found integrated in the chromosome, the prophage often exists in tandem repeats, with two or more phage genomes linked head to foot. Only lysogens in which the prophage exists in tandem repeats can be induced to make more phage (see Moyer et al., Suggested Reading). This is because the phage does not excise itself from the chromosome like λ and many other phages. Instead, the prophage replicates itself out of the chromosome by a rolling-circle mechanism to spin off more + strands, which can be packaged into phage heads, much like + strands are made from circular replicative forms in other phages. The rolling-circle replication initiates at the origin for + strand synthesis in one copy of the prophage in the chromosome and continues into the adjacent prophage DNA, terminating at the origin in the second prophage to make a complete phage genome. This explains why the prophage must exist in multiple tandem repeats in the chromosome. If only a single copy of the prophage existed in the chromosome, the phage would be able to replicate only part of its genome when it is induced.

The cholera toxin genes, *ctxA* and *ctxB*, encode another AB-type toxin in which the B subunit helps the A subunit, an enzyme, into the eukaryotic cell. The apparatus which secretes the subunits of the cholera toxin from the *V. cholerae* cell is an example of a type II secretion system and is mentioned in chapter 2 and discussed in chapter 14. It is interesting that the cholera toxin genes on the prophage are regulated by a chromosomal gene, *toxR*, which also regulates the synthesis of the pili

Figure 8.19 Genome map of a *Corynebacterium diphtheriae* prophage containing the diphtheria toxin gene (*tox*). Selected genes are annotated. The prophage is bracketed by tRNA genes of the host. Other genes involved in lysogenic conversion may be on the other end of the prophage.

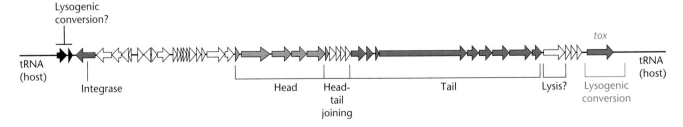

that serve as the receptor sites for the phage. These pili are also important virulence determinants because they enable the bacteria to adhere to the intestinal mucosa. The regulation of cholera toxin genes is also discussed in chapter 13.

Botulism and Tetanus

Other striking examples of diseases caused by toxins sometimes encoded by lysogen-forming phages are botulism and (probably) tetanus. Botulinum toxin causes a flaccid paralysis, in which the muscles are unable to contract, while tetanus toxin causes a rigid paralysis, in which the muscles remain flexed. Because of its ability to relax muscles, botulinum toxin (Botox) is currently used to treat involuntary muscle spasms such as facial "tics" and for the cosmetic purpose of removing facial lines associated, for example, with aging. Recent evidence indicates that these two toxins work by a common mechanism. They both cleave the same neuronal protein, synaptobrevin, in exactly the same position in the amino acid sequence, although the routes of entry of the toxins into the host and the symptoms they cause are very different (see Schiavo et al., Suggested Reading).

While some types of botulism toxin are encoded by the chromosome or plasmids, some types do seem to be encoded by phages, although these form unstable lysogens with their host, members of the gram-positive genus *Clostridium*. Botox is made by recombinant DNA techniques from the gene cloned from the prophage. However, the sequencing and characterization of these phages has been slow, due to a lack of easily manipulated techniques.

Synopsis

It is becoming increasingly apparent that some of the genes which make bacteria virulent are carried on prophages. We have listed only a few examples here; however, many other classical diseases which have plagued humankind since recorded history are turning out to have prophage involvement. It has been estimated that prophages make up as much as 10 to 20% of the DNA of some types of bacteria. They also are often the major contributor to strain diversity within a species. Within the same species, bacterial strains that cause very different diseases are often found to differ mostly in the prophages they harbor. But why are toxin genes and other virulence factors often encoded by prophages instead of being normal genes in the chromosome? The argument is similar to that used to explain why genes are carried on plasmids (see chapter 4). Having toxin and other virulence genes on a movable DNA element like a phage may allow the bacterium to adapt to being pathogenic without all the members of the population having

to carry extra genes. Furthermore, many virulence proteins are also strong antigens, thus allowing a nonlysogenic bacterium to colonize the host without alerting the host immune system and to become pathogenic only if it is infected by the phage. We have also discussed other potential advantages of carrying virulence genes on phages. They can be induced and then infect other nonpathogenic bacteria, forming a lysogen. The toxin can then be expressed from the nonpathogenic strain, often contributing to the pathogenicity.

Genetic Experiments with Phage λ

Earlier we discussed how the interaction of phage λ with its host has been one of the major contributors to our present concepts of how cells function, but we have not gone into detail about the types of experiments which contributed to these concepts. These are great examples of the uses of selectional genetics and genetic analysis in general. The following are some examples of how selectional genetics has been used to analyze the interaction of λ with its host, *E. coli*, and has led to many of the conceptual advances we have described. We do not always credit individuals for these experiments but, rather, review the types of experiments that were done and reference their work at the end. We recommend that you read these references, not just for their historical importance but also for the way they illustrate the role played by creative thinking in the advancement of scientific concepts.

Genetics of λ Lysogeny

Our understanding of how λ forms lysogens was formed by a genetic analysis. Because phage λ is capable of lysogeny, the plaques of λ are cloudy in the middle due to the growth of immune lysogens in the plaque (Figure 8.7). Mutants of λ that cannot form lysogens do not contain these immune lysogens and so are easily identified by their clear plaques. These "clear-plaque mutants," called C-type mutants, have mutations in λ genes whose products are required for the phage to form lysogens.

Complementation tests can reveal how many genes are represented by clear-plaque mutants. Now, however, rather than asking whether two mutants can help each other to multiply, we are asking whether two mutants can help each other to form a lysogen, since this is the function of the genes represented by clear-plaque mutants. Cells are infected by two different clear-plaque mutants simultaneously, and the appearance of lysogens is monitored. Lysogens can be recognized by their immunity to infection by the phage, which allows them to form colonies in the presence of the phage. One way to perform this test is to mix one of the mutant phages with the bacteria and streak the mixture on a plate. The other mutant phage is then streaked at right angles to the first streak. Very few bacterial colonies grow in the streaks because the individual mutants cannot form lysogens. However, if bacterial colonies grow in the region where the two streaks cross, immune lysogens form due to infection of some cells by both mutant phages and complemention between the two mutations to form a lysogen. Such complementation tests revealed three complementation groups of genes to which the clear-plaque mutations of λ belonged: *c*I, *c*II, and *c*III. In addition to mutations in the clear-plaque genes, mutations in the *int* gene can prevent the formation of stable lysogens, although Int⁻ mutants make somewhat cloudy plaques. In this case, the λ DNA, while not integrated into the chromosome, may make the cells transiently immune.

Further genetic tests revealed different roles for CI, CII, and CIII in lysogeny. Mutations in the *c*II and *c*III genes can be complemented to form lysogens, and these lysogens can harbor a single prophage with a *c*II or *c*III mutation. However, lysogens harboring a single prophage with a *c*I mutation are never seen. Apparently, the *c*I mutation in the phage must be complemented by another mutation to maintain the phage in the lysogenic state. This observation led to the idea that CII and CIII are required to form lysogens but are not necessary to maintain the lysogenic state once a lysogen has formed. The CI protein, on the other hand, is required to form a lysogen and to maintain the lysogenic state.

Because they can be complemented, the *c*I, *c*II, and *c*III mutations must affect *trans*-acting functions, either proteins or RNAs required to form lysogens. Another set of mutations, called the *vir* mutations, also prevent lysogeny and cause clear-plaque formation but cannot be complemented and so are *cis* acting (see chapter 3). These *vir* mutations allow the mutant phage to multiply and form clear plaques even on λ lysogens. DNA sequencing has revealed that phage with *vir* mutations are multiple mutants with mutations in the o_R1 and o_R2 sequences as well as o_L1. These mutations change the operators so that they can no longer bind the CI repressor, thereby preventing lysogeny (see "Regulation of Repressor Synthesis in the Lysogenic State" above).

Genetics of the CI Repressor

Many proteins are now known to be assembled from "modules" or domains with separable functions. One of the goals of modern proteomics is to identify the domains in proteins in an attempt to guess the function of the protein. The λ repressor product of the *c*I gene was the first protein shown to have separable domains. One region of the protein binds to the operator sequences on the DNA, and the other region binds to another

repressor monomer to form an active repressor dimer. The first indication that the CI repressor has separable domains came from genetic experiments that demonstrated intragenic complementation between temperature-sensitive mutations in the *c*I gene (see Lieb, Suggested Reading). As discussed in chapter 3, complementation usually occurs only between mutations in different genes and intragenic complementation is possible only if the protein product of the gene is a multimer composed of more than one identical polypeptide encoded by that gene.

Figure 8.20 illustrates the experiments that demonstrated intragenic complementation by some temperature-sensitive mutations in the *c*I gene. Lysogenic cells containing a prophage with one *c*I temperature-sensitive mutation were heated to the nonpermissive temperature and infected with a phage carrying a different *c*I temperature-sensitive mutation. At this temperature, infection by one or the other mutant phage alone invariably kills the cell because the repressor is inactivated so that λ cannot form a lysogen. However, if the two mutations complement each other to form an active repressor, a few cells may become lysogens and survive.

The results clearly demonstrated intragenic complementation between some of the mutations in the *c*I gene. In particular, some mutations in the amino (N)-terminal part of the polypeptide, which we now know to be involved in DNA binding, complement some mutations in the carboxyl (C)-terminal part, which we now know to be involved in dimer formation. Apparently, in some cases, dimers can form if only one of the two polypeptides has a mutation in the C-terminal domain. Furthermore, the dimer can sometimes bind to DNA if only one of the two polypeptides in the dimer has a mutation in the N-terminal domain. The ability to form active repressor out of two mutant polypeptides is what leads to intragenic complementation.

Isolation of λ *nut* Mutations

Some of the most elegant genetic experiments with phage λ involved the isolation and mapping of mutations in the λ *nut* sites (see Salstrom and Szybalski, Suggested Reading). These experiments also illustrate some basic principles of genetic selections and analysis, and so we go into them in some detail. As discussed earlier in this chapter, the existence of the *nut* sites was predicted from a model for how N antiterminates transcription. The *nut* sites (for N *ut*ilization sites) are the sites on the mRNA to which N must bind before it can bind to the RNA polymerase and prevent termination (Figure 8.4), allowing transcription to proceed into the downstream genes. Therefore, mutations in the DNA coding sequence for one of these *nut* sites could prevent the binding of the

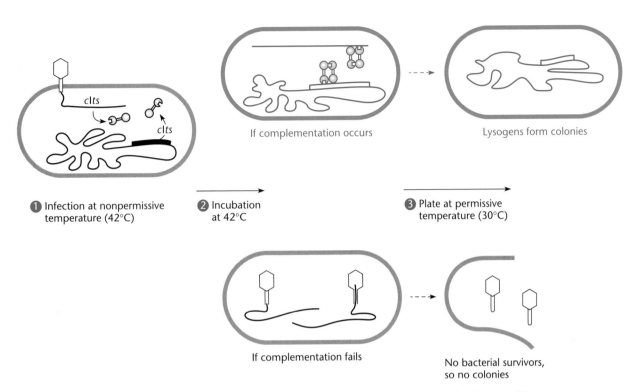

1 Infection at nonpermissive temperature (42°C)

2 Incubation at 42°C

3 Plate at permissive temperature (30°C)

If complementation occurs

Lysogens form colonies

If complementation fails

No bacterial survivors, so no colonies

Figure 8.20   An experiment to show intragenic complementation in the cI gene of λ. See the text for details.

N protein to the mRNA, thereby causing transcription to stop at the next transcription termination site and preventing transcription of downstream genes.

They had to choose whether to isolate mutants with *nutL* or with *nutR* mutations first. The investigators wisely decided that it would be easier to isolate *nutL* mutants than to isolate *nutR* mutants because *nutL* mutations would prevent the transcription of genes to the left of *cI*, including the *gam* and *red* genes, all of which are nonessential, while *nutR* mutations would prevent the transcription of essential genes to the right of the *cI* gene, including the *O* and *P* genes required for replication (see the λ map in Figure 8.2). Therefore, phages with mutations that completely inactivate *nutL* should still be viable but mutations that inactivate *nutR* should be lethal, which would preclude the isolation of *nutR* mutant phages. However, even though *nutL* mutations should not be lethal, they might be very rare. For all the investigators knew, the *nut* sequences in DNA may be very short, consisting of only a few base pairs, and only mutations that changed one of these base pairs would inactivate the *nut* site. Selecting rare mutations requires a positive selection. It meant finding conditions under which phages with a mutation that inactivates the *nutL* site can form plaques whereas wild-type λ cannot.

The positive selection used to isolate *nutL* mutations is illustrated in Figure 8.21. The selection is based on the observation that for unknown reasons, wild-type λ cannot multiply in *E. coli* lysogenized by another phage, P2, because the products of the *gam* and *red* genes (the *red* gene later turned out to be two genes, *exo* and *bet* [see chapter 10]) of the infecting λ interact somehow with the *old* gene product of the P2 prophage and kill the cell. Mutations in the predicted *nutL* site should prevent the transcription of both the *gam* and *red* genes and result in phage able to form plaques on *E. coli* lysogenic for P2. Isolating *nutL* mutants of λ should therefore be easy: just plate millions of mutagenized λ on a P2 lysogen, and some of the plaques that form may be due to λ with *nutL* mutations.

Unfortunately, *nutL* mutants are not the only type of mutant that can form plaques under these conditions. As shown in the figure, double mutants of λ with point mutations in both the *gam* and *red* genes or with a deletion mutation that simultaneously inactivates both the *red* and *gam* genes also multiply and form plaques on a P2 lysogen. Fortunately, double mutants should not be much more common than *nutL* single mutants since the chance of getting two mutations is the product of the chances of getting either single mutation. Deletion

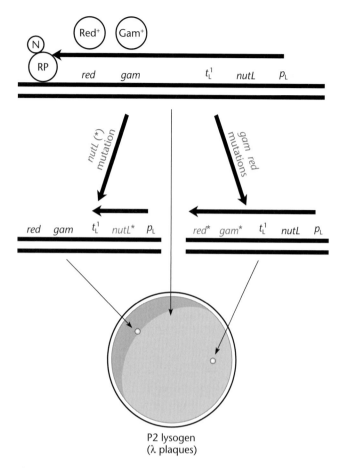

Figure 8.21 Positive selection for λ *nutL* mutations. See the text for details. RP, RNA polymerase.

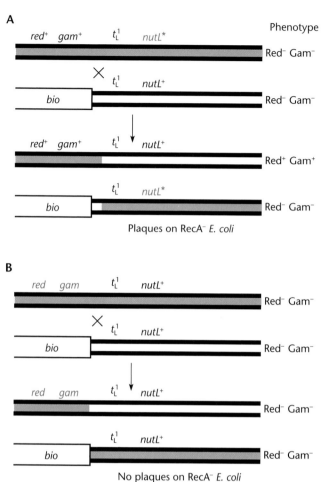

Figure 8.22 Use of λp*bio*-substituted phage to map *nutL* mutations. (A) If the mutation that makes λ phenotypically Gam⁻ Red⁻ maps outside the substituted region, e.g., a *nutL* mutation, Gam⁺ Red⁺ recombinants arise that can multiply in RecA⁻ *E. coli*. (B) If the mutations lie with the substituted region, e.g., a *red gam* double mutant, no Red⁺ Gam⁺ recombinants should arise. Only the region of the *nutL* region and the *gam* and *red* genes is shown.

mutations that include both the *red* and *gam* genes would be more frequent. However, by inducing mutations using a specific mutagen that causes only point mutations, it is possible to lower the percentage of all the mutations that will be deletions (see chapters 3 and 7). Nevertheless, in spite of all precautions, some of the λ mutants that multiply on P2 lysogens have both *gam* and *red* mutations, and these must be distinguished from any *nutL* mutants.

One way to distinguish *nutL* mutants from the other types of mutants that can form plaques on P2 lysogens is by genetic mapping, since *gam* and *red* mutations should map to the left of the t_L^1 terminator while *nutL* mutations should map to its right (Figure 8.2). Genetic mapping in phage λ is facilitated by the collection of λp*bio* specialized transducing phage in which *E. coli* genes have replaced some of the λ genes. The endpoints of the substitutions in many of these phages have been precisely mapped. The way in which they can be used to map mutations is illustrated in Figure 8.22. In the example, a mutant of λ that is phenotypically Red⁻ Gam⁻ and grows on a P2 lysogen is crossed with a phage λ in which the *bio* substitution includes the *red* and *gam* genes. The

appearance of Red⁺ Gam⁺ recombinants indicates that the mutation that makes the phage Red⁻ Gam⁻ must lie outside the substituted region. As discussed in chapter 10 in connection with the discovery of *chi* (χ) sites, only Red⁺ Gam⁺ recombinants of λ can form concatemers and hence form plaques on RecA⁻ *E. coli*. Therefore, even very rare Red⁺ Gam⁺ recombinants can be detected by plating the progeny of the cross on RecA⁻ *E. coli*.

Even if the mutation which allows the phage to multiply in a P2 lysogen maps to the right of the t_L^1 terminator, it is not necessarily a *nutL* mutation. Other types of mutations have the potential to reduce *gam* and *red* transcription. For example, leaky *N* mutations might reduce the transcription of *gam* and *red* enough to allow plaques to form on a P2 lysogen but allow sufficient *O* and *P* transcription for a plaque to form. However, such *N* mutations, as well as many other such types of mutations, can be distinguished from *nutL* mutations because they are *trans* acting rather than *cis* acting. To determine if the mutation is *cis* acting, a P2 lysogen is infected simultaneously with two different mutant λ phages, one of which has one of the potential *nutL* mutations and the other of which has a *gam red* deletion mutation. If the potential *nutL* mutation is *cis* acting, the phage should still multiply on the P2 lysogen, since the *nutL* mutant phage cannot make the *gam* and *red* gene products even if it is furnished with all the other λ gene products in *trans*. Note that *nut* mutations should behave as though they are *cis* acting even though they affect a site on the RNA (a diffusible molecule) rather than a site on the DNA, because the mutation affects transcription termination only from the same DNA.

As mentioned earlier, once the *nutL* mutations had been genetically mapped, investigators could locate the base pair change in the mutation through DNA sequencing and comparison of the sequence of the mutant DNA to the known sequence of wild-type λ DNA in this region. They then found identical sequences to the right of the *cI* gene and assumed that this was *nutR* (Figure 8.5).

Isolation of Host *nus* Mutations

The last genetic analysis with λ phage to be discussed is the isolation of host *nus* mutations, because they illustrate some additional important concepts in selectional genetics. As mentioned, host *nus* mutations affect host gene products that are required for N antitermination (see Friedman et al., Suggested Reading). The λ N gene product does not act alone, and host proteins are required for efficient N antitermination. A host chromosomal mutation that affects one of these proteins may reduce antitermination by the λ N gene product. However, host *nus* mutations are rare, and a positive selection was required to isolate them. Also, it was necessary to reduce the frequency of other types of mutants which might be much more frequent than *nus* mutants.

The selection for *nus* mutations was based on the fact that while induction of a wild-type λ prophage invariably kills the host, the cell can survive induction of an N⁻ mutant prophage. The reason they survive is that antitermination by N is required to synthesize the *P* gene product and other λ gene products that kill the cell. A *nus* mutant of the host should also survive induction of the prophage since the *nus* mutation should also prevent N antitermination. Therefore, induction of the prophage should provide a positive selection for *nus* mutations. If an *E. coli* lysogen containing a λ prophage with a temperature-sensitive mutation in its *cI* gene is induced by raising the temperature and the culture is later plated at the lower (permissive) temperature, at least some of the bacteria that survive and grow up to form a colony should have a *nus* mutation in their chromosome. However, as with the isolation of *nut* mutations in the phage, it is necessary to do the selection in a way that reduces the frequency of other types of mutants which are much more frequent than *nus* mutants. For example, cells cured of the prophage would also survive the heat treatment. To reduce the frequency of cured cells, the investigators used a deleted prophage that contained the *P* gene and could kill the cell but could not be induced to excise from the chromosome. There are also a myriad of mutations in the prophage itself that allow the cell to survive the induction, for example N mutations. The frequency of surviving mutants with N mutations or other types of mutations in the prophage could be reduced by using a double lysogen, with two copies of the prophage in the chromosome. The investigators reasoned that a mutation in the N gene of just one of the two copies of the prophage would not save the cell and that two N mutations, one in each copy of the prophage, would be required, greatly reducing the frequency of this type of survivor among the mutants. Once surviving mutants were selected, they were mapped by Hfr crosses and transduction, using methods described in chapter 3. The investigators were not interested in mutations that mapped to where the prophage was inserted in the chromosome and were presumably in the prophage; they were interested only in mutations elsewhere in the *E. coli* genome, and these mutations defined the chromosomal *nus* genes. In this way, they found the *E. coli* genes *nusA*, *nusB*, etc., whose products are involved in transcription termination and antitermination in the uninfected host.

SUMMARY

1. Some phages are capable of lysogeny, in which they persist in the cell as prophages. A bacterium harboring a prophage is called a lysogen. In the lysogen, most of the phage gene products made are involved in maintaining the prophage state. The prophage DNA can either be integrated into the chromosome or replicate autonomously as a plasmid.

2. The *E. coli* phage λ (lambda) is the prototype of a lysogen-forming phage. It was the first such phage to be discovered and the one to which all others are compared.

3. Phage λ regulates its early transcription through antitermination proteins N and Q. These proteins bind to the RNA

(continued)

polymerase and allow it to transcribe through transcription termination sites.

4. The N protein must first bind to a sequence, *nut*, in the mRNA before it can bind to the RNA polymerase. Other host proteins called the Nus proteins help it bind. The RNA polymerase with N, NusA, NusB, NusE, and NusG bound can then transcribe through transcription terminators into the *O*, *P*, and *Q* genes on the right and the *red*, *gam*, and *int* genes on the left. At least some of the Nus proteins are involved in transcription termination and antitermination in the host.

5. The Q protein allows the RNA polymerase to transcribe through a termination site into the late genes, including the head, tail, and lysis genes of the phage. The RNA polymerase first makes a short RNA and then stops. The *Q* gene product must bind to a *qut* sequence in the DNA close to the promoter before it can bind to the stalled RNA polymerase and allow it to transcribe into the late genes of the phage.

6. Phage λ DNA is linear in the phage head, with short complementary single-stranded 5′ ends called the *cos* or *cohesive* ends. Because they have complementary sequences, the *cos* ends can base pair with each other after the DNA enters the cell to form a circle. The phage DNA then replicates as a circle a few times before it enters the rolling-circle mode of replication, which leads to the formation of long concatemers in which many genome-length DNAs are linked end to end.

7. Phage λ DNA can be packaged only from concatemers and not from unit-length genomes. This is because λ begins filling the head at one *cos* site and stops only when it gets to the next *cos* site in the concatemer. The concatemers from which λ DNA is packaged can be formed either by rolling-circle replication or by recombination between single-length circles.

8. Whether λ enters the lysogenic state depends on the outcome of a race between the *cII* gene product and the products of the lytic genes of the phage. The product of the *cII* gene is a transcriptional activator that activates the transcription of the *cI* and *int* genes after infection. The *cI* gene product is the repressor that blocks transcription of most of the genes of λ in the prophage state, and the *int* gene product is a site-specific recombinase that integrates λ DNA into the bacterial chromosome by promoting recombination between the *attP* site on the phage DNA and the *attB* site in the chromosome.

9. The CI repressor protein of λ is a homodimer made up of two identical polypeptides encoded by the *cI* gene. The repressor blocks transcription by binding to operators o_R and o_L on both sides of the *cI* gene, preventing the utilization of two promoters, p_R and p_L, which are responsible for the transcription of genes to the right and to the left of the *cI* gene, respectively.

10. The repressor regulates its own synthesis in the lysogenic state through its binding to three repressor binding sites within o_R. These sites are named $o_R{}^1$, $o_R{}^2$, and $o_R{}^3$ in the order of their affinity for repressor. Repressor bound at $o_R{}^1$ blocks transcription from p_R. At higher concentrations, repressor also binds at $o_R{}^2$ and activates transcription of the *c1* gene. At yet higher concentrations, repressor binds to $o_R{}^3$ and forms a tetramer with repressor bound at $o_L{}^3$, bending the DNA and preventing synthesis of more repressor. The ability of a gene product to regulate its own synthesis is called autoregulation.

11. Damage to the DNA of its host can cause the λ prophage to be induced and produce phage. Single-stranded DNA that accumulates after DNA damage binds to host RecA protein, activating its coprotease activity. The activated RecA protein of the host causes the λ repressor protein to cleave itself between the DNA binding and dimerization domains so that it can no longer form dimers and be active. The process of excision of λ DNA is essentially the reverse of integration, except that excision requires both the *int* and *xis* gene products of λ because it requires recombination between the hybrid *attP-attB* sites flanking the prophage.

12. Very rarely, when λ DNA excises, it picks up neighboring bacterial DNA and becomes a transducing particle. This type of transduction is called specialized transduction, because only bacterial genes close to the insertion site of the prophage can be transduced.

13. Lysogen-forming phages are often useful as cloning vectors. A bacterial gene cloned into a prophage exists in only two copies, one in its normal site in the chromosome and another in the prophage at its *attB* site. This can be important in complementation tests or in other applications. If the prophage is induced, the cloned DNA can be recovered in large amounts.

14. Lysogen-forming phages often carry genes for bacterial toxins. Examples include the toxins that cause hemolytic-uremic syndrome (HUS), diphtheria, botulism, scarlet fever, cholera, and toxic shock syndrome.

QUESTIONS FOR THOUGHT

1. Why do you suppose the λ prophage uses different promoters to transcribe the *c*I repressor gene immediately after infection and in the lysogenic state?

2. Why is only one protein, Int, required to integrate the phage DNA into the chromosome while two proteins, Int and Xis, are required to excise it? Why not just make one different Int-like protein that excises the prophage?

3. How do you suppose morons containing toxin and other virulence genes move onto a phage? What is the selective pressure for a phage to pick up a moron? Where do morons come from?

4. Why do you suppose some types of prophage can be induced only if another phage of the same type infects the lysogenic cell containing them? What purpose does this serve?

5. Is P4 a phage or a genetic island? What distinguishes these two types of DNA elements?

PROBLEMS

1. Lambda (λ) *vir* mutations cause clear plaques because they change the operator sequences so that they no longer bind repressor. How would you determine if a clear plaque mutant you have isolated has a *vir* mutation instead of a mutation in any one of the three genes *c*I, *c*II, or *c*III?

2. Lambda (λ) integrates into the bacterial chromosome in the region between the galactose utilization (*gal*) genes and the biotin biosynthetic (*bio*) genes on the other side. Outline how you would isolate an HFT strain carrying the *bio* operon of *E. coli*. Would you expect your transducing phage to form plaques? Why or why not?

3. A *vir* mutation changes the operator sequences. Would you expect λ with *vir* mutations in the o_R^1 and o_L^1 sites to form plaques on λ lysogens? Why or why not?

4. The DNA of a λ specialized transducing particle usually integrates next to or into a preexisting prophage. Draw the structures you would expect from these two types of integration. Make sure you show the structures of the *att* sites at the ends of both phages and consider whether you would need both Int and Xis to excise the phages in both structures. Also, what kinds of experiments could you do to determine which kind of structure has formed in a particular dilysogen?

5. Why can you sometimes get intragenic complementation between two temperature-sensitive mutations in the *c*I gene of λ phage but never between two amber (UAG) mutations?

6. The λ CI repressor must dimerize to function. Outline how you would use this fact to identify the regions of another protein, e.g., LacZ, required for its dimerization.

7. You have isolated a relative of P2 phage from sewage, using *E. coli* as the indicator bacterium. How would you determine if P4 phage can parasitize (i.e., can be a satellite virus of) your P2-like phage?

8. You have isolated a phage from a strain of *Staphylococcus aureus* known to cause food poisoning owing to production of a toxin. Outline how you would go about determining if the toxin is encoded by a prophage. Assume that you can detect the toxin by its ability to kill human cells in culture.

SUGGESTED READING

Brussow, H., C. Canchaya, and W. D. Hardt. 2004. Phages and the evolution of bacterial pathogens: from genomic rearrangements to lysogenic conversion. *Microbiol. Mol. Biol. Rev.* 68:560–602.

Cairns, J., M. Delbrück, G. S. Stent, and J. D. Watson. 1966. *Phage and the Origins of Molecular Biology.* Cold Spring Harbor Laboratory Press, Cold Spring Harbor, N.Y.

Calendar, R. L. (ed.). 2005. *The Bacteriophages*, 2nd ed. Oxford University Press, New York, N.Y.

Casjens, S. 2003. Prophages and bacterial genomics: what have we learned so far? *Mol. Microbiol.* 49:277–300.

Court, D. L., A. B. Oppenheim, and S. L. Adhya. 2007. A new look at bacteriophage λ genetic networks. *J. Bacteriol.* 189:298–304.

Echols, H., and H. Murialdo. 1978. Genetic map of bacteriophage lambda. *Microbiol. Rev.* 42:577–591.

Freeman, V. J. 1951. Studies on the virulence of bacteriophage-infected strains of *Corynebacterium diphtheriae*. *J. Bacteriol.* 61:675–688.

Friedman, D. I., M. F. Baumann, and L. S. Baron. 1976. Cooperative effects of bacterial mutations affecting λ N gene expression. *Virology* 73:119–127.

Gottesman, M. 1999. Bacteriophage λ: the untold story. *J. Mol. Biol.* 293:177–180.

Heilpern, A. J., and M. K. Waldor. 2003. pIIICTX, a predicted CTXφ minor coat protein, can expand the host range of coliphage fd to include *Vibrio cholerae*. *J. Bacteriol.* 185:1037–1044.

Hendrix, R. W., J. W. Roberts, F. W. Stahl, and R. A. Weisberg (ed.). 1983. *Lambda II.* Cold Spring Harbor Laboratory Press, Cold Spring Harbor, N.Y.

Johnson, L. P., M. A. Tomai, and P. M. Schlievert. 1986. Bacteriophage involvement in group A streptococcal pyrogenic exotoxin A production. *J. Bacteriol.* **166:**623–627.

Juhala, R. J., M. E. Ford, R. L. Duda, A. Youlton, G. F. Hatfull, and R. W. Hendrix. 2000. Genomic sequences of bacteriophages HK97 and HK022: pervasive mosaicism in the lambdoid phages. *J. Mol. Biol.* **299:**27–51.

Kahn, M. L., R. Ziermann, G. Deho, D. W. Ow, M. G. Sunshine, and R. Calendar. 1991. Bacteriophage P2 and P4. *Methods Enzymol.* **204:**264–280.

Kobiler, O., A. Rokney, N. Friedman, D. L. Court, J. Stavans, and A. B. Oppenheim. 2005. Quantitive kinetic analysis of the bacteriophage λ genetic network. *Proc. Natl. Acad. Sci. USA* **102:**4470–4475.

Li, J., R. Horwitz, S. McCracken, and J. Greenblatt. 1992. NusG, a new *Escherichia coli* elongation factor involved in transcriptional antitermination by the N protein of phage λ. *J. Biol. Chem.* **267:**6012–6019.

Lieb, M. 1976. Lambda *cI* mutants: intragenic complementation and complementation with a *cI* promoter mutant. *Mol. Gen. Genet.* **146:**291–297.

Livny, J., and D.I. Friedman. 2004. Characterizing spontaneous induction of Stx encoding phages using a selectable reporter system. *Mol. Microbiol.* **51:**1691–1704.

Lwoff, A. 1953. Lysogeny. *Bacteriol. Rev.* **17:**269–337.

McLeod, S. M., H. H. Kimsey, B. M. Davis, and M. K. Waldor. 2005. CTXφ and *Vibrio cholerae*: exploring a newly recognized type of phage-host cell relationship. *Mol. Microbiol.* **57:**347–356.

Moyer, K. E., H. H. Kimsey, and M. K. Waldor. 2001. Evidence for a rolling circle mechanism of phage DNA synthesis from both replicative and integrated forms of CTXφ. *Mol. Microbiol.* **41:**311–323.

Nickels, B. E., C. W. Roberts, H. I. Sun, J. W. Roberts, and A. Hochschild. 2002. The σ^{70} subunit of RNA polymerase is contacted by the λQ antiterminator during early elongation. *Mol. Cell* **10:**611–622.

Ptashne, M. 2004. *A Genetic Switch: Phage Lambda Revisited*, 3rd ed. Cold Spring Harbor Laboratory Press, Cold Spring Harbor, N.Y.

Salstrom, J. S., and W. Szybalski. 1978. Coliphage λ *nutL*: a unique class of mutations defective in the site of N product utilization for antitermination of leftward transcription. *J. Mol. Biol.* **124:**195–222.

Schiavo, G., F. Benfenati, B. Poulain, O. Rossetto, P. Polverino de Laureto, B. R. DasGupta, and C. Montecucco. 1992. Tetanus and botulinum-B neurotoxins block neurotransmitter release by proteolytic cleavage of synaptobrevin. *Nature* (London) **359:**832–835.

Sunagawa, H., T. Obyama, T. Watanabe, and K. Inoue. 1992. The complete amino acid sequence of the *Clostridium botulinum* type D neurotoxin, deduced by nucleotide sequence analysis of the encoding phage d-16φ genome. *J. Vet. Med. Sci.* **54:**905–913.

Svenningsen, S. L., N. Constantine, D. L. Court, and S. Adhya. 2005. On the role of Cro in λ prophage induction. *Proc. Natl. Acad. Sci. USA* **102:**4465–4469.

Wagner, P. L., M. N. Neely, X. Zhang, D. W. K. Acheson, M. K. Waldor, and D. L. Friedman. 2001. Role for a phage promoter in Shiga toxin 2 expression from a pathogenic *Escherichia coli* strain. *J. Bacteriol.* **183:**2081–2085.

Waldor, M. K., D. I. Friedman, and S. L. Adhya (ed.). 2005. *Phages: Their Role in Bacterial Pathogenesis and Biotechnology.* ASM Press, Washington,. D.C.

Waldor, M. K., and J. J. Mekalanos. 1996. Lysogenic conversion by a filamentous phage encoding cholera toxin. *Science* **272:**1910–1914.

CHAPTER **9**

Transposition, Site-Specific Recombination, and Families of Recombinases

Recombination is the breaking and rejoining of DNA in new combinations. In homologous recombination, which accounts for most recombination in the cell, the breaking and rejoining occur only between regions of two DNA molecules that have similar or identical sequences. Homologous recombination requires that the two DNAs pair through complementary base pairing, which requires that the two DNAs have the same sequence (see chapter 10). However, other types of recombination, known as **nonhomologous recombination,** also occur in cells. As the name implies, these types of recombination do not depend on homology between the two DNA sequences involved in the recombination. Some types of nonhomologous recombination are due only to the mistaken breaking and rejoining of DNA by enzymes such as topoisomerases (see chapter 1). Other types are not mistakes and have specific purposes in the cell. These types depend on specific enzymes that promote recombination between different regions in DNA, which may or may not have sequences in common. This chapter addresses some of these examples of nonhomologous recombination in bacteria and the mechanisms involved, including transposition by transposons, the site-specific recombination that occurs during the integration and excision of prophages and other DNA elements, the inversion of invertible sequences, and the resolution of cointegrates by resolvases. Recent evidence shows that the enzymes that perform these various functions have much in common.

Transposition

Transposons are DNA elements that can hop, or **transpose,** from one place in DNA to another. Transposable DNA elements were first discovered in corn by Barbara McClintock in the early 1950s and about 20 years later in bacteria by others. Transposons are now known to exist in all organisms on Earth, including humans. In fact, from the human genome project it is apparent that almost half of our DNA may be transposons! The movement by a transposon is called **transposition,** and the enzymes that promote transposition are called **transposases.** The transposon itself usually encodes its own transposases, so that it carries with it the ability to hop each time it moves. For this reason, transposons have been called "jumping genes."

Not all DNA elements that can move are true transposons. For example, "homing" DNA elements, which include some types of moveable RNA and protein introns, move by means of endonucleases that make a specific double-strand break in the DNA at a given site. Then, through homologous recombination aimed at repairing the double-strand break, the DNA element is inserted at that site. No specific transposases are required for the movement of homing DNA, and these DNA elements can move only into the same sequence in other DNA molecules that lack them. Homing endonucleases are discussed in more detail in chapter 10 (see Box 10.1).

True transposons should also be distinguished from **retrotransposons,** so named because they behave like RNA retroviruses with a DNA intermediate. An RNA copy of a region is made and then copied into DNA by a reverse transcriptase. The DNA intermediate then integrates elsewhere by various mechanisms that may or may not be analogous to transposition. Although only a few examples of retrotransposons are known in bacteria, such elements are well known in fungi.

Although transposons probably exist in all organisms on Earth, they are best understood in bacteria, where they obviously play an important role in evolution. Transposons may offer a way of introducing genes from one bacterium into the chromosome of another bacterium to which it has little DNA sequence homology. Transposons found in different bacterial genera may be more closely related to each other than are the bacteria in which they are found. This suggests that transposons move among different genera of bacteria with some regularity. As mentioned in previous chapters, transposons may enter other genera of bacteria during transfer of promiscuous plasmids or via transducing phage. Some transposons are themselves conjugative or can be induced to form phage, as discussed later in this chapter.

Overview of Transposition

The net result of transposition is that the transposon appears at a place in DNA different from where it was originally. Many transposons are essentially cut out of one DNA and inserted into another (Figure 9.1), whereas other transposons are copied and then inserted

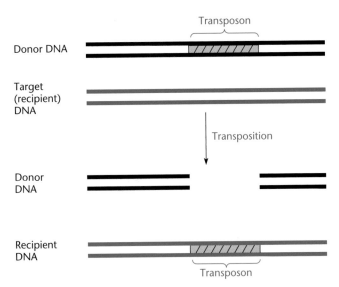

Figure 9.1   Overview of transposition. See the text for details.

elsewhere. Regardless of the type of transposon, however, the DNA from which the transposon originated is called the **donor DNA** and the DNA into which it hops is called the **target** or **recipient DNA.**

In all transposition events, the transposase enzyme cuts the donor DNA at the ends of the transposon and then inserts the transposon into the target DNA. However, the details of the mechanism can vary. Some types of transposons may exist free of other DNA during the act of transposition, but many transposons, before and after they hop, remain contiguous with other flanking DNA molecules. Later in this chapter, we discuss more detailed models for the various types of transposition.

Transposition must be tightly regulated and occur only rarely; otherwise the cellular DNA would become riddled with the transposon, which would have many deleterious effects. Transposons have evolved elaborate mechanisms to regulate their transposition so that they hop very infrequently and do not often kill the host cell. We mention some of these mechanisms later in this chapter when we discuss individual examples of transposons. The frequency of transposition varies from about once in every 10^3 cell divisions to about once in every 10^8 cell divisions, depending on the type of transposon. Thus, the chance of a transposon hopping into a gene and inactivating it is not much higher than the chance that a gene will be inactivated by other types of mutations (see chapter 3).

Structure of Bacterial Transposons

There are many different types of bacterial transposons. Some of the smaller ones are about 1,000 bp long and carry only the genes for the transposases that promote their movement in DNA and the genes that regulate this movement. Larger transposons may also contain one or more other genes, such as those for resistance to an antibiotic.

One distinguishing feature of bacterial transposons is that all those identified so far, with the exception of rolling-circle transposons, contain repeats at their ends, which are usually **inverted repeats** (Figure 9.2). As discussed in chapter 1, two regions of DNA are inverted repeats if the sequence of nucleotides on one strand in one region, when read in the 5′-to-3′ direction, is the same or almost the same as the 5′-to-3′ sequence of the opposite strand in the other region.

Another feature common to all but rolling-circle transposons is the presence of short **direct repeats** of the target DNA that bracket the transposon (Figure 9.2). Direct repeats have the same or almost the same 5′-to-3′ sequence of nucleotides on the same strand. As shown in Figure 9.2, the target DNA originally contains only one copy of the sequence at the place where the transposons insert. During the insertion of the transposon, this sequence is duplicated. Most transposons can insert into many places in DNA and so have little or no target specificity. Thus, the duplicated sequence varies with the sequence at the site in the target DNA into which the transposon inserted. However, even though the duplicated sequences differ, the number of duplicated base pairs is characteristic of each transposon. Some duplicate as few as 3 bp, and others duplicate as many as 9 bp. The molecular models for transposition discussed later in the chapter offer an explanation for the duplicated sequences.

Types of Bacterial Transposons

Each type of bacterium carries its own unique transposons, although many transposons are related across species as though they had been only recently exchanged. In this section, we describe some of the common types of transposons.

INSERTION SEQUENCE ELEMENTS

The smallest bacterial transposons are called **insertion sequence (IS) elements.** These transposons are usually only about 750 to 2,000 bp long and encode little more than the transposase enzymes that promote their transposition.

Because IS elements carry no selectable genes, they were discovered only because they inactivate a gene if

Figure 9.2 Structure of the insertion sequence element IS3 and its related family members. (A) The inverted repeats are shown as arrows, and the 3-bp target sequence that is duplicated after transposition is boxed. ORFA and ORFB encode the N terminus and C terminus of the transposase, which are translated in different reading frames and are not active by themselves. (B) A programmed −1 frameshift puts both ORFA and ORFB in the same frame and makes the active transposase. The C terminus of the IS3 transposase contains the DDE motif characteristic of this type of transposase.

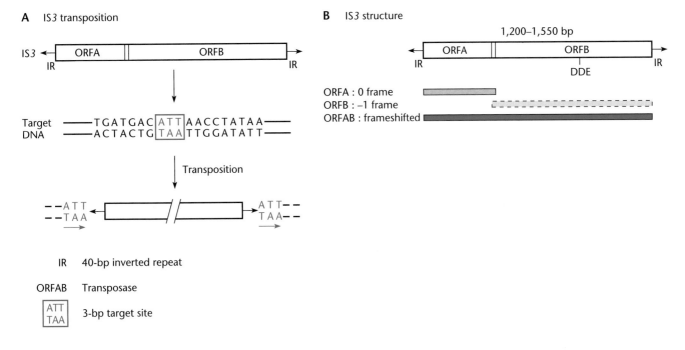

they happen to hop into it. The first IS elements were detected as a type of *gal* mutation that was unlike any other known mutations. This type of mutation resembled deletion mutations in that it was nonleaky; however, unlike deletion mutations, it could revert, albeit at a lower frequency than base pair changes or frameshifts. Such anomalous *gal* mutations were also very polar and could prevent the transcription of downstream genes (see chapter 2). Later work showed that these mutations resulted from insertion of about 1,000 bp of DNA into a *gal* gene. Moreover, they were due to insertion of not just any piece of DNA but one of very few sequences.

Originally, four different IS elements were found in *Escherichia coli*: IS1, IS2, IS3, and IS4. Most strains of *E. coli* K-12 contain approximately six copies of IS1, seven copies of IS2, and fewer copies of the others. Almost all bacteria carry IS elements, with each species harboring its own characteristic IS elements, although sometimes related IS elements can be found in different bacteria. To date, thousands of different IS elements have been found in bacteria. Plasmids also often carry IS elements, which are important in the assembly of the plasmid itself (see Figure 9.7) and in the formation of Hfr strains (see chapter 5).

Figure 9.2 also shows the structure of the IS element IS3 and how the transposase is encoded. In addition to the inverted repeats at its ends, it consists of two open reading frames (ORFA and ORFB). The reading frame of ORFB is shifted −1 relative to the reading frame of ORFA, but a programmed −1 frameshift (see Box 2.4) causes the synthesis of a fusion protein, ORFAB, which is the active transposase. The smaller protein made from ORFA when the frameshifting does not occur regulates transcription of the transposase gene. The target site sequence that is duplicated in the target DNA on insertion of IS3 is 3 bp long. As mentioned above, the length of such direct repeats is characteristic of each type of transposon.

Although the original IS elements were discovered only because they had hopped into a gene, causing a detectable phenotype, IS elements are now more often discovered during hybridization experiments with cloned regions of bacterial DNA as probes or in genomic sequencing.

COMPOSITE TRANSPOSONS

Sometimes two IS elements of the same type form a larger transposon, called a **composite transposon,** by bracketing other genes. Figure 9.3 shows the structures of three composite transposons, Tn5, Tn9, and Tn10. Tn5 consists of genes for kanamycin resistance (Kan^r) and streptomycin resistance (Str^r) bracketed by copies of an IS element called IS50. Tn9 has two copies of IS1 bracketing a chloramphenicol resistance gene (Cam^r). In Tn10, two copies of IS10 flank a gene for tetracycline

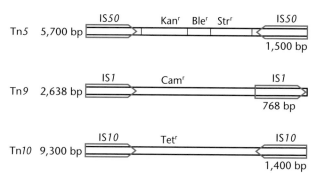

Figure 9.3 Structures of some composite transposons. The commonly used genes for kanamycin resistance, Kan^r, and the gene for chloramphenicol resistance, Cam^r, come from Tn5 and Tn9, respectively. The active transposase gene is in one of the two IS elements. Note that the IS elements can be in either the same or opposite orientation (arrows). Str^r, gene encoding streptomycin resistance; Tet^r, gene encoding tetracycline resistance; Ble^r, gene encoding bleomycin resistance.

resistance (Tet^r). Some composite transposons, such as Tn9, have the bracketing IS elements in the same orientation, whereas others, including Tn5 and Tn10, have them in opposite orientations.

Outside-End Transposition

Figure 9.4 illustrates transposition of a composite transposon. Each IS element can transpose independently as long as the transposase acts on both of its ends. However, because all the ends of the IS elements in a composite transposon are the same, a transposase encoded by one of the IS elements can recognize the ends of either IS element. When such a transposase acts on the inverted repeats at the farthest ends of a composite transposon, the two IS elements transpose as a unit, bringing along the genes between them. These two inverted repeats are called the "outside ends" of the two IS elements because they are the *farthest* from each other.

The two IS elements that form composite transposons are often not completely autonomous, because of mutations in the transposase gene of one of the elements. Thus, only one of the IS elements encodes an active transposase. However, this transposase can act on the outside ends to promote transposition of the composite transposon.

Inside-End Transposition

The transposase encoded by one IS element in a composite transposon can also act on the "inside ends" of both IS elements, that is, the two ends that are closest to each other. Inside-end transposition presumably happens as often as outside-end transposition but has very different consequences.

A IS element

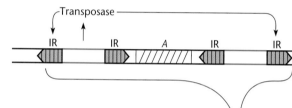

B Composite: 2 IS + gene *A*

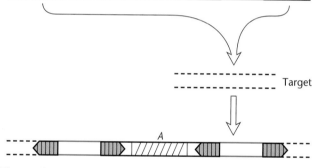

Figure 9.4 Two IS elements can transpose any DNA between them. (A) Action of the transposase at the ends of an isolated IS element causes it to transpose. (B) Two IS elements of the same type are close to each other in the DNA. Action of the transposase on their outside ends causes them to transpose together, carrying along the DNA between them. *A* denotes the DNA between the IS elements (hatched bar), and arrows indicate the inverted repeat (IR) sequences at the ends of the IS elements. Dashed lines represent the target DNA.

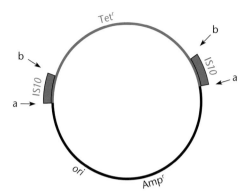

Figure 9.5 Either the outside or inside ends of the IS elements in a composite transposon can be used for transposition. Outside-end transposition (a) transposes Tn*10* (purple), including the gene encoding tetracycline resistance (Tet^r), whereas inside-end transposition (b) transposes the plasmid, including the origin of replication and the gene for ampicillin resistance (Amp^r).

One possible outcome of inside-end transposition is the creation of a new composite transposon, which was first demonstrated with Tn*10* (see Foster et al., Suggested Reading). In these experiments, Tn*10* was inserted into a small plasmid with an origin of replication (*ori*) and an ampicillin resistance gene (Amp^r). The plasmid served as the donor DNA. As shown in Figure 9.5, transposition with the outside ends of the IS*10* element would move Tn*10*, with the tetracycline resistance gene, Tet^r, to another DNA. However, transposition from the inside ends would create a new composite transposon carrying the Amp^r gene and the plasmid origin of replication (*ori*) to another DNA. If this new composite transposon hops into a target DNA that does not have a functional origin of replication, it may confer on that DNA the ability to replicate. In the experiment, a λ phage with amber mutations in its replication genes *O* and *P* was used to infect amber-suppressing cells containing the donor plasmid with the transposon. The progeny phage were then plated on a non-amber-suppressor host. In this host, the phage could not replicate from the λ origin of replication because of the amber mutations in their replication genes. However, any phage into which the new composite transposon had hopped would be able to replicate by using the plasmid origin of replication. As expected, the few phages that formed plaques did contain the new composite transposon. The inside ends of the IS*10* elements of Tn*10* must have been used for its transposition.

Deletions and inversions can also be caused by inside-end transposition of a composite transposon to a nearby target on the same DNA (Figure 9.6). The neighboring sequences between the original site of insertion of the transposon and the site into which it is trying to transpose will be either deleted or inverted. Whether a deletion or inversion is created depends on how the inside ends of the IS elements in the transposon are attached to the target DNA. If the inside ends cross over each other before they attach, the neighboring sequences will be inverted; if they do not cross over each other, the neighboring sequences will be deleted. As shown in Figure 9.6, the DNA between the two IS elements in the composite

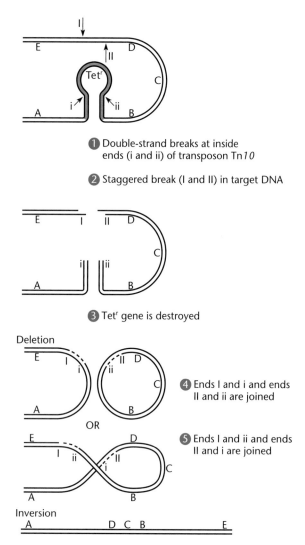

1 Double-strand breaks at inside ends (i and ii) of transposon Tn*10*

2 Staggered break (I and II) in target DNA

3 Tet^r gene is destroyed

4 Ends I and i and ends II and ii are joined

5 Ends I and ii and ends II and i are joined

Figure 9.6 Rearrangements of DNA caused by composite transposons. Attempts to transpose by the inside ends of a composite transposon to a neighboring target sequence can cause either a deletion or an inversion of the intervening sequences, depending on how the ends are attached. Antibiotic resistance genes carried by the transposon are deleted. For an explanation of steps 1 and 2, see Figure 9.14.

transposon will also be deleted, independent of what happens to the neighboring DNA. Therefore, these rearrangements are usually accompanied by the loss of any resistance gene on the composite transposon, which is how they are usually selected. For example, methods have been developed to select tetracycline-sensitive derivatives of *E. coli* harboring the Tn*10* transposon. Most of these tetracycline-sensitive derivatives have deletions or inversions of DNA next to the site of insertion of the Tn*10* element. Presumably, inside-end transposition is responsible for most of the often-observed instability of

DNA caused by composite transposons. To avoid such extensive rearrangments, some composite transposons have mechanisms to avoid inside-end transposition. One example is transposon Tn*5* which methylates adenines in the inverted repeats in the inside ends so that they will be recognized less well by the tranposase.

Assembly of Plasmids by IS Elements

Any time two IS elements of the same type happen to hop close to each other on the same DNA, a composite transposon is born. These transposons have not yet evolved a defined structure such as the named transposons (e.g., Tn*10*) described above. Nevertheless, the two IS elements can transpose any DNA between them. In this way, "cassettes" of genes bracketed by IS elements can be moved from one DNA molecule to another.

Many plasmids seem to have been assembled from such cassettes. Figure 9.7 shows a naturally occurring plasmid carrying genes for resistance to many antibiotics. Such plasmids are historically called R-factors, because they confer resistance to so many different antibiotics (see chapter 4). Notice that many of the resistance genes on the plasmid are bracketed by the same IS element. IS*3* flanks the tetracycline resistance gene, and IS*1* brackets the genes for resistance to many other antibiotics. Apparently, the plasmid was assembled in nature by resistance genes hopping onto the plasmid from some other DNA via the bracketing IS elements. In

Figure 9.7 R-factors, or plasmids containing many resistance genes, may have been assembled, in part, by IS elements. The tetracycline resistance (Tet^r) gene is bracketed by IS*3* elements, and the region containing the other resistance genes (the r determinant outlined in purple) is bracketed by IS*1* elements.

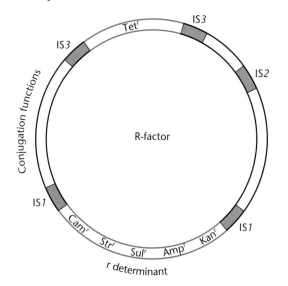

principle, any two transposons of the same type can move other DNA lying between them by a similar mechanism, but because IS elements are the most common transposons and often exist in more than one copy per cell, they probably play the major role in the assembly of plasmids.

NONCOMPOSITE TRANSPOSONS

Composite transposons are not the only ones to carry resistance genes. Such genes can also be an integral part of transposons known as **noncomposite transposons** (Figure 9.8). They are bracketed by short inverted repeats, but the resistance gene is part of the minimum transposable unit. Noncomposite transposons can also cause rearrangment of adjoining chromosomal DNA even though they do not have complete copies of IS elements at their ends and so do not have inside ends to participate in transposition events. This is because many of them transpose by a replicative mechanism (see below) in which the ends of the transposon remain attached to the flanking chromosomal DNA during the transposition event. Transposition by such a tranposon into a nearby site on the same DNA can then rearrange the DNA between the donor and target sites. This may be one reason why such transposons often exhibit target immunity (see below).

Noncomposite transposons seem to belong to a number of families in which the members are related to each other by sequence and structure. Interestingly, different members of transposon families, notably the Tn21 family, often carry different resistance genes, even though they are almost identical otherwise (Figure 9.8). Often this is the result of the resistance genes having integrated into the transposon as a cassette carrying one or more resistance genes. The cassettes, which existed elsewhere in the genome, had excised to form a circle that then integrated into an *att* site on the transposon, using a integrase much like those used to integrate lysogenic phages (see "Integrases of Transposon Integrons" below). These cassettes integrating into integrons provide one of the major ways by which bacteria can achieve resistance to a variety of antibiotics. In fact, the first known example of multiple drug resistance in pathogenic bacteria in Japan in the early 1950s was due to transposon Tn21 with multiple drug resistance cassettes acquired by integrons. This phenomenon of cassette insertion into transposons such as Tn21 is part of a more general phenomenon in which numerous gene cassettes in superintegrons (SIs) are stored in the chromosome, from which individual cassettes can then insert into mobile elements such as plasmids (see "Integrons" below).

Figure 9.8 Some examples of noncomposite transposons. The open reading frames encoding the proteins are boxed. The terminal inverted repeat ends are shown as hatched arrows. *A* is the transposase; *R* is the resolvase and repressor of *A* transcription; *res* is the site at which resolvase acts; Mer^r is the mercury resistance region; and *merR* is the regulator of mercury resistance gene transcription. In2 is an integron of the type described in Figure 9.25. Tn3 was originally found on the broad-host-range plasmid pR1drd19, Tn501 was found on the *Pseudomonas* plasmid pUS1, Tn21 was found on the *Shigella flexneri* plasmid R100, and γδ is found on the chromosome of *E. coli* and on the F plasmid. The arrows indicate the sites at which the proteins act.

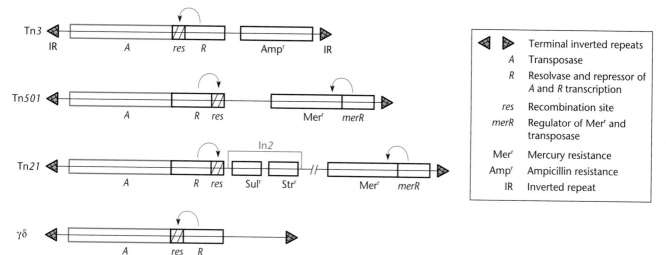

Assays of Transposition

To study transposition, we must have assays for it. As mentioned above, insertion elements were discovered because they create mutations when they hop into a gene. However, this is usually not a convenient way to assay transposition, because transposition is infrequent and it is laborious to distinguish insertion mutations from the myriad other mutations that can occur. If the transposon carries a resistance gene, the job of assaying transposition is easier. But how do we know if a transposon has hopped in the cell? The cells are resistant to the antibiotic no matter where the transposon is inserted in the cellular DNA, so hopping from one place to another makes no difference in the level of resistance of the cell. Obviously, detecting transposition requires special methods.

SUICIDE VECTORS

One way to assay transposition is with **suicide vectors.** Any DNA, including plasmid or phage DNA, that cannot replicate (i.e., is not a replicon) in a particular host can be used as a suicide vector. These DNAs are called suicide vectors because, by entering cells in which they cannot replicate, they essentially kill themselves. To assay transposition with a suicide vector, we use one to introduce a transposon carrying an antibiotic resistance gene into an appropriate host. The way in which the suicide vector itself is introduced into the cells depends on its source. If it is a phage, the cells could be infected with that phage. If it is a plasmid, it could be introduced into the cells through conjugation. However, whatever method is used, it should be very efficient since transposition is a rare event.

Once in the cell, the suicide vector remains unreplicated and eventually is lost. The only way the transposon can survive and confer antibiotic resistance on the cells is by hopping to another DNA molecule that is capable of autonomous replication in those cells, for example a plasmid or the chromosome. Therefore, when the cells under study are plated on antibiotic-containing agar and incubated, the appearance of colonies, as a result of the multiplication of antibiotic-resistant bacteria, is evidence for transposition. These cells have been mutagenized by the transposon, since the transposon has hopped into a cellular DNA molecule—either the chromosome or a plasmid, causing insertion mutations.

Phage Suicide Vectors

Some derivatives of phage λ are designed to be used as suicide vectors in *E. coli*. These phage have been rendered incapable of replication in nonsuppressing hosts by the presence of nonsense codons in their replication genes *O* and *P* (see chapter 8). They have also been rendered incapable of integrating into the host DNA by deletion of their attachment region, *attP*. Such a λ phage can be propagated on an *E. coli* strain carrying a nonsense suppressor. However, in a nonsuppressor *E. coli*, it cannot replicate or integrate. Because of the narrow host range of λ, these suicide vectors can normally be used only in strains of *E. coli* K-12.

Plasmid Suicide Vectors

Plasmid cloning vectors can also be used as suicide vectors, provided that the plasmid cannot replicate in the cells in which transposition is occurring. The plasmid containing the transposon with a gene for antibiotic resistance can be propagated in a host in which it can replicate and is then introduced into a cell in which it cannot replicate. In principle, any plasmid with a conditional-lethal mutation, nonsense or temperature sensitive, in a gene required for plasmid replication can be used as a suicide vector. The plasmid could be propagated in the permissive host or under permissive conditions and then introduced into a nonpermissive host or into the same host under nonpermissive conditions, depending on the type of mutation. Alternatively, a narrow-host-range plasmid could be used. It can be propagated in a host in which it can replicate and introduced into a different species in which it cannot.

Many general methods for assaying transposition are based on promiscuous self-transmissible plasmids because the most efficient way to introduce a plasmid into cells is by conjugation, which can approach 100% under some conditions. If the plasmid containing the transposon contains a *mob* region, it can be mobilized into the recipient cell by using the Tra functions of a self-transmissible plasmid (see chapter 5). This technique is most highly developed for gram-negative bacteria. Taking advantage of the extreme promiscuity of some self-transmissible plasmids of gram-negative bacteria, the plasmid might be mobilized into almost any gram-negative bacterium. If the mobilizable plasmid has a narrow host range, it might not be able to replicate in the host into which it has been mobilized and will eventually be lost. ColE1-derived plasmids into which a *mob* site has been introduced are often the suicide vectors of choice in such applications because they can replicate only in some enteric bacteria including *E. coli* and so can be used as suicide vectors in any other gram-negative bacterium. The hopping of the transposon can then be assayed if it carries a selectable gene, such as for antibiotic resistance, that is expressed in the recipient host. The cells become resistant only if the transposon has hopped into another replicon, e.g., a plasmid or the chromosome, in that host. Some types of transposons are also very broad in their host range for transposition, so that the transposon may hop in almost any recipient bacterium. Such methods are discussed in more detail later in this chapter (see "Transposon Mutagenesis").

THE MATING-OUT ASSAY FOR TRANSPOSITION

Transposition can also be assayed by using the "mating-out" assay, which is also based on conjugation. In this assay, a transposon in a nontransferable plasmid or the chromosome is not transferred into other cells unless it hops into a plasmid that is transferable. Figure 9.9 shows a specific example of a mating-out assay with *E. coli*. In the example shown, transposon Tn*10* carrying tetracycline resistance has been inserted into a small plasmid that is neither self-transmissible nor mobilizable. This small plasmid is used to transform cells containing F, a larger, self-transmissible plasmid. While the cells are growing, the transposon may hop from the smaller plasmid into the F plasmid in a few of the cells. Later, when these cells are mixed with streptomycin-resistant recipient cells, any F plasmid into which the transposon hopped will carry the transposon when it transfers to a new cell, thus conferring tetracycline resistance on that cell. Transposition can be detected by plating the mating mixture on agar containing tetracycline and counterselecting the donor with streptomycin.

The appearance of antibiotic-resistant transconjugants in a mating-out assay is not by itself definitive proof of transposition. Some transconjugants could become antibiotic resistant by means other than transposition of the transposon into the larger, self-transmissible plasmid. The smaller plasmid containing the transposon could have been somehow mobilized by the larger plasmid, or the smaller plasmid could have been fused to the larger plasmid by recombination or by cointegrate formation (see below). A few representative transconjugants should be tested, for example, by using restriction digests and Southern hybridizations (see chapter 1) to verify that they contain only the larger plasmid with the transposon inserted.

Mechanisms of Transposition

The process of figuring out how transposons move followed the usual course in molecular genetics. First, genetic analyses were done with certain selected transposons to identify the gene products and DNA sequences involved and obtain an overview of the process. Then the studies became more molecular, identifying the detailed molecular reactions required and determining the actual structures of the molecules involved and the ways in which these structures contribute to the transposition process. These studies of certain select transposons have revealed that transposons move by a number of different mechanisms, which are nevertheless conceptually related and often use related molecules. We first review some of the earlier genetic studies on how some transposons move and then address the molecular details of these processes.

Genetic Requirements for Transposition of Tn*3*

The first analysis of the genetic requirements for transposition used the transposons Tn*3* and Mu, which happen to transpose by similar mechanisms. We will use Tn*3* as our primary example (see Gill et al., Suggested Reading, and Figure 9.8 for a diagram of Tn*3*). The questions are essentially the same as for any other genetic analysis (see chapter 3). How many gene products are required for transposition of Tn*3*, and where are the sites at which they act? Where do the genes for these gene products lie on the transposon? Do any intermediates of transposition accumulate when one or more of these gene products is inactivated? Obtaining answers to these questions was the first step in developing a molecular model for transposition of Tn*3*.

ISOLATION OF MUTATIONS IN THE TRANSPOSON

As in any genetic analysis, the first step in analyzing the genetic requirements for transposition was to isolate mutations in the transposon. A plasmid containing the transposon was cut randomly with deoxyribonuclease (DNase), and then DNA linkers containing the recognition sequence for a restriction endonuclease were ligated into the cut plasmid. This creates random insertion

Figure 9.9 Example of a mating-out assay for transposition. See the text for details.

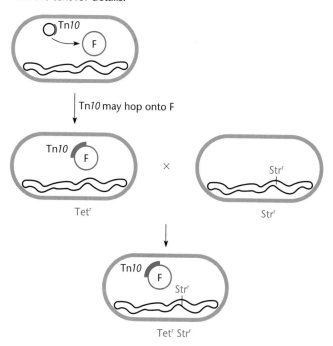

mutations around the plasmid including the transposon, and the presence of the restriction site on the inserted DNA makes the site of the mutation easy to map physically (see chapter 1). Insertion of a DNA linker also disrupts the DNA sequence; if the linker is inserted into a translated ORF, it usually causes a frameshift or creates an in-frame nonsense codon. Such methods have now been largely supplanted by more convenient methods of introducing mutations at known sites, for example, by polymerase chain reaction (PCR) or recombineering (see chapter 1 and Box 10.3).

Once a number of insertion mutations had been isolated that were scattered in various places around the transposon, they were tested for their effects on transposition by the mating-out assay. As illustrated in Figure 9.10,

cells containing both a small, nonmobilizable plasmid carrying the mutant Tn3 and a larger, self-transmissible plasmid were mixed with recipient cells and tranconjugants resistant to ampicillin were selected. Because the smaller plasmid was not mobilizable, ampicillin-resistant transconjugants could be produced only by donor cells in which the transposon had hopped from the smaller plasmid into the larger, self-transmissible one, which was then transferred into the recipient cell. When no ampicillin-resistant transconjugants were observed, the mutation in the transposon must have prevented transposition into the self-transmissible plasmid. When larger than normal numbers of ampicillin-resistant transconjugants were observed, the mutation must have increased the frequency of transposition.

Figure 9.10 Molecular genetic analysis of transposition of the replicative transposon Tn3. The transposon is in purple. The asterisk marks the position of the mutation. Amp^r, ampicillin resistance. See the text for details.

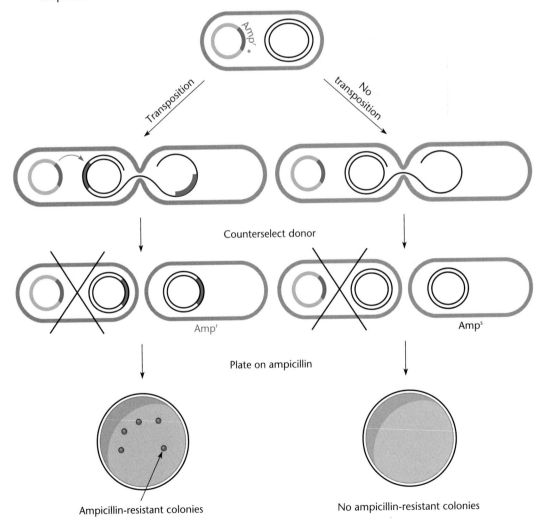

As expected, the effect of a mutation depended on its position in the transposon. As illustrated in Figure 9.11, mutations in the inverted-repeat (IR) sequences and mutations that disrupt the *tnpA* ORF can totally prevent transposition. In contrast, mutations that disrupt the *tnpR* ORF result in higher than normal rates of transposition and the formation of cointegrates, in which the self-transmissible plasmid and the smaller plasmid, which originally contained the transposon, are now joined and are transferred together into the recipient strain. The cointegrate contains two copies of the transposon bracketing the smaller plasmid as shown. Mutations in the short sequence called *res* (for *resolution* sequence) also give rise to cointegrates, but unlike *tnpR* mutations, they result in normal, not elevated, rates of transposition.

COMPLEMENTATION TESTS WITH TRANSPOSON MUTATIONS

The next step was to do complementation tests to determine which mutations in transposon Tn*3* disrupt *trans*-acting functions and which disrupt *cis*-acting sequences or sites. The complementation tests used the same mating-out assay illustrated in Figure 9.11, except that the cell in which the transposition was to occur also contained another Tn*3*-related transposon inserted into its chromosome (Figure 9.12). This other transposon is capable of transposition but lacks an ampicillin resistance gene (Amp^r) so that its own transposition does not create ampicillin-resistant transconjugants and confuse the analysis. The data are interpreted as in any other complementation test. If the mutation in the transposon in the plasmid inactivates a *trans*-acting function, it will be

Figure 9.11 Effects of mutations in different genes required for transposition of Tn*3*. In the left-hand pathway, a *tpnA* or *IR* mutation prevents transposition, and so no Amp^r transconjugants form. In the right-hand pathway, transposition by a *tnpR* or *res* mutant leads to the formation of Amp^r transconjugants that contain the mobilizable and self-transmissible plasmids fused to each other in a cointegrate. The asterisk indicates a mutation.

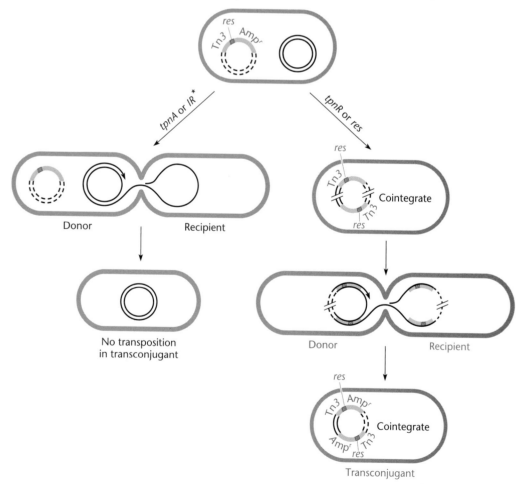

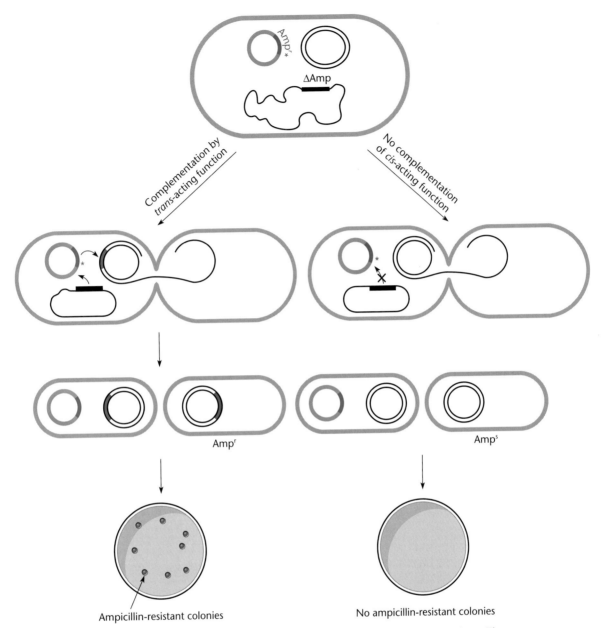

Figure 9.12 Complementation tests of transposition-defective Tn3 mutations. The mutant Tn3 transposon being complemented is in purple, and the asterisk indicates the mutation. See the text for details.

complemented by the corresponding gene in the transposon in the chromosome and the transposon should now be able to transpose. However, if the mutation in the plasmid transposon inactivates a *cis*-acting site, it will not be complemented and will not transpose properly, even in the presence of the chromosomal transposon, since mutations that inactivate *cis*-acting sites cannot be complemented. The complementation tests revealed that mutations that inactivate either the ORF called *A* or the ORF called *R* (Figure 9.8) could be complemented to give normal transposition. However, neither mutations

in the IR sequences at the ends of the transposon nor those in the sequence called *res* could be complemented to give normal transposition. Mutations in an IR sequence prevented transposition altogether, even in the presence of the complementing copy of Tn3, while mutations in *res* permitted transposition but still gave rise to cointegrates. The investigators concluded that *tnpA* and *tnpR* encode *trans*-acting proteins while IR and *res* are *cis*-acting sites on the transposon DNA.

These genetic data prompted the formulation of a model for replicative transposition of Tn3 and other

Tn3-like transposons. Briefly, mutations in the *tnpA* gene prevent transposition because the *tnpA* gene encodes the transposase TnpA, which promotes transposition. Mutations in the IR elements at the ends of the transposon also prevent transposition, because these are the sites at which the TnpA transposase acts to promote transposition.

The behavior of mutations in *tnpR* and *res* was more difficult to explain. To reiterate, *tnpR* mutations are *trans* acting and not only cause higher than normal rates of transposition but also cause the formation of cointegrates. Mutations in *res* also cause cointegrates to form but are *cis* acting and do not affect the frequency of transposition. To explain these results, the investigators proposed that *tnpR* encodes a protein with two functions. First, the TnpR protein acts as a repressor (see chapters 2 and 12), which represses the transcription of the *tnpA* gene for the transposase. By inactivating the repressor, *tnpR* mutations cause higher rates of transposition by allowing more TnpA synthesis. In addition to its role as a repressor, however, the TnpR protein acts as a recombinase that resolves cointegrates by promoting site-specific recombination between the *res* sequences in the two copies of the transposon in the cointegrate (Figure 9.13). This explains why both *tnpR* and *res* mutations cause the accumulation of cointegrates but only *tnpR* mutions can be complemented. Either type prevents the site-specific recombination that resolves the cointegrates, causing cointegrates to accumulate, but only *tnpR* mutations can be complemented because only *tnpR* encodes a diffusible gene product.

A Molecular Model for Transposition of Tn3 and Mu

The first detailed model to be developed for transposition attempted to explain all of what was known about Tn3 transposition and the transposition of other transposons such as phage Mu (see Shapiro, Suggested Reading, and Box 9.1). This model continues to be generally accepted for those types of transposons. The model incorporates the following observations, some of which have already been mentioned.

1. Whenever a transposon such as Tn3 hops into a site, a short sequence of the target DNA is duplicated. The number of bases duplicated is characteristic of each transposon. (For Tn3, 5 bp is duplicated; for IS1, 9 bp is duplicated.)
2. The formation of a **cointegrate**, in which the donor and target DNAs have become fused and encode two copies of the transposon, is an intermediate step in the transposition process.
3. Once the cointegrate has formed, it can be resolved into separate donor and target DNA molecules either by the host recombination functions or by a

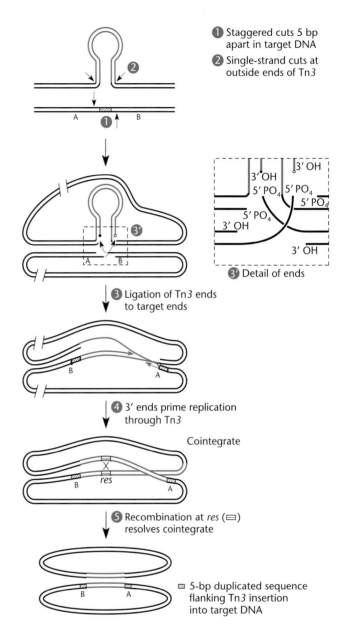

1 Staggered cuts 5 bp apart in target DNA
2 Single-strand cuts at outside ends of Tn3

3′ Detail of ends

3 Ligation of Tn3 ends to target ends

4 3′ ends prime replication through Tn3

Cointegrate

5 Recombination at *res* (▱) resolves cointegrate

▱ 5-bp duplicated sequence flanking Tn3 insertion into target DNA

Figure 9.13   Replicative transposition of Tn3 (purple) and the formation and resolution of cointegrates. At label (1 and 2) Breaks are made in the target DNA and at the ends of the transposon, respectively. (3) The 3′ OH ends of the transposon (dots) are ligated to 5′ PO₄ ends of the target DNA. The inset (3′) shows details of the ends. (4) The free 3′ ends of target DNA prime replication in both directions over the transposon to form the cointegrate. (5) The cointegrate is resolved by recombination promoted by the resolvase TnpR at the *res* sites. The A and B in the target DNA illustrate how the target DNA is reversed in step 3 for ease of drawing.

transposon-encoded **resolvase** that promotes recombination at internal *res* sequences.

4. The donor DNA and target DNA molecules both have a copy of the transposon after resolution of the cointegrate. Therefore, the transposon does not

---

BOX 9.1

Phage Mu: a Transposon Masquerading as a Phage

Phage Mu is a lysogenic phage that can integrate into the bacterial DNA after infection. However, the phage was known to be different from most lysogenic phages almost from the time it was first discovered. One unusual thing about it is that it causes random mutations when forming lysogens, giving it its name Mu, for "mutator phage." The phage causes detectable mutations because, unlike λ and other known phages, it has no unique bacterial attachment site but inserts almost randomly into the chromosome. When it happens to insert into a gene, it inactivates the gene and can cause a mutant phenotype. In contrast, λ almost always inserts at a unique site in a nonessential region between the *gal* and *bio* genes (see chapter 8); therefore, lysogenization by λ seldom causes a mutant phenotype. Other unusual properties of phage Mu were discovered when the strands of DNA in the phage heads were separated by heating, renatured, and observed under the electron microscope. These experiments were undertaken to determine whether phage Mu DNA has unique ends like T7 and λ or is cyclically permuted like T4 and P22. If the DNA has unique ends, the single-stranded DNAs will find partners that are complementary to them over their entire length, so that the renatured DNAs will be double-stranded molecules with no single-stranded ends. However, the renatured, cyclically permuted DNA molecules have a very different appearance. Each single-stranded DNA molecule usually pairs with another molecule that has different ends; most of the molecules have single-stranded ends which can pair with complementary regions in the single-stranded ends of other molecules to give very complicated, branched structures. Surprisingly, renatured Mu DNA gives neither of these patterns. Instead, the renatured Mu DNA molecules look as though they are having a bad-hair day, with single-stranded "split ends" from 500 to 2,000 bases in length. These single-stranded ends are made up of host DNA from various regions of the chromosome which are not complementary to each other. Mu has random host DNA attached to

its ends because of the way it replicates and is packaged (see Ljungquist and Bukhari, below). Mu DNA replicates by replicative transposition without resolution of the concatemers. First, it replicatively transposes to another place in the chromosome, so that it now exists in two copies in the chromosome, the original site and the new site. Rather than be resolved, each of these two Mu DNAs then replicatively transposes to yet other sites and so forth until the entire chromosome of the bacterium is riddled with hundreds of copies of the Mu DNA. These multiple copies of Mu DNA are then packaged into phage heads by making cuts in the surrounding host DNA 500 to 2,000 bp from the ends of the Mu DNA, leaving the adjacent host DNA attached to the ends. Since each Mu DNA was inserted at a different place in the chromosome, each of the packaged Mu DNA molecules will have different host DNA sequences at its ends, giving rise to "split ends" after denaturation and renaturation.

Mu uses transposition both to integrate its prophage into the host DNA to form a lysogen and to replicate its DNA during lytic development. However, it uses different transposition mechanisms for these two processes. Integration of the prophage requires a single "cut-and-paste" transposition event, while replication requires repeated rounds of replicative transposition. It is still a mystery how Mu can use these two distinct transposition mechanisms, since both require the same two Mu proteins, MuA and MuB, which make up the transposase (see the figure). Some hypotheses being tested are that the initial DNA which infects the cell has a protein attached to its ends, which allows the DNA to integrate but blocks further transposition. Another is that different regions of the MuB protein might be involved in the two different processes (see Roldan and Baker, below).

Another unusual feature of renatured Mu DNA is that it often has an unpaired region of about 3,000 bp in the middle, which forms a "bubble" when the DNA is renatured. This region of Mu DNA, called the G-segment (see the figure),

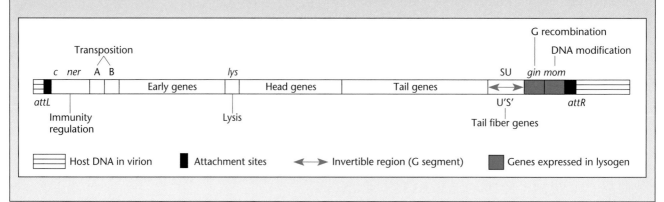

BOX 9.1 (continued)

Phage Mu: a Transposon Masquerading as a Phage

is an invertible sequence which flips around or inverts at a high frequency. If two single-stranded Mu DNAs, which have the G-segment in opposite orientations, attempt to pair, this region will loop out and form a single-stranded bubble that is clearly visible under the electron microscope. Like phage P1, which has a similar invertible segment (see the text), the host range of phage Mu is partially determined by the orientation of the G-segment, which encodes elements of the tail fibers. Recall from chapter 7 that the tail fibers help determine the host range of a phage, so that the ability of the phage to infect a particular host depends on the orientation of the G-segment. It is ironic that the G-segment may not have been observed if the phage from which the DNA had been prepared had been grown lytically in *E. coli* K-12 for many generations rather than having been induced from a lysogen. Only phage Mu with the G-segment in one orientation is able to infect *E. coli* K-12; therefore, phage with the G-segment in the other orientation is selected against when the phage is grown

in *E. coli* K-12. However, both orientations accumulate when the phage is grown as a prophage in the lysogen. When the prophage is induced, phage particles with both orientations of the G-segment are produced equally and form the bubble when the DNAs are denatured and renatured.

Because it transposes randomly into the chromosome, Mu is useful for transposon mutagenesis and for creating random gene fusions. Besides its usefulness in bacterial genetics and its general interest as a phage, Mu is an effective tool for biochemical studies of transposition because it transposes so frequently. The uses of Mu in making random gene fusions and in vivo cloning are discussed elsewhere in this chapter.

References

Ljungquist, E., and A. I. Bukhari. 1977. State of prophage Mu upon induction. *Proc. Natl. Acad. Sci. USA* **74:**3143–3147.

Roldan, L. A. S., and T. A. Baker. 2001. Differential role of MuB protein in phage Mu integration vs replication: mechanistic insights into two transposition pathways. *Mol. Microbiol.* **40:**141–156.

actually move but duplicates itself, and a new copy appears somewhere else: hence the name "replicative transposition."

5. Neither transposition nor, for some transposons, resolution of cointegrates requires the normal recombination enzymes or extensive homology between the transposon and the target DNA. Special site-specific recombinases such as TnpR resolve the cointegrate by promoting recombination between specific sites in DNA such as the *res* sites in Tn3.

Figure 9.13 shows the molecular details of the model for replicative transposition. In the first step, the transposase makes single-strand breaks at each junction between the transposon and the donor DNA and a double-strand break in the target DNA. The break in the target DNA is staggered so that the nicks in the two strands are separated by the same number of base pairs as will be duplicated in the target DNA during insertion of the transposon, as explained below. The cutting leaves two 5′ ends and two 3′ ends in the target DNA and a 5′ end and a 3′ end at each junction between the transposon and the donor DNA. The 5′ ends in the target DNA are then joined (ligated) to the 3′ ends of the transposon. Replication then proceeds in both directions over the transposon, with the free 3′ ends of the target DNA serving as primers. After replication over the transposon, the 3′ ends of the newly synthesized strands are ligated to the

remaining free 5′ ends of the donor DNA to form the cointegrate. The last step, **resolution** of the cointegrate, results from recombination between the two *res* sites in the cointegrate promoted by the resolvase of the transposon (see "S Recombinases: Mechanism" below). Resolution of the cointegrate gives rise to two copies of the transposon, one at the former (or donor) site and a new one at the target site.

This model explains why cointegrates are obligate intermediates in replicative transposition. After replication has proceeded over the transposon in both directions, the donor DNA and the target DNA are fused to each other, separated by copies of the transposon, as shown.

This model also explains why, after transposition, a short target DNA sequence of defined length has been duplicated at each end of the transposon. Because it makes a staggered break in the target DNA, the transposase causes a short region of the target DNA to be duplicated when replication proceeds from this staggered break over the transposon. The number of base pairs of target DNA duplicated at the ends of the transposon is the same as the number of base pairs between the nicks in the two strands in the staggered break and is characteristic of the transposase enzyme for each type of transposon.

Finally, this model explains why replicative transposition is independent of most host functions including DNA ligase and the normal recombination functions such as RecA (see chapter 10). The transposase cuts the target and donor DNAs and promotes ligation of the ends. Also, the normal recombination system is not needed to resolve the cointegrate into the original replicons, because the resolvase specifically promotes recombination between the *res* elements in the cointegrates. Although cointegrates can also be resolved by homologous recombination anywhere within the repeated copies of the transposon, the resolvase greatly increases the rate of resolution by actively promoting recombination between the *res* sequences.

Not all transposons that replicate by this mechanism resolve the cointegrates after they form. For example, when the Mu phage replicates itself it inserts itself around the chromosome of its bacterial host by a replicative mechanism similar to that used by Tn3 (Box 9.1). However, it does not resolve the cointegrates that form, and soon the chromosome becomes riddled with Mu genomes. These genomes are then packaged directly from the chromosomal DNA into the phage head, discarding the bacterial chromosomal DNA between the inserted Mu genomes.

Transposition by Tn*10* and Tn*5*

Further evidence indicated that not all transposons transposed by the same mechanism as Tn3 and Mu. Other transposons, represented by the composite transposons, Tn*10* and Tn*5*, transpose by a **cut-and-paste mechanism** (also known as conservative mechanism), in which the transposon is removed from one place and inserted into another as illustrated in the simplified model in Figure 9.14. In this simplified mechanism, the transposase makes double-strand breaks at the ends of the transposon, cutting it out of the donor DNA, and then pastes it into the target DNA at the site of a staggered break. When the single-strand gaps created by the nicks in the target DNA are filled in, a short sequence in the target DNA will be duplicated. For most types of transposons that replicate by a cut-and-paste mechanism, removal of the transposon from the donor DNA probably leaves breaks in the donor DNA, which is consequently degraded, as shown in the figure.

GENETIC EVIDENCE FOR CUT-AND-PASTE TRANSPOSITION

In the next few sections, we describe in detail some of the early evidence for cut-and-paste transposition by Tn*10*, and how it can be contrasted from replicative transposition.

No Cointegrate Intermediate

In Tn*10* transposition, cointegrates do not form as a necessary intermediate, as they do in the replicative mechanism. This conclusion was supported by indirect evidence. For example, there are no mutants of the transposon Tn*10* and Tn*5* that accumulate cointegrates as there are for Tn3. Moreover, even if cointegrates are formed artificially by

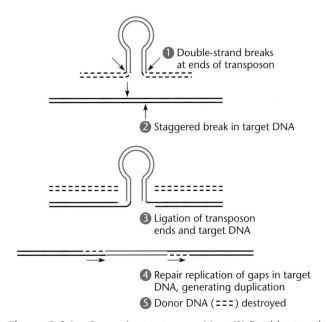

Figure 9.14 Cut-and-paste transposition. (1) Double-strand breaks are made at the ends of the transposon (arrows). (2) Staggered breaks are made in the target DNA (arrows). (3) The free 3′ ends of the transposon are ligated to the 5′ ends of the target DNA. The dashed lines represent the donor DNA, which is degraded. (4) DNA polymerase fills in the gaps of the target DNA, producing the short duplication of target DNA at the ends of the transposon. (5) Donor DNA is destroyed.

recombinant DNA techniques, there is no evidence that the cointegrates can be resolved except by the normal host recombination system. Therefore, these transposons do not seem to encode their own resolvases, which they would be likely to do if cointegrates were a normal intermediate in their transposition process.

Both Strands of the Transposon Transpose

The primary difference between replicative and cut-and-paste transposition is that in the latter, both strands of the transposon move to the target DNA. The results of genetic experiments with Tn*10* (outlined in Figure 9.15) supported this conclusion (see Bender and Kleckner, Suggested Reading).

The first step in these experiments was to introduce different versions of transposon Tn*10* into a λ suicide vector. Both of the Tn*10* derivatives contained a copy of the *lacZ* gene as well as the Tet^r gene usually carried by Tn*10*. However, one of the Tn*10* derivatives carried three missense mutations in the *lacZ* gene to inactivate it. The DNA of the two λ::Tn*10* derivatives were mixed, and the strands of the two λ DNAs were separated and reannealed. Some of the strands would reannneal with a strand of the DNA of the other derivative to make

A   Prepare λ::Tn*10*

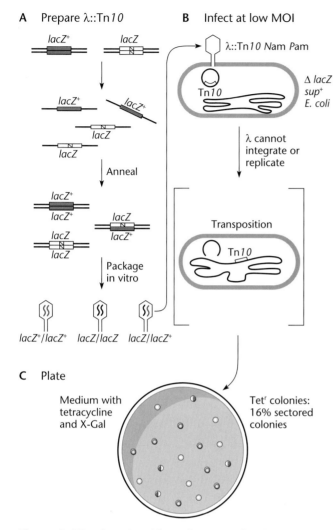

B   Infect at low MOI

C   Plate

Figure 9.15 Genetic evidence for nonreplicative transposition by Tn*10*. (A) Preparation of λ::Tn*10 lacZ/lacZ*⁺ heteroduplex DNA. (B) The λ::Tn*10* infects a nonsuppressor (sup⁺) *E. coli* host. Because λ contains *N*am and *P*am mutations, it cannot integrate or replicate, and the cells become Tet^r only if the transposon hops. If the transposon replicates during transposition, the bacteria in the Tet^r colonies will get only one or the other strand of DNA in the heteroduplex, and the colonies will be all blue or all colorless. If both strands are transferred, some colonies will be sectored, part blue and part white. See the text for further details. MOI, multiplicity of infection.

heteroduplex DNA, in which each of the strands came from a λ phage carrying a different derivative of Tn*10*. Consequently, these heteroduplex DNAs had one strand with a good copy of *lacZ* and another strand with the mutated copy of *lacZ*. In the next step, the heteroduplex DNA was packaged into λ heads in vitro (see chapter 8) and used to infect Lac⁻ *E. coli* cells. Because this λ was a suicide vector, the only cells that became Tet^r were ones

in which the Tn*10* derivatives had hopped into the chromosome. If the transposition had occurred by a replicative mechanism, the Tet^r colonies would have contained either Lac⁺ or Lac⁻ bacteria (Figure 9.16), since the information in only one of the two strands could have been transferred. If, however, the transposition had occurred by a nonreplicative cut-and-paste mechanism, both strands of the Tn*10* from a heteroduplex would have hopped into the chromosome some of the time, so that one of the strands would have the good copy of the *lacZ* gene and the other would have the *lacZ* gene with the mutations. When these heteroduplex DNAs replicated, they would give rise to both Lac⁺ and Lac⁻ bacteria in the same colony, making "sectored" blue-and-white colonies on 5-bromo-4-chloro-3-indolyl-β-D-galactopyranoside (X-Gal) plates as shown in Figure 9.15. In the experiment, about 16% of the colonies were sectored blue and white, supporting the conclusion that both strands were transferred.

Transposon Leaves the Donor DNA

A major difference between replicative and simple cut-and-paste transposition is the number of copies of the transposon created by transposition. The replicative mechanism creates two copies while the cut-and-paste mechanism creates only one, but in a different place, and the transposon is lost from its original location. It might seem easy to determine whether a copy of the transposon still exists in the donor DNA after transposition; however, it is usually difficult. For example, if the donor DNA containing the transposon is a multicopy plasmid and the transposition occurs by a cut-and-paste mechanism, only one copy of the plasmid loses its transposon, leaving many plasmids intact with the transposon. Even if the transposition occurs from a DNA that normally exists in a single copy, such as the chromosome or a single-copy plasmid, a copy of the transposon remaining in the donor DNA could be attributed to replication of the donor DNA prior to transposition.

For much the same reasons, it is difficult to determine whether the donor DNA is resealed after the transposon is cut out of it during cut-and-paste transposition. It might seem that the donor DNA must be resealed after the transposon hops; otherwise, transposition would leave a double-strand break in the donor DNA, which would be a lethal event in many cases. However, cells usually carry more than one copy of a plasmid, and even the chromosome is usually in a partial state of replication, so that many regions exist in more than one copy per cell. After transposition, the double-strand break left in the donor DNA could then be repaired by recombination with the other daughter DNA in a process called double-strand break repair (see chapter 10). Ironically,

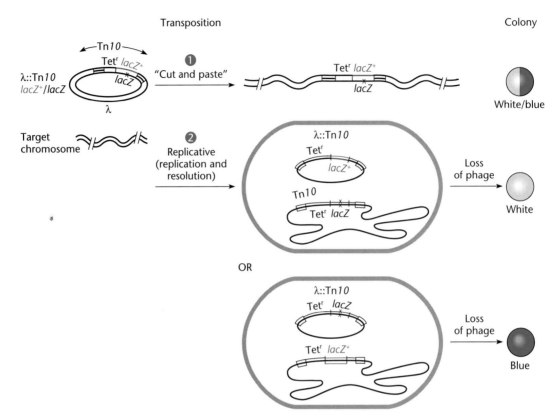

Figure 9.16   Comparison of the results predicted in the experiment in Figure 9.15 if transposition of Tn*10* is by a cut-and-paste mechanism (1) or by a replicative mechanism (2). See the text for details.

since the daughter DNA still contains the transposon, the double-strand break repair restores the transposon to its original donor site, making the transposition appear replicative even though it is not. Even if the transposon were to be cut out of an unreplicated region of the chromosome and this donor DNA were left unrepaired and the cell died, this would be still be difficult to detect. Transposition events are infrequent, and a few dead cells would go undetected in a large population of bacteria.

The best evidence that the donor DNA is not resealed or, at least, is not resealed correctly after transposition by cut-and-paste transposons such as Tn*10* and Tn*5* comes from reversion studies. In a revertant, the sequence of the DNA has returned to its original sequence (see chapter 3). If the original mutation is due to insertion of a transposon, for example into a *his* gene, the gene will be disrupted and the cell will be His⁻, unable to grow without histidine in the medium. Revertants can be easily detected by plating on media without histidine, on which only revertants multiply to form a colony. For a transposon insertion mutation to revert, the transposon must be

completely removed from the DNA in a process called **precise excision**. Not a trace of the transposon can remain, including the duplication of the short target sequence, or the gene would probably remain disrupted and nonfunctional.

If the transposon were precisely excised and the donor DNA were resealed every time a transposon hopped by a cut-and-paste mechanism, transposon insertion mutations would revert every time the transposon hopped. However, reversion of insertion mutations occurs at a much lower rate than does transposition itself, suggesting that the insertion mutation does not revert every time the transposon hops. Moreover, mutations in the transposon itself that inactivate the transposase and render the transposon incapable of transposition do not further lower the reversion frequency, as might be expected if the few revertants that are seen resulted from a transposition event. Presumably, the rare precise excisions that cause transposon insertion mutations to revert are due to homologous recombination between the short duplicated target sequences bracketing the transposon and are unrelated to transposition itself. Therefore, for cut-and-paste transposons such as Tn*10* and Tn*5*, the donor DNA is apparently left broken after the transposon is cut out of it, as shown in Figure 9.14.

Details of Transposition by the DDE Transposons

All of the transposons we have discussed so far are considered **DDE transposons,** because their transposases all have two aspartates (D) and one glutamate (E) (see inside cover) that are essential for their activity. These acidic amino acids are not next to each other in the polypeptide, but they are together in the active center when the protein is folded. Their job is to hold (by chelation) two magnesium ions (Mg^{2+}) that participate in the cleavage of phosphodiester bonds in the DNA during the transposition event. A similar structure is found for some other related enzymes such as the human immunodeficiency virus integrase, the RAG-1 protein responsible for generating antibody diversity in vertebrates, and RuvC, the enzyme that cuts Holliday junctions during recombination (see chapter 10). However, the details of how many DNA strands are cut and the fate of the ends are different for the different enzymes.

Details of the Mechanism of Transposition by Tn*5* and Tn*7*

The mechanism of transposition of the DDE transposon Tn*5* has been studied extensively and is illustrated in Figure 9.17 (see Reznikoff, Suggested Reading). The first step is the binding of one copy (monomer) of the transposase to each of the ends of the transposon in the donor DNA. The two monomers then bind each other through dimerization domains in their carboxy termini to bring the two ends of the transposon together (synapsis). Then the transposase bound to one end of the transposon cuts the DNA at the other end and vice versa to leave 3′ OH ends at each end of the transposon. These activated 3′ OH ends attack the phosphodiester bond on the other strand, forming 3′-5′ phosphodiester hairpins, as shown. This cuts the transposon out of the donor DNA. When the transposase binds to the target DNA, it cuts the two hairpin ends again and the 3′ OH ends attack phosphodiester bonds 9 bp apart in the target DNA, cutting them, and the 5′ phosphate ends in the target DNA are joined to the 3′ OH ends in the transposon, inserting the transposon into the target DNA. The 9-bp single-stranded gaps on each side of the transposon are then filled in by DNA polymerase to make the 9-bp repeats in the target DNA, characteristic of the Tn*5* transposon.

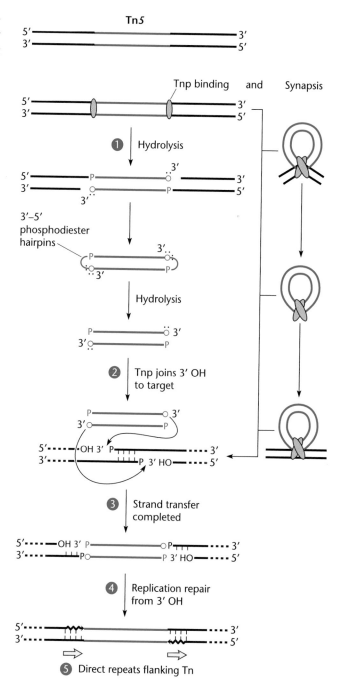

Figure 9.17 Mechanism of transposition by Tn*5*. Single copies of the transposase (TnP) bind to each end of the tranposon and then bind to each other, bringing the two ends of the transposon together (synapsis). Cutting and rejoining reactions then cut the transposon out of the donor DNA and join the 3′ and 5′ ends of the DNA to form hairpins. These hairpins are cut and simultaneously attack phosphodiester bonds 9 bp apart in the target DNA. Replication and rejoining reactions insert the transposon into the target DNA (for details, see the text).

RELATIONSHIP BETWEEN REPLICATIVE AND CUT-AND-PASTE TRANSPOSITION

Even though replicative transposition and cut-and-paste transposition by DDE transposons seem different, they are actually mechanistically related. When one compares Figures 9.13 and 9.14, the major difference is in the number of strand cuts made by the transposase enzyme in the junction between the transposon and the donor DNA. A cut-and-paste transposase makes cuts in both strands in the junction, whereas a replicative transposase cuts only one strand at the junction. Otherwise, the two mechanisms are similar. In both, the cut 5′ ends of the target DNA are joined to the free 3′ ends of the transposon. In both mechanisms, the free 3′ ends of the target DNA are then used as primers for replication that proceeds until a free 5′ end in the donor DNA is reached. Then the newly replicated DNA is joined to the target DNA. The only difference is whether the replication has to proceed over the entire transposon (replicative) or whether it has to proceed only over the short region of the target DNA that is duplicated (cut and paste). If the DNA has to replicate over the entire transposon, the transposon is duplicated and a cointegrate is created; otherwise, only a short region in the target DNA is duplicated and no cointegrate is created.

A dramatic confirmation of the similarity between the cut-and-paste and replicative mechanisms of transposition by DDE transposons came with the demonstration that the cut-and-paste transposon Tn7 can be converted into a replicative transposon by a single amino acid change in one subunit of the transposase (see May and Craig, Suggested Reading). Transposon Tn7 normally transposes by a cut-and-paste mechanism in which different subunits of the transposase make the cuts in the opposite strands of DNA at the ends of the transposon. This is illustrated in Figure 9.18. If the TnsA subunit that makes the cut that leaves the 5′ hydroxyl end is altered by a mutation, the transposase will cut only the other strand, leaving a free 3′ OH like a replicative transposase. The Tn7 transposon with such an altered transposase then transposes by the replicative mechanism, forming a cointegrate. Apparently, the transposase need only make the appropriate cuts and joinings, and the replication apparatus of the cell does the rest. It is somewhat surprising that the same transposase enzyme can support both types of transposition, since the transposon presumably needs different cellular replication machineries for each. Replication of tens of thousands of base pairs during replicative transposition would be a much more involved process than replication of a few base pairs during cut-and-paste transposition.

Other DDE transposons use a mechanism of transposition that cannot be described as either strictly

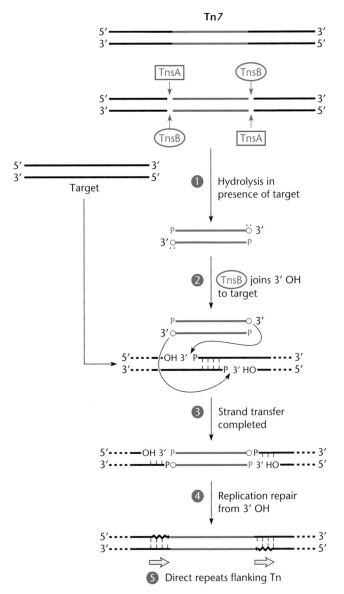

Figure 9.18 Transposition by Tn7. The TnsA and TnsB proteins are required for cleavage at the ends of the transposon. TnsA cuts at the 5′ end, and TnsB cuts at the 3′ end. They cut the donor DNA only in the presence of the target. For details, see the text.

replicative or cut-and-paste mechanisms. For example, the mechanism of transposition of the IS elements IS2 and IS3, as well as IS911, has features of both replicative and cut-and-paste transposition. Basically, one strand of the transposon is cut out of the donor DNA. The ends of this strand are then joined to form a single-stranded circle (not illustrated), and the circular strand is replicated to form the double-stranded circular transposon. This double-stranded circular transposon then attacks the target DNA, integrating itself. The strand cut out of the donor DNA is also replaced by replication, leaving a copy of the transposon in the donor DNA. Thus this transposition is replicative because a copy of the transposon appears in the target DNA but the donor DNA retains the transposon. However, a cointegrate does not form and the transposon is, in a sense, cut out of the donor DNA and pasted into the target DNA.

As an aside, while DDE transposons use a number of different mechanisms to transpose, they all share one feature: they all have some way of protecting the 5′ ends of the transposon DNA after it is cut out of the donor DNA. For the replicative ones, this is not a problem because only the 3′ OH ends of the transposon are exposed. In Tn5 (and Tn10), a hairpin is formed at the ends of the cut-out transposon, so that the 5′ ends are not exposed. In IS2, IS3, and IS911, a circular DNA is formed from the single strand of the transposon after it has been cut out, thereby protecting the ends. The exception would seem to be Tn7, where the linear transposon is cut out of the donor DNA and no hairpin seems to form at the ends (Figure 9.18). However, Tn7 does not cut itself out of the donor DNA unless the target DNA is already bound to the transposase, so that the cutting and joining reactions are coordinated and the 5′ ends are not left exposed for long (Figure 9.18). All these mechanisms might reflect the necessity of protecting the 5′ ends of the free transposon from degradation by the RecBC nuclease, which degrades linear DNA from the 5′ ends, as discussed in chapter 10.

Rolling-Circle Transposons

Not all transposons transpose by a strand exchange mechanism like that used by the DDE transposons. Other transposons, represented by IS91, use a rolling-circle mechanism to transpose themselves into a target DNA. Rather than having the motif DDE, their transposase has two essential tyrosines in its active center, hence the name **Y2 transposons** or **rolling-circle transposons**. We have encountered this form of replication in previous chapters as the mechanism of replication of some plasmids (see chapter 4), phage DNAs (chapters 7 and 8), and the mechanism of strand displacement during DNA transfer in conjugation (see chapter 5). In all of these cases, the responsible protein has a tyrosine to which the 5′ phosphate at the end of the DNA is covalently joined during the replication or transfer process. Figure 9.19 illustrates the difference between DDE transposons and rolling-circle or Y2 transposons.

The structure of Y2 transposons is also very different from that of DDE transposons, reflecting their very different mechanism of transposition. They do not have inverted repeated sequences at their ends, nor do they duplicate a target DNA sequence during integration. The details of how they transpose are not completely known, but they basically cut one strand of the DNA close to one end of the transposon (called the *ori* end in analogy to the *ori* sequence of RC plasmids) and attach the 5′ phosphate at the cleavage site to one of the tyrosines in the active center of the transposase. The free 3′ OH end then serves as a primer to replicate over the transposon, ending at the other end of the transposon called the *ter* end. The displaced old strand of the transposon enters the target DNA, and its complementary strand is synthesized in the target DNA so that both the donor and target DNAs end up with a copy of the transposon. This is therefore a form of replicative transposition. It is not clear when the target DNA is cut and invaded in this process, nor is the exact role of the two tyrosines known, since the other examples of rolling-circle replication mentioned above only require one tyrosine in the active center. The free 5′ phosphate ends created at the two ends of the transposon are presumably shuttled between the two tyrosines to allow replication back over the transposon to create two copies of the transposon. Note however, that the free 5′ ends of the transposon are being protected by being attached to one of the tyrosines in the transposon during this process, as they are during the various forms of DDE transposition.

It is becoming increasingly clear that a number of antibiotic resistance genes found on integrons are carried on Y2 transposons related to IS91. They were first identified because of common sequence elements, which led them to be called **ISCR elements** for IS common regions. Therefore, rolling-circle transposons are another common way in which antibiotic genes move from one bacterium to another (see Toleman et al., Suggested Reading).

Y and S Transposons

Other transposon-like DNA elements exist that use neither a DDE transposase nor an RC transposase to transpose. These are sometimes called Y and S transposons because they have either a single essential tyrosine (Y) or serine (S) in the active center of their transposase. However, these transposases are more akin to integrases than they are to transposases, even though they often show less specificity in their integration sites. Accordingly, they are discussed along with integrases and other recombinases in the next section on site-specific recombinases. Figure 9.19 gives an overview of all of the known types of transposons, and Table 9.1 summarizes some of their distinctive properties.

General Properties of Transposons

There are some properties that are shared by many types of transposons, even if they differ in their mechanism of transposition.

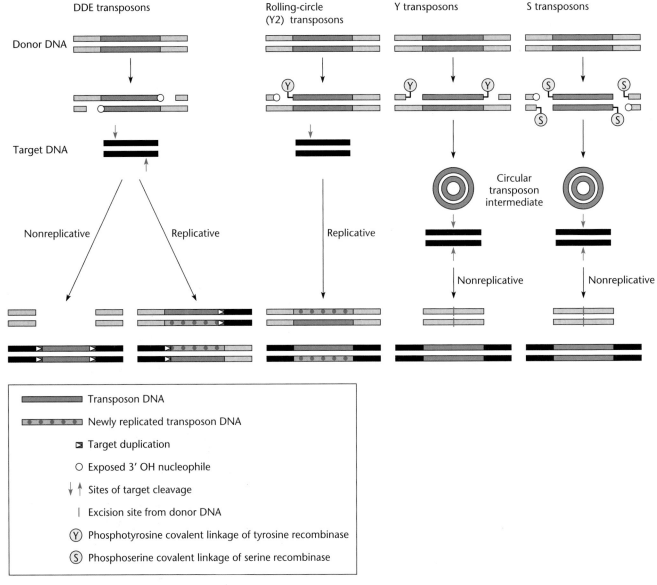

Figure 9.19 Comparison of the known mechanisms of transposition in bacteria. They differ in in the initial strand cleavage, whether and how the transposon DNA is transiently attached to the transposase, the role of DNA replication, the existence of circular intermediates, whether a target site is duplicated, and the fate of the donor DNA. The Y transposons and S transposons use mechanisms more akin to Y and S recombinases.

Target Site Specificity

While the transposition of some elements seems almost totally random, no transposable element inserts completely randomly into target DNA. Most transposable elements show some target specificity, hopping into some sites more often than into others. Even Tn5 and Mu, which are famous for hopping almost at random, prefer some sites to others, although the preference is weak.

Tn7 is the extreme case of a transposon with target specificity. It transposes with a high frequency into only

one site in the *E. coli* DNA, called *att*Tn7. Recent studies have provided insights into how this selectivity is achieved. The Tn7 transposition machinery consists of five proteins, TnsA, TnsB, TnsC, TnsD, and TnsE. Of these, TnsA and TnsB make up the transposase that cuts and joins the DNA strands (Figure 9.18) and the other proteins play ancillary roles. The role of TnsD may be to direct the transposon to its target sequence, *att*Tn7. By binding to the *att*Tn7 sequence, TnsD may induce changes representative of triple-stranded structures in

TABLE 9.1	Characteristics of transposon families			
	Characteristic			
Family	**Active-site category**	**Protein-DNA covalent linkage**	**Target duplication**	**Examples**
DDE transposons	DDE	No	Yes	Tn3
				Tn5
				Tn7
				Tn10
				Mu (see Box 9.1)
Rolling-circle/Y2 transposons	YY (related to φX174 A protein and to conjugative plasmid relaxases)	Yes, to 5'-P	No	IS91
Y transposons	Y recombinase	Yes, to 3'-P	No, but flanking "coupling sequences" are transferred to one side of target	Tn916 (see Box 5.4)
S transposons	S recombinase	Yes, to 5'-P	No	IS607 from Helicobacter pylori

the attTn7 site. This directs TnsC to stimulate transposition into the site. In the absence of TnsD, TnsE stimulates transposition into other sites in the chromosome. This transposition is inefficient but random. It is possible that this random transposition is occurring at the replication fork where replication intermediates on the lagging strand serve as recognition sites for the TnsD subunit of the Tn7 transposase. Having two options might serve Tn7 well. It transposes with high efficiency into its normal site just downstream of the glmS gene when it enters a cell where this site is unoccupied. It can also find its site in many different types of cells, since the glmS gene is highly conserved, its product performing an important step in cell wall biosynthesis. Moreover, insertion into this site downstream of glmS has no effect on the cell since it does not disrupt the gene iself, only its transcription termination site, a function that the transposon then provides. However, the mobility of Tn7 would be restricted if it could hop only into this site. In order to move from one cell to a cell that lacks a natural transformation system, it would have to hop into a conjugative element, e.g., a self-transmissible or mobilizable plasmid; however, plasmids do not normally contain the glmS gene with the attTn7 site. By having an alternative mechanism that allows it to transpose more randomly, albeit with low efficiency, it is able sometimes to hop into conjugative elements and transfer itself into other bacteria.

Effects on Genes Adjacent to the Insertion Site

Most insertion element and transposon insertions cause polar effects if they insert into a gene transcribed as a polycistronic mRNA. The inserted element contains transcriptional stop signals and may also contain long stretches of sequence that are transcribed but not translated. The latter may cause Rho-dependent transcriptional termination.

Some insertions may enhance the expression of a gene adjacent to the insertion site. This expression can result from transcription that originates within the transposon. For example, both Tn5 and Tn10 contain outward-facing promoters near their termini, and these promoters can initiate transcription into neighboring genes.

Regulation of Transposition

Transposition of most transposons occurs rarely, as discussed above, because transposons self-regulate their transposition (see Gueguen et al., Suggested Reading). The regulatory mechanisms used by various transposons differ greatly. We have discussed how, in Tn3, the TnpR protein represses the transcription of the transposase gene. For some transposons such as Tn10, transposition occurs very rarely and then primarily just after a replication fork has passed through the element. Newly replicated E. coli DNA is hemimethylated at GATC sites (see chapter 1), and hemimethylated DNA not only activates the transposase

Tn5

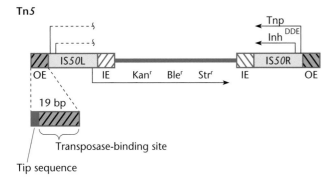

Figure 9.20 Regulation of Tn5 transposition. Two similar IS50 elements flank the antibiotic resistance genes. Only IS50R encodes the transposase Tnp and the inhibitor which is an N terminally truncated version of itself. Also, Dam methylation of the inside ends (IE) of the IS50 elements prevents the transposase from cutting these ends and transposing the individual IS50 elements. For details, see the text. OE, outer ends; Inh, inhibitor of transposase Tnp. The dashed lines indicate that Tnp and Inh made from Is50L are defective.

promoter of Tn10 but also increases the activity of the transposon ends. Also, the translation of the transposase gene of Tn10 is repressed by an antisense RNA. The transposase of Tn5, which is not very active anyway, uses a truncated version of the transposase to inhibit the active transposase. As illustrated in Figure 9.20, the translation of the truncated transposase is initiated at an internal translational initiation region (TIR) so that it lacks the N terminus but has the C terminus involved in dimer formation. When this defective transposase pairs with the normal transposase, transposition is inhibited. Most transposons employ these and similar mechanisms to modulate the level of transposase transcription and/or translation as well as the level of catalysis.

Target Immunity

Another feature of some transposons is that they prefer not to hop close in the DNA to another transposon of the same type. This is called target site immunity because DNA sequences close to a transposon in the DNA are relatively immune to insertion of another copy of the same transposon. The immunity can extend over 100,000 bp of DNA, although its reach varies among different types of transposons. The advantages of target site immunity to the cell, and therefore to the transposon, seem obvious. If two transposons were to insert close to each other, the resolution of the two copies by the transposon resolvase or homologous recombination between the two copies of the transposon would cause large deletions and often lead to death of the cell. Also,

the presence of two transposons close to each other can cause instability in the chromosome due to mechanisms like those described above in the section on inside-end transposition.

Target site immunity is limited to only some transposons of the types we have discussed. Only the Mu, Tn3 (Tn21), and Tn7 families of transposons are known to exhibit target site immunity. While not completely understood, it does seem to be related to the binding of other proteins in the transposase complex to the transposase. In Mu, where target site immunity has been most extensively studied, the MuB protein seems to be indirectly responsible for the immunity. The binding of MuB to a DNA makes it a target for the MuA transposase, which then promotes transposition into that DNA. The binding of MuA then causes MuB to dissociate from the DNA. Once a transposon has inserted, a copy of MuA may remain bound to the end of the inserted transposon. This bound MuA may then prevent the binding of other MuB to the same target DNA and prevent other transposition into that DNA. A similar mechanism may explain target site immunity by Tn7, but now the responsible proteins are TnsB and TnsC rather than MuA and MuB.

Transposon Mutagenesis

One of the most important uses of transposons is in transposon mutagenesis. This is a particularly effective form of mutagenesis because a gene that has been marked with a transposon is relatively easy to map by genetic crosses or by physical mapping with restriction endonucleases or PCR. Furthermore, genes marked with a transposon are also relatively easy to clone by using plate hybridizations or by selecting for selectable genes carried on the transposon.

Not all types of transposons are equally useful for mutagenesis. A transposon used for mutagenesis should have the following properties:

1. It should transpose at a fairly high frequency.
2. It should not be very selective in its target sequence.
3. It should carry an easily selectable gene, such as one for resistance to an antibiotic.
4. It should have a broad host range for transposition if it is to be used in several different kinds of bacteria.

Transposon Tn5 is ideal for random mutagenesis of gram-negative bacteria because it embodies all of these features. Not only does Tn5 transpose with a relatively high frequency, but also it has almost no target specificity and transposes in essentially any gram-negative bacterium. It also carries a kanamycin resistance gene that is expressed in most gram-negative bacteria. Figure 9.21A illustrates a popular method for transposon mutagenesis of gram-negative bacteria other than *E. coli* (see Simon et al. 1983, Suggested Reading). In addition to the broad host range of Tn*5* and the promiscuity of RP4, this method takes advantage of the narrow host range of ColE1-derived plasmids, which replicate only in *E. coli* and a few other closely related species. Phage Mu is another transposon-like element that can hop in many types of gram-negative bacteria and shows little target specificity (Box 9.1). Equally universal methods are not available for transposon mutagenesis in gram-positive bacteria. No transposons of gram-positive bacteria have

been identified that fulfill all the criteria above, although Tn*917* hops fairly randomly in some gram-positive bacteria. Also, Tn*916* has the advantages that it transfers itself from one cell to another and shows little target specificity although it does not integrate in many gram-positive bacteria. As mentioned, these are not strictly transposons but, rather, integrate by integrases much like phage genomes. However, they do hop fairly randomly. With recently developed techniques such as in vitro transposon mutagenesis, it is now possible to perform transposon mutagenesis in many bacteria for which useful transposons are not available (Box 9.2).

Figure 9.21 Transposon Tn*5* mutagenesis. (A) A standard protocol for transposon mutagenesis of gram-negative bacteria. A suicide ColE1-derived plasmid containing a *mob* site whose relaxase recognizes the coupling protein of the promiscuous plasmid RP4 and contains transposon Tn*5* is mobilized into the bacterium by the products of the RP4 transfer genes, which are inserted in the chromosome. The transposon hops into the chromosome of the recipient cell, and the ColE1 plasmid is lost because it cannot replicate. The Tn*5* transposon is shown in purple. (B) Random transposon mutagenesis of a plasmid. In step 1, transposon Tn*5* is introduced into cells on a suicide vector. In step 2, the culture is incubated, allowing the Tn*5* time to hop, either into the chromosome (large circle) or into a plasmid (small circle). Plating on kanamycin-containing medium results in the selection of cells in which a transposition has occurred. In step 3, plasmid DNA is prepared from Kan^r cells and used to transform a Kan^s recipient. In step 4, selection for Kan^r allows the identification of cells that have acquired a Tn*5*-carrying plasmid.

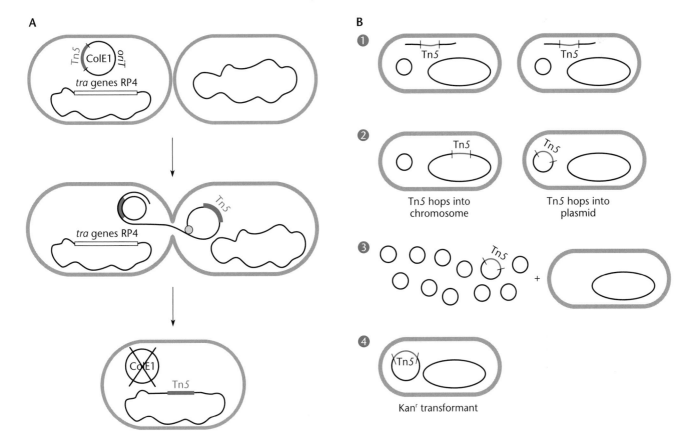

BOX 9.2

Transposon Mutagenesis In Vitro

While in vivo transposon mutagenesis is a very useful technology, it does have some limitations. One of the limitations is that it is necessary to introduce the transposon on a suicide vector, which may give some residual false-positive results for transposon insertion mutants if the suicide vector is capable of limited replication. Another limitation is that it is not very efficient and requires powerful positive selection techniques to isolate the mutants. Another limitation occurs if a specific plasmid or other smaller DNA sequence is to be mutated. There is no target specificity to the insertion mutants, and so most of the time the transposon hops into the chromosome; those few with transposon insertions in the smaller target DNA must be found among the myriad of mutants in the chromosome. Yet another limitation is that the target DNA to be mutated must be a replicon in the cell in which the mutagenesis is performed. There is also the possibility of multiple transposition events. Finally, useful transposons have not been found for most bacteria.

In vitro transposon mutagenesis avoids many of these limitations. This technology is made possible by the fact that the transposase enzyme by itself performs most of the reactions of the "cut-and-paste" transposition reaction. It cuts the DNA both at the outside ends of the inverted repeats on the transposon and in the target DNA and joins the ends to each other. Therefore, if the target DNA is mixed with a donor DNA containing the transposon and the purified transposase is added, the transposon will insert into the target DNA. Such techniques have been developed for derivatives of transposon Tn5, Tn552, Mu, Tn7, *mariner*, and others. Each of these transposons has its own advantages. For example, mutants of the Tn5 transposase are available which enhance the transposition frequency, which is necessary because the wild-type Tn5 transposase is essentially inactive. Also, only the sequences at the ends of the inverted repeats of Tn5 are needed; these are only 19 bp long. One disadvantage of Tn5 is that the transposase remains attached to the transposon after transposition and must be removed by denaturation with phenol or detergent. Apparently, a host enzyme is required to remove the transposon from the DNA which is not present in the purified system. The *mariner* transposon, which comes from a horn fly, not a bacterium, has the advantage that it requires no host functions, making it very popular to mutagenize a wide variety of cells. Many derivatives have been made, including one obtained by incorporating an *E. coli* origin of replication to facilitate the cloning of genes mutated by *mariner*. Once it is mutagenized, the target DNA can be introduced into cells by whatever means are available, and those with transposon insertions can be selected for just as in in vivo transposon mutagenesis. It helps if the transposon has been engineered to lack a transposase gene so that it does not hop in subsequent generations or cause genetic instability such as deletions once it is in the chromosome. The target DNA can be either a replicon, such as a plasmid that replicates in the recipient cell, or random linear pieces of the chromosomal DNA of the recipient if it is being introduced into cells that can be transformed with linear DNA (see chapter 6). The linear pieces recombine with the chromosome and replace the chromosomal sequence with the sequence mutated with the transposon. This offers a way of doing random chromosomal transposon mutagenesis of bacteria for which no transposon mutagenesis system is available (see, for example, Gering et al., below). Alternatively, the transposase can be expressed in the cells to be mutagenized and the mutagenesis can be done essentially in vivo (see Zhang et al., below).

Another variation of this method for doing transposon mutagenesis, which can be applied to mutagenize the DNA of almost any bacterium and even eukaryotic cells, is to use "transpososomes" (see Goryshin et al., below). A transpososome is a transposon to which the transposase protein is already attached so that it does not have to be made in the cell. This latter feature is important because the transposase gene on the transposon might not be expressed in a distantly related bacterium and certainly not in a eukaryotic cell. As in other methods of transposon mutagenesis, the transposon should carry a selectable gene that is expressed in the cell to be mutagenized. Transpososomes based on Tn5 are made by running the transposition reaction in vitro in the absence of magnesium ions. Under these conditions, the ends of the transposon in the donor DNA are not cut but if the transposon has already been cut out of the donor DNA by some other process (for example, with restriction nucleases), the transposon binds to the ends and remains attached, forming the transpososome. When the transpososome is electroporated into the cells, the transposase attached to the transposon catalyzes the DNA strand exchanges required for transposition of the transposon into the chromosome or other cellular DNA. The transposase enzyme introduced with the transposon is soon degraded, preventing further transposition.

References

Gering, A. M., J. N. Nodwell, S. M. Beverley, and R. Losick. 2000. Genomewide insertional mutagenesis in *Streptomyces coelicolor* reveals additional genes involved in morphological differentiation. *Proc. Natl. Acad. Sci. USA* 97:9642–9647.

Goryshin, I. Y., J. Jendrisak, L. M. Hoffman, R. Mais, and W. S. Reznikoff. 2000. Insertional transposon mutagenesis by electroporation of released Tn5 transposition complexes. *Nat. Biotechnol.* 18:97–100.

Zhang, J. K., M. A. Pritchett, D. J. Lampe, H. M. Robertson, and W. W. Metcalf. 2000. *In vivo* transposon mutagenesis of the methanogenic archaeon *Methanosarcina acetivorans* C2A using a modified version of the insect *mariner*-family transposable element *Himar1*. *Proc. Natl. Acad. Sci. USA* 97:9665–9670.

Transposon Mutagenesis of Plasmids

One common use of transposon mutagensis is to identify genes on large clones on a plasmid. If the transposon hops into a gene on the plasmid, it will disrupt the gene. Mapping the site of insertion of the transposon then allows the location of the gene to be determined. The relatively small size of plasmids makes physical mapping easier, and it is often fairly easy to isolate large numbers of transposon insertions in a plasmid.

Figure 9.21B illustrates the steps in the selection of plasmids with transposon insertions in *E. coli*. A suicide vector containing the transposon (Tn*5* in the example) is introduced into cells harboring the plasmid. Cells in which the transposon has hopped into cellular DNA, either the plasmid or the chromosome, are then selected by plating on medium containing the antibiotic to which a transposon gene confers resistance, in this case kanamycin. Only the cells in which the transposon has hopped to another DNA become resistant to the antibiotic, since the transposons that remain in the suicide vector are lost with the suicide vector. In most of the antibiotic-resistant bacteria, the transposon will have hopped into the chromosome rather than into the plasmid, simply because the chromosome is the larger target. The plasmids in these bacteria are normal. However, the plasmids can be isolated from the few bacteria in which the transposon has hopped into the plasmid, by mating the plasmid into another *E. coli* strain and selecting the antibiotic resistance on the transposon if the plasmid being mutagenized is self-transmissible. Alternatively, the antibiotic-resistant colonies that have the transposon either in the plasmid or in the chromosome can be pooled and the plasmids can be isolated from them by one of the procedures outlined in chapter 4. This mixture of plasmids, most of which are normal, is then used to transform another strain of *E. coli*, selecting for the antibiotic resistance gene on the transposon (the kanamycin resistance gene in the example). The antibiotic-resistant transformants should contain the plasmid with the transposon inserted somewhere in it. Voilà, in a few simple steps, plasmids with transposon insertion mutations have been isolated. This method can be used to randomly mutagenize a DNA cloned in a plasmid or to mutagenize the plasmid itself.

PHYSICAL MAPPING OF THE SITE OF TRANSPOSON INSERTIONS IN A PLASMID

Once plasmids with transposon insertion mutations have been isolated, it is fairly easy to map the sites of the insertions relative to the positions of known restriction sites on the plasmid. This mapping relies on the fact that the size and number of DNA fragments obtained with a restriction endonuclease are dependent on the number and location of the sites for that restriction endonuclease in the DNA being cut. Insertion of the transposon into the plasmid introduces new restriction sites and changes the sizes of some of the fragments.

As an example, consider the small plasmid pAT153 and the transposon Tn*5* (Figure 9.22). Use of a small plasmid such as pAT153 makes interpretation of the physical mapping data easier, but the same general methods are applicable to much larger plasmids.

To locate the site of insertion, we need to use two different restriction endonucleases, both of which cut in the transposon and the original plasmid. The restriction endonucleases PstI and HindIII fulfill this requirement (Figure 9.22). Plasmid pAT153 is 3.6 kb long and has only one HindIII site and one PstI site. Tn*5* is 5.6 kb long and has two HindIII sites and four PstI sites. Note that both HindIII and PstI cut equal distances from the ends of the transposon. HindIII cuts 1.15 kb from both ends, while PstI cuts 0.6 kb from the ends. They cut equal distances from the ends because the ends of the transposon have almost identical copies of the IS*50* element in the inverse orientation (Figure 9.3).

Cutting the original pAT153 plasmid without the transposon insertion with either HindIII or PstI should give one band of 3.6 kb. However, the plasmid with the transposon inserted is much larger (3.6 kb + 5.6 kb = 9.2 kb) and also contains restriction sites introduced on the transposon; therefore, cutting this larger molecule with HindIII should yield three bands and cutting it with PstI should yield five bands, because of the HindIII and PstI sites in the transposon. With each restriction endonuclease, the fragments should add up to 9.2 kb. Some of these fragments will always be the same size, regardless of where the transposon has inserted in the plasmid. These are called the **internal fragments** because they come from within the transposon and contain only transposon DNA. Cutting with HindIII leaves one internal fragment of 3.3 kb, while cutting with PstI leaves internal fragments of 1.08, 0.92, and 2.40 kb. However, two fragments in each case extend from a restriction site in the transposon to one in the plasmid and so contain both transposon and plasmid DNA. The sizes of these fragments, called the **junction fragments,** vary depending on where the transposon has inserted in the plasmid. From the sizes of the junction fragments, we can determine where the transposon inserted in the plasmid.

The representative data in Figure 9.23 illustrate how the site of insertion of the Tn*5* transposon can be determined for a particular insertion mutation. To obtain the pattern in Figure 9.23A, the plasmid with the transposon insertion was digested separately with HindIII and PstI and the fragments were applied to an agarose gel,

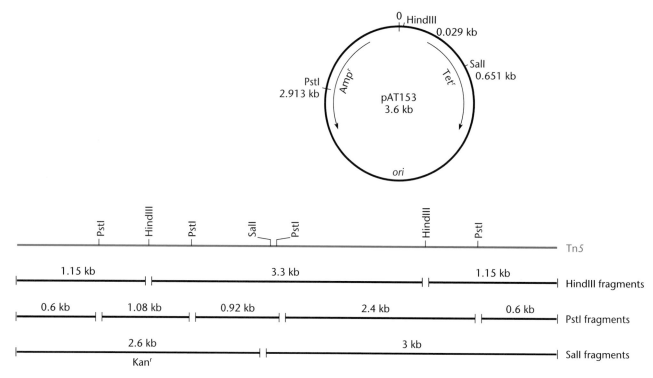

Figure 9.22 Restriction maps of plasmid pAT153 and transposon Tn5. The positions of the PstI, SalI, and HindIII sites are shown.

along with marker DNA fragments of known size. During electrophoresis, smaller fragments move faster than larger fragments and the rate of migration decreases exponentially with size (see chapter 1). Therefore, on semilog paper, a plot of distance versus size for the marker DNAs should give a standard curve approximating a straight line in the region of interest, as shown in Figure 9.23B. From this standard curve, we can estimate the sizes of the unknown fragments by plotting the distance they travel and reading over to obtain their sizes. Thus, according to Figure 9.23B, the three HindIII fragments are approximately 4.60, 3.45, and 1.40 kb, adding up to 9.45 kb, which is close enough to the expected 9.2 kb. We estimate the five PstI fragments to be 3.40, 2.20, 1.60, 1.00, and 0.87 kb, which add up to 9.07 kb, again close enough to 9.2 kb. Therefore, all the restriction fragments seem to be present and accounted for.

Our next step is to identify the internal fragments in each digestion, since the remaining fragments are the junction fragments. After HindIII digestion, the internal fragment of 3.30 kb is probably the HindIII fragment, which we have estimated to be 3.45 kb. That leaves the 4.60- and 1.40-kb fragments to be the junction fragments. Similarly, the internal PstI fragments that should be 1.08, 0.92, and 2.40 kb are probably the fragments

that we have estimated to be 1.00, 0.87, and 2.20 kb, leaving the 3.40- and 1.60-kb fragments to be the PstI junction fragments.

The next step is to examine the junction fragments to determine the distance between the insertion site for the transposon and one of the restriction sites on the plasmid. Let us start with the HindIII site on the plasmid. The smaller of the two HindIII junction fragments is about 1.40 kb, which must be the distance from the HindIII site in the original plasmid to the nearest HindIII site in the transposon. Since 1.15 kb of the junction fragment is taken up by transposon DNA, the actual site of insertion of the transposon is $1.4 - 1.15 = 0.25$ kb from the plasmid HindIII site. However, from the HindIII data alone, the transposon could be inserted either 0.25 kb clockwise or 0.25 kb counterclockwise of the HindIII site on the plasmid as the plasmid map is drawn in Figure 9.24. To determine the side of the HindIII site on which the transposon is inserted, we need to refer to the size of the PstI fragments. From the size of the smallest PstI junction fragment, which we estimated to be 1.60 kb, the transposon must be inserted $1.60 - 0.60 = 1.00$ kb from the plasmid PstI site, since 0.60 kb of this junction fragment is taken up with transposon DNA. The only way the transposon could be inserted both approximately 0.25 kb

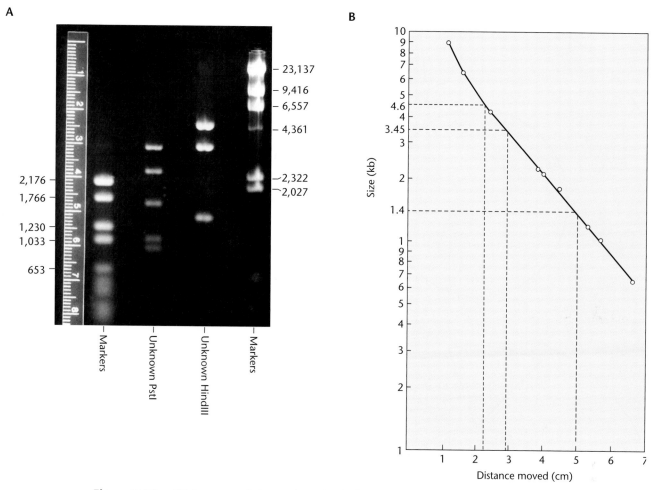

Figure 9.23 (A) Agarose gel electrophoresis of a Tn*5*-mutagenized pAT153 plasmid cut with HindIII and PstI. The sizes of the marker fragments are also shown. The DNA fragments on the gel were stained with ethidium bromide and photographed under UV illumination. (B) Standard curve of the sizes of the known fragments plotted against the distance moved in the gel shown in panel A. The sizes of the unknown fragments can be estimated from the standard curve derived from the marker fragments by plotting the distances that the unknown fragments moved on the gel (dotted lines). Only the positions of the HindIII fragments are shown.

from the HindIII site and 1.00 kb from the PstI site is to be inserted approximately 0.25 kb clockwise of the HindIII site, as shown in Figure 9.24.

Transposon Mutagenesis of the Bacterial Chromosome

The same methods used to mutagenize a plasmid with a transposon can also be used to mutagenize the chromosome. A gene with a transposon insertion is much easier to map or clone than a gene with another type of mutation, making this a popular method for mutagenesis of chromosomal genes. Transposons have been engineered to carry a bacterial origin of replication (*oriV*) or an origin of transfer (*oriT*), which facilitate cloning of the

region of the transposon insertion or genetic mapping by conjugation, as outlined in chapter 3.

The major limitation of transposon mutagenesis is that transposon insertions usually inactivate a gene, a lethal event in a haploid bacterium if the gene is essential for growth. Therefore, this method can generally be used only to mutate genes that are nonessential or essential under only some conditions. However, it can still be used to map essential genes by isolating transposon insertion mutations that are not in the gene itself but close to it in the DNA. If the transposon is inserted close enough, it might be used to map or clone the gene. It is also important to remember that insertion of some transposons may increase the expression of genes nearby to the insertion site (see above).

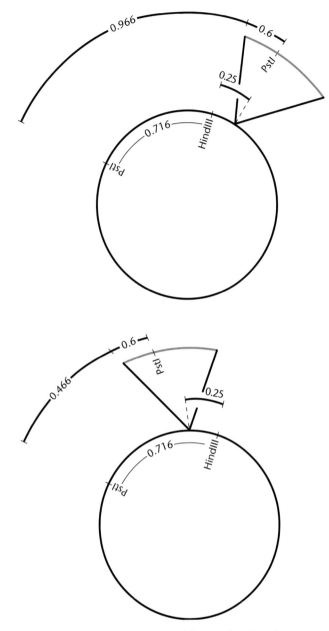

Figure 9.24 Use of the size of the smallest junction fragments from Figure 9.23 to determine the site of insertion of the transposon in the plasmid. The size of the smallest HindIII junction fragment indicates that the transposon must have inserted 0.25 kb from the HindIII site in the plasmid. The smallest PstI junction fragment indicates that the transposon must be positioned clockwise of the plasmid HindIII site. The units in the figure are kilobases.

PHYSICAL MAPPING OF CHROMOSOMAL TRANSPOSON INSERTIONS

Physical mapping of transposon insertion mutations in the chromosome is more difficult than mapping of insertions in plasmids, because the chromosome is so large.

Usually, eight-hitter restriction nucleases must be used instead of six-hitter enzymes, so that there are larger but fewer fragments. However, the fragments obtained with eight-hitter restriction endonucleases are usually too large to resolve on normal agarose gels, and techniques such as pulsed-field gel electrophoresis (see chapter 1) must be used to separate them.

Even if the fragments can be separated, fragments containing an inserted transposon are often difficult to identify because the few thousand base pairs added by a transposon usually do not make a significant difference in the size of such large fragments. However, if the transposon itself has a restriction site for the eight-hitter enzyme, any fragment containing the transposon will have a new site for the eight-hitter enzyme. The eight-hitter cuts the fragment containing the transposon into two pieces, making it easy to identify. Fortuitously, transposon Tn5 has a site for the eight-hitter NotI. Wherever transposon Tn5 has inserted in the chromosome, a new NotI site is introduced, which can be used in the physical mapping of the transposon insertion. Once the eight-hitter fragment containing the transposon has been identified, the site of insertion of the transposon in this somewhat smaller piece can be further localized by using six-hitters and more standard techniques.

Transposon Mutagenesis of All Bacteria

One of the most useful features of transposon mutagenesis is that it can be applied to many types of bacteria, even ones which have not been extensively characterized. Methods have been developed to perform transposon mutagenesis of almost all gram-negative bacteria as well as many gram-positive bacteria. All that is needed is a way of introducing a transposon into the bacterium, provided that the transposon can hop in the bacterium. The transposon should also carry a gene that can be selected in the bacterium. Some of these methods were mentioned earlier and are outlined in more detail below.

USE OF PROMISCUOUS PLASMIDS

One common method of mutagenizing gram-negative bacteria uses the transfer system of self-transmissible plasmids such as the IncP plasmids, which are very promiscuous for transfer and which transfer themselves or mobilize other plasmids into essentially any gram-negative bacterium (Figure 9.21A; see Simon et al., 1983, Suggested Reading). These plasmids are used to mobilize a smaller plasmid containing a compatible *mob* site and the transposon Tn5, which transposes in most gram-negative bacteria and contains a gene for kanamycin resistance that can be selected in most gram-negative bacteria. This smaller plasmid also has the replication origin of the ColE1 plasmid, which is narrow host range and capable only of replicating in *E. coli* and a few close relatives; this makes it a suicide vector in most gram-negative bacteria. The bacterium to be mutagenized is mixed with *E. coli* strains carrying these two plasmids. The larger IncP plasmid then mobilizes the smaller Tn*5*-containing plasmid into the cells, and the transposon hops. Cells in which the transposon has hopped can be selected on plates containing kanamycin and irgasan, an antibiotic to which *E. coli* is sensitive but most other gram-negative bacteria are resistant.

CLONING GENES MUTATED WITH A TRANSPOSON INSERTION

Genes that have been mutated by transposon insertion are usually relatively easy to clone by cloning the easily identified antibiotic resistance gene in the transposon. Since some antibiotic genes, for example the kanamycin resistance gene in Tn*5*, are expressed in many types of bacteria, this method can even be used to clone genes from one bacterium in a cloning vector from another. This is particularly desirable because most cloning vectors and recombinant DNA techniques have been designed for *E. coli*. To clone a gene mutated by a transposon from a bacterium distantly related to *E. coli*, the DNA from the mutagenized strain is cut with a restriction endonuclease that does not cut in the transposon and is ligated into an *E. coli* plasmid cloning vector cut with the same or a compatible enzyme. The ligation mixture is then transformed into *E. coli*, and the transformed cells are spread on a plate containing the antibiotic to which the transposon confers resistance. Only cells containing the mutated, cloned gene multiply to form a colony.

Transposons can be engineered to make cloning of transposon insertions even more efficient by introducing an origin of replication into the transposon so that the DNA containing the transposon need only cyclize to replicate autonomously in *E. coli*. The use of such a transposon for cloning transposon insertions is illustrated in Figure 9.25. In the example, the transposon carrying a plasmid origin of replication has inserted into the gene to be cloned. The chromosomal DNA is isolated from the mutant cells and cut with a restriction endonuclease that does not cut in the transposon, EcoRI in the example. When the cut DNA is religated, the fragment containing the transposon becomes a circular replicon with the plasmid origin of replication. If the ligation mixture is used to transform *E. coli* and the ampicillin resistance gene on the transposon is selected, the chromosomal DNA surrounding the transposon will have been cloned. Since the restriction endonuclease cuts outside the transposon, any clones of the transposon cut from the chromosome also include sequences from the gene of interest into which the transposon had inserted.

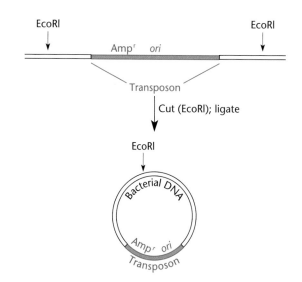

Figure 9.25 Cloning genes mutated by insertion of a transposon. A transposon used for mutagenesis of a chromosome contains a plasmid origin of replication (*ori*), and the chromosome is cut with the restriction endonuclease EcoRI and religated. If the ligation mix is used to transform *E. coli*, the resulting plasmid in the Amp^r transformants will contain the sequences that flanked the transposon insertion in the chromosome. Chromosomal sequences are shown in black, and transposon sequences are shown in purple.

Once the gene containing the transposon insertion has been cloned, it can be used in several ways. We may want to directly sequence the gene with primers complementary to the ends of the transposon. Alternatively, we may need a clone of the wild-type gene without the transposon insertion; in this case, we could use the clone with the transposon insertion as a probe to identify the wild-type gene by screening a library of wild-type DNA by plate hybridization (see chapter 1). This method allows the cloning of genes about which nothing is known except the phenotype of mutations that inactivate the gene, and it can be easily adapted to clone genes from any bacterium in which the transposon can hop to create the original chromosomal mutation. Once mutants are obtained with the transposon inserted in the gene of interest, the remaining manipulations are performed in *E. coli*. This can be a particular advantage if the bacterium being studied is difficult to grow or maintain in a laboratory situation.

Using Transposon Mutagenesis To Make Random Gene Fusions

The ability of some transposons to hop randomly into DNA has made them very useful for making random gene fusions to reporter genes (see chapter 2). Fusing a gene to a reporter gene can make regulation of the gene much easier to study or can be used to identify genes subject to a certain type of regulation or those localized to certain cellular compartments. Once a gene subject to a certain type of regulation has been identified in this way, it can be easily cloned and studied using methods such as those described above.

Transposons have been engineered to make either transcriptional or translational fusions. As discussed in chapter 2, in a transcriptional fusion, one gene is fused to the promoter for another gene so that the two genes are transcribed into mRNA together. In a translational fusion, the ORFs for the two proteins are fused to each other in the same reading frame so that translation, initiated at the TIR for one protein, continues into the ORF for the other protein, making a fusion protein.

Transposons engineered to make random gene fusions include Tn3HoHo1, Tn5lac, and Tn5lux. These transposons carry a reporter gene at one end that either has its own TIR, if the transposon is to make transcriptional fusions, or lacks a TIR, if the transposon is to make translational fusions. When the transposon hops into the chromosome, it fuses its reporter gene to whatever gene it hops into, provided that it has hopped into the gene in the right orientation.

Mud(AMP^r, *lac*)

The prototype transposon for making random gene fusions is the Mud(Amp^r, *lac*) transposon (see Casadaban and Cohen, Suggested Reading). This transposon has been largely supplanted by more elaborate constructions with specialized uses, but we can use it to illustrate the basic principles involved. The Mud(Amp^r, *lac*) transposon is derived from phage Mu, which transposes with almost no target specificity (see Box 9.1). Phage Mu also has quite a broad host range and productively infects and transposes in many gram-negative bacteria, including *Erwinia carotovora*, *Citrobacter freundii*, and *E. coli*.

Figure 9.26 shows the essential features of random gene fusions created by the Mud(Amp^r, *lac*) transposon. Most of the phage Mu DNA is removed, except the ends

Figure 9.26 Structure of random gene fusions created by the original Mud(Amp^r, *lac*) transposon. The essential transposon elements are the ends of Mu, the reporter gene *lacZ*, and a gene for ampicillin resistance (Amp^r). The *lacZ* gene lacks its own promoter and so is transcribed only from a promoter outside the transposon (*p*). The sequences outside the transposon are shown as dashed lines, and the Mu sequences of the transposon are shown in purple.

and the transposase, which is the product of Mu genes A and B. The transposase and the ends of phage Mu are sufficient for transposition of the phage DNA into the chromosome after infection. Close to one of its ends, the Mud(Amp^r, *lac*) transposon carries the *lacZ* gene of *E. coli* as its reporter gene. The *lacZ* gene has no promoter of its own and so is not normally transcribed. However, if the transposon has hopped into a target DNA in such a way that the *lacZ* gene is positioned in the correct orientation downstream of a promoter, the *lacZ* gene is turned on and expresses β-galactosidase, which is easily detected by colorimetric assays with dyes such as X-Gal or *o*-nitrophenyl-β-D-galactopyranoside (ONPG). The Mud(Amp^r, *lac*) transposon also has an ampicillin resistance gene that can be used to select only the cells that have the transposon inserted somewhere in their DNA.

The procedure for using the Mud(Amp^r, *lac*) transposon to make random gene fusions is illustrated in Figure 9.27. The first step is to prepare a lysate by inducing a Mu prophage in cells that also contain the Mud(Amp^r, *lac*) transposon. The cell must also contain the normal phage Mu (a helper phage) to furnish all the proteins needed to make a phage, including the head and tail proteins, which the Mud(Amp^r, *lac*) transposon cannot make for itself. Since Mu cannot distinguish its own DNA from the Mud(Amp^r, *lac*) transposon DNA, which has the same ends, the replicating phage Mu sometimes packages the Mud(Amp^r, *lac*) transposon instead of its own DNA. Therefore, when the cells lyse, some of the phage particles that are released contain the Mud(Amp^r, *lac*) transposon. This phage lysate is then used to infect cells at a low multiplicity of infection. Any ampicillin-resistant transductants, selected on ampicillin plates, will have the Mud(Amp^r, *lac*) transposon inserted somewhere in their chromosome. If the Mud(Amp^r, *lac*) transposon has inserted downstream of an active promoter, the *lacZ* gene on the transposon will be transcribed from the promoter and the cells will make β-galactosidase.

The fact that the Mud(Amp^r, *lac*) transposon expresses *lacZ* only if the gene into which it hops is transcribed has been used to identify genes that are turned on only under certain conditions. A classical example is the identification of the *din* (damage-inducible) genes of *E. coli*, which are turned on only after DNA damage (see Kenyon and Walker, Suggested Reading). To use Mud(Amp^r, *lac*) to identify *din* genes, these investigators first used the method described above to isolate ampicillin-resistant (Amp^r) transductants with Mud(Amp^r, *lac*) inserted somewhere in their genome. They then needed to identify transductants with the transposon inserted in the correct orientation into a gene that is induced when the DNA is damaged. To accomplish this, they replicated the plates containing the Amp^r transductants onto two other sets of

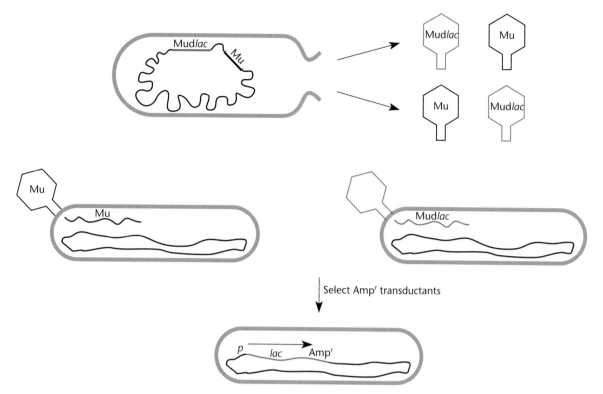

Figure 9.27 Isolating random gene fusions with Mud(Amp^r, *lac*). The Mu prophage is induced in a cell containing the Mud(Amp^r, *lac*) transposon (in purple), which will be packaged into some of the phage. All Amp^r transductants have Mud(Amp^r, *lac*) somewhere in their chromosome; if the transduced cells were originally Lac$^-$, in cells that form blue colonies on X-Gal plates, the Mud(Amp^r, *lac*) transposon has hopped downstream of a promoter in such a way that the *lacZ* gene is transcribed.

plates, one containing only X-Gal and the other containing X-Gal plus a DNA-damaging agent such as mitomycin. The colonies that were blue on the set of plates containing the DNA-damaging agent but not on the plates with X-Gal alone were known to have the transposon inserted in a *din* gene in the right orientation. By mapping the ampicillin resistance gene on the transposon, the investigators could map a number of *E. coli din* genes that are turned on after DNA damage. The *din* genes and their induction are discussed in more detail in chapter 11.

In Vivo Cloning

Transposons can also be used for in vivo cloning. Like other cloning procedures, in vivo cloning requires a library of recombinant DNA. However, these libraries do not need to be made in vitro with restriction endonucleases and DNA ligase as outlined in chapter 1. Instead, in vivo cloning relies on genetic methods and lets bacteriophages

and transposons do most of the work of making the library.

Most in vivo cloning procedures are also based on phage Mu and the fact that Mu replicates by replicative transposition without resolving the cointegrates. After Mu replicates, copies of the phage genome are inserted all over the bacterial chromosome (see Box 9.1). During the normal infection cycle, these phage genomes are then packaged into phage heads by using specific *pac* sites at the ends of the phage DNA.

CONSTRUCTING MINI-Mu ELEMENTS
Because the phage Mu packaging system recognizes the ends of phage DNA, it can, in principle, recognize the outside ends of two phage genomes lying close to each other in the chromosome and package both phage Mu DNAs plus any chromosomal DNA between them. This does not normally happen, because the phage Mu head is large enough to hold only a single Mu genome. However, if a shortened version of Mu, called a mini-Mu

element, is used, there is extra room in the phage head for chromosomal DNA (see Groisman and Casadaban, Suggested Reading). These mini-Mu elements lack all of the functions of Mu except the ends required for transposition and packaging. In addition, to make them useful for in vivo cloning, the mini-Mu elements have a plasmid origin of replication and an antibiotic resistance gene cloned into them, the significance of which will become apparent later.

Figure 9.28 illustrates how mini-Mu elements may be used for in vivo cloning. A mini-Mu element that has been introduced into a cell on a plasmid is induced to replicate (transpose) by infection of the cells with a helper phage, wild-type Mu, which contributes the transposase

Figure 9.28 In vivo cloning with mini-Mu. In the first cell, a mini-Mu (in purple) on a plasmid is induced to replicate by infection with a helper phage Mu, allowing the mini-Mu to hop randomly around the chromosome as it replicates. In the second cell, pairs of mini-Mu elements, along with chromosomal DNA lying between them, are packaged into some phage Mu heads. The sites of cleavage for packaging the two mini-Mu elements are slightly outside the ends of the mini-Mu elements (arrows). On infection of the third cell, the two mini-Mu elements can recombine to form a circular DNA that can replicate because of the origin of replication on the mini-Mu elements. The Amp^r gene also carried by the mini-Mu allows selection of transductants.

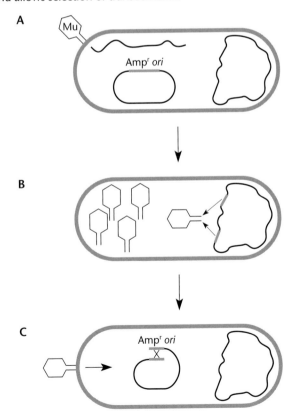

functions needed for the mini-Mu to transpose. Both the wild-type Mu and the mini-Mu elements then transpose around the chromosome, making many copies of themselves. Later in the infection, while the normal phage DNAs are being cut out and packaged, some pairs of mini-Mu elements, which happen to be separated by a phage Mu length of DNA, are also cut out and packaged along with the chromosomal DNA between them. When the cells lyse, the released phage is a mixture of phage containing normal Mu DNA and others containing chromosomal DNA packaged between copies of the mini-Mu element.

The plasmid origin of replication and the selectable gene on the mini-Mu now come into play. When new cells are infected with the phage lysate at a low multiplicity of infection, some of the cells are infected by phage containing the mini-Mu elements and chromosomal DNA. The DNA injected into these cells can then cyclize by recombination between the identical mini-Mu sequences at the ends to form a circular DNA that can replicate from the plasmid origin of replication. Because the mini-Mu element also contains the gene for ampicillin resistance, transductants containing autonomously replicating plasmids can be selected by plating on ampicillin-containing plates. If enough phage in the lysate carried chromosomal DNA, every region of the bacterial DNA is represented in the collection of plasmids, constituting a library of the bacterial DNA. Once a library has been obtained, a clone containing the desired gene can be found in the library through complementation or hybridization techniques, such as those discussed in chapters 1 and 3. This general method has been used in a variety of gram-negative bacteria, including *E. coli*, *Klebsiella* spp., and *Erwinia* strains (for an example, see Van Gijsegem and Toussaint, Suggested Reading).

Site-Specific Recombination

Another type of nonhomologous recombination, **site-specific recombination,** occurs only between specific sequences or sites on DNA. It is promoted by enzymes called **site-specific recombinases,** which recognize two specific sites in DNA and promote recombination between them. Even though the two sites generally have short sequences in common, the regions of homology are usually too short for normal homologous recombination to occur efficiently. Therefore, efficient recombination between the two sites requires the presence of a specific recombinase enzyme. We have already mentioned some site-specific recombination systems in connection with the resolution of chromosome and ColE1 plasmid dimers by the XerCD site-specific recombinase. The integrase of λ phage is another example, as are resolvases such as

TnpR of Tn*3* that resolve cointegrates formed during replicative transposition. In this section, we discuss some other examples of site-specific recombination in bacteria and phages and the recent discovery that all site-specific recombinases can be placed into two groups, the S and the Y recombinases, based on their mechanism of action.

Developmentally Regulated Excision of Intervening DNA

One type of site-specific recombination occurs during terminal differentiation in some types of bacteria. We have already mentioned examples of genes carrying sequences that are not present in the final RNA or protein product of the gene. The extra or intervening sequences in genes are often removed from the RNA or the protein product by splicing after the RNA or protein is synthesized (see Box 2.6). However, sometimes the intervening sequences are cut out of the DNA itself before the gene is expressed. Site-specific recombinases remove these intervening DNA sequences from genes. Note that this type of gene rearrangement is possible only in cell lines that do not need to reproduce themselves, that is, those that are terminally differentiated. Otherwise, the intervening DNA sequences would be lost to subsequent generations.

The classic example of DNA rearrangements during terminal differentiation is in the vertebrate immune cells that make antibodies. In this case, site-specific recombination during the differentiation of the immune cells removes different lengths of sequences from the few germ line genes that encode antibodies, thereby creating hundreds of thousands of new genes encoding antibodies of different specificities. Because sequences have been cut out of these genes, the genes can never be restored from the somatic cells of the organisms. However, the antibody-encoding genes in the germ line cells (eggs and sperm) remain intact, so that no DNA sequences are lost to future generations and the genes can undergo similar rearrangements in subsequent generations.

Most bacteria multiply by cell division, and none of the cells undergo terminal differentiation. Therefore, irreversible DNA rearrangements are generally not possible in bacteria. Nevertheless, bacteria do manifest a few examples of irreversible DNA rearrangements during terminal differentiation.

One type of terminal differentiation in bacteria occurs during sporulation in *Bacillus subtilis*. Some types of bacteria sporulate by forming endospores, a process during which the cell differentiates into a mother cell and an internal spore (see chapter 14). Both the mother cell and the spore get a copy of the chromosome to make the gene products required in that compartment for sporulation, but only the spore need retain the full complement of the

bacterial DNA. The mother cell is required only to make the spore and will eventually lyse. Thus, the formation of the mother cell is an example of terminal differentiation, like our somatic cells; therefore, the cell can undergo irreversible DNA rearrangements needed for the sporulation process. In *B. subtilis* sporulation, an intervening sequence of 42 kb is cut out of the gene for the sigma factor σ^K, which is required for transcription of some genes in the mother cell. Unless this intervening sequence is removed, σ^K is not synthesized, some of the gene products required for sporulation are not expressed, and sporulation is blocked (see Kunkel et al., Suggested Reading).

Terminal differentiation also occurs in the formation of highly specialized cells responsible for fixing atmospheric nitrogen in some types of filamentous cyanobacteria (see Golden et al., Suggested Reading). These cells, called heterocysts, appear periodically in the filaments when the bacteria are growing under conditions of nitrogen starvation. The heterocysts never divide to form new cells and disappear when the cells are provided with a good source of nitrogen. Because they never divide into new cells, they do not need a full complement of DNA and can undergo irreversible DNA rearrangements. As occurs in *B. subtilis* sporulation, intervening sequences are cut out of some genes while the cells differentiate. In the cyanobacteria, these sequences are cut out of at least two genes. One is the *nifD* gene, which encodes a product required for nitrogen fixation in the heterocysts after an intervening sequence of 11 kb is cut out. The product of the *nifD* gene is not required in the other cells in the filament that do not fix nitrogen, and so the intervening sequence need not be cut out in these cells.

The process of removing the intervening sequences seems to be similar in both developmental systems. Recombination between directly oriented sites bracketing the intervening sequence results in its excision. In both cases, the site-specific recombinase required to promote recombination between the bracketing sites is encoded within the intervening sequence itself and is expressed only during differentiation. Therefore, the recombinase is expressed only in the cells undergoing terminal differentiation, and the intervening sequence is not lost from normally replicating cells.

The function of the intervening sequences, in both *B. subtilis* and cyanobacteria, is unclear. Even though they are removed only during development, their removal seems to play no important regulatory role in either sporulation or heterocyst differentiation. Permanently removing the intervening sequences from the chromosome, so that the σ^K and *nifD* genes are not interrupted, does not adversely affect the ability of the bacteria to sporulate or to develop heterocysts that can fix nitrogen. In fact, some close relatives of these strains do not have intervening sequences in these genes. One possibility is that these intervening sequences are parasitic DNAs like phages or transposons. If they integrate themselves into an essential gene, the cell cannot easily delete them without killing itself. As with transposons, unless the deletion event precisely removes the intervening sequence, the gene will be inactive, disrupting an important cellular function. By excising themselves during differentiation, however, the parasitic DNA elements allow sporulation or nitrogen fixation by their hosts and so have no deleterious effects—the mark of a good parasite.

Integrases

Integrases are another type of site-specific recombinase. They also recognize two sequences in DNA and promote recombination between them; therefore, they are no different in principle from the site-specific recombinases that resolve cointegrates or remove intervening sequences. However, rather than remove a DNA sequence by promoting recombination between two directly repeated sequences on the *same* DNA, integrases act to integrate one DNA into another by promoting recombination between two sites on *different* DNAs.

PHAGE INTEGRASES

The best-known integrase is the Int enzyme of λ phage, which is responsible for the integration of circular phage DNA into the DNA of the host to form a prophage (see chapter 8). Briefly, the λ phage integrase specifically recognizes the *attP* site in the phage DNA and the *attB* site on the bacterial chromosome and promotes recombination between them. Usually, phage integrases are extremely specific. Only the *attP* and *attB* sites are recognized, so that the DNA integrates only at one or at most a few places in the bacterial chromosome. Other integrases seem to be somewhat less specific, including the integrase of the integrative conjugative element Tn*916* (see Box 5.4) and the integrases of integrons (see below), where there seems to be some flexibility in the sequence of the *attB* site. In a reversal of this reaction performed by the integrases, a combination of the integrase (Int) and another enzyme, often called the excisase (Xis), promotes recombination between the hybrid *attP-attB* sites flanking the integrated DNA to excise the integrated DNA, although the integrase is again the enzyme that performs the site-specific recombination.

Because of their specificity, phage integrases have a number of potential uses in molecular genetics. For example, the reaction performed by the phage λ Int and Xis has been capitalized on in cloning technologies (see, for example, Gateway Technology in the Invitrogen catalogue). In these applications, a PCR fragment containing the gene of interest is cloned into a plasmid vector called the entry vector so that the clone is flanked by one of the hybrid *attB-attP* sites which flank integrated prophage λ DNA in the chromosome. If this is mixed with a destination vector containing the other *attP-attB* hybrid site and λ integrase and excisase are added, site-specific recombination between the sites moves the cloned gene into the destination vector, where it becomes flanked by *attB* sites. While this technology does not remove the requirement for the initial cloning, which can be laborious, once a clone is made in the entry vector, it can be transferred quickly into a number of different destination vectors. This could be important if, for example, one wished to determine the effect on the solubility of a protein of fusing it to a number of different affinity tags which are encoded on a number of different destination vectors.

INTEGRASES OF TRANSPOSON INTEGRONS

Integrases are also important in the evolution of some types of transposons. The first clue was that transposons seemed to have picked up antibiotic resistance genes so that related members of some families of transposons, such as the Tn*21* family, have different resistance genes inserted in about the same place in the transposon (see Figure 9.8). Figure 9.29 shows a more detailed structure of this region, called an **integron,** and how it presumably recruits antibiotic resistance genes. A basic integron consists of an integrase gene (called *intI1* in the figure) next to a site called *attI* (for *att*achment site *i*ntegron). A gene in the *attI* site will be transcribed from the promoter p_c. The transposon originally has no antibiotic resistance gene inserted at the *attI* site. Elsewhere in the cell, there was an antibiotic resistance cassette that consisted of an antibiotic resistance gene and a site *attC* (for *att*achment site *c*assette) recognized by the integrase. This cassette excised and formed a circle, and the integrase then integrated it into the *attI* site on the integron by promoting site-specific recombination between the *attC* site on the cassette and the *attI* site in the integron. At another time, and in another cell, a similar cassette carrying a different resistance gene could integrate next to this one, adding another antibiotic resistance gene to the transposon. The advantages of this system are obvious. By carrying different resistance genes, the transposon allows the cell, and thereby itself, to survive in environments containing various toxic chemicals. However, where the resistance gene cassettes originally came from remains a mystery.

One clue to the origin of antibiotic resistance cassettes may come from the discovery of what are called chromosomal **superintegrons** in a number of different types of bacteria (see Rowe-Magnus et al., Suggested Reading). The structure of one of these from *Vibrio cholerae* is shown in Figure 9.30. It consists of 179 cassettes carrying

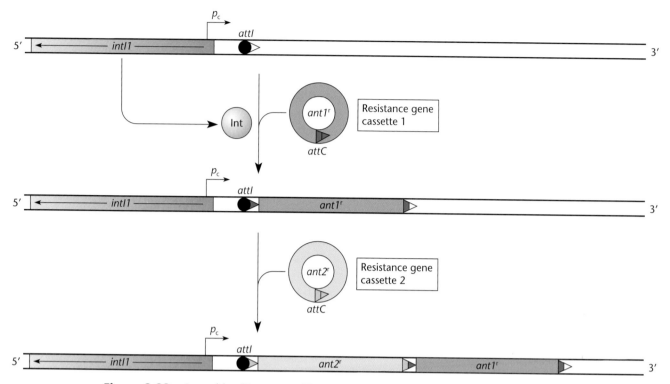

Figure 9.29 Assembly of integrons. The primary transposon carries an integron with a gene *intI1* encoding an integrase and a site *attI* transcribed by a strong promoter, p_C. A cassette carrying resistance to one antibiotic, *ant1*^{r}, has excised from elsewhere and is integrated by the integrase by recombination between its *attC* site and the *attI* site on the integron. The antibiotic resistance gene is transcribed from the promoter on the integron. Later, the integrase integrates another cassette carrying a different antibiotic resistance gene, *ant2*^{r}, at the same place. In this way, a number of different antibiotic resistance genes can be assembled by the integron on the transposon. The *attC* sites indicated by triangles contain conserved regions related to the *attC* sites between the cassettes of superintegrons shown in Figure 9.30.

Figure 9.30 Example of a superintegron from *Vibrio cholerae*. More than 100 cassettes encoding resistance to different antibiotics and other functions are associated with partially homologous *attC* sites next to an integrase gene *intIA* and an *attI* attachment site. Regions between the cassettes corresponding to possible *attC* sites are shown as arrows. Regions of sequence conservation are shown. R, purine; Y, pyrimidine.

Chromosomal superintegron (*V. cholerae*)

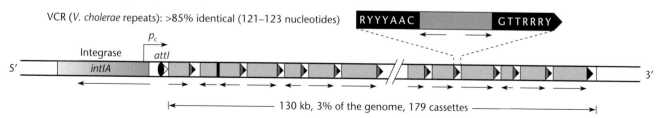

ORFs of largely unknown function, separated by partially conserved sequences, which might be *attC* sites. Presumably integrons in transposons will turn out to be one example of a larger phenomenon in which useful genes can be recruited as needed from storage areas carrying large numbers of such cassettes.

GENETIC (PATHOGENICITY) ISLANDS

Integrases and nonhomologous recombination also play a role in the integration of at least some types of genetic islands into the chromosome. Like plasmids and prophages, genetic islands often carry clusters of genes that allow the bacterium to occupy specific environmental niches. Genetic islands can be hundreds of thousands of deoxynucleotides long and carry hundreds of genes. The **pathogenicity island** (PAI) is a type of genetic island that carries genes required for pathogenicity. PAIs carry genes for resistance to multiple antibiotics in *Shigella flexneri*, for alpha-hemolysin and fimbriae in pathogenic *E. coli*, for scavenging and storing iron in *Yersinia*, and for a type III secretion system in *Helicobacter pylori*, to give just some examples. We have already mentioned the PAI SaPI1, which carries the gene for the toxin involved in toxic shock syndrome in *Staphylococcus aureus* (see Box 8.3). This PAI has its own integrase and so can integrate itself into the chromosome of new strains of *S. aureus*, making them pathogenic. It can also move from strain to strain of *S. aureus* by behaving like a satellite virus of phage 80α and allowing itself to be transduced by the phage. If a cell carrying the PAI is infected by phage 80α, the island will excise and replicate with the help of phage replication proteins. It will then be packaged into phage heads, from where it can be injected into and integrate into the chromosome of another bacterium that does not carry it, converting the new bacterium into a pathogen.

The demonstration that SaPI1 can move makes it the exception. Most PAIs have not been demonstrated to move from one bacterium to another or even to integrate into a DNA that lacks them. Often the only evidence that they have moved recently is that they are not found in all the strains of a type of bacterium and that the base composition (G+C content) of their DNA and their codon usage are often different from those of the chromosomal DNA as a whole. These characteristics are often taken as evidence of recent horizontal transfer of genes from one strain of bacterium to another. PAIs also often carry vestigial integrase genes that are broken up by nonsense and other types of mutations, suggesting that they once encoded functional integrases which are no longer functional. The PAI elements are often flanked by short repeated sequences, either direct or inverted, which may be the sites at which the integrase acted to integrate the PAI. Apparently, the PAIs did move into the strain some time ago, perhaps on a promiscuous plasmid, and integrated into the chromosome, but their DNA has mutated over time so that they are no longer capable of moving. Interestingly, many PAIs are integrated into tRNA genes in the chromosome. Part of the tRNA gene is duplicated on the PAI, so that the tRNA product of the gene is still functional (see Box 8.1). Chromosomal tRNA genes may be preferred sites of integration because they have almost the same sequence in different species of bacteria. By using a highly conserved tRNA gene as its bacterial attachment (*attB*) site, the PAI can integrate into the chromosome of any bacterial strain in which it finds itself.

Even though most PAIs cannot integrate, some researchers decided to make one that could do so (see Rakin et al., Suggested Reading). More accurately, they constructed a plasmid that could integrate using the integrase of a PAI. They accomplished this by using parts of a PAI from *Yersinia*. As discussed in chapter 4, different strains of *Yersinia* differ greatly in their pathogenicity depending on the DNA elements they carry (Box 4.1). The most pathogenic species of *Yersinia* are *Y. pestis*, which causes bubonic plague, and *Y. enterocolitica* and *Y. pseudotuberculosis*, which cause mild intestinal upsets. These strains carry a PAI called HPI (high-pathogenicity island). This PAI, of 40 kb, is integrated into one of the asparagine transfer RNA (asparagine-tRNA) genes and encodes enzymes to make small molecules called siderophores, which help scavenge for iron, which is in limited supply in the eukaryotic host. The authors reasoned that even though this PAI from different strains has not been demonstrated to move, perhaps the functional parts of each PAI could be assembled into a plasmid cloning vector of *E. coli*, which would then be able to integrate into a corresponding asparagine-tRNA gene of *E. coli*. As mentioned, the tRNA genes are very similar in different species, and *E. coli* has four asparagine-tRNA genes, all of which are very similar in the region where the PAI is integrated in *Yersinia*.

In order to integrate, the PAI needs both a functional integrase and functional *attP* and *attB* sites, by analogy to phage λ. First, the investigators reasoned that the integrase itself from the PAI in *Y. pestis* might be functional, and so they cloned this integrase gene into a plasmid cloning vector so that it would be transcribed at a high level from a phage T7 promoter (see the discussion of T7-based expression vectors in chapter 7). However, they needed to reconstruct a site that would be recognized by the integrase. The PAI itself might not have such a site. They guessed that the situation might be analogous to what happens when phage λ integrates (see chapter 8). The *attP* site on the phage recombines with the *attB* site in the chromosome, resulting in *attP-attB* hybrid sites flanking the integrated prophage, which are no longer recognized by the integrase alone. Such hybrid sites might exist at the ends of the PAI, and, if so, they might also not be recognized by the integrase. To reconstruct the original site the integrase recognizes from the hybrid sites, the authors needed to know where in the hybrid sites the "attP" sequence ends and the "attB" sequence in the asparagine-tRNA gene begins. To determine this, they compared the sequences at the ends of the PAI to the sequence of asparagine-tRNA genes in strains that do not have the PAI integrated to determine which sequences are due to the attB sequence in the tRNA gene. They then constructed the original attP sequence by using PCR and cloned this into the plasmid that already contained the integrase. When this plasmid was introduced into *E. coli* and the integrase gene on the plasmid was induced, it integrated into an asparagine-tRNA gene of *E. coli*, showing that all the features needed to integrate the PAI were still present and active.

Resolvases

The resolvases of transposons such as the TnpR protein of Tn3 discussed above are another type of site-specific recombinase. In fact, the resolvase of transposon γδ, a close relative of Tn3, is one of the best studied site-specific recombinases and has been crystallized bound to its DNA substrate. These enzymes promote the resolution of cointegrates by recognizing the *res* sequences that occur in one copy in the transposon but in two copies in direct orientation in cointegrates. Recombination between the two *res* sequences in a cointegrate excises the DNA between them, resolving the cointegrate into the donor DNA and the target DNA, both containing the transposon.

Other resolvases already mentioned resolve dimers of plasmids. Dimer formation by plasmids reduces their stability, especially if they have a low copy number, because each dimer is treated as one plasmid molecule by the partitioning system and segregated to the same side of the cell (see chapter 4). Because mutations in the resolvase gene can affect the segregation of the plasmid, some of these plasmid resolution systems were originally mistaken for partitioning systems and given the name Par (for "partitioning"). Examples of resolvases involved in resolving plasmid dimers are the Cre recombinase, which resolves dimers of the P1 prophage plasmid by promoting recombination between repeated *loxP* sites on the dimerized plasmid, and the XerC,D recombinase, which resolves dimers of the ColE1 and pSC101 plasmids by promoting recombination between repeated *cer* and *psi* sites, respectively. The XerCD recombinases also resolve the chromosome dimers formed during recombination

repair of stalled chromosome replication forks by promoting recombination between repeated *dif* sites in the dimers during cell division, and they double as integrases to integrate the single-stranded DNA of the cholera toxin-producing phage (see chapter 8), illustrating how the ability to promote recombination between specific sequences can be put to many uses.

DNA Invertases

The DNA invertases are like resolvases in that they promote site-specific recombination between two sites on the same DNA. The main difference between the reactions promoted by DNA invertases and those catalyzed by resolvases is that two sites recognized by invertases are in reverse orientation with respect to each other whereas the sites recognized by resolvases are in direct orientation. As discussed in chapter 3 in the section on types of mutations, recombination between two sequences that are in direct orientation will delete the DNA between the two sites whereas recombination between two sites that are in inverse orientation with respect to each other will invert the intervening DNA.

The sequences acted on by DNA invertases are called **invertible sequences**. These short sequences may carry the gene for the invertase or may be adjacent to it. Therefore, the invertible sequence and its specific invertase form an inversion cassette that sometimes plays an important regulatory role in the cell, some examples of which follow.

PHASE VARIATION IN *SALMONELLA* SPECIES
The classic example of an invertible sequence is the one responsible for **phase variation** in some strains of *Salmonella*. Phase variation was discovered in the 1940s with the observation that some strains of *Salmonella* can change their surface antigens. They do this by shifting from making flagella composed of one flagellin protein, H1, to making flagella composed of a different flagellin protein, H2. The shift can also occur in reverse, i.e., from making H2-type flagella to making H1-type flagella. The flagellar proteins are the strongest antigens on the surface of many bacteria, and periodically changing their flagella may help these bacteria escape detection by the host immune system.

Two features of the *Salmonella* phase variation phenomenon suggested that the shift in flagellar type was not due to normal mutations. First, the shift occurs at a frequency of about 10^{-2} to 10^{-3} per cell, much higher than normal mutation rates. Second, both phenotypes are completely reversible—the cells switch back and forth, exhibiting first one type of flagella and then the other.

Figure 9.31 outlines the molecular basis for the *Salmonella* antigen phase variation (see Simon et al. 1980, Suggested Reading). As mentioned, the two types of flagella are called H1 and H2. A DNA invertase called the Hin invertase causes the phase variation by inverting an invertible sequence upstream of the gene for the H2 flagellin by promoting recombination between two sites, *hixL* and *hixR*. The invertible sequence contains the invertase gene itself and a promoter for two other genes:

Figure 9.31 Regulation of the *Salmonella* phase variation and some other members of the family of Hin invertases. (A) Invertible sequences, bordered by purple arrows, of *Salmonella* and several phages. The recombination sites are designated *hixL* and *hixR*, etc. (B) Hin-mediated inversion. In one orientation, the H2 flagellin gene, as well as the repressor gene, are transcribed from the promoter *p* (in purple). In the other orientation, neither of these genes is transcribed and the H1 flagellin is synthesized instead. The invertase Hin is made constitutively from its own promoter.

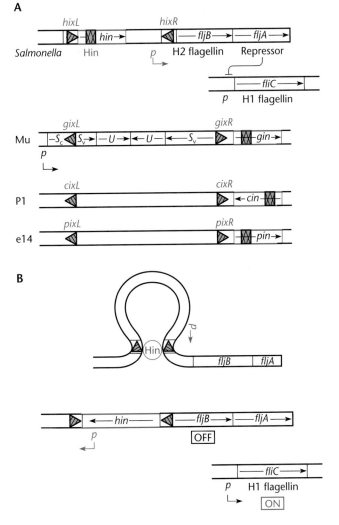

A

B

one called *fljB* encodes the H2 flagellin, and one called *fljA* encodes a repressor of H1 gene transcription. With the invertible sequence in one orientation, the promoter transcribes the H2 gene and the repressor gene, and only the H2-type flagellum is expressed on the cell surface. When the sequence is in the other orientation, neither the H2 gene nor the repressor gene is transcribed, because the promoter is facing backward. Now, however, without the repressor, the H1 gene called *fliC* can be transcribed; therefore, in this state, only the H1-type flagellum is expressed on the cell surface. Clearly, the Hin DNA invertase that is encoded in the invertible sequence is expressed in either orientation, or the inversion would not be reversible.

OTHER INVERTIBLE SEQUENCES

There are a few other known examples of regulation by invertible sequences in bacteria. For example, fimbria synthesis in some pathogenic strains of *E. coli* is regulated by an invertible sequence. Fimbriae are required for the attachment of the bacteria to the eukaryotic cell surface and may also be important targets of the host immune system.

Invertible sequences also exist in some phages. An example is the invertible region of phage Mu discussed in Box 9.1. Both phage P1 and the defective prophage e14 also have invertible regions (shown in Figure 9.31). These phages use invertible sequences to change their tail fibers. The tail fibers made when the invertible sequences are in one orientation differ from those made when the sequences are in the other orientation, broadening the host range of the phage. In phage Mu, the tail fiber genes expressed when the invertible sequence is in one orientation allow the phage to adsorb to *E. coli* K-12, *Serratia marcescens*, and *Salmonella enterica* serovar Typhi. In the other orientation, the phage is able to adsorb to other strains of *E. coli*, *Citrobacter*, and *Shigella sonnei*.

Not only do these phage invertases perform similar functions, but also they are closely related to each other. Dramatic evidence for their relationship came from the discovery that the Hin invertase inverts the Mu, P1, and e14 invertible sequences and vice versa (see Van De Putte et al., Suggested Reading). Apparently, inversion cassettes can be recruited for many puposes, much like antibiotic resistance cassettes are recruited by integrons.

Probably the most dramatic example of invertases playing a role in bacterial variability is in the **shufflons** of some self-transmissible plasmids, including R64 and Col1b-P9 (see Komano, Suggested Reading). In plasmid R64, an invertase named Rci inverts a number of short sequences encoding the C terminus of a minor pilin protein in the thin sex pilus of these plasmids. Apparently, different versions of the pilin protein created by multiple inversions bind with different affinities to the various lipopolysaccharides found on the surfaces of gram-negative bacteria, contributing to the promiscuity of these plasmids.

Y and S Recombinases

As mentioned above, many site-specific recombinases, whether they be integrases, resolvases, or invertases, are closely related to each other. This is not surprising considering that they all must perform the same basic reactions. First they must cut a total of four strands of DNA, both strands in two recognition sequences, whether these recognition sequences are on the same DNA (resolvases and invertases) or on different DNAs (integrases). Then they must join the cut end of each strand to the cut end of the corresponding strand from the other recognition sequence. We can anticipate some of the features that a site-specific recombinase must have in order to perform these reactions. First, they must somehow hold the DNA ends after they cut them so that they are not free to flop around and join with the end of any strand they happen to encounter. Second, after the strands are cut, either the DNA or at least part of the recombinase must rotate in a defined way to put the cut ends from different strands in juxtaposition with each other so that the correct rejoining can occur. If they rejoin the ends of the same strands that were cut originally, there will be no recombination and they will be back where they started.

Evidence has accumulated that all site-specific recombinases fall into two families, the Y (tyrosine) family and the S (serine) family, based on which of these amino acids, called the catalytic amino acid, plays the crucial role in their active center. The feature shared by these amino acids is a hydroxyl group in their side chain to which phosphates can be attached. After the DNA is cut, the hydroxyl group on the catalytic amino acid forms a covalent bond with the free phosphate end on the DNA. This protects the end of the strand and holds it while the recombinase moves it into position to be joined to the end of a different cut strand, the essence of recombination. However, when the strands are cut, the end of DNA that is joined to the catalytic amino acid and the structures that form after this cutting differ between the two groups of recombinases. Each of these pathways is outlined in the following sections.

Y Recombinases: Mechanism

The Y recombinases seem to be the most varied group and include recombinases that perform the most complex reactions. Some examples of Y recombinases are listed in Table 9.2, and the structures of some of them are shown in Figure 9.32. Some of these were mentioned in this and earlier chapters; they include the Cre recombinase, which resolves dimers of the P1 plasmid prophage by recombination between *loxP* sequences, and the XerC,D recombinases, which resolve chromosome dimers by recombination between repeated *dif* sites and resolve plasmid ColE1 dimers by recombination between repeated *cer* sites, as well as integrating the cholera toxin-producing phage into one of the *V. cholerae* chromosomes. The λ phage integrase discussed in chapter 8 is also a Y recombinase, as are the integrases of integrons and of the conjugative transposons (integrative conjugative elements [Box 5.4]). The Rci invertase, which creates shufflons in the R64 plasmid, is a Y recombinase. Although the reaction they perform is somewhat different, the terminases that resolve the circular dimerized plasmids created during replication of linear plasmids including those in *Borrelia* (see chapter 4) also belong to this family and use a similar mechanism. The Y recombinases are also not limited to eubacteria and include some resolvases of lower eukaryotes such as the Flp recombinase, which inverts a short sequence in the 2μm circle of yeast (Table 9.2), and they are related to some type I topoisomerases of eukaryotes, suggesting that they might have had a common origin.

Details of the molecular basis of recombination by the Y recombinases are outlined in Figure 9.33, and the structures of the sites recognized by some of them are shown in

TABLE 9.2	Examples of tyrosine (Y) recombinases
Source	**Enzyme**
Resolvases	
E. coli	Plasmid/Phage P1 Cre
	Plasmid F ResD
	Phage N15 telomere resolvase
Borrelia spp.	Telomere resolvase
Streptomyces fradiae	Tn*4556*
Shigella sonnei	Plasmid ColIb-P9 shufflon
Agrobacterium spp.	Plasmid Ti
Eubacteria	XerC,D
Saccharomyces spp.	Flp
Integrases	
E. coli	Phage λ
Eubacteria	Integrons
Superfamily	
Eukaryotes	Topoisomerases
Viruses, e.g., vaccinia virus	Topoisomerase

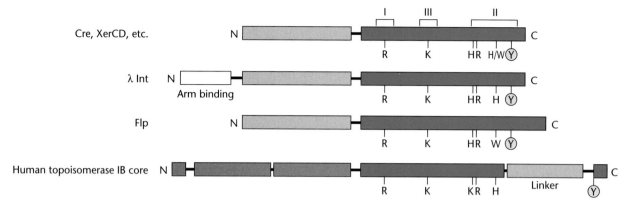

Figure 9.32   Domain structure of tyrosine recombinases (Cre, XerCD, etc., λ Int, and Flp) and eukaryotic type IB topoisomerases. The conserved C-terminal catalytic domain of the proteins is shaded in purple. Brackets show the position of three conserved regions of the catalytic domain: boxes I, II, and III. Residues of the catalytic signature of the family are indicated, and the tyrosine nucleophile is circled Y. Other protein regions are shown in different shades of gray to indicate that they are structurally unrelated. Integrases, such as λ Int, have an additional DNA-binding domain at the N terminus to bind the arm-site sequences of the recombination site. In the human type IB topoisomerase core enzyme, residues 215 to 765, the catalytic domain is interrupted by a linker region spanning the region between the active-site histidine and the tyrosine nucleophile.

Figure 9.34. Much of what we know about how Y recombinases work comes from studies of the structure of the relatively simple Cre recombinase, which has been crystallized bound to various forms of its *loxP* DNA substrate. In the absence of evidence to the contrary, we may assume that at least most features of this reaction can be extrapolated to other Y recombinases, even ones that perform more complex reactions. The *loxP* site recognized by the Cre recombinase consists of a short sequence of 8 bp, where the crossover occurs. It also has two almost identical flanking sequences of 11 bp in inverse orientation that are recognized by the recombinase. In the first step, four copies of the Cre resolvase bind to two *loxP* recognition sites (two to each site) and hold them together in a large complex. Then one strand of each of the recognition sequences is cut in the crossover region by an attack by the active-site tyrosine creating 5′ OH ends. As they are cut, the 3′ phosphate ends are transferred to the side chain of the active-site tyrosine in two of the bound recombinase molecules to form tyrosyl-3′-phosphate bonds. This holds the 3′ phosphate ends and protects them. The free 5′ OH ends then attack the 3′ tyrosyl phosphate bond in the other DNA, rejoining the 5′ OH ends to 3′ phosphate ends on the corresponding strand of the other DNA rather than on their own DNA. This causes two of the strands to cross over and hold the two DNAs together in what is called a Holliday junction. Holliday structures also form during homologous recombination; they are discussed in more detail in chapter 10. The crystal structure reveals that the Holliday junction is being held very flat by the resolvase, with the four DNA branches coming out of the complex in the same plane.

Figure 9.33   Model for the reaction promoted by the Cre tyrosine (Y) recombinase. Four Cre recombinase molecules bind two *loxP* sites, bringing them together. RBE, recombinase-binding element. (1) The active-site tyrosines in two of the Cre molecules, indicated by Y, cleave two of the strands in a phosphoryltransferase reaction that forms 3′ phosphotyrosyl bonds (boxed) and 5′ OH ends (arrows). (2) Each 5′ OH end then attacks the opposite 3′ phosphoryltyrosine bond, switching the strands and causing a 3- to 4-nucleotide swap in the complementary region. (3) The nucleophilic 5′ OH ends attack the phosphotyrosyl bonds, rejoining the ends to form a Holliday junction that isomerizes. The next steps are essentially a repeat of steps 1 to 3, but the other two Cre molecules cut the other two strands by a phosphotransferase reaction and the strands are exchanged and rejoined to form the two recombinant DNA molecules.

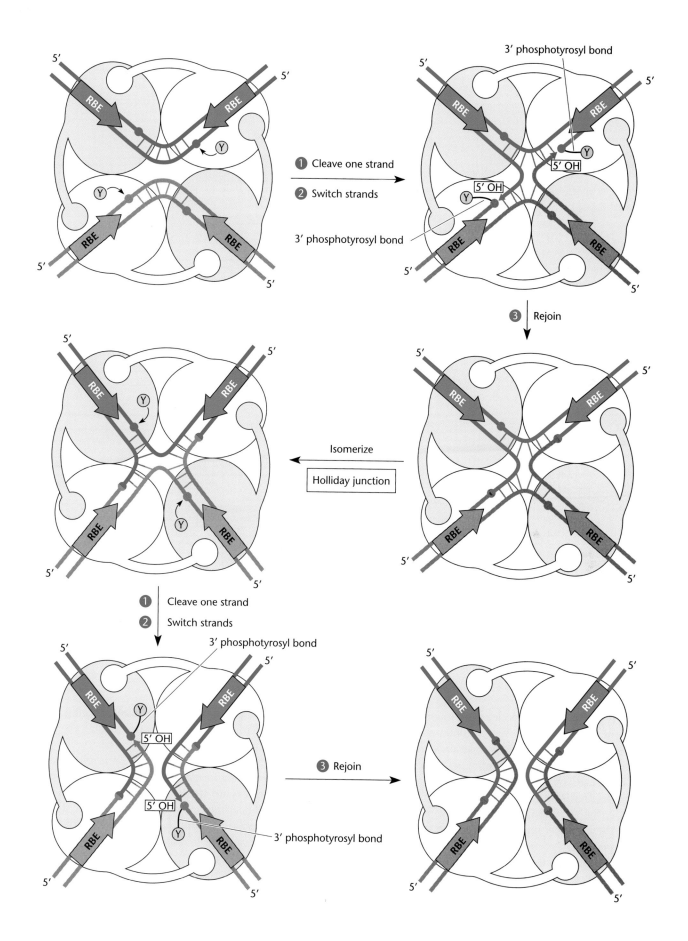

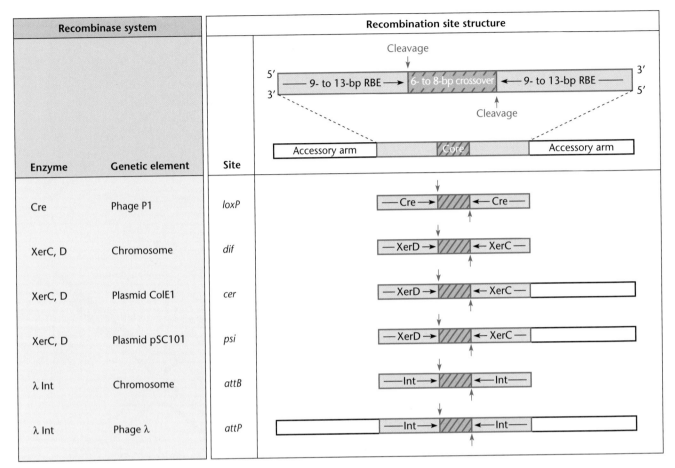

Recombinase system		Recombination site structure	
Enzyme	**Genetic element**	**Site**	
Cre	Phage P1	*loxP*	
XerC, D	Chromosome	*dif*	
XerC, D	Plasmid ColE1	*cer*	
XerC, D	Plasmid pSC101	*psi*	
λ Int	Chromosome	*attB*	
λ Int	Phage λ	*attP*	

Figure 9.34 Structure of some sites recognized by tyrosine (Y) recombinases. The top part of the figure shows (in purple) the basic structure with a core of a 6- to 8-bp crossover region flanked by two 9- to 13-bp sequences required for recombinase binding. These are referred to as recombinase-binding elements (RBE). Many sites also have flanking accessory protein-binding sites called the accessory arms. Proteins bind to these sites and help orient the recombinase and give it specificity. The bottom part of the figure shows the variations on the theme exhibited by some of the sites described in the text. The black arrows indicate inverted repeated sequences.

What happens next is less clear and might differ somewhat in different types of Y recombinases. To achieve recombination, the noncrossover strands must also be cut and joined to the corresponding strand on the other DNA. If the crossover strands are cut again and rejoined to their original strand, no recombination will occur. Therefore, the recombinase has to know which strands to cut in the Holliday junction and which strands to join them to. Holliday junctions can do a number of wonderful things, as discussed in chapter 10. They can isomerize so that the crossed strands become the uncrossed strands and vice versa. They can also migrate so that the position at which the strands are crossed over can move on the DNA, provided that the sequences of the two DNAs are almost the same in the region of migration. One possibil-

ity is that part of the resolvase rotates, forcing the isomerization of the Holliday junction so that the correct strands enter the active center of the other two copies of the resolvase to repeat the cutting and rejoining reaction. This seems unlikely, considering that major changes are not seen in the crystal structure of the complex in its various states. The other possibility is that the Holliday junction migrates, rotating the DNA so that the correct strands to be cut come in contact with the active centers of the other two copies of the resolvase and can be cut and rejoined. However, it is hard to imagine how the Holliday junction could migrate very far since that would mean either drastically rotating the DNA arms that are emerging from the complex or severely distorting the DNA in the complex.

Further complicating the models, some apparent Y resolvases, including the integrases of the so-called conjugative transposon Tn*916* and the integrases of integrons, do not require extensive homology in the crossover region of the two sites being recombined. The Tn*916* integrative element seems to integrate in many places in the chromosome, suggesting that it can use many different sequences as bacterial *att* sites, although it may prefer some sites over others. The *attC* sites in resistance cassettes seem to have very little homology either to each other or to the *intI* site on the integron. Extensive homology in the crossover region should be required for Holliday junction formation and for branch migration. Current research addresses these and other questions about the reactions performed by Y recombinases.

Other proteins besides the Y recombinases themselves are often involved in the recombination reactions (Figure 9.34). These other proteins bind close to the core region, referred to as the accessory arm regions, and help stabilize the recombinase-DNA complexes and/or orientate the recombinase proteins on the DNA. For example, the XerC,D recombinase requires two proteins, ArgR and PepA, to promote recombination at *cer* sites in plasmid ColE1 and resolve dimers (see chapter 4). These proteins also play other very different roles in the cell: one is the repressor of the arginine biosynthetic operon, and the other is an aminopeptidase. It is not clear why the ColE1 plasmid recruits these particular proteins for dimer resolution, but they may happen to have structural and/or sequence features that make them easy to adapt to this role. Whatever the reason, such situations are a nightmare for geneticists trying to deduce the function of a gene product. For example, mutations in the *argR* gene cause constitutive synthesis of the arginine biosynthetic enzymes but also reduce the stability of the ColE1 plasmid for a completely unrelated reason. The λ integrase

TABLE 9.3	Examples of serine (S) recombinases
Source	**Enzyme**
Resolvases	
E. coli	TnpR of Tn*3*
	Resolvase of γδ
	ParA of RP4/RK2
Enterococcus spp.	Tn*917*
Invertases	
S. enterica serovar Typhimurium	Hin flagellar invertase
E. coli phage Mu	Gin tail fiber invertase
Superfamily	
Streptomyces coelicolor	Integrase of phage φC31
Bacillus subtilis	SpoIVCA recombinase
Anabaena spp.	Heterocyst recombinase

also requires that other host proteins, including integration host factor, be bound close to the *attP* site for recombination to occur. This protein is bound at many places in the chromosomal DNA, where it plays multiple roles, making this less surprising.

S Recombinases: Mechanism

The S recombinases are also a large family, comprising many of the plasmid resolution and invertase systems, in both gram-positive and gram-negative bacteria. Some S recombinases are listed in Table 9.3. The TnpR cointegrate resolution systems of Tn*3*-like transposons, including γδ, are members of this family, as are the dimer resolution systems of some plasmids including the gram-negative promiscuous plasmids RK2 and RP4 and the gram-positive *Enterococcus faecalis* plasmid pAMβ1, to list some plasmids mentioned elsewhere in the book. The

Figure 9.35 Domain structure of serine (S) recombinases. The conserved catalytic domain (ca. 120 amino acids) is shown in purple. The conserved amino acids that play major roles in the catalysis are indicated and the active site serines are encircled.

Resolvase/invertase subgroup (e.g., Tn*3* and γδ resolvases and Gin and Hin invertases)

"Large" serine recombinase subgroup (e.g., ΦC31 integrase, Tn*4451* and Tn*5397* transposases, and SpoIVCA)

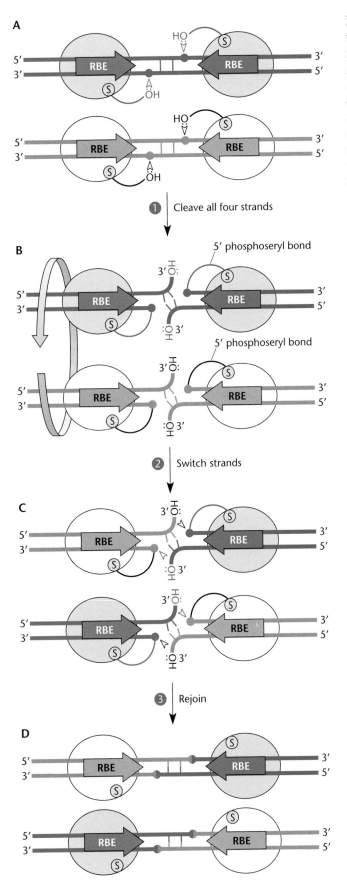

Hin invertase in *Salmonella* and its relatives that invert tail fiber genes in some phages are also members of this family, as are the excisases, mentioned above, which remove the intervening sequences during sporulation in *B. subtilis* and heterocyst formation in *Anabaena*. Some conjugative transposons (integrative conjugative elements) use S integrases rather than Y integrases to integrate into the recipient cell DNA. So far, counterparts of the S recombinases have not been found in eukaryotes, with the possible exception of some in small plasmids in marine diatoms.

The domain structure of some S recombinases is shown in Figure 9.35, and a model of the reaction they perform shown in Figure 9.36. Again, much of what we know comes from studies with one S recombinase, the resolvase of the transposon γδ, which has been crystallized with its DNA substrate. Superficially, the molecular mechanism of site-specific recombination by S recombinases is similar to that of recombination by Y recombinases. The recognition sites have a short crossover region bracketed by copies of recombinase binding sites. Four copies of the recombinase bind to two recognition sequences to form a complex in which the recombination occurs. This is where the similarities end, however. Rather than having the active-site tyrosines in two recombinase molecules make a nucleophilic attack on the same phosphodiester bond in the two DNAs, the active-site serines in all four recombinases make nucleophilic attacks a few nucleotides apart to create staggered breaks in both strands, a total of four breaks. Also, the staggered breaks leave 5′ phosphates and 3′-hydroxyl-end overhangs. The nucleophilic attacks leave the 5′ phosphate ends joined to the hydroxyl group in the side chain of the active-site serines on all four recombinase molecules to form 5′ phosphorylseryl bonds to the ends of the DNA, rather than 3′ phosphoryltyrosyl bonds, as in the Y recombinases;

Figure 9.36 Model for the reaction promoted by the γδ recombinase. (A) Four recombinase molecules bind to two copies of the recombination sites, holding them together, and the active-site serines in all four recombinases attack phosphodiester bonds a few nucleotides apart to leave staggered breaks, with the 5′ phosphate overhangs forming phosphorylserine bonds with the active-site serines. (B) A rotation of a domain of the recombinase brings the corresponding ends of the two different recombination sites together. (C) The free 3′ OH ends then attack the 5′ phosphorylserine bond on the corresponding strand on the other DNA. (D) The nicks are sealed to form the recombinant product. RBE, recombinase-binding element.

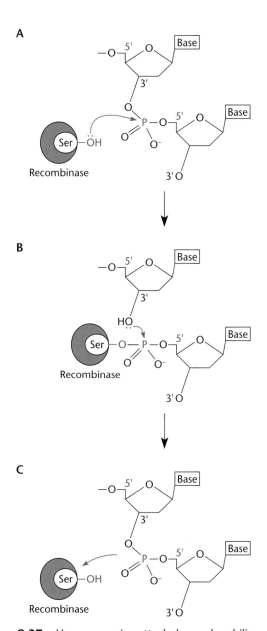

Figure 9.37   How successive attacks by nucleophilic hydroxyl groups of serine (S) recombinases can create a recombinant DNA product. (A) A nucleophilic attack by the hydroxyl group on the side chain of serine in the active center of a serine recombinase forms a 5′ phosphorylserine bond and breaks the phosphodiester bond in the DNA. (B) The free 3′ OH group can attack another 5′ phosphorylserine bond, breaking the bond. (C) This attack results in the re-formation of a phosphodiester bond to a different DNA strand.

Figure 9.37 shows these nucleophilic attacks and how they leave 5′ phosphoryseryl bonds. The nucleophilic 3′ OH on each cut strand then attacks the phosphorylseryl bond in the corresponding strand of the other recognition sequence. Rejoining of the nicks then leaves the recombinant product, without formation of a Holliday junction. Again, as with Y recombinases, it is not clear how the correct ends move and find each other. One possibility is that part of each recombinase rotates to bring its bound end into justaposition with the correct end of the other cut DNA.

Like the Y recombinases, the S recombinases often also use other proteins bound close to the recognition site to help stabilize the complex during recombination. In some cases, these are extra copies of the recombinase itself.

Importance of Transposition and Site-Specific Recombination in Bacterial Adaptation

One of the most important conclusions from studies of transposons and other types of moveable elements in bacteria is the contribution they make to bacterial adaptation. We have seen how integrons can integrate antibiotic resistance genes from large storage areas into transposons that can then tranpose into other DNAs. Conjugative elements including self-transmissible plasmids and conjugative integrative elements can move these transposons from bacterium to bacterium and even into other genera of bacteria. This presumably allows bacterial genomes to remain small but still have access to many types of genes that increase their ability to adapt to different environments. One way this impacts humans directly is in the acquisition of antibiotic resistance by bacteria. Table 9.4 gives a summary of the moveable elements discussed so far that are known to carry antibiotic resistance genes in both gram-negative and gram-positive bacteria. If we are to continue to be able to treat bacterial infections effectively, we must have a clear idea of how these moveable elements can contribute to widespread antibiotic resistance and how to combat it.

TABLE 9.4	Characteristics of genetic elements involved in the spread of antibiotic resistance genes	
Genetic element	**Characteristics**	**Role in spread of resistance genes**
Self-transmissible plasmid	Circular, autonomously replicating element; carries genes needed for conjugal DNA transfer	Transfers resistance genes; mobilizes other elements that carry resistance genes
Conjugative transposon	Integrated element that can excise to form a nonreplicating circular transfer intermediate; carries genes needed for conjugal DNA transfer	Same as self-transmissible plasmid: highly promiscuous, transferring between gram-positive and gram-negative genera and species
Mobilizable plasmid	Circular, autonomously replicating element; carries site and genes that allow it to use the conjugal apparatus provided by a self-transmissible plasmid	Transfer of resistance genes
Transposon	Moves from one DNA segment to another within the same cell	Carries resistance genes from chromosome to plasmid or vice versa
Gene cassette	Circular, nonreplicating DNA segment containing only open reading frames; integrates into integrons	Carries resistance genes
Integron	Integrated DNA segment that contains an integrase, a promoter, and an integration site for gene cassettes	Forms clusters of resistance genes that are transcribed under control of the integron promoter

SUMMARY

1. Nonhomologous recombination is the recombination between specific sequences on DNA that occurs even if the sequences are mostly dissimilar.

2. Transposition is the movement of certain DNA sequences, called transposons, from one place in DNA to another. The smallest known bacterial transposons are IS elements, which contain only the genes required for their own transposition. Other transposons carry genes for resistance to substances such as antibiotics and heavy metals. Transposons have played an important role in evolution and are useful for mutagenesis, gene cloning, and random gene fusions.

3. Composite transposons are composed of DNA sequences bracketed by IS elements. Inside-end transposition by composite transposons can cause the deletion and inversion of neighboring DNA sequences.

4. Most known transposons are DDE transposons, because their transposases have three amino acids, DDE (aspartate-aspartate-glutamate), that hold two magnesium ions which play an essential role in transposition. They are usually characterized by having inverted repeat sequences at their ends and duplicating a short sequence in the target DNA on entry. Other transposons, called Y2 or rolling-circle transposons, essentially replicate themselves into a target DNA by using a free 3' hydroxyl as primer. They do not have inverted repeat sequences on their ends. Other transposons called Y transposons and S transposons are more like integrative elements and lysogenic phages in that they use integrases to integrate into the target DNA, although they generally have less target specificity than do integrases.

5. Bacterial DDE transposons transpose by two distinct mechanisms: replicative transposition and cut-and-paste, or conservative, transposition. In replicative transposition, which is used by such transposons as Tn3 and Mu, the entire transposon is replicated, leading to formation of a cointegrate. In cut-and-paste transposition, which is used by many IS elements and composite transposons, the transposon is cut out of the donor DNA and inserted somewhere else. However, these two mechanisms are closely related and differ only in whether the donor DNA is cut in both strands at the ends of the transposon. The cut-and-paste DDE transposon Tn7 has been converted into a replicative transposon by a single mutation in part of its transposition system so that it now cuts only one strand at the end of the transposon.

6. In transposon mutagenesis, a gene is disrupted by insertion of a transposon, which can introduce a selectable marker and additional restriction sites at the site of insertion which can be used for genetic and physical mapping of the gene.

7. Especially engineered transposons carrying reporter genes can be used to make random gene fusions. Insertion of the transposon into a gene can lead to expression of the reporter gene on the transposon from the promoter or TIR of the disrupted gene, depending on whether the fusion is transcriptional or translational.

8. Genes that have been mutated by insertion of a transposon are often easy to clone in *E. coli* if an antibiotic resistance gene is present on the transposon. Some transposons have been engineered to contain an *E. coli* plasmid origin of

SUMMARY (continued)

replication so that the restriction fragment containing the transposon need not be cloned in another cloning vector but need only be cyclized after ligation to form a replicon in *E. coli.*

9. Site-specific recombinases are enzymes that promote recombination between certain sites on the DNA. Examples of site-specific recombinases include resolvases, integrases, and DNA invertases. The genes for many of these site-specific recombinases have sequences in common, suggesting they may have been derived from a common ancestor.

10. Resolvases are site-specific recombinases encoded by replicative transposons that resolve cointegrates by promoting recombination between short *res* sequences in the copies of the transposon in the cointegrate.

11. Integrases promote nonhomologous recombination between specific sequences on a DNA element such as a phage DNA and the chromosome, integrating the phage DNA into the chromosome to form lysogens. They also integrate antibiotic resistance gene cassettes into transposons. Transposons that encode an integrase which allows then to accept these antibiotic gene cassettes are called integrons. Integrases also play a role in integrating PAIs and other types of genetic islands into the chromosome. These

are large DNA elements (50,000 to 100,000 bp), which carry genes, including genes for pathogenicity, which allow the bacterium to occupy unusual ecological niches.

12. DNA invertases promote nonhomologous recombination between short inverted repeats, thereby changing the orientation of the DNA sequence between them. The sequences they invert, invertible sequences, are known to play an important role in changing the host range of phages and the bacterial surface antigens to avoid host immune defenses.

13. Recombinases can be divided into two types, Y (or tyrosine) recombinases and S (or serine) recombinases. Recombination by both types involves nucleolytic attacks by the hydroxyl group of the side chain of the amino acid on the phosphodiester bond in DNA, forming a phosphoryl bond to the amino acid. They differ in that Y transposons form a 3' phosphoryltyrosine bond whereas S transposons form a 5' phosphorylserine bond; other differences are that Y recombinases cut two strands simultaneously and form a Holliday junction which can then isomerize whereas S recombinases cut all four strands, not necessarily in any order, and do not form a Holliday junction. Rather, S recombinases may depend on rotation of part of the recombinase to bring the different strands into juxtaposition.

QUESTIONS FOR THOUGHT

1. For transposons that transpose replicatively (e.g., Tn3), why are there not multiple copies of the transposon around the genome?

2. How do you think that transposon Tn3 and its relatives have spread throughout the bacterial kingdom?

3. Transposons are not just parasitic DNAs, and they serve useful purposes for the host! List some of these purposes.

4. Where do you suppose the genes that were inserted into integrons in the evolution of transposons came from?

5. If the DNA invertase enzymes are made continuously, why do the invertible sequences invert so infrequently?

6. In the experiments of Bender and Kleckner on Tn10 transposition, why were only 16% of the colonies sectored and not 50%?

PROBLEMS

1. Outline how you would use an HFT λ strain to show that some *gal* mutations of *E. coli* are due to insertion of an IS element.

2. In the experiments shown in Figures 9.15 and 9.16, what would have been observed if Tn10 transposed by a replicative mechanism?

3. List the advantages and disadvantages of transposon mutagenesis over chemical mutagenesis to obtain mutations.

4. Outline how you would use transposon mutagenesis to mutagenize plasmid pBR322 in *E. coli* with transposon Tn5. Determine what size of junction fragments would be obtained

with PstI and HindIII if the transposon hopped into the 1-kb position clockwise of 0 kb on the plasmid. See chapter 4 for a map of pBR322. The plasmid is 4.36 kb, and the HindIII and PstI sites are at 0.029 and 3.673 kb, respectively. See Figure 9.22 for a map of Tn5.

5. How would you determine whether a new transposon you have discovered integrates randomly into DNA?

6. In the example shown in Figures 9.22 to 9.24, digestion with the restriction endonuclease SalI gives two fragments of 2.972 and 6.228 kb. In what orientation has the transposon inserted? Draw a picture of the transposon inserted in the

plasmid, showing the position of the kanamycin resistance gene in the transposon.

7. You have isolated a strain of *Pseudomonas putida* that can grow on the herbicide 2,4-dichlorophenoxyacetic acid (2,4-D) as the sole carbon and energy source. Outline how you would clone the genes for 2,4-D utilization (i) by transposon mutagenesis and (ii) by complementation in the original host.

8. Propose a mechanism whereby the same two proteins, MuA and MuB, could integrate Mu by a single cut-and-paste transposition event to form a lysogen after infection but replicatively transpose Mu DNA hundreds of times after induction of the lysogen.

9. The Int protein of conjugative transposons such as Tn*916* must integrate the transposon into the chromosome of the recipient cell after transfer. How would you show whether the Int protein of the transposon must be synthesized in the recipient cell or can be transferred in with the transposon during conjugation, much like primases are transferred during some types of plasmid conjugation?

10. How would you show whether the G segment of Mu can invert while it is in the prophage state or whether it inverts only after the phage is induced?

11. The defective prophage e14 resides in the *E. coli* chromosome and has an invertible sequence. How would you determine if its invertase can invert the invertible sequences of Mu, P1, and *Salmonella enterica* serovar Typhimurium?

12. The red color of *Serratia marcescens* is reversibly lost with a high frequency. Outline how you would attempt to determine if the change in pigment is due to an invertible sequence.

13. How would you prove that the Mu phage transposon probably does not encode a resolvase?

SUGGESTED READING

Bender, J., and N. Kleckner. 1986. Genetic evidence that Tn*10* transposes by a nonreplicative mechanism. *Cell* **45:**801–815.

Casadaban, M. J., and S. N. Cohen. 1979. Lactose genes fused to exogenous promoters in one step using a Mu-lac bacteriophage: in vivo probe for transcriptional control sequences. *Proc. Natl. Acad. Sci. USA* **76:**4530–4533.

Collis, C. M., G. D. Recchia, M.-J. Kim, H. W. Stokes, and R. M. Hall. 2001. Efficiency of recombination reactions catalyzed by class I integron integrase IntI1. *J. Bacteriol.* **183:**2535–2542.

Craig, N. L. 2002. Tn*7*, p. 423–456. *In* N. L. Craig, R. Craigie, M. Gellert, and A. M. Lambowitz (ed.), *Mobile DNA II.* ASM Press, Washington, D.C.

Derbyshire, K. M., and N. D. F. Grindley. 2005. DNA transposons: different proteins and mechanisms but similar rearrangements, p. 467–497. *In* N. P. Higgins (ed.), *The Bacterial Chromosome.* ASM Press, Washington, D.C.

Foster, T. J., M. A. Davis, D. E. Roberts, K. Takeshita, and N. Kleckner. 1981. Genetic organization of transposon Tn*10*. *Cell* **23:**201–213.

Gill, R., F. Heffron, G. Dougan, and S. Falkow. 1978. Analysis of sequences transposed by complementation of two classes of transposition-deficient mutants of Tn*3*. *J. Bacteriol.* **136:**742–756.

Golden, J. W., S. G. Robinson, and R. Haselkorn. 1985. Rearrangement of nitrogen fixation genes during heterocyst differentiation in the cyanobacterium *Anabaena*. *Nature* (London) **327:**419–423.

Groisman, E. O., and M. J. Casadaban. 1986. Mini-Mu bacteriophage with plasmid replicons for in vivo cloning and *lac* gene fusion. *J. Bacteriol.* **168:**357–364.

Gueguen, E., P. Rousseau, G. Duval-Valentin, and M. Chandler. 2005. The transpososome: control of transposition at the level of catalysis. *Trends Microbiol.* **13:**543–549.

Hallet, B., V. Vanhooff, and F. Cornet. 2004. DNA site-specific resolution systems, p. 145–180. *In* B. E. Funnell and G. J. Phillips (ed.), *Plasmid Biology.* ASM Press, Washington, D.C.

Hughes, K. Y., and S. M. Maloy (ed.). 2007. *Methods in Enzymology*, vol. 421. *Advanced Bacterial Genetics: Use of Transposons and Phage for Genomic Engineering.* Elsevier, London, United Kingdom.

Kenyon, C. J., and G. C. Walker. 1980. DNA-damaging agents stimulate gene expression at specific loci in *Escherichia coli*. *Proc. Natl. Acad. Sci. USA* **77:**2819–2823.

Komano, T. 1999. Shufflons: multiple inversion systems and integrons. *Annu. Rev. Genet.* **33:**171–191.

Kunkel, B., R. Losick, and P. Stragier. 1990. The *Bacillus subtilis* gene for the developmental transcription factor σ^K is generated by excision of a dispensable DNA element containing a sporulation recombinase gene. *Genes Dev.* **4:**525–535.

Martin, S. S., E. Pulido, V. C. Chu, T. S. Lechner, and E. P. Baldwin. 2002. The order of strand exchanges in Cre-LoxP recombination and its basis suggested by the crystal structure of the Cre-LoxP Holliday junction complex. *J. Mol. Biol.* **319:**107–127.

May, E. W., and N. L. Craig. 1996. Switching from cut-and-paste to replicative Tn*7* transposition. *Science* **272:**401–404.

Rakin, A., C. Noelting, P. Schropp, and J. Heesemann. 2001. Integrative module of the high-pathogenicity island of *Yersinia*. *Mol. Microbiol.* **39:**407–415.

Reznikoff, W. S. 2003. Tn*5* as a model for understanding DNA transposition. *Mol. Microbiol.* **47:**1199–1206.

Rowe-Magnus, D. A., A.-M. Guerout, P. Ploncard, B. Dychinco, J. Davies, and D. Mazel. 2001. The evolutionary history of chromosomal superintegrons provides an ancestry for multiresistant integrons. *Proc. Natl. Acad. Sci. USA* **98:**652–657.

Shapiro, J. A. 1979. Molecular model for the transposition and replication of bacteriophage Mu and other transposable elements. *Proc. Natl. Acad. Sci. USA* **76:**1933–1937.

Siguier, P., J. Filée, and M. Chandler. 2006. Insertion sequences in prokaryotic genomes. *Curr. Opin. Microbiol.* **9:**526–531.

Simon, M., J. Zeig, M. Silverman, G. Mandel, and R. Doolittle. 1980. Phase variation: evolution of a controlling element. *Science* **209:**1370–1374.

Simon, R., U. Preifer, and A. Puhler. 1983. A broad host range mobilization system for in vivo genetic engineering: transposon mutagenesis in gram negative bacteria. *Bio/Technology* **1:** 784–790.

Toleman, M. A., P. M. Bennett, and T. R. Walsh. 2006. ISCR elements: novel gene-capturing systems of the 21st century. *Microbiol. Mol. Biol. Rev.* **70:**296–316.

van de Putte, P., R. Plasterk, and A. Kuijpers. 1984. A Mu *gin* complementing function and an invertible DNA region in *Escherichia coli* K-12 are situated on the genetic element *e*14. *J. Bacteriol.* **158:**517–522.

Van Gijsegem, F., and A. Toussaint. 1983. In vivo cloning of *Erwinia carotovora* genes involved in catabolism of hexuronates. *J. Bacteriol.* **154:**1227–1235.

CHAPTER **10**

Molecular Mechanisms of Homologous Recombination

E ven two organisms belonging to a species that must reproduce sexually are not genetically identical. As chromosomes are assembled into germ cells (sperm and eggs), one chromosome of each pair of homologous chromosomes is chosen at random. Consequently, it is highly unlikely that two siblings will be alike, because each of their parents' germ cells contains a mixture of chromosomes originally derived from the siblings' grandparents. In addition, the chance of two siblings being the same is even lower because of **genetic recombination.** Because of recombination, genetic information that was previously associated with one DNA molecule may become associated with a different DNA molecule, or the order of the genetic information in a single DNA molecule may be altered. As mentioned at the beginning of chapter 9, the two general types of recombination are nonhomologous (site-specific) recombination and homologous recombination. Site-specific recombination occurs only in specific situations and requires special proteins that recognize specific sequences and promote recombination between them. When we talk about recombination, we are usually referring to **homologous recombination,** which occurs more generally. This type of recombination can occur between any two DNA sequences that are the same or very similar, and it usually involves the breaking of two DNA molecules in the same region, where the sequences are similar, and the joining of one DNA to the other. The result is called a **crossover.** Depending on the organism, homology-dependent crossovers can occur between homologies as short as 23 bases, although longer homologies produce more frequent crossovers.

All organisms on Earth probably have some mechanism of homologous recombination, suggesting that recombination is very important for species survival. The new combinations of genes obtained through recombination allow the species to adapt more quickly to the environment and speed up the process of evolution. Recombination can allow an organism to change the order of its own genes or move genes to a different replicon, for example, from the chromosome to a plasmid. Recombination genes also play an important role in the repair of damage to DNA and in mutagenesis, topics covered in the next chapter. Probably most important, however, is the role of recombination in replication restarts when the replication has stalled at damage in the DNA. Recombination allows an error-free way to reassemble the replication fork so that replication can continue. This subject is also treated in more detail in the next chapter.

Because of its importance in genetics, homologous recombination has already been mentioned in previous chapters, for example in discussions of deletion and inversion mutations and genetic analysis. Determination of recombination frequencies allows us to measure the distance between mutations and thus can be used to map mutations with respect to each other, as we discussed in chapters 3 and 7, among others. Moreover, clever use of recombination can take some of the hard work out of cloning genes and making DNA constructs, and we have already referred to some of these technologies.

When we discussed the use of recombination for genetic mapping and many other types of applications in previous chapters, we used a simplified description of recombination: the strands of two DNA molecules break at a place where they both have the same sequence of bases and then the strands of the two DNA molecules join with each other to form a new molecule. This model is obviously too simple, but we could use recombination without knowing its molecular mechanisms in any detail. People have been using recombination for 80 years or more without knowing the actual molecular mechanisms involved. In fact, as we will see, models of recombination are still being debated and recombination can proceed by different mechanisms depending on the situation. In this chapter, we focus on the actual mechanisms of recombination—what actually happens to the DNA molecules involved—discussing some molecular models and some of the genetic evidence that supports or contradicts those models. We also discuss the proteins involved in recombination, mostly in *Escherichia coli*, the bacterium for which recombination is best understood and which has served as the paradigm for recombination in all other organisms.

Overview of Recombination

Recombination is a remarkable process. Somehow, two enormously long DNA molecules in a cell link up and exchange sequences. Moreover, recombination usually occurs only at homologous regions of two DNAs. Thus,

these regions must line up so that they can be broken and rejoined, and the long DNA molecules on either side of the point of recombination must change their configuration with respect to each other. This complicated process clearly involves many functions. But before presenting detailed models for how recombination might occur and the functions involved, we consider the basic requirements any recombination model must satisfy and what functions to expect of gene products directly involved in recombination.

Requirement 1: Pairing between Identical or Very Similar Sequences in the Crossover Region

The distinguishing feature of homologous recombination is that the deoxynucleotide sequence in the two regions of DNA where a crossover occurs must be the same or very similar, and all recombination models must start with this requirement. This prerequisite serves a very practical function in recombination. The sequence of nucleotides in molecules of DNA from different individuals of the same species are usually almost identical over the entire lengths of the molecules, and by extension, two DNA molecules generally share identical sequences only in the same regions. Thus, recombination usually occurs only between sites on the two DNAs that are in the same place with respect to the entire molecule. Recombination between different regions in two DNA molecules does sometimes occur because the same or similar sequences occur in more than one place in the DNAs. This type of recombination, sometimes called **ectopic** or **homeologous recombination**, gives rise to deletions, duplications, inversions, and other gross rearrangements of DNA (see chapter 3). Not surprisingly, cells have evolved special mechanisms to discourage ectopic recombination, one of which is discussed in the next chapter in the section on mismatch repair.

Requirement 2: Complementary Base Pairing between Double-Stranded DNA Molecules

Logic dictates that the DNA molecules find each other by complementary base pairing. Mandatory complementary base pairing between strands of the two DNA molecules ensures that recombination will occur only between sequences at the same **locus,** that is, the same place on the molecules. The point at which two double-stranded DNA molecules are held together by complementary base pairing between their strands is called a **synapse.** All recombination models must involve synapse formation. However, in double-stranded DNA, the bases are usually hidden inside the helix, where they are not available for base pairing. Separating the strands in various regions to expose the bases would be too slow to allow efficient recombination. We know that after DNA enters a bacterial cell by transduction, transformation, or conjugation, the incoming DNA finds its complementary sequence very quickly and recombination occurs within minutes. Therefore, the cell must contain some function(s) that allows the incoming DNA quickly to scan the enormously long bacterial chromosome to locate and pair with its complementary sequence.

Requirement 3: Cutting and Rejoining by Recombination Enzymes

Once the complementary sequences have found each other and paired, for a crossover (or true recombination) to occur between the two DNA molecules, the strands of each molecule must be broken and rejoined to the corresponding strands of the other DNA molecule. Therefore, DNA endonucleases and ligases—enzymes that break DNA strands and rejoin them, respectively—are required for recombination.

Requirement 4: Heteroduplex Formation Involving All Four Strands

The regions of complementary base pairing between the two DNA molecules in a synapse are called **heteroduplexes**, because the strands in these regions come from different DNA molecules. In principle, for a synapse to form, heteroduplexes need form only between two of the strands of the DNA molecules being recombined.

However, evidence indicates that all strands of the two DNA molecules are involved in heteroduplex formation, an observation that must be explained by the models.

Molecular Models of Recombination

Several models have been proposed to explain recombination at the molecular level. These models include the required features of recombination discussed above and also account for additional experimental evidence. However, no single model of recombination can make an exclusive claim to the truth, and recombination may occur by different pathways in different situations. Nevertheless, these models serve as a framework for forming hypotheses that can be tested through experimentation and will help focus thinking about recombination at the molecular level.

The Holliday Double-Strand Invasion Model

The first widely accepted model for recombination was proposed by Robin Holliday in 1964. Figure 10.1 illustrates the basic steps of the **Holliday model**. According to this model, recombination is initiated by two single-strand breaks made simultaneously at exactly the same place in the two DNA molecules to be recombined. Then the free ends of the two broken strands cross over each other, each pairing with its complementary sequence in the other DNA molecule to form two heteroduplexes. The ends are

Figure 10.1 The Holliday model for genetic recombination. One strand of each DNA molecule is cut at the same position and then pairs with the other molecule to form a heteroduplex (region of purple paired with black). The strands are then ligated to form the Holliday junction. This DNA structure can isomerize between forms I and II. Cutting and ligating resolves the Holliday junction. Depending on the conformation of the junction, the flanking markers A, B, a, and b recombine or remain in their original conformation. The product DNA molecules contain heteroduplex patches.

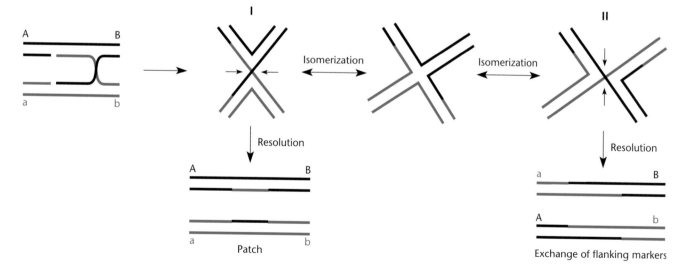

then ligated to each other, resulting in a cruciform-like structure called a **Holliday junction** in which the two double-stranded molecules are held together by their crossed-over strands (Figure 10.1). The formation of Holliday junctions is central to all the recombination models discussed here and to the site-specific recombination by Cre and other Y recombinases (see chapter 9).

Once formed, Holliday junctions can undergo a rearrangement that changes the relationship of the strands to each other. This rearrangement is called an **isomerization** because no bonds are broken. As shown in the figure, isomerization causes the crossed strands in configuration I to uncross. A second isomerization occurs to create configuration II, where the ends of the two double-stranded DNA molecules are in the recombinant configuration with respect to each other. Notice that the strands that were not crossed before are now the crossed strands and vice versa. It may seem surprising that the strands that have crossed over to hold the two DNA duplexes together can change places so readily, but experiments with models show that the two structures I and II are actually equivalent to each other and the Holliday junction can change from one form to the other without breaking any hydrogen bonds between the bases. Hence, flipping from one configuration to the other requires no energy and should occur quickly, so that each configuration should be present approximately 50% of the time.

Once formed, the crossed strands in a Holliday junction can be cut and religated, or **resolved**, as shown in Figure 10.1. Whether or not recombination occurs depends on the configuration of the junction at resolution, in other words which of the strands are the crossed strands. If the Holliday junction is in configuration I when the crossed strands are cut, the flanking DNA sequences will not be recombined and the two DNAs will return to the way they were. However, if the Holliday junction is in configuration II when the crossed strands are cut, the flanking DNA sequences will be exchanged between the two molecules and recombination will have occurred. This is indicated by recombination of the flanking markers shown in Figure 10.1.

Holliday junctions can also move up and down the DNA by breaking and re-forming the hydrogen bonds between the bases. This process is called **branch migration** (Figure 10.2). The same number of hydrogen bonds are broken and re-formed as the cross-connection moves, so that no energy is required. However, without the expenditure of some energy, hydrogen bonds may not be broken fast enough for efficient branch migration. Also, the rate of migration is decreased if the Holliday junction encounters a mismatch in the DNA, such as the one shown in Figure 10.2. Specific ATP-hydrolyzing proteins speed up the branch migration on Holliday junctions and

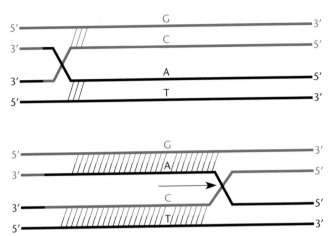

Figure 10.2 Migration of Holliday junctions. By breaking the hydrogen bonds holding the DNAs together in front of the branch and re-forming them behind, the junction migrates and extends the regions of pairing (i.e., the heteroduplexes) between the two DNAs. The heteroduplex region is hatched. In the example, two mismatches, GA and CT, form in the heteroduplex regions because one of the DNA molecules has a mutation in this region.

direct it in one direction or the other. In *E. coli*, these proteins include the Ruv proteins, RecG and and RecA as we discuss below.

As shown in Figure 10.2, branch migration in Holliday junctions can have the effect of increasing the length of the heteroduplex regions. If a heteroduplex extends to include a region of differing sequence, a mismatch will occur, possibly leading to gene conversion (discussed later in the chapter). The other models we discuss all invoke branch migration of Holliday junctions to explain the experimental evidence concerning gene conversion and the length and distribution of heteroduplexes.

The Holliday model is called a double-strand invasion model because one strand from each DNA molecule invades the other DNA molecule (Figure 10.1), explaining how heteroduplexes can form on both molecules of DNA during recombination. However, one problem with this model is that the two DNA molecules must be simultaneously cut at almost the same place to initiate recombination. But how could the two like DNA molecules line up for pairing before they are cut when the bases are hidden inside the double-stranded DNA helix and so are not free to pair with other DNA molecules? And, if the two DNA molecules were not aligned, how could they be cut at exactly the same place? To answer these questions, Holliday proposed the existence of certain sites on DNA that are cut by special enzymes to initiate the recombination. However, there is no evidence for such sites, and recombination seems to occur more or less randomly over the entire DNA.

Despite these objections, the Holliday double-strand invasion model has served as the standard against which all other models of recombination are compared. All the most favored models involve the formation of Holliday junctions, isomerization, and branch migration. They differ mostly in the earlier stages, before Holliday junctions have formed.

Single-Strand Invasion Models

One way to overcome the objections to the Holliday model is to modify it with the proposal of a single-strand invasion model, such as the one shown in Figure 10.3. In this model, a strand of one of the two DNA molecules is cut at random and then the exposed 3′ hydroxyl end invades another double-stranded DNA until it finds its complementary sequence. If it finds such a sequence, it displaces the noncomplementary strand. This displaced strand can then be cut and joined to the corresponding

Figure 10.3 A single-strand invasion model. (1) A single-stranded end invades a homologous double-stranded DNA molecule. (2) The displaced strand on the double-stranded DNA molecule is degraded (signified by black hatching). After strand exchange, DNA polymerase fills the gap. Initially, a heteroduplex (represented by purple-black hatching) forms on only one of the two DNA molecules. (3) Branch migration also causes a heteroduplex to form on the other DNA molecule. Isomerization (not shown) can recombine the flanking DNA molecules, as in the Holliday model.

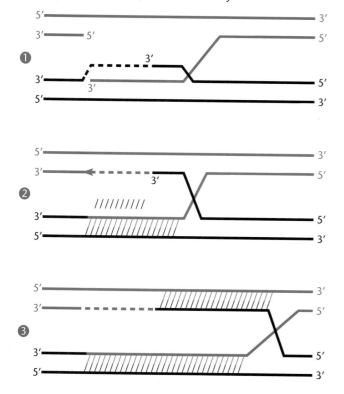

strand from the other DNA, creating a Holliday junction. The Holliday junction can then migrate to create heteroduplexes on both strands, a requirement of any model. It can also isomerize and be resolved to create recombinant products. Such models do not require the coordinated breaking of the two DNA molecules at exactly the same place; they are consistent with the action of the RecBCD enzyme in processing ends for recombination and the role of RecA protein in forming synapses. The RecBCD enzyme leaves a 3′ hydroxyl single-stranded end that can invade another DNA, and the RecA protein can bind to single-stranded DNA and allow it to invade double-stranded DNA (see below).

The Double-Strand Break Repair Model

In both single-strand and double-strand invasion models, a single-strand break in one or both of the two DNA molecules, respectively, initiates the recombination event. If recombination is initiated with a single-strand break in DNA, the other strand can still hold the molecule together. However, if recombination initiated with a double-strand break in one of the two DNAs, the two parts of the DNA should fall apart, a lethal event. Therefore, a priori, it seemed unlikely that a double-strand break in DNA could initiate recombination. Thus, double-strand break models were at first ruled out as being counterintuitive. However, it now seems clear that recombination is often initiated by a double-strand break in one of the two DNA molecules, at least in some situations.

The first evidence that double-strand breaks in DNA can initiate recombination came from genetic experiments with the yeast *Saccharomyces cerevisiae* (see Szostak et al., Suggested Reading). These experiments were aimed at analyzing the ability of recombination between plasmids and chromosomes to repair double-strand breaks and gaps in the plasmids by inserting the corresponding sequence from the chromosomes. However, the initiation of recombination by double-strand breaks is now known to be a general mechanism. For example, homing DNA endonucleases in bacteria, phages, and lower eukaryotes initiate recombination by making a double-strand break (Box 10.1).

Figure 10.4 illustrates the model for this type of recombination (see Szostak et al., Suggested Reading). Both strands of one of the two DNA molecules participating in the recombination are broken, and the 5′ ends from each break are digested by an exonuclease, leaving a gap with exposed single-stranded 3′ tails. One of these tails invades the other double-stranded DNA until it finds its complementary sequence. Then DNA polymerase extends the tail along the complementary sequence, displacing the other strand with the same sequence, until it reaches the free 5′ end of the invading strand and is joined to it by

Breaking and Entering: Introns and Inteins Move by Double-Strand Break Repair or Retrohoming

Introns and inteins are parasitic DNA elements that are sometimes found in genes, both in eukaryotes and in prokaryotes (see Box 2.6). Like all good parasites, they have as little effect on the health of their host as possible. This is a matter of self-interest, because if their host dies, they die with it. When they hop into a DNA, they could disrupt a gene, which could inactivate the gene product and be deleterious to their host. As discussed in Box 2.6, they can avoid inactivating the product of the gene in which they reside by splicing their sequences out of the mRNA before it is translated (in the case of introns) or out of the protein product of the gene after it is made (in the case of inteins). Sometimes this splicing requires other gene products of the intron or the host, and sometimes it occurs spontaneously, in a process called self-splicing. Self-splicing introns were one of the first-known examples of RNA enzymes or ribozymes.

Many introns and inteins are able to move from one DNA to another. When they move, they usually move from one gene into exactly the same location in the same gene of a member of the same species which previously lacked them. In that way, they can move through a population until almost all of the individuals in the population have the intron in that location. Because they always return to the same site, this process is called **homing.** There is a good reason why they choose to move into exactly the same position in the same gene rather than into other places in the same gene or even other places in the genome. The sequences in the gene around the intron or intein, called exon or extein sequences, respectively, also participate in the splicing reaction; therefore, if they find themselves somewhere else where these flanking exon sequences are different, they will not be able to splice themselves out of the RNA or protein. Homing allows the intron or intein to spread through a population by parasitizing other DNAs that lack it but never disrupting the product of an essential gene and disabling or killing the new host as it moves.

There are two basic mechanisms by which introns home. Some introns, called group I introns, and inteins, home by double-strand break repair. To move by double-strand break repair, the intron or intein first makes a double-strand break in the homing site of the target gene into which it must move. To accomplish this, the intron or intein encodes a specific DNA endonuclease called a homing nuclease, which makes a break only in this particular sequence. In group I introns, this homing endonuclease is usually encoded by an open reading frame on the intron. In inteins, the intein iself becomes the homing endonclease after it is spliced out of the protein. After the double-strand cut is made by the specific endonuclease,

the corresponding gene containing the intron repairs the cut by double-strand break repair, replacing the sequence without the intron with the corresponding sequence containing the intron by gene conversion (see Figure 10.4). After repair, both DNAs now contain the intron in exactly the same position. Other DNA elements, including the mating-type loci of yeast, are known to move by a similar mechanism of double-strand break repair.

Other introns, called the group II introns, move by a process called **retrohoming.** These introns essentially splice themselves into one strand of the target DNA by a process analogous to splicing the intron out of the mRNA but in reverse (see Box 2.6). The intron also encodes an endonuclease that makes a cut in the other strand, and the exposed 3' hydroxyl end then primes an intron-encoded reverse transcriptase that makes a DNA copy of the intron, which is then joined to the target site DNA by host DNA repair enzymes. The intron is homed to its target site by homologous sequences in the intron and the target site. The most important sequence in the intron for this complementary base pairing with the homing site is only 15 bp long. Other shorter complementary sites flanking this region are recognized by the nuclease that cuts the homing site, allowing integration of the intron.

Because group II introns recognize their homing site almost exclusively by complementary base pairing, it is possible to redirect the introns to other sites merely by changing the sequence of the 15-bp sequence on the intron so that it is now complementary to a different region on the chromosome. The efficiency of insertion into the new target site increases the more the other DNA sequences flanking the new complementary sequence resemble the sequences recognized by the intron nuclease in the original homing site. The ability of group II introns to be redirected to new sites, merely by changing the sequence of part of the intron, has allowed their development as site-specific mutagenesis systems. This system, marketed as TargeTron by Sigma-Aldrich, in theory allows the insertion of the intron into essentially any gene in any organism. Basically, PCR is used to make a mutated version of the 15-bp region of the transposon that is complementary to the sequence into which the intron is to be inserted. Software is provided that identifies the best region in the gene into which to insert the intron, based on which region requires the fewest changes in the other flanking complementary sequences, which can then also be changed by using multiple PCR primers. Once this region of the transposon is PCR amplified, it is cloned into a vector containing the rest of the transposon, plus a kanamycin resistance cassette that is

BOX 10.1 (continued)

Breaking and Entering: Introns and Inteins Move by Double-Strand Break Repair or Retrohoming

expressed only if the intron has excised from the vector. When the RNA is made on the intron by using the T7 phage RNA polymerase promoter and T7 RNA polymerase expressed from a different DNA, the RNA nuclease and reverse transcriptase encoded by the intron are expressed and the intron integrates itself specifically into the selected target site. Such integrations can be selected on kanamycin-containing plates. If another gene has been cloned into the intron on the plasmid, this other gene will be integrated at the new site. A variation of this method uses a mutant intron to introduce small changes such as base pair changes close to the new homing site. While less efficient than recombineering (Box 10.3), this

method has the advantage of being more easily adapted to bacteria other than *Escherichia coli* and its close relatives, by expressing the T7 RNA polymerase from a different vector in the bacterium to be mutagenized or by using a different promoter to transcribe the intron.

References
Belfort, M., M. E. Reaban, T. Coetzee, and J. Z. Dalgaard. 1995. Prokaryotic introns and inteins: a panoply of form and function. *J. Bacteriol.* **177**:3897–3903.

Karberg, M., H. Guo, J. Zhong, R. Coon, J. Perutka, and A. M. Lambowitz. 2001. Group II introns as controllable gene targeting vectors for genetic manipulation of bacteria. *Nat. Biotechnol.* **19**:1162–1167.

Figure 10.4 The double-strand break repair model. (1) A double-strand break in one of the two DNAs initiates the recombination event. The arrows indicate the degradation of the 5′ ends at the break. (2) One 3′ end, or tail, then invades the other DNA, displacing one of the strands. (3) This 3′ end serves as a primer for DNA polymerase, which extends the tail until it can eventually be joined to a 5′ end (black arrow). Meanwhile, the displaced strand (in black) serves as a template to fill the gap left in the first DNA (dashed lines). Two Holliday junctions form and may produce recombinant flanking DNA, depending on how they are resolved.

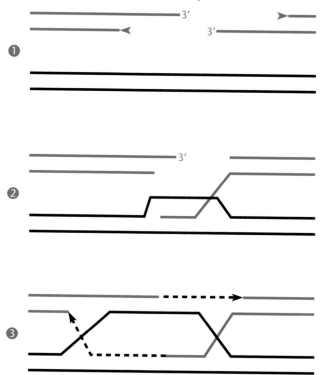

DNA ligase. The other free 3′ end can then be used as a primer to fill in the remaining gap by using the displaced strand as a template before being joined to the free 5′ end in its own strand by DNA ligase. This causes two Holliday junctions to form (Figure 10.4).

Whether recombination occurs depends on the configuration of the two Holliday junctions when resolved. If both are in the same configuration—either I or II (Figure 10.1)—when they are resolved, no crossovers, and thus no recombination, will result. However, if the two junctions are in different configurations at resolution, recombination will occur.

In this model, heteroduplex DNA initially forms in only one DNA molecule, between the strand that does the initial invading and its complementary sequence. However, as in the single-strand invasion models, the migration of the Holliday junctions can lead to heteroduplex formation on both strands. Each of these models was proposed to overcome objections to earlier models and to accommodate experimental evidence. However, as discussed below, recombination can occur by a number of different pathways, determined mostly by the DNA structures that initiate the recombination and the proteins involved.

The Molecular Basis for Recombination in *E. coli*

As with many cellular phenomena, much more is known about the molecular basis for recombination in *E. coli* than in any other organism. At least 25 proteins involved in recombination have been identified in *E. coli*, and specific roles have been assigned to many of these (Table 10.1).

TABLE 10.1	Some genes encoding recombination functions in *E. coli*		
Gene	Mutant phenotype	Enzymatic activity	Probable role in recombination
recA	Recombination deficient	Enhanced pairing of homologous DNAs	Synapse formation
recBC	Reduced recombination	Exonuclease, ATPase, helicase, χ-specific endonuclease	Initiates recombination by separating strands, degrading DNA up to a χ site
recD	Rec⁺ χ independent	Stimulates exonuclease	Degrading 3′ ends
recF	Reduced plasmid recombination	Binds ATP and single-stranded DNA	Substitutes for RecBCD at gaps
recJ	Reduced recombination in RecBC⁻	Single-stranded exonuclease	Substitutes for RecBCD at gaps
recN	Reduced recombination in RecBC⁻	ATP binding	Substitutes for RecBCD at gaps
recO	Reduced recombination in RecBC⁻	DNA binding and renaturation	Substitutes for RecBCD at gaps
recQ	Reduced recombination in RecBC⁻	DNA helicase	Substitutes for RecBCD at gaps
recR	Reduced recombination in RecBC⁻	Binds double-stranded DNA	Substitutes for RecBCD at gaps
recG	Reduced Rec in RuvA⁻B⁻C⁻	Branch-specific helicase	Migration of Holliday junctions
ruvA	Reduced recombination in RecG⁻	Binds to Holliday junctions	Migration of Holliday junctions
ruvB	Reduced recombination in RecG⁻	Holliday junction-specific helicase	Migration of Holliday junctions
ruvC	Reduced recombination in RecG⁻	Holliday junction-specific nuclease	Resolution of Holliday junctions
priA, priB, priC, dnaT	Reduced recombination	Helicase?	Reload replication forks

chi (χ) Sites and the RecBCD Nuclease

The first analysis of the genetic requirements for recombination in *E. coli* used Hfr crosses (see Clark and Margulies, Suggested Reading, and chapter 3). The *recB* and *recC* genes were among the first *rec* genes found because their products are required for recombination after Hfr crosses (see Clark and Margulies, Suggested Reading, and "Genetic Analysis of Recombination in Bacteria" below). Their products were later shown to also be required for transductional crosses. The products of these genes and the product of another gene, *recD*, form a heterotrimer, accordingly named the RecBCD nuclease. The *recD* gene product is not required for recombination after such crosses, and so this gene had to be found in other ways (see below). We now know that the RecBCD enzyme is required for the recombination that occurs after conjugation and transduction because the small pieces of DNA transferred into cells by Hfr crosses or by transduction are the natural substrates for the RecBCD enzyme. The RecBCD enzyme processes the ends of these pieces to form single-stranded 3′ ends, which can then invade the chromosomal DNA to form recombinants (see below). If these original searches had used a different selection that favored another class of *rec* genes, for example studying the requirements for recombination initiating at single-stranded gaps in the DNA, other genes would have been found first, in this case genes encoding enzymes of the RecFOR pathway (see below).

HOW RecBCD WORKS

The RecBCD protein is a remarkable enzyme with many enzyme activities. It has single-stranded DNA endonuclease and exonuclease activities as well as DNA helicase and DNA-dependent ATPase activities (see Taylor and Smith, Suggested Reading). To put all these activities in perspective, it is useful to think of the RecBCD protein as a DNA helicase with associated nuclease activities. Its job is to put single-stranded 3′ tails on DNA that can invade other DNAs to initiate recombination. To perform this job, it loads on one end of a double-stranded DNA and unwinds the DNA, looping out the 3′-to-5′ strand of DNA as it goes (Figure 10.5). These loops are cut into small pieces by its 3′-to-5′ nuclease activity, leaving the 5′-to-3′ strand mostly intact. This process continues for up to 30,000 bp or until the RecBCD protein encounters a sequence on the DNA called a *chi* (χ) site, which in *E. coli* has the sequence 5′GCTGGTGG3′ but in other types of bacteria has somewhat different sequences. These sites were first found because they stimulate recombination in phage λ (see "Discovery of χ Sites" below), so they were given the Greek symbol χ, which looks like a crossover. When RecBCD encounters a χ site, its 3′-to-5′ nuclease activity is inhibited but its 5′-to-3′ nuclease activity may be stimulated, leading to formation of the free 3′ single-stranded tail, as shown in Figure 10.5. Note that the χ sequence does not have twofold rotational symmetry like the sites recognized by many restriction

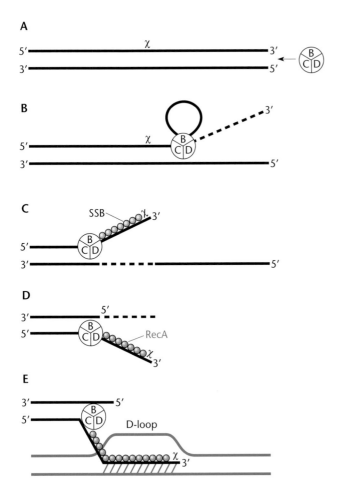

A

B

C

D

E

Figure 10.5   Model for promotion of recombination initiation at a *chi* (χ) site by the RecBCD enzyme. (A) RecBCD loads onto a double-stranded end. (B) Its helicase activity separates the strands, and its 3′-to-5′ nuclease activity degrades one strand until it encounters a χ site. (C) The χ site inhibits the 3′-to-5′ nuclease and stimulates the 5′-to-3′ nuclease, leaving a single-stranded 3′ end to which SSB protein (gray circles) binds. (D) The RecBCD protein helps RecA (in purple) displace SSB and load on the single-stranded end, forming it into an extended helical structure. (E) This helical RecA nucleoprotein filament can then invade a complementary double-stranded DNA, forming a D-loop or a triple-stranded structure (see Figure 10.7).

helping another to bind. In more physicochemical terms, the incoming RecA protein makes contact both with the DNA and with the RecBC protein already on the DNA, helping stabilize RecA binding to the DNA. This cooperative binding is necessary, because another protein called single-stranded DNA-binding (SSB) protein is already bound to the single-stranded DNA and RecA is not able to displace it by itself. The RecD protein may inhibit RecA binding so that it does not bind until RecD has been inhibited by encountering a χ site. More RecA proteins then bind cooperatively to the first RecA protein to form a helical nucleoprotein filament to prepare for the next step in recombination, synapse formation, as discussed in more detail below. This dependence on RecBCD for binding of RecA may help ensure that RecA does not bind to just any single-stranded DNA in the cell, only those that have been created by the RecBCD enzyme.

The discovery of χ sites and their role in recombination came as a complete surprise, and it took many years and a lot of clever experimentation to figure out how they work (see "Discovery of χ Sites" below). A detailed model consistent with much of the available information has emerged from this work. According to this detailed model, the RecD subunit of the RecBCD enzyme does not itself have any exonuclease activity but stimulates the 3′-to-5′ nuclease activity of the RecB subunit. The χ sites work by inhibiting the RecD subunit and thereby indirectly inhibiting the 3′-to-5′ exonuclease activity and stimulating the 5′-to-3′ nuclease activity of the remainder of the enzyme. Thus, as the RecBCD enzyme moves along the DNA, opening the strands, it degrades the 3′-to-5′ strand until it encounters the sequence 5′GCTGGTGG3′ (the χ sequence) on the strand being degraded. The DNA with this sequence can bind to the RecD subunit and inhibit it. The RecBCD enzyme continues to move on past the site, still degrading the 5′-to-3′ strand but leaving the 3′-to-5′ strand intact. The end result is a single-stranded 3′-ended tail that contains the χ site sequence at its end, as shown in Figure 10.5.

Much of the evidence for this detailed model of χ site action is genetic. First, the RecBC enzyme does have some 3′-to-5′ exonuclease activity in the absence of the RecD subunit, but it is greatly stimulated by the RecD subunit. Because of this, a linear DNA can be transformed into a RecD⁻ mutant of *E. coli* and is not degraded. Another prediction of the model, which is fulfilled, is that RecD⁻ mutants are proficient for recombination, and this recombination does not require χ sites. This is predicted by the model, since the RecBC enzyme lacking the RecD subunit still has its helicase activity to separate the strands and these single-stranded ends are not degraded even if they do not contain a χ site, so they are available to invade other DNAs and promote recombination. This property

endonucleases. This means that the sequence will be recognized on only one strand of the DNA and the RecBCD nuclease will pass over the sequence if it occurs on the other strand, making the orientation of the χ site all important, as we discuss below.

After a RecBC enzyme has formed a 3′ single-stranded tail on the DNA, it directs a RecA protein molecule to bind to the DNA next to where it is bound (see below). This is called **cooperative binding,** because one protein is

of RecD⁻ mutants, of not degrading linear DNA but still being proficient for recombination, is what makes RecD⁻ mutants of *E. coli* useful for gene replacements with linear DNA (see chapter 3).

This model also explains why recombination is stimulated only on the 5′ side of a χ site. Until the RecBCD enzyme reaches a χ site, the displaced strand is degraded and so is not available for recombination. Only after the RecBCD protein passes completely through a χ site will the strand survive to invade another DNA, so that only DNA on the 5′ side of the site survives. This model is also supported by the known enzymatic activities of the RecBCD protein, as well as by electron microscopic visualization of the RecBCD protein acting on DNA. It has also received experimental support from the results of in vitro experiments with purified RecBCD nuclease and DNA containing a χ site (see Dixon and Kowalczykowski, Suggested Reading).

WHY χ?

The hardest question to answer in biology is often why? To answer this question with certainty, we must know everything about the organism and every situation in which it might find itself, both past and present. Nevertheless, it is tempting to ask why *E. coli* and other bacteria use such a complicated mechanism involving χ sites for their major pathway of recombination initiated by ends of DNA and double-strand breaks. Adding to the mystery is the fact that they would not need χ sites at all if they were willing to dispense with the RecD subunit of the RecBCD nuclease. As mentioned above, in the absence of RecD, recombination proceeds just fine without χ sites and the cells are viable.

One idea is that the self-inflicted dependency on χ sites allows the RecBCD nuclease to play a dual role in recombination and in defending against phages and other foreign DNAs. Small pieces of foreign DNA such as a phage DNA entering the cell are not apt to have a χ site, since 8-bp sequences like χ occur by chance only once in about 65,000 bp, longer than many phage DNAs. The RecBCD nuclease degrades a DNA until it encounters a χ site, and if it does not encounter a χ site it degrades the entire DNA. *E. coli* DNA, by contrast, has many more of these sites than would be predicted by chance. In support of the idea that RecBCD is designed to help defend against phages is the extent to which phages go to avoid degradation by this enzyme. Many phages and transposons avoid degradation by RecBCD by attaching proteins to the ends of their DNA or by making proteins that inhibit RecBCD in more direct ways, such as the Gam protein of λ (see chapter 8 and Box 10.3).

Another possible reason for χ sites is that they might help direct recombination to regions in the DNA that are better for triple-strand DNA formation (see below). The χ sequence itself is relatively GT rich (note that seven of the eight bases in the χ sequence are G's or T's), and χ sites are also often surrounded by many other G's and T's. Sequences rich in G's and T's are preferred sites for binding RecA. When χ is used, GT-rich sequences more often end up on the 3′ single-stranded tail, where they can enhance the invasion of other DNAs.

It is also possible that χ sites exist to help with replication restarts, when the replication fork collapses leaving a broken end. After all, there is reason to think that the major role of recombination in bacteria is to promote replication restarts (Box 10.2). This argument is lent support from the fact that most χ sites in *E. coli* are oriented on the leading strand to help with replication restarts.

Whatever the purpose of χ sites, they seem to be universal among bacteria. Gram-positive bacteria, including *Bacillus subtilis*, contain an enzyme of similar function to RecBCD called AddAB. The *B. subtilis* enzyme is known to function similarly, except that the 5′-to-3′ strand might be degraded even before a χ site is encountered, and the χ site has a different sequence in *B. subtilis*.

The RecFOR Pathway

The other major pathway used to prepare single-stranded DNA for invading another DNA in *E. coli* is the RecFOR pathway (Figure 10.6). This pathway is so named because it requires the products of the *recF*, *recO*, and *recR* genes as well as the *recQ* and *recJ* genes. This pathway is used under different circumstances from the RecBCD pathway because it cannot prepare DNA ends for recombination and is used mostly to initiate recombination at single-stranded gaps in DNA, as shown. These gaps may be created during repair of DNA damage or by the replication fork proceeding past a lesion in the lagging strand of DNA, leaving a single-stranded gap. The RecFOR proteins can then prepare the single-stranded DNA at the gap to invade a sister DNA. This structure can be used to restart replication, making the RecFOR pathway important in recombination repair of DNA damage (see chapter 11).

The RecFOR pathway does not have a "superstar" like RecBCD that can do it all, and so it needs a number of proteins to perform the tasks which the RecBCD nuclease can perform alone (see Morimatsu and Kowalczykowski, Suggested Reading). The RecQ protein is a helicase like RecBCD but lacks a nuclease activity to degrade the strands it displaces. RecJ may provide

The Three R's: Recombination, Replication, and Repair

One of the most gratifying times in science comes when phenomena that were originally thought to be distinct are discovered to be but different manifestations of the same process. Such a discovery is usually followed by rapid progress as the mass of information accumulated on the different phenomena is combined and reinterpreted. This is true of the fields of recombination, replication, and DNA repair, sometimes called the three R's. Replication can be initiated in a number of different ways, many of which involve recombination; recombination often requires normal replication mechanisms, and some types of DNA repair require recombination as well as normal replication.

Appreciation of the role of recombination in replication was slow in coming. It was known that some phages, such as T4, need the recombination functions to initiate replication (see chapter 7 and Mosig, below). In these phages, recombination intermediates function to initiate DNA replication later in infection. However, this was thought to be unique to these phages. Normally, initiation of chromosomal replication in bacteria does not require recombination functions. However, after extensive DNA damage due to irradiation or other agents, a new type of initiation, which does require recombination, comes into play in bacteria. This type of initiation was originally called stable DNA replication (SDR) because it continued even after protein synthesis stopped (see Kogoma, below). Normally, initiation of DNA replication at the chromosomal oriC site requires new protein synthesis, and so, in the absence of protein synthesis, replication continues only until all the ongoing rounds of replication are completed, and no new rounds are initiated (see chapter 1). However, after extensive DNA damage due to irradiation, etc., initiation of new rounds of replication occurs even in the absence of protein synthesis. Interestingly, this SDR often initiates close to the oriC site, although other sites can also be used. To initiate SDR, a double-strand break is made in the DNA close to the oriC site and recombination functions cause the invasion of one daughter DNA by the other daughter DNA. The PriA, PriB, PriC, and DnaT proteins and other replisome-loading functions, maybe including DnaC, then reload the replication apparatus on the DNA, and replication is under way again (see below).

Recombination also plays a role in restarting replication forks after the replication apparatus has been derailed as a result of encountering a break or other type of damage in the DNA template, in a process analogous to stable DNA replication. In fact, some authors have gone so far as to propose that this is the primary role of recombination in bacteria: to restart replication after the replication apparatus has been derailed

(see Kuzminov and Stahl, below). The pathway of recombination that is used depends on the type of damage encountered. If a single-strand break is encountered in either the leading or lagging strand of the DNA, a double-strand break will ensue. The RecBCD enzyme then degrades in from the break until it encounters a χ site. The 3′ single-stranded end formed as a result of this degradation can then bind RecA to invade the sister DNA to form a three-stranded structure (see the text). The PriA, PriB, and DnaT proteins can bind DnaB at these structures, and replication can continue (see below). However, a gap may form if the leading strand has DNA damage over which the replication fork cannot replicate. The lagging strand may continue to replicate, but the leading-strand replication is blocked, forming a single-stranded gap. Then the RecFOR pathway is responsible for loading RecA. Once such a branched structure has formed, the PriC protein helps reassemble DnaB and the other replication fork proteins at the replication fork at the branch as above and replication continues. The Pri proteins are required to restart the fork because DnaB (with the help of DnaC alone) normally loads on the DNA only at the oriC site (see the text).

The discovery that the Pri proteins participate in recombination-mediated replication restarts also has a long and interesting history. There are three Pri proteins, PriA, PriB, and PriC, as well as another protein, DnaT, that help reload replication forks. These proteins were first discovered because they are required for the initiation of replication of the DNA of some single-stranded DNA phages (see chapter 7), so it was assumed that they also played a role in E. coli DNA replication. However, mutations that inactivate the products of these genes are not lethal, although they are defective in recombination and are more sensitive to DNA-damaging agents. Later work showed that double mutants with both priA and priC or both priB and priC mutations are dead, indicating that PriA and PriB (as well as DnaT) participate in a different pathway than PriC. Genetic and biochemical evidence indicates that the PriA PriB DnaT pathway is required to load DnaB at the types of recombination intermediates created by the RecBCD pathway while PriC is required to load DnaB at recombination intermediates created by the RecFOR pathway (see above and Lovett, below).

It is not quite clear how PriA and the other proteins reload the replication apparatus at recombination intermediates. The current view is that PriA binds to the three-stranded junction created by the recombination functions and then, in a complex with PriB, DnaC, and DnaT, directs the DnaB helicase to bind. PriC, on the other hand, preferentially binds to structures in which a single-stranded DNA at a gap has invaded a

(continued)

BOX 10.2 (continued)

The Three R's: Recombination, Replication, and Repair

double-stranded DNA, the structures created by RecFOR, and helps DnaC load DnaB. The known role of DnaC was to help DnaB helicase to bind at the normal *oriC* origin of replication, and it was not known to help load DnaB at other places, during replication restarts. A role for DnaC in replication restarts was inspired by the discovery that some *dnaC* mutants bypass the need for PriA. It is interesting that some bacteria do not encode PriA proteins and may use DnaC for both processes. The details of replication restarts are discussed in more detail in chapter 11.

A role for replication in recombination completes the circle. As described in the text, most models show recombination proceeding through one or more Holliday junctions, which are then resolved by being cut with an X-phile such as RuvC. Depending on how the resolvase cuts the Holliday junctions, a recombinant can ensue. However, recombinants need not be created in this way. If the recombination functions form a branch between two different DNAs in the

cell and the replication apparatus loads on this branch as described above, then the replication apparatus that started out replicating one DNA will have switched to replicating the other DNA and a recombinant will ensue. Such models of recombination, which used to be called "copy choice," have now come back into favor and may account for at least some of the recombination products that are observed as well as dimer formation during chromosome replication.

References

Kogoma, T. 1997. Stable DNA replication: interplay between DNA replication, homologous recombination, and transcription. *Microbiol. Mol. Biol. Rev.* **61:**212–238.

Kuzminov, A., and F. W. Stahl. 1999. Double-strand end repair via the RecBCD pathway in *Escherichia coli* primes DNA replication. *Genes Dev.* **13:**345–356.

Lovett, S. T. 2005. Filling the gaps in replication restart pathways. *Mol. Cell* **17:**751–752.

Mosig, G. 1987. The essential role of recombination in T4 phage growth. *Annu. Rev. Genet.* **21:**347–371.

the exonuclease activity that RecQ lacks, helping extend the single-stranded gaps. The RecQ protein also lacks the ability to displace the SSB protein and therefore to load the RecA protein on the single-stranded DNA it creates with its helicase activity. As we have seen, the RecBCD protein solves this problem by helping load the first RecA protein on the DNA. More RecA can then bind cooperatively, displacing SSB in the process. The RecF, RecO, and RecR proteins may help the RecA protein to bind to and coat the single-stranded DNA created by RecQ and RecJ. They do this by helping the first RecA protein bind to one end of the single-stranded gap and then displacing the SSB protein from the single-stranded DNA in the gap as shown. The RecFOR proteins may also stop RecA from invading the neighboring double-stranded DNA before the synapse with another DNA occurs (see below).

Synapse Formation and the RecA Protein

Once a single-stranded DNA is created by the RecBCD or RecFOR pathway, it must find and invade another DNA with the complementary sequence. The joining of two DNAs in this way is called a **synapse,** and the process by which an invading strand can replace one of the two strands in a double-stranded DNA is called **strand exchange.** This is a remarkable process. Somehow the single-stranded DNA must find its complementary sequence by scanning all the double-stranded DNA in the

cell, which even in a simple bacterium can be more than 1 mm long! But how could the single-stranded DNA know when the sequence is complementary, especially if it can only scan the outside of the DNA double helix and the bases of the double-stranded DNA are on the inside? Not only is synapse formation remarkably fast, but also it is remarkably efficient. Once an incoming single-stranded DNA enters the cell, for example during an Hfr cross, it finds and recombines with its complementary sequence in the chromosome almost 100% of the time.

Searching for complementary DNA is the job of the RecA protein, whose role in recombination is outlined in Figure 10.7. As the single-stranded DNA is created by RecBC or RecFOR, it is coated by RecA to form a nucleoprotein filament. As mentioned above and shown in Figures 10.5 and 10.6, the RecBCD and RecFOR proteins help RecA displace SSB protein from the single-stranded DNA, which is a prerequisite to forming this filament. The DNA in the RecA filament is also helical but much extended relative to the normal helix in DNA: it takes about twice as many nucleotides to complete a helical turn. The helical nucleoprotein filament then scans the double-stranded DNA in the cell to find its homologous sequence. Some evidence suggests that it might scan double-stranded DNA through its major groove (see chapter 1) and can base pair with its complementary sequence in the major groove once it finds it without transiently disrupting the helix. Once it finds its complementary

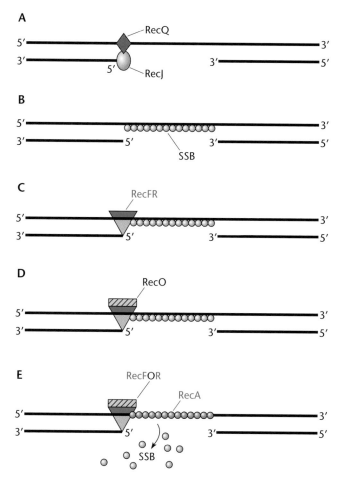

Figure 10.6   Model for recombination initiation by the RecFOR pathway. (A) RecQ, a helicase, and RecJ, an exonuclease, process gapped DNA. (B) SSB protein (gray circles) coats the gap. (C) RecF and RecR bind to the SSB-coated gap. (D) RecO binds to the RecFR DNA complex. (E) RecFOR complex nucleates RecA nucleoprotein filament assembly (purple circles), displacing SSB protein.

number of observations have been made that may shed light on this process. One observation is that a single-stranded RecA nucleoprotein filament may somehow change, or "activate," double-stranded DNAs merely by binding to them, even if the two DNAs are not complementary. This activation presumably has something to do with the way in which the RecA nucleoprotein filament scans DNA looking for its complementary sequence. Once the strands of the double-stranded DNA are activated, even by a noncomplementary RecA nucleoprotein filament, a complementary single-stranded DNA not bound to RecA can invade it and exchange with one of its strands. This was named *trans* activation because the RecA nucleoprotein filament which activated the DNA is not necessarily the one that invades it (see Mazin and Kowalczykowski, Suggested Reading). It is not clear what happens to a double-stranded DNA when it is activated by a RecA nucleoprotein filament, but the helix may be transiently extended and the strands may be partially separated to allow the single-stranded DNA to search for its complementary sequence. This is an area that needs further investigation.

The RecA protein initially forms a nucleofilament only on single-stranded DNA, either an end or a gap, and does not invade the neighboring double-stranded DNA. However, after the nucleofilament has invaded a double-stranded DNA to form what we are assuming is a triple-stranded structure, the RecA protein can continue to polymerize on the same strand, extending the nucleofilament into the neighboring double-stranded DNA. This causes the double-stranded DNA adjoining the single-stranded gap to exchange strands with the invaded DNA and eventually creates a Holliday junction. Unlike the first binding of RecA to single-stranded DNA, extension of the filament into double-stranded DNA requires energy in the form of ATP cleavage. We already mentioned that Holliday junctions can migrate spontaneously. However, unlike spontaneous migration, which is blocked at mismatches between the invading DNA and the invaded DNA, the RecA protein can drive the junction over such mismatches, creating heteroduplexes. Migration driven by RecA is also much faster than spontaneous migration and unidirectional, unlike spontaneous migration, in which the Holliday junction merely wanders slowly back and forth randomly. While the filament grows in the 5′ to 3′ direction on the strand, eventually invading the neighboring double-stranded DNA in a process called "spooling" (Figure 10.7), it can also apparently depolymerize on the other end, leaving a single-stranded gap. The single-stranded gap can then be repaired by other cellular enzymes. The Holliday junction that has formed between the two double-stranded DNAs can then be resolved by functions discussed in the next section.

sequence, it pairs with it. There is still some question of what happens next. Either it displaces one strand of the double-stranded DNA to form a **D-loop** as shown in Figure 10.5, or, as some evidence suggests, it actually forms a **triple-stranded structure** as shown in Figure 10.7. Other evidence calls into question the formation of triple-stranded structures, and it has been proposed instead that RecA can somehow approach DNA through its minor groove and somehow flip the bases out to test for complementarity. For present purposes, we often use D-loops to schematize strand displacement because they are easier to draw.

While the details of how a single-stranded DNA–RecA filament and a double-stranded DNA find each other and what kind of structures are formed remain obscure, a

RecA-DNA nucleoprotein filament

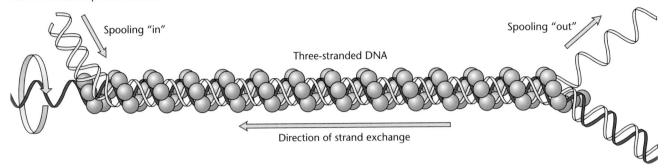

Spooling "in"

Spooling "out"

Three-stranded DNA

Direction of strand exchange

Figure 10.7 Model for synapse formation and strand exchange between two homologous DNAs by RecA protein. The RecA protein (in purple) bound to a single-stranded end, formed as in Figure 10.5 or 10.6 and forced into an extended helical structure, pairs with a homologous double-stranded DNA in its major groove to form a three-stranded structure. RecA can drive this three-stranded structure into adjacent double-stranded DNA by a "spooling" mechanism, forming a four-stranded Holliday junction (not shown).

The Ruv and RecG Proteins and the Migration and Cutting of Holliday Junctions

As discussed earlier in this chapter, Holliday junctions are remarkable structures that can do many things. The branches on the Holliday junction can **isomerize** so that the crossed strands become the uncrossed strands. This process should occur spontaneously since no hydrogen bonds need to be broken or reformed for isomerization. Holliday junctions can also **migrate** when hydrogen bonds break on one side of the Holliday junction and reform on the other side, so that the Holliday junction moves from one place on the DNA to another. Alternatively, two of the strands of the Holliday junction can be cut to resolve the Holliday junction. Depending on which two strands are cut, the two DNAs return to their original configuration or take up a new configuration to form recombinants.

Once a Holliday junction has formed as the result of the concerted action of RecBCD and RecA or the Rec-FOR pathway and RecA, at least two different pathways can resolve the junctions to make recombinant products. One pathway uses the three Ruv proteins, RuvA, RuvB, and RuvC, which are encoded by adjacent genes. Another pathway uses the RecG protein, as well, presumably, as at least one other protein whose identity is unknown. Which of these pathways is used depends on the situation. The Ruv pathway is discussed first.

RuvABC

Recent work on the crystal structures of the Ruv proteins has shown that they form interesting structures that give

clues to how they function in the migration and resolution of Holliday junctions (Figure 10.8) (see West, Suggested Reading). The RuvA protein is a specific Holliday junction binding protein whose role is to force the Holliday junction into a certain structure amenable to the subsequent steps of migration and resolution. Four copies of the RuvA protein come together to form a flat structure like a flower with four petals. The Holliday junction lies flat on the flower and thus is forced into a flat (planar) configuration. Binding of the RuvA flower also creates a short region in the middle of the Holliday junction where the strands are not base paired and the single strands form a sort of square. Another tetramer of RuvA may then bind to the first to form a sort of turtle shell, with the four arms of double-stranded DNA emerging from the "leg holes." The RuvB protein then forms a hexameric (six-member) ring encircling one arm of the DNA, as shown in Figure 10.8. The DNA is then pumped through the RuvB ring, using ATP cleavage to drive the pump, thereby forcing the Holliday junction to migrate.

After the RuvA and RuvB proteins have forced Holliday junctions to migrate, they can be cut (resolved) by the RuvC protein. The RuvC protein cuts only Holliday junctions that are being held by RuvA and RuvB. The RuvC protein is a specialized DNA endonuclease which cuts the two crossed strands of a Holliday junction simultaneously. Such enzymes are often called **X-philes** because they cut the strands of DNA crossed in a Holliday junction or branch which look like the letter X and phile means "having a tendency toward." Like many enzymes that make double-strand breaks in DNA, two

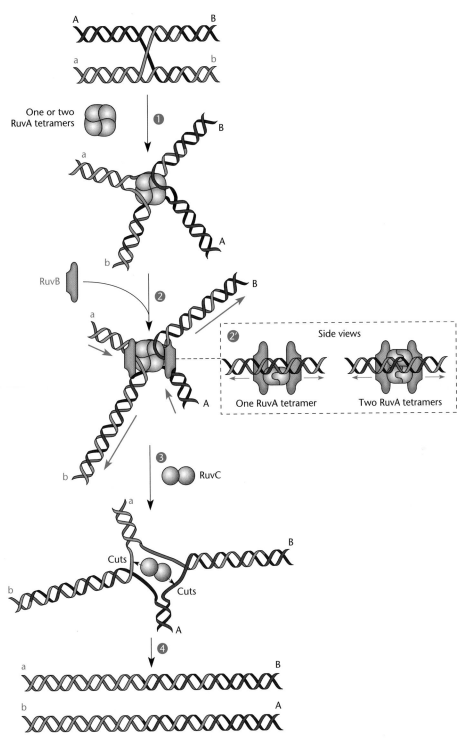

Figure 10.8   Model for the mechanism of action of the Ruv proteins. Step 1: One or two tetramers of the RuvA protein bind to a Holliday junction, holding it in a planar (flat) configuration. Note that at the beginning, the DNA has only one turn of heteroduplex (purple-gray). Step 2: Two hexamers of the RuvB protein bind to the RuvA complex, each forming a ring around one strand of the DNA. Step 2': Side view of the complex with one and two tetramers. Step 3: RuvC binds to the complex and cuts two of the strands. Step 4: The Holliday junction has been resolved into a different configuration because of the way the strands were cut. Note that there are now more turns of heteroduplex (purple-gray).

identical polypeptides encoded by the *ruvC* gene come together to form a homodimer, which is the active form of the enzyme. Because the enzyme has two copies of the polypeptide, it has two DNA endonuclease active centers, each of which can cut one of the DNA strands to make a double-strand break.

The evidence that RuvC can cut only Holliday junctions that are bound to RuvA and RuvB is mostly genetic: mutants with either a *ruvA,* a *ruvB,* or a *ruvC* mutation are indistinguishable in that they are all defective in the resolution of Holliday junctions (see below). To a geneticist, this means that RuvC cannot act to resolve a Holliday junction without RuvA and RuvB being present and bound to the Holliday junction. However, this leads to an apparent conflict with the structural information about RuvA and RuvB discussed above, indicating that RuvA forms a turtle shell-like structure over the Holliday junction. How could RuvC enter the turtle shell formed by RuvA to cleave the crossed DNA strands in the inside? Perhaps a tetramer of RuvA is bound to only one face of the Holliday junction, leaving the other face open for RuvC to bind and cut the crossed strands. However, it seems unlikely that the Holliday junction could be held tightly enough in this way to not be dislodged by the RuvB pump. Another idea is that the RuvA shell opens up somehow to allow RuvC to enter and cut the crossed strands.

Support for this model of the concerted action of RuvA, RuvB, and RuvC has come from observations of purified Ruv proteins acting on artificially synthesized structures that resemble Holliday junctions (see Parsons et al., Suggested Reading). These junctions are constructed by annealing four synthetic single-stranded DNA chains that have pairwise complementarity to each other (Figure 10.9). These synthetic structures are not completely analogous to a real Holliday junction in that they are not made from two naturally occurring DNAs with the same sequence. Rather, four single strands that are complementary to each other in the regions shown and therefore form a pairwise cross are synthesized. A Holliday junction made in this way is much more stable than a natural Holliday junction because the branch cannot migrate spontaneously. Real Holliday junctions are too unstable for these experiments; they quickly separate into two double-stranded DNA molecules.

Experiments performed with such synthetic Holliday junctions indicated that purified Ruv proteins act sequentially on a synthetic Holliday junction in a manner consistent with the above model. First, RuvA protein bound specifically to the synthetic Holliday junctions, and then a combination of RuvA and RuvB caused a disassociation of the synthetic Holliday junctions, simulating branch migration in a natural DNA molecule. The

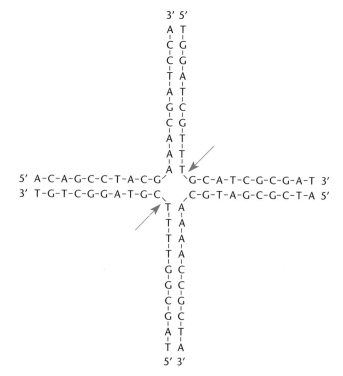

Figure 10.9 A synthetic Holliday junction with four complementary strands. The junction cannot migrate but can be disrupted by RuvA and RuvB. It can also be cut by RuvC (arrows) and other Holliday junction resolvases such as the gene *49* and gene *3* products of phages T4 and T7, respectively.

dissociation required the energy in ATP to break the hydrogen bonds holding the Holliday junction together, as predicted. The RuvC protein could also specifically cut the synthetic Holliday junction in two of the four strands.

The RuvC protein is a member of a large group of specialized DNA endonucleases called the X-philes. They are also sometimes called cruciform-cutting enzymes because the structure they cut can also be drawn as a cross. RuvC has some sequence specificity at the site where it cuts. It always cuts just downstream of two T's in the DNA preceded by either an A or a T and followed by either a G or a C (as shown in Figure 10.9). At first, it seems impossible that two T's could be opposite each other in the DNA until we remember that the crossed strands of DNA are not the complementary strands but the strands with the same sequence. If one strand has the sequence recognized by RuvC, 5'(A/T)TT(G/C)3', the other strand also has the same sequence at that position. Presumably, RuvA and RuvB cause the Holliday junction to migrate until this sequence is encountered and RuvC then can cut the crossed strands. The alternate resolvase in *E. coli*, RusA, which is encoded by a prophage, also has sequence specificity and cuts upstream of two G's in the DNA. It has been speculated that Holliday junction resolvases may have sequence specificity to distinguish them from nucleases that can cut branches, such as those that arise from single-strand invasion of a double-stranded DNA. Holliday junction resolvases such as RuvC and RusA have been shown to cut such branches if the branch point has their recognition sequence. However, such branches cannot migrate, so that the crossed-over strand always has the same sequence, which is not apt to be the sequence recognized by the Holliday junction resolvase; therefore, they are seldom cut by RuvC or RusA. Holliday junctions, on the other hand, can migrate, and so eventually they migrate to the sequence recognized by the resolvase and are cut.

Most types of bacteria have a RuvC-like Holliday junction resolvase. However, some, like *Bacillus subtilis*, have a resolvase, called YggF, which is more closely related to RusA in *E. coli* (see above), whose gene is on a prophage where it is not normally expressed. Many phages, including T4 and T7, often encode their own X-phile, which also usually cuts branches as well as Holliday junctions (see below). It will obviously take some effort to sort out the contributions of the various X-philes to recombination, replication restarts, and other aspects of DNA metabolism.

RecG

Another helicase in *E. coli* which can help junctions to migrate is RecG (Table 10.1). For a long time, the function of this helicase was unknown. It has very little effect on recombination when RuvABC is present, which suggested that its role is redundant with respect to RuvABC. The idea was that RecG could play the role of RuvAB and promote the migration of Holliday junctions. Then another resolvase could play the role of RuvC and resolve the Holliday junction, providing a backup for RuvABC. However, it now seems that the roles of these proteins are not redundant and that they each have their own unique role to play (see Bolt and Lloyd, Suggested Reading). For one, they move in opposite directions on single-stranded DNA. The RuvAB helicase moves in the 5′-to-3′ direction on single-stranded DNA, while the RecG helicase moves in the 3′-to-5′ direction. Also, unlike RuvAB, which seems to promote the migration of Holliday junctions only after they have formed, the RecG protein can turn blocked replication forks into a form of Holliday junction (see below) or bind to three-strand junctions like those at a branch. The Holliday junctions formed by RecG protein also seem to be cut by the X-phile RusA rather than by RuvC (see Bolt and Lloyd, Suggested Reading),

although, as mentioned above, this X-phile is not normally expressed. While the details of how these two pathways interact is not yet clear, it seems likely that RecG represents a completely different pathway from RuvABC for dealing with DNA branches, but the two pathways can lead to the same result, a recombinant.

As mentioned, one role of RecG might occur when the replication fork stalls as the result of encountering an obstacle such as RNA polymerase transcribing ahead of it or damage in the DNA. The RecG protein may then cause the stalled replication fork to back up, much like a train backs up to allow the track to be repaired, before it moves on (Box 10.2). Alternatively, backing up of the replication fork combined with strand switching by recombination functions could cause the formation of a type of Holliday junction called a "chicken foot" (see chapter 11). Replication could then proceed over the site of the damage (as discussed in detail in chapter 11), and replication could restart, presumably with the help of the PriA proteins and DnaC to load the DnaB helicase back on the DNA. Alternatively, if the damage is irreparable, the Holliday junction formed by the backing up of the replication fork by RecG could be cut by an unknown X-phile and the cut ends of the Holliday junction could then invade the other daughter DNA to form a branched structure, which could be acted on by the Pri proteins to restart the replication fork. Such mechanisms could increase the survival of cells after DNA damage or could increase the frequency of recombination by a copy choice mechanism as described in Box 10.2; this would explain why *recG* mutants are sensitive to DNA damage and are defective in recombination under some circumstances. There are at least 12 helicases in *E. coli*, and more work is also needed to determine the roles of the various helicases in recombination and replication. This is an active current area of research.

As discussed in chapter 3, recombination in bacteria has some differences from recombination in most other organisms. In bacteria, generally only small pieces of incoming donor DNA recombine with the chromosome, whereas in other organisms, two enormously long double-stranded DNA molecules of equal size usually recombine. Nevertheless, the requirements for recombination in bacteria are similar to those in other organisms. In fact, accumulating evidence supports the existence of proteins analogous to many of the recombination proteins of bacteria in both eukaryotes and archaeae. For example, the Rad51 proteins of yeast and humans are analogous to RecA and can form helical nucleoprotein filaments similar to those formed by the bacterial RecA protein. A protein called CCE1 in yeast mitochondria specifically cuts Holliday junctions and is related to RuvC.

Phage Recombination Pathways

Many phages encode their own recombination functions, some of which can be important for the multiplication of the phage. As discussed in chapter 7, some phages use recombination to make primers for replication and concatemers for packaging. Also, phage recombination systems may be important for repairing damaged phage DNA and for exchanging DNA between related phages to increase diversity. Phages may encode their own recombination systems to avoid dependence on host systems for these important functions.

Many phage recombination functions are analogous to the recombination proteins of the host bacteria (Table 10.2), and in many cases, the phage recombination proteins were discovered before their host counterparts. As a result, studies of bacterial recombination systems have been heavily influenced by simultaneous studies of phage recombination systems.

Rec Proteins of Phages T4 and T7

Phages T4 and T7 depend on recombination for the formation of DNA concatemers after infection (see chapter 7). Therefore, recombination functions are essential for the multiplication of these phages. Many of the T4 and T7 Rec proteins are analogous to those of their hosts. For example, the gene *49* protein of T4 and the gene *3* protein of T7 are X-phile endonucleases that resolve Holliday junctions and are representative of phage proteins discovered before their host counterparts, in this case RuvC. The gene *46* and *47* products of phage T4 may perform a reaction similar to the RecBCD protein of the host, although no evidence indicates the presence of χ-like sites associated with this enzyme. The UvsX protein of T4 and the Bet protein of λ are analogous to the RecA protein of the host.

The RecE Pathway of the *rac* Prophage

Another classic example of a phage-encoded recombination pathway is the RecE pathway encoded by the *rac* prophage of *E. coli* K-12. The *rac* prophage is integrated at 29 min in the *E. coli* genetic map and is related to λ. This defective prophage cannot be induced to produce infective phage, since it lacks some essential functions for multiplication.

The RecE pathway was discovered by isolating suppressors of *recBCD* mutations, called *sbcA* mutations (for suppressor of *BC*) that restored recombination in conjugational crosses. The *sbcA* mutations were later found to activate a normally repressed recombination function of the defective prophage. Apparently, *sbcA* mutations inactivate a repressor that normally prevents the transcription of the *recE* gene, as well as other prophage genes. When the repressor gene is inactivated, the RecE protein is synthesized and can then substitute for the RecBCD nuclease in recombination.

The Phage λ *red* System

Phage λ also encodes recombination functions. The best characterized is the *red* system, which requires the products of adjacent λ genes *exo* and *bet*. The product of the *exo* gene is an exonuclease that degrades one strand of a double-stranded DNA from the 5′ end to leave a 3′ single-stranded tail. The *bet* gene product is known to help the renaturation of denatured DNA and to bind to the λ exonuclease. Unlike many of the other recombination systems that we have discussed, the λ *red* recombination pathway does not require the RecA protein since it has its own synapse forming protein, Bet. The λ *red* system is the basis for a very useful gene replacement technique called recombineering (Box 10.3).

Interestingly, the RecE protein of *rac* and the λ *exo* exonuclease may be similar. The RecE protein needs RecA to promote *E. coli* recombination, but it does not need RecA to promote λ recombination. It is not too surprising that λ and *rac* encode similar recombination functions, since *rac* and λ are related phages.

Besides the *red* system, phage λ encodes another recombination function that can substitute for components of the *E. coli* RecF pathway (see Sawitzke and Stahl, Suggested Reading, and Table 10.2). Apparently, phages can carry components for more than one recombination pathway.

Genetic Analysis of Recombination in Bacteria

The major reason we understand so much more about recombination in *E. coli* than in most other organisms is because of the relative ease of doing genetic experiments

TABLE 10.2	Analogy between phage and host recombination functions
Phage function	**Analogous *E. coli* function**
T4 UvsX	RecA
T4 gene *49*	RuvC
T7 gene *3*	RuvC
T4 genes *46* and *47*	RecBCD
λ ORF[a] in *nin* region	RecO, RecR, RecF
Rac *recE* gene	RecJ, RecQ
λ *gam*	Inhibits RecBCD
λ *exo*	RecBCD, RecJ
λ *bet*	RecA
rusA (DLP12 prophage)	RuvC

[a] ORF, open reading frame.

Recombineering: Gene Replacements in *E. coli* with Phage λ Recombination Functions

One of the major advantages of using bacteria and other lower organisms for molecular genetic studies is the relative ease of doing gene replacements with some of these organisms (see the discussion of gene replacements in chapter 3). To perform a gene replacement, a piece of the DNA of an organism is manipulated in the test tube to change its sequence in some desired way. The DNA is then reintroduced into the cell, and the recombination systems of the cell cause the altered sequence of the reintroduced DNA to replace the normal sequence of the corresponding DNA in the chromosome. Because it depends on homologous recombination, gene replacement requires that the sequence of the reintroduced DNA be homologous to the sequence of the DNA it replaces. However, the homology need not be complete, and minor changes such as base pair changes can be introduced into the chromosome in this way as a type of site-specific mutagenesis. Also, the reintroduced DNA need not be homologous over its entire length; homology is needed only where the recombination occurs. This makes it possible to use gene replacement to make large alterations such as construction of an in-frame deletion to avoid polarity effects and insertion of an antibiotic resistance gene cassette into the chromosome. If the sequences on both sides of the alteration (the flanking sequences) are homologous to sequences in the chromosome, recombination between these flanking sequences and the chromosome will insert the alteration. Methods for gene replacement in *E. coli* have usually relied on the RecBCD-RecA recombination pathway since this is the major pathway for recombination in *E. coli*. We mentioned some of these methods in this chapter and chapter 3.

A method called recombineering has recently been developed for performing site-specific mutagenesis and gene replacements in *E. coli* by using the phage λ Red pathway or the RecET pathway of the Rac prophage in *E. coli* (Table 10.2). The λ Red system has many advantages over the RecBCD-RecA pathway for such manipulations. This method makes it possible to use single-stranded DNA oligonucleotides as short as 30 bases, although those 60 bases long or longer work better. This is important because the synthesis of single-stranded DNAs of these lengths has become routine for making PCR primers and oligonucleotides of any desired sequence can be purchased for a reasonable cost. Other methods of site-specific mutagenesis for making specific changes in a sequence are more tedious and require a certain amount of technical skill. Probably most important, recombineering is very efficient. Minor changes, such as single-amino-acid changes in a protein, usually offer no positive selection, and

most methods require the screening of thousands of individuals to find one with the replacement.

The figure outlines the original procedure for using the λ Red system for gene replacements. Panel A shows the structure of the *E. coli* strain required. It carries a defective prophage in which most of the λ genes have been deleted except the recombination (*red*) genes *gam-bet-exo* (Table 10.2; see also Figure 8.1 [the λ genetic map]). Panel B shows the replacement of a sequence in the plasmid by the corresponding region on another plasmid, in which the sequence has been disrupted by introduction of an antibiotic resistance cassette (Ab^r). This region of the plasmid has been amplified by PCR to produce a double-stranded DNA fragment carrying the antibiotic resistance cassette and some of the flanking sequences. First the cells are heated to inactivate the λ repressor and induce transcription of the *red* genes of the prophage. Then cells are made competent for electroporation and the PCR fragment is electroporated into them. The *gam* gene product, Gam, inhibits the RecBCD enzyme, so that the linear DNA fragment is not degraded as soon as it enters the cell. The *exo* gene product, Exo, then processes the fragment for recombination. Exo is an exonuclease that plays the role of RecBCD, degrading one strand of a double-stranded DNA from the 5′ end, thereby exposing a 3′ overhang single strand for strand invasion. The *bet* gene product, Bet, then plays the role of RecA, binding to the single-stranded DNA exposed by Exo and promoting synapse formation and strand exchange with a complementary DNA in the cell. The cells in which the PCR fragment has recombined with the cellular DNA so that the sequence containing the antibiotic resistance gene has replaced the corresponding sequence in the cellular DNA are then selected on plates containing the antibiotic.

To determine the effect on the cell of inactivating a gene product, it is best to delete the entire gene and replace it with an antibiotic resistance cassette. This can be accomplished by using PCR to amplify the cassette with primers whose 5′ sequences are complementary to sequences flanking the gene to be deleted. Recall that the 5′ sequences on a PCR primer need not be complementary to the sequences being amplified. When this amplified fragment is electroporated into the cells, the antibiotic resistance cassette replaces the entire gene.

Introducing an antibiotic resistance cassette into a gene simplifies the task of selecting the gene replacement and inactivating the gene. However, sometimes we want to introduce a small change into the gene for which there is no direct selection, for example a specified change in one amino acid

(continued)

BOX 10.3 (continued)

Recombineering: Gene Replacements in *E. coli* with Phage λ Recombination Functions

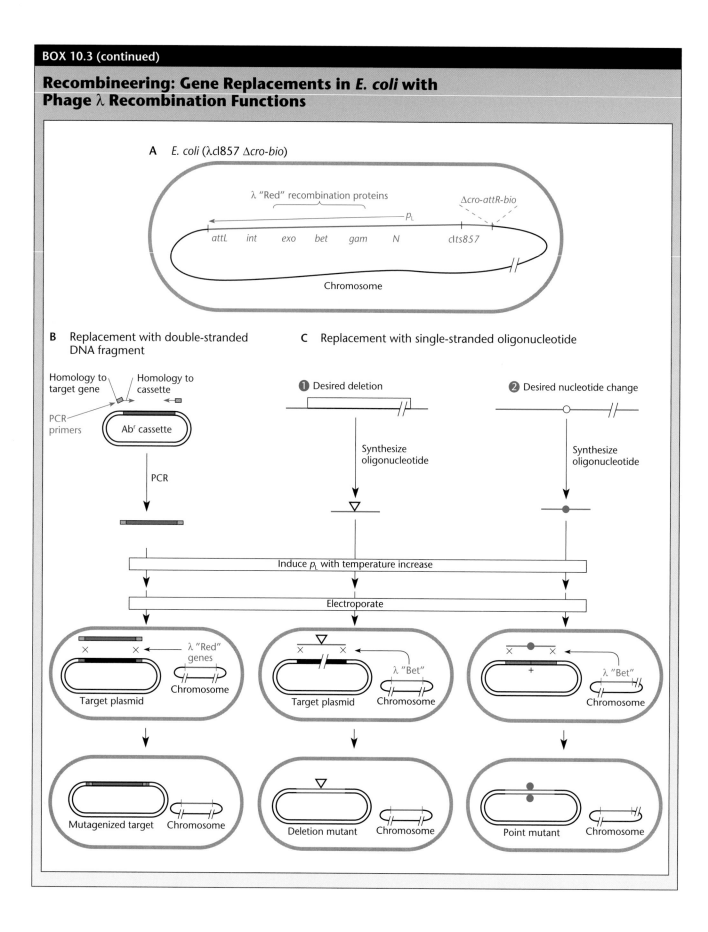

A *E. coli* (λcI857 Δcro-bio)

λ "Red" recombination proteins

Δcro-attR-bio

*p*_L

attL   int   exo   bet   gam   N   cIts857

Chromosome

B Replacement with double-stranded DNA fragment

Homology to target gene   Homology to cassette

PCR primers

Ab^r cassette

PCR

C Replacement with single-stranded oligonucleotide

❶ Desired deletion

❷ Desired nucleotide change

Synthesize oligonucleotide

Synthesize oligonucleotide

Induce *p*_L with temperature increase

Electroporate

λ "Red" genes

Target plasmid   Chromosome

λ "Bet"

Target plasmid   Chromosome

λ "Bet"

Chromosome

Mutagenized target   Chromosome

Deletion mutant   Chromosome

Point mutant   Chromosome

BOX 10.3 (continued)

Recombineering: Gene Replacements in *E. coli* with Phage λ Recombination Functions

that we think may play an important role in the protein product of the gene. A variation on this method allows the selection of recombinants that have a single defined base pair change or some other small change. It depends on having a cassette that has both a gene that can be selected by a positive selection and a gene whose product is toxic under some conditions. This allows us to select both for acquisition of the cassette and, later, for its loss. An example of a toxic gene would be a gene for sucrase (*sac*), whose product kills some bacteria in the presence of sucrose (see chapter 3). First a DNA cassette carrying both an antibiotic resistance gene and the toxic gene, flanked by sequences for the region of the gene to be replaced, are introduced into the cell by electroporation, selecting for the antibiotic resistance as above. Then another DNA fragment, identical to the region of the DNA but carrying the desired base pair change is introduced, selecting for loss of the toxic gene. Most of the surviving bacteria have the sequence with the base pair change replacing the original sequence in the gene, although this should be verified for at least one of them by sequencing.

The usefulness of this method to make specific changes in a gene increased dramatically when it was discovered that single-stranded DNA can also be used for the electroporation (see Ellis et al., below). Single-stranded DNAs with a defined sequence are easily obtainable since this is how DNA is chemically synthesized for PCR primers, etc. Using single-stranded DNAs also makes it possible to dispense with Gam and Exo since single-stranded DNA is not degraded by RecBCD and the DNA does not need Exo to make it single-stranded, since it is already single-stranded. Panel C shows the replacement if a single-stranded oligonucleotide is used to introduce either an in-frame deletion or a single-base-pair change into the target DNA. The procedure is similar, except that only *bet* need be expressed from the prophage. The Bet protein promotes pairing between the introduced single-stranded DNA and the chromosomal DNA and strand exchange. Then replication or repair replaces the normal sequence within the mutant sequence in both strands as shown.

Interestingly, gene replacement with single-stranded DNA shows a strong strand bias, meaning that a single-stranded oligonucleotide complementary to one strand is more apt to replace the corresponding chromosomal sequence than is an oligonucleotide complementary to the other strand in any particular region. Which strand is preferred correlates with the direction of replication in the region. The *E. coli* chromo-

some replicates bidirectionally from the origin (see chapter 1), so that on one side of the *oriC* region the replication fork moves in one direction while on the other side it moves in the opposite direction; the sequence to which it binds corresponds to the lagging strand. Presumably the single-stranded gaps that are produced on the lagging strand at the replication fork are sites of strand invasion promoted by Bet, which, unlike RecA, is not able to help a single-stranded DNA invade a completely double-stranded DNA. Bet apparently needs a single-stranded gap in the invaded DNA to pair with before it can get its "foot in the door" and promote the invasion of adjacent double-stranded DNA. This requirement for single-stranded gaps in the DNA might also explain why it is easier to mutagenize multicopy plasmids if they are introduced by electroporation along with the oligonucleotide. On entering the cell, the plasmid replicates to its copy number, exposing extensive single-stranded gaps in its lagging strand, which can pair with single-stranded DNA coated with Bet.

It also increases the efficiency of mutagenesis considerably to do the recombineering in a strain that is deficient in mismatch repair, or to deliberately create a C:C mismatch that will not be recognized by the mismatch repair system. For some reason, recognition of a mismatch by the mismatch repair system greatly lowers the recombination frequency, perhaps by interfering with filament formation by RecA or Bet. All of these improvements greatly increase the frequency of progeny with the desired mutation, often up to close to 50%, but it is still necessary to screen the survivors for ones in which the sequence of the introduced DNA has replaced the corresponding sequence in a plasmid or the chromosome. So far, this method has been adapted only to *E. coli* and some of its relatives such as *Salmonella* and *Yersinia*, but adaptations to other bacteria should be forthcoming because of its usefulness. One promising approach might be to identify prophages with recombination functions in other bacteria through genome sequences and try to adapt these for recombineering in these other bacteria.

References

Costantino, N., and D. L. Court. 2003. Enhanced levels of λ Red-mediated recombinants in mismatch repair mutants. *Proc. Natl. Acad. Sci. USA* **100:**15748–15753.

Ellis, H. M., D. Yu, T. DiTizio, and D. L. Court. 2001. High efficiency mutagenesis, repair and engineering of chromosomal DNA using single-stranded oligonucleotides. *Proc. Natl. Acad. Sci. USA* **98:** 6742–6746.

with this organism. In this section, we discuss some of the genetic experiments that have led to our present picture of the mechanisms of recombination in *E. coli*.

Isolating Rec⁻ Mutants of *E. coli*

As in any genetic analysis, the first step in studying recombination in *E. coli* was to isolate mutants defective in recombination. Such mutants are called **Rec⁻ mutants** and have mutations in the *rec* genes, whose products are required for recombination. Two very different approaches were used in the first isolations of Rec⁻ mutants of *E. coli*.

Some of the first Rec⁻ mutants were selected directly, based on their inability to support recombination (see Clark and Margulies, Suggested Reading). The idea behind this selection was that an *E. coli* strain with a mutation that inactivates a required *rec* gene should not be able to produce recombinant types when crossed with another strain. In one experiment, a Leu⁻ strain of *E. coli* was mutagenized with nitrosoguanidine. The mutagenized bacteria, some of which might now also have a *rec* mutation in addition to their *leu* mutation, were then crossed separately with an Hfr strain. A Rec⁻ mutant should produce no Leu⁺ recombinants when crossed with the Hfr strain.

To cross thousands of the mutagenized bacteria separately with the Hfr strain to find the few that had *rec* mutations and gave no recombinants would have been too laborious, so the investigators used replica plating to facilitate the identification of the mutants. When a plate containing colonies of individual mutagenized bacteria was replicated onto another plate lacking leucine on which an Hfr strain had been spread, the few Leu⁺ colonies that arose within the replicated colony were due to recombinants. A few colonies gave no Leu⁻ recombinants when crossed with the Hfr strain and were therefore candidates for Rec⁻ mutants. We discussed bacterial genetic techniques such as replica plating and Hfr crosses in chapter 3.

However, just because the mutants give no recombinants in a cross does not mean that they are necessarily Rec⁻ mutants. For instance, the mutants might have been normal for recombination but defective in the uptake of DNA during conjugation. This possibility was ruled out by crossing the mutants with an F′-containing strain instead of an Hfr strain. As discussed in chapter 5, apparent recombinant types can appear without recombination in an F′ cross because the F′ factor can replicate autonomously in the recipient cells; that is, it is a replicon. However, the DNA must still be taken up during transfer of the F′ factor, so that if mutants are defective in DNA uptake, no apparent recombination types would appear in the F′ cross. Normal frequencies of apparent recombinant types appeared when some of the mutants

were crossed with F′ strains; therefore, these mutants were not defective in DNA uptake during conjugation but, rather, had defects in recombination. These and other criteria were used to isolate several recombination-deficient Rec⁻ mutants.

The approach used by others to isolate recombination-deficient mutants of *E. coli* was less direct (see Howard-Flanders and Theriot, Suggested Reading). Their isolations depended on the fact that some recombination functions are also involved in the repair of UV-damaged DNA. Therefore, using methods described in chapter 11, Howard-Flanders and Theriot isolated several repair-deficient mutants and tested them to determine if any were also deficient in recombination. Some, but not all, of these repair-deficient mutants could also be shown to be defective in recombination in crosses with Hfr strains.

COMPLEMENTATION TESTS WITH *rec* MUTATIONS

Once Rec⁻ mutants had been isolated, the number of *rec* genes could be determined by complementation tests (see chapter 3). The original *rec* mutations defined three genes of the bacterium: *recA*, *recB*, and *recC*. The *recB* and *recC* mutants were less defective in recombination and repair than were the *recA* mutants. In fact, the RecA function is the only gene product absolutely required for recombination in *E. coli* and many other bacteria.

MAPPING *rec* GENES

The next step was to map the *rec* genes. This task may seem impossible, since a *rec* mutation causes a deficiency in recombination, but it is the frequency of recombination with known markers that reveals map position (see chapter 3). Crosses with Rec⁻ mutants can be successful only if the donor, but not the recipient, strain has the *rec* mutation. After DNA transfer, phenotypic lag ensures that even recipient cells which get the *rec* mutation by recombination, and which therefore eventually become Rec⁻, will retain recombination activity at least long enough for recombination to occur. Other markers are selected, and the presence or absence of the *rec* mutation is scored as an unselected marker either by the UV sensitivity of the recombinants or some other phenotype due to the *rec* mutation.

Using crosses such as those described above, investigators found that *recA* mutations mapped at 51 min and *recB* and *recC* mutations mapped close to each other at 54 min on the *E. coli* genetic map. Later studies showed that the products of the *recB* and *recC* genes, as well as of the adjacent *recD* gene, comprise the RecBCD nuclease, which initiates the major recombination pathway required after Hfr crosses (see above). The *recD* gene was not found in the original selection, because its product is not essential for recombination under these conditions. In fact, as discussed above, mutations in the *recD* gene can stimulate recombination by preventing degradation of the displaced strand and making recombination independent of χ sites.

Isolating Mutants with Mutations in Other Recombination Genes

In addition to the *recA* and *recBCD* genes, other genes whose products participate in recombination in *E. coli* have been found (Table 10.1). Many of these genes were not found in the original selections, because inactivating them alone does not sufficiently reduce recombination after Hfr or transductional crosses in wild-type *E. coli*.

THE RecFOR PATHWAY

As discussed, the RecFOR pathway in *E. coli* involves the products of the *recF*, *recJ*, *recN*, *recO*, *recQ*, and *recR* genes. The genes of the RecFOR pathway of recombination were not discovered in the original selections of *rec* mutations because they are not normally required for recombination after Hfr crosses. By themselves, the *recB* and *recC* mutations reduce recombination after an Hfr cross to about 1% of its normal level. Mutations in any of the genes of the RecFOR pathway can prevent the residual recombination that occurs in a *recB* or *recC* cell, which suggested that these genes are responsible for only a minor pathway of recombination in *E. coli*. However, further evidence suggested that the RecFOR pathway was just as important as the RecBCD pathway but was initiated at single-strand gaps in DNA rather than at free ends such as occur during Hfr crosses.

The RecFOR pathway was discovered because mutations in two other genes suppressed the requirement for RecBC after Hfr crosses. These genes were therefore named *sbcB* and *sbcC* (because they suppress the recombination deficiency in *recB* and *recC* mutants). We now know that these suppressor mutations act by allowing the RecFOR pathway to function after Hfr crosses, because the extra recombination in a *recB sbcB* mutant was eliminated by a third mutation in a gene of the RecFOR pathway, for example, *recQ*. In other words a *recB sbcB recO* triple mutant was as almost as defective in recombination as a *recA* mutant. Apparently, the products of the *sbcB* and *sbcC* genes normally interfere with the ability of the enzymes of the RecFOR pathway to process the donor DNA during an Hfr cross.

THE *ruvABC* AND *recG* GENES

As discussed above, the products of the *ruv* and *recG* genes are involved in the migration and cutting of Holliday junctions. There are three adjacent *ruv* genes: *ruvA*, *ruvB*, and *ruvC*. The *ruvA* and *ruvB* genes are transcribed into a polycistronic mRNA (see chapter 2), and the *ruvC* gene is adjacent but independently transcribed. The *recG* gene lies elsewhere in the genome.

The discovery of the role of the *ruv* genes in recombination involved some interesting genetics. The *recG* and *ruvABC* genes were not found in the original selections for recombination-deficient mutants because, by themselves, mutations in these genes do not significantly reduce recombination. The *ruv* genes were found only because mutations in them can increase the sensitivity of *E. coli* to killing by UV irradiation. The *recG* gene was found because double mutants with mutations in the *recG* gene and one of the *ruv* genes are severely deficient in recombination (see Lloyd, Suggested Reading). In genetic analysis, this is usually taken to mean that the RecG and Ruv proteins perform overlapping functions in recombination and so can substitute for each other. If any one of the *ruv* genes is inactive, the whole Ruv pathway is inactive, but the *recG* gene product is still available to migrate Holliday junctions and vice versa. The enzyme that resolves Holliday junctions in a *ruv* mutant is not clear. Redundancy of function is a common explanation in genetics for why some gene products are nonessential. We discuss other examples of redundant functions elsewhere in this book.

Once the Ruv proteins had been shown to be involved in recombination, it took some clever intuition to find that their role is in the migration and resolution of Holliday junctions (see Parsons et al., Suggested Reading). The Ruv proteins were first suspected to be involved in a late stage of recombination because of a puzzling observation: even though *ruv* mutants give normal numbers of transconjugants when crossed with an Hfr strain, they give many fewer transconjugants when crossed with strains containing F′ plasmids. As mentioned above, mating with F′ factors should result in, if anything, more transconjugants than crosses with Hfr strains, because F′ factors are replicons, which can multiply autonomously and do not rely on recombination for their maintenance.

One way to explain this observation is to propose that the Ruv proteins function late in recombination. If the Ruv proteins function late, recombinational intermediates might accumulate in cells with *ruv* mutations, thus having a deleterious effect on the cell. If so, *recA* mutations, which block an early step in recombination by preventing the formation of synapses, might suppress the deleterious effect of the *ruv* mutations on F′ crosses. This was found to be the case. A *ruv* mutation had no effect on the frequency of transconjugants in F′ crosses if the recipient cells also had a *recA* mutation. Once genetic evidence indicated a late role for the Ruv proteins in recombination, the process of Holliday junction resolution became the candidate for that role, since this is the last step in recombination. Biochemical experiments were then used to show that the Ruv proteins help in the migration and resolution of synthetic Holliday junctions, as described above.

DISCOVERY OF χ SITES

The discovery on DNA of χ sites that stimulate recombination by the RecBCD nuclease also required some interesting genetics and was a triumph of genetic reasoning. Many sites that are subject to single- or double-strand breaks are known to be "hot spots" for recombination. In some cases, such as recombination initiated by homing enzymes (Box 10.1), it is clear that breaks at specific sites in DNA can initiate recombination. In general, however, the frequency of recombination seems to correlate fairly well with physical distance on DNA, as though recombination occurs fairly uniformly throughout DNA molecules.

It came as some surprise, therefore, to discover that the major recombination pathway for Hfr crosses in *E. coli*, the RecBCD pathway, does occur through specific sites on the DNA—the χ sites. Like many important discoveries in science, the discovery of χ sites started with an astute observation. This observation was made during studies of host recombination functions using λ phage (see Stahl et al., Suggested Reading). The experiments were designed to analyze the recombinant types that formed when the phage *red* recombination genes were deleted, which are *exo* and *bet*. Without its own recombination functions, the phage requires the host RecBCD nuclease. In addition, if the phage is also a *gam* mutant, it does not form a plaque unless it can recombine. Therefore, plaque formation by a *red gam* mutant phage λ is an indication that recombination has taken place.

The reason that *red gam* mutant phage λ cannot multiply to make a plaque without recombination is somewhat complicated. As discussed in chapter 8, phage λ cannot package DNA from genome-length DNA molecules but only from concatemers in which the λ genomes are linked end to end. Normally, the phage makes concatemers by rolling-circle replication. However, if the phage is a *gam* mutant, it cannot switch to the rolling-circle mode of replication because the RecBCD nuclease, which is normally inhibited by Gam, somehow blocks the switch. Therefore, the only way a *gam* mutant phage λ can form concatemers is by recombination between the circular λ DNAs formed via θ replication (Figure 8.6). If the phage is also a *red* mutant (i.e., lacks its own recombination functions), the only way it can form concatemers is by RecBCD recombination, the major host pathway. Therefore, *red gam* mutant phage λ requires RecBCD recombination to form plaques, and the formation of plaques can be used as a measure of RecBCD recombination under these conditions.

The first χ mutations were discovered when large numbers of *red gam* mutant phage λ were plated on RecBCD⁺ *E. coli*. Very few phage were produced, and the plaques that formed were very tiny. Apparently, very little RecBCD recombination was occurring between the circular phage DNAs produced by θ replication. However, λ mutants that produced much larger plaques sometimes appeared. The circular λ DNAs in these mutants were apparently recombining at a much higher rate. The responsible mutations were named χ **mutations** because they increase the frequency of crossovers (crossed-over chromosomes are called *chi*asma in eukaryotes). Once the mutations were mapped and the DNA was sequenced, χ mutations in λ were found to be base pair changes that created the sequence 5'GCTGGTGG3' somewhere in the λ DNA. The presence of this sequence appears to stimulate recombination by the RecBCD pathway. Since wild-type λ does not need RecBCD and therefore has no such sequence anywhere in its DNA, recombination by RecBCD is very infrequent unless the χ sequence is created by a mutation.

Further experimentation with χ sites revealed several interesting properties. For example, they stimulate crossovers only to one side of themselves, the 5' side. Very little stimulation of crossovers occurs on the 3' side. Also, if only one of the two parental phages contains a χ site, most of the recombinant progeny will not have the χ site, so that the χ site itself is preferentially lost during the recombination. These and other properties of χ sites eventually led to the model for χ site action presented earlier in this chapter.

Gene Conversion and Other Manifestations of Heteroduplex Formation during Recombination

As discussed earlier in this chapter, models for recombination are based in part on the evidence concerning the formation of heteroduplexes during recombination. Such evidence indicated that heteroduplexes must form on both strands of DNA during recombination.

GENE CONVERSION

The first such evidence for heteroduplex formation during recombination came from studies of gene conversion in fungi. Understanding this process requires a little knowledge of the sexual cycles of fungi. Some fungi have long been favored organisms for the study of recombination because the spores that are the products of a single meiosis are often contained in the same bag, or ascus (see any general genetics textbook). When two haploid fungal cells mate, the two cells fuse to form a diploid zygote. Then the homologous chromosomes pair and replicate once to form four chromatids that recombine with each other before they are packaged into spores. Therefore, each ascus contains four spores (or eight in fungi such as *Neurospora crassa*, in which the chromatids replicate once more before spore packaging).

Since both haploid fungi contribute equal numbers of chromosomes to the zygote, their genes should show up in equal numbers in the spores in the ascus. In other words, if the two haploid fungi have different alleles of the same gene, two of the four spores in each ascus should have an allele from one parent and the other two spores should have the allele from the other parent. This is called a 2:2 segregation. However, the two parental alleles sometimes do not appear in equal numbers in the spores. For example, three of the spores in a particular ascus might have the allele from one parent while the remaining spore has the allele from the other parent—a 3:1 segregation instead of the expected 2:2 segregation. In this case, an allele of one of the parents appears to have been converted into the allele of the other parent during meiosis, hence the name **gene conversion**.

Gene conversion is caused by repair of mismatches created on heteroduplexes during recombination, and Figure 10.10 shows how such mismatch repair can convert one allele into the other when the DNA molecules of the two parents recombine. In the illustration, the DNAs of two parents are identical in the region of the recombination except that a mutation has changed a wild-type AT pair into a GC in one of the DNA molecules. Hence, the parents have different alleles of this gene. When the two individuals mate to form a diploid zygote and their DNAs recombine during meiosis, one strand of each DNA may pair with the complementary strand of the other DNA in this region. A mismatch will result, with a G opposite a T in one DNA and an A opposite a C in the other DNA (Figure 10.10). If a repair system changes the T opposite the G to a C in one of the DNAs, then after meiosis three molecules will carry the mutant allele sequence, with GC at this position, but only one DNA will have the wild-type allele sequence, with AT at this position. Hence, one of the two wild-type alleles has been "converted" into the mutant allele.

MANIFESTATIONS OF MISMATCH REPAIR IN HETERODUPLEXES IN PHAGES AND BACTERIA

Gene conversion is more difficult to detect in crosses with bacteria and phages than with fungi, since the products of a single recombination event do not stay together in a bacterial or phage cross. However, the existence of heteroduplexes formed during recombination in bacteria and phages, as well as the repair of mismatches in these structures, is manifested in other ways.

Map Expansion

In prokaryotes, mismatch repair in heteroduplexes can increase the apparent recombination frequency between

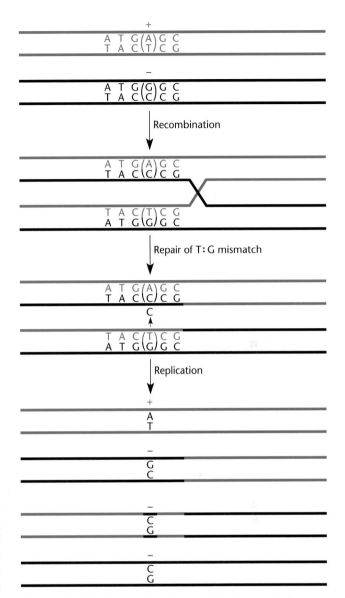

Figure 10.10 Repair of a mismatch in a heteroduplex region formed during recombination can cause gene conversion. A plus sign indicates the wild-type sequence, and a minus sign indicates the mutant sequence. See the text for details.

two closely linked markers, making the two markers seem farther apart than they really are. This manifestation of mismatch repair in heteroduplexes formed during recombination is called **map expansion** because the genetic map appears to increase in size.

Figure 10.11 shows how mismatch repair can affect the apparent recombination frequency between two markers. In the illustration, the two DNA molecules participating in the recombination have mutations that are very close to each other, so that crossovers between the two mutations

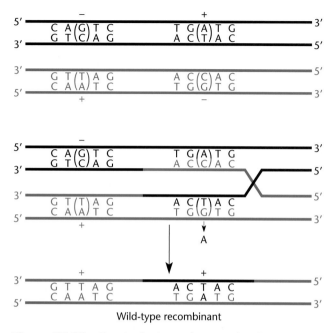

Figure 10.11   Repair of mismatches can give rise to recombinant types between two mutations. A plus sign indicates the wild-type sequence, and a minus sign indicates the mutant sequence. The position of the mutations is shown in parentheses. See the text for details.

to give wild-type recombinants should be very rare. However, a Holliday junction occurs nearby, and the region of one of the two mutations is included in the heteroduplex, creating mismatches that can be repaired. If the G in the GT mismatch in one of the DNA molecules is repaired to an A, the progeny with the DNA will appear to be a wild-type recombinant. Therefore, even though the potential crossover that caused the formation of heteroduplexes did not occur in the region between the two mutations, apparent wild-type recombinants resulted. The apparent recombination due to mismatch repair might occur even when the Holliday junction is resolved so that the flanking sequences are returned to their original configuration; thus, a true crossover does not result. Therefore, although gene conversion and other manifestations of mismatch repair are generally associated with recombinant DNA molecules, they do not appear only in DNA molecules with crossovers.

Marker Effects
Mismatch repair of heteroduplexes also can cause **marker effects**, phenomena in which two different markers at exactly the same locus show different recombination frequencies when crossed with the same nearby marker. For example, two different transversion mutations might change a UAC codon into UAA and UAG codons in different strains. However, when these two strains are crossed with another strain with a third nearby mutation, the recombination frequency between the ochre mutation and the third mutation might appear to be much higher than the recombination frequency between the amber mutation and the third mutation, even though the amber and ochre mutations are exactly the same distance on DNA from the third mutation. Such a difference between the two recombination frequencies can be explained because the amber and ochre mutations are causing different mismatches to form during recombination and one of these may be recognized and repaired more readily by the mismatch repair system than the other. In *E. coli* at least, CC mismatches are not repaired by the mismatch repair system. Note that one of the heteroduplexes that the original sequence UAC forms with the amber mutation at this site contains a CC mismatch but the heteroduplex formed with the ochre mutation does not.

Marker effects also occur because the lengths of single DNA strands removed and resynthesized by different mismatch repair systems vary (see chapter 11), and the chance that mismatch repair will lead to apparent recombination depends on the length of these sequences, or patches. As is apparent in Figure 10.11, a wild-type recombinant occurs only if mismatches due to both mutations are not removed on the same repair patch. If the patch that is removed in repairing one mismatch also removes the other mismatch, one of the parental DNA sequences will be restored and no apparent recombination will occur. In *E. coli*, mismatches due to deamination of certain methylated cytosines are repaired by very short patches (VSP repair; see chapter 1).

High Negative Interference
Another manifestation of mismatch repair in heteroduplexes is **high negative interference**. This has the reverse effect of interference in eukaryotes, a phenomenon in which the stiffness of the chromatids can cause one crossover to reduce the chance of another crossover nearby. In high negative interference, one crossover greatly *increases* the apparent frequency of another crossover nearby.

High negative interference is often detected during three-factor crosses with closely linked markers. In chapters 3 and 7, we discussed how three-factor crosses can be used to order three closely linked mutations. Briefly, if one parent has two mutations and the other parent has a third mutation, the frequency of the different types of recombinants after the cross will depend on the order of the sites of the three mutations in the DNA. Twice as many apparent crossovers are required to produce the rarest recombinant type as are needed for the more frequent recombinant types.

However, owing to mismatch repair of heteroduplexes, the rarest recombinant type can occur much more frequently than expected. Figure 10.12 shows a three-factor cross between markers that are created by three mutations. Wild-type sequences are marked with a plus, and mutant sequences are marked with a minus. With the molecules pictured, the formation of a wild-type recombinant should require two crossovers: one between mutations 1 and 2 and another between mutations 2 and 3. Theoretically, if the two crossovers were truly independent, the frequency of wild-type recombinants in the three-factor cross should equal the frequency when mutation 1 is crossed with mutation 2 times the frequency when mutation 2 is crossed with mutation 3, in separate crosses. Instead, the frequency of the wild-type recombinants in the three-factor cross is often much higher than this product. As shown in Figure 10.12, if a crossover occurs between two of the mutations, a heteroduplex formed by the Holliday junction might include the region of the third mutation. Then repair of the third mutant site mismatch in the heteroduplex would give the appearance of a second nearby crossover, greatly increasing the apparent frequency of that crossover.

It is important to realize that repair of mismatches can occur wherever two DNAs are recombining. However,

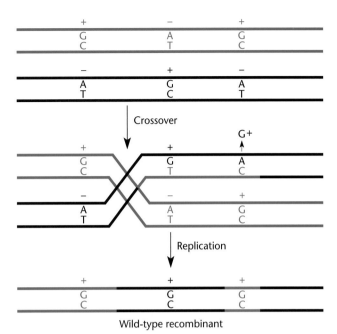

Figure 10.12 High negative interference due to mismatch repair. Inclusion of the region of the third mutation in the heteroduplex followed by repair of the mismatch can give the appearance of a second crossover.

only if two markers are close enough together that normal crossovers are infrequent will mismatch repair in the heteroduplexes contribute significantly to apparent recombination frequencies and cause effects such as map expansion, high negative interference, and marker effects.

SUMMARY

1. Recombination is the joining of DNA strands in new combinations. Homologous recombination occurs only between two DNA molecules that have the same sequence in the region of the crossover.

2. Models of recombination can be divided into one- or two-strand invasion models and double-strand break repair models.

3. All recombination models involve the formation of Holliday junctions. The Holliday junctions can migrate and isomerize so that the crossed strands uncross and then recross in a different orientation. Holliday junctions can then be resolved by specific DNA endonucleases to give recombinant DNA products.

4. The region over which two strands, originating from different DNA molecules, are paired in a Holliday junction is called a heteroduplex.

5. Repair of mismatches on the heteroduplex DNA molecules formed as intermediates in recombination can give rise to such phenomena as gene conversion, map expansion, high negative interference, and marker effects.

6. In *E. coli*, the major pathway for recombination during conjugation and transduction is the RecBCD pathway. The RecBCD protein loads on the DNA at a double-stranded break and moves along the DNA, looping out a single strand and degrading one strand from the 3′ end. If a sequence in the DNA called the χ sequence is encountered as the protein moves along the DNA, the 3′-to-5′ nuclease activity on RecBCD is inhibited and a 5′-to-3′ activity is stimulated, leaving a free 3′ end that can invade other double-stranded DNAs.

7. The RecFOR pathway is another recombination pathway in *E. coli*. It is required for recombination between
(continued)

SUMMARY (continued)

single-stranded gaps in DNA and double-stranded DNA. The RecFOR pathway requires the products of the *recF*, *recJ*, *recN*, *recO*, *recQ*, and *recR* genes. Because it functions only at single-stranded gaps in the DNA, this pathway can function in Hfr crosses only if the *sbcB* and *sbcC* genes have been inactivated. Apparently, the SbcB and SbcC enzymes destroy intermediates created during recombination by the RecFOR pathway.

8. The RecA protein promotes synapse formation and strand displacement and is required for recombination by both the RecBCD and RecFOR pathways. The RecA protein forces a single-stranded DNA into a helical nucleoprotein filament, which can then scan double-stranded DNAs, looking for its complementary sequence. If it finds its complementary sequence, it invades it, forming a D-loop or a three-strand structure which can then migrate to form a Holliday junction.

9. Holliday junctions can migrate by at least two separate pathways in *E. coli*, the RuvABC pathway and the RecG pathway. In the first pathway, the RuvA protein binds to Holliday junctions and then the RuvB protein binds to

RuvA and promotes branch migration with the energy derived from cleaving ATP. The RuvC protein is a Holliday junction-specific endonuclease that cleaves Holliday junctions to resolve recombinant products. The RecG protein is also a Holliday junction-specific helicase, but it moves Holliday junctions in the opposite direction from RuvB and can substitute for the RuvABC system.

10. Many phages also encode their own recombination systems. Sometimes, phage recombination functions are analogous to host recombination functions. The gene 49 product of T4 phage and the gene 3 product of T7 resolve Holliday junctions. A phage λ recombination system, the *red* system, is encoded by two genes, *exo* and *bet*. The *exo* gene product is analogous to RecBCD in that it degrades one strand of a double-stranded DNA to make a single-stranded DNA for strand invasion. The *bet* gene product is analogous to RecA in that it promotes synapse formation between two DNAs. These λ genes have become the basis for a very useful way of doing site-specific mutagenesis and gene replacements, sometimes called recombineering.

QUESTIONS FOR THOUGHT

1. Why do you suppose that essentially all organisms have recombination systems?

2. Why do you suppose the RecBCD protein promotes recombination through such a complicated process?

3. Why are there overlapping pathways of recombination that can substitute for each other?

4. Why do you think the cell encodes the *sbcB* and *sbcC* gene products that interfere with the RecFOR pathway?

5. Why do some phages encode their own recombination systems? Why not rely exclusively on the host pathways?

6. Propose a model for how the RecG protein could substitute for RuvABC when it has only a helicase activity and not an X-phile resolvase activity.

PROBLEMS

1. Outline how you would determine if recombinants in an Hfr cross have a *recA* mutation. Note: *recA* mutations make the cells very sensitive to mitomycin and UV irradiation.

2. How would you determine whether the products of other genes participate in the *recG* pathway of migration and resolution of Holliday junctions? How would you find such genes?

3. Describe the recombination promoted by homing double-stranded nucleases to insert an intron by using the double-strand break repair model.

4. Design an experiment to determine whether recombination due to the RecBC nuclease without the RecD subunit is still stimulated by χ.

SUGGESTED READING

Baharoglu, Z., M. Petranovic, M.-J. Flores, and B. Michel. 2006. RuvAB is essential for replication forks reversal in certain replication mutants. *EMBO J.* **25:**596–604.

Bolt, E. L., and R. G. Lloyd. 2002. Substrate specificity of RusA resolvase reveals the DNA structures targeted by RuvAB and RecG in vivo. *Mol. Cell* **10:**187–198.

Clark, A. J., and A. D. Margulies. 1965. Isolation and characterization of recombination-deficient mutants of *Escherichia coli* K12. *Proc. Natl. Acad. Sci. USA* **62:**451–459.

Cox, M. M. 2003. The bacterial RecA protein as a motor protein. *Annu. Rev. Microbiol.* **57:**551–577.

Dixon, D. A., and S. C. Kowalczykowski. 1993. The recombination hotspot Chi is a regulatory sequence that acts by attenuating the nuclease activity of the *E. coli* RecBCD enzyme. *Cell* **73:**87–96.

Holliday, R. 1964. A mechanism for gene conversion in fungi. *Genet. Res.* **5:**282–304.

Howard-Flanders, P., and L. Theriot. 1966. Mutants of *Escherichia coli* defective in DNA repair and in genetic recombination. *Genetics* **53:**1137–1150.

Jones, M., R. Wagner, and M. Radman. 1987. Mismatch repair and recombination in *E. coli*. *Cell* **50:**621–626.

Lloyd, R. G. 1991. Conjugal recombination in resolvase-deficient *ruvC* mutants of *Escherichia coli* K-12 depends on *recG*. *J. Bacteriol.* **173:**5414–5418.

Mazin, A. V., and S. C. Kowalczykowski. 1999. A novel property of the RecA nucleoprotein filament: activation of double-stranded DNA for strand exchange in *trans*. *Genes Dev.* **13:**2005–2016.

Meselson, M. S., and C. M. Radding. 1975. A general model for genetic recombination. *Proc. Natl. Acad. Sci. USA* **72:**358–361.

Morimatsu, K., and S. C. Kowalczykowski. 2003. RecFOR proteins load RecA protein onto gapped DNA to accelerate DNA strand exchange, a universal step in recombinational repair. *Mol. Cell* **11:**1337–1347.

Parsons, C. A., I. Tsaneva, R. G. Lloyd, and S. C. West. 1992. Interaction of *Escherichia coli* RuvA and RuvB proteins with synthetic Holliday junctions. *Proc. Natl. Acad. Sci. USA* **89:**5452–5456.

Radding, C. M. 1991. Helical interactions in homologous pairing and strand exchange driven by the RecA protein. *J. Biol. Chem.* **266:**5355–5358.

Renzette, N., N. Gumlaw, and S. J. Sandler. 2007. DinI and RecX modulate RecA-DNA structures in *Escherichia coli* K-12. *Mol. Microbiol.* **63:**103–115.

Sawitzke, J. A., and F. W. Stahl. 1992. Phage λ has an analog of *Escherichia coli recO*, *recR* and *recF* genes. *Genetics* **130:**7–16.

Sharples, G. J. 2001. The X philes: structure-specific endonucleases that resolve Holliday junctions. *Mol. Microbiol.* **39:**823–834.

Stahl, F. W., M. M. Stahl, R. E. Malone, and J. M. Crasemann. 1980. Directionality and nonreciprocality of Chi-stimulated recombination in phage λ. *Genetics* **94:**235–248.

Szostak, J. W., T. L. Orr-Weaver, R. J. Rothstein, and F. W. Stahl. 1983. The double-strand-break repair model for recombination. *Cell* **33:**25–35.

Taylor, A. F., and G. R. Smith. 2003. RecBCD enzyme is a DNA helicase with fast and slow motors of opposite polarity. *Nature* **423:**889–893.

West, S. C. 1998. RuvA gets X-rayed on Holliday. *Cell* **94:**699–701.

CHAPTER 11

DNA Repair and Mutagenesis

The continuity of species from one generation to the next is a tribute to the stability of DNA. If DNA were not so stable and were not reproduced so faithfully, there could be no species. Before DNA was known to be the hereditary material and its structure determined, a lot of speculation centered around what types of materials would be stable enough to ensure the reliable transfer of genetic information over so many generations (see, for example, Schrodinger, Suggested Reading in the introductory chapter). Therefore, the discovery that the hereditary material is DNA—a chemical polymer no more stable than many other chemical polymers—came as a surprise.

Evolution has resulted in a design for the DNA replication apparatus that minimizes mistakes (see chapter 1). However, mistakes during replication are not the only threats to DNA. Since DNA is a chemical, it is constantly damaged by chemical reactions. Many environmental factors can damage this molecule. Heat can speed up spontaneous chemical reactions, leading, for example, to the deamination of bases. Chemicals can react with DNA, adding groups to the bases or sugars, breaking the bonds of the DNA, or fusing parts of the molecule to each other. Irradiation at certain wavelengths can also chemically damage DNA, which can absorb the energy of the photons. Once the molecule is energized, bonds may be broken or parts may be fused. DNA damage can be very deleterious to cells because their DNA may not be able to replicate over the damaged area and so the cells could not multiply. Even if the damage does not block replication, replicating over the damage can cause mutations, many of which may be deleterious or even lethal. Obviously, cells need mechanisms for DNA damage repair.

To describe DNA damage and its repair, we need first to define a few terms. Chemical damage in DNA is called a **lesion**. Chemical compounds or treatments that cause lesions in DNA can kill cells and can also increase the frequency of mutations in DNA. Such treatments that generate mutations are called **mutagenic treatments** or **mutagens.** Some mutagens, known as **in vitro mutagens,** can be used to damage DNA in the test tube and then produce mutations when this DNA is introduced into cells. Other mutagens damage DNA only in the cell, for example by interfering with base pairing during replication. These are called **in vivo mutagens.**

In this chapter, we discuss the types of DNA damage, how each type of damage to DNA might cause mutations, and how bacterial cells repair the damage to their DNA. Many of these mechanisms seem to be universal and are shared by higher organisms including humans.

Evidence for DNA Repair

Before discussing specific types of DNA damage, we should make some general comments on the outward manifestations of DNA damage and its repair. The first question is how we even know a cell has the means to repair a particular type of damage to its DNA. One way is to measure killing by a chemical or by irradiation. The chemical agents and radiation that damage DNA also often damage other cellular constituents, including RNA and proteins. Nevertheless, cells exposed to these agents usually die as a result of chemical damage to the DNA. The other components of the cell can usually be resynthesized and/or exist in many copies, so that even if some molecules are damaged, more of the same type of molecule will be there to substitute for the damaged ones. However, a single chemical change in the enormously long chromosomal DNA of a cell can prevent the replication of that molecule and subsequently cause cell death unless the damage is repaired.

To measure cell killing—and thereby demonstrate that a particular type of cell has DNA repair systems—we can compare the survival of the cells exposed intermittently to small doses of a DNA-damaging agent with that of cells that receive the same amount of treatment continuously. If the cells have DNA repair systems, more cells survive the short intervals of treatment because some of the damage is repaired between treatments. Consequently, at the end of the experiment, fewer intermittently treated cells have been killed than cells exposed to continuous treatment. In contrast, if DNA is not repaired during the rest periods, it should make no difference whether the treatment occurs at intervals or continuously. The cells accumulate the same amount of damage regardless of the different treatments, and the same fraction of cells should survive both regimes.

Another indication of repair systems comes from the shape of the killing curves. A killing curve is a plot of the number of surviving cells versus the extent of treatment

by an agent that damages DNA. The extent of treatment can refer to the length of time the cells are irradiated or exposed to a chemical that damages DNA or to the intensity of irradiation or the concentration of the damaging chemical.

The two curves in Figure 11.1 contrast the shapes of killing curves for cells with and without DNA repair systems. In the curve for cells without a repair system for the DNA-damaging treatment, the fraction of surviving cells drops exponentially since the probability that each cell will be killed by a lethal "hit" to its DNA is the same during each time interval. This exponential decline gives rise to a straight line when plotted on semilog paper as shown.

The other curve shows what happens if the cell has DNA repair systems. Rather than dropping exponentially with increasing treatment, this curve extends horizontally first, forming a "shoulder." The shoulder appears because repair mechanisms repair lower levels of damage, allowing many of the cells to survive. Only with higher treatment levels, when the damage becomes so extensive that the repair systems can no longer cope with it, will the number of surviving cells drop exponentially with increasing levels of treatment.

Among the survivors of DNA-damaging agents, there may be many more mutants than before. However, it is very important to distinguish DNA damage from mutagenesis. In particular types of cells, some types of DNA

Figure 11.1 Survival of cells as a function of the time or extent of treatment with a DNA-damaging agent. The fraction of surviving cells is plotted against the duration of treatment. A shoulder on the survival curve indicates the presence of a repair mechanism.

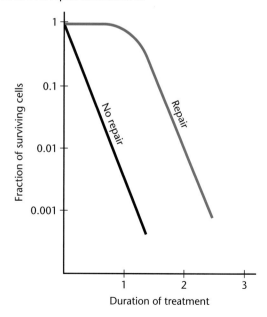

damage are mutagenic while others are not, independent of their effect on cell survival. Recall from chapter 3 that mutations are permanent heritable changes in the sequence of nucleotides in DNA. The damage to DNA caused by a chemical or by irradiation is not by itself a mutation because it is not heritable. A mutation might occur because the damage is not repaired and the replication apparatus must proceed over the damage, making mistakes because complementary base pairing does not occur properly at the site of the damage. Alternatively, mistakes might be made during attempts to repair the damage, causing changes in the sequence of nucleotides at the site of the damage. In the following sections, we discuss some types of DNA damage, how they might cause mutations, and the repair systems that can repair them. Most of what we describe is best known for *Escherichia coli*, for which these systems are best understood, but the universality of many of them is already clear.

Specific Repair Pathways

Different agents damage DNA in different ways, and different repair pathways operate to repair the various forms of damage. Some of these pathways repair only a certain type of damage, whereas others are less specific and repair many types. We first discuss examples of damage repaired by specific repair pathways.

Deamination of Bases

One of the most common types of damage to DNA is the deamination of bases. Some of the amino groups in adenine, cytosine, and guanine are particularly vulnerable and can be removed spontaneously or by many chemical agents (Figure 11.2). When adenine is deaminated, it becomes **hypoxanthine.** When guanine is deaminated, it becomes **xanthine.** When cytosine is deaminated, it becomes **uracil.**

Deamination of DNA bases is mutagenic because it results in base mispairing. As shown in Figure 11.2, hypoxanthine derived from adenine will pair with the base cytosine instead of thymine, and uracil derived from the deamination of cytosine will pair with adenine instead of guanine.

The type of mutation caused by deamination depends on which base is altered. For example, the hypoxanthine that results from deamination of adenine will pair with cytosine during replication, incorporating C instead of T at that position. In a subsequent replication, the C will pair with the correct G, causing an AT-to-GC transition in the DNA. Similarly, a uracil resulting from the deamination of a cytosine will pair with an adenine during replication, causing a GC-to-AT transition.

DEAMINATING AGENTS

Although deamination often occurs spontaneously, especially at higher temperatures, some types of chemicals react with DNA and remove amino groups from the bases. Treatment of cells or DNA with these chemicals, known as **deaminating agents,** can greatly increase the rate of mutations. Which deaminating agents are mutagenic in a particular situation depends on the properties of the chemical.

Hydroxylamine

Hydroxylamine specifically removes the amino group of cytosine and consequently causes only GC-to-AT transitions in the DNA. However, hydroxylamine, an in vitro mutagen, cannot enter cells, so it can be used only to mutagenize purified DNA or viruses. Mutagenesis by hydroxylamine is particularly effective when the treated DNA is introduced into cells deficient in repair by the uracil-N-glycosylase enzyme (see chapter 1), for reasons discussed below.

Bisulfite

Bisulfite can also deaminate only cytosine, but these cytosines must be in single-stranded DNA. This property of bisulfite has made it useful for **site-directed mutagenesis.** If the region to be mutagenized in a clone is made single stranded, the bisulfite preferentially limits mutagenesis to this single-stranded region. Use of bisulfite for site-directed mutagenesis of cloned DNAs has been largely supplanted by oligonucleotide-directed and PCR mutagenesis as well as recombineering (see chapters 1 and 10).

Nitrous Acid

Nitrous acid not only deaminates cytosines but also removes the amino groups of adenine and guanine (Figure 11.2). It also causes other types of damage. Because it is less specific, nitrous acid can cause both GC-to-AT and AT-to-GC transitions as well as deletions. Nitrous acid can enter some types of cells and so can be used as a mutagen both in vivo and in vitro.

REPAIR OF DEAMINATED BASES

Because base deamination is potentially mutagenic, special enzymes have evolved to remove deaminated bases from DNA. These enzymes, **DNA glycosylases,** break the glycosyl bond between the damaged base and the sugar in the nucleotide. A unique DNA glycosylase exists for each type of deaminated base and removes only that particular base. Specific DNA glycosylases discussed in later sections and chapter 13 remove bases damaged in other ways. There are at least a dozen specific *N*-glycosylases that remove damaged bases in *E. coli*, and they all work by basically the same mechanism.

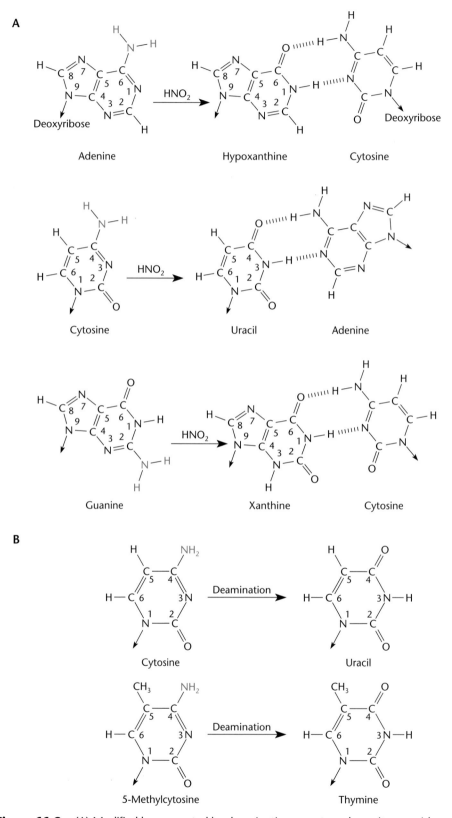

Figure 11.2 (A) Modified bases created by deaminating agents such as nitrous acid (HNO$_2$). Some deaminated bases pair with the wrong base, causing mutations. (B) Spontaneous deaminations. Deamination of 5-methylcytosine produces a thymine that is not removable by the uracil-*N*-glycosylase.

Figure 11.3 illustrates the removal of damaged bases from DNA by DNA glycosylases. There are two types of glycosylases: ones which remove just the base and others, called **AP lyases,** which both remove the base and cut the DNA backbone on the 3′ side of the damaged base. If just the base has been removed by the specific glycosylase, nucleases called **AP endonucleases** cut the sugar-phosphate backbone of the DNA on the 5′ side of the missing base, leaving a 3′ OH group. These enzymes can cut either next to the spot from which a pyrimidine (C or T) has been removed (an *apy*rimidinic site) or next to where a purine (A or G) has been removed (an *ap*urinic site). After the cut is made and processed, the free 3′-hydroxyl end is used as a primer by the repair DNA polymerase (DNA polymerase I in *E. coli*) to synthesize more DNA, while the 5′ exonuclease activity associated with the DNA polymerase degrades the strand ahead of the DNA polymerase. In this way, the entire region of the DNA strand around the deaminated base is resynthesized and the normal base is inserted in place of the damaged one.

VERY-SHORT-PATCH REPAIR OF DEAMINATED 5-METHYLCYTOSINE

Most organisms have some 5-methylcytosine bases instead of cytosines at specific sites in their DNA. These bases are cytosines with a methyl group at the 5 position on the pyrimidine ring instead of the usual hydrogen (Figure 11.2B). Specific enymes called **methyltransferases** transfer the methyl group to this position after the DNA is synthesized. The function of these 5-methylcytosines is often obscure, but we know that they sometimes help protect DNA against cutting by restriction endonucleases and may help regulate gene expression in higher organisms.

The sites of 5-methylcytosine in DNA are often hot spots for mutagenesis, because deamination of 5-methylcytosine yields thymine rather than uracil (Figure 11.2B), and thymine in DNA is not recognized by the uracil-*N*-glycosylase since it is a normal base in DNA. These thymines in DNA are located opposite guanines and so could in principle be repaired by the methyl-directed mismatch repair system (see below and chapter 1). However, in a GT mismatch created by a replication mistake, the mistakenly incorporated base can be identified because it is in a newly replicated strand, as yet unmethylated by Dam methylase, whereas the GT mismatches created by the deamination of 5-methylcytosine are generally not found in newly synthesized DNA. Repairing the wrong strand causes a GC-to-AT transition in the DNA.

In *E. coli* K-12 and some other enterics, most of the 5-methylcytosine in the DNA occurs in the second C of the sequence 5′CCWGG3′/3′GGWCC5′, where the middle base pair (W) is generally either AT or TA. The second C in this sequence is methylated by an enzyme called DNA cytosine methylase (Dcm) to give C^mCAGG/GGTC^mC. Because of the mutation potential, *E. coli* K-12 has evolved a special repair mechanism for deaminated 5-methylcytosines that occur in this sequence. This repair system specifically removes a thymine whenever it appears as a TG mismatch in this sequence.

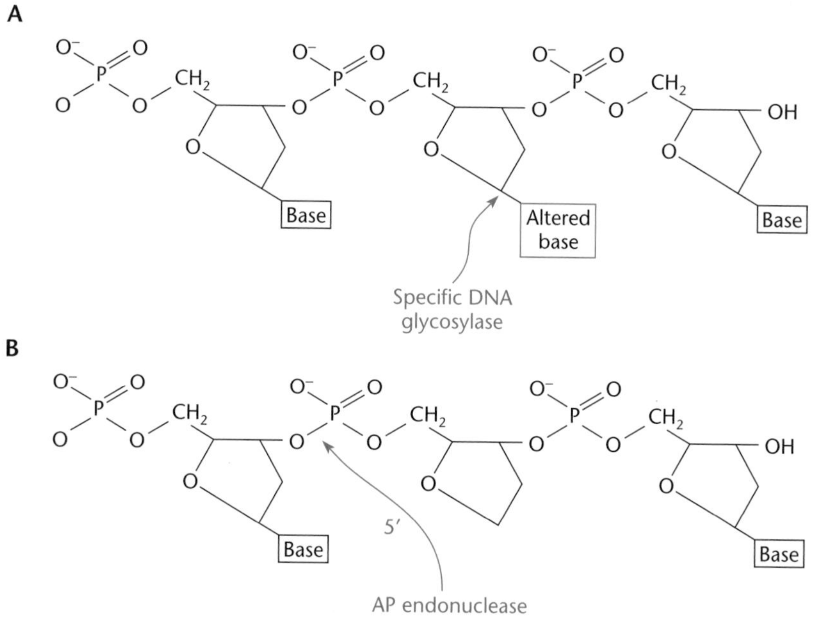

A

Specific DNA glycosylase

B

AP endonuclease

Figure 11.3 Repair of altered bases by DNA glycosylases. (A) The specific DNA glycosylase removes the altered base. (B) An apurinic or apyrimidinic (AP) endonuclease cuts the DNA backbone on the 5′ side of the apurinic or apyrimidinic site. The strand is degraded and resynthesized, and the correct base is restored (not shown).

Because a small region, or "patch," of the DNA strand containing the T is removed and resynthesized during the repair process, the mechanism is called **very-short-patch repair**. In VSP repair, the Vsr endonuclease, the product of the *vsr* gene, binds to a TG mismatch in the CmCWGG/GGWTC sequence and makes a break next to the T. The T is then removed and the strand is resynthesized by the repair DNA polymerase (DNA polymerase I), which inserts the correct C (see Lieb et al., Suggested Reading).

The Vsr repair system is very specialized and usually only repairs TG mismatches in the sequence shown above. Therefore, this repair system would have only limited usefulness if methylation did not occur in this sequence. The *vsr* gene is immediately downstream of the gene for the Dcm methylase, ensuring that cells that inherit the gene to methylate the C in CmCWGG/GGWCmC also usually inherit the ability to repair the mismatch correctly if it is deaminated. While only enterics like *E. coli* have been shown to have this particular repair system, many other organisms have 5-methylcytosine in their DNA, and we expect that similar repair systems will be found in these organisms.

Damage Due to Reactive Oxygen

Although molecular oxygen (O_2) is not very damaging to DNA and most other macromolecules, more reactive forms of oxygen, in which oxygen has acquired an extra electron, are very damaging. These more reactive forms of oxygen include superoxide radicals, hydrogen peroxide, and hydroxyl radicals, which may be produced by normal cellular reactions (Box 11.1). Alternatively, they can arise as a result of environmental factors, including UV irradiation and chemicals such as the herbicide paraquat.

Because the reactive forms of oxygen normally appear in cells, all aerobic organisms must contend with the resulting DNA damage and have evolved elaborate mechanisms to remove these chemicals from the cellular environment. In bacteria, some of these systems are induced by the presence of the reactive forms of oxygen, and these genes encode enzymes such as superoxide dismutases, catalases, and peroxide reductases, among others, that help destroy the reactive forms. These systems also include genes that encode repair enzymes which help repair the oxidative DNA damage caused by the reactive forms. The accumulation of this type of damage may be responsible for the increase in cancer rates with age and for many age-related degenerative diseases (Box 11.1).

8-oxoG

One of the most mutagenic lesions in DNA caused by reactive oxygen is the oxidized base **7,8-dihydro-**8-oxoguanine (8-oxoG or GO) (Figure 11.4). This base appears frequently in DNA because of damage caused by internally produced free radicals of oxygen, and, unless repair systems deal with the damage, DNA polymerase III often mispairs it with adenine, causing spontaneous mutations. Because of the mutagenic potential of 8-oxoG, *E. coli* has evolved many mechanisms for avoiding the resultant mutations, and we discuss these immediately below.

MutM, MutY, AND MutT

The products of the *mut* genes of an organism reduce the normal rates of spontaneous mutagenesis; therefore, organisms with a mutation that inactivates the product of a *mut* gene will suffer higher than normal rates of spontaneous mutagenesis. This is how they are selected, and we discuss the isolation of *mut* mutations below (see "Genetic Analysis of Repair Pathways" below). We have already discussed some *mut* genes of *E. coli*, selected as mutations that increase the spontaneous mutation rate, which include the genes of the mismatch repair system and the *dnaQ* (*mutD*) gene encoding the editing function ε. Other *mut* genes of *E. coli* include *mutM*, *mutY*, and *mutT*. The products of these *mut* genes are exclusively devoted to preventing mutations due to 8-oxoG. The generally high rate of spontaneous mutagenesis in *mutM*, *mutT*, and *mutY* mutants is testimony to the fact that internal oxidation of DNA is an important source of spontaneous mutations and that 8-oxoG, in particular, is a very mutagenic form of damage to DNA (Box 11.1).

The discovery that these three *mut* genes were dedicated to relieving the mutagenic effects of 8-oxoG in DNA, as well as the role played by each of them, was the result of some clever genetic experiments (see Michaels et al., Suggested Reading). First, there was the evidence that the functions of these three *mut* genes are additive to reduce spontaneous mutations, since the rate of spontaneous mutations is higher if two or all three of the *mut* genes are mutated than if only one of them is mutated. There was also evidence that mutations in each of the *mut* genes increased the frequency of some types of spontaneous mutations but not others. Below we discuss the roles of the products of each of these *mut* genes and then discuss how the genetic evidence is consistent with each of these roles.

MutM

The MutM enzyme is an *N*-glycosylase that specifically removes the 8-oxoG base from the deoxyribose sugar in DNA (Figure 11.4). This repair pathway functions like other *N*-glycosylase repair pathways discussed in this volume except that the depurinated strand is cut by the AP endonuclease activity of MutM itself, degraded by an

BOX 11.1

Oxygen: the Enemy Within

To respire, all aerobic organisms, including humans, must take up molecular oxygen (O_2). At normal temperatures, molecular oxygen reacts with very few molecules. However, some of it is converted into more reactive forms such as superoxide radicals (O_2^-), hydrogen peroxide (H_2O_2), and hydroxyl radicals (·OH). Hydrogen peroxide is probably formed inadvertently by flavoenzymes during oxidative respiration (see Seaver and Imlay, below). It is also produced deliberately in the liver to help detoxify recalcitrant molecules and by lysozomes to kill invading bacteria. Iron in the cell can then catalyze the conversion of hydrogen peroxide to hydroxyl radicals, which may be the form in which oxygen is most damaging to DNA.

Cells have evolved many enzymes to help reduce this damage. These include catalases that reduce reactive oxygen molecules as they form and repair enzymes such as exonucleases and glycosylases that remove the damaged bases from DNA before they can cause mutations (see the text and Park et al., below).

Accumulation of DNA damage due to reactive oxygen has been linked to many degenerative diseases such as cancer, arthritis, cataracts, and cardiovascular disease. It has been estimated that a rat at 2 years of age has about 2 million DNA lesions per cell, and some types of human cells have been shown to accumulate DNA damage with age. The synthesis of these active forms of oxygen helps explain why some compounds (such as asbestos) or chronic infections that are not themselves mutagenic can increase the rate of cancer. The reactive forms of oxygen that are synthesized in response to these conditions by macrophages may be the real mutagens.

The importance of internally generated reactive oxygen in cancer has received dramatic confirmation recently (see Chmiel et al., below). These authors report that a genetic disease characterized by increased rates of colon cancer is due to mutations in the human repair gene *MYH*, which is analogous to the *mutY* gene of *E. coli*. Siblings who have inherited this predisposition to cancer, called familial adenomatous polyposis, are heterozygous for different mutant alleles of the *MYH* gene. The *mutY* gene product of *E. coli* is

a specific *N*-glycosylase which removes adenine bases that have mistakenly paired with 8-oxoG in the DNA (see the text), and the human enzyme is known to have a similar activity. Also, mice which have had their *Ogg-1* and *Myh* genes inactivated (so-called knockout mice) are much more prone to lung and ovarian tumors, as well as lymphomas. The Ogg-1 gene product of mice is functionally analogous to MutM of *E. coli*. Furthermore the primary type of mutation in these knockout mice are GC-to-TA transversions, the same as they are in *E. coli* (see the text and Xie et al., below).

Obviously, any mechanism for reducing the levels of these active forms of oxygen should increase longevity and reduce the frequency of many degenerative diseases. Fruits and vegetables produce antioxidants, including ascorbic acid (vitamin C), tocopherol (vitamin E), and carotenes such as β-carotene (found in large amounts in carrots), that destroy these molecules and thereby protect the DNA in their seeds and the photosynthetic apparatus in their leaves from damage due to oxygen free radicals produced by UV irradiation. Some evidence suggests that consumption of adequate amounts of fruits and vegetables that contain these compounds may reduce the rate of cancer and many degenerative diseases.

References

Ames, B. N., M. K. Shigenaga, and T. M. Hagen. 1993. Oxidants, antioxidants and the degenerative diseases of aging. *Proc. Natl. Acad. Sci. USA* **90:**7915–7922.

Chmiel, N. H., A. L. Livingston, and S. S. David. 2003. Insight into the functional consequences of inherited variants of the *hMYH* adenine glycosylase associated with corectal cancer: complementation assays with *hMYH* variants and pre-steady state kinetics of the corresponding mutated *E. coli* enzymes. *J. Mol. Biol.* **327:**431–443.

Park, S., X. You, and J. A. Imlay. 2005. Substantial DNA damage from submicromolar intracellular hydrogen peroxide detected in Hpx⁻ mutants of *Escherichia coli. Proc. Natl. Acad. Sci. USA* **102:** 9317–9322.

Seaver, L. C., and J. A. Imlay. 2001. Alkyl hyperoxide reductase is the primary scavenger of endogenous hydrogen peroxide in *Escherichia coli. J. Bacteriol.* **183:**7173–7181.

Xie, Y., H. Yang, C. Cunanan, K. Okamoto, D. Shibata, J. Pan, D. E. Barnes, T. Lindahl, M. McIlhatton, R. Fishel, and J. H. Miller. 2004. Deficiencies in mouse *Myh* and *Ogg1* result in tumor predisposition and G to T mutations in codon 12 of the K-ras oncogene in lung tumors. *Cancer Res.* **64:**3096–3102.

exonuclease, and resynthesized by DNA polymerase I (Figure 11.3). The MutM protein is present in larger amounts in cells that have accumulated reactive oxygen, because the *mutM* gene is part of a regulon induced in response to oxidative stress. We discuss regulons in more detail in chapter 13.

MutY

The MutY enzyme is also a specific *N*-glycosylase. However, rather than removing 8-oxoG directly, the MutY *N*-glycosylase specifically removes adenine bases that have been mistakenly incorporated opposite an 8-oxoG in DNA (Figure 11.4). Repair synthesis by DNA polymerase I

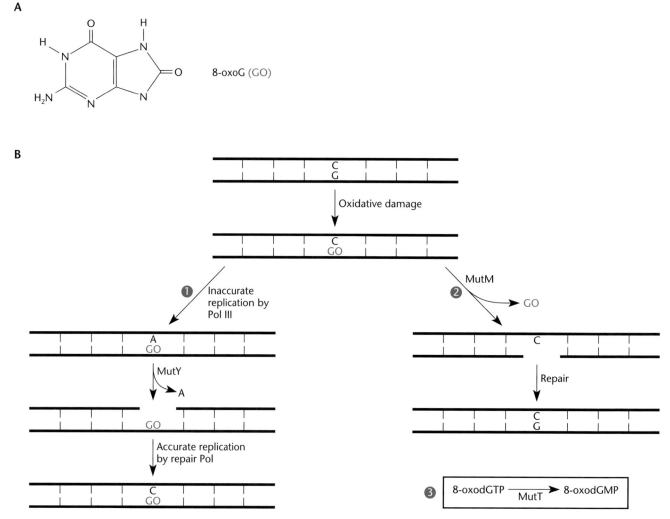

Figure 11.4 (A) Structure of 8-oxoG. (B) Mechanisms for avoiding mutagenesis due to 8-oxoG (GO). In pathway 1, an A mistakenly incorporated opposite 8-oxoG is removed by a specific glycosylase (MutY), and the strand is degraded and resynthesized with the correct C. In pathway 2, the 8-oxoG is itself removed by a specific glycosylase (MutM), and the strand is degraded and resynthesized with a normal G. In a third pathway, the 8-oxoG is prevented from entering DNA by a specific phosphatase (MutT) that degrades the triphosphate 8-oxodGTP to the monophosphate 8-oxodGMP.

then usually introduces the correct C to prevent a mutation, as with other *N*-glycosylase-initiated repair pathways.

The MutY enzyme will also apparently recognize a mismatch that results from accidental incorporation of an A opposite a normal G and will remove the A. However, its major role in avoiding mutagenesis in the cell seems to be to prevent mutations due to 8-oxoG. As evidence, mutations that cause the overproduction of MutM completely suppress the mutator phenotype of *mutY* mutants. The interpretation of this result is as follows. If a significant proportion of all spontaneous mutations in a *mutY*

mutant resulted from misincorporation of A's opposite normal G's, excess MutM should have no effect on the mutation rate, because removal of 8-oxoG should not affect this type of mispairing. However, the fact that excess MutM almost completely suppresses the extra mutagenesis in *mutY* mutants suggests that very few of the extra spontaneous mutations in a *mutY* mutant are due to A:G mispairs and most are due to A:8-oxoG mispairs.

MutT

The MutT enzyme operates by a very different mechanism (Figure 11.4): it prevents 8-oxoG from entering the DNA in the first place. The reactive forms of oxygen can oxidize not only guanine in DNA to 8-oxoG but also the base in dGTP to form 8-oxodGTP. Without MutT, 8-oxodGTP is incorporated into DNA by DNA polymerase III, which cannot distinguish 8-oxodGTP from normal dGTP. The MutT enzyme is a phosphatase that specifically degrades 8-oxodGTP to 8-oxodGMP so that it cannot be used in DNA synthesis.

GENETICS OF 8-oxoG MUTAGENESIS

There are a number of ways in which the genetic evidence obtained with *mutM*, *mutT*, and *mutY* mutants is consistent with these functions for the products of the genes. First, these activities explain why the effects of mutations in these genes are additive. If *mutT* is mutated, more 8-oxodGTP will be present in the cell to be incorporated into DNA, increasing the spontaneous mutation rate. If MutM does not remove these 8-oxoG's from DNA, spontaneous mutation rates will increase even further. If MutY does not remove some of the A's that mistakenly pair with the 8-oxoG's, the spontaneous mutation rate will be higher yet.

Mutations in the *mutM*, *mutY*, and *mutT* genes also increase the frequency of only some types of mutations, which again can be explained by the activities of these enzymes. For example, only the frequency of GC-to-TA transversion mutations is increased in *mutM* and *mutY* mutants. This is meaningful because, in general, transversions are less common than transition mutations (see chapter 1). The fact that mutations in both these genes increase the frequency of the same type of relatively uncommon mutation first suggested that they function in the same pathway and also makes sense considering the functions of the gene products. If MutM does not remove 8-oxoG from DNA, mispairing of the 8-oxoG with A can lead to GC-to-TA transversions. Moreover, GC-to-TA transversions will occur if MutY does not remove the mispaired A's opposite the 8-oxoG's in the DNA. By contrast, while *mutT* mutations can increase the frequency of relatively uncommon GC-to-TA transversion mutations, they can also increase the frequency of TA-to-GC transversions. This is possible because an 8-oxodGTP molecule, which owes its existence to the lack of MutT to degrade it, may cause mutations in two different ways. It may incorrectly enter the DNA by pairing with an A and then, once in the DNA, correctly pair with a C to result in an AT-to-CG transversion, or it can enter the DNA correctly as a G by pairing with a C but then, once in the DNA, pair wrongly with an A to result in a GC-TA transversion.

Damage Due to Alkylating Agents

Alkylation is another common type of damage to DNA. Both the bases and the phosphates in DNA can be alkylated. The responsible chemicals, known as **alkylating agents,** usually add alkyl groups (CH_3, CH_3CH_2, etc.) to the bases or phosphates in DNA, although any electrophilic reagent that reacts with DNA could be considered an alkylating agent. For example, the anticancer drug cisplatin is an alkylating agent that reacts with guanines in the DNA. Other examples of alkylating agents are ethyl methanesulfonate (EMS; nitrogen mustard gas), methyl methanesulfonate (MMS), and *N*-methyl-*N*'-nitro-*N*-nitrosoguanidine (nitrosoguanidine, NTG, or MNNG). Some of these alkylate DNA directly, whereas others react with cellular constituents such as gluthionine that are supposed to inactivate them but instead convert them into alkylating agents for DNA and worsen their effect. Many alkylating agents are known mutagens and carcinogens, and some are used as chemotherapeutic agents for the treatment of cancer. Not all alkylating agents are artificially synthesized; some are produced normally in cells or in the environment. For example, methylchloride produced in large quantities by marine algae is a DNA-alkylating agent, as are *S*-adenosylmethionine and methylurea, produced as normal cellular metabolites. Obviously, the cell needs repair systems to deal with alkylation damage to DNA.

Many reactive groups of the bases can be attacked by alkylating agents. The most reactive are N^7 of guanine and N^3 of adenine. These nitrogens can be alkylated by EMS or MMS to yield methylated or ethylated bases such as N^7-methylguanine and N^3-methyladenine, respectively. Alkylation of the bases at these positions can severely alter their pairing with other bases, causing major distortions in the helix.

Some alkylating agents, such as nitrosoguanidine, can also attack other atoms in the rings, including the O^6 of guanine and the O^4 of thymine. The addition of a methyl group to these atoms makes O^6-methylguanine (Figure 11.5) and O^4-methylthymine, respectively. Altered bases with an alkyl group at these positions are particularly mutagenic because the helix is not significantly distorted, so that the lesions cannot be repaired by the more general repair systems discussed below. However, the altered base often mispairs, producing a mutation. In this section, we discuss the repair systems specific to these types of alkylated bases.

SPECIFIC *N*-GLYCOSYLASES

Some types of alkylated bases can be removed by specific *N*-glycosylases. The repair pathways involving these enzymes work in the same way as other *N*-glycosylase repair pathways in that first the alkylated base is removed by the specific *N*-glycosylase and then the apurinic or apyrimidinic DNA strand is cut by an AP endonuclease. Exonucleases degrade the cut strand, which is then resynthesized by DNA polymerase I. In *E. coli*, two

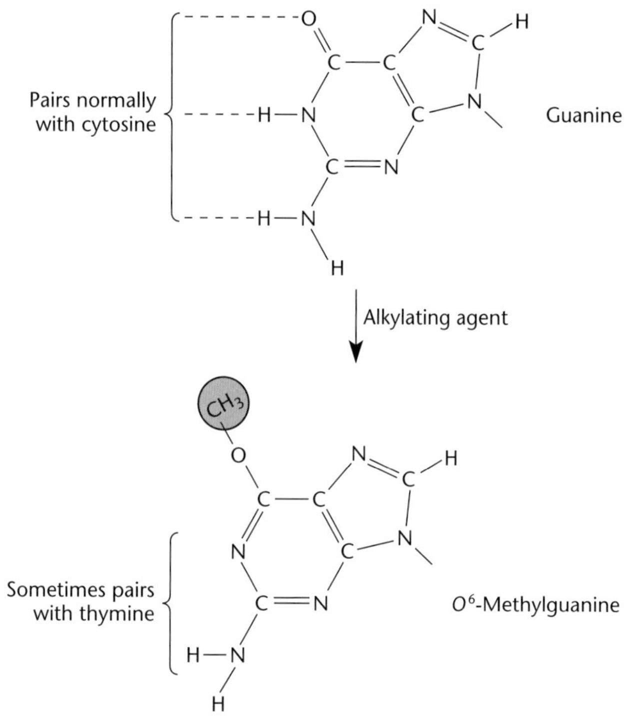

Figure 11.5 Alkylation of guanine to produce O^6-methylguanine. The altered base sometimes pairs with thymine, causing mutations.

N-glycosylases that remove methylated and ethylated bases from the DNA have been identified. One of these, TagA (for "three methyladenine glycosylase A") removes the base 3-methyladenine and some related methylated and ethylated bases. Another, AlkA, is less specific. It not only removes 3-methyladenine from the DNA but also removes many other alkylated bases including 3-methylguanine and 7-methylguanine. This enzyme, which is encoded by the *alkA* gene, is induced as part of the adaptive response (see below).

METHYLTRANSFERASES

Other repair systems for alkylated bases act by repairing the damaged base rather than removing it and resynthesizing the DNA. These proteins, called methyltransferases, directly remove the alkyl group from the base by transferring the methyl or other alkyl group from the altered base in the DNA to themselves. They are not true enzymes, because they do not catalyze the reaction but, rather, are consumed during it. Once they have transferred a methyl or other alkyl group to themselves, they become inactive and are eventually degraded. The two major methyltransferases in *E. coli* are Ada and Ogt, sometimes called alkyltransferases I and II, respectively. Both of these proteins

repair bases damaged from alkylation of the O^6 carbon of guanine and the O^4 carbon of thymine. Ogt plays the major role when the cells are growing actively, but when the cells reach stationary phase and quit growing or if the cell is exposed to an external methylating agent, Ada is induced as part of the adaptive response and then becomes the major methyltransferase (see below). That the cell is willing to sacrifice an entire protein molecule to repair a single O^6-methylguanine or O^4-methylthymine lesion is a tribute to the mutagenic potential of these lesions.

AlkB AND AidB

Two other enzymes that repair damage induced by alkylating agents are **AlkB** and AidB. The enzyme AlkB basically oxidizes the methyl groups on 1-methyladenine and 3-methylcytosine to formaldehyde, releasing them and restoring the normal base. More precisely, it is an α-ketoglutarate-dependent dioxygenase that couples the decarboxylation of α-ketoglutarate to the hydroxylation of the methyl group on 1-methyladenine or 3-methylcytosine to release formaldehyde (see Trewick et al., Suggested Reading). Its cofactor, α-ketoglutarate, is an intermediate in the tricarboxylic acid cycle, with many uses in nitrogen assimilation, etc.; therefore, it is always available in large quantities. The function of AidB, in contrast, is unknown; however, determining this function should be a high priority since AidB can be assumed to play a role in protecting against alkylation damage to DNA, and many such enzymes are important in protecting against cancer (see Box 11.5).

THE ADAPTIVE RESPONSE

Many of the genes whose products, including the specific *N*-glycosylases and methytransferases that repair alkylation damage in *E. coli*, are part of the **adaptive response.** The products of these genes are normally synthesized in small amounts, but they are produced in much greater amounts if the cells are exposed to an alkylating agent. The name "adaptive response" comes from early evidence suggesting that *E. coli* "adapted" to damage caused by alkylating agents. If *E. coli* cells are briefly treated with an alkylating agent such as nitrosoguanidine (NTG), they will be better able to survive subsequent treatments with this and other alkylating agents. We now know that the cell adapts to the alkylating agents by inducing a number of genes whose products are involved in repairing alkylation damage to DNA. The adaptive response genes seem to be most important for conferring resistance to alkylating agents that transfer methyl (CH_3) groups to DNA. Resistance to alkylating agents that transfer longer groups, such as ethyl (CH_3CH_2) groups, to DNA seems to be due mostly to excision repair (see below).

Regulation of the Adaptive Response

Treatment of *E. coli* cells with an alkylating agent causes the concentration of some of the proteins involved in repairing alkylation damage to increase from a few to many thousands of copies. The genes induced as part of the adaptive response include *ada*, *aidB*, *alkA*, and *alkB*, discussed above. The regulation is achieved through the state of methylation of one of the alkylation-repairing proteins, the Ada protein (Figure 11.6). The *ada* gene is part of an operon with *alkB*, while the *aidB* and *alkA* genes are separately transcribed, as shown in the figure. The Ada

protein can regulate the transcription of itself as well as the other genes under its control because, in addition to its role in repairing alkylation damage to DNA, it is a transcriptional activator. However, the Ada protein becomes a transcriptional activator only if the alkylation damage is quite extensive. It can discern the level of damage in the DNA by having two amino acids to which methyl groups can be transferred, cysteine-321 (the 321st amino acid from its N terminus) and cysteine-38 (the 38th amino acid from its N terminus). After mild alkylation damage has occurred, most of the methylation is confined to the bases; these methyl groups can be transferred from either O^6-methylguanine or O^4-methylthymine to cysteine-321 close to the C terminus. This removes the methyl groups from the damaged bases but inactivates the Ada protein as far as transfer of more methyl groups from the bases is concerned. At higher levels of damage, some of the phosphates in the DNA background also become methylated in the form of phosphomethyltriesters, and these methyl groups can be transferred only to cysteine-38 close to the N terminus. The presence of a methyl group on cysteine-38 converts Ada into a transcriptional activator that activates transcription of its own as well as the other genes under its control. Transcriptional activators are discussed in detail in chapter 12.

Interestingly, some of the adaptive response genes are also turned on during stationary phase when the cells stop growing. The transcription of genes during stationary phase requires σ^s rather than σ^{70} (see chapter 13), and the methylated Ada protein can also activate transcription by RNA polymerase containing σ^s at these promoters. However, transcription only of the *ada-alkB* operon and the *aidB* gene is activated by methylation in stationary phase; *alkA* gene transcription is not activated but instead is actually repressed by methylated Ada protein at this stage. It seems possible that the purpose of inducing these genes in stationary phase is to prevent DNA damage due to the natural accumulation of nitrosoamines, which could accumulate when the cells run out of oxygen. Nitrosoamines such as nitrosourea can accumulate in cells that have run out of oxygen and are using nitrate as a terminal electron acceptor for anaerobic respiration. Accepting an electron converts nitrate to nitrite, which is chemically reactive and can react with other cellular constituents to form alkylating agents (see Taverna and Sedgwick, Suggested Reading). This could also explain why the *alkA* gene is repressed in the stationary phase. The AlkA protein is a specific glycosylase that cuts the DNA if it is alkyated but then requires DNA replication to finish the job of resynthesizing the deleted stretch of DNA (Figure 11.3). Very little replication of DNA occurs in stationary-phase cells, so these cuts could become lethal in stationary phase.

Figure 11.6   (A) The adaptive response. (B) Regulation of the adaptive response. Only a few copies of the Ada protein normally exist in the cell. After damage due to alkylation, the Ada protein, a methyltransferase, transfers alkyl groups from methylated DNA phosphates to an amino acid in the N terminus (N) of itself, converting itself into a transcriptional activator (1), or from a methylated base to an amino acid in its C terminus (C), inactivating itself (2). See the text for details.

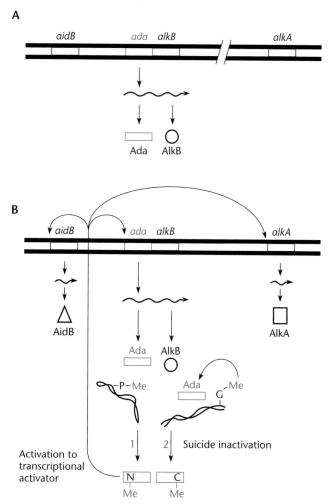

Damage Due to UV Irradiation

One of the major sources of natural damage to DNA is ultraviolet (UV) irradiation due to sun exposure. Every organism that is exposed to sunlight must have mechanisms to repair UV damage to its DNA. The conjugated-ring structure of the bases in DNA causes them to strongly absorb light in the UV wavelengths. The absorbed photons energize the bases, causing their double bonds to react with other nearby atoms and hence to form additional chemical bonds. These chemical bonds result in abnormal linkages between the bases in the DNA and other bases or between bases and the sugars of the nucleotides.

One common type of UV irradiation damage is the **pyrimidine dimer,** in which the rings of two adjacent pyrimidines become fused (Figure 11.7). In one of the two possible dimers, the carbon atoms at positions 5 and 6 of two adjacent pyrimidines are joined to form a **cyclobutane ring.** In the other type of dimer, the carbon at position 6 of one pyrimidine is joined to the carbon at position 4 of an adjacent pyrimidine to form a **6-4 lesion.**

PHOTOREACTIVATION OF CYCLOBUTANE DIMERS

Because the cyclobutane-type pyrimidine dimer due to UV irradiation is so common, a special type of repair system called **photoreactivation** has evolved to repair it. The photoreactivation repair systems separate the fused bases of the cyclobutane pyrimidine dimers rather than replacing them. This mechanism is named photoreactivation because this type of repair occurs only in the presence of visible light (and so used to be called light repair).

In fact, photoreactivation was the first DNA repair system to be discovered. In the 1940s, it was observed that the bacterium *Streptomyces griseus* was more likely to survive UV irradiation in the light than in the dark. Photoreactivation is now known to exist in most organisms on Earth, with the important exception of placental mammals such as humans. Do not make the mistake of thinking that when you are soaking up rays the damaging effects of UV irradiation are being repaired just because you are also being exposed to visible light. Humans (and other placental mammals) do not seem to have a photoreactivation system, although a repair system derived from it may exist.

The mechanism of action of photoreactivation is shown in Figure 11.8. The enzyme responsible for the repair is called **photolyase.** This enzyme, which contains a reduced flavin adenine dinucleotide ($FADH_2$) group that absorbs light of wavelengths between 350 and 500 nm, binds to the fused bases. Absorption of light then gives photolyase the energy it needs to separate the fused bases. A different but related enzyme has evolved to repair 6-4 lesions in eukaryotes.

There is some evidence that the photoreactivating system may also help repair pyrimidine dimers even in the dark, by cooperating with the excision repair system. By binding to pyrimidine dimers, it may help make them more recognizable by the nucleotide excision repair system discussed below.

N-GLYCOSYLASES SPECIFIC TO PYRIMIDINE DIMERS

There are also specific *N*-glycosylases that recognize and remove pyrimidine dimers. This repair mechanism is similar to the mechanisms for deaminated and alkylated bases discussed above and involves AP endonucleases or lyases and the removal and resynthesis of strands of DNA containing the dimers.

Figure 11.7 Two common types of pyrimidine dimers caused by UV irradiation. In the top panel, two adjacent thymines are linked through the 5- and 6-carbons of their rings to form a cyclobutane ring. In the bottom panel, a 6-4 dimer is formed between the 6-carbon of a cytosine and the 4-carbon of a thymine 3′ to it.

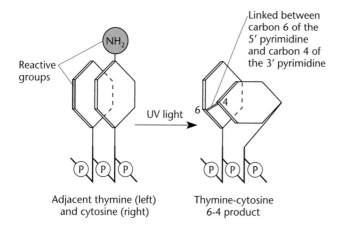

Cyclobutane ring

Linked across carbons 5 and 6 of each pyrimidine ring

Reactive double bonds

UV light

Adjacent thymines

Thymine dimer (T̂T)

Reactive groups

Linked between carbon 6 of the 5′ pyrimidine and carbon 4 of the 3′ pyrimidine

UV light

Adjacent thymine (left) and cytosine (right)

Thymine-cytosine 6-4 product

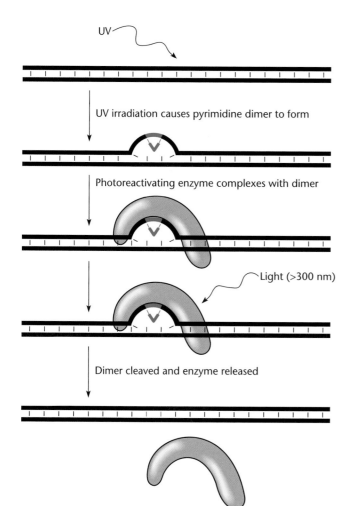

Figure 11.8 Photoreactivation. The photoreactivating enzyme (photolyase) binds to cyclobutane pyrimidine dimers (purple) even in the dark. Absorption of light by the photolyase causes it to cleave the bond between the two pyrimidines, restoring the bases to their original form.

General Repair Mechanisms

As mentioned, not all repair mechanisms in cells are specific for a certain type of damage to DNA. Some types of repair systems can repair many different types of damage. Rather than recognizing the damage itself, these repair systems recognize distortions in the DNA structure caused by improper base pairing and repair them, independent of the type of damage that caused the distortion.

The Methyl-Directed Mismatch Repair System

One of the major pathways for avoiding mutations in *E. coli* is the **methyl-directed mismatch repair system.** The mismatch repair system is mentioned in chapter 1, in connection with lowering the rate of mistakes made during the replication of DNA, and in chapter 10, in connection

with gene conversion. The methyl-directed mismatch repair system is not specific and can repair essentially any damage that causes a slight distortion in the DNA helix, including mismatches, frameshifts, incorporation of base analogs, 8-oxoG, and some types of alkylation damage that cause only minor distortions in the DNA helix. The types of alkylation that cause more significant distortions are repaired by other systems including nucleotide excision repair (discussed below). In general, DNA damage that causes only minor distortions of the helix is repaired by the methyl-directed mismatch repair system whereas damage that causes more significant distortions is repaired by other pathways.

BASE ANALOGS

Base analogs are chemicals that resemble the normal bases in DNA. Because they resemble the normal bases, these analogs are sometimes converted into a deoxynucleoside triphosphate and enter DNA. Incorporation of a base analog can be mutagenic because the analog often pairs with the wrong base, leading to base pair changes in the DNA. Figure 11.9 shows two base analogs, 2-aminopurine (2-AP) and 5-bromouracil (5-BU). 2-AP resembles adenine, except that it has the amino group at the 2 position rather than at the 6 position. The other base analog pictured,

Figure 11.9 Base analogs 2-aminopurine (2-AP) and 5-bromouracil (5-BU). The amino groups that are at different positions in adenine (A) and 2-AP are circled in purple, as are the methyl group in thymine (T) and the bromine group in 5-BU.

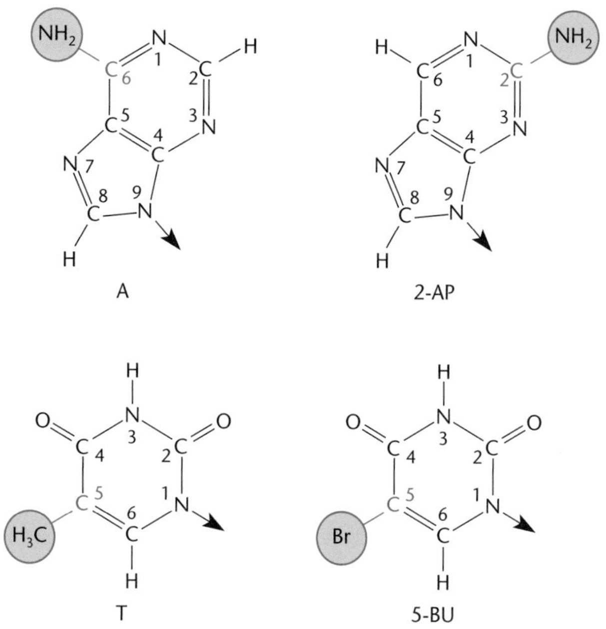

5-BU, resembles thymine except for a bromine atom instead of a methyl group at the 5 position.

Figure 11.10 shows how mispairing by a base analog might cause a mutation. In the illustration, the base analog 2-AP has entered a cell and been converted into the

Figure 11.10 Mutagenesis by incorporation of the adenine analog 2-AP into DNA. The 2-AP is first converted to the deoxyribose nucleoside triphosphate and then inserted into DNA. (A) Analog incorrectly pairs with a C during its incorporation into the DNA strand. (B) Analog is incorporated correctly opposite a T but mispairs with C during subsequent replication. The mutation is circled in both panels.

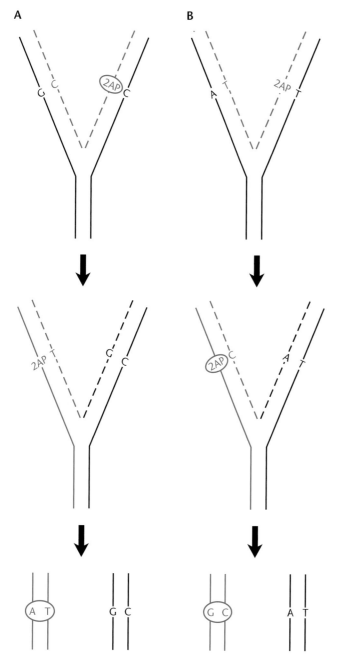

nucleoside triphosphate. The deoxyribose 2-aminopurine triphosphate is then incorporated during synthesis of DNA sometimes pairing with cytosine in error. Which type of mutation occurs depends on when the 2-AP mistakenly pairs with cytosine. Even if the 2-AP enters the DNA by mistakenly pairing with C instead of T, in subsequent replications it usually pairs correctly with T. This causes a GC-to-AT transition mutation. However, if it is incorporated properly by pairing with T but pairs mistakenly with C in subsequent replications, an AT-to-GC transition mutation is the result. Similar arguments can be made for 5-BU, which sometimes mistakenly pairs with cytosine instead of adenine.

FRAMESHIFT MUTAGENS

Another type of damage repaired by the methyl-directed mismatch repair system is the incorporation of frameshift mutagens, which are usually planar molecules of the acridine dye family (see chapters 1 and 7). These chemicals are mutagenic because they intercalate between bases in the same strand of the DNA, increasing the distance between the bases and preventing them from aligning properly with bases on the other strand. The frameshift mutagens include acridine dyes such as 9-aminoacridine, proflavine, and ethidium bromide, as well as some aflatoxins made by fungi.

Figure 11.11 illustrates a model for mutagenesis by a frameshift mutagen. Intercalation of the dye forces two of the bases apart, causing the two strands to slip with respect to each other. One base is thus paired with the base next to the one with which it previously paired. This slippage is most likely to occur where a base pair in the DNA is repeated, for example in a string of AT or GC base pairs. Whether a deletion or addition of a base pair occurs depends on which strand slips, as shown in Figure 11.11. If the dye is intercalated into the template DNA prior to replication, the newly synthesized strand might slip and incorporate an extra nucleotide. However, if it is incorporated into the newly synthesized strand, the strand might slip backward, leaving out a base pair in subsequent replication.

MECHANISM OF METHYL-DIRECTED MISMATCH REPAIR

As discussed briefly in chapter 1, the methyl-directed mismatch repair system requires the products of the *mutS, mutL,* and *mutH* genes. Like the *mut* genes whose products are involved in avoiding mutagenesis due to 8-oxoG, these *mut* genes were found in a search for mutations that increase spontaneous mutation rates (see below). Another gene product that participates in the methyl-directed mismatch repair system is the product of the *dam* gene. The *dam* gene encodes the Dam methylase, which methylates adenine in the sequence GATC/CTAG. This enzyme

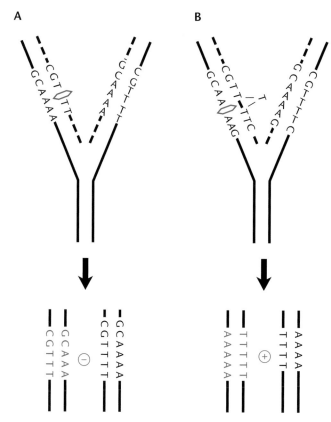

Figure 11.11 Mutagenesis by a frameshift mutagen. Intercalation of a planar acridine dye molecule between two bases in a repeated sequence in DNA forces the bases apart and can lead to slippage. (A) The dye inserts itself into the new strand, resulting in deletion of a base pair (–). (B) The dye comes into the old strand, adding a base pair.

methylates the DNA at this sequence only after the DNA is synthesized, so that the new strand of DNA is temporarily unmethylated after the replication fork has passed through. By having the *mut* gene products repair only the newly synthesized strand, the cell usually avoids mutations because most mistakes are made during synthesis of new strands. If the old strand were degraded and resynthesized with the new strand as a template, the damage would be fixed as a mutation and would be passed on to future generations.

It is somewhat mysterious how the mismatch repair system can use the state of methylation of the GATC/CTAG sequence to direct itself to the newly synthesized strand, even though the nearest GATC/CTAG sequence is probably some distance from the alteration. Figure 11.12 presents a model that has been used to explain this mechanism. First, a dimer of the MutS protein binds to the alteration in the DNA that is causing a minor distortion in the helix (marked with an X in the figure). A nearby GATC sequence is still unmethylated on the newly synthesized strand. Then two copies of MutL bind to the MutS and a

copy of MutH binds. This binding activates the MutH nuclease to cut the nearest hemimethylated GATC sequence in the newly synthesized unmethylated strand and exonucleases degrade the DNA past the site of the original mismatch. DNA polymerase III then fills the gap, and ligase seals the break. This model is supported by experiments showing that the mismatch repair system preferentially repairs mismatches on hemimethylated DNA by correcting the sequence on the unmethylated strand of DNA to match the sequence on the methylated strand (see below and Pukkila et al., Suggested Reading). But DNA is degraded in either the 5′-to-5′ or 5′-to-3′ direction depending on which side of the nearest GATC sequence the mismatch has occurred. According to the model, *E. coli* solves this problem by using four different exonucleases, two of which, ExoVII and RecJ, can degrade in only the 5′-to-3′ direction and two of which, ExoI and ExoX, can degrade in only the 3′-to-5′ direction. Also, these exonucleases can degrade only single-stranded DNA, and so a helicase is needed to separate the strands. This is the job of the UvrD helicase, which is a general helicase also used by the excision repair system (see below). After the strand containing the mismatch has been degraded, DNA polymerase III resynthesizes a new strand, using the other strand as the template and removing the cause of the distortion.

While this model more or less explains the genetic and biochemical evidence concerning the mismatch repair system (see below), evidence is accumulating that it may not be the whole story. For one, it is somewhat surprising that DNA polymerase III, the replication DNA polymerase, is used for this repair. Most other repair reactions use DNA polymerase I or II, the normal repair enzymes. How does the DNA polymerase III load itself on the DNA at the gap created by the exonucleases? It normally loads on DNA only at the origin of replication or with the help of Pri proteins at recombinational intermediates at blocked replication forks. If this repair was going on ahead of the replication fork, it could create a replication block that could be corrected by a replication restart mechanism (see Figure 11.16), but this repair is behind the fork. Suspicions are also raised by the fact that other organisms seem to be able to identify the new strand for mismatch repair, without the help of Dam methylation. Mismatch repair systems are universal, as is the ability to distinguish the newly synthesized strand from the old strand, even though most organisms, even most types of bacteria, do not have a Dam methylase and so could not use hemimethylation to identify the new strand. There is also evidence that the mismatch repair system is in much closer contact with the replicating polymerase than indicated by the model. MutS may bind to the β clamp which holds the replicating DNA polymerase on the DNA, and MSH, the eukaryotic equivalent of MutS (Box 11.2), binds to the proliferating-cell nuclear antigen protein, which is the eukaryotic equivalent

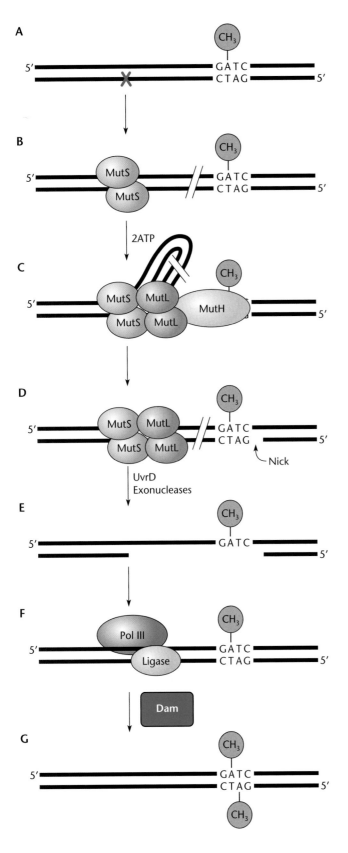

of the β-clamp protein in bacteria. Finally, what role might SeqA play in this process? SeqA was first discovered because it binds to hemimethylated GATC sequences of DNA in the origin of replication and delays methylation of the unmethylated strand to help prevent new initiations of rounds of replication (see chapter 1). However, evidence is accumulating that clusters of SeqA are associated with the replication fork throughout replication rather than staying with the origin. Does SeqA help maintain contact between hemimethylated DNA and the replicating polymerase, allowing the cell to distinguish the new from the old strand, even after the fork has passed? It will be interesting to see how new evidence obtained with the bacterial systems furthers our understanding of this most interesting and important of DNA repair mechanisms, especially considering its role in preventing cancer in humans (Box 11.2).

VSP Repair

In addition to their role in general mismatch repair, the MutS and MutL proteins participate in very-short-patch repair, which occurs at the site of methylated cytosines in *E. coli* (see above and Lieb et al., Suggested Reading). The MutS protein may bind to the T in the T-G mismatch created by the deamination of the methylated C at this position, thereby attracting the attention of the Vsr endonuclease to the mismatch. The MutL protein may then recruit the UvrD helicase and exonucleases to degrade the strand causing the mismatch, consistent with the roles of these proteins in general mismatch repair.

GENETIC EVIDENCE FOR METHYL-DIRECTED MISMATCH REPAIR

Models like the one presented in Figure 11.12 are based on biochemical and genetic evidence. The results of these

Figure 11.12 MutSLH DNA repair in *E. coli*. (A) One arm of a replication fork is shown at the top of the figure with methylated and unmethylated GATC (*dam*) sequences and a replication mistake "X" generating a base-base or deletion/insertion mismatch. (B) The mismatch is bound by MutS. (C) In an ATP-dependent reaction, a ternary complex is formed with MutS, MutL, and MutH proteins. (D) Incision by activated MutH occurs in the newly synthesized strand at the unmethylated GATC sequence. (E) The nick is extended into a gap by excision in either the 3'-to-5' direction or the 5'-to-3' direction. Only the 5'-to-3' direction is depicted in the figure. The gap is formed by the action of exonucleases, including exonuclease I (ExoI), ExoVII, ExoX, and RecJ, and the direction of excision is determined by the UvrD helicase. (F) Resynthesis is accomplished by DNA polymerase III holoenzyme, and the nick is sealed by DNA ligase. (G) Subsequent methylation by Dam completes the process.

BOX 11.2

Cancer and Mismatch Repair

Cancer is a multistep process that is initiated by mutations in oncogenes and other tumor-suppressing genes. It is not surprising, therefore, that DNA repair systems form an important line of defense against cancer. All organisms probably have mismatch repair systems that help reduce mutagenesis. However, unlike bacteria, which have only one or two different MutS proteins and one MutL protein, humans have at least five MutS analogs and four MutL homologs (see Kang et al., below). Attention was focused on the role of the mismatch repair system in cancer with the discovery that people with a mutation in a mismatch repair gene are much more likely to develop some types of cancer, including cancer of the colon, ovary, uterus, and kidney. One such genetic predisposition, called HNPCC (for *hereditary nonpolyposis colon cancer*), results from mutations in a gene called *hMSH2* (for *human mutS homolog 2*). This gene was first suspected to be involved in mismatch repair because people who inherited the mutant gene showed a higher frequency of short insertions and deletions that should normally be repaired by the mismatch repair system (see the text). People with this hereditary condition were found to have inherited a mutant form of a gene homologous to the *mutS* gene of *E. coli*. Moreover, the *hMSH2* gene can cause increased spontaneous mutations when expressed in *E. coli*, probably because it interferes with the normal mismatch repair system. Like its analog, MutS, the

hMSH2 gene product may bind to mismatches but then does not interact properly with MutL and MutH.

Another reason why the mismatch repair system is relevant to cancer research in humans is the role it plays in making cells sensitive to cancer therapeutic agents. Apparently, much of the toxicity of some antitumor agents such as cisplatin and other alkylating agents is due to the mismatch repair system. When tumor cells become resistant to the drug, it is often because they have acquired a mutation in a human *mut* gene (see Karran, below).

References

Fishel, R., M. K. Lescoe, M. R. S. Rao, N. G. Copeland, N. A. Jenkins, J. Garber, M. Kane, and R. Kolodner. 1993. The human mutator gene homolog MSH2 and its association with hereditary nonpolyposis colon cancer. *Cell* **75**:1027–1038.

Gradia, S., D. Subramanian, T. Wilson, S. Acharya, A. Makhov, J. Griffith, and R. Fishel. 1999. hMSH2-hMSH6 forms a hydrolysis-independent sliding clamp on mismatched DNA. *Mol. Cell* **3**: 255–261.

Harfe, B. D., and S. Jinks-Robertson. 2000. DNA mismatch repair and genetic instability. *Annu. Rev. Genet.* **34**:359–399.

Kang, J., S. Huang, and M. J. Blaser. 2005. Structural and functional divergence of MutS2 from bacterial MutS1 and eukaryotic MSH4-MSH5 homologs. *J. Bacteriol.* **187**:3528–3537.

Karran, P. 2001. Mechanisms of tolerance to DNA damaging therapeutic drugs. *Carcinogenesis* **22**:1931–1937.

experiments support the conclusion that the state of methylation of GATC sequences in the DNA help direct the mismatch repair system to the newly synthesized strand of DNA. Some of this evidence is briefly reviewed in this section.

Isolation of *mut* Mutants

The *mut* genes of *E. coli* were discovered because mutations in these genes increase spontaneous mutation rates. The resulting phenotype is often referred to as the "mutator phenotype." Other mutations with this phenotype, including the *mutT*, *mutY*, and *mutM* mutations in the genes for reducing mutations due to oxygen damage to DNA and the *mutD* mutations in the gene for the editing function of DNA polymerase, are discussed in other sections. Mutations in the *uvrD*, *vsp*, and *dam* genes of *E. coli* also increase the rate of at least some spontaneous mutations, although these genes were found in other ways and so were not named *mut*.

A common method for detecting mutants with abnormally high mutation rates is colony papillation. This scheme is based on the fact that all of the descendants of a

bacterium with a particular mutation are mutants of the same type. Colonies grow from the middle out, so that if a mutation occurs in a growing colony, the mutant descendants of the original mutant stay together to form a sector, or papilla, as the colony grows. A growing colony contains many papillae composed of mutant bacteria of various types. A *mut* mutation increases the frequency of papillation for many types of mutants, so that the phenotype used in a colony papillation test is purely a matter of convenience.

Lac$^+$ revertants of *lac* mutations in *E. coli* are an obvious choice for papillation tests, because the revertants form conspicuous blue papillae on 5-bromo-4-chloro-3-indolyl-β-D-galactopyranoside (X-Gal) plates. Figure 11.13 shows a papillation test using reversion of a *lac* mutation as an indicator of mutator activity. If the bacteria forming a colony are *mut* mutants with a higher than normal spontaneous mutation frequency, the colony will have more blue papillae than normal. This is how many of the *mut* genes we have discussed were detected.

Once a number of *mut* genes were detected, they were classified into complementation groups and arbitrarily

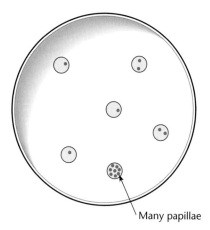

Many papillae

Figure 11.13 Colonies due to *mut* mutants have more papillae. A *lacZ* mutant was plated on X-Gal-containing medium. Revertants of the *lacZ* mutation produce blue sectors or papillae (shown in purple). A *mut* mutant gives more blue papillae than normal owing to increased spontaneous mutation frequencies (arrow).

lettered (see chapter 3). They were also combined into multiple mutants to see how they interacted. The products of three of the *mut* genes, *mutS*, *mutL*, and *mutH*, were predicted to participate in the same repair pathway because of the observation that double mutants did not exhibit higher spontaneous mutation rates than cells with mutations in each of the single genes alone. In other words, the effect of inactivating two of these *mut* genes by mutation is not additive. Generally, mutations that affect steps of the same pathway do not have additive effects.

Other experiments established the role of Dam methylation in mismatch repair. Probably the most convincing evidence for the role of methylation in directing the repair system came from experiments with heteroduplexes of λ DNA (see Pukkila et al., Suggested Reading). We have discussed the synthesis and experimental use of heteroduplex DNAs in other chapters. For example, in chapter 9 we showed how they can be used to distinguish replicative from cut-and-paste transposition.

To determine the role of methylation in directing the mismatch repair system, heteroduplex λ DNAs in which the two strands came from phage with different mutations were prepared. We shall call them phage mutant 1 and phage mutant 2. The DNA from phage mutant 2 was unmethylated because it was propagated on *E. coli* with a *dam* mutation. After the DNA from both mutants was purified, the DNA strands of both were separated by heating and the two complementary strands were separated on the basis that one of the complementary strands of λ DNA binds more RNA homopolymers such as poly(rG) than the other, making it denser on CsCl equilibrium gradients. The purified complementary strands from each of the

phage were then mixed and rehybridized to create a mismatch at the site of the mutation, with one strand unmethylated at its GATC sequences and the other strand methylated. These heteroduplex DNAs were then transfected into cells, the phage progeny were plated, and their genotypes were tested to see which mutation they had inherited and which of the two strands had been preferentially repaired. The progeny phage exhibited predominantly the genotype of mutant 1, which had the methylated DNA. In the reverse experiment, mutant 1 was propagated on the *dam* mutant so that it would have the unmethylated DNA. In this case, the progeny phage that arose from the heteroduplex DNA had the genotype of mutant 2; therefore, once again, the methylated DNA sequence was preferentially preserved. These results indicate that the sequence of the unmethylated strand is usually the one that is repaired to match the sequence of the methylated strand.

Other evidence supporting the role of methylation in directing the mismatch repair system came from genetic studies of 2-AP sensitivity in *E. coli* (see Glickman and Radman, Suggested Reading). These experiments were based on the observation that a *dam* mutation makes *E. coli* particularly sensitive to killing by 2-AP. This is expected from the model since, as mentioned, cells incorporate 2-AP indiscriminately into their DNA because they cannot distinguish it from the normal base adenine. The cellular mismatch repair system repairs DNA containing 2-AP because the incorporated 2-AP causes a slight distortion in the helix. Since the 2-AP is incorporated into the newly synthesized strand during the replication of the DNA, the strand containing the 2-AP is normally transiently unmethylated, so that the 2-AP-containing strand is repaired, removing the 2-AP. In a *dam* mutant, however, neither strand is methylated, and so the mismatch repair system cannot tell which strand was newly synthesized and may try to repair both strands. The rationale was that if two 2-APs are incorporated close enough to each other, the mismatch repair system may try to simultaneously remove the two mismatches by repairing the opposite strands. However, cutting two strands at sites opposite each other causes a double-strand break in the DNA, which may kill the cell. Whether this is the real mechanism of killing is unclear, but it is supported by the fact that recombination systems that repair double-strand breaks in the DNA relieve the toxicity of *dam* mutations.

One prediction based on this proposed mechanism for the sensitivity of *dam* mutants to 2-AP is that the toxicity of 2-AP in *dam* mutants should be reduced if the cells also have a *mutL*, *mutS*, or *mutH* mutation. Without the products of the mismatch repair enzymes, the DNA is not cut on either strand, much less on both strands simultaneously. The cells may suffer higher rates of mutations, but at least they will survive. This prediction was fulfilled. Double mutants that have both a *dam* mutation and a mutation in one of the three *mut* genes were much less sensitive to 2-AP than were mutants with a *dam* mutation alone. Furthermore, *mutL*, *mutH*, and *mutS* mutations can be isolated as suppressors of *dam* mutations on media containing 2-AP. In other words, if large numbers of *dam* mutant *E. coli* cells are plated on medium containing 2-AP, the bacteria that survive are often double mutants with the original *dam* mutation and a spontaneous *mutL*, *mutS*, or *mutH* mutation. A similar situation seems to prevail in human cells (Box 11.2).

ROLE OF THE MISMATCH REPAIR SYSTEM IN PREVENTING HOMEOLOGOUS AND ECTOPIC RECOMBINATION

As discussed in chapter 3, some DNA rearrangements, such as deletions and inversions, are caused by recombination between similar sequences in different places in the DNA. This is called **ectopic recombination**, or "out-of-place" recombination. Many sequences at which ectopic recombination occurs are similar but not identical. Also, recombination between DNAs from different species often occurs between sequences that are similar but not identical. In general, recombination between similar but not identical sequences is called **homeologous recombination.**

The mismatch repair system helps reduce ectopic and other types of homeologous recombination. As evidence, recombination between similar but unrelated bacteria such as *E. coli* and *Salmonella enterica* serovar Typhimurium is greatly enhanced if the recipient cell has a *mutL*, *mutH*, or *mutS* mutation. Also, the frequency of deletions and other types of DNA rearrangements is enhanced among bacteria with a *mutL*, *mutH*, or *mutS* mutation, since these rearrangements often depend on recombination between similar but not identical sequences. The frequency of site-specific mutations caused by recombineering is also enhanced by *mutS* mutations (see Box 10.3) since this also depends on homeologous recombination between the DNA electroporated into the cell and the endogenous DNA. Evidence suggests that the mismatch repair system may inhibit homeologous and ectopic recombination by interfering with the ability of RecA to form synapses where there are extensive mismatches (see Worth et al., Suggested Reading). Because the sequences at which the homeologous recombination occurs are not identical, the heteroduplexes formed during recombination between different regions in the DNA or between the same region from different species often contain many mismatches that bind MutS and the other proteins of the mismatch repair system and interfere with the binding of RecA.

Nucleotide Excision Repair

One of the most important general repair systems in cells is **nucleotide excision repair,** so named because the entire damaged nucleotides are cut out of the DNA and replaced. This type of repair is very efficient and seems to be common to most types of organisms on Earth. It is also relatively nonspecific and repairs many different types of damage. Because of its efficiency and relative lack of specificity, the nucleotide excision repair system is very important to the ability of the cell to survive damage to its DNA.

The nucleotide excision repair system is relatively nonspecific because, like the mismatch repair system, it recognizes the distortions in the normal DNA helix that result from damage, rather than the chemical structure of the damage itself. This makes it capable of recognizing and repairing damage to DNA as diverse as most types of alkylation and almost all the types of damage caused by UV irradiation, including cyclobutane dimers, 6-4 lesions, and base-sugar cross-links. Nucleotide excision repair also collaborates with recombination repair to remove cross-links formed between the two strands by some chemical agents such as psoralens, *cis*-diamminedichloroplatinum (cisplatin), and mitomycin (see below). However, because nucleotide excision repair recognizes only major distortions in the helix, it does not repair lesions such as base mismatches, O^6-methylguanine, O^4-methylthymine, 8-oxoG, or base analogs, all of which result in only minor distortions and must be repaired by other repair systems.

The excision repair system can be distinguished from most other types because the DNA containing the damage is actually excised from the DNA and ends up outside the cell. For example, after cells are irradiated with UV light, short pieces of DNA (oligodeoxynucleotides) containing pyrimidine dimers and the other types of damage induced by UV appear in the medium, due to the nucleotide excision repair system.

MECHANISM OF NUCLEOTIDE EXCISION REPAIR
Because nucleotide excision repair is such an important line of defense against some types of DNA-damaging agents, including UV irradiation, mutations in the genes whose products are required for this type of repair can make cells much more sensitive to these agents. In fact, mutants defective in excision repair were identified because they are killed by much lower doses of irradiation than is the wild type. Table 11.1 lists the *E. coli* genes whose products are required for nucleotide excision repair. Comparative genomic analysis has found UvrA, UvrB, and UvrC orthologs in all eubacterial species, as well as in some members of the Archaea. The products of some of these genes, such as *uvrA*, *uvrB*, and *uvrC*, are involved only in excision repair, while the products of others,

TABLE 11.1	Genes involved in the UvrABC endonuclease repair pathway
Gene	**Function of gene product**
uvrA	DNA-binding protein
uvrB	Loaded by UvrA to form a DNA complex; nicks DNA 3′ of lesion
uvrC	Binds to UvrB-DNA complex; nicks DNA 5′ of lesion
uvrD	Helicase II; helps remove damage-containing oligonucleotide
polA	Polymerase I; fills in single-strand gap
lig	Ligase; seals single-strand nick

including the *polA* and *uvrD* genes, are also required for other types of repair.

How these gene products participate in excision repair is illustrated in Figure 11.14. The products of the *uvrA*, *uvrB*, and *uvrC* genes interact to form what is called the **UvrABC endonuclease.** The function of these gene products is to make a nick close to the damaged nucleotide, causing it to be excised. In more detail, two copies of the UvrA protein and one copy of the UvrB protein form a complex that binds nonspecifically to DNA even if it is not damaged. This complex then migrates up and down the DNA until it hits a place where the helix is distorted because of DNA damage (in the illustration, because of a thymine dimer). The complex then stops, the UvrB protein binds to the damage, and the UvrA protein leaves, being replaced by UvrC. The binding of UvrC protein to UvrB causes UvrB to cut the DNA about 4 nucleotides 3′ of the damage. The UvrC protein then

cuts the DNA 7 nucleotides 5′ of the damage. Once the DNA is cut, the UvrD helicase removes the oligonucleotide containing the damage and the DNA polymerase I resynthesizes the strand that was removed, using the complementary strand as a template.

INDUCTION OF NUCLEOTIDE EXCISION REPAIR

Although the genes of the excision repair system are almost always expressed at low levels, *uvrA*, *uvrB*, and *uvrD* are expressed at much higher levels after the DNA has been damaged. This is a survival mechanism which ensures that larger amounts of the repair proteins are synthesized when they are needed. Because they are inducible by DNA damage, the *uvr* genes induced by DNA damage fall into a class of genes known as the *din* genes (for "damage inducible"), which includes *recF*, *recA*, *umuC*, and *umuD*. Many *din* genes, including *uvrA*, *uvrB*, and *uvrD*, are induced because they are part of the SOS regulon (see later sections).

Damage in some regions of the DNA presents a more immediate problem for the cell than does damage in other regions. For example, pyrimidine dimers in transcribed regions of DNA block not only replication of the DNA but also transcription of RNA from the DNA, when the RNA polymerase stalls at the damage. It makes sense for the cell to first repair the damage that occurs in transcribed genes, so that these genes can be transcribed and translated into proteins. Box 11.3 describes such a system called transcription-repair coupling that helps remove RNA polymerase stalled at damage in the DNA and helps direct the nucleotide excision repair system to the damage.

DNA Damage Tolerance Mechanisms

In all of the repair systems discussed above, the cell removes the damage from the DNA, often using the information in the complementary strand to restore the correct DNA sequence. Hopefully, the damage is repaired before the replication apparatus arrives on the scene and tries to replicate over the damage, causing the chromosome to break or mistakes to be made. Cells have

Figure 11.14   Model for nucleotide excision repair by the UvrABC endonuclease. See the text for details. A, UvrA; B, UvrB; C, UvrC; D, UvrD; I, DNA polymerase I.

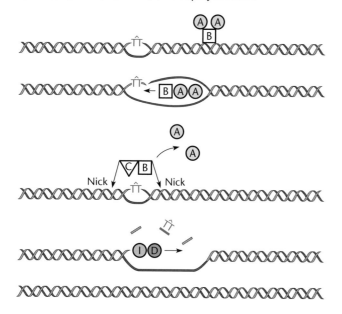

BOX 11.3

Transcription-Repair Coupling

DNA damage in transcribed regions presents a special problem for the cell because the RNA polymerase can stall at the damage, interfering with both expression of the gene and repair of the damage. Predictably, there is a special mechanism to deal with RNA polymerase that has stalled at damage to the DNA. This is called transcription-repair coupling, and the factor involved in bacteria is called Mfd protein (for *m*utation *f*requency *d*ecline). The gene for this protein, *mfd*, was discovered more than 40 years ago by Evelyn Witkin as mutations which prevent the decrease in mutations if protein synthesis is inhibited immediately following DNA damage. Similar systems exist in eukaryotes, where they are called Rad26 in yeast and Csb in humans. Some of the early evidence suggesting the existence of transcription-repair coupling was that DNA damage that occurs within transcribed regions of the DNA and in the transcribed strand is repaired preferentially by the nucleotide excision repair system. Also, mutations occur more frequently when the nontranscribed strand of DNA in a particular region is the one damaged, as expected if damage in this strand were not repaired by the relatively mistake-free excision repair system. It was some time before the Mfd protein was linked to these phenomena.

RNA polymerase stalled at damage in the DNA creates two potential problems for the cell. One is that the stalled RNA polymerase can block access of the nucleotide excision repair system to the damage and thereby prevent its repair. Another is that the stalled RNA polymerase can block the passage of replication forks. The Mfd protein overcomes these potential problems through its translocase activity. When RNA polymerase stalls on the DNA, it backtracks, shoving the 3' hydroxyl end of the growing RNA out of the active center of the enzyme into the secondary channel (see chapter 2). It would sit there like that indefinitely unless something came along that could push the RNA polymerase forward, forcing the 3' hydroxyl end of the RNA back into position. The Mfd protein accomplishes this by binding to the DNA behind the stalled RNA polymerase and translocating (moving) itself forward, pushing the RNA polymerase ahead of it, like the pushers who push commuters onto crowded subways. This has one of two effects. If the damage is still there, the RNA polymerase cannot move forward and the forward movement will disrupt the RNA-DNA transcription bubble holding the RNA polymerase on the DNA. This causes the RNA polymerase to be released, getting it out of the way of repair and replication. If the RNA polymerase can move forward, the block is relieved and the RNA polymerase can continue to make the RNA. Some evidence also indicates that the Mfd protein binds to UvrA, part of the nucleotide excision system, and so it might also help direct the repair system to the DNA damage after the RNA polymerase has left. Translocation by the Mfd protein, with perhaps other proteins, could be a general way to move RNA polymerases out of the way of the much faster replication apparatus.

References

Park, J. S., M. T. Marr, and J. W. Roberts. 2002. *E. coli* transcription repair coupling factor (Mfd protein) rescues arrested complexes by promoting forward translocation. *Cell* **109**:757–767.

Witkin, E. M. 1966. Radiation induced mutations and their repair. *Science* **152**:1345–1353.

---

mechanisms to delay replication after extensive damage to prevent this from happening. But what happens if the damage is not repaired in time before a replication fork arrives? In all the instances where the damage is not repaired, the cell has no choice but to tolerate it by replicating over it. Mechanisms that allow the cell to tolerate DNA damage without ever repairing it are called **damage tolerance** mechanisms.

Recombination Repair of Damaged Replication Forks

One type of damage tolerance mechanism is **recombination repair of replication forks.** This type of repair uses the recombination functions to basically allow the replication fork to bypass the damage to the DNA rather than repair it. After the replication fork has moved on, the damage is still there; it can be repaired by other systems or remain to be a problem for another later-arriving fork. Such damage could involve single- or double-strand breaks in the DNA or damage to the bases so that they cannot base pair properly. It could also involve other molecules such as RNA polymerase stalled on the DNA (Box 11.3). When the replication fork encounters a block in the DNA that it cannot replicate over, it stalls or collapses. The recombination functions allow the leading and lagging strands to exchange information so that the replication apparatus can reassemble on the other side of the damage and replication can continue.

When recombination was first discovered in bacteria, it was assumed that its only purpose was to exchange genes between bacteria and increase their diversity, by analogy to sexual reproduction in higher organisms. Now it is generally assumed that the major role of recombination is to reset replication forks after they have become derailed or have collapsed. Investigation of the role of recombination in restoring replication forks has a long history and is still under way. Early on, it was recognized that mutations in the *rec* genes of *E. coli* made the cell more sensitive to DNA-damaging agents (see Kato et al., Suggested Reading, and below). Mutations in genes in either the RecBCD or RecFOR pathway can make the cell more sensitive to killing by DNA damage, as can mutations in the *ruvABC* genes or *recG* that migrate and resolve Holliday junctions.

We can imagine scenarios in which the recombination functions can help reset replication forks after they have been disrupted by damage to the DNA. Such scenarios must take the following facts into account. Basically, a fork is reset when the DnaB helicase has been loaded on the DNA, as well as the DNA polymerase holoenzyme, which includes the two polymerase molecules that replicate the leading and lagging strands and all of their accessory proteins. There must also be a free 3′ hydroxyl end to provide a primer from which the leading strand can elongate. Somehow the activities of the recombination pathways must help achieve this end. As discussed in chapter 10, the RecBCD pathway is used at double-strand breaks or ends of DNA and creates a 3′ hydroxyl-ended single-stranded DNA that can bind RecA and invade another double-stranded DNA with a complementary sequence to form a fork. These forks can then migrate into neighboring double-stranded DNA to form four-stranded Holliday junctions. The RecFOR pathway, in contrast, is used primarily at gaps in the DNA, with no free double-stranded ends. It allows RecA protein to bind to the single-stranded DNA in the gap and invade another double-stranded DNA with the complementary sequence, basically creating two branches that can migrate into the neighboring double-stranded DNA to create two Holliday junctions. The Holliday junctions created by either pathway can migrate and be cut with the aid of RecA, RuvABC, and RecG.

There is also the problem of reloading the replication apparatus on the DNA after the damage has been bypassed. Recall from chapter 1 that the DnaB helicase normally loads on only at the origin of replication (*oriC*). The DnaC protein helps with this loading, and then the DNA polymerase III and all of its accessory proteins are reassembled. However, damage can occur anywhere on the DNA (not just at the *oriC* region) and can block a replication fork. Somehow, the DnaB helicase must be able to be loaded on the DNA at any site where damage has derailed a replication fork. Other proteins, named the Pri proteins (for "primer proteins"), help reload the DnaB helicase at these sites. They do this not by recognizing specific sequences in the DNA but by recognizing structures in the DNA, namely, intermediates in recombination pathways. There are three Pri proteins, PriA, PriB, and PriC, along with other proteins (DnaT, DnaC, and possibly Rep) that play this role. They form two pathways, one which uses PriA, PriB, DnaT, and probably DnaC and is called the PriA pathway, and another which uses PriC, DnaC, and maybe Rep and is called the PriC pathway. Which of these pathways are used depends on the type of recombinational intermediate that has formed at the site of the stalled or collapsed fork. The PriA pathway loads the DnaB helicase onto recombinational intermediates created by invasion of a free single-stranded 3′ OH end into a double-stranded DNA, an intermediate in recombination promoted by the RecBCD enzyme. In contrast, the PriC pathway loads the DnaB helicase on DNA invaded by a single-stranded DNA at a gap in the DNA, and these recombination intermediates are created by the RecFOR system (see Box 10.2 and Heller and Marians, Suggested Reading). The discovery of the Pri proteins and the genetic and biochemical evidence for their role in replication restarts are discussed in Box 10.2.

At present, there is no proof for any particular detailed model for how the recombination functions might restore replication forks. Large gaps still exist in our knowledge of the functions of some of the proteins involved, and the replication fork is a very complex structure, involving many proteins. Nevertheless, a number of models have been proposed that are more or less consistent with what we know from various sources of the capabilities of the various proteins involved, the effect of mutations that inactivate one or more of the pathways, and biochemical studies of the behavior of the proteins involved when presented with different types of DNA structures. Some models also illustrate how the recombination functions could help replication forks stalled at damage get out of the way so that the damage can be repaired by other systems; however, this discussion is restricted for models in which the DNA damage is bypassed but not necessarily repaired. The bad news is that because we do not know everything the recombination proteins are capable of, we cannot propose a definitive model for how replicative bypass works. The good news is that because we do not know everything they are capable of, we can invent models for how the bypass occurs, admittedly with an occasional black box.

One thing shared by all models for how the recombination functions can help the DNA replication apparatus bypass damage to the DNA is that they all invoke recombination between a good daughter DNA and the damaged DNA at the site of the damage. This means that one of the two strands must be able to replicate past the damage even though the other one has stopped. Therefore, the replication of the leading strand and lagging strand must be temporarily uncoupled, allowing the replication of one strand to proceed even though the replication of the other strand is blocked. Such uncoupling is somewhat unexpected, considering that the leading- and lagging-strand DNA polymerases are bound to each other through the τ protein (see chapter 1). However, there is some experimental support for this uncoupling, so we will assume that it happens.

LAGGING-STRAND DAMAGE

One imagines that the consequences of encountering damage in the lagging strand, the strand running in the 5′-to-3′ direction, would be very different from the consequences of encountering it in the leading strand, running in the 3′-to-5′ direction. If the damage is on the lagging strand, the replication fork might be able to pass right over the damage, provided that the DnaB helicase, which is encircling the lagging strand (see Figure 1.12) can proceed over the damage. The lagging strand is replicating discontinuously anyway, and so the worst that could happen is that a gap would be left opposite the damage. This gap could be moved opposite a good strand of DNA by the RecFOR pathway, and the gap could be filled, after the replication fork has moved on.

LEADING-STRAND DAMAGE

The consequences of encountering damage on the leading strand could be much more severe than those of encountering damage on the lagging strand. Take the scenario shown in Figure 11.15. In this case the leading-strand replicating polymerase encounters a block in the DNA that it cannot replicate over (in the example, the block is due to a cyclobutane thymine dimer). The leading-strand DNA polymerase stops, but the DnaB helicase continues, separating the strands for a short distance, and the lagging-strand polymerase continues to replicate the lagging strand. This leaves single-stranded leading-strand DNA containing the damage opposite good double-stranded DNA in the same region. The

Figure 11.15   Model for recombination-mediated bypass of DNA damage in the leading strand. DNA polymerase III stalls at the thymine dimer (purple) but the helicase can continue on the lagging strand, allowing lagging-strand synthesis to continue. The RecFOR proteins help RecA bind to the single-stranded DNA (ssDNA) at the gap, and the RecA nucleoprotein filament invades the sister DNA. Note that in subsequent steps, the order of the purple strands is reversed for convenience. The gap is now opposite a good strand and can be filled in, leaving two Holliday junctions, which are resolved by RuvABC or RecG. The PriC protein then helps reload the DnaB helicase and the replicative DNA polymerase III holoenzyme on the DNA, and replication continues (dashed black line). The DNA in which the thymine dimer remains depends on how the Holliday junctions are resolved.

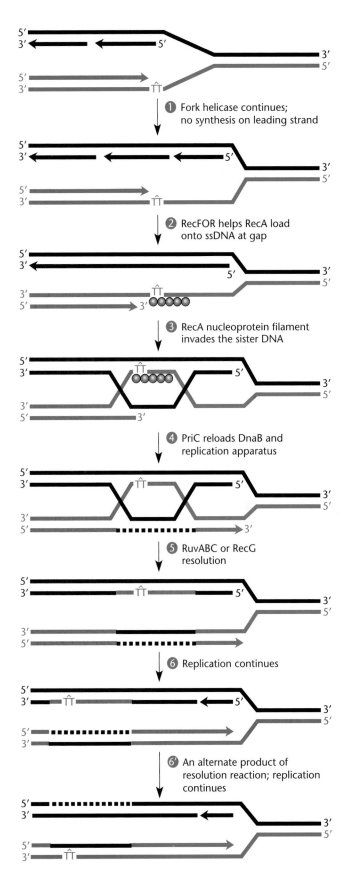

RecFOR proteins could load RecA onto the single-stranded DNA and the filament could invade the double-stranded lagging-strand DNA. This is the preferred substrate for the PriC pathway, which might be able to reload DnaB and the replication apparatus, which could then replicate past the damage (see Heller and Marians, Suggested Reading). The thymine dimer is now opposite a good lagging strand and the gap is opposite a good leading strand, so it can be filled in by repair functions.

A somewhat more satisfying model proposes that the replication fork can regress, as shown in Figure 11.16. The model starts out the same as the one discussed above, in that single-stranded DNA forms when replication of the leading strand is blocked at the damage. The newly synthesized strands might then pair with each other in a strand exchange reaction promoted by the RecFOR and RecA proteins. This creates a branch, which can migrate backward (fork regression) to form a type of Holliday junction called a "chicken foot" structure, in which the two new strands are now hybridized to each other, with the strand created by replicating the lagging stand past the damage being somewhat longer than the blocked leading strand. The shorter new strand due to blocked replication of the leading strand could then furnish a 3' hydroxyl primer for a repair polymerase (DNA polymerase I?) to extend this shorter strand until the two strands are the same length. On the right side of the figure, some arbitrary base pairs are shown to illustrate how this allows replication past the site of the thymine dimer. The chicken foot Holliday junction could then migrate forward, possibly with the help of RecG or some other helicase, past the site of the damage. We now have a bona fide replication fork on the other side of the damage with a free 3' hydroxyl group to prime replication of the leading strand. The DnaB helicase and other replication proteins could then be reloaded, perhaps by the PriA-PriB-DnaT-DnaC pathway, and replication is again under way.

These are just some scenarios to explain the genetic evidence for how the recombination functions can help tolerate damage to the DNA template. Final proof of any model requires more detailed studies of what happens when the extremely complex replication fork encounters damage on the DNA.

REPAIR OF INTERSTRAND CROSS-LINKS IN DNA

The recombination functions may also collaborate with other repair functions to repair damage to DNA. One example might be in the repair of chemical cross-links in the DNA (Figure 11.17). Many chemicals such as light-activated psoralens, mitomycin, cisplatin, and ethyl methanesulfonate can form **interstrand cross-links** in the DNA, in which two bases in the opposite strands of the DNA are covalently joined to each other (hence the prefix "inter," or "between"). Interstrand cross-links present special problems and cannot be repaired by either nucleotide excision repair or recombination repair alone. Cutting both strands of the DNA by the UvrABC endonuclease would cause double-strand breakage and death of the cell. Also, recombination repair by itself cannot repair DNA cross-links, because the cross-link prevents the replication fork from separating the strands.

Although DNA cross-links cannot be repaired by either excision or recombination repair alone, they can be repaired by a combination of recombination repair and nucleotide excision repair, as shown in Figure 11.17. In the first step, the UvrABC endonuclease makes nicks in one strand on either side of the interstrand cross-link, as though it were repairing any other type of damage. This leaves a gap opposite the DNA damage, as shown in the figure. In the second step, the gap is widened by the 5' exonuclease activity of DNA polymerase I. In the third step, recombination repair replaces the gap with a good strand from the other daughter DNA in the cell. The DNA damage is now confined to only one strand and is opposite a good strand; therefore, it can be repaired by the nucleotide excision repair pathway. Notice that this repair is possible only when the DNA has already replicated and there are already two copies of the DNA in the cell, since the helicase cannot open the cross-linked strands. However, in fast-growing cells, most of the DNA does exist in more than one copy (see chapter 1).

Figure 11.16 A fork regression model for recombination-mediated replicative bypass of a thymine dimer in DNA when the damage is on the leading strand. The leading strand replicating polymerase could stall, but the lagging strand replicating polymerase keeps going, making good double-stranded DNA opposite the damage. The fork could then back up (regress) to form a Holliday junction (HJ)-like "chicken foot" in which the newly synthesized strands have now paired with each other, maybe with the help of the RecFOR pathway and RecA. The free 3' end due to truncated synthesis of the leading strand could then serve as a primer for the synthesis of DNA past the original site of the damage. The right side of the figure illustrates this. The junction could then migrate back over the damage, perhaps with the help of RecG, and the replication fork machinery could be reloaded by the PriA pathway. The site of the thymine dimer is in bold type.

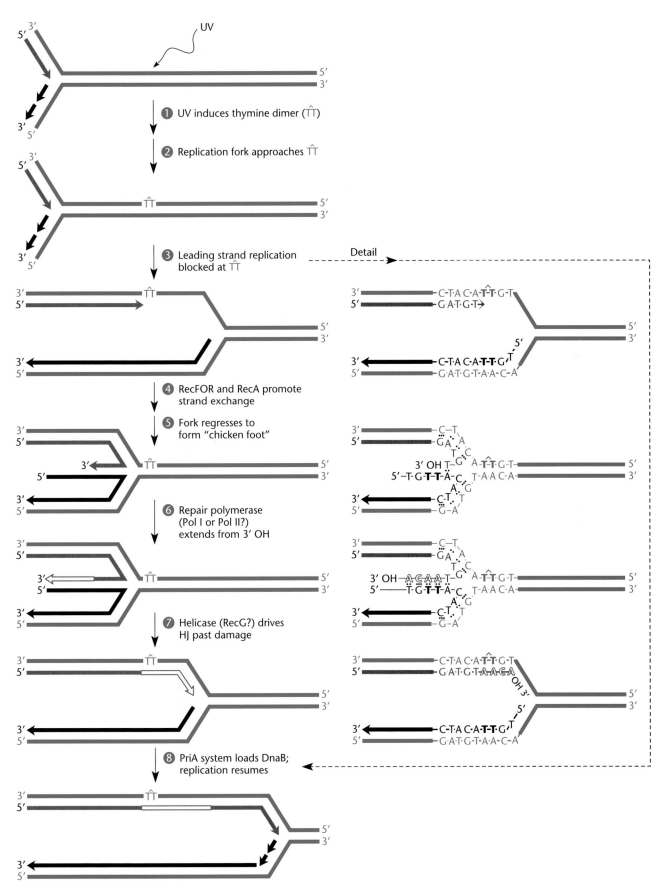

1. UV induces thymine dimer ($\widehat{TT}$)

2. Replication fork approaches $\widehat{TT}$

3. Leading strand replication blocked at $\widehat{TT}$

4. RecFOR and RecA promote strand exchange

5. Fork regresses to form "chicken foot"

6. Repair polymerase (Pol I or Pol II?) extends from 3' OH

7. Helicase (RecG?) drives HJ past damage

8. PriA system loads DnaB; replication resumes

Detail

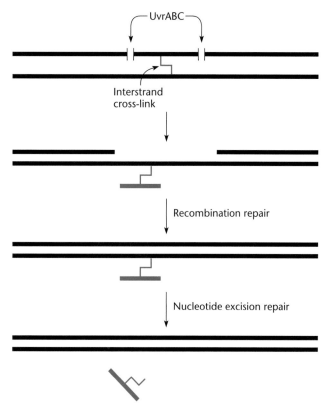

Figure 11.17 Repair of a DNA interstrand cross-link through the combined action of nucleotide excision repair and recombination repair. See the text for details.

SOS Inducible Repair

As mentioned above, DNA damage leads to the induction of genes whose products are required for DNA repair. In this way, the cell is better able to repair the damage and survive. The first indication that repair systems for UV damage are inducible came from the work of Jean Weigle on the reactivation of UV-irradiated λ phage (see Weigle, Suggested Reading). He irradiated phage λ and their hosts and tested the survival of the phage by their ability to form plaques when they were plated on *E. coli*. He observed that more irradiated phage survived when plated on *E. coli* cells that had themselves been irradiated than when plated on unirradiated *E. coli* cells. Because the phage were restored to viability, or reactivated, by being plated on preirradiated *E. coli* cells, this phenomenon was named **Weigle reactivation** or **W-reactivation**. Apparently, a repair system was being induced in the irradiated cells that could repair the damage to the λ DNA when this DNA entered the cell.

THE SOS RESPONSE

Earlier in the chapter, we mentioned a class of genes induced after DNA damage, called the *din* genes (for *damage inducible*), which includes genes that encode products that are part of the excision repair and recombination repair pathways. The products of other *din* genes help the cell survive DNA damage in other ways. For example, some *din* gene products transiently delay cell division until the damage can be repaired and some allow the cell to replicate past the DNA damage (see below).

Many *din* genes are regulated by the **SOS response,** so named because this mechanism rescues cells that have suffered severe DNA damage. Genes under this type of control are called **SOS genes.** Originally, classical genetic analysis uncovered some 31 SOS genes. More recent microarray analyses have found a few more, not all of which may be directly regulated by the SOS system. For example, some of the new ones are genes on cryptic prophages that are being induced by DNA damage, indirectly increasing the expression of the gene (see below and Courcelle et al., Suggested Reading).

Figure 11.18 illustrates how genes under SOS control are induced. The SOS genes are normally repressed by a protein called the LexA repressor, which binds to sequences called the **SOS box** upstream of the SOS genes and prevents their transcription. The SOS box is the operator sequence that binds the LexA repressor by analogy to other operator sites such as *lacO* and the operator sites that bind the λ repressor. Any gene directly regulated by LexA would therefore have one of these SOS boxes close to its promoter. Quite often, genes in the same regulon have a common upstream sequence that binds the regulatory protein, and these are often referred to as boxes, with the name of the regulon. In fact, this is often how the genes of a regulon are first identified, by the presence of one of these boxes close to their promoter. Examples of regulons and their boxes are discussed in chapter 13.

The LexA repressor remains bound to the SOS boxes upstream of the promoters for the genes, repressing their transcription, until the DNA is extensively damaged by UV irradiation or other DNA-damaging agents. This causes the LexA repressor to cleave itself, an action known as **autocleavage,** thereby inactivating itself and allowing transcription of the SOS genes. This is reminiscent of the cleavage of the λ repressor during induction of phage λ described in chapter 8, which also cleaves itself following DNA damage. The two mechanisms are in fact remarkably similar, as discussed below.

The reason why the LexA repressor no longer binds to DNA after it is cleaved is also well understood. Each LexA polypeptide has two separable domains. One, the **dimerization domain,** binds to another LexA polypeptide to form a dimer (hence its name), and the other, the **DNA-binding domain,** binds to the DNA operators upstream of the SOS genes and blocks their transcription.

The LexA protein will bind to DNA only if it is in the dimer state. The point of cleavage of the LexA polypeptide is between the two domains, and autocleavage separates the DNA binding domain from the dimerization domain. The DNA binding domain cannot dimerize and so by itself cannot bind to DNA and block transcription of the SOS genes.

Figure 11.18 also illustrates the answer to the next question: why does the LexA repressor cleave itself only after DNA damage? After DNA damage, single-stranded DNA appears in the cell, probably owing to blockage of replication forks at the damage. This single-stranded DNA binds RecA protein to form RecA nucleoprotein filaments, which then bind to LexA and cause it to autocleave. This feature of the model—that LexA cleaves itself in response to RecA binding rather than being cleaved by RecA—is supported by experimental evidence. When heated under certain conditions, even in the absence of RecA, LexA eventually cleaves itself. This result indicates that RecA acts as a **coprotease** to facilitate LexA autocleavage, rather than doing the cleaving itself like a standard protease.

The RecA protein thus plays a central role in the induction of the SOS response. It senses the single-stranded DNA that accumulates in the cell as a by-product of attempts to repair damage to the DNA and then causes LexA to cleave itself, inducing the SOS genes. We have already discussed another activity of RecA, the recombination function involved in synapse formation; later, we discuss yet another, in translesion synthesis.

We can speculate why the RecA protein might play two different roles, one in recombination and the other in repair. It must bind to single-stranded DNA to promote synapse formation during recombination, and so it must be a sensor of single-stranded DNA in the cell. Also, even though it itself is encoded by an SOS gene and is induced following DNA damage, it is always present in large enough amounts to bind to all the LexA repressor and quickly promote autocleavage of all the repressor.

As mentioned, the regulation of the SOS response through cleavage of the LexA repressor is strikingly similar to the induction of λ through autocleavage of the CI repressor (see chapter 8). Like the LexA repressor, the

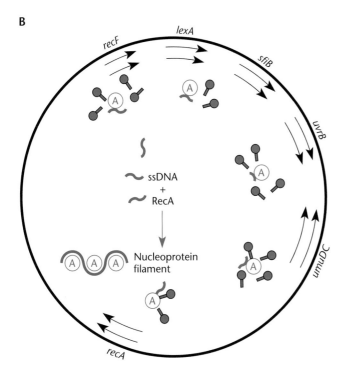

Figure 11.18 Regulation of the SOS response regulon in *E. coli*. (A) About 30 genes around the *E. coli* chromosome are normally repressed by the binding of a LexA dimer (barbell structure) to their operators; only a few of these genes are shown. Some SOS genes are expressed at low levels, as indicated by single arrows. (B) After DNA damage, the single-stranded DNA (ssDNA) that accumulates in the cell binds to RecA (circled A), forming a RecA nucleoprotein filament, which binds to LexA, causing LexA to cleave itself. The cleaved repressor can no longer bind to the operators of the genes, and the genes are induced as indicated by two arrows. The approximate positions of some of the genes of the SOS regulon are shown.

λ repressor must be a dimer to bind to the operator sequences in DNA and repress λ transcription. Each λ repressor polypeptide consists of an N-terminal DNA binding domain and a C-terminal dimerization domain. Autocleavage of the λ repressor, stimulated by the activated RecA nucleoprotein filaments, also separates the DNA-binding domain and dimerization domain, preventing the binding of the λ repressor to the operators and allowing transcription of λ lytic genes. The sequences of amino acids around the sites at which the LexA and λ repressors are cleaved are also similar. By modeling its own repressor after the LexA repressor, the λ prophage allows the SOS regulatory system of the host to induce it following DNA damage, thereby allowing λ to escape a doomed host cell. The activated RecA coprotease also promotes the autocleavage of the UmuD protein involved in SOS mutagenesis (see below).

GENETICS OF SOS INDUCIBLE MUTAGENESIS

It is well known that many types of DNA damage, including UV irradiation, are mutagenic and increase the number of mutations in cells. This is true of all organisms, from bacteria to humans (Box 11.4). This implies that one or more of the repair mechanisms used to repair damage are mistake prone. Early evidence suggested that at least one of the repair systems encoded by the SOS genes appears to be very mistake prone. As we show, this system, known as **translesion synthesis** (**TLS**), allows the replication fork to proceed over damaged DNA so that the molecule can be replicated and the cell can survive.

BOX 11.4

Translesion Synthesis and Cancer

Conversion of a normal cell to a cancer cell requires multiple mutations. To grow out of control without the normal constraints, a cancer cell must have mutations in genes whose products normally control the normal cell cycle checkpoints, control communication with surrounding cells, and cause the cells to kill themselves by apoptosis if they get out of hand, among others. Therefore, it is not surprising that conditions that increase the number of mutations in cells would increase the frequency of their transformation into cancer cells. We have already discussed how people with mutations in the human equivalent of *mutY*, which removes A's opposite GOs, and *mutS* of the mismatch repair system have an increased frequency of some types of cancer (Boxes 11.1 and 11.2).

As mentioned in the text, a number of genes related to *E. coli umuC* and *dinB* have been found in other organisms including humans. Once UmuC was discovered to be a mutagenic polymerase, the products of some of these genes were purified and found also to be DNA polymerases. The family was named the Y polymerases. One of the more universal of these is Rev1, which is found in all eukaryotes. The gene for Rev1 was originally found in yeast by using a selection similar to the one used to isolate *umuC* and *umuD* mutations in *E. coli*, but in this case by selecting mutants with increased mutation frequency after DNA damage. The Rev1 polymerase always inserts C's opposite of damage, independent of the type of damage, and so is very mutagenic. Interestingly, it may also play a role in selecting which Y polymerase is used for translesion synthesis over a particular type of damage (see Guo et al., below). One of these is DNA polymerase η, the product of the yeast gene called *RAD30*. The *RAD* genes of yeast were isolated because they make yeast more sensitive to UV light. This DNA polymerase seems to be smarter than most of the other more mutagenic polymerases, however, in that it incorporates A's opposite the T's in a cyclobutane thymine dimer, the most common form of UV damage, unlike UmuC, which randomly incorporates nucleotides opposite such a dimer. By incorporating two A's opposite the thymine dimer, DNA polymerase η does not make mistakes and restores the original sequence. By choosing η to replicate over a thymine dimer, the Rev1 polymerase can avoid mutations, explaining why Rev1 mutations are so mutagenic. A gene encoding a DNA polymerase homologous to polymerase η has been found in humans; it is mutated in one type of hereditary skin disease called xeroderma pigmentosum (XP). People with xeroderma pigmentosum are very sensitive to sunlight, and even limited exposure to UV light can cause them to develop a type of skin cancer called basal cell carcinoma. Most types of xeroderma pigmentosum are due to defects in excision repair; however, this type, called XPV (for XP variant, because it was known to be different), is due to a mutant DNA polymerase η. Apparently, the thymine dimers which accumulate in skin cells exposed to UV light are much more mutagenic in the absence of DNA polymerase η. If DNA polymerase η is not there to accurately replicate over the thymine dimers, they must be dealt with in other ways which are much more mutagenic, leading to cancers.

References

Guo, C., P. L. Fischhaber, M. Luk-Paszye, Y. Masuda, J. Zhou, K. Kisker, and E. C. Friedberg. 2003. Mouse Rev1 protein interacts with multiple DNA polymerases involved in translesion DNA synthesis. *EMBO J.* **22**:6621–6630.

Masutani, C., R. Kusumoto, A. Yamada, N. Dohmae, M. Yokoi, M. Yuasa, M. Araki, S. Iwai, K. Takio, and F. Hanaoka. 1999. The XPV (*Xeroderma pigmentosum* variant) gene encodes human DNA polymerase eta. *Nature* **399**:700–704.

This mechanism seems to be a last resort that operates only when the DNA damage is so extensive that it cannot be repaired by other, less mistake-prone mechanisms.

The first indications that a mistake-prone pathway for UV damage repair is inducible in *E. coli* came from the same studies that showed that repair itself is inducible (see Weigle, Suggested Reading). In addition to measuring the survival of UV-irradiated λ plated on UV-irradiated *E. coli*, Weigle counted the number of clear-plaque mutants among the surviving phage. (Recall from chapter 8 that lysogens form in the center of wild-type phage λ plaques, making the plaques cloudy, but mutants that cannot lysogenize form clear plaques that can be easily identified.) There were more clear-plaque mutants among the surviving phage if the bacterial hosts had been UV irradiated prior to infection than if they had not been irradiated. Therefore, the increased mutagenesis was due to induction of a mutagenic repair system after DNA damage. This inducible mutagenesis was named **Weigle mutagenesis** or just **W-mutagenesis**. Later studies showed that the inducible mutagenesis results from induction of one or more of the SOS genes; it does not occur without RecA and cleavage of the LexA repressor. Thus, the inducible mutagenesis Weigle observed is now often called **SOS mutagenesis**.

Determining Which Repair Pathway Is Mutagenic

Although Weigle's experiments showed that one of the inducible UV damage repair systems in *E. coli* is mistake prone and causes mutations, they did not indicate which system was responsible. A genetic approach was used to answer this question (see Kato et al., Suggested Reading). To detect UV-induced mutations, the experimenters used the reversion of a *his* mutation. Their basic approach was to make double mutants with both a *his* mutation and a mutation in one or more of the genes of each of the repair pathways. The repair-deficient mutants were then irradiated with UV light and plated on medium containing limiting amounts of histidine, so that only His⁺ revertants could multiply to form a colony. Under these conditions, each reversion to *his*⁺ results in only one colony, making it possible to measure directly the number of *his*⁺ reversions that have occurred. This number, divided by the total number of surviving bacteria, gives an estimation of the susceptibility of the cells to mutagenesis by UV light.

The results of these experiments led to the conclusion that recombination repair of blocked replication forks does not seem to be mistake prone. While *recB* and *recF* mutations reduced the survival of the cells exposed to UV light, the number of *his*⁺ reversions per surviving cell was no different from that of cells lacking mutations in their *rec* genes. Also, nucleotide excision repair does not seem to be mistake prone. Addition of a *uvrB* mutation to the *recB* and *recF* mutations made the cells even more susceptible to killing by UV light but also did not reduce the number of *his*⁺ reversions per surviving cell.

However, *recA* mutations did seem to prevent UV mutagenesis. While these mutations made the cells extremely sensitive to killing by UV light, the few survivors did not contain additional mutations. We now know that two genes, *umuC* and *umuD*, must be induced for mutagenic repair and that *recA* mutations prevent their induction. Thus, the UmuC and UmuD proteins act in the opposite way to most repair systems. Rather than repairing the damage before mistakes in the form of mutations are made, the UmuC and UmuD proteins actually make the mistakes themselves. If they were not present, the cells which survive UV irradiation and some other types of DNA damage would have fewer mutations. The payoff, however, is that more cells survive. Besides its role in inducing the SOS genes, the RecA protein is directly involved in SOS mutagenesis (see below).

Isolation of *umuC* and *umuD* Mutants

Once it was established that most of the mutagenesis after UV irradiation can be attributed to the products of SOS-inducible genes and that these gene products are not the known ones involved in recombination and nucleotide excision repair, the next step was to identify the unknown genes. Note that the products of these genes seem to work in the opposite way to most repair pathways. Mutations that inactivate a *mut* gene or another repair pathway gene increase the rate of spontaneous mutations because normally the gene products repair damage before it can cause mistakes in replication. However, as noted above, mutations that inactivate gene products of a mutagenic or mistake-prone repair pathway have the opposite effect, decreasing the rate of at least some induced mutations. Because the repair pathway itself is mutagenic, mutations should be less frequent if the repair pathway does not exist than if it does exist. Cells with a mutation in a gene of the mistake-prone repair pathway may be less likely to survive DNA damage, but there should be a lower percentage of newly generated mutants among the survivors.

Mutations which reduce the frequency of mutants after UV irradiation were found to fall in two genes, named *umuC* and *umuD* (for "ultraviolet-induced mutations C and D"). The first *umuC* and *umuD* mutants were isolated in two different laboratories by essentially the same method—reversion of a *his* mutation that makes cell growth require histidine to measure mutation rates—but we describe only the one used by Kato and Shinoura (Suggested Reading). The basic strategy was to treat the *his* mutant with a mutagen that induces DNA damage similar to that caused by UV irradiation and to identify mutant bacteria in which fewer his^+ revertants occurred. These mutants would have a second mutation that inactivated the mutagenic repair pathway and reduced the rate of reversion of the *his* mutation.

Figure 11.19 illustrates the selection in more detail. The *his* mutant of *E. coli* was heavily mutagenized, so that some of the bacteria would have mutations in the putative mutagenic repair genes. Individual colonies of the bacteria were then patched with a loop onto plates with medium containing histidine, and the plates were incubated until patches due to bacterial growth first appeared. Each plate was then replicated onto another plate containing

Figure 11.19 Detection of a mutant defective in mutagenic repair. Colonies of mutagenized *his* bacteria are picked individually from a plate and patched onto a new plate containing histidine. This plate (plate 1) is then replicated onto a plate containing 4NQO (plate 2) to induce DNA damage similar to that induced by UV irradiation. The 4NQO-containing plate is then replicated onto another plate with limiting amounts of histidine (plate 3). After incubation, mottling of a patch caused by many His$^+$ revertants indicates that the bacterium that made the colony on the original plate was capable of mutagenic repair. The colony circled in purple on the original plate may have arisen from a mutant deficient in mutagenic repair, because it gives fewer His$^+$ revertants when replicated onto plate 3.

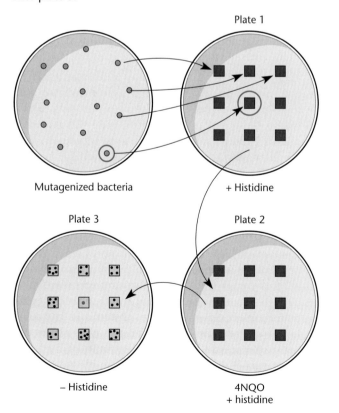

Plate 1

Mutagenized bacteria

+ Histidine

Plate 3

Plate 2

− Histidine

4NQO
+ histidine

4-nitroquinoline-1-oxide (4NQO), which causes DNA damage similar to that caused by UV irradiation. After the patches developed, this plate was replicated onto a third plate containing limiting amounts of histidine. Most bacteria formed patches with a few regions of heavier growth due to His$^+$ revertants, indicating that they were being mutagenized by the 4NQO. However, a few bacteria formed patches with very few areas of heavier growth and therefore had fewer His$^+$ revertants. The bacteria in the patches were candidates for descendants of mutants that could not be mutagenized by 4NQO or, presumably, by UV irradiation, since, as mentioned, the types of damage caused by the two mutagens are similar.

The next step was to map the mutations in some of the mutants. Some of the mutations that prevented mutagenesis by UV irradiation mapped to either *recA* or *lexA*. These mutations could have been anticipated because mutations in these genes could prevent the induction of all the SOS genes including those for mutagenic repair. The *recA* mutations presumably inactivate the coprotease activity of the RecA protein, preventing it from causing autocleavage of the LexA repressor and thereby preventing induction of the SOS genes. The *lexA* mutations in all probability change the LexA repressor protein so that it cannot be cleaved, presumably because one of the amino acids around the site of cleavage has been altered. This is a special type of *lexA* mutation called a (Ind$^-$) mutation. If the LexA protein is not cleaved following UV irradiation of the cells, the SOS genes, including the genes for mutagenic repair, will not be induced.

Some of the mutations that prevented UV mutagenesis mapped to a previously unknown locus, distant from *recA* or *lexA*, on the *E. coli* map. Complementation tests between mutations at this locus revealed two genes at this site required for UV mutagenesis, later named *umuC* and *umuD*. Later experiments also showed that these genes are transcribed into the same mRNA and so form an operon, *umuDC*, in which the *umuD* gene is transcribed first. Experiments with *lacZ* fusions revealed that the *umuDC* operon is inducible by UV light and is an SOS operon, since it is under the control of the LexA repressor (see Bagg et al., Suggested Reading).

Experiments Showing that Only *umuC* and *umuD* Must Be Induced for SOS Mutagenesis
The fact that *umuC* and *umuD* are inducible by DNA damage and are required for SOS mutagenesis does not mean that they are the *only* genes that must be induced for this pathway. Other genes that must be induced for SOS mutagenesis might have been missed in the mutant selections. Some investigators sought an answer to this question (see Sommer et al., Suggested Reading). Their experiments used a *lexA*(Ind$^-$) mutant, which, as mentioned, should

permanently repress all the SOS genes. They also mutated the operator site of the *umuDC* operon so that these genes would be expressed constitutively and would no longer be under the control of the LexA repressor. Under these conditions, the only SOS gene products that should be present are those of the *umuC* and *umuD* genes, since all other such SOS genes are permanently repressed by the LexA(Ind⁻) repressor. In addition, a shortened form of the *umuD* gene was used. This altered gene synthesizes only the carboxyl-terminal UmuD′ fragment that is the active form for SOS mutagenesis (see below). With the altered *umuD* gene, the RecA coprotease is not required for UmuD to be autocleaved to the active form.

The experiments showed that UV irradiation is mutagenic if UmuC and UmuD′ are expressed constitutively, even if the cells are *lexA*(Ind⁻) mutants, indicating that *umuC* and *umuD* are the only genes that need to be induced by UV irradiation for SOS mutagenesis. However, this result does not entirely eliminate the possibility that other SOS gene products are involved. As discussed below, the RecA protein is also directly required for UV mutagenesis. The *recA* gene is induced to higher levels following UV irradiation but apparently is present in large enough amounts for UV mutagenesis even without induction. The GroEL and GroES proteins are also required for UV mutagenesis, presumably because they help fold mutagenic repair proteins, but the *groEL* and *groES* genes are not under the control of the LexA repressor and so are expressed even in the *lexA*(Ind⁻) mutants.

Experiments Showing that RecA Has a Role in UV Mutagenesis in Addition to Its Role as a Coprotease

Similar experiments were performed to show that RecA has a required role in UV mutagenesis in addition to its role in promoting the autocleavage of LexA and UmuD. A sufficient explanation for why mutations in the *recA* gene can prevent UV mutagenesis was that they prevent the induction of all the SOS genes including *umuC* and *umuD* and that they also prevent the autocleavage of UmuD to UmuD′, which is required for TLS (see below). However, this did not eliminate the possibility that RecA plays another role in TLS besides its role as a LexA and UmuD coprotease. The mutants that express UmuC and the cleaved form of UmuD′ constitutively could also be used to answer this question. If the coprotease activity of RecA alone is required for TLS, *recA* mutations should not affect UV mutagenesis in this genetic background. However, it was found that *recA* mutations still prevented UV mutagenesis even if UmuC and UmuD′ were made constitutively, indicating that RecA also participates directly in mutagenesis. This is what inspired the model that the UmuD′₂C mutagenic polymerase could

only replicate over DNA that was in a RecA nucleoprotein filament (see below and Figure 11.21).

Mechanism of Induction of SOS Mutagenesis

Dramatic progress has been made in understanding how the UmuC and UmuD proteins promote mutagenesis and allow an *E. coli* cell to tolerate damage to its DNA. As is often the case, these discoveries have implications far beyond the UmuC and UmuD proteins and DNA repair in *E. coli* (Box 11.4). The UmuC protein was found to be a DNA polymerase which, in contrast to the normal replicative DNA polymerase, can replicate right over some types of damage to the DNA, including the thymine cyclobutane dimers and cytosine-thymine 6–4 dimers induced by UV irradiation. It can also replicate over abasic sites in which the base has been removed by a glycosylase (see above); therefore, obviously it does not require correct base pairing before it can move on. Accordingly, UmuC was renamed **DNA polymerase V;** because it is capable of replicating over DNA damage or lesions in the template, it was said to be capable of TLS. This answered the question of why UmuC was so mutagenic. Because thymine dimers and other types of damaged bases cannot pair properly, DNA polymerase V must incorporate bases almost at random opposite the damaged bases, causing mutations.

The fact that *umuC* is an SOS gene, and is therefore induced only after extensive damage, would have been sufficient to explain why UV mutagenesis is inducible. However, the regulation of UV mutagenesis is more complicated and refined than this, presumably because it is in the best interest of the cell to not induce this mutagenic system unless the damage cannot be repaired in other ways. Figure 11.20 illustrates why SOS mutagenesis occurs only after very extensive damage to the DNA. After the DNA has been extensively damaged so that the LexA repressor has been cleaved and the *umuDC* operon has been induced, along with the other SOS genes, the newly synthesized UmuC and UmuD proteins come together to form a heterotrimer complex with two copies of the UmuD protein and one copy of the UmuC protein (UmuD₂C). This complex is inactive as DNA polymerase V, although it might bind to the DNA polymerase III and temporarily arrest replication to create a "checkpoint" (see below). However, as the single-stranded DNA-RecA nucleoprotein filaments accumulate, they also cause UmuD to cleave itself (to form UmuD′), in much the same way that they cause LexA and the λ repressor to cleave themselves. The autocleavage of UmuD to UmuD′ requires a higher concentration of RecA nucleoprotein filaments than does the autocleavage of LexA; therefore, rather than happening immediately, it occurs only if the damage is so extensive that it cannot be immediately repaired by other SOS

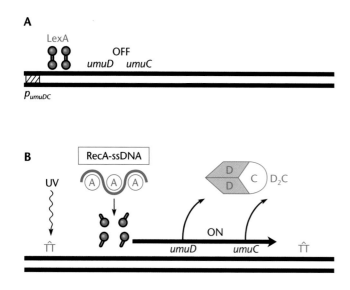

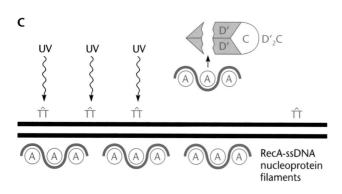

Figure 11.20 Regulation of SOS mutagenesis in *E. coli.*
(A) Before DNA damage occurs, the LexA protein represses the transcription of SOS genes including the *umuDC* operon. (B) After limited DNA damage, the RecA protein binds to the single-stranded DNA (ssDNA), which accumulates, forming RecA nucleoprotein filaments. These filaments bind to LexA, promoting its autocleavage and inducing the SOS genes including the *umuDC* operon. The UmuC and UmuD proteins form a heterotrimer composed of two copies of UmuD and one copy of UmuC (UmuD$_2$C). (C) More damage causes more RecA nucleoprotein filaments to accumulate, eventually promoting the autocleavage of UmuD to form UmuD′$_2$C. (D) The UmuC protein bound to UmuD′$_2$ is an active mutagenic polymerase that can replicate right over the damage, often making mistakes in the process; some wrong bases are shown mistakenly incorporated opposite thymine dimers.

functions. However, once UmuD is cleaved to UmuD′, the UmuC in the UmuD′$_2$C complex is active as a translesion DNA polymerase and replicates over the damage in the DNA, making mistakes in the form of mutations. This allows replication to continue past the damage and permits the cell to survive, but at the price of increasing the frequency of mutations among the survivors.

As mentioned, the trimer complex UmuD$_2$C containing uncleaved UmuD may also play a role even before the UmuD is cleaved. Overproduction of these proteins inhibits DNA replication at low temperatures, even in the absence of DNA damage (see Sutton et al., Suggested Reading). This has been interpreted to mean that the UmuC and UmuD proteins might inhibit replication after they are first induced, creating a checkpoint and allowing more time for the other repair systems to work before the replication forks can move again and encounter the damage. Some evidence suggests that the UmuD$_2$C complex inhibits replication by interfering with the editing functions of the DNA polymerase III since overproduction of the β-clamp protein and the editing function of DNA polymerase affect the inhibition.

Mechanism of Translesion Synthesis by the UmuD′$_2$C Complex

Figure 11.21 shows a recent model for how the UmuD′$_2$C complex performs TLS and causes mutations. This model attempts to explain a number of observations concerning SOS mutagenesis, including the observation that RecA protein is directly involved in TLS in addition to its role in cleaving LexA and UmuD (see above and Sommer et al., Suggested Reading). It also includes roles for the β clamp and clamp-loading proteins of DNA polymerase III in TLS, since some evidence indicates that these functions are also directly required for TLS. In the first step of the model, the DNA polymerase III holoenzyme encounters damage to the DNA that has not yet been repaired (in the example, a cyclobutane thymine dimer). The editing functions on DNA polymerase III does not let it polymerize over the damage, and so it stalls. However, the DnaB helicase which is separating the strands of the DNA ahead of the replicating polymerase keeps going for a short time on the lagging strand, opening the DNA at the damage. The RecA protein then polymerizes on the single-stranded DNA ahead of the stalled leading-strand polymerase including the damaged region to form a helical RecA nucleoprotein filament. The coating of the single-stranded DNA by RecA is required for two reasons. The UmuD′$_2$C complex may replicate DNA efficiently only if it is coated in a RecA nucleoprotein filament, and the coating may also make it less mistake prone when not replicating directly opposite the damage. Also, the RecA protein in the filament may attract UmuD′$_2$C complex to the site, since it is known that RecA binds strongly to UmuD'$_2$C. Once the UmuD'$_2$C complex is bound to the DNA, the β clamp is somehow transferred from the stalled DNA polymerase III to the UmuD'$_2$C complex. Some evidence suggests that the clamp is required to hold the UmuD'$_2$C complex on the DNA. The UmuD'$_2$C polymerase then replicates over the damage, inserting deoxynucleotides essentially at random opposite the damaged bases, thereby

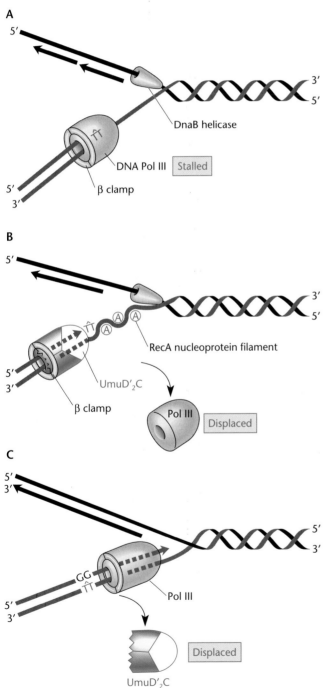

causing mutations. The UmuD'$_2$C polymerase makes as short a DNA as possible, often not much more than 5 nucleotides long, before DNA polymerase III is reloaded on the DNA and normal replication continues. There is a good reason for limiting replication by UmuD'$_2$C. It not only makes mistakes opposite DNA lesions but also makes more mistakes in general, even when replicating nondamaged DNA; these are called untargeted mutations (see below).

This model for TLS will almost certainly have to be revised as more experimental evidence becomes available. We already know some observations about TLS that are difficult to reconcile with the model. One is an apparent continued requirement for DNA polymerase III in TLS in addition to providing the clamp for the UmuD'$_2$C polymerase complex. There is some evidence that DNA polymerase III activity is directly required for SOS mutagenesis, and DNA polymerase III must be added to some in vitro systems for TLS by UmuD'$_2$C in the test tube. On the surface, it does not seem to make much sense that DNA polymerase III would be directly involved in TLS. Why would TLS need another DNA polymerase when UmuC has its own DNA polymerase activity? Perhaps the very short DNA chains made by UmuD'$_2$C over the damage are too unstable unless they are immediately extended by DNA polymerase III. It is also not clear how the DNA polymerase III reloads on the DNA after TLS. PriA may play a role, as it does in replication restarts after recombination repair (see above), but PriA has been postulated to bind only to D-loops to recruit the DnaB helicase and restart replication. There is no evidence that D-loops form during TLS; in fact, as mentioned above, Kato et al. had shown that the recombination functions required for D-loop formation are not required for SOS mutagenesis (Kato et al., Suggested Reading). Perhaps DNA polymerase III never leaves the DNA during TLS, and so it does not need to be reloaded later. There is also some evidence that the role of RecA in TLS is not to form single-stranded

Figure 11.21   A detailed model for TLS by the UmuD'$_2$C complex. (A) In the first step, the DnaB helicase proceeds on the lagging strand past the damage, separating the strands, but the DNA polymerase III stalls, leaving a single-stranded gap. (B) The RecA protein binds to the single-stranded DNA in the gap, forming a RecA nucleoprotein filament. The bound RecA protein attracts the UmuD'$_2$C complex, which "commandeers" the β clamp from the stalled DNA polymerase III and uses it to hold itself on the DNA as it replicates over the damage, often making mistakes in the process. (C) Having done its job, the UmuD'$_2$C complex is replaced by the normal replicative DNA polymerase III and the replication continues.

DNA-RecA nucleprotein filaments for PolV to replicate over, since there is experimental evidence that only two copies of RecA are required.

ROLE OF MUTAGENIC REPAIR

In the discussion above, we assume that SOS mutagenic repair is a last-ditch attempt by the cell to survive extensive damage to its DNA. However, the actual function of SOS mutagenic repair remains a mystery. If this explanation for the existence of mutagenic repair were correct, *umuC* and *umuD* mutants that lack mutagenic repair should be much more sensitive than wild-type bacteria to being killed by agents that extensively damage DNA. However, mutations that inactivate the *umuC* and/or *umuD* gene make *E. coli* only slightly more sensitive to killing by UV irradiation and other DNA-damaging agents. One way to explain this puzzling observation is to propose that UmuC and UmuD protect against types of DNA-damaging agents other than those that have been tested. Another possibility is that UmuC and UmuD offer more protection against DNA damage under conditions different from those that exist in the laboratory.

Other Specialized Polymerases

The UmuC polymerase is a member of a large group of polymerases called the **Y polymerases**. Structural studies have shown that the active center of Y polymerases is more open than those of the replicative polymerases and has fewer contacts between the polymerase and DNA template, allowing the Y polymerases to replicate over many types of lesions in the DNA. They also lack editing functions that would cause them to stall at mismatches. Most types of bacteria, archaea, and eukaryotes are known to have Y DNA polymerases related to UmuC (Box 11.4). In addition, some naturally occurring plasmids carry analogs of the *umuC* and *umuD* genes. The best studied of these genes are the *mucA* and *mucB* genes of plasmid R46, of which pKM101 is a derivative (see chapter 5). The products of these plasmid genes can substitute for UmuC and UmuD' in mutagenic repair of *E. coli* and *Salmonella* spp. Because the pKM101 plasmid makes *Salmonella* spp. more sensitive to mutagenesis by many types of DNA-damaging agents, it has been introduced into the *Salmonella* strains used for the Ames test (Box 11.5).

Besides UmuC, there are two other Y-type polymerases in *E. coli*, polymerase II and polymerase IV, that play specialized roles. Both of these polymerases are induced after DNA damage and are part of the SOS response, presumably because they help replicate over some types of damage. Polymerase II also seems to help replication restarts after UV irradiation, maybe by extending 3' OH ends that the polymerase III holoenzyme cannot. If the first oligonucleotides made by UmuCD' past damage are too short to be stable, the editing function of the polymerase III holoenzyme may reject them. Polymerase II, which is less

---

BOX 11.5

The Ames Test

It is now well established that cancer is initiated by mutations in genes including oncogenes and tumor-suppressing genes. Therefore, chemicals that cause mutations are often carcinogenic for humans. Many new chemicals are being used as food additives or otherwise come into contact with humans, and each of these chemicals must be tested for its carcinogenic potential. However, such testing in animal models is expensive and time-consuming. Because many carcinogenic chemicals damage DNA, they are mutagenic for bacteria as well as humans. Therefore, bacteria can be used in initial tests to determine if chemicals are apt to be carcinogenic. The most widely used of these tests is the Ames test, developed by Bruce Ames and his collaborators. This test uses revertants of *his* mutations of *Salmonella* spp. to detect mutations. The chemical is placed on a plate lacking histidine and on which has been spread a His⁻ mutant of *Salmonella*. If the chemical can revert the *his* mutation, a ring of His⁺ revertant colonies will appear around the chemical on the plate. A number of different *his* mutations must be used because different mutagens cause different types of mutations and they all have preferred sites of mutagenesis (hot spots). Also, the test is made more sensitive by eliminating the nonmutagenic nucleotide excision repair system with a *uvrA* mutation and introducing the pKM101 plasmid containing the *mucA* and *mucB* genes. These genes are analogs of *umuC* and *umuD* and so increase mutagenesis (see the text). Some chemicals are not mutagenic themselves but can be converted into mutagens by enzymes in the mammalian liver. To detect these precursors of mutagens, we can add a liver extract from rats to the plates and spot the chemical on the extract. If the extract converts the chemical into a mutagen, His⁺ revertants will appear on the plate.

Reference

McCann, J., and B. N. Ames. 1976. Detection of carcinogens as mutagens in the *Salmonella*/microsome test assay of 300 chemicals: discussion. *Proc. Natl. Acad. Sci. USA* **73**:950–954.

fussy and lacks an editing function, might need to extend them further before polymerase III can take over. This role may be played by a different DNA polymerase, polymerase I, in *Bacillus subtilis* (see Duigou et al, Suggested Reading).

The function of polymerase IV is even more mysterious. It causes spontaneous mutations when it is overproduced, even without DNA damage. Many of these are frameshifts, but only -1 frameshifts. While it is induced following DNA damage, it is always present in the cell. It is largely responsible for the spontaneous mutations that occur during stationary phase, when the cells are not growing. Why the cell would tolerate a DNA polymerase that does not seem to increase its resistance to DNA-damaging agents and only increases its spontaneous mutation rate is something of a mystery.

Eukaryotes have a large number of different mutagenic polymerases related to UmuC and polymerase IV

(Box 11.4). Each type of mutagenic DNA polymerase may be required to replicate over a particular type of damage to DNA and thereby play a role in avoiding mutations due to a particular type of DNA damage. Mutations in some of these genes have been implicated in increased cancer risk. In the absence of the specialized Y-type polymerase for a particular type of lesion in the DNA, another one will take over, causing the mutations that lead to the increased risk of cancer.

Summary of Repair Pathways in *E. coli*

Table 11.2 shows the repair pathways that have been discussed and some of the genes whose products participate in each pathway. Some pathways, such as photoreactivation and most base excision pathways, have evolved to repair specific types of damage to DNA. Some, such as VSP repair, mend damage only in certain sequences.

TABLE 11.2	Genetic pathways for damage repair and tolerance	
Repair mechanism	**Genetic loci**	**Function**
Methyl-directed mismatch repair	*dam*	DNA adenine methylase
	mutS	Mismatch recognition
	mutH	Endonuclease that cuts at hemimethylated sites
	mutL	Interacts with MutS and MutH
	uvrD (*mutU*)	Helicase
Very-short-patch repair	*dcm*	DNA cytosine methylase
	vsr	Endonuclease that cuts at 5′ side of T in TG mismatch
"GO" (guanine oxidizations)	*mutM*	Glycosylase that acts on GO
	mutY	Glycosylase that removes A from A:GO mismatch
	mutT	8-OxodGTP phosphatase
Alkylation/adaptive response	*ada*	Alkyltransferase and transcriptional activator
	alkA	Glycosylase for alkylpurines
	alkB	α-Ketoglutarate-dependent dioxygenase
Nucleotide excision	*uvrA*	Component of UvrABC
	uvrB	Component of UvrABC
	uvrC	Component of UvrABC
	uvrD	Helicase
	polA	Repair synthesis
Base excision	*xthA*	AP endonuclease
	nfo	AP endonuclease
Photoreactivation	*phr*	Photolyase
Recombination repair	*recA*	Strand exchange
	recBCD	Helicase and nuclease at double-strand breaks
	recFOR	Recombination function
	ssb	Single-stranded DNA-binding protein
SOS system	*recA*	Coprotease
	lexA	Repressor
	umuDC	Translesion synthesis (Pol V)
	dinB	Mutagenic polymerase (Pol IV)
	polB	Replication restarts (Pol II)

Others, such as mismatch repair and nucleotide excision repair, are much more general and repair essentially any damage to DNA, provided that it causes a distortion in the DNA structure.

The separation of repair functions into different pathways is in some cases artificial. Some repair genes are inducible, and the repair enzymes themselves can play a role in their induction as well as in the induction of genes in other pathways. For example, the RecBCD nuclease is involved in recombination repair but can also help make the single-stranded DNA that activates the RecA coprotease activity after DNA damage to induce SOS functions. The RecA protein is required for both recombination repair and induction of the SOS functions, and it is directly involved in SOS mutagenesis. Needless to say, the overlap of the functions of the repair gene products in different repair pathways has complicated the assignment of roles in these pathways.

Bacteriophage Repair Pathways

The DNA genomes of bacteriophages are also subject to DNA damage, either when the DNA is in the phage particle or when it is replicating in a host cell. Not surprisingly, many phages encode their own DNA repair enzymes. In fact, the discovery of some phage repair pathways preceded and anticipated the discovery of the corresponding bacterial repair pathways. By encoding their own repair pathways, phages avoid dependence on host pathways and repair proceeds more efficiently than it might with the host pathways alone.

The repair pathways of phage T4 are perhaps the best understood. This large phage encodes at least seven different repair enzymes that help repair DNA damage due to UV irradiation, and some of these enzymes are also involved in recombination (see chapter 10). Table 11.3 lists the functions of some T4 gene products and their

TABLE 11.3	Bacteriophage T4 repair enzymes
Repair enzyme	**Host analog**
DenV	UV endonuclease of *M. luteus*
UvsX	RecA
UvsY	RecOR
UvsW	RecG
gp46/47 exonuclease	RecBCD recombination repair
gp49 resolvase	RuvABC recombination repair
gp59	PriA

homologous bacterial enzymes. The phage also uses some of the corresponding host enzymes to repair damage to its DNA.

One of the most important functions for repairing UV damage in phage T4 is the product of the *denV* gene, which is an AP lyase (see above) having both *N*-glycosylase and DNA endonuclease activities (see Dodson and Lloyd, Suggested Reading). The *N*-glycosylase activity specifically breaks the bond holding one of the pyrimidines to its sugar in pyrimidine cyclobutane dimers of the *cis-syn* type (Figure 11.7). The endonuclease activity then cuts the DNA just 3′ of the pyrimidine dimer, and the dimer is removed by the exonuclease activity of the host cell DNA polymerase I. As mentioned above, AP lyases thus work very differently from the UvrABC endonuclease of *E. coli* and, in a sense, combine both the *N*-glycosylase and AP endonuclease activities of some other repair pathways. The bacterium *Micrococcus luteus* has a similar enzyme. The endonuclease activity of the DenV protein also functions independently of the *N*-glycosylase activity and cuts apurinic sites in DNA as well as heteroduplex loops caused by short insertions or deletions. Because it cuts next to pyrimidine dimers in DNA, the purified DenV endonuclease of T4 is often used to determine the persistence of pyrimidine dimers after UV irradiation.

SUMMARY

1. All organisms on Earth probably have mechanisms to repair damage to their DNA. Some of these repair systems are specific to certain types of damage, while others are more general and repair any damage that makes a significant distortion in the DNA helix.

2. Specific DNA glycosylases remove some types of damaged bases from DNA. Specific DNA glycosylases are known that remove uracil, hypoxanthine, some types of alkylated bases, 8-oxoG, and any A mistakenly incorporated opposite 8-oxoG in DNA. After the damaged base is

removed by the specific glycosylase, the DNA is cut by an AP endonuclease and the strand is degraded and resynthesized to restore the correct base.

3. The positions of 5-methylcytosine in DNA are particularly susceptible to mutagenesis, because deamination of 5-methylcytosine produces thymine instead of uracil, and thymine is not removed by the uracil-*N*-glycosylase. *E. coli* has a special repair system, called VSP repair, that recognizes the thymine in the thymine-guanine mismatch at the site of 5-methylcytosine and removes it, preventing mutagenesis. The products of

the *mutS* and *mutL* genes of the mismatch repair system also make this pathway more efficient, perhaps by helping attract the Vsr endonuclease to the mismatch.

4. The photoreactivation system uses an enzyme called the photolyase to specifically separate the bases of one type of pyrimidine dimer created during UV irradiation. The photolyase binds to pyrimidine dimers in the dark but requires visible light to separate the fused bases.

5. Methyltransferase enzymes remove the methyl group from certain alkylated bases and phosphates and transfer it to themselves. Others are dioxygenases that oxidize the methyl group, converting it into formaldehyde and removing it. In *E. coli*, some of these alkylation defense proteins are inducible as part of the adaptive response. Methylation of the Ada protein converts it into a transcriptional activator for its own gene as well as for some other repair genes involved in repairing alkylation damage. Some genes of the adaptive response are also turned on when the cells reach stationary phase because they are transcribed from σ^s promoters.

6. The methyl-directed mismatch repair system recognizes mismatches in the DNA and removes and resynthesizes one of the two strands, restoring the correct pairing. The products of the *mutL*, *mutS*, and *mutH* genes participate in this pathway in *E. coli*. The Dam methylase product of the *dam* gene helps select the strand to be degraded. The region including the mismatch in the newly synthesized strand is degraded and resynthesized because the A in neighboring GATC sequences on that strand has not yet been methylated by the Dam methylase.

7. The nucleotide excision repair pathway encoded by the *uvr* genes of *E. coli* removes many types of DNA damage that cause gross distortions in the DNA helix. The UvrABC endonuclease cuts on both sides of the DNA damage, and the entire oligonucleotide including the damage is removed and resynthesized.

8. Recombination repair does not actually repair the damage but helps the cell tolerate it. Replication of the lagging strand proceeds past the damage, leaving a gap opposite the damaged bases. Recombination with the other strand can put a good strand opposite the damage, and the replication can continue. This type of repair in *E. coli* requires the recombination functions RecBCD, RecFOR, RecA, RecG,

and RuvABC, as well as PriA, PriB, PriC, DnaC, and DnaT to restart replication forks.

9. A combination of excision and recombination repair may remove interstrand cross-links in DNA. The excision repair system cuts one strand, and the single-stranded break is enlarged by exonucleases to leave a gap opposite the damage. Recombination repair can then transfer a good strand to a position opposite the damage. Excision repair can then remove the damage, since it is now confined to one strand. Interstrand cross-links can be repaired only if there are two or more copies of that region of the chromosome in the cell.

10. The SOS regulon includes many genes that are induced following DNA damage. The genes are normally repressed by the LexA repressor, which cleaves itself (autocleavage) after extensive DNA damage. The autocleavage is triggered by single-stranded DNA-RecA nucleoprotein filaments that accumulate following DNA damage.

11. SOS mutagenesis is due to the products of the genes *umuC* and *umuD*, which are induced following DNA damage. Immediately after induction, these proteins form a heterotrimer, UmuD$_2$C, which is not active until the RecA single-stranded DNA nucleoprotein filament promotes the autocleavage of UmuD to UmuD'.

12. The UmuD'$_2$C heterotrimer is a mutagenic DNA polymerase, called polymerase V, that can replicate over abasic sites and the two forms of pyrimidine dimers formed by UV irradiation as well as some other types of DNA damage, inserting bases randomly opposite the damage and causing mutations. This polymerase is a member of a large family of translesion DNA polymerases called the Y polymerases, which are found in all the kingdoms of life, and mutations in the genes for these polymerases are known to be the cause of some types of genetic susceptibility to cancer.

13. Other proteins besides polymerase V play a role in translesion synthesis. Translesion synthesis also requires at least the β clamp and clamp-loading functions of the replicative DNA polymerase III, probably to hold polymerase V on the DNA. The RecA protein also plays a direct role in SOS mutagenesis in addition to its role in promoting the autocleavage of LexA and UmuD. Polymerase V may be able to replicate only DNA in the form of RecA nucleoprotein filaments.

QUESTIONS FOR THOUGHT

1. Some types of organisms, for example, yeasts, do not have methylated bases in their DNA and so would not be expected to have a methyl-directed mismatch repair system. Can you think of any other ways besides methylation that a mismatch repair system could be directed to the newly synthesized strand during replication?

2. How would you reconcile the results of Pukkila et al., using heteroduplex DNA to show that the unmethylated strand is preferentially repaired with the speculation that GATC sequences stay in contact with the DNA polymerase III holoenzyme, perhaps through SeqA, after the replication fork has passed?

3. Why do you suppose so many pathways exist to repair some types of damage in DNA?

4. Why do you think the SOS mutagenesis pathway exists if it contributes so little to survival after DNA damage?

PROBLEMS

1. Outline how you would determine if a bacterium isolated from the gut of a marine organism at the bottom of the ocean has a photoreactivation-based DNA repair system.

2. How would you show in detail that the mismatch repair system preferentially repairs the base in a mismatch on the strand unmethylated by Dam methylase? Hint: Make heteroduplex DNA of two mutants of your choice.

3. Outline how you would determine if the photoreactivating system is mutagenic.

4. Outline how you would determine if the nucleotide excision repair system of *E. coli* can repair damage due to aflatoxin B. Hint: Use a *uvr* mutant.

5. How would you determine if the RecA protein plays a direct role in SOS mutagenesis independent of its role in cleaving LexA and UmuD?

6. Outline how you would determine if the *recN* gene is a member of the SOS regulon.

SUGGESTED READING

Bagg, A., C. J. Kenyon, and G. C. Walker. 1981. Inducibility of a gene product required for UV and chemical mutagenesis in *Escherichia coli. Proc. Natl. Acad. Sci. USA* **78:**5749–5753.

Courcelle, J., A. Khodursky, B. Peter, P. O. Brown, and P. C. Hanawalt. 2001. Comparative gene expression profiles following UV exposure in wild-type and SOS-deficient *Escherichia coli* cells. *Genetics* **158:**41–64.

Deaconescu, A. M., A. L. Chambers, A. J. Smith, B. E. Nickels, A. Hochschild, N. J. Savery, and S. A. Darst. 2005. Structural basis for bacterial transcription-coupled DNA repair. *Cell* **124:**507–520.

Dodson, M. L., and R. S. Lloyd. 1989. Structure-function studies of the T4 endonuclease V repair enzyme. *Mutat. Res.* **218:**49–65.

Duigou, S., S. D. Ehrlich, P. Noirot, and M.-F. Noirot-Gros. 2005. DNA polymerase I acts in translesion synthesis mediated by the Y-polymerases in *Bacillus subtilis. Mol. Microbiol.* **57:**678–690.

Friedberg, E. C., G. C. Walker, W. Siede, R. D. Wood, R. A. Schultz, and T. Ellenberger. 2006. *DNA Repair and Mutagenesis*, 2nd ed. ASM Press, Washington, D.C.

Friedberg, E. C., A. R. Lehmann, and R. P. P. Fuchs. 2005. Trading places: how do DNA polymerases switch during translesion DNA synthesis? *Mol. Cell* **18:**499–505.

Glickman, B. W., and M. Radman. 1980. *Escherichia coli* mutator mutants deficient in methylation instructed DNA mismatch correction. *Proc. Natl. Acad. Sci. USA* **77:**1063–1067.

Heller, R. C., and K. J. Marians. 2005. The disposition of nascent strands at stalled replication forks dictates the pathway of replisome loading during restart. *Mol. Cell* **17:**733–743.

Kato, T., R. H. Rothman, and A. J. Clark. 1977. Analysis of the role of recombination and repair in mutagenesis of *Escherichia coli* by UV irradiation. *Genetics* **87:**1–18.

Kato, T., and Y. Shinoura. 1977. Isolation and characterization of mutants of *Escherichia coli* deficient in induction of mutations by ultraviolet light. *Mol. Gen. Genet.* **156:**121–131.

Landini, P., and M. R. Volkert. 2000. Regulatory responses of the adaptive response to alkylation damage: a simple regulon with complex regulatory features. *J. Bacteriol.* **182:**6543–6549.

Lieb, M., S. Rehmat, and A. S. Bhagwat. 2001. Interaction of MutS and Vsr: some dominant-negative *mutS* mutations that disable methyladenine-directed mismatch repair are active in very-short-patch repair. *J. Bacteriol.* **183:**6487–6490.

Michaels, M. L., C. Cruz, A. P. Grollman, and J. H. Miller. 1992. Evidence that MutY and MutM combine to prevent mutations by an oxidatively damaged form of guanine in DNA. *Proc. Natl. Acad. Sci. USA* **89:**7022–7025.

Nohni, T., J. R. Battista, L. A. Dodson, and G. C. Walker. 1988. RecA-mediated cleavage activates UmuD for mutagenesis: mechanistic relationship between transcriptional derepression and posttranslational activation. *Proc. Natl. Acad. Sci. USA* **85:**1816–1820.

Pandya, G. A., I. Y. Yang, A. P. Grollman, and M. Moriya. 2000. *Escherichia coli* responses to a single DNA adduct. *J. Bacteriol.* **182:**6598–6604.

Pukkila, P. J., J. Petersson, G. Herman, P. Modrich, and M. Meselson. 1983. Effects of high levels of adenine methylation on methyl directed mismatch repair in *Escherichia coli*. *Genetics* **1044:**571–582.

Rupp, W. D., C. E. I. Wilde, D. L. Reno, and P. Howard-Flanders. 1971. Exchanges between DNA strands in ultraviolet irradiated *E. coli*. *J. Mol. Biol.* **61:**25–44.

Sommer, S., K. Knezevic, A. Bailone, and R. Devoret. 1993. Induction of only one SOS operon, *umuDC*, is required for SOS mutagenesis in *E. coli*. *Mol. Gen. Genet.* **239:**137–144.

Sutton, M. D., M. F. Farrow, B. M. Burton, and G. C. Walker. 2001. Genetic interactions between the *Escherichia coli* *umuDC* gene products and the β processivity clamp of the replicative DNA polymerase. *J. Bacteriol.* **183:**2897–2909.

Tang, M., X. Shen, E. G. Frank, M. O'Donnell, R. Woodgate, and M. F. Goodman. 1999. UmuD′(2)C is an error-prone DNA polymerase *Escherichia coli* pol V. *Proc. Natl. Acad. Sci. USA* **96:**8919–8924.

Taverna, P., and B. Sedgwick. 1996. Generation of endogenous DNA-methylating agent by nitrosation in *Escherichia coli*. *J. Bacteriol.* **178:**5105–5111.

Teo, I., B. Sedgwick, M. W. Kilpatrick, T. V. McCarthy, and T. Lindahl. 1986. The intracellular signal for induction of resistance to alkylating agents in *E. coli*. *Cell* **45:**315–324.

Trewick, S. C., T. F. Henshaw, R. P. Hausinger, T. Lindahl, and B. Sedgwick. 2002. Oxidative demethylation by *Escherichia coli* AlkB directly reverts DNA base damage. *Nature* **419:**174–178.

Weigle, J. J. 1953. Induction of mutation in a bacterial virus. *Proc. Natl. Acad. Sci. USA* **39:**628–636.

Worth, L., Jr., S. Clark, M. Radman, and P. Modrich. 1994. Mismatch repair proteins MutS and MutL inhibit RecA-catalyzed strand transfer between diverged DNAs. *Proc. Natl. Acad. Sci. USA* **91:**3238–3241.

CHAPTER **12**

Regulation of Gene Expression: Operons

The DNA of a cell contains thousands to hundreds of thousands of genes depending on whether the organism is a relatively simple single-celled bacterium or a complex multicellular eukaryote like a human. All of the features of the organism are due, either directly or indirectly, to the products of these genes. However, all the cells of an organism do not always look or act the same, even though they share essentially the same genes. Even the cells of a single-celled bacterium can look or act differently depending on the conditions under which the cells find themselves, because the genes of a cell are not always expressed at the same levels. The process by which the expression of genes is turned on and off at different times and under different conditions is called the **regulation of gene expression.**

Cells regulate the expression of their genes for many reasons. A cell may express only the genes that it needs in a particular environment so that it does not waste energy making RNAs and proteins which are not needed at that time. Or the cell may turn off genes whose products might interfere with other processes going on in the cell at the time. Cells also regulate their genes as part of developmental processes, such as embryogenesis and sporulation.

As described in chapter 2, the expression of a gene moves through many stages, any one of which offers an opportunity for regulation. First, RNA is transcribed from the gene. Even if RNA is the final gene product, that molecule may require further processing to be active. If the final product of the gene is a protein, the mRNA synthesized from the gene might have to be processed before it can be translated into protein. Then the protein might have to be further processed or transported to its final location to be active.

Even after the gene product is synthesized in its final form, its activity might be modulated under certain environmental conditions.

By far the most common type of regulation occurs at the first stage, when RNA is made. Genes that are regulated at this level are said to be **transcriptionally regulated.** This form of gene regulation seems the most efficient, since synthesizing mRNA that will not be translated seems wasteful. However, not all genes are transcriptionally regulated, at least not exclusively. Examples abound in which the expression of a gene is regulated even after RNA synthesis.

Any regulation that occurs after the gene is transcribed into mRNA is called **posttranscriptional regulation.** There are many types of posttranscriptional regulation; the most common is **translational regulation.** If a gene is translationally regulated, the mRNA may be continuously transcribed from the gene but its translation is sometimes inhibited.

Transcriptional Regulation in Bacteria

Thanks to the relative ease of doing genetics with bacteria, transcriptional regulation in these organisms is better understood than in other organisms and has served as a framework for understanding transcriptional regulation in eukaryotic organisms. There are important differences between the mechanisms of transcriptional regulation in bacteria and higher organisms, many of which relate to the presence of a nuclear membrane. Nevertheless, many of the strategies used are similar throughout the biological world, and many general principles have been uncovered through studies of bacterial transcriptional regulation.

As discussed in chapter 2, most transcriptional regulation occurs at the level of transcription initiation at the promoter. Transcriptional regulation occurs through proteins called **transcriptional regulators,** which usually bind to DNA through helix-turn-helix motifs (Box 12.1). Regulation of transcription initiation can be either negative or positive or both. If it is negative, it is controlled by a repressor that binds to an operator sequence in the DNA and prevents initiation of transcription by RNA polymerase. A repressor can perform this role of preventing initiation of transcription in a number of ways. The operator sequence may overlap with the promoter sequence so that binding of the repressor prevents binding of the RNA polymerase to the promoter. Alternatively, or in addition, the repressor might bend the promoter so that RNA polymerase can no longer bind. The repressor might also hold the RNA polymerase on the promoter so that it cannot leave as it begins to make RNA (see Rojo, Suggested Reading). If the regulation is positive, initiation of transcription is controlled by an activator that is required for initiation by RNA polymerase at the promoter. Activators may work by increasing the tightness of binding of the RNA polymerase to the promoter, by allowing it to open the strands of DNA at the promoter, by rotating and bending the promoter to bring the recognition sites together, or even by allowing RNA polymerase to move from (escape) the promoter into the first gene and begin making RNA.

Some regulatory proteins can be either repressors or activators, depending on where they bind to the promoter region. If the binding site overlaps the binding site for RNA polymerase to the promoter, the protein might sterically inhibit (i.e., physically get in the way of) the binding of RNA polymerase to the promoter and repress transcription. However, if it binds even slightly further upstream, it might make contact with the RNA polymerase, stabilizing the binding of RNA polymerase to the promoter and activating transcription. There is even one case of a phage ϕ29 regulatory protein that can activate transcription from some promoters and repress transcription from others even though it binds in approximately the same position in both types of promoters and makes contact with the same region of RNA polymerase bound to the promoters. In the former case, it increases the strength of binding of RNA polymerase to the promoter enough to activate transcription. In the latter case, the RNA polymerase already binds tightly to the promoter. The regulatory protein then stabilizes the binding too much, essentially "tying" the RNA polymerase to the promoter so that it cannot escape the promoter and begin making RNA. The details of repressor and activator action are discussed later in this chapter.

Genetic Evidence for Negative and Positive Regulation

Negatively and positively regulated operons behave very differently in genetic tests. One difference is in the effect of mutations that inactivate the regulatory gene for the operon. If an operon is negatively regulated, a mutation that inactivates the regulatory gene allows transcription of the operon genes, even in the absence of the inducer. If the regulation is positive, mutations that inactivate the regulatory gene prevent transcription of the genes of the operon, even in the presence of the inducer. A mutant in which the genes of an operon are always transcribed, even in the absence of the inducer, is called a **constitutive mutant.** Constitutive mutations are much more common with negatively than with positively regulated operons because any mutation that inactivates the repressor will result in the constitutive phenotype. With positively regulated operons, a constitutive phenotype can be caused only by mutational changes that do not inactivate the activator protein but

BOX 12.1

The Helix-Turn-Helix Motif of DNA-Binding Proteins

Proteins that bind to DNA, including repressors and activators, often share similar structural motifs determined by the interaction between the protein and the DNA helix. One such motif is the helix-turn-helix (HTH) motif. A region of approximately 7 to 9 amino acids forms an α-helical structure called helix 1. This region is separated by about 4 amino acids from another α-helical region of 7 to 9 amino acids called helix 2. The two helices are at approximately right angles to each other, hence the name helix-turn-helix. When the protein binds to DNA, helix 2 lies in the major groove of the DNA double helix while helix 1 lies crosswise to the DNA, as shown in the figure. Because they lie in the major groove of the DNA double helix, the amino acids in helix 2 can contact and form hydrogen bonds with specific bases in the DNA. Thus, a DNA-binding protein containing an HTH motif recognizes and binds to specific regions on the DNA. Many DNA-binding proteins exist as dimers and bind to inverted repeated DNA sequences. In such cases, the two polypeptides in the dimer are arranged head to tail so that the amino acids in helix 2 of each polypeptide can make contact with the same bases in the inverted repeats.

In the absence of structural information, the existence of an HTH motif in a protein can often be predicted from the amino acid sequence, since some sequences of amino acids cause the polypeptide to assume an α-helical form and the bent region between the two helices usually contains a glycine. The presence of an HTH motif in a protein helps identify it as a DNA-binding protein.

A variant on the HTH domain is the winged HTH (wHTH) domain, in which the "winged turn" is 10 amino acids or more in length, longer than the 3 or 4 amino acids of the "turns" of other HTH domains (see Kenney, below).

References
Kenney, L. J. 2002. Structure/function relationships in OmpR and other winged-helix transcription factors. *Curr. Opin. Microbiol.* **5:**135–141.

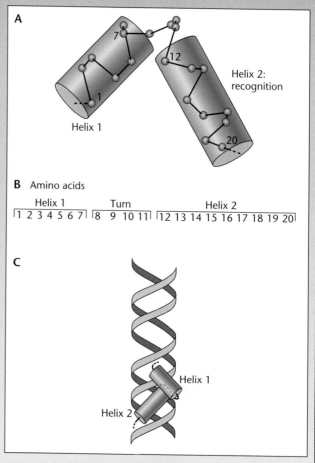

HTH motif of DNA-binding proteins. (A) The structure of an HTH domain; (B) the number of amino acids in the HTH domain of the CAP protein; (C) the interactions of helix 1 and helix 2 with double-stranded DNA.

Steitz, T. A., D. H. Ohlendorf, D. B. McKay, W. F. Anderson, and B. W. Matthews. 1982. Structural similarity in the DNA-binding domains of catabolite gene activator and *cro* repressor proteins. *Proc. Natl. Acad. Sci. USA* **79:**3097–3100.

alter it so that it can activate transcription without binding to the inducer. Such changes tend to be rare.

Complementation tests reveal another difference between negatively and positively regulated operons. Constitutive mutations of a negatively regulated operon are often recessive to the wild type (see chapter 3 for genetic definitions). This is because any normal repressor protein in the cell encoded by a wild-type copy of the gene binds to the operator and blocks transcription, even

if the repressor encoded by the mutant copy of the gene in the same cell is inactive. In contrast, constitutive mutations in a solely positively regulated operon are often dominant to the wild type. A mutant activator protein that is active without inducer bound might activate transcription even in the presence of a wild-type activator protein. In the next sections we describe the regulation of some operons and how genetic evidence contributed to this knowledge.

Negative Transcriptional Regulation

The *E. coli lac* Operon

The classic example of negative regulation is regulation of the *Escherichia coli lac* operon, which encodes the enzymes responsible for the utilization of the sugar lactose. The experiments of François Jacob and Jacques Monod and their collaborators on the regulation of the *E. coli lac* genes are excellent examples of the genetic analysis of a biological phenomenon in bacteria (see Jacob and Monod, Suggested Reading). Although these experiments were performed in the late 1950s, only shortly after the discovery of the structure of DNA and the existence of mRNA, they still stand as the framework with which all other studies of gene regulation are compared.

MUTATIONS OF THE *lac* OPERON

When Jacob and Monod began their classic work, it was known that the enzymes of lactose metabolism are **inducible** in that they are expressed only when the sugar lactose is present in the medium. If no lactose is present, the enzymes are not made. From the standpoint of the cell, this is a sensible strategy since there is no point in making the enzymes for lactose utilization unless lactose is available for use as a carbon and energy source.

To understand the regulation of the lactose genes, Jacob and Monod first isolated many mutations affecting lactose metabolism and regulation, which fell into two fundamentally different groups. Some mutants were unable to grow with lactose as the sole carbon and energy source and so were called Lac⁻ mutants. Other mutants made the lactose-metabolizing enzymes whether or not lactose was present in the medium and so were called constitutive mutants.

COMPLEMENTATION TESTS WITH *lac* MUTATIONS

To analyze the regulation of the *lac* genes, Jacob and Monod needed to know which of the mutations affected *trans*-acting gene products—either protein or RNA—involved in the regulation and how many different genes these mutations represented. They also wished to know if any of the mutations were *cis* acting (affecting sites on the DNA involved in regulation).

To answer these questions, they needed to perform complementation tests, which require that the organisms be diploid, with two copies of the genes being tested. Bacteria are normally haploid, with only one copy of each of their genes, but are "partial diploids" for any genes carried on an introduced prime factor. Recall that a prime factor is a plasmid into which some of the bacterial chromosomal genes have been inserted (see chapter 5). By introducing prime factors carrying various mutated *lac* genes into cells with different mutations in the chromosomal *lac* genes, Jacob and Monod performed complementation tests on each of their *lac* mutations. Their methods depended on the type of mutation being tested.

Whether a particular *lac* mutation is dominant or recessive was determined by introducing an F′ factor carrying the wild-type *lac* region into a strain with the *lac* mutation in the chromosome. If the partial diploid bacteria are Lac⁺ and can multiply to form colonies on minimal plates with lactose as the sole carbon and energy source, the *lac* mutation is recessive. If the partial diploid cells are Lac⁻ and cannot form colonies on lactose minimal plates, the *lac* mutation is dominant. Jacob and Monod discovered that most *lac* mutations are recessive to the wild type and so presumably inactivate genes whose products are required for lactose utilization.

The question of how many genes are represented by recessive *lac* mutations could be answered by performing pairwise complementation tests between different *lac* mutations. Prime factors carrying the *lac* region with one *lac* mutation were introduced into a mutant strain with another *lac* mutation in the chromosome (Figure 12.1). In this kind of experiment, if the partial diploid cells are Lac⁺, the two recessive mutations can complement each other and are members of *different* complementation groups or genes. If the partial diploid cells are Lac⁻, the two mutations cannot complement each other and are members of the *same* complementation group or gene. Jacob and Monod found that most of the *lac* mutations sorted into two different complementation groups, which they named *lacZ* and *lacY*. We now know of another gene,

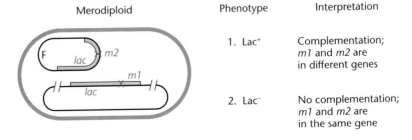

Figure 12.1 Complementation of two recessive *lac* mutations. One mutation is in the chromosome, and the other is in a prime factor. If the two mutations complement each other, the cells will be Lac⁺ and will grow with lactose as the sole carbon and energy source. The mutations will not complement if they are in the same gene or if one affects a regulatory site or is polar. See the text for more details.

Merodiploid   Phenotype   Interpretation

1. Lac⁺   Complementation; *m1* and *m2* are in different genes

2. Lac⁻   No complementation; *m1* and *m2* are in the same gene

lacA, which was not discovered in their original selections because its product is not required for growth on lactose.

cis-Acting *lac* Mutations

Not all Lac⁻ mutants have *lac* mutations that affect diffusible gene products and can be complemented. Immediately adjacent to the *lacZ* mutations are other *lac* mutations that are much rarer and have radically different properties. These mutations cannot be complemented to allow the expression of the *lac* genes on the same DNA, even in the presence of good copies of the *lac* genes. Mutations that cannot be complemented are *cis* acting and presumably affect a site on DNA rather than a diffusible gene product like an RNA or protein (see chapter 3).

To show that a *lac* mutation is *cis* acting, i.e., affects only the expression of genes on the same DNA where it occurs, we could introduce an F′ factor containing the potential *cis*-acting *lac* mutation into cells containing either a *lacZ* or a *lacY* mutation in the chromosome (Figure 12.2). Any *trans*-acting gene products encoded by the F′ factor *lacZ* or *lacY* genes would complement the chromosomal *lacY* or *lacZ* mutations, respectively. However, if the resulting phenotypes are Lac⁻, the *lac* mutation in the F′ factor must prevent expression of both LacZ and LacY proteins from the F′ factor. The mutation in the F′ factor is therefore *cis* acting.

As discussed below, Jacob and Monod named one type of the *cis*-acting *lac* mutations "*lacp* mutations" and hypothesized that they affect the binding of RNA polymerase to the beginning of the gene, in other words are mutations in the promoter region. We now know that many of these mutations are strong polar mutations in the beginning of the *lacZ* gene that also prevent the transcription of the downstream *lacY* gene (see below).

Lac⁻ Mutants with Dominant Mutations

Some Lac⁻ mutants have mutations that affect diffusible gene products but are dominant rather than recessive. A dominant *lac* mutation makes the cell Lac⁻ and unable to use lactose even if there is another good copy of the lactose operon in the cell, either in the chromosome or in the F′ factor. These dominant *lac* mutations are called *lacI*ˢ mutations, for "superrepressor mutations." As shown below, these mutants have mutations that change the repressor so that it can no longer bind the inducer.

COMPLEMENTATION TESTS WITH CONSTITUTIVE MUTATIONS

As mentioned, some *lac* mutations do not make the cells Lac⁻ but rather make them constitutive, so that they express the *lacZ* and *lacY* genes even in the absence of the inducer lactose. In complementation tests between constitutive mutations, partial diploids are made in which either the chromosome or the F′ factor, or both, carry a constitutive mutation. The partial diploid cells are then tested to determine whether they express the *lac* genes constitutively or only in the presence of the inducer. If the partial diploid cells express the *lac* gene in the absence of the inducer, the constitutive mutation is dominant. However, if the partial diploid cells express the *lac* genes only in the presence of the inducer, the mutations are recessive. Using this test, Jacob and Monod found that some of the constitutive mutations, which could be recessive or dominant, affect a *trans*-acting gene product, either protein or RNA. Complementation between the recessive constitutive mutations revealed they are all in the same gene, which these investigators named *lacI* (Figure 12.3A).

cis-Acting *lacO*ᶜ Mutations

A rarer constitutive mutation is *cis* acting, allowing constitutive expression of the *lacZ* and *lacY* genes from the DNA that has the mutation, even in the presence of a wild-type copy of the *lac* DNA. Jacob and Monod named these *cis*-acting constitutive mutations *lacO*ᶜ mutations, for *lac* operator-constitutive mutations. Figure 12.3B shows the partial diploid cells used in these complementation tests.

trans-Acting Dominant Constitutive Mutations

Some *lacI* mutations, called *lacI*⁻ᵈ mutations, are *trans* dominant, making the cell constitutive for expression of

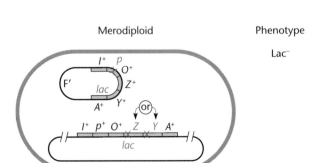

Merodiploid	Phenotype	Interpretation
	Lac⁻	No complementation; the *cis*-acting *lacp* mutation prevents expression of *lacZ*, *lacY*, and *lacA*

Figure 12.2 The *lacp* mutations cannot be complemented and are *cis* acting. A *lacp* mutation in the prime factor prevents the expression of any of the other *lac* genes on the prime factor, so that a *lac* mutation in the chromosome will not be complemented. Partial diploid cells will be Lac⁻. Mutated *lac* regions are shown in purple.

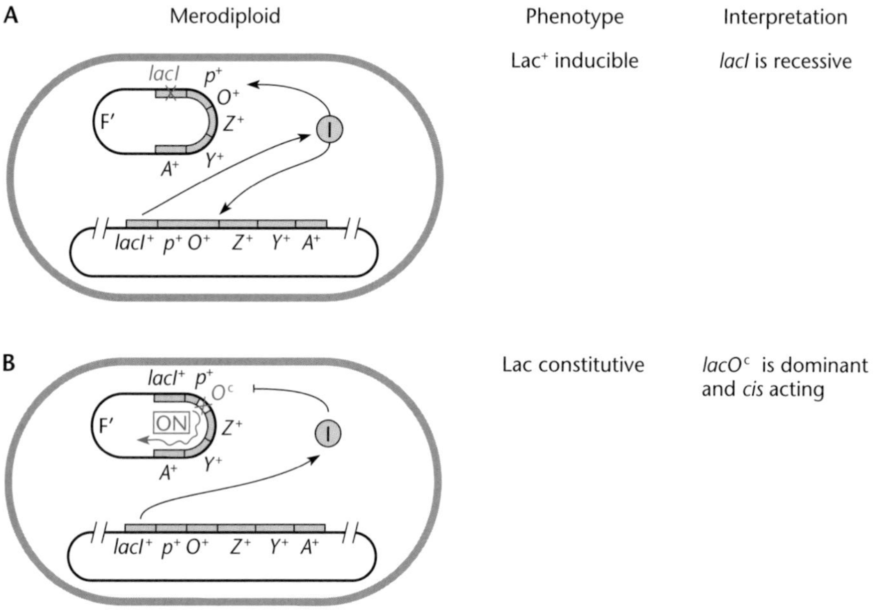

A   Merodiploid   Phenotype   Interpretation

Lac⁺ inducible   *lacI* is recessive

Figure 12.3 Complementation with two types of constitutive mutations. (A) The *lacI* mutation can be complemented, and so other genes on the prime factor will be inducible in the presence of a wild-type copy of the *lacI* gene. (B) In contrast, *lacO^c* mutations cannot be complemented by a wild-type *lacO* region in the chromosome and so are *cis* acting. Mutated regions are shown in purple.

B   Lac constitutive   *lacO^c* is dominant and *cis* acting

the *lac* operon even in the presence of a good copy of the *lacI* gene. As explained below, these *lacI*⁻ᵈ mutations are possible because the LacI polypeptides form a homotetramer. A mixture of normal and defective subunits can be nonfunctional, causing the constitutive LacI⁻ phenotype. Hence, the *lacI*⁻ᵈ mutations are *trans* dominant. Table 12.1 summarizes the behavior of the various *lac* mutations in complementation tests.

JACOB AND MONOD OPERON MODEL

On the basis of this genetic analysis, Jacob and Monod proposed their **operon model** for *lac* gene regulation (Figure 12.4). The *lac* operon includes the genes *lacZ* and *lacY*. These genes, known as the **structural genes** of the operon, encode the enzymes required for lactose utilization. The *lacZ* gene product is a β-galactosidase that cleaves lactose to form glucose and galactose, which can then be used by other pathways. The *lacY* gene product

is a permease that allows lactose into the cell. The operon also includes the *lacA* gene. The *lacA* gene product is a transacetylase, whose function is unknown. This enzyme was originally thought to be also encoded by the *lacY* gene.

Their operon model explained why the structural genes are expressed only in the presence of lactose. The product of the *lacI* gene is a repressor protein. In the

Figure 12.4 The Jacob and Monod model for negative regulation of the *lac* operon. In the absence of the inducer, lactose, the LacI repressor binds to the operator region, preventing transcription of the other genes of the operon by RNA polymerase (RNA Pol). In the presence of lactose, the repressor can no longer bind to the operator, allowing transcription of the *lacZ*, *lacY*, and *lacA* genes. It was later determined that lactose is not the true inducer but, rather, that allolactose, which is a metabolite of lactose, is the inducer.

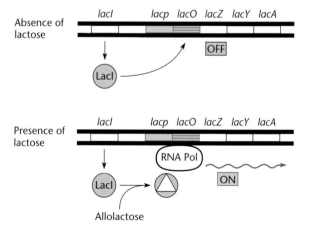

TABLE 12.1	Genetic behavior of *lac* mutations	
Mutation	**Inducibility**	**Complementation behavior**
lacZ	Noninducible	Recessive; *trans* acting
lacI	Constitutive	Recessive; *trans* acting
lacI^s	Noninducible	Dominant; *trans* acting
lacI⁻ᵈ	Constitutive	Dominant; *trans* acting
lacI^q	Inducible^a	Dominant; *trans* acting
lacO^c	Constitutive	Dominant; *cis* acting
lacP	Noninducible	Recessive; *cis* acting

ᵃ Tighter on-off control.

absence of lactose, this repressor binds to the operator sequence (*lacO*) close to the promoter and thereby prevents the RNA polymerase from binding to the promoter and blocks the transcription of the structural genes. In contrast, when lactose is available, the inducer binds to the repressor and changes its conformation so that it can no longer bind to the operator sequence. The RNA polymerase can then bind to *lacp* and transcribe the *lacZ, lacY,* and *lacA* genes. The LacI repressor is very effective at blocking transcription of the structural genes of the operon. In the absence of repressor, transcription is about 1,000 times more active than in its presence.

It is worth emphasizing how the Jacob and Monod operon model explains the behavior of mutations that affect the regulation of the *lac* enzymes. Mutants with *lacZ* and *lacY* mutations are Lac⁻ because they do not make an active β-galactosidase or permease, respectively, both of which are required for lactose utilization. These mutations are clearly *trans* acting, because they are recessive and can be complemented. An active β-galactosidase or permease made from another DNA in the same cell can provide the missing enzyme and allow lactose utilization.

The behavior of *lacp* mutations is also explained by the model. Jacob and Monod proposed that *lacp* mutations change the sequence on the DNA to which the RNA polymerase binds. The RNA polymerase normally first binds to *lacp* before it moves through *lacZ, lacY,* and *lacA,* making an RNA copy of these genes. This explains why *lacp* mutations are *cis* acting; if the site on DNA at which RNA polymerase initiates transcription is changed by a mutation so that it no longer binds RNA polymerase, the *lacZ, lacY,* and *lacA* genes on that DNA are not transcribed into mRNA, even in the presence of a good copy of the *lac* region elsewhere in the cell.

Their model also explains the behavior of the two constitutive mutations: *lacI* and *lacO*ᶜ. The *lacI* mutations affect a *trans*-acting function because they inactivate the repressor protein that binds to the operator and prevents transcription. The LacI repressor made from a functional copy of the *lacI* gene anywhere in the cell can bind to the operator sequence and block transcription in *trans*. However, the *lacO*ᶜ mutations change the sequence on DNA to which the LacI repressor binds to block transcription. The LacI repressor cannot bind to this altered *lacO* sequence, even in the absence of lactose. Therefore, the RNA polymerase is free to bind to the promoter and transcribe the structural genes. The *lacO*ᶜ mutations are *cis* acting because they allow constitutive expression of the *lacZ, lacY,* and *lacA* genes from the same DNA, even in the presence of a good copy of the *lac* operon elsewhere in the cell.

The existence of superrepressor *lacI*ˢ mutations is also explained by their model. These are mutations that

change the repressor molecule so that it can no longer bind the inducer. The mutated repressor binds to the operator even in the presence of inducer, making the cells permanently repressed and phenotypically Lac⁻. The fact that this type of mutation is dominant over the wild type is also explained. The mutated repressors repress the transcription of any *lac* operon in the same cell, even in the presence of inducer, and so they make the cell Lac⁻ even in the presence of a good *lac* operon, either in the chromosome or in an F′ factor.

The *lac* genes provide a good example of what is meant by "operon." As stated earlier, an operon includes all the genes that are transcribed into the same mRNA plus any adjacent *cis*-acting sites that are involved in the transcription or regulation of transcription of the genes. The *lac* operon of *E. coli* consists of the three structural genes, *lacZ, lacY,* and *lacA,* which are transcribed into the same mRNA, as well as the *lac* promoter from which these genes are transcribed. It also includes the *lac* operator, since this is a *cis*-acting regulatory sequence involved in regulating the transcription of the structural genes. However, the *lac* operon does *not* include the gene for the repressor, *lacI*. The *lacI* gene is adjacent to the *lacZ, lacY,* and *lacA* genes and regulates their transcription, but it is not transcribed onto the same mRNA as the structural genes. Moreover, its product is *trans* acting rather than *cis* acting.

UPDATE ON THE REGULATION OF THE *lac* OPERON

The operon model of Jacob and Monod has survived the passage of time. In 1965, it earned them a Nobel Prize, which they shared with Andre Lwoff. Because of its elegant simplicity, the operon model for the regulation of the *lac* genes of *E. coli* serves as the paradigm for understanding gene regulation in other organisms. The *lac* genes and *cis*-acting sites also have many uses in the molecular genetics of all organisms. They have been introduced into many other types of organisms, where they are used to study many aspects of gene regulation and developmental and cell biology. Some of these uses are discussed later in the section.

While still largely intact, the operon model has undergone a few refinements over the years. As mentioned, Jacob and Monod did not know of the existence of the *lacA* gene and thought that *lacY* encoded the transacetylase rather than the permease, which was unknown at the time. Also, most of the mutations that Jacob and Monod defined as *lacp* were not promoter mutations but, rather, strong polar mutations in *lacZ* that prevent the transcription of all three structural genes (*lacZ, lacY,* and *lacA*). Later studies also revealed that the true inducer that binds to the LacI repressor is not lactose itself but, rather, allolactose, a metabolite of lactose. In most experiments, an analog of allolactose called isopropyl-β-D-thiogalactopyranoside (IPTG) is used as the inducer because it is not metabolized by the cells.

The most significant refinement of the Jacob and Monod operon model came from the discovery that the LacI repressor can bind to not just one but three operators, called o_1, o_2, and o_3 (Figure 12.5). The operator closest to the promoter, o_1, seems to be the most important for repressing the transcription of *lac* and acts by sterically interfering with binding of RNA polymerase to the promoter. However, deleting both o_2 and o_3 diminishes repression as much as 50-fold.

Why does the *lac* operon have more than one operator, especially since one of the operators (o_3) is so far upstream of the promoter that it seems unlikely that it could block binding of the RNA polymerase to that site? The purpose of having more than one operator is so that the same LacI repressor molecule can bind to two operators simultaneously, bending the DNA—and the promoter—between them (Figure 12.5). The bent promoter might not be able to bind RNA polymerase or might not undergo the changes in structure required for the initiation of transcription. As discussed below, many other operons including the *gal* and *ara* operons also contain multiple operators that may also bend the DNA in the promoter region.

CATABOLITE REPRESSION OF THE *lac* OPERON

In addition to being under the control of its own specific repressor, the *lac* operon is regulated through **catabolite repression**. **Catabolites** are carbon-containing molecules that are used to build other molecules. The catabolite repression system ensures that the genes for lactose utilization are not expressed if a better carbon and energy source such as glucose is available. The name "catabolite repression" is a misnomer, at least in *E. coli*, since the expression of operons under catabolite control in *E. coli* requires a transcriptional activator, the catabolite activator protein (CAP), and the small-molecule effector, cyclic AMP (cAMP). Many operons are under the control of CAP, and we defer a detailed discussion of the mechanism of catabolite repression until chapter 13, since this is a type of global regulation. It is ironic that, even though catabolite repression by CAP was first studied in detail with the *lac* operon, it is more effective in many other operons and most of the inhibitory effect of glucose on the expression of the *lac* operon is due to inducer exclusion, keeping the inducer out of the cell, as discussed in chapter 13.

STRUCTURE OF THE *lac* CONTROL REGION

Figure 12.6 illustrates the structure of the *lac* control region in detail, showing the nucleotide sequences of the

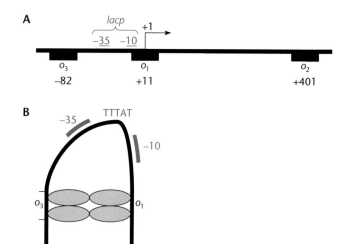

Figure 12.5 Locations of the three operators in the *lac* operon (A) and a model for how binding of the tetrameric LacI repressor to two of these operators may help repress the operon. (B) Repressor (solid ellipses) bound to o_1 and o_3, or o_2 and o_3 (not shown), could bend the DNA in the promoter region and help prevent RNA polymerase binding to the promoter. The AT-rich region may facilitate bending.

lac promoter and one of the operators (o_1) as well as the region to which CAP binds. The *lac* promoter is a typical σ^{70} bacterial promoter with the characteristic -10 and -35 regions (see chapter 2). One of the operators to which the LacI repressor binds (o_1) actually overlaps the mRNA start site ($+1$ in Figure 12.6A) for transcription of *lacZ*, *lacY*, and *lacA*. Although their position is not shown, the other *lacO* operator sequences lie nearby, and their sequence is shown in Figure 12.6B. Each symmetrical half of an operator binds a LacI monomer. The CAP-binding site that enhances initiation by RNA polymerase in the absence of glucose is just upstream of the promoter, as shown.

Fine-Structure Analysis of the *lacI* Gene of *E. coli*

Some of the most elegant early experiments in bacterial genetics involved analysis of the structure of the *lacI* gene of *E. coli* (see Miller and Schmeissner; and Schmeissner et al., Suggested Reading). These experiments demonstrated the domain structure of proteins, with different regions of the protein being dedicated to different functions of the protein.

As already outlined, the LacI repressor is a tetramer, formed from four polypeptides encoded by the *lacI* gene, and binds to the operator regions to prevent transcription of the structural genes of the operon. However, when the inducer allolactose or one of its analogs binds to the LacI

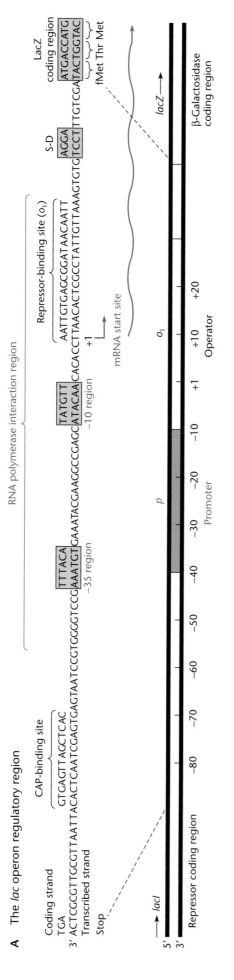

Figure 12.6 (A) DNA sequence of the promoter and operator regions of the *lac* operon. The entire region is only 100 bp long. Only the position of the o₁ (operator) sequence is shown. (B) Alignment of the three natural *lac* operator sequences. Nucleotides o₃ and o₂ that do not match o₁ are shown in lowercase.

repressor, LacI can no longer bind to the operators, and transcription ensues. Because of the many functions of *lacI*, missense mutations in *lacI* can cause various phenotypes, some of which are discussed above (Table 12.2). To reiterate, a *lacI* mutation may inactivate the repressor by preventing binding to the operators or preventing tetramer formation or both. Alternatively, it might prevent binding of the inducer. The *lacI* mutations can also be subdivided into two groups: recessive and dominant. Any *lacI* mutation that completely inactivates the repressor so that it no longer binds to the operators is recessive to the wild type and is called simply a *lacI* mutation. However, some *lacI* mutations inactivate the operator-binding domain of the repressor but do not affect the tetramerization domain and so can lead to the formation of tetramers. As mentioned above, these *lacI* mutations can be dominant to the wild type if a mixed tetramer, made up of both wild-type and mutant subunits, cannot bind to the operators. They are therefore called *lacI*$^{-d}$ mutations, where the "d" stands for "dominant." Other dominant *lacI* mutations, called *lacI*^{s} mutations, result in permanent repression, even in the presence of inducer. These mutations prevent binding of the inducer to the repressor. With some mutations, called *lacI*^{rc}, the situation is reversed. These mutations change the repressor so that it binds to the operators only in the presence of inducer.

These investigators reasoned that the position in the *lacI* gene of missense mutations that cause a particular phenotype may reveal information about which regions of the LacI protein are involved in its various functions. Accordingly, they isolated numerous different types of *lacI* missense mutations and mapped them to determine their position in the *lacI* gene.

ISOLATING DELETION MUTATIONS IN THE *lacI* GENE

The most efficient way to map a large number of mutations in a particular gene is by deletion mapping. We have already discussed deletion mapping of *rII* mutations in phage T4 (see chapter 7) and *thyA* mutations by marker rescue with M13 phage clones (see chapter 3),

and the principles are the same here. No wild-type recombinants appear in crosses between a mutant with a point mutation and a mutant with a deletion mutation if the point mutation lies within the deleted region.

The first step in deletion mapping is to obtain a large number of deletion mutations extending into different regions of the gene. One property of deletion mutations is that they can inactivate more than one gene simultaneously (see chapter 3). Therefore, one way to select *lacI* deletions is to select mutations in a nearby gene and then screen them to identify those that are also *lacI*. Most of these are deletions that extend from the nearby gene into *lacI*. The investigators decided to use the *tonB* gene to select such deletions because they had a positive selection for mutations in the *tonB* gene. We have already discussed the *tonB* gene in chapter 3 in connection with the classic experiments of Luria and Delbrück and of Newcombe on inheritance in bacteria; these investigators used *E. coli* resistance to phage T1 as the mutant phenotype in their selections. The *tonB* gene product is part of the receptor for phage T1 and some types of bacteriocins, and so mutations that inactivate the *tonB* gene make the cells resistant to these agents. Therefore, any mutant bacteria that multiply to form a colony in the presence of phage T1 have a *tonB* mutation. A subset of these *tonB* mutations are deletions, and some of these deletions could extend into the *lacI* gene if it is nearby.

Actually, there was an additional complication. The *lacI* gene is not normally near the *tonB* gene, and so it had to be moved there, by integrating a prophage carrying the *lacI* gene nearby. It also had to be moved in such a way that there were no essential genes between *tonB* and *lacI* or the deletions would also remove this intervening gene and be lethal. The genomic structure of the lysogenic strain they constructed, *E. coli* X7800, with the *lacI* gene close to *tonB*, is diagrammed in Figure 12.7. Mutant bacteria with *tonB* mutations were then selected by plating with phage T1, and those with deletions extending into *lacI* were detected by plating the *tonB* mutants on 5-bromo-4-chloro-3-indolyl-β-D-galactopyranoside (X-Gal) plates without inducer. Only constitutive *lac* mutants make blue colonies on X-Gal plates in the absence of the inducer. As illustrated in Figure 12.7, any *tonB* mutants that were also constitutive for *lac* expression probably had deletion mutations extending through *tonB* and into *lacI*, inactivating both genes simultaneously, since deletions are much more frequent than double point mutations that inactivate both genes. The deletions must end in the *lacI* gene, however, and not extend into *lacZ*, because the product of the *lacZ* gene, β-galactosidase, is required to cleave X-Gal and turn the colonies blue. The end points of the deletions in the *lacI* gene were then mapped by crossing them with a few *lacI* point

TABLE 12.2	Types of *lacI* mutations	
Mutation	**Function affected**	**Phenotype**
lacI	Operator binding or tetramer formation	Constitutive; recessive
lacI^{-d}	Operator binding	Constitutive; dominant
lacI^s	Inducer binding	Permanently repressed; dominant
lacI^{rc}	Conformational change after inducer binding	Repressed only with inducer bound; dominant

mutations that had been mapped previously by three-factor crosses.

ISOLATION OF *lacI* MISSENSE MUTATIONS

The next step in their analysis was to isolate a large number of the different types of *lacI* missense mutations so they could be mapped. To facilitate mapping, these *lacI* mutations were isolated in an F′ factor rather than in the chromosome. F′ *lac-pro*, the F′ factor used, contains the *lac* genes as well as the wild-type *proB* gene as a selectable marker. This prime factor was maintained in a strain that had the chromosomal *lac* genes deleted, so that any *lacI* mutations would occur in the F′ factor.

Isolation of different types of *lacI* missense mutations required different selection procedures. To isolate mutants with constitutive *lacI* mutations, the bacteria containing the F′ factor were mutagenized and mutants that formed blue colonies on X-Gal plates in the absence of inducer were selected. As discussed above, under these conditions only constitutive *lacI* mutants express the *lacZ* gene and form blue colonies.

Isolation of strains with *lacI*s mutations was somewhat more difficult. These mutants form colorless colonies on X-Gal plates, even with inducer, but *lacZ* mutants also form colorless colonies under these conditions and are much more common. However, unlike most other mutations that make the colonies colorless, *lacI*s mutations are dominant over the wild type. In other words, they make a strain colorless even if there is a wild-type *lac* operon in the chromosome. Therefore, the mutagenized F′ factor was mated into a strain with a functional *lac* operon in the chromosome and the transconjugants were plated on X-Gal plates with the inducer. Under these conditions, many of the cells that form colorless colonies will have a *lacI*s mutation in the F′ factor.

MAPPING OF *lacI* MISSENSE MUTATIONS

Figure 12.8 shows the process they used to map their collection of *lacI* mutations. The F′ factor containing the particular point mutation to be mapped, as well as the wild-type *proB* gene, was crossed into derivatives of *E. coli* X7800 that each contained one of the chromosomal deletions extending into the *lacI* gene, selecting for the ability to grow without proline (Pro+). The partial diploid strains containing the F′ factor were grown to allow possible recombination between the mutant *lacI* gene on the F′ factor and the partially deleted *lacI* gene in the chromosome and plated. The presence of any *lacI*+ recombinants in the population was evidence that the *lacI* mutation in the F′ factor lies outside the deleted region in the chromosome. However, any *lacI*+ recombinants, even if they exist in a particular cross, are rare and must be selected. The selection method used also depends on the type of *lacI* mutation being mapped.

Figure 12.8   Deletion mapping of mutations in the *lacI* gene of *E. coli*. Step 1: cells lacking a chromosomal *lac* region but with an F′ factor containing this region are mutagenized. Step 2: the F′ factor containing a *lacI* mutation (*lacI**) is mated into a strain with one of the *lacI* deletions isolated as in Figure 12.7, and Pro+ bacteria are selected. A few *lacI*+ recombinants appear if the deleted region in the chromosome does not extend into the region of the point mutation in the prime factor. How these *lacI*+ recombinants are selected depends on the nature of the original *lacI* mutation.

Figure 12.7   Selecting *tonB* deletions that extend into the *lacI* gene in *E. coli* X7800. Mutant bacteria that have *tonB* deletion mutations extending into *lacI* form blue colonies on X-Gal plates in the absence of inducer.

E. coli X7800

trp	tonB	lacI	lacPOZYA

Select Δ *tonB* mutations by plating on bacteriocins or phage

trp	tonB	lacI	lacPOZYA

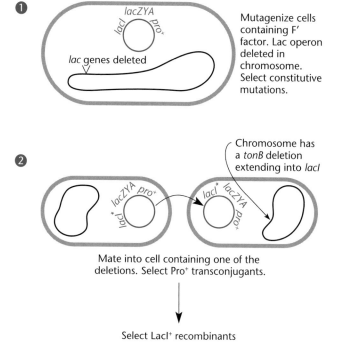

① Mutagenize cells containing F′ factor. Lac operon deleted in chromosome. Select constitutive mutations.

lacZYA

lacI

pro+

lac genes deleted

② Chromosome has a *tonB* deletion extending into *lacI*

Mate into cell containing one of the deletions. Select Pro+ transconjugants.

Select LacI+ recombinants

Mapping of *lacI* Constitutive Mutations

To detect *lacI*⁺ recombinants in crosses with mutants that had constitutive *lacI* point mutations, the investigators used the fact that *galE* mutants are killed by galactose (in other words, they are galactosemic [see chapter 3 and below]). The *lac* deletion mutant strains were made mutant for *galE* by crossing in a *galE* mutation. The investigators could then use phenyl-β-D-galactoside (P-Gal) to select *lacI*⁺ recombinants. The selection was based on the fact that the β-galactosidase product of the *lacZ* gene can cleave the galactose off P-Gal and kill the *galE* mutant strains but P-Gal itself is not an inducer of the *lac* operon. In the absence of inducer, only strains that are still constitutive for the expression of the *lacZ* gene cleave P-Gal and kill themselves whereas any *lacI*⁺ recombinants do not make β-galactosidase and hence survive. As a consequence, if P-Gal is added to the partial-diploid *galE* mutant strain, all the bacteria that are constitutive for *lac* gene expression are killed and only *lacI*⁺ recombinants multiply to form a colony. Therefore, when the transconjugants in step 2 of Figure 12.8 were plated on agar containing P-Gal, the appearance of colonies on the plates was evidence that the *lacI* point mutation lies outside the deleted region.

Mapping of *lacI*ˢ Mutations

The selection method outlined above could not be used to detect *lacI*⁺ recombinants with *lacI*ˢ mutant on their F′ plasmid. The *lacI*ˢ mutations are dominant, so the *lacZ* gene is not expressed to make β-galactosidase to cleave P-Gal and kill the *galE* mutant cells, even in partial diploids with a *lacI* deletion in the chromosome. Therefore, the transconjugants grow on the P-Gal plates regardless of whether recombination with the chromosome occurs, rendering the detection of rare *lacI*⁺ recombinants impossible.

One way to detect possible *lacI*⁺ recombinants in crosses with *lacI*ˢ mutations depends on the fact that the partial diploids with a *lacI*ˢ mutation in the F′ factor and a *lacI* deletion in the chromosome cannot use lactose because the *lacZ* and *lacY* genes cannot be induced. Therefore, if the transconjugants are plated on minimal plates with lactose as the sole carbon and energy source, any *lacI*⁺ cells in which the *lacI*ˢ gene in the F′ factor has recombined with the *lacI* deletion in the chromosome become inducible and multiply to form a colony. However, another type of recombinant also exhibits the Lac⁺ phenotype. In this recombinant, both the F′ factor and the chromosome have the *lacI* deletion, because the *lacI* deletion has been transferred to the F′ factor by recombination. The latter type of recombinant can occur even if the *lacI*ˢ mutation lies in the deleted region, but it can be distinguished from the other type because it is constitu-

tive. Therefore, to determine whether the *lacI*ˢ mutation lies outside the deleted region, only Lac⁺ recombinants that are *lacI*⁺ and not constitutive should be counted. In the experiment, the two types of Lac⁺ recombinants were distinguished by replicating the colonies onto X-Gal plates in the presence and absence of the inducer IPTG. Only the *lacI*⁺ recombinants should form blue colonies in the presence of IPTG and colorless colonies in the absence of IPTG. The *lacI* constitutive recombinants should form blue colonies on both types of plates and so can be subtracted from the total.

LOCATION OF THE VARIOUS REGIONS IN THE THREE-DIMENSIONAL STRUCTURE OF THE LacI REPRESSOR

Later, when the *lacI* gene had been sequenced, the exact amino acid changes in the *lac* repressor due to some of their mutations could be identified. The locations of the amino acid changes that cause the various phenotypes should give clues to the regions of the LacI protein that are involved in the various functions of the repressor. For example, the *lacI*ˢ mutations should be largely confined to the site on the protein where the inducer binds, allowing this site to be identified. Likewise, *lacI*⁻ᵈ mutations should not inactivate the regions involved in dimer or tetramer formation since they must form mixed dimers or tetramers with the wild-type polypeptide to be dominant. We would expect that constitutive *lacI*⁻ mutations would be scattered around the protein, since they could affect almost any activity of the protein, including its ability to bind to DNA and its ability to fold into its final conformation. However, it was found that the various mutations are often scattered around the gene and are not always concentrated in a certain region. To make sense of their distribution, it was necessary to view the protein in three dimensions since amino acids that are not in the same region in the linear polypeptide might be close to each other in the folded protein.

After many years of trying, the LacI protein was finally crystallized and its three-dimensional structure was determined by X-ray diffraction (see Lewis et al., Suggested Reading). Nuclear magnetic resonance spectroscopy was also used to determine its interaction with the *lac* operators and how this structure changes when the inducer, IPTG, is bound. Figure 12.9 shows the structure of the LacI repressor and the regions affected by amino acid changes due to some types of mutations. For example, the amino acid changes due to *lacI*ˢ mutations are often in the inducer-binding pocket (see Pace et al., Suggested Reading). The spacing between the DNA-binding N-terminal domains changes when inducer is bound, preventing its binding to the operator (not shown). The fact that inducer binding in one region of the protein can effect changes in

A

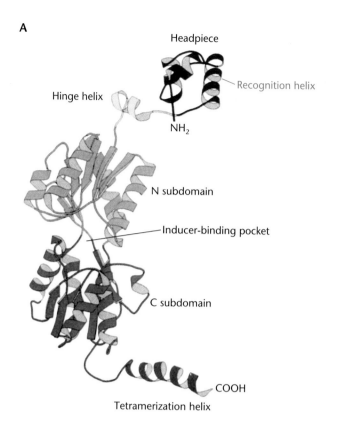

Headpiece

Recognition helix

Hinge helix

NH₂

N subdomain

Inducer-binding pocket

C subdomain

COOH

Tetramerization helix

B

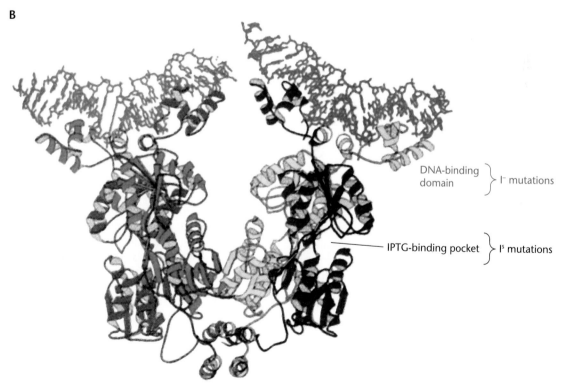

DNA-binding domain   } I⁻ mutations

IPTG-binding pocket   } Iˢ mutations

Figure 12.9   Three-dimensional structure of the LacI protein, showing both the monomer and some regions devoted to its various functions as well as the tetramer formed by two dimers and how it interacts with *lac* operators on the DNA. Also shown are the sites of some types of mutational changes mentioned in the text.

another region is called an **allosteric interaction**. Some *lacI*[s] mutations also change amino acids in the allosteric signaling region, the region which signals that inducer is bound in the inducer-binding pocket to the DNA-binding domains. This region doubles as the dimerization domain, which helps hold two monomers together, and changing the orientation of the two monomers may be part of the signal. The way in which some other repressors interact with their operators and how this interaction changes with the binding of inducers are discussed later in this chapter in the sections on the *gal* and *trp* repressors.

EXPERIMENTAL USES OF THE *lac* OPERON

The *lac* genes and regulatory regions have found many uses in molecular genetics. For example, the *lacZ* gene is probably the most widely used reporter gene and has been introduced into a wide variety of different organisms ranging from bacteria to fruit flies to human cells. It is so popular because its product, β-galactosidase, is easily detected by colorimetric assays using substrates such as X-Gal and ONPG (*o*-nitrophenyl-β-D-galactopyranoside) It is also an unusually stable protein and can be translationally fused to almost any other protein and still retain its activity. Also, the N-terminal portion of the protein is nonessential for activity, making it easier to make translation fusions. The only disadvantage of *lacZ* as a reporter gene is that its polypeptide product is very large, which can be a disadvantage in some types of expression systems.

The *lac* promoter or its derivatives are used in many expression vectors (see chapter 2). This promoter offers many advantages in these expression vectors. It is fairly strong, allowing high levels of transcription of a cloned gene. It is also inducible, which makes it possible to clone genes whose products are toxic to the cell. The cells can be grown in the absence of the inducer IPTG, so that the cloned gene is not transcribed. Only when the cells reach a high density is the inducer IPTG added and the cloned gene transcribed. Even if the protein is toxic and kills the cell when made in such large amounts, enough of the protein is usually synthesized before the cell dies.

There are many derivatives of the *lac* promoter in use. These derivatives retain some of the desirable properties of the wild-type *lac* promoter but have additional features. For example, the mutated *lac* promoter, *lacUV5*, is no longer sensitive to catabolite repression and so is active even if glucose is present in the medium (see chapter 13).

A hybrid *trp-lac* promoter called the *tac* promoter has also been widely used. The *tac* promoter has the advantages that it is even stronger than the *lac* promoter and is insensitive to catabolite repression but still retains its inducibility by IPTG.

Because it binds so tightly to its operator sequences, the LacI repressor protein also has many uses. One current use is to locate regions of DNA in the cell in bacterial cell biology. In one application, the LacI protein is translationally fused to green fluorescent protein, which can be detected in the cell by fluorescence microscopy. If this fusion protein is then expressed in a cell in which multiple copies of the operator sequence have been introduced into a region of the chromosome, for example close to the origin of replication, the fusion protein binds in multiple copies to the origin region of the chromosome. The location of the origin region of the chromosome can then be tracked in the cell as it goes through its cell cycle by monitoring the fluorescence given off by the fusion protein. We discuss such experiments in chapter 1.

PROSPECTUS

The *lac* operon is one of the simplest regulatory systems known, and so it is fortunate that it was one the first to be chosen to study. The relatively simple regulation of the *lac* operon encouraged attempts to understand other types of regulation. As discussed later in this chapter, regulation of most other operons is more complicated and would have been even more difficult to understand if the *lac* operon had not been available as a point of reference. In the next sections, we discuss the regulation of some other representative bacterial operons.

The *E. coli gal* Operon

The operon of *E. coli* involved in the utilization of the sugar galactose, the *gal* operon, is another classic example of negative regulation. Figure 12.10 shows the organization of the genes in this operon. The products of three structural genes, *galE*, *galT*, and *galK*, are required for the utilization of galactose and convert galactose into glucose, which can then enter the glycolysis pathway.

The specific reactions catalyzed by each of the enzymes of the *gal* pathway appear in Figure 12.11. The *galK* gene product is a kinase that phosphorylates galactose to make galactose-1-phosphate. The product of the *galT* gene is a transferase that transfers the galactose-1-phosphate to UDPglucose, displacing the glucose to make UDPgalactose. The released glucose can then be used as a carbon and energy source. The GalE gene product is an epimerase that converts UDPgalactose to UDPglucose to continue the cycle. It is not clear why *E. coli* cells use this seemingly convoluted pathway to convert galactose to glucose so that the latter can be used as a carbon and energy source. However, many organisms, including both plants and animals, use this pathway.

Unlike the genes for lactose utilization, not all the genes for galactose utilization are closely linked in the *E. coli* chromosome. The *galU* gene, whose product synthesizes UDPglucose, is located in a different region of

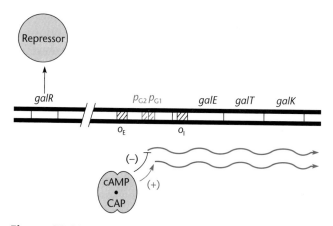

Figure 12.10 Structure of the galactose operon of *E. coli*. The *galE, galT,* and *galK* genes are transcribed from two promoters, p_{G1} and p_{G2}. CAP with cAMP bound turns on p_{G1} and turns off p_{G2}, as shown. There are also two operators, o_E and o_I. The repressor genes are some distance away, as indicated by the broken line. Only the *galR* repressor gene is shown.

Galactose + ATP $\xrightarrow[\text{GalK}]{}$ galactose-1-PO_4

Galactose-1-PO_4 + UDPglucose $\xrightarrow[\text{GalT}]{}$ UDPgalactose + glucose

UDPgalactose $\xrightarrow[\text{GalE}]{}$ UDPglucose

UTP + glucose-1-PO_4 $\xrightarrow[\text{GalU}]{}$ UDPglucose

Figure 12.11 Pathway for galactose utilization in *E. coli*. See the text for details.

the chromosome. Also, the genes for the galactose permeases, which are responsible for transporting galactose into the cell, are not part of the *gal* operon. Another difference with the *lac* operon is that not only are there two repressor genes, *galR* and *galS*, but also they are nowhere near the operon, unlike the *lacI* gene, which is adjacent to the *lac* operon. This scattering of the genes for galactose metabolism reflects the fact that galactose not only serves as a carbon and energy source but also plays other roles. For example, the UDPgalactose synthesized by the *gal* operon donates galactose to make polysaccharides for lipopolysaccharide and capsular synthesis.

REGULATION OF THE *gal* OPERON

Everything about the *gal* operon comes in twos. There are two promoters and two operators. There are even two repressors, encoded by different genes. Either of these repressors can repress the *gal* operon, although one is more effective than the other.

Two *gal* Repressors: GalR and GalS

The two repressors in control of the *gal* operon are GalR and GalS, encoded by the *galR* and *galS* genes, respectively. GalR was discovered first because mutations in *galR* cause constitutive expression of the *gal* operon. However, it was apparent that the *gal* operon is also subject to other regulation. If the GalR repressor were solely responsible for regulating the *gal* operon, mutations that inactivate the *galR* gene should result in the same level of *gal* expression whether galactose is present or not. Yet some regulation of the *gal* operon could be observed,

even in *galR* mutants. When galactose was added to the medium in which *galR* mutant cells were growing, more of the enzymes of the *gal* operon were made than if the cells were growing in the absence of galactose. The product of another gene, *galS*, was responsible for the residual regulation. As evidence, double mutants with mutations that inactivate both *galR* and *galS* are fully constitutive. Later studies showed that the product of the *galS* gene is also a repressor that negatively regulates the *gal* operon.

The GalS and GalR repressor proteins are closely related, and they both bind the inducer galactose. Even so, they may play somewhat different roles. The GalR repressor is responsible for most of the repression of the *gal* operon in the absence of galactose. The GalS repressor plays only a minor role in regulating the *gal* operon but solely controls the genes of the galactose transport system, which transports galactose into the cell. The reason for this two-tier regulation is unclear but also may be related to the diverse roles of galactose in the cell (see above).

Two *gal* Operators

There are also two operators in the *gal* operon. One is upstream of the promoters, and the other is internal to the first gene, *galE* (Figure 12.10). The two operators are named o_E and o_I for operator *external* to the *galE* gene and operator *internal* to the *galE* gene, respectively. The discovery of the o_I operator involves some interesting genetics, so we discuss it in some detail.

Isolating gal *operator mutants.* The first mutant with an o_I mutation was isolated as part of a collection of constitutive mutants of the *gal* operon (see Irani et al., Suggested Reading). These mutants are easier to isolate in strains with superrepressor *galR^s* mutations than in wild-type *E. coli*. The *galR^s* mutations are analogous to *lacI^s* mutations. The superrepressor mutation will make a strain Gal⁻ and uninducible because galactose cannot bind to the mutated repressor. Therefore, *E. coli* with a *galR*ˢ mutation cannot multiply to form colonies on plates containing only galactose as the carbon and energy source. However, a constitutive mutation that inactivates the GalRˢ repressor or changes the operator sequence will prevent the mutant repressor from binding to the operator and allow the cells to use the galactose and multiply to form a colony. Thus, if bacteria with a *galR*ˢ mutation are plated on medium with galactose as the sole carbon and energy source, only constitutive mutants multiply to form a colony. However, most of the constitutive mutants isolated this way have mutations in the *galR* gene that inactivate the GalRˢ repressor rather than operator mutations, since the operator is by far the smaller target. Many *galR* mutants would have to be screened before a single operator mutant was found. Therefore, to make this method practicable for isolating constitutive mutants with operator mutations, the frequency of *galR* mutants must be decreased until it is not too much higher than that of constitutive mutants with mutations in the operator sequences.

One way to reduce the frequency of *galR* mutants is to use a strain that is a partial diploid for (has two copies of) the *galR*ˢ gene. Then, even if one *galR*ˢ gene is inactivated by a mutation, the other *galR*ˢ gene continues to make the GalRˢ protein, making the cell phenotypically Gal⁻. Only two independent mutations, one in each *galR*ˢ gene, can make the cell constitutive. Since the frequency of two independent mutations is the product of the frequency of each of the single mutations, the presence of two independent *galR* mutations should be very rare, probably no more frequent than single operator mutations, making cells with operator mutations a significant fraction of the total constitutive mutants and easier to identify. Moreover, constitutive mutants with operator mutations can be distinguished from the double mutants with mutations in both the *galR*ˢ genes by the locations where they map. Operator mutations map in the *gal* operon, unlike mutations in the *galR*ˢ genes, which map elsewhere in the genome.

Accordingly, a partial diploid that had one copy of the *galR*ˢ gene in the normal position and another copy in a specialized transducing λ phage integrated at the λ attachment site was constructed (see chapter 8). When this strain was plated on medium containing galactose as the sole carbon and energy source, a few Gal⁺ colonies arose due to constitutive mutants. The mutations in two of these constitutive mutants mapped in the region of the *gal* operon and so were presumed to be operator mutations. When the DNA of the two mutants was sequenced, it was discovered that one mutation had changed a base pair in the known operator region (o_E, for operator *external*), just upstream of the promoters as expected. However, the other operator mutation had changed a base pair

downstream in the *galE* gene, suggesting that a sequence in that gene also functions as an operator. This operator was named o_I, for operator *internal* to the *galE* gene. Furthermore, this mutation occurred in a sequence homologous to the 15 bp making up the known operator o_E. In fact, 12 bp of this sequence is identical in o_I and o_E. Moreover, the mutation in the *galE* gene was *cis* acting for constitutive expression of the *gal* operon, one of the criteria for an operator mutation.

Escape synthesis of the Gal enzymes. As mentioned, the o_I mutations in the *galE* gene cause *cis*-acting constitutive expression of the *gal* genes, suggesting that the o_I sequence functions as an operator. However, the constitutive mutants could be explained in other ways.

If the sequence in the *galE* gene truly is an operator, it would bind the GalR repressor protein. To test this, the experiment illustrated in Figure 12.12 was performed. First, DNA containing the *galE* gene is cloned into a multicopy plasmid. When this multicopy plasmid is transformed into a cell, that cell contains many copies of the *galE* gene. Then, if the sequence in the *galE* gene does bind the repressor, these extra copies should bind most of the GalR protein in the cell, leaving too little to completely repress expression of the *gal* operon. Thus, the cells would appear to be constitutive mutants. This general method is called **titration**, and the enzymes of the *gal* operon are synthesized through **escape synthesis,** so named since the operon is "escaping" the effects of the repressor. In the actual experiment, cells containing many copies of the *galE* gene region did exhibit a partially constitutive phenotype. In contrast, multiple copies of mutant DNA with the putative operator mutation (i.e., o_I^C) do not cause escape synthesis of the Gal enzymes. This result was interpreted to confirm the presence in *galE* of a second binding site for repressor, which is inactivated by the mutation.

Why does the gal *operon have two operators?* There are two general hypotheses for why the *gal* operon has two operators. According to one, the two operators function independently to block transcription of the *gal* operon. The other proposes that the operators cooperate to block transcription. Genetic evidence supports the second general hypothesis—the two operators cooperate to block transcription. If the two operators functioned independently, the effect of mutations in both operators would be additive; in other words, the level of expression of the genes in the operon when both operators were mutated would be the sum of the levels of operon expression when each of the operators was mutated separately. However, genetic experiments showed that the level of expression in the double mutant is lower than the

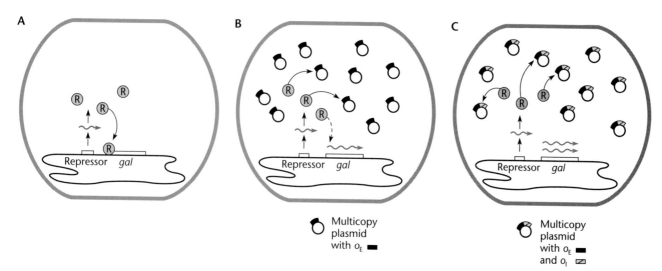

Figure 12.12 Escape synthesis of the enzymes of the galactose operon caused by additional copies of the operator regions. Clones of the operator regions in a multicopy plasmid dilute out the repressor, inducing the operon even in the absence of galactose. (A) The cell contains the normal number of operators, and the operon is not induced. (B) The multicopy plasmid contains only o_E, and the operon is only partially induced. (C) The multicopy plasmid contains both o_E and the *galE* gene containing o_I, and the operon is fully induced.

sum of the expression levels in each single mutant. This observation indicated some cooperation between the two operators.

Figure 12.13 shows a model for how the two operators cooperate to block transcription in the absence of galactose. Repressor molecules bound to the two operators interact with each other to bend the DNA of the promoter that is between the two operators. The bent promoter does not bind RNA polymerase, and so there is no initiation of RNA synthesis at the promoter. Repressor bound to only one of the two operators might still interfere with transcription to some extent, but the promoter would not be bent and the repression would be much less severe. A similar model had been proposed for the *ara* operon, and another was proposed later for the extra *lac* operators (Figure 12.5).

This model leads to a specific prediction: the spacing in the DNA between the two operators should be important for the repression. Double-stranded DNA is quite stiff over short distances, making it difficult to twist. Therefore, the two repressor molecules in the dimer must be able to bind to each other and to bind to the operator sites without twisting the DNA significantly. The more they have to twist the DNA between them to bind, the less strongly they bind and the less severe the repression. However, to calculate how far apart they should be to minimize twisting and optimize repression, we must make a number of assumptions. First we have to make the assumption of how the repressor molecules bind

Figure 12.13 Repressor molecules (circled Rs) bound to the two operators of the *gal* operon can interact more easily if they are bound on opposite sides of the helix. (A) Multiples of 10 bp separating o_E and o_I allow GalR binding to opposite sides of the helix. (B) Arrows denote twisting of the molecule that is necessary when the operators are a multiple of 10 plus 5 bp apart. See the text for details.

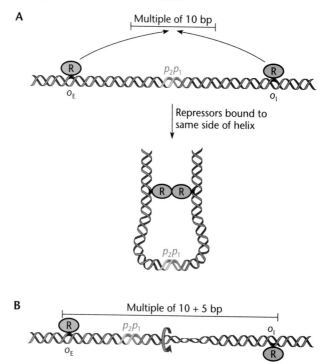

to the DNA. Presumably, they specifically recognize the sequence of DNA in the operator and bind to it. However, the operator sequences are partially palindromic, meaning that the sequence reads the same in the 5′-to-3′ direction on both strands with a few exceptions (see chapter 1). This means that the repressor sees both faces of the double-stranded DNA helix as looking the same in the operator regions but with opposite polarity. Do both repressor molecules in the dimer bind to the same side of the two operators (in the parallel orientation) or to the opposite sides (the antiparallel orientation)? We will assume that the repressor molecules bind to the operators in an antiparallel fashion, meaning that one copy of the repressor binds to the operator on one face of the helix and the other copy of the repressor in the dimer binds to the other operator on the other face (otherwise why would the operators be palindromic?). The same face of the double helix appears every multiple of 10 bp (actually 10.5 bp) on average, and the opposite face appears every multiple of 10 plus 5 bp. Therefore, the opposite faces of the operator should show up on the same side of the helix, and repression should be optimal if they are a multiple of 10 plus 5 bp apart (in other words, 15, 25, 35, etc., bp apart). Figure 12.13 shows a simple picture to illustrate this bending and why one would predict this spacing.

In an experiment to determine the effect of the spacing on repression, the two operators were moved farther apart by inserting extra DNA sequences between them. The results were dramatic and strongly supported the general model that the spacing mattered but the predicted optimal spacing was wrong. The two operators still functioned optimally if they were moved farther apart, but only if the spacing was increased in multiples of 10 bp. If the spacing was changed to a multiple of 10 plus 5 bp, the cells were partially constitutive, suggesting that the DNA was being twisted so that the two repressor molecules in the dimer could not bind as well. Also, the two operators in the *gal* operon are normally a multiple of 10 bp apart, before their spacing is altered. The riddle of why the optimal spacing is a multiple of 10 bp was solved with the discovery that another protein, called HU (for histone-like), binds between the two operators and introduces a 180° turn in the DNA, which puts the two operators in the antiparallel configuration in a multiple of 10 bp, so that the repressor dimers can be bound to opposite sides of them without twisting. Further work has confirmed this model and has contributed some details to the looping structure (see Geanacopoulos et al., Suggested Reading) (Figure 12.14). Also, the repressor bound is actually a tetramer, with two dimers of the repressor bound to each other as shown. This large structure with the loop of promoter DNA held by the repressor was named the repressosome.

Two Gal Promoters and Catabolite Repression of the *gal* Operon

As mentioned, the *gal* operon also has two promoters called p_{G1} and p_{G2} (Figure 12.10). The *gal* operon may have two promoters because, unlike *lac*, the enzymes are needed even when a better carbon source is available, since they are involved in making polysaccharides as well as in utilizing galactose (see above). One of these promoters is like the *lac* promoter in that it is regulated by catabolite repression so that it is repressed if a better carbon source such as glucose is available. However, the other promoter is active even in the presence of glucose and continues making the Gal enzymes so that other cellular constituents containing galactose can continue to be made. The differential regulation of the two *gal* promoters is discussed in more detail in chapter 13 under catabolite repression.

Negative Regulation of Biosynthetic Operons: Aporepressors and Corepressors

The enzymes encoded by the *lac* and *gal* operons are involved in degrading compounds to obtain catabolites in order to build other molecules. Consequently, these

Figure 12.14 Formation of the *gal* operon repressosome. Two dimers of the *galR* repressor gene product bind to each other and to the operators o_E and o_I (A) to bend the DNA between the operators. (B) A histone-like DNA binding protein, HU, introduces a 180° twist in the DNA (shown by the arrow), so that the repressor can bind to the two operators in an antiparallel configuration. Bending of the DNA in the promoter regions inactivates the promoters. See the text for details.

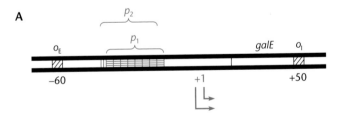

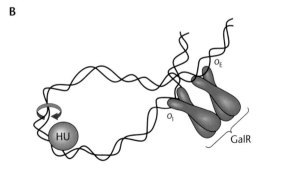

operons are called **catabolic operons** or **degradative operons.** Not all operons are involved in degrading compounds, however. The enzymes encoded by some operons synthesize compounds needed by the cell, such as nucleotides, amino acids, and vitamins. These operons are called **biosynthetic operons.**

The regulation of a biosynthetic operon is essentially opposite to that of a degradative operon. The enzymes of a biosynthetic pathway should *not* be synthesized in the presence of the end product of the pathway, since the product is already available and energy should not be wasted in synthesizing more. However, the mechanisms by which degradative and biosynthetic operons are regulated are often similar. Biosynthetic operons can also be regulated negatively by repressors. If the genes of a biosynthetic operon are constitutively expressed in the absence of the regulatory gene product, even if the compound is present in the medium, the biosynthetic operon is negatively regulated.

The terminology used to describe the negative regulation of biosynthetic operons differs somewhat from that used for catabolic operons, despite shared principles. The effector that binds to the repressor and allows it to bind to the operators is called the **corepressor.** A repressor that negatively regulates a biosynthetic operon is not active in the absence of the corepressor and in this state is called the **aporepressor.** However, once the corepressor is bound, the protein is able to bind to the operator and so is now called the repressor.

THE *trp* OPERON OF *E. COLI*

The tryptophan (*trp*) operon of *E. coli* is the classic example of a biosynthetic operon that is negatively regulated by a repressor. The enzymes encoded by the *trp* operon (Figure 12.15) are responsible for synthesizing the amino acid L-tryptophan, which is a constituent of most proteins and so must be synthesized if none is available in the medium. The products of five structural genes in the operon are required to make tryptophan from chorismic acid. These genes are transcribed from a single promoter, p_{trp}, shown in Figure 12.15. The *trp* operon is negatively regulated by the TrpR repressor protein, whose gene, like the *gal* repressor gene, is unlinked to the rest of the operon. Also, like the *galR* gene, this may reflect the fact that TrpR regulates more than one operon. In addition to the *trp* operon, it regulates the *aroH* operon to make chorismate and it is an autoregulator; i.e., it regulates its own gene, *trpR* (see below).

Figure 12.16 shows the model for the regulation of the *trp* operon by the TrpR repressor. By binding to the operator, the TrpR repressor can prevent transcription from the p_{trp} promoter. However, the TrpR repressor can bind to the operator only if the corepressor tryptophan is present in the medium. The tryptophan binds to the TrpR aporepressor protein and changes its conformation so that it can bind to the operator.

The TrpR repressor has been crystallized and its structure has been determined in both the aporepressor form, when it cannot bind to DNA, and the repressor form with tryptophan bound, when it can bind to the operator. These structures have led to a satisfying explanation of how tryptophan corepressor binding changes the TrpR protein so that it can bind to the operator (Figure 12.17). The TrpR repressor is a dimer, and each copy of the *trpR* polypeptide has α-helical structures (shown as cylinders). Helices D and E form the helix-turn-helix (HTH) DNA-binding domain (see Box 12.1). Helix D corresponds to helix 1, the nonspecific DNA-binding helix, and helix E corresponds to helix 2, the DNA sequence specific recognition helix. In the aporepressor state, the conformations of the two HTH domains in the dimer do not allow proper interactions with successive major

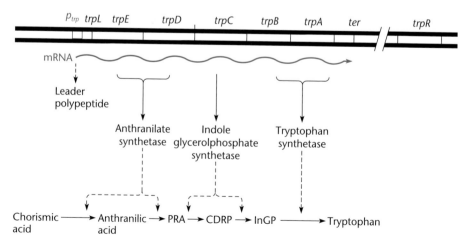

Figure 12.15 Structure of the tryptophan biosynthetic (*trp*) operon of *E. coli*. The structural genes *trpEDCBA* are transcribed from the promoter p_{trp}. Upstream of the structural genes is a short coding sequence for the leader peptide called *trpL*. The *trpR* repressor gene is unlinked, as shown by the broken line. PRA, phosphoribosyl anthranilate; CDRP, 1-(*o*-carboxyphenylamino)-1-deoxyribulose-5-phosphate.

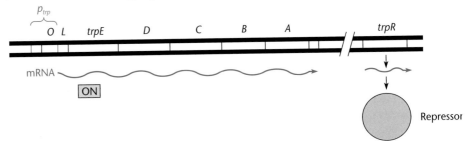

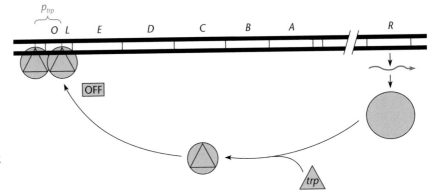

Figure 12.16 Negative regulation of the *trp* operon by the TrpR repressor. See the text for details.

grooves appearing on one side of the DNA helix. Binding of the typtophan corepressor alters the HTH conformations, allowing repressor to bind to the operators.

Autoregulation of the *trpR* Gene

As mentioned, the TrpR repressor negatively regulates not only the transcription of the *trp* operon but also the transcription of its own gene, *trpR*. In the absence of the TrpR protein, the transcription of the *trpR* gene is about five times higher than in the presence of TrpR. When the product of a gene regulates the expression of its own gene, it is called **autoregulation.**

It is perhaps surprising that the TrpR repressor would negatively regulate the transcription of its own gene, since there would be less repressor in the cell when tryptophan is present than when tryptophan is absent. It would seem advantageous to have more repressor present when tryptophan is present to better repress the *trp* operon. One possible answer to this riddle is that through negative autoregulation of transcription of the *trpR* gene, the cell ensures that repression can be established more quickly if tryptophan suddenly appears in the medium.

Isolation of *trpR* Mutants

Like other negatively regulated operons, constitutive mutations of the *trp* operon are quite common and most map in *trpR*, inactivating the product of the gene.

Mutants with constitutive mutations of the *trp* operon can be obtained by selecting for mutants resistant to the tryptophan analog 5-methyltryptophan in the absence of tryptophan. This tryptophan analog also binds to the TrpR repressor and acts as a corepressor. However, 5-methyltryptophan cannot be used in place of tryptophan to make active proteins. Therefore, in the presence of the analog, the *trp* operon is not induced even in the absence of tryptophan and the cells will starve for this amino acid. Only constitutive mutants that continue to express the genes of the operon in the presence of 5-methyltryptophan can multiply to form colonies on plates with this analog but without tryptophan.

Other Types of Regulation of the *trp* Operon

The *trp* operon is also subject to a completely different type of regulation called **attenuation.** This type of regulation is discussed later in the chapter. Also, as in many biosynthetic pathways, the first enzyme of the *trp* pathway is subject to **feedback inhibition** by the end product of the pathway, tryptophan. We also return to feedback inhibition later in the chapter.

Positive Regulation

The first part of the chapter covered the classic examples of negative regulation by repressors, but many operons are regulated positively by activators. An operon under

A Aporepressor dimer

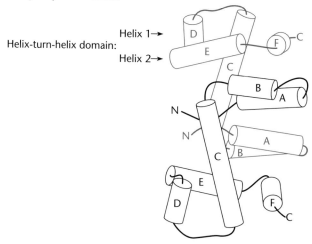

Helix-turn-helix domain:

Helix 1→

Helix 2→

B Aporepressor-repressor conformational change

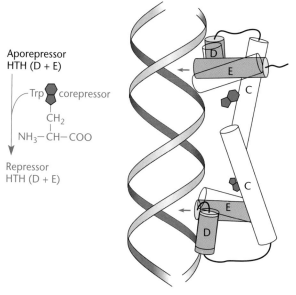

Aporepressor
HTH (D + E)

Trp ⬡ corepressor

CH₂
|
NH₃−CH−COO

Repressor
HTH (D + E)

Figure 12.17 Structure of the TrpR repressor and an illustration of how tryptophan binding allows it to convert from the aporepressor to the repressor that binds to the operator. (A) The helices (shown as cylinders) of the aporepressor dimer in the inactive state with no tryptophan bound. (B) The purple cylinders represent the HTH domains of the active repressor with tryptophan bound.

the control of an activator protein is transcribed only in the presence of that protein with the inducer bound. The next portion of the chapter is devoted to positive regulation in bacteria.

The *E. coli* L-*ara* Operon

The L-*ara* operon was the first example of positive regulation in bacteria to be discovered. The L-*ara* operon, usually called the *ara* operon, is responsible for the

utilization of the five-carbon sugar L-arabinose. The genes of this operon are responsible for converting L-arabinose into D-xylulose-5-phosphate, which can be used by other pathways. *E. coli* can also utilize D-arabinose, an isomer of L-arabinose, but the enzymes for D-arabinose utilization are encoded by a different operon, which lies elsewhere in the chromosome.

Figure 12.18A illustrates the structure of the *ara* operon. Three structural genes in the operon, *araB*, *araA*, and *araD*, are transcribed from a single promoter, p_{BAD}. Upstream of the promoter is the activator region, *araI*, where the activator protein AraC binds to activate transcription in the presence of L-arabinose, and the CAP site, at which CAP binds. There are also two operators, $araO_1$ and $araO_2$, at which the AraC protein binds to repress transcription. The *araC* gene, which encodes the regulatory protein, is also shown. This gene is transcribed from the promoter p_C in the opposite direction from *araBAD*, as shown by the arrows in the figure. As described below, the AraC protein is a positive activator of transcription. As such, it is a member of a large family of activator proteins (Box 12.2).

GENETIC EVIDENCE FOR POSITIVE REGULATION OF THE *ara* OPERON

Early genetic evidence indicated that the *lac* and *ara* operons are regulated by very different mechanisms (see Englesberg et al., Suggested Reading). One observation was that loss of the regulator proteins results in very different phenotypes. For example, deletions and nonsense mutations in the *araC* gene—mutations that presumably inactivate the protein product of the gene—lead to a "superrepressed" phenotype in which the genes of the operon are not expressed, even in the presence of the inducer arabinose. Recall that deletion or nonsense mutations in the regulatory gene of a negatively regulated operon such as *lac* result in a constitutive phenotype, not a superrepressed phenotype. Another difference between *ara* and negatively regulated operons is in the frequency of constitutive mutants. Mutants that constitutively express a negatively regulated operon are relatively common because any mutation that inactivates the repressor gene causes constitutive expression. However, mutants that constitutively express *ara* are very rare, which suggests that mutations that result in the constitutive phenotype do not merely inactivate AraC.

Isolating Constitutive Mutations of the *ara* Operon

Because constitutive mutations of the *ara* operon are so rare, special tricks are required to isolate them. One method for isolating rare constitutive mutations in *araC* uses the anti-inducer D-fucose. This anti-inducer binds to the AraC protein and prevents it from binding

L-arabinose, thereby preventing induction of the operon. As a consequence, wild-type *E. coli* cannot multiply to form colonies on agar plates containing D-fucose with L-arabinose as the sole carbon and energy source. Only mutants that constitutively express the genes of the *ara* operon can form colonies under these conditions.

Another selection for constitutive mutations in the L-arabinose operon cleverly plays off the operon responsible for the utilization of its isomer, D-arabinose. The enzymes produced by the L- and D-*ara* operons cannot use each other's intermediates, with one exception. The product of the *araB* gene (the ribulose kinase enzyme of the L-*ara* operon pathway, which phosphorylates L-ribulose as the second step of the pathway) can also phosphorylate D-ribulose, so that the L-*ara* kinase can substitute for that of the D-*ara* operon. Nevertheless, *E. coli* mutants that lack the D-*ara* kinase cannot multiply to form a colony on plates containing only D-arabinose as a carbon and energy source, because D-arabinose is not an inducer of the L-*ara* operon. Only constitutive mutants of the L-*ara* operon can grow if D-*ara* kinase-deficient mutants are plated on agar plates containing D-arabinose as the sole carbon and energy source.

A MODEL FOR THE POSITIVE REGULATION OF THE *ara* OPERON

The contrast in phenotypes between mutations that inactivate the *lacI* and *araC* genes led to an early model for the regulation of the *ara* operon. According to this early model, the AraC protein can exist in two states, called P1 and P2. In the absence of the inducer, L-arabinose, the AraC protein is in the P1 state and inactive. If L-arabinose is present, it binds to AraC and changes the protein conformation to the P2 state. In this state, AraC binds to the DNA at the site called *araI* (Figure 12.18A) in the promoter region and activates transcription of the *araB*, *araA*, and *araD* genes.

This early model explained some, but not all, of the behavior of the *araC* mutations. It explained why mutations in *araC* that cause the constitutive phenotype are rare but do occur at a very low frequency. According to this model, these mutations, called *araC*[c] mutations, change AraC so that it is permanently in the P2 state, even in the absence of L-arabinose, and thus the operon is always transcribed. Such mutations would be expected to be very rare because only a few amino acid changes in the AraC protein could specifically change the conformation of the AraC protein to the P2 state.

Figure 12.18 (A) Structure and function of of the L-arabinose operon of *E. coli*. (B) Binding of the inducer L-arabinose converts the AraC protein from an antiactivator P1 form to an activator P2 form. See the text for details.

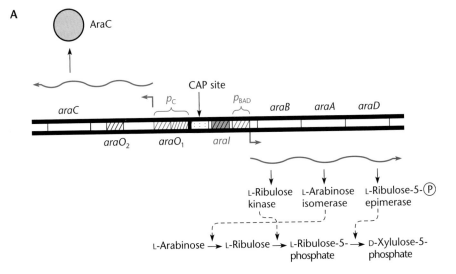

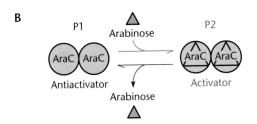

BOX 12.2

Families of Regulators

The techniques of comparative genomics have made it possible to identify repressor and activator genes in a wide variety of bacteria. These transcriptional regulatory proteins belong to a limited number of families, based on sequence and structural conservation, even though they regulate operons of very different functions and respond to different effectors. Transcriptional regulators can be assigned to a family based on sequence and structural homology, the organization of their motifs, and whether they use the same motif to bind to DNA. Many of them use a helix-turn-helix (HTH) motif or a winged HTH motif (Box 12.1), or they can use a looped-hinge helix or a zinc finger motif, among others. There are at least 15 different families of activators, many of which, such as AraC, can also function as repressors. Some of these families are quite large, with dozens of known members. Some families consist of only repressors, others consist of only activators, and some consist of both repressors and activators. Most of the families are named after the first member of the family to be studied. For example, the LacI family includes GalR and consists mostly of repressors that regulate operons involved in carbon source utilization and generally also respond to CAP and cAMP. These repressors usually function as homotetramers, have an HTH DNA-binding motif to bind to DNA at their N terminus, an effector-binding motif in the middle, and dimerization and tetramerization motifs at their C terminus.

The TetR family of repressors, named after the repressor that regulates the tetracycline resistance gene in Tn*10,* is even larger, with almost 100 known members so far, including LuxR, which regulates light emission in chemiluminescent bacteria (see chapter 13). These family members are found in both gram-negative and gram-positive bacteria, and they repress operons involved in a variety of functions including antibiotic resistance and synthesis, osmotic regulation, efflux pumps for multidrug resistance, and virulence genes in pathogenic bacteria. The TetR repressor itself is widely used to regulate gene expression in eukaryotic cells because it binds very tightly to its operators and because tetracycline readily diffuses into eukaryotic cells. TetR has an HTH motif that binds to DNA, but only when the repressor has not bound tetracycline, and partially unwinds the operator DNA through its major groove. Presumably this can change the structure of the DNA at the promoter. Interestingly, a member of this family that regulates a multidrug efflux pump has a very broad effector-binding pocket, allowing it to bind a variety of antibiotics.

The AraC family of activators is also very large, with more than 100 known probable members, based on sequence similarities. These activators seem to fall into at least two subfamilies, those that regulate carbon source utilization, like AraC and the XylS activator protein of the Tol plasmid of *Pseudomonas putida*, described in this chapter, and function as dimers; and those that respond to stress responses, like SoxS, and function as monomers. A signature of this family is that the HTH motif that binds to DNA is in the C-terminal part of the protein, not in the N terminus like many activators.

Other large classes are the LysR activators and the NtrC activators. The NtrC activators are particularly interesting in that they activate transcription only from σ^{54} promoters (the nitrogen sigma [see chapter 13]). They include the XylR activator of the *tol* operon, discussed in this chapter. These activators are organized into distinguishable domains (see also Box 13.3); the domain of the activator protein that either binds the inducer or is phosphorylated is located at the N terminus, and the DNA-binding domain is located at the C terminus. The middle region of the polypeptide contains a region that interacts with RNA polymerase and has an ATPase activity required for activation.

Experiments with hybrid activators, made by fusing the C-terminal DNA-binding domain of one activator protein to the N-terminal inducer-binding domain of another activator protein from the same family, provide a graphic demonstration that members of a family of regulators all use the same basic strategy to activate transcription of their respective operons (see Parek et al., below). Sometimes, such hybrid activators can still activate the transcription of an operon, but the operon that is activated by the hybrid activator depends on the source of the C-terminal DNA-binding domain, while the inducer that induces the operon depends on the source of the N-terminal inducer-binding domain. This leads to a situation where an operon is induced by the inducer of a different operon. It is intriguing to think that all activator proteins may have evolved from a single precursor protein through simple changes in its effector-binding and, to some extent, its DNA-binding regions, yet they continue to activate the RNA polymerase by the same basic mechanism.

Regulatory proteins use almost every conceivable mechanism to regulate trancription. Repressors act on essentially every step required for initiation of transcription, although many affect more than one step. Some repressors act, at least in part, by preventing the binding of RNA polymerase to the promoter either by getting in the way (steric hindrance) or by bending the DNA at the promoter. They can also act by preventing RNA polymerase from separating the strands of DNA at the promoter (open-complex formation) or even hindering the ability of RNA polymerase to move out of the promoter and begin making RNA (promoter escape). In one particularly

(continued)

BOX 12.2 (continued)

Families of Regulators

illustrative case, shown in Figure 1, the repressor/activator protein P4 of a *Bacillus subtilis* phage represses an already strong promoter by making it so strong that the RNA polymerase cannot escape it and begin transcription. Figure 1A shows how its binding to a sequence at −82 relative to the start site of a promoter, A3, activates transcription from that promoter (shown as ON in the figure), while its binding to a sequence at −71 relative to the start site of another promoter, A2c, inhibits transcription from that promoter (shown as OFF in the figure).

Activators can also act at any of the steps of transcription initiation. Many of them "recruit" the RNA polymerase to the promoter by binding both to the DNA close to the promoter and to the RNA polymerase, thereby stabilizing the binding of the RNA polymerase to the promoter. Figure 2 gives some examples of how activators and sequences around a promoter can recruit RNA polymerase to promoters. Figure 2A shows the RNA polymerase binding to the −35 and −10 regions of a σ⁷⁰ promoter. Figure 2B shows how binding of the αCTD domains (the C-terminal domains of the α subunits of RNA polymerase) to a sequence upstream of the promoter called an UP element can stabilize the binding. Figure 2C and D shows how CAP (CRP in the figure) bound at different sites upstream of a promoter can make contact with different regions of the RNA polymerase and stabilize its binding. In Figure 2C, CAP is bound further upstream and makes contact with one αCTD domain. In Figure 2D, it is bound closer to the start site and can make contact with one of the αNTDs

(N-terminal domains of the α subunit) as well as the αCTDs. Figure 2E shows how the C1 protein of λ phage (shown as dumbbells [see chapter 8]) can activate transcription from the p_{RM} promoter by binding cooperatively to the operators o^2_R and o^1_{R}, from where they can contact both σ⁷⁰ and the αCTDs. Some activators such as SoxS may even bind to the RNA polymerase before it binds to the promoter (Figure 3A). Only after the SoxS activator has bound to it can the RNA polymerase bind to the promoter (Figure 3B). Others, such as the NtrC-type activators, do not recruit the RNA polymerase but allow an RNA polymerase already bound at the promoter to open the DNA to form an open complex (see chapter 13). In one newly discovered mechanism, illustrated in Figure 4, an activator of the MerR family actually remodels the promoter by "scrunching" it to make it stronger and activate transcription (see Huffman and Brennan, below). Normally the spacing between the −35 sequence and the −10 sequence in a σ⁷⁰-type promoter is 17 bp, but this promoter has a 19-bp spacing (Figure 4), which rotates the two elements by about 70°, making it difficult for the σ⁷⁰ subunit of the RNA polymerase to contact both of them (see chapter 2). The activator binds to the promoter even without the inducer bound, but when the inducer binds, the activator bends and compacts the DNA between the two promoter elements, even breaking a hydrogen bond in the process. This brings the −35 and −10 elements closer together and rotates them so that their orientation and spacing more closely resemble those in a normal σ⁷⁰ promoter (Figure 4).

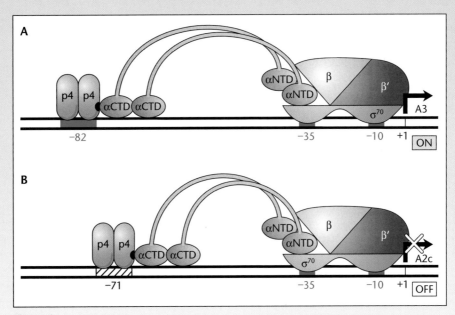

Figure 1

BOX 12.2 (continued)

Families of Regulators

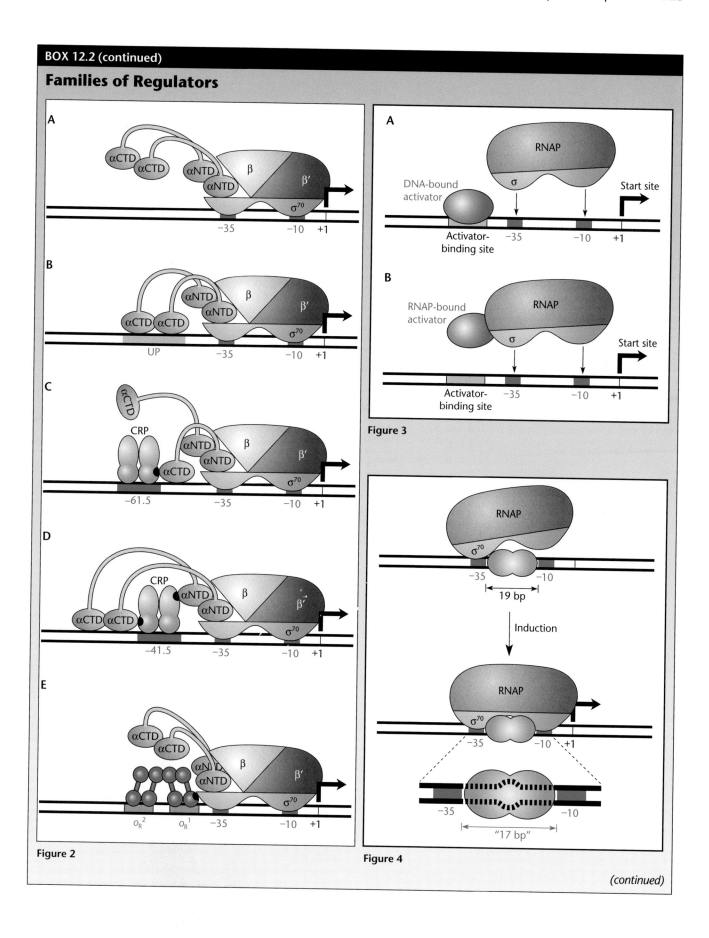

Figure 2

Figure 3

Figure 4

(continued)

BOX 12.2 (continued)	

Families of Regulators

References

Dove, S. L., and A. Hochschild. 2005. How transcription initiation can be regulated in bacteria, p. 297–310. *In* N. P. Higgins (ed.), *The Bacterial Chromosome*. ASM Press, Washington, D.C.

Egan, S. 2002. Growing repertoire of AraC/XylS activators. *J. Bacteriol.* **184:**5529–5532.

Huffman, J. L., and R. G. Brennan. 2002. Prokaryotic transcription regulators: more than just the helix-turn-helix motif. *Curr. Opin. Struct. Biol.* **12:**98–106.

Parek, M. R., S. M. McFall, D. L. Shinabarger, and A. M. Chakrabarty. 1994. Interaction of two LysR-type regulatory proteins CatR and ClcR with heterologous promoters: functional and evolutionary implications. *Proc. Natl. Acad. Sci. USA* **91:**12393–12397.

Ramos, J. L., M. Martinez-Bueno, A. J. Molina-Henares, W. Teran, K. Watanabe, X. Zhang, M. T. Gallegos, R. Brennan, and R. Tobes. 2005. The TetR family of transcriptional repressors. *Microbiol. Mol. Biol. Rev.* **69:**326–356.

AraC IS NOT JUST AN ACTIVATOR

One prediction of this model for regulation of the *ara* operon is that *araC^c* mutations should be dominant over the wild-type allele in complementation tests. If AraC acts solely as an activator, partial diploid cells that have both an *araC^c* allele and the wild-type allele would be expected to constitutively express the *araB, araA,* and *araD* genes. In other words, *araC^c* mutations should be dominant over the wild type, since the mutant AraC in the P2 state should activate transcription of *araBAD,* even in the presence of wild-type AraC protein in the P1 state.

The prediction of the model was tested with complementation. An F′ factor carrying the wild-type *ara* operon was introduced into cells with an *araC^c* mutation in the chromosome. Figure 12.19 illustrates that the partial diploid cells were inducible, not constitutive, indicating that *araC^c* mutations were recessive rather than dominant. This observation was contrary to the prediction of the model. Therefore, the model had to be changed.

Figure 12.20 illustrates a more detailed model to explain the recessiveness of *araC^c* mutations. In this model, the P1 form of the AraC protein that exists in the absence of arabinose is not simply inactive but takes on a new identity as an antiactivator (Figure 12.20A). The P1 state is called an antiactivator rather than a repressor because it does not repress transcription like a classical repressor but, rather, acts to prevent activation by the P2 state of the protein. In the P1 state, the AraC protein preferentially binds to the operator *araO₂* and another site, *araI₁*, bending the DNA between the two sites like the GalR repressor. Because AraC in the P1 form preferentially binds to *araO₂*, it cannot bind to *araI₂* and activate transcription from the *p*$_{BAD}$ promoter. In the presence of L-arabinose, however, the AraC protein changes to the P2 form and now preferentially binds to *araI₁* and *araI₂*, activating the RNA polymerase to transcribe the operon (Figure 12.20B).

This model explains why the *araC^c* mutations are recessive to the wild type in complementation tests,

because AraC^c in the P2 form can no longer bind to *araI₁* and *araI₂* to activate transcription as long as wild-type AraC in the P1 state is already bound to *araO₂* and *araI₁*. It also explains the behavior of certain deletion mutations, known as the Englesberg deletions, which we have not mentioned yet. These deletions remove the *araO₂* region but leave the *araI₁* and *araI₂* regions intact. In this case, *araC^c* mutations are no longer recessive to the wild-type allele of *araC*. Without *araO₂* to bind to, the AraC protein in the P1 form seems unable to antiactivate transcription of the operon.

AUTOREGULATION OF AraC

The AraC protein not only regulates the transcription of the *ara* operon but also negatively autoregulates its own transcription. Like TrpR, the AraC protein seems to repress its own synthesis, so that less AraC protein is synthesized in the absence of arabinose than in its presence. However, if the concentration of AraC becomes too high, its synthesis will again be repressed.

Figure 12.20 also shows a model for the autoregulation of AraC synthesis. In the absence of arabinose, the interaction of two AraC monomers bound at *araO₂* and *araI₁* bend the DNA in the region of the *araC* promoter *p*$_C$, thereby inhibiting transcription from this promoter (Fig 12.20A). In the presence of arabinose, the AraC protein is no longer bound to *araO₂*, and so the *p*$_C$ promoter is no longer bent and transcription from *p*$_C$ occurs. However, if the AraC concentration becomes too high, the excess AraC protein binds to the operator *araO₁*, preventing further transcription of *araC* from the *p*$_C$ promoter (Fig 12.20C).

CATABOLITE REGULATION OF THE L-*ara* OPERON

The *ara* operon is also regulated through catabolite repression, so the genes for arabinose utilization are not expressed if the medium contains a better carbon source. CAP, which regulates the transcription of genes subject to

Absence of arabinose

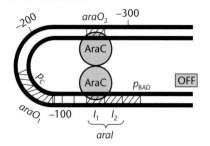

P1 form binds to operators of both operons, preventing activation of *araBAD* expression by the AraC in the P2 state

Figure 12.19   Recessiveness of *araC*c mutations. The presence of a wild-type copy of the *araC* gene prevents activation of transcription of the operon by AraCc. See text for details.

A   Absence of L-arabinose

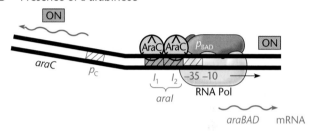

B   Presence of L-arabinose

C   Excess of AraC

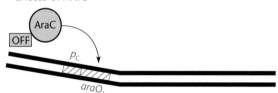

Figure 12.20   A model to explain how AraC can be a positive activator of the *ara* operon in the presence of L-arabinose and an antiactivator in the absence of L-arabinose, as well as how AraC can negatively autoregulate transcription of its own gene. (A) In the absence of arabinose, AraC molecules in the P1 state preferentially bind to *araI*$_1$ and *araO*$_2$, preventing any AraC in the P2 state from binding to *araI*$_2$. No transcription occurs because AraC must bind to *araI*$_2$ to activate transcription from p_{BAD}. Bending of the DNA between the two sites may also inhibit transcription of the *araC* gene itself by inhibiting transcription from the p_C promoter. The bend in the DNA is also facilitated by the binding by the CAP protein (see chapter 13). (B) In the presence of arabinose, AraC shifts to the P2 state and preferentially binds to *araI*$_1$ and *araI*$_2$. AraC bound to *araI*$_2$ activates transcription from p_{BAD}. (C) If the AraC concentration becomes very high, it will also bind to *araO*$_1$, thereby repressing transcription from its own promoter, p_C.

catabolite repression, is also a positive activator, like the AraC protein. By binding to the CAP binding site shown in Figure 12.18A, CAP may help open the loop of DNA created when AraC binds to *araO*$_2$ and *araI*$_1$. Opening the loop may prevent AraC from binding to *araO*$_2$ and *araI*$_1$, facilitating the binding of AraC to *araI*$_1$ and *araI*$_2$ and the activation of transcription from p_{BAD}. Thus, the absence of glucose or another carbon source better than arabinose enhances the transcription of the *ara* operon.

USES OF THE L-*ara* OPERON

Besides its historic importance in the pioneering studies of positive regulation, the *ara* operon has many uses in biotechnology. The p_{BAD} promoter, from which the genes of the *ara* operon are transcribed, is often used instead of the *lac* promoter in expression vectors because it is more tightly regulated than the *lac* promoter. Because of its combination of positive and negative regulation by the AraC activator, very little transcription occurs from the promoter unless L-arabinose is present in the medium. The p_{BAD} promoter is also more tightly regulated by catabolite repression than the *lac* promoter, making it possible to accumulate large amounts of a toxic gene product by first growing the cells in medium containing glucose and then washing out the glucose and adding L-arabinose. A widely used series of *E. coli* expression vectors use the p_{BAD} promoter and have other desirable features (Guzman et al., Suggested Reading). They have a variety of antibiotic resistance genes for selection, and some have the p15A plasmid origin of replication so that they can coexist with the more standard cloning vectors that have the compatible ColE1 origin (see chapter 4).

The *E. coli* Maltose Operons

Other well-studied and heavily used positively regulated operons in bacteria include those for the utilization of the sugar maltose and polymers of maltose in *E. coli*, shown in Figure 12.21. Rather than being organized in one operon, the genes for maltose transport and metabolism are organized in four clusters at 36, 75, 80, and 91 min on the *E. coli* genetic map. The operon at 75 min has two genes, *malQ* and *malP*, whose products are involved in converting maltose and polymers of maltose into glucose and glucose-1-phosphate. This cluster also includes the regulatory gene *malT*. The *malS* gene at 80 min encodes an enzyme that breaks down polymers of maltose such as amylase. The other cluster, at 91 min, has two operons whose gene products can transport maltose into the cell. An operon at 36 min encodes enzymes that degrade polymers of maltose.

Although they allow the cell to use maltose as a carbon source, the more significant function of the products of these operons is probably to enable the cell to transport and degrade polymers of maltose called **maltodextrins**. These compounds are products of the breakdown of starch molecules, which are very long polysaccharides stored by cells to conserve energy. The sugar maltose is itself a disaccharide composed of two glucose residues with a 1-4 linkage, and the enzymes of the *malP-malQ* operon can break the maltodextrins down into maltose and then into glucose-1-phosphate, which can enter other pathways. Some bacteria, including species of *Klebsiella*, excrete extracellular enzymes that degrade long starch molecules and allow the bacteria to grow on starch as the

sole carbon and energy source. *E. coli* lacks some of the genes needed to degrade starch to maltodextrins; therefore, in nature it probably depends on neighboring microorganisms to break the starch down to the smaller maltodextrins that it can use.

THE MALTOSE TRANSPORT SYSTEM

Most of the protein products of the *mal* operons are involved in transporting maltodextrins and maltose through the outer and inner membranes into the cell (Figure 12.22). Five different proteins make up the transport system. The product of the *lamB* gene resides in the outer membrane, where it can bind maltodextrins in the medium. This protein forms a large channel in the outer membrane through which the maltodextrins can pass. LamB is not required for growth on maltose, probably because maltose is small enough to pass through the outer membrane without its help. The LamB protein in the outer membrane also serves as the cell surface receptor for phage λ, and so the gene name is derived from the phage name (*lamB* from lambda). Mutants of *E. coli* resistant to λ have mutations in the *lamB* gene and lack the receptor for λ in the outer membrane.

Once maltodextrins are through the outer membrane, the MalS protein in the periplasm may degrade them into smaller polymers before they can be transported through

Figure 12.21   The maltose operons in *E. coli*. The MalT activator protein regulates both operons at 75 min and the operons at 80 and 91 min on the *E. coli* map. Another operon at 36 min is also induced by maltose. If maltose is not being transported, MalK, a part of the transport system, binds to MalT, inactivating it (see the text for details).

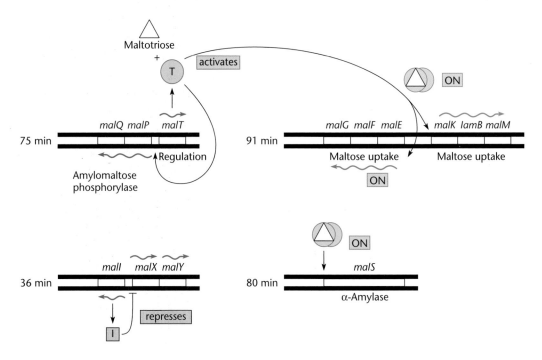

the inner membrane. The smaller polymers of maltose bind to MalE in the periplasm between the outer and inner membranes. The MalF, MalG, and MalK proteins then transport the maltodextrin through the inner membrane. The MalF and MalG proteins are membrane-bound permeases, while the MalK protein is the ATPase that provides the energy. All together these proteins form what is called a high-affinity ATP-binding cassette (ABC) transporter, based on the presence of an ATP-binding motif. There are many examples of related ABC transporters involved in transporting substances into or out of the cell, some of which are part of protein-secreting systems (see chapter 14).

REGULATION OF THE *mal* OPERONS

The regulation of the *mal* operons is also illustrated in Figure 12.21. The inducer of the *mal* operons is **maltotriose,** which is composed of three molecules of glucose held together by the maltose linkage. Maltotriose can be synthesized from maltose brought into the cell by some of the enzymes encoded by the operons. Also, the cell normally contains polymers of maltose that were synthe-

Figure 12.22 Function of the genes of the maltose regulon in the transport and processing of maltodextrins in *E. coli.* The LamB protein binds maltodextrins and transports them across the outer membrane. The MalE protein in the periplasmic space then passes them through a pore in the cytoplasmic or inner membrane formed by the MalF, MalB, and MalK proteins, an ABC transporter. The MalK protein binds the MalT transcriptional activator and releases MalT only if maltose is being transported. Once in the cytoplasm, the maltodextrins and maltose are degraded by MalP and MalQ to glucose-1-phosphate and glucose, respectively. These compounds can then be converted into glucose-6-phosphate for use as energy and carbon sources.

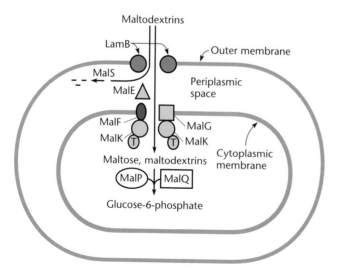

sized in the cell from glucose (and so do not need to be transported in) and can be broken down into the inducer, maltotriose. The enzymes that degrade maltose polymers therefore play an indirect role in regulating the operons.

The genes in all three clusters are regulated by a single activator, encoded by the *malT* gene in the first cluster. The MalT activator is a member of a large family of activators that includes SoxS, which regulates genes that relieve oxygen toxicity (Box 12.2). The MalT activator protein specifically binds the inducer, maltotriose, and activates transcription of the operons. Activation of the genes of the *mal* regulon by MalT involves the DNA wrapping around many copies of the protein (oligomerization), thereby changing the DNA conformation. As with many catabolic operons, however, transcription of the *mal* operons occurs only if glucose, a better carbon source than maltose, is not present. Glucose regulates the ability of MalT to activate transcription of the *mal* operons in two ways, one through repressing the synthesis of MalT and the other through regulating its activity. Glucose represses the synthesis of the MalT activator by acting through a repressor protein, Mlc (for *makes large colonies*), that represses transcription of the *malT* gene as well as many other genes involved in using alternative carbon sources. How Mlc represses the transcription of *malT* and other genes is interesting but complicated. Whether or not Mlc can repress transcription of the *malT* gene depends on the state of the system that transports glucose and other sugars into the cell; this system is called the phosphotransferase system. If glucose is present in the medium, the transporter of glucose in the membrane, PtsG, is transporting glucose into the cell by transferring phosphate to glucose. PtsG therefore tends to be unphosphorylated if there is glucose; however, if there is no glucose, PtsG tends to be phosphorylated. The Mlc repressor binds specifically to unphosphorylated PtsG, sequestering it so that it is unavailable to repress transcription of *malT* and the other genes under Mlc control. The other way glucose prevents MalT function is by acting through CAP, which binds cAMP and is required for activation of many promoters. It is known that CAP bound to cAMP helps MalT oligomerize on the *malE* promoter and presumably the other *mal* promoters. The dependence of MalT on CAP for its oligomerization ensures that the *mal* operons will not be turned on when a better carbon source (e.g., glucose) is available and cAMP levels are low.

The genetic analysis of the regulation of the *mal* operons was complicated by the fact that maltose polymers are natural components of the cell and play many roles including protecting the cell against high osmolarity. In such situations, caution must be exercised in concluding that constitutive mutations are located in regulatory genes. For example, preliminary genetic evidence suggested that the MalK protein is a repressor of the *mal* operons, since *malK* mutants appear to be constitutive for the expression of the other genes of the operons. It now appears that MalK, which is normally part of the transport system (Figure 12.22), may bind to MalT and hold it in an inactive state if maltose is not present. This may be another way of coupling the regulation of the *mal* genes to transport (see Joly et al., Suggested Reading). There are other known examples of where the specific transport system affects expression of the transcriptional regulator for utilization of a sugar, although they work in many different ways. Another example is addressed in the Problems section in a question on the *bgl* operon.

EXPERIMENTAL USES OF THE *mal* GENES

The proteins of the *mal* operons have many properties that have made them very useful in molecular genetics. For example, the MalE protein, sometimes called the maltose-binding protein (MBP), binds very tightly to polymers of maltose, making it very useful as an affinity tag (see chapter 2). Cloning vectors which translationally fuse the *malE* gene to the coding sequence of a cloned gene have been constructed. The MalE portion of the fusion protein then binds very tightly to a column made of amylose, a polymer of maltose, making it possible to bind the fusion protein on amylose affinity columns, wash out the other proteins, and then elute the purified MalE fusion protein with maltose.

The maltose operon genes have also been very useful in studying transport through the membrane. Because the products of the *E. coli mal* operons are involved in transporting long molecules into the cell, the *mal* operons have been of particular interest in studies of large-molecule transport systems. Also, many of the proteins encoded by the operons are themselves localized in the inner and outer membranes or the periplasmic space. These proteins must be transported into or through the inner membrane to get to their final destination, and so they serve as models for the study of protein transport through cellular membranes. The use of the *mal* genes in genetic analysis of protein transport is discussed in chapter 14.

The *tol* Operons

Many of the bacterial operons of soil bacteria involved in the degradation of cyclic hydrocarbons are also subject to positive regulation. Cyclic hydrocarbons are based on the conjugated ring structure of benzene. Many do not exist naturally and are pesticides and other industrially important manufactured chemicals. Moreover, many of the manufactured chemicals are also chlorinated. Chlorinated hydrocarbons were very rare in nature until the advent of the modern chemical industry, yet despite this short time interval, some types of bacteria have evolved enzymatic pathways to degrade some of these compounds. **Bioremediation** is the use of bacteria and other microorganisms to remove contaminating chemicals from the environment. Understanding the regulation of operons involved in degrading cyclic compounds may lead to more rational approaches to bioremediation of toxic waste.

The *tol* operons in plasmid pWWO, originally isolated from the soil bacterium *Pseudomonas putida*, encode enzymes that degrade toluene and the closely related compound xylene. Toluene itself is not chlorinated and has presumably always existed in nature, but operons related to *tol* have been discovered that degrade similar compounds that are chlorinated, including chlorinated catechols.

Figure 12.23 diagrams the *tol* operons of plasmid pWWO. The pathway consists of two operons, separated by a few thousand base pairs of DNA. The first operon encodes the enzymes of the "upper pathway," which converts toluene into benzoate; the other operon encodes enzymes of the "lower pathway," which breaks the ring of benzoate and degrades it to intermediates of the tricarboxylic acid cycle to be used for energy and to make carbon-containing compounds.

REGULATION OF THE *tol* OPERONS

The regulation of the *tol* operons is also illustrated in Figure 12.23. The regulation of the upper and lower *tol* operons is both coordinated and independent. Both operons must be coordinately turned on if toluene is present in the medium, since both operons are required to degrade toluene to tricarboxylic acid cycle intermediates. However, only the lower operon should be turned on if only benzoate is present in the medium, since there is no need to induce the upper pathway under these conditions.

The coordinate regulation of the upper and lower *tol* operons in the presence of toluene is achieved through two activators, one of which activates transcription of the other's gene. The activator for the upper operon is XylR, a member of the NtrC family of activators, and the activator for the lower pathway is XylS, a member of the LysR family of activators (Box 12.2). In the presence of toluene (or xylene), the XylR activator activates transcription of the upper operon from the promoter called p_U. Toluene is thus degraded into benzoate, but not enough benzoate is produced to bind to XylS and activate transcription of the lower operon. However, the XylR activator also activates transcription of the *xylS* gene from the promoter p_S. This activates transcription of the lower operon because at higher concentrations, XylS can activate transcription of the lower operon even without benzoate being bound. If, however, high levels of benzoate are present in the medium, the benzoate binds

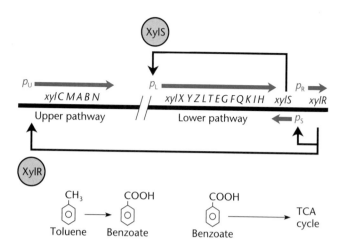

Figure 12.23 Structure of the *tol* operons of the Tol plasmid pWWO of *Pseudomonas putida*. The upper pathway converts toluene (or xylene) to benzoate. The lower encodes a meta-cleavage pathway that converts the benzoate to intermediates of the TCA cycle. The two operons are separated by a few thousand base pairs of DNA, as indicated by the broken line. The promoters activated by each of the activator proteins XylS and XylR are indicated by the arrows, and the direction of mRNA transcription is indicated by the purple arrows. p_U is the promoter for the upper operon; p_L is the promoter for the lower operon.

to XylS and induces the lower pathway; much less XylS is required for this type of activation. This regulatory interaction ensures that both the upper and lower pathways are induced if toluene is present but only the lower pathway is induced in the presence of benzoate alone.

The regulation of the *tol* operons and their regulatory genes is similar to the regulation of the genes of other catabolic pathways in bacteria. For instance, the XylR protein also regulates its own transcription; that is, it is transcriptionally autoregulated, like AraC. Therefore, in the presence of toluene, more XylR is present to activate transcription of the other genes. Another similarity is that the promoters p_U and p_S are recognized by an RNA polymerase containing the alternate sigma factor σ^{54}, the same sigma factor used to transcribe the nitrogen-regulated genes discussed in chapter 13. Why these promoters should use an alternate sigma factor is not known, but the expression of operons in bacteria as distantly related as the nitrogen-regulated genes in *E. coli* and the toluene-degrading operons in *P. putida* share this feature.

GENETICS OF THE *tol* OPERONS

The above picture of the organization of the genes of the *tol* operons came from molecular genetic experiments. These studies of the organization and regulation of the *tol* operons were greatly aided by the fact that plasmid pWWO carrying the *tol* operons is a broad-host-range,

self-transmissible plasmid that can transfer itself from the original *P. putida* strain into *E. coli*, where more sophisticated genetic tests have been developed. Once in *E. coli*, the plasmid could be easily mutagenized with transposon Tn*5* and then transferred back into *P. putida* to determine whether a particular transposon insertion inactivates a gene required for growth on toluene. Insertions that inactivate *tol* genes could then be located by restriction endonuclease mapping as discussed in chapter 9. In this way, the maps of the *tol* operons shown in Figure 12.23 were obtained.

Molecular genetic tests also yielded the picture for the regulation of the *tol* operons outlined above. The XylS protein was first identified as a positive activator of the lower pathway because clones that did not include the *xylS* gene failed to express the genes of the lower pathway, even in the presence of benzoate. Similarly, clones of the upper pathway that excluded the *xylR* gene did not express the upper-pathway operon, and *xylS* was transcribed at a lower rate if *xylR* was missing. These observations implicated XylR as a positive activator of the upper operon and the *xylS* gene. Clones that overexpress XylS turn on the lower pathway, even in the absence of benzoate, leading to the model that XylS can activate transcription of the lower pathway in the absence of benzoate, provided that XylS is present at a high concentration.

USING SELECTIONAL GENETICS TO BROADEN THE RANGE OF INDUCERS OF THE *tol* LOWER OPERON

A promising avenue of research is to use genetic selections to alter known pathways so that they can use alternate substrates, which is undoubtedly how this occurs in nature. For example, although the *tol* lower pathway can degrade some substituted forms of benzoate, including 3-methylbenzoate and 4-methylbenzoate, and allow the cell to use them as carbon and energy sources, it cannot use other derivatives of benzoate such as 4-ethylbenzoate. This particular substrate cannot be used because the second enzyme of the pathway cannot use it as a substrate and because it does not function as an inducer of the operon. Even if the enzymes encoded by an operon can degrade a derivative of the normal substrates for the operon, the enzymes are not present if the derivative does not function as an inducer of the pathway.

Gene fusion techniques were used to select *xylS* mutants that can use 4-ethylbenzoate and other derivatives of benzoate as an inducer of the lower *tol* operon (see Ramos et al., Suggested Reading). These studies could be performed with *E. coli* because the XylS protein can activate transcription of the lower operon even in *E. coli*. In these experiments, the promoter p_L for the lower operon was fused to a tetracycline resistance gene on a plasmid so that the tetracycline resistance gene would not be transcribed unless the p_L promoter was activated. This plasmid was then used to transform *E. coli* containing a second compatible plasmid expressing the XylS protein. When a particular derivative of benzoate caused XylS to activate transcription from the p_L promoter, the cells became tetracycline resistant (Tet^r) and grew on plates containing tetracycline. However, when the derivative did not induce the operon, the cells remained tetracycline sensitive (Tet^s). As expected, benzoate made the cells Tet^r. Moreover, some derivatives of benzoate, including 2-chlorobenzoate, functioned well as inducers, making the cells Tet^r. However, other derivatives, such as 4-ethylbenzoate and 2,4-dichlorobenzoate, were not inducers and the cells remained Tet^s.

Selectional genetics was then used to try to isolate mutants with altered XylS proteins in which 4-ethylbenzoate or similar noninducing derivatives could function as inducers. The bacteria described above containing the two plasmids were mutagenized, and large numbers were spread on plates containing tetracycline and a potential inducer, for example, 4-ethylbenzoate. Most of the bacteria did not multiply to form a colony, but a few colonies of Tet^r mutants appeared. Some of these were constitutive mutants that were Tet^r even in the absence of inducer, and so they were discarded. However, some had mutations changing the XylS protein so that it could use the new inducer. Mutants with *xylS* mutations that allowed induction by 4-ethylbenzoate were separated from other, unwanted types by isolating the *xylS*-containing plasmid and transforming it into new bacteria containing the *tet* fusion. Only if the mutation was in the plasmid containing the *xylS* gene, and therefore presumably in the *xylS* gene itself, were the transformants Tet^r in the presence of ethylbenzoate. The mutation was shown to be in the *xylS* gene and not somewhere else in the plasmid by recloning the *xylS* gene into a new plasmid and showing that this plasmid also confers the Tet^r phenotype in the presence of 4-ethylbenzoate. Finally, the mutated *xylS* gene could be sequenced to determine which amino acid changes in the XylS protein can allow it to use 4-ethylbenzoate as an inducer. The success of these experiments with 4-ethylbenzoate and other benzoate derivatives revealed that the inducer specificity of activator proteins can sometimes be changed by simple mutations. Presumably this is the origin of families of activators (Box 12.2).

Regulation by Attenuation of Transcription

In the above examples, the transcription of an operon is regulated through the initiation of RNA synthesis at the promoter of the operon. However, this is not the only

known means of regulating operon transcription. Another mechanism is the **attenuation of transcription.** Unlike repressors and activators, which turn transcription from the promoter on or off, the attenuation mechanism works by allowing transcription to begin constitutively at the promoter but then terminating it before the RNA polymerase reaches the first structural gene of the operon, but only if the gene products of the operon are not needed. The classic examples of regulation by attenuation are the *his* and *trp* operons of *E. coli*. Closely related mechanisms regulate such *E. coli* biosynthetic operons as leucine (*leu*), phenylalanine (*phe*), threonine (*thr*), and isoleucine-valine (*ilv*) and the *Bacillus subtilis trp* operon. In all of these cases, the availability of metabolites can change the secondary structure of the first RNA made from the operon, the leader RNA, affecting termination of transcription. Some types of riboswitch regulation due to the binding of small molecules directly to the leader RNA also act through antitermination; these are discussed later in the chapter. In this section, we discuss the regulation of the *E. coli* and *B. subtilis trp* operons by attenuation.

Regulation of the *E. coli trp* Operon by Attenuation

The archetype of attenuation control of transcription is the *trp* operon of *E. coli*. As discussed earlier in the chapter, the *trp* operon is like the *lac* operon in that it is negatively regulated, in this case by the TrpR repressor protein. However, early genetic evidence suggested that this is not the only type of regulation for *trp*. If the *trp* operon were regulated solely by the TrpR repressor, the levels of the *trp* operon enzymes in a *trpR* mutant would be the same in the absence and the presence of tryptophan. However, even in a *trpR* null mutant, the expression of these enzymes is higher in the absence of tryptophan than in its presence, indicating that the *trp* operon is subject to a regulatory system in addition to the TrpR repressor.

Early evidence suggested that tRNA^{Trp} plays a role in the regulation of the *trp* operon in the absence of TrpR (see Morse and Morse, Suggested Reading). Mutations in the tryptophanyl transfer RNA (tryptophanyl-tRNA) synthetase (the enzyme responsible for transferring tryptophan to tRNA^{Trp}) and mutations in the structural gene for the tRNA^{Trp}, as well as mutations in genes whose products are responsible for modifying the tRNA^{Trp}, increase the expression of the operon. All these mutations presumably lower the amount of aminoacylated tRNA^{Trp} in the cell, suggesting that this other regulatory mechanism is sensing not the amount of free tryptophan in the cell but, rather, the amount bound to the tRNA^{Trp}.

Other evidence suggested that the region targeted by this other type of regulation is not the promoter but a region downstream of the promoter called the **leader** region, or *trpL* (Figures 12.15 and 12.24). Deletions in this region, which lies between the promoter and *trpE*, the first gene of the operon, eliminate the regulation, so that double mutants, with both a deletion mutation of the leader region and a *trpR* mutation, are completely constitutive for expression of the *trp* operon. Deletions of the leader region are also *cis* acting and affect only the expression of the *trp* operon on the same DNA. Later evidence indicated that transcription terminated in this leader region in the presence of tryptophan because of an excess of aminoacylated tRNA^Trp. Because the regulation seemed to be able to stop, or attenuate, transcription that had already initiated at the promoter, it was called attenuation of transcription, in agreement with an analogous type of regulation already discovered for the *his* operon.

MODEL FOR REGULATION OF THE *trp* OPERON BY ATTENUATION

Figures 12.24 and 12.25 illustrate a current model for regulation of the *trp* operon by attenuation (see Yanofsky and Crawford, Suggested Reading). According to this model, the percentage of the tRNA^Trp that is aminoacylated (i.e., has tryptophan attached) determines which of several alternative secondary-structure hairpins will form in the leader RNA. Recall from chapter 2 that that the secondary structure of an RNA, or a hairpin, results from complementary pairing between the bases in RNA transcribed from inverted repeated sequences.

Whether transcription termination occurs depends on whether the attenuation mechanism senses relatively low or high levels of tryptophan. The *trpL* region, which contains two consecutive *trp* codons, provides the signal. The *trp* codons are there to allow the ribosome to "test the water" before the RNA polymerase is allowed to plunge into the structural genes of the operon. If levels of tryptophan are low, the levels of tryptophanyl-tRNA^Trp (tRNA^Trp with tryptophan attached) will also be low. When a ribosome encounters one of the *trp* codons, it temporarily stalls, unable to insert a tryptophan. This stalled ribosome in the *trpL* region therefore communicates that the tryptophan concentration is low and that transcription should continue (Figure 12.25).

Figures 12.24 and 12.25 also show how the hairpins operate in attenuation. Four different regions in the *trpL* leader RNA, regions 1, 2, 3, and 4, can form three different hairpins, 1:2, 2:3, and 3:4, as shown in Figure 12.24. The formation of hairpin 3:4 causes RNA polymerase to terminate transcription because this hairpin is part of a factor-independent transcription termination signal (see chapter 2). Whether hairpin 3:4 forms is determined by the dynamic relationship between ribosomal translation of the *trp* codons in the *trpL* region and the progress of RNA polymerase through the *trpL* region (also illustrated

Figure 12.24 Structure and relevant features of the leader region of the *trp* operon involved in regulation by attenuation. UGGUGG (in purple) indicates the two *trp* codons in the leader region. See the text for details.

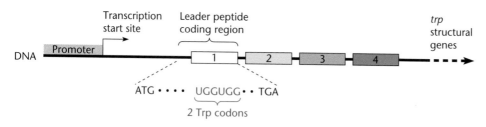

in Figure 12.25). After RNA polymerase initiates transcription at the promoter, it moves through the *trpL* region to a site located just after region 2, where it pauses. The hairpin formed by mRNA regions 1 and 2 is an important part of the signal to pause. The pause is short, probably less than 1 s, but it ensures that a ribosome has time to load on the mRNA before the RNA polymerase proceeds to region 3. The moving ribosome may help release the paused RNA polymerase by catching up with it and colliding with it.

The progress of ribosome translation through the *trp* codons of *trpL* then determines whether hairpin 3:4 will form, causing termination, or whether region 2 will pair instead with region 3, preventing formation of the 3:4 hairpin. Region 3 will pair with region 2 if the ribosome stalls at the *trp* codons because of low tryptophan concentrations (Figure 12.25D). If the ribosome does not stall at the *trp* codons, it will continue until it reaches the UGA stop codon at the end of *trpL*. By remaining at the stop codon while region 4 is synthesized, the ribosome prevents hairpin 2:3 from forming. Therefore, hairpin 3:4 can form and terminate transcription (Figure 12.25C).

GENETIC EVIDENCE FOR THE MODEL
No model is satisfactory unless it is supported by direct experimental evidence. The existence and in vivo functioning of hairpin 2:3 were supported by the phenotypes produced by mutation *trpL75*. This mutation, which changes one of the nucleotides and prevents pairing of two of the bases holding the hairpin together, should destabilize the hairpin. In the *trpL75* mutant, transcription terminates in the *trpL* region, even in the absence of tryptophan, consistent with the model that formation of hairpin 2:3 normally prevents formation of hairpin 3:4.

That translation of the leader peptide from the *trpL* region is essential to the regulation is supported by the phenotypes of mutation *trpL29*, which changes the AUG start codon of the leader peptide to AUA, preventing initiation of translation. In *trpL29* mutants, termination also occurs even in the absence of tryptophan. The model also explains this observation as long as we can assume that the RNA polymerase paused at hairpin coding sequence

1:2 will eventually move on, even without a translating ribosome to nudge it, and will eventually transcribe the 3:4 region. Without a ribosome stalled at the *trp* codons, however, hairpin 1:2 will persist, preventing the formation of hairpin 2:3. If hairpin 2:3 does not form, hairpin 3:4 will form and transcription will terminate.

One final prediction of the model is that stopping translation at codons other than those for tryptophan in *trpL* should also relieve attenuation. The codon immediately downstream of the second tryptophan codon in the *trpL* region is for arginine. Starving the cells for arginine also prevents attenuation of the *trp* operon. The *trp* operon is now being regulated by the availability of arginine, dramatically fulfilling this prediction of the model.

Regulation of the *trp* Operon of *B. subtilis* by Attenuation

It is always interesting to compare the regulation of the same operon in different types of bacteria. We often find that the same regulation can be achieved in different ways. The only other *trp* operon for the biosynthesis of tryptophan that has been studied extensively is the *trp* operon of *B. subtilis*, which consists of seven genes whose products are required to make tryptophan from chorismic acid. Interestingly, although *B. subtilis* uses different mechanisms from *E. coli* to regulate its *trp* operon, the result is the same: the operon can respond both to the amount of free tryptophan in the cell and to the amount of tRNA^{Trp} that does not have tryptophan attached (unaminoacylated tRNA^{Trp}).

While *B. subtilis* uses antitermination to regulate transcription of its *trp* operon in response to limiting tryptophan in the medium, the mechanism is very different, as illustrated in Figure 12.26. Rather than depending on pausing by the ribosome at tryptophan codons to sense limiting tryptophan and alter the secondary structure of the mRNA, the *B. subtilis trp* operon uses a protein called TRAP (for *trp* RNA-binding attenuation protein). This protein has 11 subunits, each of which can bind tryptophan. The subunits are arranged in a wheel, and each subunit binds to a repeated 3-base (triplet) sequence (either GAG or UAG) in the leader mRNA, but only if the

Figure 12.25 Details of regulation by attenuation in the *trp* operon of *E. coli*. (A) RNA polymerase pauses after transcribing regions 1 and 2. (B) A ribosome has time to load on the mRNA and begin translating, eventually reaching the RNA polymerase and bumping it off the pause site. (C) In the presence of tryptophan, the ribosome translates through the *trp* codons and prevents the formation of hairpin 2:3, thereby allowing the formation of hairpin 3:4, which is part of a transcription terminator. Transcription terminates. (D) In the absence of tryptophan, the ribosome stalls at the *trp* codons, and hairpin 2:3 forms, preventing the formation of hairpin 3:4 and allowing transcription to continue through the terminator. See the text for more details.

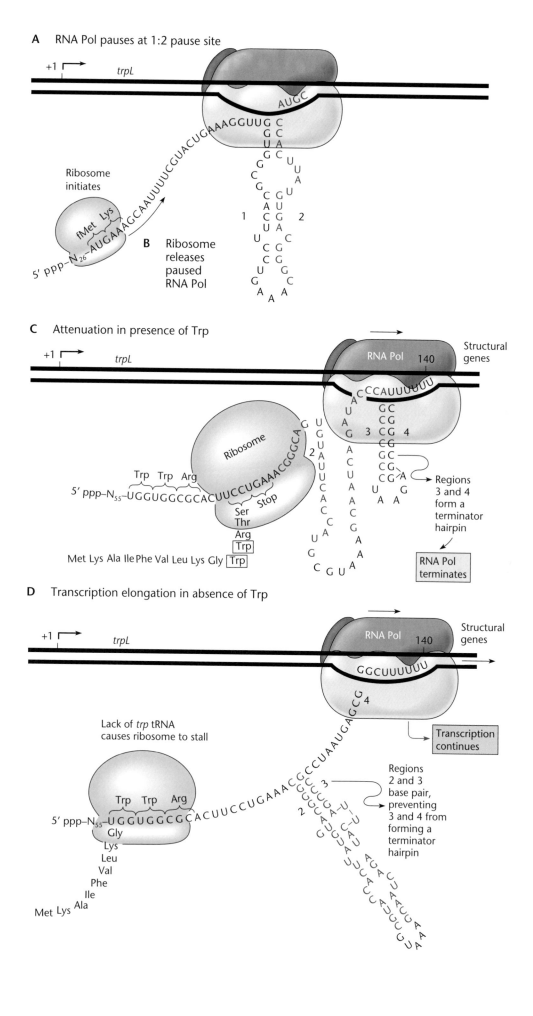

A RNA Pol pauses at 1:2 pause site

Ribosome initiates

B Ribosome releases paused RNA Pol

C Attenuation in presence of Trp

RNA Pol 140   Structural genes

Ribosome

Trp Trp Arg

5′ ppp–N₅₅–UGGUGGCGCACUUCCUGAAACGGGCA

Ser
Thr
Arg
Trp

Met Lys Ala Ile Phe Val Leu Lys Gly Trp

Regions 3 and 4 form a terminator hairpin

RNA Pol terminates

D Transcription elongation in absence of Trp

RNA Pol 140   Structural genes

GGCUUUUUU

Transcription continues

Lack of *trp* tRNA causes ribosome to stall

Trp Trp Arg

5′ ppp–N₅₅–UGGUGGCGCACUUCCUGAAACGCCUAAUGAGCG

Gly
Lys
Leu
Val
Phe
Ile
Met Lys Ala

Regions 2 and 3 base pair, preventing 3 and 4 from forming a terminator hairpin

subunit is bound to a tryptophan. The optimal spacing between the triplets is 2 bp. This causes the leader RNA to wrap around TRAP as shown, preventing formation of an antiterminator hairpin. If the antiterminator hairpin does not form, a less stable downstream terminator hairpin, which shares some sequences with it, can form, and transcriptional termination occurs. Therefore, transcription termination blocks transcription of the operon only if tryptophan is present in the medium. The fact that there are 11 sites on TRAP for binding tryptophan may allow the regulation to be fine-tuned in response to intermediate levels of tryptophan when only some of the 11 sites are occupied.

TRAP, when bound to tryptophan, can also directly block translation of the first gene of the *trpE* operon by binding to similar 3-base repeats just upstream of the *trpE* Shine-Dalgarno (S-D) sequence. It also binds to the same region in the mRNAs for other genes in other operons, including the one that includes the gene for anti-TRAP (see below). TRAP binding close to the S-D sequence prevents the ribosome from binding and blocks translation of the gene. Therefore, tryptophan in the medium can inhibit both the transcription and translation of the genes of the *trp* operon.

Also, like the *E. coli trp* operon, the *B. subtilis trp* operon is regulated in response to the state of aminoacylation of tRNA^{Trp}. The first indication was that mutations in the tryptophan aminoacyl synthetase gene showed enhanced *trp* operon expression. These mutations should increase the amount of unaminoacylated tRNA^{Trp} (tRNA^{Trp} without tryptophan attached) in the cell. However, the state of aminoacylation of tRNA^{Trp} was not regulating the *trp* operon directly. Rather, the trail led to another operon, containing two genes of unknown function. Not only did this operon contain a series of 3-base repeats characteristic of a TRAP-binding site, but also the leader sequence contained a T-box much like those that regulate aminoacyl-tRNA synthetase genes (see below), suggesting how it could be regulated by the state of aminoacylation of tRNA^{Trp}. Later work showed that the product of one of the genes, *yczA*, would cause constitutive expression of the *trp* operon if overproduced and that it binds directly to TRAP, inactivating it. The protein was therefore named anti-TRAP and binds to TRAP even if tryptophan is bound to TRAP. The transcription of the operon containing the gene for anti-TRAP is regulated by antitermination of transcription in response to the state of aminoacylation of tRNA^{Trp}. The aminoacylated tRNA^{Trp} binds directly to the T-box and causes termination of transcription. Regulation due directly to the binding of a small effector to the leader RNA is called riboswitch regulation and is discussed in a later section.

Translation of the gene for anti-TRAP is also inhibited if tRNA^{Trp} is unaminoacylated. There are three codons for tryptophan (UGG) in the leader RNA; if these codons are translated, the ribosomes activate translation from the downstream gene for anti-TRAP, a form of translational coupling (see chapter 2). Thus, even though superficially the *trp* operon of *B. subtilis* behaves much like the *trp* operon of *E. coli* in that transcription is antiterminated in the leader sequence if tryptophan is limiting or if most of the tRNA^{Trp} in the cell is not charged with tryptophan, the actual molecular mechanisms used are very different.

Comparative genomics has been used to determine whether other types of bacteria use TRAP and a similar mechanism to regulate their *trp* operon (see Gutierrez-Preciado et al., Suggested Reading). Some other gram-positive bacilli closely related *B. subtilis* do use TRAP, but many do not. Also, many of those that use TRAP do not have anti-TRAP. Most gram-positive bacteria, however, even some related to *B. subtilis*, do not use TRAP, including *Bacillus*

Figure 12.26 TRAP regulation of the *trp* operon in *B. subtilis*. (A) Model for transcription attenuation of the *trp* operon. When tryptophan is limiting (−Tryptophan), TRAP is not activated. During transcription, antiterminator formation (A and B) prevents formation of the terminator (C and D), which results in transcription of the *trp* operon structural genes. When tryptophan is in excess (+Tryptophan), TRAP is activated. Tryptophan-activated TRAP can bind to the (G/U)AG repeats and promote termination by preventing antiterminator formation. The overlap between the antiterminator and terminator structures is shown. (B) Translational control of *trpE* by TRAP. Under tryptophan-limiting conditions, TRAP is not activated and is unable to bind to the *trp* leader transcript. In this case, the *trp* leader RNA adopts a structure such that the *trpE* S-D sequence is single stranded and available for translation. Under excess-tryptophan conditions, TRAP is activated and binds to the (G/U)AG repeats. As a consequence, the *trpE* S-D blocking hairpin forms, which prevents ribosome binding and translation. The overlap between the two alternative structures is shown. Numbering is from the start of transcription.

A Transcription attenuation

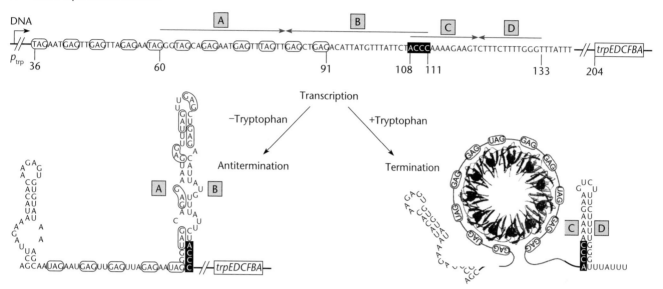

B Translational control

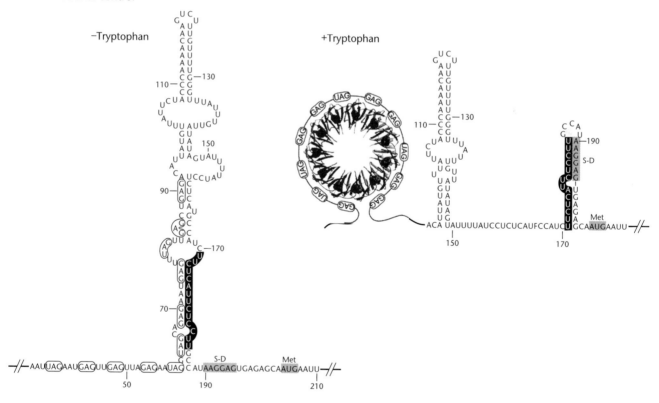

anthracis (the cause of anthrax); instead, they use a T-box riboswitch mechanism to antiterminate transcription of their *trp* operon (see below).

Regulation of the *bgl* Operon of *E. coli*

Sometimes an antiterminator protein can act to stabilize an antiterminator hairpin rather than to destabilize one as in the case of TRAP. This is the strategy used by the *bgl* operon of *E. coli*, whose gene products degrade β-glucosides for use as carbon and energy sources (see Fux et al., Suggested Reading). In this case, an antiterminator named BglG binds to and stabilizes two antiterminator hairpins that share sequences with two terminator hairpins. If the BglG protein binds to the antiterminator hairpins and allows them to form, the terminators do not form and transcription continues into the operon, expressing the genes for β-glucoside degradation.

As with other catabolic pathways, the *bgl* operon is induced only when β-glucosides are present in the medium. However, rather than the β-glucosides binding directly to the antiterminator protein, BglG, the activity of BglG is coupled to the transport of β-glucosides into the cell. The active form of the BlgG protein that can bind to the antiterminator loops is a homodimer made up of two identical polypeptide products of the *bglG* gene. The BglG protein forms a dimer only if it is not phosphorylated. If it is phosphorylated, it will be in the monomer form, composed of only one subunit, and does not bind. The state of phosphorylation of another protein, BglF, determines whether BglG is phosphorylated and not active. BglF is part of the PTS transport system in the membrane that specifically transports β-glucosides into the cell. If β-glucosides are not being transported, BglF transfers a phosphate to BglG, inactivating it. It also binds the inactive phosphorylated form of BglG, so BglG is not longer free in the cytoplasm, much like the phosphorylated MalK transporter binds MalT in the case of the maltose operon (see above). This ensures that the genes for β-glucoside utilization are turned on only if β-glucosides are being transported into the cell.

Regulation by Secondary-Structure Changes in the mRNA

In most of the examples cited so far, proteins play a role in the regulation of operons, either independently by binding the promoter region or to the leader sequence of the messenger RNA (mRNA) or as part of the ribosome translating the leader sequence. However, there are more and more examples where the leader RNA itself can play the regulatory role. It can do this either through antitermination or perhaps by functioning as a ribozyme to cleave its own RNA in response to environmental conditions. These leader sequences in the mRNA that can regulate genes are

upstream or 5' of the initiation codon for the first gene on the mRNA. This part of the mRNA is usually not translated into a protein and so is often referred to as the 5' untranslated sequence (5'UTR); this terminology is used below.

REGULATION BY MELTING SECONDARY STRUCTURE IN THE mRNA

A 5'UTR can affect the expression of a gene directly through the effects of temperature on its secondary structure. Base pairing between complementary sequences on the mRNA can cause secondary structures to form in the RNA in the form of hairpins, pseudoknots, etc. (see chapter 2). Secondary structures are less stable at higher temperatures because the base pairing that holds them together can melt at these temperatures. One way in which temperature can regulate the expression of a gene is if the secondary structures that have formed in the 5'UTR block access of the ribosome to the translational initiation region (TIR) of the mRNA, for example if they include the S-D sequence and/or the initiator codon. When the temperature rises, these secondary structures could melt, exposing the TIR so that the ribosomes can bind and initiate translation of the mRNA. Temperature regulation of the translation of a gene through the secondary structure of the 5'UTR was first discovered in the gene for the heat shock sigma, σ^{32}, in *E. coli*. After *E. coli* is exposed to an abrupt increase in temperature called a heat shock, some genes are transcribed from promoters that use σ^{32} instead of the normal σ^{70}. The level of σ^{32} is increased following an abrupt increase in temperature, and the melting of secondary structure in the 5'UTR of its mRNA is one of the reasons. Regulation by heat shock is a form of global regulation and is discussed in chapter 13.

Many pathogenic bacteria use a mechanism involving temperature-dependent melting of mRNA secondary structure to help turn on their virulence genes. Our body temperature, and that of most other warm-blooded hosts, is much higher than that of the outside environments usually inhabited by bacteria. Pathogenic bacteria often use temperature as one of the clues that they are in a eukaryotic host, and a rise in temperature tells them that it is time to turn on the virulence genes that allow them to survive and multiply in the host. By having the 5'UTR for a regulatory gene contain secondary structure that obstructs the TIR at lower temperatures but melts at the higher temperatures characteristic of their host, the bacterium begins to express the virulence genes under its control only when it is in the host. One example of such temperature regulation is in the expression of *lcrF* (low calcium response gene F) in *Yersinia pestis*, the bacterium that causes bubonic plague. The product of this gene is the transcriptional regulator that turns on virulence genes in rats (and humans).

A secondary structure in the 5′UTR of the mRNA for LcrF normally blocks the S-D sequence, preventing translation, but this secondary structure melts at higher temperature, allowing translation of the LcrF transcriptional regulator. As in most such cases, temperature is not the only clue that the bacterium is in a eukaryotic host, and transcription of the virulence genes of *Y. pestis* is turned on only if calcium ions are also limiting, another characteristic of the environment within a eukaryotic host.

Riboswitch Regulation

Sometimes regulation is achieved by altering the secondary structure of the mRNA through the binding of a small effector molecule to the RNA rather than through heat-induced melting. This can then affect antitermination of transcription or act in other ways to affect regulation of the gene. This mechanism has been named **riboswitch regulation** because the binding of the small molecule is affecting a switch in the structure of the mRNA directly, with no proteins involved. More and more examples of riboswitch regulation are accumulating, both in bacteria and in eukaryotes; some of them are discussed in this section.

RIBOSWITCH REGULATION BY
ANTITERMINATION OF TRANSCRIPTION
OF THE AMINOACYL-tRNA SYNTHETASE
GENES OF *B. SUBTILIS*
The first example to be discovered of an effector of gene expression acting directly on the mRNA is the regulation of the transcription of the genes for the aminoacyl-tRNA synthetase genes in *B. subtilis*. Depriving the cell of an amino acid causes much of the tRNA for that amino acid to be unaminoacylated. The bacteria respond by synthesizing higher levels of the aminoacyl-tRNA synthetase for that tRNA. Higher levels of the synthetases presumably allow more efficient attachment of the limiting amino acids to their cognate tRNAs.

The expression of the aminoacyl-tRNA synthetase genes in *B. subtilis* is regulated through antitermination of transcription. If the tRNA for that amino acid is mostly unaminoacylated, with no amino acid attached, transcription terminates less often in the leader sequence and more synthetase is made. If most of the tRNA for that amino acid has the amino acid attached, i.e., is aminoacylated, transcription of the synthetase gene often terminates in the leader region. Therefore, whether transcription terminates in the leader sequence of each synthetase gene is determined by the relative levels of the unaminoacylated cognate tRNA for that synthetase.

Extensive research led to the remarkable conclusion that no proteins were involved in this antitermination mechanism and that the mRNA could do it all by itself in response to the level of aminoacylated tRNA. It also led

to a remarkably clear understanding of how the antitermination regulation works for at least some synthetase genes (Figure 12.27) (see Yousef et al., Suggested Reading). According to the dictates of thermodynamics, an RNA folds to form the most stable secondary structure overall, i.e., the one that allows the most hydrogen bonds to form. In this case, the base pairing of regions in the tRNA to regions of the 5′UTR of the mRNA for a synthetase enzyme can contribute additional base pairing and cause a different secondary structure to become the most stable one. To ensure that only the correct tRNA binds to a particular 5′UTR, the anticodon of the tRNA must form complementary base pairs with a strategically placed codon for that amino acid in an exposed region called the specifier loop in a hairpin at the 5′ end of the mRNA (Figure 12.27). The other end of the tRNA, the amino acid acceptor end, can then form complementary base pairs with a downstream exposed region (called the bulge in Figure 12.27). The bulge lies in the sequence that forms an antiterminator hairpin, and this base pairing helps stabilize the antiterminator hairpin. By stabilizing the antiterminator hairpin, the tRNA prevents the formation of a less stable third hairpin, the terminator hairpin, which shares some of its sequences. However, if the tRNA is aminoacylated, the attached amino acid interferes with base pairing between the acceptor end of the tRNA and the bulge and the antiterminator hairpin does not form. Instead, the terminator hairpin forms and the transcription terminates before the initiator codon for the synthetase. In vitro experiments have confirmed many aspects of this model, including the fact that binding of tRNA to the 5′UTR can affect its secondary structure and stabilize the antiterminator hairpin (see Yousef et al., Suggested Reading). The antitermination occurs when tRNA is added to a purified system, which has only the RNA polymerase and the DNA template, confirming that no other factors, including translation, are required. By having the secondary structure of the 5′UTR depend on specific base paring with the tRNA, many different synthetase genes can all be regulated by the same mechanism but each synthetase gene responds only to levels of its own cognate amino acid. This was named riboswitch regulation because the regulation was due to binding of the effector molecule directly to the mRNA, causing a "switch" in the secondary structure of the mRNA.

OTHER EXAMPLES OF RIBOSWITCH
REGULATION
Once riboswitch regulation of some *B. subtilis* aminoacyl-tRNA synthetase genes was discovered, many other examples of riboswitch regulation by binding of small molecules directly to the mRNA, both in bacteria and in eukaryotes, were forthcoming. Many of these involve the

A Limiting amino acid

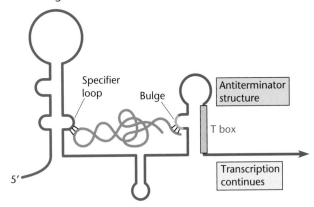

B Excess amino acid

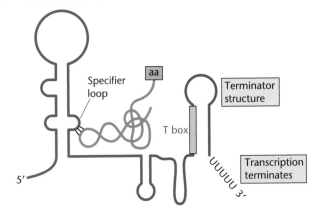

Figure 12.27 Riboswitch regulation by binding of tRNA to the T-box in a leader RNA for aminoacylsynthetases in *B. subtilis*. The specific tRNA base-pairs to the specifier loop through a codon-anticodon interaction. Whether or not the acceptor end can also pair with the antiterminator bulge and stabilize the antiterminator loop depends on whether the tRNA is aminoacylated. (A) The amino acid is limiting, and the tRNA is not aminoacylated. The acceptor end can pair with the bulge, stabilizing the antiterminator loop. The terminator cannot form, and transcription continues into the downstream gene. (B) There is excess amino acid (aa), and the tRNA is aminoacylated. The attached amino acid interferes with the pairing with the bulge. The antiterminator loop does not form, and so the terminator loop can form, terminating transcription. The T-box sequence shared between the antiterminator loop and the terminator loop is boxed.

binding of small metabolites, either the end product or one of the intermediates of a biosynthetic pathway, to the mRNA, affecting its secondary structure and regulation of expression of genes of the pathway. So far, examples of riboswitch regulation include amino acids (lysine and glycine), vitamins (B_{12} and thiamine pyrophosphate [B_1]), nucleic acid bases (guanine and adenine), as well as flavin

mononucleotide and glucosamine-6-phosphate. The regulation is sometimes through antitermination of transcription, as with the *B. subtilis* aminoacylsynthetase genes, but also can occur through translation, by blocking the TIR on the mRNA, or perhaps even through converting the 5′UTR into a ribozyme that cleaves itself, although the latter has been demonstrated less unambiguously. In eukaryotes, which have the option of using RNA splicing as a regulatory mechanism, the secondary-structure changes caused by binding the metabolite can inhibit expression of a gene by inhibiting splicing of the mRNA.

One well-studied example of riboswitch regulation by binding a small metabolite is the regulation by antitermination of genes involved in methionine metabolism (see McDaniel et al., Suggested Reading). If methionine is limiting, the genes for methionine are induced and antitermination of transcription through a riboswitch plays a role in the induction. Methionine plays a special role in the cell, besides being one of the 20 amino acids in proteins (actually 22 [see Box 2.3]). Methionine is also converted into *S*-adenosylmethionine, which is the donor of methyl groups in many biochemical reactions in the cell. We have mentioned some of these reactions in earlier chapters, including the methylation of DNA by restriction endonucleases, but there are many more, some of which are essential to the functioning of the cell. Reflecting this role, methionine is not the effector that binds to the 5′UTRs of the methionine biosynthetic operons to effect antitermination; instead, *S*-adenosylmethionine does so. In this case, the effector *S*-adenosylmethionine binds to a region of the 5′UTR called an S-box, which is highly conserved in methionine biosynthetic genes in gram-positive bacteria related to *B. subtilis*.

The binding of *S*-adenosylmethionine to the S-box in the 5′UTR of a gene involved in methionine biosynthesis and how this binding causes termination of transcription is illustrated in Figure 12.28. This regulation is reminiscent of transcription antitermination in the *trp* operon in *E. coli* in that it involves not simply an antiterminator and a terminator hairpin but uses an anti-antiterminator, an antiterminator, and a terminator hairpin. The binding of *S*-adenosylmethionine stabilizes the anti-antiterminator, which prevents the formation of the antiterminator hairpin. If the antiterminator hairpin does not form, the terminator hairpin forms and transcription terminates. As with the T-box regulation of the aminoacyltransferase genes, the S-box-mediated termination occurs when *S*-adenosylmethionine is added to a purified transcription system consisting only of RNA polymerase and template DNA, showing that no other factors are required. Also, the importance of some of the hairpin loops has been confirmed by mutation studies (see McDaniel et al., Suggested Reading).

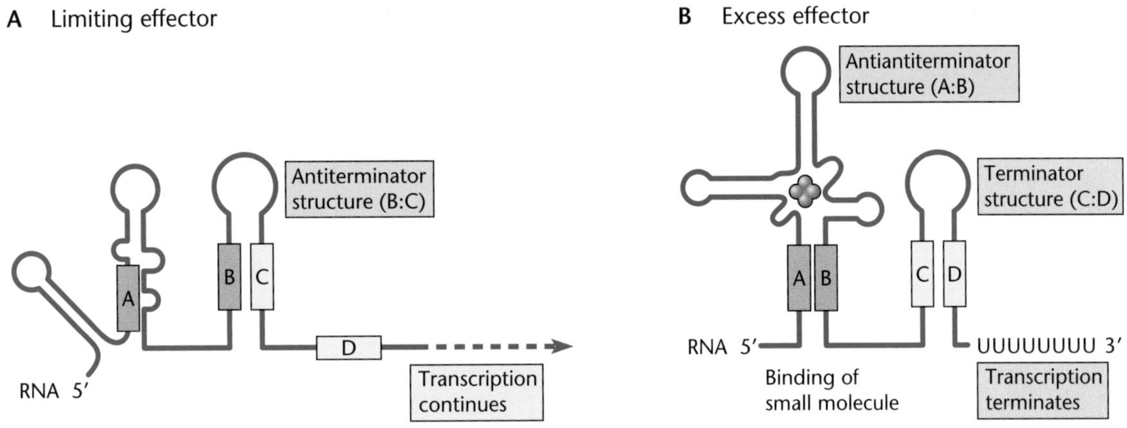

A Limiting effector

Antiterminator
structure (B:C)

Transcription
continues

RNA 5'

B Excess effector

Antiantiterminator
structure (A:B)

Terminator
structure (C:D)

RNA 5'

UUUUUUUU 3'

Binding of
small molecule

Transcription
terminates

Figure 12.28 Riboswitch regulation by binding of a small effector molecule to the leader mRNA of an operon. Regions A, B, C, and D indicate mRNA sequences that can bind to form alternative secondary structures. (A) Limiting effector. The antiterminator structure forms, allowing transcription to continue. (B) Excess effector. Binding of effector molecule causes the anti-antiterminator structure to form. Consequently, a terminator structure can form, and transcription terminates.

Posttranslational Regulation: Feedback Inhibition

Biosynthetic pathways are not regulated solely through transcriptional regulation of their operons and translational regulation of their mRNAs; they are also often regulated by feedback inhibition of the enzymes once they are made. In **feedback inhibition,** the end product of a pathway binds to the first enzyme of the pathway, inhibiting its activity. Feedback inhibition is common to many types of biosynthetic pathways and is a more sensitive and rapid mechanism for modulating the amount of the end product than are transcriptional regulation and translational regulation, which respond only slowly to changes in the concentration of the end product of the pathway.

Feedback Inhibition of the Tryptophan Operon

The tryptophan biosynthetic pathway of *E. coli* is subject to feedback inhibition. Tryptophan binds to the first enzyme of the tryptophan synthesis pathway, anthranilate synthetase, and inhibits it, thereby blocking the synthesis of more tryptophan. The tryptophan analog 5-methyltryptophan has been used to study this process. At high concentrations, 5-methyltryptophan binds to anthranilate synthetase in lieu of tryptophan and inhibits the activity of the enzyme, starving the cells for tryptophan. Only mutants defective in feedback inhibition because of a missense mutation in the *trpE* gene that pre-

vents the binding of tryptophan (and 5-methyltryptophan) to the anthranilate synthetase enzyme can multiply to form a colony in the absence of tryptophan.

A similar method was described earlier for isolating constitutive mutants with mutations of the *trp* operon, but selection of constitutive mutants requires lower concentrations of 5-methyltryptophan. If the concentration of this analog is high enough, even constitutive mutants will be starved for tryptophan.

Feedback Inhibition of the Isoleucine-Valine Operon

Feedback inhibition is also responsible for the valine sensitivity of *E. coli*. If *E. coli* cells are presented with high concentrations of valine, they will die as long as isoleucine is not provided in the medium. The reason is that valine and isoleucine are synthesized by the same pathway, encoded by the *ilv* (isoleucine-valine) operon. The first enzyme of the pathway, acetohydroxy acid synthase, is feedback inhibited by valine, so that if the concentration of valine is high, the cells can make neither valine nor isoleucine. The cells then starve for isoleucine unless this amino acid is provided in the medium. Such a situation seldom occurs in nature, since degraded proteins are the usual source of amino acids and since isoleucine and valine are two of the most common amino acids and so are both present in most proteins.

While most *E. coli* strains are valine sensitive, valine-resistant mutants are easily isolated by plating *E. coli* in the presence of high concentrations of valine with no isoleucine. Any colonies that arise are due to the multiplication of valine-resistant mutants. These mutants are about as frequent as mutants resistant to the tryptophan analog 5-methyltryptophan, and a priori one might assume that they had the same molecular basis, in this case an altered acetohydroxy acid synthase that is still active but is no longer feedback inhibited by valine. However, mutations to valine resistance are often revertants of a mutation that normally inactivates a gene for another acetohydroxy acid synthase that is not feedback inhibited by valine and so performs the first step in the synthesis of isoleucine, regardless of the valine concentration.

Posttranslational Modifications of Enzymes

In some cases, regulation is achieved by the reversible covalent modification of an enzyme after it has been translated. One example is the adenoribosylation (addition of AMP) to the glutamine synthetase enzyme, which synthesizes glutamine from glutamate. At high concentrations of glutamine, this enzyme is adenoribosylated, which temporarily inhibits its activity until glutamine levels drop. The AMP groups are then removed, and the activity is restored. This and other examples of reversible covalent modification of proteins are discussed in chapters 13 and 14.

Operon Analysis for Sequenced Genomes

The operons discussed so far were all discovered in genetic analyses that used forward genetics; that is, they were discovered by isolating and analyzing mutations that changed the wild-type sequence to a mutant sequence, i.e., changed a wild-type allele to a mutant allele. Moreover, the mutations were found because mutant strains could be identified on the basis of phenotypic changes that were related directly to the biological role of the operon studied. For example, *lac* mutations could be isolated on color indicator plates, such as MacConkey agar plates, which use a color change to indicate a perturbation in β-galactoside metabolism. Also, both mutant selections and mutant screens could be used, and creative exploitation of the metabolic properties of the operon studied yielded robust collections of mutations. It is because of studies of the variety of mutant alleles in both the structural genes and the regulatory genes and sequences (consider the *araC* alleles) that we now know as much as we do about operons and their regulation.

Alleles of Operon Genes

Mutant alleles of structural genes may alter the function of a gene so as to eliminate, change, or increase its activity.

A mutation that eliminates gene function creates a null allele. The term "loss-of-function allele" generally is interchangeable with "null allele." The term "knockout"—such as a mutation due to a transposon insertion—implies a null mutation but is often used too loosely because some activity could remain. Although many valuable collections of mutants have been obtained by transposon mutagenesis, there is no assurance that all of these are null mutations. Also, such mutations can be polar and prevent expression of other genes downstream in the same operon. In some cases, they might even provide promoter activity and express genes upstream that are transcribed in the opposite direction of the gene in which the insertion occurred. Furthermore, the methods used to obtain these collections are often laborious and difficult to standardize. After the mutations are obtained, they must be characterized by PCR.

USING REVERSE GENETICS TO CONSTRUCT NULL, NONPOLAR ALLELES

When the sequence of the DNA of a bacterium is known, it is often possible to make null mutations of genes, using systematic methods that avoid many of the complications discussed above. Figure 12.29 illustrates one such method for *E. coli* (see Baba et al., Suggested Reading). This procedure is designed to delete the entire gene but leave the TIR at the beginning of the gene and the terminator codon at the end of the gene plus some sequences upstream of the terminator codon in case the coding sequence for the gene being deleted includes the S-D sequence for the downstream gene. Overlap of the terminator codon for one gene with the TIR for the next gene in an operon often occurs and can cause translational coupling. In this method, upstream and downstream PCR primers are designed to amplify an antibiotic resistance gene from a plasmid and introduce sequences at one end of the amplified fragment for the TIR of the gene being deleted and at the other end of the fragment for the terminator codon plus a short upstream sequence from the gene. This fragment is then electroporated into an *E. coli* strain that expresses the Red functions of λ phage or the recombination functions of the Rac prophage to promote recombineering between the amplified fragment and the gene in the chromosome (see Box 10.3). Recombination between the sequences flanking the antibiotic resistance gene in the amplified fragment and the corresponding sequences in the gene in the chromosome deletes most of the gene, replacing it with the antibiotic resistance gene. As an additional feature, the antibiotic resistance gene can be later removed if the antibiotic resistance gene on the plasmid is flanked by sequences for a site-specific recombinase, in the example the FLP recombinase of yeast. When another plasmid expressing this recombinase is introduced into the *E. coli* organism, the antibiotic resistance gene is excised, leaving behind a scar.

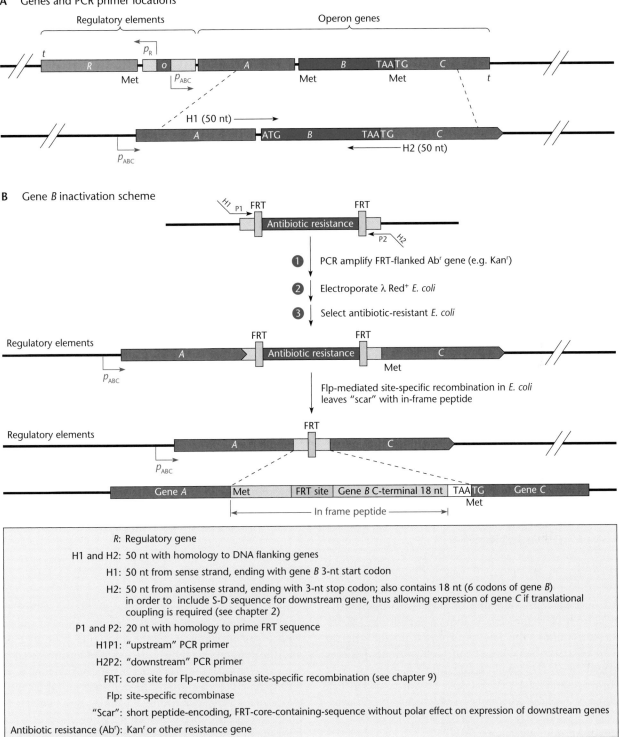

A Genes and PCR primer locations

B Gene *B* inactivation scheme

① PCR amplify FRT-flanked Ab^r gene (e.g. Kan^r)

② Electroporate λ Red$^+$ *E. coli*

③ Select antibiotic-resistant *E. coli*

Flp-mediated site-specific recombination in *E. coli* leaves "scar" with in-frame peptide

R: Regulatory gene

H1 and H2: 50 nt with homology to DNA flanking genes

H1: 50 nt from sense strand, ending with gene *B* 3-nt start codon

H2: 50 nt from antisense strand, ending with 3-nt stop codon; also contains 18 nt (6 codons of gene *B*) in order to include S-D sequence for downstream gene, thus allowing expression of gene *C* if translational coupling is required (see chapter 2)

P1 and P2: 20 nt with homology to prime FRT sequence

H1P1: "upstream" PCR primer

H2P2: "downstream" PCR primer

FRT: core site for Flp-recombinase site-specific recombination (see chapter 9)

Flp: site-specific recombinase

"Scar": short peptide-encoding, FRT-core-containing-sequence without polar effect on expression of downstream genes

Antibiotic resistance (Ab^r): Kan^r or other resistance gene

Figure 12.29 Method for constructing null mutations in sequenced genes of *E. coli* that are in-frame deletions so they are not polar and do not affect translational coupling. (A) An operon with genes A, B, and C. Gene B is to be inactivated. PCR primers are constructed to introduce sequences upstream of and including the initiating AUG codon for gene B and sequences just upstream of the gene B terminator codon, UAA in the example, to include possible S-D sequences for gene C. When these primers are used to PCR amplify an antibiotic resistance cassette from a plasmid, the amplified fragment has the sequences from the chromosome at its ends. (B) When this PCR fragment is electroporated into *E. coli* expressing the λ Red functions, recombination between the sequences at the ends of the fragment replaces most of gene B with the fragment. If the PCR primers also included sites for a site-specific recombinase, the recombinase can be expressed in the cells and excises the cassette, leaving behind a short "scar" including one site for the recombinase as well as other sequences. If this scar is a multiple of three bases of the sequence that was replaced, gene B now expresses a short polypeptide to prevent polarity on downstream genes. The process can then be repeated on other genes to test the effect of inactivating multiple genes. R, regulatory gene; Met, initiating methionine. Purple boxes show sequences flanking the antibiotic resistance cassette in the plasmid that were used to amplify the cassette.

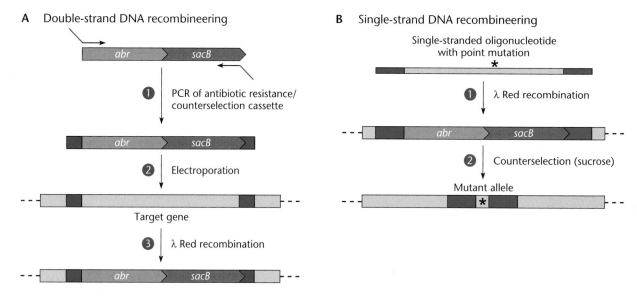

Figure 12.30 Method for introducing a base pair change or other small point mutation into a sequenced gene in *E. coli*. The first step is similar to the method shown in Figure 12.29 except that the cassette contains both a selectable gene for antibiotic resistance (*abr*) and a counterselectable gene for sensitivity to sucrose (*sacB*). Also, the complementary sequences introduced into the amplified fragment are closer together in the gene, on either side of the site of the desired mutation (*). The amplified fragment is electroporated into λ Red-expressing cells, and cells in which recombination between the sequences at the ends of the amplified fragment and complementary sequences in the chromosome have replaced the sequence in the chromosome can be selected by selection for antibiotic resistance on the cassette. An oligonucleotide containing the desired change can then be electroporated into these cells, and those in which the sequence of the oligonucleotide has replaced the cassette can be selected by plating on sucrose. Complementary sequences in the chromosome and at the ends of the amplified PCR fragment are shown in purple. See the text for details.

If this scar has a multiple of 3 bp of the number of base pairs deleted from the gene, a short polypeptide is translated that contains too few sequences from the original gene to be active but whose expression prevents polarity or effects of translational coupling on the expression of the downstream gene. The same procedure can be then repeated to replace other genes with the antibiotic resistance gene to assess the effects of multiple gene knockouts on the bacterium (see chapter 14).

USING REVERSE GENETICS TO CONSTRUCT A COLLECTION OF CHANGE- OR GAIN-OF-FUNCTION ALLELES

Recombineering can also be used to create a variety of allele types in a gene of *E. coli*. Figure 12.30 illustrates one such method, which is briefly described in Box 10.3. This method depends on having both a gene that can be selected, e.g., for antibiotic resistance (*abr*), and a gene that can be counterselected, e.g., for sucrose sensitivity

(*sacB*). It also requires two rounds of recombineering recombination, one to replace a short region of the gene with a cassette containing the antibiotic resistance gene and the counterselectable gene and the second to replace the cassette with the displaced region of the gene with a base pair change or some other minor mutation whose effect you wish to determine. In step A, two PCR primers that have been constructed amplify the cassette from a plasmid and introduce sequences that are complementary to sequences on either side of the site to be mutagenized in the chromosomal gene. The introduced sequences should not be too far apart in the gene, since later they, as well as the sequences between them in the gene, must be included in a single oligonucleotide, and excessively long single-stranded deoxyoligonucleotides are prohibitively expensive. These primers are then used to PCR amplify the cassette from a plasmid, and the amplified fragment is introduced into λ Red-expressing *E. coli*, selecting for the antibiotic resistance on the cassette. In these cells, recombination between the sequences at the ends of the cassette and their complementary sequences in the gene has replaced the sequence between the complementary sequences with the cassette as shown. In step B, a single-stranded PCR primer that includes both complementary sequences and spans the region of the gene between them but includes the desired point mutation is electroporated into the cells, this time selecting for loss of the counterselectable gene by growing the cells in sucrose. In many of the surviving cells, the cassette has been replaced by the sequence of the oligonucleotide and, depending on where the recombination occurred, the desired point mutation has been introduced into the gene sequence without otherwise disrupting the sequence of the gene. This region should be sequenced in some of the sucrose-resistant strains to find one in which the desired change has occurred.

Alleles of Regulatory Genes and Elements

The methods discussed above can also be applied to assess a gene to determine if it has a regulatory function. For example, a gene that is adjacent to an operon in a genome sequence, e.g., *lacI* or *araC*, is a candidate for encoding a regulator, especially if it contains a helix-turn-helix motif. Determination of the null phenotype of the gene can show whether the gene product is a positive or negative regulator. Generation of other allele types can add to understanding of how the regulator functions. However, it is still very difficult to predict which amino acid changes will elicit a particular phenotype, for example, a superrepressor phenotype, even if the regulator falls into one of the families discussed above. Imagine how naive our view of regulation would be if some regulatory genes had not been intensively studied, using selectional genetics to identify the amino acids important for the various phenotypes.

SUMMARY

1. Regulation of gene expression can occur at any stage in the expression of a gene. If the amount of mRNA synthesized from the gene differs under different conditions, the gene is transcriptionally regulated. If the regulation occurs after the mRNA is made, the gene is posttranscriptionally regulated. A gene is translationally regulated if the mRNA is made but not always translated at the same rate.

2. In bacteria, more than one gene is sometimes transcribed into the same mRNA. Such a cluster of genes, along with their adjacent *cis*-acting controlling sites, is called an operon.

3. The regulation of operon transcription can be negative, positive, or a combination of the two. If a protein blocks the transcription of the operon, the operon is negatively regulated and the regulatory protein is a repressor. If a protein is required for transcription of an operon, the operon is positively regulated and the regulatory protein is an activator.

4. If an operon is negatively regulated, mutations that inactivate the regulatory gene product will result in constitutive mutants in which the operon genes are always expressed. If the operon is positively regulated, mutations that inactivate the regulatory protein will cause permanent repression of the expression of the operon. In general, constitutive mutations are much more common with negatively regulated operons than with positively regulated operons.

5. Sometimes the same protein can be both a repressor and an activator in different situations, which complicates the genetic analysis of the regulation.

6. The regulation of transcription of bacterial operons is often achieved through small molecules called effectors, which bind to the repressor or activator protein, changing its conformation. If the presence of the effector causes the operon to be transcribed, it is called an inducer; if its presence blocks trancription of the operon, it is called a corepressor. The substrates of catabolic operons are usually inducers, whereas the end products of biosynthetic pathways are usually corepressors.

7. The regions on DNA to which repressors bind are called operators. Some repressors act by physically interfering with the binding of the RNA polymerase to the promoter (closed-complex formation). Others allow repressor binding but prevent opening of the DNA at the promoter (open complexes). Yet others prevent the RNA polymerase from escaping the promoter to begin RNA synthesis (promoter clearance). Some repressors act by binding to two operators on either side of the promoter simultaneously, bending the DNA between them and inactivating the promoter.

8. The regions to which activator proteins bind are called activator sequences. Some activator proteins recruit RNA polymerase to the promoter by binding both to a region on the DNA close to the promoter and to an exposed region of the RNA polymerase, thereby stabilizing the binding of the RNA polymerase to the promoter. Others interact with RNA polymerase already at the promoter and allow it to form an open complex. Yet others remodel the promoter by scrunching the DNA, thereby making the spacing and orientation of the -10 and -35 regions more optimal.

9. Some regulatory proteins seem to bind to different regions on DNA depending on whether the effector is present, in this way acting as both repressors and activators.

(continued)

SUMMARY (continued)

10. Some operons are transcriptionally regulated by a mechanism called attenuation. In operons regulated by attenuation, transcription begins on the operon but then terminates after a short leader sequence has been transcribed if the enzymes encoded by the operon are not needed.

11. Attenuation of biosynthetic and degradative operons is sometimes determined by whether certain codons in the leader sequence are translated. Pausing of the ribosome at these codons can cause secondary-structure changes in the leader mRNA, leading to termination of transcription by RNA polymerase before it reaches the first gene of the operon. In other operons, attenuation is mediated by antiterminator proteins that can bind to antiterminator hairpin loops and either stabilize or destabilize them.

12. Other operons are regulated solely through changes in the 5′ untranslated sequence (5′UTR) on the mRNA. A temperature change can melt a secondary structure in an mRNA, allowing translation of the mRNA. Another type of regulation through the mRNA, called riboswitch regulation, occurs when binding of another small effector molecule to the mRNA alters a secondary structure, affecting translation or transcription of the operon.

13. The activity of biosynthetic operons is often not regulated solely through the transcription or translation of the genes of the operon. It may also be regulated through reversible regulation of the activity of the enzymes of the pathway. This reversible regulation can occur by feedback inhibition, which results from binding of the end product of the biosynthetic pathway to the first enzyme of the pathway, or it can occur by reversible covalent modification of an enzyme of the pathway.

QUESTIONS FOR THOUGHT

1. Why do you suppose both negative and positive mechanisms of transcriptional regulation are used to regulate bacterial operons?

2. Why are regulatory protein genes sometimes autoregulated?

3. Why do you suppose the genes for the biosynthesis of most amino acids such as arginine, tryptophan, and histidine are arranged together in an operon while the genes for methionine biosynthesis are scattered around the genome?

4. What advantages or disadvantages are there to regulation by attenuation? Would it not be less wasteful to regulate all operons through initiation of RNA synthesis at the promoter by repressors or activators?

PROBLEMS

1. Outline how you would isolate a *lacI*ˢ mutant of *E. coli*.

2. Is the AraC protein in the P1 or P2 state with D-fucose bound?

3. What would the phenotype of the following merodiploid *E. coli* cells be with one form of the operon region in the F′ factor and the other in the chromosome?

a. F′ *lac*⁺/*lacI*ˢ

b. F′ *lac*⁺/*lacO*ᶜ *lacZ*(Am) [the *lacZ*(Am) mutation is an N-terminal polar nonsense mutation]

c. F′ *ara*⁺/*araC* (the *araC* mutation is inactivating)

d. F′ *ara*⁺/*araI*

e. F′ *ara*⁺/*araP*₍BAD₎

4. The *phoA* gene of *E. coli* is turned on only if phosphate is limiting in the medium. What kind of genetic experiments would you do to determine whether the *phoA* gene is positively or negatively regulated?

5. Outline how you would use 5-methyltryptophan to isolate constitutive mutants of the *trp* operon and then use these to isolate feedback inhibition mutants.

6. The MalQ protein is involved in degrading polymers of maltose. Why are MalQ⁻ mutants constitutive for the expression of the *mal* operons?

7. Would you expect BglG⁻ mutants to be constitutive or superrepressed? BglF⁻ mutants?

SUGGESTED READING

Baba, T., T. Ara, M. Hasegawa, Y. Takai, Y. Okumura, M. Baba, K. A. Datsenko, M. Tomita, B. L. Wanner, and H. Mori. 2006. Construction of *Escherichia coli* K-12 in-frame, single-gene knockout mutants: the Keio collection. *Mol. Syst. Biol.* 2:2006.0008. [Online.] www.molecularsystemsbiology.com.

Barker, A., S. Oehler, and B. Müller-Hill. 2007. "Cold-sensitive" mutants of the Lac repressor. *J. Bacteriol.* 189: 2174–2175.

Bell, C. E., and M. Lewis. 2001. The Lac repressor: a second generation of structural and functional studies. *Curr. Opin. Struct. Biol.* 11:19–25.

Boos, W., and A. Bohm. 2000. Learning new tricks from an old dog: MalT of the *Escherichia coli* maltose system is part of a complex regulatory network. *Trends Genet.* 16: 404–409.

Busby, S., and R. H. Ebright. 1999. Transcription activation by catabolite activator protein (CAP). *J. Mol. Biol.* 293: 199–213.

Englesberg, E., C. Squires, and F. Meronk. 1969. The arabinose operon in *Escherichia coli* B/r: a genetic demonstration of two functional states of the product of a regulator gene. *Proc. Natl. Acad. Sci. USA* 62:1100–1107.

Fux, L., A. Nussbaum-Shochat, L. Lopian, and O. Amster-Choder. 2004. Modulation of monomer formation of the BglG transcriptional antiterminator from *Escherichia coli*. *J. Bacteriol.* 186:6775–6781.

Geanacopoulos, M., G. Vasmatzis, V. B. Zhurkin, and S. Adhya. 2001. Gal repressosome contains an antiparallel DNA loop. *Nat. Struct. Biol.* 8:432–436.

Gutierrez-Preciado, A., R. A. Jensen, C. Yanofsky, and E. Merino. 2005. New insights into regulation of the tryptophan biosynthetic operon in gram-positive bacteria. *Trends Genet.* 21:432–436.

Guzman, L. M., D. Belin, M. J. Carson, and J. Beckwith. 1995. Tight regulation, modulation, and high-level expression by vectors containing the arabinose P$_{BAD}$ promoter. *J. Bacteriol.* 177:4121–4130.

Irani, M. H., L. Orosz, and S. Adhya. 1983. A control element within a structural gene: the *gal* operon of *Escherichia coli*. *Cell* 32:783–788.

Jacob, F., and J. Monod. 1961. Genetic regulatory mechanisms in the synthesis of proteins. *J. Mol. Biol.* 3:318–356.

Joly, N., A. Bohm, W. Boos, and E. Richet. 2004. MalK, the ATP-binding cassette component of the *Escherichia coli* maltodextrin transporter, inhibits the transcriptional activator MalT by antagonizing inducer binding. *J. Biol. Chem.* 279:33123–33130.

Lewis, M., G. Chang, N. C. Horton, M. A. Kercher, H. C. Pace, M. A. Schumacher, R. G. Brennan, and P. Lu. 1996. Crystal structure of the lactose operon repressor and its complexes with DNA and inducer. *Science* 271: 1247–1254.

McDaniel, B. A., F. J. Grundy, and T. M. Henkin. 2005. A tertiary structural element in S box leader RNAs required for S-adenosylmethionine-directed transcription termination. *Mol. Microbiol.* 57:1008–1021.

Merino, E., and C. Yanofsky. 2005. Transcription attenuation: a highly conserved regulatory strategy used by bacteria. *Trends Genet.* 21:260–264.

Miller, J. H., and U. Schmeissner. 1979. Genetic studies of the *lac* repressor. X. Analysis of missense mutations in the *lacI* gene. *J. Mol. Biol.* 131:223–248.

Morse, D. E., and A. N. C. Morse. 1976. Dual control of the tryptophan operon is mediated by both tryptophanyl-tRNA synthetase and the repressor. *J. Mol. Biol.* 103:209–226.

Oxender, D. L., G. Zurawski, and C. Yanofsky. 1979. Attenuation in the *Escherichia coli* tryptophan operon. Role of RNA secondary structure involving the tryptophan codon region. *Proc. Natl. Acad. Sci. USA* 76:5524–5528.

Pace, H. C., M. A. Kercher, P. Lu, P. Markiewicz, J. H. Miller, G. Chang, and M. Lewis. 1997. Lac repressor genetic map in real space. *Trends Biochem. Sci.* 22:334–339.

Ramos, J. L., C. Michan, F. Rojo, D. Dwyer, and K. Timmis. 1990. Signal-regulator interactions: genetic analysis of the effector binding site of *xylS*, the benzoate-activated positive regulator of *Pseudomonas* Tol plasmid *meta*-cleavage pathway operon. *J. Mol. Biol.* 211:373–382.

Rojo, F. 1999. Repression of transcription initiation in bacteria. *J. Bacteriol.* 181:2987–2991.

Roy, S., S. Garges, and S. Adhya. 1998. Activation and repression of transcription by differential contact: two sides of a coin. *J. Biol. Chem.* 273:14059–14062.

Schlegal, A., O. Danot, E. Richet, T. Ferenci, and W. Boos. 2002. The N-terminus of the *Escherichia coli* transcription activator MalT is the domain of interaction with MalY. *J. Bacteriol.* 184:3069–3077.

Schleif, R. 2000. Regulation of the L-arabinose operon of *Escherichia coli*. *Trends Genet.* 16:559–565.

Schmeissner, U., D. Ganem, and J. H. Miller. 1977. Genetic studies of the *lac* repressor. II. Fine structure deletion map of the *lacI* gene, and its correlation with the physical map. *J. Mol. Biol.* 109:303–326.

Snyder, D., J. Lary, Y. Chen, P. Gollnick, and J. L. Cole. 2004. Interaction of the *trp* RNA-binding attenuation protein (TRAP) with antiTRAP. *J. Mol. Biol.* 338:669–682.

Trinh, V., M.-F. Langelier, J. Archambault, and B. Coulombe. 2006. Structural perspective on mutations affecting the function of multisubunit RNA polymerases. *Microbiol. Mol. Biol. Rev.* **70:**12–36.

Wray, L. V., Jr., and S. V. Fisher. 2007. Functional analysis of the carboxy-terminal region of *Bacillus subtilis* TnrA, a MerR family protein. *J. Bacteriol.* **189:**20–27.

Yanofsky, C., and I. P. Crawford. 1987. The tryptophan operon, p. 1453–1472. *In* F. C. Neidhardt, J. L. Ingraham, K. B. Low, B. Magasanik, M. Schaechter, and H. E. Umbarger (ed.), Escherichia coli *and* Salmonella typhimurium: *Cellular and Molecular Biology*, vol. 2. American Society for Microbiology, Washington, D.C.

Yousef, M. R., F. J. Grundy, and T. M. Henkin. 2005. Structural transitions induced by the interaction between tRNA^{Gly} and the *Bacillus subtilis glyQS* T box leader RNA. *J. Mol. Biol.* **349:**273–287.

CHAPTER 13

Global Regulation: Regulons and Stimulons

Bacteria must be able to adapt to a wide range of environmental conditions to survive. Nutrients are usually limiting, so bacteria must be able to protect themselves against starvation until an adequate food source becomes available. Different environments also vary greatly in the amount of water or in the concentration of solutes, so bacteria must also be able to adjust to desiccation and differences in osmolarity. Temperature fluctuations are also a problem for bacteria. Unlike humans and other warm-blooded animals, bacteria cannot maintain their own cell temperature and so must be able to function over wide ranges of temperature.

Mere survival is not enough for a species to prevail, however. The species must compete effectively with other organisms in the environment. Competing effectively might mean being able to use scarce nutrients efficiently or taking advantage of plentiful ones to achieve higher growth rates and thereby become a higher percentage of the total population of organisms in the environment. Moreover, different compounds may be available for use as carbon and energy sources. The bacterium may need to choose the carbon and energy source it can use most efficiently and ignore the rest, so that it does not waste energy making extra enzymes.

Not only do conditions vary in the environment, but their changes can be abrupt. The bacterium may have to adjust the rate of synthesis of its cellular constituents quickly in response to the change in growth conditions. For example, different carbon and energy sources allow different rates of bacterial growth. Different growth rates require different rates of synthesis of cellular macromolecules such as DNA, RNA, and proteins, which in turn require different concentrations of the components of the cellular macromolecular synthesis machinery such as ribosomes, tRNA, and RNA polymerase. Moreover, the relative rates of synthesis of the different cellular components must be coordinated so that the cell does not accumulate more of some components than it needs.

Adjusting to major changes in the environment requires regulatory systems that simultaneously regulate numerous operons. These systems are called **global regulatory mechanisms.** Often in global regulation, a single regulatory protein controls a large number of operons, which are then said to be members of the same **regulon.** Most genes are part of some regulon, and some regulons are very large. Regulons are often overlapping in their response to changing conditions. The collection of regulons that respond to the same environmental conditions is called a **stimulon.** Microarray analysis can be used to find essentially all of the genes of a regulon and stimulon. Such studies reveal that only seven regulators control almost one half of all the genes of *Escherichia coli.*

Table 13.1 lists some global regulatory mechanisms known to exist in *E. coli.* If the genes are under the control of a single regulatory gene (and so are members of the same regulon), the regulatory gene is also listed. Some examples of regulons are discussed in previous chapters. For example, all the genes under the control of the TrpR repressor, including the *trpR* gene itself, are part of the TrpR regulon. The Ada regulon comprises the adaptive response genes, including those encoding the methyltransferases that repair alkylation damage to DNA (see chapter 11); all of these genes are under the control of the Ada protein. Similarly, the SOS genes induced after UV irradiation and some other types of DNA-damaging treatments are all under the control of the same protein, the LexA repressor, and so are part of the LexA regulon. In other cases, the molecular basis of the global regulation is less well understood and may involve a complex interaction between several cellular signals.

In this chapter, we discuss what is known about how some global regulatory mechanisms operate on the molecular level and describe some of the genetic experiments that have contributed to this knowledge. The ongoing studies of the molecular basis of global regulatory mechanisms represent one of the most active areas of research involving bacterial molecular genetics; therefore, new developments will probably have occurred in many of these areas by the time you read this book.

Catabolite-Sensitive Operons

One of the largest global regulatory systems in bacteria coordinates the expression of genes involved in carbon and energy source utilization. All cells must have access to high-energy, carbon-containing compounds, which they degrade to generate ATP for energy and smaller molecules needed for cellular constituents. Smaller molecules resulting from the metabolic breakdown of larger molecules are called **catabolites.**

In times of plenty, bacterial cells may be growing in the presence of several different carbon and energy sources, some of which can be used more efficiently than others. Energy must be expended to synthesize the enzymes needed to metabolize the different carbon sources, and the utilization of some carbon compounds requires more enzymes than does the utilization of others. By making only the enzymes for the carbon and energy source that yields the highest return, the cell gets the most catabolites and energy, in the form of ATP, for the energy it expends. The mechanism for ensuring that the cell preferentially uses the best carbon and energy source available is called **catabolite repression,** and operons subject to this type of regulation are **catabolite sensitive.** The name "catabolite repression" originates from the fact that cells growing in better carbon sources have more catabolites, which seem to repress the transcription of operons for the utilization of poorer carbon sources. However, as we shall see, the name "catabolite repression" is sometimes a misnomer because, at least in some of the regulatory systems in *E. coli,* the genes under catabolite control are *activated* when poorer carbon sources are the only ones available. Catabolite repression is also sometimes called the **glucose effect** because glucose, which yields the highest return of ATP per unit of expended energy, usually strongly represses operons for other carbon sources. To use glucose, the cell need only convert it to glucose-6-phosphate, which can enter the glycolytic pathway. Thus, glucose is the preferred carbon and energy source for most types of bacteria.

Figure 13.1 illustrates what happens when *E. coli* cells are growing in a mixture of glucose and galactose. The cells first use the glucose, and only after it is depleted do they begin using the galactose. When the glucose is gone, the cells stop growing briefly while they synthesize the enzymes for galactose. Growth then resumes but at a slightly lower rate.

cAMP and the cAMP-Binding Protein

Most bacteria and lower eukaryotes are known to have systems for catabolite repression. The best understood is the **cyclic AMP (cAMP)**-dependent system of *E. coli* and other enteric bacteria. cAMP is like AMP (see chapter 2), with a single phosphate group on the ribose sugar, but the phosphate is attached to both the 5′-hydroxyl and the 3′-hydroxyl groups of the sugar, thereby making a circle out of the phosphate and sugar. Only *E. coli* and other closely related enterics seem to use this cAMP-dependent system. Some bacteria have an entirely different system, which does not involve cAMP. Even *E. coli*

TABLE 13.1	A sampling of *E. coli* global regulatory systems			
System	Response	Regulatory gene(s) (protein[s])	Category of mechanism	Some genes, operons, regulons, and stimulons
Nutrient limitation				
Carbon	Catabolite repression	*crp* (CAP, also called CRP)	DNA-binding activator or repressor	*lac, ara, gal, mal*, and numerous other C source operons
	Control of fermentative vs. oxidative metabolism	*cra* (*fruR*) (CRA)	DNA-binding activator or repressor	Enzymes of glycolysis, Krebs cycle
Nitrogen	Response to ammonia limitation	*rpoN* (NtrA)	Sigma factor (σ^{54})	*glnA* (GS) and operons for amino acid degradation
		ntrBC (NtrBC)	Two-component system	
Phosphorus	Starvation for inorganic orthophosphate (P$_i$)	*phoBR* (PhoBR)	Two-component system	>38 genes, including *phoA* (bacterial alkaline phosphatase) and *pst* operon (P$_i$ uptake)
Growth limitation				
Stringent response	Response to lack of sufficient aminoacylated-tRNAs for protein synthesis	*relA* (RelA), *spoT* (SpoT)	(p)ppGpp metabolism	rRNA, tRNA, ribosomal proteins
Stationary phase	Switch to maintenance metabolism and stress protection	*rpoS* (RpoS)	Sigma factor (σ^S)	Many genes with σ^S promoters; complex effects on many operons
Oxygen	Response to anaerobic environment	*fnr* (Fnr)	CAP family of DNA-binding proteins	>31 transcripts, including *narGHJI* (nitrate reductase)
	Response to presence of oxygen	*arcAB* (ArcAB)	Two-component system	>20 genes, including *cob* (cobalamin synthesis)
Stress				
Osmoregulation	Response to abrupt osmotic upshift	*kdpDE* (KdpD, KdpE)	Two-component system	*kdpFABC* (K$^+$ uptake system)
	Adjustment to osmotic environment	*envZ/ompR* (EnvZ/OmpR)	Two-component system	OmpC and OmpF outer membrane proteins
		micF	Antisense RNA	*ompF* (porin)
Oxygen stress	Protection against reactive oxygen species	*soxS* (SoxS)	AraC family of DNA-binding proteins	Regulon, including *sodA* (superoxide dismutase) and *micF* (antisense RNA regulator of *ompF*)
		oxyR (OxyR)	LysR family of DNA-binding proteins	Regulon, including *katG* (catalase)
Heat shock	Tolerance of abrupt temperature increase	*rpoH* (RpoH)	Sigma factor (σ^{32})	Stimulon, Hsps (heat shock proteins), including *dnaK, dnaJ*, and *grpE* (chaperones), and *lon, clpP, clpX*, and *hflB* (proteases)
Envelope stress	Misfolded Omp proteins	*rpoE* (RpoE)	Sigma factor (σ^E)	>10 genes, including *rpoH* (σ^{32}) and *degP* (encoding a periplasmic protease)
	Misfolded pilus	*cpxAR* (CpxAR)	Two-component system	Overlap with RpoE regulon
pH shock	Tolerance of acidic environment	Many	Many	Complex stimulon

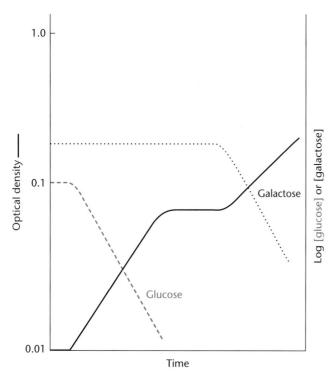

Figure 13.1 Growth of *E. coli* in a mixture of glucose and galactose. The concentration of the sugars in the medium is shown as the dashed lines. The optical density, a measure of cell density, is shown in black. Only after the cells deplete all the glucose do they begin to use the galactose, and then only after a short lag while they induce the *gal* operon (plateau in optical density). They then grow more slowly on the galactose. See the text for details.

has a second, cAMP-independent system for catabolite repression, which is discussed in Box 13.1.

REGULATION OF cAMP SYNTHESIS

Catabolite regulation in *E. coli* is achieved through fluctuation in the levels of cAMP, which vary inversely with the levels of cellular catabolites. In other words, cellular concentrations of cAMP are higher when catabolite levels are lower, the situation that prevails when the bacteria are growing in a relatively poor source of carbon such as lactose or maltose. The synthesis of cAMP is controlled through the regulation of the activity of adenylate cyclase. This enzyme, which makes cAMP from ATP, is more active when cellular concentrations of catabolites are low and less active when catabolite concentrations are high. The adenylate cyclase enzyme is associated with the inner membrane and is the product of the *cya* gene.

Figure 13.2 outlines the current picture of the regulation of adenylate cyclase activity. An important factor in the regulation is the phosphoenolpyruvate (PEP)-dependent sugar phosphotransferase system (PTS), which,

as the name implies, is responsible for transporting certain sugars, including glucose, into the cell. We mentioned PTS systems in connection with the regulation of the *mal* and *bgl* operons in chapter 12. One of the protein components of the PTS, named IIA^{Glc}, can exist in either an unphosphorylated (IIA^{Glc}) or a phosphorylated ($IIA^{Glc}{\sim}P$) form. The $IIA^{Glc}{\sim}P$ form activates adenylate cyclase to make cAMP. However, $IIA^{Glc}{\sim}P$ forms in low concentration when glucose or another of the sugars it is involved in transport of is present in the medium. Thus, little of the $IIA^{Glc}{\sim}P$ form then exists to activate the adenylate cyclase, and cAMP levels drop.

The ratio of $IIA^{Glc}{\sim}P$ to IIA^{Glc} is determined largely by the ratio of PEP to pyruvate in the cell. When a rapidly metabolizable substrate such as glucose is present in the medium, the PEP/pyruvate ratio is low; when only poorer carbon sources are available, this ratio is high (Box 13.1). The PEP transfers its phosphate to another protein called Hpr (for *h*istidine *p*rotein, which is the amino acid in this protein to which it is transferred) and becomes pyruvate. The phosphate from Hpr~P is then transferred to IIA^{Glc} to make $IIA^{Glc}{\sim}P$. Therefore, the higher the PEP/pyruvate ratio, the higher the Hpr~P/Hpr ratio and the higher the $IIA^{Glc}{\sim}P/IIA^{Glc}$ ratio. This is called a **phosphorylation cascade** because phosphates are transferred from one molecule to another, much like water is transferred down a cascade of waterfalls. We give other examples of phosphorylation cascades later in this chapter.

The unphosphorylated form of IIA^{Glc} also inhibits other sugar-specific permeases that transport sugars such as lactose (Fig. 13.2). Therefore, less of these other sugars enters the cell if glucose or another better carbon source is available, and hence less inducer is present to induce synthesis from their respective operons (see chapter 12). This effect is called **inducer exclusion** (Figure 13.2). In fact, it is difficult to distinguish effects of inducer exclusion on operon induction from effects of cAMP on the promoter (see Inada et al., Suggested Reading, and below).

CATABOLITE ACTIVATOR PROTEIN

The mechanism by which cAMP turns on catabolite-sensitive operons in *E. coli* is quite well understood and has served as a model for transcriptional activation. The cAMP binds to the protein product of the *crp* gene, which is an activator of transcription of catabolite-sensitive operons. This activator protein goes by two names, CAP (for *c*atabolite gene *a*ctivator *p*rotein) and CRP (for *c*AMP *r*eceptor *p*rotein). We refer to it both ways, but we most often call it CAP. The activator CAP with cAMP bound (CAP-cAMP) functions like other activator proteins discussed in chapter 12 in that it interacts with

BOX 13.1

cAMP-Independent Catabolite Repression

Not all catabolite repression in bacteria is mediated by cAMP. In fact, most gram-positive bacteria, including *B. subtilis,* do not even have cAMP. Even in *E. coli* and other enteric bacteria, there is a mechanism of catabolite repression that does not depend on cAMP. This mechanism involves the Cra protein, named for its function as a *catabolite repressor/activator* (sometimes called FruR, for *fructose repressor*). The Cra protein is encoded by the *cra* gene and is a DNA-binding protein similar to LacI and GalR. Cra was discovered during the isolation of mutants with mutations that suppress *ptsH* mutations of *E. coli* and *S. enterica* serovar Typhimurium. The *ptsH* gene encodes the Hpr protein, which phosphorylates many sugars during transport that can be transported by the PTS system, including glucose (Figure 13.2). Therefore, *ptsH* mutants cannot use these sugars, because phosphorylation is the first step in the glycolytic pathway. The *cra* mutations suppress *ptsH* mutations and allow growth on PTS sugars by allowing the constitutive expression of the fructose catabolic operon, one of whose genes encodes a protein that can sub-

stitute for Hpr. The *cra* mutants were found to be pleiotropic in that they are unable to synthesize glucose from many substrates, including acetate, pyruvate, alanine, and citrate. However, they demonstrate elevated expression of genes involved in glycolytic pathways.

The pleiotropic phenotype of *cra* mutants suggested that Cra functioned as a global regulatory protein, activating the transcription of some genes and repressing the synthesis of others. As for other regulatory proteins, whether Cra activates or represses transcription depends on where it binds to the operon. If its binding site is upstream of the promoter, it activates transcription of the operon; if its binding site overlaps or is downstream of the promoter, it represses transcription. In either case, Cra comes off the DNA if it binds the sugar catabolites fructose-1-phosphate or fructose-1,6-bisphosphate (FBP) that are present in high concentrations during growth in the presence of sugars such as glucose. The effect of this on the transcription of a particular operon depends on whether Cra functions as a repressor or activator of that

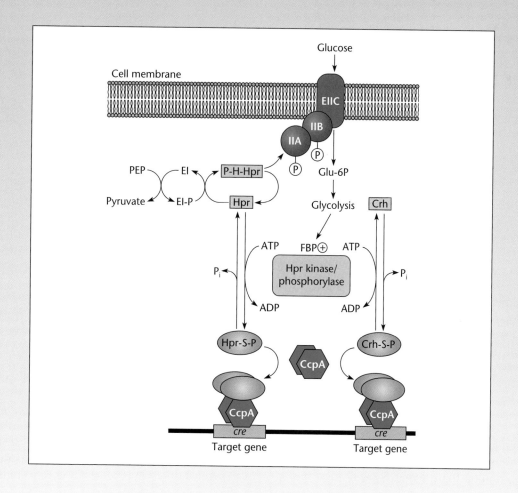

(continued)

BOX 13.1 (continued)

cAMP-Independent Catabolite Repression

operon. If it functions as a repressor, the transcription of the operon increases; if it functions as an activator, the transcription of the operon decreases—a process called antiactivation. In general, the Cra protein functions as a repressor of operons whose products are involved in alternate pathways for sugar catabolism, such as the Embden-Meyerhof and Entner-Doudoroff pathways, so that the transcription of these genes increases when glucose and other good carbon sources are available. In contrast, it usually activates operons whose products are involved in synthesizing glucose from acetate and other metabolites (gluconeogenesis), and so it does not activate the transcription of these operons if glucose is present. Why make glucose if some is already available in the medium?

It is in *B. subtilis* where cAMP-independent catabolite repression has been most extensively studied. This pathway is illustrated in the figure. As mentioned, this bacterium does not have cAMP and so depends exclusively on cAMP-independent pathways to regulate its carbon source utilization pathways. In fact, diauxic growth (Figure 13.1) was first discovered in this bacterium. In some ways, the pathway for catabolite repression in *B. subtilis* is similar to the cAMP-independent pathway in *E. coli*, in that there is a repressor protein called CcpA (for *c*atabolite *c*ontrol *p*rotein *A*), which is a member of the LacI and GalR family of regulators. The CcpA repressor binds to operator sites called *cre* sites (for *c*atabolite *re*pressor sites) in the promoters of many catabolite-sensitive genes, repressing their transcription. Almost 100 genes in *B. subtilis* are under the control of this repressor. However, whether the CcpA repressor binds to the *cre* operators depends on the state of phosphorylation of another protein

called Hpr (for *h*istidine *p*rotein). If cells are growing in a good carbon source such as glucose, high levels of intermediates in the glycolytic pathway accumulate, including FBP. High levels of FBP stimulate the phosporylation of Hpr on one of its serines, which converts it into a corepressor that binds to the CcpA and allows it to bind to the *cre* operators, repressing transcription. Interestingly, the same Hpr protein which serves as a corepressor with CcpA when it is phosphorylated on the serine also serves as the phosphate donor in the PTS for sugar transport when it is phosphorylated on a histidine. Apparently, phosphorylation of Hpr at the serine can inhibit phosphorylation at the histidine and inhibit the transport of sugars that use this system. This allows the close coordination of sugar transport and the regulation of catabolite-sensitive operons. Another corepressor of CcpA called Crh, which is also phosphorylated on one of its serines in the presence of high FBP levels, also exists in *B. subtilis* and acts on some of the catabolite-sensitive operons. However, this second corepressor may exist only in strains of *Bacillus*, and its significance is unclear.

References

Deutscher, J., C. Francke, and P. W. Postma. 2006. How phosphotransferase system-related protein phosphorylation regulates carbohydrate metabolism in bacteria. *Microbiol. Mol. Biol. Rev.* **70:**939–1031.

Lorca, G. L., Y. J. Chung, R. D. Barabote, W. Weyler, C. H. Schilling, and M. H. Saier, Jr. 2005. Catabolite repression and activation in *Bacillus subtilis*: dependency on CcpA, HPr, and HprK. *J. Bacteriol.* **187:**7826–7839.

Warner, J. B., and J. S. Lolkema. 2003. CcpA-dependent catabolite repression in bacteria. *Microbiol. Mol. Biol. Rev.* **67:**475–490.

the RNA polymerase to activate transcription from promoters for operons under its control, including *lac*, *gal*, *ara*, and *mal*. These operons are all members of the **CAP regulon,** or the catabolite-sensitive regulon (Table 13.1). However, the mechanism of CAP-cAMP regulation varies. As discussed below, CAP can function not only as an activator but also as a repressor, depending on where it binds relative to the promoter (see below).

REGULATION OF *lac* BY CAP-cAMP
The mechanism by which CAP activates transcription varies from promoter to promoter. Some of these mecha-

nisms are shown in Figure 13.3. Upstream of the promoter is a short sequence called the **CAP-binding site,** which is similar in all catabolite-sensitive operons and so can be easily identified. CAP can bind to this site only when it is bound to cAMP, so that this site is occupied only when cAMP levels are high. CAP functions like other activators to make contact with the RNA polymerase at the promoter and stimulate one or more of the steps in the initiation of transcription (see Busby and Ebright, Suggested Reading). Transcriptional activators are discussed in chapter 12. CAP is interesting in that it can contact different regions of the RNA polymerase and

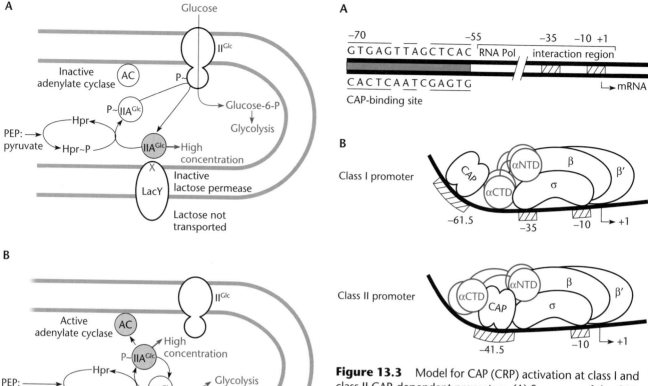

A

Figure 13.2 Exogenous glucose inhibits both cAMP synthesis and the uptake of other sugars, such as lactose. (A) In the presence of glucose, the ratio of IIA^{Glc} to $IIA^{Glc}\sim P$ is high, as glucose is phosphorylated on transport by the glucose transporter, II^{Glc}. Unphosphorylated IIA^{Glc} inhibits the lactose permease, resulting in "inducer exclusion." (B) In the absence of glucose, the $IIA^{Glc}\sim P$ concentration is high, and it activates adenylate cyclase. Also, lactose transport is permitted. This is an oversimplified model of the way in which this complex regulation occurs in different bacteria. For more complete accounts, see Deutscher et al., cited in Box 13.1, and Pompeo et al., Suggested Reading.

Figure 13.3 Model for CAP (CRP) activation at class I and class II CAP-dependent promoters. (A) Sequence of the CAP binding site upstream of the class I *lac* promoter. (B) The binding and location interactions of CAP with the C-terminal and N-terminal domains of the α subunit with class I and class II promoters, respectively.

stimulate different steps in initiation, depending on where it is bound relative to the promoter. This is illustrated in Figure 13.3. At so-called class I CAP-dependent promoters, for example, the *lac* promoter, a dimer of CAP with cAMP, binds upstream of the promoter and contacts the C-terminal end of the α subunit of RNA polymerase (αCTD). This contact strengthens the

binding of RNA polymerase to the promoter (closed complex). In other promoters, i.e., the so-called class II CAP-dependent promoters such as the *gal* promoter p_{G1}, the CAP dimer-binding site slightly overlaps that of RNA polymerase, and it contacts a region in the N terminus of the α subunit (αNTD). In this position, it stimulates the opening of the DNA at the promoter (the open complex). There are even promoters in which more than one CAP dimer binds and that stimulate both steps of initiation. CAP also bends the DNA when it binds to the CAP-binding site, although the significance of this bending is unknown.

The positioning of the CAP-binding sequence relative to the promoter can be very different from that of *lac* and *gal*. In the *ara* operon, the CAP-binding site is far upstream, with the AraC-binding site between it and the promoter (Figure 13.4). Nevertheless, it can still make contact with the C terminus of the α subunit of RNA polymerase, which can reach far up the DNA, as shown. In cases such as these, CAP can also stimulate transcription by interacting with the activator or can stimulate transcription by preventing the binding of a repressor.

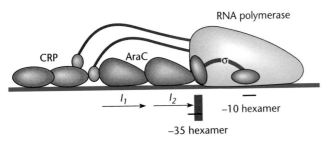

Figure 13.4 Summary of the polymerase-promoter and activator-promoter interactions at p_{BAD}. The σ^{70} subunit of RNA polymerase contacts the –35 and –10 hexamers. Occupancy of the I_1 and I_2 half-sites by AraC activates transcription with the aid of CAP, most probably utilizing the α-subunit–activator interactions as shown. The binding sites of σ^{70} and AraC overlap by 4 bp at p_{BAD}. The nucleotides in the –35 hexamer that lie outside the region of overlap are shaded.

Some operons in the CAP regulon, such as *gal*, are less sensitive to catabolite repression than others. As discussed in chapter 12, the *gal* operon is not totally repressed by glucose in the medium because it has two promoters, p_{G1} and p_{G2} (see Figure 12.10). p_{G2} does not require CAP for its activation. In fact, CAP-cAMP represses this promoter, probably because it binds to the –35 sequence of the promoter. The p_{G2} promoter permits some expression of the *gal* operon even in the presence of glucose. This low level of expression is necessary to allow the synthesis of cell wall components that include galactose, since the UDP-galactose synthesized by the operon serves as the donor of galactose in biosynthetic reactions. However, the level of expression of the *gal* operon from p_{G2} is not high enough for the cells to grow well on galactose as a carbon and energy source.

RELATIONSHIP OF CATABOLITE REPRESSION TO INDUCTION

An important point about CAP regulation of catabolite-sensitive operons is that it occurs in addition to any other regulation to which the operon is subject (Figure 13.5). Two conditions must be met before catabolite-sensitive operons can be transcribed: better carbon sources such as glucose must be absent, and the inducer of the operon must be present. Take the example of the *lac* operon. If a carbon source better than lactose is available, cAMP levels are low and so CAP-cAMP does not bind upstream of the *lac* promoter to activate transcription. Also, the *lac* transport system is inhibited, excluding the inducer from the cell. However, even at high cAMP levels, the *lac* operon is not transcribed unless the inducer, allolactose, is also present. In the absence of inducer, the LacI repressor is bound to the operator and prevents the RNA poly-

merase from binding to the promoter and transcribing the operon (see chapter 12).

Genetic Analysis of Catabolite Regulation in *E. coli*

The above model for the regulation of catabolite-sensitive operons is supported by both genetic and biochemical analysis of catabolite repression in *E. coli*. This analysis has involved the isolation of mutants defective in the global regulation of all catabolite-sensitive operons, as well as mutants defective in the catabolite regulation of specific operons.

ISOLATION OF *crp* AND *cya* MUTATIONS
According to the model presented above, mutations that inactivate the *cya* and *crp* genes for adenylate cyclase and CAP, respectively, should prevent transcription of all the catabolite-sensitive operons. In these mutants, there is no CAP with cAMP attached to bind to the promoters. In other words, *cya* and *crp* mutants should be Lac⁻, Gal⁻, Ara⁻, Mal⁻, and so on. In genetic terms, *cya* and *crp* mutations are **pleiotropic**, because they cause many phenotypes, i.e., the inability to use many different sugars as carbon and energy sources.

The fact that *cya* and *crp* mutations should prevent cells from using several sugars was used in the first isolations of *crp* and *cya* mutants (see Schwartz and Beckwith, Suggested Reading). The selection was based on the fact that colonies of bacteria turn tetrazolium salts red while they multiply, provided that the pH remains high. However, bacteria that are fermenting a carbon source give off organic acids such as lactic acid that lower the pH, preventing the conversion to red. As a consequence, wild-type *E. coli* cells growing on a fermentable carbon source form white colonies on tetrazolium-containing plates whereas mutant bacteria that cannot use the fermentable carbon source utilize a different carbon source and so form red colonies. Some of these red-colony-forming mutants might have *crp* or *cya* mutations, although most would have mutations that inactivate a gene within the operon for the utilization of the fermentable carbon source. Thus, without a way to increase the frequency of *crp* and *cya* mutants among the red-colony-forming mutants, many red-colony-forming mutants would have to be tested to find any with mutations in either *cya* or *crp*.

For these experiments, the investigators reasoned that they could increase the frequency of *crp* and *cya* mutants by plating heavily mutagenized bacteria on tetrazolium agar containing *two different* fermentable sugars, for example, lactose and galactose or arabinose and maltose. Then, to prevent the utilization of both sugars, either two mutations, one in each operon, or a single mutation,

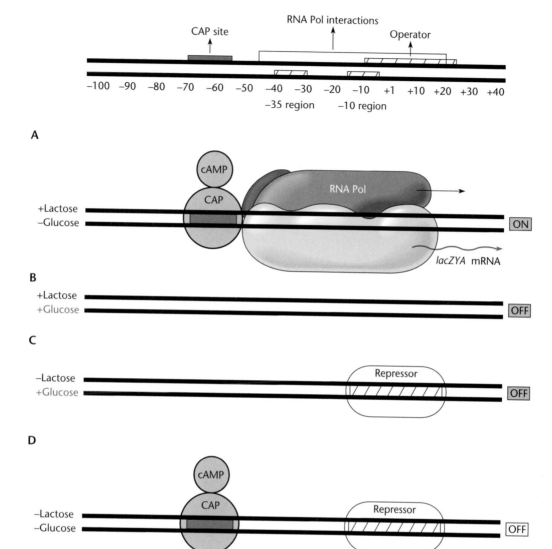

Figure 13.5 Regulation of the lactose operon by both glucose and the inducer lactose. (A) The operon is on only in the absence of glucose and the presence of lactose, which is converted into the inducer, allolactose. (B and C) The operon is off in the presence of glucose whether or not lactose is present, because the CAP-cAMP complex is not bound to the CAP site. (D) The operon is also off if lactose is not present, even if glucose is also not present, because the repressor is bound to the operator. The relative positions of the CAP-binding site, operator, and promoter are shown. The entire regulatory region covers about 100 bp of DNA.

in *cya* or *crp*, would have to occur. Since mutants with single mutations should be much more frequent than mutants with two independent mutations, the *crp* and *cya* mutants should be a much higher fraction of the total red-colony-forming mutants growing on two carbon and energy sources. Indeed, when the red-colony-forming mutants that could not use either of the two sugars provided were tested, most of them were found to be deficient in adenylate cyclase activity or to lack a protein,

now named CAP, later shown to be required for the activation of the *lac* promoter in vitro.

PROMOTER MUTATIONS THAT AFFECT CAP ACTIVATION

Genetic experiments with the *lac* promoter also contributed to the models of CAP activation. Three classes of mutations have been isolated in the *lac* promoter. Those belonging to class I change the CAP-binding site

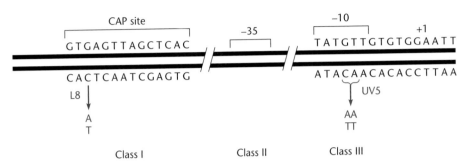

Figure 13.6 Mutations in the *lac* regulatory region that affect activation by CAP. The class I mutation L8 changes the CAP-binding site so that CAP can no longer bind and the promoter cannot be turned on even in the absence of glucose (high cAMP). The class III mutation UV5 changes 2 bp in the −10 sequence of the promoter so that the promoter no longer requires activation by CAP and the operon can be induced even in the presence of glucose (low cAMP). No class II mutations are shown. The changes in the sequence in each mutation appear in purple.

so that CAP can no longer bind to it. The *lac* promoter mutation L8 is an example (Figure 13.6). By preventing the binding of CAP-cAMP upstream of the promoter, this mutation weakens the *lac* promoter. As a result, the *lac* operon is expressed poorly, as measured by β-galactosidase activity, even when cells are growing in lactose and cAMP levels are high. However, the low level of expression of the *lac* operon is less strongly affected by the carbon source and is not reduced much more if glucose is added and cAMP levels drop.

Other promoter mutations called class II mutations change the −35 region of the RNA polymerase-binding site so that the promoter is less active even when cAMP levels are high. However, with this type of mutation, the residual expression of the *lac* operon is still sensitive to catabolite repression. Consequently, the amount of β-galactosidase synthesized when cells are growing in the presence of lactose plus glucose while cAMP levels are low is smaller than the amount synthesized when the cells are growing in the presence of a poorer carbon source plus lactose when cAMP levels are high.

A third, very useful mutated *lac* promoter, class III, was found by isolating apparent Lac⁺ revertants of class I mutations such as p_{L8} or of *cya* or *crp* mutations. One such mutation is called *placUV5*. This mutant promoter is almost as strong as the wild-type *lac* promoter but no longer requires cAMP-CAP for activation. As shown in Figure 13.6, the *lacUV5* mutation changes a 2-bp stretch of the −10 region of the *lac* promoter, so that the sequence reads TATAAT instead of TATGTT. This mutant −10 sequence more closely resembles the sequence of a consensus σ⁷⁰ promoter (see chapter 2), perhaps explaining why it no longer requires CAP. Some expression vectors use the *lacUV5* promoter rather than the wild-type *lac* promoter, so that the promoter can be

induced even if the bacteria are growing in glucose-containing media.

INTERACTION OF RNA POLYMERASE WITH CAP
Biochemical experiments first identified the carboxyl terminus of the α subunit as a region of the RNA polymerase that contacts CAP for some promoters (see Igarashi and Ishihama, Suggested Reading). In these experiments, the α subunits of RNA polymerase were synthesized in vitro by using a clone of the *rpoA* gene, the gene for the α subunit. If the region of the *rpoA* gene encoding the carboxyl terminus of the α subunit is deleted from the clone, an RNA polymerase containing the truncated α subunits can be assembled and is still active for transcription from most promoters. However, the defective RNA polymerase cannot initiate transcription from catabolite-sensitive promoters, even in the presence of CAP-cAMP. These experiments therefore suggested that CAP interacts with the carboxyl-terminal portion of the α subunit of RNA polymerase to activate transcription from class I catabolite-sensitive promoters. Genetic experiments confirmed this role of the carboxyl-terminal region of the α subunit and identified some of the amino acids important for the interaction (see Savery et al., Suggested Reading).

Uses of cAMP in Other Organisms

The use of cAMP to regulate catabolite-sensitive operons seems unique to enteric bacteria such as *E. coli*. Other bacteria use different mechanisms to regulate catabolite-sensitive operons (Box 13.1). Many bacteria do not appear to make cAMP at all, whereas in others the levels of cAMP do not vary depending on the carbon sources available. Furthermore, the CAP-cAMP-mediated catabolite regulatory system is not the only system used by *E. coli* and other enteric bacteria (Box 13.1).

In eukaryotes, cAMP does not regulate carbon source utilization but does have many other uses, including the regulation of G proteins and roles in cell-to-cell communication. In view of the general importance of cAMP in various biological systems, it is somewhat surprising that, among the bacteria, only the enteric bacteria seem to use this nucleotide to regulate carbon source utilization.

A Bacterial Two-Hybrid System Based on Adenylate Cyclase

A two-hybrid system named the BACTH (for *b*acterial *a*denylate *c*yclase *t*wo-*h*ybrid) system, for detecting protein-protein interactions is based on interaction-mediated reconstitution of the cAMP signaling cascade (see Karimova et al., 1998, Suggested Reading) (Figure 13.7A).

Figure 13.7 A cAMP-based bacterial two-hybrid system for detecting protein-protein interactions. (A) Synthesis of cAMP by the *B. pertussis* adenylate cyclase requires that the two regions T25 and T18 of the cyclase bind to each other. The coding regions for the two domains are fused in frame to the coding regions of potentially interacting proteins or regions of proteins (P1 and P2 are shown in purple), and introduced into a Δ*cya* mutant of *E. coli*. If the two domains P1 and P2 interact, they bring the domains of adenylate cyclase together, cAMP is synthesized, and the *mal* and *lac* operons are expressed. (B) This system can be used to investigate the interaction of membrane proteins. If X and Y interact, cAMP is synthesized.

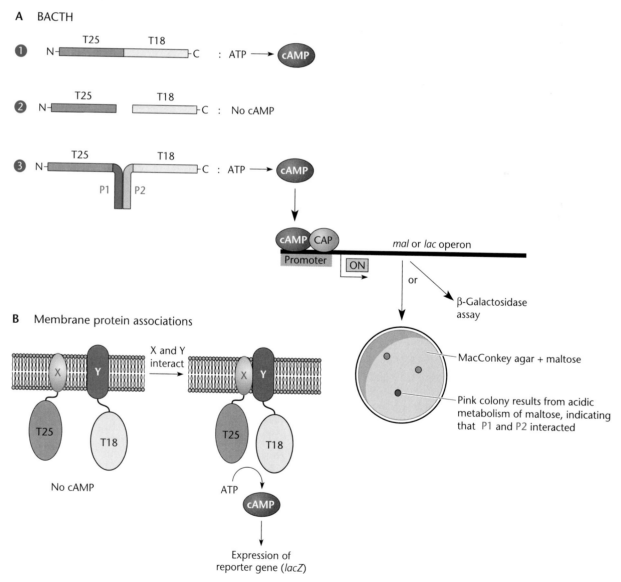

Based on their experience, some investigators think that this two-hybrid system is best used for determining whether two specific proteins interact rather than for "fishing expeditions" to find proteins that interact, an application for which the yeast two-hybrid system is preferred. The *Bordetella pertussis* adenylate cyclase enzyme has been separated into two complementing fragments: T25, from amino acids 1 to 224, and T18, from amino acids 225 to 399. Functional complementation of these two fragments results in cAMP synthesis and so allows transcription of cAMP-dependent catabolic operons. Clones expressing the two *B. pertussis* fragments, T25 and T18, can be introduced into an *E. coli* mutant with an adenylate cyclase mutation, e.g., Δ*cya*, and the *Bordetella* adenylate cyclase complements the *E. coli* mutation provided the two fragments come together to form the functional adenyl cyclase. Thus functional complementation of T25 and T18 can be monitored by culture assay of β-galactosidase or by plating colonies on MacConkey plus maltose indicator plates. The essence of the use of these T25 and T18 gene fragments is that functional complementation occurs only if the fragments are fused to proteins that interact with each other since the two fragments do not interact by themselves.

A notable advantage of the BACTH system is that it does not require that interactions between hybrid proteins take place in association with the transcriptional complex, unlike the yeast two-hybrid system. For this reason, it seems to work well for membrane proteins. For example, in one application, protein-protein interactions of the *E. coli* membrane-associated septum assembly proteins were investigated (Figure 13.7B) (see Karimova et al., 2005, Suggested Reading). Another example of its use was to identify a dimerization motif internal to a *Streptomyces* developmental regulator (see Hudson and Nodwell, Suggested Reading).

Regulation of Nitrogen Assimilation

Nitrogen is a component of many biological molecules, including nucleotides, vitamins, and amino acids. Thus, all organisms must have a source of nitrogen atoms for growth to occur. For bacteria, possible sources include ammonia (NH_3) and nitrate (NO_3^-), as well as nitrogen-containing organic molecules such as the bases in nucleosides and amino acids. Some bacteria can even use atmospheric nitrogen (N_2) as a nitrogen source, a capability that makes them apparently unique on Earth (Box 13.2).

Whatever the source of nitrogen, all biosynthetic reactions ultimately involve the addition of nitrogen either by transferring NH_3 directly or by transferring nitrogen in the form of an NH_2 group from glutamate and glutamine, which in turn are synthesized by directly adding NH_3 to α-ketoglutarate and glutamate, respectively. Thus, because NH_3 is directly or indirectly the source of nitrogen in biosynthetic reactions, most other forms of nitrogen must be reduced to NH_3 before they can be used in these reactions. This process is called **assimilatory reduction** of the nitrogen-containing compounds, because the nitrogen-containing compound converted into NH_3 is introduced, or assimilated, into biological molecules. In another type of reduction, **dissimilatory reduction**, oxidized nitrogen-containing compounds such as NO_3^- are reduced when they serve as electron acceptors in anaerobic respiration (in the absence of oxygen). However, the compounds are generally not reduced all the way to NH_3 in this process, and the nitrogen is not assimilated into biological molecules. Here, we discuss only the assimilatory uses of nitrogen-containing compounds. The genes whose products are required for anaerobic respiration are members of a different regulon, the FNR regulon, which is turned on only in the absence of oxygen, when other, less efficient, electron acceptors are required (Table 13.1).

Pathways for Nitrogen Assimilation

Enteric bacteria use different pathways to assimilate nitrogen depending on whether NH_3 concentrations are low or high (Figure 13.8). When NH_3 concentrations are low, for example when the nitrogen sources are amino acids, which must be degraded for their NH_3, an enzyme named **glutamine synthetase**, the product of the *glnA* gene, adds the NH_3 directly to glutamate to make glutamine. About 75% of this glutamine is then converted to glutamate by another enzyme, **glutamate synthase,** sometimes called GOGAT, which removes an -NH_2 group from glutamine and adds it to α-ketoglutarate to make two glutamates. These glutamates can, in turn, be converted into glutamine by glutamine synthetase. Because the NH_3 must all be routed by glutamine synthetase to glutamine when NH_3 concentrations are low, the cell needs a lot of the glutamine synthetase enzyme under these conditions. This pathway requires a lot of energy, but it is necessary if nitrogen availability is limited. The significance of this is addressed later.

If NH_3 concentrations are high because the medium contains NH_3 in the dissolved form of NH_4OH, but carbon sources are limited, the nitrogen is assimilated through a very different pathway. This pathway requires less energy but is possible only if NH_3 concentrations are high. In this case, the enzyme **glutamate dehydrogenase** adds the NH_3 directly to α-ketoglutarate to make glutamate. Some of the glutamate is subsequently converted into glutamine by glutamine synthetase. Much less glutamine is required for protein synthesis and biosynthetic

BOX 13.2

Nitrogen Fixation

Some bacteria can use atmospheric nitrogen (N_2) as a nitrogen source by converting it to NH_3 in a process called nitrogen fixation, which appears to be unique to bacteria. However, N_2 is a very inconvenient source of nitrogen. The bond holding the two nitrogen atoms together must be broken, which is very difficult, and 16 mol of ATP must be cleaved to cleave 1 mol of dinitrogen. Bacteria that can fix nitrogen include the cyanobacteria and members of the genera *Klebsiella*, *Azotobacter*, *Rhizobium*, and *Azorhizobium*. These organisms play an important role in nitrogen cycles on Earth.

Some types of nitrogen-fixing bacteria, including members of the genera *Rhizobium* and *Azorhizobium*, are symbionts that fix N_2 in nodules on the roots or stems of plants and allow the plants to live in nitrogen-deficient soil. In return, the plant furnishes nutrients and an oxygen-free atmosphere in which the bacterium can fix N_2. This symbiosis therefore benefits both the bacterium and the plant. An active area of biotechnology is the use of N_2-fixing bacteria as a source of natural fertilizers.

The fixing of N_2 requires the products of many genes, called the *nif* genes. In free-living nitrogen-fixing bacteria such as *Klebsiella* spp., the fixing of nitrogen requires about 20 *nif* genes arranged in about eight adjacent operons. Some of the *nif* genes encode the nitrogenase enzymes directly responsible for fixing N_2. Others encode proteins involved in assembling the nitrogenase enzyme and in regulating the genes. Plant-symbiotic bacteria also require many other genes whose products produce the nodules on the plant (*nod* genes) and whose products allow the bacterium to live and fix nitrogen in the nodules (*fix* genes).

Because atmospheric nitrogen is such an inconvenient source of nitrogen, the genes involved in N_2 fixation are part of the Ntr regulon and are under the control of the NtrC activator protein. In *Klebsiella pneumoniae*, in which the regulation of the *nif* genes has been studied most extensively, the phosphorylated form of NtrC does not directly activate all eight operons involved in N_2 fixation. Instead, the phosphorylated form of NtrC activates the transcription of another activator gene, *nifA*, whose product is directly required for the activation of the eight *nif* operons. The nitrogenase enzymes are very sensitive to oxygen, and in the presence of oxygen, the *nif* operons are negatively regulated by the product of the *nifL* gene. The NifL protein is able to sense oxygen because it is a flavoprotein with a bound FAD group, which is oxidized in the presence of oxygen. The NifL protein then forms a stable complex with NifA and inactivates it so that the *nif* genes are not transcribed. Because the nitrogenase enzymes are so sensitive to oxygen, the bacteria can fix nitrogen only in an oxygen-free (anaerobic) environment, such as exists in the nodules on the roots of plants or in other anaerobic environments in the soil or in the heterocysts of filamentous cyanobacteria.

References

Margolin, W. 2000. Differentiation of free-living rhizobia into endosymbiotic bacteroids, p. 441–466. *In* Y. V. Brun and L. J. Shimkets (ed.), *Prokaryotic Development*. ASM Press, Washington, D.C.

Martinez-Argudo, I., R. Little, N. Shearer, P. Johnson, and R. Dixon. 2004. The NifL-NifA system: a multidomain transcriptional regulatory complex that integrates environmental signals. *J. Bacteriol.* **186:**601–610.

reactions than for assimilation of limiting nitrogen from the medium. Therefore, cells need much less glutamine synthetase when growing in high concentrations of NH_3 than when growing in low concentrations.

The reaction catalyzed by glutamate dehydrogenase when NH_3 concentrations are high is very efficient and requires little energy. The rapid assimilation of nitrogen under these conditions allows bacteria to multiply rapidly. However, some bacteria, including *Bacillus subtilis*, lack this pathway and probably have no efficient way of using inorganic NH_3 as a nitrogen source. Why some bacteria would prefer organic sources of nitrogen, such as glutamine or other amino acids, is not known but might reflect the lack of NH_3 in environments in which they normally live. For example, acidic environments have very little NH_3, since it is very volatile at low pH and quickly evaporates.

REGULATION OF NITROGEN ASSIMILATION PATHWAYS BY THE Ntr SYSTEM

The operons for nitrogen utilization are part of the **Ntr system**, for *ni*trogen *r*egulated. Ntr regulation ensures that the cell does not waste energy-making enzymes for the use of nitrogen sources such as amino acids or nitrate when NH_3 is available. Transport systems for alternative nitrogen sources are also part of this regulon. However, it may also be a sort of stress response system that allows the cell to cope with the slower growth associated with nitrogen limitation. In this section, we discuss what is known about how the Ntr global regulatory system

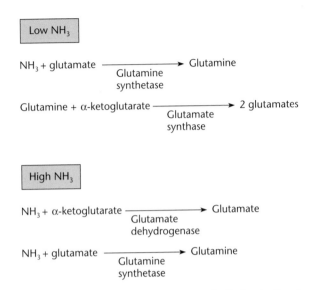

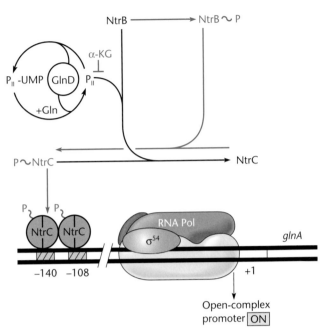

Figure 13.8 Pathways for nitrogen assimilation in *E. coli* and other enteric bacteria. With low NH_3 concentrations, the glutamine synthetase enzyme adds NH_3 directly to glutamate to make glutamine. Glutamate synthase can then convert the glutamine plus α-ketoglutarate into two glutamates, which can reenter the cycle. In the presence of high NH_3 concentrations, the NH_3 is added directly to α-ketoglutarate by the glutamate dehydrogenase to make glutamate, which can be subsequently converted to glutamine by glutamine synthetase.

Figure 13.9 Regulation of Ntr operons by a signal transduction pathway in response to NH_3 levels. At low NH_3 concentrations, the reactions shown in purple predominate. At high NH_3 concentrations, the reactions shown in black predominate. α-KG, α-ketoglutarate. The effect of α-ketoglutarate on P_{II} is indirect. See the text for details.

works. As usual, geneticists led the way, by identifying the genes whose products are involved in the regulation, so that a role could eventually be assigned to each one. The Ntr regulatory system is remarkably similar in all genera of gram-negative bacteria, including *Escherichia*, *Salmonella*, *Klebsiella*, and *Rhizobium*, but with important exceptions, some of which are pointed out here. Early indications are that this system may be different in gram-positive bacteria, although work on Ntr in these bacteria is in its early stages.

Regulation of the *glnA-ntrB-ntrC* Operon by a Signal Transduction Pathway

Since cells need more glutamine synthetase while growing at low NH_3 concentrations than when growing at high NH_3 concentrations, the expression of the *glnA* gene, which encodes glutamine synthetase, must be regulated according to the nitrogen source that is available. This gene is part of an operon called *glnA-ntrB-ntrC*. The products of the two other genes in the operon, NtrB and NtrC (*nitrogen regulator B and C*), are involved in regulating the operon. (These proteins are also called NR_{II} and NR_I, respectively, but we use the Ntr names in this chapter.) Because the *ntrB* and *ntrC* genes are part of

the same operon as *glnA*, their genes are autoregulated and their products are also synthesized at higher levels when NH_3 concentrations are low.

Figure 13.9 illustrates the regulation of the *glnA-ntrB-ntrC* operon and other Ntr genes in detail. In addition to NtrB and NtrC, the proteins GlnD and P_{II} participate in regulation of the operon. These four proteins form a **signal transduction pathway,** in which news of the available nitrogen source is passed, or transduced, from one protein to another until it gets to its final destination, the transcriptional regulator, NtrC, which activates genes in the Ntr regulon.

The availability of nitrogen is sensed through the level of glutamine in the cell. If the cell is growing in a nitrogen-rich environment, the levels of glutamine are high, whereas if the cell is growing in limiting nitrogen, the levels of glutamine are low. How the levels of glutamine affect the regulation of the Ntr genes involved in using alternative nitrogen sources is probably best explained by working backward from the last protein in the signal transduction pathway, NtrC. The NtrC protein can be phosphorylated to form NtrC~P, the form in which it is a transcriptional activator that activates transcription of the Ntr genes, which are turned on in limiting nitrogen

(Figure 13.9). The penultimate protein in the pathway is NtrB. Together, NtrB and NtrC form a two-component phosphorelay system (see below) in which the NtrB protein is the sensor kinase and NtrC is the response regulator. NtrB can phosphorylate itself to form NtrB~P and transfer this phosphate to NtrC to form NtrC~P and activate transcription of the Ntr genes, but it does this only if nitrogen is limiting. Whether NtrB can phosphorylate itself depends on the state of modification of another protein called P_{II}. The P_{II} protein can be modified not by phosphorylation but by having UMP attached to it (P_{II}~UMP). If nitrogen is limiting, making the glutamine level low, most of this protein exists as P_{II}~UMP. However, if nitrogen is in excess and the glutamine level is high, the glutamine stimulates an enzyme called GlnD to remove the UMP from P_{II}. The unmodified P_{II} protein binds to NtrB and inhibits its autokinase activity so that it cannot phosphorylate itself to form NtrB~P. If it cannot phosphorylate itself, it cannot transfer a phosphate to NtrC to activate transcription of the Ntr genes involved in using alternate nitrogen sources.

NtrB and NtrC: a Two-Component Sensor-Response Regulator System

As mentioned, the NtrB and NtrC proteins form a two-component system in which NtrB is a **sensor kinase** and NtrC is a **response regulator**. Such protein pairs are now known to be common in bacteria, and the corresponding members of such pairs are remarkably similar to each other (see below). Typically, one protein of the pair "senses" an environmental parameter and phosphorylates itself at a histidine. This phosphoryl group is then passed on to an aspartate in the second protein. The activity of this second protein, the response regulator, depends on whether it is phosphorylated. Many response regulators are transcriptional regulators. In chapter 6, we discuss another example of a sensor-response regulator pair, ComP and ComA, involved in the development of transformation competence in *Bacillus subtilis*, and later in this chapter we describe other pairs.

Regulation of Other Ntr Operons

The other operons, besides *glnA-ntrB-ntrC*, that are activated by NtrC~P depend on the type of bacteria and the other nitrogen sources they can use. In general, operons under the control of NtrC~P are those involved in using poorer nitrogen sources. For example, genes for the uptake of the amino acids glutamine in *E. coli* and histidine and arginine in *Salmonella enterica* serovar Typhimurium are under the control of NtrC~P. An operon for the utilization of nitrate as a nitrogen source in *Klebsiella pneumoniae* is activated by NtrC~P, but

neither *E. coli* nor *S. enterica* serovar Typhimurium has such an operon.

In some bacteria, the Ntr genes are not regulated directly by NtrC~P but are under the control of another gene product whose transcription is activated by NtrC~P. For example, operons for amino acid degradative pathways in *Klebsiella aerogenes* use σ^{70} promoters that do not require activation by NtrC~P. However, they are indirectly under the control of NtrC~P because transcription of the gene for their transcriptional activator, *nac*, is activated by NtrC~P. The nitrogen fixation genes of *K. pneumoniae* are similarly under the indirect control of NtrC~P because this protein activates transcription of the gene for their activator protein, *nifA* (Box 13.2).

Another Ntr Regulatory Pathway?

The need for so many steps in the regulation of the Ntr system is not clear, but it may reflect the central role of nitrogen utilization in the physiology of the cell. Other pathways may intersect with the various steps of the signal transduction pathway for nitrogen utilization, so that expression of the genes can be coordinated with many other cellular functions. In support of this idea, some Ntr regulation occurs even in *ntrB* mutants (see the section on genetics of Ntr regulation, below). At least one other pathway besides the one involving NtrB must lead to the phosphorylation and dephosphorylation of NtrC in response to changes in the nitrogen source.

TRANSCRIPTION OF THE *glnA-ntrB-ntrC* OPERON

The *glnA-ntrB-ntrC* operon is transcribed from three promoters, only one of which is NtrC~P dependent. The positions of the three promoters and the RNAs made from each are shown in Figure 13.10. Of the three promoters, only the p_2 promoter is activated by NtrC~P and is responsible for the high levels of glutamine synthetase and NtrB and NtrC synthesis under conditions of low NH_3. The other two promoters, p_1 and p_3, are discussed later.

The p_2 promoter is immediately upstream of the *glnA* gene, and RNA synthesis initiated at this promoter continues through all three genes, as shown in Figure 13.10. However, some transcription terminates at a transcriptional terminator located between the *glnA* and *ntrB* genes, so that much less NtrB and NtrC is made than glutamine synthetase.

The "Nitrogen Sigma," σ^{54}

The p_2 promoter and other Ntr-type promoters activated by NtrC~P are unusual in terms of the RNA polymerase holoenzyme that recognizes them. Most promoters are recognized by the RNA polymerase holoenzyme with σ^{70}

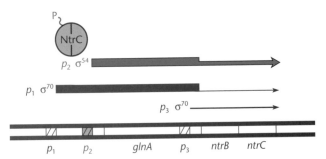

Figure 13.10 The *glnA-ntrB-ntrC* operon of *E. coli*. Three promoters service the genes of the operon. The arrows show the mRNAs that are made from each promoter. The purple arrow indicates the mRNA expressed from the nitrogen-regulated promoter, p_2. The thickness of the lines indicates how much RNA is made from each promoter in each region. See the text for details.

attached, but the Ntr-type promoters are recognized by holoenzyme with σ^{54} attached (Box 13.3). As shown in Figure 13.11, promoters recognized by the σ^{54} holoenzyme look very different from promoters recognized by the σ^{70} holoenzyme. Unlike the typical σ^{70} promoter, which has RNA polymerase binding sequences at -35 and -10 bp upstream of the RNA start site, the σ^{54} promoters have very different binding sequences at -24 and -12 bp. Because promoters for the genes involved in Ntr regulation are recognized by RNA polymerase with σ^{54}, this sigma factor was first named the "nitrogen sigma" and the gene for σ^{54} was named *rpoN* (Table 13.1). However, σ^{54}-type promoters have been found in many operons unrelated to nitrogen utilization, including in the flagellar genes of *Caulobacter* spp. and some promoters of the toluene-biodegradative operons of the *Pseudomonas putida* Tol plasmid (see chapter 12). Interestingly, all of the known σ^{54}-type promoters require activation by an activator protein (Box 13.3).

The Transcription Activator NtrC

The polypeptide chains of NtrC-type activators have the basic structure shown in Figure 13.12. A DNA-binding domain that recognizes the σ^{54}-type promoter lies at the carboxyl-terminal end of the polypeptide. A regulatory domain that either binds inducer or is phosphorylated is present at the amino-terminal end. The region of the polypeptide responsible for transcriptional activation is in the middle. This region has an ATP-binding domain and an ATPase activity that cleaves ATP to ADP. As mentioned in Box 12.2 (see chapter 12), the N-terminal domain somehow masks the middle domain for activation unless the N-terminal domain has bound inducer or has been phosphorylated.

The mechanism of activation by NtrC-type activators has been studied in some detail. The NtrC-activated promoters, including p_2 of the *glnA-ntrB-ntrC* operon, are unusual in that NtrC binds to an **upstream activator sequence** (UAS), which lies more than 100 bp upstream of the promoter. For most prokaryotic promoters, the activator protein binding sequences are adjacent to the site at which RNA polymerase binds. Some evidence suggests that the function of these upstream activator sequences is merely to increase the local concentration of NtrC. If NtrC is overproduced in the cell, the UASs become nonessential for the activation. Activation at a distance, such as occurs with the NtrC-activated promoters, is much more common in eukaryotes, where many examples are known.

Figure 13.12 shows a detailed model for how NtrC~P activates transcription from p_2 and other Ntr-activated promoters. The RNA polymerase can bind to the promoter even when nitrogen is not limiting and NtrC is not phosphorylated. Also, NtrC can bind to the UAS even if it is not phosphorylated. However, when nitrogen is limiting and NtrC becomes phosphorylated, oligomers (probably octamers) of NtrC~P bound at the UAS activate transcription from the promoter, perhaps because the phosphorylated NtrC molecules bind more tightly to each other and this stimulates their ATPase activity. Activation of the promoter must involve bending of the DNA to bring the NtrC activators in contact with the RNA polymerase bound at the promoter. Activation of transcription at a σ^{54} promoter requires cleavage of ATP by NtrC for open-complex formation, unlike transcription from other promoters where RNA polymerase can do the job itself. The eight subunits in an octamer of NtrC might form a ring structure around which the DNA wraps. It is interesting that NtrC-type activators belong to a larger class of ATPases called mechanochemical ATPases, which form ring-shaped complexes that convert the cleavage of ATP to mechanical energy, in this case the separation of strands of DNA.

Function of the Other Promoters of the *glnA-ntrB-ntrC* Operon

As mentioned, the p_2 promoter is only one of three promoters that service the *glnA-ntrB-ntrC* operon (Figure 13.10). The other promoters are p_1 and p_3. The p_1 promoter is further upstream of *glnA* than is p_2, and most of the transcription that initiates at the p_1 promoter terminates at the transcription termination signal just downstream of *glnA*, as shown in Figure 13.10. The p_3 promoter is between the *glnA* and *ntrB* genes and services the *ntrB* and *ntrC* genes.

The p_1 and p_3 promoters use σ^{70} and so do not require NtrC~P for their activation. In fact, they are repressed by

BOX 13.3

Sigma Factors

$\mathcal{S}$igma (σ) factors seem to be unique to bacteria and their phages and are not found in eukaryotes. They are discussed in previous chapters. These proteins cycle on and off RNA polymerase and help direct it to specific promoters. They also help RNA polymerase melt the promoter to initiate transcription, and they help in promoter clearance after initiation. They also often contain the contact points of activator proteins that help these proteins stabilize the RNA polymerase on the promoter and activate transcription (see chapter 12). Promoters are often identified by the σ factor they use; for example, a "σ⁷⁰ promoter" is one which uses the RNA polymerase holoenzyme with σ⁷⁰ attached while a σˢ promoter uses σˢ RNA polymerase. A caveat: different σ factors are often given the same name in different bacteria. For example, σᴴ in *E. coli* and *B. subtilis* refers to very different σ factors. The σᴴ in *E. coli* is the heat shock sigma, while the σᴴ in *B. subtilis* is involved in sporulation.

Sigma factors can be found in the genomic sequences of bacteria based on their sequence conservation. This has revealed that the number of different sigma factors varies widely from one bacterial type to another. The bacterium with the least known so far, *Helicobacter pylori*, has only 3 different types, while the current record holder, *Streptomyces coelicolor*, has 63. In general, bacteria that are free living have more sigma factors than do obligate parasites, probably reflecting the greater environmental challenges faced by free-living bacteria.

There are two major classes of sigma factors in bacteria: the σ⁷⁰ class, which includes most of the sigma factors discussed in this book, including σˢ, σ³², σ²⁸, σᴱ, etc., and another class, σ⁵⁴, which seems to form a class by itself. While all sigma factors in the σ⁷⁰ class have some sequence and functional homology, there is no sequence similarity between members of this class and members of the σ⁵⁴ class. The two classes also seem to differ fundamentally in their mechanism of action (see below). Members of the σ⁷⁰ class are found in all bacteria and play many diverse roles, some of which are discussed in previous chapters. The σ⁵⁴-type promoters are also widely distributed among both gram-positive and gram-negative bacteria, but they are not universal. This sigma factor was originally named the "nitrogen sigma," σᴺ, because the promoters it uses were first found in the genes for Ntr regulation that were turned on during nitrogen-limited growth in *E. coli* (see the text). However, it is now known that there is no common theme for σ⁵⁴-expressed genes. In some soil bacteria, this sigma factor is used to express biodegradative genes, for example to degrade toluene (see chapter 12), and in other species, including some pathogens, it is used by some of the flagellar genes, to make components of type III secretion systems, as well as to make alginate in *Pseudomonas aeruginosa* (see Kazmierczak et al., below).

One major distinction between the two classes of sigma factors is in the way their promoters are activated. For example, many promoters that use a sigma factor of the σ⁷⁰

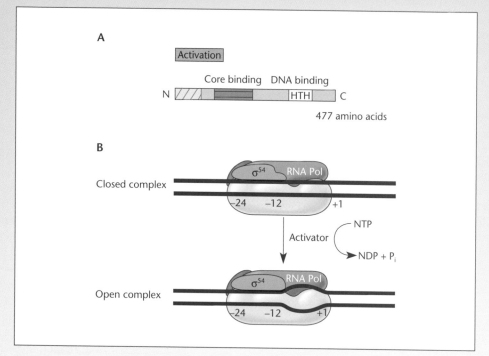

(continued)

Sigma Factors

family can initiate transcription without the help of an activator protein. If a σ^{70} promoter requires an activator, it generally binds adjacent to the promoter and helps recruit RNA polymerase to the promoter (see chapter 12). However, all σ^{54} promoters studied thus far absolutely require a specialized activator protein with ATPase activity, which binds to an enhancer sequence that can be hundreds of base pairs upstream of the promoter (see Box 12.2 and below). In some ways, this makes σ^{54} promoters more like the RNA polymerase II promoters of eukaryotes. They also differ from σ^{70} promoters in their mechanism of activation. The activation of a σ^{54} promoter is illustrated in the figure. Panel A shows the functionally important domains of σ^{54} that allow binding to core RNA polymerase and DNA. The N-terminal domain allows σ^{54} to respond to activators. Panel B shows that the σ^{54}-RNA polymerase forms a stable but closed complex with the promoter, even in the absence of the activator bound to the upstream enhancer. The activator has a latent ATPase activity, which becomes activated by phosphorylation that is often passed down from a regulatory cascade. Many of the activators of σ^{54} promoters are at the end of a phosphorylation cascade, with NtrC being the prototype (see the text). Phosphorylation of the N terminus of the activator alters its affinity for enhancer sites. Multimerization and formation of a DNA-bound complex activates the ATPase activity in its central domain. Once activated, the ATPase can cause the σ^{54} to undergo a conformational change that overcomes its inhibition of open-complex formation by the σ^{54}-RNA polymerase and allows initiation at the promoter (see also Box 12.2).

References

Buck, M., M.-T. Gallegos, D. J. Studholme, Y. Guo, and J. D. Gralla. 2000. The bacterial enhancer-dependent σ^{54} (σ^{N}) transcription factor. *J. Bacteriol.* **182:**4129–4136.

Kazmierczak, M. J., M. Weidmann, and K. J. Boor. 2005. Alternate sigma factors and their roles in bacterial virulence. *Microbiol. Mol. Biol. Rev.* **69:**527–543.

Paget, M. S. B., and J. D. Helmann. 2003. The sigma 70 family of sigma factors. *Genome Biol.* **4:**203–215.

NtrC~P, so they are not used if NH_3 concentrations are low. The function of these promoters is presumably to ensure that the cell has some glutamine synthetase and NtrB and NtrC when NH_3 concentrations are high. Unless glutamine is provided in the medium, the cell must have glutamine synthetase to make glutamine for use as the $-NH_2$ group donor in some biosynthetic reactions and in protein synthesis since it is one of the amino acids. The cell must also have some NtrB and NtrC in case conditions change suddenly from a high-NH_3 to a low-NH_3 environment, in which the products of the Ntr genes are suddenly needed. Other genes of the Ntr regulon also have σ^s promoters, which allow them to be turned on at times of stress (see below). This is one of the reasons for suspecting that the Ntr response is also a stress response.

ADENYLYLATION OF GLUTAMINE SYNTHETASE
Regulating the transcription of the *glnA* gene is not the only way the activity of glutamine synthetase is regulated

Figure 13.11 Sequence comparison of the promoters recognized by the RNA polymerase holoenzyme carrying the normal sigma factor (σ^{70}), the nitrogen sigma factor (σ^{54}), and the heat shock sigma factor (σ^{32}). Instead of consensus sequences at −10 and −35 bp with respect to the RNA start site, the σ^{54} promoter has consensus sequences at −12 and −24 bp. The σ^{32} promoter has consensus sequences at approximately −10 and −35 bp, but they are different from the consensus sequences of the σ^{70} promoter. X indicates that any base pair can be present at this position. +1 is the start site of transcription.

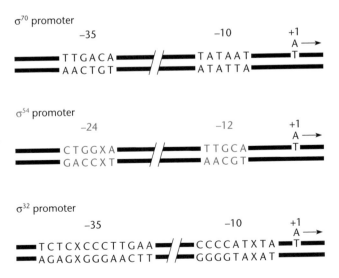

A

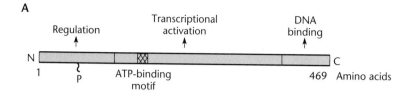

B

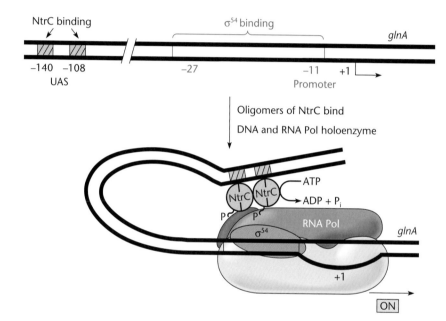

Figure 13.12   Model for the activation of the p_2 promoter by phosphorylated NtrC protein. (A) Functions of the various domains of NtrC. P denotes a phosphate. (B) Oligomers (shown as two dimers) of phosphorylated NtrC bind to the inverted repeats in the upstream activator sequence (UAS). The DNA is bent between the UAS and the promoter, allowing contact between NtrC and the RNA polymerase bound at the promoter more than 100 bp downstream. Cleavage of ATP due to the NtrC ATPase activity is required for the RNA polymerase to form open complexes at the promoter.

in the cell. This activity is also modulated by the adenylylation of (transfer of AMP to) a specific tyrosine in the enzyme by an adenylyltransferase (ATase) enzyme when NH_3 concentrations are high. The adenylylated form of the glutamine synthetase enzyme is less active and is also much more susceptible to feedback inhibition by glutamine than is the unadenylylated form (see chapter 12 for an explanation of feedback inhibition). This makes sense, considering how glutamine synthetase plays different roles when NH_3 concentrations are high and when they are low. When NH_3 concentrations are high, the primary role of the glutamine synthetase is to make glutamine for protein synthesis, which requires less enzyme; the little enzyme activity that remains should be feedback inhibited to ensure that the cells do not accumulate too much glutamine. When NH_3 concentrations are low, more enzyme is required. In this situation, the enzyme should not be feedback inhibited, because its major role is to assimilate nitrogen.

The state of adenylylation of the glutamine synthetase enzyme is also regulated by GlnD and the state of the P_{II} protein. When the cells are in low NH_3 concentrations so that the P_{II} protein has UMP attached (Figure 13.9), the adenylyltransferase removes AMP from glutamine synthetase. When the cells are in high NH_3 concentrations, so that the P_{II} protein does not have UMP attached, the P_{II} protein binds to and stimulates the adenylyltransferase to add more AMP to the glutamine synthetase.

The P_{II} protein may not be the only way to stimulate the deadenylylation of glutamine synthetase in *E. coli*, however. Some work has suggested that another protein, GlnK, may form GlnK-UMP in response to ammonia deprivation. The GlnK-UMP protein can also stimulate the adenylyltransferase to adenylylate glutamine synthetase when NH_3 concentrations are high (see van Heeswijk et al., Suggested Reading). Clearly, many pathways interact to regulate nitrogen source utilization by bacteria.

Coordination of Catabolite Repression, the Ntr System, and the Regulation of Amino Acid-Degradative Operons

Not only must bacteria sometimes use one or more of the 20 amino acids as a nitrogen source, but also they must sometimes use amino acids as carbon and energy sources. The type of amino acids that different bacteria can use varies. For example, *E. coli* can use almost any amino acid as a nitrogen source except tryptophan, histidine, and the branched-chain amino acids such as valine. However, it can use only alanine, tryptophan, aspartate, asparagine, proline and serine as carbon sources. *Salmonella* can also use many amino acids as nitrogen sources but can use only alanine, cysteine, proline, and serine as carbon sources. Like the sugar-utilizing and biosynthetic operons, the amino acid-utilizing operons not only have their own specific regulatory genes, so that they are transcribed only in the presence of their own inducer, but often are also part of larger regulons. As discussed in this section, they are often under Ntr regulation and so are not induced in the presence of their inducer while a better nitrogen source such as NH_3 is in the medium. In some bacteria, these operons are also under the control of the catabolite regulon and are not expressed in the presence of better carbon sources such as glucose.

Sometimes multiple levels of operon control can be a disadvantage to the bacterium. For example, *S. enterica* serovar Typhimurium cells are starved for nitrogen if they are growing in the presence of histidine plus glucose but no other nitrogen source. The glucose apparently prevents CAP from activating the promoter for the *hut* operon, so that the cells cannot use the histidine as a nitrogen source. Glucose itself contains no nitrogen, and so, although the cells can use it as a carbon source, they are starved for nitrogen.

In addition to their potential as nitrogen, carbon, and energy sources, amino acids are necessary for other purposes, including protein synthesis and, in the case of proline, osmoregulation. Therefore, the use of amino acids as carbon and nitrogen sources can present strategic problems for the cell. The way in which all these potentially conflicting regulatory needs are resolved is bound to be complicated.

Genetic Analysis of Nitrogen Regulation in Enteric Bacteria

The present picture of nitrogen regulation in bacteria began with genetic studies. Most of this work was first done with *K. aerogenes*, although some genes were first found in *S. enterica* serovar Typhimurium or *E. coli*.

The first indications of the central role of glutamine and glutamine synthetase in Ntr regulation came from the extraordinary number of genes that, when mutated, could affect the regulation of the glutamine synthetase or give rise to an auxotrophic growth requirement for glutamine (see Magasanik, Suggested Reading). The *gln* genes were originally named *glnA*, *glnB*, *glnD*, *glnE*, *glnF*, *glnG*, and *glnL* (Table 13.2). These genes are not lettered consecutively because, as often happens in genetics, genes presumed to exist because of a certain phenotype were later found to not exist or to be the same as another gene, and so their letters were retired. Sorting out the various contributions of the *gln* gene products to arrive at the model for nitrogen regulation outlined above was a remarkable achievement. It took many years and required the involvement of many people. In this section, we describe how the various *gln* genes were first discovered and how the phenotypes caused by mutations in the genes led to the model.

TABLE 13.2	Genes for nitrogen regulation		
Gene	Alternate name	Product	Function
glnA		Glutamine synthetase	Synthesize glutamine
glnB		P_{II}, P_{II}-UMP	Inhibit phosphatase of NtrB, activate adenylyltransferase
glnD		Uridylyltransferase (UTase)/ Uridylyl-removing enzyme (UR)	Transfer UMP to and from P_{II}
glnE		Adenylyltransferase (ATase)	Transfer AMP to glutamine synthetase
glnF	*rpoN*	σ^{54}	RNA polymerase recognition of promoters of Ntr operons
ntrC	*glnG*	NtrC, NtrC-PO_4	Activate promoters of Ntr operons
ntrB	*glnL*	NtrB, NtrB-PO_4	Autokinase, phosphatase; phosphate transferred to NtrC

THE *glnB* GENE

The *glnB* gene encodes the P_{II} protein. The first *glnB* mutations were found among a collection of *K. aerogenes* mutants with the Gln⁻ phenotype, the inability to multiply without glutamine in the medium. Mutations in *glnB* apparently can prevent the cell from making enough glutamine for growth. However, early genetic evidence indicated that these *glnB* mutations do not exert their Gln⁻ phenotype by inactivating the P_{II} protein. As evidence, transposon insertions and other mutations that should totally inactivate the *glnB* gene (null mutations) do not result in the Gln⁻ phenotype. In fact, null mutations in *glnB* are intragenic suppressors of the Gln⁻ phenotype of the original *glnB* mutations (see chapter 3 for a discussion of the different types of suppressors).

We now know that the original *glnB* mutations do not inactivate P_{II} but, rather, change the binding site for UMP so that UMP cannot be attached to it by GlnD. This should have two effects. P_{II} without UMP binds to NtrB~P and prevents phosphorylation of NtrC. Therefore, even under low NH_3 concentrations, little glutamine is synthesized. By itself, however, this effect does not explain the Gln⁻ phenotype, since the *glnA* gene can also be transcribed from the p_1 promoter, which does not require NtrC~P for activation (see above). It is the second function of P_{II}—the stimulation of the adenylyltransferase—that causes the Gln⁻ phenotype. In the mutants, enough P_{II} without UMP attached accumulates to stimulate the adenylyltransferase to the extent that glutamine synthetase is too heavily adenylated to synthesize enough glutamine for growth. This also explains why null mutations in *glnB* do not cause the Gln⁻ phenotype. By inactivating P_{II} completely, null mutations prevent the P_{II} protein from stimulating the adenylyltransferase, so that less glutamine synthetase is adenylylated and enough glutamine is synthesized for growth.

THE *glnD* GENE

The *glnD* gene was also discovered in a collection of mutants with mutations that cause the Gln⁻ phenotype. However, in this case, experiments showed that null mutations in *glnD* cause the Gln⁻ phenotype. Furthermore, null mutations in *glnB* suppress the Gln⁻ phenotype of null mutations in *glnD*. These observations are consistent with the above model. Since the GlnD protein is the enzyme that transfers UMP to P_{II}, null mutations in *glnD* should behave like the original *glnB* mutations and prevent UMP attachment to P_{II} but leave the P_{II} protein intact. This makes the cells Gln⁻ for the reasons given above. Null mutations in *glnD* are suppressed by null mutations in *glnB* because the absence of GlnD protein has no effect if the cell contains no P_{II} protein to bind to the adenylyltransferase and stimulate the attachment of

AMP to glutamine synthetase. The glutamine synthetase without AMP attached remains active and synthesizes enough glutamine for growth.

THE *glnL* GENE

The *glnL* gene was later renamed *ntrB* to better reflect its function. Mutations in this gene were first discovered not because they cause the Gln⁻ phenotype but because they are extragenic suppressors of *glnD* and *glnB* mutations. These mutations do not inactivate NtrB completely; instead, they leave NtrB intact and prevent the binding of P_{II} so that NtrB transfers its phosphate to NtrC regardless of the presence of P_{II}-UMP.

THE *glnG* GENE

The *ntrC* gene, originally called *glnG*, was discovered because mutations in it can suppress the Gln⁻ phenotype of *glnF* mutations. The *glnF* gene encodes the nitrogen sigma, σ^{54} (see below). Further work showed that the suppressors of *glnF* are null mutations in *ntrC* because transposon insertions and other mutations that inactivate the *ntrC* gene also suppress *glnF* mutations. Null mutations in *ntrC* do not cause the Gln⁻ phenotype, because the *glnA* gene can also be transcribed from the p_1 promoter, which does not require NtrC for activation. They do, however, prevent the expression of other Ntr operons that require NtrC~P for their activation, many of which do not have alternative promoters.

Some mutations in *ntrC* do cause the Gln⁻ phenotype, however. These presumably are mutations that change NtrC so that it can no longer activate transcription from p_2, but it retains the ability to repress transcription from p_1 because it can be phosphorylated.

THE *glnF* GENE

As mentioned, the *glnF* gene, now renamed *rpoN*, encodes the nitrogen sigma, σ^{54}. The gene was also discovered in a collection of Gln⁻ mutants. Without σ^{54}, the p_2 promoter cannot be used to transcribe the *glnA-ntrB-ntrC* operon. By itself, inactivation of p_2 would not be enough to cause the Gln⁻ phenotype, since the *glnA* gene can also be transcribed from the p_1 promoter, which does not require σ^{54}. However, the NtrC~P form of NtrC represses the p_1 promoter if the cells are growing in low NH_3 concentrations (see above). In high NH_3 concentrations, the p_1 promoter is not repressed, but the small amount of glutamine synthetase synthesized is heavily adenylylated, preventing the synthesis of sufficient glutamine for growth. This interpretation of the Gln⁻ phenotype of *rpoN* mutations is consistent with the fact that null mutations in *ntrC* suppress the Gln⁻ phenotype of *rpoN* mutations as discussed above. Without NtrC~P present to repress the p_1 promoter, sufficient glutamine synthetase is synthesized from the p_1 promoter, even in low NH_3 concentrations.

Stress Responses in Bacteria

Many of the global regulons in bacteria are designed to deal with stress. In order to survive, all organisms must be able to deal with abrupt changes in the conditions in which they find themselves. The osmolarity, temperature, or pH of their surroundings might abruptly increase or decrease, or they might be suddenly deprived of growth requirements and have to enter a dormant state. If they have invaded a eukaryotic host, they might suddenly be exposed to reactive forms of oxygen or nitric oxide (NO) as part of the host defense. Not only must they be able to respond quickly to such changes, but also their response must be flexible enough to deal with a variety of different stresses, or even more than one stress at a time. As expected, bacteria and other organisms have evolved complicated interactive pathways to deal with such changes, and this is a subject of active current research. In this section we discuss what has been learned about some major pathways and how they interact.

Heat Shock Regulation

The regulation of gene expression following a heat shock is one of the most extensively studied global regulatory pathways in bacteria and other organisms. One of the major challenges facing cells is to survive abrupt changes in temperature. To adjust to abrupt increases, cells induce at least 30 different genes encoding proteins called the **heat shock proteins (Hsps)**. The concentrations of these proteins quickly increase in the cell after a temperature upshift and then slowly decline, a phenomenon known as the **heat shock response**. Besides being induced by abrupt increases in temperature, the heat shock genes are induced by other types of stress that damage proteins, such as the presence of ethanol and other organic solvents in the medium. Therefore, the heat shock response is more of a general stress response rather than a specific response to an abrupt increase in temperature. Nevetheless, the name heat shock has stuck, and we retain it for this discussion.

Unlike most shared cellular processes, the heat shock response was observed in cells of higher organisms long before it was seen in bacteria. Some of the Hsps are remarkably similar in all organisms and presumably play similar roles in protecting all cells against heat shock. The mechanism of regulation of the heat shock response may also be similar in organisms ranging from bacteria to higher eukaryotes (see Craig and Gross, Suggested Reading).

HEAT SHOCK REGULATION IN *E. COLI*

The molecular basis for the heat shock response was first understood in *E. coli* and is illustrated in Figure 13.13. In this bacterium, about 30 genes encoding 30 different Hsps are turned on following a heat shock. The functions of many of these Hsps are known (Table 13.1) (see chapter 2). Most of these Hsps play roles during the normal growth of the cell and so are always present at low concentrations, but after a heat shock, their rate of synthesis increases markedly and then slowly declines to normal levels.

Some Hsps, including GroEL, DnaK, DnaJ, and GrpE, are chaperones that direct the folding of newly translated proteins (see chapter 2). The names of these proteins do not reflect their function but, rather, how they were orginally discovered. For example, DnaK and DnaJ were found because they affect the assembly of a protein complex required for λ phage DNA replication, but they are not themselves involved directly in replication. Chaperones may help the cell survive a heat shock by binding to proteins denatured by the abrupt rise in temperature and either helping them to refold properly or targeting them for destruction. As mentioned, the chaperones are among the most highly conserved proteins in cells, being largely unchanged from bacteria to humans.

Other Hsps, including Lon and Clp, are proteases, which may degrade proteins that are so badly denatured by the heat shock that they are irreparable and so are best degraded before they poison the cell. Some other Hsps are proteins normally involved in protein synthesis, including special aminoacyltransferases that are induced after a heat shock. The function of this type of Hsp in protecting the cell after a temperature rise is not clear.

Knowing that many Hsps are involved in helping proteins fold properly or in destroying denatured proteins helps explain the transient nature of the heat shock response. Immediately after the temperature increases, the concentrations of salts and other cellular components, which were adjusted for growth at lower temperatures, are not appropriate for protein stability at the higher temperature; this leads to massive protein unfolding. Later, after the temperature has been elevated for some time, the internal conditions have had time to adjust, and so proteins are no longer denatured and the increased number of chaperones and other Hsps is no longer necessary. Hence, the synthesis of the Hsps declines.

Genetic Analysis of Heat Shock in *E. coli*

As with other regulatory systems, the analysis of the heat shock response was greatly aided by the discovery of mutants with defective regulatory genes. The first of

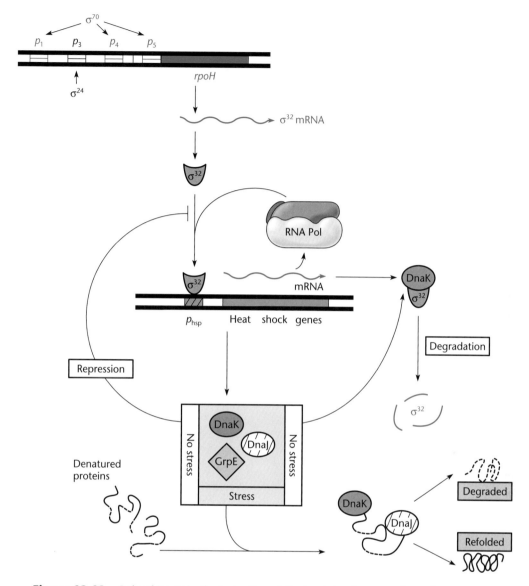

Figure 13.13 Role of DnaK in the induction of the heat shock genes after cells are exposed to an abrupt increase in temperature or other type of stress. The σ^{32} with DnaK bound is susceptible to a protease and is quickly degraded. In addition, σ^{32} with DnaK bound is less active for the initiation of transcription from the heat shock promoters. After an abrupt increase in temperature, many other proteins are denatured, and DnaK, with the help of GrpE and DnaJ (see chapter 2), binds to these to help them refold properly. This action frees σ^{32}, stabilizing it and making it more active for transcription initiation. When the cell adjusts to the higher temperature and DnaK accumulates to the point where some is again available to bind to σ^{32}, the activity of σ^{32} in the cell again drops and the transcription of the heat shock genes returns to basal levels.

such mutants was found in a collection of temperature-sensitive mutants. It was later shown to be unable to induce the Hsps after a shift to high temperature (see Zhou et al., Suggested Reading). This mutant, which failed to make the regulatory gene product, made it possible to clone the regulatory gene by complemen-

tation. A library of wild-type *E. coli* DNA was introduced into the mutant strain, and clones that permitted the cells to survive at high temperatures were isolated. When the sequence of the cloned gene for the regulatory Hsp was compared with those of genes encoding other proteins, the gene was found to encode a new type of sigma factor that was only 32 kDa long; therefore, it was named σ^{32}, heat shock sigma, or σ^H. The RNA polymerase holoenzyme with σ^{32} attached recognizes promoters for the heat shock genes that are different from the promoters recognized by the normal σ^{70} and the nitrogen sigma, σ^{54} (Figure 13.11). The gene for the heat shock sigma was named *rpoH* (for RNA *po*lymerase subunit *h*eat shock).

Regulation of σ^{32} Synthesis

Normally, very few copies of σ^{32} exist in the cell. However, immediately after an increase in temperature from 30 to 42°C, the amount of σ^{32} in the cell increases 15-fold. This increase in concentration leads to a significant rise in the rate of transcription of the heat shock genes, since they are transcribed from σ^{32}-type promoters. Understanding how the heat shock genes are turned on after a heat shock requires an understanding of how this increase in the amount of σ^{32} occurs.

An abrupt increase in temperature might increase the amount of σ^{32} through several mechanisms. One possibility is that the *rpoH* gene for σ^{32} is transcriptionally autoregulated. According to this hypothesis, the cell would normally contain a small amount of σ^{32}, which is somehow activated after a heat shock, and this activated σ^{32} would direct more of the RNA polymerase to the *rpoH* gene, leading to the synthesis of more σ^{32}, and so forth. For this hypothesis to be correct, the *rpoH* gene would have to be strongly transcriptionally regulated. However, although the rate of transcription of the *rpoH* gene increases slightly after a heat shock, it does not increase enough to explain the large rise in σ^{32} levels. Moreover, if the *rpoH* gene is transcriptionally autoregulated, at least one of the promoters servicing the *rpoH* gene should be of the type that uses the heat shock sigma. However, of the four different promoters from which *rpoH* is transcribed, none are recognized by σ^{32}. Three promoters are used by RNA polymerase with the normal σ^{70}, and another is probably used by RNA polymerase with another type of sigma factor, σ^E (also called σ^{24}), which is more active at higher temperatures, perhaps explaining the slight increase in *rpoH* transcription after a heat shock.

Because the transcription of the *rpoH* gene does not increase significantly after heat shock, the amount of σ^{32} in the cell must be posttranscriptionally regulated. In fact, immediately after the temperature upshift, σ^{32} stability increases markedly and the translation rate of the *rpoH* mRNA increases 10-fold.

DnaK: the Cellular Thermometer?

At least some of the posttranscriptional regulation of σ^{32} levels after a heat shock is due to the protein chaperone DnaK. In other words, DnaK is the "cellular thermometer" that senses the change in temperature and induces the transcription of itself and the other heat shock genes (see Craig and Gross, Suggested Reading). It can play this role because it normally binds to nascent proteins in the process of being synthesized and helps them to fold properly (see chapter 2). Under heat shock conditions, the DnaK chaperone can also bind to the denatured proteins and help them refold.

Figure 13.13 shows a model for how the protein-binding ability of the DnaK chaperone indirectly regulates the synthesis of itself as well as all the other heat shock proteins. One of the proteins to which DnaK binds is σ^{32} (see Liberek et al., Suggested Reading). By binding to σ^{32}, the DnaK protein regulates the transcription of the heat shock genes in two ways. First, it affects the stability of σ^{32}. The σ^{32} protein with DnaK bound is more susceptible to a cellular protease called FtsH than is free σ^{32} (see Blaszczak et al., Suggested Reading). The σ^{32} bound to DnaK is degraded almost as soon as it is made, keeping the amount of σ^{32} at low levels and therefore reducing the transcription of the heat shock genes. Second, the binding of DnaK inhibits the activity of σ^{32}. With DnaK protein bound, σ^{32} may be less active in transcription, either because it is less able to bind to RNA polymerase or because the complex of RNA polymerase, σ^{32}, and DnaK is less able to bind to the heat shock promoters. By inhibiting the activity of σ^{32}, the DnaK protein lowers the transcription of the heat shock genes even more.

How, then, do we reconcile the fact that an increase in temperature, which increases the synthesis of DnaK, leads to increased stability of σ^{32} and induction of the heat shock genes? The answer lies in the chaperone role of DnaK. In addition to binding to the σ^{32} protein, DnaK binds to denatured proteins to help them refold properly. After the heat shock, many denatured proteins appear in the cell and most of the DnaK protein is commandeered to help renature these unfolded proteins. This leaves less DnaK available to bind to σ^{32}. The σ^{32} protein is then more stable and accumulates in the cell. It is also more active, increasing the transcription of the heat shock genes including the *dnaK* gene itself.

This model also explains the transient nature of the heat shock response, in which the concentration of the Hsps slowly declines after a sharp increase following a temperature rise. When enough DnaK has accumulated to bind to all the unfolded proteins and internal conditions have adjusted so that no more proteins are denatured at the higher temperature, extra DnaK once again becomes available to bind to and inhibit σ^{32}, leading to the observed drop in the rate of synthesis of the Hsps.

The σ^{32} protein may also be **translationally autoregulated,** in other words, able to repress its own translation.

Generally, such proteins bind to their own translational initiation region (TIR) in the mRNA, thereby blocking access by ribosomes. After an upshift in temperature, the translational repression of σ^{32} synthesis is less complete but then returns, suggesting that DnaK may be involved. The translational repression of σ^{32} synthesis also involves secondary structures that are formed in the σ^{32} mRNA at lower temperatures but melt at higher temperatures (see "Regulation by Secondary-Structure Changes in the mRNA" in chapter 12 and Nagai et al., Suggested Reading). In fact, temperature regulation by the secondary structure of the σ^{32} mRNA was the first such regulation to be discovered.

Another alternative sigma factor, σ^{E}, is responsible for transcription of some heat shock genes at high temperature, including the *rpoH* gene encoding σ^{32} (see above). Because this sigma factor is activated by damage to the outer membrane of the cell by heat and other agents, it is discussed below in connection with extracytoplasmic stress responses.

HEAT SHOCK REGULATION IN OTHER BACTERIA

Once heat shock regulation in *E. coli* was fairly well understood, it was of of interest to see whether other bacteria used the same mechanism. Surprisingly, most other bacteria, including *B. subtilis*, in which it has been studied in the greatest detail, use a very different mechanism. Rather than using a heat shock sigma homologous to σ^{32}, *B. subtilis* and many other types of bacteria use the normal sigma and a repressor protein named HrcA to repress transcription from heat shock genes during growth at lower temperatures. The HrcA repressor binds to an operator sequence called CIRCE, which is highly conserved among bacteria as diverse as *B. subtilis* and cyanobacteria, suggesting that this type of regulation may be very ancient. Bacteria that use HrcA do use a chaperone as a cellular thermometer, but rather than using DnaK, they seem to use the chaperonin, GroEL (see chapter 2). GroEL may be required to fold HrcA; when the temperature increases abruptly and GroEL is recruited to help other proteins fold, HrcA may misfold. This would then cause derepression of the heat shock genes under its control.

General Stress Response in Gram-Negative Bacteria

In addition to the heat shock sigma σ^{32}, which responds to an abrupt increase in temperature, *E. coli* has another sigma factor, called the stationary-phase sigma (σ^{s}), which is used to transcribe genes that are involved in the general stress response (Box 13.3). The gene for this sigma factor, *rpoS*, is turned on following many different types of stress including nutritional deprivation, oxidative damage, and acidic conditions. σ^{s} is very closely related to the normal vegetative sigma factor, σ^{70}, which transcribes most genes in *E. coli* and recognizes very similar promoters. The only difference is that the -10 sequence recognized by σ^{s} may be somewhat extended, and, in fact, some promoters may be recognized by both σ^{s} and σ^{70}.

DNA microarray analysis has been used in attempts to identify all of the *E. coli* genes whose transcription is affected by RNA polymerase containing σ^{s} (see Weber et al., Suggested Reading). A total of 481 genes, more than 10% of the total number of genes in *E. coli*, are affected, either positively or negatively, by the absence of σ^{s}. Of these, 140 are affected under all the conditions tested in early stationary phase while the remaining 341 are transcribed under only some conditions, such as low pH or high osmolarity in the medium. Many of these are activator genes that are then activated only under a certain set of conditions. Besides genes whose products are obviously involved in stress relief, many genes involved in central energy metabolism such as glycolysis are also affected. The products of some of these genes probably play a role in switching the cell from aerobic metabolism to anaerobic metabolism as the cell enters stationary phase. Others are efflux pumps that may help to remove toxic compounds from the cell or to scavenge for rare nutrients. The picture thus arises of σ^{s} as the master regulator at the top of a large regulatory pyramid, turning on, in stationary phase, a number of genes for more specialized activators that then respond to more individualized stress conditions.

With so much of the fate of *E. coli* in its hands, σ^{s} must be able to respond quickly to a number of different environmental signals. Transcriptional regulation is efficient but rather slow. While different stress conditions do affect *rpoS* transcription (see Magnusson et al., Suggested Reading), the levels of *rpoS* mRNA remain high throughout exponential phase, indicating that most of the regulation seems to occur posttranscriptionally, either in the ability of the *rpoS* mRNA to be translated or in the stability of σ^{s} itself; we therefore concentrate on these.

One way in which translation of *rpoS* mRNA is regulated is through small noncoding RNAs (see Box 13.5). The best understood of these is DsrA, a small RNA whose synthesis increases at low temperatures. It then stimulates *rpoS* mRNA translation by binding to the 5′ untranslated region with the help of a protein, Hfq, and possibly opening a secondary structure, exposing the TIR for translation initiation. Interestingly, DsrA also regulates H-NS, a DNA-binding histone-like protein that also binds to the 5′UTR of *rpoS* mRNA and inhibits *rpoS* translation, suggesting some sort of feedback loop between DNA structure and RpoS induction. Another small RNA, RprA, seems to be able to substitute for DsrA under some conditions, suggesting redundancy in the regulation.

As mentioned, RpoS levels are also regulated through stability of the sigma factor σ^s after it is made. While cells are growing exponentially, σ^s is being continuously made but very little of it accumulates since it has a half-life of only 1 to 2 min. However, when the cells run out of energy or are subjected to some other stress, the half-life of σ^s increases markedly, leading to its rapid accumulation. The reason for this is quite well understood. During normal growth, the σ^s protein is being degraded by the ClpXP protease, which consists of two proteins, a barrel-shaped chaperone, made up of six copies of the ClpX protein that unfolds proteins, and the ClpP protease, which then degrades the unfolded protein (see chapter 2). However, ClpXP cannot degrade σ^s directly but can do so only if it is bound to another protein, RssB. The RssB protein is a response regulator much like others we have discussed in that is dephosphorylated in response to stress, presumably by the phosphatase activity of a sensor kinase that senses the stress, although the phosphatase partner of the sensor kinase has yet to be identified. However, RssB differs from most of the other response regulators already discussed in that it does not bind to DNA but binds only to the σ^s protein. Only the unphosphorylated form of RssB can bind to σ^s and promote its degradation, explaining why σ^s is unstable only when the cells are growing exponentially. Interesting, the *rssB* gene itself is transcribed from a σ^s promoter, so that the σ^s is essentially regulating itself, keeping its concentration low under exponential growth conditions, when it is not needed.

General Stress Response in Gram-Positive Bacteria

Many gram-negative bacteria are known to have a general stress response based on σ^s and probably similar to that of *E. coli*. However, gram-positive bacteria, including *B. subtilis*, use another sigma factor, σ^B, to transcribe stress-induced genes. This sigma factor is quite different from σ^s in sequence and is activated by a very different pathway. Its mechanism of activation is reminiscent of the activation of σ^E in *E. coli* (see below) and σ^F in *B. subtilis* sporulation (see chapter 14) in that it depends on inactivating an **anti-sigma factor**. The signaling pathway that activates σ^B is fairly long and complicated, using a phosphorelay system involving serine-threonine kinases and phosphatases that is more reminiscent of eukaryotes than it is of eubacteria. Also, there seems to be one pathway to sense energy deficiency when the cells run out of a carbon and energy source and a different pathway to sense an environmental stress such as heat shock or a pH change. This section gives an overview of the latter signaling pathway as it is presently understood, as well as some of the genetic techniques that have been used to elucidate it.

The stress sigma factor σ^B is normally held quiescent in the membrane because it is bound to an anti-sigma factor in the membrane. The ball starts rolling when, after an environmental stress, sensing systems in the membrane (see below) remove the phosphate from an anti-anti-anti-sigma. The anti-anti-anti-sigma factor that is used depends on whether the signal is energy depletion or environmental stress. They activate the phosphatase activity of the anti-anti-anti-sigma so that it removes the phosphate from an anti-anti-sigma. The anti-anti-sigma can then bind to the anti-sigma and release it from σ^B. Free σ^B can then bind to RNA polymerase and direct it to the stress response genes. If this seems unnecessarily complicated, it is probably because we do not know all the pathways that must interact with this pathway. With more steps, other pathways that respond to different environmental conditions can interact at different steps and influence the outcome.

Recent research has concentrated on how environmental signals are communicated to the first step of the pathway, causing the the anti-anti-anti-sigma factor to become active (see Kim et al., Suggested Reading). This research used a genetic approach to determine **epistasis**. A general definition of genetic epistasis is a genetic interaction between alleles in different genes in which the phenotype due to one allele predominates over that due to another allele. Epistasis tests in genetics are often used to determine whether the function of one gene product is dependent on the previous functioning of another gene product. If so, the latter gene is epistatic over the former gene. If the two gene products can act independently of one another, neither gene is epistatic over the other.

A total of five periplasmic proteins called the Rsb proteins were known to be involved in the negative control of activation of the anti-anti-anti-sigma factor and therefore induction of the stress response. They all seem to be part of the same large complex (the stressosome) that senses external stresses and induces the stress response by activating the signal transduction system that activates σ^B. All five Rsb proteins are similar in their carboxyl termini, but one of them, RsbS, is shorter, consisting of only the shared carboxyl terminus. These proteins are all serine or threonine kinases with phosphate groups being added to or removed from them in response to external stresses. Preliminary evidence had suggested that the longer Rsb proteins (called RsbRA, RsbRB, RsbRC, and RsbRD) all act through RsbS, i.e., are epistatic to RsbS. As evidence, null mutations in the gene for RsbS cause constitutive expression of the stress response, but three or more of the longer RsbR proteins must be inactivated to have the same effect. It was hypothesized that the longer Rsb proteins each respond to a different, but overlapping, external stress. In the presence of their particular stress, they allow the phosphorylation of RsbS. Under normal conditions, RsbS is not phosphorylated and binds to the RsbT kinase protein, preventing it from activating the phosphatase activity of the anti-anti-anti-sigma factor. A stress causes RsbS to be phosphorylated on serine 59 (the 59th amino acid from the N terminus is the serine that is phosphorylated). With phosphate bound to its serine 59, RsbS cannot bind to RsbT, and the anti-anti-anti-sigma factor becomes active and begins the process of activating the stress response. This model explained why all five of the Rsb proteins negatively regulated the stress response. In the presence of any of the other Rsb proteins, the RsbS protein can be phosphorylated on its serine 59 and the stress response is induced. In the absence of RsbS, the stress response is also induced, but for a different reason: it is not there to bind RsbT and inactivate the anti-anti-anti-sigma.

The two possibilities for how the five proteins act are illustrated in Figure 13.14. In one model, shown in Figure 13.14A, the larger Rsb proteins all act through RsbS and are required for its phosphorylation and the release of RsbT. In other words, RsbS is epistatic over the other proteins in that the others must all act through it. In the second model, shown in Figure 13.14B, at least one of the R proteins is required to bind RsbT and lack of phosphorylation of RsbS by itself is insufficient. To test the epistatic hypothesis genetically, the investigators introduced a reporter gene fusion into the cells, with the *lacZ* reporter gene fused to a σ^B-dependent promoter (see chapter 2). If σ^B is activated, the *lacZ* gene is transcribed and β-galactosidase, which is easy to assay, is induced. They also used another trick. Rather than depending on the state of phosphorylation of the serine 59, in RsbS, they used site-specific mutagenesis to replace the serine with a different amino acid, aspartate. The side chain of aspartate mimics phosphate attached to serine or threonine because it is acidic and about the same size. However, unlike phosphate, it is not removed by phosphatases in response to cellular conditions, which would complicate the analysis. Consistent with the hypothesis, replacing serine 59 of RsbS with aspartate led to constitutive expression of the reporter gene fusion, a situation that simulates constitutive phosphorylation of RsbS. If hypothesis A (Figure 13.14A) is correct, then replacing serine 59 with alanine, which neither resembles

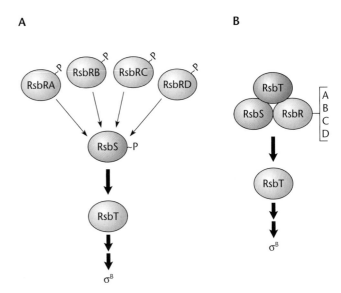

Figure 13.14   Genetic test for epistasis. Phosphorylation of RsbS releases it from RsbT, leading to elevated levels of σ^B. If the four RsbR proteins act only to phosphorylate RsbS, as shown in panel A, then substituting the serine that is phosphorylated with alanine, which cannot be phosphorylated, should lead to repressed levels of σ^B even if all four RsbR genes are inactivated. However, at least one of the RsbR proteins is still required for repression, indicating that the model in panel B is correct.

phosphorylated serine nor can be phosphorylated, should lead to repression of the levels of σ^B even if all the larger RsbR proteins are inactivated. However, deleting three or more of the *rsbR* genes in the presence of this mutated RsbS caused constitutive expression of the reporter gene, showing that they do not work through *rsbS*, at least not exclusively. At least one of them must act with unphosphorylated RsbS to bind to RsbT and negatively regulate the stress response, as shown in Figure 13.14B. Apparently, unphosphorylated RsbS plus at least one of the RsbR proteins is required to prevent the activation of σ^B under nonstress conditions, and each of them responds to a somewhat different stress. How these proteins actually sense the stress, and what they sense, as well as why there seem to be redundant sensors for the same signals, remains unknown (see Reeves and Haldenwang, Suggested Reading).

Extracytoplasmic (Envelope) Stress Responses

The membranes of bacteria are the first line of defense against external stresses. They are also particularly sensitive to abrupt changes in osmolarity, damaging agents such as hydrophobic toxins, heat shock, pH changes, etc. Not surprisingly, many stress responses are dedicated to preserving the integrity of the bacterial membranes. These are often referred to as extracytoplasmic stress responses because they respond to changes outside of the cytoplasm.

Regulation of Porin Synthesis

One of the changes often faced by bacteria is a change in osmolarity due to changing solute concentrations outside the cell. The osmotic pressure is normally higher inside the cell than outside it. This pressure would cause water to enter the cell and the bacterium to swell, but the rigid cell wall keeps the cell from expanding. Even the cell wall is not invincible, however, and bacteria must keep the difference in osmotic pressure inside and outside the cell from becoming too great. The ability to monitor osmolarity can be important to bacteria for a second reason: bacteria also sometimes sense changes in their external environment by detecting changes in osmolarity. In fact, one way in which pathogenic bacteria sense that they are inside a host, and induce their virulence genes, is by the much higher osmolarity inside the host (see the section on global regulation in pathogenic bacteria, below). The systems by which bacterial cells sense these changes in osmolarity and adapt are global regulatory mechanisms, and many genes are involved.

Much is known about how bacteria respond to media of different osmolarities. One way they regulate the differences in osmotic pressure across the membrane is by excreting or accumulating K^+ ions and other solutes such as proline and glycine betaine. Gram-negative bacteria such as *E. coli* have the additional problem of maintaining an equal osmotic pressure across the outer membrane. They achieve this in part by synthesizing oligosaccharides in the periplasmic space to balance solutes in the external environment.

One of the major mechanisms by which gram-negative bacteria balance osmotic pressure across the outer membrane is by synthesizing pores to let solutes into and out of the periplasmic space. These pores are composed of outer membrane proteins called **porins**. To form pores, three of the polypeptide products of these genes come together (trimerize) in the outer membrane to form what are called β barrels with central channels that selectively allow hydrophilic molecules through the very hydrophobic outer membrane.

The two major porin proteins in *E. coli* are OmpC and OmpF. Pores composed of OmpC are slightly smaller than those composed of OmpF, and the size of the pores can determine which solutes can pass through the pores and thus confer protection under some conditions. For example, the smaller pores, composed of OmpC, may prevent the passage of some toxins such as

the bile salts in the intestine. The larger pores, composed of OmpF, may allow more rapid passage of solutes and so confer an advantage in dilute aqueous environments. Accordingly, *E. coli* cells growing in a medium of high osmolarity, such as the human intestine, have more OmpC than OmpF, whereas *E. coli* cells growing in a medium of low osmolarity, such as dilute aqueous solutions, have less OmpC than OmpF.

Other environmental factors besides osmolarity can alter the ratio of OmpC to OmpF. This ratio increases at higher temperatures or pHs or when the cell is under oxidative stress due to the accumulation of reactive forms of oxygen (see Box 11.1). The ratio also increases when the bacterium is growing in the presence of organic solvents such as ethanol or some antibiotics and other toxins. Presumably, the smaller size of OmpC pores limits the passage of many toxic chemicals into the cell. Many of these abrupt changes occur when the *E. coli* bacterium leaves the external environment and passes through the stomach into the intestine of a warm-blooded vertebrate host, its normal habitat. It then must synthesize mostly OmpC-containing pores to keep out toxic materials such as bile salts, as mentioned above. Other conditions cause a decrease in both OmpC and OmpF concentrations. To respond to all of these other changes, the *ompC* and *ompF* genes are in a number of different regulons, which respond to different external stresses. We first discuss one of these pathways in *E. coli*, the regulation in *ompC* and *ompF* expression by EnvZ and OmpR in response to high osmolarity. While not yet completely understood, this system has served as a model for two-component signal transduction systems that allow the cell to sense the external environment and adjust its gene expression accordingly; therefore, we discuss this subject in some detail.

GENETIC ANALYSIS OF PORIN REGULATION BY OSMOLARITY

As in the genetic analysis of any regulatory system, the first step in studying the osmotic regulation of porin synthesis in *E. coli* was to identify the genes whose products are involved in the regulation. The isolation of mutants defective in the regulation of porin synthesis was greatly aided by the fact that the some of the porin proteins also serve as receptors for phages and bacteriocins, so that mutants which lack a particular porin are resistant to a given phage or bacteriocin. This fact offers an easy selection for mutants defective in porin synthesis. Only mutants that lack a certain porin in the outer membrane are able to form colonies in the presence of the corresponding phage or bacteriocin.

Using such selections, investigators isolated mutants that had reduced amounts of the porin protein OmpF in their outer membrane. These mutants were found to have mutations in two different loci, which were named *ompF* and *ompB*. Mutations in the *ompF* locus can completely block OmpF synthesis, whereas mutations in *ompB* only partially prevent its synthesis. The quantitative difference in the effect of mutations in the two loci suggested that *ompF* is the structural gene for the OmpF protein and that an *ompB*-encoded protein(s) is required for the expression of the *ompF* gene. Complementation experiments showed that the *ompB* locus actually consists of two genes, *envZ* and *ompR*. Using *lacZ* fusions to *ompF* to monitor the transcription of the *ompF* gene (see chapter 2), investigators confirmed that EnvZ and OmpR are required for optimal transcription of the *ompF* gene (see Hall and Silhavy, Suggested Reading).

EnvZ and OmpR: a Sensor Kinase and Response Regulator Pair of Proteins

The *envZ* and *ompR* genes were cloned and sequenced by methods such as those discussed in chapter 1. Similarities in amino acid sequence between EnvZ and OmpR and other sensor kinase and response regulator pairs, including NtrB and NtrC, suggested that these proteins are also a sensor kinase and response regulator pair of proteins. Like many sensor proteins, the EnvZ protein is an inner membrane protein, with its N-terminal domain in the periplasm and its C-terminal domain in the cytoplasm (Box 13.4, figure). The N-terminal domain of EnvZ apparently senses an unknown signal in the periplasm which indicates that the osmolarity is low and transfers this information to the cytoplasmic domain. The information is then transferred to the OmpR protein, a transcriptional regulator that regulates transcription of the porin genes. Like the NtrB protein, the EnvZ protein is known to be autophosphorylated, and its phosphate is also known to be transferred to OmpR. This led to the simple model that EnvZ was more heavily phosphorylated at high osmolarity and that this phosphate was then transferred to OmpR. If OmpR is phosphorylated, it activates transcription of *ompC*; if it is unphosphorylated, it activates transcription of *ompF*. However, further evidence did not support this simple model, since phosphorylated OmpR was found to be required for the transcription of both *ompC* and *ompF*. While we still do not completely understand how this regulation works, it is worth reviewing this genetic evidence.

1. Mutant phenotypes. Table 13.3 lists several relevant phenotypes of *envZ* and *ompR* mutations. According to the simple model, mutations that totally inactivate the EnvZ protein (*envZ* null mutations) should completely prevent the phosphorylation of OmpR under any conditions, freezing it in the form that activates

transcription of *ompF*. Therefore, *envZ* null mutations would be predicted to completely prevent the transcription of *ompC* but allow high-level transcription of *ompF*. However, the evidence shows that while *envZ* null mutations do prevent the transcription of *ompC*, they also allow only very limited transcription of *ompF* (shown as OmpF^{+-} in Table 13.3; see Slauch et al., Suggested Reading). The EnvZ protein is apparently required to activate the transcription of *both* porin genes, perhaps because some OmpR~P is required to activate the transcription of both genes.

2. Constitutive mutations in *ompR*. One type of constitutive mutation, called *ompR2*(Con) in Table 13.3, prevents the expression of *ompC* but causes the constitutive expression of *ompF*, even when coupled with a null mutation in *envZ*. A second type of constitutive mutation, called *ompR3*(Con), causes the constitutive expression of *ompC* but prevents the expression of *ompF*. The existence of these constitutive mutations could have been predicted from the simple model. In analogy to the AraC activator (see chapter 12), the OmpR activator may exist in two forms (see chapter 12), one when it is phosphorylated and another when it is not. The *ompR3*(Con) mutations could change OmpR into the form in which it normally exists when phosphorylated, even without phosphorylation. It could then activate the transcription of *ompC* but not *ompF*. However, the behavior of the *ompR*(Con) mutations in complementation tests is hard to reconcile with the simple model. According to the simple model, in partial diploids carrying both a constitutive allele and a wild-type allele [*envZ*$^+$ *ompR3*(Con) / *envZ*$^+$ *ompR*$^+$ in Table 13.3], both *ompC* and *ompF* would be predicted to be expressed at high levels, even at low osmolarity, because the mutant OmpR protein should activate the transcription of *ompC* whereas the wild-type OmpR protein should activate the transcription of *ompF*. In genetic terms, the constitutive mutations should be dominant over the wild-type allele for the expression of *ompC* but recessive for the expression of *ompF* under conditions of low osmolarity. Instead, as shown in Table 13.3, *ompF* is not expressed in the partial diploids, even in media of low osmolarity, although *ompC* is constitutively expressed.

The Affinity Model

A more current model for how EnvZ and OmpR regulate the transcription of the porin genes incorporated these and other observations about the regulation. The EnvZ protein is known to have both phosphotransferase and phosphatase activities, allowing it to both donate a

BOX 13.4

Signal Transduction Systems in Bacteria

Many regulation mechanisms require that the cell sense changes in the external environment and change the expression of its genes or the activity of its proteins accordingly. The sensors can be of many types (see the figure, panel A). Some are threonine serine kinases-phosphatases (STYK) such as the Rsb proteins that activate the stress sigma factor, σ^B, of *B. subtilis* and many other bacteria (see the text). Others already discussed are adenyl cyclases (ACyc), which make cAMP in enteric bacteria such as *E. coli* in response to nutritional conditions such as a relatively poor carbon source. An interesting, recently discovered type of signaling molecule is cyclic diGMP (c-diGMP), which consists of two GMP molecules linked to each other's 3' carbons through their 5' phosphates to form a sort of circle of phosphate-ribose sugar groups. Specific enzymes called diguanylate cyclases make this small-molecule effector, and specific phosphodiesterases destroy it. These enzymes were discovered primarily through genomic analysis because the cyclases have the domain GGDEF and the phosphodiesterases have the domain EAL (see inner cover for amino acid assignments). Quite often, signaling proteins have both a diguanylate cyclase and a diguanylate phosphodiesterase domain. They are widespread, having been found in many types of bacteria, and play diverse roles in the attachment of bacteria to surfaces, in the formation of biofilms, in the regulation of photosynthesis, and in motility. However, how this effector acts is not well understood, as indicated by the question mark in the figure (see Ryan et al., below).

Some of the most common and widely studied sensor systems are the so-called two-component signal transduction systems, which consist of a sensor histidine kinase (His Kin) which autophosphorylates (transfers phosphates to itself) and a response regulator (RR) that transfers the phosphate to itself and then performs a specific action in the cell. Two-component systems have been found in all bacteria and some plants but, at least in this form, are absent from animals. Large bacteria can have hundreds of these systems. As the name implies, they usually consist of two proteins, but in some cases the sensor kinase and the response regulator activities are domains of the same protein. In only a few cases is the stimulus to which the sensor kinase responds known (see Mascher et al., below). The output responses of the systems also vary (see Galperin, below). To name just a few, the response regulator is quite often a DNA-binding transcriptional regulator with a helix-turn-helix (HTH) domain or antiterminator, but we also discuss cases where it destabilizes a protein by targeting it for proteolysis. The cellular functions that enlist two-component signal transduction systems also vary widely, including involvement in motility in response to chemical attractants, i.e., the methylated chemotaxis proteins (MCP in the figure, panel A) (see Box 14.1); the induction of pathogenesis operons after entry into a suitable host; and the activation of extracellular stress responses.

The way in which these two-component sensor kinase and response regulator systems operate in general is illustrated in the figure, panel B. Part I shows that sensor kinases are often integral membrane proteins responsive to external signals. Part II shows the functions of the protein domains. The C-terminal domain of a sensor kinase has the conserved histidine that is phosphorylated (step 1). The response regulators are similar in their N-terminal region, which includes the phosphorylated aspartate (step 2). The remainder of the

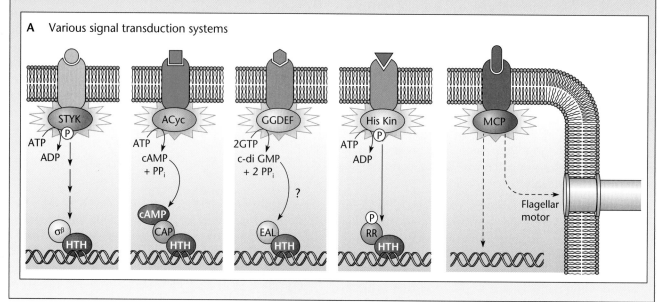

A   Various signal transduction systems

BOX 13.4 (continued)

Signal Transduction Systems in Bacteria

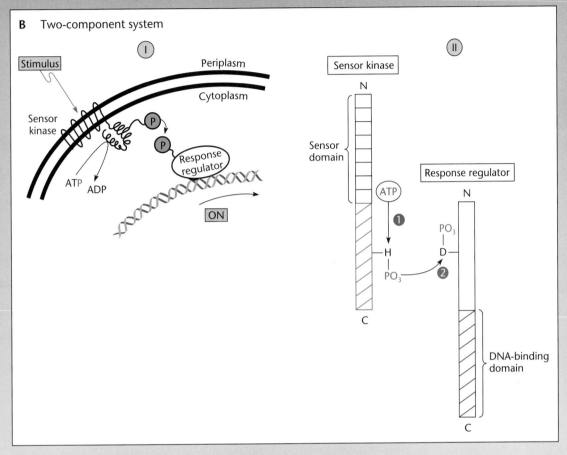

protein differs depending on its function, although different subfamilies of response regulators show regions of high homology in other parts of the protein, including the helix-turn-helix motif of many transcriptional regulators.

Not only do many sensor kinase and response regulator pairs of proteins work in remarkably similar ways, but also they have considerable amino acid sequence homology, as though they were evolutionarily derived from each other. Panel C of the figure illustrates sequence conservation in histidine kinases. The key shows that the most highly conserved amino acids are indicated by letters and the less highly conserved ones are indicated by diamonds. Nonconserved amino acids are indicated by dots. Structural studies showed that the conserved region consists of two separate domains, the kinase domain HisKA (PF00512 in the Pfam database [http://pfam.wustl.edu]) and HATPase (PF02518 in Pfam). The sequence conservation in the two domains in part III is in

a WebLogo (http://weblogo.berkeley.edu) representation, which reflects the statistical importance of every given position.

References

Bourret, R. B., N. W. Charon, A. M. Stock, and A. H. West. 2002. Bright lights, abundant operons—fluorescence and genomic technologies advance studies of bacterial locomotion and signal transduction: review of the BLAST meeting, Cuernavaca, Mexico, 14 to 19 January 2001. *J. Bacteriol.* **184:**1–17.

Galperin, M. Y. 2006. Structural classification of bacterial response regulators: diversity of output domains and domain combinations. *J. Bacteriol.* **188:**4169–4182.

Mascher, T., J. D. Helmann, and G. Unden. 2006. Stimulus perception in bacterial signal-transducing histidine kinases. *Microbiol. Mol. Biol. Rev.* **70:**910–938.

Ryan, R. P., Y. Fouhy, J. F. Lucey, and J. M. Dow. 2006. Cyclic di-GMP signaling in bacteria: recent advances and new puzzles. *J. Bacteriol.* **188:**8327–8334.

(continued)

BOX 13.4 (continued)

Signal Transduction Systems in Bacteria

C Histidine kinases

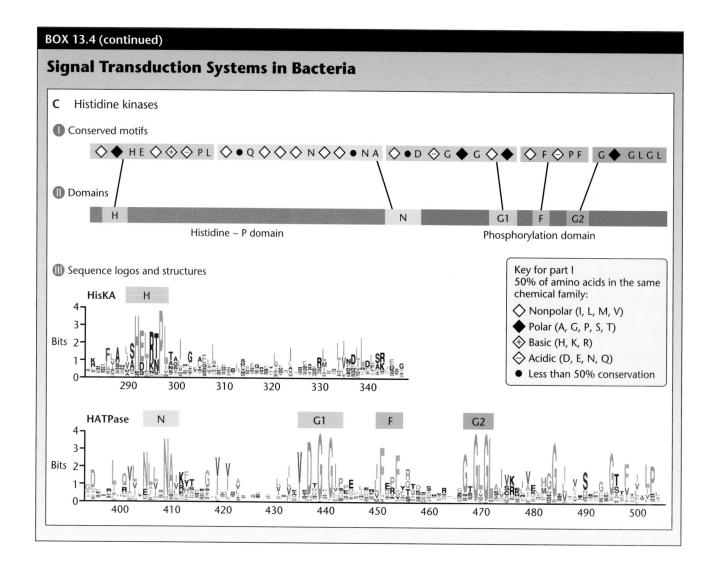

I Conserved motifs

II Domains

Histidine ~ P domain

Phosphorylation domain

III Sequence logos and structures

HisKA

HATPase

Key for part I
50% of amino acids in the same chemical family:
◇ Nonpolar (I, L, M, V)
◆ Polar (A, G, P, S, T)
⊕ Basic (H, K, R)
◇ Acidic (D, E, N, Q)
● Less than 50% conservation

TABLE 13.3	Phenotypes of *envZ* and *ompR* mutations
Genotype	**Phenotype**
envZ⁺ *ompR*⁺	OmpC⁺ OmpF⁺
envZ⁺ *ompR1*	OmpC⁻ OmpF⁻
envZ(null) *ompR*⁺	OmpC⁻ OmpF⁺⁻ᵃ
envZ⁺ *ompR2*(Con)	OmpC⁻ OmpF⁺ (low osmolarity)
	OmpC⁻ OmpF⁺ (high osmolarity)
envZ(null) *ompR2*(Con)	OmpC⁻ OmpF⁺ (low osmolarity)
	OmpC⁻ OmpF⁺ (high osmolarity)
envZ⁺ *ompR3*(Con)	OmpC⁺ OmpF⁻ (low osmolarity)
	OmpC⁺ OmpF⁻ (high osmolarity)
envZ⁺ *ompR3*(Con)/	OmpC⁺ OmpF⁻ (low osmolarity)
envZ⁺ *ompR*⁺	OmpC⁺ OmpF⁻ (high osmolarity)

ᵃ+ − indicates that OmpF levels are reduced but not eliminated.

phosphoryl group to and remove one from OmpR. Many sensor kinases have both activities. Whether the phosphorylated or unphosphorylated form of OmpR predominates depends on whether the phosphotransferase or phosphatase activity of EnvZ is most active. Under conditions of high osmolarity, the phosphotransferase activity predominates and the levels of phosphorylated OmpR (OmpR~P) are high. Under conditions of low osmolarity, the phosphatase activity predominates, and most of the OmpR is unphosphorylated.

To explain how *envZ* null mutations can prevent the optimal transcription of both genes, we must propose that the phosphorylated form of OmpR is required to activate transcription of both the *ompC* and *ompF* genes. But how can the relative levels of phosphorylated OmpR regulate the transcription of *ompC* and *ompF* differently? In other words, how could higher levels of OmpR~P favor the transcription of *ompC* while lower levels favor the transcription of *ompF* even though both require OmpR to be phosphorylated? One solution to this question was to propose the existence of multiple binding sites for OmpR~P upstream of the promoters for the *ompF* and *ompC* genes (see Pratt et al., Suggested Reading). Some of these sites bind OmpR~P more tightly than others; in other words, some are high-affinity sites whereas others are low-affinity sites. By binding to these sites, the OmpR~P protein can either activate or repress transcription from the promoters, depending on the position of the binding site relative to the promoter (see chapter 2). However, subsequent evidence suggested that the binding affinities of phosphorylated OmpR to *ompF* and *ompC* are too similar to account for the regulation, so things were back where they started. It is important not to give up trying to understand the relatively simple EnvZ-OmpR two-component sensor kinase and response regulator system, however. The EnvZ protein is one of the better defined transmembrane proteins that can communicate information about what is happening outside the cell to the internal regulatory pathways. Therefore, studies with this system should further our understanding of how cells sense the external environment. How the EnvZ protein achieves this feat should tell us much about how information is transferred across cellular membranes in general. Also, understanding the novel form of regulation by phosphorylated OmpR could reveal a new mechanism of transcription activation.

REGULATING OmpF BY A SMALL RNA: MicF

As mentioned, the OmpC/OmpF ratio increases not only when the osmolarity increases but also when the temperature or pH increases or when toxic chemicals including active forms of oxygen, nitric oxide (NO), or organic solvents such as ethanol are in the medium. In general, under conditions where nutrients and toxins are high, such as in the vertebrate intestine, the OmpC levels are high. Since OmpC forms narrower channels, fewer toxins can get in. Fewer nutrients can get in as well, but since their concentration in the intestine is high, this is not a problem. If nutrient levels are low, such as in water outside of the vertebrate host, OmpF levels are higher, because allowing the available nutrients to get into the cell becomes more important than keeping toxins out. The rationale for regulating the porins with temperature or pH is less obvious. One possibility is that the bacterium uses temperature and pH as sensors to indicate that the water in which it lives has just been drunk by a vertebrate and the bacterium is about to pass into the vertebrate intestine. The

porins would then need to be regulated to combat the onslaught of toxins that will be faced by the bacterium, including oxidative bursts by macrophages, bile salts, etc.

Much of the synthesis of the OmpF porin in response to these other forms of stress is through the small noncoding RNA, MicF. There are many such small regulatory RNAs in bacteria (Box 13.5) including DsrA, the small RNA that regulates the translation of the stress sigma factor, σ^s, in *E. coli* (see above). The MicF RNA was one of the first to be discovered, and it was found by chance because it is adjacent to the gene for OmpC but transcribed in the opposite direction (divergently transcribed). When the region of the chromosome containing the *ompC* gene was cloned into a high-copy-number plasmid and introduced into *E. coli* cells, the synthesis of MicF was inhibited. At first it was assumed that the OmpC protein was somehow inhibiting the synthesis of the OmpF protein. However, it was not the OmpC protein that was doing the inhibiting but, rather, a small RNA encoded just upstream of the *ompC* gene. The RNA was named MicF for *m*ulticopy *i*nhibitor of Omp*F*. Later it was shown to have partial homology to the 5'UTR of the mRNA for *ompF* and can pair with it and inhibit its translation. Because it is only partially complementary to the mRNA for OmpF, however, it cannot pair with it very strongly, and another protein called Hfq is required for the pairing. This seems to be true of most regulatory small RNAs (Box 13.5). By pairing with the TIR of the mRNA for *ompF*, MicF can prevent access by ribosomes to the TIR and hence prevent translation of OmpF. As a consequence, *E. coli* cells containing higher concentrations of MicF RNA will make less OmpF.

The cellular levels of MicF RNA increase under certain conditions because the promoter for the *micF* gene contains binding sites for many transcriptional activators of the AraC family. Because they are from the same family, they might share DNA-binding specificity and the ability to activate the RNA polymerase at the *micF* promoter and might differ only in the effector-binding pocket (Box 12.2). The activators seem to work independently, and each activates transcription from the *micF* promoter under its own particular set of conditions. For example, the transcriptional activator SoxS activates transcription from the *micF* promoter when the cell is under oxidative stress. Another, MarA, activates the transcription of *micF* when weak acids or some antibiotics are present. A third activator, Rob, may induce transcription of *micF* in the presence of cationic peptide antibiotics. Even OmpR, which binds the promoter of *ompC*, which lies in the same region as the promoter of *micF*, may also activate the transcription of *micF*, allowing another level of regulation of OmpF synthesis by osmolarity.

Regulatory RNAs

It is becoming increasingly evident that much of the work of regulation in cells is done by small regulatory RNAs (sRNAs) (see Altuvia and Wagner, below). This mode of regulation is sometimes called "riboregulation" and can occur at many different levels. Some regulatory RNAs are involved in transcriptional regulation. For example, an sRNA plays a role in the pheromone-responsive plasmid transfer in *Enterococcus faecalis*, where the 200-nucleotide mD RNA enhances transcriptional termination in *trans* (see Tomita and Clewell, below). Other sRNAs might bind to proteins and regulate them; for example, a small 6S RNA binds to σ^{70} RNA polymerase in stationary phase and inhibits it (see Trotochaud and Wassarman, below). In addition, sRNAs can regulate gene expression posttranscriptionally as antisense RNAs. These genes overlap those of the mRNAs they regulate, and their activities depend on complementary base-pairing interactions with their targets. We have already discussed some antisense RNAs in connection with the regulation of plasmid replication and transposition. For antisense RNAs, the complementarity between the antisense and target RNAs is exact and may extend as far as 100 nucleotides. Because of their extensive complementarity, the antisense RNAs are usually made as the complement of their target—hence the name antisense—and only one target RNA is regulated by a particular antisense RNA.

The genes for many chromosomal regulatory RNAs do not overlap their target genes, however, and may even be distantly located on the chromosome. Some of these small RNAs are mentioned in the text: the DsrA, MicF, and Ryh RNAs, which regulate *rpoS*, *ompF*, and the genes for iron-containing proteins, respectively. They are usually highly regulated and allow another level of regulation of the genes they control. Because these small RNAs regulate genes from a distance, they are often said to regulate in *trans*. By binding to their RNA targets, they can regulate gene expression in many ways. For example, they can inhibit translation by binding close to the TIR of an mRNA and blocking access by the ribosome to the TIR, or by targeting the mRNA for degradation by a cellular RNase. Alternatively, they can stimulate translation by binding close to the TIR and melting a secondary structure that includes the TIR or by stabilizing the mRNA. For these *trans*-acting RNAs, the regions of complementarity are short, often less than 12 bp, and scattered around the RNA. Because of such short interactions, a single antisense RNA may be able to regulate more than one target gene, with different regions of the sRNA base pairing with the various target sequences. Also, the RNAs usually require the Hfq protein to help them pair with their target mRNA (see below).

The DsrA and OxyS sRNAs are examples of *trans*-acting small RNAs that regulate multiple targets. The DsrA and OxyS RNAs can either increase or decrease the expression of the target gene, depending on the specific situation (see the figure). The DsrA RNA has at least two targets: *hns*, whose product is a histone-like silencer of genes, and *rpoS*, which encodes the stationary-phase sigma factor σ^s (Table 13.1). Short base-pairing interactions of DsrA with its target mRNAs in their Shine-Dalgarno regions either inhibit or allow translation of the mRNA (see Lease and Belfort, below). Interaction of DsrA with its target mRNA also affects the stability of the target mRNA, either decreasing it, as in the case of *hns* mRNA stability, or (as shown in the figure) increasing it, as in the case of *rpoS* mRNA. Whether DsrA inhibits or stimulates translation depends on how it binds to and affects different regions on the mRNA.

The OxyS sRNA, which is induced in response to oxidative stress, also presumably has multiple targets of regulation. One well-characterized target is *fhlA*, a transcriptional activator of formate metabolism. The mechanism of regulation of *fhlA* expression is similar to that of other sRNAs in that OxyS binds to the *fhlA* translation initiation region, inhibiting ribosome binding (see the figure).

As mentioned above, many sRNAs require a protein, Hfq, to bind to their target RNAs (see the figure and Valentin-Hansen et al., below). This protein was first found as a host-encoded protein in the Qβ replicase that is required to replicate the genomic RNA of this small phage, hence its name Hfq (for *host factor* Qβ). This protein is ubiquitous in bacteria and is even homologous to the Sm RNA-binding proteins involved in RNA splicing in eukaryotes. Six polypeptide products of the *hfq* gene form a ring (a hexameric ring). The protein helps the sRNA bind to a specific region of the target RNA, even though there is very little complementary base pairing to hold them together. Presumably, the Hfq protein recognizes sequences in both the sRNA and the target RNA, although these sequences have not yet been identified.

The first RNAs were found by accident, for example when they inhibited the synthesis of a gene product when they were overproduced from a multicopy plasmid. None of them were found in classical genetic analyses, perhaps because these genes are small and therefore are small targets for mutagenesis or because they often have redundant functions, so that inactivating mutations have no obvious phenotypes. Now, however, new sRNAs can be found by analysis of the genomic sequences of bacteria. Using such methods, over 60 sRNAs have been found in *E. coli*, i.e.,

BOX 13.5 (continued)

Regulatory RNAs

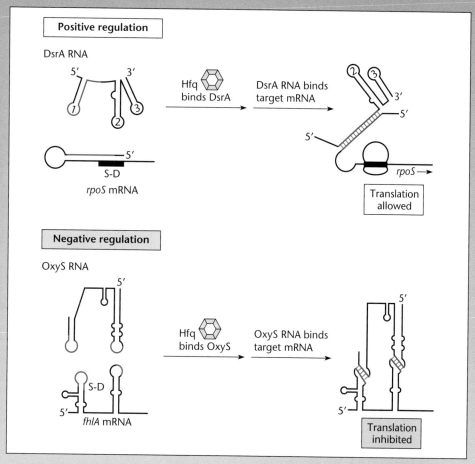

almost 2% of the coding capacity of the genome. The sRNAs might be in intergenic regions (the regions between genes) and be noticed because they have consensus promoters or transcription terminators that are "orphans" because there is no obvious ORF encoding a protein between them. Alternatively, they might be recognized as conserved sequences in intergenic regions by a comparison of the genomes of closely related bacteria or by microarray analysis, especially with Affymetrix chips. For example, sRNAs have been found for *E. coli* by comparing the *E. coli* genome to the genomes of other closely related bacteria, *Klebsiella pneumoniae* and *Salmonella* spp., using a combined computational and microarray approach (see Storz et al., below). The reasoning is that the sequences of intergenic regions, which normally differ, are conserved if they encode important regulatory sRNAs. In a more recent approach, the Hfq protein was used to fish sRNAs out of the RNA pool from cells, since many of them bind to Hfq. They could then be amplified by reverse transcription-PCR and characterized.

References

Altuvia, S., and G. H. Wagner. 2000. Switching on and off with RNA. *Proc. Natl. Acad. Sci. USA* **97:**9824–9826.

Gottesman, S. 2005. Micros for microbes: non-coding regulatory RNAs in bacteria. *Trends Genet.* **21:**399–404.

Lease, R., and M. Belfort. 2000. A *trans*-acting RNA as a control switch in *Escherichia coli*: DsrA modulates function by forming alternative structures. *Proc. Natl. Acad. Sci. USA* **97:**9919–9924.

Storz, G., S. Altuvia, and K. M. Wassarman. 2005. An abundance of RNA regulators. *Annu. Rev. Biochem.* **74:**199–217.

Tomita, H., and D. B. Clewell. 2000. A pAD1-encoded small RNA molecule, mD, negatively regulates *Enterococcus faecalis* pheromone response by enhancing transcription termination. *J. Bacteriol.* **182:**1062–1073.

Trotochaud, A. E., and K. M. Wassarman. 2004. 6S RNA function enhances long-term cell survival. *J. Bacteriol.* **186:**4978–4985.

Valentin-Hansen, P., M. Eriksen, and C. Udesen. 2004. The bacterial Sm-like protein Hfq: a key player in RNA transactions. *Mol. Microbiol.* **51:**1525–1533.

Wassarman, K. M., F. Repoila, C. Rosenow, G. Storz, and S. Gottesman. 2001. Identification of novel small RNAs using comparative genomics and microarrays. *Genes Dev.* **15:**1637–1651.

The MicF RNA inhibits the translation of OmpF but not OmpC. We might expect that a similar regulation would apply to OmpC, so that less of it would be made under conditions that favor OmpF. In fact, OmpC translation is inhibited by its own small RNA, named MicC, in analogy to the MicF small RNA. The MicC RNA was discovered because it has some sequence homology to the 5'UTR of *ompC* mRNA and was shown to inhibit OmpC trasnlation in high copy. Interestingly, the MicC gene is also adjacent to a gene for a porin, in this case OmpN, which is very poorly expressed in *E. coli*, at least under laboratory conditions, but might be expected to form pores with sites similar to those of OmpF. It is likely that OmpA, yet another porin, might be regulated by a small RNA. The regulation of porin synthesis by osmolarity and other environmental factors is obviously central to cell survival, which is why it is so complicated and involves so many interacting systems and regulatory molecules.

Regulation of the Envelope Stress Response by CpxA-CpxR: a Two-Component Sensor-Kinase Response-Regulator System

Another way that *E. coli* senses stress to the outer membrane is via the two-component system, CpxA and CpxR (Figure 13.15). This two-component system works like many two-component systems (Box 13.4) in that CpxA is a sensor kinase, which transfers a phosphate to the response regulator CpxR, a transcriptional activator that activates the transcription of more than 100 genes under its control. The CpxA sensor phosphorylates itself when it senses that proteins to be secreted, such as pili or curli fibers that play a role in attaching the bacterium to surfaces such as eukaryotic cells, are piling up in the periplasm as a result of some defect in transport due to damage to the outer membrane. The accumulation of these proteins in the periplasm is toxic, and many of the genes regulated by CpxR encode proteases and chaperones in the periplasm that help fold and degrade these proteins. The synthesis of the proteins that make up cellular appendages such as pili and curli fibers is also inhibited by phosphorylated CpxR, perhaps so that their intermediates will not accumulate in the periplasm.

The finding that the CpxA-CpxR two-component system also regulates the ratio of OmpC to OmpF came from a genetic screen for mutations that affect the ratio of OmpF to OmpC (see Batchelor et al., Suggested Reading). Since this analysis used some of the techniques described in previous chapters, it is discussed in some detail here. First, the investigators needed an easy way to determine if the ratio of OmpF to OmpC has been altered, since they would have to screen many mutants. To do this, they used transcriptional fusions that fused

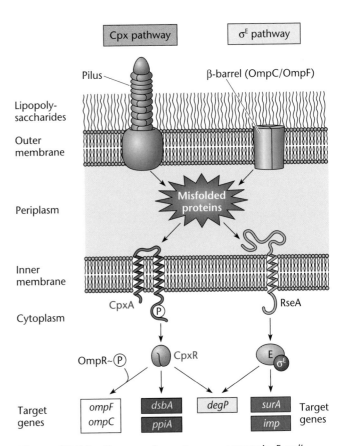

Figure 13.15 Two envelope stress responses in *E. coli* that respond to different stress signals in the periplasm. The Cpx pathway responds to the accumulation of pilin subunits in the periplasm, while the σ^E pathway responds to the accumulation of OmpA. DsbA is a periplasmic chaperone and oxidoreductase that makes disulfide bonds in exported proteins (see chapter 2), PpiA is a peptidyl-prolyl *cis-trans* isomerase, DegP is a protease and chaperone, SurA is a chaperone, and Imp is an usher protein for outer membrane proteins.

derivatives of the *gfp* reporter gene, called *yfp* and *cfp*, to the promoters for *ompF* and *ompC*. Fluoresence by the original Gfp protein emits green light, while these mutant forms emit different colors, yellow and cyan (blue), respectively. If the *ompF* promoter is more active, the colonies fluoresce more yellow, whereas if the *ompC* promoter is more active, the colonies fluoresce more blue. They constructed a strain containing both fusions by making the fusions in lysogenic phages and then integrating the phages into the chromosome so that the genes would be present in only one copy.

The next step was to mutagenize their strain to try to obtain mutants that were altered in the relative expression of *ompF* and *ompC*. They used transposon mutagenesis because it would be easier to characterize the resulting mutations. They used a transposon called miniTn5 that lacks its own tranposase so that it cannot transpose again when the strains are propagated and also does not cause chromosomal instability around its site of insertion (see Box 9.2). About 5,000 transposon insertion mutants were plated, and the fluorescence of their colonies was observed. One mutant in particular exhibited a dramatic decrease in OmpF transcription and a modest increase in OmpC transcription, especially in glucose minimal medium. When the DNA flanking the site of transposon insertion in this mutant was sequenced, the transposon was determined to be in the *cpxA* gene.

To be certain that the transposon insertion was inactivating the *cpxA* gene product and that this was the basis for the relative effects on *ompC* and *ompF* expression, the investigators used recombineering (see Box 10.3) to replace the entire *cpxA* gene in a different strain with a chloramphenicol resistance cassette and observed the same phenotypes with this strain. They also directly measured the amounts of OmpF and OmpC protein in the mutant cells to ensure that the amount of fluorescence accurately reflected the amount of protein product made on the genes. It is easy to detect OmpF and OmpC on stained sodium dodecyl sulfate-polyacrylamide gels because these are two of the most abundant proteins in *E. coli*, making their bands clearly visible among all the other protein bands.

They also wished to know the effect of inactivating both the *cpxR* gene and the *cpxA* gene in the same bacterium. We might expect that inactivating CpxR when CpxA had already been inactivated would have no effect since the only known role of CpxA is to transfer its phosphate to CpxR. However, when they also inactivated the *cpxR* gene, the phenotypes returned to normal and OmpF and OmpC were made in normal amounts. They explained this by assuming that the altered phenotypes in a *cpxA* mutant were due to a high level of phosphorylation of CpxR, rather than to no phosphorylation of CpxR. Like EnvZ and other sensor kinases, the CpxA protein is both a phosphotransferase, which attaches a phosphate to CpxR, and a phosphatase, which removes a phosphate from CpxR. In the absence of the phosphatase, phosphates might accumulate on CpxR as a result of another phosphorylating enzyme that transfers phosphates to many proteins from the universal donor, acetyl phosphate. They concluded that high levels of CpxR with phosphate attached (CpxR~P) repress *ompF* transcription but stimulate *ompC* transcription. Apparently, when some toxins enter the cell, CpxA is phosphorylated, perhaps because the toxins damage proteins in the periplasm or prevent the assembly of structures on the surface and cause their subunits to accumulate in the periplasm. The phosphorylated CpxA then transfers its phosphate to CpxR to form CpxR~P. Somehow CpxR~P then represses *ompF* and activates *ompC*, probably acting through phosphorylated OmpR (OmpR~P) at the promoters since both *ompF* and *ompC* require OmpR~P for their activation. The net result is that fewer toxins can get into the periplasm because more of the OmpC porin is made, which has a smaller pore than OmpF. In this way, the composition of the porins can respond both to changes in osmolarity (EnvZ and OmpR) and to the presence of toxins or proteins in the periplasm (CpxA, CpxR, and OmpR).

The Extracytoplasmic Function: *E. coli* Sigma Factor σ^E

In *E. coli*, extracytoplasmic stress is also sensed by another system which uses a specialized sigma factor to transcribe genes in the stress response (see Alba and Gross, Suggested Reading). This alternative sigma factor is called the extracytoplasmic function sigma, σ^E, because it is used mostly to express genes that function in the periplasm, such as proteases that degrade defective proteins in the periplasm and chaperones that help fold proteins as they pass through the periplasm. It was first discovered because one of the promoters that transcribes the *rpoH* gene for the heat shock sigma at high temperatures is recognized by this sigma factor.

As mentioned, the activation of σ^E in *E. coli* is reminiscent of the activation of σ^B in *B. subtilis* in that it involves inactivation of an anti-sigma factor. However, rather than being displaced by another anti-anti-sigma, the σ^E anti-sigma is degraded by proteases in response to envelope stress. This aspect of the induction of the envelope stress response is quite well understood and is also illustrated in Figure 13.15. The anti-sigma factor RseA is an inner membrane protein with domains in both the periplasm and the cytoplasm. The cytoplasmic domain binds σ^E, inactivating it and sequestering it to the membrane. When the outer membrane is damaged, Omp proteins, including OmpC, accumulate in the periplasm because they cannot be assembled into porins in the damaged outer membrane. The carboxyl-terminal domain of the Omp proteins binds to the carboxyl-terminal domain of a protease called DegS in the periplasm. The carboxyl terminus of DegS may normally inhibit its protease activity, and binding of the carboxyl terminus of an Omp protein to it activates the DegS protease, which then cleaves off the periplasmic domain of RseA. Another protease named YaeL then degrades the transmembrane domain of RseA, releasing σ^E to bind RNA polymerase and transcribe genes in the envelope stress response.

There is experimental evidence for each of these steps. Mutations that inactivate the *rpoE* gene for σ^E are lethal at any temperature, showing that the σ^E protein is essential for viability at any temperature. It is not clear why σ^E is essential, even at low temperatures, but maybe some periplasmic proteins under its control are essential, such as protein chaperones in the periplasm that help insert other proteins into the outer membrane. Mutations that inactivate *rseA* cause constitutive induction of the σ^E response as expected, since there is then no anti-sigma to inhibit σ^E even in the absence of stress. Mutations that inactivate DegS and YaeL are also lethal, but double mutants with both an *rseA* mutation and a *degS* or *yaeL* mutation are viable. In genetic terms, *degS* and *yaeL* mutations are suppressors of *rseA* mutations. This shows that the only essential role for DegS and YaeL is to degrade RseA and, by extension, induce the σ^E stress response. The role of the carboxyl terminus of DegS in inhibiting its own protease activity was found when the region of the *degS* gene encoding the carboxyl terminus was deleted, which led to constitutive activation of σ^E. Finally, the role of the carboxyl terminus of the Omp proteins in activating the protease activity of DegS was discovered when it was shown that just overproducing the carboxyl terminus of these proteins was sufficient to induce σ^E but did not increase the induction further if the carboxyl terminus of DegS had been deleted. It is possible that other pathways for releasing the RseA anti-sigma from σ^E exist, since *degS* mutants still show some activation of σ^E following envelope damage.

Iron Regulation in *E. coli*

Iron is an important nutrient, both for bacteria and for humans (see Kadner, Suggested Reading). Many enzymes use iron as a catalyst in their active center; many transcriptional regulators such as FNR, which regulates genes for anaerobic metabolism, use it as a sensor of oxygen levels; and hemes use it as an oxygen carrier. However, too much iron can also can be very damaging to cells and requires a stress response. It catalyzes the conversion of hydrogen peroxide and other reactive forms of oxygen to hydroxyl free radicals, the most mutagenic form of oxygen (see Box 11.1). Because of its Dr. Jekyll and Mr. Hyde properties, iron storage proteins such as ferritin store any excess iron in the cell to reduce the damage it can do, while at the same time making it available if it is needed.

Iron exists in two states in the environment, the ferric (Fe^{3+}) state, with a valence of three, and the ferrous (Fe^{2+}) state, with a valence of two. Iron in the ferric state forms largely insoluble compounds that are not easily used by bacteria and other organisms. Because oxygen quickly converts iron in the ferrous state to iron in the ferric state, most iron in the environment exists in the insoluble ferric state. Accordingly, many bacteria secrete proteins called siderophores that bind ferric ions and transport them back into the cell, where they can be converted to ferrous ions in the reducing atmosphere of the cytoplasm. As expected, these siderophores are made and secreted only if iron is limiting, to avoid the damaging effects of too much iron in the cell. There are three basic mechanisms in *E. coli* and most other bacteria for regulating genes involved in iron metabolism: the Fur regulon based on a protein repressor and transcriptional regulation; a small regulatory RNA called the RyhB RNA, which is regulated by Fur and promotes mRNA degradation; and the aconitase enzyme of the tricarboxylic acid (TCA) cycle, which doubles as a translational repressor. We discuss the Fur regulon first.

The Fur Regulon

The Fur repressor is a classical repressor with a helix-turn-helix DNA-binding domain in its N terminus, a dimerization domain in its C terminus, and an effector-binding pocket in the middle. Figure 13.16 illustrates regulation by Fur, which is much like the regulation of the *trp* operon by TrpR repressor. When ferrous iron is in excess, it acts as a corepressor by binding to the Fur aporepressor, changing its conformation to the repressor form, which can bind to an operator sequence called, in this case, a Fur box. By binding to the Fur box, which overlaps the −10 sequence of the σ^{70} promoters it regulates, it blocks access of the RNA polymerase to the promoters and represses transcription of the operons of the Fur regulon. If ferrous ions are in short supply and limiting, the repressor is in the aporepressor state and cannot bind to the operator sequences, turning on the transcription of the genes under its control, including genes for the siderophores and iron transporters in the membrane (called iron assimilation proteins in the figure). The Fur regulation system is highly conserved among gram-negative bacteria. Gram-positive bacteria have a similar system based on repressors related to the DtxR repressor of *Corynebacterium diphtheriae* (see below), which has little sequence similarity to Fur but is structurally similar and probably works in similar ways.

The RyhB RNA

While many genes are turned off when iron is in excess, other genes are turned on. These include the ferritin-like storage proteins and many other proteins that contain iron, including aconitase A (AcnA), an iron-containing enzyme of the TCA cycle, and an iron-containing superoxide dismutase that destroys peroxides in the cell before they can be converted into hydroxyl radicals by the iron and damage DNA and other cellular constituents. In the presence of excess iron, the concentration of these

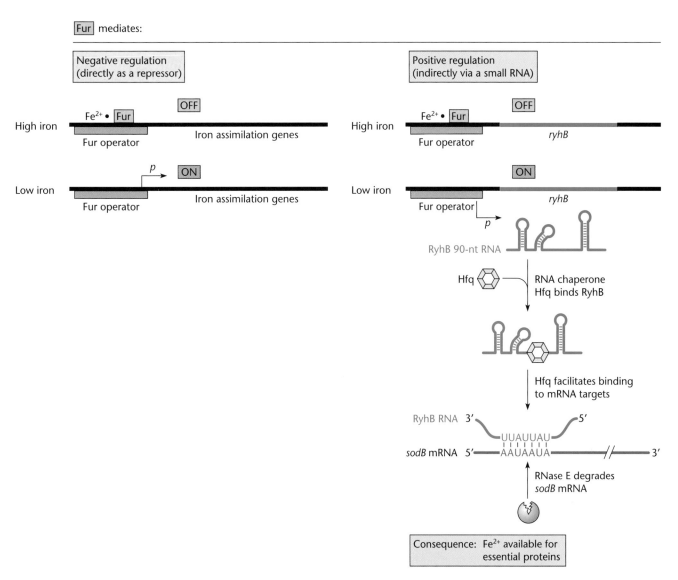

Figure 13.16   Regulation of operons in the Fur regulon. (Left) Negative regulation. Fur is a repressor that binds to the Fur operator in the presence of iron, blocking transcription of operons under its control. (Right) Indirect positive regulation by Fur. One gene repressed by Fur in the presence of iron encodes the small RNA, RyhB. If this small RNA is made when iron is limiting, it will base-pair, with the help of the Hfq protein, to the 5′ end of SodB mRNA and the mRNAs of other genes that are turned on by iron. The binding of the small RNA creates double-stranded RNAs which are the targets for degradation by a cellular RNase called RNase E. This prevents expression of the genes in the absence of iron but allows their expression in its presence, when Fur represses the transcription of the RyhB small RNA. Turning off some genes when iron is limiting makes more iron available for essential iron-containing proteins.

proteins increases rather than decreases, but only in Fur⁺ cells. Very little of them is made in a Fur⁻ mutant, suggesting that Fur is activating their transcription. Microarray analysis, comparing the levels of mRNA for all genes of *E. coli* in the presence and absence of iron,

revealed that the transcription of 53 genes increases when iron is in excess while that of 48 genes decreases. Accordingly, it was proposed that Fur could act either as a repressor or as an activator, like many other transcriptional regulators, repressing some genes in the presence of iron and activating others. However, Fur was being given credit for something it cannot do—it is only a repressor—and the apparent transcriptional activation by Fur was explained by its repression of the synthesis of a small regulatory RNA named RyhB. If this small RNA is made, it inhibits the expression of iron-responsive genes under its control. In the presence of excess iron, Fur represses the transcription of this small RNA, causing the apparent activation of transcription of these iron-responsive genes.

The way in which the RyhB RNA regulates the genes under its control is also shown in Figure 13.16. Like many such small regulatory RNAs, the RyhB RNA sequence is partially complementary to sequences close to the 5′ end of the mRNA of genes under its control. This complementarity allows the RyhB to pair with the mRNA, with the help of the Hfq protein (Box 13.5). However, rather than blocking translation of the mRNAs, like other small regulatory RNAs discussed so far, this binding creates a partially double-stranded RNA that is the substrate of the ribonuclease (RNase E) enzyme, one of the major RNases in *E. coli*, which degrades the mRNA. It has been proposed that prevention of the synthesis of many iron-containing proteins by the RyhB RNA when iron is limiting reserves the available iron for the most essential iron-containing enzymes in *E. coli*, including ribonucleotide reductase, which is required to make deoxynucleotides and hence DNA.

The Aconitase Translational Repressor

The more we learn about cells, the more we find out about how many functions are coordinated in the cell. These discoveries are often serendipitous, as was the case when aconitases were discovered to also play a role in iron regulation in bacteria and eukaryotes. This dual role was first discovered in eukaryotes during studies of proteins called iron-responsive proteins (IRPs). When iron is limiting, IRPs bind to the mRNA for proteins involved in iron metabolism. They can either inhibit the translation of an mRNA by binding to the 5′UTR and preventing ribosome access to the initiator codon or increase translation by binding to the other end of the mRNA, the 3′UTR, and stabilizing the mRNA. In general, they inhibit the translation of proteins that are needed only in the presence of an excess of iron, such as ferritins, and stimulate the translation of proteins that are needed when iron is limiting, such as iron transport proteins or transferrins. The sequences to which they bind are highly conserved in evolution and are called iron-responsive elements (IREs). However, an IRP can bind to an IRE sequence and inhibit or stimulate translation only if iron is limiting, which is why these proteins were named iron-responsive proteins.

It came as a complete surprise when IRPs were purified, partially sequenced, and discovered to be aconitases. Aconitases are enzymes that function in the TCA cycle to convert citrate into isocitrate. The TCA cycle produces essential carbon-containing compounds (including α-ketoglutarate), used in many biochemical reactions, and generates reducing power in the form of NAPDH to feed electrons into the electron transport system to make ATP. Aconitases contain iron in the form of an iron-sulfur cluster, often written $[4Fe-4S]^{2+}$. Many iron-containing enzymes have the iron in this prosthetic group, and it usually makes them sensitive to oxygen. In fact, much of the sensitivity of cells to oxygen may be due to the oxygen sensitivity of the iron-sulfur clusters in some of their essential enzymes, and proteins sometimes use their iron-sulfur cluster as a sensor of oxygen (see Sutton et al., Suggested Reading).

Bacteria and mitochondria have aconitases related to the cytoplasmic aconitases of eukaryotes. *E. coli* has two aconitases, which are similar but differ in their regulation and their sensitivity to oxygen. Aconitase A (AcnA) is induced following stress and in the stationary phase. We have already mentioned it in connection with genes for iron-containing proteins whose expression is regulated by the small RNA, RyhB. The other aconitase, aconitase B (AcnB), is the major aconitase synthesized during exponential growth and is the most sensitive to oxygen. AcnB also regulates the translation and stability of mRNAs involved in iron metabolism in response to iron deficiency, much like the aconitases from eukaryotes. A similar dual role has been found for the aconitases of other bacteria, including *B. subtilis* and some pathogenic bacteria, so we can assume that this phenomenon is universal. Also, some pathogenic bacteria may use their aconitase to sense the availability of iron as part of the signal that they are inside a eukaryotic host and to adjust their physiology accordingly.

It is not clear why aconitase plays the dual role of sensing iron levels and performing an essential step in the TCA cycle. One possibility relates to the extreme toxicity and mutagenic properties of hydroxyl radicals (see Box 11.1). These are produced from hydrogen peroxide in the presence of high iron concentrations. If the cellular iron concentration is high, the TCA cycle may run at full capacity, to increase the reducing power in the cell and reduce the amount of such dangerous reactive oxygen species. If the cellular iron concentration is low, the TCA cycle can run at a lower rate, just fast enough to produce essential intermediates and electron donors for the electron transport system.

Regulation of Virulence Genes in Pathogenic Bacteria

Many of the stress responses and other types of regulons discussed above have relevance to the induction of virulence genes in bacterial pathogens and their ability to survive in a eukaryotic host. The virulence genes of pathogenic bacteria represent a type of global regulon. Virulence genes allow pathogenic bacteria to adapt to their eukaryotic hosts and cause disease. Most pathogenic bacteria express their virulence genes only in the eukaryotic host; somehow, the conditions inside the host turn on the expression of these genes. Virulence genes can be identified because mutations that inactivate them render the bacterium nonpathogenic but do not affect its growth outside the host. Some examples of the regulation of virulence regulons are discussed in this section.

Diphtheria

Diphtheria is caused by the bacterium *Corynebacterium diphtheriae*, a gram-positive bacterium that colonizes the human throat. It is spread from human to human through aerosols created by coughing or sneezing. The colonization of the throat by itself results in few symptoms. However, strains of *C. diphtheriae* that harbor a prophage named β (see chapter 8) produce diphtheria toxin, which is responsible for most of the symptoms. Excreted from the bacteria in the throat, the toxin enters the bloodstream, where it does its damage.

DIPHTHERIA TOXIN

The diphtheria toxin is a member of a large group of A-B toxins, so named because they have two subunits, A and B. In most A-B toxins, the A subunit is an enzyme that damages host cells and the B subunit helps the A subunit enter the host cell by binding to specific cell receptors. The two parts of the diphtheria toxin are first synthesized from the *tox* gene as a single polypeptide chain, which is cleaved into the two subunits A and B as it is excreted from the bacterium. These two subunits are held together by a disulfide bond until they are translocated into the host cell, where the disulfide bond is reduced and broken, releasing the individual A subunit into the cell.

The action of the diphtheria toxin A subunit on eukaryotic cells is quite well understood and is widely used in studies of the eukaryotic cell. The A subunit enzyme specifically ADP-ribosylates (adds ADP-ribose to) a modified histidine amino acid of the translation elongation factor EF-2 (called EF-G in bacteria; see chapter 2). The ADP-ribosylation of the translation factor blocks translation and kills the cell. Interestingly, the opportunistic pathogen *Pseudomonas aeruginosa* makes a toxin that is identical in action to the diphtheria toxin, although it has a somewhat different sequence.

Regulation of the *tox* Gene of *C. diphtheriae*

Iron limitation presents a problem for bacteria in general and for pathogenic bacteria in particular. All of the iron in the human body is tied up in other molecules, such as transferrins and hemoglobin. Thus, to multiply in a eukaryotic host, pathogenic bacteria must extract the iron from the transferrins and other proteins to which it is bound and transport it into its own cell. For this purpose, *C. diphtheriae* and many other pathogenic bacteria synthesize very efficient siderophores, much like those of free-living bacteria (see above). These small siderophores are excreted from the bacterial cells into the host, where they bind Fe^{2+} more tightly than do other molecules and so can wrest Fe^{2+} from them. The siderophores then are transported back into the bacterial cell with their "catch" of Fe^{2+}. As in free-living bacteria, the genes for making the siderophores and a high-efficiency transport system for iron are also synthesized only when iron is limiting.

Not only is iron an essential nutrient for bacteria, but also its limitation is often used as a signal that the bacterium is in the internal eukaryotic environment and the virulence genes should be turned on. The way iron turns on the virulence genes is illustrated in Figure 13.17. In *C. diphtheriae* the virulence genes, including the *tox* gene and iron uptake genes, are under the control of the same global regulator, DtxR (for *d*iphtheria *tox*in *r*egulator). The DtxR protein of *C. diphtheriae* is a repressor which functions similarly to the Fur repressor protein of gram-negative bacteria including *E. coli*. Like Fur, the DtxR protein binds to the operators of genes under its control only if it is bound to iron. Interestingly, even though the *tox* gene encoding the toxin of *C. diphtheriae* is carried on the lysogenic phage, it is regulated by the DtxR repressor, which is encoded on the chromosome (Figure 3.17) (see Schmidt and Holmes, Suggested Reading). Most other genes controlled by DtxR are chromosomal genes. This is just one of many examples of the contribution of lysogenic phages to the pathogenicity of bacteria. A more detailed map of phage β is shown in Figure 8.19.

Cholera and Quorum Sensing

The disease cholera is another well-studied example of the global regulation of virulence genes. *Vibrio cholerae*, the causative agent, is a gram-negative bacterium that is spread through water contaminated with human feces. The disease continues to be a major health problem worldwide, with periodic outbursts, especially in countries with

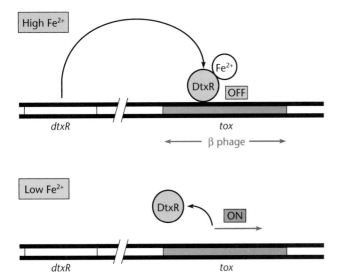

Figure 13.17 Regulation of the *tox* gene of the *Corynebacterium diphtheriae* prophage β. The DtxR repressor protein binds to the operator for the *tox* gene only in the presence of ferrous ions (Fe^{2+}).

poor sanitation. When ingested by a human, *V. cholerae* colonizes the small intestine, where it synthesizes cholera toxin, which acts on the mucosal cells to cause a severe form of diarrhea. Other virulence determinants are the flagellum, which allows the bacterium to move in the mucosal layer of the small intestine, and pili called TCP (*t*oxin-*c*oregulated *p*ili) that allow it to stick to the mucosal surface. Other virulence determinants are quorum sensing, which allows the bacterium to sense when other *V. cholerae* cells are nearby, and biofilm formation, in which the bacteria band together and surround themselves with an impregnable layer of polymers to keep from being washed out of the intestine and to resist host defense systems.

CHOLERA TOXIN
The mechanism of action of the cholera toxin has been the subject of intense investigation, in part because of what it reveals about the normal action of eukaryotic cells. The cholera toxin is composed of two subunits, CtxA and CtxB, which are secreted from the bacterial cell by a type II secretion system. The two subunits are secreted through the inner membrane by the SecYEG channel and then assemble in the periplasm before being secreted to the outside of the cell through a large structure called a secretin in the outer membrane. The functioning of type II secretion systems is discussed in chapter 14. Once outside the cell, the CtxB subunit helps the CtxA subunit enter the eukaryotic cell. Like diphtheria toxin, the CtxA subunit of cholera toxin is an ADP-ribosylating enzyme. However, rather than ADP-ribosylating an elongation

factor for translation, the cholera toxin ADP-ribosylates a mucosal cell membrane protein called Gs, which is part of a signal transduction pathway that regulates the activity of the adenylcyclase enzyme that makes cAMP. The ADP-ribosylation of Gs causes cAMP levels to rise and alters the activity of transport systems for sodium and chloride ions. The osmotic pressure caused by loss of sodium and chloride ions from the cells releases water from the cells, resulting in severe diarrhea and dehydration. Treatment is to force feed water until the condition of the patient improves.

Regulation of the Synthesis of Cholera Toxin and Other Virulence Determinants
The *ctxA* and *ctxB* genes encoding the cholera toxin are part of a large regulon containing as many as 20 genes. In addition to the *ctx* genes, the genes of this regulon include those encoding pili, colonization factors, and outer membrane proteins related to osmoregulation. Although some of these genes, including the *ctx* genes, are carried on the prophage, others, including the pilin genes, are carried on the bacterial chromosome. The transcription of the genes of this regulon is activated only under conditions of high osmolarity and in the presence of certain amino acids, conditions that may mimic those in the small intestine. Here we describe the cascade of genes involved in the regulation of transcription of *ctx* and other virulence genes.

ToxR-ToxS. The cholera virulence regulon was first found to be under the control, either directly or indirectly, of the activator protein ToxR, the product of the *toxR* gene (see Yu and DiRita, Suggested Reading). The ToxR protein combines in the same polypeptide elements that are normally part of different proteins of the two-component sensor and response regulator type of system. The ToxR polypeptide traverses the inner membrane so that the carboxy-terminal part of the protein is in the periplasm, where it may sense the external environment. The amino-terminal part is in the cytoplasm, where it contains an OmpR-like DNA-binding domain that can activate the transcription of genes under its control.

While the ToxR protein resembles other response regulators in some respects, it is unlike orthodox response regulator proteins in that it is not activated by phosphorylation. Also, it is not known to bind to any small-molecule effectors. A clue to its mechanism of activation could be found in another protein, ToxS. The *toxS* gene is immediately downstream of *toxR* in the same operon, and the gene was discovered because mutations that inactivate it also prevent expression of the genes of the ToxR regulon. The ToxS protein is also anchored in the inner membrane, but a large domain sticks out into the periplasm.

One model for how ToxS might activate ToxR came from the experiments designed to investigate the membrane topology of ToxR. The purpose of these experiments was to determine if all of ToxR was in the cytoplasm or if part of ToxR traversed the inner membrane into the periplasm. A method for determining the membrane topology of proteins in gram-negative bacteria uses translational fusions of various regions of the protein to PhoA (see chapter 2). The PhoA protein is an alkaline phosphatase enzyme that cleaves X-P, which is like X-Gal except that the dye is fused to phosphate instead of galactose. If PhoA cleaves the phosphate off of X-P, the colonies turn blue. However, PhoA must be in the periplasm to be active, probably because it must form dimers that are held together by disulfide bonds between cysteines in its subunits. However, these disulfide bonds form only in the periplasm. Therefore, if a region of a protein is fused to PhoA, the alkaline phosphate will be active and make blue colonies on X-P plates only if that region of the protein is in the periplasm (see chapter 14). To use this method to determine which parts of ToxR are in the periplasm, fusion proteins composed of the reporter gene PhoA fused to both the N-terminal and C-terminal portions of ToxR were created. The results revealed that the C-terminal portion of the ToxR protein is in the periplasm but the N-terminal portion is in the cytoplasm, because bacteria containing fusions to the N-terminal portion did not form blue colonies on X-P plates while fusions to the C-terminal portion did. A surprising result was that some ToxR-PhoA fusions exhibited ToxR activation of transcription, even when *toxS* was inactivated by a mutation. It was hypothesized that by dimerizing in the periplasm, the PhoA part of these fusions was driving the dimerization of the ToxR portion and activating transcription. This suggested that the normal function of ToxS is to dimerize ToxR, since ToxS could be dispensed with if PhoA drove the dimerization.

Other evidence suggests that dimerization, by itself, may not be enough to activate the ToxR protein. These results suggested that some feature related to the membrane anchoring of ToxR may also be required. If dimerization were sufficient, attaching other dimerization domains, such as the ones from the CI repressor protein of λ phage, to the cytoplasmic domain of the ToxR protein should cause ToxR to dimerize in the cytoplasm and activate transcription. However, fusion proteins composed of ToxR and other dimerization domains are still inactive unless the ToxR protein retains its transmembrane domain. Perhaps the ToxR protein must be at least partly in the membrane to be active and ToxS plays some other role in stabilizing ToxR. Perhaps by being anchored in the membrane, ToxR can be activated directly by external signals and ToxS somehow helps in this activation.

It is known that ToxR can be activated by membrane-damaging agents such as bile.

ToxT and TcpP. More recent work indicated that the regulation of virulence genes in *V. cholerae* is much more complicated than a single activator, ToxR, turning on virulence genes. A general outline of the various regulatory pathways used to turn on the genes required for *V. cholerae* pathogenicity is given in Figure 13.18. As mentioned, *V. cholerae* has many virulence genes besides the toxin genes that are also considered part of the ToxR regulon, since *toxR* mutations prevent their expression. However, while ToxR can activate the transcription of the toxin genes directly, most of the ToxR regulon genes are not directly controlled by the ToxR protein but, rather, are controlled by an intermediary, the ToxT protein, which is a member of the AraC family of regulators (see Box 12.2). The transcription of the *toxT* gene is activated by ToxR, and the transcription of the other genes is then activated by ToxT.

Activation of the *toxT* gene also requires the activator TcpP. The activity of the TcpP activator somehow requires the activity of another protein, TcpH. Both TcpP and TcpH are inner membrane proteins, but it is not clear how TcpH regulates TcpP. Transcription of the *tcpP-tcpH* operon responds to environmental cues, but we do not understand how these two genes, together with the *toxR-toxS* genes, transduce environmental signals that the bacterium is in the intestine of its host into activation of ToxT. This type of serial activation, where one regulator activates another regulator which activates another regulator, is a called a **regulatory cascade**. Obviously, much more needs to be done before we can begin to understand this important model system of bacterial pathogenicity.

Interestingly, the *toxT, tcpP,* and *tcpH* genes are all located on a DNA element in *V. cholerae* called VPI (*V. cholerae p*athogenicity *i*sland). The cholera toxin is also encoded by a DNA element, in this case a lysogenic phage related to the single-stranded DNA phage M13 (see chapter 8). Nevertheless these genes are regulated by ToxR, encoded by a chromosomal gene. This is yet another example of virulence traits in pathogenic bacteria being encoded on exchangeable DNA elements but interacting with the products of normal chromosomal genes, as well as being another example of pathogenic bacteria that are derived from their free-living relatives by acquisition of virulence genes encoded on interchangeable DNA elements.

CLONING THE *toxR* GENE

Our current level of understanding of the virulence regulon in *V. cholerae* would not have been possible without clones of the *toxR* gene (see Miller and Mekalanos,

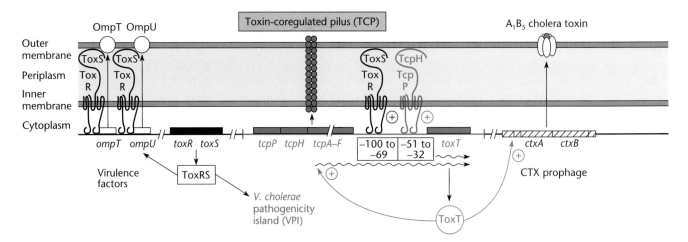

Figure 13.18 Regulatory cascade for *Vibrio cholerae* virulence factors. The ToxR-ToxS activator-effector pair directly regulates *omp* virulence factors and the *toxT* regulatory gene located on a *V. cholerae* pathogenicity island (VPI) (indicated in purple). The VPI-encoded TcpP-TcpH activator-effector pair also regulates *toxT* transcription. ToxT activates the Ctx prophage-encoded *ctxAB* toxin genes and the toxin-coregulated pilus (TCP) genes. ToxT also positively regulates its own expression from the promoter for *tcpA* to *tcpF* transcription.

Suggested Reading). When this work was begun, little was known except that mutations in a *V. cholerae* gene called *toxR* caused the cells to produce little toxin. This suggested that the *toxR* gene encoded a positive regulatory protein that was required to activate transcription of the toxin genes.

For convenience, the cloning was attempted in *E. coli* rather than in *V. cholerae*. As discussed in chapters 1 and 4, many cloning vectors have been developed for *E. coli*, and cloning techniques are much more advanced for this species than for other bacteria. The investigators had reason to hope that the *toxR* gene would be expressed and function in *E. coli*, since *V. cholerae* and *E. coli* are fairly closely related and since most *V. cholerae* genes that had been tested are expressed well in *E. coli*.

Their strategy was to screen a library (see chapter 1) for clones that expressed a protein which could activate transcription from the promoter for the toxin genes, the *ctxA* promoter, in *E. coli*. They first made a partial Sau3A library of wild-type *V. cholerae* DNA in an *E. coli* plasmid cloning vector. This library was then transformed into a strain of *E. coli* harboring a lysogenic phage carrying a transcriptional fusion of *ctxA* to a reporter gene, *lacZ*. The *ctxA* and *lacZ* genes were arranged so that the *lacZ* gene would be transcribed only from the *ctxA* promoter. The investigators then plated the transformants on agar containing X-Gal and looked for blue colonies due to synthesis of β-galactosidase, the product of the *lacZ* gene. Most transformants made only faintly blue colonies. However, 2 of about 5,000

transformants made deep blue colonies. The clones in these transformants were presumed to contain the *toxR* gene, which was activating transcription from the *ctxA* promoter.

Further confirmation that the clones contained the *toxR* gene came from complementation tests in which plasmids containing the clones were mobilized into *toxR* mutants of *V. cholerae* with a ColE1-derived plasmid cloning vector, so that it could be mobilized by the F-plasmid Tra functions (see chapter 5). The narrow-host-range ColE1-derived plasmids can nevertheless replicate to some extent in *V. cholerae*, since, as mentioned, *V. cholerae* is closely related to *E. coli*. Because the plasmid clones complemented *toxR* mutations in *V. cholerae* to allow normal synthesis of the toxin, they presumably contained the *toxR* gene. With clones of the *toxR* gene, it was also possible to construct PhoA fusions to investigate the membrane topology of the ToxR protein (see above).

QUORUM SENSING

As mentioned above, quorum sensing is an important contributor to virulence in *V. cholerae*. For a long time, it was thought that single-celled bacteria such as *V. cholerae* lived a lonely life, with no way of telling if other members of the same species were nearby. However, with the discovery of quorum sensing, it became apparent that bacteria do have ways of communicating with each other. They do this by giving off small molecules that can be taken up by other members of the same species. When the concentration of bacteria is high, the concentration of these secreted small molecules also becomes high, signaling that the bacterium is in the presence of many other bacteria of the same type. In response, they induce certain types of genes, to adapt them for a more communal existence such as in a biofilm.

Quorum sensing was first discovered in the marine bacterium *Photobacterium fischeri (Vibrio fischeri)*, which is a facultative symbiont that can live either free in the ocean or in the light organ of some fishes and squids. These bacteria are concentrated in the light organ of the fish or squid, where they give off light due to chemiluminescence. The light is due to induction of the *lux* genes, which are turned on only when a bacterium is close to other bacteria of the same type. The squid and fish with which these bacteria form a symbiosis live deep in the ocean, where there is little light. By emitting light when they are concentrated in the light organ, the bacteria help the marine organisms find each other. Recent work has concentrated on a free-living marine bacterium, *Vibrio harveyi*, which also gives off light when the bacteria are concentrated in certain regions of the ocean, but it is not known to be a symbiont of any marine animal. A likely explanation is that *V. harveyi* also forms such a symbiosis, but with an unknown marine organism. This explanation is supported by the fact that some of the genes induced with *lux* form a type III secretion system. In other types of bacteria, such systems are used to inject effectors into eukaryotic cells and are important in avoiding host defenses and establishing infections or symbioses (see chapter 14).

Extensive research over a long period revealed that the chemiluminescence of *V. harveyi* is induced because the bacteria give off two small molecules, called autoinducers AI-1 and AI-2. AI-1 is a homoserine lactone, and AI-2 is a furanosyl borate diester. Autoinducer AI-2 is made by the product of the *luxS* gene, which is found in many gram-negative and gram-positive bacteria. Its structure is shown in Figure 13.19B. Because AI-2 is made by so many different types of bacteria, it has been proposed that it is a universal autoinducer allowing different types of bacteria to communicate with each other, for example in the formation of biofilms. Both AI-1 and AI-2 act through two-component sensor kinase and response regulators pairs that determine the state of phosphorylation of LuxO, a transcriptional regulator. LuxO is a member of the NtrC family of activators and, like NtrC, is active only if it is phosphorylated. It also activates transcription only from σ^{54}-type promoters, like the other members of this family of regulators (Box 13.3). Genes under the control of the LuxO activator include four small noncoding sRNAs (Box 13.5). These small RNAs inhibit the translation of an activator named

LuxR by binding to and destabilizing its mRNA so that its mRNA is degraded almost as fast as it is made. If LuxR is not made, the *lux* gene is not transcribed and the cells do not give off light. At a low concentration of cells, the kinase activity of the sensor kinase is active, which leads to the phosphorylation of LuxO so that it is active as a transcriptional activator. The four small RNAs are made, LuxR is not made, the *lux* gene is not transcribed, and the cells do not give off light. At a high concentration of cells, and therefore at a high concentration of the autoinducer, the binding of the autoinducer to the sensor kinase activates its phosphatase activity so that it is not phosphorylated. It compensates by taking the phosphate from the response regulator protein in the next step of the pathway until eventually LuxO loses its phosphate and is unphosphorylated. If it is not phosphorylated, the small RNAs are not made, the LuxR activator is made, the *lux* operon is transcribed, and the cells give off light.

Quorum Sensing in *V. cholerae*

An understanding of the pathways of quorum sensing in *V. harveyi* led the way to an understanding of them in its pathogenic relative, *V. cholerae*, which was also found to have quorum-sensing systems, as do many types of pathogenic bacteria including plant pathogens such as *Agrobacterium*. While it lacks the AI-1 system, *V. cholerae* has the AI-2 system and two other systems with unknown autoinducers named CA-1 and VarS-VarA, which also function through sensor kinases and response regulators to finally determine the state of phosphorylation of LuxO.

A comparison of quorum sensing in *V. cholerae* and *V. harveyi* is presented in Figure 13.19. Interestingly, a total of seven small noncoding RNAs (Box 13.5) are used by the pathways for quorum-sensing signaling in *V. cholerae* (see Lenz et al., 2005, Suggested Reading). Not only do they have the four small RNAs whose transcription is activated by LuxO in *V. harveyi*, but also they have three others that are used in the VarS-VarA pathway. These other small RNAs bind to and inhibit the activity of a regulator named CsrA, which in turn affects the state of phosphorylation of LuxO by an unknown mechanism. Instead of an activator, LuxR, which activates the transcription of the *lux* gene and other genes appropriate for a free-living organism, *V. cholerae* has an activator, HapR, which activates virulence genes and genes involved in biofilm formation, etc., that are important for pathogenesis.

Genetic Experiments That Led to the Detection of the Small RNAs That Regulate LuxR

It took many investigators a long time to unravel the signal transduction pathways of quorum sensing in *V. harveyi*, and these efforts are still under way. We discuss

A

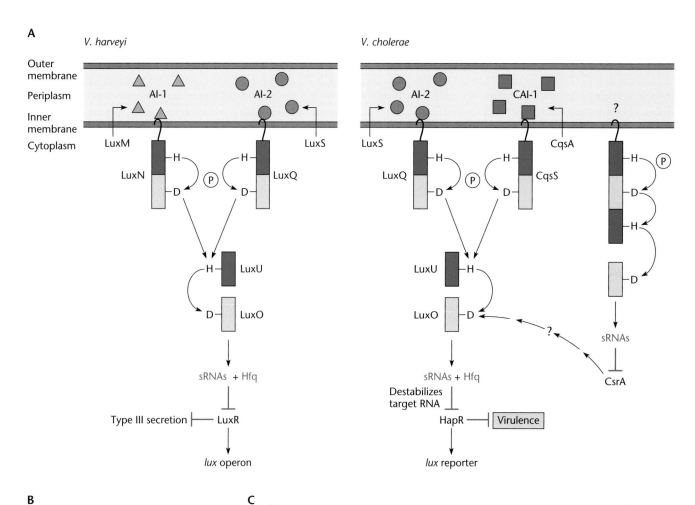

B

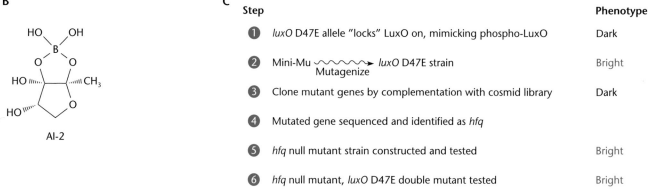

AI-2

C

Step		Phenotype
1	*luxO* D47E allele "locks" LuxO on, mimicking phospho-LuxO	Dark
2	Mini-Mu ⟿ *luxO* D47E strain Mutagenize	Bright
3	Clone mutant genes by complementation with cosmid library	Dark
4	Mutated gene sequenced and identified as *hfq*	
5	*hfq* null mutant strain constructed and tested	Bright
6	*hfq* null mutant, *luxO* D47E double mutant tested	Bright

Conclusion: Hfq is required for quorum sensing

Inference: Hfq requirement suggests small RNA involvement

Figure 13.19 Quorum sensing in *Vibrio harveyi* and *Vibrio cholerae*. (A) *V. cholerae* shares the AI-2 quorum-sensing system with *V. harveyi* and contains two additional phosphorelay systems to regulate virulence. AI-1 (triangles), AI-2 (circles), and CAI-1 (squares) are small molecules (see the text). LuxS synthesizes AI-2, while LuxM and Cqs synthesize AI-1 and CAI-1, respectively. LuxN, LuxQ, LuxS, LuxU, and LuxO form a phosphorelay (see Box 14.2). The flow of phosphate depicts a low-cell-density state. High cell density would reverse the flow. (B) Structure of AI-2. (C) Steps in identifying quorum-sensing regulatory small RNAs.

only one set of experiments here, the ones that led to the discovery of the small RNAs that bind to and destabilize the mRNA for the LuxR activator (see Lenz et al., 2004, Suggested Reading). These experiments illustrate some principles of genetics and some of the methods already described. They also illustrate the power of comparing two closely related bacteria to find mechanisms that are common to both.

It was hypothesized that in *V. harveyi*, LuxO did not act directly to activate the transcription of the *lux* operon but, rather, activated the transcription of a repressor of LuxR, because LuxO null mutations are constitutive for light production. If the LuxO activator is not made, the putative repressor would not be made, the LuxR activator would be made constitutively, and the cells should give off light, even at low concentrations. This suggested a way of isolating mutants with mutations in the putative repressor, since they should have the same effect as *luxO* mutations. A mutation that inactivated the putative repressor should also cause constitutive light production, even when the cells were present at low concentrations. The problem is that it is difficult to see the light given off by individual cells when the cells are present at low concentrations. It is much easier to screen colonies for mutants, but the cells in a colony are effectively at high concentrations. To overcome this difficulty, the investigators introduced a mutated form of the *luxO* gene into the cells by site-specific mutagenesis (see chapter 1). This mutated form of the gene produces a LuxO protein in which the aspartate to which the phosphate normally attaches is changed to a glutamate, whose side chain more nearly resembles a phosphate group. We mentioned this approach earlier in the chapter in connection with the analysis of the activation pathway for σ^B, the general stress response sigma factor of gram-positive bacteria. The *V. harveyi* strain with this mutated form of LuxO makes the putative repressor constitutively and is permanently dark, even when the cells are present at high concentration in colonies. The investigators then mutagenized these cells with mini-Mu (see chapter 9) and looked for mutants that formed constitutive light-producing colonies. They found *luxO* mutants as expected, since LuxO is required to make the putative repressor, as well as mutants with mutations in *rpoN*, the gene for σ^{54}, which is required for LuxO to act. However, they also found mutants in *hfq*, the gene for the ubiquitous small-RNA binding protein that promotes the binding of small regulatory RNAs to their targets (Box 13.5). This led them to suspect that the repressor was a small RNA rather than a protein. In other systems, including *E. coli*, small RNAs inhibit translation. The reason that none of the constitutive mutants had mutations in the gene for this small RNA could merely be

because the gene is small and small genes are difficult to inactivate by a mutation. Figure 13.19C shows the steps that implicated a small RNA in this regulation.

Because it is hard to identify genes for RNAs by a classical genetic approach, the investigators searched for the genes directly, using bioinformatics and the fact that the genome of *V. cholerae* had been sequenced, a prerequisite for such studies. They thought they could use *V. cholerae* for this search because the two species are very similar and *hfq* mutants were also constitutive for light production in *V. cholerae* into which the *lux* operon and the *luxR* gene had been introduced. In particular, they were interested in regions that contained a σ^{54} promoter, since this is what LuxO recognizes, and a factor-independent transcriptional terminator, which is the type of terminator used by most small regulatory RNA genes. They found four candidates for small RNAs that negatively regulate the translation of the LuxR activator in *V. harveyi* and the HapR regulator in *V. cholerae*. The transcription of these RNAs was under the control of the LuxO activator. Mutations in one, two, or even three of these genes had no effect on the translation of LuxR or HapR. It was necessary to inactivate all four of the small RNAs to obtain constitutive expression of the *lux* operon. Further experiments showed that these small RNAs acted by base pairing to the mRNA for LuxR or HapR and apparently created the double-stranded RNA substrate for a cellular RNase which then degraded the mRNA and also the small RNA in the process.

Whooping Cough

Another well-studied disease used to illustrate global regulation of virulence genes is whooping cough, caused by the gram-negative bacterium *Bordetella pertussis*. Whooping cough is mainly a childhood disease and is characterized by uncontrolled coughing, hence the name. The bacteria colonize the human throat and are spread through aerosols resulting from the coughing. Effective vaccines have been developed, but the disease continues to kill hundreds of children worldwide, mainly in areas where the vaccines are not available. This bacterium infects humans only through the aerosols created by coughing and cannot survive long outside the human throat.

Despite their very different symptoms, the diseases caused by *V. cholerae* and *B. pertussis* have similar molecular bases. *B. pertussis* also makes a complex A-B toxin (pertussis toxin), which is in some ways remarkably similar to the cholera toxin. The pertussis toxin has six subunits, although only two of them are identical. One of the subunits (S1) is the enzyme, while the others (S2 to S5) are involved in adhesion to the mucosal surface of the throat. They make a type IV secretion system that is capable of secreting the toxin only from the periplasm to the outside of the cell. The SecYEG translocase must get it through the inner membrane first (see chapters 2 and 14). Once outside the bacterial cell, the B domains of the toxin bind to receptors on the ciliated epithelial cells and transfer the A domains into the epithelial cells, where they perform their enzymatic reaction, the ADP-ribosylation of the G protein. This type IV secretion system is unusual in that most type IV secretion systems transfer effector proteins directly through both membranes into eukaryotic cells (see chapters 5 and 14). The pertussis toxin is also similar to the cholera toxin in that it ADP-ribosylates a G protein in a signal transduction pathway involved in deactivating the adenylate cyclase, leading to elevated cAMP levels. However, rather than causing a loss of water from the cells, the elevated cAMP levels seem to cause an increase in mucus production, presumably because the pertussis toxin attacks cells of the throat rather than of the small intestine.

In addition to pertussis toxin, *B. pertussis* synthesizes a number of other toxins and other virulence proteins. One toxin is an adenylate cyclase enzyme that enters host cells and presumably directly increases intracellular cAMP levels by synthesizing cAMP. This observation supports the importance of increased cAMP levels to the pathogenesis of the bacterium, although the contribution of this adenylate cyclase to the symptoms is unknown. Other known toxins include one that causes necrotic lesions on the skin of mice and a cytotoxin that is not a protein but a peptidoglycan fragment that kills ciliated cells of the throat. Other virulence factors are involved in the adhesion of the bacterium to the mucosal layer, fimbriae, factors required to survive nutrient deprivation, motility, etc.

REGULATION OF PERTUSSIS VIRULENCE GENES
Like the virulence genes of *C. diphtheriae* and *V. cholerae*, many of the virulence genes of *B. pertussis* are presumably expressed only when the bacterium enters the eukaryotic host at the same time that others, expressed in the free-living state, are repressed. The regulation of the virulence genes of *B. pertussis* is achieved by a sensor kinase and response regulator pair of proteins encoded by linked genes, *bvgA* and *bvgS* (for *bordetella virulence genes*); *bvgS* encodes the sensor, and *bvgA* encodes the transcriptional regulator.

The BvgS-BvgA system is similar to many other sensor kinase and response regulator pairs in that the BvgS protein is a transmembrane protein, with its N terminus in the periplasm and its C terminus in the cytoplasm, allowing it to communicate information from the external environment across the membrane to the inside of the cell. It also exists as a dimer in the inner membrane. Furthermore, like many other sensor proteins that work in two-component systems, the BvgS protein is autophosphorylated in response to some signal from the external environment and donates this phosphate to the BvgA protein, which then regulates transcription of the virulence genes.

Attempts have been made to determine the signals to which BvgS responds to phosphorylate itself. In vitro (or, as a medical person would say, ex vivo [outside the body]), the signal transduction pathway to transcribe the pertussis toxin gene and other virulence genes can be activated by growing the bacteria in medium with low nicotinamide and magnesium concentrations and possibly low iron concentrations, conditions that presumably mimic those in the throat. Also, expression of the virulence genes is highest at 37°C, the temperature of the human body.

Once phosphorylated, the BvgA response regulator can be either an activator or a repressor. It can activate the transcription of genes required in the host and repress genes that allow it to survive outside the host. This is particularly clear for *Bordetella bronchiseptica*, a close relative of *B. pertussis* which can infect other mammals besides humans and can survive for longer times outside the host. This more versatile relative has a BvgS-BvgA system very closely related to that of *B. pertussis*; under conditions that mimic those inside one of its hosts, the phosphorylated BvgA protein represses a number of genes whose products are required in the free-living state (see Cummings et al., Suggested Reading).

One difference between BvgS and many other sensor kinases is that it has more than one amino acid that can accept phosphates on signal activation. As with other sensor kinases, one histidine in the periplasmic N terminus of the protein phosphorylates itself in response to conditions that mimic entrance into the host bronchial tubes. This phosphate group can then be transferred to an aspartate closer to the C terminus of the same polypeptide and then transferred again to another histidine closer yet to the C terminus before it is transferred to an aspartate in the BvgA response regulator. Thus, in successive steps, the phosphate is being transferred closer and closer to the C terminus of the sensor kinase and therefore closer to the BvgA response regulator in the cytoplasm. Sensor kinases like BvgS, which transfer phosphates within themselves in a phosphorelay, have been called polydomain sensors because they transfer phosphates in a phosphorelay within the same polypeptide as opposed to phosphorelays from one protein to another (Box 13.4).

There is speculation about why *Bordetella* should use a polydomain sensor to signal that it is in a mammalian throat, rather than just having a single site of phosphorylation in the sensor kinase or a multiprotein phosphorelay as in many other signal transduction systems. This speculation centers on recent research indicating that changes in gene expression occur in more than one stage. When *Bordetella* enters conditions like those in the host, the virulence genes are not just turned on but, rather, proceed through a least one intermediate (see Cotter and Jones, Suggested Reading). The three stages have been named Bvg⁻ for outside the host, Bvg⁺ for inside the host, and Bvgⁱ for an intermediate state when the bacterium has just entered the host and neither has established an infection nor is in a state where it can be spread to another host in an aerosol by coughing. Genes expressed in the Bvg⁻ state include those that allow it to survive in a free-living state, such as genes for carbon site utilization, nutrient deficiency, and growth at low temperatures. Genes expressed in the Bvgⁱ state are those that promote attachment to the epithelial cells in the throat, and those expressed in the Bvg⁺ state are the toxin genes and others that should be expressed only when the bacterium has already established an infection and wants the host to start coughing so that the bacterium may be spread to other hosts. The polydomain sensor of BvgS may allow intermediate states of activation, allowing the sensor kinase and response regulator system to function more like a "rheostat" than just an on-off switch.

CLONING THE *bvgA-bvgS* OPERON

As with the ToxR regulon, the results described above would not have been obtained without clones of the *bvg* gene region. The *bvg* locus, where the two genes lie, was first found by isolating Tn*5* transposon insertion mutations that prevent the synthesis of many of the virulence factors of *B. pertussis* (see Stibitz et al., Suggested Reading). By cloning the kanamycin resistance gene on Tn*5* and the flanking chromosomal sequences, the investigators constructed a clone of the *bvg* region. This clone was then used as a probe in colony hybridizations to screen a library made from wild-type *B. pertussis* DNA to identify bacteria containing the wild-type *bvg* locus.

When the *bvg* locus was sequenced, it was found to contain an operon with two open reading frames (ORFs), one of which encoded a protein with sequence homology to sensor autokinases and so was presumably a gene encoding a sensor autokinase. The other ORF encoded a protein with a helix-turn-helix DNA-binding motif and other features of a transcriptional activator and so presumably encoded one of these proteins. The genes were later named *bvgS* and *bvgA*, respectively.

Clones of the *bvg* genes facilitated the genetic analysis of the activation mechanism. Some mutations in *bvgS* cause constitutive expression of the virulence genes, even

at high concentrations of nicotinamide and magnesium. Also, deletion of the N-terminal periplasmic domain of the BvgS protein forestalled the requirement for conditions of low nicotinamide and magnesium concentrations to activate the regulon. Apparently, these mutations change BvgS so that it no longer requires a signal from the environment to be autophosphorylated.

Regulation of Ribosome and tRNA Synthesis

To compete effectively in the environment, cells must make the most efficient use possible of the available energy. One of the major ways in which cells conserve energy is by regulating the synthesis of their ribosomes and tRNAs, so that they make only enough to meet their needs. More than half of the RNA made at any one time comprises rRNA and tRNA. Moreover, each ribosome is composed of about 50 different proteins, and there are about as many different tRNAs.

The number of ribosomes and tRNA molecules needed by the cell varies greatly, depending on the growth rate. Fast-growing cells require many ribosomes and tRNA molecules to maintain the high rates of protein synthesis required for fast cellular growth. Cells growing more slowly, either because they are using a relatively poor carbon and energy source or because some nutrient is limiting, need fewer ribosomes and tRNAs. As a consequence, a rapidly growing *E. coli* cell contains as many as 70,000 ribosomes but a slow-growing cell has fewer than 20,000. As with most of the global regulatory systems we have discussed, the regulation of ribosome and tRNA synthesis is much better understood for *E. coli* than any other organism. Even in *E. coli*, however, some major questions remain unanswered. In this section, we confine our discussion to *E. coli*, with occasional references to other bacteria when information is available.

Ribosomal Proteins

Ribosomes are composed of both proteins and RNA (see chapter 2). The ribosomal proteins are designated by the letter L or S, to indicate whether they are from the large (50S) or small (30S) subunit of the ribosome, respectively, followed by a number for the particular protein. Thus, protein L11 is protein number 11 from the large 50S subunit, whereas protein S9 is protein number 9 from the small 30S subunit. The gene names begin with *rp*, for ribosomal protein, followed by a lowercase *l* or *s* to indicate whether the protein product resides in the large or small subunit. Another capital letter designates the specific gene. For example, the gene *rplK* is ribosomal protein gene K encoding the L11 protein; note that

K is the 11th letter of the alphabet. Similarly, *rpsL* encodes the S12 protein; *L* is the 12th letter of the alphabet.

MAPPING OF RIBOSOMAL PROTEIN GENES

A total of 54 different genes encode the 54 polypeptides that compose the *E. coli* ribosome, and mapping these genes was a major undertaking. Some ribosomal protein genes were mapped simply by mapping mutations that caused resistance to antibiotics such as streptomycin, which binds to the ribosomal protein S12, blocking the translation of other genes (see chapter 2). More complex techniques involving specialized transducing phages and DNA cloning (see chapters 1 and 8) were needed to map the other ribosomal protein genes. Often, clones containing these genes were identified because they synthesize a particular ribosomal protein in coupled transcription-translation systems (see chapter 2). Nowadays, annotation makes it relatively easy to place these genes in the sequenced genome of one of the bacteria that has been sequenced, due to their high degree of conservation throughout evolution.

The final map revealed some intriguing aspects to the organization of the ribosomal protein genes in the chromosome of *E. coli*. Rather than being randomly scattered around the chromosome, the 54 genes are organized into large clusters of operons, with the largest at 73 and 90 min in the *E. coli* genome. Furthermore, these operons also often contain genes for other components of macromolecular synthesis, including subunits of RNA polymerase, tRNAs, and genes for proteins of the DNA replication apparatus.

In the cluster shown in Figure 13.20, four genes for tRNAs, *thrU*, *tyrU*, *glyT*, and *thrT*, are followed by *tufB*, a gene for the translation elongation factor EF-Tu. These five genes constitute one operon; they are all cotranscribed into one long precursor RNA, from which the individual tRNAs are cut out later. The next operon in the cluster contains two genes for ribosomal proteins, *rplK* and *rplA*, and a third operon has four genes, *rplJ* and *rplL* (encoding two more ribosomal proteins) and

rpoB and *rpoC* (encoding the β and β′ subunits of the RNA polymerase, respectively).

Several hypotheses have been proposed to explain why genes involved in macromolecular synthesis are clustered in *E. coli* DNA. First, the products of these genes must all be synthesized in large amounts to meet the cellular requirements. Some clusters are near the origin of replication, *oriC*, and cells growing at high growth rates have more than one copy of the genes near this site (see chapter 1), which allows higher rates of synthesis of the gene products. Other possible reasons have to do with the structure of the bacterial nucleoid (see chapter 1). Clustered genes are probably on the same loop of the nucleoid. A loop for the macromolecular synthesis genes might be relatively large and extend out from the core of the nucleoid to allow RNA polymerase and ribosomes to gain easier access to the genes. A third possible explanation is that the genes for macromolecular synthesis must be coordinately regulated with the growth rate and that their assembly in clusters of operons facilitates their coordinate regulation, although why this should be the case is not clear.

REGULATION OF THE SYNTHESIS OF RIBOSOMAL PROTEINS

The regulation of the synthesis of ribosomal proteins and ribosomal RNA (rRNA) presents an interesting case of coordinate regulation. Even though the ribosomal proteins and rRNAs are synthesized independently and only later assembled into mature ribosomes, there is never an excess of either free ribosomal proteins or free rRNA in the cell, suggesting that their synthesis is somehow coordinated. Either the rate of ribosomal protein synthesis is adjusted to match the rate of synthesis of the rRNA or vice versa. As it turns out, the rate of ribosomal protein synthesis is adjusted to match the rate of rRNA synthesis, evidence for which is discussed below.

The regulation of ribosomal protein synthesis is best understood in *E. coli*. However, there is every reason to believe that the regulation is similar in all bacteria and probably, to some extent, in eukaryotes as well. Like the synthesis of the heat shock sigma (σ^{32}), the synthesis of the ribosomal proteins is translationally autoregulated. The ribosomal proteins bind to TIRs in the mRNA and repress their own translation. However, rather than having each ribosomal gene of the operon translationally regulate itself independently, the protein product of only one of the genes of the operon is designated as responsible for regulating the translation of all the ribosomal proteins encoded by the operon. Basically, this protein translationally regulates the first gene in the operon and the translation of the other proteins on the mRNA is then translationally coupled to the translation of this

Figure 13.20 Arrangement of a gene cluster in *E. coli* encoding ribosomal proteins and other gene products involved in macromolecular synthesis. The cluster contains three operons, transcribed in the direction shown by the arrows. The *thrU*, *tyrU*, *glyT*, and *thrT* genes all encode tRNAs. The *tufB* gene encodes translation elongation factor EF-Tu. The *rplK*, *rplA*, *rplJ*, and *rplL* genes encode proteins of the large subunit of the ribosome. The *rpoB* and *rpoC* genes encode subunits of the RNA polymerase.

thrU | tyrU | glyT | thrT | tufB | rplK | rplA | rplJ | rplL | rpoB | rpoC

protein (see chapter 2 for an explanation of translational coupling). This designated protein can also bind to free rRNA in the cell, which, as discussed later, is what coordinates the synthesis of the ribosomal proteins with the synthesis of rRNA.

Figure 13.21 illustrates this regulation with the relatively simple operon *rplK-rplA*. Of the two ribosomal proteins encoded by the operon, L1 is the one thought to function in regulation of the operon. The L1 protein can bind to both free rRNA and the TIR for the *rplK* gene on the mRNA. By binding to the *rplK* TIR and thereby inhibiting the translation of this gene, the L1 protein simultaneously inhibits the translation of its own gene, since the two genes are translationally coupled. However, if the cell contains free rRNA, the L1 protein preferentially binds to it, releasing L1 protein from the *rplK* TIR and allowing translation of L11 and L1 to resume. When there is no longer any free rRNA in the cell—because it is all taken up by the ribosomes—the L1 protein begins to accumulate, binds again to the TIR for the *rplK* gene, and once again represses the translation of the L11 protein and thereby of itself. The same thing happens to the other ribosomal proteins. Their designated regulatory protein accumulates when rRNA is not freely available because it is all taken up by ribosomes. The free designated protein binds to the TIR for the first gene in the operon and represses the synthesis of all the genes in the operon. The synthesis resumes when more free rRNA accumulates. In this way, the cell ensures that there will be neither an excess of free ribosomal proteins nor an excess of free rRNA.

The protein in each operon designated to be the regulator may have been chosen because it normally binds to rRNA during the assembly of the ribosome and so already has an rRNA-binding ability. In at least some cases, the sequence to which the designated protein binds in the rRNA may be mimicked in the TIR for the first gene in the mRNA.

Experimental Support for the Translational Autoregulation of Ribosomal Proteins

Evidence that the synthesis of ribosomal proteins is translationally autoregulated came from a series of experiments called **gene dosage experiments,** in which the number of copies, or dosage, of a gene in the cell is increased and the effect of this increase on the rate of synthesis of the protein product of the gene is determined (see Yates and Nomura, 1980 and 1981, Suggested Reading). Figure 13.22 illustrates the principle behind a gene dosage experiment. If the gene is not autoregulated, the rate of synthesis of the gene product should be approximately proportional to the number of copies of the gene. However, if the gene is autoregulated, the product of the gene should repress its own synthesis and the rate of synthesis of the gene product should not increase, regardless of the number of gene copies.

Figure 13.21 Translational autoregulation of the ribosomal proteins, as illustrated by the *rplK-rplA* operon shown in Figure 13.20. See the text for details.

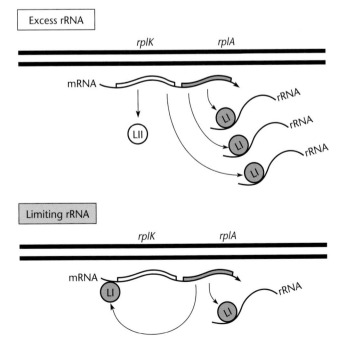

Figure 13.22 A gene dosage experiment to determine if a gene is autoregulated. The number of copies of the gene for the A protein is doubled. If the gene is not autoregulated, twice as much of protein A is made. If gene *A* is autoregulated, the same amount of protein A is made.

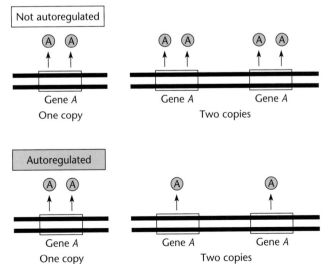

To introduce additional copies of the genes, the experiments were performed with cloning vectors based on phage λ, which is capable of lysogeny (see chapter 8). The phage carrying operons encoding one or more ribosomal proteins were then integrated at the normal phage attachment site. As a result, the cell contained two copies of these ribosomal protein genes, one at the normal position and another at the site of integration of the phage DNA. It was found that the same amounts of the ribosomal proteins were synthesized with two copies of the ribosomal protein genes as with one, indicating that their synthesis is autoregulated.

The next step was to determine if the autoregulation occurs at the level of transcription or translation. If the synthesis of the ribosomal proteins were transcriptionally autoregulated, the rate of transcription of the ribosomal genes would not increase when the number of copies of the genes increased. However, if the autoregulation occurred only at the level of translation, the rate of gene transcription would increase. The investigators found that the rate of transcription approximately doubled when the gene dosage was doubled, indicating that the autoregulation did not occur transcriptionally. If genes are not transcriptionally autoregulated, by a process of elimination they must be translationally autoregulated. Therefore, the ribosomal proteins are capable of repressing their own translation.

The next step was to determine whether all the proteins in the operon independently repress their own translation or whether some of the proteins in each operon are responsible for regulating their own translation as well as that of the others. To answer this question, the investigators systematically deleted genes for some of the proteins encoded by each operon cloned in the phage and evaluated the effect of their absence on the synthesis of the other proteins. Deleting most of the genes in each operon had no effect on the synthesis of the proteins encoded by the other genes. However, when one particular gene in each operon was deleted, the rate of synthesis of the other proteins doubled. Therefore, this one protein represses the translation of itself and the other proteins encoded by the same operon.

Regulation of rRNA and tRNA Synthesis

If the translation of the ribosomal proteins is coupled to the synthesis of the rRNAs under different growth conditions, then how is the synthesis of rRNA regulated? As discussed in chapter 2, the 16S, 23S, and 5S rRNAs are synthesized together as a long precursor RNA, often with intermingled tRNA sequences. After synthesis, the long precursor RNA is processed into the individual rRNAs and tRNAs. Every ribosome contains one copy of each of the three types of rRNAs, and synthesizing all

three rRNAs as part of the same precursor RNA ensures that all three are made in equal amounts.

Each cell has tens of thousands of ribosomes, requiring the synthesis of large amounts of rRNA. To meet this need, bacteria have evolved many ways to increase the output of their rRNA genes. For example, many bacteria have more than one copy of the genes for rRNA. *E. coli* and *S. enterica* serovar Typhimurium have seven copies of the genes for the rRNAs. Many bacteria also have very strong promoters for their rRNA genes. The RNA polymerase molecules initiate transcription and start down the operon, one immediately after another, so that as many as 50 RNA polymerase molecules can be transcribing each rRNA operon simultaneously, almost as many as a DNA of this length will hold. The rRNA promoters are so strong because they have a sequence called the UP element upstream of the promoter (see Figure 2.13B). This sequence enhances initiation of transcription from the promoter by interacting with the carboxyl terminus of the α subunit of RNA polymerase, much like CAP and some other activator proteins, which interact with the polymerase to activate transcription.

The rate of rRNA gene transcription is also increased because rRNA operons have antitermination sequences lying just downstream of the promoter and in the spacer region between the 16S and 23S coding sequences. These antitermination sequences reduce pausing by RNA polymerase and prevent termination at ρ-dependent transcription termination sites which would occur otherwise since rRNA is not translated (see Figure 2.19). These antitermination sequences thereby allow the synthesis of more complete rRNAs and allow them to be completed in a shorter time (see Condon et al., Suggested Reading).

The rRNA and tRNA genes in *E. coli* seem subject to at least two types of regulation. In the first type, **stringent control,** rRNA and tRNA synthesis ceases when cells are starved for an amino acid. In the second, **growth rate regulation,** the rate of rRNA and tRNA synthesis decreases when cells are growing slowly on a poor carbon source or when some essential nutrient is limiting and increases when they are growing in a good carbon source with ample nutrients. As discussed next, these two types of regulation have some features in common in that they both seem to involve a small molecule, ppGpp.

STRINGENT CONTROL

Protein synthesis requires all 20 amino acids, which must be either made by the cell or furnished in the medium. A cell is said to be **starved** for an amino acid when the amino acid is missing from the medium and the cell cannot make it. The ribosomes of the cell then stall whenever they encounter a codon for the missing amino acid because it is not available for insertion into the growing polypeptide chain.

In principle, rRNA synthesis can continue in cells starved for an amino acid, since RNA does not contain amino acids. However, in *E. coli*, and probably other types of cells, the synthesis of rRNA and tRNA ceases when an amino acid is lacking. This coupling of the synthesis of rRNA and tRNA to the synthesis of proteins is called stringent control. Stringent control saves energy; there is no point in making ribosomes and tRNA if one or more of the amino acids are not available for protein synthesis.

Synthesis of ppGpp during Stringent Control

The stringent control of rRNA synthesis results from the accumulation of an unusual nucleotide, **guanosine tetraphosphate (ppGpp)**. This nucleotide is first made as **guanosine pentaphosphate (pppGpp)** by transferring two phosphates from ATP to the 3' hydroxyl of GTP but is then quickly converted to ppGpp by a phosphatase. These nucleotides were originally called magic spots I and II (MSI and MSII) because they show up as distinct spots during some types of chromatography (see Cashel and Gallant, Suggested Reading).

Figure 13.23 shows a model for how amino acid starvation stimulates the synthesis of ppGpp. The nucleotide is made by an enzyme called RelA (for *relaxed control gene A*), which is bound to the ribosome. When *E. coli* cells are starved for an amino acid (lysine in the example), the tRNAs for that amino acid, tRNA^Lys, are uncharged. The uncharged tRNA does not bind EF-Tu (see chapter 2) and so does not enter the ribosome. Consequently, when a ribosome moving along an mRNA encounters a codon for that amino acid (the codon AAA in the example), the ribosome stalls. If the ribosome stalls long enough, an uncharged tRNA may eventually enter the A site of the ribosome even though it is not bound to EF-Tu. Uncharged tRNA entering the A site stimulates the RelA protein on the ribosome to synthesize ppGpp.

The intracellular levels of ppGpp during amino acid starvation are also regulated by the SpoT (for "magic spot") protein, which is the product of the *spoT* gene. The SpoT protein normally degrades ppGpp, but its degradation activity is inhibited after amino acid starvation, leading to more accumulation of ppGpp. Therefore, after amino acid starvation, the cellular concentration of ppGpp is determined both by the activation of the ppGpp synthesis activity of RelA and by the inhibition of the ppGpp-degrading activity of SpoT. The *spoT* gene product not only can degrade ppGpp but also may synthesize ppGpp during growth rate regulation (see

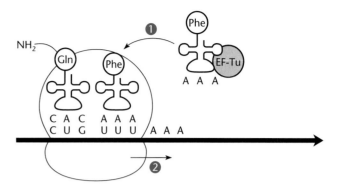

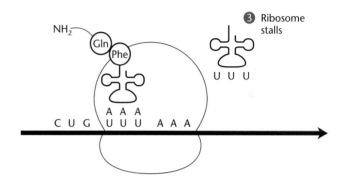

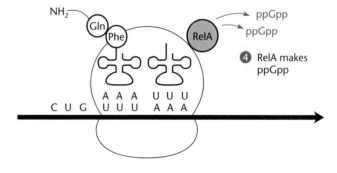

Figure 13.23 Model for synthesis of ppGpp after amino acid starvation. Cells are starved for the amino acid lysine. The tRNA^Lys then has no lysine attached, and EF-Tu cannot bind to a tRNA that is not aminoacylated. A ribosome moving along the mRNA then stops when it arrives at a codon for lysine (AAA), because it has no aminoacylated tRNA to translate the codon. However, if the ribosome remains stalled at the codon long enough, a tRNA^Lys (anticodon UUU in the example) binds to the A site of the ribosome even though EF-Tu is not bound. This binding causes RelA to synthesize ppGpp.

below). It is not clear how the different activities of SpoT are regulated under changing conditions, but some types of bacteria have only SpoT or a SpoT-RelA hybrid which seems to perform both roles.

Isolation of *relA* Mutants

Evidence that the accumulation of ppGpp is responsible for stringent control came from the behavior of *relA* mutants of *E. coli* that lack the RelA enzyme activity. These mutants do not accumulate ppGpp after amino acid starvation and also do not shut off rRNA and tRNA synthesis. Because rRNA synthesis and tRNA synthesis are not stringently coupled to protein synthesis in *relA* mutants, strains with *relA* mutations are called **relaxed strains,** which is the origin of the gene name.

As is often the case, the first *relA* mutant was isolated by chance. A mutant strain of *E. coli* was observed to have a difficult time recovering after amino acid starvation, whereas the wild-type parent could start growing almost immediately after the amino acid was restored. A later study showed that rRNA and tRNA synthesis continued after this mutant was starved for an amino acid, and so this mutant was called a relaxed mutant.

The relative inability of the original *relA* mutant to recover from amino acid starvation suggested a way of enriching for *relA* mutants (see Fiil and Frieson, Suggested Reading). This procedure is based on the fact that growing cells are killed by ampicillin but cells that are not growing will survive (see chapter 3 for a description of the methods). *E. coli* cells that were auxotrophic for an amino acid were mutagenized, washed, and resuspended in a medium without the amino acid, so that they stopped growing. After a long period of incubation, the amino acid was restored and ampicillin was added to the medium. The ampicillin was removed shortly thereafter, and the process was repeated. After a few cycles of this treatment, the bacteria were plated. A high percentage of the bacteria that survived and could multiply to form colonies were *relA* mutants, as evidenced by the fact that most of the surviving strains continued to synthesize rRNA and tRNA after amino acid starvation. In this case, their relative inability to recover from amino acid starvation and begin growing contributed to the survival of the *relA* mutants.

GROWTH RATE REGULATION OF rRNA AND tRNA SYNTHESIS

As mentioned above, cells growing fast in rich medium have many more ribosomes and a higher concentration of tRNA than do cells growing slowly in poor medium. This regulation of rRNA and tRNA synthesis is called growth rate regulation.

An interesting model has been proposed to explain the growth rate dependence of rRNA synthesis (see Barker and Gourse, Suggested Reading). According to this model, more than for most types of promoters, the rate of initiation at rRNA promoters is dependent on the concentration of the initiating nucleotide, which in turn depends on the energy state of the cell and the growth rate. Like most transcription, rRNA synthesis begins with ATP or GTP, so one or the other of these initiates transcription of each of the rRNA genes. According to the model, the rRNA promoters are unusual in that they have a high affinity for RNA polymerase but form very short-lived open complexes (see chapter 2 for a detailed discussion of closed and open complexes and initiation). The high affinity for RNA polymerase allows them to form closed complexes at a high rate and compete very effectively for RNA polymerase. They then quickly enter the open complex. However, because these open complexes are short-lived, they are quickly reversed unless the initiating nucleotide immediately enters through the secondary channel and transcription initiates. The higher the concentration of the initiating nucleotide, the more likely this is to happen in time before the open complex is reversed. When growth rates are high, the concentration of ATP and GTP increases, leading to more frequent initiations at the rRNA promoters and therefore more synthesis of rRNAs and ribosomes. When growth rates are low, for whatever reason, the concentration of ATP and GTP is usually lower, leading to less frequent initiation of transcription from these promoters and less rRNA and tRNA.

In support of this model, the *rrn* promoters that are least sensitive to initiating nucleoside triphosphate concentrations in vitro are also least sensitive to growth rate regulation in vivo. Also, transcription from promoters that initiate with GTP tend to be more sensitive to GTP concentrations than they are to ATP concentrations, while the reverse is true for promoters that initiate with ATP. The cellular level of ppGpp is also higher when cells are growing more slowly, and this also seems to affect growth rate regulation. One hypothesis is that ppGpp competes with the initiating nucleotides in the secondary channel and therefore further interferes with initiation at rRNA and tRNA promoters (see below).

ROLE OF ppGpp IN GROWTH RATE REGULATION, AFTER STRESS, AND IN THE STATIONARY PHASE

Not only are ppGpp levels higher when *E. coli* cells are growing more slowly in poorer medium, but also they increase when cells run out of nutrients and begin to reach stationary phase. Many stress conditions can also cause ppGpp levels to increase. Under these conditions, the rate of rRNA and tRNA synthesis is reduced, but other genes are turned on whose products are required in stationary phase or for stress responses. This suggests that ppGpp is a sort of general "alarmone," signaling that major changes in gene expression are soon to be required (see Magnusson et al., Suggested Reading).

Surprisingly, SpoT, which plays a role in degrading ppGpp in the stringent response, seems to play the major role in synthesizing it during growth rate regulation and during stress responses. However, mutations in both genes are required to completely block ppGpp synthesis, and a *relA spoT* double mutant completely lacks ppGpp.

Because *relA spoT* double mutants lack ppGpp, many experiments have been done comparing these mutants with the wild type to determine the effect of ppGpp on cells. However, in spite of many years of such experimentation, it is not clear how ppGpp affects transcription or even whether more than one mechanism is involved. The ppGpp nucleotide seems to have both positive and negative effects on gene transcription. For example, it has been known for some time that *relA spoT* mutants do not grow in minimal medium without added amino acids, i.e., are effectively auxotrophic for some amino acids. They require nine different amino acids, suggesting that ppGpp serves as an inducer for the transcription of the operons to make these amino acids when they are lacking in the medium. Also, *relA spoT* double mutants, which completely lack ppGpp, show phenotypes similar to mutants that lack σ^s (the stationary-phase σ), which responds to many types of stresses in the cell (see above). The activity of other alternate "stress" sigmas including σ^H (the heat shock sigma) and σ^E (the extracytoplasmic stress sigma) are also enhanced by ppGpp. Apparently, σ^E can be activated by ppGpp in exponentially growing cells, via a pathway that is independent of the degradation of its antisigma, RseA (see above and Costanzo and Ades, Suggested Reading).

One factor that complicates the interpretation of these experiments and necessitates the use of careful controls is the possibility of indirect effects due to competition for RNA polymerase. Genes that seem to be positively regulated by ppGpp might merely be those that do not bind RNA polymerase as tightly as the promoters for rRNA and tRNA genes; they would therefore require higher concentrations of free RNA polymerase to be active. At any one time, fully half of all the RNA polymerase in the cell is devoted to making rRNA and tRNA. If the initiation of synthesis of the rRNAs and tRNAs is blocked by ppGpp, more RNA polymerase becomes available to transcribe other genes. However, recent discoveries from in vitro experiments have indicated that competition for RNA polymerase could not be the sole explanation for the apparent positive effect of ppGpp on the transcription of some genes (see below).

MECHANISM OF ppGpp ACTION

Recent discoveries promise to help clear up some of the ambiguities surrounding the action of ppGpp. One of these was the crystallization of RNA polymerase bound to ppGpp, revealing the whereabouts of ppGpp in the complex (see Artsimovitch et al., Suggested Reading). It had been known for some time that ppGpp binds to RNA polymerase, but it was not known where it binds. Knowing its exact location in the structure of RNA polymerase could help explain its mechanism of action. Surprisingly, these structural studies showed that ppGpp is bound in two orientations in the active center, which are almost mirror images of one another. In one of these orientations it is in position to base pair with a −1 cytosine in the nontranscribed strand of DNA in the open complex (see chapter 2 for the numbering of nucleotides around a promoter). This binding could further destabilize the open-complex formation and reduce transcription from the rRNA promoters. Intriguingly, rRNA promoters are characterized by a cytosine-rich sequence between the −10 region and the start site of transcription where cytosines would be available for base pairing with ppGpp in the open complex, and earlier evidence suggests that this region is important for the stringent control of the promoters. It should be determined whether both of the orientations of ppGpp are physiologically relevant, but it is possible that one orientation might occur when the RNA polymerase is bound to negatively regulated promoters with strategically located cytosines and the other orientation occurs when it is bound to positively regulated promoters without such cytosines. The different orientations of ppGpp could be due to subtle changes in RNA polymerase bound to the two types of promoters. In one orientation it might base pair with a nucleotide in the untranscribed strand of the promoter and destabilize open complexes; in the other orientation, it might stimulate the binding of initiator nucleotides and/or extend open-complex formation.

DksA: a Partner in ppGpp Action?

Another important discovery is that ppGpp does not act alone but instead acts in concert with a protein named DksA. As is often the case, the gene for this protein was discovered by chance, and its role in ppGpp-mediated regulation was not suspected for some time. It was first found because in high copy it suppresses growth defects of *dnaK* mutants (the name DksA means *dnaK* suppressor A). It is not clear even in retrospect why excess DksA would have this effect. Recall from chapter 2 that DnaK is the bacterial Hsp70 protein chaperone that helps fold some proteins as they emerge from the ribosome and also plays the role of the cellular thermometer that induces heat shock (see above). The reason why *dnaK* null mutants are sick is that DnaK normally binds to the heat shock sigma σ^{32} and targets it for degradation. In the absence of DnaK, σ^{32} accumulates and the heat shock genes are expressed constitutively, slowing cell growth.

Apparently, excess DksA relieves the toxicity of the constitutive induction of the heat shock response, but how it does this is not clear.

The role of DksA in ppGpp-mediated regulation was not discovered until later, when it was noticed that the phenotypes of deletion mutations of the *dksA* gene (Δ*dksA*) are similar to the phenotypes of *relA spoT* double mutants that do not make ppGpp. The Δ*dksA* mutants are effectively auxotrophic for some (but not all) of the same amino acids; they show increased rates of rRNA transcription, and they seem to lack stringent control and growth rate regulation of rRNA synthesis. In vitro experiments in which the effect was studied of adding ppGpp to RNA polymerase, with or without purified DksA, showed that DksA markedly enhances the ability of ppGpp to inhibit transcription from rRNA promoters (see Paul et al., Suggested Reading). It also greatly increases the dependence of initiation from rRNA promoters on the concentration of the initiator nucleotides (ATP or GTP) and seems to further shorten the half-life of the open complexes on these promoters. Later, the same group of investigators showed that it enhances the ppGpp stimulation of transcription from the promoters for some amino acid biosynthetic operons, indicating that this effect of ppGpp is not all due to competition for RNA polymerase (see above).

Clues to how the DksA protein has these effects came from the observation that DksA bears a remarkable structural similarity to GreA (see Peredina et al., Suggested Reading). Recall from chapter 2 that GreA is a transcription factor that inserts into the secondary channel in RNA polymerase through which deoxynucleotide triphosphates enter (see Figure 2.17). It has a long extended coiled-coil probe that extends all of the way into the channel to the active center and has two acidic amino acids on the end of the probe (aspartate and glutamate) that may bind the Mg^{2+} in the active center, which plays a role in the polymerization reaction. From this position, GreA can degrade backtracked RNAs and unstall stalled transcription complexes. DksA also has an extended coiled-coil structure of about the same length, with two conserved acidic amino acids in the end (in this case both aspartates), although there is little or no amino acid sequence similarity otherwise. This striking similarity in structure suggests that DksA may also enter RNA polymerase through the secondary channel, like GreA. In its position in the secondary channel, it might help ppGpp enter the channel and/or help direct the binding of ppGpp to one of the two orientations in the active center (see above). It might also affect the binding of the initiator nucleotides and might further shorten the half-life of open complexes at some promoters and lengthen it at others. However, it differs from GreA in that it does not

seem to degrade backtracked RNA or relieve stalled transcription complexes. While many questions remain, it is to be hoped that these advances may help break what has been essentially an impasse in understanding the mechanism of this important regulatory molecule and its role in global regulation.

Microarray and Proteomic Analysis of Regulatory Networks

Each new genome sequence tempts us with a renewed challenge: to pursue the quest for complete understanding of cellular regulation. The (almost) complete gene content of an organism can be displayed on a computer screen, usually in multicolor. The ever-growing sequence databases allow the assignment of presumptive functions, enzymatic or structural, to most genes of an organism.

Transcriptome Analysis

RNA profiling can be efficiently accomplished with high-density DNA microarrays (see Rhodius and LaRossa, Suggested Reading). These microarrays can be produced in several ways. The least expensive methods involve spotting ORF-specific PCR products representing more than 95% of the ORFs on nylon membranes or glass slides. Only three slides are needed for the entire *E. coli* genome. Glass slides containing tens of thousands of oligonucleotide probes require the synthesis of "long" oligomers. These are more expensive but allow more sensitive measurements. Competitive hybridizations between two culture samples can distinguish the relative RNA abundances in the two samples. The above two types of microarrays are called two-color arrays (Figure 13.24).

TWO-COLOR MICROARRAYS

The color representations are actually false-color representations of the fluorescence data. Green and red have been popular because, when the hybridizations of two samples are approximately equal, the "spots" can be visualized as red-yellow-green. Blue and yellow are also popular because red-green color-blind people cannot distinguish the red and green spots.

The number of microarray experiments published in the literature has mushroomed in recent years. Most of the regulons and stimulons discussed in this book have been analyzed by microarrays, and it would be impossible to provide a listing of even some of the better analyses. Nevertheless, we have chosen a recent example in which the authors published their data in the blue-yellow color scheme.

Figure 13.25 illustrates an experiment based on a topic discussed in chapter 6: regulation of *B. subtilis* competence (DNA uptake) by cell density peptides and the

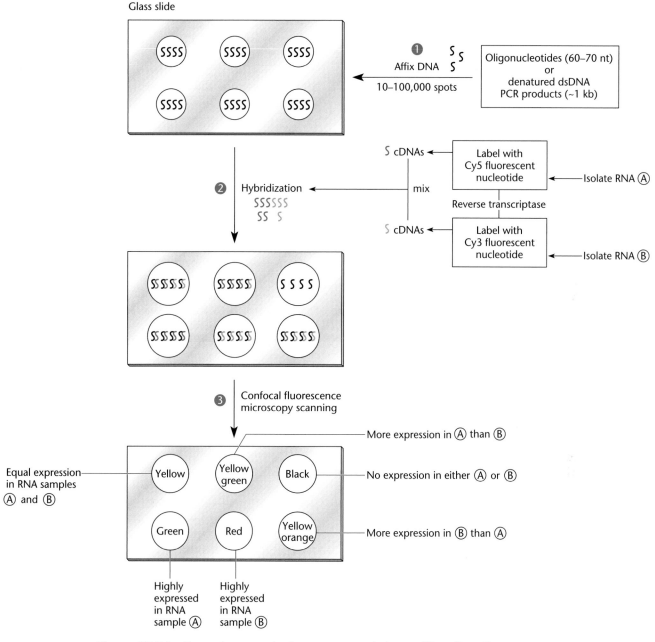

Figure 13.24 Two-color spotted microarray transcription profiling. (Step 1) DNA probes representing the entire genome sequence are affixed to a glass slide. (Step 2) Fluorescently labeled cDNAs, produced by reverse transcriptase primed by random hexamers from template RNA isolated from different cultures (A and B), are hybridized to the glass slide. (Step 3) Scanning of the slide produces dots that vary in color from red to green to yellow to intermediate shades. The colors are interpreted as shown. dsDNA, double-stranded DNA; nt, nucleotides.

ComA transcription regulator (see Figure 6.3). Recall that the PhrC peptide stimulates ComA phosphorylation and thus ComA activates the transcription of competence genes. Figure 13.25 illustrates a DNA microarray experi-

ment that accesses the ComA-dependent genes when several peptide-encoding genes named *phrC*, *phrF*, and *phrK* are deleted. In this experiment, ComA-dependent mRNAs were affected by each of the *phr* deletions. The

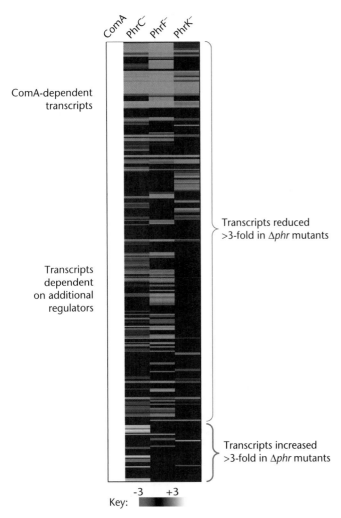

ComA-dependent transcripts

Transcripts dependent on additional regulators

Transcripts reduced >3-fold in Δ*phr* mutants

Transcripts increased >3-fold in Δ*phr* mutants

Key: -3   +3

Figure 13.25 Two-color microarray data showing the effect of *phrC*, *phrF*, or *phrK* deletion on the expression of genes activated by ComA. A threefold or greater decrease in gene expression is shaded light purple, and a threefold or greater increase is shaded gray. Genes whose expression does not change are shaded black. The boxes to the left of the visualization indicate genes whose expression was previously shown to be regulated by the response regulator ComA.

results illustrate that *B. subtilis*, like the *Vibrio* species discussed earlier in this chapter, use multiple quorum-sensing signals.

Determining a value for "significant" change in RNA expression between two RNA samples has been rather arbitrary, but it has been conventionally agreed to be approximately twofold or greater. It goes without saying that biological and technical replicates should be included. Additionally, all experiments must be assessed for inherent biases. For example, Cy3 and Cy5, the two fluorescent nucleotides, are not incorporated with equal efficiencies, and so it is advisable to perform experiments that include reversal of Cy3 and Cy5 incorporation.

Data normalization should also include a reference sample. Choice of what will best serve as such a sample is not obvious, nor is it consistent among different research groups. A good approach is for various research groups to use a reference that is standard with respect to both sample preparation procedures and the strains used as the "wild type."

Finally, microarray experiments measure only steady-state RNA levels. Often, proteomic analyses provide important complementary information (see below).

Single-Gene Analysis

Once a comparative microarray analysis has been done, differences in expression of a particular gene should be further studied. An emerging method for monitoring transcription of a specific gene is Q-RT-PCR (for *q*uantitative *r*eal-*t*ime *PCR*). This method first uses the enzyme reverse transcriptase to produce a cDNA. The primers used to prime the reverse transcriptase reaction are random hexamers. Then PCR amplifications of the cDNA template are performed, using specific primers to amplify the cDNA corresponding to the RNA being quantitated. Where this differs from reverse transcriptase-PCR is that an attempt is made to estimate the amount of cDNA template, and hence the amount of original RNA, by measuring the rate of accumulation of PCR product rather than the amount of end product, which is measured in standard PCR amplifications. The assumption is that the rate of accumulation of the PCR product depends, at least initially, on the concentration of the cDNA template. The direct way to make these measurements is to begin a number of parallel reactions and stop them at various times to determine how much product has been made by running the products on agarose gels. Data points are chosen when the accumulation of product is still linear with respect to the number of cycles of amplification. To ensure that the measurements are more or less quantitative, known amounts of a different template can be added to some of the reaction mixtures, ideally the same gene with a short deletion or insertion, so that a PCR product of a somewhat different size will also accumulate. The rate of accumulation of this other product relative to its known concentration allows a more confident statement about the concentration of the unknown RNA in the original sample.

A more sensitive and less laborious, although more expensive, way to measure the rate of accumulation of the product is to carry out the PCR in a specially designed fluorimeter that measures the accumulation of product based on the fluorescence of a dye that fluoresces only when bound to a double-stranded DNA. To minimize the background due to hybridized primers, etc., the measurements are made just before the denaturation step in each cycle, at a temperature when the primers are usually denatured since they are shorter.

GENE CHIP ARRAYS

A very sensitive but very expensive method for transcriptome analysis uses **gene chips** (Figure 13.26). Besides measuring the transcription of protein-coding sequences,

this technique can be used to evaluate short transcripts from intergenic regions.

In a now classic study analyzing intergenic regions (see Zheng et al., 2001a and 2001b, Suggested Reading), the response of *E. coli* to hydrogen peroxide stress revealed the induction of at least 140 mRNAs. Because previous genetic analysis had identified the OxyR protein as a peroxide response regulator, comparison of isogenic wild-type and *oxyR* deletion strains not only confirmed the

Figure 13.26 Gene chip transcription profiling. (Step 1) A computer algorithm designs 10 to 20 oligonucleotides to represent each known or predicted gene. Intergenic regions can also be included. A mismatch oligonucleotide is a control for cross-hybridization. Masked photolithography produces the oligonucleotides on the chip. (Steps 2 and 3) Following isolation from a culture (step 2), RNA is transcribed into cDNA by RNA-dependent reverse transcription (step 3). (Step 4) Biotinylated nucleotides are added to the cDNA termini. (Step 5) The cDNA is hybridized to the silicon chip. (Step 6) A streptavidin-phycoerythrin conjugate binds to biotin where DNA-RNA hybridization has occurred. Fluorescence imaging provides data. If the mismatch control gives a signal as strong or almost as strong, the data point is discarded. nt, nucleotide.

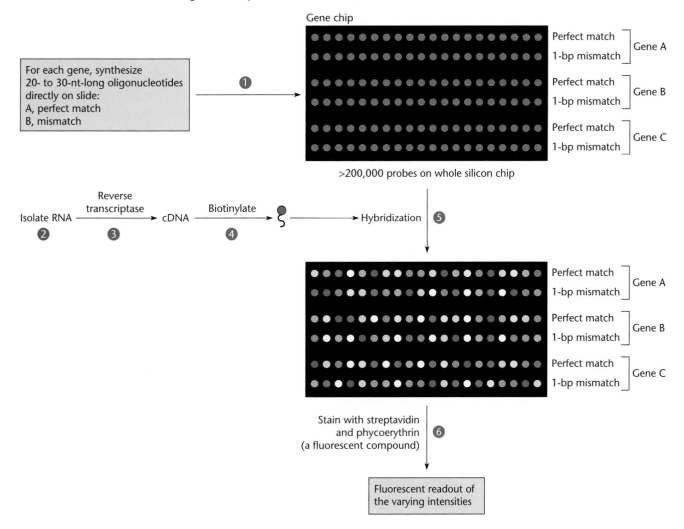

known OxyR-regulated genes but also identified new candidates for the OxyR regulon. The addition of a bioinformatics computational search for OxyR-binding sites revealed several more OxyR-regulated genes.

Proteome Analysis

The techniques of **proteomics** can be used to identify proteins in order to analyze the levels of proteins in cells, to determine relative changes in protein levels of regulons or stimulons, to evaluate protein-protein interactions, and to study subcellular localization of proteins.

MASS SPECTROMETRY

Mass spectrometry (MS) is an important tool of proteomics. It is used both to identify proteins and to quantitate protein expression (see Han and Lee, Suggested Reading). Protein MS involves the ionization of proteins and peptides and subsequent measurements of mass-to-charge (*m/z*) ratios. Figure 13.27 illustrates the steps involved. Once proteins have been obtained from a biological sample, they are fragmented into peptides, often by trypsin digestion because it produces small peptides that are suitable for the first step in MS: ionization. Two popular ionization methods are **MALDI** (matrix-assisted laser desorption ionization) and **ESI** (electrospray ionization) (see Kolker et al., Suggested Reading). Figure 13.27A shows that a mass spectrometer is composed of three chambers. Ionization of peptides, which transfers a peptide from the solid phase to a gaseous ionic phase, takes place in the first chamber. In the second chamber, the separation chamber, the charge and mass of the ionized peptides determine their relative **time of flight (TOF)**. In a third

Figure 13.27 Mass spectrometry in proteomics. (A) The mass spectrometer contains several chambers which first ionize peptides and then separate them based on their mass-to-charge ratio and allow the user to collect fractions containing the different peaks. (B) For tandem mass spectrometry, a peptide fraction collected from the first round of spectrometry is subjected to fractionation into even smaller peptides, which, when separated in the final detection chamber, allow very high resolution of mass-to-charge ratios such that the sequence of the fragmented peptide can be deduced.

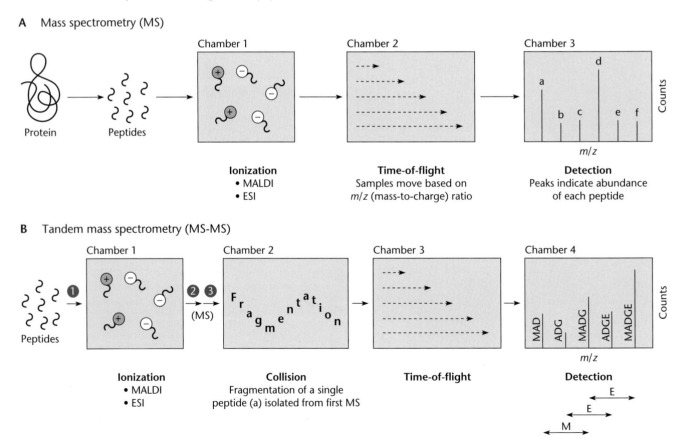

chamber, detection of the ionized peptide fragments occurs and the spectrum readout resolves the different species and provides peaks that show the abundance of each peptide.

TANDEM MASS SPECTROMETRY

Tandem mass spectrometery (MS-MS) allows proteins to be identified from a biological sample. The process is illustrated in Figure 13.27B. After a protein has been subjected to MS, a single peptide is isolated and shunted through a "collision" chamber in which nitrogen or argon gas breaks the peptide into subfragments. These fragments then continue through a time-of-flight chamber and go on to the detection chamber determined by their *m/z*. The MS-MS spectrum produced can be analyzed such that the sequence of the peptide can be deduced. Computerized comparison to databases can identify the sequence of the peptide and, if the genome sequence is known, the protein and gene from which the peptide came.

PROTEIN SAMPLE PREPARATION

Two-dimensional protein electrophoresis (**2D-PAGE**) can separate many of the proteins of the cell into individual spots on a membrane, and the spots can be identified by MS as above. Because sample complexity limits the ability to identify individual proteins, non-gel protein separation techniques such as liquid chromatography (LC) are often used. This method separates the protein samples into subsamples that are less complex.

PROTEIN ANNOTATION

LC-MS-MS has been used to identify more than 1,000 *E. coli* proteins; popular database search programs for peptide identification are SEQUEST and X!Tandem.

From Genes to Regulons to Networks to Genetic Analysis

Microarray and proteomic analyses are most useful if seen as hypothesis-generating tools rather than as an end in themselves. Regulon and stimulon components that are identified by whole-genome or high-throughput technologies can be studied further by the use of genetic techniques.

Gene knockout experiments, performed so as to avoid polar effects (see chapter 12), are necessary to determine the loss-of-function phenotypes of individual genes. However, in genetics, unlike in boxing, a knockout is not the end of the bout. As we can see in many of the examples of genetic analysis discussed in this textbook, an understanding of gene function often requires the study of a variety of allelic forms that result from base pair changes. With genome sequences, we can better direct base pair changes to specific regions of proteins. Regions of proteins or RNAs with presumptive functions or predicted regulatory sequences can be altered at will by using recently developed methods of site-specific mutagenesis which involve the λ phage Red system (see Box 10.3). This type of reverse genetic approach is not applicable to all bacteria. One limitation is that the genetic constructs are usually made in *E. coli* and electroporation or transformation of bacteria that are not naturally transformable can also be a limiting factor.

COMBINING TRANSCRIPTOME, PROTEOME, AND GENETIC ANALYSES: A CASE STUDY

An example of combining genomic and genetic techniques is provided by studies of tuberculosis, a disease caused by the bacterium *Mycobacterium tuberculosis*. For a cautionary tale regarding the limitation of more than 50 microarray experiments performed on bacteria grown in cultures, in mice, and in humans, see Kendall et al., Suggested Reading. Developments in *M. tuberculosis* genetic analysis are described by Murray and Rubin, Suggested Reading. The understanding of regulation has not reached a plateau for geneticists but is just beginning.

SUMMARY

1. The coordinate regulation of a large number of genes is called global regulation. Operons that are regulated by the same regulatory protein are part of the same regulon.

2. In catabolite repression, the operons for the use of alternate carbon sources cannot be induced when a better carbon and energy source, such as glucose, is present. In *E. coli* and other enteric bacteria, this catabolite regulation is achieved, in part, by cAMP, which is made by adenylate cyclase, the product of the *cya* gene. When the bacteria are growing in a poor carbon such as lactose or galactose, the adenylate cyclase is activated and cAMP levels are high. When the bacteria are growing in a good carbon source such as glucose, cAMP levels are low. The cAMP acts through a protein called CAP, also called CRP, the product of the *crp* gene. CAP is a transcriptional activator, which, with cAMP bound, activates the transcription of catabolite-sensitive operons such as *lac* and *gal*.

3. Bacterial cells induce different genes depending on the nitrogen sources available. Genes that are regulated through the nitrogen source are called Ntr genes. Most bacteria,

(continued)

including *E. coli*, prefer NH_3 as a nitrogen source and do not transcribe genes for using other nitrogen sources when growing in NH_3. Glutamine concentrations are low when NH_3 concentrations are low. A signal transduction pathway is then activated, culminating in the phosphorylation of NtrC. This signal transduction pathway begins with the GlnD protein, a uridylyl transferase, which is the sensor of the glutamine concentration in the cell. At low concentrations of glutamine, GlnD transfers UMP to the P_{II} protein, inactivating it. However, at high concentrations of glutamine, the GlnD protein removes UMP from P_{II}. The P_{II} protein without UMP attached can bind to NtrB, somehow preventing the transfer of phosphate to NtrC and causing the removal of phosphates from NtrC. The phosphorylated NtrC protein activates the transcription of the *glnA* gene, the gene for glutamine synthetase, as well as the *ntrB* gene and its own gene, *ntrC*, since they are part of the same operon as *glnA*. It also activates the trancription of operons for using other nitrogen sources.

4. The NtrB and NtrC proteins form a sensor and response regulator pair and are highly homologous to other sensor kinase and response regulator pairs in bacteria.

5. NtrC-regulated promoters of *E. coli* and other enterics require a special sigma factor called σ^{54}, which is also used to transcribe the flagellar genes and some biodegradative operons in other types of bacteria.

6. The cell also regulates the activity of glutamine synthetase by adenylating the glutamine synthetase enzyme. The enzyme is highly adenylated at high glutamine to α-ketoglutarate ratios, which makes it less active and subject to feedback inhibition.

7. Bacteria induce a set of proteins called the heat shock proteins in response to an abrupt increase in temperature. Some of the heat shock proteins are chaperones, which assist in the refolding of denatured proteins; others are proteases, which degrade denatured proteins. The heat shock response is common to all organisms, and some of the heat shock proteins have been highly conserved throughout evolution.

8. In *E. coli*, the promoters of the heat shock genes are recognized by RNA polymerase holoenzyme with an alternative sigma factor called the heat shock sigma, or σ^{32}. The amount of this sigma factor markedly increases following a heat shock, leading to increased transcription of the heat shock genes. The increase in σ^{32} following a heat shock involves DnaK, a chaperone that is one of the heat shock proteins. The DnaK protein normally binds to σ^{32}, targeting the sigma factor for degradation. Immediately after a heat shock, DnaK binds to other denatured proteins, making less DnaK available to bind to σ^{32} so that the sigma factor is stabilized and more of it accumulates.

9. In addition to the vegetative sigmas, bacteria have stress sigmas that are activated by a wide variety of different stresses.

10. Bacteria also have ways of detecting stress to their membranes including osmotic stress and damage to the outer membrane. These are called extracytoplasmic stresses.

11. One of the ways that bacteria adjust to changes in the osmolarity of the medium is by changing the ratio of their porin proteins, which form pores in the outer membrane through which solutes can pass to equalize the osmotic pressure on both sides of the membrane. The major porins of *E. coli* are OmpC and OmpF, which make pores of different sizes, thereby allowing the passage of different-sized solutes. The relative amounts of OmpC and OmpF change in response to changes in the osmolarity of the medium. The *ompC* and *ompF* genes in *E. coli* are regulated by a sensor and response regulator pair of proteins, EnvZ and OmpR, which are similar to NtrB and NtrC. The EnvZ protein is an inner membrane protein with both kinase and phosphatase activities that, in response to a change in osmolarity, can transfer a phosphoryl group to or remove one from OmpR, a transcriptional activator. The state of phosphorylation of OmpR affects the relative rates of transcription of the *ompC* and *ompF* genes.

12. The ratio of OmpF to OmpC porin proteins is also affected by an antisense RNA named MicF. A region of the MicF RNA can base pair with the translation initiation region of the OmpF mRNA and block access by ribosomes, thereby inhibiting OmpF translation. The *micF* gene is regulated by a number of transcriptional regulatory proteins, including SoxS, which induces the oxidative stress regulon, and MarA, which induces genes involved in excluding toxic chemicals and antibiotics from the cell.

13. Bacteria detect damage to their outer membrane by detecting the accumulation of outer membrane proteins in the periplasm. The two systems in *E. coli* are Cpx and σ^E, which respond to the accumulation in the periplasm of pilin subunits and Omp proteins, respectively.

14. The virulence genes of pathogenic bacteria are also global regulons and are normally transcribed only when the bacterium is in its vertebrate host.

15. The diphtheria toxin gene, *dtxR*, encoded by a prophage of *C. diphtheriae*, is turned on only when iron is limiting, a condition mimicking that in the host. The *dtxR* gene is regulated by a chromosomally encoded repressor protein, DtxR, which is similar to the Fur protein involved in regulating the genes of iron availability pathways in *E. coli* and other enteric bacteria.

16. The toxin genes of *V. cholerae* are also carried on a prophage and are regulated by a regulatory cascade that begins with a transcriptional activator, ToxR. The ToxR protein traverses the inner membrane and is activated by a second protein, ToxS. ToxR and ToxS act in concert with another gene pair, TcpP-TcpH, to activate the transcription of *toxT*, whose gene product in turn activates the transcription of virulence genes.

17. The virulence genes of *B. pertussis* are regulated by a sensor and response regulator pair of proteins, BvgS and BvgA. The regulation goes through more than two stages as the bacterium enters its host.

18. The genes encoding the ribosomal proteins, rRNAs, and tRNAs are part of the largest regulon in bacteria, with hundreds of genes that are coordinately regulated. A large proportion of the cellular energy goes into making the rRNAs, tRNAs, and ribosomal proteins; therefore, regulating the expression of these genes saves the cell considerable energy.

19. The synthesis of ribosomal proteins is coordinated by coupling the translation of the ribosomal protein genes to the amount of free rRNA that is not yet in a ribosome. The ribosomal protein genes are organized into operons, and one ribosomal protein of each operon plays the role of translational repressor. The same protein also binds to free rRNA, so that when there is excess rRNA in the cell, all of the repressor protein binds to the free rRNA and none is available to repress translation.

20. The synthesis of rRNA and tRNA following amino acid starvation is inhibited by guanosine tetraphosphate (ppGpp), synthesized by an enzyme associated with the ribosome called RelA. All types of bacteria contain ppGpp,

and so the regulation may be universal. However, it is not yet clear how higher levels of ppGpp inhibit transcription of the genes for rRNA and tRNA and stimulate transcription of others. A protein named DksA may enter the secondary channel and help ppGpp regulate transcription.

21. Cells contain fewer ribosomes when they are growing more slowly in poorer media. This is called growth rate control and may be due to the lower concentration of the initiating ribonucleosides, ATP and GTP, in slower-growing cells. RNA polymerase forms short-lived open complexes on the promoters for the rRNA genes, and these may have to be stabilized by immediate initiation of transcription with high concentrations of ATP and GTP. Guanosine tetraphosphate also plays a role in growth rate control, perhaps by reducing the concentration of ATP and GTP or by competing with ATP and GTP for the initiating complex.

22. Many small RNAs play important roles in gene regulation in bacteria. These small RNAs can pair with complementary sequences in mRNA and block translation or target the mRNA for degradation by RNases.

23. Techniques such as microarrays have made it possible to monitor the expression of many genes simultaneously. This has led to the discovery of many more genes belonging to the same regulon. In the techniques of proteomics, proteins can be isolated and identified by tandem mass spectrometry. If the genome sequence of the organism is known, the gene and protein can be identified.

QUESTIONS FOR THOUGHT

1. Why do you suppose that proteins involved in gene expression (i.e., transcription and translation) are among the heat shock proteins?

2. Why do you think genes for the utilization of amino acids as a nitrogen source are not under Ntr regulation in *Salmonella* spp. but are under Ntr regulation in *Klebsiella* spp.?

3. Why are the corresponding sensor kinase and response regulator genes of the various two-component systems so similar to each other?

4. Why is the enzyme responsible for ppGpp synthesis in response to amino acid starvation different from the one responsible for ppGpp synthesis during growth rate control? Why might SpoT be used to degrade ppGpp made by RelA after amino acid starvation but be used to synthesize it during growth rate control?

PROBLEMS

1. You have isolated a mutant of *E. coli* that cannot use either maltose or arabinose as a carbon and energy source. How would you determine if your mutant has a *cya* or *crp* mutation or whether it is a double mutant with mutations in both the *ara* operon and a *mal* operon?

2. What would you expect the phenotypes of the following mutations to be?

a. a *glnA* (glutamine synthetase) null mutation

b. an *ntrB* null mutation

c. an *ntrC* null mutation

d. a *glnD* null mutation that inactivates the UTase so that P_{II} has no UMP attached

e. a constitutive *ntrC* mutation that changes the NtrC protein so that it no longer needs to be phosphorylated to be active

f. a *dnaK* null mutation

g. a *dtrR* null mutation of *C. diphtheriae*

h. a *relA spoT* double null mutant

3. How would you show that the toxin gene of a pathogenic bacterium is not a normal chromosomal gene but is carried on a prophage not common to all the bacteria of the species?

4. Explain how you would use gene dosage experiments to prove that the heat shock sigma (σ^{32}) gene is not transcriptionally autoregulated.

5. Explain how you would show which of the ribosomal proteins in the *rplJ-rplL* operon is the translational repressor.

SUGGESTED READING

Alba, B. M., and C. A. Gross. 2004. Regulation of the *Escherichia coli* σ^E-dependent envelope stress response. *Mol. Microbiol.* **52**:613–620.

Artsimovitch, I., V. Patlan, S. I. Sekine, M. N. Vassylyeva, T. Hosaka, K. Ochi, S. Yokoyama, and D. G. Vassylyev. 2004. Structural basis for transcription regulation by alarmone ppGpp. *Cell* **117**:299–310.

Auchtung, J. M., C. A. Lee, and A. D. Grossman. 2006. Modulation of the ComA-dependent quorum response in *Bacillus subtilis* by multiple Rap proteins and Phr peptides. *J. Bacteriol.* **188**:5273–5285.

Baba, T., T. Ara, M. Hasegawa, Y. Takai, Y. Okumura, M. Baba, K. A. Datsenko, M. Tomita, B. L. Wanner, and H. Mori. 2006. Construction of *Escherichia coli* K-12 in-frame single-gene knockout mutants: the Keio collection. *Mol. Syst. Biol.* Epub 2006 Feb 21.

Barabote, R. D., and M. H. Saier, Jr. 2005. Comparative genomic analysis of the bacterial phosphotransferase system. *Microbiol. Mol. Biol. Rev.* **69**:608–634.

Barker, M. M., and R. L. Gourse. 2001. Regulation of rRNA transcription correlates with nucleoside triphosphate sensing. *J. Bacteriol.* **183**:6315–6323.

Batchelor, E., D. Walthers, L. J. Kenney, and M. Goulian. 2005. The *Escherichia coli* CpxA-CpxR envelope stress response system regulates expression of the porins OmpF and OmpC. *J. Bacteriol.* **187**:5723–5731.

Blaszczak, A., C. Georgopoulos, and K. Liberek. 1999. On the mechanism of FtsH-dependent degradation of the sigma 32 transcriptional regulator of *Escherichia coli* and the role of the DnaK chaperone machine. *Mol. Microbiol.* **31**:157–166.

Bothwell, D., and J. Sambrook (ed.). 2003. *DNA Microarrays: a Molecular Cloning Manual.* Cold Spring Harbor Laboratory Press, Cold Spring Harbor, N.Y.

Boucher, P. E., M.-S. Yang, D. A. Schmidt, and S. Stibitz. 2001. Genetic and biochemical analyses of BvgA interaction with the secondary binding region of the *fha* promoter of *Bordetella pertussis*. *J. Bacteriol.* **183**:536–544.

Busby, S., and R. H. Ebright. 1999. Transcription activation by catabolite activator protein (CAP). *J. Mol. Biol.* **293**:199–213.

Cashel, M., and J. Gallant. 1969. Two compounds implicated in the function of the RC gene in *Escherichia coli*. *Nature* (London) **221**:838–841.

Condon, C., C. Squires, and C. L. Squires. 1995. Control of rRNA transcription in *Escherichia coli*. *Microbiol. Rev.* **59**:623–645.

Costanzo, A., and S. E. Ades. 2006. Growth phase-dependent regulation of the extracytoplasmic stress factor σ^E by guanosine 3′, 5′-bispyrophosphate (ppGpp). *J. Bacteriol.* **188**: 4627–4634.

Cotter, P. A., and A. M. Jones. 2003. Phosphorelay control of virulence gene expression in *Bordetella*. *Trends Microbiol.* **11**:367–373.

Craig, E., and C. A. Gross. 1991. Is Hsp70 the cellular thermometer? *Trends Biochem. Sci.* **16**:135–140.

Cummings, C. A., H. J. Bootsma, D. A. Relman, and J. F. Miller. 2006. Species- and strain-specific control of a complex flexible regulon by *Bordetella* BvgAS. *J. Bacteriol.* **188**:1775–1785.

Eisen, M. B., and P. D. Brown. 1999. DNA arrays for analysis of gene expression. *Methods Enzymol.* **303**:179–205.

Fiil, N., and J. D. Friesen. 1968. Isolation of relaxed mutants of *Escherichia coli*. *J. Bacteriol.* **95**:729–731.

Fisher, S. H. 1999. Regulation of nitrogen metabolism in *Bacillus subtilis*: vive la difference! *Mol. Microbiol.* **32**:223–232.

Gall, T., M. S. Bartlett, W. Ross, C. L. Turnbough, Jr., and R. L. Gourse. 1997. Transcription initiation by initiating NTP concentration: rRNA synthesis in bacteria. *Science* **278**: 2092–2097.

Gibson, G., and S. V. Muse. 2004. *A Primer of Genome Science*, 2nd ed. Sinauer Associates, Inc., Sunderland, Mass.

Graves, P. R., and T. A. J. Haystead. 2002. Molecular biologist's guide to proteomics. *Microbiol. Mol. Biol. Rev.* **66**: 39–63.

Hall, M. N., and T. J. Silhavy. 1981. Genetic analysis of the *ompB* locus in *Escherichia coli* K-12. *J. Mol. Biol.* **151**:1–15.

Han, M.-J. and S. Y. Lee. 2006. The *Escherichia coli* proteome: past, present, and future prospects. *Microbiol. Mol. Biol. Rev.* **70**:362–439.

Hernandez, V. J., and H. Bremer. 1991. *Escherichia coli* ppGpp synthetase II activity requires *spoT*. *J. Biol. Chem.* **266**: 5991–5999.

Hirvonen, C. A., W. Ross, C. E. Wozniak, E. Marasco, J. R. Anthony, S. E. Aiyer, V. H. Newburn, and R. L. Gourse. 2001.

Contributions of UP elements and the transcription factor FIS to expression from the seven *rrn* P1 promoters in *Escherichia coli*. *J. Bacteriol.* **183**:6305–6314.

Hudson, M. E., and J. R. Nodwell. 2004. Dimerization of the RamC morphogenetic protein of *Streptomyces coelicolor*. *J. Bacteriol.* **186**:1330–1336.

Igarashi, K., and A. Ishihama. 1991. Bipartite functional map of *E. coli* RNA polymerase α subunit: involvement of the C-terminal region in transcription activation by cAMP-CRP. *Cell* **65**:1015–1022.

Inada, T., K. Kimata, and H. Aiba. 1996. Mechanisms responsible for glucose-lactose diauxie in *Escherichia coli*: challenge to the cAMP model. *Genes Cells* **1**:293–301.

Kadner, R. J. 2005. Regulation by iron: RNA rules the rust. *J. Bacteriol.* **187**:6870–6873.

Karimova, G., J. Pidoux, A. Ullmann, and D. Ladant. 1998. A bacterial two-hybrid system based on a reconstituted signal transduction pathway. *Proc. Natl. Acad. Sci. USA* **95**:5752–5756.

Karimova, G., N. Dautin, and D. Ladant. 2005. Interaction network among *Escherichia coli* membrane proteins involved in cell division as revealed by bacterial two-hybrid analysis. *J. Bacteriol.* **187**:2233–2243.

Kazmierczak, M. J., M. Wiedmann, and K. J. Boor. 2005. Alternative sigma factors and their roles in bacterial virulence. *Microbiol. Mol. Biol. Rev.* **69**:527–543.

Kendall, S. L., S. C. G. Rison, F. Movahedzadeh, R. Frita, and N. G. Stoker. 2004. What do microarrays really tell us about *M. tuberculosis*? *Trends Microbiol.* **12**:537–544.

Kim, T.-J., T. A. Gaidenko, and C. W. Price. 2004. A multicomponent protein complex mediates environmental stress signaling in *Bacillus subtilis*. *J. Mol. Biol.* **341**:135–150.

Kolker, E., R. Rigdon, and J. M. Hogan. 2006. Protein identification and expression analysis using mass spectrometry. *Trends Microbiol.* **14**:229–235.

Krukonis, E. S., R. R. Yu, and V. J. DiRita. 2000. The *Vibrio cholerae* ToxR/Tcp/ToxT virulence cascade: distinct roles for the two membrane-localized transcriptional activators on single promoter. *Mol. Microbiol.* **38**:67–84.

Lee, S.-J., W. Boos, J.-P. Bouche, and J. Plumbridge. 2000. Signal transduction between a membrane-bound transporter, PtsG, and a soluble transcription factor, Mlc, of *Escherichia coli*. *EMBO J.* **19**:5353–5361.

Lenz, D. H., M. B. Miller, J. Zhu, R. V. Kulkarni, and B. L. Bassler. 2005. CsrA and three redundant small RNAs regulate quorum sensing in *Vibrio cholerae*. *Mol. Microbiol.* **58**:1186–1202.

Lenz, D. H., K. C. Mok, B. N. Lilley, R. V. Kulkarni, N. S. Wingreen, and B. L. Bassler. 2004. The small RNA chaperone Hfq and multiple small RNAs control quorum sensing in *Vibrio harveyi* and *Vibrio cholerae*. *Cell* **118**:69–82.

Liberek, K., T. P. Galitski, M. Zyliez, and C. Georgopoulos. 1992. The DnaK chaperon modulates the heat shock response of *E. coli* by binding to the σ^{32} transcription factor. *Proc. Natl. Acad. Sci. USA* **89**:3516–3520.

Magasanik, B. 1982. Genetic control of nitrogen assimilation in bacteria. *Annu. Rev. Genet.* **16**:135–168.

Magnusson, L. U., A. Farewell, and T. Nystrom. 2005. ppGpp: a global regulator in *Escherichia coli*. *Trends Microbiol.* **13**:236–242.

Mattison, K., R. Oropeza, N. Byers, and L. J. Kenney. 2002. A phosphorylation site mutant of OmpR reveals different binding conformations at *ompF* and *ompC*. *J. Mol. Biol.* **315**:497–511.

Merkel, T. J., C. Barros, and S. Stibitz. 1998. Characterization of the *bvgR* locus of *Bordetella pertussis*. *J. Bacteriol.* **180**:1682–1690.

Miller, V. L., and J. J. Mekalanos. 1984. Synthesis of cholera toxin is positively regulated at the transcriptional level by *toxR*. *Proc. Natl. Acad. Sci. USA* **81**:3471–3475.

Mount, D. W. 2004. *Bioinformatics: Sequence and Genome Analysis*, 2nd ed. Cold Spring Harbor Laboratory Press, Cold Spring Harbor, N.Y.

Murray, J. P., and E. J. Rubin. 2005. New genetic approaches shed light on TB virulence. *Trends Microbiol.* **13**:366–372.

Nagai, H., H. Yuzawa, and T. Yura. 1991. Interplay of two *cis*-acting mRNA regions in translational control of σ^{32} synthesis during the heat shock response of *Escherichia coli*. *Proc. Natl. Acad. Sci. USA* **88**:10515–10519.

Nomura, M., J. L. Yates, D. Dean, and L. E. Post. 1980. Feedback regulation of ribosomal protein gene expression in *Escherichia coli*: structural homology of ribosomal RNA and ribosomal protein mRNA. *Proc. Natl. Acad. Sci. USA* **77**:7084–7088.

Paul, B. J., M. M. Barker, W. Ross, D. A. Schneider, C. Webb, J. W. Foster, and R. L. Gourse. 2004. DksA: a critical component of the transcription initiation machinery that potentiates the regulation of rRNA promoters by ppGpp and the initiating NTP. *Cell* **118**:311–322.

Peredina, A., V. Setlov, M. N. Vassylyeva, T. H. Tahirov, S. Yokoyama, I. Artsimovitch, and D. G. Vassylyev. 2004. Regulation through the secondary channel-structural framework for ppGpp-DksA synergism during transcription. *Cell* **118**:297–309.

Pompeo, F., J. Luciano, and A. Galinier. 2007. Interaction of GapA with HPr and its homologue, Crh: novel levels of regulation with a key step of glycolysis in *Bacillus subtilis*? *J. Bacteriol.* **189**:1154–1157.

Pratt, L. A., W. Hsing, K. E. Gibson, and T. J. Silhavy. 1996. From acids to *osmZ*: multiple factors influence synthesis of the OmpF and OmpC porins in *Escherichia coli*. *Mol. Microbiol.* **20**:911–917.

Reeves, A., and W. G. Haldenwang. 2007. Isolation and characterization of dominant mutations in the *Bacillus subtilis* stressosome components RsbR and RsbS. *J. Bacteriol.* **189**:1531–1541.

Reitzer, L., and B. L. Schneider. 2001. Metabolic context and possible physiological themes of σ^{54}-dependent genes in *Escherichia coli*. *Microbiol. Mol. Biol. Rev.* **65**:422–444.

Rhodius, V. A., and R. A. LaRossa. 2003. Uses and pitfalls of microarrays for studying transcriptional regulation. *Curr. Opin. Microbiol.* **6**:114–119.

Savery, N. J., G. S. Lloyd, S. J. W. Busby, M. S. Thomas, R. H. Ebright, and R. L. Gourse. 2002. Determinants of the C-terminal domain of the *Escherichia coli* RNA polymerase α subunit important for transcription at class I cyclic AMP receptor protein-dependent promoters. *J. Bacteriol.* **184:** 2273–2280.

Schmidt, M., and R. K. Holmes. 1993. Analysis of diphtheria toxin repressor-operator interactions and the characterization of mutant repressor with decreasing binding activity for divalent metals. *Mol. Microbiol.* **9:**173–181.

Schwartz, D., and J. R. Beckwith. 1970. Mutants missing a factor necessary for the expression of catabolite-sensitive operons in *E. coli*, p. 417–422. *In* J. R. Beckwith and D. Zipser (ed.), *The Lactose Operon.* Cold Spring Harbor Laboratory Press, Cold Spring Harbor, N.Y.

Skorupski, K., and R. K. Taylor. 1997. Control of the ToxR virulence regulon in *Vibrio cholerae* by environmental stimuli. *Mol. Microbiol.* **25:**1003–1009.

Slauch, J. M., S. Garrett, D. E. Jackson, and T. J. Silhavy. 1988. EnvZ functions through OmpR to control porin gene expression in *Escherichia coli* K-12. *J. Bacteriol.* **170:**439–441.

Stibitz, S., A. A. Weiss, and S. Falkow. 1988. Genetic analysis of a region of *Bordetella pertussis* chromosome encoding filamentous hemagglutinin and the pleiotropic regulatory locus *vir*. *J. Bacteriol.* **170:**2904–2913.

Sutton, V. R., E. L. Mettert, H. Beinert, and P. J. Kiley. 2004. Kinetic analysis of the oxidative conversion of the [4Fe-4S]$^{2+}$ cluster of FNR to a [2Fe-2S]$^{2+}$ cluster. *J. Bacteriol.* **186:**8018–8025.

Tang, Y., J. R. Guest, P. J. Artymiuk, and J. Green. 2005. Switching aconitase B between catalytic and regulatory modes involves iron-dependent dimer formation. *Mol. Microbiol.* **56:**1149–1158.

Tu, K. C., and B. Bassler. 2007. Multiple small RNAs act additively to integrate sensory information and control quorum sensing in *Vibrio harveyi*. *Genes Dev.* **21:**221–233.

van Heeswijk, W. C., S. Hoving, D. Molenaar, B. Stegeman, D. Kahn, and H. V. Westerhoff. 1996. An alternative P$_{II}$ protein in the regulation of glutamine synthetase in *Escherichia coli*. *Mol. Microbiol.* **21:**133–146.

Weber, H., T. Polen, J. Heuveling, V. F. Wendisch, and R. Hengge. 2005. Genome-wide analysis of the general stress response network in *Escherichia coli*: σ^s-dependent genes, promoters, and sigma factor selectivity. *J. Bacteriol.* **187:** 1591–1603.

Yates, J. L., and M. Nomura. 1980. *E. coli* ribosomal protein L4 is a feedback regulatory protein. *Cell* **21:**517–522.

Yates, J. L., and M. Nomura. 1981. Localization of the mRNA binding sites for ribosomal proteins. *Cell* **24:**243–249.

Yu, R. R., and V. J. DiRita. 2002. Regulation of gene expression in *Vibrio cholerae* by ToxT involves both antirepression and RNA polymerase stimulation. *Mol. Microbiol.* **43:**119–134.

Zheng, M., X. Wang, L. J. Templeton, D. R. Smulski, R. A. LaRossa, and G. Storz. 2001a. DNA microarray-mediated transcriptional profiling of the *Escherichia coli* response to hydrogen peroxide. *J. Bacteriol.* **183:**4562–4570.

Zheng, M., X. Wang, B. Doan, K. A. Lewis, T. D. Schneider, and G. Storz. 2001b. Computation-directed identification of OxyR DNA binding sites in *Escherichia coli*. *J. Bacteriol.* **183:**4571–4579.

Zhou, Y. N., N. Kusukawa, J. W. Erickson, C. A. Gross, and T. Yura. 1988. Isolation and characterization of *Escherichia coli* mutants that lack the heat shock sigma factor σ^{32}. *J. Bacteriol.* **170:**3640–3649.

Zimmer, D. P., E. Soupene, H. L. Lee, V. F. Wendisch, A. B. Khodursky, B. J. Peter, R. A. Bender, and S. Kustu. 2000. Nitrogen regulatory protein C-controlled genes of *Escherichia coli*: scavenging as a defense against nitrogen limitation. *Proc. Natl. Acad. Sci. USA* **97:**14674–14679.

CHAPTER **14**

Bacterial Cell Compartmentalization and Sporulation

The cells of higher eukaryotes are generally very large and contain a number of organelles and compartments that are in communication with each other. Larger organisms also consist of many cells, arranged in tissues and organs, which also communicate with each other. They also undergo seemingly miraculous embryonic development processes, in which a single cell divides and eventually becomes a complete organism consisting of thousands or even billions of cells, depending on the organism, each playing a defined role. Many of these features of higher organisms have their counterparts in bacteria in a relatively simple form.

Bacterial cells are simple only in relation to eukaryotic cells; they are still unimaginably complex. Even relatively simple single-celled bacteria consist of many compartments, which are in communication with each other. The cytoplasmic compartment contains the DNA, the ribosomes and other components of the translational machinery, and many of the enzymes that perform biosynthetic and metabolic functions. The cytoplasmic compartment is surrounded by a bilipid cytoplasmic membrane, a compartment containing many proteins that perform sensory roles, communicating information from outside the cell to the cytoplasmic compartment. It also contains the components of the electron transport system, the major generator of ATP, and has channels such as SecYEG that selectively allow molecules in and out of the cell. Surrounding this cytoplasmic membrane is a rigid cell wall, which gives the bacterium its shape and protects it from osmotic shock under rapidly changing solute conditions. In gram-negative bacteria, the cell wall is surrounded by a second bilipid membrane called the outer membrane, which creates a space between the two compartments. This space is called the periplasm and creates another compartment, containing enzymes that degrade larger molecules so they can be taken up to be used as energy sources,

613

chaperones that help fold proteins and protect them from degradation as they pass through the compartments, sensor proteins that detect stress and damage to the outer membrane, and proteases that degrade abnormal proteins released from the outer membrane, among others. The outer membrane has channels that allow the selective passage of proteins and small molecules and receptors to which these molecules bind before they are taken up; it also contains anchoring points for extracellular organelles such as pili and flagella. Gram-positive bacteria do not have an outer membrane but have a thicker cell wall that performs many of the same functions. They may also have a rudimentary periplasmic space between this cell wall and the cytoplasmic membrane, and this space performs some of the same functions as the periplasmic space in gram-negative bacteria.

Recent discoveries have revealed that macromolecules, including proteins, do not drift freely among these compartments but are directed on a subcellular matrix, much as they are in larger multicellular organisms. In earlier chapters we have discussed how these matrices are composed of the same types of filaments as they are in eukaryotes, i.e., actin, tubulin, and intermediate filaments. The enzymes that synthesize the cell wall are directed on similar intracellular matrices to those for the proteins that segregate the chromosomes during cell division.

Bacteria also undergo developmental processes, and some types are multicellular, with different cells playing clearly distinguishable roles. These developmental processes are rigorously programmed and require communication between different cells in the developing structure as development proceeds.

In this chapter we discuss a selection of the best-studied examples of cellular compartmentalization, communication, and development in bacteria and explain how molecular genetic analysis has contributed to our understanding of these phenomena. Some of these topics were touched on in earlier chapters; in this chapter we provide more details about the processes involved and the experiments which contributed to our understanding. These are certainly some of the most important areas of research in biological science, and the relatively malleable and simple bacteria offer the best hope of furthering our understanding of these striking manifestations of living organisms.

Analysis of Protein Transport in *Escherichia coli*

The process of how proteins leave the cytoplasm and move into the surrounding compartments has been studied most extensively in *Escherichia coli*, a gram-negative bacterium. As discussed in chapter 2, about one-fifth of the proteins made in a bacterium do not remain in the cytoplasm but are transported or exported into or through the surrounding membranes. Proteins that remain in either the inner or outer membrane are membrane proteins, while those that remain in the periplasmic space are periplasmic proteins. Proteins that are passed out of the cell into the surrounding environment are secreted proteins.

A number of exported proteins are discussed in other chapters. For example, the LamB protein resides in the outer membrane, where it binds polymers of maltose and serves as the receptor for λ phage. The β-lactamase enzyme that makes the cell resistant to penicillin resides in the periplasm and so must be exported through the inner membrane. The protein disulfide isomerases also reside in the periplasm and form disulfide linkages in some periplasmic or extracellular enzymes as these other proteins pass through the periplasm. The maltose-binding protein, MalE, also resides in the periplasm, where it can help transport maltodextrins into the cell, while MalF is in the inner membrane and helps form the channel through which maltodextrins pass into the cytoplasm. The *tonB* gene product of *E. coli* must also pass through the inner membrane to its final destination in the outer membrane, where it participates in transport processes and serves as a receptor for some phages and colicins. Sensor kinases such as EnvZ have regions that reside in all three inner compartments, with part in the periplasm, where it can sense the external environment, and a transmembrane region that passes through the inner membrane to another region in the cytoplasm, where they can communicate this information to response regulators in the cytoplasm.

The overall process of protein transport is outlined in chapter 2; therefore, it is only briefly reviewed here. Most proteins that are transported into the other compartments surrounding the cytoplasm use the SecYEG channel, sometimes called the translocon, to enter or pass through the inner membrane. However, the pathway that is used to target a protein to this channel depends on its final destination. Proteins which are transported into the periplasmic space and beyond to the outer membrane or to the outside of the cell generally use the SecB-SecA targeting pathway to target proteins to the SecYEG channel in the membrane. The SecB protein is a chaperone that binds to a short, somewhat hydrophobic signal sequence on the N terminus of the protein and prevents the protein from folding prematurely. The SecA protein then binds the protein to the SecYEG channel and, through cleavage of ATP, drives the unfolded protein into the channel. The short signal sequence is cleaved off by proteases as the unfolded protein passes through the SecYEG channel.

However, most proteins that are destined for the inner membrane of bacteria (and hence do not continue further into the surrounding compartments) are targeted by a different system, the signal recognition particle (SRP) system, which includes the Ffh protein, a small 4.5S RNA, and a membrane receptor or "docking protein," FtsY. At least some inner membrane proteins may also use SecA as well as another protein, YidC, to direct them to the SecYEG channel although their roles are not completely understood. Proteins targeted by the SRP system generally lack a cleavable signal sequence but have long hydrophobic transmembrane domains which traverse the membrane one or more times. Inner membrane proteins are not usually translated in their entirety before they enter a SecYEG channel. Usually the emergence of the most N-terminal of the transmembrane domains, and therefore the first to emerge from the ribosome, identifies the protein being translated as an inner membrane protein to bind to the FtsY docking protein and be targeted by the SRP system to the SecYEG channel. The protein is then translated as it is inserted into the membrane, a process called cotranslation. This cotranslation probably serves two purposes. The energy of translation due to cleavage of GTP to GDP probably drives the polypeptide into the SecYEG channel, obviating the need for ATP cleavage by SecA, except perhaps for the some of the largest proteins, which might get stuck otherwise. More importantly, cotranslation is necessary to prevent precipitation of these proteins in the cytoplasm. They are so hydrophobic that they would irreversibly precipitate if they were completely translated in the acqueous environment of the cytoplasm. However, by translating them as they are inserted into the membrane, the SRP system ensures that they remain soluble and then fold properly in the hydrophobic environment of the inner membrane.

It seems likely that the need for cotranslation of inner membrane proteins explains why there are two types of targeting systems in *E. coli*. The SecB-SecA system can target the much less hydrophobic outer membrane and secreted proteins after they are synthesized. They are much less hydrophobic than the inner membrane proteins and so do not irreversibly precipitate if they are translated in their entirety in the cytoplasm. However, the SRP system is needed to target the very hydrophobic inner membrane proteins and force their cotranslation to keep them from precipitating in the cytoplasm. We would predict that a bacterium could exist with only the SRP targeting system, since this system could target both types of proteins, even ones which are less hydrophobic, to the SecYEG channel. In fact, many types of bacteria and eukaryotes do seem to have only an SRP system and to lack the equivalent of the SecB-SecA targeting system.

It should be mentioned, however, that in spite of extensive research, there is still no direct evidence for obligatory cotranslation of SRP-targeted proteins in bacteria. It is not clear whether translation stops after the first transmembrane domain emerges from the ribosome and can continue only after the transmembrane domain binds to the FtsY docking protein on the SecYEG channel in the membrane or whether the binding is so rapid that it takes place anyway before the protein has been completely translated. The situation is clearer in eukaryotes, where cotranslation is known to be enforced by the SRP binding to the signal sequence (see Wild et al., Suggested Reading). The 54-kDa protein, which is analogous to Ffh in *E. coli* (see above), binds to the signal sequence as it emerges from the ribosome. Other proteins then bind to the 54-kDa protein, making the SRP large enough to extend all the way to the A site of the ribosome, where it can block incoming translation factor EF-1α, the eukaryotic equivalent of EF-Tu. The SRP can thus block translation until the ribosome binds to the docking protein on the Sec channel in the membrane. It is then released, and the protein is cotranslated with its insertion into the channel. It seems possible that something similar happens in bacteria but that the smaller SRP cannot extend all the way to the ribosome A site. Other proteins, perhaps with additional roles in the cell, may bind to Ffh and extend the effective size of the bacterial SRP; this allows it to bind to the hydrophobic N terminus of the protein when it first emerges from the channel in the ribosome and still reach all the way to the A site of the ribosome so that it can block further translation of the protein. It is known that archaea use a similar strategy to force cotranslation of their membrane proteins.

Use of the *mal* Genes To Study Protein Transport: Signal Sequences, *sec*, and SRP

Identification of the genes and proteins of the protein transport systems in *E. coli* involved some very elegant genetic experiments. These experiments used the genes of the maltose transport operons (*mal* operons) that transport maltose and maltodextrins, polymers of maltose, into the cell (see chapter 12). To fulfill their roles, the products of many of the genes of the *mal* operons are membrane or periplasmic proteins and so must be transported into or through the inner membrane. They had also been the subject of extensive earlier studies, making them a good basis for these studies. These studies continue to be very important in informing our view of how protein transport occurs in all organisms; the relevance of their conclusions to the mechanisms of protein transport (as discussed in chapter 2) is addressed here.

ISOLATION OF MUTATIONS THAT AFFECT THE SIGNAL SEQUENCE OF THE MalE AND LamB PROTEINS

As mentioned, the products of the *lamB* and *malE* genes reside in the outer membrane and periplasm of *E. coli*, respectively (see Figure 12.22). To reach their final destinations, these proteins must be exported through the inner membrane. Like most exported proteins, LamB and MalE are first made as precursor polypeptides with approximately 25 extra amino acids at their N-terminal ends, called the signal sequence (see chapter 2). The signal sequence consists of three domains, a short N-terminal domain with some positively charged (basic) amino acids, a middle or H domain with mostly hydrophobic amino acids, and a C-terminal domain that contains the consensus sequence for cleavage by the signal peptidase which cleaves the signal peptide off the protein as it traverses the inner membrane.

Gene fusion techniques and selectional genetics were used to determine which amino acids in the signal sequence of MalE and LamB were important for the secretion of these proteins (see Bassford and Beckwith, Suggested Reading). Mutations that cause amino acid changes in a short sequence such as a signal sequence are rare and require a positive selection. The method capitalized on translational fusions that joined the N terminus of the MalE or LamB protein, including the signal sequence, to the LacZ protein (β-galactosidase), which is normally a cytoplasmic protein (see chapter 2 for an explanation of translational fusions). The signal sequence on the N terminus of the fusion protein directs the fusion protein to the membrane, and the transport machinery attempts to export it. However, for unknown reasons, the large fusion protein becomes trapped in the membrane, killing the cells, perhaps by causing them to lyse.

The sensitivity of cells to transport of the fusion proteins offers a means of positively selecting transport-defective mutants (Figure 14.1). Since the fusion protein is made in large amounts only in the presence of the inducer maltose, which turns on transcription from the *mal* promoter (see Figure 12.21), bacteria containing these fusions are maltose sensitive (Mal^s) and are killed by the addition of maltose to the growth medium. However, bacteria with mutations that prevent transport of the fusion proteins are resistant to maltose (Mal^r) and survive. Therefore, any colonies that form when bacteria containing the gene fusions are plated on maltose-containing plates may be due to mutant bacteria that no longer can transport the fusion proteins, provided that they still make the fusion protein in large amounts. This can be checked by Western blotting using antibodies to the LacZ protein.

Some of the Mal^r mutants may have mutations that change the coding region for the signal sequence in the

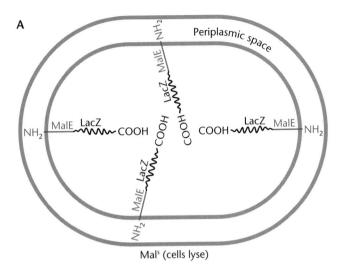

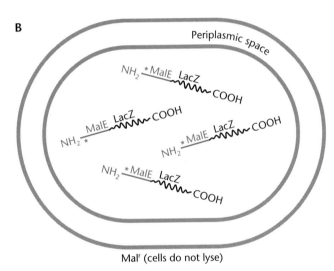

Figure 14.1 Model for the maltose sensitivity (Mal^s) of cells containing a *malE-lacZ* gene fusion. (A) In Mal^s cells, the presence of maltose induces the synthesis of the fusion protein, which cannot be transported completely through the membrane and so lodges in the membrane, causing the cell to lyse. (B) In Mal^r cells, a mutation in the region encoding the signal sequence (asterisk) prevents transport of the fusion protein into the membrane.

MalE or LamB portion of the fusion protein so that the signal sequence no longer functions in transport. Then the fusion protein no longer enters the membrane and kills the cell, and the cell becomes resistant to maltose. Determining which amino acids in the signal sequence have been changed in these Mal^r mutants should reveal which amino acids in the signal sequence are required for signal sequence function. These mutations can be distinguished by their map positions from other mutations that might be in genes that encode proteins required for secretion. Signal sequence mutations should map in the *malE-lacZ* fusion gene, while mutations that affect other proteins required for transport should map elsewhere in the *E. coli* genome. Furthermore, mutations that change the signal or some other sequence in the MalE or LamB part of the fusion protein should specifically affect the transport of the fusion protein and should not affect the transport of other transported proteins.

On the basis of this selection, a number of signal sequence mutations were isolated and later sequenced to determine which amino acid changes could prevent the function of the signal sequence. Many of these were changes from hydrophobic to charged amino acids in the H region of the signal peptide. These changes presumably interfere with the insertion of the signal sequence into the SecYEG channel in the hydrophobic membrane or its recognition by the SecB-SecA targeting system.

ISOLATION OF MUTATIONS IN *sec* GENES

The *mal* genes were also used to select mutants with mutations in genes whose products are part of the protein transport machinery (see Oliver and Beckwith, Suggested Reading). These were named the *sec* genes (for protein *sec*retion) because their products are required for the transport of some or maybe even all of the membrane, periplasmic, and exported proteins. Therefore, unlike signal sequence mutations, which affect the transport of only one protein, *sec* mutations should cause defects in the transport of many proteins into or through the membranes and map in many places on the genome.

A different, somewhat more sensitive selection was needed to isolate *sec* mutants compared with the selection used to isolate signal sequence mutations. This is because, unlike signal sequence mutations that affect the transport of only the fusion protein, *sec* mutations should affect the transport of many proteins, some of which might be essential. Therefore, only mutants with very leaky *sec* mutations might be viable, requiring a more sensitive selection. Their selection is illustrated for the selection of *secB* mutants in Figure 14.2A and is based on the observation that cells containing a particular *malE-lacZ* fusion do make some MalE-LacZ fusion protein, even in the absence of maltose in the medium, but do not make enough to kill the cells. Nevertheless, even though they do seem to make sufficient fusion protein to make the cells Lac⁺, the cells are Lac⁻ and unable to multiply with lactose as the sole carbon and energy source. The investigators reasoned that the cells may not exhibit the β-galactosidase activity because the fusion protein is being transported to the periplasmic space (Figure 14.2A), where the normally cytoplasmic β-galactosidase may be inactivated by the formation of disulfide bonds between its cysteines by disulfide isomerases in the periplasm.

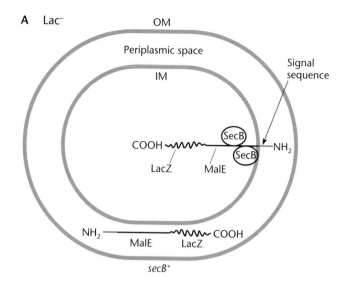

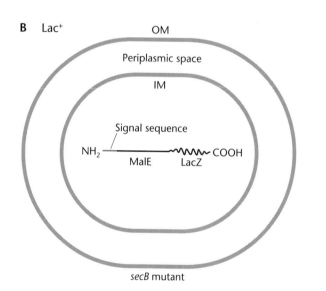

Figure 14.2 Function of SecB in protein transport and how *secB* mutations can be selected. (A) The SecB protein (black) prevents premature folding of a protein, keeping the signal sequence (purple) exposed so that it can enter the membrane after protein synthesis is complete. The fusion protein can then enter the periplasm, where the LacZ portion is not active. (B) In a *secB* mutant, the signal sequence folds into the interior of the protein and so the protein cannot be transported. The fusion protein stays in the cytoplasm, where the β-gal portion of the fusion is active. OM, outer membrane; IM, inner membrane.

The fact that transport of the fusion protein inactivates its β-galactosidase activity, thereby preventing growth on lactose, offers a positive selection for *sec* mutations that prevent its transport. Any mutation that prevents transport of the fusion protein through the membrane, for example a *secB* mutation, should make the cells Lac⁺ and able to form colonies on lactose minimal plates, because at least some of the MalE-LacZ fusion proteins should remain in the cytoplasm and retain their β-galactosidase activity (Figure 14.2B). Therefore, to isolate *sec* mutants, these investigators could merely plate cells containing the *malE-lacZ* fusion on minimal plates containing lactose but no maltose. They identified six different *sec* genes in this way and named them *secA*, *secB*, *secD*, *secE*, *secF*, and *secY*. As discussed above and in chapter 2, the *secY* and *secE* gene products are required to form a channel in the membrane through which other proteins pass. The *secG* gene product also forms part of this channel but is not absolutely required for protein transport through the channel and so was not selected by this procedure. The other *sec* genes encode the chaperones required to target such signal sequence-containing proteins to the membrane channel.

ISOLATION OF MUTATIONS IN THE SRP PATHWAY FOR INNER MEMBRANE PROTEINS

The *mal* genes were also used to isolate mutations in genes of the SRP pathway, which, as discussed, targets proteins to the inner membrane (see Tian and Beckwith, Suggested Reading). The method used to isolate mutations in genes that target proteins to the inner membrane is similar to that used above to isolate mutations in genes required for transport and signal sequence mutations. However, rather than using the MalE protein, which resides in the periplasm, they used the MalF protein, which resides in the inner membrane with some domains in the periplasm and some in the cytoplasm (see Figure 12.22) and therefore presumably uses the SRP system for its transport into the membrane. They fused the N-terminal coding region for MalF, including the first transmembrane domain and first periplasmic domain, to *lacZ* and introduced this gene fusion into a phage λ vector, which they then integrated into the chromosome. Now, however, rather than selecting for sensitivity to maltose or ability to grow on lactose as a sole carbon source, they looked for blue colonies on 5-bromo-4-chloro-3-indolyl-β-D-galactopyranoside (X-Gal) plates. Cells containing this fusion normally make white colonies on X-Gal plates because the MalF portion of the fusion protein is transported into the inner membrane, dragging the β-galactosidase portion of the fusion along with it, perhaps as far as the periplasm. Again, it is not clear why the β-galactosidase portion of the fusion is not active. If it is lodged in the membrane, it might not form the dimers required for activity; if it is in the periplasm, the β-galactosidase might be inactive because its cysteines become cross-linked, as discussed above. Whatever the

reason, the transport of the β-galactosidase in the fusion protein somehow makes it inactive, so that it cannot cleave the X-Gal on the plates and turn the colonies blue, and therefore they remain white.

These investigators reasoned that they might be able to use this system to isolate mutants of *E. coli* with mutations in genes required to transport proteins into the inner membrane. If a mutation in such a gene prevents the transport of the MalF portion of the fusion protein into the inner membrane so that it remains in the cytoplasm, the β-galactosidase to which it is fused also remains in the cytoplasm, where it remains active, making the mutant colonies blue on X-Gal plates. The mutants they isolated in this way had mutations in the gene encoding the 4.5S RNA part of the SRP particle, in the *ffh* gene, which encodes the *f*ifty-*f*our *h*omolog, so named because it is homologous to a 54-kDa protein in the eukaryotic SRP (see above), and in the *ftsY* gene, which encodes the docking protein. Surprisingly, the only mutations they found in a *sec* gene were in *secM*, whose product plays a role in regulating SecA. Why they did not find mutations in the *secY* or *secE* genes is unclear, since we now think that the products of these genes form the channel that also plays a role in transporting proteins targeted by the SRP system into the inner membrane.

SUPPRESSORS OF SIGNAL SEQUENCE MUTATIONS

Just as it was possible to isolate mutations that are defective in transport, it was also possible to isolate extragenic suppressors of these mutations, and these have been very informative as to mechanisms of protein secretion. In chapter 3 we gave some examples of suppressing mutations and how they are analyzed genetically. An extragenic suppressing mutation is a mutation in another gene that alleviates the effect of the first mutation. Generally, mutants with suppressing mutations are easy to isolate, especially if the original mutation causes a lethal phenotype under some conditions. Large numbers of the original mutant are plated under conditions where it is lethal; any colonies that arise either are revertants of the original mutation or have suppressing mutations. The two can be distinguished because mutants with a suppressing mutation still retain the original mutation. The kinds of mutations that can suppress the original mutation often give clues to the function of the original gene product and the other cellular proteins with which it interacts to perform its function. This information can complement structural information and give insights into pathways by which proteins and regions of proteins interact that can not easily be obtained from structural information alone. Such is the case for studies of suppressors of mutations that affect protein translocation in *E. coli*. These studies complemented the conclusions from the structure of the SecYEG channel (see Figure 2.40).

Isolation of Suppressors of Signal Sequence Mutations

The first suppressors to be isolated were those that suppressed mutations in the signal sequence of the *lamB* gene (see Emr et al., Suggested Reading). The hope was that these suppressor mutations would be in the *sec* genes and would reveal something about how the signal sequences interact with components of the Sec apparatus. Recall that these signal sequence mutations were selected because they prevented export of the LamB-LacZ fusion protein to the outer membrane and thereby prevented this fusion from being lethal in the presence of maltose, offering a positive selection for signal sequence mutations. Suppressors of these signal sequence mutations should restore export of the fusion protein and thereby restore the lethality of the fusion protein and make the cells Mal^s. However, this would be a negative selection, and they needed a positive selection for suppressors of these mutations, which were apt to be rare.

Fortunately, there was a way to select for mutations in which the export of LamB to the outer membrane is restored. The function of LamB is to bind maltodextrins (polymers of maltose) in the outer membrane and direct them to the maltose transport system. Therefore, in the absence of LamB in the outer membrane, the cells cannot grow on maltodextrins as a sole carbon and energy source (the Dex$^-$ phenotype). Consequently, if they crossed the signal sequence mutations into the normal *lamB* gene and used this mutated gene to replace the normal *lamB* gene in the chromosome, the LamB protein with the mutated signal sequence would not be exported, the cells would be Dex$^-$, and they would not grow on minimal plates with dextrins. Any mutants that formed colonies would be candidates for having suppressor mutations that allowed the export to the outer membrane of the LamB with the mutated signal sequence. However, the investigators still had another potential problem to overcome. Revertants of the orginal signal sequence mutation could also grow on maltodextrins, and these were apt to be more common. To eliminate revertants, the investigators used short in-frame deletion mutations of the signal sequence coding region rather than point mutations. Even very short deletion mutations seldom revert, so that suppressing mutations would be much more common than reversions.

Using this selection and others, a number of suppressing mutations of signal sequence mutations, called *prl* mutations, have been isolated and mapped. Some of the most interesting of these turned out to be in the *secY* gene (*prlA*) and the *secE* gene (*prlG*). These suppressor mutations could suppress even deletions of the entire signal sequence coding region, allowing the export of some LamB outer membrane protein that lacked a signal sequence altogether. This suggested that these mutations partially open the channel so that proteins can pass through the channel, even without a signal sequence. The existence of these suppressor mutations, combined with structural data on the SecYEG channel, contributed to the more detailed current models for how the SecYEG channel is opened to allow passage of proteins to be exported through the channel (see Figure 2.4). The model is that part of the SecY protein lies in the channel and forms a "plug" that normally blocks the channel. When the signal sequence binds to the channel, the plug moves over and binds to a region of SecE, which holds it as the protein passes through. Apparently, the suppressing mutations in *secY* and *secE* hold the plug open so that the signal sequence is no longer required to open the channel. The location of the suppressing mutations within *secY* and *secE* also helped identify the region in SecY that forms the plug and the region in SecE to which it binds to keep the channel open as the protein is coming through.

Constructing Double Mutants with *prl* Mutations: Synthetic Lethality

Combinations of suppressing mutations can also contribute information about the interactions of proteins within a structure that are not easily revealed by structural information alone. One prediction we could make is that putting different *prl* mutations together in the same double mutant could be lethal (see Smith et al., Suggested Reading). If *prl* mutations displace the plug and cause the SecYEG channel to remain partially open, even if no protein is being exported, we would predict that they could not open the channel very much, otherwise they would let too many molecules through, including small metabolites and proteins that should not be exported. However, putting two *prl* mutations together, both of which have the effect of partially opening the channel, could open the channel even further and be lethal. Putting different mutations together to see if the combination is lethal is called testing for **synthetic lethality**. Tests for synthetic lethality are discussed in chapter 1 in connection with isolation of mutants defective in nucleoid occlusion that are lethal only in the presence of another mutation, in that case one in a *min* gene.

Normally, double mutants are made by starting with a strain containing one mutation and then introducing the other mutation by site-specific mutagenesis. But how can you put two mutations together in the same strain if the combination is lethal? Generally, methods of site-specific mutagenesis mutagenize only a subset of the population, so how would we know that this subset had died? Screening a large number of strains with only the original mutation, until the absence of the double mutant becomes statistically significant, is not practical. A much easier way, which depends on the original mutation being dominant, is to start with a clone of the gene in an inducible expression vector, for example a pBAD plasmid vector, in which the cloned gene is expressed from the L-arabinose inducible promoter (see Figure 12.18). Because the *prl* mutation is dominant, the cell shows the Prl phenotype when expression of the clone is induced, even though the corresponding wild-type gene is in the chromosome. We would expect most *prl* mutations to be dominant because if any of the SecYEG channels are formed with a SecY (or SecE) subunit with the *prl* mutation, these channels would be partially open and allow protein through, even though the other channels composed of the wild-type SecY (or SecG) subunit are closed. The second mutation can then be introduced into the cloned gene in the plasmid or into a different chromosomal gene without inducing the expression of the cloned gene. If the cells die when the expression of the cloned gene is induced, by adding L-arabinose in this example, the double mutant is synthetically lethal. In this way it was determined that some combinations of *prlA* (*secY*) mutations are lethal when combined with other *prlA* mutations, as were some combinations of *prlA* and *prlG* (*secE*) mutations. It was concluded that the regions of these mutations interact in the SecYEG channel, and this was later confirmed by the structural studies.

The Tat Secretion Pathway

Not all proteins that are transported into or through the cytoplasmic membrane use the SecYEG channel, which is quite narrow. Only proteins that have not yet folded and so are still long, slender polypeptides can be transported by this narrow channel. Once folded into their final three-dimensional structure, proteins are much too wide to fit through this channel. However, some proteins must fold in the cytoplasm before they can be transported. Often these are membrane proteins that contain redox factors such as molybdopterin and FeS clusters, which are synthesized in the cytoplasm and can be inserted into the protein only after it has folded. These cofactors are not available in the membrane to bind to the proteins. Other examples of proteins that are transported after they are folded are some heterodimers in which only one member has a signal sequence, so the other partner is left behind if the signal sequence-containing polypeptide is transported before the two polypeptides have combined and folded.

At least most eubacteria and archaea as well as the chloroplasts of plants (remember that chloroplasts are descended from cyanobacteria [see Introduction]) have a Tat secretion system, although it might differ in the number of subunits. The Tat system from *E. coli* has three subunits, TatA, TatB, and TatC, that of while *Bacillus subtilis* has only two, with TatA and TatB seemingly combined into one larger subunit. Much less is known about how the Tat system works than about how the SecYEG translocase works. All three Tat subunits are membrane bound. The TatC subunit has a number of transmembrane domains and seems to form the channel in the membrane, perhaps with the assistance of TatB. The TatA subunit may be recruited only after the protein to be exported has bound to the channel. The electric field provided by the proton gradient across the membrane may provide the only energy for the transport.

THE Tat SIGNAL SEQUENCE

Proteins that are to be transported by the Tat pathway also have a signal sequence that is cleaved off while they are transported. This signal sequence is somewhat longer and less hydrophobic than the sequences of proteins transported by the SecYEG channel. They also have a characteristic motif at the N terminus, S-R-R-X-, where the X can be any amino acid. The presence of the two arginines (R) in this motif give the transport system its name, the *t*win *a*rginine *t*ransport system, although the first of the twin arginines is sometimes a lysine (K). This sequence is followed by two hydrophobic amino acids, usually F and L, and then often by a K.

The presence of this particular signal sequence at the N terminus of a newly synthesized protein signals that this is a protein to be transported by the Tat system rather than by the SecYEG channel. This raises an interesting question. How does the system know the protein has already folded properly and contains all the needed cofactors, etc., so that it is time to transport it? If anything, the signal sequence should be more accessible in the polypeptide before rather than after it has folded. The Tat system needs a "quality control" system to ensure that it transports only properly folded proteins and does not transport proteins that are unfolded or only partially folded. This quality control system should also be specific for each protein to be transported, since each type of folded protein has a unique structure (see chapter 2).

The *E. coli* cell seems to solve this problem by encoding proteins that specifically bind to the Tat signal sequence of only one type of protein and come off only when that protein has folded properly (see Palmer et al., Suggested Reading). For example, the HyaE protein may play this role for HyaA, the small subunit of hydrogenase 1, and a different protein, HybE, may interact with the signal sequence of the small subunit HybO of hydrogenase 2. Interestingly, in the latter case the HybE protein may also interact with the large subunit of hydrogenase 2, HybC, which lacks a signal sequence. The hydrogenases are heterodimers composed of the two subunits and an NiFe cofactor. When the protein folds, the interaction with the newly acquired large subunit in the proper position may cause the HybE protein to come off the signal sequence, indicating that the two subunits of hydrogenase 2 have already come together, folded, and bound the NiFe cofactor and that hydrogenase 2 is ready to be transported.

Genetic Analysis of Transmembrane Domains of Inner Membrane Proteins in Gram-Negative Bacteria

As mentioned, most proteins transported out of the cytoplasm are destined for the cytoplasmic membrane, surrounding the cytoplasm. Rather than being completely buried in the membrane, most cytoplasmic membrane proteins have regions exposed to the cytoplasm. They also have regions exposed on the other side of the membrane, which is the periplasm in gram-negative bacteria and the external environment in gram-positive bacteria. Having exposed regions on both sides of the membrane allows the membrane protein to pass information from the external environment to the interior of the cell or from one cellular component to another. Proteins that are exposed at both surfaces of a membrane are called transmembrane proteins, and the regions of the polypeptide that traverse the membrane from one surface to the other are called the transmembrane domains. Some transmembrane proteins traverse the membrane many times. The transmembrane domains alternate with those exposed to the cytoplasm (cytoplasmic domains) and the periplasm (periplasmic domains). The term **membrane topology** refers to the way the different sections of the protein are distributed in the membrane and in the external and internal compartments.

The transmembrane domains of a polypeptide can often be distinguished from the periplasmic or cytoplasmic domains merely by the primary sequence of the gene. Because the transmembrane domains are embedded in the hydrophobic membrane, they are composed mostly of hydrophobic amino acids such as phenylalanine, leucine, and methionine (see inside front cover), which makes them more soluble in hydrophobic environments. The periplasmic and cytoplasmic domains have a larger number of charged and polar amino acids, such as arginine, glutamic acid, or asparagine, which make them more soluble in the ionic environments on either side of the membrane. Sometimes it is possible to guess which regions are in the various compartments just by counting the transmembrane domains, if you know that one domain has to

be either in the cytoplasm or outside the membrane. However, whether a domain is in the periplasm or the cytoplasm cannot be absolutely determined from the amino acid sequence alone. For example, the plug domain of the SecY subunit of the SecYEG channel was originally thought to reside in the periplasm, based on such considerations. However, later it was found to extend into the relatively hydrophilic channel to form the plug.

Translational fusions to the alkaline phosphatase gene (*phoA*) of *E. coli* have been used to identify genes that encode transmembrane proteins and to study the membrane topology of inner membrane proteins of some gram-negative bacteria (see Hoffman and Wright and San Milan et al., Suggested Reading). These methods can be made to work with *E. coli* and some other closely related gram-negative bacteria. The alkaline phosphatase product of the *phoA* gene is a scavenger enzyme, which cleaves phosphates off larger molecules so that the phosphates can be transported into the cell to be used in cellular reactions. To fulfill this role, the alkaline phosphatase resides in the periplasmic space, where it can obtain phosphates from molecules, even if they are too large or too ionic to be easily transported.

The property of *E. coli* alkaline phosphatase which makes it so useful for studying the membrane topology of proteins is that it is active only in the periplasm and not in the cytoplasm. To be active, the enzyme must form a homodimer of two identical polypeptide products of the *phoA* gene and the two monomers in the dimer must be held together by disulfide bonds between their cysteines. Disulfide bonds form only in the oxidizing environment of the periplasm, where the protein disulfide isomerase enzymes that form disulfide bonds reside, and not in the reducing environment of the cytoplasm (see chapter 2). Therefore, the PhoA enzyme is active only if it is in the periplasm. The PhoA enzyme is also easy to assay, and bacteria which synthesize active PhoA make blue colonies on plates containing the chromogenic compound 5-bromo-4-chloro-3-indolylphosphate (XP), which turns blue when the phosphate is cleaved off by alkaline phosphatase.

The way *phoA* translational fusions can be used to determine the domains of a transmembrane protein that are in the periplasm is illustrated in Figure 14.3. Briefly, the carboxy-terminal coding region for PhoA, without its signal sequence, is fused to various lengths of the coding sequence for the N terminus of the transmembrane protein. If the region of the protein to which PhoA is fused is in the periplasm, i.e., is a periplasmic domain, the PhoA part of the fusion is also in the periplasm and active so that cells containing the fusion make blue colonies on XP plates. However, if the region of the protein is a cytoplasmic domain, the colonies are colorless.

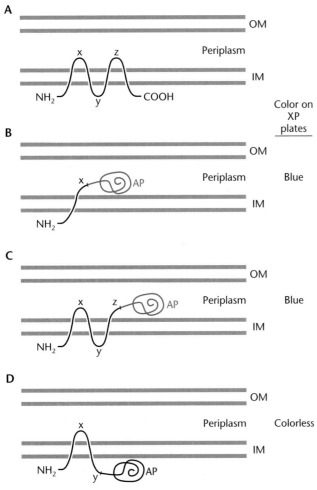

Figure 14.3 Using *phoA* fusions to determine the membrane topology of a transmembrane protein. (A) Transmembrane protein, showing both periplasmic and cytoplasmic domains. (B and C) A fusion that joins the transmembrane protein to alkaline phosphatase (AP) at x or z leaves the alkaline phosphatase in the periplasm, where it is active. The bacteria form blue colonies on XP plates. (D) The transmembrane protein is fused at y to alkaline phosphatase, leaving the alkaline phosphatase in the cytoplasm, where it is inactive. The bacteria form colorless colonies on XP plates.

Identification of Genes for Transported Proteins by Random *phoA* Fusions

The fact that PhoA is active only in the periplasm allows it to be used to identify inner membrane proteins in gram-negative bacteria. Some transposons have been engineered to generate random gene fusions by transposon mutagenesis (see, for example, the discussion of Mud*lac* in chapter 9). These transposons contain a reporter gene that is expressed only if the transposon hops into an expressed gene in the correct orientation. One such transposon, Tn*phoA*, was developed to identify genes whose

protein products are transported into or through the inner membrane (see Gutierrez et al., Suggested Reading). Tn*phoA* has the *phoA* reporter gene inserted so that it does not have its own promoter or translational initiation region and lacks its own signal sequence coding region. A fusion protein with PhoA fused to another protein is synthesized whenever the transposon hops into an expressed open reading frame (ORF) in such a way that PhoA is translated in the right reading frame. However, the PhoA part of the fusion protein has alkaline phosphatase activity and turns colonies blue on XP plates only if the Tn*phoA* has integrated into a gene whose protein product is translocated and the *phoA* gene happens to be fused to a periplasmic domain of the protein. Obviously, such insertions are rare, but blue colonies can be easily spotted, even on plates with crowded colonies.

Protein Secretion

So far we have discussed the transport of proteins into the cytoplasmic membrane or through the membrane into the surrounding structures or environment. All types of cells have a cytoplasmic membranes, and their functions are highly conserved. For example, the SecYEG channel is found in a related form in all cells—eubacterial, archaeal, and eukaryotic—and is one of the most highly evolutionarily conserved functions in cells. The Tat transport system seems to be somewhat less widespread but exists in chloroplasts and at least most gram-negative and gram-positive eubacteria. However, some proteins are transported through the membranes to the outside of the cell, where they can remain attached, enter the surrounding medium, or even enter another cell. As mentioned, this process is called protein secretion and differs between gram-negative and gram-positive bacteria because of the different structures that surround their cells. The gram-negative bacteria are surrounded by an outer membrane outside their cell wall, creating a periplasmic space, while the gram-positive bacteria have only one membrane, with a thicker cell wall layer outside the membrane. As mentioned, it is a long-running dispute whether gram-positive bacteria have something resembling a periplasmic space between the cell wall and the cytoplasmic membrane; however, if such a space exists, it is much narrower than the periplasmic space of gram-negative bacteria and is not surrounded by a highly hydrophobic lipid bilayer outer membrane.

Because of their different surrounding structures, gram-negative and gram-positive bacteria face different challenges when secreting a protein. In gram-negative bacteria, once a protein has been translocated through the inner membrane it is only in the periplasm and still faces the challenge of getting through the extremely hydrophobic outer membrane. In gram-positive bacteria, once it is through the inner membrane it is essentially outside the cell. Because of the additional challenge created by the outer membrane, gram-negative bacteria have developed elaborate specialized structures to get secreted proteins through the outer membrane. These structures are discussed in the next section.

Protein Secretion Systems in Gram-Negative Bacteria

The protein secretion systems of gram-negative bacteria come in five basic types, imaginatively named types I to V. All of these secretion systems rely on channels in the outer membrane (called β-barrels) formed from β-sheets organized in a ring. It is a curiosity that hydrophobic regions of outer membrane proteins are often arranged in β structures while hydrophobic transmembrane domains in the inner membrane are usually organized in α-helices (see Figure 2.23 for an explanation of protein secondary structures). The β-barrels are assembled so that the side chains of charged and polar amino acids tend to be in the center of the barrel, where they are in contact with hydrophilic proteins that are passing through, while the side chains of hydrophobic amino acids are on the outside of the barrel in contact with the very hydrophobic surrounding membrane. The barrels are normally closed at one or both ends, but they open to allow passage of the secreted protein (see below).

Having channels in the outer membrane presents some of the same problems associated with having channels, such as the SecYEG channel, in the cytoplasmic membrane. For example, how do they select some proteins to go through and others to keep in, as well as keeping small molecules out? This process is called **channel gating**: the gate is open only when the protein being secreted passes through. They also have other unique problems. Where does the energy come from to secrete a protein through the outer membrane? There is no ATP or GTP in the periplasmic space to provide energy, and the outer membrane is not known to have a proton gradient across it to create an electric field. Also, how do they themselves get through the inner membrane to reach the outer membrane? And once there, can they assemble themselves into channels, or are other proteins involved? Not all of these questions have been completely answered, but in this section we try to address possible mechanisms used by the various secretion systems for solving these and other problems. We also mention some examples of proteins secreted by each of the systems.

TYPE I SECRETION (T1S) SYSTEMS

Type I secretion systems secrete a protein directly from the cytoplasm to the outside of the cell (Figure 14.4).

They are different from members of the other types of secretion systems and more closely related to a large family of ATP-binding cassette (ABC) transporters that export small molecules, including antibiotics and toxins, from the cell. These ABC transporters tend to be more dedicated, exporting only certain molecules from the cell. The dedicated part of the system consists of two proteins: an ABC protein in the inner membrane and an integral membrane protein that bridges the inner and outer membranes. They then use a multiuse protein, TolC, that forms the β-barrel channel in the outer membrane to get molecules through the outer membrane. Because the TolC channel has other uses and also exports other molecules including toxic compounds from the cell, it is recruited to this system only when the specific protein is to be secreted. When the molecule to be secreted binds to the ABC protein, the integral membrane protein recruits the third protein, TolC, which forms a β-barrel in the outer membrane through which the molecule can pass. The cleavage of ATP by the ABC protein presumably provides the energy to push the molecule all the way through the TolC channel to the outside of the cell.

The classical example of a protein secreted by a type I secretion system is the HylA hemolysin protein of pathogenic *E. coli*. This toxin inserts itself into the plasma membrane of eukaryotic cells, creating pores that allow the contents to leak out. It also has its own dedicated type I secretion system composed of HylB (the ABC protein) and HylD (the integral membrane protein), which secretes it through the membranes. Because HylA is not transported through the inner membrane by either the Sec system or the Tat system, it does not contain a cleavable N-terminal signal sequence. Instead, like all proteins secreted by type I systems, it has a sequence at its carboxyl terminus that is recognized by the ABC transporter but is not cleaved off as the protein is secreted.

Another well-studied protein secreted by a type I secretion system is the adenylate cyclase toxin of *Bordetella pertussis*. This toxin enters eukaryotic cells and makes cyclic AMP, thereby disrupting their signaling pathways. The use of the pertussis adenyl cyclase in bacterial two-hybrid selections is discussed in chapter 13.

The TolC channel has been crystallized and has had its structure determined (see Koronakis et al., Suggested Reading). This structure has provided interesting insights into the structure of β-barrels and how they can be gated and opened to transport specific molecules. Briefly, three TolC polypeptides come together to form the channel through the outer membrane. Each of these monomers contributes four transmembrane domains to form a β-barrel that is always open on one side of the outer membrane, the side on the outside of the cell. In addition,

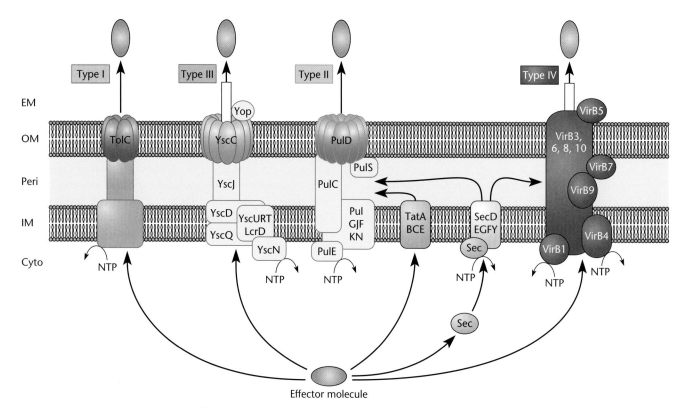

Figure 14.4 Schematic representation of the type I, II, III, and IV protein secretion systems. The examples shown are for type I (hemolysin A [HlyA] of *E. coli*), type II (pullulanase of *Klebsiella oxytoca*), type III (Yop of *Yersinia*), and type IV (*vir* of *Agrobacterium tumefaciens*). EM, extracellular milieu; OM, outer membrane; Peri, periplasmic space; IM, inner membrane; Cyto, cytoplasm. Arrows indicate which pathways use the Sec and Tat pathways through the inner membrane. Purple indicates the secretin-formed channels.

each monomer has four longer α-helical domains that are long enough to extend all the way across the periplasm. These four α-helical domains contribute to the formation of a second channel that is aligned with the first channel and traverses the periplasm. Because of these two channels, the secreted protein can be transported all the way from the inner membrane to the outside of the cell. In addition, the channel in the periplasm can open and close and therefore "gate" the channel. The periplasmic channel remains closed, perhaps because the α-helical domains are twisted, until the molecule to be exported binds to the ABC protein. When the molecule to be exported binds to the ABC transporter in the cytoplasm, the TolC channel is recruited and the α-helical domains of the periplasmic channel may rotate, which untwists them and opens the gate on the periplasmic side. The molecule is then secreted all the way through both channels to the outside of the cell.

TYPE II SECRETION (T2S) SYSTEMS
We have already mentioned **type II secretion systems** because of their relationship to some competence systems

(see chapter 6). They are also closely related to the systems that assemble type IV pili on the cell surface (see below). Type II secretion was originally called the main terminal branch of the Sec secretion pathway because it was thought that all gram-negative bacteria have them. However, it is now known that, even though very common, they are not universally shared among gram-negative bacteria.

Some examples of proteins secreted by type II secretion systems are the pullulanase of *Klebsiella oxytoca* and the cholera toxin of *Vibrio cholerae*. The pullulanase degrades starch, and the cholera toxin is responsible for the watery diarrhea associated with the disease cholera, illustrating the variety of proteins secreted by these systems. In the case of the cholera toxin, after transport by the SecYEG channel, one A and five B subunits of the toxin assemble in the periplasm, from where they are secreted through the secretin channel into the intestine. The associated B subunit then assists the A subunit into mucosal cells, where the A subunit ADP-ribosylates a protein that regulates the adenylate cyclase (see chapter 13). This disupts the signaling pathways and causes diarrhea. Because the B subunits of the toxin must help the A subunit to enter the eukaryotic cell, the subunits must be associated with each other before they are secreted from the bacterium, otherwise they would be unable to find each other.

Type II secretion systems are very complex, consisting of as many as 15 different proteins (Figure 14.4). Most of these proteins are in the inner membrane and periplasm, and only one is in the outer membrane, where it forms a β-barrel that is the channel through which secreted proteins pass. This outer membrane protein is a member of a family of outer membrane proteins called **secretins,** which are related, among others, to the protein that forms the β-channel in type III secretion systems (see below). It is thought that 12 of the secretin proteins come together to form a large β-barrel with a pore large enough to pass already folded proteins. The formation of this channel is not spontaneous but requires the participation of normal cellular lipoproteins that might become part of the structure. The secretin protein has a long N terminus that might extend all the way through the periplasm to make contact with other proteins of the type II secretion system in the inner membrane. This periplasmic portion of the secretin may also gate the channel, as with the TolC channel.

Even though many of the components of the type II secretion systems are in the inner membrane, they use either the SecYEG channel or the Tat pathway to get their substrates through the inner membrane. Therefore, proteins secreted by this system have cleavable signal sequences at their N terminus, either the Sec type or the Tat type. Once in the periplasm, the proteins usually fold, if they have not already, before they are secreted through the outer membrane. Some of the periplasmic and inner membrane proteins of the secretion system are related to components of pili and have been called pseudopilin proteins even though they do not normally appear outside the cell (see chapter 6). It has been proposed that the formation and retraction of these pseudopili works like a piston to push the protein through the secretin channel in the outer membrane to the outside of the cell. In this way the energy for the secretion could come from the inner membrane or the cytoplasm, as shown in the figure since, as mentioned, there is no source of energy in the periplasm. In support of this model, the pseudopili have been seen to produce pili outside an *E. coli* cell when the gene for a pilin-like protein was cloned and overproduced in *E. coli*.

TYPE III SECRETION (T3S) SYSTEMS

The **type III secretion systems** are probably the most dramatic of the secretion systems in gram-negative bacteria. They are composed of about 20 proteins that form a syringe-like structure which takes up virulence proteins called effectors from the cytoplasm of the bacterium and injects them directly through both membranes into a eukaryotic cell (Figure 14.4). For this reason, they are sometimes called **injectisomes.** One striking feature of type III secretion systems is how similar they are in both animal and plant pathogens. They exist in almost the same form in many gram-negative animal pathogens, including *Salmonella* and *Yersinia*, but are also found in many plant pathogens including *Erwinia* and *Xanthomonas*. In all these bacteria, the parts of the secretion systems involved in getting the secreted protein through the bacterial membranes are very similar. Where they differ is in the protuberance called the needle that penetrates the eukaryotic cell wall to allow injection through the wall into the host cell cytoplasm. Animal and plant cells are surrounded by very different cell walls, and so the needle of a syringe that can penetrate the membrane of a mammalian cell would be expected to be very different from a needle that can penetrate a plant cell wall.

Type III secretion systems are usually encoded on pathogenicity islands, and their genes are induced only when the bacterium encounters its vertebrate host or under comparable conditions. They then induce the genes for the injectisome and assemble it through the cell membranes. The proteins they inject, called effectors, are also encoded by the same DNA element, and their genes are turned on at the same time. The part of the injectisome that traverses the outer membrane is composed of a secretin protein related to those of type II secretion systems. It also forms a β-barrel composed of about a dozen secretin subunits. Like the secretins of type II systems, these might require normal bacterial lipoproteins to assemble the channel in the membrane, but these lipoproteins might not remain as part of the barrel like they do in type II systems. They might also require other components of the secretion machinery to assemble. Type III secretion systems are related to the flagellar motor that drives bacterial movement in liquid media (Box 14.1).

The identifying mark of proteins to be secreted by at least some type III systems is a short sequence located on the N terminus of the protein, as it is for the *sec* and Tat systems, but this signal is not cleaved off when the protein is injected. Some of them may even use a sequence at the 5′ end of the messenger RNA (mRNA) encoding the protein to drag it to the injectisome to be secreted as it is translated (see Anderson and Schneewind, Suggested Reading), although this is still controversial.

Many of the proteins secreted into eukaryotic cells by type III secretion systems are involved in subverting the host defenses against infection by bacteria. This can be illustrated by *Yersinia pestis*, the bacterium that causes bubonic plague and in which type III secretion systems

BOX 14.1

Secretion Systems and Motility

Type III secretion systems are structurally related to the flagellar motors that provide motility to many bacteria in liquid environments (see Blocker et al., below). Not only are they structurally related, but also some flagellar systems may play a role in secreting some virulence proteins. It makes sense since these two systems are superficially similar. Flagellar motors have appendages that extrude from the cell surface but in the form of flagella rather than the needle of the type III secretion apparatus. They also consist of a motor buried in the membranes that rotates the flagella, and these proteins are structurally related to the syringe-forming proteins of the type III secretion systems. However, the major functions of these systems are very different, as reflected in their structures. A number of flagella are clustered on one end of the cell. If they rotate in one direction, counterclockwise, the flagella wrap around each other (bundle), they all turn in the same direction, and the cell moves forward in a straight line. If they rotate in the other direction, clockwise, the flagella separate and the cell moves in circles without any defined direction (tumble). The direction in which the flagella rotate depends on the state of methylation of a number of protein receptors in the inner membrane called MCPs (methyl-accepting chemotaxis proteins), which bind attractants such as amino acids that might provide a food source (see Bray, below). The state of methylation of these receptors is determined by how much attractant is bound, and this level is adjusted every 3 or 4 s. If the level of methylation is not consistent with how much attractant is bound, it means that the cell is moving up or down a gradient of the attractant. This information is communicated down a phosphorelay system called the Che proteins to a component of the flagellar motor, which then continues rotating counterclockwise (if it is moving up a gradient and should continue in the same direction) or clockwise (if is moving down a gradient and should try another direction).

The relationship between flagella and type III secretion systems is not the only known example of a secretion system that has been adapted to provide motility to the cell; type II secretion systems have been also been adapted to assemble type IV pili on the cell surface (see the text). These pili provide motility on solid surfaces to some bacteria, including *Myxococcus xanthus*, by extending and contracting and thus pulling the cell along. They are on the front end of the cell, where they can fulfill their pulling role, while flagella are on the rear end, where they can push. Thus, secretion systems, which have extensions that extend through the outer membrane, seem particularly adaptable to providing motility.

References

Blocker, A., K. Komoriya, and S. Aizawa. 2003. Type III secretion systems and bacterial flagella: insights into their function from structural similarities. *Proc. Natl. Acad. Sci. USA* **100**:3027–3030.

Bray, D. 2002. Bacterial chemotaxis and the question of gain. *Proc. Natl. Acad. Sci USA* **99**:7–9.

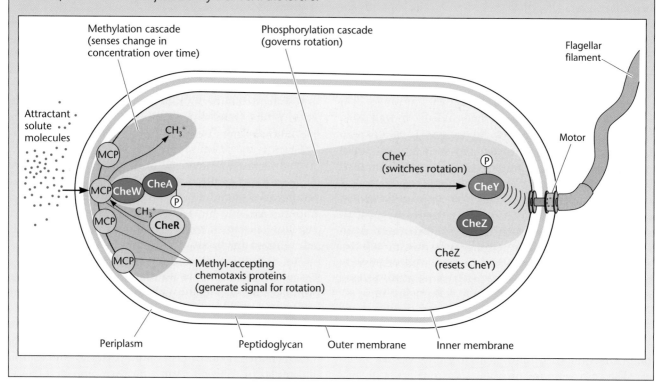

were first discovered. In animals, one of the first lines of defense against infecting bacteria are the macrophages, phagocytic white blood cells that engulf invading bacteria and destroy them by emitting a burst of oxidizing compounds. However, when a macrophage approaches a *Yersinia* cell, the bacterium injects effectors called Yop proteins into the macrophage cell before it can be engulfed. Once in the eukaryotic cell, these effectors can disarm the cell by interfering with its signaling systems and thus diverting the macrophage from its purpose of engulfing the bacterium. The effector proteins that are injected have a remarkably intimate understanding of how the eukaryotic cell works. For example, one of the Yop proteins is a tyrosine phosphatase, which removes phosphates from proteins in a signal transduction system in the macrophage, blocking the signal to take up the bacterium and preventing the burst of oxidizing compounds that kill the bacterium. Some bacteria even inject proteins called intimins that provide receptors on the cell surface to which the bacterium can absorb in order to enter the eukaryotic cell. This allows them to open the door into the cell from the inside, if the eukaryotic cell has not been nice enough to provide a receptor on its surface to which the bacterium can bind.

Plants use very different defense mechanisms against bacteria, and so plant pathogens have to adapt their strategy accordingly. Plants defend themselves against infection by inducing necrosis or destruction of the infected tissue and inducing phenolic compounds which destroy the bacterium. This is called a hypersensitive response, and it is induced by proteins called Avr (for *avirulence*) proteins that are injected into the plant cell by the type III secretion system. In a susceptible plant, these Avr proteins do not elicit the hypersensitive response.

TYPE IV SECRETION (T4S) SYSTEMS

Type IV secretion systems are discussed in chapters 5 and 6 because they are also involved in DNA transfer during conjugation and transformation. Like type III systems, they can inject proteins through both membranes directly into eukaryotic cells, although there could be some exceptions that use the Sec system to get through the inner membrane (see below).

As discussed in chapter 5, the T-DNA transfer system of *Agrobacterium tumefaciens* has served as the prototype of type IV secretion systems and is the one about which the most is known. Accordingly, the genes and proteins of other type IV systems are numbered after their counterparts in the T-DNA transfer system, named the *vir* genes because of their virulence in plants. Recall that this system transfers part of the Ti plasmid, called the T-DNA, directly into plant cells. The T-DNA is attached to a relaxase protein that directs the T-DNA

into the nucleus of the plant cell, where it integrates into the plant DNA. The T-DNA has plant-like genes that encode plant hormones which cause growth of the plant cell, leading to the formation of tumors called crown galls. In addition to the T-DNA, this system directly injects proteins into the plant cell, which makes it a bona fide protein secretion system.

The structure of the T-DNA transfer system is also shown in Figure 14.4. VirB9 is a secretin-like protein that forms a β-barrel channel in the outer membrane and extends into the periplasm, where it makes contact with proteins in the inner membrane. However, unlike true secretins, it seems to require another outer membrane protein, VirB7, to make a channel. The VirB9 protein is covalently attached to the VirB7 protein, which is, in turn, covalently attached to the lipid membrane, making the structure very stable.

Type IV secretion systems work through a coupling protein, named VirD4 in the T-DNA transfer system, that binds proteins to be secreted. The coupling protein then allows them into the channel. Therefore, to be secreted, a protein must bind to this coupling protein, ensuring that only certain proteins are secreted. These proteins presumably have a short domain that specifically binds the coupling protein but has been identified in only a few cases, including some relaxases and the VirB-VirD4 secretion system of *Bartonella* spp. (see below). The energy of secretion probably comes from the cleavage of ATP or GTP in the cytoplasm by some channel-associated proteins (Figure 14.4).

Even though type IV secretion systems are related to conjugation systems of plasmids, some are more closely related than others. For example, the type IV secretion system of *Helicobacter pylori* has extensive homology and the same order of genes as the *tra* genes of pKM101 (Box 5.2 in chapter 5). However, the winner so far of the lookalike contest is the Trw system of *Bartonella* (see Schroder and Dehio, Suggested Reading). This bacterium, the causative agent of trench fever and some other diseases, has two type IV secretion systems: one, VirB-VirD4, mentioned above, which is highly homologus to the T-DNA transfer system of *Agrobacterium*, and another, Trw, which is carried on a pathogenicity island and is highly homologous to the *tra* system of the R388 plasmid of *E. coli*. In fact, the latter two systems, one from a type IV secretion system involved in pathogenesis and the other involved in plasmid transfer, are so similar that their genes are given the same names. The order of genes in the two systems is almost the same and can be as much as 80% identical in some cases. Even the regulatory genes *korA* and *korB* are very similar in the two systems. One difference is that some genes are duplicated many times, in slightly different forms, in the Trw type IV secretion system of *Bartonella*. Interestingly, these are genes whose products probably form extracellular pilus components. Expressing so many different variations of these pili may allow *Bartonella* to bind to different tissues or to evade the host immune systems. Another difference is that the *Bartonella* Trw system seems to lack a coupling protein, raising concerns that this system might be inactive. However, the Trw system is required for pathogenicity, which is presumably because of an active role in secreting effector proteins. Perhaps it can use the coupling protein of the VirB-VirD4 system much like mobilizable plasmids can use the coupling protein of self-transmissible plasmids (see chapter 5). Another possibility is that it uses a different system such as the Sec system to get through the inner membrane, which is also thought to be the case for the pertussis toxin.

TYPE V SECRETION (T5S) SYSTEMS: AUTOTRANSPORTERS

All of the secretion systems discussed above use some sort of structure formed of β-sheets assembled into a ring called a β-barrel to get them through the outer membrane. Some of these are part of the secretion apparatus itself, while some, like TolC, are recruited from other functions in the cell. However, some secreted proteins do not take it for granted that they will find a β-barrel in the outer membrane to allow them through to the outside of the cell when they get there. They carry their own β-barrel with them in the form of a domain of the protein that can create a β-barrel when it gets to the outer membrane. These proteins are called **autotransporters** because they transport themselves and do not depend on preexisiting secretion systems. The prototypical autotransporter is the immunoglobulin A protease of *Neisseria gonorrhoeae*, which is typical of most autotransporters. It is involved in evading the host immune system by cleaving antibodies. Most known autotransporters are large virulence proteins, such as toxins and intimins, that perform various roles in bacterial pathogenesis or in helping evade the host immune system.

The mechanism used by autotransporters is illustrated in Figure 14.5, which also shows the basic structure of most autotransporters. Autotransporters consist of four domains, the translocator domain at the C terminus that forms a β-barrel in the outer membrane, an adjacent flexible linker domain (not shown) that may extend into the periplasm, a passenger domain that contains the functional part of the secreted protein, and sometimes a protease domain that may cleave the passenger domain off of the translocator domain after it passes through the transporter channel.

Autotransporters are typically transported through the SecYEG channel, so they have a signal sequence that is cleaved off as they pass into the periplasm. Their translocator domain then enters the outer membrane, where it forms a β-barrel. The flexible linker domain then guides the passenger domain into and through the channel to the outside of the cell. The passenger domain can then be cleaved off by its own protease domain or remain attached to the translocator domain and protrude outside the cell, depending on the function of the passenger domain.

In spite of this simple picture, some questions remain about autotransporters. The first is the question of where the passenger domain folds. This question is related to the size of the pore formed by the transport domain; the channel formed by a single translocator domain would be too small to accommodate a folded passenger domain, much less the linker domain, if this is guiding it into the pore. Perhaps the passenger domain folds on the outside of the cell once it is secreted. However, this would require that it have the capacity to fold spontaneously in the hostile environment outside the cell without the help of other cellular constituents. One way out of this dilemma is to propose that a number of translocator domains come together to form a larger pore, with each monomer contributing some β-structure domains to the larger β-barrel. Passenger domains could then pass through this shared larger β-barrel even if they had already folded. There is some structural evidence for such shared pores, at least for some autotransporters. There is also the question of where the energy comes from for autotransportation, since, as mentioned, there is no ATP or GTP in the periplasm and the outer membrane does not have a membrane potential. Perhaps the autotransporter arrives at the periplasm in a "cocked" or high-energy state that drives its own transport. Such an explanation has been proposed for the chaperone usher transport of some types of pilin proteins (see below).

Two-Partner Secretion

In a variation on autotransporters, sometimes the β-barrel-forming domains and the passenger domains are on different polypeptides. This has been called **two-partner secretion** (TPS) and is found in a large variety of the gram-negative proteobacteria, where it is largely responsible for secreting large toxins, much like the AT pathway. The two-partner polypeptides, called TpsA and TpsB, are transported separately through the inner membrane and the TpsB protein forms a β-barrel in the outer membrane. The TpsA protein, the equivalent of the passenger domain in autotransporters, interacts with the TpsB protein at the periplasmic side of the channel and is secreted through the channel, where it can either remain associated with the cell or enter the surrounding medium. The TpsB protein is highly specific for its TpsA

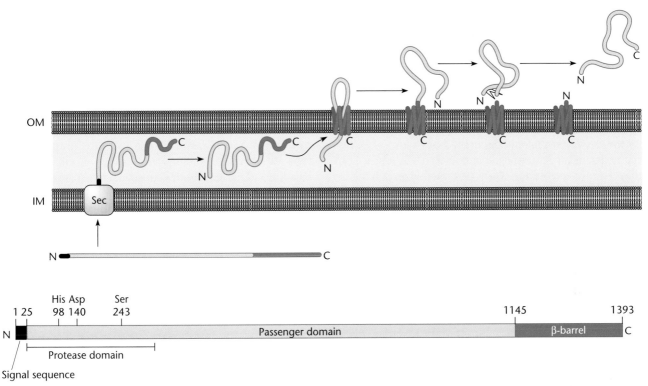

Figure 14.5 Structure and function of a typical autotransporter. Shown is a *Haemophilus influenzae* adhesin; the length in amino acids of each domain, where known, is indicated by the number above the structure, as are some of the important amino acids in the protease domain. The transporter domain at the C terminus that forms a β-barrel in the outer membrane is shown in dark purple; the passenger domain and the protease domain that cleaves the passenger domain off the transporter domain outside the cell are shown in light purple. The flexible linker domain is not indicated. The signal sequence that is cleaved off when the protein passes through the SecYEG (Sec) channel in the inner membrane is shown in black. OM, outer membrane; IM, inner membrane.

partner and secretes no other partner proteins. It also contains motifs that may participate in the processing and folding of its partner protein. This is another case where the source of the energy for secretion is not clear.

Chaperone-Usher Secretion

Another type of secretion related to type V secretion is **chaperone-usher secretion**. This type of secretion is usually used to assemble some pilins on the cell surface, such as the P pilus of uropathogenic *E. coli*. The secretion system consists of three proteins, a β-barrel-forming protein in the outer membrane called the usher, a periplasmic protein called the chaperone, and the pilin subunit to be assembled on the cell surface. The pilin protein is transported through the inner membrane by the SecYEG channel and therefore has a cleavable signal sequence. Once in the periplasm, the pilin protein is bound by the

dedicated periplasmic chaperone, which helps it fold properly and prevents it from prematurely associating with other pilin subunits. The chaperone then makes contact with the usher at the outer membrane and releases the pilin subunit in an orderly process as assembly of the pilus occurs. In spite of many studies, this process is still not well understood. The pilus assembles from the end, with the adhesin at the end of the pilus added first, followed by other subunits. Apparently, the usher knows what subunit to put on next, depending on where it is in the pilus. Also, two usher channels seem to cooperate with each other, forming twin channels that may alternate somehow in adding subunits. Again, there is the problem of where the energy for pilus assembly comes from since the assembly of the pilus is occurring at the inner face of the outer membrane, after the pilin protein has been transported through the inner membrane and cytoplasm, which are the sources of energy. One idea is that the periplasmic chaperone holds the pilin protein in a high-energy state and its eventual folding at the usher drives the assembly process.

Protein Secretion in Gram-Positive Bacteria

So far, we have limited our discussion of protein secretion to the mechanisms used by gram-negative bacteria. The protein secretion systems discussed above are, of necessity, restricted to gram-negative bacteria; without a lipid bilayer outer membrane, gram-positive bacteria have no need for them. However, some of our best friends (and worst enemies) are gram-positive bacteria, so we must not neglect them. To give some examples, the lactobacilli that are used to make food products, including yogurt and cheese, are gram positive, as are the biodegradable-insecticide-producing *Bacillus thuringiensis* and the *Streptomyces* species that make most of the known antibiotics. But so are *Staphylococcus aureus,* the agent of many serious infections, *Bacillus anthracis*, the cause of anthrax, and *Streptococcus mutans*, which causes dental plaque. In this section, we discuss some features of secretion systems that are unique to gram-positive bacteria.

INJECTOSOMES OF GRAM-POSITIVE BACTERIA
While pathogenic gram-positive bacteria lack the type I, III, and IV secretion systems described above, they require

mechanisms to translocate virulence effectors into eukaryotic cells. Some AB-type toxins such as diphtheria toxin and clostridial neurotoxins such as botulinum toxin are self-translocating (see chapter 8). The B subunit binds to the surface receptors on eukaryotic cells and helps the A-subunit toxin enter the cell. But some gram-positive bacteria, such as *Streptococcus pyogenes*, inject a virulence effector into the mammalian target cell by a mechanism functionally analogous to the gram-negative type III secretion system (see Madden et al., Suggested Reading). This has been named an injectosome to distinguish it from the gram-negative type III injectisome (note the different spellings). As shown in Figure 14.6, the function of the injectosome requires translocation of the effector by the Sec-dependent secretion system across the bacterial membrane before it translocates the effector directly across the membrane of the eukaryotic target cell. The figure illustrates the functional analogy of the gram-positive injectosome and the gram-negative injectisome.

Sortases

Since gram-positive bacteria lack a bilipid outer membrane, their cell wall is not surrounded by a membrane and so is available to the external surface of the cell. The gram-positive bacteria can therefore attach proteins to their cell wall and have them exposed on the cell surface. In

Figure 14.6 The gram-positive injectosome (A) compared to the gram-negative type III injectisome (B). See the text for details.

A   Gram-positive injectosome

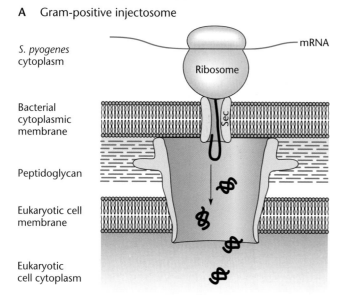

B   Gram-negative type III injectisome

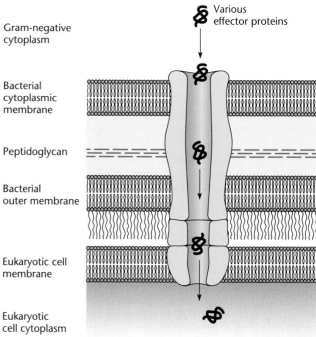

gram-positive bacteria, proteins destined for covalent attachment to the outer cell surface are the targets of a type of cell wall-sorting enzyme called a **sortase**. A sortase is able to create covalent attachments by catalyzing a transpeptidation reaction. The transpeptidation substrate can be the cell wall peptidoglycan or a specific polypeptide. Sortases can direct proteins to unique positions on the outer surface of a cell and can also function in gram-positive bacterial pilus assembly. Surface proteins that are sortase targets include an N-terminal signal peptide and a 30- to 40-residue C-terminal sorting signal, which is composed of a pentapeptide cleavage site, commonly LPXTG, and a hydrophobic domain (Figure 14.7A).

Figure 14.7B illustrates a typical sortase pathway. The N-terminal signal sequence of the sortase target protein directs the protein to the membrane translocase, where the signal sequence is removed. After the protein has been translocated across the cytoplasmic membrane, the sorting signal is processed by a sortase. The sortase cuts in the pentapeptide motif and then covalently links the C terminus to an amino group in a peptidoglycan cross-bridge, often a glycine.

Five sortase subfamilies are currently defined, differing in their taxonomic distribution in the gram-positive genera but also on the basis of differences in the pentapeptide cleavage sites.

Figure 14.7 The sortase pathway. (A) Typical sortase substrate. The protein is composed of an N-terminal signal peptide and a C-terminal cell wall sorting signal (Cws). The Cws contains a conserved LPXTG motif followed by a hydrophobic stetch of amino acids and positively charged residues at the C terminus. (B) A model for the cell wall sortase pathway. (1) The full-length surface protein precursor is secreted through the cytoplasmic membrane via an N-terminal signal sequence. (2) A charged tail at the C terminus of the protein may serve as a stop transfer signal. Following cleavage of this secretion signal, a sortase enzyme cleaves the protein between the threonine and glycine residues of the LPXTG motif, forming a thioacyl-enzyme intermediate. The free amine of the cell wall cross-bridge of lipid II is deprotonated in the SrtA site. The Pro/Gly/Ser/Thr region may help it through the cell wall.

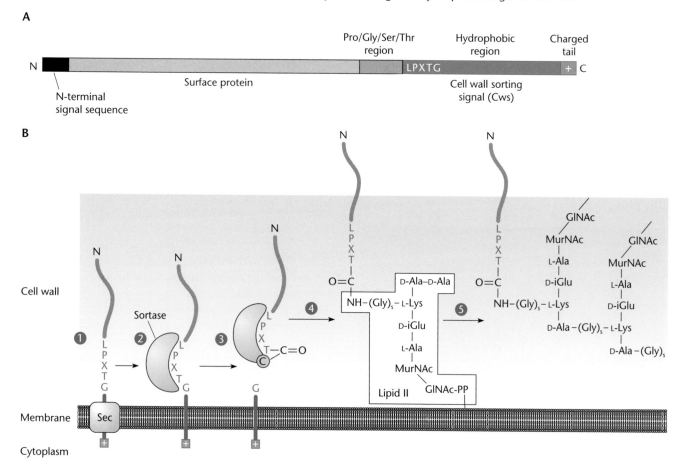

Example of a Sortase-Dependent Pathway: Sporulation in *Streptomyces coelicolor*

A common soil bacterium, *Streptomyces coelicolor*, uses sortase enzymes in a specific phase of growth: morphological differentiation. These bacteria grow as filaments (hyphae), and morphological differentiation produces colonies that are reminiscent of filamentous fungi in that a feeding mycelium of hyphae "morphs" into a fluffy-surfaced colony covered with specialized spore-forming hyphae (Figure 14.8) (see Chater and Horinouchi and also Flardh, Suggested Reading). In plate-grown cultures, the hyphae that will differentiate into spore chains are often referred to as "the aerial mycelium" because the prespore hyphae grow away from the feeding mycelium, which would mean that they grow vertically into the air. In their natural soil habitat, growth of the sporulating hyphae from one soil particle-water droplet to another allows the dispersal of spores into new soil niches.

In both growth environments—plate and soil—the growing, feeding mycelial clump acquires a hydrophobic surface, thus allowing spore-producing hyphae to alter their spatial orientation at an air-water interface (i.e., on a culture plate surface) or at a water-soil particle interface. The mycelial hydrophobicity is due, in part, to the production of a layer of hydrophobic proteins. A mature, sporulating colony is extremely hydrophobic (Figure 14.9).

One type of colony surface hydrophobic protein is called a **chaplin** (for *c*oelicolor *h*ydrophobic *a*erial *p*rotein). The streptomycete chaplin genes, *chp*, comprise a **multigene family** (Figure 14.10) with eight members, all sharing a hydrophobic domain of ~40 residues (the chaplin domain). All also have a secretion signal, and many

Figure 14.9 Hydrophobicity of *Streptomyces coelicolor*. The aerial mycelia on a confluently streaked plate are so hydrophobic that water droplets form. The color illustrates the blue coloration of one of the antibiotics produced by this bacterium.

also have a "cell wall sorting signal," which allows them to be attached to the cell surface by the sortase cell wall localizing pathway.

In *S. coelicolor*, the sortase enzyme cleaves its target between the threonine and glycine residues of an LAXTG motif (Figure 14.10). The sortase then covalently links the target protein to the cell wall peptidoglycan at a cross-bridge of lipid II as shown in Figure 14.7B. Thus, the sortase localizes the *S. coelicolor* chaplins to a colony surface, allowing the aerial mycelium to develop and the sporulation process to proceed.

GENETIC ANALYSIS OF A MULTIGENE FAMILY
Genetic analysis of the chaplin-encoding multigene family demonstrated that the presence of the chaplins on the colony surface is essential for the morphological differentiation process. Because the chaplins are a multigene family, the chaplin genetic analysis required a sophisticated approach for constructing and analyzing mutant strains (see Elliot et al., Suggested Reading). In fact, this analysis provides a model for how to deal with **genetic redundancy**. Although the entire *S. coelicolor* genome sequence was known and methods for systematic gene knockout were available, the analysis required that multiple mutant

Figure 14.8 Cross section of a mature, differentiated, sporulating *S. coelicolor* colony, as visualized by scanning electron microscopy.

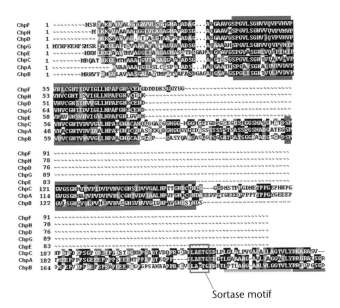

Sortase motif

Figure 14.10   Alignment of the eight chaplin proteins. The thick grey lines mark the chaplin domains. For ChpA to ChpC, the LAXTG sortase recognition motif is boxed in purple, the hydrophobic regions following are underlined in light purple, and the positively charged C-terminal tails are underlined in dark purple.

strains be made. In fact, knockout of one chaplin gene gave no phenotype, knockout of two genes gave no phenotype, knockout of three genes gave no phenotype, and it was only knockout of a fourth gene that produced a strain with a phenotype!

Construction of a Multiple-Mutant Strain

Figure 14.11 illustrates that some of the *chp* genes are adjacent on the *S. coelicolor* chromosome. This gene arrangement, which is not unusual for a multigene family, was extremely helpful in constructing the quadruple-mutant strain. Figure 14.12 illustrates the scheme used for generation of the multiple-knockout mutants. The

Figure 14.11   Organization of the chaplin genes (purple arrows). Unrelated intervening genes are shown as grey arrows. The numbers indicate ORF designations in the *S. coelicolor* genome sequence.

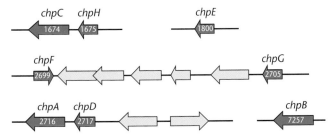

figure also shows that some mutations were insertions of antibiotic resistance genes, some were in-frame deletions, and the quadruple mutant contained a combination constructed by a series of sequential insertions and deletions as illustrated in Figure 14.12.

The mutant phenotype of the quadruple *chp* mutant is shown in Figure 14.13. Encouraged that this mutant showed a morphogenetic phenotype, the researchers went on to make a quintuple mutant! Their effort was rewarded by the observation that the defect in morphogenesis was now even more severe. Indeed, subsequent knockout of all eight *chp* genes has generated a strain unable to produce any aerial hyphae.

Localization and Spatial Expression of the *chp* Genes

In additional experiments, isolation of surface proteins and analysis with matrix-assisted laser desorption ionization–time-of-flight (MALDI-TOF) mass spectrometry confirmed that the chaplins could indeed be identified on the cell surface. Finally, transcriptional expression of the chaplin genes in the aerial mycelium could be seen by using green fluorescent protein transcriptional fusions (Figure 14.14).

Example of a Functional Genomic Analysis

We have not yet discussed how the chaplin genes were first identified. In fact, this process illustrates the power of a functional genomics analysis that begins with a hypothesis-based microarray experiment, continues with single-gene transcriptional analysis, and then finally exploits the known genome sequence of the bacterium for reverse genetics.

First, microarrays were used to identify regulon candidates for an extracytoplasmic sigma factor (see chapter 13), σ^N, which was known to be required for differentiation of a colony into an aerial hypha-producing colony. The essential role of σ^N in development was evident because loss-of-function σ^N mutants produced no aerial hyphae; instead, they displayed a smooth-colony phenotype named Bld for "bald," rather than the "hairy" colony phenotype characteristic of wild-type *S. coelicolor*. Thus, genes that were σ^N dependent were identified as transcripts that were down-regulated in microarrays that used RNA isolated from a *bldN* strain (i.e., a σ^N-minus mutant) and its "wild-type" (i.e., **congenic**) parent.

The set of genes that was down-regulated in the σ^N mutant was further tested by **single-gene transcriptional analysis**. Among the genes that showed greater than a twofold difference—17 of the 7,071 genes represented on the microarrays—were several secreted proteins. Sequence analysis (e.g., by **BLAST**) showed that a

A Gene disruption scheme

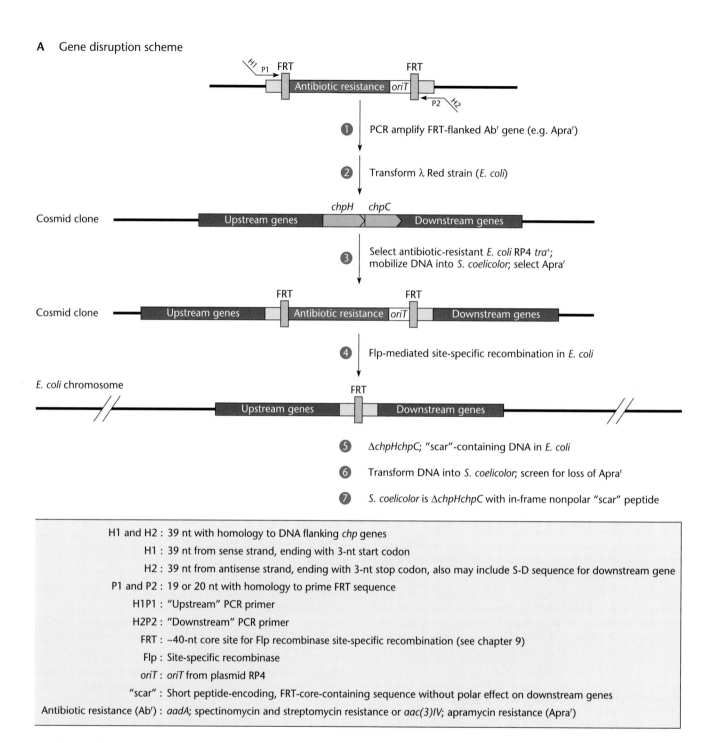

① PCR amplify FRT-flanked Ab^r gene (e.g. Apra^r)

② Transform λ Red strain (*E. coli*)

Cosmid clone

③ Select antibiotic-resistant *E. coli* RP4 *tra*⁺; mobilize DNA into *S. coelicolor*; select Apra^r

Cosmid clone

④ Flp-mediated site-specific recombination in *E. coli*

E. coli chromosome

⑤ Δ*chpHchpC*; "scar"-containing DNA in *E. coli*

⑥ Transform DNA into *S. coelicolor*; screen for loss of Apra^r

⑦ *S. coelicolor* is Δ*chpHchpC* with in-frame nonpolar "scar" peptide

H1 and H2 : 39 nt with homology to DNA flanking *chp* genes

H1 : 39 nt from sense strand, ending with 3-nt start codon

H2 : 39 nt from antisense strand, ending with 3-nt stop codon, also may include S-D sequence for downstream gene

P1 and P2 : 19 or 20 nt with homology to prime FRT sequence

H1P1 : "Upstream" PCR primer

H2P2 : "Downstream" PCR primer

FRT : ~40-nt core site for Flp recombinase site-specific recombination (see chapter 9)

Flp : Site-specific recombinase

oriT : *oriT* from plasmid RP4

"scar" : Short peptide-encoding, FRT-core-containing sequence without polar effect on downstream genes

Antibiotic resistance (Ab^r) : *aadA*; spectinomycin and streptomycin resistance or *aac(3)IV*; apramycin resistance (Apra^r)

B Quintuple-mutant construction

① Construct Δ*chpAD*::*aac(3)IV*; *S. coelicolor* (Apra^r)

② Flp-out Apra^r with FRT/Flp; *S. coelicolor* is Δ*chpAD*

③ Construct Δ*chpCH*::*aadA*; strain is Δ*chpAD* Δ*chpCH*::*aadA*

④ Construct Δ*chpB*::*aac(3)IV*; strain is Δ*chpAD* Δ*chpCH*::*aadA* Δ*chpB*::*aac(3)IV*

A

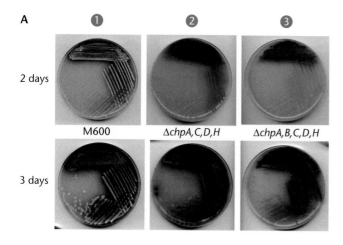

B

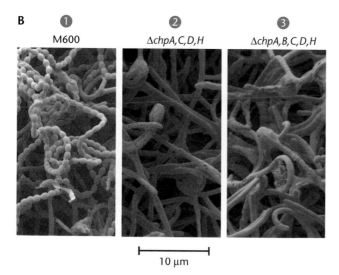

Figure 14.13 Phenotypes of *chp* mutants. (A) Plate-grown cultures: (1) "wild-type" strain (named M600); (2) quadruple *chp* mutant, which displays a delay in morphogenesis; (3) quintuple *chp* mutant, which displays a more severely delayed morphogenetic capacity. Purple on plates indicates purple-blue-red antibiotic synthesis in a colony. (B) Scanning electron microscope images of the strains after 4 days of growth of the plate cultures.

subset of these defined the new multigene family that encodes the chaplins. These were then further analyzed as described above.

In summary, this analysis demonstrates how microarrays can be used to both identify the genes in a regulatory network and identify which gene(s) in the regulon contributes to the phenotype, even if the genes are redundant, by systematic gene knockouts. Note that "forward mutations" can identify single genes that cause a Bld phenotype, e.g., for σ^N, but it would have been difficult to identify the *chp* genes using forward genetics alone because the genes are redundant, and so strains with mutant phenotypes would have been relatively rare, probably requiring a change-of-function mutation in a *chp* gene.

Genetic Analysis of Sporulation in *Bacillus subtilis*

As mentioned in the introductory chapter, many bacteria undergo complex developmental cycles. In their development, some bacteria perform many functions reminiscent of higher organisms: they undergo regulatory cascades; their cells communicate with each other and differentiate and form complex multicellular structures; the cells in these multicellular structures often perform different distinct functions which require compartmentalization and cell-cell communication; and the cells use phosphorelays to respond to changes in communication with other cells and with the external environment. Because of the relative ease of molecular genetic analysis with some bacteria, some of these developmental processes have been extensively investigated as potential model systems for even more complex developmental processes in higher organisms (see Brun and Shimkets, Suggested Reading).

The best understood bacterial developmental system is sporulation in *B. subtilis*. When starved, *B. subtilis* cells undergo genetically programmed developmental changes. They first attempt to obtain nutrients from neighboring

Figure 14.12 Construction of an *S. coelicolor* strain containing multiple knockout mutations. (A) (1) PCR is used to amplify an antibiotic resistance gene and introduce flanking sites for a site-specific recombinase, Flp, for yeast (see chapter 9), as well as sequences complementary to the gene to be knocked out. (2) Recombineering (see chapter 10) is used to insert this sequence into the gene cloned in an *E. coli* plasmid vector, deleting much of the gene. (3) It is then conjugatively introduced into *S. coelicolor,* selecting for the antibiotic resistance and thus replacing the indigenous gene. (4 and 5) If desired, an in-frame nonpolar deletion can be made by introducing the Flp recombinase into *E. coli* (above) containing the gene with the insert which excises the antibiotic resistance cassette, leaving a "scar." If this is mobilized into *Streptomyces*, using the transfer system of the promiscuous IncP plasmid RP4 (see chapter 5) with screening for loss of the antibiotic resistance, the scar will have replaced the antibiotic resistance cassette. (B) Construction of a quintuple mutant involves stepwise insertion and deletion of antibiotic resistance genes as above.

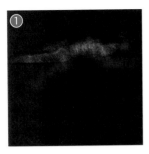
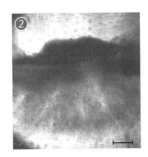
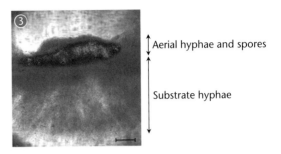

Aerial hyphae and spores

Substrate hyphae

Figure 14.14 Confocal microscope image of a cross section of a wild-type colony carrying a *chpH-gfp* transcriptional fusion. 1, Fluorescent image; 2, phase-contrast image; 3, overlay of panel 1 on panel 2, illustrating the localization of fluorescence to the aerial structures. GFP in purple in 1 and 3.

organisms by producing antibiotics and extracellular degradative enzymes. If starvation conditions persist, the cells sporulate, producing endospores that are metabolically dormant and highly resistant to environmental stresses.

The process of sporulation starts with an asymmetric division that produces two cell types with different morphological fates. The larger cell, which is called the mother cell, engulfs the smaller forespore and then nurtures it. Eventually the mother cell lyses, releasing the endospore.

Many of the changes that occur in the sporulating cell can be visualized by electron microscopy. These are schematized in Figure 14.15. The figure also shows the proteins that have been identified as key regulators of specific stages of development. We describe the experiments that identified these key regulators below.

Identification of Genes That Regulate Sporulation

Isolation of mutants was crucial to the process of identifying the important regulators of sporulation. Many mutants were isolated on the basis of a phenotype referred to as Spo⁻ (for *spo*rulation minus). Such mutants could be identified as nonsporulating colonies because plate-grown cultures of the wild type develop a dark brown spore-associated pigment whereas the nonsporulators remain unpigmented.

Spo⁻ mutants were phenotypically characterized by electron microscopy and then grouped according to the stage at which development was arrested (Figure 14.16). Some of the key regulatory genes defined by analysis of the mutants are listed in Table 14.1. The names of *B. subtilis* sporulation genes reflect three aspects of the genetic analysis of these genes. The roman numerals refer to the results of phenotypic categorization of the mutant stains, with the numbers 0 through V indicating the stages of sporulation at which mutants were found to be

blocked. The gene names also contain one or two letters. The first letter designates the separate loci that mutated to cause similar phenotypes. Each such locus was defined by the set of mutations that caused the same morphological block and that were genetically closely linked. The second letter in the names indicate the individual ORFs that were found when DNA sequencing revealed that a locus contained several ORFs.

Regulation of Initiation of Sporulation

Much of what we understand about the mechanism of sporulation initiation is based on studies of the class of sporulation-minus mutants, which were designated *spo0* because of their failure to begin the sporulation process. Many of these mutants have pleiotropic phenotypes, meaning that they are altered in several characteristics. Besides being unable to sporulate, they fail to produce the antibiotics or degradative enzymes that are characteristically produced by starving cultures, and they do not develop competence for transformation (see chapter 6).

Two of the *spo0* genes, *spo0A* and *spo0H*, encode transcriptional regulators. *spo0A* encodes a "two-component system" response regulator that is responsible for regulating the cellular response to starvation. The product of *spo0H* is a sigma factor (σ^H). Many of the genes that are targets for Spo0A regulation are transcribed by the σ^H-containing RNA polymerase holoenzyme (see Britton et al., Suggested Reading).

Like most response regulators, Spo0A must be phosphorylated in order to carry out its transcriptional regulatory functions. Phosphorylation of Spo0A involves a "phosphorelay" system (Figure 14.17) that includes another two of the *spo0* gene products, Spo0F and Spo0B. The phosphorelay also involves at least five protein kinases: these each phosphorylate Spo0F under certain growth conditions. Spo0B is a phosphotransferase enzyme that transfers phosphoryl groups from Spo0F~P to Spo0A. Spo0A~P then regulates its target genes by

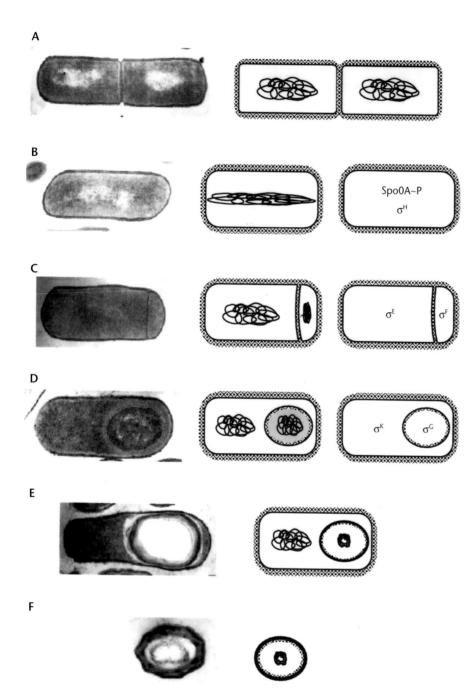

Figure 14.15 Stages of sporulation. The left of each panel shows an electron micrograph of the stage of sporulation, and the right shows, in cartoon form, the disposition of the chromosomes and the time and site of action of the principal regulatory proteins that govern sporulation gene expression. (A) Vegetative cells. (B through E) Sporangia at entry into sporulation (stage 0) (B), at polar division (also called polar septation) (stage II) (C), at engulfment (stage III) (D), and at cortex and coat formation (stages V to VI) (E). (F) A free spore.

binding to their promoter regions, activating some and repressing others (Box 14.2).

The regulatory effect of Spo0A on a given target gene depends on the amount of Spo0A~P in the cell, reminiscent of OmpR~P regulation of porin genes and *Bordetella* BvgA~P regulation of virulence genes (see chapter 13). At low levels, Spo0A~P positively regulates genes involved in the synthesis of antibiotics and degradative enzymes as well as competence and biofilm formation.

This positive regulation may result from what is actually a "double-negative" series of events, in which the direct effect of Spo0A~P action is repression of a gene called *abrB*, which itself encodes a repressor that acts on the antibiotic and degradative enzyme genes. At higher levels, Spo0A~P directly activates several sporulation operons, including *spoIIA*, *spoIIE*, and *spoIIG*. Activation of these genes irreversibly commits a cell to the sporulation process.

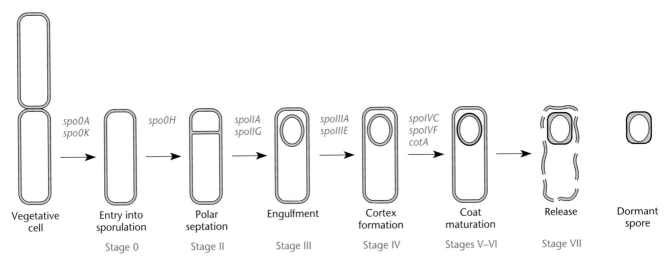

Figure 14.16 Morphological events of *B. subtilis* sporulation and some of the genes required for sporulation. Mutations in the genes above the arrows give a mutant phenotype similar to that shown immediately before the arrow. Stage I is skipped in the figure because it is less clearly defined.

REGULATION OF THE Spo0A PHOSPHORELAY SYSTEM

Numerous genes participate in regulating the amount of Spo0A~P produced in a cell. Several of these encode the kinases mentioned above, which phosphorylate Spo0F and therefore increase Spo0A~P levels. Two of these kinases, KinA and KinB, phosphorylate Spo0F and con-

sequently Spo0A to high levels in response to severe extended starvation and commit the cell to sporulation; the others, KinC, KinD, and KinE, phosphorylate Spo0F only to low levels and commit the cell only to competence and biofilm formation. As mentioned above, the signals that activate these respective kinases are unknown but are the subject of active investigation. Other signals

TABLE 14.1	*Bacillus subtilis* sporulation regulators	
Stage of mutant arrest	**Gene**	**Function**
0	*spo0A*	Transcription regulator
	spo0B	Phosphorelay component
	spo0F	Phosphorelay component
	spo0E	Phosphatase
	spo0L	Phosphatase
	spo0H	Sigma H
II	*spoIIAA*	Anti-anti-sigma
	spoIIAB	Anti-sigma[a]
	spoIIAC	Sigma F
	spoIIE	Phosphatase
	spoIIGA	Protease
	spoIIGB	Pro-sigma E
III	*spoIIIG*	Sigma G
IV	*spoIVCB-spoIIIC*[b]	Pro-sigma K
	spoIVF (operon)[c]	Regulator of sigma K

[a] Loss-of-function mutations in *spoIIAB* do not produce a Spo⁻ phenotype; rather, they cause lysis.

[b] In *B. subtilis* and some other bacilli, two gene fragments undergo recombination to produce the complete σ^K coding region.

[c] Subsequent work has further defined two genes, *spoIVFA* and *spoIVFB* (see the text).

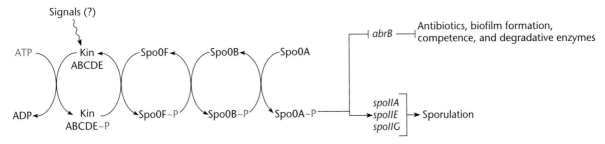

Figure 14.17 The phosphorelay activation (phosphorylation) of the transcription factor Spo0A. The phosphorelay (see Box 14.2) is initiated by at least five histidine kinases, which autophosphorylate on a histidine residue in response to unknown signals. Kinases A and B phosphorylate to high levels and initiate sporulation and kinases C, D, and E phosphorylate only to low levels for competence, biofilm formation, antibiotic synthesis, and degradative enzymes. The phosphate is transferred to Spo0F, to Spo0B, and finally to Spo0A. Spo0A~P regulates transcription as described in the text.

such as DNA damage could also activate kinases A and B and commit the cell to sporulation unless there is intervention by a checkpoint protein, Sda. This protein binds to the kinases and inhibits them, preventing sporulation in response to DNA damage (see Ruvolo et al., Suggested Reading). As discussed in Box 4.3, *B. subtilis* also encodes an addiction module that kills some cells in the population in response to starvation, allowing the killed cells to be cannibalized by other cells to delay or prevent their sporulation. Obviously, the cell sporulates only if it is absolutely necessary.

Other genes encode phosphatases that can dephosphorylate Spo0F~P, thereby draining phosphate out of the phosphorelay and diminishing Spo0A~P levels, or dephosphorylate Spo0A~P directly (see below). These phosphatases also respond to physiological and environmental signals, only a few of which are known.

Negative Regulation of the Phosphorelay by Phosphatases

Genetic analysis of two of the *spo0* loci revealed that their gene products functioned as negative regulators of sporulation. The combined results from sequencing mutant alleles, constructing gene knockouts, obtaining enhanced expression from many copies of the genes, and isolating suppressor mutations were all important to deciphering the gene functions (See Perego et al., Suggested Reading). We list these regulators below.

1. For the *spo0E* locus, sequence analysis determined that two mutant alleles in Spo⁻ strains contained nonsense mutations. This result would usually be interpreted as an indication that the *spo0E* gene product plays a positive role in initiating sporulation, since nonsense mutations usually inactivate the gene product and a requirement of the regulatory gene product

for expression is the genetic definition of positive regulation (see chapter 2). However, deletion analysis of *spo0E* was contradictory, since Δ*spo0E* strains were capable of sporulation—in fact, they hypersporulated. Furthermore, multiple cloned copies of the *spo0E* gene inhibited sporulation. The last two observations exemplify behavior typical of negative regulators. A resolution of this paradox came from a clue provided by one aspect of the Δ*spo0E* mutant phenotype: Δ*spo0E* mutant strains had a tendency to segregate Spo⁻ papillae that were visible as translucent patches on the surface of sporulating colonies. Genetic analysis of these Spo⁻ papillae showed that they contained suppressor mutations, several of which mapped to the *spo0A* gene. This result suggested the hypothesis that cells lacking *spo0E* experienced an especially strong pressure to sporulate because increased expression or activity of the phosphorelay components produced an exceptionally high level of Spo0A~P. The *spo0E* mutants were found to have no alteration in phosphorelay gene transcription, but a biochemical study of the Spo0E protein showed that it functioned as a specific phosphatase of Spo0A~P. Such an activity would indeed provide a negative regulatory function.

What would explain the finding that nonsense *spo0E* mutants had a Spo⁻ phenotype? Both of these mutations were found to affect the C terminus of Spo0E, leaving most of the protein intact. One hypothesis is that the Spo0E C terminus is a regulatory domain, perhaps one that binds a signal molecule which regulates the phosphatase activity. If so, the mutations may prevent signal binding and hence lock Spo0E into the phosphatase mode, thereby preventing Spo0A~P accumulation and sporulation.

BOX 14.2

Phosphorelay Activation of the Transcription Factor Spo0A

Some bacterial pathways respond to multiple signal inputs. An example is the sporulation phosphorelay of *B. subtilis*. As illustrated in panel A of the figure, the histidine kinase and the phosphorylated aspartate domains can be found on separate polypeptides. Of the five kinases, only KinA is shown. The phosphoryl group is transferred from one protein to the next,

as shown in steps 1 through 4. Panel B illustrates a cocrystal structure of a Spo0B dimer containing the conserved histidine residues (H in purple) with two Spo0F polypeptides, which contain the aspartate residues (D in purple). The close proximity of the histidine- and aspartate-containing active sites allows phosphoryl group transfer (shown as arrow). Panel C

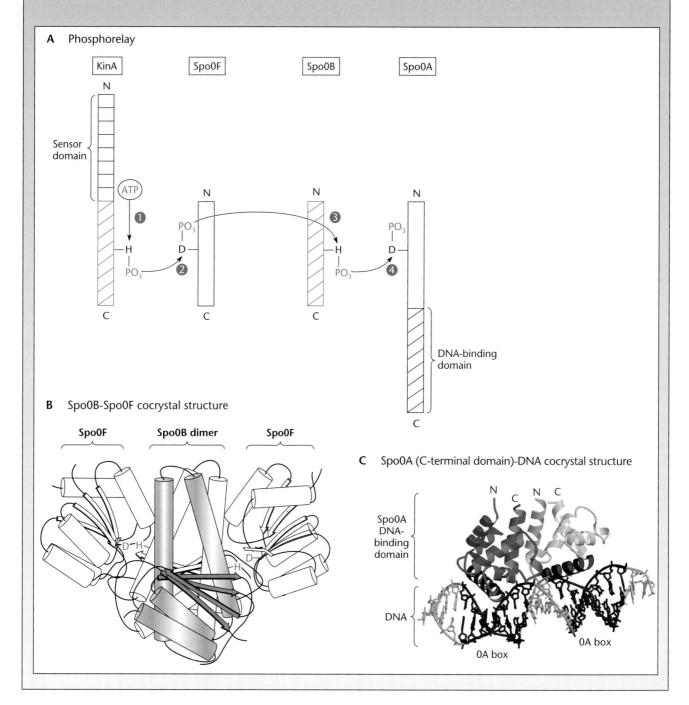

A   Phosphorelay

B   Spo0B-Spo0F cocrystal structure

C   Spo0A (C-terminal domain)-DNA cocrystal structure

BOX 14.2 (continued)

Phosphorelay Activation of the Transcription Factor Spo0A

shows a cocrystal structure of the transcription regulator Spo0A with its DNA-binding site. Only the C-terminal domain of Spo0A was used in the crystallization experiment. In this response regulator, the unphosphorylated N-terminal domain inhibits C-terminal domain binding to DNA. When the phosphorelay is operating, phosphorylation of the N-terminal domain relieves the N-terminal domain-mediated inhibition.

References

Bourret, R. B., N. W. Charon, A. M. Stock, and A. H. West. 2002. Bright lights, abundant operons—fluorescence and genomic technologies advance studies of bacterial locomotion and signal transduction: review of the BLAST meeting, Cuernavaca, Mexico, 14 to 19 January 2001. *J. Bacteriol.* **184:**1–17.

Hoch, J. A., and K. I. Varughese. 2001. Keeping signals straight in phosphorelay signal transduction. *J. Bacteriol.* **183:**4941–4949.

2. The *spo0L* locus shared some genetic properties with *spo0E*: multiple copies of *spo0L* caused a sporulation deficiency, and Δ*spo0L* mutations caused hypersporulation, as well as accumulation of Spo⁻ segregants. Like Spo0E, Spo0L behaved like a negative regulator of the phosphorelay. Because the *spo0L* mutants isolated on the basis of their Spo⁻ phenotype contained missense mutations of *spo0L*, it was reasoned that isolation of suppressors of these mutations might identify the target of Spo0L activity. Accordingly, a plan was made to mutagenize a Spo⁻ *spo0L* mutant strain and then look for Spo⁺ colonies. One problem that arose was that the most frequent class of sporulating mutants contained null mutations of *spo0L*. To overcome this problem, a strain containing two copies of the *spo0L* missense allele was constructed; this strain was mutagenized for isolation of extragenic suppressor mutations. The result of this experiment was isolation of a suppressor mutation in the *spo0F* gene, one of the phosphorelay components. When tested for phosphatase activity, Spo0L proved to be a phosphatase of Spo0F~P.

An additional phosphatase of Spo0F~P was found in the *B. subtilis* genome sequence as a Spo0L homolog. Named Spo0P, this homologous protein inhibited sporulation when hyperexpressed from multicopies of the gene and encoded a phosphatase with 60% identical amino acid residues to Spo0L. The recent renaming of Spo0L and Spo0P to RapA and RapB, respectively, reflects their roles as *r*esponse regulator *a*spartyl-phosphate *p*hosphatases.

Figure 14.18 illustrates inhibition of the phosphorelay by the RapA, RapB, and Spo0E phosphatases. Since these phosphatases function to reduce the accumulation of Spo0A~P, their activities must be inhibited under conditions that promote antibiotic synthesis and sporulation. A regulator of Spo0E has been hypothesized, as discussed above. For RapA and RapB, the known regulatory signals are peptide molecules named PhrA and

competence-stimulating factor (CSF). Produced by the bacilli themselves, they may function as indicators of population density or "quorum sensors." The quorum sensors of gram-negative bacteria are typically homoserine lactones, while those of gram-positive bacteria are more typically peptides.

Regulation of the phosphorelay involves additional signals, besides those mentioned above. Starvation, cell density, metabolic states, cell cycle events, and DNA damage are all known to influence Spo0A~P levels. Many of the signals involved and the mechanisms by which they affect Spo0A~P are poorly understood and are the subjects of active investigation.

Compartmentalized Regulation of Sporulation Genes

The mother cell and the forespore are genetically identical, but certain proteins must be made specifically in the developing spore and others (such as those that form the sturdy spore coat) must be made in the surrounding mother cell cytoplasm. Thus, the set of genes transcribed from the mother cell DNA must differ from the set

Figure 14.18 Regulation of the phosphorelay by phosphatases. RapA and RapB are phosphatases for SpoF~P, and Spo0E is a phosphatase for Spo0A~P (see Stephenson and Perego, Suggested Reading). RapA and RapB are inhibited by the PhrA and CSF (PhrC) pentapeptides, respectively. It is not known what controls the activity of the phosphatase, Spo0E.

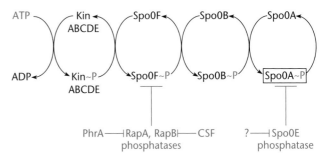

transcribed from the forespore DNA. What mechanisms account for transcription of different sporulation genes in the two compartments?

REGULATION OF SPORULATION GENES BY SEQUENTIAL AND COMPARTMENT-SPECIFIC ACTIVATION OF RNA POLYMERASE SIGMA FACTORS

The entire collection of sporulation genes can be sorted into a handful of classes on the basis of transcription by a specific sigma factor. The sporulation sigma factors replace the principal vegetative cell sigma factor A (σ^A) in RNA polymerase holoenzyme, possibly by outcompeting σ^A for RNA polymerase. The σ^A of *B. subtilis* plays a role similar to the σ^{70} of *E. coli* (see chapter 2). As shown in Table 14.2, there are five distinct sigma factors associated with sporulation, sigma H (σ^H), sigma E (σ^F), sigma F (σ^F), sigma G (σ^G), and sigma K (σ^K); these associate with RNA polymerase to transcribe the sporulation genes. Each of the sigma factors is active at a specific time during sporulation. Four of the sigmas are regulated such that they are active in only one of the two developing cell compartments: σ^E and σ^K are sequentially active in the mother cell, and σ^F and σ^G are sequentially active in the forespore.

Analysis of the Role of Sigma Factors in Sporulation Regulation

Four kinds of information have been important to understanding gene regulation in *B. subtilis*.

TEMPORAL PATTERNS OF REGULATION

Measurements of the times of expression of the sporulation genes indicated that many of the genes underwent dramatic increases in expression at specific times after the sporulation process started. Use of gene fusions allowed large-scale comparisons of the complete set of sporulation genes. The most commonly used reporter genes were *lacZ* and *gus* from *E. coli* (Figure 14.19) (see chapter 2). The product of the *lacZ* gene, β-galactosidase, and the product of the *gus* gene, β-glucuronidase, could be assayed by adding "artificial" substrates (such as o-nitrophenyl-β-D-galactopyranoside [ONPG] or methylumbelliferyl-β-glucuronide [MUG]) to samples of the

test culture at various times after induction of sporulation by a nutritional downshift. The appearance of β-galactosidase or β-glucuronidase activity indicated the onset of gene expression. In addition, direct measurements of mRNA of various sporulation genes correlated well with the results of *lacZ* and *gus* fusion experiments. Therefore, the use of such fusions became widespread, because of the relative ease and convenience of fusion assays.

A significant outcome of comprehensive fusion experiments was the extensive assessment and comparison of the times of expression of many sporulation genes. Moreover, the timing of *lacZ* expression could be correlated with the timing of morphological changes visible as sporulation progressed.

DEPENDENCE PATTERNS OF EXPRESSION

Fusions with *lacZ* were also used to determine whether the expression of one gene depended on the activity of a second gene. If the expression of one gene depends on a second gene, the second gene may encode a direct or indirect regulator of the first. The use of *spo* mutations in combination with *spo-lacZ* fusions (Figure 14.20) allowed the testing of many regulatory dependencies. An example of a set of experimental data is shown in Figure 14.20B. In this example, expression of a *spoIIA::lacZ* fusion was dependent on all of the *spo0* loci but not on any of the "later" loci. Results of tests of many pairwise combinations of *spo* mutations and gene fusions are also summarized in Table 14.3. Besides the data for the genes shown, data for many dozens of additional genes have contributed to our understanding of gene regulation.

TRANSCRIPTION FACTOR DEPENDENCE

Once *spo* genes had been cloned and sequenced, it was possible to determine the functions of some of the proteins because of their amino acid sequence similarities to known families of regulatory proteins such as sigma factors, which share characteristic amino acid motifs. The *spo0H* gene could be seen to encode a sigma factor, as could ORFs of the *spoIIA*, *spoIIG*, *spoIIIG*, and *spoIVC* loci (Table 14.2). In vitro experiments confirmed the functions of these proteins as transcription factors.

TABLE 14.2	Sporulation sigma factors and their genes		
Sigma factor	**spo gene name**	**sig gene name**	**Compartment of activity**
σ^H	*spo0H*	*sigH*	Predivisional sporangium
σ^F	*spoIIAC*	*sigF*	Forespore
σ^E	*spoIIGB*	*sigE*	Mother cell
σ^G	*spoIIIG*	*sigG*	Forespore
σ^K	*spoIVCB-spoIIIC*	*sigK*	Mother cell

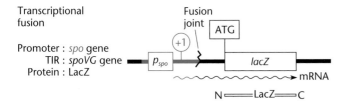

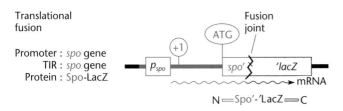

Figure 14.19 Reporter gene *lacZ* fusions to sporulation genes. Translational and transcriptional gene fusions are both transcribed from a *B. subtilis spo* promoter. In translational fusions, a fusion protein is expressed from the translation initiation region (TIR) of the gene being studied. In a transcriptional fusion, *lacZ* is translated from the TIR of a *B. subtilis spo* gene, often that of *spoVG*.

It was possible to infer the sigma factor dependence of many of the other sporulation genes on the basis of sequence comparisons around the transcription start sites. In some cases, allele-specific suppressors of promoter mutations could be isolated in a particular sigma factor gene. Remember from chapter 3 that an allele-specific suppressor can suppress only a particular type of mutation in a gene. Another important type of experiment involved in vitro transcription studies with RNA polymerase containing specific sigma factors.

CELLULAR LOCALIZATION

Several methods have been used to determine the cellular location of expression of sporulation genes. For example, expression of β-galactosidase in the forespore can be distinguished from that in the mother cell on the basis that the forespore is more resistant to lysozyme. Immunoelectron microscopy has been useful for visualizing the expression of β-galactosidase, and, more recently, the use of green fluorescent protein fusions has allowed a determination of the cellular locations of numerous sporulation proteins.

From studies like these, it could be seen that all of the genes turned on after septation were expressed in only one compartment. The genes transcribed by RNA polymerase with σ^F and σ^G were expressed only in the forespore compartment, and the genes transcribed by RNA polymerase with σ^E and σ^K were expressed only in the mother cell.

Intercompartmental Regulation during Development

When the observations on timing, dependence relationships, and localization of sporulation gene expression are combined, a complex pattern of regulation that includes a cascade of sigma factors and signaling between the developing compartments is revealed.

Figure 14.21 shows that after septation, gene expression in the forespore depends at first on σ^F and later on σ^G. An early σ^F-dependent transcript, *gpr*, encodes a protease that is important during spore germination. Besides its dependence on σ^F, *gpr* requires functional *spo0* genes, because σ^F expression and activity depends on them. Another σ^F-transcribed operon is *spoIIIG*, which encodes the late forespore sigma, σ^G. Transcription of *spoIIIG* differs from that of *gpr* in that it occurs later and, although confined to the forespore, requires functioning of the *spoIIG* locus (Table 14.3), which encodes the mother cell-specific sigma, σ^E.

Once σ^G is produced in the forespore, it transcribes a set of *ssp* genes, which encode spore-specific proteins that condense the nucleoid. As shown in Table 14.3, *ssp* transcription is blocked in *spoIIIG* mutants as well as in mutants with mutations in all of the genes discussed above, such as the *spoIIG* gene, because they are involved in σ^G production.

Gene expression in the mother cell also reveals intercompartmental regulation. Figure 14.21 shows that one gene transcribed relatively early in the mother cell by σ^E RNA polymerase is the *gerM* gene, which encodes a germination protein. Later, σ^E RNA polymerase transcribes the gene for σ^K. σ^K RNA polymerase then transcribes *cotA*, one of a set of *cot* genes that encode proteins incorporated into the spore coat. Table 14.3 shows that *cotA* transcription also requires the activity of the late forespore sigma, σ^G.

Activation of the sigma factors alternates between the two developing compartments. As shown in Figure 14.22, each successive activation step requires intercompartmental communication. The critical information is whether morphogenesis and/or gene expression in the other compartment has progressed beyond a "checkpoint." The way in which the two compartments communicate their status to each other is a fascinating area of current research and is discussed in the next sections.

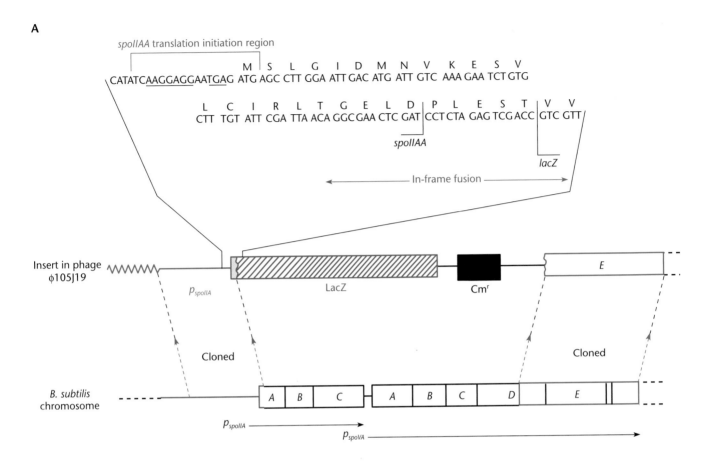

A

spoIIAA translation initiation region

```
                              M   S   L   G   I   D   M   N   V   K   E   S   V
CATATCAAGGAGGAATGAG ATG AGC CTT GGA ATT GAC ATG ATT GTC  AAA GAA TCT GTG
```

```
        L   C   I   R   L   T   G   E   L   D   P   L   E   S   T   V   V
       CTT TGT ATT CGA TTA ACA GGC GAA CTC GAT CCT CTA GAG TCG ACC GTC GTT
                                                    spoIIAA                lacZ
```

In-frame fusion

Insert in phage φ105J19 p_{spoIIA} LacZ   Cm^r E

Cloned   Cloned

B. subtilis chromosome   A B C   A B C D   E

p_{spoIIA} p_{spoVA}

B

| Mutation | β-Galactosidase activity (units ml^{-1}) | | No. of determinations |
	$t_{1.5}$	t_4	
spo0			
spo0A43	0.016	0.037	2
spo0B136	0.023	0.054	2
spo0E11	0.018	0.015	2
spo0F221	0.021	0.050	2
spo0H17	<0.01	0.018	2
spo0J93	0.048	0.25	2
spo0K141	0.013	0.050	2
spoII			
spoIIAA562	0.41	0.38	2
spoIIAC1	0.44	0.51	2
spoIIB131	0.36	0.17	3
spoIID298	0.43	0.11	2
spoIIE48	0.67	0.77	3
spoIIG55	0.69	0.58	3

| Mutation | β-Galactosidase activity (units ml^{-1}) | | No. of determinations |
	$t_{1.5}$	t_4	
spoIII			
spoIIIA65	0.18	0.22	4
spoIIIB2	0.65	0.13	2
spoIIIC94	0.24	0.23	4
spoIV			
spoIVA67	0.29	0.080	3
spoIVB165	0.12	0.19	3
spoIVC23	0.29	0.15	3
spoV			
spoVA89	0.44	0.34	2
spoVB91	0.49	0.21	2
spoVC134	0.61	0.46	2

Figure 14.20 Testing the regulatory dependencies of *spo* genes. (A) Use of a *B. subtilis* transducing phage to create translational *lacZ* fusions to a chromosomal *spo* gene. Shown is the structure of phage φ105J19 carrying the *spoIIAA-lacZ* gene fusion. The lower part of the figure shows the region of the *B. subtilis* chromosome containing the *spoIIA* operon (three genes) and the adjacent *spoVA* operon (five genes). The phage contains a cloned fragment of chromosomal DNA covering these operons, but the central portion of the insert has been replaced by the *E. coli lacZ* gene (purple hatches) and a chloramphenicol

TABLE 14.3	Timing and dependence patterns of gene expression						
Gene or operon fusion[a]	Time of expression (min)	Expression in mutants of gene or operon:					
		spo0	spoIIA (σ^F)	spoIIE	spoIIG (σ^E)	spoIIIG (σ^G)	spoIVC (σ^K)
spoIIA	40	−	+	+	+	+	+
spoIIE	30–60	−	+	+	+	+	+
spoIIG	0–60	−	+	+	+	+	+
gpr	80–120	−	−	−	+	+	+
spoIIIG	120	−	−	−	−	+	+
ssp	>120	−	−	−	−	−	+
spoIVC	150	−	−	−	−	+	+
cotA	240	−	−	−	−	−	−

[a] The functions of these are described in the following sections.

TEMPORAL REGULATION AND COMPARTMENTALIZATION OF σ^E AND σ^F

Both the *sigE* and *sigF* genes, encoding σ^E and σ^F, respectively (Table 14.2), are transcribed in the developing cell before the sporulation septum divides off the forespore compartment (Figure 14.21). However, neither sigma factor starts to transcribe its target genes until after the septum forms, and then, as mentioned above, each sigma becomes active in only one compartment: σ^E in the mother cell and σ^F in the forespore. Before septation, the sigma factors are held in inactive states, with a different inhibitory mechanism acting on each sigma. In σ^E, the active form of the protein must be proteolytically released from an inactive precursor, Pro-σ^E. In σ^F, the active protein must be released from a complex that contains an inhibitory "anti-sigma" factor, SpoIIAB (Figure 14.23). Once the sporulation septum forms, σ^F becomes active in the forespore. Subsequently, σ^E becomes active in the mother cell.

Activation of σ^F in the forespore requires the interplay of a set of proteins. Two of these are binding partners and are named SpoIIAA and SpoIIAB. It is SpoIIAB that is the above-mentioned anti-sigma factor that binds to and inactivates σ^F. SpoIIAA is an anti-anti-sigma factor which nullifies the anti-sigma activity of SpoIIAB; it does this because of its own ability to bind SpoIIAB.

A cycle of phosphorylation and dephosphorylation of SpoIIAA modulates the binding of SpoIIAA to SpoIIAB. Only when it is unphosphorylated can SpoIIAA bind to SpoIIAB. Before septation, SpoIIAA is in a phosphorylated state and so does not bind to SpoIIAB in the preseptational sporangium. Unbound by SpoIIAA, SpoIIAB is free to bind to and inactivate σ^F. After septation, SpoIIAA is in the unphosphorylated state in the forespore; hence, it binds SpoIIAB, releasing σ^F.

The enzymes that phosphorylate and dephosphorylate SpoIIAA are SpoIIAB and SpoIIE, respectively. Their opposing activities, before and after septation, determine the balance between the two forms of SpoIIAA. Before septation, SpoIIAB kinasing of SpoIIAA predominates, whereas once the spore septum has formed, SpoIIE phosphatase activity predominates in the forespore. Regulation of the SpoIIE phosphatase by septum formation is an area of current investigation.

Regulation of σ^F

Important progress in understanding the regulation of σ^F activity came from studying the two genes, *spoIIAA* and

resistance gene (Cm^r) (in black). The insertion is arranged so that the region encoding the *lacZ* gene is fused in frame to the N terminus of the *spoIIAA* gene. (B) Effect of *spo* mutations on the production of β-galactosidase by the *spoIIAA-lacZ* gene fusion during sporulation. Phage φ105J19 (A) was transduced into a series of isogenic strains carrying *spo* mutations. Sporulation was induced, and samples were taken for assay of β-galactosidase. The results shown are mean activities in samples at 1.5 and 4 h after induction of sporulation. Not shown are the control values. The wild-type Spo$^+$ strain containing the phage produced 0.58 unit of β-galactosidase per ml at *t* = 1.5 h and 0.17 unit at *t* = 4 h. With no *lacZ* fusion, the Spo$^+$ strain produced 0.013 unit per ml at *t* = 1.5 h and 0.032 unit per ml at *t* = 4 h.

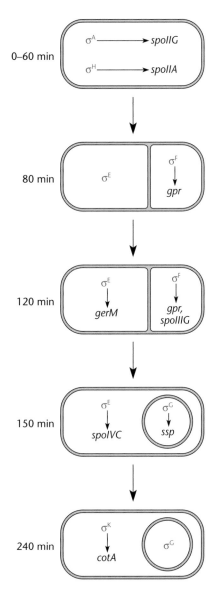

Figure 14.21 Compartmentalization of sigma factors and temporal regulation of transcription within compartments. The genes for σ^E and σ^F are transcribed before polar septation. σ^F is active in the forespore compartment and is required for transcription of the gene for σ^G, which succeeds it. σ^E is active in the mother cell and is required for transcription of the gene for its successor, σ^K. Within their compartments, σ^F and σ^E are required for transcription of their target genes at various times.

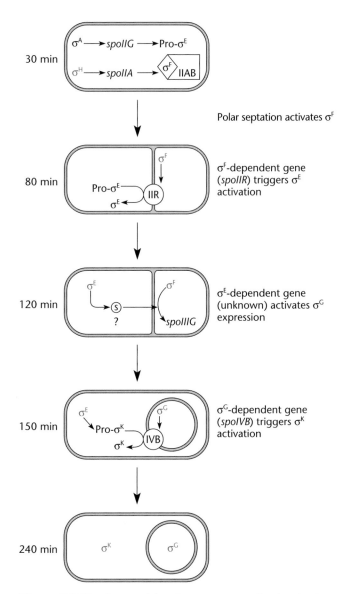

Figure 14.22 Sequential and compartmentalized activation of the *B. subtilis* sporulation sigma factors (purple). A series of signals allows communication between the two developing compartments, as described in the text, at approximately the times shown after induction of sporulation.

spoIIAB, that are cotranscribed with the *sigF* gene in the same operon (Figure 14.24A). Two key findings were that the *spoIIAB* gene product is a protein that inhibits σ^F activity and is, in turn, inhibited by SpoIIAA. These conclusions were drawn from experiments on the effects of *spoIIAA* and *spoIIAB* mutations on σ^F activity (see Schmidt et al., Suggested Reading).

The test for σ^F activity in these experiments was to measure the expression of genes with σ^F-dependent promoters. Two such genes are *spoIIIG* and *gpr* (Table 14.3). The use of *lacZ* fusions to these genes allowed gene expression to be monitored by β-galactosidase assays. Control experiments showed that *sigF* transcription and translation were normal, ensuring that differences in β-galactosidase activity from the σ^F-dependent *lacZ* fusions reflected the activity of σ^F and not its expression.

The comparisons of β-galactosidase activity levels in *spoIIAA* and *spoIIAB* mutant cultures and wild-type

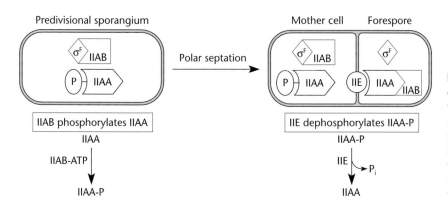

Predivisional sporangium

Mother cell   Forespore

Polar septation

IIAB phosphorylates IIAA

IIAA

IIAB-ATP

IIAA-P

IIE dephosphorylates IIAA-P

IIAA-P

IIE   P$_i$

IIAA

Figure 14.23   Model for the regulation of σ^F activity. SpoIIAB holds σ^F in an inactive state in both the predivisional sporangium and the mother cell. SpoIIAB phosphorylates SpoIIAA. SpoIIAA~P cannot bind SpoIIAB. The SpoIIE phosphatase controls the generation of SpoIIAA, which complexes with SpoIIAB. These reactions allow the release of σ^F.

cultures showed that *spoIIIG::lacZ* and *gpr::lacZ* expression was substantially higher in the *spoIIAB* mutant strain than in the wild type. Conversely, the tester fusions were not expressed at all in the *spoIIAA* mutant cultures. Outcomes like these could occur if SpoIIAA function is required for σ^F activity and SpoIIAB function inhibits σ^F activity (Figure 14.24B). An additional genetic experiment showed that SpoIIAA is required because it is needed to counteract SpoIIAB. This experiment was performed to assay the tester fusions in double *spoIIAA spoIIAB* mutant cultures; these cultures overexpressed the tester fusions, like the *spoIIAB* mutant had. Thus, SpoIIAA is required only if SpoIIAB is active.

Further study showed that *spoIIAB* inhibition of σ^F is an essential event in the normal course of sporulation. One observation was that the *spoIIAB* mutant strains could not survive sporulation and could be maintained only in media that suppressed sporulation. The proposed explanation for this phenotype was that unregulated transcription by σ^F is lethal to the cell. It is important to note here that the *spoIIAB* mutant strain used in the experiments above was not isolated in a mutant hunt for Spo$^-$ strains but, rather, was a constructed deletion mutant. Lethality caused by deregulated σ^F activity could

also be observed if *sigF* was artificially induced in vegetative cells from the p_{spac} promoter (Figure 14.25).

Another important advance in our understanding of σ^F regulation came from biochemical studies of SpoIIAA and SpoIIAB. The SpoIIAB amino acid sequence suggested that it might have protein kinase activity. It was indeed able to phosphorylate a protein: its substrate turned out to be SpoIIAA! Additional studies examined the binding interactions of the three proteins SpoIIAA, SpoIIAB, and σ^F and found that (i) SpoIIAA could bind to SpoIIAB, but only if SpoIIAA was not phosphorylated, and (ii) SpoIIAB could bind to σ^F. Genetic evidence included site-directed mutagenesis of the SpoIIAA phosphorylation target, a serine residue, changing it to aspartate or alanine, and so mimicking phosphorylated and nonphosphorylated states, respectively (see Diederich et al., Suggested Reading, and chapter 13). Together, these observations formed the basis for the following model for σ^F regulation in a sporulating cell (Figure 14.23). As soon as the three *spoIIA* operon genes are expressed in the preseptation cell, SpoIIAB binds to and therefore inactivates σ^F. SpoIIAB also phosphorylates SpoIIAA and so prevents SpoIIAA-SpoIIAB binding. Thus, the SpoIIAB-σ^F complex is stable and σ^F cannot

Figure 14.24   The *spoIIA* operon and its gene products. (A) The three genes that are cotranscribed. (B) The inhibitory effects of SpoIIAA and SpoIIAB, as inferred from genetic experiments described in the text.

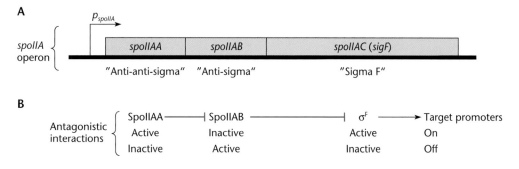

A

P_{spoIIA}

spoIIA operon

spoIIAA   *spoIIAB*   *spoIIAC (sigF)*

"Anti-anti-sigma"   "Anti-sigma"   "Sigma F"

B

Antagonistic interactions

	SpoIIAA	SpoIIAB	σ^F	Target promoters
	Active	Inactive	Active	On
	Inactive	Active	Inactive	Off

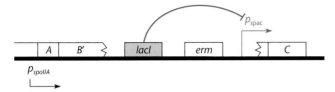

Figure 14.25 Induction of *spoIIAC* (*sigF*) from the p_{spac} promoter, which consists of RNA polymerase recognition sequences of *B. subtilis* phage SPO1 and *lac* operator sequences. Insertion of *lacI* accompanies insertion of p_{spac}, and so *spoIIAC* transcription is inducible with IPTG.

direct transcription. However, σ^F could become active in the forespore if SpoIIAB released it after septation. Thus, the model proposed that SpoIIAA is dephosphorylated in the forespore, so that it can bind SpoIIAB and cause the release of active σ^F.

The necessity that SpoIIAA be dephosphorylated in order to release σ^F activity predicted that the sporulating cell must express a SpoIIAA~P phosphatase. The collection of *spo* mutants was evaluated for the possibility that one of the known genes might encode the hypothesized phosphatase. A candidate for such a phosphatase was the *spoIIE* gene product, because *spoIIE* mutants had a phenotype consistent with a defect in σ^F activation: they expressed the *spoIIA* (*sigF*) operon but failed to express σ^F-dependent genes (Table 14.3). This prediction was borne out by in vitro studies in which SpoIIE dephosphorylated SpoIIAA~P.

The SpoIIE protein associates with the polar septum. This and other incompletely understood factors contribute to the limitation of σ^F activity to the forespore (see Hilbert and Piggot, Suggested Reading).

Regulation of σ^E

Mother cell transcription depends on σ^E, which is encoded by the *spoIIGB* gene of the *spoIIG* operon. The primary product of *spoIIGB* is an inactive precursor, named Pro-σ^E, which is processed to form the active sigma factor. The protease that cleaves Pro-σ^E is the product of the first gene in the *spoIIG* operon, *spoIIGA*. Although the *spoIIG* operon is expressed in the predivisional sporangium, the *spoIIGA* product does not process SpoIIGB immediately but waits about an hour, until after septation has occurred. Then, notification from the forespore that development is proceeding and that σ^F has become active comes via the messenger SpoIIR, which is the product of a σ^F-transcribed gene. SpoIIR is secreted into the spaces between the cell membranes, where it signals to SpoIIGA to process Pro-σ^E.

An important clue to the explanation for the time lag in Pro-σ^E processing was the observation that mutants

which lacked σ^F activity failed to process Pro-σ^E to σ^E. Could a σ^F-transcribed gene be required for the processing mechanism? A genetic search for such a gene was undertaken (see Karow et al., Suggested Reading) with the rationale that a strain with a mutant in the hypothetical gene might have a SigF$^+$ SigE$^-$ phenotype, i.e., it would express σ^F-dependent fusions but would not express σ^E-dependent fusions. Accordingly, mutants that expressed the σ^F-dependent fusion *gpr-gus* but did not express two σ^E-dependent fusions, *spoIID-lacZ* and *spoVID-lacZ*, were sought. The use of two *lacZ* fusions was intended to reduce the likelihood of isolating Lac$^-$ strains that were mutant in *lacZ* itself rather than in the desired regulatory gene. The Gus$^+$ Lac$^-$ mutants isolated from this screen were of two types: mutants with mutations of the *spoIIG* locus, as would be expected, and mutants with mutations of a new locus, which was named *spoIIR*. Evaluation of the *spoIIR* mutants for Pro-σ^E synthesis and processing showed that Pro-σ^E was indeed synthesized but was not processed. Thus, the *spoIIR* gene seemed to have the predicted properties. Subsequently, the SpoIIR protein was found to be secreted into the spaces between the membranes, with the outcome that SpoIIGA is activated to process Pro-σ^E. There is an additional factor, forespore specific degradation of Pro-σ^E, that operates to restrict σ^E accumulation to the mother cell (Figure 14.26).

σ^G, A SECOND FORESPORE-SPECIFIC SIGMA FACTOR

The σ^G-encoding gene, *spoIIIG*, is transcribed in the forespore by σ^F RNA polymerase. Its transcription lags behind that of other σ^F-transcribed genes, evidently because it requires a signal from the mother cell sigma, σ^E. The evidence for the existence of such a signal is indirect at present—the *spoIIIG* gene is not transcribed in a *spoIIG* (σ^E) mutant—but the signal has not yet been identified. Another aspect of σ^G regulation involves a mechanism related to σ^F regulation. Indeed, the two sigmas are closely related in their amino acid sequences, and, like σ^F, σ^G is inactivated by the anti-sigma SpoIIAB. However, the mechanism that releases σ^G from SpoIIAB differs, since the phosphorylation cycle described above for σ^F does not appear to affect σ^G regulation.

σ^K, A MOTHER CELL SIGMA

The last sigma to be made, σ^K, is expressed only in the mother cell. σ^E RNA polymerase transcribes the *sigK* gene. Like σ^E, σ^K is cleaved from a precursor protein, Pro-σ^K. Also, like σ^E processing, σ^K processing depends on a signal from the forespore. In this case, the signal is expression of the *spoIVB* gene under the control of the forespore sigma, σ^G (Figure 14.22).

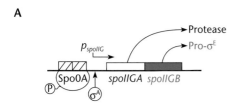

A

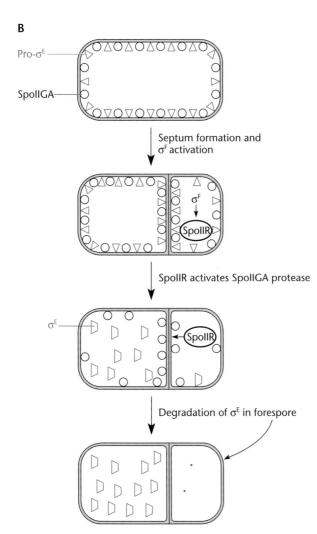

B

Figure 14.26 Model for activation of σ^E in the mother cell compartment. (A) The *spoIIG* operon is transcribed by σ^A RNA polymerase and requires activation by Spo0A~P (hatched boxes indicate Spo0A~P-binding sites [see Box 14.2]). (B) Pro-σ^E and SpoIIGA are associated with the cytoplasmic membrane in the sporangium. After septum formation, both proteins are associated with all cell membranes. Then SpoIIR is expressed in the forespore under the control of σ^F and SpoIIR activates SpoIIGA protease, which cleaves Pro-σ^E to form active σ^E, which is distributed in the cytoplasm. Finally, any Pro-σ^E or σ^E is degraded in the forespore.

σ^K Activation

The SpoIVB protein is thought to be secreted across the innermost membrane surrounding the forespore and to communicate with the Pro-σ^K processing factors across the membrane, thus activating the SpoIVFB protease that cleaves Pro-σ^K (Figure 14.27). SpoIVB activation of the σ^K-specific protease does not occur directly but, rather, by deactivation of proteins, SpoIVFA and BofA, that inhibit the protease. The complexity of this mechanism was revealed by isolation of "bypass suppressor" mutations. These were mutations that bypassed the requirement for σ^G involvement in σ^K activation.

The motivation for isolating suppressor mutations was the observation that late mother cell gene expression depends not only on the mother cell sigma, σ^K, but also on σ^G, the forespore sigma, and other forespore proteins. Moreover, the σ^G requirement is manifested at the step of Pro-σ^K processing, since Pro-σ^K protein accumulated in *spoIIIG* mutant cells. It was hypothesized that mutations that bypassed the σ^G requirement might provide information about the mechanism of σ^G involvement.

Genetic Analysis of σ^K Activation

The isolation of suppressor mutations involved a screen for mutations that allowed expression of a σ^K-dependent fusion, *cotA::lacZ*, in a ΔsigG strain (see Cutting et al.,

Figure 14.27 Model for regulation of Pro-σ^K processing based on the genetic data discussed in the text.

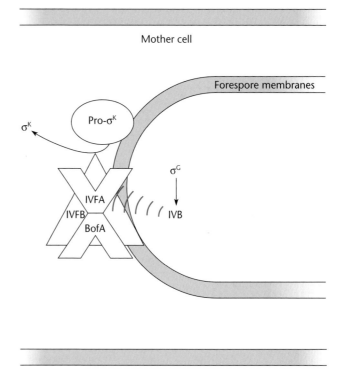

Suggested Reading). The *sigG* mutant strain was mutagenized with nitrosoguanidine to produce a broad spectrum of mutations. The *cotA::lacZ* fusion was then introduced by specialized transduction, and lysogens were screened for a blue-colony phenotype on X-Gal plates. Two classes of mutations were isolated and localized, defining loci that were named *bofA* and *bofB* (for *b*ypass *of f*orespore). Characterization of the *bof* mutants showed that *cotA* expression still required σ^K and, importantly, that Pro-σ^K processing was restored in *bof* mutants. The *bofA* mutations defined a new gene. However, the *bofB* mutations were missense or nonsense mutations in a previously discovered gene, *spoIVFA*. The second gene of this operon, *spoIVFB*, is thought to be the protease that processes Pro-σ^K, as mentioned above. Recent work has also shown that SpoIVFA and the BofA protein work together to inhibit SpoIVFB (Figure 14.27).

The identity of SpoIVB as the σ^G-dependent signal was inferred from the observation that *bof* mutations restored *cotA::lacZ* expression in a *spoIVB* mutant, just as for the *spoIIIG* mutant. The SpoIVB protein is a serine protease, and it is hypothesized that this protein, produced in the forespore as discussed above, can cross the innermost membrane surrounding the forespore to trigger the Pro-σ^K proteolysis mechanism by cleaving SpoIVFA.

Finding Sporulation Genes: Mutant Hunts, Suppressor Analysis, and Functional Genomics

The discussions in the preceding sections show that a variety of approaches have been useful for identifying *B. subtilis* sporulation genes. A very large percentage of the regulatory genes were defined by mutations that caused a Spo⁻ phenotype. However, an important class of regulatory genes was not well represented in the Spo⁻ mutant collection. These genes are the negative regulators. Loss-of-function mutations—which are the most frequent kind of mutation—in negative regulators would not cause a Spo⁻ phenotype, as discussed above for genes such as *spoIIAB*. "Special" alleles, e.g., gain-of-function

mutations, might cause a Spo⁻ phenoype, as for *spo0E* and *spo0L*, the negative regulators of the phosphorelay, but these would generally be found at a low frequency. However, suppression analysis is a powerful tool in such cases and has revealed the existence of important negative regulators. As one example, the *abrB* gene (Figure 14.17) was defined by mutations that restored degradative enzyme and antibiotic synthesis, but not sporulation, to *spo0A* mutants.

Another type of gene underrepresented in the sporulation mutant collections encodes a function that is redundant, or overlapping, with that encoded by another gene. An example is the set of *kin* genes, which encode the kinases that initiate the phosphorelay. Mutations that inactivate any one of these genes cause only a weak Spo⁻ phenotype. Only one of these, *kinA*, was found as a very leaky Spo⁻ mutant in a collection of Tn*917*-induced mutants.

Additional genes involved in sporulation have been defined by gene knockout analysis of ORFs annotated in the *Bacillus* genome sequence (http://bacillus.genome.ad.jp) and by regulon analysis (see Wang et al., Suggested Reading). Recent technological developments using gene fusions to fluorescent probes now make it possible to monitor the movement of proteins in real time during sporulation. Figure 14.28 shows a method involving the fusion of proteins to the gene for the fluorescent protein GFP to monitor the distribution of the protein during sporulation. In fact, fusions constructed in this way were used to make the illustration on the cover of this book, showing the formation of FtsZ division septa in the early stages of division in *B. subtilis*. Information obtained using methods such as these, building on the wealth of information obtained on *B. subtilis* sporulation using molecular genetic and biochemical analyses, promises to deliver new insights into how multicompartmental communication occurs during this relatively simple developmental process. It will be surprising if the same principles are not used in more complex multicellular development.

Figure 14.28 Method for constructing reporter gene fusions to reporter genes and integrating them into the chromosome of *B. subtilis*. (A) Structure of the cloning vector pMUTIN-GFP (see Figure 4.26) showing the restriction sites that allow C-terminal translational fusions to the gene for GFP. (B) A PCR product including the gene to be fused to GFP, in the example the gene for FtsZ, is cut with the restriction endonucleases and cloned into the plasmid vector. The plasmid is then transformed into *B. subtilis*, selecting erythromycin-resistant (Erm^r) colonies. Because the pMUTIN plasmid cannot replicate in *B. subtilis*, the only way it can be maintained is if it integrates by a single crossover into the *ftsZ* gene in the chromosome. The chromosome now contains a duplication of the gene for FtsZ bracketing the plasmid vector. The cell now makes normal FtsZ plus FtsZ fused to GFP. *sbp* is a gene for a conserved membrane protein adjacent to the *ftsA-ftsZ* operon.

A Structure of pMUTIN-GFP⁺

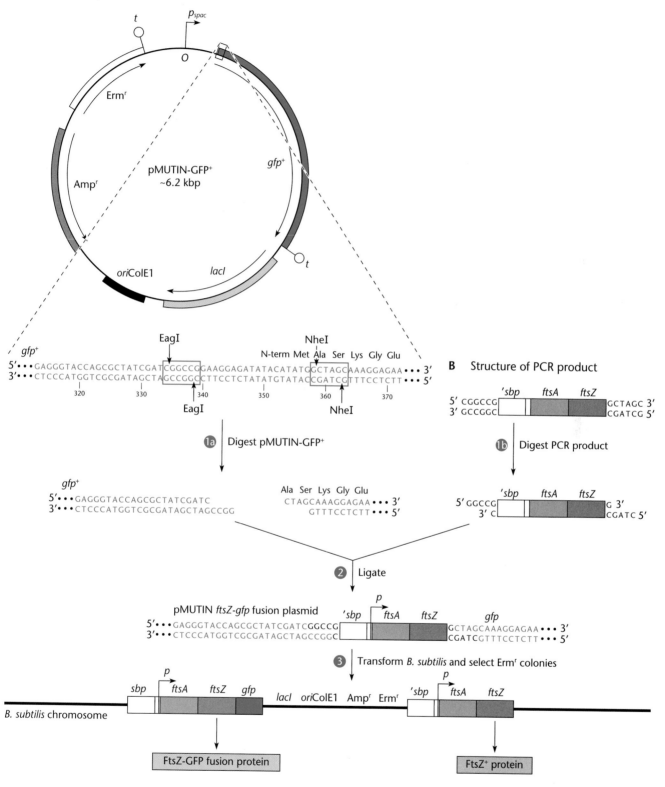

EagI

NheI

N-term Met Ala Ser Lys Gly Glu

5'•••GAGGGTACCAGCGCTATCGATCGGCCGGAAGGAGATATACATATGGCTAGCAAAGGAGAA•••3'
3'•••CTCCCATGGTCGCGATAGCTAGCCGGCTTCCTCTATATGTATACGATCGTTTCCTCTT•••5'

320   330   340   350   360   370

EagI

NheI

B Structure of PCR product

5' CGGCCG 'sbp ftsA ftsZ GCTAGC 3'
3' GCCGGC   CGATCG 5'

1a Digest pMUTIN-GFP⁺

1b Digest PCR product

gfp⁺

5'•••GAGGGTACCAGCGCTATCGATC
3'•••CTCCCATGGTCGCGATAGCTAGCCGG

Ala Ser Lys Gly Glu
CTAGCAAAGGAGAA•••3'
     GTTTCCTCTT•••5'

5' GGCCG 'sbp ftsA ftsZ G 3'
3' C   CGATC 5'

2 Ligate

pMUTIN ftsZ-gfp fusion plasmid

5'•••GAGGGTACCAGCGCTATCGATCGGCCG 'sbp ftsA ftsZ gfp GCTAGCAAAGGAGAA•••3'
3'•••CTCCCATGGTCGCGATAGCTAGCCGGC   CGATCGTTTCCTCTT•••5'

3 Transform B. subtilis and select Ermʳ colonies

B. subtilis chromosome

sbp ftsA ftsZ gfp lacI oriColE1 Ampʳ Ermʳ 'sbp ftsA ftsZ

FtsZ-GFP fusion protein

FtsZ⁺ protein

SUMMARY

1. Genetic experiments, using fusions to genes of the maltose transport system, combined with structural information, have established a model for protein export by the SecYEG channel. Part of the SecY protein forms a plug in the channel that moves over to contact the C terminus of the SecE protein on binding of the signal sequence of the protein to be exported.

2. In bacteria, the SRP pathway mostly targets proteins destined for insertion into the inner membrane while the Sec system (SecB and SecA) targets proteins to be exported to the periplasm, the outer membrane, or the outside of the cell.

3. The Tat pathway exports proteins that have already folded in the cytoplasm, such as proteins that must bind redox factors in the cytoplasm after folding.

4. Fusions to the *E. coli* alkaline phosphatase can be used to determine the periplasmic and cytoplasmic domains of transmembrane proteins in gram-negative bacteria.

5. Protein secretion systems differ between gram-negative bacteria and gram-positive bacteria because of the presence of an outer bilipid membrane in gram-negative bacteria.

6. Secretion systems of gram-negative bacteria form pores in the outer membrane that are composed of β-sheets organized into rings to form a β-barrel.

7. There are five different types of protein secretion systems in gram-negative bacteria, types I to V. Type I systems use a specific ABC transporter and the TolC channel in the outer membrane. Type II systems are related to some competence systems and the systems that assemble type IV pili; they use the SecYEG channel to get proteins through the inner membrane and then use a pseudopilus to push the protein through a secretin channel in the outer membrane. Type III

systems form a syringe-like injectisome that injects effector proteins directly through both membranes into a eukaryotic cell; they are related to the flagellar motors. Type IV systems are related to DNA conjugation systems and some competence systems and also inject proteins through both membranes into eukaryotic cells. Type V systems include autotransporters, twin-partner secretion systems, and chaperone-usher systems; they form dedicated β-channels in the outer membrane that can be part of the secreted protein itself. Chaperone-usher systems use a dedicated periplasmic chaperone and an outer membrane protein called an usher to assemble some types of pilins. The chaperone helps assemble the pilin subunits on the periplasmic side of an outer membrane-usher channel, and then the pili are passed through the channel as they are assembled.

8. Gram-positive bacteria also make syringe-like structures called injectosomes, that secrete effector proteins directly into eukaryotic cells.

9. Because their cell wall is not surrounded by an outer membrane, gram-positive bacteria can attach proteins to their cell wall and have them exposed to the external environment. Sortases are enzymes that specifically attach proteins to the cell wall.

10. While forward genetics is unsurpassed at identifying regulatory genes, it is sometimes less good at identifying the gene products responsible for their phenotypes. The use of microarrays combined with genome sequencing information can sometimes identify genes responsible for phenotypes, even if they are redundant.

11. *Bacillus subtilis* sporulation is the best understood developmental system. It involves a regulatory cascade of sigma factors and communication between cellular compartments vis-à-vis phosphatases, kinases, and proteases.

QUESTIONS FOR THOUGHT

1. Why do all types of cells including eukaryotes have an SRP system and cotranslate proteins to be inserted into the SecYEG channel but only bacteria have SecB and SecA to secrete proteins after they have been translated?

2. Why do type II secretion systems have so many proteins in the inner membrane when they use the SecYEG or Tat channels to get proteins through the inner membrane and use their

own channel only to get the proteins through the outer membrane?

3. Why are *B. subtilis* cells so averse to sporulating that they sporulate only after prolonged starvation and even induce genes to kill some cells so they can be cannibalized to feed the other cells and delay sporulation? What does this say about the purpose of sporulation?

PROBLEMS

1. Why did the selections for mutations in the Sec and SRP pathways not yield mutations in genes for the Tat pathway? How would you design a selection for genes in the Tat pathway?

2. The product of the *envZ* gene is a transmembrane protein in the inner membrane of *E. coli* (see chapter 13). Describe how you would determine which of the regions of the EnvZ protein are in the periplasm and which are in the cytoplasm.

3. Outline how you would isolate suppressors of mutations in the signal sequence coding region of the *malE* gene.

4. How were *B. subtilis* regulatory genes identified?

5. Contrast the similarities and differences of the *B. subtilis* phosphorelay and typical two-component systems (see Box 13.4).

6. What types of *spo0E* and *spo0L* mutations suggested that the genes were positive regulators? Negative regulators? How did suppression analysis help in understanding that Spo0E and Spo0L are actually negative regulators?

7. What is the difference between *spo-lacZ* transcriptional and *spo-gfp* translational fusions? What different questions do they answer?

8. The *tlp* and *cotD* genes encode proteins that are structural components of the endospore. Their dependence patterns of expression are similar to those of the *ssp* genes and *cotA*, respectively. On which sigma factor is *tlp* transcription dependent? *cotD*? Which compartments would you predict these genes are expressed in?

SUGGESTED READING

Anderson, D. M., and O. Schneewind. 1997. A mRNA signal for the type III secretion of Yop proteins by *Yersinia enterocolitica*. *Science* **278:**1140–1143.

Bassford, P., and J. Beckwith. 1979. *Escherichia coli* mutants accumulating the precursor of a secreted protein in the cytoplasm. *Nature* (London) **277:**538–541.

Bayan, N., I. Guilvout, and A. P. Pugsley. 2006. Secretins take shape. *Mol. Microbiol.* **60:**1–4.

Britton, R. A., P. Eichenberger, J. E. Gonzalez-Pastor, P. Fawcett, R. Monson, R. Losick, and A. D. Grossman. 2002. Genome-wide analysis of the stationary-phase sigma factor (sigma-H) regulon of *Bacillus subtilis*. *J. Bacteriol.* **184:**4881–4890.

Broder, D. H., and K. Pogliano. 2006. Forespore engulfment mediated by a ratchet-like mechanism. *Cell* **126:**917–928.

Brun, Y. V., and L. J. Shimkets (ed.). 2000. *Prokaryotic Development*. ASM Press, Washington, D.C.

Chater, K. F., and S. Horinouchi. 2003. Signaling early developmental events in two highly diverged *Streptomyces* species. *Mol. Microbiol.* **48:**9–15.

Cutting, S., V. Oke, A. Driks, R. Losick, S. Liu, and L. Kroos. 1990. A forespore checkpoint for mother cell gene expression during development in *B. subtilis*. *Cell* **62:**239–250.

Diederich, B., J. F. Wilkinson, T. Magnin, S. M. A. Najafi, J. Errington, and M. D. Yudkin. 1994. Role of interactions between SpoIIAA and SpoIIBB in regulating cell-specific transcription factor σ^F of *Bacillus subtilis*. *Genes Dev.* **8:**2653–2663.

Elliot, M. A., N. Karoonuthaisiri, J. Huang, M. J. Bibb, S. N. Cohen, C. M. Kao, and M. J. Buttner. 2003. The chaplins: a family of hydrophobic cell-surface proteins involved in aerial mycelium formation in *Streptomyces coelicolor*. *Genes Dev.* **17:**1727–1740.

Emr, S. D., S. Hanley-Way, and T. J. Silhavy. 1981. Suppressor mutations that restore export of a protein with a defective signal sequence. *Cell* **23:**79–88.

Errington, J. 1993. *Bacillus subtilis* sporulation: regulation of gene expression and control of morphogenesis. *Microbiol. Rev.* **57:**1–33.

Errington, J., and J. Mandelstam. 1986. Use of a *lacZ* gene fusion to determine the dependence pattern of sporulation operon *spoIIA* in *spo* mutants of *Bacillus subtilis*. *J. Gen. Microbiol.* **132:**2967–2976.

Feucht, A., T. Magnin, M. D. Yudkin, and J. Errington. 1996. Bifunctional protein required for asymmetric cell division and cell-specific transcription in *Bacillus subtilis*. *Genes Dev.* **10:**794–803.

Flardh, K. 2003. Growth polarity and cell division in *Streptomyces*. *Curr. Opin. Microbiol.* **6:**564–571.

Guest, B., G. L. Challis, K. Fowler, T. Kieser, and K. F. Chater. 2003. Gene replacement by PCR targeting in *Streptomyces* and its use to identify a protein domain involved in the biosynthesis of the sesquiterpene odour geosmin. *Proc. Natl. Acad. Sci. USA* **100:**1541–1546.

Gutierrez, C., J. Barondess, C. Manoil, and J. Beckwith. 1987. The use of transposon Tn*phoA* to detect genes for cell envelope proteins subject to a common regulatory stimulus. *J. Mol. Biol.* **195:**289–297.

Hilbert, D. W., and P. J. Piggot. 2004. Compartmentalization of gene expression during *Bacillus subtilis* spore formation. *Microbiol. Mol. Biol. Rev.* **68:**234–262.

Hoch, J. A., and K. I. Varughese. 2001. Keeping signals straight in phosphorelay signal transduction. *J. Bacteriol.* **183:**4941–4949.

Hoffman, C. S., and A. Wright. 1985. Fusions of secreted proteins to alkaline phosphatase: an approach for studying protein secretion. *Proc. Natl. Acad. Sci. USA* 82:5107–5111.

Hueck, C. J. 1998. Type III protein secretion systems in bacterial pathogens of animals and plants. *Microbiol. Mol. Biol. Rev.* 62:379–433.

Karow, M., P. Glaser, and P. J. Piggot. 1995. Identification of a gene, *spoIIR*, that links the activation of σ^E to the transcriptional activity of σ^F during sporulation in *Bacillus subtilis*. *Proc. Natl. Acad. Sci. USA* 92:2012–2016.

Koronakis, V., J. Eswaran, and C. Hughes. 2004. Structure and function of TolC: the bacterial exit duct for proteins and drugs. *Annu. Rev. Biochem.* 73:467–489.

Kostakioti, M., C. L. Newman, D. G. Thanassi, and C. Stathopoulos. 2005. Mechanisms of protein export across the bacterial outer membrane. *J. Bacteriol.* 187:4306–4314.

Lazazzera, B., T. Palmer, J. Quisel, and A. D. Grossman. 1999. Cell density control of gene expression and development in *Bacillus subtilis*, p. 27–46. *In* G. M. Dunny and S. C. Winans (ed.), *Cell-Cell Signaling in Bacteria*. ASM Press, Washington, D.C.

Levin, P. A., and R. Losick. 2000. Asymmetric division and cell fate during sporulation in *Bacillus subtilis*, p. 167–190. *In* Y. V. Brun and L. J. Shimkets (ed.), *Prokaryotic Development*. ASM Press, Washington, D.C.

Madden, J. C., N. Ruiz, and M. Caparon. 2001. Cytolysin-mediated translocation (CMT): a functional equivalent of type III secretion in gram-positive bacteria. *Cell* 104:143–152.

Marraffini, L. A., A. C. DeDent, and O. Schneewind. 2006. Sortases and the art of anchoring proteins to the envelopes of gram-positive bacteria. *Microbiol. Mol. Biol. Rev.* 70:192–221.

Oliver, D. B., and J. Beckwith. 1981. *E. coli* mutant pleiotropically defective in the export of secreted proteins. *Cell* 25: 2765–2772.

Palmer, T., F. Sargent, and B. C. Berks. 2005. Export of complex cofactor-containing proteins by the bacterial Tat pathway. *Trends Microbiol.* 13:175–180.

Perego, M., C. Hanstein, K. M. Welsh, T. Djavakhishvili, P. Glaser, and J. A. Hoch. 1994. Multiple protein-aspartate phosphatases provide a mechanism for the integration of diverse signals in the control of development in *B. subtilis*. *Cell* 79:1047–1055.

Ruvolo, M. V., K. E. Mach, and W. F. Burkholder. 2006. Proteolysis of the replication checkpoint protein Sda is necessary for the efficient initiation of sporulation after transient replication stress in *Bacillus subtilis*. *Mol. Microbiol.* 60:1490–1508.

San Milan, J. L., D. Boyd, R. Dalbey, W. Wickner, and J. Beckwith. 1989. Use of *phoA* fusions to study the topology of the *Escherichia coli* inner membrane protein leader peptidase. *J. Bacteriol.* 171:5536–5541.

Schatz, P. J., and J. Beckwith. 1990. Genetic analysis of protein export in *Escherichia coli*. *Annu. Rev. Genet.* 24:215–248.

Schmidt, R., P. Margolis, L. Duncan, R. Coppolecchia, C. P. Moran, Jr., and R. Losick. 1990. Control of developmental transcription factor σ^F by sporulation regulatory proteins SpoIIAA and SpoIIAB in *Bacillus subtilis*. *Proc. Natl. Acad. Sci. USA* 87:9221–9225.

Schroder, G., and C. Dehio. 2005. Virulence-associated type IV secretion systems of *Bartonella*. *Trends Microbiol.* 13: 336–342.

Smith, M. A., W. M. Clemons, Jr., C. J. DeMars, and A. M. Flower. 2005. Modeling the effects of *prl* mutations on the *Escherichia coli* SecY complex. *J. Bacteriol.* 187: 6454–6465.

Sonenshein, A. L., J. A. Hoch, and R. Losick (ed.). 2001. Bacillus subtilis *and Its Closest Relatives: from Genes to Cells*. ASM Press, Washington, D.C.

Stephenson, S. J., and M. Perego. 2002. Interaction surface of the Spo0A response regulator with the Spo0E phosphatase. *Mol. Microbiol.* 44:1455–1467.

Sun, C., S. L. Rusch, J. Kim, and D. A. Kendall. 2007. Chloroplast SecA and *Escherichia coli* SecA have distinct lipid and signal peptide preferences. *J. Bacteriol.* 189:1171–1175.

Tian, H., and J. Beckwith. 2002. Genetic screen yields mutations in genes encoding all known components of the *Escherichia coli* signal recognition particle pathway. *J. Bacteriol.* 184:111–118.

Wang, S. T., B. Setlow, E. M. Conlon, J. L. Lyon, D. Imamura, T. Sato, P. Setlow, R. Losick, and P. Eichenberger. 2006. The forespore line of gene expression in *Bacillus subtilis*. *J. Mol. Biol.* 358:16–37.

Wild, K., K. R. Rosendal, and I. Sinning. 2004. A structural step into the SRP particle. *Mol. Microbiol.* 53:357–363.

Answers to Questions for Thought and Problems

Chapter 1

Questions for Thought

1. The two strands of bacterial DNA probably are not replicated in the 3′-to-5′ direction simultaneously because replicating a DNA as long as the chromosome in this manner would leave single-stranded regions that were so long that they would be unstable or susceptible to nucleases.

2. DNA molecules may be very long because if cells contained many short pieces of DNA, each one would have to be segregated individually into the daughter cells.

3. DNA may be the hereditary material instead of RNA because double-stranded DNA has a slightly different structure than double-stranded RNA. The B-form structure of DNA may have advantages for replication, etc. The use of DNA instead of RNA may allow the primer, which is made of RNA, to be more easily identified and removed from a DNA molecule by the editing functions. The editing functions do not operate when the 5′ end is synthesized. By removing and resynthesizing any RNA regions, using upstream DNA as primer, mistakes can be minimized.

4. A temperature shift should cause the rate of DNA synthesis to drop but not too abruptly. Each cell would complete the rounds of replication that were under way at the time of the shift but would not begin another round. The synthesis rate would drop exponentially. If cells are growing rapidly, the drop would be less steep because a number of rounds of DNA replication would be under way in each molecule and each would have to complete the cycle.

5. The gyrase of *Streptomyces sphaeroides* might be naturally resistant to novobiocin. You could purify the gyrase and test its ability to introduce supercoils into DNA in vitro in the presence of novobiocin.

6. How chromosome replication and cell division are coordinated in bacteria like *Escherichia coli* is not well understood. One hypothesis is that a protein required for cell division might be encoded by a gene located close to the termination region and this gene is transcribed into RNA only when it replicates. Cell division could then begin only when the termination region has replicated. You could move the terminus of replication somewhere else in the chromosome and see if this affects the timing of cell division.

7. No known answer, but maybe this allows for more genome rearrangements such as inversions and duplications, which may play an important role in evolution. If chromosome replication obligatorily stopped at a certain *ter* sequence, a major chromosome rearrangement may make it impossible to replicate the entire DNA because then the *ter* sequence could be encountered before the entire DNA had replicated. With some readthrough, a few of the cells survive and the new gene arrangement might prove more effective.

Problems

1. 5′GGATTA3′

2. 5′GGAddT3′

   5′GGATddT3′

   5′GGATTACGGddT3′

   5′GGATTACGGTAAGGddT3′

3. $I = 25$ min, $C = 40$ min, $D = 20$ min

4. $I = 90$ min, $C = 40$ min, $D = 20$ min

5. The *topA* mutant lacks topoisomerase I, which removes negative supercoils, so there should be more negative supercoils in the DNA of the mutant.

6. 5′NNGAATTCATTAAGATCG3′, where the N's could be any deoxynucleotide. Extra nucleotides are added at the end so that it can be cut by EcoRI, which does not efficiently cut sites right at the end of the DNA. Preferably, G's or C's rather than A's or T's should be put at the end to minimize "fraying" of the ends during amplification since G's and C's form more stable base pairs than A's and T's.

7. The cloning vector primer is 5′NNCTCTTTCGTACGTCGC3′. The insert primer is 5′NNCTCTTCTACGAAGCTCT3′, where the N's can be any deoxynucleotide but preferably are G's or C's for the reasons given in the answer to problem 6.

Chapter 2
Questions for Thought

1. Some people think RNA came first because the peptidyltransferase that links amino acids to make protein is the 23S rRNA, and some other enzymes are also RNA. However, the question remains open to speculation.

2. The genetic code may be universal because once the code was established, too many components—aminoacyl-tRNA synthetases, etc.—were involved in translating the code to change all of them.

3. Why eukaryotes rarely have polycistronic mRNAs is open to speculation. It might have something to do with the necessity for exporting mRNA from the nucleus before it can be translated.

4. The genetic code of mitochondrial genes differs from the chromosomal code of eukaryotes because mitochondria were once bacteria with their own simple translation apparatus, ribosomes, etc. This translation apparatus has remained independent of that for the chromosomal genes, so they have gone their separate ways.

5. Selenocysteine may be a relic of what was once a useful process in an earlier organism from which all other organisms evolved. In some proteins, selenocysteine in the active center does enhance the reaction rate.

6. The translation apparatus is very highly conserved evolutionarily, so that an antibiotic that inhibits the translation apparatus of one type of bacteria is apt to inhibit the translation apparatus of all bacteria. This is less true of amino acid biosynthetic pathways.

7. The two-chamber structure of chaperonins is very mysterious and is made even more so by the fact that the chaperonins of archaea and the eukaryotic cytoplasm have a two-cylinder structure similar to those of eubacteria, even though the sequences of the two types of chaperonins are not related and hence apparently evolved independently. One idea is that this structure has something to do with regulation and the taking up of proteins that must be folded in a chaperonin chamber, such as actin. One chamber plays a regulatory role, recognizing sequences on the protein as the protein emerges from the exit pore of the ribosome, and the other chamber then takes up the protein from the N terminus, which could be some distance away. The presence of two chambers means that the ends are far enough apart to play both roles without having to make a longer structure.

8. The *sec* system transports proteins only through the inner membrane of gram-negative bacteria, and so secreted proteins must have their own system to get through the outer membrane. It might be best to have the transport system through the outer membrane directly coupled to the transport system through the inner membrane, because all the ATP for energy is in the cytoplasm or in the form of a membrane potential in the inner membrane. It might be more difficult to coordinate both processes if the SecYEG channel is being used. Also, it might be more efficient to have their own system because they do not have to compete for transport with other transported proteins. Another possibility is that the presence of their own specialized transport system makes secretion of a protein encoded by a prophage less dependent on the bacterium in which the prophage is located. The protein can secrete itself from the cell even if the cell has an incompatible *sec* system.

9. How this is achieved is not known and is the subject of active investigation. One possibility is that other proteins such as YidC might somehow open up the SecYEG channel to let transmembrane domains of inner membranes out into the inner membrane as they are traversing the channel. Maybe the extreme hydrophobicity of these domains helps open the channel.

10. One reason is to save energy by not making a particular protein unless it is needed. At least two GTPs and one ATP are required to translate each amino acid, not to mention the hundreds of nucleoside triphosphates required to transcribe a gene. Another reason is to prevent interference between intermediates of pathways. For example, the intermediates in some degradative pathways might inhibit other degradative pathways. A third reason is that it might help in replication of the chromosome. Transcribing RNA polymerase can interfere with the replication fork, and so the transcription of only a few genes at any one time may help speed up chromosome replication.

Problems

1. 5'CUAACUGAUGUGAUGUCAACGUCCUACUCUA-GCGUAGUCUAA3'

2. It is likely that translation begins in the second triplet GUG because this is followed by a long open reading frame, although this hypothesis would have to be tested.

3. The answer is (a). Both sequences have a string of A's, but only in (a) is this preceded by an inverted repeat that could form a hairpin loop in the mRNA.

4. A good primer would be 5'CC GGA TCC ATG TTG CGA TTT3'. A good way to think of it is that the BamHI restriction site in the cloning vector is read GGA TCC, so that the BamHI site in the fragment being cloned will also be read this way after it is cloned. Reading the BamHI site in this way in the primer puts the downstream coding region in the correct frame. You also have to add at least two random deoxynucleotides to the 5' end before the amplified fragment can be cut with BamHI. We added two C's because G's and C's are preferred on the end (see problem 6, chapter 1). Some restriction endonucleases require more than two deoxynucleotides at the 5' end before they cut at a site.

5. a. Expression of the operon is constitutive and the genes are expressed even if inducer is not added.

b. Expression of the operon is turned off, and the operon cannot be induced.

6. It is usually best to PCR amplify from the cDNA rather than from the chromosomal DNA because eukaryotic genes often have large introns and these introns will have been spliced out of the mRNA from which the cDNA is made. To make the cDNA, it is necessary to use reverse transcriptase and a primer that is complementary to the region immediately upstream and including the terminator codon but read in the 3'-to-5' direction. This cDNA is then used as a template for PCR with two primers: one which has, at its 3' end, the same sequence as the mRNA immediately downstream of the AUG initiator codon for the human gene and the other which has, at its 3' end, the complement of the sequence immediately upstream of the terminator codon, read in the 3'-to-5' direction. At the 5' end of the primers, sequences for restriction sites are added to facilitate cloning into an E. coli expression vector and fusing of the gene sequence in frame with an affinity tag encoded by the cloning vector on either the N or C terminus of the protein. The vector

containing the cloned gene can then be introduced into E. coli, and the promoter on the expression vector can be induced to express the human protein in E. coli. It can then be purified on the affinity column to which the affinity tag binds. It may be desirable to cut the affinity tag off the protein after it is purified, using a sequence-specific protease and the fact that a site for the protease has been introduced on the cloning vector.

Chapter 3
Questions for Thought

1. Why the genetic maps of *Salmonella* and *Escherichia* spp. are similar is unknown. Perhaps there is an optimal way to arrange genes on a chromosome, with genes that are expressed at high levels closest to the origin of replication and transcribed in the same direction in which they are replicated. Inversions would then be selected against.

2. Duplication mutations would allow the number of genes of an organism to increase. The duplicated genes could then evolve so that their products could perform novel functions. Sometimes organisms with a duplication of a particular region may have a selective advantage in a particular environment, and so the duplication would be preserved.

3. The cells of higher organisms may be more finely tuned because runthrough proteins resulting from translation through the ends of genes may create more problems for eukaryotic cells, which, being more complicated, may be less tolerant of aberrant proteins. Alternatively, they may be less efficient at degrading aberrant proteins, or they may trigger an immune response in those organisms that have one.

4. It is not known how directed, or adaptive, mutations might occur. In their purest form, adaptive mutations would require that the cell somehow sense that a mutation would be desirable and change the DNA sequence accordingly. There would have to be some flow of information from the protein product back to the DNA.

Problems

1. The cultures with the largest number of mutants probably had the earliest mutation.

2. Arginine auxotrophy would have a higher mutation rate because many genes encode enzymes to make arginine and any mutation that inactivates the product of one of these genes would make the cell Arg⁻. Rifampin resistance, however, can be caused by only a few mutations in the gene for the β subunit of RNA polymerase because very few amino acids can be changed and have the RNA polymerase no longer bind rifampin but still be active for transcription.

3. Approximately 8×10^{-10}

4. 5.5×10^{-8}

5. a. *arg-1* is probably a leaky missense or another type of base pair change mutation since it seems to retain some activity of the gene product and it reverts.

b. *arg-2* is probably a deletion mutation since it is not leaky and does not revert.

c. *arg-1* could be a frameshift mutation close to the end of the gene encoding the carboxyl end of the protein, so that the gene product is not totally inactivated. The mutant with *arg-2* could be a double mutant with two missense mutations in *arg* genes, so that both of them would seldom revert simultaneously.

6. You could make a double mutant with a *dam* mutation and, for example, an *arg* mutation that makes the cell Arg⁻. Then you could compare the reversion frequency to Arg⁺ of this double mutant with that of the single *arg* mutant without the *dam* mutation.

7. If the mutants are isolated from the same culture, they could be siblings with the same original mutation and so they would be the same and not representative of all the mutations that can cause the phenotype.

8. a. Plate large numbers of the bacteria on plates containing the antibiotic coumermycin and all the other needed growth supplements.

b. Plate large numbers of the Trp⁻ bacteria on minimal plates lacking tryptophan but containing all the other needed growth supplements.

c. Plate large numbers of the *dnaA*(Ts) mutant on plates at the high, nonpermissive temperature to isolate mutants that can multiply to form a colony at the high temperature. Test by crossing with the wild type to show that there are temperature-sensitive recombinant types but that the two mutations are closely linked.

d. Plate large numbers of the *araD* mutant on plates containing L-arabinose plus another carbon source. Test mutants that can multiply by crossing with the wild type to show that there are some arabinose-sensitive recombinants.

e. Make a partial diploid that has a polar *hisC* mutation in one copy of the operon and a *hisB* mutation in the other copy. Then plate large numbers of the partial diploid on minimal plates lacking histidine but having all the other necessary growth supplements. Mutants that can grow to form a colony could have a polarity-suppressing mutation in *rho* that allows *hisB* expression from the first copy of the operon, complementing the *hisB* mutation in the other copy.

9. Grow large numbers of the cells at the low, permissive temperature and then shift to the high, nonpermissive temperature before adding ampicillin. After incubating the cells for a period, wash out the ampicillin and plate the cells at the lower temperature. Repeat the enrichment once or twice. Test colonies which arise at the low temperature to find ones due to mutant bacteria which cannot form colonies at the high temperature.

10. Nonsense suppressors are dominant. The mutant tRNA still inserts an amino acid at the nonsense codon even in the presence of the normal tRNA.

11. The mutation is probably a nonsense suppressor in a tRNA gene. You could test it to see if the strain propagates a phage with a nonsense mutation in an essential gene or if it suppresses nonsense mutations in other genes.

12. The *trpA* mutation is the selected marker, and the *argH* and *hisG* mutations are the unselected markers.

13. Very close to the *thyA* gene.

14. The selected marker is the *metB1* mutation, and the unselected markers are the *leuA5* mutation and the Tn*5* insertion mutation. A little before 44 min.

15. Close to *argG*.

16. The cotransduction frequency of the *argH* and *metB* markers is 37%, and that of the *argH* and *rif* markers is 20%. The order is probably *metB1–argH5–rif-8*, which is also consistent with the three-factor cross data. The order *argH5–metB1–rif-8* is consistent with some of the data, but the *argH* marker is probably in the middle since most of the Rif* transductants are not also Met⁺, as they would be if they were on the same side of the *argH* marker.

17. If the *metA15* mutation had been suppressed, you would not expect any Met⁺ transductants because both the donor and recipient have the *metA15* mutation. About 86% would be Arg⁻, and 14% would be Arg⁺ (the cotransduction frequency between the *arg* and *met* markers).

18. First, transduce the Tn*10* transposon insertion mutation into your *hemA* mutant, selecting the tetracycline resistance gene on the transposon by plating on plates containing tetracycline. Identify a tetracycline-resistant transductant that still has the *hemA* mutation because it does not grow on plates lacking δ-aminolevulinic acid. This strain can be used to move the *hemA* mutation, selecting Tet*.

19. Many more than 4,000.

20. The clone probably does not contain all of the *hemA* gene. Recombination between the clone and the *hemA* gene in the chromosome seems to be required to make a functional *hemA* gene, explaining why not all the bacteria containing the clone are HemA⁺. If the clone contained the entire *hemA* gene, it would complement the mutation in the chromosome, and all the bacteria would be HemA⁺ and would grow without δ-aminolevulinic acid.

21. Introduce the gene with the Cm* cassette cloned in a plasmid cloning vector into the cell under conditions where the cloning vector cannot replicate (i.e., is a suicide vector), and plate the cells on plates containing chloramphenicol. Test any Cm* cells for the presence of the cloning vector. For example, if the cloning vector carries ampicillin resistance (Ap*), you could screen for cells which are Cm* but not Ap*. These are presumably strains which have had two crossovers, replacing the normal *arg* gene with the *arg* gene containing the Cm* cassette. You would expect the strain to be Arg⁻.

Chapter 4
Questions for Thought

1. If essential genes were carried on a plasmid, cells that were cured of the plasmid would die. By having nonessential genes on plasmids, the chromosome can be smaller so that the cells can multiply faster; the species can adapt to a wide range of environments because the cells can exchange plasmids and sudden selection for cells with a particular plasmid will allow some members of the population to survive. You would not expect genes encoding enzymes of the tricarboxylic acid cycle, such as isocitrate dehydrogenase, or genes for proteins involved in macromolecular synthesis, such as RNA polymerase, to be carried on plasmids. Genes involved in using unusual carbon sources such as the herbicide 2,4-D or genes required for resistance to antibiotics such as ampicillin might be expected to be on plasmids.

2. Why some but not all plasmids have a broad host range is unknown. A broad-host-range plasmid can parasitize more species of bacteria. A narrow-host-range plasmid can develop a better commensal relationship with its unique host.

3. Perhaps a single copy of the plasmid binds to each of two sites on the ends of filaments formed by the ParA protein. When the filaments grow, they push these two copies of the plasmid apart toward opposite poles of the cell. This may be why the plasmid must finish replicating before the filaments can grow.

4. If the genes required for replication of the plasmid are not all closely linked to the *ori* site, you could find them by isolating temperature-sensitive mutants of the plasmid that cannot replicate at high temperature and then looking for pieces of plasmid DNA that can help the mutant plasmid replicate at the high temperature when introduced into the cell in a cloning vector.

5. You could determine which of the replication genes of the host *E. coli* (*dnaA*, *dnaC*, etc.) are required for replication of a given plasmid by introducing the plasmid into cells with temperature-sensitive mutations in each of the genes and then determining whether the plasmid can replicate at the high nonpermissive temperature for the mutant. Plasmid replication could be determined by adding radioactive precursors for DNA synthesis after the temperature has been raised and determining if the radioactive DNA hybridizes to plasmid DNA.

6. One advantage to having the leader sequence degraded, rather than the mRNA for the RepA protein itself, is that it may help prevent the synthesis of defective RepA protein that could interfere with replication. If the mRNA for the RepA protein is itself degraded, defective translation products of RepA due to running off the end of the degraded mRNA might compete with normal RepA for replication.

7. It actually seems more consistent with the pairing model. The daughter plasmids may remain paired until just before cell division. The ParA filaments then begin to form, and they push each of the plasmids in the pair to opposite poles, ensuring that each cell gets at least one copy of the plasmid.

8. The DNA polymerase can replicate all the way to the end of the leading strand but cannot replicate the inverted repeat at the end of the lagging strand. The newly synthesized inverted repeat on the leading strand could flip over and substitute for the inverted repeat on the lagging strand since they have the same sequence in the 5′-to-3′ direction. The inverted repeat on the leading strand could then be resynthesized using the upstream DNA as primer. Make a diagram of your model.

Problems

1. You introduce your plasmid into cells with the other plasmid and grow the cells in the presence of streptomycin and/or sulfonamide but in the absence of kanamycin. After a few generations, you plate the cells and test the colonies for kanamycin resistance. If more of the cells are kanamycin sensitive than would be the case if your plasmid were replicating in the same type of cell without the other plasmid, your plasmid is probably an IncQ plasmid and so is incompatible with the other plasmid.

2. 1/2,048

3. You take the colonies due to the transformants on ampicillin plates and test them on tetracycline plates to determine whether they are tetracycline resistant. The plasmids in the cells that are ampicillin resistant but tetracycline sensitive probably have an insert in the BamHI site, since the BamHI site of pBR322 is in the Tet^r gene.

4. Make plates containing higher and higher concentrations of ampicillin until bacteria containing the RK2 plasmid can no longer multiply to form a colony. Plate large numbers of bacteria containing the plasmid on this concentration of ampicillin. Any bacteria which form colonies may contain high-copy-number mutants of RK2. The copy number of the plasmid in these bacteria could be determined, and the plasmid could be introduced into new bacteria to show that the new bacteria are also made resistant to higher concentrations of ampicillin by the plasmid, thereby showing that the mutation is in the plasmid itself and not in the chromosome. The *repA* gene in the mutant plasmid could then be sequenced to ensure that this is the region of the responsible mutation.

5. The plasmid should have an easily selectable gene such as for resistance to an antibiotic. Cells containing the plasmid are grown through a number of generations in the absence of the antibiotic and plated without the antibiotic. The number of cells cured of the plasmid is then determined by replicating these plates onto a plate containing the antibiotic. Any colonies that do not transfer onto the new plate are due to bacteria that have been cured of the plasmid. If this number is much smaller than predicted from the normal distribution based on the copy number of the plasmid, the plasmid may have a partitioning system.

6. You can purchase the RNA polymerases of the phages. You mix one of these with the pBAC DNA containing the clone along with the nucleoside triphosphates, one of which is radioactively labeled. To be safe, it might be helpful to cut the clone on the other side with a restriction endonuclease to prevent transcription all the way through the clone and into the cloning vector on the other side. The RNAs that are synthesized are hybridized to other pBAC clones to find ones which hybridize and are therefore overlapping.

7. Changing one of the complementary sequences prevents pseudoknot formation, and therefore the Shine-Dalgarno sequence of *repZ* mRNA is masked by structure III. No RepZ is made whether or not Inc antisense RNA is made, and the plasmid does not replicate.

Chapter 5

Questions for Thought

1. If plasmids of the same Inc group all have the same Tra functions, the exclusion functions prevent a plasmid from transferring into a cell that already contains a plasmid of the same Inc group, where one or the other would be lost.

2. If the *tra* genes and the *oriT* site on which they act are close to each other, recombination between the *oriT* site and the genes for the Tra functions seldom occurs. Separating the Tra functions from the *oriT* site on which they act renders them nonfunctional.

3. Plasmids with certain *mob* sites can be transferred only by certain corresponding Tra functions because the relaxase of the mobilizable plasmid must bind to the coupling protein of the self-transmissible plasmid.

4. Self-transmissible promiscuous plasmids may encode their own primases, so they can transfer themselves into a distantly related host cell with an incompatible primase and still synthesize their complementary strand.

5. The absence of an outer membrane in gram-positive bacteria may make a pilus unnecessary. The question is open to speculation.

6. The answer is not known. Perhaps the role of the Mpf structure, including the pilus, in helping transmit DNA into the cell can be easily subverted by the phage to transfer its own DNA into the cell. Alternatively, the fact that the pilus is extended outside the cell may make it an attractive absorption site, much like it makes it an antigen.

7. Plasmids are generally either self-transmissible or mobilizable because if they were neither, they would not be able to move to cells that did not already contain them. By being promiscuous, plasmids can expand their host range and parasitize other cells.

8. They can be mobilized by plasmids of a number of different Inc groups. Apparently, they are designed so that their Dtr system can communicate with a number of different coupling factors. A self-transmissible plasmid lacking its Mpf functions can be mobilized only by members of its own Inc group.

Problems

1. The recipient strain is the one that becomes recombinant and retains most of the characteristics of the original strain. If the transfer is due to a prime factor, the apparent recombinants become donors of the same genes.

2. Determining which of the *tra* genes of a self-transmissible plasmid encodes the pilin protein is not easy. For example, phage-resistant mutants do not necessarily have a mutation in the pilin gene. They could also have a mutation in a gene whose product is required to assemble the pilus on the cell surface. You could purify the pili and make antibodies to them. Then *tra* mutants with mutations in the pilin gene do not make

an antigen that will react with the antibody. Similarly, to determine which *tra* mutants do not make the DNase that nicks the DNA at the *ori* region or the helicase that separates the DNA strands, you may have to develop assays for these enzymes in crude extracts and determine which *tra* mutants do not make the enzyme in your assay.

3. You can show that only one strand of donor DNA enters a recipient cell by using a recipient that has a temperature-sensitive mutation in its primase gene. The plasmid DNA should remain single stranded after transfer into such a strain, provided, of course, that the plasmid cannot make its own primase. Single-stranded DNA is more sensitive to some types of DNases and behaves differently from double-stranded DNA during gel electrophoresis.

4. If the tetracycline resistance gene is in a plasmid, the tetracycline-resistant recipient cells should have acquired the plasmid. If it is in a conjugative transposon, no transferred plasmid will be in evidence. Also, Southern blots should reveal that the conjugative transposon now has different flanking sequences.

5. Male-specific phage cannot infect cells containing only a mobilizable plasmid because mobilizable plasmids do not encode a pilus, which serves as the adsorption site for the phage.

6. The protein product of the *eex* gene prevents the entry of a plasmid of the same Inc group into the cell. You could mutagenize the plasmid randomly with a transposon such as Tn*5* and isolate a number of insertion mutants with mutations in the plasmid by mating it into another strain, selecting the kanamycin resistance on the Tn*5*. Cells containing the plasmid with different insertion mutants could then be patched on a plate on which have been spread cells containing a plasmid of the same Inc group but carrying a different antibiotic resistance gene. After incubation, this plate could be replicated onto another plate carrying the second antibiotic. Any patches containing transconjugants that have become resistant to the second antibiotic are candidates for having contained the plasmid with the Tn*5* transposon in its *eex* gene, and they could be tested directly for plasmid exclusion. It might be necessary to have kanamycin on the first plate and have the potential donor cells also be resistant to kanamycin, in addition to being resistant to the second antibiotic carried on the plasmid, if the background due to cells cured of the mutagenized plasmid is too high.

7. If a plasmid is self-transmissible, it also often mobilizes at least some type of mobilizable plasmid. A collection of different mobilizable plasmids, containing easily selectable markers for antibiotic resistance, could be tried. They could be individually introduced into cells containing the indigenous plasmid by electroporation, selecting the antibiotic resistance on the mobilizable plasmid. These cells could then be mixed with related cells that lack the indigenous plasmid, selecting for antibiotic resistance on the mobilizable plasmid and counterselecting the donor strain. If one of the mobilizable plasmids is mobilized into this strain, it will become resistant to the antibiotic carried on the mobilizable plasmid. If you cannot introduce the mobilizable plasmids into the cells by electroporation, you could try

a triparental mating to see if the indigenous plasmid can mobilize any of the mobilizable plasmids into a third strain.

Chapter 6
Questions for Thought

1. The chapter lists some possible reasons for the development of competence in bacteria. They include the ability to try combinations of alleles to enhance fitness, to repair damage to DNA, or even to provide nutrition, or a combination of these. At this time, we do not know the correct answer or even whether the same reason is true for all competent bacteria.

2. To discover whether the competence genes of *Bacillus subtilis* are turned on by UV irradiation and other types of DNA damage, you could make a gene fusion with a reporter gene such as *lacZ* to one of the competence factor-encoding genes and see if the reporter gene is induced following UV irradiation.

3. To determine whether antigenic variation in *Neisseria gonorrhoeae* results from transformation between bacteria or recombination within the same bacterium, you could introduce a selectable gene for antibiotic resistance into one of the antigen genes and see if it is transferred naturally under conditions where antigenic variation occurs.

Problems

1. To determine whether a given bacterium is naturally competent, you would isolate an auxotrophic mutant, such as a Met⁻ mutant, and mix it with DNA extracted from the wild-type bacterium. The mixture would then be plated on medium without methionine. The appearance of colonies due to Met⁺ recombinants would be evidence of transformation.

2. To isolate mutants defective in transformation, you would take your Met⁻ mutant, mutagenize it, and repeat the above test on individual isolates. Any mutants that do not give Met⁺ recombinants when mixed with the wild-type DNA might be mutants with a second mutation in a competence gene.

3. To discover whether a naturally transformable bacterium can take up DNA of only its own species or any DNA, you could make radioactive DNA and mix it with your competent bacteria. Any DNA taken up by the cells would become resistant to added DNase, and the radioactivity would be retained with the cells on filters. Try this experiment with radioactive DNA from the same species as well as from different species.

4. If the bacterium can take up DNA of only the same species, it must depend on uptake sequences from that species. The experiment should be done as in problem 3 but with only known pieces of DNA instead of the entire molecule. If a known piece of DNA is taken up, the responsible uptake sequence could be determined by trying overlapping fragments to see what region they must have in common to be taken up.

5. If the DNA of a phage successfully transfects competent *E. coli*, plaques appear when the transfected cells are plated with bacteria sensitive to the phage.

Chapter 7
Questions for Thought

1. If phages made the proteins of the phage particle at the same time as they made DNA, the DNA might be prematurely packaged into phage heads, leaving no DNA to replicate.

2. Phages that make their own RNA polymerase can shut off host transcription by inactivating their host RNA polymerase without inactivating their own RNA polymerase. However, phages that use the host RNA polymerase can take advantage of the ability of the host molecule to interact with other host proteins, allowing more complex regulation.

3. In this way, they can infect a wider range of hosts and still replicate their DNA.

4. No known answer. Perhaps again it has something to do with the range of hosts they can infect. In one range of hosts, the Pri proteins may be more compatible, while in another range of hosts, the RNA polymerases may be more compatible.

5. No known answer, but if the phage injects a protein with the DNA that is intended to be used early, this protein could present problems for the phage which is in later states of development. Also, the DNA of the second infecting phage would be in a different stage of replication, which might interfere with the replication of the DNA of the first infecting phage.

Problems

1. You could mix a known amount of the virus with the cells and then measure the fraction of survivors. From the Poisson distribution, you could then measure the effective multiplicity of infection (MOI). e^{-MOI} = fraction of surviving bacteria. The ratio of the effective MOI to the actual MOI is the fraction of the viruses that actually infected the cells.

2. The regulatory gene is probably gene M, because mutations in this gene can prevent the synthesis of many different gene products. The other genes probably encode products required for the assembly of tails and heads.

3. Amber mutations introduce a nonsense UAG codon into the coding sequence of an mRNA, stopping translation and leading to synthesis of a shortened gene product. The *ori* sequence does not encode a protein, and so an amber mutation could not be isolated in it.

4. The order is *A-Q-M*, because with this order most of the Am⁺ recombinants with a crossover between *amA* and *amQ* would have the Ts mutation in gene *M*. A second crossover between *Q* and *M* would be required to give the wild-type recombinant.

5. T1 has a linear genetic map, which is expanded for the genes at the end of the linear DNA.

6. A phage with a mutation in its lysozyme gene does not lyse the cells and release phage unless egg white lysozyme is added. Infect the cells and allow the infection to proceed long enough for the cells to lyse if they had been infected by the wild-type

phage. Then divide the culture in half, and add lysozyme to one of the divided cultures. Plate both cultures with indicator bacteria. If the culture with the added lysozyme yields many more plaques than the culture without lysozyme, the phage that had infected the cells contained a mutation in its lysozyme gene. Addition of the lysozyme caused the cells to lyse, releasing their phage and making many more plaques rather than just one plaque where the original infected cell was located.

7. It should lyse cells infected with an antiholin mutant because this mutant would still make the holin and allow the lysozyme to destroy the cell wall, allowing access of the $CHCl_3$ to the cytoplasmic membrane. It should not lyse a holin mutant for the same reason. It should also not lyse a lysozyme mutant for the same reason.

8. If you infect at a high MOI, the phage you isolate that is displaying the peptide on the surface of its capsid, and is thus being "panned," may not be the one that encoded it. This used to be called phenotypic mixing in phage genetics.

Chapter 8
Questions for Thought

1. Perhaps the λ prophage uses different promoters to transcribe the cI repressor gene immediately after infection and in the lysogenic state because this may allow the repressor gene to be transcribed from a strong, unregulated promoter immediately after infection but then be transcribed from a weaker, regulated promoter in the lysogenic state.

2. By making two proteins, the cell can use the Int protein to promote recombination for both integration and excision. Then the smaller Xis protein need only recognize the hybrid *att* sites at the ends of the prophage.

3. Morons might have been integrated in a process much like integrons pick up gene cassettes, using an integrase encoded by the phage or even by a different DNA element in the cell. It is difficult to identify the *att* sites on such cassettes since they show a lot of variability. The selection could vary from one moron to another. If the prophage carries the moron, cells in which it has formed a lysogen might have some selective advantage over lysogenic cells containing the prophage without it.

4. It is not known why some types of prophage can be induced only if another phage of the same type infects the lysogenic cell containing them. Perhaps there is some way of inducing them that has not been tried.

5. There is not much to distinguish them. P4 encodes very few of its own gene products to make a phage and depends on P2 for its head and tail proteins and most of its other functions. Perhaps the major difference is that P4 phage DNA can replicate in the cell by itself.

Problems

1. If the clear mutant has a *vir* mutation, it forms plaques on a λ lysogen.

2. A specialized transducing phage carrying the *bio* operon of *E. coli* would be isolated in the same way as the λ*dgal* phage in the text, except that a Bio⁻ mutant would be infected and plated on medium without biotin. The Bio⁺ bacteria would be isolated, and the phage would be induced. It might be necessary to add a wild-type helper phage before induction since *bio* substitutions extend into the *int* and *xis* genes. The λ*pbio* phage should form plaques because no replication genes should be substituted.

3. A λ phage with *vir* mutations in the o_1 sites of o_L and o_R should form plaques on a λ lysogen because the repressor must bind first to the o_1 sites of the operators.

4. Both Int and Xis are required to integrate phage λ transducing particles next to an existing prophage because the recombination occurs between two hybrid *attP-attB* sites.

5. Amber mutations lead to termination of translation, so you cannot make the remainder of the protein carrying the other domain for intragenic complementation.

6. Infect a P4 lysogen with your phage, and see if the lysate contains any P4 phage. They can be detected because their head is smaller and they can form plaques on a P2 lysogen but not on a nonlysogen.

7. To determine whether the *Staphylococcus aureus* toxin is encoded by a prophage, you could take the strain of *S. aureus* and try to induce the prophage from it and cure the cells by UV irradiation, etc. One way to do this would be to introduce a selectable gene, for example for antibiotic resistance, into the prophage and screen for cells that are no longer resistant after UV irradiation. To test for toxin production, the cured cells could be filtered out and the cell-free medium could be tested for the presence of toxin by the ability to kill human cells in culture. If the induced prophage forms plaques, the medium could be plated on a closely related *S. aureus* strain to see if plaques form. You could then use the phage in the plaques to try to isolate lysogens of the related *S. aureus* strain to determine whether the lysogens of this strain now produce the toxin.

Chapter 9
Questions for Thought

1. Perhaps replicative transposons do not occur in multiple copies around a genome because their resolution functions cause deletions between repeated copies of the transposon, resulting in the death of cells with more than one copy. Also, a poorly understood phenomenon called target immunity inhibits transposition of a transposon into a DNA that already contains the same transposon.

2. The transposon Tn3 and its relatives may have spread throughout the bacterial kingdom on promiscuous plasmids.

3. Transposons sometimes, but not always, carry antibiotic resistance or other traits of benefit to the host. They may also help the host move genes around, as in the construction of plasmids carrying multiple drug resistance.

4. The origin of the gene cassettes within integrons is not known, but they may come from superintegrons, like the one found in *Vibrio cholerae*.

5. It is possible that invertible sequences rarely invert because very little of the invertase enzyme is made or because the enzyme works very inefficiently on the sites at the ends of the invertible DNA sequence.

Problems

1. You would integrate a λd*gal* at the normal λ attachment site close to the *gal* operon with one of the *gal* mutations by selecting for Gal⁺ transductants. Sometimes the *gal* genes in the chromosome would recombine with the *gal* genes on the integrated λ and the cell would become Gal⁻ by gene conversion (homogenoting). When the λd*gal*s are induced, their DNA should be longer and the phage should be denser, owing to the inserted DNA making the λ genome longer.

2. The colonies should not be sectored because only one strand, either the *lac* or *lac*⁺ strand, of the original heteroduplex transposon has been inserted and the other strand has been copied from it.

3. The advantages are that transposon insertions almost always inactivate the gene and are not leaky, so that the phenotypes of a null mutant can be known. They also mark the site of the mutation both genetically and physically, so that the site of the mutation is easier to determine by either genetic or physical mapping techniques. The disadvantage is that they almost always inactivate the gene, so that the effects of other types of mutations in the gene cannot be studied. You also cannot get transposon insertion mutations in an essential gene in a haploid organism because they are lethal.

4. Follow the procedure for the mutagenesis and mapping of pAT153 outlined in the text. Only the sizes of the junction fragments will be different.

5. Perform Southern blot analyses using a sequence within the transposon as a probe. Isolate a number of strains with independent transposon insertions. Digest DNA with a restriction endonuclease that has no sites in the transposon. If different-sized fragments always light up in the different strains, the junction fragments must all be different and the transposon integrates randomly.

6. The Kan^r gene should be on the side closest to the SalI site on the plasmid. The transposon is inserted 0.372 kb from the SalI site on the plasmid, so that, with this orientation, the smallest SalI site should be 2.6 + 0.372 = 2.972 kb and the largest should be 9.2 − 2.972 = 6.228 kb.

7. a. Use a promiscuous self-transmissible plasmid such as RP4 to mobilize a plasmid suicide vector containing Tn*5* into the strain of *Pseudomonas putida*. Select the Kan^r transconjugants on rich medium containing kanamycin, and then replicate them onto minimal medium with 2,4-D as the sole carbon source, looking for transposon insertion mutants that cannot use 2,4-D as a sole carbon and energy source and so cannot

grow to form colonies. Pick the corresponding colonies from the kanamycin plates, and isolate the DNA from such mutants. Cut the DNA with a restriction endonuclease that does not cut in the transposon, and ligate the pieces of the DNA into an *E. coli* plasmid cloning vector. Use the ligation mix to transform *E. coli*, selecting Kan^r transformants. These should contain a plasmid clone with at least part of a gene whose product is required to use 2,4-D. The corresponding wild-type gene could be found by using the clone with the transposon as a probe to screen a wild-type library of the *P. putida* strains by plate hybridizations.

b. Make a library of the DNA of the *P. putida* strain in a broad-host-range mobilizable plasmid such as RSF1010. Mobilize the plasmid library into mutants of the *P. putida* strain that cannot use 2,4-D, and select transconjugants that can form colonies on minimal plates containing 2,4-D as the sole carbon source. The plasmid cloning vector in the bacteria in these colonies should contain the gene that was mutated to prevent growth on 2,4-D, and the plasmid gene is complementing the chromosomal mutation.

8. Perhaps MuB or some other protein remains attached to the end of the Mu DNA when it enters the cell, allowing only one round of a cut-and-paste transposition event.

9. You could introduce an amber mutation into the *int* gene of the conjugative transposon and use an amber suppressor strain for the donor cell and a nonsuppressor for the recipient cell. If the transposon can still integrate into the chromosome of the recipient, it must have transferred the Int protein made in the donor cell since it cannot make Int protein in the recipient.

10. Start a culture from a single colony, isolate the DNA from the culture, and perform a Southern hybridization after cutting with a restriction nuclease that cuts off center in the G segment and using a probe complementary to the G segment. If you get two bands, the segment has inverted. Another way might be to map transposon insertion mutations to antibiotic resistance in the prophage G segment with respect to markers in the neighboring chromosomal DNA by three-factor crosses. If you get a consistent order, the G segment is not inverting in the prophage. An even better way would be to insert a promoterless reporter gene next to the invertible element so that it is transcribed from a promoter of the invertible sequence only if the invertible sequence flips into the other orientation. If the reporter gene is expressed in some cells, the invertible element is inverting.

11. You can use mutants of the phages that lack a functional DNA invertase of their own and propagate them in isogenic cells that are lysogenic for e14 and cured of e14. Pick plaques, and test the host range of the phage in the plaques to determine if their invertible sequence has inverted and changed their host range. To test for *Salmonella* phase shift, again use a *Salmonella* mutant which lacks the invertase and introduce the e14 prophage. Test whether it can now shift from one cell surface antigen to the other.

12. By methods such as those outlined in chapter 1, you could clone the pigment gene and use it as a probe in Southern blot analyses to see whether the flanking sequences around the gene change when it is in the pigmented as opposed to the nonpigmented form.

13. You could make a plasmid that has two copies of the Mu phage by cloning a piece of DNA containing Mu into a plasmid cloning vector also containing Mu. You could then see if the two repeated Mu elements could resolve themselves in the absence of the host recombination functions.

Chapter 10
Questions for Thought

1. Recombination is required to restart replication forks that have stalled at damage to the DNA. It might also help speed up evolution by allowing new combinations of alleles to be tried, and it helps in the repair of DNA damage.

2. The reason that RecBCD recombination is so complicated is perhaps because its real role might be to repair double-strand breaks caused when replication encounters damage in the DNA. The free ends can then invade the other daughter DNA and, with the help of the Pri proteins, re-form a replication fork. This interpretation is supported by the observation that *chi* sites are mostly arranged so that they can help re-form replication forks.

3. The different pathways of recombination may function under different conditions for recombination between short and long DNAs or at breaks and gaps, etc.

4. The RecF pathway is preferred under conditions different from the ones normally used in the laboratory to measure recombination. The SbcB and SbcC functions may interfere with the RecF pathway only under the conditions normally used in laboratory crosses such as transductional, transformational, or Hfr crosses, where the RecBCD pathway is preferred.

5. By encoding their own recombination functions, phages can increase their rate of recombination. Also, some phages use the recombination functions for replication and, by encoding their own functions, can inhibit the host recombination functions to prevent them from interfering with phage replication, as in the case of the RecBCD function and λ rolling-circle replication.

6. There could be another as yet undetected X-phile in the cell that cuts the Holliday junctions migrated by the RecG helicase. Alternatively, the RecG helicase may allow replication restarts by backing up the replication fork to form a type of Holliday junction called a chicken foot and then replicating the remainder of the DNA to remove the Holliday junction rather than resolving it by using an X-phile.

Problems

1. A common way to determine if recombinants in an Hfr cross have a *recA* mutation is to take the individual recombinants and streak them across a plate. Then half of each streak is covered with a glass plate (glass is opaque to UV radiation), and the plate is irradiated before incubation. If the recombinant is RecA⁻, it grows only in the part of the streak that was covered by the plate because RecA⁻ mutants are much more sensitive to killing by UV.

2. To determine which other genes, if any, participate in the *recG* pathway, you could set up a synthetic lethal screen based on the fact that RecG is required only in the absence of RuvABC. You could construct a strain in which the *ruv* genes are transcribed from an inducible promoter on a plasmid in a cell with *ruvABC* deleted in the chromosome, and isolate transposon insertion mutants in the presence of the inducer. The mutants you are interested in do not grow in the absence of inducer. You can then see if any of the mutants have mutations in genes other than the *recG* gene.

3. The recombination promoted by homing double-stranded nucleases to insert an intron occurs in the same manner as in Figure 10.4, except that the double-strand break that initiates the recombination occurs at the site in the target DNA into which the intron will home. The invading DNA then pairs with the homologous flanking DNA on one side of the transposon in the donor DNA and replicates over the transposon until it meets the other 5′ end, inserting the intron.

4. In a RecB⁺C⁺D⁻ host, compare recombination between the same two markers in λ DNAs, one with a *chi* mutation and another without. If there is no difference, *chi* sites stimulate recombination only in the presence of the RecD function.

Chapter 11
Questions for Thought

1. The answer is unknown. Perhaps the newly synthesized strand of DNA somhow remains bound to the replication apparatus for some distance behind the replication fork. This is suggested by the observation that the SeqA protein bound to hemimethylated DNA in *E. coli* apparently travels with the replication fork. Perhaps a protein or RNA remains bound to the newly synthesized DNA strand for a period after it has been synthesized.

2. Different repair pathways work better depending on where the damage occurs, the type of damage, or the extent of the damage. For example, some common types of damage have their own dedicated repair system. If damage is so extensive that lesions occur almost opposite each other in the two strands of the DNA, it might be easier to repair the lesions with excision repair than with recombination repair. Alternatively, if the damage is irreparable, it might be better just to replicate over it.

3. The SOS mutagenesis pathway might exist to allow the cells to survive damage other than that due to UV irradiation, or it might be more effective under culture conditions different from those used in the laboratory.

Problems

1. Assuming that you can grow the organism in the laboratory (you might have to grow it under high pressure), you could irradiate it in the dark and then divide the culture in half and expose half of the cells to visible light before diluting and counting the surviving bacteria. If more of the cells survive after they have been exposed to visible light, the bacterium has a photoreactivation system.

2. The procedure is explained in the text. Briefly, to show that the mismatch repair system preferentially repairs the unmethylated strand, you could make heteroduplex DNA of λ phage. One strand should be unmethylated and heavier than the other because the λ phages from which this strand was derived were propagated on Dam⁻ *E. coli* cells grown in heavy isotopes. The two λ phages used to make the heteroduplex DNA should also have mutations in different genes, so that there are mismatches at these positions. After the heteroduplex λ DNA is transfected into cells, test the progeny phage to determine which genotype prevails: the genotype of the phage from which the unmethylated DNA was prepared or the genotype of the phage with methylated DNA. Then reverse the two DNAs so that the other DNA has the heavy isotope to eliminate marker effects.

3. To determine whether the photoreactivating system is mutagenic, perform an experiment similar to that in problem 1 but with a *umuCD* mutant of *E. coli*. Measure the frequency of mutations (such as reversion of a *his* mutation) among the survivors of UV irradiation in the dark as opposed to those that have been exposed to visible light after UV irradiation. More cells should survive if they are exposed to visible light, but a higher *frequency* of these survivors should be His⁺ revertants if photoreactivation is mutagenic. If photoreactivation is not mutagenic, a lower frequency should be His⁺ revertants because the photoreactivation system will have removed some of the potentially mutagenic lesions. That is why it may be better to do this experiment with a *umuCD* mutant, to lower the background mutations due to SOS mutagenesis.

4. To find whether the nucleotide excision repair system can repair damage due to aflatoxin B, treat wild-type *E. coli* cells and a *uvrA*, *uvrB*, or *uvrC* mutant with aflatoxin B. Dilute and plate. Compare the survivor frequencies of the *uvr* mutant and the wild type.

5. Express *umuC* and *umuD* from a clone that has a constitutive operator mutation so that the cloned genes are not repressed by LexA. Also, be sure that part of the *umuD* gene has been deleted so that UmuD′, rather than the complete UmuD, is synthesized. This clone can be put into isogenic RecA⁻ and RecA⁺ strains of *E. coli* that have a *his* mutation. After UV irradiation, the frequency of His⁺ revertants among the surviving bacteria can be compared for each strain. If the RecA⁺ strain shows a higher frequency of His⁺ revertants, the RecA protein may have a role in UV mutagenesis other than inducing *umuCD* by cleaving LexA and then cleaving UmuD.

6. You could make a transcriptional fusion of the *recN* gene to a reporter gene such as *lacZ* and then determine whether more of the reporter gene product is synthesized after UV irradiation, as it should be if *recN* is an SOS gene. A strain that also has a *lexA*(Ind⁻) mutation should not show this induction if *recN* is induced because it is an SOS gene and not for some other reason.

Chapter 12
Questions for Thought

1. Why operons are regulated both positively and negatively is not clear. However, there may be different advantages to the two types of regulation. For example, negative regulation might require more regulatory protein but might allow more complete repression, while it might be easier to achieve intermediate levels of expression with positive regulation. There may be less interaction between negative regulatory systems than between positive regulatory systems, in which a regulatory protein might inadvertently turn on another operon. Also, constitutive mutants are rarer with positive regulation.

2. The genes for regulatory proteins may be autoregulated to save energy. If they are autoregulated, only the amount of regulatory protein needed is synthesized. Also, in the case of positive regulators, more can be made after induction to further increase the expression of the operons under their control.

3. These other amino acids have only one role in the cell: to be incorporated into proteins. Methionine has other roles, including being converted into *S*-adenosylmethionine, which is the donor of methyl groups in biosynthetic reactions. Since the *met* gene products must interact with many other pathways, its genes are separately subject to regulation by the other pathways so that there is no advantage to having its biosynthetic genes together in the same operon on the chromosome.

4. Regulation by attenuation of transcription of amino acid biosynthetic operons offers the advantage that the ability of the cell to translate codons for that amino acid can be exploited to regulate the operon. The regulatory system can be designed so that transcription continues into the structural genes of the operon if ribosomes stall in the leader region at codons for the amino acid because not enough of the amino acid is available. For degradative operons, it might have the advantage that the expression of the operon can be turned on more quickly if the substrate becomes available. A major disadvantage of this type of regulation is that it is wasteful. A short RNA is always made from the operon, even if it is not needed.

Problems

1. To isolate a *lacI*ˢ mutant, take advantage of the fact that *lacI*ˢ mutations, while rare, are dominant Lac⁻ mutations and the β-galactosidase, which is synthesized when the *lac* operon is induced, cleaves P-Gal, furnishing galactose that kills *galE* mutants. Mutagenize a *galE* mutant that contains an F′ factor with the *lac* operon, and plate it on P-Gal medium containing another carbon source such as maltose. The survivors that form colonies are good candidates for *lacI*ˢ mutants because inactivation of the *lacZ* genes in both the chromosome and the F′ factor requires two independent mutations, which should be even rarer than single *lacI*ˢ mutations. The mutants can be further tested by mating the F′ factor from them into other strains whose chromosome contains a wild-type *lac* operon. If the F′ factor makes the other strain Lac⁻, the F′ factor must contain a *lacI*ˢ mutation.

2. AraC must be in the P1 state, since it represses the *ara* operon.

3. a. Lac⁻ permanently repressed. The LacIˢ repressor binds to the operators of both operons, even in the presence of the inducer, and prevents transcription of the *lacZYA* structural genes.

b. Inducible Lac⁺. In other words, it is wild type for the *lac* operon. The *lacOᶜ* mutation in the chromosome makes the *lac* operon on the chromosome constitutive, but LacZ and LacY are not made anyway because of the polar mutation in *lacZ*.

c. Inducible Ara⁺. The inactivating *araC* mutation is recessive to the wild-type *araC* allele, and the cell is wild type for the *ara* operon and inducible by L-arabinose.

d. Inducible Ara⁺. The *araI* mutation prevents transcription of the operon in the chromosome, but the mutation is *cis* acting and so does not prevent transcription of the operon on the F′ factor.

e. Inducible Ara⁺. Same reason as (d). The *cis*-acting p_{BAD} promoter mutation prevents transcription of the chromosomal operon but not the operon on the F′ factor.

4. To determine whether *phoA* is negatively or positively regulated, you could first isolate constitutive mutants to determine how frequent they are. Since PhoA turns XP blue, you could mutagenize cells and isolate mutants that form blue colonies on XP-containing medium even in the presence of excess phosphate in the medium. If *phoA* is negatively regulated, constitutive mutants should be much more frequent than if it is positively regulated. Also, at least some of these constitutive mutants should have null mutations, deletions, etc., that inactivate the regulatory gene.

5. Plate wild-type *E. coli* cells in the presence of low concentrations of 5-methyltryptophan in the absence of tryptophan. Only constitutive mutants can multiply to form colonies under these conditions, because 5-methyltryptophan is a corepressor of the *trp* operon but cannot be used for protein synthesis, so that the nonmutant wild-type *E. coli* cells starve for tryptophan. To isolate mutants defective in feedback inhibition of tryptophan synthesis, plate a constitutive mutant in the presence of higher concentrations of 5-methyltryptophan in the absence of tryptophan. Even constitutive mutants cannot multiply to form colonies under these conditions, because the first enzyme of tryptophan synthesis is feedback inhibited by the 5-methyltryptophan. Only mutants that are defective in feedback inhibition form colonies.

6. The MalQ protein degrades polymers to glucose, which is not the inducer of the operon; maltotriose is the inducer. In a MalQ⁻ mutant, maltotriose accumulates, inducing the operon.

7. BglG⁻ mutants should be permanently repressed (superrepressed) because the BglG protein binds to antiterminator hairpins, stabilizing them. In the absence of BglG, the antiterminator hairpins do not form, the terminator hairpins form, and transcription termination occurs. BglF⁻ mutants, on the other hand, should be constitutive. BglF transfers phosphates to BglG when β-glucosides are not being transported, inactivating BglG. It also binds to BglG, sequestering it. In the absence of BglF, BglG cannot be phosphorylated and is active and free to bind to the antiterminator hairpins, even in the absence of β-glucosides.

Chapter 13
Questions for Thought

1. Perhaps it is important to make more of the proteins involved in synthesizing new proteins so that the rate of protein synthesis increases after heat shock, allowing more rapid replacement of the proteins irreversibly denatured as a result of the shock.

2. Perhaps *Salmonella* species, which are normal inhabitants of the vertebrate intestine, are usually in an environment where amino acids are in plentiful supply but NH₃ is limiting. *Klebsiella* species may usually be free-living, where NH₃ is present but amino acids are not.

3. Perhaps the genes for corresponding sensor and response regulator genes are similar to allow cross talk between regulatory pathways. If the genes are similar, a signal from one pathway can be passed to the other pathway, allowing coordinate regulation in response to the same external stimulus. However, there is no good evidence for the importance of cross talk. Another possible explanation is that the genes had a common ancestor in evolution and still retain many of the same properties.

4. The enzymes responsible for ppGpp synthesis during amino acid starvation and during growth rate control may be different because the enzymes involved in stringent control and growth rate regulation must be in communication with different cellular constituents. The RelA protein works in association with the ribosome, where it can sense amino acid starvation, while the enzyme involved in synthesizing ppGpp during growth rate regulation might have to sense the level of energy in ATP and GTP. SpoT might be involved both in synthesizing and degrading ppGpp if the equilibrium of the reaction is somehow shifted. All enzymes function by lowering the activation energy, and so in a sense they catalyze both the forward and backward reactions. However, because the equilibrium usually favors the forward reaction, this is the reaction that predominates. If the equilibrium were shifted, perhaps by sequestering the ppGpp as it is made, SpoT could synthesize ppGpp rather than degrade it.

Problems

1. Determine if the mutant can use other carbon sources such as lactose and galactose. If it has a *cya* or *crp* mutation, it should not be able to induce other catabolite-sensitive operons and so cannot grow on these other carbon sources.

2. a. Gln⁻ (glutamine requiring); Ntr constitutive (express Ntr operons even in presence of NH₃)

b. Ntr⁻ (cannot express Ntr operons even at low NH_3 concentrations); make intermediate levels of glutamine synthetase independent of the presence or absence of NH_3

c. Gln⁻, Ntr⁻

d. Gln⁻, Ntr⁻

e. Ntr constitutive

f. Grows slowly because it constitutively synthesizes heat shock proteins

g. Constitutive expression of diphtheria toxin and other virulence determinants, even in the presence of Fe^{2+}

h. No ppGpp; grows very slowly and is auxotrophic for some amino acids

3. As described in chapter 1, you could clone the gene for the toxin and then perform Southern blot analysis to show that the gene is carried on a large region of DNA that is not common to all the members of the species. Using the methods described in chapter 8, you could also try to induce a phage from the cells and show that production of the toxin requires lysogeny by the phage.

4. If the *rpoH* gene for σ³² is transcriptionally autoregulated, the same amount of RNA should be made from the gene when it exists in two or more copies as is made when it exists in only one copy. Introduce a clone of *rpoH* in a multicopy plasmid into cells, and measure the amount of RNA made on the gene by DNA-RNA hybridization. If more RNA is made from *rpoH* under these conditions, the gene is not transcriptionally autoregulated.

5. To show which ribosomal protein is the translational repressor, introduce an in-frame deletion into the *rplJ* gene and determine if the synthesis of L12 increases. Similarly, introduce a mutation (any inactivating mutation will do) into *rplL* and determine if L10 synthesis increases.

Chapter 14
Questions for Thought

1. It might be because eukaryotic cells are much larger, making it impractical to keep proteins unfolded after they are translated but before they are translocated through the membrane.

2. It might have to do with providing energy to get the proteins through both membranes. Since the only known sources of energy are in the cytoplasm in the form of ATP and GTP and in the inner membrane in the form of membrane potential, they must have some source of energy to get proteins through the outer membrane. Presumably, all of these extra proteins allow them to transport the proteins through both membranes if the proteins are to be secreted.

3. The bacteria might be slow to commit to sporulation because it limits their options. Once cells are committed to sporulation, they must go all the way, making it harder to reverse course and begin growing if nutrients later become

available. It may be that the real role of sporulation is to disseminate the bacteria to new locations and that other means are used to survive nutrient deprivation.

Problems

1. Because only proteins that have a specific Tat signal sequence and that have already folded are transported by the Tat pathway. Unlike the SecYEG transported proteins, which need only the signal sequence, the Tat transported proteins must have folded, and their folded state must be recognized by other specific proteins. You might try fusing *lacZ* to an entire protein transported by the Tat pathway and hope that the Tat transported protein would fold properly and be recognized by the Tat system.

2. You would fuse the *phoA* gene to regions of the *envZ* gene in such a way that variable-length N-terminal fragments of EnvZ are translationally fused to the alkaline phosphatase product of *phoA*. You would then express these fusions in an *E. coli* strain in which the chromosomal *phoA* gene is deleted. The fusion proteins in which the alkaline phosphatase protein is fused to a region of the EnvZ protein in the periplasm will make the colony blue on XP plates.

3. A signal sequence mutation in *malE* makes the cells Mal⁻ and unable to transport maltose for use as a carbon and energy source. A suppressor of the signal sequence mutation makes them Mal⁺ and able to grow on minimal plates with maltose as the sole carbon and energy source. Spread millions of bacteria with the *malE* signal sequence mutations on minimal plates with maltose. Any colonies which arise are Mal⁺ and may be revertants or may have suppressors of the signal sequence mutation. To distinguish those which have suppressors from the true revertants, you could use some of them as donors to transduce the *mal* operon into another strain, selecting for a nearby marker. If any of the transductants are Mal⁻, the *mal* operon in that Mal⁺ apparent revertant had a mutation somewhere else in the chromosome which was suppressing the signal sequence mutation.

4. By isolating mutants blocked in sporulation. The regulatory genes could then be identified because mutations in these genes blocked the expression of many other genes as determined by *lacZ* fusions. Also, some were similar in sequence to other known regulators, including sigma factors and transcriptional activators.

5. Both involve histidine kinases, which phosphorylate on histidine residues. Transfer of phosphoryl groups occurs between histidine and aspartate residues, and transcriptional activators are regulated by phosphorylation. The *Bacillus subtilis* phosphorelay contains a series of proteins that carry out discrete steps. Each of these proteins is regulated by other phosphatases and kinases.

6. The original mutations in *spo0E* and *spo0L* both conferred the Spo⁻ phenotype, suggesting that the products of the genes were positive regulators. However, these mutations did not inactive the gene products, and deletion mutations of the genes caused hypersporulation, suggesting that they were in fact negative regulators. Also, multiple copies of the genes inhibited sporulation, as expected of negative regulators. Spo⁻ suppressors of *spo0E* deletion mutations were in the gene for Spo0A, a phosphorylated protein. Also, Spo⁺ suppressors of *spo0L* missense mutations were in the gene for Spo0F, another phosphorylated protein. This suggested that Spo0E and Spo0L reduced the phosphorylation of these proteins by acting as phosphatases, and this was subsequently confirmed.

7. In a *spo-lacZ* transcriptional fusion, the *lacZ* gene has its own TIR and so is translated independently of the upstream *spo* sequences. Transcriptional fusions can be used to determine when the *spo* gene is transcribed. In the *spo-lacZ* translational fusions, the coding region of the *spo* gene is fused to the *lacZ* gene to encode a fusion protein with β-galactosidase activity. These fusions can be used to determine when the *spo* gene product is made.

8. The *tlp* gene is dependent on σ^G, and *cotD* is dependent on σ^K. The *tlp* gene is expressed in the forespore, and *cotD* is expressed in the mother cell.

Glossary

Activator. A protein that regulates transcription of an operon by interacting with RNA polymerase at the promoter and allowing RNA polymerase to begin transcribing the operon. Regulation of transcription by an activator is said to be positive because transcription of the operon is enhanced when the activator is active.

Activator site. A sequence in DNA upstream of the promoter to which the activator protein binds.

Adaptive mutation. *See* Directed-change (adaptive) mutation hypothesis.

Adaptive response. Activation of transcription of the genes of the Ada regulon, which is involved in the repair of some types of alkylation damage to DNA.

Adenine (A). One of the two purine (two-ringed) bases in DNA and RNA.

Affinity tag. A polypeptide that binds tightly to some other molecule. If the polypeptide coding sequence is translationally fused to the coding sequence of another protein, it allows the protein to be purified more easily.

Alkylating agent. A chemical that reacts with DNA and thereby forms a carbon bond to one of the atoms in DNA.

Allele. One of the forms of a gene, e.g., the gene with a particular mutation. Can refer to the wild-type or mutant form.

Allele-specific suppressor. A second-site suppressor mutation that alleviates the effect of other mutations, but only certain mutations or certain types of other mutations.

Allelism test. A complementation test to determine if two mutations are in the same gene, i.e., if they create different alleles of the same gene.

Allosterism. A change in the conformation of a domain of a protein as a result of a change in a different domain, e.g., when binding of the allolactose inducer to the inducer-binding pocket of the LacI repressor changes the angle of the DNA-binding domain.

Amber codon. The nonsense codon UAG.

Amber mutation. A mutation that causes the nonsense codon UAG to appear in frame in the protein-coding region of an mRNA.

Amber suppressor. A nonsense suppressor (usually a mutant tRNA) that inserts an amino acid for the nonsense UAG codon.

Amino group. The NH_2 chemical group.

Amino terminus. *See* N terminus.

Antibiotic. Generally, a substance—often a natural microbial product or its semisynthetic derivative—that kills (i.e., is bacteriocidal) or inhibits the growth of (i.e., is bacteriostatic) bacteria. Some antibacterial substances are chemically synthesized.

Antibiotic resistance gene cassette. A fragment of DNA, usually bracketed by restriction sites for ease of cloning, that contains a gene whose product confers resistance to an antibiotic for easy selection.

Anticodon. The 3-nucleotide sequence in a tRNA that pairs with the codon in mRNA by complementary base pairing.

Antiparallel. A configuration in which, moving in one direction along a double-stranded DNA or RNA, the phosphates in one strand are attached 3′ to 5′ to the sugars while the phosphates in the other strand are attached 5′ to 3′.

Antisense RNA. RNA that contains a sequence complementary to a sequence in an mRNA.

Anti-sigma factor. A protein that binds to a sigma factor, reversibly inactivating it as the next to last step of a signal transduction pathway.

Antitermination. A regulatory process in which changes in the RNA polymerase allow it to transcribe through transcription termination signals in DNA.

AP endonuclease. A DNA-cutting enzyme that cuts on the 5′ side of a deoxynucleotide that has lost its base, usually due to a DNA glycosylase, i.e., an apurinic or apyrimidinic site. This cutting allows the DNA strand to

be degraded and resynthesized, replacing the apurinic or apyrimidinic site with a normal nucleotide.

AP lyase. A DNA-cutting activity, usually associated with an *N*-glycosylase, that cuts on the 3′ side of the apurinic or apyrimidinic site created by the *N*-glycosylase activity of the enzyme.

Aporepressor. A protein that can be converted into a repressor by undergoing a conformational change if a small molecule called the corepressor is bound to it.

Archaea. A separate kingdom of prokaryotic single-celled organisms that share some of the features of both eukaryotes and prokaryotes and usually inhabit extreme environments.

A site. The site on the ribosome to which the incoming aminoacylated tRNA binds.

Assimilatory reduction. Addition of electrons to nitrogen-containing compounds to reduce them to NH_3 for incorporation into cellular constituents.

Attenuation. Regulation of an operon by premature termination of transcription, under conditions where less of the gene product(s) is needed.

Autocleavage. The process by which a protein cuts itself.

Autokinase. A protein able to transfer a PO_4 group from ATP to itself.

Autophosphorylation. Process by which a protein transfers a PO_4 group to itself, independent of the source of the PO_4 group.

Autoregulation. The process through which a gene product controls the level of its own synthesis.

Autotransporter. *See* Type V secretion system.

Auxotrophic mutant. A mutant that cannot make or use a growth substance that the normal or wild-type organism can make or use.

Backbone. The chain of phosphates alternating with deoxyribose sugars that holds the DNA chain together.

Bacterial artificial chromosome (BAC) vector. An *Escherichia coli* plasmid cloning vector that can accept very large clones of DNA (>300 kb) because it is derived from the single-copy F plasmid. Often used to make DNA libraries in genome-sequencing projects.

Bacterial lawn. The layer of bacteria on an agar plate that forms when many bacteria are plated and the bacterial colonies grow together.

Bacteriophage. A virus that infects bacteria.

Base. Carbon-, nitrogen-, and hydrogen-containing chemical compounds with structures composed of one or two rings that are constituents of the DNA or RNA molecule.

Base analog. A chemical that resembles one of the bases and so is mistakenly incorporated into DNA or RNA during synthesis.

Base pair. Each set of opposing bases in the two strands of double-stranded DNA or RNA that are held together by hydrogen bonds and thereby help hold the two strands together. Also used as a unit of length.

Base pair change. A mutation in which one type of base pair in DNA (e.g., an AT pair) is changed into a different base pair (e.g., a GC pair).

Basic local alignment search tool (BLAST). A genome annotation tool that uses bioinformatics to find similar regions in DNA sequences. It can compare DNA or amino acid sequences.

Binding. Process by which molecules are physically joined to each other by noncovalent bonds.

Bioinformatics. A repertoire of technologies that allow prediction of open reading frames, prediction of pIs and molecular weights of proteins, prediction of posttranslational modifications of proteins, prediction of subcellular localization of proteins, prediction of the level of expression of genes, and prediction of the function of gene products.

Bioremediation. The removal of toxic chemicals from the environment by microorganisms.

Biosynthesis. The synthesis of chemical compounds by living organisms.

Biosynthetic operon. An operon composed of genes whose products are involved in synthesizing compounds, such as amino acids or vitamins, rather than degrading them. *See* Degradative operon.

Blot. The filter to which DNA, RNA, or protein has been transferred by blotting.

Blotting. The process of transferring DNA, RNA, or protein from a gel or agar plate to a filter.

Blunt end. A double-stranded DNA end in which the 3′ and 5′ termini are flush with each other, that is, with no overhanging single strands.

Branch migration. The process by which the site at which two double-stranded DNAs held together by crossed-over strands (such as in a Holliday junction) moves, changing the regions of the two DNAs that are paired in heteroduplexes.

Broad host range. The ability of a phage, plasmid, or other DNA element to enter and/or replicate in a wide variety of bacterial species.

Bypass suppression. A suppressor mutation that bypasses the need for a gene product.

cAMP. *See* Cyclic AMP.

Campbell model. The model in which λ phage forms a circle and then integrates into the chromosome by recombination between a site normally internal to the λ phage DNA and a site on the chromosome, which creates a prophage genetic map that is a cyclic permutation of the phage genetic map. Named after the person who first purposed it.

CAP. *See* Catabolite activator protein.

CAP-binding site. The sequence on DNA to which the CAP protein binds.

CAP regulon. All of the operons that are regulated, either positively or negatively, by the CAP protein.

Capsid. The protein and/or membrane coat that surrounds the genomic nucleic acid (DNA or RNA) of a virus.

Carboxyl group. The chemical group COOH.

Carboxyl terminus. *See* C terminus.

Catabolic operon. An operon composed of genes whose products degrade organic compounds.

Catabolism. The degradation of an organic compound, such as a sugar, to make smaller molecules with the concomitant production of energy.

Catabolite. A small molecule produced by the degradation of larger carbon-containing organic compounds such as sugars.

Catabolite activator protein. The DNA- and cAMP-binding protein that regulates catabolite-sensitive operons in enteric bacteria by binding to their promoter regions. Also called catabolite repressor protein (Crp).

Catabolite repression. The reduced expression of some operons in the presence of high cellular levels of catabolites due to growth on an efficiently utilized carbon source.

Catabolite-sensitive operons. Operons whose expression is regulated by the cellular levels of catabolites.

Catenenes. Structures formed when two or more circular DNA molecules are joined like links in a chain.

CCC, *See* Circular and covalently closed.

Cell division. The splitting of a mother cell into two daughter cells.

Cell division cycle. The events occurring between the time a cell is created by division of its mother cell and the time it divides.

Cell generations. The total number of times in a culture that new cells have been made by the growth and division of old cells.

Central dogma. The tenet that protein is translated from RNA that was transcribed from DNA.

Change-of-function mutation. A mutation that changes the activity of a protein rather than inactivating all or part of it, e.g., a mutation that makes an activator respond to a different inducer.

Channel gating. Blocking a membrane channel unless the substrate is being transported. This prevents other molecules from leaking into or out of the cell through the membrane channel.

Chaperone. A protein that binds to other proteins and helps them fold correctly or prevents them from folding prematurely.

Chaperone-usher secretion. *See* Type V secretion system.

Chaperonin. A protein which forms double back-to-back chambers, which alternate in taking up denatured proteins and refolding them. Represented by the GroEL (Hsp60) protein in *E. coli*.

Chaplin. Coelicolor hydrophobic aerial protein. A hydrophobic protein in Streptomycetes that contributes to colony surface hydrophobicity, allowing aerial mycelia to escape the surface of the plate.

***chi* (χ) mutation.** A mutation that causes the sequence of a *chi* site to apppear in the DNA.

***chi* (χ) site.** The sequence 5′GCTGGTGG3′ in *E. coli* DNA. Stimulates recombination by the RecBCD nuclease in *E. coli* by inhibiting the 3′-to-5′ nuclease activity of RecBCD.

Chromatography. A method to separate molecules on the basis of charge, size and shape, or affinity differences. Liquid chromatography, based on a cation exchange column or a hydrophobicity column, is useful for separating peptides prior to mass spectrometry.

Chromosome. In a bacterial cell, the DNA molecule that contains most of the genes required for cellular growth and maintenance, usually the largest DNA molecule in the cell, and the one that contains a characteristic *oriC* sequence.

Circular and covalently closed. A circular double-stranded DNA with no breaks or discontinuities in either of its strands.

CI repressor. The phage λ-encoded protein that binds to the phage operator sequences close to the p_R and p_L promoters and prevents transcription of most of the genes of the phage.

***cis*-acting mutation.** A mutation that affects only the DNA molecule in which it occurs and not other DNA molecules in the same cell.

***cis*-acting site.** A functional region on a DNA molecule that does not encode a gene product and so affects only the DNA molecule in which it resides (e.g., an origin of replication).

Clamp loader. A protein that helps the ring-like sliding clamp accessory protein of the replicative DNA polymerase load onto DNA. Represented by the δ and δ′ proteins of *E. coli*.

Classical genetics. The study of genetic phenomena by using only intact living organisms.

Clonal. A situation in which all the descendants of an organism or replicating DNA molecule remain together, as in colonies on an agar plate.

Clone. A collection of DNA molecules or organisms that are all identical to each other because they result from replication or multiplication of the same original DNA or organism.

Cloning vector. An autonomously replicating DNA (replicon), usually a phage or plasmid, into which can be introduced other DNA molecules that are not capable of replicating themselves so that the non-self-replicating DNA can be cloned.

Closed complex. The complex which forms when RNA polymerase first binds to a promoter and before the strands of DNA at the promoter separate.

Cluster of orthologous groups of genes. A compilaton of presumptive genes from a diversity of organisms (that represent the major phylogenetic lineages) that are proposed to be functionally analogous, based on sequence similarity.

Cochaperone. A smaller protein which helps chaperones fold proteins or cycle their adenine nucleotide. Represented by DnaJ and GrpE in *E. coli*.

Cochaperonin. A smaller protein which helps chaperonins fold proteins by forming the cap on the chamber once the protein has been taken up. Represented by the GroES protein (Hsp10) in *E. coli*.

Coding strand. The strand of DNA in a gene that has the same sequence as the mRNA transcribed from the gene.

Codon. A 3-base sequence in mRNA that stipulates one of the amino acids.

COG. *See* Cluster of orthologous groups of genes.

Cognate aminoacyl-tRNA synthetase. The enzyme that attaches the correct amino acid to a tRNA.

Cointegrate. A DNA molecule after a transposition event from a donor DNA into a target DNA in which the donor and target DNAs are joined, separated by copies of the transposon.

Cold-sensitive mutant. A mutant that cannot live and/or multiply in the lower temperature ranges at which the normal or wild-type organism can live and/or multiply.

Colony. A small lump or pile made up of millions of multiplying cells on an agar plate.

Colony papillation. A process leading to sectors or sections in a colony that appear different from the remainder of the colony.

Colony purification. Isolation of individual bacteria on an agar plate so that all the cells in a colony that forms after incubation will be descendants of the same bacterium.

Compatible restriction endonucleases. Restriction endonucleases that leave the same overhangs after cutting a DNA molecule. The resulting ends can pair, allowing the molecules to be ligated to each other.

Competence pheromones. Small peptides given off by bacterial cells. Required to induce competence in neighboring cells when the cells are at high concentrations.

Competent. The state during which cells are capable of taking up DNA.

Complementary. The property that two nucleotides can have that allows them to be held together by basepairing between their bases.

Complementary base pair. A pair of nucleotides that can be held together by hydrogen bonds between their bases, e.g., dGMP and dCMP or dAMP and dTMP.

Complementation. Restoration of the wild-type phenotype when two DNAs containing different mutations that cause the same mutant phenotype are in the cell together. Usually means that the two mutations affect different genes.

Complementation group. A set of mutations of which none complement any of the others. An indication that they are all in the same gene.

Composite transposon. A transposon made up of two almost identical insertion (IS) elements plus the DNA between them.

Concatemer. Two or more almost identical DNA molecules linked tail to head.

Condensation. A way of making the chromosome occupy a smaller space, for example by supercoiling or by binding condensins.

Condensins. Proteins that bind chromosomal DNA in two different places, folding it into large loops and thereby making it more condensed. Represented by the Smc protein in *Bacillus subtilis* and by the MukB protein in *E. coli.*

Conditional lethal mutation. A mutation that inactivates an essential cellular component, but only under a certain set of circumstances; for example, a temperature-sensitive mutation that inactivates RNA polymerase only at relatively high temperatures or a nonsense mutation that inactivates an essential gene product, but only in the absence of a nonsense suppressor.

Congenic. Identical genotype except for a particular allele variant. *See* isogenic.

Conjugation. The transfer of DNA from one bacterial cell to another by the transfer functions of a self-transmissible DNA element such as a plasmid.

Conjugative transposon. A transposon that encodes functions that allow it to transfer itself into other bacteria, where it can integrate almost randomly. Sometimes called an integrating conjugative element (ICE), because the known elements integrate by an integrase recombinase rather than a transposase.

Consensus sequence. A nucleotide sequence in DNA or RNA, or an amino acid sequence in protein, in which each position in the sequence has the nucleotide or amino acid that has been found most often at that position in molecules with the same function and a similar sequence.

Conservative. A reaction involving double-stranded DNA in which the molecule retains both of its original strands.

Constitutive mutant. A mutant in which the genes of an operon are transcribed whether or not the inducer of the operon is present.

Context. The sequence of nucleotides in DNA or RNA surrounding a particular sequence that affects its efficiency; e.g., the sequence around a nonsense codon that affects the efficiency of translation termination at the nonsense codon.

Cooperative binding. Process in which the binding of one protein molecule to a site (often on DNA) greatly enhances the binding of another protein molecule of the same type to an adjacent site. The proteins bound at adjacent sites interact through their multimerization domains, which stabilizes the binding.

Coprotease. A protein that binds to another protein and thereby activates the second protein's autocleavage or other protease activity.

Copy. A molecule of a particular type identical to another in the same cell. Often refers to a gene that has been moved somewhere else so that it now exists in more than one place in the genome.

Copy number. The number of copies of a plasmid per cell immediately after cell division. Also the ratio of the number of plasmids of a particular type in the cell to the number of copies of the chromosome.

Core polymerase. The part of the DNA or RNA polymerase that actually performs the polymerization reaction and functions independently of accessory and regulatory proteins that cycle on and off the protein.

Corepressor. A small molecule that binds to an aporepressor and converts it into a repressor.

Cosmid. A plasmid that carries the sequence of a *cos* site so that it can be packaged into λ phage heads.

***cos* site.** The sequence of deoxynucleotides at the ends of λ DNA in the phage head. A staggered cut in this sequence at the time the phage DNA is packaged from concatamers gives rise to complementary or cohesive ends that can base pair with each other to form circular DNA on infection of another host cell.

Cotranscribed. Two or more contiguous genes transcribed by a single RNA polymerase molecule from a single promoter.

Cotransducible. Two genetic markers that are close enough together on the DNA that they can be carried in the same phage head during transduction.

Cotransduction. A type of transduction in which transductants that were selected for being recombinant for one marker in DNA are also recombinant for a second marker.

Cotransduction frequency. The percentage of transductants selected for being recombinant for one genetic marker that have also become recombinant for another genetic marker. A measure of how far apart the markers are on DNA.

Cotransformable. As in cotransducible, but the regions of two genetic markers are close enough together to be carried on the same piece of DNA during transformation.

Cotransformation. A type of transformation in which transformants that were selected for being recombinant for one marker in DNA are also recombinant for a second marker.

Cotransformation frequency. As in cotransduction frequency, except that the percentage is of the frequency of transformants recombinant for the selected marker that are also recombinant for an unselected marker. A measure of how far apart the markers are on DNA.

Cotranslational translocation. A type of translocation in which a protein is translated as it is inserted into the membrane by the SRP system. Required if the protein is to be inserted in the inner membrane so is highly hydrophobic.

Counterselection of donor. Selection of transconjugants under conditions in which the donor bacterium cannot multiply to form colonies.

Coupling model. A model for the regulation of replication of iteron plasmids in which two or more plasmids are joined by binding to the same Rep protein through their iteron sequences. *See* Handcuffing model.

Coupling protein. A protein that is part of the Mpf system of self-transmissible plasmids. The coupling protein binds to the relaxase of the Dtr system to communicate that contact has been made with a recipient cell.

Covalent bond. A bond that holds two atoms together by sharing their electron orbits.

Covalently closed circular DNA. *See* Circular and covalently closed.

Cross. Any means of exchange of DNA between two organisms.

Crossing. Allowing the DNAs of two strains of an organism to enter the same cell so they can recombine with each other.

Crossover. Site of the breaking and rejoining of two DNA molecules during recombination.

C-terminal amino acid. The amino acid on one end of a polypeptide chain that has a free carboxyl group unattached to the amino group of another amino acid.

C terminus. The end of a polypeptide chain with the free carboxyl (COOH) group.

Cured. Loss of a DNA element such as a plasmid, prophage, or transposon by a cell.

Cut and paste. A mechanism of transposition in which the entire transposon is excised from one place in the DNA and inserted into another place.

Cyclically permuted genome. The mathematical definition of a cyclic permutation is a permutation that shifts all elements of a set by a fixed offset with the elements shifted off the end inserted back at the beginning. In a cyclically permuted genome, there are no unique ends. If the genome of such a phage is drawn as a circle, each genome starts somewhere on the circle and extends around the circle until it returns to the same place, so that the individual genomes have different endpoints but contain all of the genes.

Cyclic AMP. Adenosine monophosphate with the phosphate attached to both the 3′ and 5′ carbons of the ribose sugar.

Cyclobutane ring. A ring structure of four carbons held together by single bonds. Present in some types of pyrimidine dimers in DNA.

Cytoplasmic domain. A region of the polypeptide chain of a transmembrane protein that is in the interior or cytoplasm of the cell.

Cytosine (C). One of the pyrimidine (one-ringed) bases in DNA and RNA.

Damage tolerance mechanism. A way of dealing with damage to DNA which does not involve repairing the damage, for example replication restart or translesion synthesis.

Daughter cell. One of the cells arising from division of a mother cell.

Daughter DNA. One of the two DNAs arising from replication of another DNA.

DDE transposon. A family of transposons containing the motif DDE (aspartate-aspartate-glutamate) in their transposase. These amino acids chelate the magnesium ions required in the active center for transposase activity.

Deaminating agent. A chemical that reacts with DNA, causing the removal of amino (NH_2) groups from the bases in DNA.

Deamination. The process of removing amino (NH_2) groups from a molecule. In mutagenesis, the removal of amino groups from the bases in DNA.

Decatenation. The process performed by type II topoisomerases of passing DNA strands through each other to resolve catenenes.

Defective prophage. A DNA element in the bacterial chromosome that contains phage-like DNA sequences and presumably was once capable of being induced to form phages but has lost genes essential for lytic development.

Degenerate probe. A chemically synthesized oligonucleotide that is made to be complementary to a certain protein-coding sequence in DNA or RNA but in which the third base in some codons has been randomized to include all the codons that could encode each amino acid.

Degradative operon. Like a catabolic operon, an operon whose genes encode enzymes required for the breakdown of molecules into smaller molcules with the concomitant release of energy and/or compounds needed for other pathways. *See* Biosynthetic operon.

Deletion mapping. A convenient procedure for mapping point mutations in which mutants that have point mutations to be mapped are crossed with mutants that have deletion mutations with known endpoints. Wild-type recombinants appear only if the unknown mutation lies outside the deleted region, allowing the mutation to be localized.

Deletion mutation. A mutation in which a number of contiguous base pairs have been removed from the DNA.

Deoxyadenosine. An adenine base attached to a deoxyribose sugar.

Deoxyadenosine methylase (Dam methylase). An enzyme that attaches a CH_3 (methyl) group to the adenine base in DNA. Represented by the Dam methylase in *E. coli* that methylates the A in the sequence GATC.

Deoxycytidine. A cytosine base attached to a deoxyribose sugar.

Deoxyguanosine. A guanine base attached to a deoxyribose sugar.

Deoxynucleoside. A base (A, G, T, or C) attached to a deoxyribose sugar.

Deoxyribose. A sugar similar to the five-carbon sugar ribose but with a hydrogen (H) atom rather than a hydroxyl (OH) group attached to the 2′ carbon.

Deoxythymidine. The thymine base attached to deoxyribose sugar.

Dimer. A protein made up of two polypeptides.

Dimerization domain. The region of a polypeptide that binds to another polypeptide of the same type to form a dimer.

Dimerize. To bind two identical polypeptides to each other.

Diploid. The state of a cell containing two copies of each of its genes, which are not derived from replication of the same DNA. *See* Haploid.

Directed-change (adaptive) mutation hypothesis. The hypothesis that mutations in DNA occur preferentially when they benefit the organism or help it adapt to a new environment.

Directional cloning. Cloning a piece of DNA into a cloning vector in such a way that it can be inserted in only one orientation, for example by using incompatible restriction endonucleases to join each of the ends.

Direct repeat. A short sequence of deoxynucleotides in DNA closely followed by an almost identical sequence on the same strand.

Dissimilatory reduction. The reduction of nitrogen-containing compounds such as nitrate that occurs when they are used as terminal electron acceptors in anaerobic respiration. The reduced nitrogen-containing compounds are not necessarily incorporated into the cellular molecules.

Disulfide bonds. Covalent bonds between two sulfur atoms, such as those between the side chain sulfur atoms in two cysteine amino acids in a polypeptide.

Disulfide oxidoreductases (Dsb). Enzymes in the periplasmic space, which can form or break disulfide bonds between cysteines by reducing or oxidizing the bonds. They contain the motif CXXC, where X can be any amino acid, and exchange cysteine bonds in the protein with the cysteines in the Dsb protein.

Division septum. The cross wall that forms between two daughter cells just before they separate.

Division time. The time taken by a newborn bacterial cell to grow and divide again in a particular growth environment.

D-loop. The three-stranded structure that forms when a single strand of DNA invades a double-stranded DNA, displacing one of the strands.

DnaA box. The sequence 5'TTATCCACA3' in DNA to which the DnaA protein binds. The DnaA protein is required for the initiation of chromosome replication in *E. coli*.

DNA-binding domain. The region of a polypeptide in a DNA-binding protein that binds to DNA.

DNA box. Sequence on DNA to which a protein binds.

DNA clone. A fragment of DNA inserted in a cloning vector such that many identical copies of the fragment are made when the cloning vector replicates.

DNA glycosylase. An enzyme that removes bases from DNA by cleaving the bond between the base and the deoxyribose sugar.

DNA helicase. An enzyme that uses the energy of ATP to separate the strands of double-stranded DNA.

DNA library. A collection of clones of the DNA of an organism that together represent all the DNA sequences of that organism.

DNA ligase. An enzyme that can join the phosphate-terminated 5' end of one DNA strand to the 3' hydroxyl end of another.

DNA polymerase accessory proteins. Proteins that travel with the DNA polymerase during replication.

DNA polymerase III holoenzyme. The replicative DNA polymerase in *E. coli*, including all the accessory proteins, sliding clamp, editing functions, etc.

DNA polymerase V. The product of the *umuC* gene of *E. coli*. When bound to UmuD', the autocleaved form of UmuD, UmuC becomes a DNA polymerase capable of translesion synthesis.

DNA replication complex. The entire complex of proteins, including the DNA polymerase, that moves along the DNA at the replication fork.

DNA transfer functions (Dtr component). The *tra* gene functions of a plasmid responsible for preparing the DNA for transfer.

Domain. A region of a polypeptide with a particular function or localization.

Dominant mutation. A mutation that affects the phenotype, even in a diploid organism containing a wild-type allele of the gene.

Dominant phenotype. The phenotype exerted by a mutation or other genetic marker even in an organism that is diploid for the region because it also contains the corresponding region from the wild-type organism.

Donor allele. The form of the gene that exists in the donor strain if the donor and recipient in a cross have different forms of a gene.

Donor DNA. DNA that is extracted from the donor strain of bacteria and used to transform a recipient strain of bacteria. In transposition, the DNA in which the transposon originally resides before it transposes to the target DNA.

Donor strain. The bacterial strain that is the source of the transferred DNA in a bacterial cross. For example, in a transductional cross, the donor strain is the strain in which the phage was previously propagated; in conjugation, it is the strain harboring the self-transmissible plasmid.

Double mutant. A mutant with two mutations.

Downstream. From a given point, sequences that lie in the 3' direction on RNA or in the 3' direction on the coding strand of a DNA region from which an RNA is made.

Dtr component. DNA transfer component of a plasmid transmission system. The *tra* or *mob* genes of the plasmid involved in preparing the plasmid DNA for transfer.

Duplication junction. The point at which a crossover occurred, resulting in a tandem duplication mutation in DNA.

Early gene. A gene expressed early during a developmental process, for example during bacterial sporulation or phage infection.

Ectopic recombination. "Out-of-place" recombination: homologous recombination occurring between two, usually nonidentical, sequences in different regions of the two DNAs participating in the recombination. It is often responsible for deletions, inversions, and other types of DNA rearrangements and is sometimes called "unequal crossing over." *See* Homeologous recombination.

Editing. Process of removing and replacing a wrongly inserted deoxynucleotide during replication, for example a C inserted opposite a template A, to reduce the frequency of mutations.

Editing functions. The 3′ exonuclease activities that remove nucleotides erroneously incorporated during replication. Such activities can be part of the DNA polymerase polypeptide itself or can be accessory proteins that travel with the DNA polymerase during replication. They are represented by ε protein in *E. coli*.

Effector. A small molecule that binds to a protein and changes its properties.

EF-G. *See* Translation elongation factor G.

EF-Tu. *See* Translation elongation factor Tu.

Eight-hitter. A type II restriction endonuclease that recognizes and cuts in an 8-bp sequence in DNA.

8-OxoG. A damaged DNA base commonly caused by reactive forms of oxygen, in which an oxygen atom has been added to the 8-position of the small ring of the base guanine.

Electroporation. The introduction of nucleic acids or proteins into cells through exposure of the cells to an electric field.

Electrospray ionization. A method used for preparation of samples for mass spectrometry that produces singly and multiply charged ions from a peptide so that multiple peaks are seen in a mass spectrometric analysis. The sample is introduced into an electric field in a liquid solution. Ions are formed when the solution is sprayed from a fine needle into the electric field. As solvent evaporates, intact peptides are left with different numbers of charges, depending on the sequence of the peptide.

Elongation factor G. *See* Translation elongation factor G.

Elongation factor Tu. *See* Translation elongation factor Tu.

ELPH (Estimated locations of pattern hits). Online software that can identify motifs in a set of protein or DNA sequences. If a large set of sequences is submitted, the program can search for the most common motif(s). For an example in this book, see Box 13.4, part C of the figure. For software, consult the University of Maryland Center for Bioinformatics and Computational Biology at www.cbab.umd.edu/software/ELPH/.

Endonuclease. An enzyme that can cut phosphodiester bonds between nucleotides internal to a polynucleotide.

Enrichment. The process of increasing the frequency of a particular type of mutant in a population, often by using an antibiotic, such as ampicillin, that kills cells only if the are growing.

Epistasis. A type of interaction in which a mutation at one locus predominates over a different locus.

Escape synthesis. Induction of transcription of an operon as a result of titration of its repressor owing to an increase in the number of operators to which the repressor binds. *See* Titration.

ESI. *See* Electrospray ionization.

E site. The site on the ribosome at which the tRNA binds after it has contributed its amino acid to the growing polypeptide and just before it exits the ribosome. It may help maintain the correct reading frame.

Essential genes. Genes whose products are required for maintenance and/or growth of the cell under all known conditions.

Eubacteria. "True" bacteria: members of the kingdom of organisms characterized by a relatively simple cell structure free of many cellular organelles, the presence of 16S and 23S rRNAs, and usually a four-component core RNA polymerase, among other features.

Eukaryotes. Members of the kingdom of organisms whose cells contain a nucleus surrounded by a nuclear membrane and many other cellular organelles, including a Golgi apparatus and an endoplasmic reticulum. They have 18S and 28S rRNAs rather than the 16S and 23S of eubacteria.

E-value. A measure of the number of similarities or local alignment scores that are reported in a sequence search based on comparing a query sequence with a database, identifying matching sequences, and calculating the probability that a particular match could have occurred by chance. For example, the score of a query sequence and the same sequence in a database would be very close to 0. For genome annotation, only E-values less than $1e^{-5}$ are usually considered evidence of a reliable match.

Exons. The sequences of nucleotides in a gene encoding a protein or RNA after all the introns have been removed.

Exonuclease. A nuclease enzyme that can remove nucleotides only from the end of a polynucleotide.

Expected value. *See* E-value.

Exported proteins. Proteins which leave the cytoplasm after they are made and end up in a membrane, in the periplasmic space, or outside the cell.

Expression vector. A cloning vector in which a cloned gene can be transcribed and sometimes also translated from a vector promoter and translational initiation region, respectively.

Exteins. The sequences of nucleotides in a gene after all the inteins have been removed.

Extracellular protein. A protein that is secreted from cells after it is made.

Extragenic. Involving a different gene.

Extragenic suppressor. *See* Intergenic suppressor.

Factor-dependent transcription termination site. A DNA sequence that causes transcription termination only in the presence of a particular protein, such as the Rho protein of *E. coli*.

Factor-independent transcription termination site. A DNA sequence that causes transcription termination by RNA polymerase alone, in the absence of other proteins. In bacteria, it is characterized by a GC-rich region with an inverted repeat followed by a string of A's on the template strand.

FASTA. An early database search program. It has been largely replaced by BLAST and related search tools, but the "FASTA format" is still used to submit raw sequences for a database search. For example, amino acid sequences are submitted using standard amino acid codes (as in the inside front cover of this book) with the addition of "X" for any amino acid, "*" for translation stop and "-" for a gap of indeterminate length. Also see http://fasta.bioch.virginia.edu/fasta.

Feedback inhibition. Inhibition of synthesis of the product of a pathway resulting from binding of the end product of the pathway to the first enzyme of the pathway, thereby inhibiting the activity of the enzyme.

Ffh protein. The protein component of the signal recognition particle of eubacteria. It is related to the 54-kilodalton protein component of the signal recognition particle in eukaryotes (fifty-four homolog).

Filamentous phage. A type of phage with a long, floppy appearance. The nucleic acid genome of these phages is merely coated with protein, making the phage as long as the genome and giving the floppy appearance. In contrast, the nucleic acids of most phages are encapsulated in a rigid, almost spherical, icosahedral head.

Filter mating. A procedure in which two bacteria are trapped on a filter to hold them in juxtaposition so that conjugation can occur.

Fimbriae. Another name for pili, except for conjugative sex pili encoded by self-transmissible plasmids, which are always called pili and never fimbriae.

5' end. The end of a nucleic acid strand (DNA or RNA) at which the 5' carbon of the ribose sugar is not attached through a phosphate to another nucleotide.

5' exonuclease. A deoxyribonuclease (DNase) that degrades DNA starting with a free 5' end.

5' overhang. A short, single-stranded 5' end on an otherwise double-stranded DNA molecule.

5' phosphate end. In a polynucleotide, a 5' end that has a phosphate attached to the 5' carbon of the ribose sugar of the last nucleotide.

5'-to-3' direction. The direction on a polynucleotide (RNA or DNA) from the 5' end to the 3' end.

5' untranslated region. The untranslated sequence of nucleotides that extends from the 5' end of an mRNA to the first initiation codon for a polypeptide encoded by the mRNA.

Flanking sequences. The sequences that lie on either side of a gene or other DNA element.

Formylmethionyl-tRNA$_f^{Met}$. The special tRNA in prokaryotes that is activated by formylmethionine and is used to initiate translation at prokaryotic translational initiation regions. It binds to translation initiation factor IF2 and responds to the initiator codons AUG and GUG and, more rarely, to other codons in a translational initiation region.

Forward genetics. The classical genetic approach where genes are first identified by the phenotypes of mutations in the genes.

Forward mutation. Mutation that changes wild-type DNA sequence to mutant DNA sequence.

Four-hitter. A type II DNA restriction endonuclease that recognizes and cuts at a 4-bp sequence in DNA.

4.5S RNA. The RNA component of the signal recognition particle of eubacteria.

Frameshift mutation. Any mutation that adds or removes one or a very few (but not a multiple of 3) base pairs from DNA, whether or not it occurs in the coding region for a protein.

FtsY protein. The docking protein that binds proteins to be exported by the SRP pathway and directs them to the SecYEG channel. This term is a misnomer because mutations in this gene were isolated in a search for genes involved in cell division that cause cells to form filaments rather than to divide (filament formation Ts gene Y).

Functional domain. The region of a polypeptide chain that performs a particular function in the protein.

Functional genomics. A technique which involves all experimentation that seeks to define all functions of all genes and regulatory sequences in a genome. It includes biochemical, structural, and genetic analyses.

Fusion protein. A protein created when coding regions from different genes are fused to each other in frame so that one part of the protein is encoded by sequences from one gene and another part is encoded by sequences from a different gene.

Gain-of-function mutation. A mutation that results in gene overexpression or expression that is incorrect in time or location in the cell or creates a new activity for the gene product.

Gel electrophoresis. A procedure for separating proteins, DNA, or other macromolecules. It involves the application of the macromolecules to a gel made of agarose, acrylamide, or some other gelatinous material and then the application of an electric field, forcing the electrically charged macromolecules to move toward one or the other electrode. The speed at which the macromolecules move depends on their size and their charge.

Gene. A region on DNA encoding a particular polypeptide chain or functional RNA such as an rRNA, tRNA, or small noncoding RNA.

Gene chip. A glass slide or membrane on which dots of DNA are arranged and then used for hybridization with a nucleic acid probe.

Gene conversion. Nonreciprocal apparent recombination associated with mismatch repair on heteroduplexes that are formed between two DNA molecules during recombination. The name comes from genetic experiments with fungi in which the alleles of the two parents were not always present in equal numbers in an ascus, as though an allele of one parent had been "converted" into the allele of the other parent.

Gene disruption. An alteration of the structure or activity of a gene which is intended to inactivate a gene. *See* Null mutation.

Gene dosage experiment. An experiment in which the number of copies of a gene in a cell is increased to determine the effect on the amount of gene product synthesized or on other cellular phenotypes.

Gene ontology. Gene relationships categorized by cellular component, molecular process, and/or biological process. The Gene Ontology Consortium has developed a standard vocabulary for gene function(s) and a hierarchical framework that organizes functions from the general to the specific (see www.geneontology.org).

Generalized recombination. *See* Homologous recombination.

Generalized transduction. The transfer, via phage transduction, of essentially any region of the bacterial DNA from one bacterium to another. The transducing phage head contains only bacterial DNA.

Generation time. The time it takes for the cells in an exponentially growing culture to double in number (*see* Division time).

Gene replacement. A molecular genetic technique in which a cloned gene is altered in the test tube and then reintroduced into the organism, selecting for organisms in which the altered gene has replaced the corresponding normal gene in the organism.

Genetic code. The assignment of each mRNA nucleotide triplet to an amino acid.

Genetic linkage map. An ordering of the genes of an organism solely on the basis of recombination frequencies between mutations in the genes in genetic crosses.

Genetic marker. A difference in sequence of the DNAs of two strains of an organism in a particular region that causes the two strains to exhibit different phenotypes that can be used for genetic mapping of the region of sequence difference.

Genetic recombination. The joining of genetic markers into new combinations.

Genetic redundancy. A situation in which more than one gene can provide a needed biological function and therefore null mutations in one gene will not cause a mutant phenotype, even in a haploid organism.

Genetics. The science of studying organisms on the basis of their genetic material.

Genome. The nucleic acid (DNA or RNA) of an organism or virus that includes all the information necessary to make a new organism or virus.

Genomics. The process of using the sequence of the entire DNA of an organism to study its physiology and relationship to other organisms.

Genotype. The sequence of nucleotides in the DNA of an organism, usually discussed in terms of the alleles of its genes.

Glimmer (Gene locator and interpolated Markov models). A software program that is used to find coding regions in bacterial genomes. See http://www.cbcb.umd.edu/software/glimmer/.

Global regulatory mechanism. A regulatory mechanism that affects many operons scattered around the genome.

Glucose effect. The regulation of genes involved in carbon source utilization based on whether glucose is present in the medium. *See* Catabolite repression.

Glutamate dehydrogenase. An enzyme that adds ammonia directly to α-ketoglutarate to make glutamate. Responsible for assimilation of nitrogen in high ammonia concentrations.

Glutamate synthase. An enzyme that transfers amino groups from glutamine to α-ketoglutarate to make glutamate.

Glutamine synthetase. An enzyme that adds ammonia to glutamate to make glutamine. Responsible for the assimilation of nitrogen in low ammonia concentrations.

GO. *See* 8-OxoG.

GOGAT. *See* Glutamate synthase.

Gradient of transfer. In a conjugational cross, the decrease in the transfer of chromosomal markers the farther they are in one direction from the origin of transfer of an integrated plasmid.

Gram-negative bacteria. Bacteria characterized by an outer membrane and a thin peptidoglycan cell wall that stains poorly with a stain invented by the Danish physician Hans Christian Gram in the 19th century.

Gram-positive bacteria. Bacteria characterized by having no outer membrane and a thick peptidoglycan layer that stains well with the Gram stain.

GroEL. *See* Hsp60 chaperonin.

GroES. *See* Hsp10.

Growth rate regulation of ribosomal synthesis. The regulation of ribosomal synthesis that ensures that cells growing more slowly have fewer ribosomes. It is proposed to be at least partially due to the levels of the initiating nucleotides GTP and ATP, which affect the stability of open complexes on the promoters for rRNAs.

GS. *See* Glutamine synthetase.

Guanine (G). One of the two purine (two-ringed) bases in DNA and RNA.

Guanosine. The base guanine with a ribose sugar attached to form a nucleoside.

Guanosine pentaphosphate (pppGpp). The nucleoside guanosine with two phosphates attached to the 3′ carbon and three phosphates attached to the 5′ carbon of the ribose sugar. It is quickly converted to ppGpp, which is responsible for stringent control.

Guanosine tetraphosphate (ppGpp). The nucleoside guanosine with two phosphates attached to each of the 3′ and 5′ carbons of the ribose sugar. It is responsible for stringent control as well as being involved in many other stress responses.

Gyrase. A type II topoisomerase capable of introducing negative supercoils two at a time into DNA with the concomitant cleavage of ATP. It is apparently unique to bacteria.

Hairpin. A secondary structure formed in RNA or single-stranded DNA when one region of the polynucleotide chain folds back on itself and pairs by complementary base pairing with another region located a few nucleotides away.

Handcuffing model. A model for the regulation of replication of iteron plasmids in which two plasmid molecules are held together by binding to the same Rep protein through their iteron sequences. *See* Coupling model.

Haploid. The state of a cell containing only one copy or allele of each of its chromosomal genes. *See* Diploid.

Haploid segregant. A haploid cell or organism derived from multiplication of a partially or fully diploid or polyploid cell.

Hairpin secondary structure. A secondary structure in RNA, single-stranded DNA, or protein characterized by a region folding back on itself due to antiparallel noncovalent pairing between bases or amino acids in nearby sequences in the nucleic acid or protein, respectively.

Headful packaging. A mechanism of encapsulation of DNA in a virus head in which the concatemeric DNA is cut after uptake of a length of DNA sufficient to fill the head.

Heat shock protein (Hsp). One of a group of highly evolutionarily conserved proteins whose rate of synthesis markedly increases after an abrupt increase in temperature or certain other stresses on the cell.

Heat shock regulon. The group of *E. coli* genes under the control of σ^{32}, the heat shock sigma.

Heat shock response. The cellular changes that occur in the cell after an abrupt rise in temperature.

Helix-destabilizing protein. A protein that preferentially binds to single-stranded DNA and so can help keep the two complementary strands of DNA separated during replication or remove secondary structure from DNA. Represented by Ssb in *E. coli*.

Helper phage. A wild-type phage that furnishes gene products that a deleted form of the phage cannot make, thereby allowing the deleted form to multiply and form phage.

Hemimethylated. DNA in which only one strand is methylated at a sequence with twofold symmetry, such as when only one of the two A's in the sequence GATC/CTAG is methylated.

Heterodimer. A protein made of two polypeptide chains that are different because they are encoded by different genes. *See* Heteromultimer, Homodimer.

Heteroduplex. A double-stranded DNA region formed during recombination, in which the two strands come from different DNA molecules and so can have somewhat different sequences, leading to mismatches.

Heteroimmune. Related lysogenic phages that carry different immunity regions and therefore cannot repress each other's transcription, so they can multiply on cells lysogenic for the other phage. *See* Homoimmune.

Heterologous probe. A DNA or RNA hybridization probe taken from the same gene or region of a different organism. It is usually not completely complementary to the sequence being probed. *See* Hybridization probe.

Heteromultimer. A protein made of more than one polypeptide chain (usually more than two) that are different because they are encoded by different genes. *See* Heterodimer.

Hfr strain. A bacterial strain that contains a self-transmissible plasmid integrated into its chromosome and thus can transfer its chromosome by conjugation.

HFT lysate. The lysate of lysogenic phage containing a significant percentage of transducing phage with bacterial DNA substituted for some of the phage DNA.

Hidden Markov model. A stastical model that can be used to identify families of proteins by identifying which amino acids are conserved and which are not. It is useful in genomics because it can produce an alignment of a set of nucleotide or amino acid sequences, taking into account all possible combinations of matches, gaps, and mismatches. The term "hidden" refers to an aspect of the model that allows it to operate even though specific data are missing because a probabability distribution of all possible data can substitute. This aspect of the model is possible because the model can be "trained" on a group of known related sequences, and this "training" allows the observed sequence variations to be used to adjust the model's parameters.

High multiplicity of infection. A state of a virus or phage infection in which the number of viruses greatly exceeds the number of cells being infected so that most cells are infected by more than one virus.

High negative interference. A phenomenon in which a crossover in one region of the DNA greatly increases the probability of an apparent second crossover close by. It is caused by mismatch repair of mismatches on heteroduplexes formed at the site of the crossover.

HMM. *See* Hidden Markov model.

Holliday junction. An intermediate in homologous recombination in which one strand from each of two DNAs crosses over and is joined to the corresponding strand on the opposite DNA.

Holliday model. A model for homologous recombination developed by Robin Holliday, stating that one strand of each DNA is cut at exactly the same place and crosses over to be joined to the corresponding strand on the other DNA. The resulting structure, called a Holliday junction, can then migrate and/or isomerize and be cut in the crossed strands to recombine the flanking DNA sequences.

Holoenzyme. RNA polymerase or DNA polymerase attached to all of its accessory proteins that help it make RNA or DNA, respectively.

Homeologous recombination. Homologous recombination in which the deoxynucleotide sequences of two participating regions are somewhat different from each other, usually because they are in different regions of the DNA or because the DNAs come from different species. *See* Ectopic recombination.

Homing. The process of double-strand break repair and gene conversion, by which an intron or intein in a gene enters the same site in the same gene in a new DNA which lacks it. A double-strand break is made in the target DNA by a specific endonuclease encoded by the intron or intein, and double-strand break repair inserts the DNA element.

Homing endonuclease. The sequence-specific DNA endonuclease encoded by an intron or intein that makes a double-strand break in the target DNA to initiate the homing of an intron or intein.

Homodimer. A protein made up of two polypeptide chains that are identical, usually because they are encoded by the same gene. *See* Heterodimer, Homomultimer.

Homoimmune. Two related phages that have the same immunity region so that they repress each other's transcription; hence, one cannot multiply on a lysogen of the other. *See* Heteroimmune.

Homologous proteins. Proteins encoded by genes derived from a common ancestral gene.

Homologous recombination. A type of recombination that depends on the two DNAs having identical or at least very similar sequences in the regions being recombined because complementary base pairing between strands of the two DNAs must occur as an intermediate state in the recombination process.

Homologs. Two or more nucleotide or protein sequences that are derived from a common ancestor.

Homomultimer. A protein made up of more than one polypeptide (usually more than two), which are identical usually because they are encoded by the same gene. *See* Homodimer.

Host range. All of the types of host cells in which a DNA element, plasmid, phage, etc., can multiply.

Hot spot. A position in DNA that is particularly prone to mutagenesis by a particular mutagen.

Hsp10. Heat shock protein of 10 kDa found in eubacteria, chloroplasts, and mitochondria. Hsp10 is the cochaperonin to Hsp60 and forms a cap on the cylinder in which denatured proteins are folded. It is represented by GroES in bacteria.

Hsp60 chaperonin. A highly evolutionarily conserved heat shock-induced chaperonin of 60 kDa, found in eubacteria, chloroplasts, and mitochondria. It is represented by GroEL in bacteria.

Hsp70. A highly evolutionarily conserved heat shock-induced protein chaperone of 70 kDa, represented by DnaK in bacteria.

Hybridization. The process by which two complementary strands of DNA or RNA, or a strand of DNA and a strand of RNA, are allowed to base pair with each other and form a double helix.

Hybridization probe. A DNA or RNA that can be used to detect other DNAs and RNAs because it shares a complementary sequence with the DNA or RNA being sought and so hybridizes to it by base pairing.

Hypoxanthine. A purine base derived from the deamination of adenine.

IF2. *See* Initiation factor 2.

IMM. *See* Interpolated Markov model.

IMP. *See* Inner membrane protein.

Incompatibility. The interference of plasmids with one another's replication and/or partitioning.

Incompatibility (Inc) group. A set of plasmids that interfere with each other's replication and/or partitioning and so cannot be stably maintained together in the descendents of the same bacterium.

Induced mutations. Mutations that are caused by deliberately irradiating cells or treating cells or DNA with a mutagen such as a chemical.

Inducer. A small molecule that can increase the transcription of an operon.

Inducer exclusion. The process by which the inducer of an operon such as a sugar is kept out of the cell by inhibiting its transport through the membrane. Often, a more efficiently utilized sugar such as glucose inhibits the transport of other less efficiently used sugars such as lactose.

Inducible. The ability of an operon to have its transcription increased by an inducer.

Induction. In gene regulation, the turning on of the expression of the genes of an operon. In phage, the initiation of lytic development of a prophage.

In-frame deletion. A deletion mutation in an open reading frame that removes a multiple of 3 bp and so does not cause a frameshift. These deletions are particularly useful because they cannot be polar and can remove a specific domain without removing the rest of the protein.

Initiation codon. The 3-base sequence in an mRNA that specifies the first amino acid to be inserted in the synthesis of a polypeptide chain. In prokaryotes, it is the 3-base sequence (usually AUG or GUG) within a translational initiation region for which formylmethionine is inserted to begin translation. In eukaryotes, the AUG closest to the 5′ end of the mRNA is usually the initiation codon and methionine is inserted to begin translation.

Initiation factor 2. The protein that helps formylmethionyl-tRNA (tRNA$_f^{Met}$) bind correctly to the P site of the ribosome in response to an initiation codon that is part of a translational initiation region.

Initiation mass. The size of a bacterial cell at which initiation of a new round of chromosome replication occurs.

Initiation transcription complex. The complex formed by the RNA polymerase holoenzyme including the σ factor, the promoter, and the first nucleoside triphosphate.

Injectisome. *See* Type III secretion system.

Injectosome. A needle-like structure in gram-positive bacteria that injects proteins directly through the bacterial membrane and cell wall into eukaryotic cells.

Inner membrane protein. A protein that resides, at least in part, in the cytoplasmic (inner) membrane of gram-negative bacteria.

Insertional inactivation. Inactivation of the product of a gene by an insertion mutation; quite often this involves inactivation of the product of a gene on a plasmid cloning vector by cloning a fragment of DNA into the gene.

Insertion element. *See* Insertion sequence element.

Insertion mutation. A change in a DNA sequence due to the incorporation of another DNA sequence such as a transposon or antibiotic resistance cassette into the sequence.

Insertion sequence element. A small transposon in bacteria that carries only genes for the enzymes needed to promote its own transposition.

Integrase. A type of site-specific recombinase that promotes recombination between two defined sequences in DNA, causing the integration of one DNA into another DNA (e.g., the integration of a phage DNA into the chromosome).

Integron. A transposon that contains a gene for an integrase and an *att* site for integration of gene cassettes, often for antibiotic resistance. A promoter is also lined up on the transposon to allow transcription of cassette genes inserted into the transposon *att* site.

Intein. A parasitic DNA that encodes a polypeptide sequence that, when inserted into the gene for another polypeptide, introduces a polypeptide sequence into the other polypeptide that must be spliced out before the other polypeptide can be active. Inteins are usually self-splicing.

Intergenic. In different genes.

Intergenic suppressor. A suppressor mutation located in a gene different from that containing the mutation it suppresses. Also called extragenic suppressor.

Internal fragments. Fragments, created by cutting DNA containing a transposon or other inserted DNA element with a restriction endonuclease, that come from entirely within the DNA element and do not include the insertion junctions.

Interpolated Markov model. One type of hidden Markov model that is useful for locating genes in a particular organism because it has been trained on known sequences of that organism.

Interstrand cross-links. Covalent chemical bonds between the two complementary strands of DNA in a double-stranded DNA.

Intervening sequence. A sequence inserted into a polypeptide or polynucleotide that must be removed before the polypeptide or polynucleotide can be functional.

Intragenic. In the same gene.

Intragenic complementation. Complementation between two mutations in the same gene. It is rare and allele specific; it usually occurs only if the protein product of the gene is a homodimer or homomultimer.

Intragenic suppressor. A suppressor mutation that occurs in the same gene as the mutation it is suppressing.

Intron. A parasitic DNA that, when inserted into a gene for a protein, introduces polynucleotide sequences into the mRNA, which must be spliced out before the mRNA can be translated into functional protein.

Inversion junctions. The points where the recombination events occurred that inverted a sequence.

Inversion mutation. A change in DNA sequence as a result of flipping a region within a longer DNA so that it lies in reverse orientation. It is usually due to homologous recombination between inverted repeats in the same DNA molecule.

Inverted repeat. Two nearby sequences in DNA that are the same or almost the same when read in the 5′-to-3′ direction on the opposite strands.

Invertible sequence. A sequence in DNA that inverts often, owing to the action of a site-specific recombinase protein that promotes recombination between inverted repeats at the ends of the sequences.

In vitro mutagen. A mutagen that reacts only with purified DNA or with viruses or phage. It cannot be used to mutagenize intact cells, either because it cannot get in or because it is too reactive and is destroyed before it reaches the DNA.

In vitro packaging. The incorporation of DNA or RNA into virus or phage heads in the test tube.

In vivo mutagen. A mutagen that enters and mutagenizes the DNA of intact cells.

IS element. *See* Insertion sequence element.

Isogenic. Strains of an organism that are almost identical genetically except for one small region or gene.

Isolation of mutants. The process of obtaining a pure culture of a particular type of mutant from among a myriad of other types of mutants and the wild type.

Isomerization. Changing of the spatial conformation of a molecule without breaking any bonds. In DNA recombination, it refers to the rotating of the DNAs in a Holliday junction so the other strands are the ones which are crossed.

Iteron sequences. Short DNA sequences, often repeated many times in the origin region of some types of plasmids, that bind the Rep protein required for replication of the plasmid and play a role in the regulation of replication of the plasmid, possibly through promoting plasmid coupling.

Junction fragments. Fragments, created by cutting a DNA containing a transposon or other DNA element, that contain sequences from one of the ends of the DNA element as well as flanking sequences from the DNA into which the element has inserted.

KEGG map (Kyoto encyclopedia of genes and genomes). An automated reconstruction of the metabolic pathways of an organism. See http://www.genome.adijp/kegg2.html.

Kinase. An enzyme that transfers a phosphate group from ATP to another molecule.

Kleisins. Proteins that bind to condensins and help them bind to and condense DNA molecules.

Knockout mutation. A mutation that presumably eliminates the function of a gene, i.e., is presumably a null mutation.

Lagging strand. During DNA replication, the newly synthesized strand that must be synthesized in the direction opposite the overall movement of the replication fork, i.e., in the 3′-to-5′ direction overall.

Late gene. A gene that is expressed only relatively late in the course of a developmental process, e.g., a late gene of a phage.

Lawn. *See* Bacterial lawn.

Leader region or sequence. An RNA sequence close to the 5′ end of an mRNA that is translated but does not encode a functional polypeptide.

Leading strand. During DNA replication, the newly synthesized strand that is made in the same direction as the overall direction of movement of the replication fork, i.e., in the 5′-to-3′ direction.

Leaky mutation. A mutation in a gene that does not completely inactivate the product of the gene, hence leaving some residual activity.

Lep protease. One of the enzymes that cleaves the signal sequence off secreted proteins as they pass through the SecYEG channel, and probably also the Tat pathway.

Lesion. Any change in a DNA molecule as a result of chemical alteration of a base, sugar, or phosphate.

Linkage. Situation occurring when two genetic markers are sufficiently close together on the DNA that recombination between them is less than random.

Linked. A genetic term referring to the fact that two markers are close enough on the DNA that they are separated by recombination less often than if they sorted randomly.

Locus. A region in the genome of an organism.

Low multiplicity of infection. State of a virus or phage infection in which the number of cells almost equals or exceeds the number of viruses, so that most cells remain uninfected or are infected by at most one or very few viruses.

Lyse. To break open cells and release their cytoplasm into the medium.

Lysogen. A strain of bacterium that harbors a prophage.

Lysogenic conversion. A property of a bacterial cell caused by the presence of a particular prophage.

Lysogenic cycle. The series of events following infection by a bacteriophage and culminating in the formation of a stable prophage.

Lysogenic phage. A phage which is known to be capable of entering a prophage state in some host.

Lytic cycle. The series of events following infection by a bacteriophage or induction of a prophage and culminating in lysis of the bacterium and the release of new phage into the medium.

Macromolecule. A large molecule such as DNA, RNA, or protein.

Major groove. In the DNA double helix, the larger of the two grooves between the two strands of DNA wrapped around each other.

MALDI (matrix-assisted laser desorption ionization). A procedure that is used to measure peptide mass. A singly charged ion is produced from a peptide, resulting in one peak on a mass spectrometric analysis. This technique can be used to directly analyze complex peptide mixtures if a complete genome sequence is available.

Male bacterium. A bacterium or strain harboring a self-transmissible plasmid or other conjugative element.

Male-specific phage. A phage that infects only cells carrying a particular self-transmissible plasmid. The plasmid produces the sex pilus used by the phage as its adsorbtion site.

Maltodextrins. Short chains of glucose molecules held together by α1–4 (maltose) linkages. They are breakdown products of starch.

Maltotriose. A chain of three glucose molecules held together by α1–4 linkages.

Map distance. The distance between two markers in the DNA as measured by recombination frequencies.

Map expansion. A phenomenon that occurs in genetic linkage experiments, in which two markers appear to be farther apart than they are because of hyperactive apparent recombination. This is often due to hot spots for recombination or to preferential mismatch repair of some mismatches.

Map unit. A distance between genetic markers corresponding to a recombination frequency of 1% between the markers.

Marker effect. A difference in the apparent genetic linkage between the site of a mutation and other markers depending on the type of mutation at the site. It is due to preferential mismatch repair of some mismatches relative to others.

Marker rescue. Acquisition of a genetic marker by the genome of an organism or virus through recombination with a cloned DNA fragment containing the marker.

Markov model. A statistical tool that can be applied to a system that is represented by discrete states. For example, when used for protein annotation, a discrete state could be 1 of the 23 amino acids at a position in the protein.

Mass spectrometry. An analytical method that measures ion abundances based on their mass-to-charge (m/z) ratios. First, gas phase ions are produced from the compound of interest (*see* MALDI and Electrospray ionization). Then the ions are separated on the basis of their m/z ratios. Finally, the ions at different m/z ratios are detected (*see* Time of flight) and counted.

Maxam-Gilbert sequencing. A method for DNA sequencing that depends on the ability of certain chemicals to react with and cleave DNA at particular bases.

Membrane protein. A protein that at least partially resides in, or is tightly bound to, one of the cellular membranes.

Membrane topology. Distribution of the various regions of a membrane protein between the membrane and the two surfaces of the membrane. In the inner membrane of gram-negative bacteria, topology refers to which domains are in the cytoplasm, which are in the periplasm, and which are buried in and traverse the membrane from one side to the other.

Merodiploid. A bacterial cell that is mostly haploid but is diploid for some region of the genome due to some chromosomal genes being carried on a prophage or plasmid. *See* Partial diploid.

Messenger RNA. An RNA transcript that includes the coding sequences for at least one polypeptide.

Methionine aminopeptidase. An enzyme that removes the N-terminal methionine from newly synthesized polypeptides.

Methyl-directed mismatch repair system. The mismatch repair system in enteric bacteria that recognizes mismatches in newly replicated DNA and specifically removes and resynthesizes the new strand, which is distinguishable from the old strand because it is the strand that is not methylated at nearby hemimethylated GATC sequences.

Methyltransferase. In DNA repair, an enzyme that removes a CH_3 (methyl) or CH_3CH_2 (ethyl) group from a base in DNA by attaching the group to itself.

Microarray. A high-density array of spots of DNA probes that represent the genome of an organism. The spots may be attached to a glass slide or other solid medium such as a nylon membrane. In some methods, the DNA probes are synthesized directly on the solid support. A microarray might be used for several different types of experiments. A common application is to quantitate the mRNA transcripts present in a cell under a given culture condition. Other applications are possible, including DNA-DNA hybridization to compare genome sequences and determination of the gene content of a tumor cell line.

Migration. *See* Branch migration.

Mini-Mu. A shortened version of phage Mu DNA in which most of the phage DNA has been deleted except the inverted-repeat ends and the transposase genes, leaving it unable to replicate or be packaged into a phage head without the assistance of a helper wild-type phage Mu. Other DNAs, such as genes for antibiotic resistance and a plasmid origin of replication, can be inserted between copies of the mini-Mu.

Minor groove. In double-stranded DNA, the smaller of the two gaps between the two strands of DNA wrapped around each other in a helix.

Minus (−) strand. In a virus with a single-stranded nucleic acid genome (DNA or RNA), the strand that is complementary to the strand in the virus head.

−10 sequence. In a bacterial σ^{70}-type promoter, a short sequence that lies about 10 bp upstream of the transcription start site. The canonical or consensus sequence is TATAAT/ATATTA.

−35 sequence. In a bacterial σ^{70}-type promoter, a short sequence that lies about 35 bp upstream of the transcription start site. The canonical or consensus sequence is TTGACA/AACTGT.

Mismatch. Improper pairing of the normal bases in DNA, e.g., an A opposite a C.

Mismatch repair system. A pathway for removing mismatches in DNA by degrading a strand containing the mismatched base and replacing it by synthesizing a new strand containing the correctly paired base.

Missense mutation. A base pair change mutation in a region of DNA encoding a polypeptide that changes an amino acid in the polypeptide.

mob genes. The genes on a mobilizable DNA element that allow it to be mobilized by a self-transmissible element such as a self-transmissible plasmid. They often encode Dtr (DNA transfer) functions and a coupling protein that allows the element to communicate with the mating-pair formation (Mpf) system of the self-transmissible element.

Mobilizable DNA element. A plasmid or other DNA element that cannot transfer itself into other bacteria but can be transferred by other self-transmissible elements. Naturally occurring mobilizable plasmids usually encode Dtr (DNA transfer) functions and a coupling protein but not Mpf (mating pair formation) genes.

Mobilization. The process by which a mobilizable DNA element, incapable of self-transmission, is transferred into other cells by the conjugation functions of a self-transmissible element.

mob region. A region in DNA carrying an origin of transfer (*oriT* sequence) and often genes whose products allow the plasmid or other DNA element to be mobilized by self-transmissible elements.

MOI. *See* Multiplicity of infection.

Molecular genetic analysis. Any study of cellular or organismal functions that involves manipulations of DNA in the test tube.

Molecular genetic techniques. Methods for manipulating DNA in the test tube and reintroducing the DNA into cells.

Moron gene. A phage gene that has apparently moved into the phage DNA fairly recently from an unknown source and has its own promoter.

Mother cell. A cell that divides or differentiates to give rise to a new cell or spore.

Motif. A conserved nucleotide or amino acid sequence that is relatively short and suggests similarity of function.

Mpf component. Mating-pair formation. This component is made up of *tra* gene products of a self-transmissible plasmid involved in making the surface structures (pilus, etc.) that contact another cell and transfer the DNA during conjugation as well as the coupling protein that communicates with the Dtr component.

mRNA. *See* Messenger RNA.

MS. *See* Mass spectrometry.

Multigene family. A set of related genes, such a paralogs, which perform similar or redundant functions.

Multimeric protein. A protein that consists of more than one polypeptide chain (usually more than two).

Multimerized. A protein that has folded to include all of its subunits.

Multiple cloning site. A region of a cloning vector that contains the sequences cut by many different type II restriction endonucleases. It is also called a polyclonal site.

Multiplicity of infection. The ratio of phages or viruses to cells that initiates an infection.

Mutagen. A chemical or type of irradiation that causes mutations by damaging DNA.

Mutagenic repair. A pathway for repairing damage to DNA that sometimes changes the sequence of deoxynucleotides as a consequence.

Mutagenic treatments or chemicals. Treatments or chemicals that cause mutations by damaging DNA.

Mutant. An organism that differs from the normal or wild type as a result of a change in the sequence (mutation) of its DNA.

Mutant allele. The mutated gene of a mutant organism that makes it different from the wild type.

Mutant enrichment. A procedure for increasing the frequency of a particular type of mutant in a culture.

Mutant phenotype. A characteristic that makes a mutant organism different from the wild type.

Mutation. Any heritable change in the sequence of deoxynucleotides in DNA.

Mutation rate. The probability of occurrence of a mutation causing a particular phenotype each time a newborn cell grows and divides.

Narrow host range. A range of hosts (in which a DNA element can enter and/or replicate) that includes only a few closely related types of cells.

Natural competence. The ability of some types of bacteria to take up DNA at a certain stage in their growth cycle without chemical or other treatments.

Naturally transformable bacteria. Types of bacteria that have a growth stage during which they are naturally competent for taking up DNA.

NBU elements. *See* Nonreplicating *Bacteroides* units.

Negatively supercoiled. A DNA molecule in which the two strands of the double helix are wrapped around each other less than about once every 10.5 bp.

Negative regulation. A type of regulation in which a protein or RNA molecule, in its active form, inhibits a process such as the transcription of an operon or translation of an mRNA.

Negative selection. The process of detecting a mutant on the basis of the inability of the mutant to multiply under a certain set of conditions in which the normal or wild-type organism can multiply.

Nicked DNA. Double-stranded DNA in which one strand contains a broken phosphate-deoxyribose bond in the phosphodiester backbone.

Nonreplicating *Bacteroides* units. DNA elements found in the chromosome of some *Bacterioides* strains that are mobilizable by conjugative transposons.

Noncomposite transposon. A transposon in which the transposase genes and the inverted-repeat ends are included in the minimum transposable element and are not part of autonomous IS elements. *See* Composite transposon.

Noncovalent change. Any change in a molecule that does not involve the making or breaking of a chemical covalent bond due to shared electron orbits in the molecule.

Nonhomologous recombination. The breaking and rejoining of two DNAs into new combinations, which does not necessarily depend on the two DNAs having similar sequences in the region of recombination.

Nonpermissive conditions. Conditions under which a mutant organism or virus cannot multiply but the wild type can multiply.

Nonpermissive host. A host organism in which a mutant phage or virus cannot multiply but the wild type can multiply.

Nonpermissive temperature. A temperature at which the wild-type organism or virus but not the mutant organism or virus can multiply.

Nonselective. Conditions or media in which both the mutant and wild-type strains of an organism or virus can multiply.

Nonsense codon. A codon that does not stipulate an amino acid but, rather, triggers the termination of translation. In most organisms, the codons UAG, UGA, and UAA are nonsense codons.

Nonsense mutation. In a region of DNA encoding a protein, a base pair change mutation that causes one of the nonsense codons to be encountered in frame when the mRNA is translated.

Nonsense suppressor. A suppressor mutation that allows an amino acid to be inserted at some frequency for one or more of the nonsense codons during the translation of mRNAs.

Nonsense suppressor tRNA. A mutation in the gene for a tRNA that allows the tRNA to pair with one or more of the nonsense codons in mRNA during translation and therefore causes an amino acid to be inserted for the nonsense codon. This type of mutation usually changes the anticodon on the tRNA.

Northern blotting. Transfer of RNA from a gel to a filter for hybridization to a sequence-specific probe.

N-terminal amino acid. The amino acid on the end of a polypeptide chain whose amino (NH_2) group is not attached to another amino acid in the chain through a peptide bond.

N terminus. The end of a polypeptide chain with the free amino (NH_2) group not attached to the carboxyl group of another amino acid.

Ntr (nitrogen regulation) system. A global regulatory system that regulates a number of operons in response to the nitrogen sources available.

Nuclease. An enzyme that cuts the phosphodiester bonds in DNA or RNA polymers.

Nucleoid. A compact, highly folded structure formed by the chromosomal DNA in the bacterial cell and in which the DNA appears as a number of independent supercoiled loops held together by a core.

Nucleoid core. The center of the nucleoid of unknown composition.

Nucleoid occlusion. A process that prevents the formation of the division septum in a region of the cell still occupied by the nucleoid.

Nucleotide excision repair. A system for the repair of DNA damage in which the entire damaged nucleotide is removed rather than just the damaged base. A cut is made on either side of the damage on the same strand, and the damaged strand is removed and resynthesized.

Null mutation. A mutation in a gene that abolishes the function of the gene product.

Ochre codon. The nonsense codon UAA.

Ochre mutation. In a region of DNA encoding a polypeptide, a base pair change mutation that causes the nonsense codon UAA to appear in frame in the mRNA for the polypeptide.

Ochre suppressor. A suppressor (usually a mutant tRNA) that causes an amino acid to be inserted at some frequency wherever the nonsense codon UAA is encountered in frame during translation of a protein-coding region of an mRNA.

Okazaki fragments. The short pieces of DNA that are initially synthesized in the opposite direction of movement of the replication fork during replication of the lagging strand at the fork.

Oligopeptide. A short polypeptide only a few amino acids long.

OMP. *See* Outer membrane protein.

Opal codon. The nonsense codon UGA; also called umber codon.

Opal mutation. A base pair change that causes the nonsense codon UGA to appear in frame in the protein-coding region of an mRNA (also called an umber mutation).

Opal suppressor. A suppressor (usually a mutant tRNA) that inserts an amino acid for the UGA nonsense mutation.

Open complex. The complex of RNA polymerase and DNA at a promoter in which the strands of the DNA have been separated.

Open reading frame. A sequence on DNA, read 3 nucleotides at a time, that is unbroken by any nonsense codons.

Operator. Usually a sequence on DNA to which a repressor protein binds to block transcription. More generally, any sequence in DNA or RNA to which a negative regulator binds.

Operon. A DNA region encompassing genes that are transcribed into the same mRNA, as well as any adjacent *cis*-acting regulatory sequences.

Operon model. The model proposed by Jacob and Monod for the regulation of the *lac* operon, in which transcription of the structural genes of the operon is prevented by the LacI repressor binding to the operator region and thereby preventing access of the RNA polymerase to the promoter. In the presence of lactose, the inducer binds to LacI and changes its conformation so that it can no longer bind to the operator, and as a result, the structural genes are transcribed.

ORF. *See* Open reading frame.

***oriC*.** A sequence of DNA consisting of the site in the bacterial chromosome at which initiation of a round of replication normally occurs and all of the surrounding *cis*-acting sequences required for initiation.

Origin of replication. The site on a DNA at which replication initiates, including all of the surrounding *cis*-acting sequences required for initiation.

Orthologs. Genes in different species that are derived from a common ancestor. They may differ in function, but they usually have identical functions.

Outer membrane protein. A protein that resides, at least in part, in the outer membrane of gram-negative bacteria.

P site. The site on the ribosome to which the peptidyl-tRNA is bound.

***pac* site.** The sequence in phage DNA at which packaging of the phage DNA into heads begins.

Packaging site. *See pac* site.

PAI. *See* Pathogenicity island.

Papilla. A section of a bacterial colony with an appearance different from that of most of the colony.

Par function. A site or gene product that is required for the partitioning of a plasmid.

Paralogs. Genes that have resulted from duplication of a common ancestor. They generally have similar functions but may have distinct functions.

Parent. One of the two strains of an organism participating in a genetic cross.

Parental types. Progeny of a genetic cross that are genetically identical to one or the other of the parents.

Partial digest. A restriction endonclease digestion of DNA in which not all the available sites are cut, either because the amount of endonuclease enzyme is limiting or because the time of incubation is too short.

Partial diploid. A bacterium that has two copies of part of its genome, usually because a plasmid or prophage in the bacterium contains some bacterial DNA. Also called merodiploid.

Partitioning. An active process by which at least one copy of a replicon (plasmid, chromosome, etc.) is distributed into each daughter cell at the time of cell division.

Pathogenicity island. A DNA element integrated into the chromosome of a pathogenic bacterium which carries genes whose products are required for pathogenicity and which, based on its base composition and codon usage, shows evidence of having been acquired fairly recently in evolution and in some cases carries genes for its own integration. Pathogenicity islands form a subset of a more general class of integrated elements called genetic islands.

PCR. *See* Polymerase chain reaction.

Peptide bond. A covalent bond between the amino (NH_2) group of one amino acid and the carboxyl (COOH) group of another.

Peptide deformylase. An enzyme that removes the formyl group from the amino-terminal formylmethionine of newly synthesized polypeptides.

Peptidyltransferase. The ribozyme activity of the 23S rRNA (28S rRNA in eukaryotes) which forms a bond between the carboxyl group of the growing polypeptide and the amino group of the incoming amino acid.

Peptidyl tRNA hydrolase. An enzyme that removes polypeptides from tRNA. It is not associated with the ribosome, and so it may be a scavenger enzyme that is involved in the recycling of tRNAs bound to polypeptides that are prematurely released during translation.

Periplasm. The space between the inner and outer membranes in gram-negative bacteria.

Periplasmic domain. A region of a membrane protein located in the periplasm of the cell.

Periplasmic protein. A protein located in the periplasm.

Permissive conditions. Conditions under which a mutant organism or virus can multiply.

Permissive host. A strain of an organism which can support the multiplication of a particular mutant virus.

Permissive temperature. A temperature at which both a temperature-sensitive mutant (or cold-sensitive mutant) and the wild type can multiply.

Pfams. Protein family and domain databases that are useful for categorizing predicted genes or proteins, primarily based on compilation of protein domains.

Phage. *See* Bacteriophage.

Phage genome. The nucleic acid (DNA or RNA) that is packaged into the phage head and contains all the genes of the phage.

Phase variation. The reversible change of one or more of the cell surface antigens of a bacterium at a frequency higher than normal mutation frequencies.

Phasmid. A hybrid DNA element containing both plasmid and phage sequences.

Phenotype. Any identifiable characteristic of a cell or organism that can be altered by mutation.

Phenotypic lag. The delay between the time a mutation occurs in the DNA and the time the resulting change in the phenotype of the organism becomes apparent.

Phosphate. The chemical group PO_4.

Photolyase. An enzyme that uses the energy of visible light to split pyrimidine butane dimers in DNA, restoring the original pyrimidines.

Photoreactivation. The process by which cells exposed to visible light after DNA damage achieve greater survival rates than cells kept in the dark. It is due to the photolyase restoring pyrimidine dimers to the individual pyrimidines.

Physical map. A map of DNA showing the actual distance in deoxynucleotides between somehow identifiable sites such as restriction sites.

Pilin. A protein that makes up the structure of pili. *See* Pilus.

Pilus. A protrusion or filament composed of protein attached to the surface of a bacterial cell. *See* Sex pilus, Fimbriae.

Plaques. Clear spots in a bacterial lawn as a result of phage killing and lysing the bacteria as the bacterial lawn is forming and the phage is multiplying.

Plaque purification. Isolation of a pure strain of a phage by diluting and plating to obtain individual plaques, each of which contains descendents of only a single phage.

Plasmid. Any DNA molecule in cells that replicates independently of the chromosome and regulates its own replication so that the number of copies of the DNA molecule remains relatively constant.

Plasmid incompatibility. *See* Incompatibility.

Pleiotropic mutation. A mutation that causes many phenotypic changes in the cell.

Plus (+) strand. In a virus with a single-stranded genome (DNA or RNA), the strand packaged in the phage head.

Poisson distribution. A mathematical distribution that can be used to calculate probabilities in certain situations. It can be used to approximate a binomial distribution when the probability of success in a single trial is low but the number of trials is large. Named after the mathematician who first derived it.

Polarity. A condition in which a mutation in one gene reduces the transcription of a downstream gene that is cotranscribed into the same mRNA.

Polycistronic mRNA. An mRNA that contains more than one translational initiation region so that more than one polypeptide can be translated from the mRNA.

Polyclonal site. *See* Multiple cloning site.

Polymerase chain reaction. A technique involving a sucession of heating and cooling steps that uses the DNA polymerase from a thermophilic bacterium and two primers to make many copies of a given region of DNA occurring between sequences complementary to the primers.

Polymerization. A reaction in which small molecules are joined in a chain to make a longer molecule.

Polymerizing. The act of joining small molecules to form a chain.

Polymorphism. A difference in DNA sequence between otherwise closely related strains.

Polypeptide. A long chain of amino acids held together by peptide bonds. Polypeptides are the product of a single gene.

Porin. A protein that forms channels in the outer membrane of gram-negative bacteria by forming a β-barrel in the outer membrane.

Positive regulation. A type of regulation in which the gene is expressed only if the active form of a regulatory protein (or RNA) is present.

Positive selection. Conditions under which only a strain with the desired mutation or a particular recombinant type can multiply.

Positively supercoiled. A DNA molecule in which the two strands of the double helix are wrapped around each other more than about once every 10.5 bp.

Postreplication repair. *See* Recombination repair.

Posttranscriptional regulation. Any regulation in the expression of a gene that occurs after the mRNA has been synthesized from the gene, for example in the rate of translation of the mRNA.

Posttranslational translocation. Tat-, SecB-, and SecA-mediated translocation of proteins through the cytoplasmic membrane after they have been translated. This form of translocation is limited to proteins destined for the periplasm, the outer membrane, or the outside of the cell.

Precise excision. Removal of a transposon or other foreign DNA element from a DNA in such a way that the original DNA sequence is restored.

Precursors. The smaller molecules that are polymerized to form a polymer.

Presecretory protein. A secreted protein after it has been translated and while its signal sequence is still attached.

Primary structure. The sequence of nucleotides in an RNA or of amino acids in a polypeptide.

Primase. An enzyme that synthesizes short RNAs to prime the synthesis of DNA chains.

Prime factor. A self-transmissible plasmid carrying a region of the bacterial chromosome.

Primer. A single-stranded DNA or RNA that can hybridize to a single-stranded template DNA and provide a free 3′ hydroxyl end to which DNA polymerase can add deoxynucleotides to synthesize a chain of DNA complementary to the template DNA.

Primosome. A complex of proteins involved in making primers for the initiation of synthesis of DNA strands.

Probe. A short oligonucleotide (DNA or RNA) that is complementary to a sequence being sought and so hybridizes to the sequence and allows it to be identified from among many other sequences.

Prokaryotes. Organisms whose cells do not contain a nuclear membrane and visible nucleus or many of the other organelles characteristic of the cells of higher organisms. They include the eubacteria and archaea.

Prolyl isomerase. An enzyme, often associated with chaperones, that can catalyze the conversion of one isomer of proline to the other isomer. Proline is the only amino acid that has more than one isomer because the carbon in the carboxyl group is not free to rotate.

Promiscuous plasmid. A self-transmissible plasmid that can transfer itself into many types of bacteria, which need not be closely related to each other.

Promoter. A region on DNA to which RNA polymerase binds in order to initiate transcription.

Prophage. The state of phage DNA in a lysogen in which the phage DNA is integrated into the chromosome of the bacterium or replicates as a plasmid.

Protein disulfide isomerase. An enzyme that catalyzes the oxidation of the sulfhydryl groups of cysteines in polypeptides, cross-linking the cysteines to each other.

Protein export. The transport of proteins into or through the cellular membranes.

Protein secretion. The transport of proteins through the cellular membranes to the outside of the cell.

Proteome. The complete set of proteins expressed in an organism.

Proteomics. Global analysis of protein expression patterns and protein interactions. It includes techniques such as mass spectrometry, phage display, and two-hybrid analyses.

Pseudoknot. An RNA tertiary structure with interlocking loops held together by regions of hydrogen bonding between the bases.

PSI-BLAST. Reiterative sequence alignments are performed with the goal of defining as large a potential family of functionally related proteins as possible. This program can find a set of related sequences based on the presence of common sequence patterns.

Purine. A base in DNA and RNA with two ring structures.

Pyrimidine. A base in DNA and RNA with only one ring.

Pyrimidine dimer. A type of DNA damage in which two adjacent pyrimidines are covalently joined by chemical bonds.

Quantitative reverse transcriptase PCR (Q-RT-PCR). A way of quantitating the amount of an individual RNA in cells. First, cDNAs are made on the cellular RNAs with reverse transcriptase and then the rate of accumulation of a PCR fragment corresponding to that RNA is measured during the linear phase to determine the concentration of the cDNA for that RNA.

Quaternary structure. The complete three-dimensional structure of a protein including all the polypeptide chains making up the protein and how they are wrapped around each other.

Random gene fusion. A technique in which transposon mutagenesis is used to fuse reporter genes to different regions in the chromosome. A transposon containing a reporter gene hops randomly into the chromosome, resulting in various transposon insertion mutants that have the reporter gene on the transposon fused either transcriptionally or translationally to different genes or to different regions within each gene.

Random-mutation hypothesis. A hypothesis explaining the adaptation of organisms to their environment. It states that mutations occur randomly, free of influence from their consequences, but that mutant organisms preferentially survive and reproduce themselves if the mutations inadvertently confer advantages under the conditions experienced by the organisms.

Random shotgun sequencing. A method for sequencing a long DNA molecule, in which the longer DNA is first broken into random smaller pieces, often by mechanical shearing, and then the smaller fragments are cloned and sequenced. The random sequences are then ordered into a continuous sequence by computer-aided alignment of overlapping sequences.

RBS finder. A program that uses an algorithm to find translational initiation regions in both eubacterial and archaeal genomes. It is usually used after gene finders such as Glimmer. Once a gene is found, RBS finder looks for a probable sequence to which ribosomes bind, hence the name.

RC plasmid. *See* Rolling-circle plasmid.

Reading frame of translation. Any sequence of nucleotides in RNA or DNA read three at a time in succession, as during translation of an mRNA.

Rec⁻ (recombination-deficient) mutant. A mutant strain in which DNA shows a reduced capacity for recombination due to a mutation in a *rec* gene whose product is involved in recombination.

Recessive mutation. In complementation tests, a mutation that does not exhibit its phenotype in the presence of a wild-type allele of the gene.

Recessive phenotype. The phenotype exhibited by a mutation or other genetic marker that does not exert itself in an organism diploid for the region because it also contains the corresponding region from the wild-type organism.

Recipient allele. The sequence of a gene or allele as it occurs in the recipient bacterium.

Recipient strain. In a genetic cross between two bacterial strains, the strain of bacterium that receives DNA from another strain of bacterium.

Reciprocal cross. A genetic cross in which the alleles of the donor and recipient strain are reversed relative to an earlier cross. An example would be a transduction in which the phage was grown on the strain that had the alleles of what was previously the recipient strain and used to transduce a strain with the alleles of what was previously the donor strain. In bacterial crosses, generally what was before the selected marker now becomes an unselected marker.

Recombinant DNA. A DNA molecule derived from the sequences of two different DNAs joined to each other in a test tube.

Recombinant type. In a genetic cross, progeny that are genetically unlike either parent in the cross because they have DNA sequences that are the result of recombination between the parental DNAs.

Recombinase. An enzyme that specifically recognizes two sequences in DNA and breaks and rejoins the strands to cause a crossover within the sequences.

Recombination. The breakage and rejoining of DNA into new combinations.

Recombination frequency. In a genetic cross, the number of progeny that are recombinant types for the two parental markers divided by the total number of progeny of the cross.

Recombination repair. A DNA damage tolerance mechanism that requires the recombination functions, which function to restart replication forks stalled at the damage. In one scenario the lagging strand may replicate past the damage to leave a gap and the undamaged strand may be used to fill the gap using the RecFOR recombination functions. In another scenario, the damage may leave a double-stranded end that can then invade the other daughter DNA, using the RecBCD functions. Alternatively, the replication fork may back up to form a Holliday junction, which can then migrate past the damage. In all these pathways, the Pri proteins help the replication proteins reload on the DNA to restart the fork past the damage.

Redundancy. *See* Terminally redundant DNA.

Regulation of gene expression. Control of the rate of synthesis of the active product of a gene, so that the active gene product can be synthesized at different rates, depending, for example, on the developmental stage of the organism or the state in which the organism finds itself.

Regulatory cascade. A strategy for regulating the expression of genes during developmental processes in which the products of genes expressed during one stage of development turn on the expression of genes for the next stage of development and turn off genes from the previous stage.

Regulatory gene. A gene whose product regulates the expression of other genes, as well as, sometimes, its own expression.

Regulon. The set of operons that are all regulated by the product of the same regulatory gene.

Relaxase. The protein of a self-transmissible or mobilizable plasmid that makes a cut at the *oriT nic* site, remains attached to the 5′ end at the cut, is secreted into the recipient cell, and rejoins the cut ends in the recipient cell.

Relaxed control. Control that occurs when the synthesis of rRNA and other stable RNAs continues even after protein synthesis is blocked by starvation for an amino acid.

Relaxed DNA. A DNA that contains no supercoils.

Relaxed plasmid. A plasmid that has a high copy number, so that its replication need not be too tightly controlled.

Relaxed strain. A bacterial strain that continues to make rRNA and other stable RNAs even if starved for an amino acid. These strains have a mutation in the *relA* gene that inactivates the RelA enzyme, so that they do not synthesize ppGpp in response to amino acid starvation.

Relaxosome. The complex of proteins, including the relaxase, which is bound to the *oriT* sequence of a self-transmissible or mobilizable plasmid in the donor cell.

Release factors. Nonsense codon-specific proteins that are required, along with EF-G, for the termination of polypeptide synthesis and the release of the newly synthesized polypeptide from the ribosome when the ribosome encounters an in-frame nonsense codon in the mRNA.

Replica plating. A technique in which bacteria grown on one plate are transferred to a fuzzy cloth and then are transferred from the fuzzy cloth onto another plate so that the bacteria on the first plate are transferred to the corresponding position(s) on the second plate.

Replication fork. The region in a replicating double-stranded DNA molecule where the two strands are separating to allow synthesis of the complementary strands.

Replication restart. The process of reloading the replication apparatus on the DNA after it has been dissociated, for example at a nick or damage to the DNA.

Replicative bypass. A process in which the replication fork moves past damage to the DNA that interferes with proper base pairing, either by skipping over the damage and leaving a gap in the newly synthesized strand or by inserting deoxynucleotides at random opposite the damaged bases.

Replicative form. The double-stranded DNA or RNA that forms by synthesis of the complementary minus strand after infection by a phage or virus that has a single-stranded genome.

Replicative transposition. A type of transposition in which single-stranded nicks are made at each end of the transposon and a staggered double-strand break is made in the target DNA. The free 3′ ends at the extremities of the transposon are ligated to the free 5′ ends of the target DNA and the free 3′ ends of the target DNA are used as primers to synthesize over the transposon, giving rise to a cointegrate.

Replicon. A DNA molecule capable of autonomous replication because it contains an origin of replication that functions in the cell in which it is located.

Reporter gene. A gene whose product is stable and easy to assay and so is convenient for detecting and quantifying the expression of genes to which it is fused.

Repressor. A protein that, in its active state, binds to operator sequences close to the promoter for an operon, thereby preventing transcription of the operon. More generally, it refers to a protein or RNA that negatively regulates transcription or translation so that synthesis of the gene product is reduced when it is active.

Resolution of a cointegrate. Separation of the two DNAs joined in a cointegrate by recombination between repeated sequences in the two copies of the transposon, leaving each DNA with one copy of the transposon.

Resolution of Holliday junctions. Cutting of the two crossed strands of DNA in a Holliday junction, for example by an X-phile, so that the DNA molecules, held together by the crossed strands in the Holliday junction, are separated.

Resolvase. A type of site-specific recombinase that breaks and rejoins DNA in *res* sequences in the two copies of the transposon in a cointegrate, thereby resolving the cointegrate into separate DNAs, each with one copy of the transposon.

Response regulator protein. A protein that is part of a two-component system and that, on receiving a signal (usually in the form of a phosphoryl group) from another protein, the sensor protein, performs a regulatory function, e.g., activates transcription of operons.

Restriction fragment. A piece of DNA obtained by cutting a longer DNA with a restriction endonuclease.

Restriction fragment length polymorphism. A difference in the size of restriction fragments obtained by cutting DNA from two different strains with the same restriction endonuclease. The polymorphism reflects differences in the DNA sequences between the sites.

Restriction modification system. A complex of proteins whose members can recognize specific sequences in DNA, methylate a base in the sequence, and cut in or near the sequence if it is not methylated.

Retrohoming. The process by which a retrotransposon inserts itself into the same site in a different DNA which lacks it.

Retroregulation. Regulation occurring when the amount of an RNA in the cell is determined by an event at the 3′ end of the RNA rather than at the 5′ end.

Retrotransposon. A type of transposon that hops into the same site in a DNA that lacks it by first making an RNA copy of itself and then making a DNA copy of this RNA with a reverse transcriptase, while it inserts this DNA copy into the target DNA by a sort of reverse splicing.

Reverse genetics. The process in which the function of the product of a gene is determined by first altering the sequence in DNA in the test tube, using molecular biology techniques, and then reintroducing the DNA into a cell to see what effect the mutation has on the organism. Contrast with forward genetics, in which the mutation is first recognized because of the phenotype it causes.

Reverse transcriptase PCR. A procedure for amplifying PCR fragments from RNA. A cDNA is first made of the RNA with reverse transcriptase, and then this cDNA is used as a template for PCR.

Reversion. Restoration of a mutated sequence in DNA to the wild-type sequence.

Reversion rate. The probability that a mutated sequence in DNA will change back to the wild-type sequence each time the organism multiplies.

Revert. *See* Reversion.

Revertant. An organism in which the mutated sequence in its DNA has been restored to the wild-type sequence.

RF. *See* Replicative form.

RFLP. *See* Restriction fragment length polymorphism.

Ribonucleoside triphosphate. A base (usually A, U, G, or C) attached to a ribose sugar with three phosphate groups attached in tandem to the 5′ carbon of the sugar.

Ribonucleotide reductase. An enzyme that catalyzes the reduction of nucleoside diphosphates to deoxynucleoside diphosphates by removing the hydroxy group at the 2′ carbon of the ribonucleoside diphosphate and replacing it with a hydrogen.

Riboprobe. A hybridization probe made of RNA rather than DNA.

Ribosomal proteins. The proteins that, in addition to the rRNAs, make up the structure of the ribosome.

Ribosomal RNA. Any one of the three RNAs (16S, 23S, and 5S in bacteria) that make up the structure of the ribosome.

Ribosome. The cellular organelle, made up of about 50 different proteins and 3 different RNAs, that is the site of protein synthesis.

Ribosome-binding site. *See* Translational initiation region.

Ribosome cycle. The association and dissociation of the 30S and 50S ribosomes during initiation and termination of translation.

Ribosome release factor. *See* Release factors.

Ribozyme. An RNA that has enzymatic activity.

R-loop. A three-stranded structure formed by the invasion of a double-stranded DNA by an RNA, displacing one of the strands of the double-stranded DNA.

RNA modification. Any covalent change to RNA, such as methylation of a base, that does not involve the breaking and joining of phosphate-phosphate or phosphate-ribose bonds in the backbone of the RNA.

RNA polymerase. An enzyme that polymerizes ribonucleoside triphosphates to make RNA chains by using a DNA or RNA template.

RNA polymerase holoenzyme. The $\alpha_2\beta\beta'$ RNA polymerase with a σ factor attached.

RNA processing. Covalent changes to RNA that involve the breaking and joining of phosphate-phosphate or phosphate-ribose bonds in the backbone of the RNA.

Robust regulation. Overlapping regulatory circuits to ensure that regulation is not too sensitive to any change in conditions.

Rolling-circle plasmid. A plasmid which replicates by a rolling-circle mechanism.

Rolling-circle replication. A type of replication of circular DNAs in which a single-stranded nick is made in one strand of the DNA and the 3′ hydroxyl end is used as a primer to replicate around the circle, displacing the old strand.

Rolling-circle transposon. *See* Y2 transposon.

rRNA. *See* Ribosomal RNA.

RT-PCR. *See* Reverse transcriptase PCR and Quantitative reverse transcriptase PCR.

Round of replication. The cycle of replication of a circular DNA in which a complete copy of the DNA is made.

Sanger dideoxy sequencing. A method for DNA sequencing in which the chain-terminating property of the dideoxynucleotides is used.

Satellite virus. A naturally occurring virus that depends on another virus for its multiplication.

Screening. The process (usually streamlined) of testing a large number of organisms for a particular mutant type.

Seamless cloning. The process of inserting a PCR fragment into a cloning vector by using a restriction endonuclease that cuts outside its recognition site so that no extraneous base pairs are added between the cloned DNA fragment and the cloning vector.

SecA. A protein with ATPase activity that drives proteins to be exported into the SecYEG channel.

SecB. A chaperone that binds exported proteins and keeps them from folding prematurely before they can be taken up by the SecYEG channel.

***sec* gene.** One of the genes whose products are required for transport of proteins through the inner membrane.

Secondary structure. A structure of a polynucleotide or polypeptide chain that results from noncovalent pairing between nucleotides or amino acids in the chain.

Secretin. A protein encoded by type II and type III protein secretion systems of gram-negative bacteria that form multisubunit β-barrels in the outer membrane through which proteins are secreted.

Secreted protein. A protein which leaves the cell after it is made and moves into the outside environment.

Sec system. The general system encoded by the *sec* genes of eubacteria for secreting proteins across the cytoplasmic membrane; it consists of the targeting factors SecA and SecB and includes the components of the SecYEG channel in the inner membrane.

SecYEG channel, SecYEG translocase. The channel in the inner membrane of eubacteria, composed of the SecY, SecE, and SecG proteins, through which many proteins are translocated. *See* Translocase.

Segregation. The process by which newly replicated DNAs or genetic alleles are separated into daughter cells or spores.

Selected marker. A difference in DNA sequence between two strains in a bacterial or phage cross that is used to select recombinants. The cross is plated so that only recombinants that have received the donor sequence or allele can multiply.

Selection. A procedure in which bacteria or viruses are placed under conditions in which only the wild type or the desired mutant or recombinant can multiply, allowing the isolation of even very rare mutants and recombinants.

Selectional genetics. Genetic analysis in which selection of mutants or recombinants is used.

Selective conditions. Conditions under which only the wild type or the desired mutant can multiply.

Selective media. Media that have been designed to allow multiplication of only the desired mutant or wild type. Such media often lack one or more nutrients or contain a substance that is toxic.

Selective plate. An agar plate made with selective media.

Self-transmissible plasmid. A plasmid that encodes all the gene products needed to transfer itself to other bacteria through conjugation.

Semiconservative replication. A type of DNA replication in which the daughter DNAs are composed of one old strand and one newly synthesized strand.

Sensitive cell. A type of cell that can serve as the host for a particular type of virus.

Sensor kinase. In two-component systems, a protein that transfers the γ phosphate of ATP to itself in response to a certain environmental or cellular signal and then transfers this phosphate to a response regulator protein that performs some cellular function.

Sensor protein. The protein in a two-component system that detects changes in the environment and communicates this information to the response regulator, usually by transferring a phosphoryl group. *See* Sensor kinase.

Sequestration. The entry of the origin of replication (*oriC*) of the bacterial chromosome into an incompletely understood dormant state after a round of chromosome replication has initiated.

Serial dilution. A procedure in which an aliquot of a solution is diluted into one vessel and then an aliquot of the solution in this vessel is diluted into a second vessel, and so forth. The total dilution is the product of each of the individual dilutions.

7,8-Dihydro-8-oxoguanine. *See* 8-OxoG.

Sex pilus. A rod-like structure that forms on the surface of bacterial cells containing a self-transmissible plasmid and facilitates transfer of the plasmid or other DNAs into another bacterium, probably by holding the two cells together.

Shine-Dalgarno sequence. A short sequence, usually about 10 nucleotides upstream of the initiation codon in a bacterial translational initiation region, that is complementary to a sequence in the 3′ end of the 16S rRNA; it helps position the ribosome for initiation of translation. It is named after the persons who discovered it.

Shufflon. A region of a self-transmissible plasmid that contains many cassette sequences that integrate by a site-specific integrase to alter the carboxyl terminus of a pilin protein and change the cell surface receptors to which the pilus can bind.

Shuttle vector. A plasmid cloning vector that contains two origins of replication which function in different types of cells so that the plasmid can replicate in both types of cells.

Siblings. In microbial genetics, two cells or viruses that arose from the multiplication of the same mutant cell or virus.

Signal recognition particle. Universally evolutionarily conserved particle composed of both RNA and protein that binds the signal sequence of the cytoplasmic membrane proteins as the signal sequence emerges from the ribosome exit channel. The complex containing this particle then binds to the docking protein to direct proteins to the translocon channel. The signal recognition particle is composed of the 4.5S RNA and the Ffh protein in eubacteria.

Signal sequence. A sequence, composed of mostly hydrophobic amino acids, that is located at the N terminus of some membrane and secreted proteins and that targets the protein for transport into or through the cytoplasmic membrane. In bacterial proteins secreted by the Sec system or the Tat system, the signal sequence is removed as the protein passes into or through the membrane; in proteins targeted by the signal recognition particle to the cytoplasmic membrane, it is usually the first transmembrane domain and is not removed.

Signal transduction pathway. A set of proteins that pass a signal from one to the other by direct contact. They do this by chemically altering each other by proteolysis, by transferring a chemical group such as a phosphoryl or methyl group, or by binding to each other.

Silent mutation. A change in a DNA sequence of a gene encoding a protein that does not change the amino acid sequence of the protein, usually because it changes the last base in a codon, thereby changing it to another codon but one that encodes the same amino acid.

Single-gene transcriptional analysis. A method for evaluating the transcription of a single gene, including promoter location and quantitation (e.g., quantitative reverse transcriptase PCR).

Single mutant. A mutant organism that has only one of the two or more mutations being studied.

Single mutation. A mutation due to a single event that changed the DNA sequence, independent of how many base pairs were changed by the event.

Site-specific (site-directed) mutagenesis. One of many methods for mutagenizing DNA in such a way that the change is localized to a predetermined base pair or small region in the DNA.

Site-specific recombinases. Enzymes that recognize two specific sites on DNA and promote recombination between them.

Site-specific recombination. Recombination that occurs only between defined sequences in DNA. It is usually performed by site-specific recombinases.

6-4 lesion. A type of damage to DNA in which the carbon at the 6 position of a pyrimidine is covalently bound to the carbon at the 4 position of an adjacent pyrimidine.

Six-hitter. A type II restriction endonuclease that recognizes and cuts at a specific 6-bp sequence in DNA.

Sortase. An enzyme, in gram-positive bacteria, which cuts a protein to be displayed on the cell surface at its sorting signal and attaches it through a new peptide bond to a cross bridge in the cell wall.

SOS gene. A gene that is a member of the LexA regulon, so that its transcription is normally repressed by LexA repressor.

SOS mutagenesis. *See* Weigle mutagenesis.

SOS response. Induction of transcription of the SOS genes in response to DNA damage. It is due to stimulation of autocleavage of LexA repressor by the RecA single-stranded DNA nucleoprotein coprotease.

Southern blot hybridization. A procedure for transferring DNA from an agarose gel to a filter for hybridization. It is named after the person who developed the procedure.

Specialized transduction. A type of transduction restricted to phages capable of lysogeny in which the prophage integrates into the bacterial chromosome and in which only DNA sequences close to the attachment site of the prophage in the chromosome are transduced. It results from mistaken excision of the prophage that substitutes some flanking chromosomal DNA sequences for some of the phage DNA sequences that are normally packaged in the phage head.

Spontaneous mutations. Mutations that occur in organisms without deliberate attempts to induce them by irradiation or chemical treatment.

Sporulation. A developmental process that leads to the development of spores, which are dormant cells containing the DNA of the organism and are often resistant to desiccation and other harsh environmental conditions.

SRP. *See* Signal recognition particle.

Start point of transcription. *See* Transcription start site.

Starve. To deprive an organism of an essential nutrient that it cannot make for itself.

Sticky end. The short single-stranded DNA that sticks out from the end of the DNA molecule after it has been cut with a type II restriction endonuclease that makes a staggered break in the DNA.

Stimulon. The collection of all of the operons that are turned on by a particular environmental condition, independent of whether they are part of the same regulon. *See* Regulon.

Strain. A group of organisms that are identical to each other but differ genetically from other organisms of the same species. A strain is a subdivision of a species.

Strain typing. A useful method for classifying viruses or clinical isolates of pathogenic bacteria. Serotyping is a method based on detecting antigens. Molecular methods often involve PCR analysis of DNA.

Strand exchange. The process by which a strand of a double-stranded DNA changes partners so that it pairs with a different complementary strand of DNA, as in D-loop formation.

Strand passage. A reaction performed by topoisomerases in which one or two strands of a DNA are cut and the ends of the cut DNA are held by the enzyme to prevent rotation while other strands of the same or different DNAs are passed through the cuts.

Stringent control. Cessation of synthesis of rRNA and other stable RNAs in the cell when the cells are starved for an amino acid. It is due to the accumulation of ppGpp synthesized by the RelA protein on the ribosome.

Stringent plasmid. A plasmid that exists in only one or very few copies per cell, so that its replication must be very tightly controlled.

Structural gene. One of the genes in an operon for a pathway that encodes one of the enzymes of the pathway.

Subclone. A smaller DNA clone obtained by cutting a larger clone and cloning one of the pieces.

Sugar. A simple carbohydrate with the general formula $(CH_2O)_n$, as found in nature; n is 3 to 9.

Suicide vector. A cloning vector, usually plasmid or phage DNA, that cannot replicate in the cells into which it is being introduced.

Supercoiling. A condition in which the two strands of the DNA double helix are wrapped around each other either more or less often than predicted from the Watson-Crick helical structure of DNA, i.e., more or less than about 10.5 bp per turn.

Superinfection. Infection of cells by a virus when the same cells are already infected by the same type of virus.

Suppression. Alleviation of the effects of a mutation by a second mutation elsewhere in the DNA.

Suppressive effect. An effect that occurs if a clone, when introduced into a cell in high copy number, alleviates the effect of a mutation indirectly rather than complementing the mutation.

Suppressor mutation. A mutation elsewhere in the DNA that alleviates the effects of another mutation.

Symmetric sequence. A sequence of deoxynucleotides in double-stranded DNA that reads the same in the 5′-to-3′ direction on both strands.

Synapse. In recombination, a structure in which two DNAs are held together by pairing between their strands.

Synchronize. To treat a culture of cells so that they are all at approximately the same stage in their cell cycle at the same time.

Synteny. Conservation of gene order or genetic linkage in the genomes of different types of organisms.

Synthetic lethality screen. A selection system set up to isolate mutants with mutations in genes whose products are required for viability only in the absence of another gene product. Often the other gene is set up so that it is transcribed only from an inducible promoter. The mutations being sought are lethal only in the absence of inducer, when the other gene is not being expressed.

Synthetic phenotype. A situation in which a given mutant phenotype results from two or more mutations that individually would not cause the particular phenotype.

Tag. A sequence of amino acids added to a protein so that the protein can be purified more easily.

Tag vector. A cloning vector designed so that an amino acid sequence that is easy to purify will be added to a protein if the gene for the protein is cloned into the vector in frame with the translational initiation region on the vector and with no intervening nonsense codons.

Tandem duplication. A type of mutation that causes a DNA sequence to be followed immediately by the same sequence in the same orientation.

Tandem mass spectrometry. Two mass spectrometric analyses run sequentially, such that the first analysis allows selection of a specific peptide ion and the second analysis includes fragmentation of the selected peptide, analyzing the masses of the pieces, and thereby determining partial peptide sequences. In the fragmentation step, the bonds that break are almost exclusively along the peptide backbone, and therefore the ion species detected in the second analysis mostly represent peptide ions.

Target DNA. The DNA into which a transposon hops.

Tautomer. A (usually temporary) form of a molecule in which the electrons are distributed differently among the atoms.

Temperature-sensitive mutant. A mutant that cannot grow in the temperature range in which the wild type can multiply, usually at a higher temperature.

Template strand. The strand of DNA in a region from which RNA is synthesized that has the complementary sequence of the RNA and so serves as the template for RNA synthesis.

Terminally redundant DNA. A DNA, usually a phage genome, that has direct repeats at both ends; that is, the sequences at both ends are the same in the direct orientation.

Termination of replication. The process by which the replication apparatus leaves the DNA when DNA replication is completed and the daughter DNAs are separated.

Termination of transcription. The process by which the RNA polymerase leaves the DNA and the RNA chain is released at a transcription termination site in the DNA.

Termination of translation. The process by which the ribosome leaves the mRNA and the polypeptide is released when a nonsense codon in the mRNA is encountered in frame.

Tertiary structure. The three-dimensional structure of a polypeptide or polynucleotide.

Tetramer. A protein made up of four polypeptides.

Theta replication. A type of replication of circular DNA in which the replication apparatus initiates at an origin of replication and proceeds in one or both directions around the circle with leading and lagging strands of replication. The molecule in an intermediate state of replication resembles the Greek letter theta (θ).

Three-factor cross. A type of genetic cross used to order three closely linked mutations and in which one parent has two of the mutations and the other parent has the third mutation.

3′ end. The terminus of a polynucleotide chain (DNA or RNA) ending in the nucleotide that is not joined at the 3′ carbon of its ribose to the 5′ phosphate of another nucleotide.

3′ exonuclease. An enzyme that degrades a polynucleotide from its 3′ end by removing nucleotides one at a time.

3′ hydroxyl end. In a polynucleotide, a 3′ end that has a hydroxyl group on the 3′ carbon of the ribose sugar of the last nucleotide without a phosphate group attached.

3′ overhang. An unpaired single strand extending from the strand with a free 3′ end in a double-stranded DNA.

3′ untranslated region (3′ UTR). In an mRNA, the sequences downstream or 3′ of the nonsense codon of the last open reading frame encoding a protein.

Thymine (T). One of the pyrimidine (one-ringed) bases in DNA and some tRNAs.

Time of flight. A method used in mass spectrometry to separate ionic species. For example, each time a peptide ion has been produced by a laser pulse (*see* MALDI), a trigger starts a clock so as to measure the velocity of the ion's movement through an electric field.

TIGRFAM. An extensive database of protein families that is useful for categorizing coding sequences. The database is maintained by TIGR (The Institute for Genomic Research).

TIR. *See* Translational initiation region.

Titration. A process of increasing the concentration of one of two types of molecules that bind to each other until all of the other type of molecule is bound.

TLS. *See* Translesion synthesis.

tmRNA. A small RNA that is a hybrid between a tRNA and an mRNA. It can be aminoacylated with alanine like a tRNA and enter the A site of the ribosome if the A site is unoccupied, e.g., if the ribosome has reached the 3′ end of an mRNA without encountering a nonsense codon. A short reading frame on the tmRNA is then

translated, fusing a short peptide sequence to the C terminus of the truncated protein, which targets the protein for degradation by a protease.

TOF. *See* Time of flight.

Topo cloning. Using a topoisomerase with a sequence recognition site to insert DNA into a cloning vector. It is more efficient than cloning using DNA ligase but has the disadvantage that it requires specially designed cloning vectors.

Topoisomerase. An enzyme that can alter the topology of a DNA molecule by cutting one or both strands of DNA, passing other DNA strands through the cuts while holding the cut ends so that they are not free to rotate, and then resealing the cuts.

Topoisomerase IV. A type II topoisomerase of *E. coli* that is responsible for decatenation of daughter chromosomes after replication and for relieving positive supercoils ahead of the replication fork.

Topology of DNA. Relationship of the strands of DNA to each other in space.

Tra functions. Gene products encoded by the *tra* genes of self-transmissible DNA elements that allow the plasmids to transfer themselves into other bacteria.

***trans*-acting function.** A gene product that can act on DNAs in the cell other than the one from which it was made.

***trans*-acting mutation.** A mutation which affects a gene product that leaves the DNA from which it is made and so can be complemented.

Transconjugant. A recipient cell that has received DNA from another cell by conjugation.

Transcribe. To make an RNA that is a complementary copy of a strand of DNA.

Transcribed strand. In a region of a double-stranded DNA that is transcribed into RNA, the strand of DNA that is used as a template and so is complementary to the RNA.

Transcript. An RNA made from a region of DNA.

Transcriptional activator. A protein that is required for transcription of an operon. The protein makes contact with the RNA polymerase and allows the RNA polymerase to initiate transcription from the promoter of the operon.

Transcriptional autoregulation. The process by which a protein regulates the transcription of its own gene, by being either a repressor or an activator of its own gene.

Transcriptional fusion. Introduction of a gene downstream of the promoter for another gene or genes so that it is transcribed from the promoter for the other gene(s) into the same mRNA but is translated as a separate polypeptide from its own translational initiation region.

Transcriptional regulation. Regulation in which the amount of product of a gene that is synthesized under certain conditions is determined by how much mRNA is made from the gene.

Transcriptional regulator. Any protein that regulates the transcription of genes, e.g., a repressor, activator, or antitermination protein.

Transcription antitermination. *See* Antitermination.

Transcription bubble. The ~17-bp region in DNA during transcription in which the two strands of DNA have been separated by the RNA polymerase and within which the newly synthesized RNA forms a short RNA-DNA duplex with the transcribed strand of DNA.

Transcription start site. The nucleotide in the coding strand of DNA in a promoter that corresponds to the first nucleotide polymerized into RNA from the promoter.

Transcription termination site. A DNA sequence at which the RNA polymerase falls off the template, stopping transcription. It can be either factor independent or dependent on a transcription termination factor such as ρ.

Transcription vector. A cloning vector that contains a promoter from which a cloned DNA can be transcribed.

Transcriptome. The complete set of transcripts expressed in an organism. The transcripts actually detected depend on their abundance under the experimental conditions used.

Transducing particle. A phage whose head contains bacterial DNA instead of its own DNA.

Transducing phage. A type of phage that sometimes packages bacterial DNA during infection and introduces it into other bacteria during infection of those bacteria.

Transductant. A bacterium that has received DNA from another bacterium by transduction.

Transduction. A process in which DNA other than phage DNA is introduced into a bacterium via infection by a phage containing the DNA in its head.

Transfection. Initiation of a virus infection by introducing virus DNA or RNA into a cell by transformation rather than by infection by the virus.

Transfer RNA. The small stable RNAs in cells to which specific amino acids are attached by aminoacyl tRNA synthetases. The tRNA with the amino acid attached enters the ribosome and base pairs through its anticodon sequence with a 3-nucleotide codon sequence in the mRNA to insert the correct amino acid into the growing polypeptide chain.

Transformant. A cell that has received DNA by transformation.

Transformasomes. Globular structures that appear on the surfaces of some types of bacteria into which DNA first enters during natural transformation of the bacteria.

Transformation. Introduction of DNA into cells by mixing the DNA and the cells.

Transformylase. The enzyme that transfers a formyl (CHO) group to the amino group of methionine to make formylmethionine.

Transgenic organism. An organism that has inherited foreign DNA sequences that have been experimentally introduced into its ancestors. The introduced DNA sequences are passed down from generation to generation because they are inserted into a stably inherited DNA, such as a chromosome.

Transgenics. The process of introducing foreign DNA into the chromosome of an organism to make a transgenic organism.

Transition mutation. A type of base pair change mutation in which the purine base has been changed into the other purine base and the pyrimidine base has been changed into the other pyrimidine base (e.g., AT to GC or GC to AT).

Translated region. A region of an mRNA that encodes a protein.

Translational coupling. A gene arrangement in which the translation of one protein-coding sequence on a polycistronic mRNA is required for the translation of the second, downstream coding sequence. Often, translation of the upstream coding sequencing is required to remove secondary structure in the mRNA that blocks the translational initiation region for the downstream coding sequence.

Translational fusion. The fusion of parts of the coding regions of two genes so that translation initiated at the translational initiation region for one polypeptide on the mRNA will continue into the coding region for the second polypeptide in the correct reading frame for the second polypeptide. A polypeptide containing amino acid sequences from the two genes that were joined to each other will be synthesized.

Translational initiation region. The initiation codon, the Shine-Dalgarno sequence, and any other surrounding sequences in mRNA that are recognized by the ribosome as a place to begin translation. Also called a ribosome-binding site (RBS).

Translationally autoregulated. A protein that can affect the rate of translation of its own coding sequence on its mRNA. Usually in such cases the protein binds to its own translational initiation region or that of an upstream gene to which it is translationally coupled; hence, the protein represses its own translation.

Translational regulation. Variation, under different conditions, in the amount of synthesis of a polypeptide due to variation in the rate at which the polypeptide is translated from the mRNA.

Translation elongation factor G. The protein required to move the peptidyl-tRNA from the A site to the P site on the ribosome with the concomitant cleavage of GTP to GDP, after the peptide bond has formed.

Translation elongation factor Tu. The protein that binds to aminoacylated tRNA and accompanies it into the A site of the ribosome. It then cycles off the ribosome with the concomitant cleavage of GTP to GDP, leaving the aminoacylated tRNA behind.

Translation termination site. Any one of the nonsense codons for the organism in the frame being translated.

Translation vector. A cloning vector which contains a TIR from which a cloned DNA sequence can be translated.

Translesion synthesis. Synthesis of DNA over a template region containing a damaged base or bases that are incapable of proper base pairing.

Translocase. The evolutionarily highly conserved channel in the cytoplasmic membrane through which proteins are exported. In bacteria, it is represented by the SecYEG membrane channel.

Translocation. During translation, the movement of the tRNA with the polypeptide attached from the A site to the P site on the ribosome after the peptide bond has formed. In exported proteins, the act of moving an exported protein through the membranes.

Transmembrane domain. The region in a polypeptide between a region that is exposed to one surface of a membrane and a region that is exposed to the other surface. This region must traverse and be embedded in the membrane. Usually, transmembrane domains have a stretch of at least 20, mostly hydrophobic, amino acids that is long enough to extend from one face of a bilipid membrane to the other.

Transmembrane protein. A membrane protein that has surfaces exposed at both sides of the membrane.

Transposase. An enzyme encoded by a transposon that cuts the target DNA and the DNA at both ends of the transposon and joins the cut ends of the target DNA to the ends of the transposon DNA during transposition.

Transposition. Movement of a transposon from one place in DNA to another.

Transposon. A DNA sequence that can move from one place in DNA to a different place with the help of transposase. It should be distinguished from homing DNA elements, which usually move only into the same sequence in another DNA and depend on homologous recombination, or DNA elements that insert into other DNAs by using recombinases called integrases.

Transposon mutagenesis. A technique in which a transposon is used to make random insertion mutations in DNA. The transposon is usually introduced into the cell in a suicide vector, and so it must transpose into another DNA in the cell to become established.

TransTerm. A database that contains mRNA sequences compiled from GenBank and that allows the user to find regulatory elements such as initiation and termination regions (http://transterm.cbcb.umd.edu).

Transversion mutation. A type of base pair change mutation in which the purine in the base pair is changed into the pyrimidine and vice versa, e.g., GC to TA or GC to CG.

Trigger factor. A chaperone in *E. coli* that is closely associated with the exit pore of the ribosome and that helps proteins fold as they emerge from the ribosome. It can partially substitute for DnaK.

Triparental mating. A conjugational mating, for introducing mobilizable plasmids into cells, in which three strains of bacteria are mixed. One strain contains a self-transmissible plasmid, and the second strain contains the mobilizable plasmid, which is then mobilized into the third strain.

Triple-stranded DNA structure. Three strands of DNA held together in a triple-stranded structure, as has been hypothesized to form when a RecA nucleoprotein filament invades a double-stranded DNA.

tRNA. *See* Transfer RNA.

tRNA$_f^{Met}$. The tRNA to which formylmethionine is added and that pairs with the initiator codon in a translational initiation region to initiate translation of a polypeptide in bacteria.

Ts mutant. *See* Temperature-sensitive mutant.

Two-component regulatory system. A pair of proteins, one of which, the sensor, undergoes a change in response to a change in the environment and communicates this change, usually in the form of a phosphate, to another protein, the response regulator, which then causes the appropriate cellular response. Different two-component systems are often highly homologous to each other, so they can be identified in sequenced bacterial genomes. Also referred to as two-component signal transduction.

Two-dimensional polyacrylamide gel electrophoresis (2D-PAGE). A separation technique in which proteins are applied to a pI (isoelectric point) strip and separated by charge by using isoelectric focusing and then this strip is poured into another slab gel containing sodium dodecyl sulfate so the proteins move at right angles to the first gel and are separated by size.

Two-hybrid screen. A technique for determining if two proteins or regions of proteins bind to each other or for identifying proteins that bind to a particular protein or region of a protein. It is based on the ability of proteins that bind to each other to bring two parts of a tester protein together, restoring its activity. In yeast two-hybrid screens, it is the ability of two proteins or regions to dimerize a transcription factor that has had its dimerization domain deleted and thereby restore the activity of the transcription factor. In a bacterial two-hybrid system, it is the ability to bring two parts of an adenylate cyclase enzyme together, restoring its ability to make cyclic AMP.

Two-partner secretion. *See* Type V secretion system.

Type I secretion system. A protein secretion system in gram-negative bacteria based on a specific ATP-binding cassette (ABC) transporter. The ABC transporter binds the protein in the cytoplasm and cleaves ATP to furnish the energy to push the protein through a channel formed by a specific inner membrane protein and through the β-barrel channel formed by TolC in the outer membrane. The TolC channel, a multiuse channel, may be recruited and opened by the inner membrane protein only when the protein is to be transported. Proteins secreted by type I systems recognize a signal sequence in the carboxyl terminus of the protein that is not cleaved off during transport. The hemolysin A of *E. coli* is an example of a protein secreted by a type I system.

Type II secretion system. A protein secretion system of gram-negative bacteria that uses either the SecYEG channel or the Tat channel to transport proteins through the inner membrane. It then uses a specific secretin β-channel to secrete the protein through the outer membrane. It makes a complicated structure called a pseudopilus, which may push the protein through the inner membrane channel and through the secretin channel to the outside of the cell. Examples of proteins

secreted by type II systems are pullulanase of *Klebsiella oxytoca* and the cholera toxin of *Vibrio cholerae*. It is related to the transport systems that assemble type IV pili on the cell surface, to some transformation systems, and to the systems that secrete filamentous bacteriophages from the infected cell.

Type III secretion system. A protein secretion system of pathogenic gram-negative bacteria that forms a syringe-like structure, sometimes called an injectisome, that injects effector proteins directly through both bacterial membranes into eukaryotic cells. A short sequence on the N terminus of the protein directs it to the injectisome, but this short sequence is not cut off during transport. It forms a secretin channel through the outer membrane that is related to the secretin channel of type II secretion systems. Type III secretion systems are found in both plant and animal pathogens. An example is the type III secretion system in *Yersinia pestis* that injects Yop proteins directly into macrophages, inactivating them. It is structurally related to the flagellar motor.

Type IV secretion system. A protein secretion system of gram-negative bacteria that can inject proteins directly through both bacterial membranes into other cells, although some seem to use the SecYEG channel to transport the protein through the inner membrane. These systems form a secretin-like channel in the outer membrane. They work through a coupling protein that binds to the protein to be secreted and directs it to the channel in the membranes. Plasmid conjugation systems are essentially type IV secretion systems. In this case, the protein being secreted is the relaxase which is bound to the DNA being transferred. The T-DNA transfer system of *Agrobacterium tumefaciens* is an example of a type IV secretion system. It injects T-DNA as well as other effector proteins directly into the plant cell nucleus, where they cause tumors to form on the plant.

Type V secretion system. A group of secretion systems that includes the autotransporters, the two-partner secretion systems, and the chaperone-usher systems. These secretion systems all form a dedicated β-barrel in the outer membrane that secretes only one or a select group of proteins, and they also all use the SecYEG channel to transport the secreted protein through the inner membrane. Autotransporters transport themselves since they have a carboxyl-terminal domain that forms a β-barrel channel in the outer membrane. Another domain of the protein, the passenger domain, passes through the channel. A protease domain can then cut the passenger domain off the transporter domain, depending on whether the passenger domain is to be released from the cell or is to be displayed on the cell surface. The immunoglobulin A protease of *Neisseria gonorrhoeae* is an example of an autotransported protein. Two-partner secretion systems are similar except that the two domains are on separate proteins. Chaperone-usher systems assemble some types of pili on the cell surface. They consist of three proteins, the pilin proteins (of which there can be more than one type depending on where they are in the pilus), the usher (which forms a β-barrel in the outer membrane), and a periplasmic chaperone. The chaperone binds the pilin protein in the periplasm, preventing it from folding prematurely and keeping it in an energized "cocked" state while it delivers the pilin to the usher in the outer membrane. The usher then secretes the pilin proteins, somehow assembling them in the right order into the growing pilus on the cell surface.

UAS. *See* Upstream activator sequence.

Umber codon. The codon UGA; also called the opal codon.

Unselected marker. A difference between the DNA sequences of two bacteria or phages involved in a genetic cross that can be used for genetic mapping. It is a difference other than the difference used to select recombinants. Mapping information can be obtained by testing recombinants that have been selected for being recombinant for one marker, the selected marker, to determine if they have the sequence of the donor or the recipient for another marker, an unselected marker.

Untargeted mutations. Mutations that occur in DNA at sites other than the sites of DNA damage.

Upstream. From a given point, sequences that lie in the 5′ direction on RNA or in the 5′ direction on the coding strand of a DNA region from which an RNA is made.

Upstream activator sequence. A DNA sequence upstream of a promoter that increases transcription from the promoter by binding an activator protein. It is usually associated with NtrC family activators and σ^{54} promoters. It can be many hundreds of base pairs upstream from the promoter. Also called upstream activator site.

Uptake sequence. A short DNA sequence that allows DNA containing the sequence to be taken up by some types of bacteria during natural transformation.

Uracil (U). One of the pyrimidine (one-ringed) bases; naturally found in RNA.

Uracil-*N*-glycosylase. An enzyme that removes the uracil base from DNA by cleaving the bond between the base and the deoxyribose sugar.

UTM. *See* Untargeted mutations.

UvrABC endonuclease. A complex of three proteins that cuts on both sides of any DNA lesion causing a significant distortion of the helix, as a first step in excision repair of the damage. Also called the UvrABC excinuclease.

Very short patch (VSP) repair. A type of repair in enteric bacteria that removes the mismatched T in the sequence CT(A/T)GG/GG(T/A)CC and replaces it with a C. The C at this position is commonly methylated in enteric bacteria, and its deamination to a T is not recognized by the uracil-*N*-glycosylase. A very short stretch of the DNA strand around this mismatched T is removed and resynthesized during the repair, hence the name. The gene for the methylase that methylates this cytosine is linked to the gene for a protein, Dcm, that recognizes the T in this mismatch and directs components of the mismatch repair system to remove it.

Watson-Crick structure of DNA. The double-helical structure of DNA first proposed by James Watson and Francis Crick. The two strands of the DNA are antiparallel and held together by hydrogen bonding between the bases.

Weigle mutagenesis. Another name for SOS mutagenesis. It refers to the increase in the number of phage mutations if phage infects cells that have been preirradiated with UV. It is due to SOS induction of the *umuCD* genes as well as *recA*. It is named after Jean Weigle, who first observed it.

Weigle reactivation. The increased ability of phages to survive UV irradiation damage to their DNA if the cells they infect have been previously exposed to UV irradiation. It is due to SOS induction of repair functions. It is named after Jean Weigle, who first observed it.

Western blot. A membrane onto which proteins have been transferred from a gel.

Wild type. The normal type. Literally, the term refers to the organism as it was first isolated from nature. In a genetic experiment, it is the strain from which mutants are derived.

Wild-type allele. The form of a gene as it exists in the wild-type organism.

Wild-type phenotype. The particular outward trait characteristic of the wild type that is different in the mutant.

W-mutagenesis. *See* Weigle mutagenesis.

Wobble. The property of the genetic code in which codons for the same amino acid often differ only in the last (third) nucleotide. It reflects the fact that the base of the first nucleotide (read 3′ to 5′) in the anticodon of a tRNA can often pair with more than one base in the third nucleotide (read 5′ to 3′) of a codon in the mRNA.

W-reactivation. *See* Weigle reactivation.

Xanthine. A purine base that results from deamination of guanine.

XerC,D recombinase. The recombinanse in *E. coli* and many other bacteria that separates dimerized chromosomes by promoting recombination between repeated *dif* sequences.

X-phile. One of a group of enzymes that can cut the crossed DNA strands at a Holliday junction.

YidC protein. An inner membrane protein of unknown function that cooperates with the SecYEG channel in inserting inner membrane proteins into the inner membrane.

Y polymerases. A large family of DNA polymerases, represented by Pol II, DinB (Pol IV), and UmuC (Pol V) in *E. coli*, that are capable of translesion synthesis, perhaps because they have a more open active center and lack editing functions.

Y2 transposon. A transposon with two Y's (tyrosines) in its active center, sometimes called a rolling-circle transposon because the mechanism of transposition resembles rolling-circle replication of phages and plasmids. These transposons do not have inverted repeated sequences at their ends but, rather, have one end that serves as the origin end and the one end that is the terminus end. The transposon is nicked at its origin end to form 3′ hydroxyl and 5′ phosphate ends, and one of the tyrosines forms a phosphoryl linkage through its side chain to the 5′ phosphate. The 3′ hydroxyl end is then used as s primer to replicate over the transposon. The old strand of the transposon may then loop out as it is displaced to form a single-stranded circle that integrates into the target DNA or may integrate into the target DNA as it is displaced. The second tyrosine may then somehow play a role in replicating back over the transposon to form the complementary strand.

Zero frame. In the coding region of a gene, the sequence of nucleotides, taken three at a time, in which the polypeptide encoded by the gene is translated.

Figure and Table Credits

Chapter 1

Figure 1.17 Adapted from Figure 2, p. 179, of J. E. Camara and E. Crooke, p. 177–191, *in* N. P. Higgins (ed.), *The Bacterial Chromosome* (ASM Press, Washington, D.C., 2005), with permission.

Figure 1.22 Reprinted from Figure 1, p. 159, of D. E. Pettijohn, p. 158–166, *in* F. C. Neidhardt, R. Curtiss III, J. L. Ingraham, E. C. C. Lin, K. B. Low, B. Magasanik, W. S. Reznikoff, M. Riley, M. Schaechter, and H. E. Umbarger (ed.), Escherichia coli *and* Salmonella: *Cellular and Molecular Biology*, 2nd ed. (ASM Press, Washington, D.C., 1996), with permission.

Figure 1.23 Reprinted from Figure 4, p. 161, of D. E. Pettijohn, p. 158–166, *in* F. C. Neidhardt, R. Curtiss III, J. L. Ingraham, E. C. C. Lin, K. B. Low, B. Magasanik, W. S. Reznikoff, M. Riley, M. Schaechter, and H. E. Umbarger (ed.), Escherichia coli *and* Salmonella: *Cellular and Molecular Biology*, 2nd ed. (ASM Press, Washington, D.C., 1996), with permission.

Box 1.3 figure (B) Redrawn from Figure 4, p. 613, of J. Errington, *ASM News* 69:608–614, 2003, with permission. (C) Redrawn from Figure 9 of K. A. Michie et al., *J. Bacteriol.* 188:1680–1690, 2006. (D) Adapted from Figure 6H, p. 743, of Y.-L. Shih and L. Rothfield, *Microbiol. Mol. Biol. Rev.* 70:729–754, 2006, with permission.

Chapter 2

Figure 2.8 Adapted from Color Plate 3A (associated with K. Geszvain and R. Landick, p. 283–296) *in* N. P. Higgins (ed.), *The Bacterial Chromosome* (ASM Press, Washington, D.C., 2005), with permission.

Figure 2.13 Based on Figure 2, p. 299, of S. L. Dove and A. Hochschild, p. 297–310, *in* N. P. Higgins (ed.), *The Bacterial Chromosome* (ASM Press, Washington, D.C., 2005), with permission.

Figure 2.16 Adapted from Color Plate 3B (associated with K. Geszvain and R. Landick, p. 283–296) *in* N. P. Higgins (ed.), *The Bacterial Chromosome* (ASM Press, Washington, D.C., 2005), with permission.

Figure 2.28 (A) Reprinted from Figure 1 of M. M. Yusupov, G. Z. Yusupova, A. Baucom, K. Lieberman, T. N. Earnest, J. H. Cate, and H. F. Noller, *Science* **292**:883–896, 2001, with permission. Copyright 2001 American Association for the Advancement of Science. (B) Reprinted from Figure 5 of J. H. Cate, M. M. Yusupov, G. Yusupova, T. N. Earnest, and H. F. Noller, *Science* **285**:2100, 1999, with permission. Copyright 1999 American Association for the Advancement of Science.

Figure 2.42 Adapted from Figure 2, p. 383, of J. Collado-Vides, B. Magasanik, and J. D. Gralla, *Microbiol. Rev.* **55**:371–394, 1991, with permission.

Figure 2.45 Adapted from p. 108 of the *Novagen Catalog* (Novagen, Madison, Wis., 1997), www.novagen.com, with permission.

Table 2.2 Adapted from Tables 1 and 2 of M. C. Ganoza, M. C. Kiel, and H. Aoki, *Microbiol. Mol. Biol. Rev.* **66**:460–485, 2002, with permission.

Box 2.5 figure Based on Figure 4, p. 339, of S. Kushner, p. 327–345, *in* N. P. Higgins (ed.), *The Bacterial Chromosome* (ASM Press, Washington, D.C., 2005), with permission.

Box 2.5 table Based on Table 1, p. 328, of S. Kushner, p. 327–345, *in* N. P. Higgins (ed.), *The Bacterial Chromosome* (ASM Press, Washington, D.C., 2005), with permission.

Box 2.7 Figure 2 Adapted from Figure 1 of S. R. Gill et al., *J. Bacteriol.* **187**:2426–2438, 2005, with permission.

Chapter 3

Figure 3.30 Reprinted from Figure 2, p. 118, of B. J. Bachmann, B. Low, and A. L. Taylor, *Bacteriol. Rev.* **40**:116–167, 1976, with permission.

Figure 3.38 Adapted from Figure 1, p. 372, of M. Belfort and J. Pedersen-Lane, *J. Bacteriol.* **160**:371–378, 1984, with permission.

Chapter 4

Figure 4.4 Adapted from Figure 2, p. 66, of S. A. Khan, p. 63–78, *in* B. E. Funnell and G. J. Phillips (ed.), *Plasmid Biology* (ASM Press, Washington, D.C., 2004), with permission.

Figure 4.7 Adapted from Figure 2, p. 1620, of S. E. Luria and J. L. Suit, p. 1615–1624, *in* F. C. Neidhardt, J. L. Ingraham, K. B. Low, B. Magasanik, M. Schaechter, and H. E. Umbarger (ed.), Escherichia coli *and* Salmonella typhimurium: *Cellular*

and Molecular Biology (American Society for Microbiology, Washington, D.C., 1987).

Figure 4.11 Adapted from Figure 1D, p. 50, of S. Brantl, p. 47–62, *in* B. E. Funnell and G. J. Phillips (ed.), *Plasmid Biology* (ASM Press, Washington, D.C., 2004), with permission.

Figure 4.18 Adapted from Figure 3B and C of Y.-L. Shih and L. Rothfield, *Microbiol. Mol. Biol. Rev.* **70**:729–754, 2006, with permission.

Figure 4.22 Adapted from p. 766 of the *BRL-Gibco Catalog* (Invitrogen Corp., Carlsbad, Calif., 2002). Copyright 2002 Invitrogen Corporation, with permission.

Figure 4.26 Adapted from Figures 1 and 2 of V. Vagner, E. Dervyn, and S. D. Ehrlich, *Microbiology* **144**:3097–3104, 1998, with permission.

Box 4.3 figure (A–C) Adapted from R. M. Lacatena and G. Cesareni, *Nature* (London) **294**: 623–626, 1981, with permission.

Chapter 5

Figure 5.2 Redrawn from Figure 1 of N. Firth, K. Ippen-Ihler, and R. A. Skurray, p. 2377–2401, *in* F. C. Neidhardt, R. Curtiss III, J. L. Ingraham, E. C. C. Lin, K. B. Low, B. Magasanik, W. S. Reznikoff, M. Riley, M. Schaechter, and H. E. Umbarger (ed.), Escherichia coli *and* Salmonella: *Cellular and Molecular Biology*, 2nd ed. (ASM Press, Washington, D.C., 1996), with permission.

Figure 5.3 Redrawn from Color Plate 6 (associated with T. Lawley, B. M. Wilkins, and L. S. Frost, p. 203–226) *in* B. E. Funnell and G. J. Phillips (ed.), *Plasmid Biology* (ASM Press, Washington, D.C., 2004), with permission.

Figure 5.13 Reprinted from Figure 4, p. 406, of S. Winans and G. Walker, *J. Bacteriol.* **161**:402–410, 1985.

Figure 5.19 Redrawn from Figure 1, p. 4831, of J. R. Chandler, A. R. Flynn, E. M. Bryan, and G. M. Dunny, *J. Bacteriol.* **187**:4830–4843, 2005, with permission.

Box 5.2 figure Redrawn from Figure 3, p. 461, of P. J. Christie, p. 455–472, *in* B. E. Funnell and G. J. Phillips (ed.), *Plasmid Biology* (ASM Press, Washington, D.C., 2004), with permission.

Chapter 7

Figure 7.1 (A) (Left) Electron micrograph by Sally Burns, Michigan State University. (Right) Provided by Arthur Zachary and Lindsay Black. (B) Photograph by Kurt Stepnitz, Michigan State University.

Figure 7.5 Adapted from p. 158 of the *Novagen Catalog* (EMD Biosciences, San Diego, Calif., 2007), www.emdbiosciences.com/novagen, 2007, with permission.

Figure 7.6 Reprinted from J. D. Karam (ed.), *Molecular Biology of Bacteriophage T4* (ASM Press, Washington, D.C., 1994), with permission.

Figure 7.8 Reprinted from Figure 8 of E. S. Miller, E. Kutter, G. Mosig, F. Arisaka, T. Kunisawa, and W. Ruger, *Microbiol. Mol. Biol. Rev.* 67:86–156, 2003, with permission.

Figure 7.14 Figure and legend reprinted from Figure 1, p. 418, of A. Gupta, A. B. Oppenheim, and V. K. Chaudhary, p. 415–429, *in* M. K. Waldor, D. Friedman, and S. L. Adhya (ed.), *Phages: their Role in Bacterial Pathogenesis and Biotechnology* (ASM Press, Washington, D.C., 2005), with permission.

Figure 7.19 Adapted from Figure 1, p. 231, of K. N. Kreuzer and B. Michel, p. 229–250, *in* N. P. Higgins (ed.), *The Bacterial Chromosome* (ASM Press, Washington, D.C., 2005), with permission.

Figure 7.20 Derived from Figure 2A, p. 97, of R. Young, p. 92–127, *in* M. K. Waldor, D. I. Friedman, and S. L. Adhya (ed.), *Phages: their Role in Bacterial Pathogenesis and Biotechnology* (ASM Press, Washington D.C., 2005), with permission.

Figure 7.23 Photograph by Kurt Stepnitz, Michigan State University.

Figure 7.26 Adapted from S. Benzer, *Proc. Natl. Acad. Sci. USA* 47:403–415, 1961, with permission.

Chapter 8

Figure 8.7 Photograph by Kurt Stepnitz, Michigan State University.

Figure 8.18 Adapted from Figure 1, p. 2082, of P. L. Wagner, M. N. Neely, X. Zhang, D. W. K. Acheson, M. K. Waldor, and D. L. Friedman, *J. Bacteriol.* 183:2081–2085, 2001, with permission.

Figure 8.19 Adapted from Figure 18 of H. Brussow, C. Canchaya, and W. D. Hardt, *Microbiol. Mol. Biol. Rev.* 68:560–602, 2004, with permission.

Chapter 9

Figure 9.2 (B) Based on Figure 7, p. 322, of M. Chandler, and J. Mahillon, p. 305–366, *in* N. L. Craig, R. Craigie, M. Gellert, and A. M. Lambowitz (ed.), *Mobile DNA II* (ASM Press, Washington, D.C., 2002), with permission.

Figure 9.17 Adapted from Figure 1, p. 8241, of W. S. Reznikoff, S. R. Bordenstein, and J. Apodaca, *J. Bacteriol.* 186:8240–8247, 2004, with permission.

Figure 9.18 Adapted from Figures 7 and 8, p. 431, of N. L. Craig, p. 423–456, *in* N. L. Craig, R. Craigie, M. Gellert, and A. M. Lambowitz (ed.), *Mobile DNA II* (ASM Press, Washington, D.C., 2002), with permission.

Figure 9.19 Adapted from Figure 1, p. 468, of K. Derbyshire and N. Grindley, p. 467–497, *in* N. P. Higgins (ed.), *The Bacterial Chromosome* (ASM Press, Washington, D.C., 2005), with permission.

Figure 9.29 Based on Figure 1 of D. Mazel, *ASM News* 70:520–525, 2004, with permission.

Figure 9.30 Based on Figure 2 of D. Mazel, *ASM News* 70:520–525, 2004, with permission.

Figure 9.32 Adapted from Figure 13, p. 165, of B. Hallet, V. Vanhooff, and F. Cornet, p. 145–180, *in* B. E. Funnell and G. J. Phillips (ed.), *Plasmid Biology* (ASM Press, Washington, D.C., 2004), with permission.

Figure 9.33 Adapted from Figure 14, p. 166, of B. Hallett, V. Vanhooff, and F. Cornet, p. 145–180, *in* B. E. Funnell and G. J. Phillips (ed.), *Plasmid Biology* (ASM Press, Washington, D.C., 2004), with permission.

Figure 9.34 Adapted from Figure 1, p. 95, of G. D. Van Duyne, p. 93–117, *in* N. L. Craig, R. Craigie, M. Gellert, and A. M. Lambowitz (ed.), *Mobile DNA II* (ASM Press, Washington, D.C., 2002), with permission.

Figure 9.35 Adapted from Figure 9, p. 160, of B. Hallet, V. Vanhooff, and F. Cornet, p. 145–180, *in* B. E. Funnell and G. J. Phillips (ed.), *Plasmid Biology* (ASM Press, Washington, D.C., 2004), with permission.

Figure 9.36 Adapted from Figure 10, p. 161, of B. Hallet, V. Vanhooff, and F. Cornet, p. 145–180, *in* B. E. Funnell and G. J. Phillips (ed.), *Plasmid Biology* (ASM Press, Washington, D.C., 2004), with permission.

Figure 9.37 Adapted from Figure 8B, p. 245, of R. C. Johnson, p. 230–271, *in* N. L. Craig, R. Craigie, M. Gellert, and A. M. Lambowitz (ed.), *Mobile DNA II* (ASM Press, Washington, D.C., 2002), with permission.

Table 9.1 Based on Table 1, p. 469, of K. Derbyshire and N. Grindley, p. 467–497, *in* N. P. Higgins (ed.), *The Bacterial Chromosome* (ASM Press Washington, D.C., 2005), with permission.

Chapter 11

Figure 11.4 Adapted from Figure 1, p. 6322, of M. L. Michaels and J. H. Miller, *J. Bacteriol.* 174:6321–6325, 1992, with permission.

Figure 11.6 Courtesy of P. C. Hanawalt.

Figure 11.12 Figure and legend adapted from Figure 1, p. 415, of M. G. Marinus, p. 413–430, *in* N. P. Higgins (ed.), *The Bacterial Chromosome* (ASM Press, Washington, D.C., 2005), with permission, and from additional material provided by the author.

Chapter 12

Figure 12.5 Adapted from Figure 5, p. 1291, of H. Choy and S. Adhya, p. 1287–1299, *in* F. C. Neidhardt, R. Curtiss III, J. L. Ingraham, E. C. C. Lin, K. B. Low, B. Magasanik, W. S. Reznikoff, M. Riley, M. Schaechter, and H. E. Umbarger (ed.), *Escherichia coli* and *Salmonella: Cellular and Molecular*

Biology, 2nd ed. (ASM Press, Washington, D.C., 1996), with permission.

Figure 12.9 Reprinted from Figures 5A and 6A of M. Lewis, G. Chang, N. C. Horton, M. A. Kercher, H. C. Pace, M. A. Schumacher, R. G. Brennan, and P. Lu, *Science* **271**:1247–1254, 1996, with permission. Copyright 1996 American Association for the Advancement of Science.

Figure 12.22 Adapted from Figure 2, p. 1484, of M. Schwartz, p. 1482–1502, *in* F. C. Neidhardt, J. L. Ingraham, K. B. Low, B. Magasanik, M. Schaechter, and H. E. Umbarger (ed.), Escherichia coli *and* Salmonella typhimurium: *Cellular and Molecular Biology* (American Society for Microbiology, Washington, D.C., 1987), with permission.

Figure 12.26 Reprinted from Figures 2 and 3, p. 5797 and 5798, of P. Babitzke and P. Gollnick, *J. Bacteriol.* **183**:5795–5802, 2001, with permission.

Box 12.2 Figures 1, 2, 3, and 4 Adapted from Figures 4, 2, 6, and 5, respectively, of S. L. Dove and A. Hochschild, p. 297–310, *in* N. P. Higgins (ed.), *The Bacterial Chromosome* (ASM Press, Washington, D.C., 2005), with permission.

Chapter 13

Figure 13.3 Adapted from Figure 5, p. 2276, of N. J. Savery, G. S. Lloyd, S. J. W. Busby, M. S. Thomas, R. H. Ebright, and R. L. Gourse, *J. Bacteriol.* **184**:2273–2280, 2002, with permission.

Figure 13.4 Reprinted from Figure 1, p. 5077, of A. Dhiman and R. Schleif, *J. Bacteriol.* **182**:5076–5081, 2000, with permission.

Figure 13.7 (B) Adapted from Figure 1, p. 2234, of G. Karimova, N. Dautin, and D. Ladant, *J. Bacteriol.* **187**:2233–2243, 2005, with permission.

Figure 13.13 Adapted from Figure 1, p. 511, of W. H. Mager and A. J. J. de Kruijff, *Microbiol. Rev.* **59**:506–531, 1995, with permission.

Figure 13.15 Adapted from Figure 3 of P. A. Digiuseppe and T. J. Silhavy, *ASM News* **70**:71–79, 2004, with permission.

Figure 13.25 Reprinted from Figure 1 of J. M. Auchtung, C. A. Lee, and A. D. Grossman, *J. Bacteriol.* **188**:5273–5285, 2006, with permission.

Box 13.1 figure Adapted from Figure 1, p. 7827, of G. Lorca, Y.-J. Chung, R. Barbote, W. Weyler, C. Schilling, and M. Saier, Jr., *J. Bacteriol.* **187**:7826–7839, 2005, with permission.

Box 13.3 figure Adapted from Figure 1, p. 4130, of M. Buck, M.-T. Gallegos, D. J. Studholme, Y. Guo, and J. D. Gralla, *J. Bacteriol.* **182**:4129–4136, 2000, with permission.

Box 13.4 figure (A and C) Adapted from Figures 1 and 4 of M. Y. Galperin, and M. Gomelsky, *ASM News* **71**:326–333, 2005, with permission.

Box 13.5 figure Adapted from Figure 1, p. 9825, of S. Altuvia and G. H. Wagner, *Proc. Natl. Acad. Sci. USA* **97**:9824–9826, 2000, and from Figure 5, p. 9923, of R. Lease and M. Belfort, *Proc. Natl. Acad. Sci. USA* **97**:9919–9924, 2000, with permission. Copyright 2000 National Academy of Sciences.

Chapter 14

Figure 14.4 Adapted from Figure 1 of I. R. Henderson, F. Navarro-Garcia, M. Desvaux, R. C. Fernandez, and D. Ala'Aldeen, *Microbiol. Mol. Biol. Rev.* **68**:692–744, 2004, with permission.

Figure 14.5 Adapted from Color Plate 28 (associated with N. K. Surana, S. E. Cotter, H.-J. Yeo, G. Waksman, and J. W. St. Geme III, p. 129–148) *in* G. Waksman, M. Caparon, and S. Hultgren (ed.), *Structural Biology of Bacterial Pathogenesis* (ASM Press, Washington, D.C., 2005), with permission.

Figure 14.6 Adapted from Color Plate 49 (associated with R. K. Tweten and M. Caparon, p. 223–239) *in* G. Waksman, M. Caparon, and S. Hultgren (ed.), *Structural Biology of Bacterial Pathogenesis* (ASM Press, Washington, D.C., 2005), with permission.

Figure 14.7 Adapted from Figure 2, p. 104, of K. M. Connolly and R. T. Clubb, p. 101–127, *in* G. Waksman, M. Caparon, and S. Hultgren (ed.), *Structural Biology of Bacterial Pathogenesis* (ASM Press, Washington, D.C., 2005), with permission.

Figure 14.8 Courtesy of Andrew Davis, Marie Elliot, and Mark Buttner.

Figure 14.9 Courtesy of Kim Findlay and Mark Buttner.

Figure 14.10 Reprinted from Figure 2A, p. 1729, of M. A. Elliot, N. Karoonuthaisiri, J. Huang, M. J. Bibb, S. N. Cohen, C. M. Kao, and M. J. Buttner, *Genes Dev.* **17**:1727–1740, 2003, with permission.

Figure 14.11 Reprinted from Figure 2B, p. 1729, of M. A. Elliot, N. Karoonuthaisiri, J. Huang, M. J. Bibb, S. N. Cohen, C. M. Kao, and M. J. Buttner, *Genes Dev.* **17**:1727–1740, 2003, with permission.

Figure 14.13 Reprinted from Figure 6, p. 1733, of M. A. Elliot, N. Karoonuthaisiri, J. Huang, M. J. Bibb, S. N. Cohen, C. M. Kao, and M. J. Buttner, *Genes Dev.* **17**:1727–1740, 2003, with permission.

Figure 14.14 Reprinted from Figure 4, p. 1731, of M. A. Elliot, N. Karoonuthaisiri, J. Huang, M. J. Bibb, S. N. Cohen, C. M. Kao, and M. J. Buttner, *Genes Dev.* **17**:1727–1740, 2003, with permission.

Figure 14.15 Reprinted from Figure 1, p. 168, of P. A. Levin and R. Losick, p. 167–190, *in* Y. V. Brun and L. J. Shimkets (ed.), *Prokaryotic Development* (ASM Press, Washington, D.C., 2000), with permission, and reprinted from A. Driks, Figure 2A to F, p. 23, *in* V. E. A. Russo, D. J. Cove, L. G. Edgar, R. Jaenisch, and F. Salamini (ed.), *Development: Genetics, Epigenetics and Environmental Regulation* (Springer-Verlag, Berlin, Germany, 1999), with permission. Electron micrographs provided courtesy of A. Driks.

Figure 14.16 Adapted from Figure 1 of J. Errington, *Microbiol. Rev.* 57:1–33, 1993, with permission.

Figure 14.18 Adapted from Figure 8, p. 39, of B. Lazazzera, T. Palmer, J. Quisnel, and A. D. Grossman, p. 27–46, *in* G. M. Dunny and S. C. Winans (ed.), *Cell-Cell Signaling in Bacteria* (ASM Press, Washington, D.C., 1999), with permission.

Figure 14.20 Adapted from Figure 1 and Table 2 of J. Errington and J. Mandelstam, *J. Gen. Microbiol.* 132:2967–2976, 1986, with permission.

Figure 14.28 Redrawn from material provided by A. I. Derman.

Box 14.1 figure Adapted from Figure 13.10, p. 266, of M. Schaechter, J. L. Ingraham, and F. C. Neidhardt, *Microbe* (ASM Press, Washington, D.C., 2006), with permission.

Box 14.2 figure (B) Adapted from Figure 4E, p. 6, of R. B. Bourret et al., *J. Bacteriol.* 184:1–17, 2002, with permission. (C) Adapted from Figure 5 of J. A. Hoch and K. I. Varughese, *J. Bacteriol.* 183:4941–4949, 2001, with permission.

Index